The Geophysics of Sea Ice

NATO ASI Series

Advanced Science Institutes Series

A series presenting the results of activities sponsored by the NATO Science Committee, which aims at the dissemination of advanced scientific and technological knowledge, with a view to strengthening links between scientific communities.

The series is published by an international board of publishers in conjunction with the NATO Scientific Affairs Division

A	**Life Sciences**	Plenum Publishing Corporation
B	**Physics**	New York and London
C	**Mathematical and Physical Sciences**	D. Reidel Publishing Company Dordrecht, Boston, and Lancaster
D	**Behavioral and Social Sciences**	Martinus Nijhoff Publishers
E	**Engineering and Materials Sciences**	The Hague, Boston, and Lancaster
F	**Computer and Systems Sciences**	Springer-Verlag
G	**Ecological Sciences**	Berlin, Heidelberg, New York, and Tokyo

Recent Volumes in this Series

Volume 139—New Vistas in Nuclear Dynamics
edited by P. J. Brussaard and J. H. Koch

Volume 140—Lattice Gauge Theory: A Challenge in Large-Scale Computing
edited by B. Bunk, K. H. Mütter, and K. Schilling

Volume 141—Fundamental Problems of Gauge Field Theory
edited by G. Velo and A. S. Wightman

Volume 142—New Vistas in Electro-Nuclear Physics
edited by E. L. Tomusiak, H. S. Caplan, and E. T. Dressler

Volume 143—Atoms in Unusual Situations
edited by Jean Pierre Briand

Volume 144—Fundamental Aspects of Quantum Theory
edited by Vittorio Gorini and Alberto Frigerio

Volume 145—Atomic Processes in Electron-Ion and Ion-Ion Collisions
edited by F. Brouillard

Volume 146—The Geophysics of Sea Ice
edited by Norbert Untersteiner

Series B: Physics

The Geophysics of Sea Ice

Edited by

Norbert Untersteiner
University of Washington
Seattle, Washington

Springer Science+Business Media, LLC

Based on the proceedings of the NATO Advanced Study Institute on
Air-Sea-Ice Interaction,
held September 28-October 10, 1981,
in Acquafredda di Maratea, Italy

Library of Congress Cataloging in Publication Data

The Geophysics of sea ice.

(NATO ASI series. Series B, Physics; vol. 146)
"Published in cooperation with NATO Scientific Affairs Division."
"Based on the proceedings of the NATO Advanced Study Institute on Air-Sea-Ice Interaction held September 28-October 10, 1981, in Acquafredda di Maratea, Italy"—T.p. verso.
Includes bibliographical references and index.
1. Sea ice—Congresses. I. Untersteiner, Norbert, 1926- . II. North Atlantic Treaty Organization. Scientific Affairs Division. III. NATO Advanced Study Institute on Air-Sea-Ice Interaction (1981: Acquafredda di Maratea, Italy) IV. Series: NATO ASI series. Series B, Physics; v. 146.
GB2401.2.G46 1986 551.3′43 86-22651
ISBN 978-1-4899-5354-4 ISBN 978-1-4899-5352-0 (eBook)
DOI 10.1007/978-1-4899-5352-0

Originally published by Plenum Press, New York in 1986
Softcover reprint of the hardcover 1st edition 1986

PREFACE

The intent of this book is to present the topic of sea ice in the broad and interdisciplinary context of atmospheric and oceanographic science. It is the result of a NATO Advanced Study Institute for which the authors had provided preprints of their respective chapters. The material was presented orally in several lectures by each author, followed by discussions among the approximately fifty attendees of the Institute. By prior agreement between the editor and the authors, the material is introduced on a tutorial level, and developed toward problems currently under study.

Thanks to the willing cooperation by the authors, it was possible to present the material in a more or less structured form, although it was clear from the outset that a book written by twenty authors would not be a text in the usual sense. Even so, we hope that it will help the beginning researcher to survey the subject of sea ice and its role in the global fluid system, and will provide the expert with summaries of less familiar topics.

Instruments and methods of making observations in the field receive inadequate attention. This is, in part, due to the immense variety of methods needed to approach the interdisciplinary subject of sea ice. Also, owing to the use of microprocessors, new power sources, sophisticated means of data storage and transmission, and data acquisition by remote sensing, there is a rapid development toward automation. It seemed an inopportune time to review techniques of observation that would be outdated by the time of their publication.

Seattle, July 1986 N. Untersteiner

Note concerning literature citations

It is widely known and appreciated that researchers in the Soviet Union have played a leading role in sea ice research. They demonstrated, in 1938, that ordinary ice floes can be safely used as manned research platforms. Shortly thereafter (1943) N.N. Zubov published his monumental treatise "Arctic Ice" which, to this day, remains a source of valuable information and insight. During the past three decades, researchers at the Arctic and Antarctic Research Institute, Leningrad, have produced prodigious amounts of literature on sea ice, polar oceanography, meteorology, and climatology.

Even though much of that literature is translated eventually, typical lag times are several years. When the Soviet work becomes available in English it is no longer "fresh" and, as a result, receives inadequate attention. We trust that the sparsity of Soviet citations in this book will not be interpreted as disregard, but simply as the regrettable consequence of the language barrier.

ACKNOWLEDGMENTS

My most sincere thanks to the authors and, among them, the Editorial Committee consisting of

G. Herman, Madison, Wisconsin, USA,
P. Lemke, Hamburg, Federal Republic of Germany
G. McBean, Sidney, British Columbia, Canada, and
M. McPhee, Naches, Washington, USA,

who not only wrote their own chapters but also arranged for peer reviews of all sections of the book. All authors were extraordinarily punctual in providing pre-prints of their section for use at the Institute. Their cooperative spirit in providing the material in the requested form and level of sophistication is also greatly appreciated. That it has taken five years to produce this book is entirely my fault and the result of the erroneous assumption that such a job can be done "on the side" with part-time help. In that regard, too, the authors were generous and updated their material. Most of the chapters including their final corrections were received during 1985.

After a number of false starts, preparation of the camera-ready manuscripts finally ended up in the hands of Marcus Duke, Publications Office, School of Fisheries, University of Washington, whose competence and efficiency were instrumental in finishing that large task. Generous and invaluable help was also given by Stephen Bowen, Chief Editor of the U.S. Army Cold Regions Research and Engineering Laboratory, Hanover, N.H.

The original suggestion to hold an Advanced Study Institute on the subject of this book came from the Special Program Panel on Air-Sea Interaction of the NATO Scientific Affairs Division, especially from Professor Clayton Paulson. With that encouragement, I submitted a proposal to NATO, outlining a program of structured lectures and discussions spanning ten working days, with the idea that the Institute would result in a more or less systematic treatise on air-sea-ice interaction rather than a volume of conference proceedings.

Our collective thanks are due to the NATO Scientific Affairs Division for a grant in support of travel and subsistence for instructors, subsidies for student participants, and other expenses associated with the Institute. With advice and help from Dr. Mario di Lullo and Dr. Tilo Kester, the choice of a location for the Institute fell on Hotel Villa del Mare at Acquafredda di Maratea, about 150 km south of Naples. At the time of the Institute (28 September to 10 October 1981), we had the hotel virtually to ourselves and enjoyed superb hospitality (Mr. A. Guzzardi, Hotel Director) and conference and recreational facilities, and a collection of spirited and hard working attendees. Unfortunately, all local expenses in Italy were accountable based on the U.S. dollar, which experienced a sharp increase in value during the interval between issuance of the grant and its actual expenditure, resulting in the loss of a sizable fraction of the NATO grant in Belgian francs. Thanks to the financial rescue by the Office of Naval Research (Arctic Program) and the National Oceanic and Atmospheric Administration (Research and Development) it was possible to finance the Institute as planned.

I am grateful to the Oceanographer of the Navy and to the Applied Physics Laboratory at the University of Washington for their generous help, especially to Dr. John Harlett.

CONTENTS

Selected Special Topics

THE GEOPHYSICS OF SEA ICE: OVERVIEW

Norbert Untersteiner

Department of Atmospheric Sciences
and Polar Science Center
University of Washington
Seattle, Washington 98195 U.S.A.

Terrestrial temperatures happen to lie in a range where, in the course of a year, the sea surface at high latitudes becomes cold enough to freeze. According to paleoclimatic studies of the sea floor sediments, this has been the case for at least several million years. During that time, however, the extent of sea ice underwent large fluctuations. For instance, at the last glacial maximum 18,000 years ago, Atlantic sea ice extended as far south as the shores of France and northern Spain in Europe, and the eastern coastal states in North America. The heat of fusion released by the sea surface to form an ice layer of, say, one meter thickness is an order of magnitude smaller than the mean annual total of either short-wave or long-wave radiation at the surface. In a single year, these annual totals may easily deviate ten percent from their mean values. Thus it is not surprising that small changes in the climatic forcing are accompanied by large changes in the extent of the ice-covered area.

These circumstances make sea ice one of the most perishable constituents of the earth's surface. Wherever ice occurs, the nearly black sea surface (albedo 0.1 or less) turns highly reflective to solar radiation, and the surface can become very cold, thereby curtailing the loss of thermal infrared radiation. Being a mechanical as well as thermal barrier between ocean and atmosphere, ice suppresses the exchanges of kinetic energy, heat, water vapor, and salt between the two media, while the ocean temperature beneath the ice is held at a nearly constant value near the freezing point.

The large variations of sea ice cover are not a passive phenomenon, forced unilaterally by oceanic and atmospheric conditions. Rather, they participate in a variety of feedback effects, both positive and negative. For instance, a positive anomaly of sea ice extent enlarges the area of low albedo, reducing the net short-wave radiation and thus cooling the surface and increasing the chance of more freezing and a further expansion of the ice cover (positive feedback). At the same time, the incoming long-wave radiation remains largely unaffected while the outgoing long-wave radiation diminishes as the surface becomes colder; thus the net long-wave radiation increases. In addition, the cold surface builds a temperature inversion in the boundary layer, directing more sensible heat toward the surface (negative feedback). From these two examples it is evident that the ice is engaged in a complex interplay with its environment whose unraveling is the central theme of the geophysics of sea ice.

A standard practice in all geophysical sciences is to cast the interacting processes into a set of equations that describe (model) the system to the best possible approximation. For the atmosphere and the ocean, the application of classical fluid dynamics has been highly successful. The approach here is to use the principles of conservation of momentum and mass, the first law of thermodynamics, and equations of state, and to solve these equations for a steady state or, starting from an initial condition, for a future state of the system. Deterministic short-range weather prediction proceeds in that fashion. The main reason why the same kinds of predictions are just beginning for the ocean is that it is technically far more difficult to determine its initial conditions at the appropriate scales.

This hydrodynamic approach has been applied to sea ice. As will be outlined below, this leads to difficulties not encountered in the atmosphere and ocean: The conservation of ice mass (Nikiforov, 1957; Untersteiner, 1962) can be formally expressed by

$$\frac{\partial(\rho h)}{\partial t} = -\hat{u} \cdot grad(\rho h) - \rho h \nabla \cdot \hat{u} + \frac{1}{q} Q \qquad (1)$$

where ρ and h denote density and thickness, $\hat{u}$ the velocity, q the heat of fusion, and Q that part of the heat balance available for freezing or melting ice. This equation states that the amount of ice present at a given point is controlled by advection, divergence, and sources or sinks due to freezing or melting. However, unlike the atmosphere and the ocean whose total mass is practically invariant, the total amount of ice is extremely variable, the boundaries themselves are subject to advection, and a small imbalance in the quantity Q can induce large changes in the area covered by ice. Thus, in a physical as well as computational sense, the continuity equation (1) is not a constraint on the system that simplifies matters but, instead, imposes an increased demand on observations and computational accuracy for model output validations.

The first law of thermodynamics, implied in the last term of equation (1), is needed to account for heat energy taken up or released in melting and freezing, and in changes of the ice temperature:

$$\rho q \frac{dh}{dt} = \sum F_s + \sum F_b + \int_h \rho c \frac{\partial T}{\partial t}\, dz \qquad (2)$$

where $\sum F_s$ and $\sum F_b$ are the sums of all heat fluxes, turbulent and radiative, at the surface and at the bottom of the ice, and the last term is the change of the internal energy associated with the change of ice temperature, T, in a slab of thickness h. (Note that both ice density, ρ, an specific heat, c, are functions of ice temperature and salinity; see Chapters 1 and 5.)

The continuity equation (1) is stated for a medium that can be described by state variables at a point and by local derivatives of velocity and thickness, i.e., for a continuum. It remains to be seen whether, or under what circumstances, such a formulation is useful. The same situation holds for the momentum equation (Nansen, 1902),

$$\rho h \frac{d\hat{u}}{dt} = \tau_a + \tau_w - \rho h f \hat{k} \times \hat{u} + div\sigma - \rho h g \nabla H - h \nabla P \qquad (3)$$

The term $d\hat{u}/dt$ denotes ice acceleration, τ_a and τ_w surface stresses at the top and bottom of the ice, f the Coriolis parameter, $\hat{k}$ the unit vector in the vertical, and σ the horizontal stress in the ice resulting from floe-to-floe stress transmission. The term $-\rho hg \cdot gradH$ is a component of the gravitational acceleration, g, along a sea surface with the slope $gradH$. This slope is induced by the atmospheric pressure gradient (barometer effect) and by dynamic tilt of the sea surface. The term $-h\, gradP$ is due to the horizontal gradient of the atmospheric pressure at the surface (Rothrock, 1975). The surface stresses are generally taken to be proportional to a drag coefficient characterizing the roughness of the ice, and to the square of the relative velocity between ice-air and ice-water. The term $div\sigma$ represents the internal stress propagated through the ice. This equation, which has been used in virtually all studies of ice dynamics, presumes that ice is a continuum.

Air and water consist of material elements (molecules) so small in size and so large in number that, on a macroscopic scale, their collective statistical behavior can be reduced to, for all practical purposes, simple laws, or "equations of state," for a continuum. Pressure, volume, density, humidity, and salinity are the primary state variables that describe atmosphere and ocean in this manner.

On a scale of up to several meters, sea ice can also be considered a continuum. Much is known about the physical properties of sea ice on that scale (Chapters 1 and 2). The same is true for exchange processes at the air-ice-ocean interfaces. Chapters 3 and 4 describe the application of boundary layer fluid dynamics to a sea ice field with average properties, while Chapter 5 focuses on the exchange of thermal energy and its controlling effect on ice build-up or decay, in a fashion where it is of little relevance whether or not the ice is a continuum. In treating the dynamics of sea ice on scales greater than tens of meters, the fact that ice is not a continuum may become important under some circumstances. On those space scales, the ice is an aggregate of pieces of variable size, shape, and thickness that are separated by open water and thin ice, or tightly packed, with rubble fields and pressure ridges.

Since it is the purpose of equation (3) to explain ice acceleration as the result of all external and internal forces, one must consider the chance to observe them in nature. Velocity, air stress, water stress, and sea surface slope are, at least in principle, observable. The internal stress term poses a severe problem. The state of stress in the ice is related to deformation, in a general way, by a "constitutive law" (Coon, 1980; Hibler, 1980) of the form

$$\sigma = f\,(\varepsilon, \dot{\varepsilon}, \ldots) \tag{4}$$

where f denotes a functional relationship. Since it is not likely that this functional relationship can be derived from basic physical principles, one could attempt to find it from observations in nature. With some limitations with regard to scales (see Chapter 7) the strain rate can be measured, but measuring the stress poses seemingly insurmountable problems. To measure a force one must counterbalance it with another that is opposite and equal. Translated into a real situation in the field, this would amount to installing arrays of calibrated springs that would prevent the movement of ice floes relative to each other. Also, in principle it might be possible to install numerous strain gauges in a single floe and, with the knowledge of the continuum scale elastic properties of the ice, calculate the state of stress within that floe. It is not likely that either technique can be used to measure the internal stress in such a way that it can be applied to large-scale ice dynamics.

Even if such measurements were possible, the function f in equation (4) would have to be stated in terms of various state variables and their mutual relationships. Obvious candidate state variables are the ice thickness and its statistical distribution, the compactness of the ensemble of ice "floes" (ice concentration), the distribution of floe shapes and sizes, the statistical geometry of open water areas, possible relationships between floe size and thickness, the yield strength of an ensemble of floes and other similar parameters. Work in that direction is just beginning (Chapter 8). It should be noted that the momentum balance (equation 3) can be greatly simplified in some cases. For instance, it has been shown that the internal ice stress may be ignored, provided that the ice cover is "loose" or consists primarily of thin, first-year ice, such as is found in the arctic marginal seas and in most of the waters surrounding Antarctica.

A step toward formulating an equation of state for sea ice was the thickness distribution theory of Thorndike et al. (1975). The continuity equation (1) makes no statement about a relationship between the dynamic and thermodynamic terms. They are, of course, related because the ice thickness changes in time by melting and freezing, as well as by mechanical deformation. That relationship can be expressed by

$$\frac{\partial g}{\partial t} + \hat{u} \cdot \nabla g = -\frac{\partial}{\partial h}\left[\frac{dh}{dt} g\right] + \psi - g\, div\, \hat{u} \tag{5}$$

where g denotes the fractional area covered by ice of a certain range in thickness. The thermodynamic growth rate dh/dt, a function of ice thickness and time, is given by equation (2). The function ψ describes the mechanical redistribution of ice thickness associated with the strain rate. Some principles of formulation for the function ψ have been established, but data available to date have not allowed the choice of a specific form (see Chapter 8).

Given these problems of dealing with the large-scale state variables and material properties of sea ice, it may prove impossible to find generally applicable equations of state and constitutive laws. Undaunted by such prospects, scientists have mustered an impressive amount of ingenuity in modeling sea ice and have achieved many useful results. The general philosophy is to select, based on observations and intuition, a limited number of state variables that are likely to be dominant and important, depending on scales in space and time of the phenomena to be modeled and studied. Chapters 8 and 9 describe these problems and the results of work on ice dynamics to date.

Mechanistic fluid dynamics models are, of course, not the only way to study sea ice. The sea ice data base (see Chapter 17) is now large enough to allow statistical investigations of the air-ice-sea system. These are particularly informative on the large spatial and temporal scales of climate. Studies of that kind are described in Chapters 12 and 13.

The physical interactions between the ice and the atmosphere and the ocean are treated in Chapters 10 and 11. On time scales of decades and longer, an important relationship exists between the seasonal formation of sea ice (and its anomalies) and the formation of deep water masses. Global water mass analyses (Sarmiento and Toggweiler, 1984) have shown that three-quarters of the water masses of the World Ocean acquire their properties in an area covering only five percent of the global ocean surface, all of which is located at high latitudes. The reasons for this are the convective processes driven by surface cooling and ice formation. It is these same convective processes that bring deep waters to the surface where their rich content in nutrients combines with light to produce the most biologically productive waters.

A region with especially numerous and complicated processes is the seasonal ice zone, i.e., that part of the ocean that becomes ice-covered and ice-free in a seasonal cycle. By definition, it is that region where anomalous behavior of the sea ice cover is most apparent and where the search for its causes is most intensive. The multitude of physical processes occurring in the seasonal sea ice zone are treated in Chapter 14.

The appearance of ice on the ocean changes the surface radiative properties over the entire spectrum from visible to microwave (centimeter wavelengths) frequencies. Both passive (emitted) and active (reflected) microwave radiation from the surface is independent of daylight and passes the atmosphere nearly unattenuated. Satellite instruments obtaining images of the polar regions in these frequencies have become an important research tool (see Chapter 16), and have been accumulating valuable data about the geographical and seasonal distribution of sea ice in both hemispheres. Not only can passive and active microwave instruments sense the presence or absence of ice, but they can also obtain information about the physical state of the ice, such as its temperature, porosity, brine content, crystallography, and roughness. These parameters combine to give the ice, as seen by a satellite-borne microwave receiver, a certain "apparent brightness." It was an important discovery (Gloersen et al., 1973) that the loss of salt occurring in the surface layer of young ice during its first summer changes its microwave emissivity in such a way that, in principle, recently formed ice can be distinguished from ice that is more than one year old. From a current census, through one or more annual cycles, of the amount of first-year (FY) and multi-year (MY) ice in a large-scale region, one could evaluate the integral of the mass balance equation (1), or an approximation to it. The routine use of passive microwave data for the discrimination of ice types is still hampered by calibration problems. The spatial resolution of present passive microwave imagers on satellites is 15 to 35 kilometers. The only reliable ground truth comes from areas where all the ice melts in summer (all FY) or where the ice cover never breaks open (all MY). To calibrate regions with a mixed ice cover it would be necessary to survey the fractions of FY and MY ice present in surface elements (corresponding to pixels) measuring several hundred square kilometers, by independent methods of unquestionable interpretation. In practice, such methods are not yet available.

Even so, ways will be found to overcome that problem and FY/MY maps will be produced that offer insight into the ice mass balance. An essential ingredient for such interpretations is the continued acquisition of ice drift data by means of automatic data buoys to monitor the average field of ice velocity (Colony and Thorndike, 1984) and other parameters.

As in most other branches of quantitative earth science, the small-scale physical mechanisms of the air-sea-ice system are as well understood as classical physics, and have been for decades. The earlier difficulties of analyzing the simultaneous interactive workings of many such mechanisms, which are the hallmark of our physical environment, are becoming diminishingly serious with the emergence of computing devices of astounding power to produce numerical simulations of reality.

There is little doubt that, at least in the near future, the formal methods and the means for modeling will stay a step ahead of our ability to provide the models with a realistic description of what they are designed to simulate. For instance, the progressive inaccuracy of weather models with increasing forecast intervals is not only the fault of the models but, to a large extent, also the fault of inaccurately described initial conditions of the atmosphere (Thompson, 1957). The future acquisition of data will have to be guided by studies of the sensitivity of models to specific processes and effects (WMO-ICSU, 1984). Throughout the history of geophysics it has been the observational part of science that generated the new ideas and concepts. Given the steady evolution of ever more sophisticated satellite instruments and autonomous in situ observing devices, the prospects are excellent. At the same time, it is becoming increasingly important to direct observing programs toward a clearly defined problem, and to make the thorough analysis of the data an integral part of the plan to acquire them.

The most difficult task of the future will be to conduct our quest for knowledge in such a way that it strikes a balance between benefits and burdens to society.

REFERENCES

Colony, R. and A.S. Thorndike (1984) An estimate of the mean field of arctic sea ice motion. J. Geophys. Res., 89: 10623-10629.

Coon, M.D. (1980) A review of AIDJEX modelling. In: Sea Ice Processes and Models (R.S. Pritchard, Ed.), Univ. of Washington Press, Seattle and London, 12-27.

Gloersen, P., W. Nordberg, T.J. Schmugge and T.T. Wilheit (1973) Microwave signatures of first-year and multiyear sea ice. J. Geophys. Res., 78: 3564-3572.

Hibler, W.D. III (1980) Sea ice growth, drift, and decay. In: The Dynamics of Snow and Ice Masses (S.C. Colbeck, Ed.), Academic Press, New York, 141-209.

Nansen, F. (1902) The oceanography of the north polar basin. In: Norwegian North Pole Expedition, 1893-1896, Scientific Results, Vol. 3, No. 9, Longmans, Green, Christiania, 357-386.

Nikiforov, Ye. G. (1957) Ob izmenenii splochennosti ledianogo pokrova v sviazi s ego dinamikoi. (A change in the concentration of the ice cover in connection with its dynamics.) Probl. Arkt. 2: 59-72 (Transl. by Am. Met. Soc. NTIS No. AD 232204).

Rothrock, D.A. (1975) The mechanical behavior of pack ice. Ann. Rev. of Earth and Planet. Sci., 3: 317-342.

Sarmiento, J.L. and J.R. Toggweiler (1984) A new model for the role of the ocean in determining atmosphere PCO_2. Nature, 308: 621-623.

Thompson, P.D. (1957) Uncertainty of initial state as a factor in the predictability of large-scale atmospheric flow patterns. Tellus, 9: 275-295.

Thorndike, A.S., D.A. Rothrock, G.A. Maykut and R. Colony (1975) The thickness distribution of sea ice. J. Geophys Res., 80: 4501-4513.

Untersteiner, U. (1963) Ice budget of the Arctic Ocean. In: Arctic Basin Symposium (Hershey, PA, Oct. 1962), Arctic Inst. of N. America, Wash. D.C.: 219-230.

WMO-ICSU (1984) Report of the Meeting of Experts on Sea Ice and Climate Modelling (Geneva, Dec. 1983), WCP-77, World Meteorological Organization, Geneva, 1-34.

CHAPTER 1

THE GROWTH, STRUCTURE, AND PROPERTIES OF SEA ICE

W. F. Weeks and S. F. Ackley

Snow and Ice Branch
U.S. Army Cold Regions Research and Engineering Laboratory
72 Lyme Road
Hanover, New Hampshire 03755 U.S.A.

1. INTRODUCTION

On the geophysical scale sea ice is a thin, fragile, dynamic, solid layer that forms under the thermodynamic conditions that occur near the Poles. There it serves as a boundary between two much larger fluid bodies -- the ocean and the atmosphere. Typical scales of interest would be 10^3 to 10^6 m. In the present paper we take a more detailed view, focusing on the ice itself at scales ranging between 10^0 and 10^{-3} m, with an occasional glimpse at a scale of 10^{-10} m. It is our purpose to help the reader understand the internal structure of sea ice, how this structure develops, and how it affects the bulk properties of the ice. Although this is a subject that has received little attention in comparison to similar studies of metals and ceramics, it is, in our view, very important, as many details in the behavior of sea ice are structurally controlled. The variations in structure are, in turn, determined by the environmental conditions under which the ice has formed.

In the following, we will take many observations and theories based on studies of other materials and apply them to sea ice. This is a natural progression, in that it moves from the well studied to the little studied. The reader should, however, be warned that such transfers of ideas are, many times, far from straightforward. In fact, it is to be hoped that in the long run the study of sea ice as a material will be able to contribute as much to the study of other materials as it receives. The reasons for this optimism are several. First, the natural range of temperatures at which sea ice exists is just a few degrees off its melting point. In fact, sea ice normally is only partially solidified in that it contains melt inclusions. Second, this temperature range is one in which man can operate without special precautions or equipment. Therefore, experiments on sea ice are much easier than experiments on most metals and ceramics studied at similar near-melting temperatures. Finally, pure ice, the primary phase composing sea ice, is transparent. This makes it relatively easy to look inside the ice and to observe what is actually occurring. This is a luxury rarely available to workers in metallurgy and ceramics.

2. THE ATOMIC STRUCTURE OF ICE

One of the unusual characteristics of ordinary ice is that it is less dense than its melt. Geophysically this low density is quite important in that it causes ice to float on the surface of lakes and seas. If ice sank into its melt, as do most solids, there would be a tendency for natural water bodies to freeze completely to their beds, a situation hardly conducive to the development of aquatic life in either temperate or polar regions.

The fact that ordinary ice is such an open, low density solid also suggests that there may be other, more compact atomic arrangements of H and O that would result in H_2O polymorphs with a higher density. There are, in fact, at least nine other polymorphs of H_2O (Hobbs, 1974) that can be produced. However, ordinary ice (ice I) is the only one of these that exists in significant quantities under the physical conditions encountered at the earth's surface. In the remainder of this paper, when ice is referred to, the ice polymorph will always be ice I.

The geometric arrangement of the oxygen atoms, which in the ice structure are the principal diffracting centers for X-rays, is well understood, as ice was one of the first substances to have its structure determined (Bragg, 1922; Lonsdale, 1958; Owston, 1958; Hobbs, 1974). The results of these studies are shown in Figure 1. Each oxygen atom is located at the center of a tetrahedron with four other oxygen atoms located at each of the apices. The O-O distance is 0.276 nm at 0^oC, resulting in an open, low-density structure. The tetrahedral coordination of the oxygen atoms results in a crystal structure possessing hexagonal symmetry, a fact that will affect many of the large scale characteristics of ice. An important feature of the structure of ice I is that the oxygen atoms are concentrated close to a series of parallel planes that are referred to as the basal planes. The direction perpendicular to these planes is referred to as the principal hexagonal or c-axis. The arrangement is such that, in any unit cell which

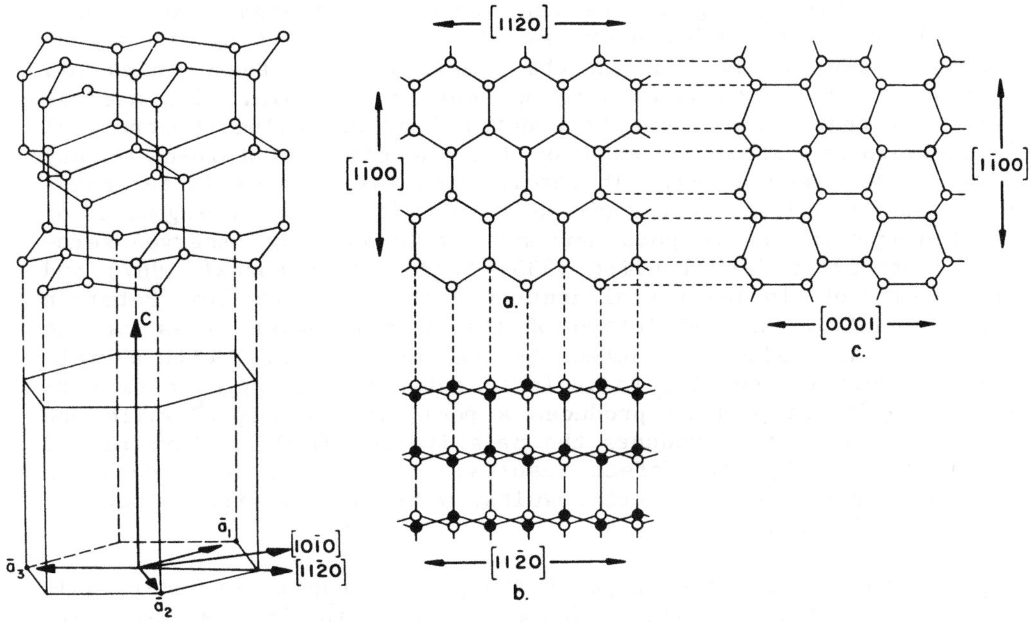

Fig. 1. Structure of ice I.

contains four oxygen atoms, fracture along the basal or (0001)* plane involves the rupture of only two bonds. On the other hand, fracture along any plane normal to this plane requires the breaking of at least four bonds. Therefore, the fact that ice glides and cleaves readily on the basal plane can be easily rationalized in terms of its atomistic structure. When the positions of the oxygen atoms are projected parallel to the [0001] direction, the resulting hexagonal array (Figure 1a) can be seen to be composed of three close-packed rows of atoms. These rows lie in the <11$\bar{2}$0> direction parallel to the three secondary or a-axes. As will be seen, the a-axis directions, which are, of course, all equivalent, correspond to the directions of the arms of a) dendritic sea ice crystals growing from the melt, b) snowflakes growing from the vapor, and c) internal melt features that form inside ice crystals as the result of absorbed solar radiation; again all are external manifestations of the atomic structure.

This highly regular net of oxygen atoms is bonded together by a series of hydrogen bonds. The position of the hydrogen atoms in these bonds are disordered and obey the so-called Bernal-Fowler

---

*The parentheses around the Miller indices signify a single plane or a set of planes; indices of a direction are given in brackets, i.e., [11$\bar{2}$0], and the full set of equivalent directions of a form are indicated by carets, i.e., <11$\bar{2}$0> (Barrett, 1952, pp. 8-12).

rules: 1) two hydrogens are near each oxygen atom, and 2) only one hydrogen atom can be on or near the line connecting two neighboring oxygen atoms. Within the constraints of these rules all configurations are considered to be equally probable. The average structure that results can be specified by assigning hypothetical "half-hydrogen" atoms to each of the 4N sites contained in any array of N oxygen atoms. Therefore each oxygen atom is tetrahedrally surrounded by four "half-hydrogens." This arrangement of "half-hydrogens" is in good agreement with both the observed zero-point entropy of ice (Pauling 1935; Nagle 1966; Suzuki 1967) and the results of single-crystal neutron diffraction studies (Peterson and Levy 1957). Any violation of the Bernal-Fowler rules can be considered as producing a defect in the ice structure, with a violation of rule 1 resulting in an ionic defect: an oxygen atom surrounded by three protons produces a positive ion $(H_3O)^+$ while one with only one proton produces the negative ion $(OH)^-$. When rule 2 is violated, a Bjerrum defect results: when two protons occur on the bond a so-called D-defect results, while a bond with no protons gives an L-defect.

Although beyond the scope of the present paper, an understanding of the nature of the hydrogen bonding in ice is certainly necessary to the study of its electrical, diffusive, thermodynamic and mechanical behavior. For instance, it is the nature of the highly directional hydrogen bond that is responsible for the anomalous physicochemical properties of ice and water. As was mentioned, the density of ice at 0°C is less than that of water at the same temperature, with the density of water increasing to a maximum at +4°C. The low density of ice is the result of the tetrahedral coordination of each water molecule by hydrogen bonding causing a very open structure. On melting, some of the hydrogen bonds are broken, allowing the molecules to assume a more closely packed structure with an associated increase in density. This process of further bond breakage and compaction of molecules continues with rising temperature until a point is reached (+4°C) when density increases are more than compensated for by thermal expansion.

3. THE SEA ICE PHASE DIAGRAM

Now that we have a general knowledge of the structure of pure ice, we next need information on the phase relations of the ice-salt and ice-seawater systems that are of concern in the study of sea ice. By phase relations we refer to information that specifies the different phases that coexist at equilibrium at different temperatures and pressures in a system with a known bulk composition. One of the first questions that must be considered is "Does the ice that forms from freezing salt solutions incorporate salt in solid solution?" Here by solid solution we refer to the process when the impurity atoms actually occupy lattice sites in the atomic structure of the ice crystal. Solid solutions form most readily when the foreign atoms have the right size to fit into the host structure, tend to form a similar type of chemical bond, and have an appropriate charge to maintain electrostatic neutrality. Impurities fulfilling these requirements for substitution into ice are rare. Because the ionic and atomic radii of fluorine and nitrogen are similar to those of oxygen, F^-, HF, NH^+_4, and NH_3 are likely candidates to replace some of the water molecules in the ice structure. Other possibilities are NH_4OH, NH_4F, and the hydrohalogen acids (HCl, HBr, and HI). In all of the above cases some limited substitution in the ice structure is observed. Maximum substitution occurs at a concentration of 10^{-2} mol/L, which corresponds to a molar ratio of about 1 in 5000. However, none of the above materials are present in significant quantities in seawater.

How much of the common components of seawater enter substitutionally into ice crystals grown from it? There is, to our knowledge, no precise information on the subject, although the amounts would appear to be extremely small. Harrison and Tiller (1963) estimate that for most solutes in ice the equilibrium solute partition coefficient $k_o(k_o = X^s_A/X_A^\ell$ where X^s_A and X_A^ℓ are the mole fraction of solute A in the solid and liquid respectively) is less than $k_o \leq 10^{-4}$. In short, although thermodynamic arguments can be made that suggest that at equilibrium a finite amount of impurity should exist in the ice structure, this amount is apparently so small that for most purposes the ice can be considered as a pure phase. Therefore, as ice forms from a salt or seawater solution essentially all the solute is rejected back into the liquid or melt. So, as more and more ice forms with gradually decreasing temperature the melt that coexists in equilibrium with the ice becomes saltier and saltier. The phase relations simply specify the number and composition of the different phases (brine, ice, solid salts) that coexist at the different temperatures and pressures. Although, in principle, it is possible to calculate the phase relations from thermodynamic information, in fact phase diagrams, and particularly those for complex systems such as seawater, are invariably determined experimentally. The results of such experimentation are then displayed in the form of a phase diagram.

For example, Figure 2 shows the water-rich portion of the H_2O - NaCl phase diagram, which is the simple two-component system that most closely approximates seawater. What happens when we start with an NaCl solution and gradually cool it down past its freezing point until it becomes completely solidified can be seen on this diagram. Consider an NaCl solution with a composition of 35°/oo NaCl at a temperature of +5°C. The phase rule is P + F = C + 2, where P is the number of phases that coexist, C is the number of components, and F is the number of degrees of freedom (for details refer to any standard physical chemistry text). However, in our case we have fixed one degree of freedom by the fact that our

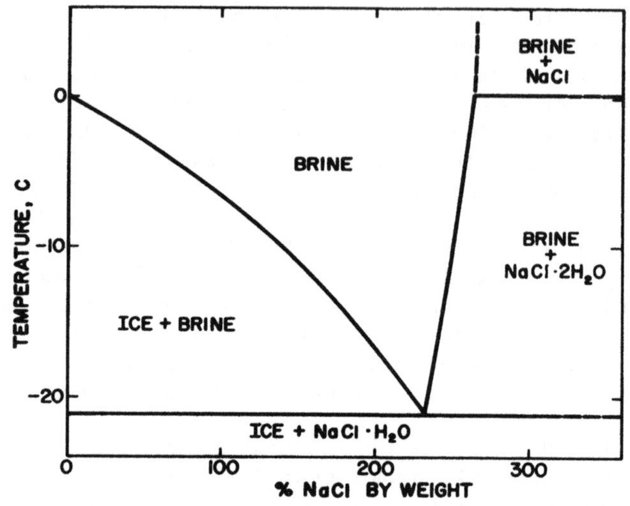

Fig. 2. A portion of the phase diagram $NaCl-H_2O$.

experiment was carried out at a constant pressure of 1 atmosphere and, therefore, the appropriate formulation of the phase rule becomes $P + F = C + 1$. We have a two-component system, H_2O and NaCl (so $C = 2$), with only one phase (the brine solution) present ($P = 1$). Therefore, $F = 2$ and two variables (temperature and bulk composition) must be specified to fix the state of the system. Figure 2 shows that experimentally ice is observed to start forming from a 35°/oo NaCl solution at -2.0°C. Once ice forms P becomes 2, C remains 2, and $F = 1$, indicating that the system is now univariate. Therefore, as long as ice and brine coexist it is only necessary to specify the temperature or the bulk composition to fix the state of the system. In our case we initially specified the composition of the system, so if we additionally specify the temperature we have also fixed the composition of the brine that can coexist in equilibrium with the ice. The liquidus curve, in fact, gives the composition of the NaCl brine that is in equilibrium with ice at different temperatures. As the system continues to cool, more and more pure ice forms from the brine, causing the remaining brine to decrease in volume and become more saline. For example, at -5°C ice coexists with a brine containing 80°/oo NaCl, while at -10°C and -20°C the brine compositions are respectively 140 and 225°/oo. If cooling is continued to -21.2°C a third phase (the solid salt $NaCl \cdot 2H_2O$) is observed to form. Now P has become 3, C remains 2, and F must equal 0. The system is now referred to as being invariate, and this point (where the curve giving the composition of the brines that are in equilibrium with ice meets the curve giving the composition of the brines that are in equilibrium with $NaCl \cdot 2H_2O$) is called the eutectic point. As the system is invariate further cooling cannot continue until one of the phases disappears. Solidification continues at the eutectic temperature, with the two solids ice + $NaCl \cdot 2H_2O$ crystallizing until all of the remaining brine has disappeared. The system is now composed of a mixture of the two solids ice + $NaCl \cdot 2H_2O$, and a further decrease in temperature can proceed.

The above can be summarized as follows. When an NaCl solution with a composition similar to seawater (35°/oo salt) is frozen, the first ice crystals start to form at -2.0°C. As freezing proceeds, more and more pure ice forms, with the remaining brine gradually becoming saltier and saltier. As long as ice and brine coexist the composition of the brine at each temperature is fixed. The formation of ice continues with further cooling, and the amount of brine becomes less and less until at a temperature of -21.2°C the eutectic point is reached and an additional solid phase, the sodium chloride dihydrate, is formed. At temperatures below the eutectic temperature all brine has solidified and only the two phases ice and $NaCl \cdot 2H_2O$ exist.

When seawater is frozen a similar but more complex series of events occur. In this case we are dealing with an eight-component system for artificial seawater and a greater-than-eight-component system for natural seawater. Therefore, these systems never become invariate exhibiting eutectic points. With cooling, ice forms and the remaining brine becomes more saline. As cooling continues, different solid salts also precipitate from the brine. However they precipitate over a temperature range as opposed to a fixed eutectic temperature. At a temperature of -70°C there is still a measurable amount of brine believed to be present in the ice (Richardson and Keller, 1966), and at least five solid salts are presumed to be present ($CaCO_3 \cdot 6H_2O$, $Na_2SO_4 \cdot 10H_2O$, $NaCl \cdot 2H_2O$, KCl, $MgCl_2 \cdot 12H_2O$).

Present phase information on sea ice is largely based on three different experimental studies (Ringer, 1906; Gitterman, 1937; Nelson and Thompson, 1954). All these contributions are very creditable, but the first two works are old and the investigators were not aware of many potential problems that must be overcome in any proper determination of the phase relations (Tsurikov and Tsurikova, 1972). Also, the Nelson and Thompson study was not focused on sea ice (they were interested in the possibility of the formation of sulfate deposits by freezing) and, therefore, certain desirable parameters of importance in the study of sea ice were not determined.

Without going into details concerning who did what and why, we feel that current deficiencies in the experimental studies of the sea ice phase diagram are as follows:

1. There has never been a direct determination of the composition of the solid salts that form in sea ice (their composition

has always been inferred from changes in the composition of the brine).

2. Although equilibrium is always assumed in such studies, there does not appear to have been any attempt to verify this assumption by approaching the same state via different cooling (and heating) paths.

3. It is very important in phase studies to assure that the liquid phase is always compositionally homogeneous and can react with all the solid phases in the system. It is almost certain that this was not true in the three studies we mentioned. The general experimental technique has been to freeze a solution to a desired temperature, remove the ice from the brine, and if necessary prepare a new solution for further freezing based on the composition of the remaining brine. This is an excellent procedure as long as the brine remains homogeneous and is in continuous contact with all the phases present. However, if isolated brine pockets form in the ice phase then the liquid and solids in these pockets can no longer react with the bulk liquid. This problem becomes particularly pronounced if there are reactions between the brine and previously formed solid phases. Several such reactions have been suggested by Gitterman and also by Savel'ev (1963), although the evidence for their occurrence cannot at present be considered conclusive.

4. All experimental work to date has ignored the carbonate content of the brine, although $CaCO_3 \cdot 6H_2O$ would appear to be the first solid salt to form during the formation of sea ice.

Considering the above problems, is our present knowledge of the phase diagram useful or just misleading? We feel that in most cases it is useful. The phase diagram as commonly utilized was worked out by Assur (1958), based on the work of Nelson and Thompson and of Ringer. Figure 3 shows the freezing point of the brine derived from standard seawater as a function of its composition (expressed as grams of salt per kilogram of H_2O). The different experimental determinations are in good agreement. Note that there are pronounced breaks in slope at temperatures where major solid salts start to crystallize. Figure 4 presents the phase diagram for standard sea ice, which is defined as sea ice of such a composition that its meltwater will have the same relative concentration of ions (to each other) as normal seawater (Assur, 1958). The idea of standard sea ice is a useful one and is an extension of the observation that although the absolute amounts of salt in seawater may vary widely, the ratios of the ions relative to each other remain surprisingly constant. How applicable this observation is to sea ice will be examined later. It should be noted here that if the ratios of the ions were to change, the crystallization temperatures of the solid salts would also be changed.

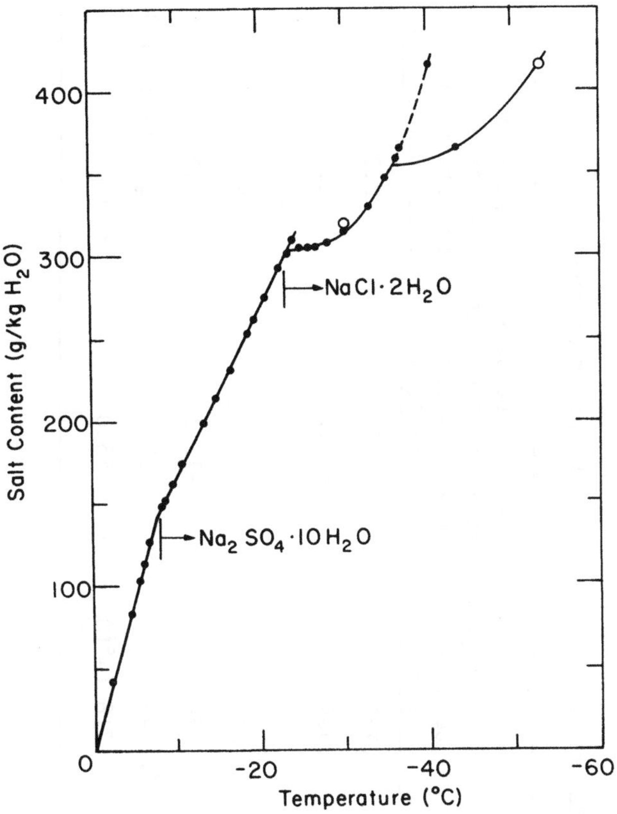

Fig. 3. Freezing point of brine as a function of the ratio of the weight of dissolved salts to pure water (linear relations used to -24°C). Solid dots computed from data of Nelson and Thompson (1954). Open circles computed from Ringer (1906). The dashed line indicates a possible path if salts precipitate in a different order (after Assur 1958).

Figure 4 shows that at -10°C 1000 g of standard sea ice with a salinity of 34.325°/oo is composed of 768 g of ice, 4.2 g of solid salt (4.0 g of $Na_2SO_4 \cdot 10H_2O$ + 0.2 g $CaCO_3 \cdot 6H_2O$) and 228 g of brine composed of 195 g H_2O and 32 g of various ions. At -30°C, on the other hand, the amount of ice and of solid salt present has risen to 917.0 g and 43.4 g, respectively (76% of the solid salt is $NaCl \cdot 2H_2O$), while the amount of brine has decreased to 39.5 g (the exact figures can be found in Assur (1958)). The initial temperatures of crystallization of the various solid salts believed to be present in sea ice are given in Table 1. Also listed are some other characteristics of these salts, including their eutectic temperatures in pure salt-H_2O solutions. The differences in these temperatures are the results of the presence of the other ions in

Table 1. Some properties of the solid salts presumed to occur in sea ice.

Salt composition	Mineral name	Crystal system	Density (Mg/m^3)	Eutectic temperature of the salt in an aqueous solution (C°)	Temperature of initial salt formation in sea-water brine (C°)
$CaCO_3 \cdot 6H_2O$		Monoclinic	1.771	?	- 2.2
$Na_2SO_4 \cdot 10H_2O$	Mirabilite	Monoclinic	1.464	- 3.6	- 8.2
$MgCl_2 \cdot 8H_2O$				-33.6	-18.0
$NaCl \cdot 2H_2O$	Hydrohalite	Monoclinic	1.630 (0°C)	-21.1	-22.9
KCl	Silvite	Cubic	1.984	-11.1	-36.8
$MgCl_2 \cdot 12H_2O$		Monoclinic	(1.24)	-33.6	-43.2 (erratic)
$CaCl_2 \cdot 6H_2O$	Antarcticite	Hexagonal	1.718 (4°C)	-55.0	<-55.0

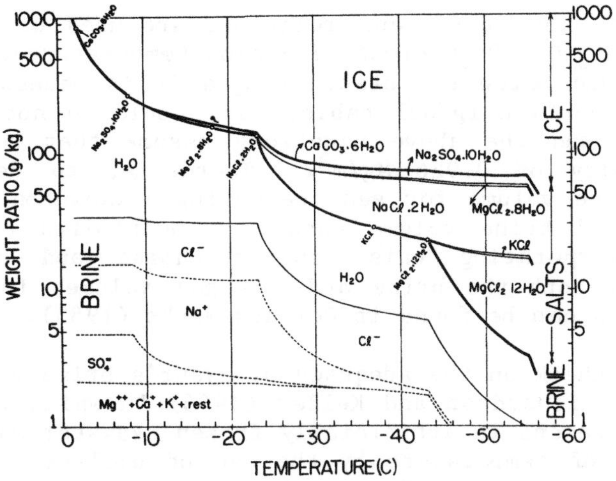

Fig. 4. Phase relations for "standard" sea ice. Circles on the brine-salt line indicate temperatures at which solid salts precipitate (Assur 1958).

seawater brine that are not present in the aqueous solutions of the individual salts.

To change precisely from measurements of the weight of brine per kilogram of sample to the volume of brine per kilogram of sample, accurate measurements must be available of the density of the seawater brine coexisting with ice at different temperatures. As Nelson and Thompson only reported brine densities at +4°C, this calculation cannot be made with the confidence desired. A table of the estimated values of the relative volume of brine (usually indicated by the symbol ν_b) for standard sea ice is given in Assur (1958) for the temperature range of -0.1 to -54°C. In sea ice with a salinity of 1°/oo only 0.075°/oo of brine remains at -54°C as compared with 24°/oo existing at -2.0°C (a 320-fold increase). In most applications where extreme accuracy is not deemed essential, the fact that ν_b can be linearized as a function of $1/\theta$ and S_i, where θ is the temperature and S_i is the salinity of the ice, is utilized, and the empirical equations developed by Frankenstein and Garner (1967)

$$\nu_b = S_i \left(\frac{45.917}{\theta} + 0.930\right) \qquad -8.2 \leq \theta \leq 2.0°C \tag{1.1}$$

and

$$\nu_b = S_i \left(\frac{43.795}{\theta} + 1.189\right) \qquad -22.9 \leq \theta \leq -8.2°C \tag{1.2}$$

are used (note that different relations are used on the opposite sides of the $Na_2SO_4 \cdot 10H_2O$ crystallization temperature). At temperatures above and below the stated ranges it is necessary to refer directly to Assur's original table. It should be noted that both Assur's table and the above equations assume that sea ice has a constant density of 0.926 Mg/m^3. Therefore, to determine the applicable brine volume for sea ice having a known bulk density of ρ the calculated brine volume should be multiplied by ($\rho/0.926$). Equations incorporating this density change and allowing the calculation of both the brine and the gas volumes in sea ice of known densities can be found in Cox and Weeks (1983).

The best check on the adequacy of Assur's brine volume diagram is a study by Richardson and Keller (1966) in which the amount of liquid water present in artificially frozen seawater was determined as a function of temperature by the use of nuclear magnetic resonance (NMR) techniques. These values were then compared with a curve of the same parameter based on Nelson and Thompson's chemical measurements. The results are shown in Figure 5. The agreement is very good at temperatures above -43^oC (the $MgCl_2 \cdot 12H_2O$ crystallization temperature). At lower temperatures systematic differences were noted, with NMR giving the higher water contents. Fortunately, natural sea ice rarely exists at temperatures below -40^oC.

Based on the NMR study can one assume that the phase relations published by Assur are adequate at temperatures above -40^oC? This is a hard question to answer at present. The problems arise in two different but similar ways. First, as was mentioned, brine pockets may form, isolating portions of the brine and solid salts, and changing the sequence of possible reactions between these phases and the ice. Second, and probably more important, it is possible that as sea ice ages and brine drainage proceeds, the solid salts will remain fixed in the ice. This process would result in the enrichment of ions that are present in the early-formed solid phases. For instance, if the upper surface of sea ice were to be maintained at a temperature of -20^oC for a long period of time, $CaCO_3 \cdot 6H_2O$ and $Na_2SO_4 \cdot 10H_2O$ would form while all the Cl^- ion would remain in the brine. If brine drainage were to proceed, this could produce changes in the $SO_4^=/Cl^-$ and $CO_3^=/Cl^-$ ratios. If these changes were large enough the standard sea ice phase diagram of Assur would no longer be applicable, with appreciably different phase relations being exhibited by ices that have had different thermal histories. As the thermal history of most pack ice is relatively unknown, we would be reduced to either performing a complete chemical analysis on each sea ice specimen or developing a technique that could easily be used to determine brine volume directly. Both of these methods would presumably be much more time consuming than the present technique of determining the salinity, density and temperature of a sample and determining the amount of brine, gas and solid salt in the ice from Assur's tables or Cox and Weeks' equations.

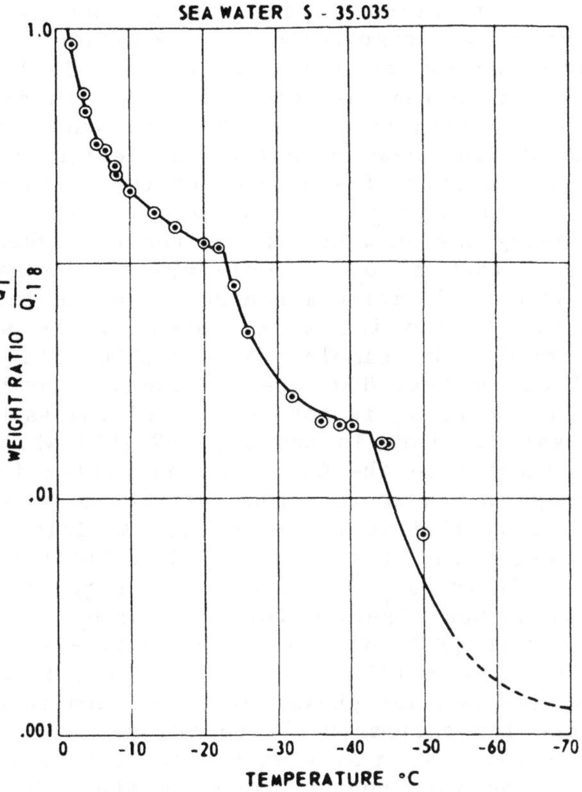

Fig. 5. Relative water content in percent versus temperature for sea ice (sea-water sample 35.035°/oo salinity). The curve is based on Nelson and Thompson's analytical data. The points are measured by NMR techniques. (After Richardson and Keller 1966).

Do significant changes occur in the ratios of the ions in sea ice? The evidence is mixed. In newly formed ice, and in first-year ice that has not been subject to low temperatures for long periods of time, most measurements (Bennington, 1963b; Blinov, 1965; Addison, 1977) suggest that major changes in the ion ratios are small. For thicker ice, and particularly for multiyear ice where the changes in ion ratios would be expected to be largest, experimental results are quite variable. The most thorough study of the problem is by Tsurikov (1974, 1976), who calculated the ratio

$$D_c = \frac{(c/Cl)_i - (c/Cl)_w}{(c/Cl)_w} \qquad (1.3)$$

where c is the concentration of the ion of interest and i and w indicate ice and water respectively. Positive values of D_c indicate that the content of ion c in the ice is higher than in the water. The chloride ion was chosen as the "standard" as it is retained completely within the brine at temperatures above -22.9°C and, therefore, should show extensive depletion via drainage. Figure 6 shows the relative frequencies of the D_c values calculated for several different ions. Although negative D_c values do occur, all the histograms show pronounced positive skews. The most interesting case is that of $D_{HCO_3^-}$; only 60% of the samples showed non-zero $D_{HCO_3^-}$ values indicating measurable changes in the relative ratios of (HCO_3^-/Cl) in the ice as compared to the seawater from which the ice formed. One sample showed a 2000-fold enrichment in HCO_3^- and several values near 800 were observed. That higher values might be observed for $D_{HCO_3^-}$ is not too surprising as $CaCO_3{\cdot}6H_2O$ is the first solid salt to form in sea ice (-2.2°C) which would presumably tend to immobilize the $CO_3^=$ ion, resulting in enrichment. However, such large values would appear difficult to explain. The other histograms show similar but much less striking trends, with near-zero values occurring in excess of 70% of the time. A 14-fold enrichment was noted for $D_{SO_4^{-2}}$, which is not surprising. However, a 30-fold enrichment was observed for D_{K^+}, which is surprising in that the first K salt (KCl) does not form until -36.8°C and therefore K^+ should be more mobile than Cl^-. The effects of these ratio changes on the sea ice phase relations, and in particular on current techniques for estimating brine volume, are not presently clear. Certainly this is an area worthy of further study. To add to the confusion, the most recent paper on the subject (Reeburgh and Young, 1982) obtained data that suggest an initial change in the $SO_4^=/Cl^-$ ratio during the formation of first-year ice but no change in the ratio for multiyear ice. As in most of the previous studies, the sampling was sparse (1 first-year and 1 multiyear site).

4. FORMATION OF THE INITIAL ICE COVER

In comparison to the rather voluminous literature devoted to the freezing of bodies of "fresh" water, little attention has been paid to the initial formation of sea ice. It is possible, however, by piecing together the available information on sea ice and comparing it with the "fresh" water observations, to arrive at a fairly complete qualitative picture of the gross features of the process. This initial crystal formation is unique in that, although the cold source is the overlying air, the velocity of crystal growth is probably largely limited by the rate of heat and solute dissipation through the liquid. It is, therefore, the only portion of the growth of a sea ice sheet that can be considered as true dendritic growth, i.e., that type of branched external crystal morphology that is characteristic of growth into "thermally" supercooled melts.

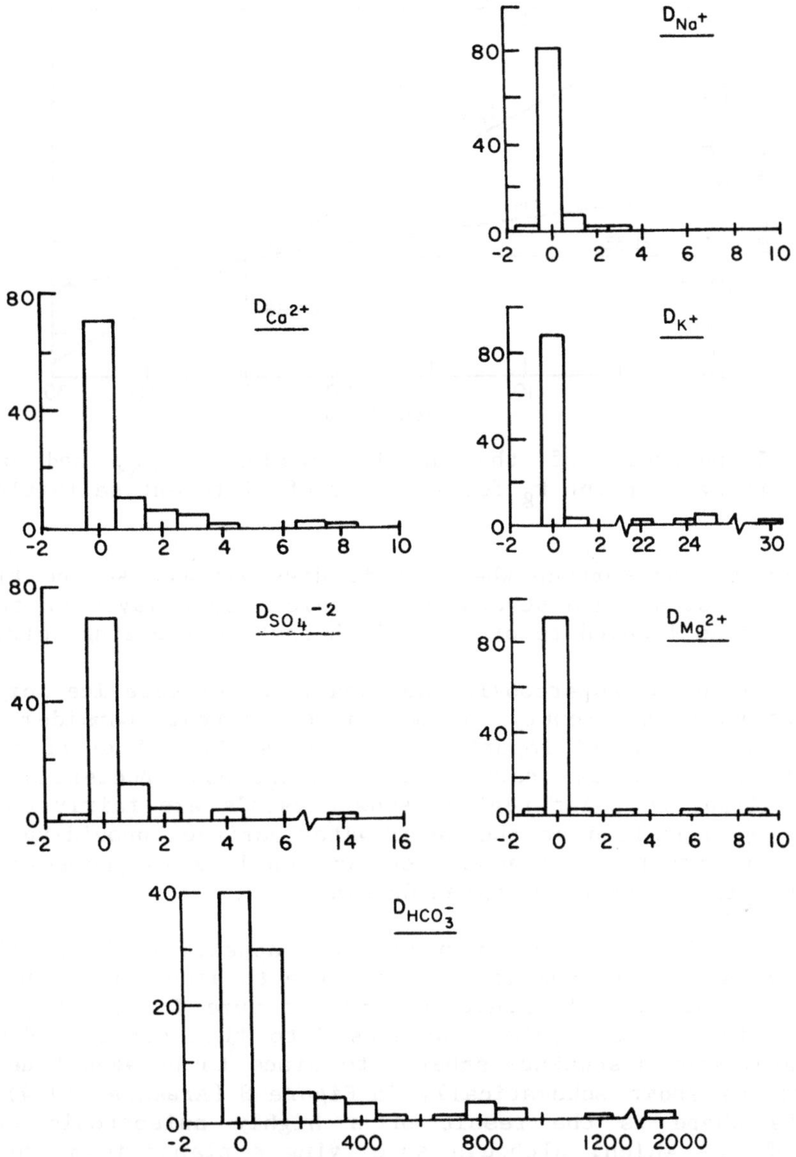

Fig. 6. Probabilities (expressed as percentage) of values for D_c for various ions (Tsurikov, 1974).

Because normal seawater has a salinity greater than 24.7°/oo, therefore possessing a freezing point higher than its temperature of maximum density (Figure 7), surface cooling creates an unstable vertical density distribution, causing convective mixing. This process transports the sensible heat stored in the lower layers of

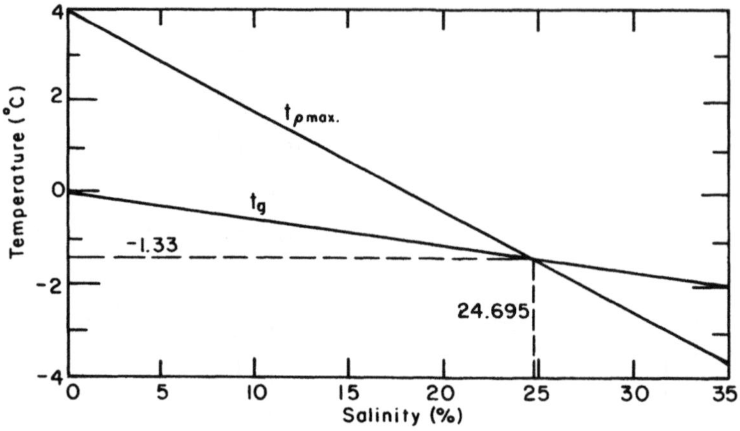

Fig. 7. Temperature of the density maximum t_{max} and of the freezing point t_g for seawater of different salinities.

the water to the surface where it is dissipated. At the time of initial ice formation a several-meter-thick upper layer of the sea will have been lowered to or slightly below the freezing point.

The amount of supercooling necessary to initiate ice formation in seawater has apparently not been investigated. Considering the large number of solid impurities in any small volume of natural seawater, it is almost certain that homogeneous nucleation never occurs. Also, in polar regions snow crystals are fairly continuously being deposited on the upper water surface, providing nuclei for further growth. The amount of supercooling is probably a few hundredths to tenths of a degree Celsius.

In general, the freezing of seawater appears to be similar to that of fresh water, and it is reasonable to assume that the first crystals to form will be minute spheres of pure ice (Hobbs, 1974). Growth rapidly changes these spheres into thin circular discs in the general growth sequence spheres to discs to hexagonal dendritic stars, as shown schematically in Figure 8 (Arakawa, 1954). The disc-like shape is the result of a highly anisotropic surface energy of ice which, although specifying a planar form, does not specify any particular growth direction in this plane. At small undercoolings, when heat and solute dissipation problems are not important, the growing disc can minimize its surface-to-volume ratio by maintaining a circular outline. That the growth rate of ice is strongly dependent on the growth direction is well known (Hillig, 1958, 1959; Macklin and Ryan, 1966). In ice discoids this plane of maximum growth is the (0001) or basal plane, which is the plane of maximum reticular atomic density in an ice crystal (Owston, 1958). Therefore, ice growth can be said to be well

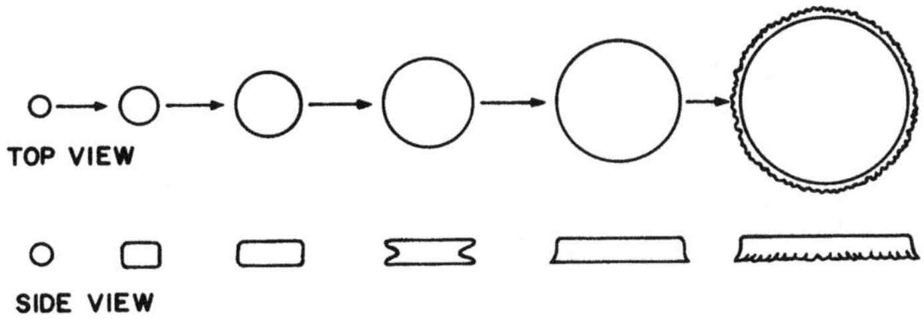

Fig. 8. Growth sequence for ice crystals in bulk water (Arakawa, 1954).

described by Bravais' law; the smaller the reticular density of a crystal surface, the faster it grows normal to itself. The final crystal consists only of those surfaces showing the slowest normal growth, the close-packed surfaces (Rosenberg and Tiller, 1957). Crystal growth parallel to the c-axis is orders of magnitude slower than growth in the (0001) plane, and the resulting kinetics are consistent with the classical Volmer-Stranski models, which are based on the assumption that two-dimensional nucleation and growth is the rate-controlling mechanism (Hillig, 1958).

The maximum diameter to which discs grow is, at least in fresh water, on the order of 2 to 3 mm, and is a function of the supercooling. A discussion of the morphological stability of such disc crystals, including a specific discussion of ice discs growing in pure water, is given by Fujioka and Sekerka (1974). Related work concerning the morphological stability of ice cylinders freezing from aqueous solutions has been published by Hardy and Coriell (1973).

Figure 9 shows a large number of such discs developing in the upper centimeter of seawater (observations were made at Thule, Greenland). Note the characteristic notched edges on several of the tilted plates. Limited experimental observations (Kumai and Itagaki, 1953) indicate that under comparable growth conditions the maximum disc size is depressed by the presence of an appreciable amount of solute in the water. Discs did, however, form in all cases when inorganic solutes were used (Arakawa and Higuchi, 1954). This decrease in maximum disc size is quite reasonable since it is necessary for an ice crystal growing in seawater to dissipate both heat and solute into the surrounding liquid. As the radius of curvature of the disc increases, the ease of heat and solute dissipation decreases, until at some critical radius -- determined by the thermal conductivity of the melt, the diffusion coefficient of the solute in the melt, and the growth velocity (Glen, 1955) -- the discoidal growth form becomes unstable, breaking up into a hexagonal dendritic star which again offers a much smaller tip radius of curvature. Although the change to a stellar

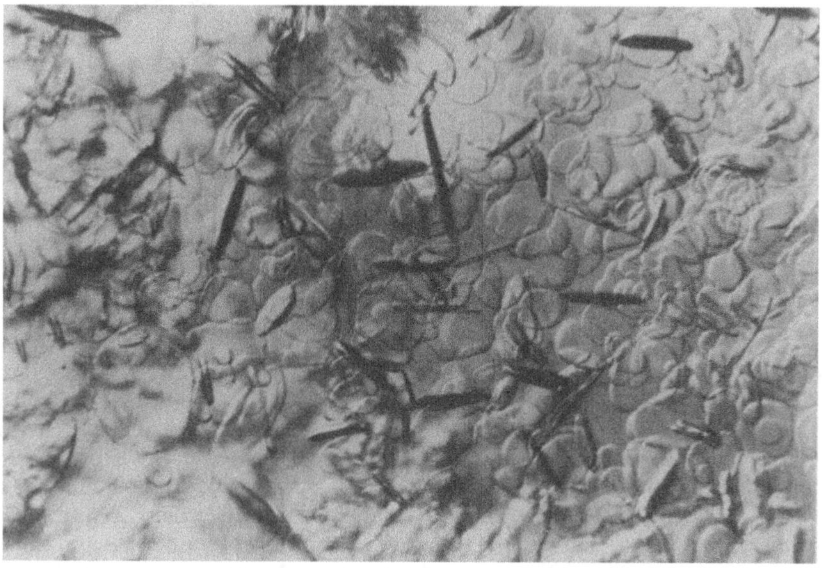

Fig. 9. Initial discs during the freezing of seawater. Disc size is approximately 1 mm (Thule, Greenland).

form causes an increase in the relative amount of surface, this is apparently compensated for by the fact that the crystal is now more readily able to dispose of heat and solute at the advancing interface (the so-called "point effect of diffusion"). The disc-to-star transition is therefore marked by an appreciable increase in the growth velocity (Kumai and Itagaki, 1953). That most, if not all, of the stellar crystals initially went through a discoidal stage can be seen by examining the centers of the stars under a high magnification (Figures 10 and 11). The arms of the stars form parallel to the a-axis direction $\langle 11\bar{2}0\rangle$ in the ice crystal, which, as was mentioned, is the shortest lattice vector in the (0001) plane.

Optically, these initial crystals, both discoidal and stellar, show a small but definite biaxiality. Optic angles of 3 to 5 degrees were observed in crystals from ice skims formed at Hopedale, Labrador, and Thule, Greenland. Golovkov (1936) reported 2V values as great as 20 degrees from crystals of Kara Sea ice. This biaxiality is undoubtedly produced by strains introduced in the crystals during their growth. Due to its low birefringence, ice is optically extremely stress-sensitive.

The star-like crystals grow rapidly across the surface of calm seawater until they overlap and freeze together, forming a continuous thin ice skim. Because of the tabular nature of both

Fig. 10. Stellar ice crystals growing in seawater.

Fig. 11. Stellar ice crystals grown from seawater.

the initial ice discoids and stars, they float with their basal planes (0001) in the plane of the water surface. Ideally this would produce an ice skim with a completely c-axis-vertical orientation. However, since water conditions are rarely completely calm a number of discoids or stars are usually caught in an intermediate position, i.e. with their c-axes inclined at some angle to the vertical. This is well shown in Figures 9 and 12 which are photomicrographs of initial ice crystal formation in seawater of 32°/oo salinity. A petrofabric diagram of such a skim will appear similar to the first fabric diagram in Figure 23, a strong c-axis-vertical maximum and a few crystals in other orientations.

During the initial freezing of fresh water at low supercoolings, surface needles are extremely common (Fujino and Suzuki, 1959; Hallett, 1960). The needles form when a disc becomes inclined at an angle to the water surface. Subsequent growth then proceeds rapidly, in the form of long, thin, needle-like crystals, from the points where the disc intersects the surface. The needles appear to be confined strictly to the thin surface layer of water that is appreciably supercooled. If crystal growth were to continue downward along the basal plane, the crystal would immediately encounter water that is not supercooled and the growth would stop. If larger supercoolings are encountered more complex morphologies develop -- simple double pyramids between -2.7 and -5.5°C and, at yet lower temperatures, complex double pyramids showing secondary and higher order non-rational growth directions (photographs of such crystals can be found in Macklin and Ryan (1966) or in Hobbs (1974)).

The limited observations available on the freezing of sea- and salt-water indicate that although surface needles are occasionally encountered, they appear to be rarer than in fresh water

Fig. 12. Initial ice crystal formation in calm seawater.

(Fukutomi et al., 1949; Arakawa and Higuchi, 1954; Suzuki, 1955; Fujino and Suzuki, 1959). This observation is readily explained by the fact that prior to the nucleation of the initial sea ice crystals, convective mixing has lowered an appreciable thickness of the water layer to the freezing point. This makes it possible for inclined discs to grow both downward as well as along the water surface. A significant downward growth component causes the disc to develop as a dendritic star instead of a needle, and also produces a moment due to buoyancy forces which tends to rotate the stellar crystal until its (0001) plane is parallel to the water surface. Possibly needle-like growth in seawater occurs only when part of the needle has frozen to or is spatially restricted by other ice crystals, so that the needle is prevented from rotating when a downward growth component develops. Figure 13 is a photomicrograph of the upper surface of a thin ice skim showing the tip of such a needle (the surrounding ice is of an essentially c-axis-vertical orientation). Figure 14 is an inked copy of a rubbing of the upper surface of a similar ice skim. This technique, which has been successfully applied to both glacier and lake ice for grain size measurements and texture analysis (Seligman, 1949; Ragle, 1962), utilizes the fact that projections on the ice surface appear as lines or "ticks" on the rubbing, while "smooth" (0001) surfaces appear as white or uniformly shaded areas. Each group of parallel lines represents a single crystal of sea ice. Figure 15 shows schematically the relation between the ice surface topography and the lines that appear on the rubbings. The majority of the surface of the ice skim in Figure 14 is composed of

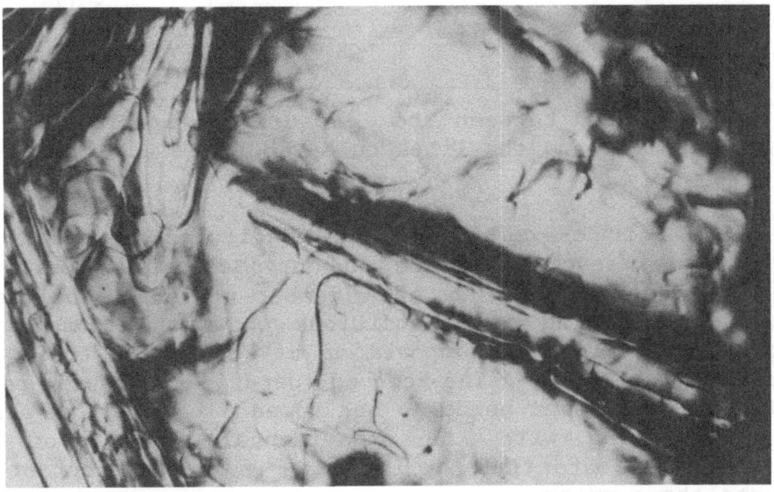

Fig. 13. A needle surrounded by crystals with their c-axes vertical (normal to the plane of the photograph) (right 2/3 of photograph).

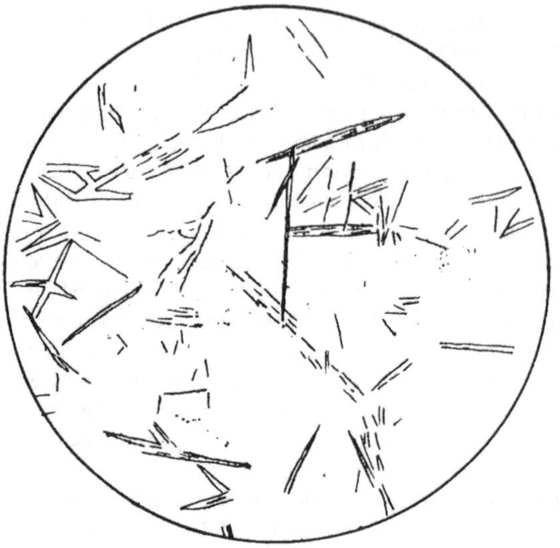

Fig. 14. Inked copy of a surface rubbing of an ice skim.

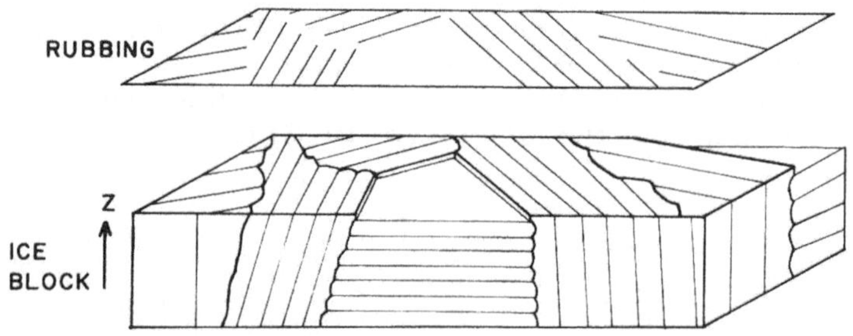

Fig. 15. The relation between the ice surface topography produced by etching and the lines on a rubbing.

crystals with their c-axes roughly perpendicular to the freezing surface, although several needles and crystals with their c-axes inclined are present. A study of a number of such rubbings from ice skims formed during calm conditions shows that commonly >50% of the upper ice surface is composed of crystals with their c-axes approximately vertical. The curved dendritic growth patterns often noted on freshwater ponds do not appear to occur in sea ice. This is in keeping with the experimental results of Knight (1962a), which indicate that an increase in solute content tends to eliminate the curvature.

Available experiments suggest that the ice crystal morphologies developed during the freezing of pure water and of aqueous solutions are similar. Also, the presence of solutes appears to reduce the supercooling necessary for crystal growth to occur in non-rational crystallographic directions (Ryan and Macklin, 1968; Ryan, 1969). Therefore, one might expect to find reports of single and double pyramidal crystals forming during the freezing of seawater. We are not aware of such reported occurrences. We can, therefore, conclude that either natural supercoolings are not sufficient for pyramidal growth to develop or that no one has bothered to look. Both are probably true.

Crystal growth processes like those described above occur during calm, cold periods when there is no turbulence in the upper layer of the sea. The resulting ice cover "presents a smooth unbroken surface on which there are no highly evident horizontal changes in the structure of the ice layer" and has been termed "sheet ice" by Wilson et al. (1954). Commonly, however, in the open ocean there is some wave-induced turbulence during initial ice formation. This mixing introduces more nuclei into the area of active freezing and provides the energy to overcome the buoyancy forces so that the initial crystals may be "stirred" throughout a depth of up to several meters (Savel'ev, 1958; Martin and Kauffman, 1981). The effective supercooling is reduced, more crystals form per unit volume, and abrasive action between crystals is increased. As a result, extensive discoidal growth is favored. Even if stars were to form, their arms would commonly be broken off. It should be mentioned that in this freezing mode it has been found that it is not necessary for each individual crystal to nucleate separately.

Wind action herds the resultant ice crystals, which are generally termed frazil, into agglomerations that take on various forms. For instance, grease ice is an agglomeration of frazil crystals into a soupy layer that can be distinguished as a result of the low reflectivity, matte-like appearance of its upper surface (Figure 16). Its major distinction from thicker and/or more compact agglomerations of frazil is in its mechanical behavior where it exhibits viscous fluid-like properties as opposed to the rafting and bending behavior of thicker or more compact floes. Martin and Kauffman (1981) describe a wave-damping experiment indicating that a general concentration of about 40% by volume of frazil ice crystals is necessary for the mechanical properties to transition to more solid characteristics. These wind-driven frazil accumulations can reach thicknesses of 1 m as opposed to the 0.01- to 0.1-m-thick frazil accumulations that occur before transitioning to columnar ice when quiescent conditions (no wind or waves) are present.

Pancake ice, roughly circular pieces of new ice with upturned edges, also results from frazil accumulations (Figure 17). During the initial stages of the development of pancakes from frazil ice the pancakes are only semi-consolidated slush. When touched by a

Fig. 16. Grease ice forming on the surface of a lead and being driven by the wind into its downwind end (Martin and Kauffman, 1981).

Fig. 17. Pancake ice forming from frazil crystals (Dunbar and Weeks, 1975).

probe this will break up into horizontally floating discoids that will not sustain their own weight when removed from the water. The areas between the pancakes are usually relatively free of crystals and act as source areas for the growth of new crystals. Wave action then causes oscillatory motions that result in repeated contacts and separations between the floes, pushing the newly formed frazil crystals up to build the characteristic raised pancake rims. Bonding between frazil particles takes place when they are brought into contact, and is also a necessary process in forming these features. These intercrystal bonds develop by deformation and by surface energy forces (described in Hobbs (1974) and Colbeck (1979)). Two effects weld the particles. The first is regelation, where the force pressing the particles tends to locally depress the melting point, leading to heat flow away from and freezing at other contacts. As well, when the stress is removed, the unfrozen contact will change its freezing point and refreeze. The second mechanism is dependent on the radius of curvature and generally takes place in the unsaturated or above-sea-level portions of the pancake ice. Small particles in contact with the water and vapor that is present are thermodynamically unstable with respect to the equilibrium pressures between solid-liquid, solid-vapor and liquid-vapor at the surfaces of larger particles. The imbalance in pressures tends to deform the ice particles, leading to a flattened solid interface between them and causing them to be welded together at a solid boundary. A distinction between grease ice and pancake ice in the formation process may be related to the ability of the pancake floes to drain somewhat and increase the likelihood of freezing contacts developing by the non-equilibrium pressures due to radius of curvature effects between particles. Generally, in the completely saturated case when the ice crystals are immersed in seawater, bond development is not as dramatic as in the unsaturated compacts. These processes are, however, invoked cautiously as speculative models since they have been developed primarily to explain wet snow properties. The specific application to seawater-ice crystal systems has not been made, and since the thermodynamics is affected by dissolved impurities a complete formulation of the problem is necessary with the seawater parameters specified. It seems clear, however, that the physical processes are qualitatively similar and that bonding between particles is necessary to build these ice forms. When such crystals are packed together into pancakes by wave motion and the movement of slush pancakes against one another, visual observations show that there is some tendency for the discoids to be stacked in a vertical position (c-axis horizontal) so that a closer lateral packing of discoids is permitted. The degree of orientation produced by this packing process is, of course, far from perfect, and in many cases the overall orientation may be close to random. Unfortunately, fabric diagrams of this type of ice are apparently lacking. The initial ice cover that results when a slush layer congeals is usually fine-grained equigranular in texture. A series of excellent vertical and horizontal photomicrographs of such ice are shown in Tabata and Ono (1957, Figures 10 and 11-1 to 11-6).

The general shape of the initial pancakes depends upon their position in relation to the wave motion and shore. In the open ocean, away from the edge effects of shore, the pancakes tend to be circular in outline due to constant abrasion by other pancakes, which removes sharp corners. Close to shore the pancakes are commonly unequidimensional, with their long axes parallel to the shoreline.

During periods of wave action, as the growing ice sheet moves out from shore there is commonly a definite sequence in its development. There is a change from open water containing frazil crystals to grease ice to agglomerating pancakes and grease ice to loosely joined pancakes and finally to a complete sheet of joined pancakes. If at any given time during this process turbulence subsides, grease ice will form between the pancakes, producing a composite ice sheet (Figure 18). Also, after the initial ice sheet has formed, sheet ice will develop beneath the pancakes and grease ice. Photographs illustrating the different aspects of the formation of both sheet and composite ice sheets can be found in Weeks and Lee (1958, 1962) and the illustrated ice glossaries (Armstrong et al., 1966; U.S. Navy Hydrographic Office, 1952; Rodhe, 1959).

Fig. 18. Composite ice sheet of pancakes and sheet ice, Gulf of St. Lawrence (Dunbar and Weeks, 1975).

In larger open-water regions, such as polynyas and open ocean-ice edge regions, cold, windy conditions cause grease ice and frazil crystals to stream out approximately parallel to the wind in long ice plumes. Figure 19 shows these plumes developing at surface temperatures of about -20°C and wind speed of 15 m/s in a polynya near Nome, Alaska. Dunbar (private communication; quoted in Martin and Kauffman (1981)) first suggested that a Langmuir circulation causes this organization of grease and pancake ice. Figure 20 schematically shows such a circulation. At the downwind head of such plumes a broadening occurs, usually consisting primarily of pancake ice. Such features have been dubbed "tadpoles" by Dunbar and Weeks (1975) because of their broad head - narrow tail shape. When formed by the offshore winds along shorelines or ice shelves these ice plumes eventually either disappear across regions of open ocean with warm conditions or form the relatively intact ice covers of larger fields of sea ice. Estimates of the

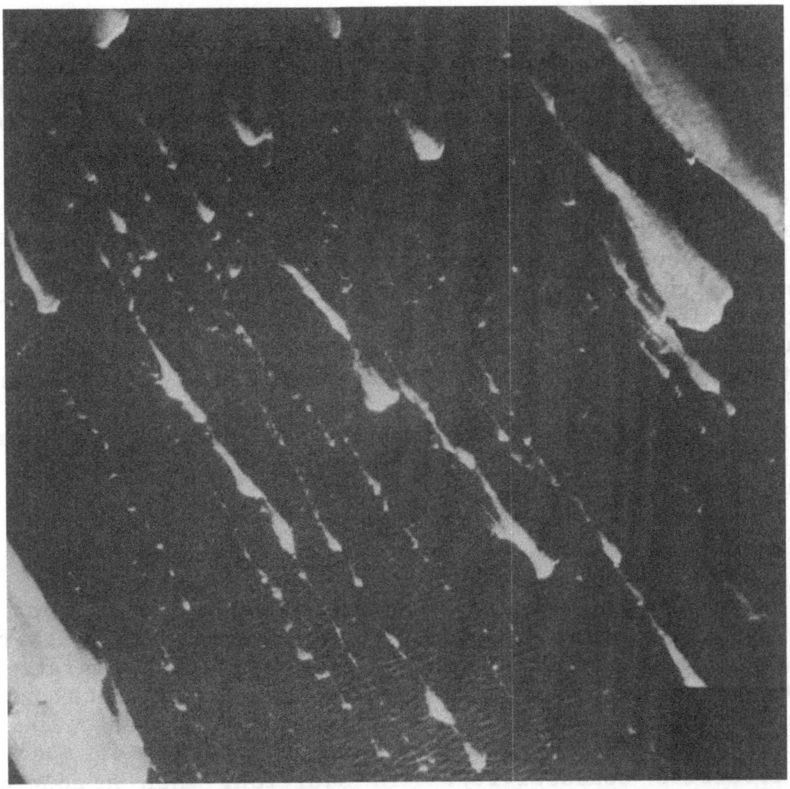

Fig. 19. Aerial photograph from elevation of 147 m of grease ice showing tadpole shapes in a large polynya in the Bering Sea (Martin and Kauffman, 1981).

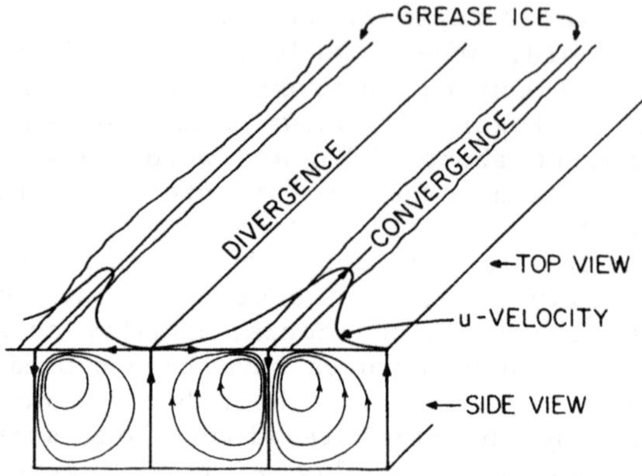

Fig. 20. Schematic drawing of the general effect of Langmuir circulation on grease and pancake ice (Martin and Kauffman, 1981).

ice production in such polynyas are large, with the possibility that an equivalent thickness of 2 m of ice can develop in as little as 20 hours, as opposed to an entire season's growth of 2 m of ice that would form under undisturbed conditions (Martin, 1981).

As will be seen later, frazil ice formation can influence deep ocean circulation. It does this through salt fluxes caused by high ice production rates and through mechanical effects like the Langmuir circulations. Based on our most recent observations of Antarctic ice covers, frazil ice appears to be much more widespread than was previously thought. The mechanisms of ice-ocean interaction that contribute to frazil ice other than the initial forms we have just discussed are just being developed. We discuss the evidence for these other features and for possible mechanisms that could lead to increased occurrences of this ice type in a later section.

5. THE TRANSITION ZONE

Once a continuous skim of ice has formed across the sea surface, the possibilities for crystal growth due to purely "thermal" supercooling are greatly reduced. An ice skim now separates the melt (seawater) from the cold source (air). The latent heat is, therefore, extracted entirely through the ice sheet, and the growth rate is determined by the temperature gradient in the sheet and its effective conductivity. In addition, when a continuous ice layer has formed, the ice crystals lose a degree of growth freedom: only if the grain boundaries are exactly perpendicular to the freezing interface can crystal growth proceed without one grain interfering with the growth of another. Any tendency for anisotropic growth will produce geometric selection, with the crystals in the favored orientation eliminating the crystals in the unfavored orientation by cutting them off from the melt. Since "free-floating" ice crystals, forming both from fresh- and seawater, show pronounced anisotropic growth (ratios as high as 100:1; Hallett 1960), with the plane of maximum growth parallel to the (0001) crystallographic plane (Hobbs, 1974), it is reasonable to suppose that geometric selection will occur. In fact, this phenomenon always does occur in natural ice sheets, and the ice layer in which the preferred orientation "wins out," producing a characteristic growth fabric, has been termed the transition layer by Perey and Pounder (1958). In sea ice, the orientation change is usually essentially completed within a vertical distance of 5 to 10 cm, although the base of the transition layer may be located at distances in excess of 1 meter below the upper ice surface if the initial ice sheet is composed of a thick layer of frazil (Pounder and Little, 1959; Weeks and Gow, 1980).

A typical rubbing of the bottom surface of a 2.5-cm ice skim in which the process of geometric selection is obviously occurring is shown in Figure 21. The blank plane-sided polygonal areas are where c-axes vertical or near-vertical crystals occur. These crystals grow appreciably slower than the surrounding c-axes

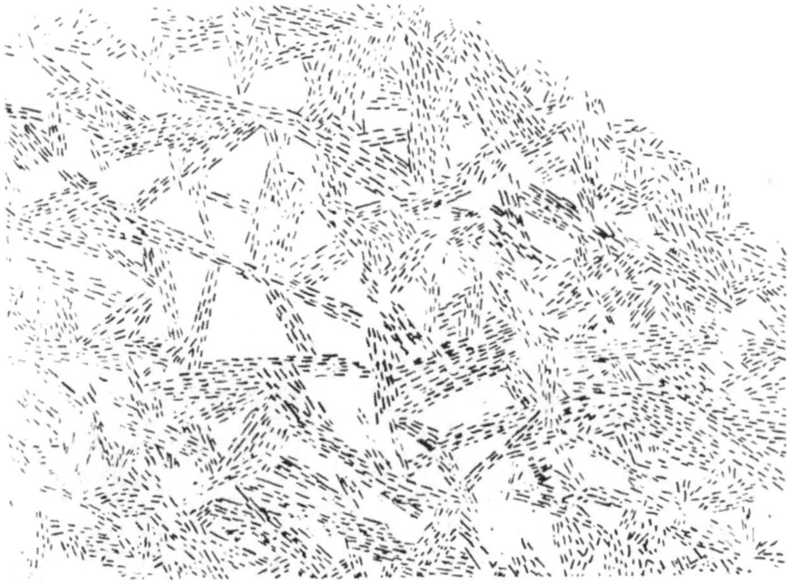

Fig. 21. Rubbing of the bottom of a 2.5-cm-thick ice skim that is undergoing geometric selection (Hopedale, Labrador, 1955).

horizontal crystals. They therefore form depressed (2- to 5-mm) areas on the bottom surface of the ice skim and do not show in the rubbings. A photomicrograph of the bottom surface of an ice skim showing such a depressed area due to the presence of a c-axis vertical crystal is shown in Figure 22. If this ice were allowed to continue growing, the crystal in the center of the photomicrograph would be rapidly eliminated by the surrounding crystals as a result of selected growth parallel to the (0001) plane. Once the c-axes vertical crystals are eliminated, rubbings from the lower levels in the ice sheet are similar to Figure 27.

A more quantitative picture of this process can be obtained by making a series of c-axis orientation measurements at different levels in the ice sheet using a 4-axis universal stage. Figure 23 shows these results plotted on a Schmidt net from a layer of sheet ice that formed during a calm night. It is quite clear that the majority of c-axes vertical crystals are eliminated in the upper 1 cm of the ice sheet. By the 5-cm level, all the remaining crystals have their c-axes within a few degrees of the horizontal. Since the azimuthal distributions of c-axes in Figure 23 and in other studied examples (Tabata and Ono, 1957) did not show any pronounced pattern, it was assumed that they are random and only the angle between the c-axis and the vertical need be considered. Figure 24 is a diagram prepared on the basis of this assumption. It clearly shows that in the initial skim, as well as in the top

Fig. 22. Photomicrograph of the lower surface of sea ice showing a triangular crystal being eliminated by geometric selection.

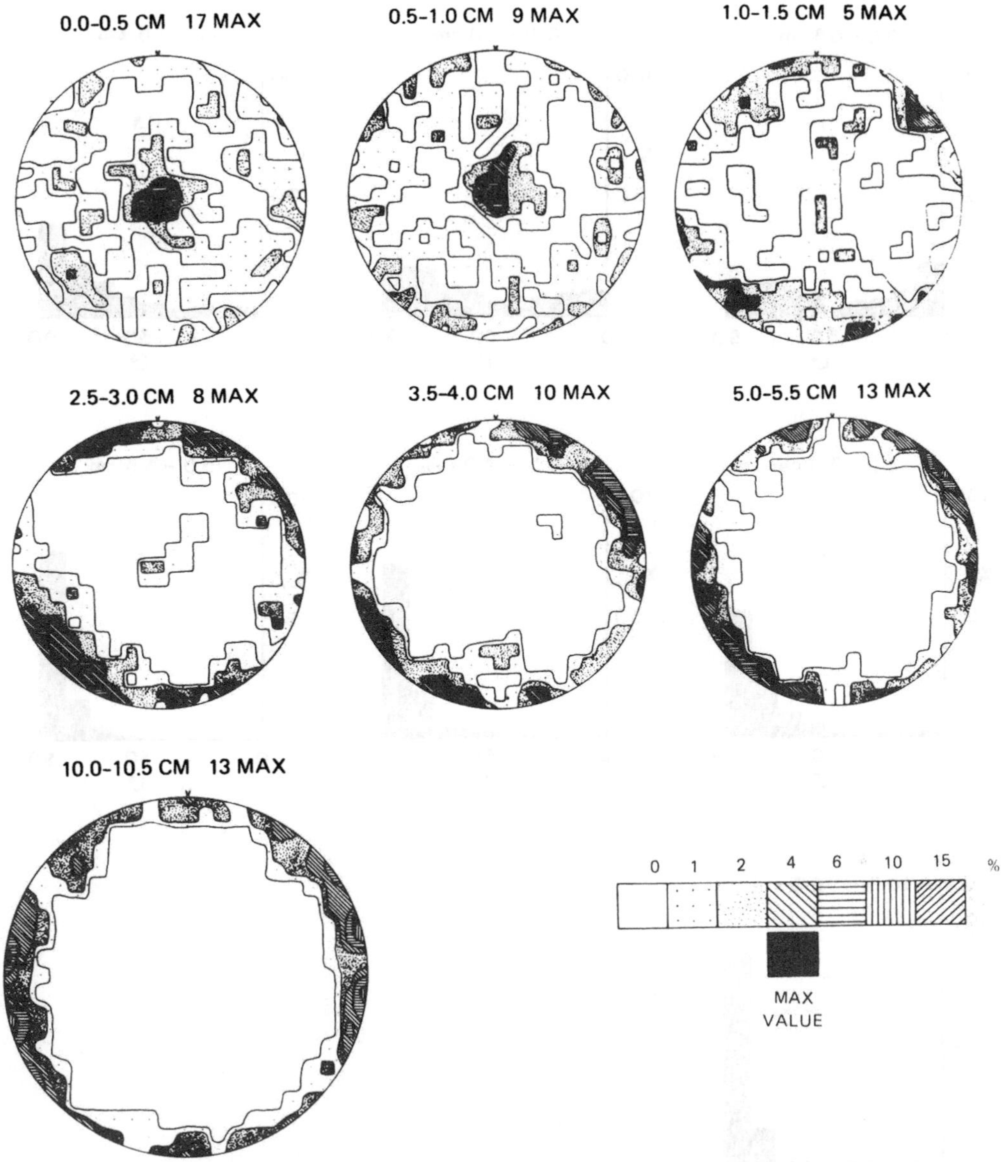

Fig. 23. Orientation of the c-axes of sea ice crystals from different depths below the upper surface of an ice sheet (Thule, Greenland, 1957) plotted and contoured on the upper hemisphere of a Schmidt net. The diagrams are in the horizontal plane. Contour intervals (per 1% circle) as indicated.

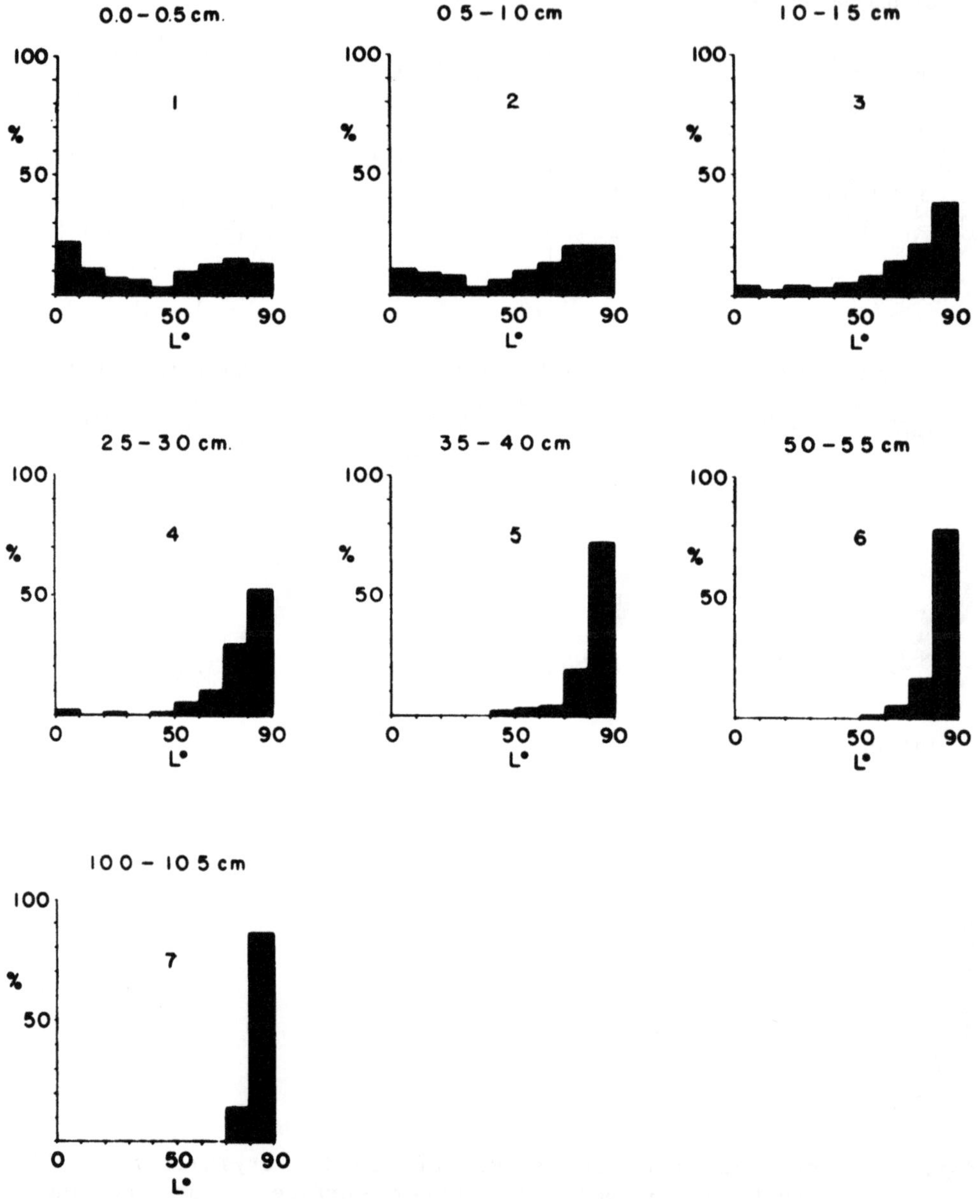

Fig. 24. Histograms showing the relative percentage of different c-axis orientations (0° = vertical, 90° = horizontal) in the ice studied in Figure 23.

1.5 cm of the ice sheet, ice crystals exist with all different orientations. Therefore, to form the characteristic c-axis orientation of the columnar zone (below the transition layer), it is not necessary to nucleate new grains with a c-axis horizontal orientation. The new orientation presumably forms by the survival of those grains with the favored orientation that are present in the initial ice skim. Similar conclusions have been reached from the study of the solidification of metal ingots (Walton and Chalmers, 1959).

The process of geometric selection in ice can easily be rationalized in terms of crystal growth theory. The theory predicts that atoms from the melt arriving at a crystal surface will have a much higher probability of remaining if they arrive at a step in the surface than if they arrive at an atomistically smooth face. Therefore, growth normal to the closest packed plane [the (0001) plane in ice] requires either the existence of step-producing defects or the nucleation of an embryo that will then grow laterally to form a new plane of atoms. The activation energy for this latter process is quite high, necessitating a larger supercooling, δT. These conclusions have been verified for ice by Hillig (1958, 1959), who found that the kinetics of ice growth parallel to the c-axis was consistent with the classical two-dimensional nucleation and growth model, while the growth kinetics parallel to the (0001) plane could be explained by a screw dislocation model (Hillig and Turnbull, 1956).

When the initial downward growth starts in the ice sheet, the crystals in the most favorable orientations rapidly grow ahead of crystals with less favored orientations since they 1) can grow more rapidly at any given supercooling -- at the same growth velocity, growth in a high index direction generally occurs at a much smaller δT than growth in a low index direction -- and they 2) have to grow a smaller linear distance to advance the same distance parallel to the axis of heat flow. In addition, the latent heat evolved and the solute rejected by the more rapidly growing crystals reduces the effective interface supercooling of their slower neighbors. This selective growth process is shown diagrammatically in Figure 25 (Perey and Pounder, 1958).

In the final growth orientation, all the (0001) planes are aligned approximately parallel to the thermal gradient (Figure 23, diagrams from depths of greater than 5 cm). This orientation has been found repeatedly in studies of both sea and NaCl ice (Shumskii, 1955; Tabata and Ono, 1957; Perey and Pounder, 1958; Weeks and Lee, 1958; Weeks and Assur, 1963; Langhorne, 1980; Weeks and Gow, 1980). In sea ice, because the heat loss into the atmosphere is unidirectional there is no obvious reason why the (0001) planes should have any particular orientation in the horizontal plane of the ice sheet. Early field observations appeared to

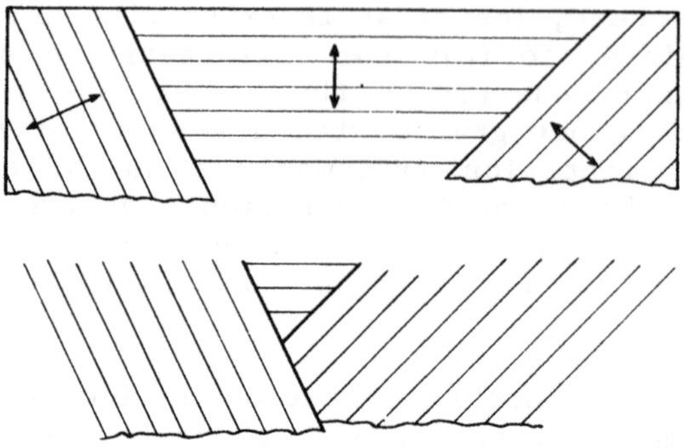

Fig. 25. Schematic diagram showing the process of geometric selection according to Perey and Pounder (1958). The arrows indicate the direction of the c-axes.

verify this, at least for relatively thin ice (Tabata and Ono, 1957; Weeks, 1958).

However, if the heat loss were two-directional, as it is during the refreezing of small vertical cracks in sea ice, then heat would be lost both to the air and to the cold ice that composes the sides of the crack. The ice crystals that form would therefore be expected to have their c-axes horizontal and parallel to the axis of the crack. Figure 26 is a diagram showing the c-axis orientations observed in healed sea ice cracks at Hopedale, Labrador. The orientations are exactly as expected. Similar observations from lake ice have been reported by Taylor and Lyons (1959, p. 13). In fact this pronounced orientation in the refrozen crack explains why once a crack heals it commonly does not refracture in tension but new cracks develop parallel to it (the

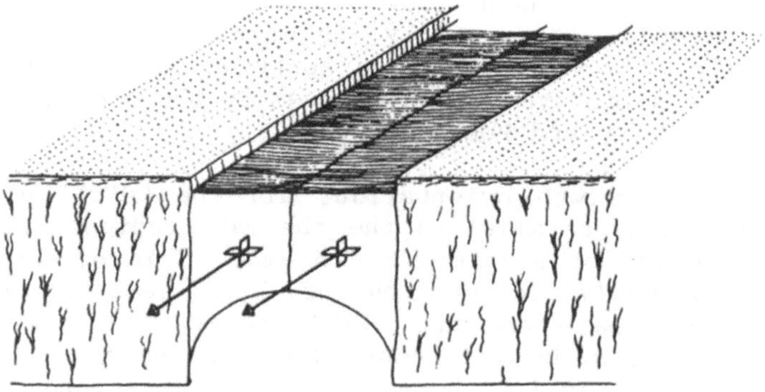

Fig. 26. c-axis orientation in a healed crack at Hopedale, Labrador.

crack has a crystal orientation that gives the highest possible tensile strength in the direction normal to the crack).

The percentage of a winter's ice sheet composed of transition zone ice is small. But the importance of this zone exceeds its size. By studying the transition one can examine the processes that control the structure of the underlying columnar zone, the zone which contains the majority of the sea ice.

6. THE COLUMNAR ZONE

Below the transition layer in sea ice, the ice has all the characteristics associated with the so-called "columnar" zone in metal ingots (i.e., a strong crystal elongation parallel to the direction of heat flow, a pronounced crystal orientation, and an increase in crystal size over crystals closer to the cold source (Walton and Chalmers, 1959)). Therefore, it is useful to apply this term to sea ice. Because the transition layer is thin and occurs at or near the ice/air interface, it commonly sublimates away or recrystallizes into the snow cover as the ice sheet grows. Therefore, the complete thickness of a one-winter sheet of ice can, to a good approximation, be considered as being in the columnar zone. Compared with the pronounced changes occurring in the transition layer, the changes in the columnar zone are considerably more subtle.

Fabric diagrams from the columnar zone are extremely monotonous, as shown by Figure 23 (diagrams below 5 cm). All c-axes are oriented within a few degrees of the horizontal plane. An inked rubbing of columnar zone ice showing the grain boundaries and the trace of the (0001) plane for each crystal is presented in Figure 27. This rubbing was made just below the transition layer on the lower surface of a 5-cm ice skim formed at Thule, Greenland. The grain boundaries are highly sutured, the details of the boundaries corresponding to the locations of individual platelets. Even with this suturing, it is usually quite easy to distinguish the boundaries of individual crystals. Suturing is usually clearly evident both in horizontal rubbings (Figure 27) and thin sections (Figure 35) and appears to be a common characteristic of ice grown with a dendritic solid/liquid interface (see Knight, 1962d, p. 323). A histogram showing the angles between the c-axes of neighboring crystals from the rubbing in Figure 27 is presented in Figure 28. If the diagram were isotropic, the theoretical frequency in each class would be 37 (11.1%). A chi-squared test shows that there is no reason to doubt that this fabric is isotropic at the 0.05 significance level. Similar results have been obtained by Tabata and Ono on thin sheets (<33 cm) from Hokkaido, Japan.

As compared with horizontal rubbings, vertical rubbings in the columnar zone are relatively uninformative. A typical example

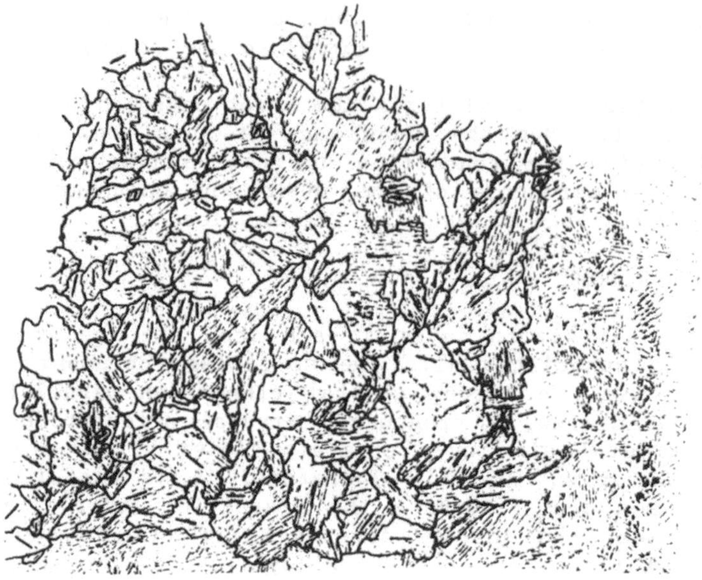

Fig. 27. Rubbing from just below the transition layer on a 5-cm-thick ice sheet at Thule, Greenland. The inked "tick" marks indicate the trace of the (0001) plane.

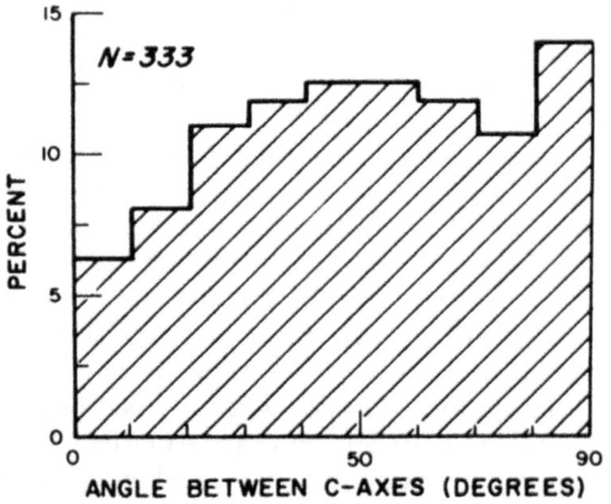

Fig. 28. Histogram giving the angles between the c-axes of neighboring ice crystals shown in Figure 27.

is shown in Figure 29. The lines reflect the vertical orientation of the plate-like sea ice structure. The change in horizontal spacing of the lines is produced by changes in both the plate spacing at different positions in the ice sheet and the angle at which the plates intersect the plane of the rubbing.

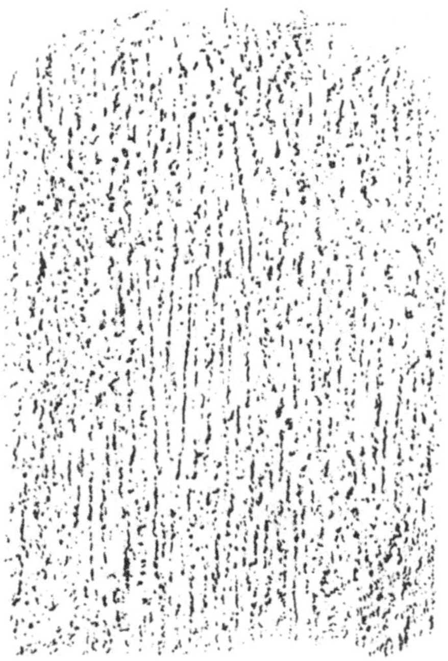

Fig. 29. Vertical rubbing of columnar zone ice.

Although the c-axis orientations of sea ice have received considerable attention, the orientations of the a-axes have not been examined, as they are harder to measure. The one exception to this is a recent study by Kawamura and Ono (1980), who studied both c- and a-axis orientations in sheets of thin (4.0 to 8.5 cm), newly formed sea ice at Barrow, Alaska. Measurements were made at two different levels in the ice. Although formally both these samples would probably be considered to be from the transition zone, the samples from near the bottom of the ice sheets showed strong c-axis-horizontal orientations (88%). The relative frequency of the a-axis orientations of the different grains from the bottom centimeter of the sheets is shown in Figure 30. The vertical angle between the horizontal and the a-axis nearest the ice surface is indicated by the symbol theta. As can be seen, the most frequent a-axis orientations are either horizontal to near-horizontal (38%) or at 30 degrees (17%). This suggests that the favored a-axis growth direction is close or equal to either $\langle 11\bar{2}0 \rangle$ or $\langle 10\bar{1}0 \rangle$.

We would like to suggest that once the columnar zone has developed and all the remaining crystals are oriented with their c-axes essentially horizontal, it is the crystal with the most favorable a-axis orientation that has the growth advantage and will ultimately dominate at the interface. We would guess that

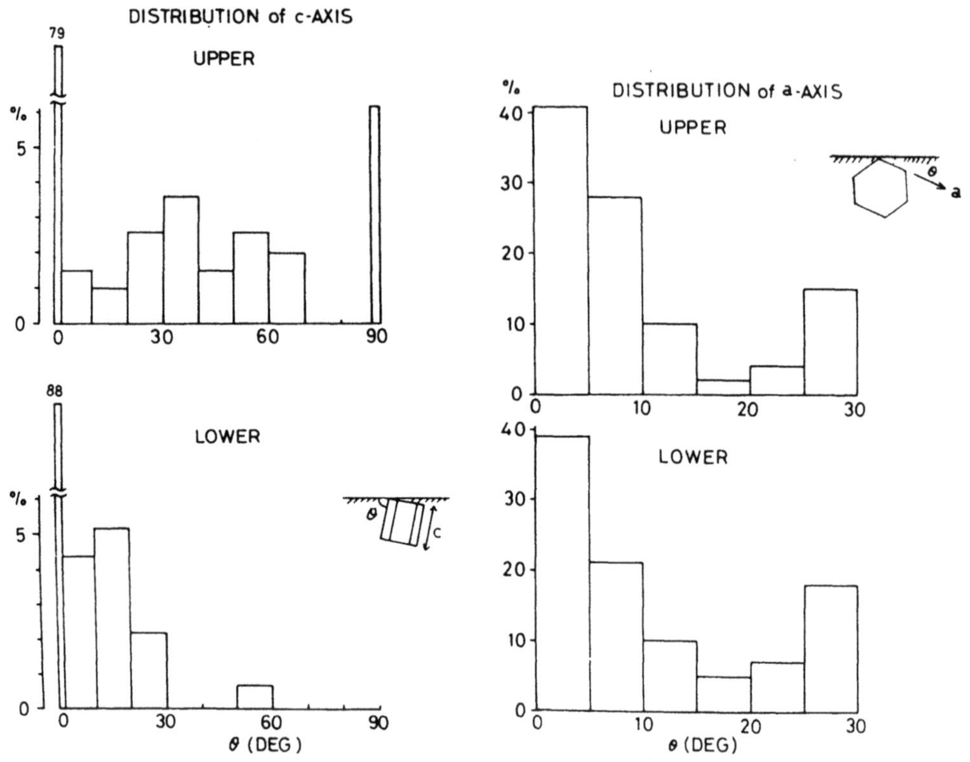

Fig. 30. The distribution of c- and a-axis orientations in the upper (4.0-7.7 cm) and lower (17-22 cm) portions of thin ice sheets examined at Barrow, Alaska (Kawamura and Ono, 1980).

the favored growth direction would be $\langle 11\bar{2}0\rangle$, that is parallel to the a-axis as opposed to normal to the first-order hexagonal prism $\langle 10\bar{1}0\rangle$. We would also guess that the growth advantage of $\langle 11\bar{2}0\rangle$ would be small, and that a strong $\langle 11\bar{2}0\rangle$ orientation in the vertical direction would take a prolonged period of ice growth before it would dominate. It should not be difficult to check these contentions by using etch techniques on samples collected from thick undeformed first-year and multiyear sea ice.

Another interesting aspect of the columnar zone is the variation in crystal size. The general relation between the maximum length L and the maximum width W of a number of crystals measured in horizontal thin sections has been studied by Weeks and Hamilton (1962) using relatively thin (31.4 cm) ice from Point Barrow, Alaska. Length measurements were always taken parallel to the (0001) plane and width measurements parallel to the c-axis. The data are reasonably well fitted by a linear regression line giving approximately a 2:1 length-to-width ratio. This indicates that the ratios of sidewise to edgewise growth of the individual crystals do not differ nearly as much as might be expected from the studies of growth anisotropy in free-floating ice crystals. Figure 31 shows a plot of the length-width values of a number of crystals measured from horizontal thin sections made at 5-cm intervals through the upper 60 cm of the fast ice at Thule, Greenland. Here <u>L</u> refers to the maximum dimensions of the crystal and <u>W</u> the value measured at right angles to <u>L</u> (i.e., the crystallographic orientation was not considered). Therefore, by definition, the L:W ratio can never be less than 1. Keeping this difference in mind, the Thule results are quite comparable to those from Point Barrow.

The most noticeable change in the upper part of the columnar zone is the marked increase in grain size with increase in depth in the ice sheet. This change is clearly shown in Figure 32 (Weeks and Hamilton, 1962), which plots the number of crystals per square centimeter at different depths in the ice sheet. Once the rapid grain size change at the base of the slush layer is

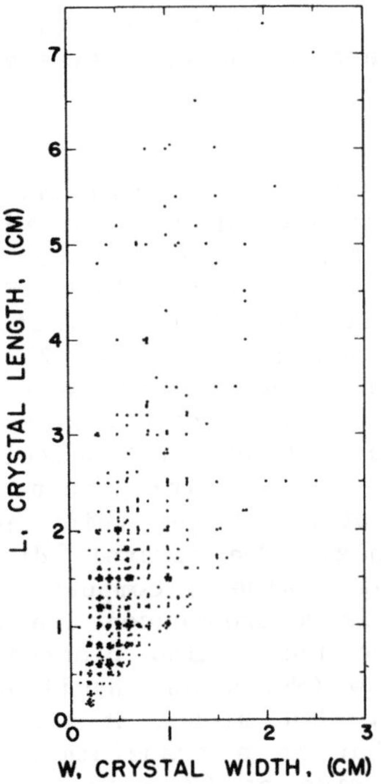

Fig. 31. Lengths and widths (in the horizontal plane) of ice crystals at different levels in a sea ice sheet at Thule, Greenland.

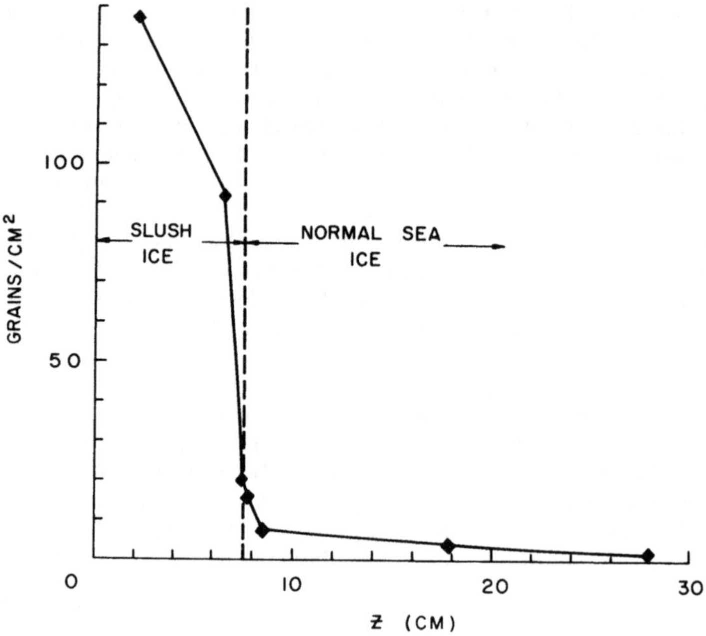

Fig. 32. Increase in crystal size with distance below the surface of an ice sheet at Barrow, Alaska (Weeks and Hamilton, 1962).

completed, there is a slow but regular increase with depth. The grain size curve in this ice sheet is quite comparable to grain-coarsening effects observed in the columnar zones of both pure and impure metals (Walton and Chalmers, 1959, Figure 11-12, p. 451). The change in the grain size characteristics has been studied in some detail from the upper 60 cm of a 125-cm-thick ice sheet at Thule, Greenland. The measurements were stopped at the 60-cm level because a large number of crystals were not completely contained in a given thin section. Fortunately, as a result of the fact that all grains in sea ice are columnar, the investigator is not faced with the problem of the thin section not cutting the "centers" of the crystals. The crystal "diameter" d ($d \equiv \sqrt{LW}$) was measured and the average value $\bar{d}$ computed for several levels in the ice sheet. There is a pronounced linear increase in d with increasing depth in the ice. Also plotted in this figure are d values from Point Barrow (Weeks and Hamilton, 1962) and Hokkaido (Tabata and Ono, 1962). The slopes of all three curves are similar. This suggests that if a transformation is made correcting for the initial grain size at (Z = 0), all these curves can be considered as identical. Figure 33 shows a plot of [d - d(Z = 0)] versus Z. The least-squares line through the origin has a slope of 0.033. The [d(Z = 0)] value for Hokkaido (0.87) was obtained

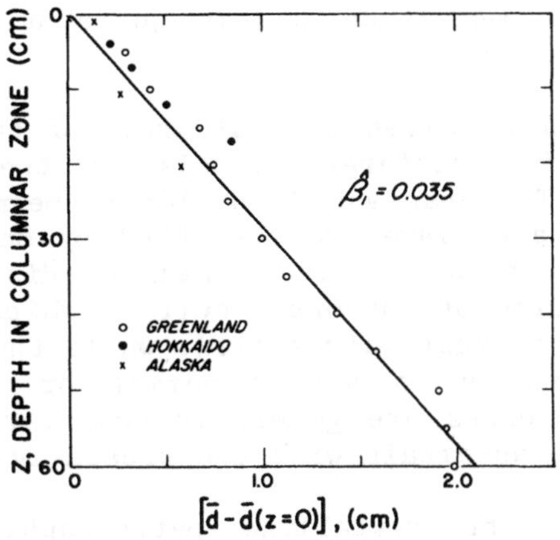

Fig. 33. Plot of change in grain size with distance below the upper ice surface. The grain size curves are shifted to zero at the ice surface.

by least squares extrapolation and the value for Point Barrow (0.22) by setting (Z = 0) at the slush - normal sea ice boundary. This preliminary examination suggests that grain size may prove to be an interesting and useful parameter. If further measurements support the hypothesis of "equal slopes," once the average grain size is determined at one level in an ice sheet it would specify the grain size for the complete sheet. This presumably could only be the case for continuously grown one-winter ice sheets since nucleation of new crystals would cause a discontinuity in the grain diameter curves. We would anticipate that the simple "equal slopes" hypothesis will not be verified and that the rate of increase in the mean grain diameter will be a function of the growth rate of the ice, with the largest values of the slope occurring at the lowest ice growth rates. More will be said later in this paper about grain size variations in sea ice with strong c-axis alignments.

There are, in fact, data already available in the literature that tend to cast doubt on a simple relation between d and the vertical position in the ice sheet. For instance, Koerner (1963) observed that sea ice cores from Jones Sound off Devon Island in the Canadian Arctic showed a smaller initial grain size and a more rapid increase in crystal size with depth than ice collected closer to shore. He also suggested that this difference might be the result of changes in growth rates. Also, in old sea ice incorporated into the ice island Arlis II, Smith (1964) observed areas as large as 10 m on a side with almost perfect c-axis alignment. Using a slope of 0.033 it would require an ice thickness of approximately 300 m to give such an average grain size. Ice thicknesses of this magnitude appear quite unreasonable at the present.

Figure 34 shows a series of histograms of the distribution of crystal diameters for different levels in the ice sheet from Thule, Greenland. The abscissa is scaled to permit ready comparison with similar data on lake ice presented by Marshall (Wilson et al., 1954). The sea ice crystal diameter distributions in sea ice, with the exception of the upper surface which would be in the transition layer, are negatively skewed, while the lake ice distributions have a tendency to appear normal or even rectangular. Whether these differences are generally true between lake and sea ice is impossible to ascertain with the presently available data.

The preceding intracrystalline petrographic parameters are best measured in horizontal thin sections. There are also several petrographic features that are only clearly seen in vertical thin sections. In contrast to the preceding parameters, which are either constant or change gradually with position in the ice sheet, these commonly appear as sharp bands that extend for large lateral distances. Small scale horizontal banding in sea ice has been noted at widely separated regions (Shumskii, 1955; Tabata and Ono, 1957, plate XVI; Langleben, 1959, Figure 3; Bennington, 1963a, Figure 1; Paige, 1966). For example, Shumskii observed a total of 58 secondary layers in a 291-cm section of old pack ice,

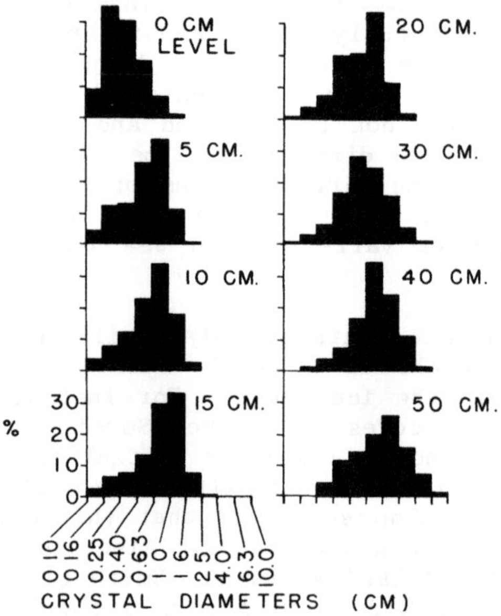

Fig. 34. Histograms of crystal diameters at different levels in an ice sheet at Thule, Greenland.

giving an average layer thickness of 5 cm. The layers are produced by changes in the amount of impurities (brine and air) trapped in the ice. Similar effects are well known in studies of metals. As far as is known, this type of layering is not accompanied by the nucleation of new crystals. Air bubble layering in lake ice is also well known (Taylor and Lyons, 1959, pp. 16-19; Swinzow, 1966; Gow and Langston, 1977). These differences in impurity content are produced by changes in the growth rate of the ice and can make convenient horizontal markers in the ice sheet throughout any region where the weather is laterally homogeneous. Langleben (1959) has, quite effectively, used this layering as a reference mark in studying the location of the transition layer throughout the growth season. A similar technique should also be useful in studying the relative amounts of ablation from the lower and upper surfaces of pack ice during the melt season. This has been done for lake ice by Gow and Langston (1977). A theoretical analysis of the effect of changing the freezing velocity on the amount of impurity incorporated into the solid phase, assuming that solute transfer in the melt takes place by diffusion only, has been published by Tiller et al. (1953) and Smith et al. (1955). Up to the present time, there have been two attempts to apply this theory to bubbles in freshwater ice grown in the laboratory (Carte, 1961; Bari and Hallett, 1974) and one example using field observations (Gow and Langston, 1977). Apparently, no work has been done on sea ice, although we would expect similar relations to hold.

There are several other important aspects of the structure of the columnar zone (e.g., variations in the width of the substructure in the ice crystals, brine drainage channels, annual layering in multiyear ice, directional c-axis alignments in the horizontal plane). These subjects will be treated in detail later in the paper.

7. SUBSTRUCTURE

One invariably identifies sea ice by the presence of a characteristic cellular substructure consisting of reasonably evenly spaced ice platelets or cells separated by small angle grain boundaries. It is along these boundaries that the salt in sea ice is present in the form of liquid and solid inclusions. This cellular substructure is clearly shown in Figures 35 and 36. It has been known for some time that the spacings of this substructure change with vertical location in the ice sheet (Fukutomi et al., 1953). In an attempt to understand these variations, observations have been carried out on natural sea ice (Tabata and Ono, 1962; Weeks and Hamilton, 1962; Paige, 1966; Gow and Weeks, 1977; Nakawo and Sinha, 1984) and experiments have been undertaken on a variety of salt ices (Weeks and Assur, 1963; Rohatgi and Adams, 1967a,b; Lofgren and Weeks, 1969). The most important parameter in

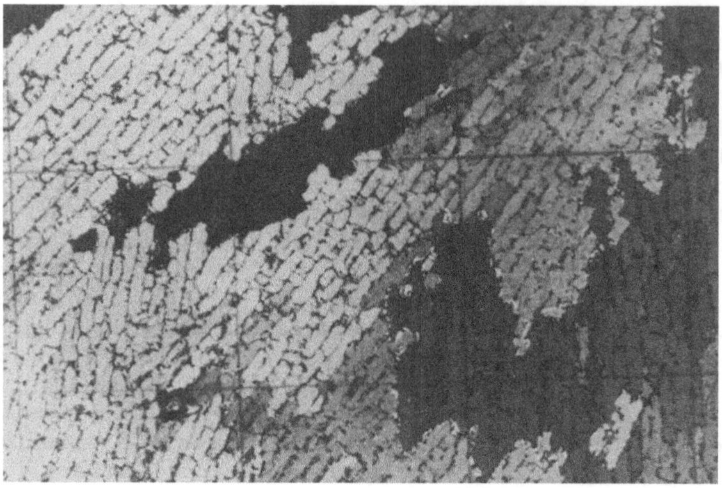

Fig. 35. Horizontal thin section of sea ice from 27.9 cm below the upper ice surface, Point Barrow, Alaska (Weeks and Hamilton, 1962). Grid is 1 cm on a side.

characterizing the substructure is the brine layer spacing a_o (also referred to as the plate width), which is the distance between adjacent layers of brine pockets measured parallel to the c-axis (see Figure 36; double-headed arrow indicates the c-axis direction).

Laboratory experiments have shown that in NaCl ice formed under conditions of unidirectional freezing, there is a gradual increase in the value of a_o with increasing distance from the upper ice surface or from the constant temperature cold plate. An analysis of the factors controlling the variation in a_o has been published by Bolling and Tiller (1960) for the case where the solute transfer in the liquid ahead of the advancing interface is by diffusion only. Although the final form of the analysis is approximate, the physical reasoning behind the analysis is clear. The exact solution to the steady-state solute distribution in the liquid ahead of an advancing plate tip consists of both plane and non-plane wave terms. These latter terms cause lateral diffusion and can be considered to extend some effective distance y into the liquid. Once lateral diffusion starts it continues until it is terminated by the advancing interface. The time allowed for diffusion is $t = (y + y^+)/v$ where y^+ is a function of interface shape and v is the interface velocity. The distance that the

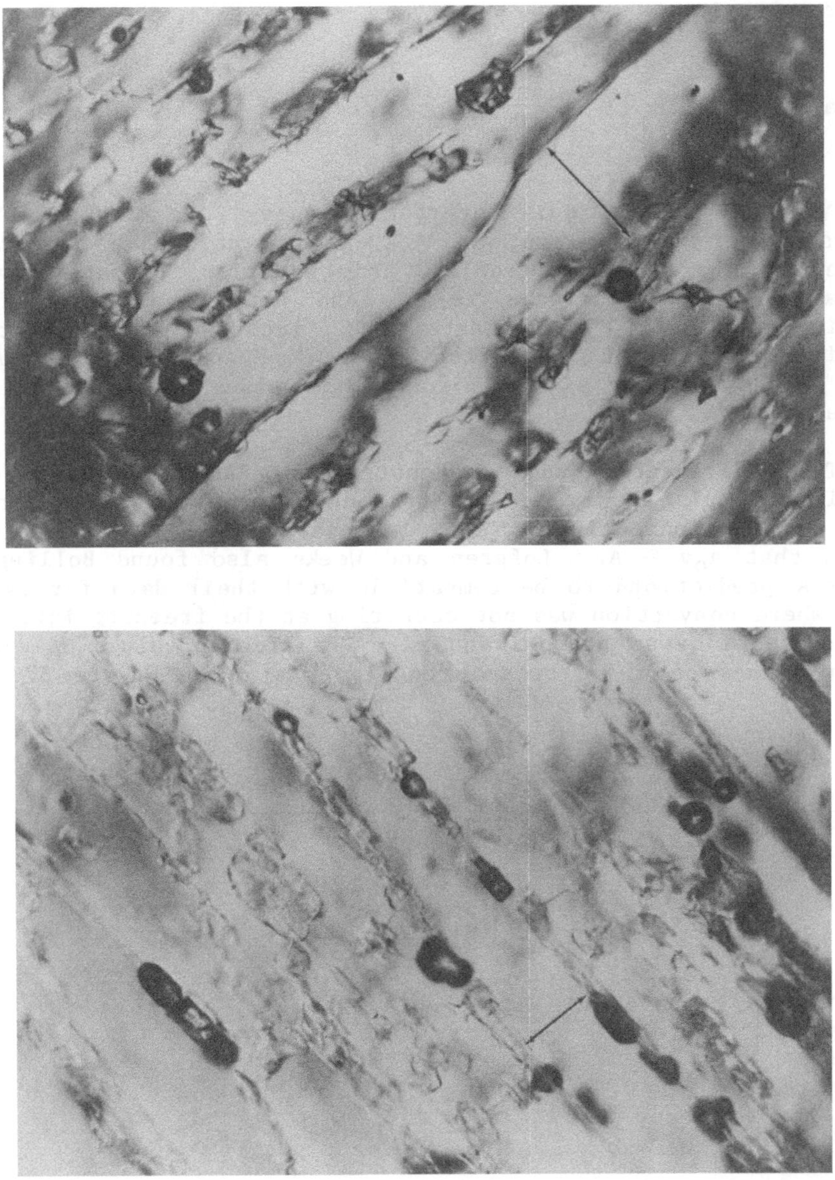

Fig. 36. Photographs of two thin sections showing cellular substructure in sea ice. The c-axis direction is in the plane of the section and normal to the platy structure. The spacing between brine layers is approximately 0.6 mm.

solute can diffuse in this time is $x \approx \sqrt{Dt}$. Assuming that $x = a_o/2$ results in

$$a_o = \frac{D}{2\pi v} \{1 + [1 + \frac{16\pi^2 v}{D} (\frac{0.6\gamma}{\Delta S\ G'_o})^{1/2}]^{1/2}\} \tag{1.4}$$

Here γ is the solid/liquid interfacial energy, ΔS is the entropy of fusion per unit volume, and G'_o is the effective temperature gradient at the tip of the cell. This equation indicates that a_o increases with an increase in C_o, the bulk composition of the freezing solution (G'_o decreases as C_o increases), and that the functional form of the relation between a_o and v is $a_o v = A$ for small values of v and $a_o \sqrt{v} = A$ for large values of v (here A is a constant).

The experimental studies (Rohatgi and Adams, 1967a, b) corresponding to this theory are in reasonable agreement (Lofgren and Weeks, 1969), indicating that a_o increases as a linear function of C_o and that $a_o v \approx A$. Lofgren and Weeks also found Bolling and Tiller's predictions to be compatible with their data for experiments where convection was not occurring at the freezing interface. When convection was predominant $a_o\sqrt{v} \approx A$ at high values of v while at low values of v, a_o became roughly constant. Figure 37 shows

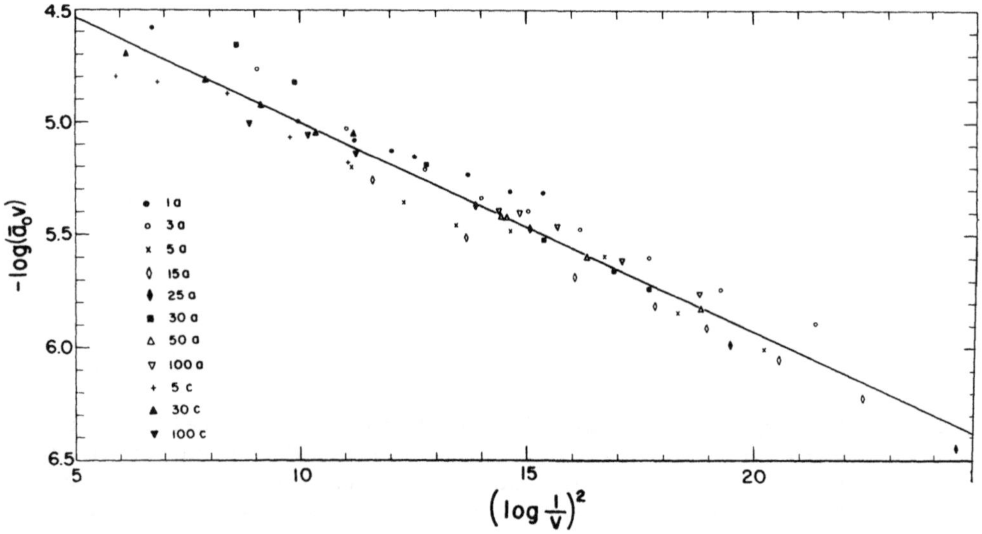

Fig. 37. Plot of $-\log(a_o v)$ versus $(\log 1/v)^2$. (From Lofgren a Weeks, 1969).

the test data; the fitted relation is of the form

$$\log (\bar{a}_o v) = \log A + n_1(\log\frac{1}{v})^2 \tag{1.5}$$

where A and n_1 are constants.

The field observations have in general shown an increase in a_o with increasing depth Z in the ice sheet (Tabata and Ono, 1962; Weeks and Hamilton, 1962; Paige, 1966). However, at Narwhal Island, Alaska, a decrease in a_o with increasing Z was noted by Gow and Weeks (1977), and at Eclipse Sound, Baffin Land, a rather complex variation was observed by Nakawo and Sinha (1984). Typical a_o variations would range from 0.4 mm near the upper ice surface to 1.0 mm at Z = 2.0 m in first-year sea ice. At a given level in the ice sheet the distribution of a_o values is approximately normal about the mean (Weeks and Hamilton, 1962), as might be expected from the central limit theorem.

The best available study of the relation between a_o and growth velocity in a natural ice cover was undertaken at Eclipse Sound, N.W.T., by Nakawo and Sinha (1984). Curve (a) in Figure 38 shows the calculated growth rate plotted as a function of Z, the vertical position (depth) in the ice sheet. Curve (b) is the mean of the growth rate for an interval of $\pm$ 50 mm centered on the 25-mm segment for which it is plotted. There is a pronounced variation in growth rate with Z, with the maxima at about 0.2, 0.5, and 1.0 m corresponding to the observed cold periods in November, December, and January. Curve (c) presents the salinity profile obtained from an adjacent core. There is clearly a positive correlation between growth rate and salinity. More will be said about this later. The curve on the right shows the corresponding values of a_o and indicates an inverse proportionality with growth rate in agreement with the work of other investigators. Figure 39 shows the same data presented as an a_o versus v plot. The dashed line is a least-squares fit of the Bolling and Tiller equation. The agreement is quite reasonable if the two a_o values from near the bottom of the ice sheet are disregarded. The ice studied here showed a pronounced c-axis alignment, not only in the horizontal plane but specifically in the NE-SW direction in that plane. At a given level in the ice sheet (i.e., at a constant growth velocity) limited data suggest that a_o is a maximum value when the c-axis is parallel to the favored direction, and that it systematically decreases when the angle to this direction increases. If this can be verified it both complicates the picture and makes studies of a_o variations very interesting indeed.

The hypothesis that slower growth corresponds to larger a_o values is quite strongly supported by field observations on very thick sea ice -- ice that in all probability grew extremely slowly.

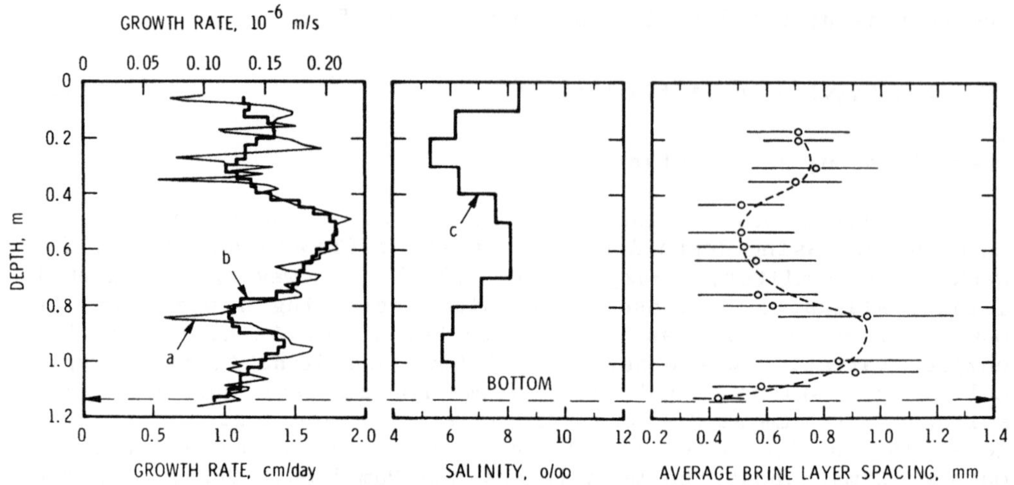

Fig. 38. Profiles of growth rate, salinity and brine layer spacing. Curve b represents the mean of calculated growth rate, curve a, for an interval of ±50 mm for every 25 mm (Nakawo and Sinha, 1984).

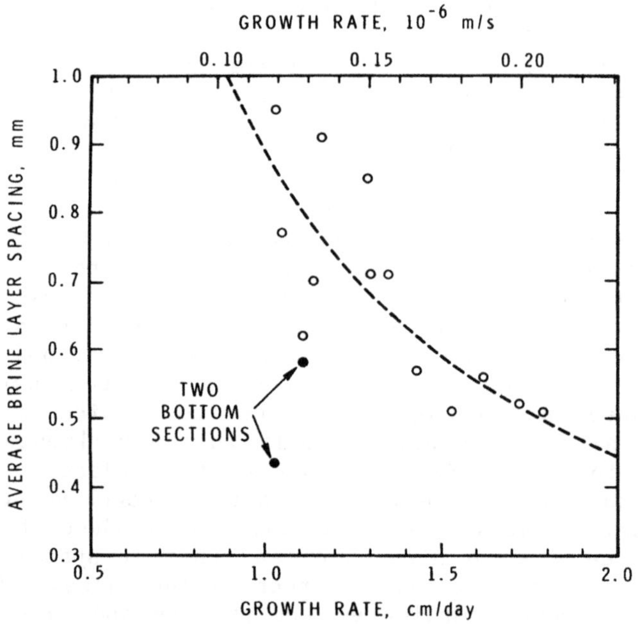

Fig. 39. Plot of average brine layer spacing versus corresponding rate of growth from curve b (Figure 38). Broken line shows a least squares fit, assuming that the spacing is inversely proportional to the growth rate (Nakawo and Sinha, 1984).

Cherepanov (1964) obtained a_o values of 1.5 mm on old sea ice 10 to 12 m thick observed on drifting station NP-6; Gow (personal communication) obtained a_o values of 1 to 1.5 mm on thick sea ice that formed as part of the Koettlitz Ice Tongue, McMurdo Sound, Antarctica; and finally Zotikov et al. (1980) obtained a_o values of 5 mm from sea ice formed on the base of the 416-m-thick Ross Ice Shelf. Figure 40 shows a vertical thin section from the bottom of this core. The core was oriented so that the plane of the thin section was in general normal to the (0001) planes containing the brine pockets. The growth velocity at the sample site (Camp J9) was estimated to be 2 cm/year, far slower than could readily be obtained in the laboratory.

Fig. 40. Vertical thin section of sea ice from the lower portion of the core from site J-9, Ross Ice Shelf, Antarctica. The scale is marked in units of 5 mm (thin section prepared by A. J. Gow). The ice shelf at site J-9 is 416 m thick.

8. CONSTITUTIONAL SUPERCOOLING

Up to the present time we have demonstrated that the ice phase that forms from seawater should be pure ice. If this were strictly true it would be simple to desalinate seawater simply by freezing it. The reason this is not done is that salt becomes entrapped in the ice phase, largely in the form of liquid inclusions located along the boundaries between the ice plates. To understand why this salt entrapment occurs one must first understand why sea ice grows with a platelet-like structure in the first place. In short, why does sea ice form with a non-planar, dendritic interface, resulting in brine entrapment, while lake ice forms with a planar interface, resulting in almost complete impurity rejection back into the melt. Both result from the freezing of impure melts; it is only a matter of degree (35°/oo as compared with <0.1°/oo). The answer is that sea ice is a classic example of a material that invariably shows pronounced constitutional supercooling.

What is constitutional supercooling (C.S.) and how does it work? Briefly stated, the answer is as follows (Rutter and Chalmers, 1953; Tiller et al., 1953; Elbaum, 1959). Consider a salt solution freezing as the result of one-dimensional cooling. For such a system the distribution coefficient k_o (defined as the ratio of the solute content of the solid to the solute content of the liquid when the two phases are at equilibrium) has a value of 10^{-4} or less (Harrison and Tiller, 1963). Next assume that:

(a) Diffusion in the solid is negligible.

(b) Convection and mixing in the liquid, due to causes other than diffusion, are negligible (for salt solutions this is a good assumption only if we freeze from the bottom up).

(c) k_o is constant.

If the origin of the coordinate system is taken to be at the freezing interface, then freezing can be viewed as moving this liquid toward the interface at a velocity v producing a net flow out of a unit volume of v (dC/dx) where C is the concentration and x is the distance normal to the interface. At the same time there is a diffusive flow of impurity into the volume element of $D(d^2C/dx^2)$, where D is the diffusion coefficient. If a steady state is achieved these two terms must be equal, leading to the differential equation

$$D \frac{d^2C}{dx^2} + v \frac{dC}{dx} = 0. \qquad (1.6)$$

The boundary conditions are

$$C_L = C_o \text{ at } x = \infty$$
$$C_L = C_o/k \text{ at } x = 0.$$

Here C_L is the concentration in the liquid and C_o is the initial concentration in the liquid (the composition far removed from the interface). The steady-state solution is then

$$C_L = C_a \exp(-(vx)/D) + C_o \tag{1.7}$$

where $C_a \equiv C_i - C_o$ and C_i is the composition of the liquid at the interface. This ultimately leads to

$$C_L = C_o \left[1 + \left(\frac{1-k}{k}\right) \exp\left(-\frac{v}{D}x\right)\right] \tag{1.8}$$

indicating that there should be a thin layer of liquid at the interface that shows an exponential decrease in C_L as one moves away from the interface. This is shown schematically in Figure 41. Because of these compositional variations there are associated changes in the equilibrium freezing temperature of the liquid (Figure 41) inasmuch as

$$T_E = T_o - m\, C_L \tag{1.9}$$

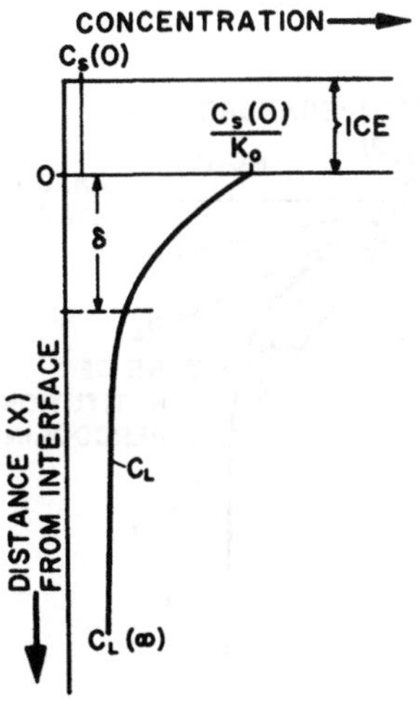

Fig. 41. Steady-state distribution of solute in the liquid ahead of an advancing solid/liquid interface.

where T_E is the equilibrium freezing temperature, T_o is the melting point of the pure solvent (ice), and m is the slope of the liquidus line from the phase diagram. This results in

$$T_E = T_o - m\,C_o\,[1 + \frac{1-k}{k}\exp\left(-\frac{v}{D}x\right)]. \qquad (1.10)$$

Next, take the actual temperature distribution in the liquid to be given by

$$T = T_o - m\left(\frac{C_o}{k}\right) + G_L X \qquad (1.11)$$

where G_L is the temperature gradient in the liquid. The first two terms on the right side of this equation give the temperature at the interface. For some compositional profiles and temperature profiles the composition is such that the liquid ahead of the interface becomes supercooled (see Figure 42). Hence the term "compositional supercooling."

When the slope of the actual temperature profile is less than the slope of the T_E curve at the interface, constitutional supercooling occurs. The criterion for this condition is

$$\frac{G_L}{v} < \frac{mC_o}{D}\left(\frac{1-k}{k}\right) \qquad (1.12)$$

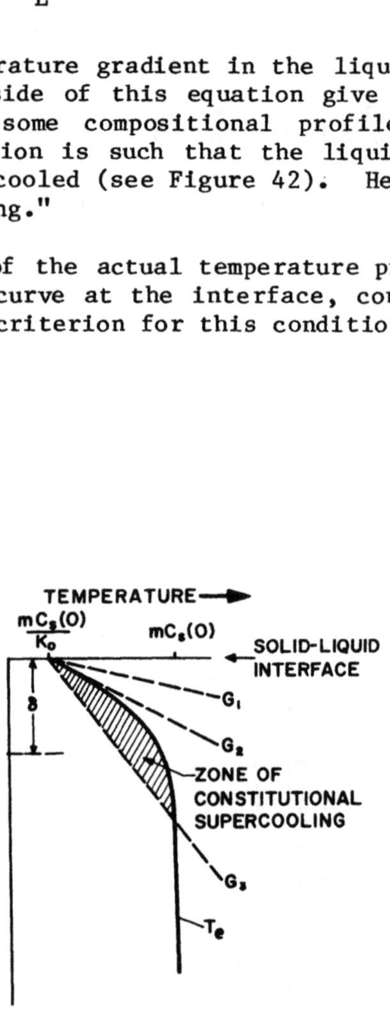

Fig. 42. Constitutionally supercooled zone of liquid ahead of an advancing interface.

a relation obtained by differentiating eqs. (1.10) and (1.11) with respect to x and evaluating them at x = 0. This relation has been found to be in good agreement with experimental observations in metal systems (Walton et al., 1955). When eq. (1.12) is applied to the freezing of both salt solutions and seawater, it is found that even when optimum conditions for the maintenance of a planar interface occur ($v \approx 10^{-6}$ cm/s, $G \approx 1.0$ C/cm, and $k_o = 10^{-4}$), the ratio $G/v = 10^6$. For this ratio C.S. should exist (Figure 43) even for salinities representative of typical lake water (≈ 100 ppm or 0.1 °/oo), not to mention the salt contents studied in this paper (24.7°/oo and greater).

Although the existence of a C.S. layer ahead of the advancing solid/liquid interface is necessary for cell formation, it is not necessarily sufficient. In addition, there is considerable uncertainty in the preceding calculation as the result of possible variations in G. This difficulty can be avoided by calculating if the growth conditions are such that the steep-walled cell boundary grooves associated with the entrapment of brine along plate boundaries are stable. The stability criterion for the formation of these grooves (Tiller, 1962) is

$$\frac{G_s}{v} < \frac{-m\,C_i\,(1-k)}{D} \qquad (1.13)$$

where G_s is the temperature gradient in the solid and C_i is the composition of the liquid at the interface. If transfer in the liquid is by diffusion only, $C_i = C_o/k$ and eq. (1.13) becomes similar to eq. (1.12) with G_s substituted for G_L. Equation (1.13)

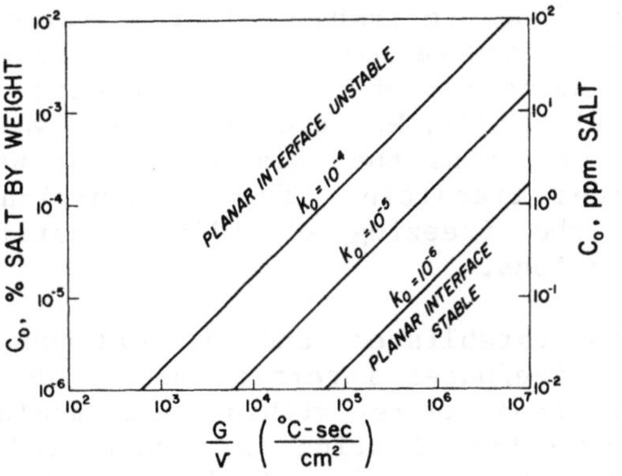

Fig. 43. A plot of the boundaries that for different k_o values separate growth conditions where a planar interface is stable and where it is unstable. Solute transfer in the liquid is assumed to be by diffusion only.

is a more stringent criterion than eq. (1.12) because for ice-solute systems $G_s >> G_L$. Now, if

$$t = \frac{\rho L h^2}{2\kappa \Delta T_o} \quad (1.14)$$

(Carslaw and Jaeger, 1959), where t = time, ρ = density of the solid, L = latent heat of fusion, ΔT = difference between the upper surface (i.e., cold plate/ice) temperature of the ice and the freezing temperature of the seawater, h = thickness of the ice, and κ = thermal conductivity of the ice, then

$$v = \frac{dh}{dt} = \frac{\kappa \Delta T_o}{\rho L h} \quad (1.15)$$

and

$$G_s = \frac{\Delta T_o}{h}. \quad (1.16)$$

Therefore, eq. (1.13) becomes

$$\frac{\rho L}{\kappa} < \frac{-m\, C_o\, (1 - k)}{D\, k} \quad (1.17)$$

in which C_o is the only parameter under the control of the experimenter. Substitution of numerical values in eq. (1.17) shows that for all seawater and brackish water salinities cell boundary grooves are stable if solute transfer in the liquid is by diffusion only. In actuality, because natural sea ice freezes from the top down, the primary mechanism of solute transfer is free convection (Farhadieh and Tankin, 1972; Wakatsuchi, 1977), which causes C_i to approach C_o. This explains the observations of Weeks and Lofgren (1967), who recorded transitions from a non-planar to a planar interface during the freezing of $1^o/oo$ (unstirred) and $3^o/oo$ (stirred) NaCl solutions.

Once C.S. is established and a cell-boundary groove is stable, the planar ice/water interface breaks down into parallel rows of cellular projections within each grain (Harrison and Tiller, 1963). Concentrated brine is accumulated in the grooves that develop between the rows of cells (Figure 44) and is eventually trapped when adjacent cells develop lateral connections. The entrapped brine leaves a permanent record of the positions of the intercellular grooves. It is the distance between the midpoints

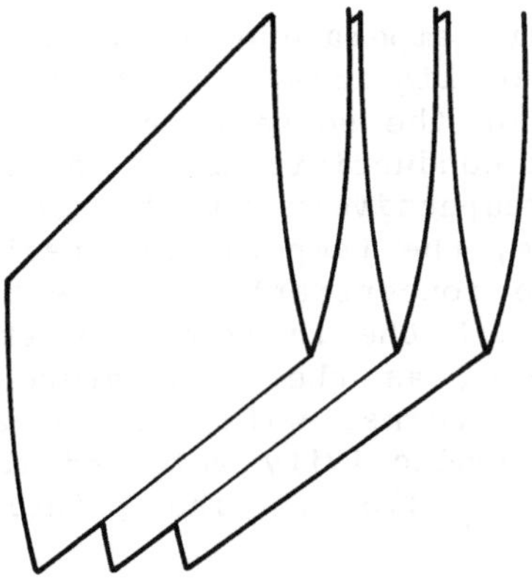

Fig. 44. Schematic of knife-edged cells (Harrison and Tiller, 1963).

of these intercellular grooves that was termed the brine layer spacing a_o earlier in this paper.

The relations we have just presented are useful in understanding the general phenomenon of C.S. However, a treatment of C.S. that is truly adequate for sea ice does not as yet exist. It will have to deal explicitly with the fact that the liquid is mixed by free convection and that the appropriate boundary layer is not a diffusive one, but a convective one.

It should also be mentioned that considerable work is currently underway to further develop and apply a more sophisticated interface stability theory, one that is based on the dynamics of the whole system rather than just the thermodynamics of the liquid and is capable of describing the time evolution of a perturbed interface and the accompanying temperature and concentration fields. A review of these developments has been published by Sekerka (1968). The diffusion equations are solved for a planar interface with a sinusoidal perturbation of a small amplitude δ superimposed upon it. The growth velocity of the perturbation δ is then calculated, a velocity which is a function of the wavelength of the perturbation λ and several thermal parameters. If δ is greater than zero, the instability grows and develops into cells. The stability criterion involves the function $S(\omega)$, which for $k \ll 1$ is

$$s(\omega) = -T\Gamma\, \omega^2 - G_\ell + mG_c + \left(\frac{k_s - k_\ell}{k_s + k_\ell}\right) G_\ell - \frac{Lv}{k_s + k\ell} \,. \qquad (1.18)$$

Here T is the melting temperature of a flat interface; $\Gamma = \gamma/L$ where γ is the specific liquid/solid interfacial free energy, and L is the latent heat of the solvent per unit volume; $\omega = 2\pi\lambda$; k_s and k_ℓ are the thermal conductivities of the solid and liquid; and G_s, G_ℓ, and G_c are respectively the temperature gradient in the solid at the interface, the temperature gradient in the liquid at the interface, and the concentration gradient in the liquid at the interface. When $S(\omega) < 0$ the interface is stable. The principal stabilizing terms are G_ℓ as the temperature gradient is positive and the first and last terms, which are usually small. Because for ice $k_s > k_\ell$, the conductivity weighted term in G_ℓ offsets the stabilizing effect of G_ℓ alone. The principal term leading to instability is mG_c.

This relation is only strictly applicable to the study of perturbations about the steady-state solution. When expressed as a stability criterion it is usually stated as

$$|m| \frac{vC_i}{D} < \frac{2\,k_\ell G_\ell}{k_s + k_\ell} + \frac{Lv}{k_s + k_\ell} \tag{1.19}$$

where C_i is the concentration in the liquid at the interface. Sekerka et al. (1967) have also expressed this relation as

$$\left[\frac{1}{k} - \left(\frac{1}{k} - 1\right) e^{-t/t_{rise}}\right] |m| \frac{vC_o}{D} < \frac{2\,k_\ell G_\ell}{k_s + k_\ell} + \frac{Lv}{k_s + k_\ell} \tag{1.20}$$

where C_o is the bulk composition of the freezing solution, t is time and $t_{rise} = D/(kv^2)$ is a characteristic time of solute build-up at the growing interface. In eq. (1.20) the term in brackets increases monotonically from a value of 1 at $t = 0$ to a value of nearly $1/k \approx 10^4$ for $t = t_{rise}$. In inserting the appropriate values for seawater, Sekerka et al. (1967) found that a seawater/ice interface is effectively instantaneously unstable with respect to the time t_{rise} associated with the buildup of salt at an advancing ice interface.

Modifications of this theory incorporating coupled convective and constitutional interface instabilities are under development. The equations are quite complicated (Coriell et al., 1980) and have not, as yet, been applied to the freezing of seawater.

It should also be mentioned that this theory can be used to estimate λ_{max}, which is presumably equivalent to the brine layer spacing a_o. The resulting equations suggest that

$$\lambda_{max} \propto v^{-1/2} \tag{1.21}$$

and that for seawater freezing with a velocity of 10^{-5} cm/s, a_o should have a value of ≈0.2 mm. In fact, a_o from 0.5 mm to 0.95 mm are encountered, in reasonable agreement.

9. VARIATIONS

9.1 Directional C-Axis Alignments

In the earlier portions of this paper we have pointed out that the crystals in the columnar zone have their c-axes oriented in the horizontal plane. We have also shown some early fabric diagrams (Figure 23) in which the c-axis alignments within the horizontal plane were either random or near-random. It was reasonable to expect that this might be the case as the heat flow was apparently one-dimensional. Also, such radially random c-axis alignments were common in the solidification of metals.

However, in the early 1960's Peyton (1963, 1966, 1968), while examining a 3x3-m block of 1.6-m-thick sea ice from Barrow, found that the bottom meter exhibited a near-constant c-axis orientation over the entire 9-m^2 cross section. In his studies of mechanical properties, Peyton utilized this ice to investigate the effect of changes in the orientation of the uniaxial stress to the c-axis direction on the failure strength. At roughly the same time Cherepanov (1964) and Smith (1964) observed that old sea ice incorporated in the ice islands SP-6 and Arlis II showed near-perfect c-axis alignment over large areas (on SP-6 the entire 80-km^2 area of the "island" showed a similar c-axis alignment). Clearly such alignments were not statistical fluctuations, and specific efforts to study this phenomenon (or these phenomena) were in order.

In 1971 a quite remarkable set of c-axis observations (Figure 45) were published by Cherepanov, who observed nearly constant c-axis alignments over thousands of square kilometers in the Kara Sea. This was followed by studies of the directional dependence of the electrical properties of sea ice from widely separated sites in the Canadian Arctic that also supported the contention that c-axis alignments might well extend over lateral distances of the order of tens of kilometers (Campbell and Orange, 1974; Kohnen, 1976). This work was followed by a series of three papers that provided a detailed look (Figure 46) at crystal alignments along the coast of arctic Alaska (Gow and Weeks, 1977; Weeks and Gow, 1978, 1980). Similar studies are now available for the Mackenzie Delta portion of the Beaufort Sea (Langhorne, 1980), for the ice surrounding artificial gravel islands that are used for

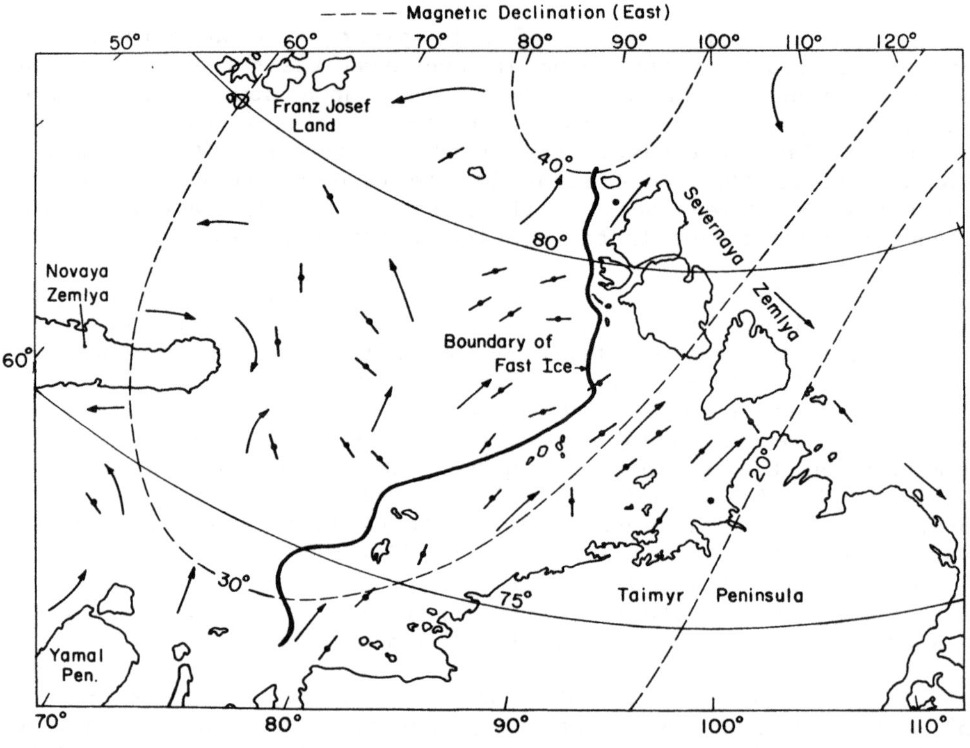

Fig. 45. The c-axis orientations as determined by Cherepanov (1971) in the Kara Sea. The arrows indicate the estimated directions of the surface currents obtained from U.S. Navy sources (Weeks and Gow, 1978).

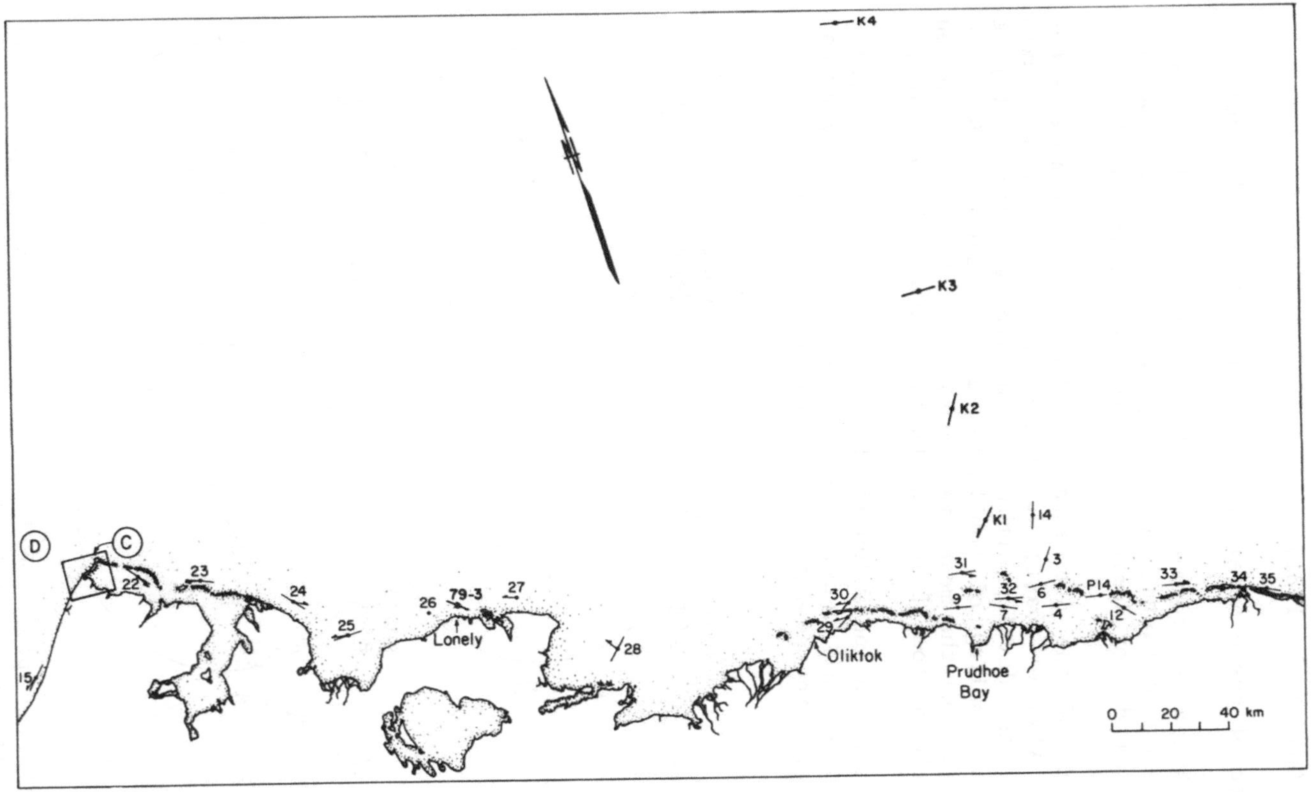

Fig. 46. Sample site locations (dots), mean c-axis alignments (bars), and instantaneous current directions (arrows) determined along the coast of the Beaufort Sea. All the samples were collected during 1978 except site 79-3 north of Lonely and the low "out-of-sequence" sites (3,4,6,7,9,12,14 and P14) north of Prudhoe Bay, which were collected during 1977 (Weeks and Gow, 1978) and are included for comparative purposes. Also shown are four orientations determined by radar (sites K1-K4) by Kovacs and Morey (1980).

drilling platforms (Vittoratos, 1979), for channels in the Canadian Arctic Islands (Nakawo and Sinha, 1984), and for the Antarctic (Serikov 1963; Gow et al. 1981).

The results of these observations indicate the following:

1. In all sea ice formed by unidirectional freezing, c-axis horizontal orientations develop rapidly after an initial ice skim forms, and they dominate the rest of the ice growth.

2. In much of the ice examined (95% of the sample sites along the Alaskan coast; Weeks and Gow 1980) strong c-axis alignments also develop within the horizontal plane. Figure 47 shows a sampling of published fabric diagrams. Standard deviations around the mean are usually <10° for ice collected from near the bottom of ice sheets in excess of 1.5 m thick.

3. Alignments may occasionally be present in ice 20 cm thick and they invariably become stronger (show less scatter around the mean) with increasing depth in the ice (Figure 48).

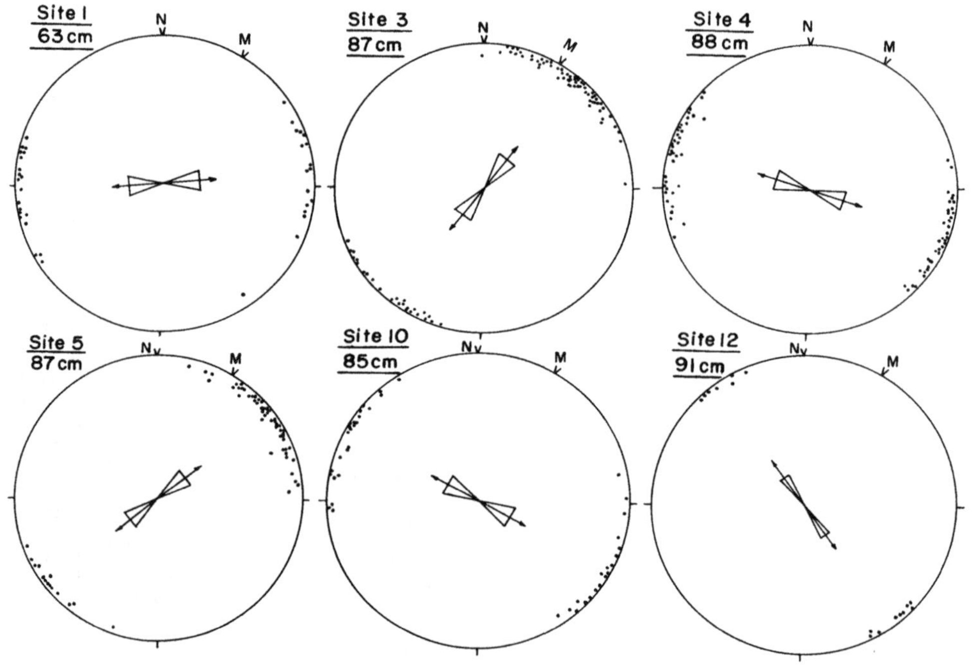

Fig. 47. Schmidt net plots of individual c-axis orientations for six different sampling sites along the coast of the Beaufort Sea (Weeks and Gow, 1978).

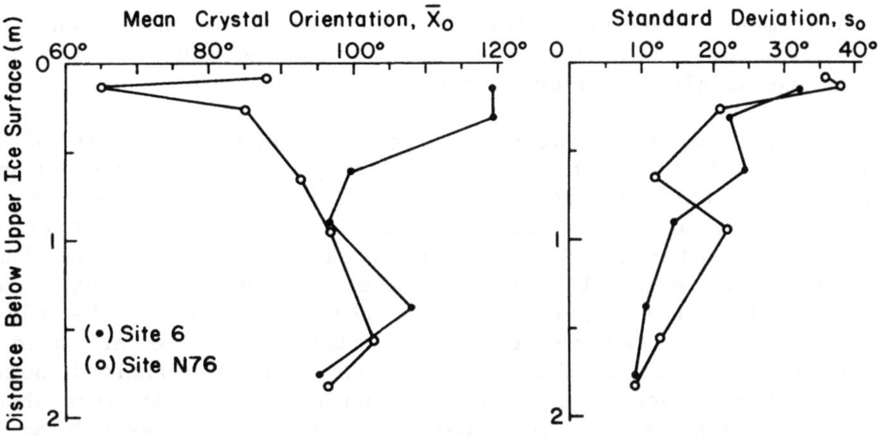

Fig. 48. Mean crystal orientation in the horizontal plane, $\bar{X}_o$, and the standard deviation, s_o, as a function of vertical location in the ice sheet. Sites along the coast of the Beaufort Sea (Weeks and Gow, 1978).

4. Although the mean c-axis direction is not constant at different levels in the ice sheet, the variations are commonly <20°.

5. Mean c-axis directions at the same site commonly appear to be similar from year to year; however, the "strength" of the alignment may vary.

6. In the nearshore regime the c-axes generally parallel the coast, swing around both natural and artificial islands in a streamlike-pattern, and are aligned parallel to the axes (throats) of passes and inlets.

7. Limited information (Figures 45 and 46) indicates that strong crystal alignments can also develop in areas of pack ice.

There have been several different suggestions concerning the cause of these alignments. Cherepanov (1971) initially suggested that some interaction between the earth's magnetic field and the electric potential produced at the freezing interface caused them. No information was given concerning how such couplings might work. Weeks and Gow (1978) rejected this suggestion, pointing out that a) there does not appear to be a theoretical reason why the sea ice crystal orientation should be affected by the earth's magnetic field, b) similar alignments have not been observed in experiments on the unidirectional freezing of NaCl solutions in magnetic fields 80,000 times stronger than the earth's, and c) Cherepanov's own measurements show many examples of crystal alignment that are not in agreement with the direction of the geomagnetic field at the time the samples were collected.

Weeks and Gow then suggest, after examining a variety of alternatives, that the mechanism most likely to produce the aligned crystals works through the control that the current has on the composition of the liquid at the dendrite tips, and thus on the interface temperature. If the current is parallel to the basal ice plates, then a stable boundary layer can build up along a dendrite tip. If, however, the flow is perpendicular to the plates (i.e., parallel to the c-axis), then mixing will be enhanced at each plate tip, and the thickness of the diffusion-limited solute boundary layer reduced. A reduced boundary layer thickness should give the crystals in this orientation a very slight growth advantage, permitting them to grow sideways at the expense of neighboring crystals. Although approximate calculations suggest that the Weeks and Gow hypothesis is plausible, its mechanisms are far from proved. It is, however, possible to test the suggested correlation between mean current direction and the direction of the c-axis alignment. These observations have been collected by Weeks and Gow (1978, 1980) and by Kovacs and Morey (1978). Figure 49 shows the relative frequency of different angular deviations between the observed "instantaneous" current direction determined just under the ice and the mean c-axis direction. The agreement is quite good, considering that short-term current measurements are known

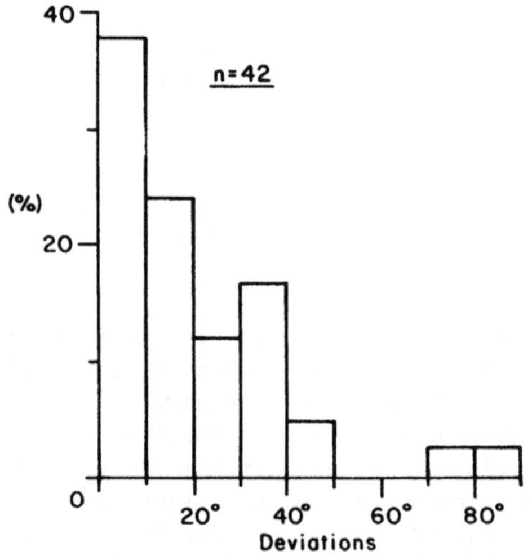

Fig. 49. Histogram showing the relative frequency of different deviations between the observed instantaneous current direction and the mean c-axis direction (Weeks and Gow, 1980).

to commonly show significant deviations from the long-term average current direction because of eddies. It should also be noted that Weeks and Gow measured a 7-hour average of the current vector at a site near Barrow and obtained an average value that differed from the mean c-axis alignment by only 4°.

Recently, Stander and Gidney (1980) have proposed a very different model to explain c-axis alignments: that the alignments are caused by the mechanical rotation of crystals as the result of stress-activated mechanisms such as grain boundary sliding and dislocation generation. Although we would agree that grain rotations could occur in sea ice sheets undergoing large strains, we doubt that this theory is applicable to c-axis alignments in sea ice. The fabrics in sea ice are the result of preferred directions of crystal growth from the melt. The orientation seen at each level in the ice appears to be the orientation that was present when the ice layer initially formed. During this initial phase of crystal growth, each crystal is composed of a series of separated platelets that extend downward into the melt (Figure 50). These platelets do not touch each other until approximately 2.5 cm above the gross solid/liquid interface. It is a bit difficult to talk about stress transfer between crystals at a level in the ice where the crystals are not in direct contact. If the Stander and Gidney hypothesis were correct, there should be evidence that significant strains

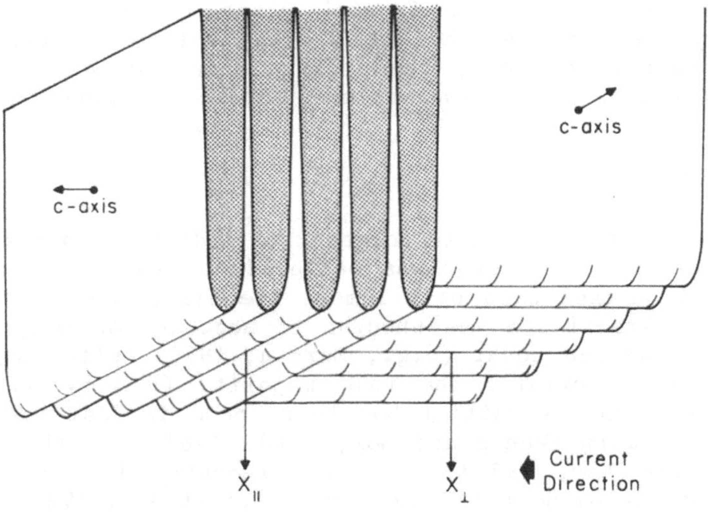

Fig. 50. Sketch showing the interface geometry of two crystals of sea ice oriented so that the c-axis in one crystal (the left) is parallel to the current direction, while in the other crystal (the right) the c-axis is oriented normal to it (Weeks and Gow, 1978).

have occurred. To date we have seen no petrographic evidence that this is the case.

It should be possible to resolve by experiments many of the questions about the mechanisms that lead to the c-axis alignments. Such work is now underway (Langhorne, 1980). If the c-axis direction is a measure of the mean current direction it might at first glance be concluded that alignments would only be found in ice that was fast when the alignment developed. Two pieces of information suggest that this is not always true. Cherepanov (1971) found alignments in first-year ice in areas of the Kara Sea that were identified as pack (Figure 45), and Kovacs and Morey (1980) obtained similar results for offshore areas of the Beaufort Sea (Figure 46). This can be explained if the current velocity at the ice/water interface remains reasonably constant relative to the ice long enough for an alignment to develop. As both these sets of observations were made in the later winter when the ice pack in the Beaufort and Kara Seas was extremely tight, this would appear to be a plausible explanation.

Present observations suggest that very significant portions of the world's fast ice show strong c-axis alignments. The percentage of the pack that shows such alignments is unknown. We would guess that pack ice that forms at locations such as the Bering Sea, Baffin Bay, and much of the Antarctic, where the ice is highly mobile and the floes are free to rotate, would show random alignments in the horizontal plane. Where the pack ice becomes very tight during the ice growth season, restricting the possible rotation of floes, then aligned ice may be the dominant form. Systematic field observations should be made to resolve these questions.

9.2 Frazil Ice

Given an initial ice skim formed from individual discs or particles of ice, the ice cover was believed to usually take on the characteristic congelation or columnar ice structures so frequently observed. Frazil ice was thought to account for only the top 5% or so of the ice cover (e.g., Martin, 1979), with congelation or columnar ice providing the vast majority (95% by volume) of most sea ice. This condition has been found in fast ice regimes in the Arctic Basin (Weeks and Gow, 1978, 1980), in the Labrador Sea (Weeks and Lee, 1958), and more recently in the fast ice region of Antarctica near McMurdo sound (Gow et al., 1981).

In ice edge regions and in large polynyas where substantial open water can be maintained, Martin and others have described rapid frazil ice generation that could conceivably result in frazil ice thicknesses of up to a meter prior to the development of a columnar congelation structure.

Work in the Weddell Sea has, however, indicated even greater potential for frazil growth. A serious revision of ideas concerning the importance and distribution of frazil ice, at least for the Antarctic pack ice regions, appears to be in order. If frazil ice is more widespread than once believed, mechanisms for ice-ocean interaction must also be modified. We first give the evidence for these new frazil structures and then discuss some preliminary ideas for the ice-ocean mechanisms leading to the formation of the observed structures.

Figure 51 shows the location of sites in the Weddell Sea sampled during February-March 1980. Complete cores were obtained from 66 individual floes (13 multiyear floes), with a total of 138 m of core being retrieved. Salinity and fluorescence measurements were also taken at 10-cm intervals over the entire length of core. The crystal structure was defined in two ways. The first was by preparing two to six horizontal thin sections per core (depending on its length). This method gave only point samples of the structure, albeit at several locations along the core length. To verify the frazil percentages initially estimated from these point samples, vertical thick (2 to 3 mm) sections were then prepared of the ice. When these sections were viewed between crossed polaroids, the differences between relatively fine-grained equiaxed

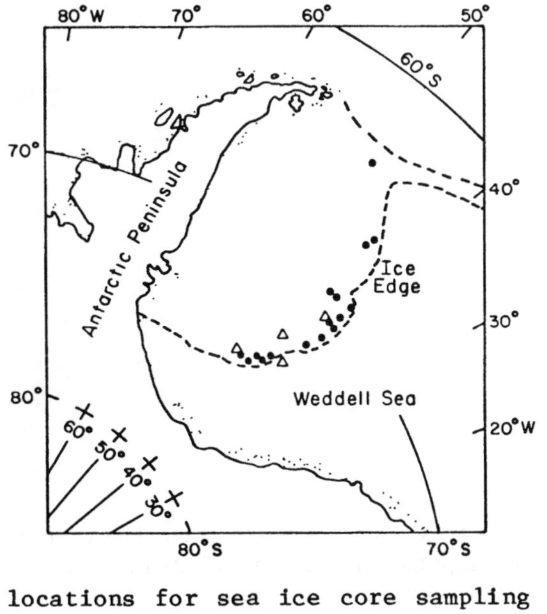

Fig. 51. Site locations for sea ice core sampling from the USCGC Polar Sea in Feb-March 1980.

crystals (frazil) or coarse-grained columnar crystals (congelation ice) could easily be distinguished. These changes were then sketched schematically (cartooned) as a function of the distance along the core.

We will now briefly describe the structure of several representative floes (after Gow et al., 1981).

Floe 42-A-1 (see Figure 52). This is an example of thick (4.8 m) multiyear ice with a maximum salinity of 9°/oo and an average salinity of 3.5°/oo. The floe was composed of 4% snow ice (designated S in the vertical structure profile), 61% frazil ice (designated F) and 35% congelation ice (designated C). The congelation ice component is characterized by a very strong alignment of the crystallographic c-axes as indicated by the arrows in the photographs of thin sections from 2.24 m and 2.93 m. In this

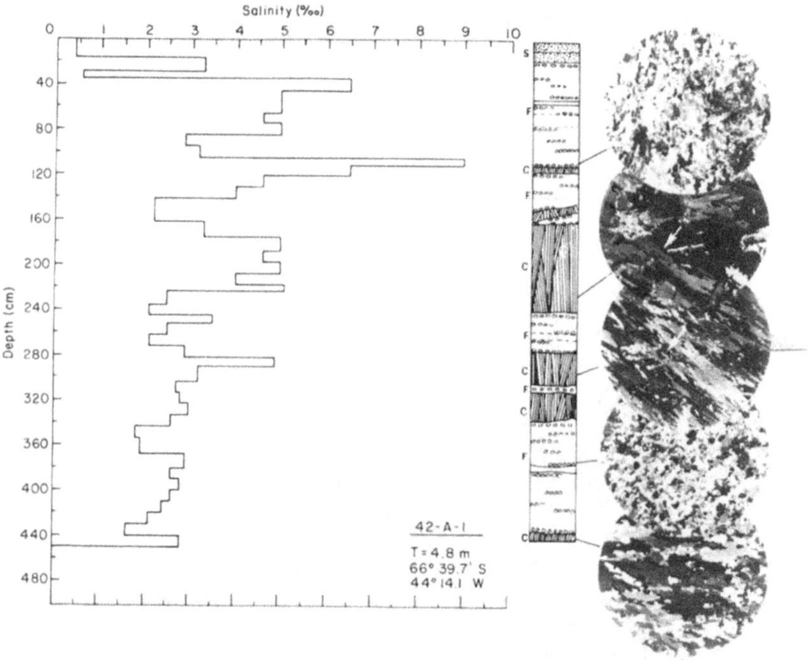

Fig. 52. Salinity structure cartoon and thin section photos from selected depths for Floe 42-A-1. S = snow ice, F = frazil, C = congelaton ice.

and other profiles presented here the much smaller crystals and the absence of brine lamellae-ice plate structure serve to distinguish frazil ice from the generally fibrous-textured congelation ice.

Floe 43-A-2 (see Figure 53). The exact age and origin of this 2.4-m-thick floe are difficult to assess. However, a maximum salinity of 5°/oo and an average value of about 3°/oo would indicate that this particular floe is multiyear. It is composed of 90% frazil ice of variable grain size with congelation ice composing less than 6% of the floe thickness.

Floe 44-G-3 (see Figure 54). This floe was located in a region of recently disintegrated sheet ice that produced floes measuring approximately 30 m in length. Identified as first-year ice with an average salinity of 5°/oo it is composed almost entirely of congelation ice (95%) with crystals showing well developed brine lamellae-ice plate structure and a very strong alignment of the crystallographic c-axes. The constancy of this alignment (indicated by arrows) would suggest either a fast ice origin or growth of ice in place in a relatively immobilized part of the winter pack.

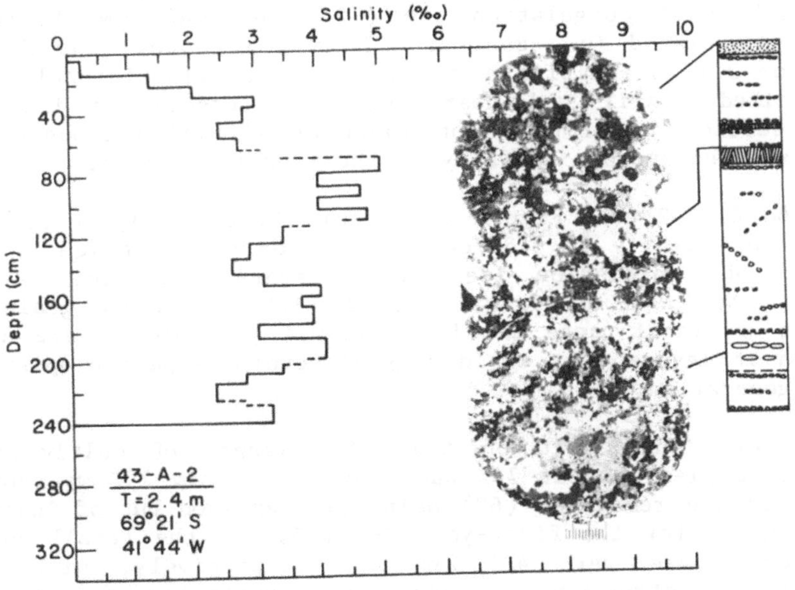

Fig. 53. Salinity, structure and thin section photos from Floe 43-A-2. A key to the ice structure diagram is given in Figure 52.

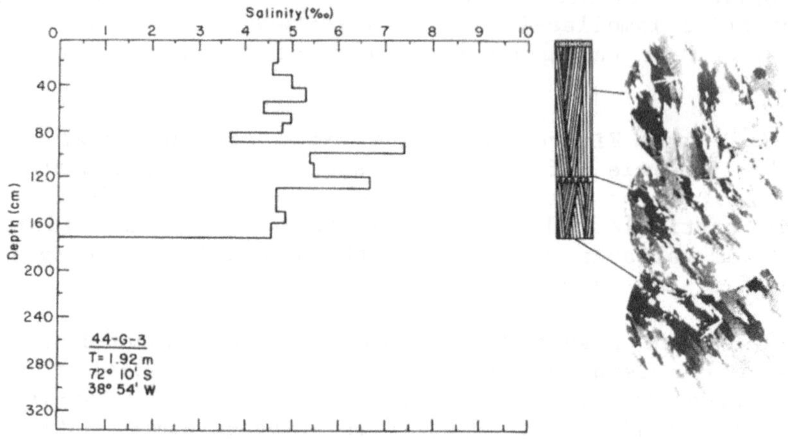

Fig. 54. Salinity, structure and thin section photos from Floe 44-G-3. A key to the ice structure diagram is given in Figure 52.

Floe 49-G-2 (see Figure 55). This first-year floe was composed mainly of congelation ice (72%) but had some frazil ice (22%) incorporated in a zone at 0.26-0.70 m depth. The thin section from 0.70 m exhibits a texture that is transitional between fine-grained frazil and coarser-grained congelation ice. The bottom section is composed of randomly aligned fibrous-textured crystals. The average salinity of the floe is 4.7‰.

Floe 61-G-2 (see Figure 56). This is a multiyear floe with an average salinity of 3.5‰. The upper meter of ice is mostly composed of congelation ice with moderately aligned crystals, underlain by 2.9 m of frazil ice of highly variable grain size. Crystals of frazil range in size from 0.4 to 4 mm and are usually arranged in layers that would suggest several separate episodes of frazil generation.

It was found that of the total length of multiyear core (35.85 m samples), about 75% was frazil and only 19% was congelation, with the remainder (6%) being fine-grained but of indeterminate origin. For the first-year ice (85.7 m) the frazil congelation percentages were 46% and 47% respectively, with 7% of indeterminate structure. These characteristics are shown in Figure 57, where congelation ice percentage is plotted as a function of ice thickness. The data indicate a slightly higher percentage of congelation ice for ice less than 1 m thick and roughly equal (but highly varied) percentages for ice between 1 and 2 m

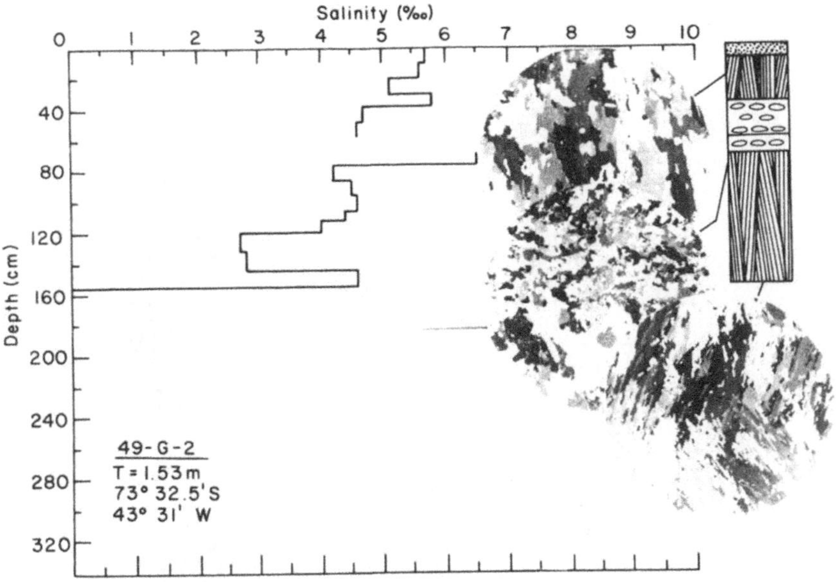

Fig. 55. Salinity, structure and thin section photos from Floe 49-G-2. A key to the ice structure diagram is given in Figure 52.

thick. However, no floe exceeding 2 m in thickness is composed of more than 50% congelation ice. Therefore the thickest ice in the Weddell Sea (>2 m) is composed primarily of ice with frazil structure.

One's first reaction in looking at these statistics is that it is the almost complete reversal of what would be presumed based on the commonly held assumptions about the growth of sea ice. Instead of the thinnest ice (<1 m) being primarily composed of frazil, it is the thickest ice that commonly contains the largest percentage of frazil.

Such frazil ice as occurs in the Weddell Sea pack ice is particularly interesting in that its origin and properties are very different in most all respects from the congelation ice upon which most work has been concentrated. This will be seen quite clearly later in this paper when the effect of ice structure on ice properties is discussed. We will now discuss some possible mechanisms for the generation of oceanic frazil ice.

Mechanisms of Frazil Generation in Oceans. To generate fine-grained frazil ice during freezing from the melt, the observations from rivers, lakes and oceans suggest one necessary condition, namely turbulence. Frazil generation originates as a result

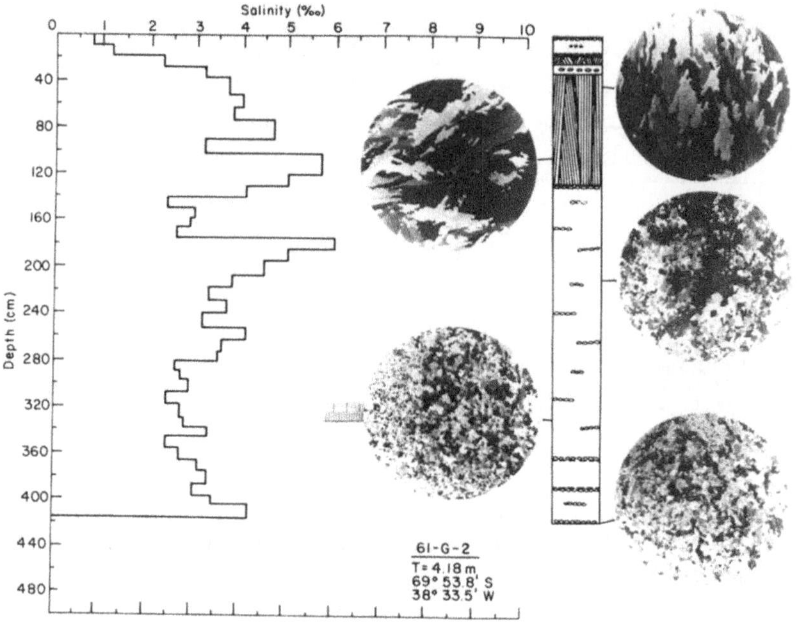

Fig. 56. Salinity, structure, and thin section photos from Floe 61-G-2. The key to the structure diagram is given in Figure 52.

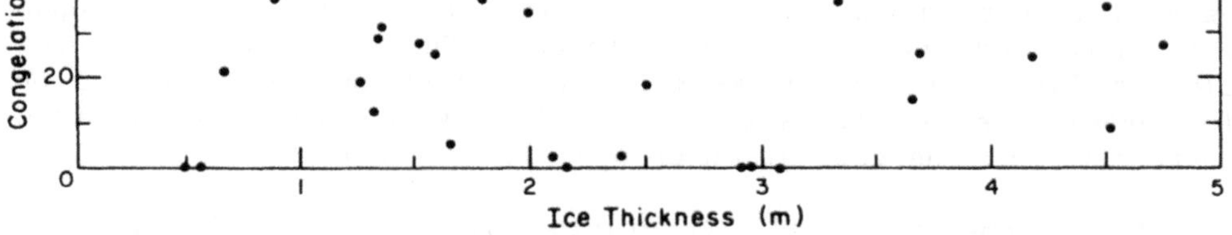

Fig. 57. Congelation ice (percentage) vs ice thickness for all cores obtained in the Weddell Sea.

of the large-scale (> cm scale) mixing of waters and during its development has been likened to a blizzard (an upside-down one) in the water column. However, some basic differences exist in the momentum and heat flow characteristics of rivers and oceans. In rivers the turbulence is nearly always purely mechanically derived from accelerations in the flow by increasing discharge, varying bed geometry (falls and rapids) or the like. Although the literature is not clear on this point, it appears that a threshold velocity of 0.6 m/s is necessary under freezing conditions to allow the onset of frazil formation (Ashton, pers. comm.). The constraint is primarily a mechanical one; while frazil generation requires efficient heat transfer between the cold atmosphere and the water, sufficient mechanical shear is necessary to mix the flow and carry the newly formed ice crystals rapidly away before a complete skim of ice can form. In the ocean, however, these mechanical influences are much less pronounced. Four possible sources of turbulence that can affect freezing conditions leading to frazil formation in the ocean are as follows.

(1) Wind- and wave-induced turbulence. As we previously described, this effect can only account for limited frazil production in wide leads and polynyas and near ice edges due to the extreme damping of these effects when a significant percentage of the sea surface is covered by ice.

(2) Water which is at a depressed freezing point due to pressure at depth and is suddenly adiabatically brought to the surface (Foldvik and Kvinge, 1974). Water of this type typically exists under large ice shelves, and the ascent occurs near the front. If this were the only mechanism producing frazil one might expect the percentage of frazil in floes to increase as ice shelves are approached. Correlations of this type have not as yet been observed. As well, the amount of ice production available by this mechanism (since possible freezing point depressions are only a few tenths of a degree) does not seen compatible with the large amount of frazil ice observed ($\approx$50% of the observed ice).

(3) Contact between two water masses of significantly different salinity but both at their freezing point. We will discuss this mechanism further in relation to multiyear ice structure since this type of ice formation occurs in under-ice melt ponds (of primarily fresh water) which drain into the ocean from melting surface ice and are stratified due to density differences. Freezing then occurs due to double diffusion (transfer of heat at a faster rate than salt) from the cold (<-1°C) seawater to the fresh water on top of it (Martin, 1974). This process could also occur associated with the drainage of river or glacier water into seawater that is colder than the fresh water freezing point. Again these effects are undoubtedly local in the regions they affect and

it is difficult to see how they could account for the observed ice conditions in the Weddell Sea.

A process of sufficient generality that occurs over a broad enough region to account for much of the observed frazil production is:

(4) Thermohaline convection initiated by surface cooling and freezing. Figure 58 indicates some differences between Arctic and Antarctic water column structure that indicate more potential for frazil ice generation in the Antarctic. In general, the mixed layer in the Antarctic is both considerably deeper than in the Arctic and is, also, totally at its local freezing point. Martin (1974), in a study of ice stalactites, indicated that the heat loss of a descending brine plume could be, when the water was at its freezing point, as much at 50% accounted for by frazil production. As the temperature of the brine plume is closer to the freezing point of the surrounding water, there is less tendency for stalactite forms to occur, implying all the heat loss could be taken out by frazil ice formation. We note, however, that in Martin's study the frazil crystals appeared to be entrained downward with the descending plume. More likely is that frazil crystals are generated in waters adjacent to the descending plume for two reasons. One effect is buoyancy. That is, the energy necessary to force an ice crystal downward with a density deficit approaching 10% compared to surrounding waters is considerable and much larger than the relative density differences that can lead to

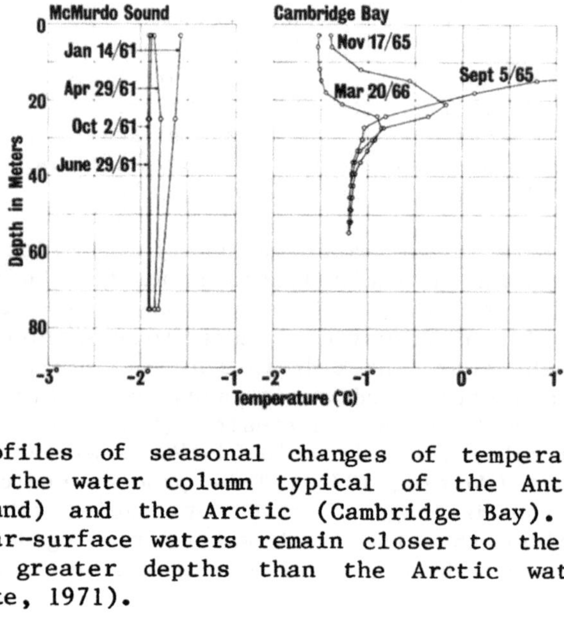

Fig. 58. Profiles of seasonal changes of temperature with depth in the water column typical of the Antarctic (McMurdo Sound) and the Arctic (Cambridge Bay). The Antarctic near-surface waters remain closer to the freezing point at greater depths than the Arctic waters (Lewis and Lake, 1971).

water column overturning. Secondly, if the waters of a descending cold brine plume are entraining surrounding waters then the mixture will also be in equilibrium at its freezing point. That is, there should not be any ice production, merely a more saline mixture at a temperature intermediate between the cold brine plume and the surrounding water. A simpler mechanism that preserves the buoyant characteristic of the ice crystals and still allows ice production at depth is to maintain the integrity of the descending brine plume and allow it to gain heat at a different rate than salt. This is known as double diffusion and has been observed in several other contexts in the ocean (e.g., Turner 1973). In this mechanism, the descending brine plume gains heat but diffuses salt at a much lower rate, thereby cooling adjacent waters to temperatures below their freezing point (at their given salinity). An ice crystal is then nucleated and rises due to buoyancy, and the remaining water also descends at a new equilibrium freezing temperature and salinity. While the process appears to be diffusion-limited by the ability to extract heat from the brine plume, the convection is continuously strengthened by the ice formation so the effect is a convection cascade which should quite effectively use the heat sink of the brine plume and establish a circulation boosting the convective process. Based on the observed structures (Figures 52-56), we also see that frazil forms under thicker ice. Advecting surface-generated frazil downward underneath this thicker ice is again a problem of overcoming strong upward buoyancy forces. A generation mechanism whereby ice crystals are formed at depth due to thermohaline convection is not constrained in this way. Descending brine plumes in leads and polynyas will set up a local circulation that could bring the ascending arm (containing some of the frazil) underneath adjoining ice. As well, substantial shear usually exists between the ice and the water column due to wind action on the ice, so motion of the ice cover would tend to bring a different piece (possibly thicker) into position over the ascending frazil crystals, if they are formed at depths greater than a few meters.

None of the above-mentioned mechanisms have, as yet, been detailed to any substantial degree. Whether they do indeed account for the observed frazil ice structures and whether frazil structures are more widespread than previously thought or are only an anomaly of the Weddell region remain subjects for future study.

Suspended Material in Frazil Ice. One very interesting aspect of frazil ice is that it may contain large amounts of fine-grained sediment and algae. Our observations in the Weddell Sea (Ackley, 1982; Buck and Garrison, 1982; Garrison and Buck, 1982) indicate that enough algae was incorporated into thin (<20 cm), young (≃1 day old) ice to give the ice significant coloration. Of primary importance is that our measurements indicated that the concentration of algae in this thin ice was several times the concentration in the water column in which the ice was growing. This observation is strongly in opposition to our past experience with growing ice sheets from the melt, where water impurities, dissolved or suspended, are for the most part rejected and lesser concentrations are generally observed in the ice than in the water. Similar evidence from the Arctic has been gathered (T. Osterkamp and J. Gosink, personal communication) where, in coastal waters, bottom sediments that are stirred up and incorporated into the growing ice are at higher concentrations than observed in the water column itself.

The short time scales (a few hours) observed in the growth of the algae-loaded ice in the Antarctic mitigates against biological growth affecting the increase in the concentration in the ice relative to the water column. Two possible mechanisms, both relying on frazil ice formation processes, can be proposed to explain these increases in concentration of suspended material in the ice relative to the water column. The first is ice nucleation by foreign particles. Concentration by ice nucleation can occur if frazil ice crystals preferentially nucleate on suspended particles. Since the small observed supercoolings limit the volume of ice formed to a small fraction of the volume of water, the effect is to have the same number of foreign particles in a much smaller volume, that is, to increase the concentration of particles in the ice phase. We note, however, that this process still requires the ice to form <u>in</u> the water column (frazil ice) rather than on <u>top</u> of the water column (congelation ice).

The second mechanism, which we call concentration by ice scavenging, occurs by collision of the frazil ice crystal with particles as it travels through the water column to the upper water surface (Garrison et al., 1983). Each crystal could collect several particles, depending upon how much the length of the path that the ice crystal travels exceeds the mean distance between particles in the water column.

We have microscopically examined thin sections of some of the Antarctic ice samples and have found evidence suggesting that for algae particles both nucleation and scavenging operate. This examination indicates that cells are found <u>within</u> individual ice crystals, suggesting that the cell nucleated the crystal, and also are found at grain boundaries <u>between</u> individual crystals, suggesting that the crystal scavenged the cell.

It appears, however, that scavenging is the dominant mechanism for incorporation of material. If nucleation were the only mechanism, a limit on the amount of material incorporated would be the number of crystals (one cell per crystal) per unit volume. Frazil ice crystal sizes observed are of the order of $\approx$1 mm diameter, implying an upper bound of $\approx 10^6$ cells/liter if nucleation is the only mechanism. Measured concentrations of algae exceed this by one to two orders of magnitude ($\approx 10^8$ cells/liter), indicating scavenging may be the more effective mechanism for incorporating algae. Another observation is that most foreign materials allow water supercoolings of the order of a few degrees Celsius to exist before they become effective as ice nuclei (Hobbs, 1974). Although supercoolings of a few tenths of a degree have been observed at the surfaces of Arctic leads (Katsaros, 1973) these values are characteristically less than required for most suspended mineral matter to serve as a nucleating agent. Biological materials, especially living cells from cold waters, have not been well studied as to their effectiveness as nucleating agents, and this behavior should be investigated. In conclusion, in the few cases studied the total particle concentration relative to the ice crystal size as well as the wide range of organic and inorganic materials incorporated in the ice support the contention that scavenging is a more important incorporation mechanism than nucleation.

9.3 Multiyear Ice

Considering that vast areas of the polar seas are covered with multiyear ice, there is an amazing lack of information on its structure and properties. In the following discussion we will use the term "multiyear ice" to refer to ice that has survived at least one summer's melt season. This rather casual usage corresponds to the way that the word "multiyear" is commonly utilized by sea ice specialists. This differs from the recommended WMO usage in which second-year ice has survived one melt season, multiyear ice has survived two or more melt seasons, and the phrase old ice is used to characterize both of these ice types.

In the Arctic, multiyear ice is distinguished by two main characteristics: 1) its rolling, hummocky surface, which is the result of differential melt, and 2) its salinity profile, which indicates values near zero for the ice above sea level and shows a gradual increase to values of 3.0 to 3.5°/oo for the ice in the lower portion of the floe. As will be discussed later, this profile appears to be primarily produced by the flushing of nearly pure surface meltwater downward through the ice sheet.

What is the structure of this ice? The limited work that has been performed on multiyear ice has largely focused on the horizontal layers that are produced by the summer melt cycle. The reason for this interest is obvious: the annual layering allows one to establish a minimum age for thick floes of pack ice. Previous to these studies age estimates were based upon subjective criteria such as the amount of pressuring and the detailed morphology of the ice surface. Age estimates have also been made from the ice thickness and from the locations of the maxima and minima in the salinity profiles (Savel'ev, 1958). As will be discussed later, although there is a general tendency to form a low salinity layer on the underside of the ice during the summer melt period, this salinity distribution is rapidly modified by brine migration and is difficult to use in estimating ice age except in extremely simple cases.

The first study of a horizontal "translucent-longitudinal" section of a thick (291 cm) pack ice floe was by Shumskii (1955) when in July of 1954 he made a brief visit to the drifting station SP-3. As a result of the above-freezing temperatures during the period of field work, very few detailed observations were made. The ice was divided into two main types, infiltered (0-38 cm) and normal sea ice (38-291 cm). The infiltered ice layer was described as resembling the firn of arctic glaciers and contained four distinguishable layers which were interpreted as annual. The normal sea ice was also classified into two types. The upper part (38 to 150 cm) contained lens-like irregular strata that were believed to be the result of thaw cycles. These strata were absent in the lower portion (150 to 291 cm) of the ice. Therefore, Shumskii reasoned that the lower portion of the sea ice (141 cm thick) was formed after the end of the thaw period while the 4-year accumulation of infiltered ice was forming at the surface of the ice sheet. The upper 112 cm of the sea ice was believed to be older. The total age of the floe was estimated at not less than 6 to 7 years. Shumskii also suggested that the presence of infiltered ice indicated that the snow line in the central Arctic is presently at sea level.

Following this preliminary study, detailed investigations of the horizontal layering in pack ice were made on SP-4 (Cherepanov, 1957) and Station Alpha (Schwarzacher, 1959). It was found that the annual layering is readily observable in the lower part of the ice sheet and less clearly defined nearer the upper surface where the effects of the summer melt are more pronounced. Cherepanov found ten recognizable annual layers in a 335-cm-thick floe (Figure 59), while Schwarzacher found either seven or eight layers in a 345-cm-thick floe. The boundaries between the annual layers were of two different types. The most common was a thin (2-5 mm) layer of milky white ice with a sharp upper boundary and an irregular lower boundary. The details of the formation of this layer are not known but it apparently develops during the period when the ice growth has stopped. The milky color of the layer may, in some way, be connected with the biological activity beneath the ice during the summer months. The formation of the milky layer is apparently not associated with either recrystallization or the nucleation of new grains, since when ice growth starts in the fall the crystals of the previous winter commonly resume growth with the same crystallographic orientation (Schwarzacher, 1959).

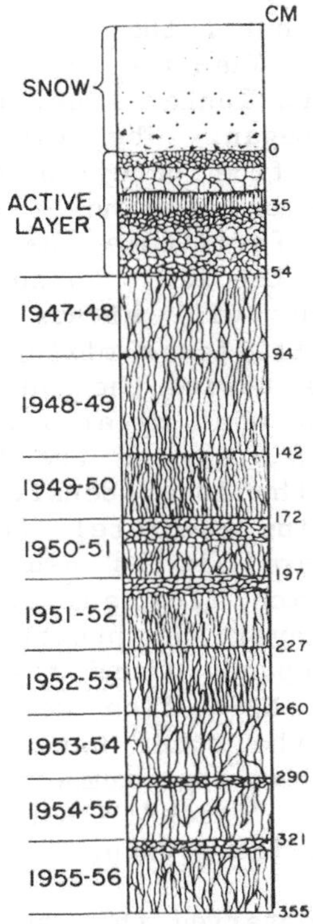

Fig. 59. Cross section of a multi-year floe (Cherepanov 1957).

The other type of summer layer is considerably thicker (1 to 10 cm) and shows a sharp decrease in grain size relative to the overlying winter ice. The grain shapes in this layer have been described as slightly elongated in the vertical direction. The c-axis orientations are in general horizontal but they show deviations up to 30 degrees (Schwarzacher, 1959). In the ice studied by Cherepanov the crystals were equi-axed and the orientation was random. This ice does not show the platy substructure so characteristic of sea ice and has a much lower salinity (1 to 1.5°/oo than the surrounding ice. The formation of such an annual layer is clearly the result of the formation during the summer of a stable layer of fresh meltwater between the lower surface of the ice and the underlying dense seawater. This process has been known for some time (Malmgren, 1927) and has been described in considerable detail on Station Alpha by Untersteiner and Badgley (1958). They found that many of the freshwater melt ponds that develop during the summer drain through core holes, forming an irregular layer of fresh water between the ice and the seawater. At the interface between the fresh (0°C) and saltwater (-1.6°C), ice crystals formed and floated upward until the fresh water was filled with a mesh of fragile crystals. When the crystals forming at the fresh/saltwater boundary could no longer rise, the formation of a solid layer of freshwater ice began. The formation of such a layer is to be expected under any floe occupied by a scientific station because the large number of core holes drilled in the floe for technical studies readily facilitate the rapid transfer of meltwater to the underside of the ice. On an unoccupied floe early in the melt period, meltwater will either be channeled into open leads where wave action will favor mixing with the seawater or it will form large melt pools toward the interior of the floe. Only later in the melt period when deterioration is far advanced do drainage holes form in all areas of the floe (Bennington, 1959). Even during this period, the lower density of the fresh meltwater inhibits it from percolating completely through the floe. When a freshwater layer does form, it is usually localized under the thinner portions of the ice floe and is particularly prevalent near open leads. This layer, if preserved by freezing, is very useful as a marker; Cherepanov found that four out of the nine annual boundaries in the NP-4 floe were marked by "fresh" ice layers (Figure 59). On the other hand, Schwarzacher (1959) has noted from examination of a large number of cores in the vicinity of Station Alpha that the "fresh" ice layers do not contribute significantly to the total ice growth.

Schwarzacher (1959) observed that the top 50 cm of the floes he examined contained ice which formed at or near the surface of the floe. This ice was predominantly the result of the freezing of meltwater pools that form during the summer and was characterized by c-axes-vertical orientations. A fine-grained granular ice may also be formed by snow falling in the meltwater pools, producing a slush which subsequently freezes. This melt and refreezing associated with the upper ice surface tends to obliterate the annua layering in the upper part of the ice sheet and makes the accurat estimation of the age of the ice in a floe difficult if only on cross section is examined. For example, of the 150 cores collected by Schwarzacher in the summer of 1958 from areas that did no show signs of old pressure ridges, only 25% consisted of undisturbed ice and only 2% were sufficiently unaltered to show a annual layering throughout the complete thickness of the flo (eight layers to 1950). Inasmuch as neither Schwarzacher Cherepanov, or Bennington observed any annual stratification c superimposed firn, this casts considerable doubt on Shumskii' suggestion that the snow line in the central Arctic is currentl at sea level. Both Cherepanov and Schwarzacher have noted th the formation of infiltered ice is strictly local and is highl dependent upon the details of the surface relief. If infilter ice forms at all it will form in small depressed areas where o heavy snowdrifts are present. In most areas on both NP-4 a Alpha, both the snow cover and 20 to 30 cm of ice thawed duri the summer (Untersteiner, 1961). It therefore seems quite like that the ice identified by Shumskii as superimposed firn was actually ice formed by melting and refreezing in the upper part of the ice sheet. If this is the case the SP-3 floe would be a minimum of 9 years old instead of 6 or 7 as suggested by Shumskii.

It is also of interest to point out that the thicknesses of the annual layers can be used to determine if the top annual layer is the layer representing the year when the ice sheet initially formed. Figure 60 shows the winter increments of ice growth plotted against the number of the annual layer measured from the top of the floe. The data are from Cherepanov (1957) and Schwarzacher (1959). For the last 7 years (SP-4) and 5 years (Alpha) it is clear that the ice thickness has approximately reached a steady state, the ice growth in the winter being the same as the expected ablation during the summer. Even the thickest annual layer measured on Alpha indicates an existing ice thickness of 260 to 280 cm at the end of the summer prior to the accumulation of the layer if the empirical relations of Schwarzacher (1959, p. 2365) and Untersteiner (1961, p. 165) are used. Inasmuch as 280 cm is the maximum one winter's fast sea ice accumulation known in the Canadian Arctic (Eureka, 1946-47; see Bilello, 1961), this indicates that although the floe studied by Schwarzacher contained 8 years' accumulation of ice, the initial floe formed a minimum of 9 years prior to the time of measurement. Because the accumulation of winter ice in the Arctic Basin is usually considerably less than 260-280 cm (160 cm at NP-2 (Petrov, 1954-55), 173 cm at NP-4 (Cherepanov, 1957, Figure 4, p. 183) and 180 cm at Alpha [Frankenstein, personal communication)] the estimate of 9 years for the age of the floe at Alpha is probably quite conservative, and the floe at NP-4 must be more than 11 years old.

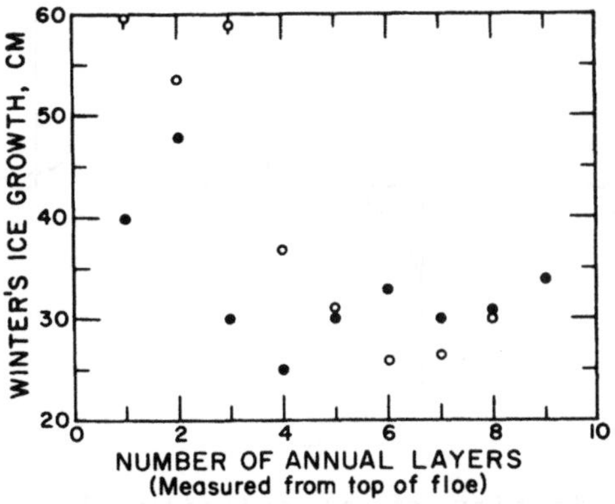

Fig. 60. Winter's increment of ice growth plotted against the number of annual layers measured from the top of the floe. Data from Cherepanov (1957) and Schwarzacher (1959). The circles and dots represent observations on Station Alpha and SP-4 respectively.

One interesting aspect of the freezing of the freshwater layers that can form beneath multiyear ice floes is that the heat transfer rates are 5 to 10 times those calculated assuming that all heat transfer is by thermal diffusion (Martin and Kauffman, 1974). What appears to happen is as follows. When the surface meltwater flows off the pack into the ocean, and ultimately beneath the ice, we have an ice layer at its melting point (≈0°C) underlain by a nearly freshwater layer at its freezing point (≈0°C), in turn underlain by seawater at its freezing point (-1.8°C). At the boundary between the fresh- and seawater a zone of water forms which, because of the rapid diffusion of heat relative to salt and the fact that the density of water with a salinity of <24.7 decreases on cooling, is both supercooled and less dense than the overlying water (Figure 61). The resulting density distribution is unstable, with a Rayleigh number of 10^3 to 10^4, and free convection results. When this supercooled water rises, it is nucleated by the overlying ice layer and forms a mesh of thin vertical interlocking ice crystals that ultimately grow down to the freshwater/seawater interface. At this time the presence of ice crystals throughout the complete thickness of the freshwater layer eliminates supercooling, constrains both temperature and salinity to lie on the freezing curve, and permits them to diffuse in the vertical direction until both quantities become horizontally uniform. The combination of these processes results in the lateral growth of the ice crystals, which continues until a horizontal ice sheet forms. Excellent photographs showing the growth of ice crystals during a laboratory simulation of this process can be found in Martin and Kauffman (1974).

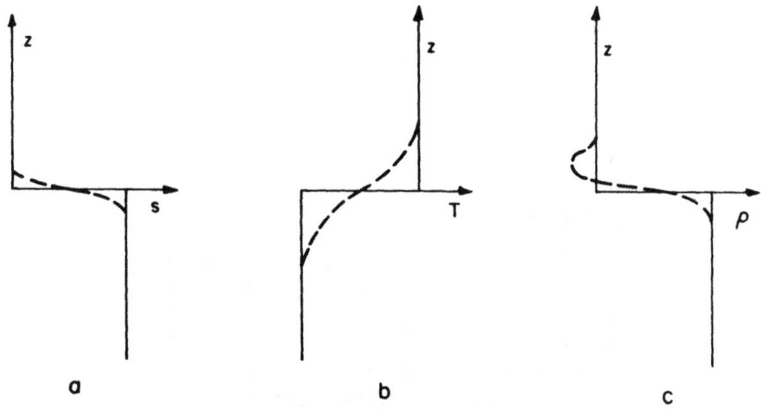

Fig. 61. The distribution of a) salinity, b) temperature and c) the resultant density caused by the diffusion of a freshwater layer over a saltwater layer, with both layers at their respective freezing points. The size of the density inversion is exaggerated (Martin and Kauffman, 1974).

Just what is the structure of multiyear ice? The only thing that can be said at present is that we don't know. As was discussed, we used to think that we did, that multiyear ice was a layer-cake of annual growth layers with a layer thickness roughly equaling the amount of summer ablation. As has been shown, this type of ice clearly exists. However, based on the limited samples (150 cores) collected by Schwarzacher in the Arctic Ocean from multiyear ice that appeared to be undeformed, only 25% of the ice actually was undeformed and only 2% showed annual layering throughout the complete thickness of the floe. These results should be coupled with the results obtained in the Weddell Sea by Gow et al. (1982) where 75% of the multiyear ice was frazil. It should also be noted (Meeks et al., 1974) that recrystallized ice has tentatively been identified in the low salinity, above-water-level portions of thick multiyear floes studied in the Beaufort Sea during the 1972 AIDJEX pilot experiment. Perhaps the multimaximum fabric reported by Savel'ev (1958, Figure 5) from a hummock on a multiyear flow is from this type of ice (similar fabrics are produced by recrystallization in glacier ice). If this interpretation is correct, we will not be surprised, as we have believed for a long time that sea ice would probably recrystallize if the brine inclusions, which presumably lock in the substructure, were drained from the ice. This drainage, in fact, does occur in the upper part of thick multiyear floes.

When all the above information is combined, we frankly must admit that we do not know what to think. However, we do know what is needed: a program that systematically obtains cores from different regions of the Arctic and Antarctic pack ice and studies their structure. Such a program is not hard to accomplish, but it requires good logistic support.

In concluding this section it should be pointed out that understanding why the structurally different types of sea ice form and being able to estimate how much of each ice type will exist at different locations at different times of the year is not only of academic interest. It is also a problem of considerable applied importance as each of these different ice types has different physical properties. At the present time we at least know what the problem is. This alone should stimulate further work in this area.

This is a convenient place to also mention that although undeformed multiyear ice commonly has a steady-state thickness of between 2.5 and 5 m (Maykut and Untersteiner, 1971) it appears to be possible to obtain undeformed multiyear sea ice with thicknesses of at least 12 m under present climatic conditions (Walker and Wadhams, 1979). Examples of such thick ice would appear to be the floe upon which NP-6 was situated (10-12 m, Cherepanov, 1964), a floe from Nansen Sound, Ellesmere Island, described by Serson

(1972, 10 m thick), the sea ice portions of the Koettlitz Glacier, McMurdo Sound, Antarctica (15 m, Gow and Epstein, 1972) and the Ward Hunt Ice Shelf off Northern Ellesmere Island (20 m, Lyons et al., 1971). The formation of such thick ice requires rather special conditions in that the oceanic heat flux must be near zero and the annual snowfall must be sufficient to maintain a high surface albedo without allowing the accumulation of snow on the ice surface. Walker and Wadhams (1979) have calculated that under such conditions a thickness of 12 m would be reached in about 65 years, with a final equilibrium thickness of roughly 20 m requiring about 200 to 300 years.

10. THE SALINITY PROFILE

10.1 Observations

There are ice salinity data scattered through the sea ice literature. However, most of this information is very spotty -- a profile here, a profile there. There are only a few exceptions where reasonably systematic salinity observations were made at a given site throughout the complete growth season. The best example of this type of study is by Nakawo and Sinha (1981), who obtained salinity profiles from a site in Eclipse Sound near Pond Inlet, Baffin Island, on a weekly basis from November 1977 to April 1978. The profiles are shown in Figure 62 (a two-week interval is used here for clarity). A plot of the variations in mean daily

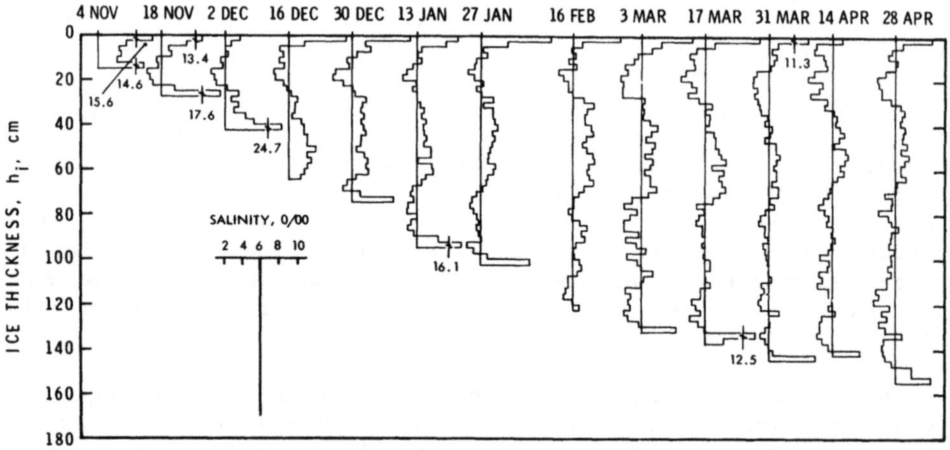

Fig. 62. Salinity profiles of ice of Eclipse Sound at intervals of two weeks during the winter of 1977-78. Scale for salinity is shown in insert. Vertical solid lines represent a value of 6°/oo and are given as reference (Nakawo and Sinha, 1981).

air temperature which can be consulted in the original reference shows values that were below -10°C for all but a very few days just after freeze-up and commonly were below -20°C. These profiles show characteristics that are similar to those of innumerable other profiles obtained from ice in the high Arctic. Note that there is initially a rapid decrease in the salinity at a given level in newly formed ice [see also Malmgren (1927) and Weeks and Lee (1958, 1962)]. After this initial rapid decrease, which appears to occur within a week after the ice forms, the salinity values attain quasi-stable values which then decrease very slowly throughout the growth season. In many ice sheets the uppermost layer of ice shows a higher salinity, as does the lowermost layer. This results in the profiles having a C-shaped appearance.

The other set of reasonably complete salinity measurements was obtained at Hopedale, Labrador, by Weeks and Lee (1958). The climate at Hopedale was very different than at Pond Inlet, with several thaw periods occurring when air temperatures went above -10°C and at times above freezing. The continuous, relatively rapid desalination that was observed (Figure 63) even in December and January is presumably caused by the higher air temperatures, which in turn produced higher ice temperatures, resulting in

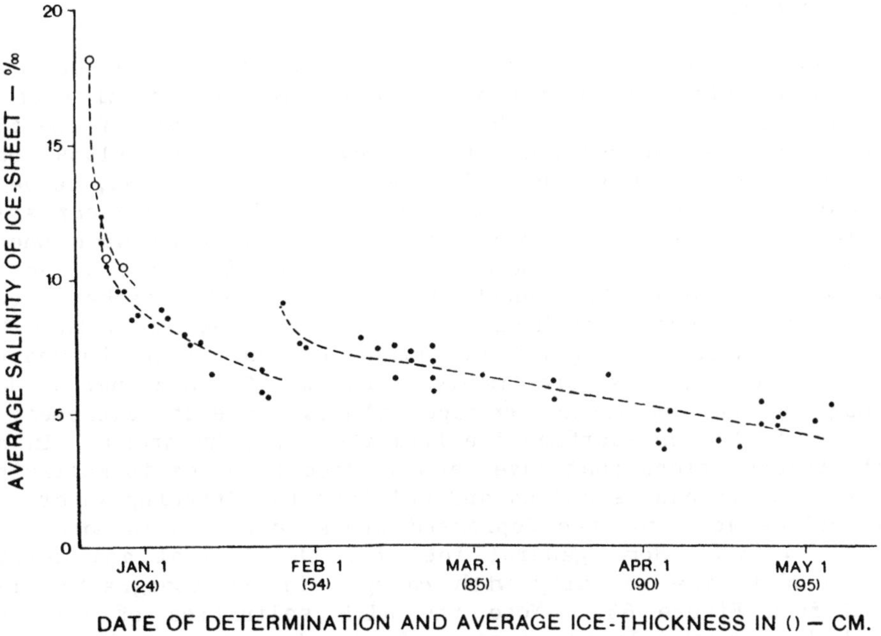

Fig. 63. Average salinity of ice sheet at Hopedale, Labrador, plotted against ice thickness and date of measurement (Weeks and Lee, 1958).

increased brine volumes and enhanced brine drainage. Nakawo and Sinha note that at Pond Inlet similar rapid desalination also occurred but not until May and June (data not shown in Figure 62).

It should be remembered in examining salinity profiles that even in the most homogeneous-appearing ice there is a small-scale, apparently random variation in the salinity. In the only available study performed to date of such variations (Thule, Greenland), a standard deviation of 0.62^{o}/oo was obtained at a given depth level in ice that would seem to be completely uniform (Weeks and Lee, 1962).

Plots of the average ice salinity for the complete ice sheet versus ice sheet thickness have been prepared by Cox and Weeks (1974) for first-year sea ice using data from a wide variety of sources and locations. The results are rather surprising in that they are not only consistent but also show only a small amount of scatter (Figure 64). Initially there is a rather rapid linear drop in the average salinity until an ice thickness of approximately 0.4 m is reached. At this and larger thicknesses the curve of average salinity versus ice thickness is also linear but with an appreciably decreased slope. Recent work by Gow et al. (1982) shows that the bulk salinity of Antarctic sea ice tends to be greater than that of sea ice of comparable age and thermal history in the Arctic.

When in the spring the upper surface of the ice starts to melt, there initially is a rapid decrease in the salinity of the above-sea-level portions of the ice. The best example of the final salinity profile for multiyear ice is by Schwarzacher (1959) based on 40 cores and 2060 salinity determinations (Figure 65, curve C). In the more recent study by Cox and Weeks (1974) multiyear salinity data were separated into two classes depending upon whether the core was selected from a hummock (Figure 65, curve A) or from a depression (curve B). Based on these data Schwarzacher's curve appears representative of hummocky ice or at least ice with a high freeboard. Note the significant amount of scatter in the measurements. The same is presumably true of Schwarzacher's data, although he only presented average values. Note the extremely low salinity of the near-surface ice from the hummocky areas. In many of the thicker floes that have been studied this ice is sufficiently pure that it can be melted and utilized for drinking water. The near-surface ice from the depressed areas (curve B) is more saline by 3 to 4°/oo. Some general feel for the lateral and vertical variations in ice salinity with changes in ice topography can be gained from Figure 66. Note the high salinities of the newly formed ice in the lower portion of the ice sheet and the great variability in the upper portion of the profile. Figure 67 shows a plot of average salinity versus average ice thickness for multiyear ice. There is an appreciable amount of scatter and a

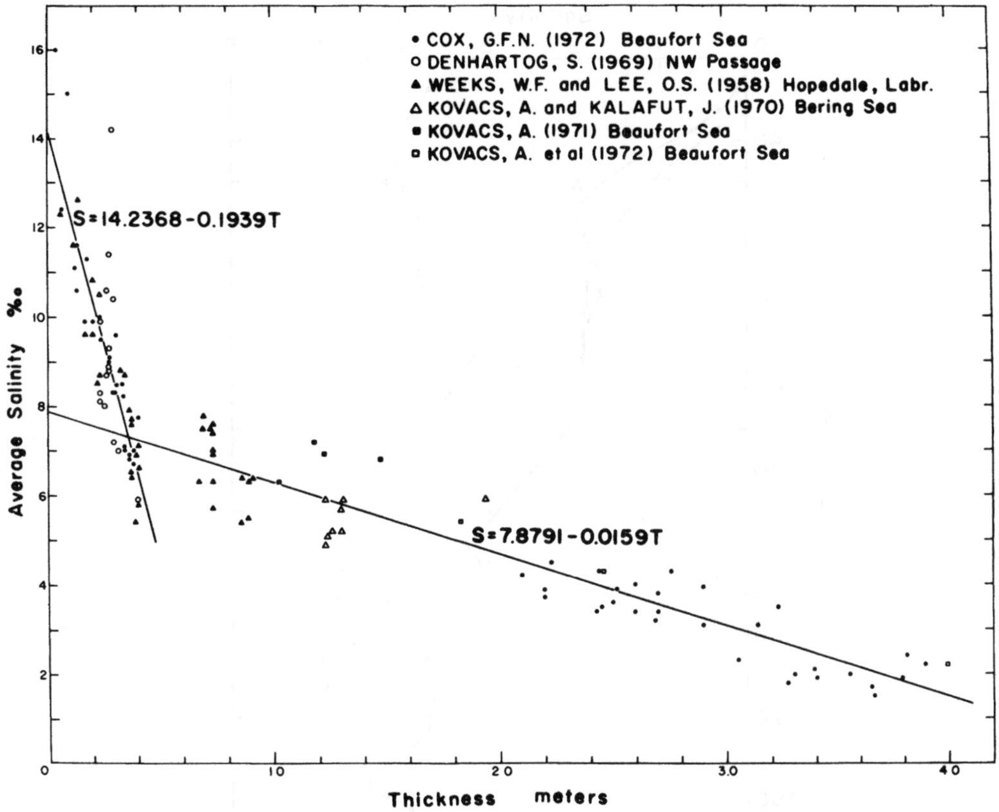

Fig. 64. Average salinity of sea ice as a function of ice thickness for cold sea ice sampled during the growth season (Cox and Weeks, 1974).

very slight increase in mean salinity with increasing ice thickness.

There appear to be several factors that contribute in different degrees to determining the salinity profile that exists at any given time:

1. The initial amount of salt entrapped in the ice.
2. The migration of liquid inclusions through the ice crystals.
3. Brine expulsion or the squeezing of brine out of the ice as the result of differential volume changes in the different phases composing the sea ice.
4. Brine drainage.
5. Flushing.

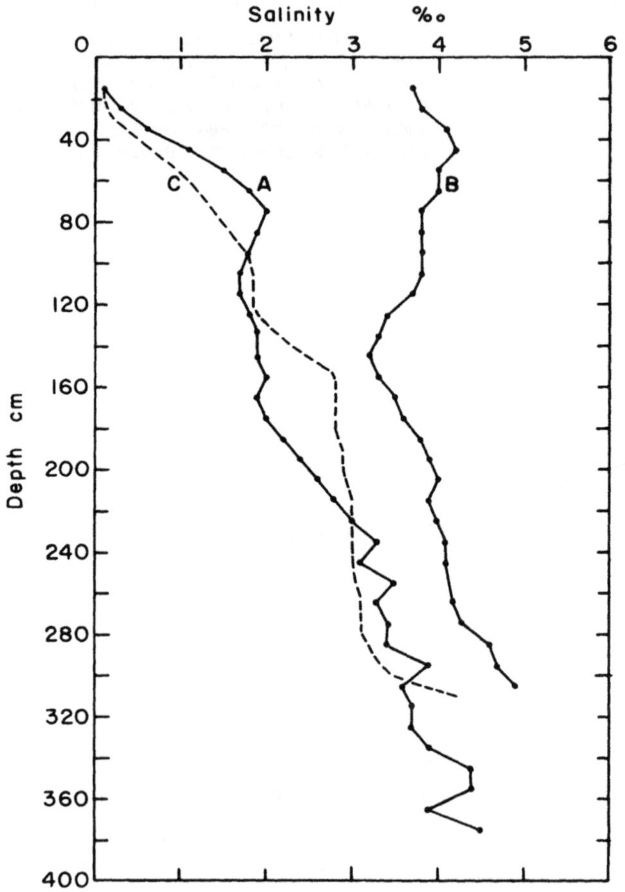

Fig. 65. Average multiyear salinity profiles for hummocked (curve A) and depressed (curve B) areas. Curve C is the average multiyear salinity profile determined on Station Alpha by Schwarzacher (1959). (From Cox and Weeks, 1975).

10.2 Initial Salt Entrapment

The first step in unraveling the variations in the salinity profile of ice is to understand how much salt is initially entrapped in the ice as a function of growth conditions and structure. Surprisingly little work has been carried out on this subject. Fortunately, there has been considerable work on this general subject for other similar materials to serve as a guide to understanding the situation for sea ice.

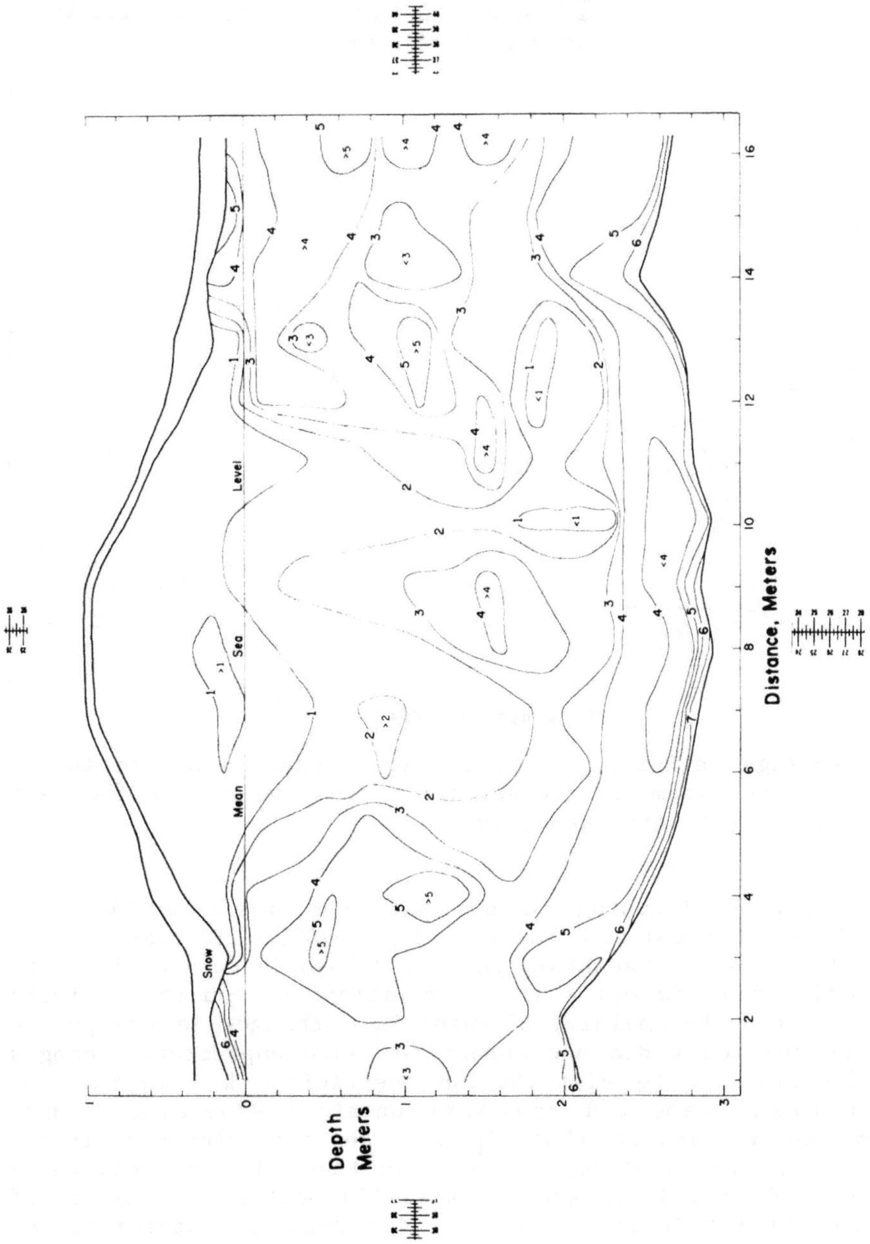

Fig. 66. A cross section of an area of multiyear ice showing the variation of salinity with topography (Cox and Weeks, 1975).

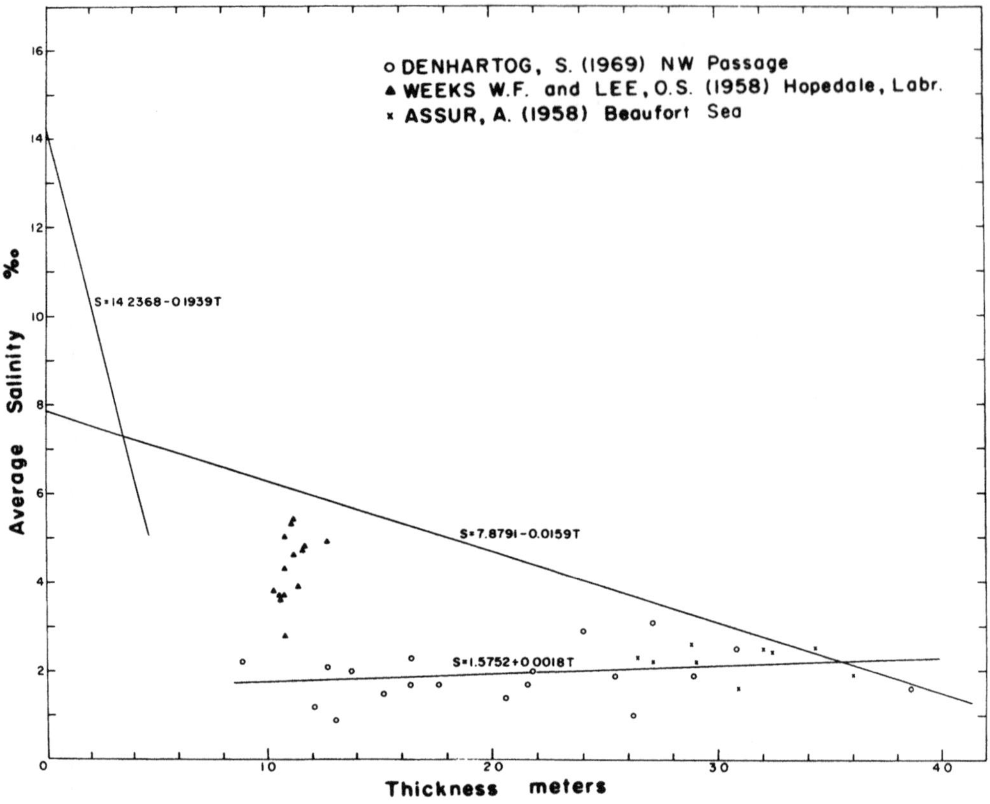

Fig. 67. Average salinity of sea ice as a function of ice thickness for warm sea ice sampled during the end of the melt season (Cox and Weeks, 1975).

Early studies focusing on the salt entrapment problem are those of Johnson (1943) and Adams et al. (1963). In some aspects these results were contradictory in that Johnson's results indicated that although there was a linear relation between the salinity of the ice S_i and the salinity of water S_w , changes in the growth velocity of the ice v did not appear to cause appreciable changes in the effective solute distribution coefficient $k \equiv S_i/S_w$. On the other hand, Adams and coworkers obtained experimental data indicating quite clearly that S_i was an approximately linear function of v. In an attempt to resolve these differences, Weeks and Lofgren (1967) and Cox and Weeks (1975) performed a number of experiments using NaCl ice as a model for related behavior in sea ice. These two sets of experiments were similar in purpose but quite different in technique. In the Weeks and Lofgren study, ice sheets up to 25 cm thick were formed during freezing runs that lasted between 48 and 110 hours. At the end of the run the ice was removed and cut into 1-cm-thick slices, and the salinity of each slice measured. Because the freezing runs were very short, brine drainage was neglected, which is, of course, a rather questionable assumption because of the rapid drop in salinity that usually occurs during initial ice growth. Water salinities were also calculated for each stage of ice growth by considering the amount of salt rejected from the ice. The values of k were then determined by combining the ice salinities with the water salinities that were appropriate for that ice level.

In the Cox and Weeks study the radioactive isotope ^{22}Na was used as a tracer and the salinity of both the ice and water was determined by sequentially measuring the gamma ray emission from different levels of the sample which, as in the previous experiment, was produced by unidirectional freezing. After considerable processing of the resulting data it was possible to obtain, via non-destructive observations, replicate salinity profiles that were in good agreement with more directly determined values. As these freezing runs took over 900 hours, it was no longer possible to neglect brine drainage. Therefore, only the salinities from the newly formed ice located just above the skeleton layer (that essentially no-strength layer at the bottom of a growing ice sheet where there are no lateral bonds between the ice platelets that compose each crystal) were used. It was assumed that the skeleton layer had a constant thickness of 3 cm.

In analyzing these results Weeks, Cox and Lofgren used the theory of Burton et al. (1953) which has been quite effective in treating experimental results in the field of metallurgy. They assumed that because of mixing the concentration C will approach a constant value of C_ℓ at some distance delta from the growing interface. For distances less than δ the velocity component normal to the interface approaches that due to crystal growth v. For a steady state at $x < \delta$, the continuity equation is

$$D \frac{d^2C}{dx^2} + v \frac{dC}{dx} = 0 \tag{1.22}$$

with the boundary conditions

$$(C_i - C_s)\, v + D \frac{dC}{dx} = 0 \quad \text{at } x = 0 \tag{1.23}$$

where C_i and C_s are the concentrations in the liquid at the interface (x = 0) and in the solid respectively and

$$C = C_\ell \text{ at } x \geq \delta.$$

The solutions to eq. (1.22) with these boundary conditions are given by

$$\frac{C - C_s}{C_\ell - C_s} = \exp\left[\frac{v}{d}(\delta - x)\right] \tag{1.24}$$

and the concentration C_i of the liquid at the interface is

$$\frac{C_i - C_s}{C_\ell - C_s} = \exp\left[\frac{\delta v}{D}\right]. \tag{1.25}$$

If δ is chosen properly this last equation can be forced to give the correct value of C_i. Because $C_s/C_i = k_o$, eq. (1.25) can be rewritten as

$$k = \frac{k_o}{k_o + (1 - k_o)\exp\left[-\frac{\delta v}{D}\right]}. \tag{1.26}$$

Here k_o can be considered to be the value of k at v = 0 (provided that a cellular interface were to remain stable). Inasmuch as $k_o < 1$ for salt-water systems, eq. (1.26) can be rearranged as

$$\ln\left(\frac{1}{k} - 1\right) = \ln\left(\frac{1}{k_o} - 1\right) - \frac{\delta v}{D} \tag{1.27}$$

which is a straight line on a plot of ln(1/k - 1) vs v with a slope of $-\delta/D$ and a zero intercept of ln $(1/k_o - 1)$. Figure 68 shows the resulting plot of Cox and Weeks' (1975) data. The data are quite linear with the exception of the values obtained at very low v values. Based on Lofgren and Weeks' observations, the drift toward higher (1/k - 1) values in the low v range is undoubtedly related to changes in the morphology of the solid/liquid interface as conditions are approached where a planar interface becomes stable. Ignoring the values at low v, least-squares gives k_o = 0.26 and δ/D = 7243 s/cm. The results obtained by Lofgren and Weeks were similar but the value obtained for δ/D was lower (5090 s/cm). Figure 69 shows the later data set presented as a plot of k versus [v/D]. Note that at higher growth velocities essentially all the salt in the solution is trapped in the ice.

How do these results compare with field observations? Figure 38 showed a plot of v as a function of depth in the ice against salinity S_i. There is clearly a positive correlation between v and S_i as predicted by the BPS theory. Replotting these results (Nakawo and Sinha, 1981) as shown in Figure 70 clearly shows the

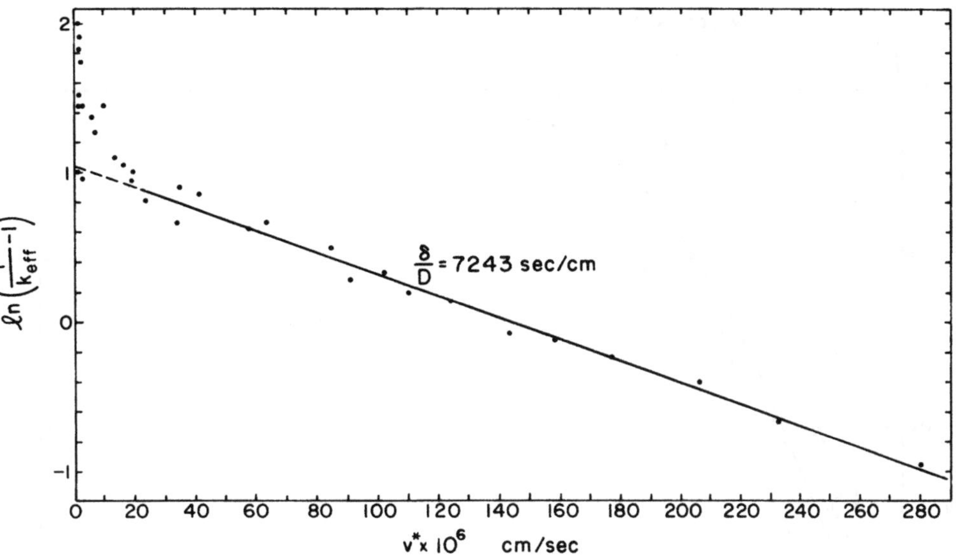

Fig. 68. Plot of ln(1/k − 1) versus v using the salinity entrapment data of Cox and Weeks (1975).

linear relation suggested by the BPS theory. Presenting these results in a more direct manner by plotting k versus v gives Figure 71. Here the ice between 5 and 25 cm consists of discontinuous columnar grains, with the length of the columns less than 1 cm. Below 25 cm the ice consists of long columnar grains. Note that there does not appear to be any appreciable difference between discontinuous columnar grains and continuous columnar grains. The BPS theory appears to work quite satisfactorily. Also shown in Figure 71 are data points obtained by Cox and Weeks. These latter observations consistently show more salt present in the ice, a not unexpected result inasmuch as they were determined by nondestructive testing and did not experience the brine drainage that commonly occurs during normal sampling procedures. Therefore, at the present time the best way to estimate the amount of salt initially entrapped in sea ice is to use the relation

$$S_i = S_w \left[\frac{k^*}{k^* + (1 - k^*) \exp\left(-\frac{\delta v}{D}\right)} \right] \tag{1.28}$$

with $k^* = 0.12$ and $\delta/D = 4.2 \times 10^4$ s/cm (Nakawo and Sinha, 1981). Here we have replaced k_o with k^* to distinguish it from the true equilibrium k_o value.

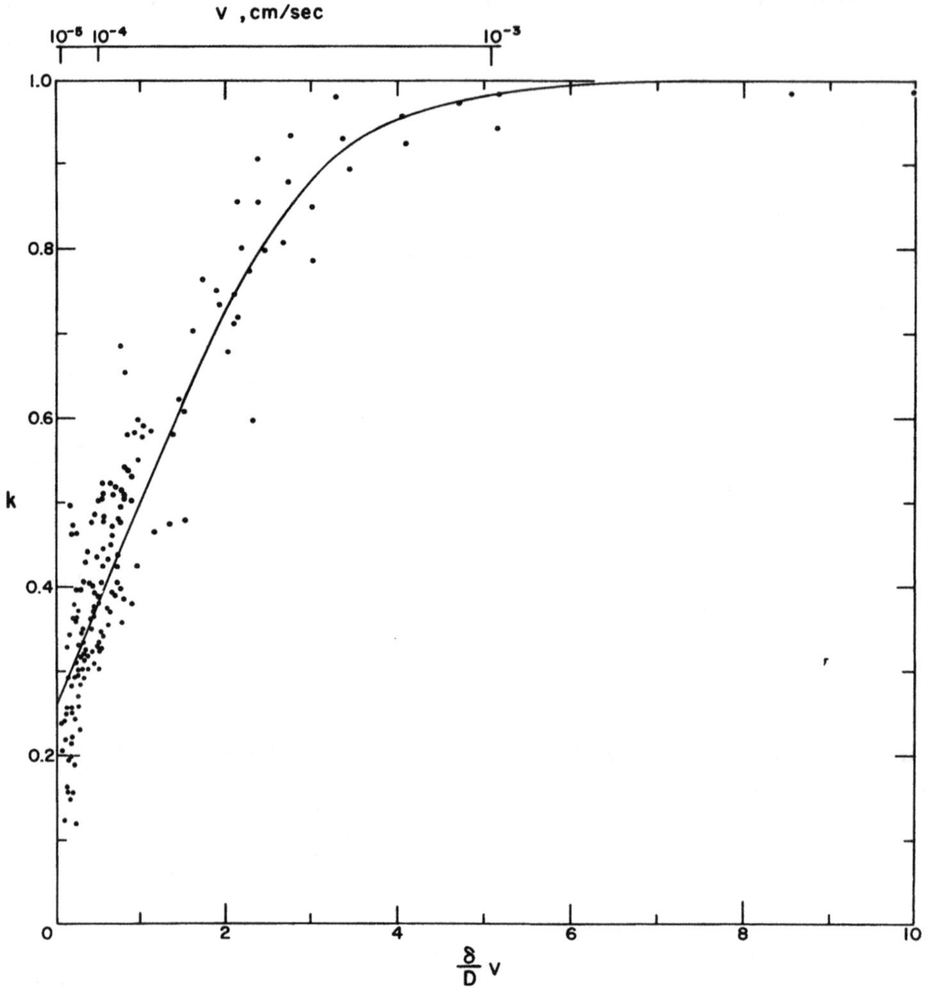

Fig. 69. Plot of k versus v/D for the salinity entrapment data of Weeks and Lofgren (1967). The curve is a least-squares fit of eq. (26).

Although it would appear that our knowledge of initial salt entrapment is in reasonable shape, there are some disturbing aspects. Although the BPS theory appears to work it is based on a series of assumptions that strictly apply only to impurities that are incorporated in solid-solution in the solid phase. In the sea ice case we are clearly dealing with bulk entrapment of the melt at a dendritic interface. There have been some attempts to analyze this situation (Edie and Kirwan, 1973; Myerson and Kirwan, 1977a,b; Ozum and Kirwan, 1976). In the last reference this approach was applied to the freezing of stirred NaCl-H_2O solutions

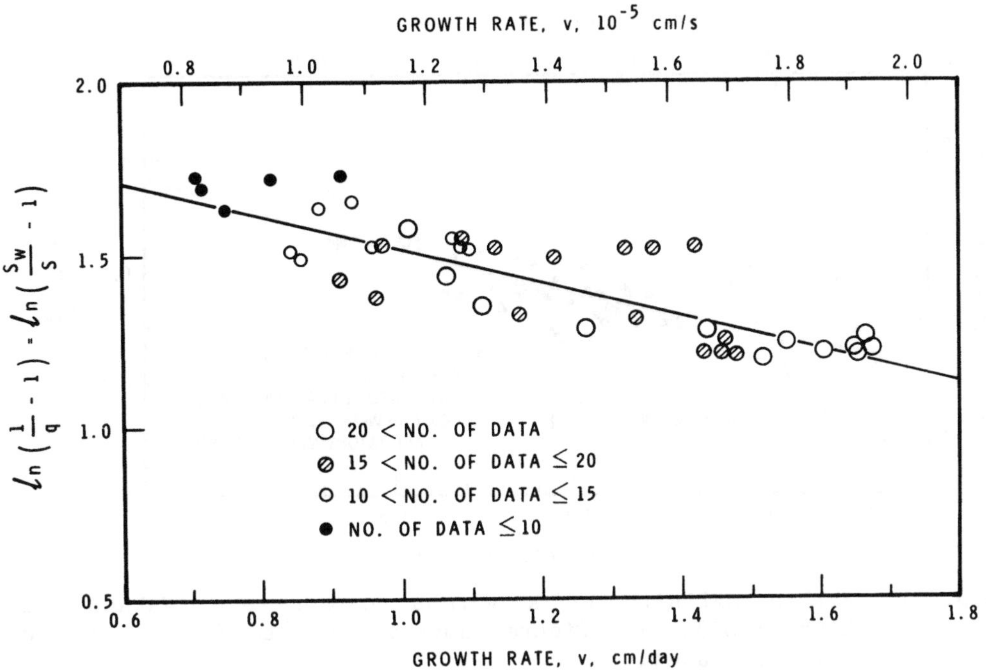

Fig. 70. Plot of ln(1/k-1) versus v for columnar-grained sea ice from depths between 25 and 125 cm, Eclipse Sound, NWT. The solid line is a least-squares fit of eq. (1.27). (After Nakawo and Sinha 1981).

with considerable success. It would be interesting to modify these relations in order to treat free convection as the mixing process and then apply this analysis to the case of sea ice.

10.3 Brine Drainage Mechanisms

Brine Pocket Migration. In the first published discussion of a possible mechanism for removing salt from sea ice, Whitman (1926) pointed out that a temperature gradient in the ice establishes a concentration gradient in a brine pocket if phase equilibrium is to be maintained. This, in turn, during the winter causes the diffusion of solute from the cold, saline upper end of the brine pocket to the warmer, less saline lower end of the pocket. Associated with this diffusion, the ice at the warm end of the pocket dissolves while freezing occurs at the cold end, resulting in the migration of the brine pocket toward the warm side of the ice.

Since this early paper the phenomenon of liquid inclusions or liquid zones migrating through solid crystals has received

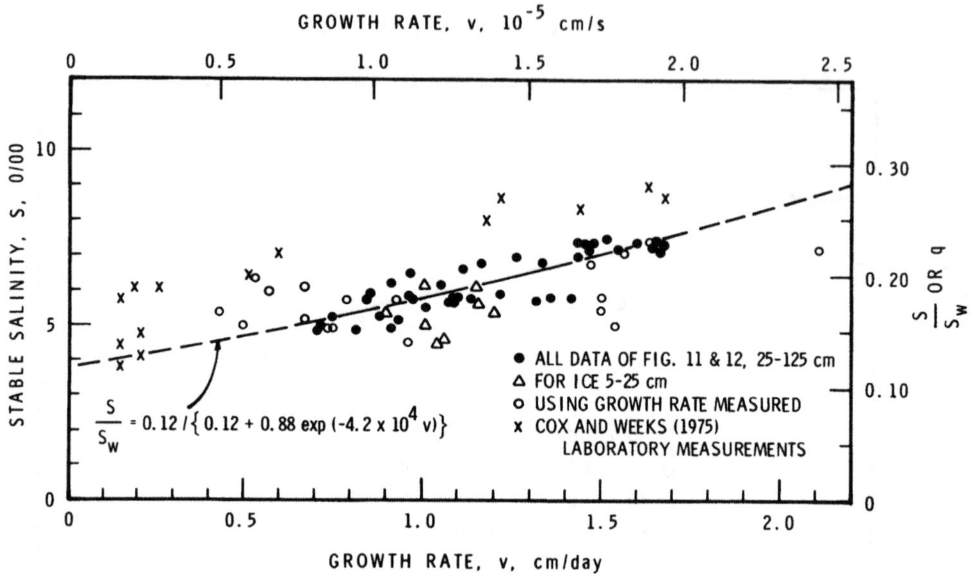

Fig. 71. Stable salinity or k versus growth rate (Nakawo and Sinha 1981). The figure numbers refer to figures in the original paper.

considerable attention in the crystal growth and crystal purification literature (Wernick, 1956; Pfann, 1958; Tiller, 1963) where it is referred to as temperature gradient zone melting (TGZM).

The theory is as follows. Because for salt-water systems the equilibrium value of k is very small, for the migration of a small inclusion such as a brine pocket effectively all the salts can be considered to be excluded from the ice phase. Therefore J, the flux of solute at the freezing interface of a brine pocket, is

$$J = Cv\,(1-k) \approx Cv. \tag{1.29}$$

For the steady state, eq. (1.29) can be written in terms of the concentration gradient dC/dx as

$$Cv = -D\frac{dC}{dx} \tag{1.30}$$

where D is the diffusion coefficient of the salt (or salts) in water. Then substituting

$$\frac{dC}{dx} = \frac{dC}{d\theta} \cdot \frac{d\theta}{dx} \tag{1.31}$$

where θ = temperature and (dC/dθ) is specified by the phase diagram, it follows that

$$v = -\frac{D}{C}\left[\frac{dC}{d\theta} \cdot \frac{d\theta}{dx}\right]. \tag{1.32}$$

Here D, C, dC/dθ and v are all dependent on the temperature θ. In eq. (1.32) the appropriate temperature gradient is the gradient in the liquid G_ℓ.

For a spherical droplet a solution of Laplace's equation provides a relation between G_ℓ and the temperature gradient in the solid G_s which is specified

$$\frac{G_\ell}{G_s} = \frac{3\kappa_s}{(2\kappa_s + \kappa_\ell)} . \tag{1.33}$$

Here κ indicates the thermal conductivity. As κ_{ice} is approximately 4 times κ_{brine}, the gradient ratio is fairly insensitive to small changes in κ_{brine} caused by variations in the brine composition. Also, as the values of κ change slowly with temperature, the G_ℓ/G_s ratio is roughly constant. Substituting appropriate thermal conductivity values Seidensticker (1965) obtained G_ℓ = 1.34 G_s. Making an additional correction caused by the density (ρ) differences between the ice and the brine Seidensticker finally obtained

$$v = \left(\frac{D}{mC}\right)\left(\frac{2\kappa_s}{(2\kappa_s + \kappa_\ell)}\right)\left(\frac{\rho_\ell}{\rho_s}\right) = 1.46 \frac{D}{mC} \tag{1.34}$$

as the equation giving the velocity of migration in a unit thermal gradient. In eq. (1.34) m = dθ/dC and is the slope of the liquidus curve from the phase diagram (note that the units of C do not matter as long as m and C are in compatible units).

How do the predictions of eq. (1.34) agree with observed brine pocket migration rates? There have been a number of studies of the migration of brine inclusions through ice crystals (Kingery and Goodnow, 1963; Hoekstra et al., 1965; Harrison, 1965; Jones, 1973, 1974). That the theory and observations are in reasonable agreement is shown by Figures 72 and 73. Figure 72 shows the relation between migration velocity and ice temperature at a constant temperature gradient of 1°C/cm in the ice. The general shapes of the curves are in good agreement although the theory predicts velocities that are ≈20% higher than observed. That v is

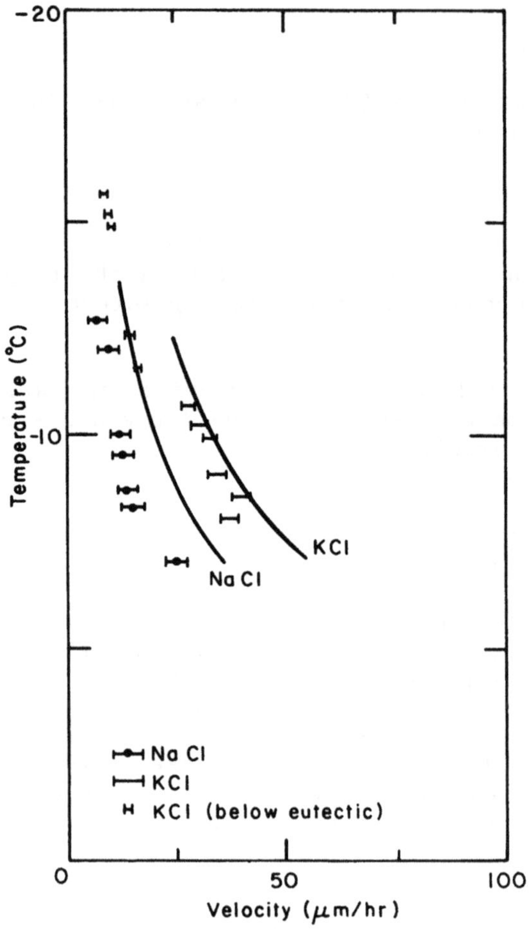

Fig. 72. Relation between brine pocket migration velocity and ice temperature at a constant temperature gradient of 1°C/cm. The curves are calculated using eq. (1.34) (Hoekstra et al., 1965; Seidensticker, 1965).

a linear function of G_s, the temperature gradient in the ice, is clearly shown in Figure 73. Analysis of the experimental data shows that the rate of droplet migration is, indeed, almost completely controlled by the rate of transport of solute through the liquid. At the present time the results suggest that in some, if not all, of the cases studied droplet migration depends on the presence of crystal defects (presumably screw dislocations) in the solid ahead of the high temperature interface (Jones, 1973).

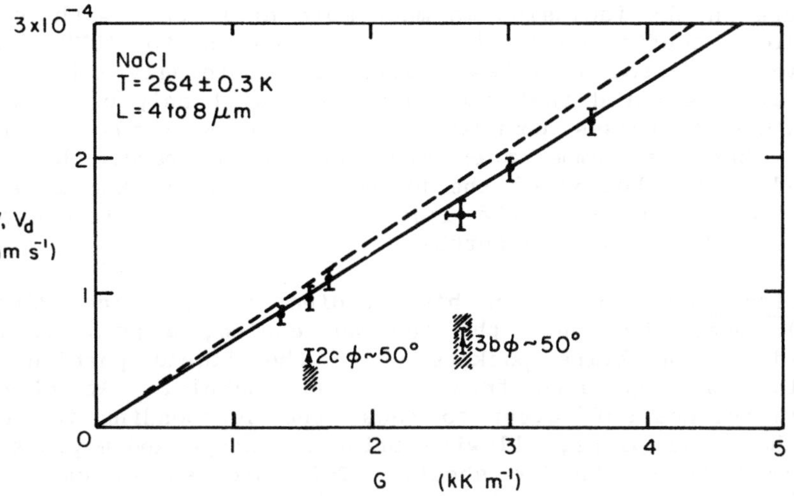

Fig. 73. Relation between the temperature gradient in the ice and the brine pocket migration rate (Jones 1973). Each brace marked with a solid circle gives the mean and mean deviation of the velocities of about 10 separate droplets; the braces marked by solid triangles comprise rather smaller samples. The dashed line indicates, to an accuracy of $\pm 10\%$, the maximum (diffusion-limited) values of V determined from calculated values of D. The cross-hatched areas give values of $V\cos^2$ for those droplets migrating diagonally.

The obvious problem with brine pocket migration as an effective contributor to desalination is the fact that brine pockets appear to migrate extremely slowly. In fact Untersteiner (1968) has calculated that, if representative values for the temperature distribution in multiyear ice are used, a brine pocket between 10 and 20 cm below the ice surface would migrate 2 cm downward between August and April and then between May and July experience an essentially equal migration upward for a net year's travel of near zero. As an example of an extreme case of brine pocket migration consider a 30°C temperature difference across a 1-m-thick piece of first-year ice giving a temperature gradient of 0.3°C/m. If the temperature at the brine pocket location is -6°C, eq. 1.34 gives a migration velocity of 14.0 μm/hr, which corresponds to almost exactly 1 cm migration per month.

In closing this discussion, we note that the brine pockets that have been studied have all been very small (4 to 80 μm). In inclusions this small there is little chance for convective motions to occur in the brine. In much larger inclusions, with diameters on the order of 1 mm, convection becomes possible, and the transfer of solute in the inclusion is much more efficient, with effective "diffusion" coefficients of 10^{-3} as opposed to 10^{-6} for pure diffusion. Migration rates should be accordingly enhanced. It should also be noted that the convection will only occur when the temperature increases downward (during the winter). During the summer, when the temperature gradient is reversed, the salt is produced at the bottom of the brine pocket, resulting in a stable density distribution in the liquid. It would be interesting to see this problem explored further.

Brine Expulsion. In his studies of sea ice Bennington (1963a) noted that when the ice was cooled, a pressure buildup occurred in the brine pockets where the liquid portion of the inclusion had separated from the vapor bubble. At times this pressure becomes sufficient to cause the surrounding ice to fail along the basal plane, allowing brine to escape and migrate toward the warm side of the ice sheet. This process is called brine expulsion, and its contribution to the overall removal of brine from sea ice has been explored by Untersteiner (1968) and by Cox and Weeks (1975). Untersteiner has used a simple model which gives S, the decrease in the bulk ice salinity, from its initial value of S_o when the temperature drops from θ_o to θ, as

$$S = S_o \ (\theta_o/\theta)^{\Delta\rho/1-\Delta\rho} \qquad (1.35)$$

where $\Delta\rho \approx 0.1$ is the difference in density between water and ice. This relation only holds while the temperature is continuously dropping from θ_o to θ. Therefore the salinity of the ice at any time $t > 0$ depends on the initial salinity of the sample and on all subsequent periods of cooling that the ice has undergone in its history. Eq. (1.35) is then applied to estimating the salinity profile of a multiyear floe. The resulting salinity values indicate that the near-surface ice should have a salinity of a few tenths of a part per thousand (as observed), and that below the near-surface ice the salinity profile in all but the lowest part of the ice is very slightly concave upwards (which is not observed).

Cox and Weeks develop a more complex model and utilize it via the application of a finite difference scheme to calculate the change in salinity that should occur in first-year ice. A portion of the results are shown in Figure 74, where the solid curves give the initial and final experimental salinity profiles and the broken curves were calculated from the model. From this, combined with a number of other comparisons, it was concluded that brine expulsion plays only a small role in the desalination of first-year ice. However, the change in salinity is significant and cannot be neglected. This is especially true during periods of rapid ice growth when the rate of change of temperature at each

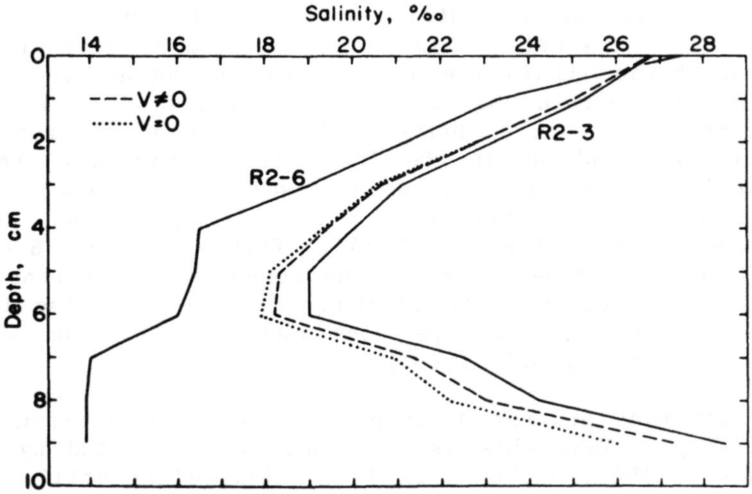

Fig. 74. Comparison between experimental salinity curves and theoretical salinity curves determined from a brine expulsion model. R2-3 and R2-6 are the initial and final observed curves. The other curves were calculated. The dashed curve considers the effect of brine velocity; the dotted curve does not (Cox and Weeks, 1975).

level in the ice sheet is high. In a way it is unfortunate that brine expulsion is not the dominant desalination mechanism as it is independent of the ice permeability in that the volume changes rupture the ice, allowing the brine to migrate (no permeability is required as the process makes its own). Therefore a detailed knowledge of the brine pocket and brine channel structure is not required. Photomicrographs showing the effects of brine expulsion in multiyear ice can be examined in Knight (1962c).

Gravity Drainage. If brine pocket migration and brine expulsion are not sufficiently vigorous processes to explain the observed changes in the salinity profiles of growing sea ice then gravity drainage must be the culprit (unless there is something "undiscovered" taking place). Here by gravity drainage we refer to all processes where brine, under the influence of gravity, drains out of the ice sheet into the underlying seawater. As an ice sheet grows, its surface gradually rises higher above sea level to maintain isostatic equilibrium. This in turn produces a pressure head in the interconnected brine system, driving the underlying brine out of the ice (Eide and Martin, 1975). In addition, because the density of the brine in equilibrium with ice is determined by the temperature distribution, during the period of time when the temperature within the ice increases downward an unstable vertical density distribution exists within the brine channels in an ice sheet. This should produce a convective overturn of the brine within the ice as well as an exchange between the denser brine within the ice and the underlying seawater. In such a process the rate of brine drainage should be a function of the temperature gradient in the ice, which determines the brine density gradient and the brine volume, which is presumably related to the permeability. Whether or not this is true in an ice sheet was examined by Cox and Weeks (1975). Figures 75 and 76 indicate that as either the brine volume or the temperature gradient in the ice sheet increases, the rate of change of salinity also increases. In fact gravity drainage does appear to be the dominant mechanism for removing salt from growing sea ice.

The fact that gravity drainage is the dominant mechanism is not surprising. That this is the case was suggested by Kingery and Goodnow in 1963, although in 1968 Untersteiner still remained skeptical that gravity drainage was important in ice thicker than a few tens of centimeters. What is surprising is how gravity drainage works. Although it might be reasonable to expect that gravity drainage would be a process that occurs in a uniform manner at any given level in the ice sheet, this clearly does not appear to be the case. Knight (1962b) and Lake and Lewis (1970) first noted that in natural sea ice large vertical tubular brine drainage

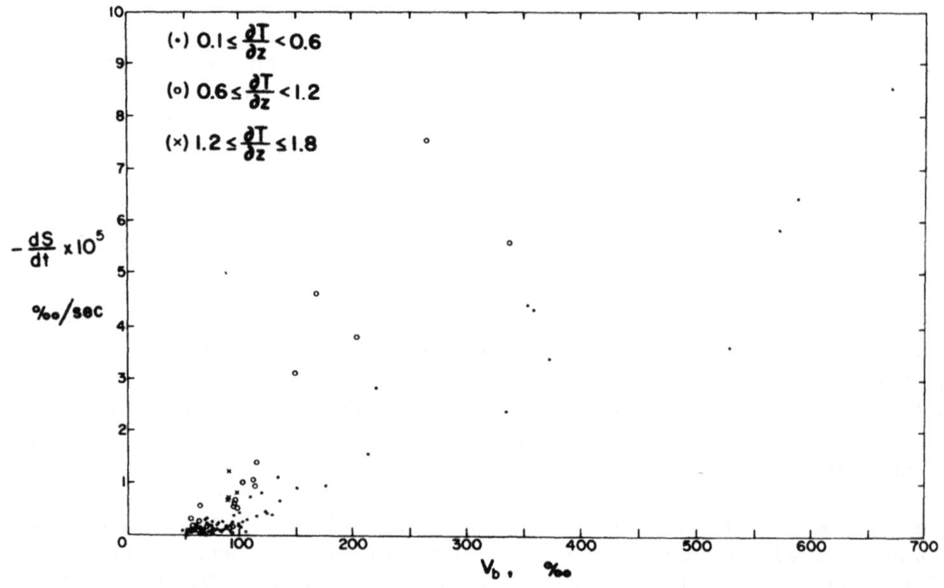

Fig. 75. Plot of rate of change of salinity due to gravity drainage ds/dt versus brine volume for different temperature gradients $\partial t/\partial z$ ($^{\circ}C/m$). (After Cox and Weeks 1975).

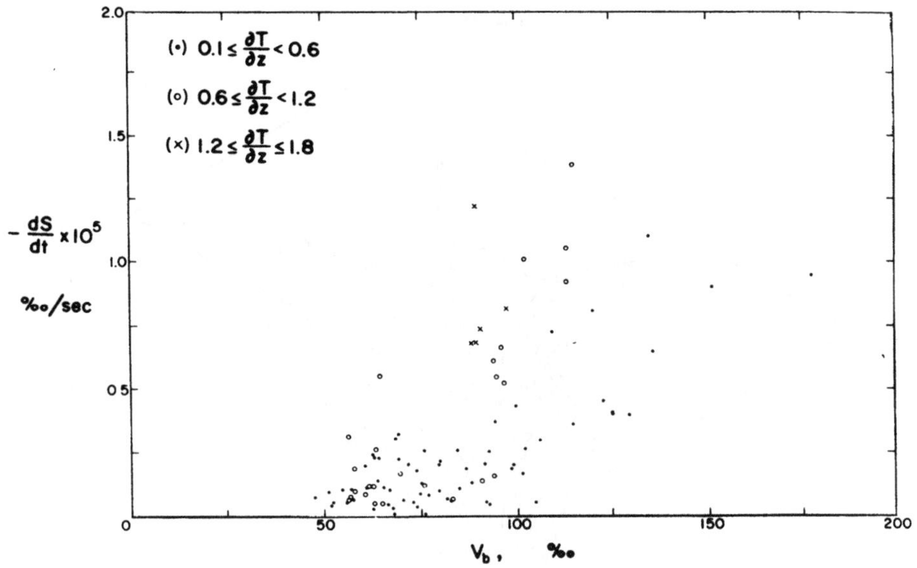

Fig. 76. Plot of rate of change of salinity ds/dt due to gravity drainage against brine volume (at low brine volumes) for different temperature gradients $\partial t/\partial z$ (°C/m). (After Cox and Weeks 1975).

structures are attended by smaller tributary tubes, much as in a vertically oriented, radially symmetric river system. Figure 77 shows a schematic drawing of such a brine drainage tube (Lake and Lewis, 1970). Representative diameters of the tubes at the bottom of a 1.55-m-thick ice sheet were 0.4 cm, and there was one large channel every 180 cm^2. In a given horizontal plane the large channels show a starburst pattern, with the arms of the star generally following the crystal boundaries. In thin ice (<10 cm) it has been found (Saito and Ono, 1980) that the number of brine drainage channels is related to the average growth rate (Figure 78) but not to the average grain size. Even at the lowest growth velocities studied there were three channels per 100 cm^2 or 33 cm^2 per channel as compared to the 180 cm^2 per channel noted in thicker ice by Lake and Lewis. Presumably, some of the channels in the thin ice merge in the same manner as river systems join as the ice thickens.

The nature of brine drainage channels has been investigated in a series of papers by Martin and his coworkers (Dayton and Martin, 1971; Martin, 1974; Eide and Martin, 1975; Niedrauer and Martin, 1979). The flow from the channels is oscillatory, with the duration of the downward flow being typically shorter than the duration of the upward flow. Oscillatory periods of approximately 1 hour have been described, with an 8- to 15-minute inflow followed

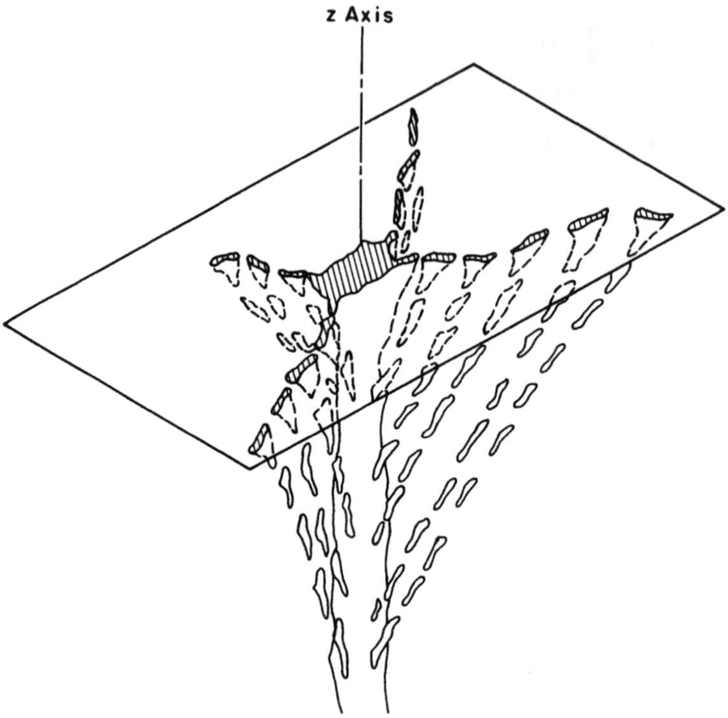

Fig. 77. Schematic drawing of a cut through a brine drainage channel (Lake and Lewis, 1970).

by a 45-minute outflow (Eide and Martin, 1975). Dayton and Martin (1971) have described the drainage of brine from such channels at a rate of approximately 1 liter/minute. They have also noted channel diameters of the order of 10 cm although most channels appear to have diameters ranging between 0.1 and 1.0 cm.

When the cold, saline brine streamers leave the base of the ice sheet, hollow tubes of ice form around them. These ice stalactites (Paige, 1970) can be very large. Dayton and Martin (1971) present excellent photographs of such features and note stalactite lengths of 1.5 to 6.0 m (lengths of 20 to 50 cm are probably more representative). Once the ice tube is formed around the brine streamer, the horizontal transfer of heat between the brine and the surrounding seawater occurs through the ice wall. However, lateral transfer of salt cannot occur. Immediately after a tube forms, the "relatively" warm brine is in contact with the inner ice wall. For phase equilibrium to be maintained, this brine must be both cooled and diluted back to the composition specified by the eutectic curve. This is accomplished by melting the inner tube wall, with the heat of melting coming from ice accretion on

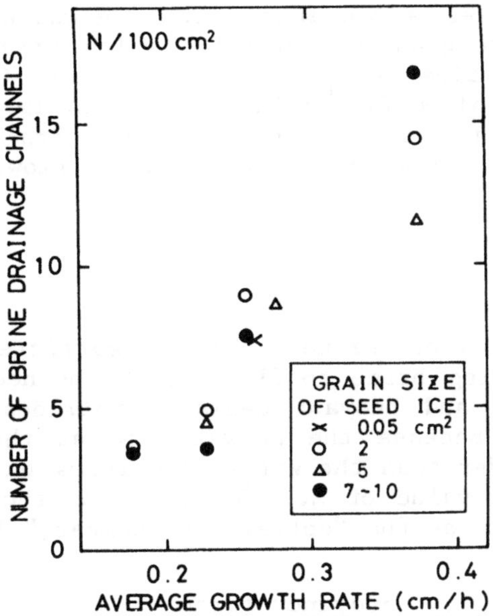

Fig. 78. Number of brine drainage channels developed in thin ice sheets versus the average growth rate (Saito and Ono, 1980).

the outer wall. Therefore, the stalactite both accretes on the outside and ablates on the inside while at the same time lengthening.

A similar enlargement appears to take place in brine channels within the ice sheet. Let a flow of cold, dense brine occur through a brine channel from the upper (cold) part of the ice to the lower (warm) part. To warm the cold brine there is a lateral flow of heat (through the ice) to the brine channel. As in the stalactite the brine warms above its equilibrium temperature. To reestablish equilibrium the channel wall melts, cooling the ice and diluting the brine. Therefore, the results of pumping brine down through the ice are both the cooling of the adjacent ice and an increase in the diameter of the brine channel (Martin, 1974).

Because the seawater beneath growing sea ice is at its freezing point, there is an important difference between brine drainage channels and river systems. A river system usually continuously widens as its mouth is approached. Brine channels, on the other hand, neck just above the ice/water interface (a similar neck also occurs at the tips of growing stalactites; Eide and Martin, 1975). Two factors appear to determine this necking. Not only is the seawater below the ice sheet at its freezing point, but it is also less dense than the brine flowing from the ice. This density difference $\Delta\rho$ creates a buoyancy force $\Delta\rho g$ which acts upward and opposes the pressure gradient force F_p, which drives the brine out of the ice. The value of F_p can be calculated (Martin, 1974) by assuming that the volume flux q in the neck is imposed on the neck by the nature of the upper portion of the drainage channel and that the flow through the neck is Poisseuille flow. Then

$$F_p = \frac{8\rho_p \nu_b q}{\pi a^4} \tag{1.36}$$

where a is the radius of the neck. For a sufficiently small value of a, $F_p > \Delta\rho g$, causing brine to flow out of the neck. For a larger neck radius $F < \Delta\rho g$, and buoyant seawater intrudes into the neck. When this occurs, because the seawater is at the freezing point and the ice is colder than the water, ice forms on the side of the neck, reducing the value of a. The ratio of these two forces E, which is referred to as the "entrainment number," is given by

$$E = \frac{\pi \Delta\rho g a^4}{8 \rho_b \nu_b q} = \frac{\text{buoyancy force}}{\text{pressure gradient force}} . \tag{1.37}$$

For stalactite tips the tip radius is nearly predicted by $E = 1$ (Martin, 1974). The sequence of adjustments in the tip radius is as follows. First, the brine flows uniformly out of the tip, causing the tip radius a to increase so that $E > 1$. The flow then decreases and seawater intrudes into the neck, freezing to the walls and reducing the radius back to the subcritical value ($E < 1$), at which time the whole process starts anew.

For very slow ($q \rightarrow 0$) flows salt diffusion is important and the above criterion is not valid. In this situation Lake and Lewis (1970) have argued that a Rayleigh number criterion

$$\frac{g \frac{1}{\rho} \frac{\partial\rho}{\partial x} a^4}{D} \approx 68 \tag{1.38}$$

(where $\frac{\partial\rho}{\rho z}$ is the density gradient inside the tube caused by the temperature field in the ice, μ is the viscosity, and D is the salt diffusivity) determines the minimum tube radius. Substituting appropriate values into eq. (1.38) we obtain $a = 0.3$ mm, which is of the observed order, indicating that in the limit of $q = 0$ the channels will remain open.

The reason that oscillations occur is that the brine level inside the ice oscillates between two positions of hydrostatic equilibrium. When cold, dense, saline brine from the upper levels of the ice fills the drainage channel the equilibrium brine level is lower than when warmer, less saline seawater fills the channel. When the tip radius has increased until E>1 then any small mass perturbation in the stalactite results in a large pressure imbalance in the brine drainage tube, which accelerates seawater up the tube until the second higher equilibrium level is reached. At the same time, ice growth associated with the influx of seawater is decreasing a, causing E<1. In addition, the seawater is being cooled to a brine and the process starts over again. The important factor in governing the oscillations would appear to be the geometry of the brine channels (Martin, 1974).

In conclusion, although this has not been investigated, it would appear that the brine drainage channel - stalactite combinations, in addition to providing an efficient mechanism for removing brine that is within the channel, also facilitate desalination by three other mechanisms:

1. The increase in the diameter of the brine channel both entrains brine pockets and reduces the viscous drag on the downward-flowing brine.
2. The pressure drop associated with the formation of the stalactites facilitates the drainage of brine into the channel.
3. The cooling of the ice surrounding the channel causes the expulsion of brine into the channel (Martin, 1972).

There are many aspects of brine drainage that have not been explored at all. In particular, we refer to the effects of the formation of ice in the brine drainage system and the effective permeability of such systems. We would expect that features similar to brine drainage tubes develop during the solidification of other materials.

Flushing. Although flushing is actually a particular type of gravity drainage, it is considered separately in that the pressure head necessary to overcome capillary retention is provided by the hydrostatic head resulting from surface meltwater. Therefore, flushing can only occur in the spring and summer when surface melt is possible. The other requirement for flushing is that the ice be permeable. The presence of significant amounts of surface meltwater also suggests that the ice is either at or near the pressure melting point. Such near-melting ice temperatures result in very high brine volumes for a given salinity, and high brine volumes generally correlate with large permeabilities.

It is generally believed that flushing is the most effective mechanism for desalination. The reason for this belief is the fact that the time when flushing starts corresponds to the time during the spring and early summer when major changes occur in the salinity of sea ice. It is also reasonable to believe that the continual percolation of fresh water down and through sea ice should result in significant decreases in the salinity. Considering its probable importance as well as the fact that flushing is undoubtedly similar to (and probably simpler than) processes that have been extensively studied in soils and other permeable materials it is surprising that the only discussion of flushing is by Untersteiner (1968).

In this study, it is assumed that a steady-state salinity profile will be achieved when the downward flux of salt F_s is balanced by the upward flux $\omega \cdot s$. Here ω is the velocity of upward ice displacement (in the Arctic Ocean $\omega \approx 30$ cm/yr) and s is the ice salinity in g/cm^3. To specify F_s it is assumed that $\partial F_s/\partial z$ is a function f of the largest brine volume ν_m attained during the period when melt is occurring at the upper surface of the ice floe. Letting +z be the vertical axis (positive downward) the steady-state case is given by

$$\frac{\partial s}{\partial t} = -\omega \left(\frac{\partial s}{\partial z}\right) + f\,(\nu_m) \equiv 0 \cdot \qquad (1.39)$$

Then, following Assur (1958), by letting $\nu_m = c(s/\theta_m)$ where $c \approx$ constant ≈ -55 and θ_m and ν_m are the highest temperature and highest brine volume attained during the year, Untersteiner obtains

$$w\,\frac{\partial s}{\partial z} = f\left(c\,\frac{s}{\theta_m}\right) \qquad (1.40)$$

Untersteiner then presents field data obtained on drifting station Alpha showing vertical profiles of ν_m and (ds/dz). Both relations are similar, supporting the suggestion that the greatest salt loss occurs at the level where ν_m reaches its highest value. Based on the observed salinity profile, the values of salt advection necessary to maintain the profile are then calculated and plotted against ν_m (Figure 79). Based on this plot, it is assumed that $F_s \approx 0$ at $\nu_m \approx 0.05$ (5%). The steady-state case then becomes

$$\omega\left(\frac{\partial s}{\partial z}\right) \equiv \omega\left(\frac{ds}{dz}\right) \equiv \frac{ds}{dt} = 0.04\,\nu_m^3 \qquad (1.41)$$

a relation that can be used to recalculate the salinity profile. The resulting salinity profile shows a very reasonable agreement with the observed salinity profile for multiyear ice (see Untersteiner, 1968, Figure 1). However, as was also pointed out by Untersteiner, this apparent agreement between "theory" and observation is probably misleading in that the reasoning is somewhat circular: first a salinity profile characteristic of multiyear sea ice was selected, then a relation was selected between ($\partial s/\partial t$)

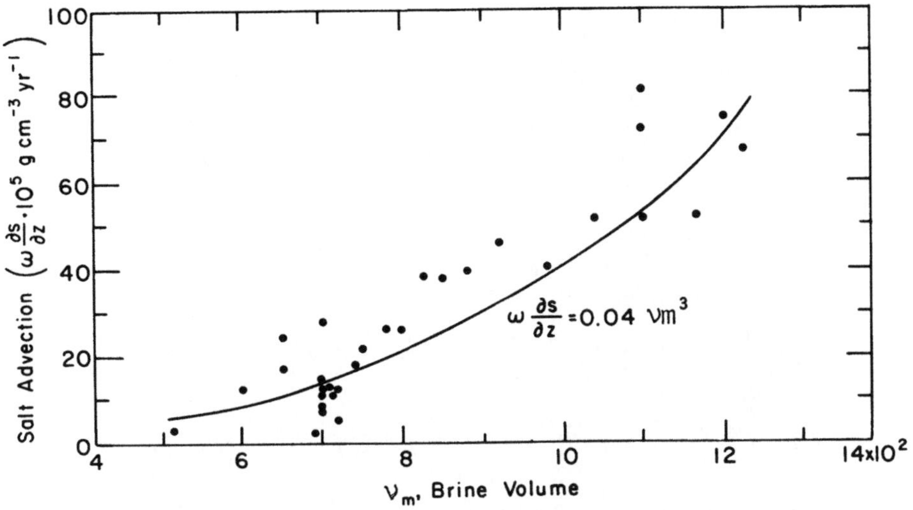

Fig. 79. Relationship of maximum brine volume and salt advection required to maintain the steady-state salinity profile shown as curve C in Figure 65 (Untersteiner, 1968).

and ν_m that would maintain the steady-state salinity profile. This relation (eq. 1.41) then allowed the recalculation of the original salinity profile. Although encouraging, the agreement can hardly be considered proof that the multiyear salinity profile is the result of flushing. However, flushing certainly does appear to be the most likely (and almost the only) candidate.

10.4 Solid Salt Crystals

There is only one paper that examines the nature of the solid salt crystals in sea ice (Sinha, 1977). In this work a microtoming and replicating technique was used to examine the nature of the brine pockets and the precipitation pattern of the salt crystals at low temperatures (-30°C, which is below the crystallization temperature of $CaCO_3 \cdot 6H_2O$, $Na_2SO_4 \cdot 10H_2O$, $MgCl_2 \cdot 8H_2O$ and $NaCl \cdot 2H_2O$). Figure 80 shows vertical scanning electron micrographs of a brine pocket. Although the individual crystals were not positively identified the majority of the crystals are undoubtedly $NaCl \cdot 2H_2O$. Note the random distribution of the salt crystals. Most crystals appear to be loosely packed in the cavities and can be removed either by preparing successive replicates or by washing the microtomed surface with kerosene before the replicates are prepared.

Some brine pocket replicates were also made at -10°C, a temperature at which only $Na_2SO_4 \cdot 10H_2O$ should be present in the

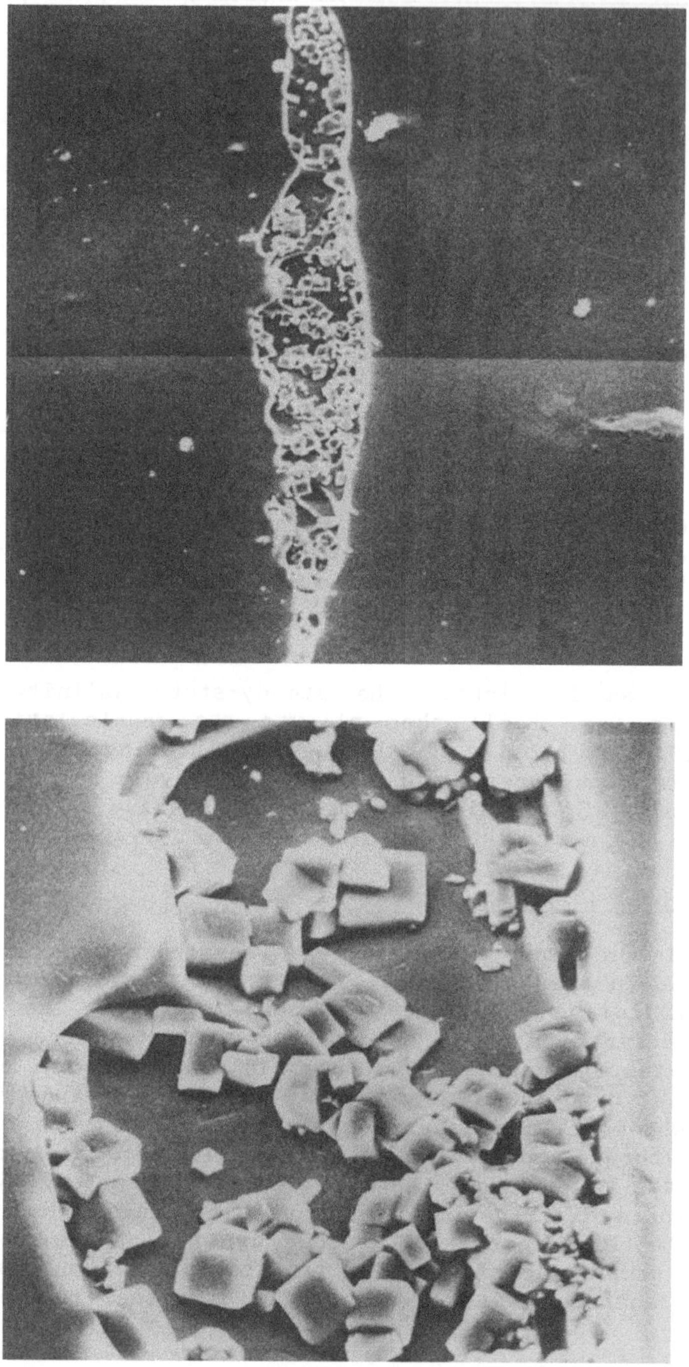

Fig. 80. Scanning electron micrographs of a vertical section of brine pocket at -30°C, second replica (Sinha 1977).

ice in appreciable quantities. In this case the walls of the pockets were quite smooth and it appears that many of the salt crystals remained in the liquid brine. If this is the case it is difficult to see how the presence of $Na_2SO_4 \cdot 10H_2O$ could have a significant effect on the strength of sea ice as suggested by Peyton (1966).

Clearly more work using these or related techniques to examine the solid salt crystals is needed. It certainly would also be useful to positively identify the different crystals that are observed.

11. RELATIONSHIPS BETWEEN SEA ICE STRUCTURE AND PROPERTIES

In the above we have discussed a number of aspects of our current understanding of the internal structure of sea ice. We will now review selected examples of how the physical properties of sea ice are affected by changes in its internal structure. Here we will start with the mechanical properties, then discuss the thermal properties, and finally treat our limited understanding of the electromagnetic properties of sea ice. In this last section we will stress electromagnetic properties in the microwave range in that these properties are important in developing and interpreting several different types of remote sensing observations that are currently used in sea ice research.

11.1 Models for the Variation in the Mechanical Properties of Sea Ice

Failure Strength. Early studies of the mechanical properties of sea ice revealed that the fracture surface was commonly controlled by the pronounced sea ice substructure. For example, Figure 81 (Anderson and Weeks, 1958) clearly shows the fracture surface in a sample subjected to tension to tend to run parallel to the (0001) planes of the ice crystals. Additional rubbings showing similar trends have been published by Tabata (1960) and by Paige and Kennedy (1967). Clearly the breaks are following the planes in which the brine pockets and air bubbles are concentrated. That these planes, if correctly oriented, should be planes of weakness is reasonable in that the gas and brine inclusions reduce the percentage of the failure plane that is solid (ice), allowing failure to occur at lower bulk stress levels. Therefore it would be expected that sea ice samples with low porosities would, being essentially pure ice, have strengths similar to that of freshwater ice. Also that sea ice with higher porosities would have very low strengths. Experimental observations (Schwarz and Weeks, 1977; Weeks and Assur, 1967, 1969) support this reasoning.

What was needed was a model that could be used to extrapolate and interpolate strength values at intermediate values of porosity

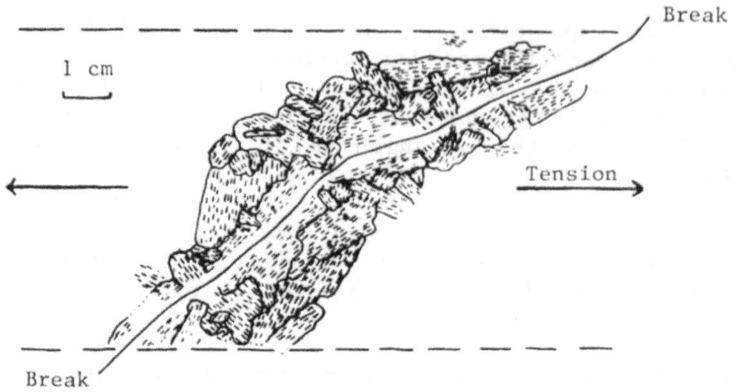

Fig. 81. Rubbing of a broken segment of sea ice (Anderson and Weeks, 1958).

and that was based on a reasonable model of the pore geometry of real sea ice. Such a model was developed by Assur (1958), Anderson and Weeks (1958) and Anderson (1958, 1960). They assumed that it should be possible to express the variation in the failure strength σ_f for sea ice produced by variations in its porosity in the general form

$$\frac{\sigma_f}{\sigma_o} = 1 - \psi \qquad (1.42)$$

where σ_o is the basic strength of sea ice (i.e., the strength of an imaginary material that contains no brine, but still possesses the sea ice substructure and fails as the result of the same mechanism(s) that cause failure in natural sea ice) and Ψ is the "plane porosity" or relative reduction in the area of the failure plane as the result of the presence of brine and air inclusions. The critical value of Ψ in the failure plane is

$$\Psi = f(\nu) = f(\nu_a + \nu_b) \qquad (1.43)$$

where ν is the void volume or porosity and ν_a and ν_b are the volume of air and brine respectively in the sea ice. Usually it is considered that $\nu_b \gg \nu_a$ (that the volume of brine is sufficiently larger than the volume of air so that ν_a can be neglected).

It is now necessary to express Ψ in terms of ν_b through a simplified model of the brine geometry in real sea ice. Figure 82 shows such a model given by Assur (1958). Here the relative brine volume is

$$\nu_b = \frac{F_g}{a_o b_o g_o} \qquad (1.44)$$

where F_g is the average area of a brine inclusion in the BG plane and defining

$$\beta_o \equiv \frac{b_o}{a_o} \qquad (1.45)$$

$$\gamma \equiv \frac{g}{g_o} \qquad (1.46)$$

the reduction in the cross-sectional area caused by the presence of the brine pockets is

$$\psi = \frac{2r_b\, g}{b_o\, g_o} = \frac{2r_b\, \gamma}{\beta_o\, a_o}\,. \qquad (1.47)$$

The question now is: How do the geometric parameters r_b and Ψ vary with ν_b? A number of assumptions are possible; however, two are commonly used. The simplest is to assume that geometric similarity is maintained along the B-axis (see Figure 82) and that the width as well as the relative length (γ) of the brine pockets remain constant. In this instance r_b changes proportionally to ν_b and an equation of the form

$$\frac{\sigma_f}{\sigma_o} = 1 - c\,\nu_b \qquad (1.48)$$

is obtained. The other assumption commonly made is that the average length and spacing of the brine pockets remain constant and that changes in ν_b are reflected only in the BC cross section. In this case the resulting equation is

$$\frac{\sigma_f}{\sigma_o} = 1 - c\,\nu_b^{1/2}. \qquad (1.49)$$

These two models can be represented as straight lines in σ_f, ν^k coordinates where k is 1 and 1/2, respectively. The σ axis intercept is σ_o and

$$c = \nu_o^{-k} \qquad (1.50)$$

where ν_o is the volume of brine required to cause the ice to have zero strength.

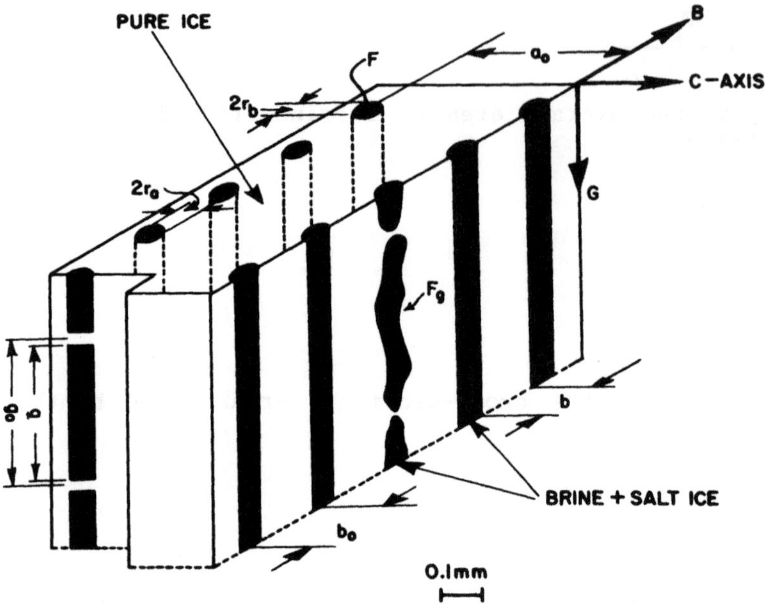

Fig. 82. Idealized diagram of the shape of the brine inclusions in sea and NaCl ice (Assur, 1958).

Two specific models that have been utilized in discussing strength results are the constant width and elliptical cylinder models. In the constant width model

$$F = 4\ r_a\ r_b \tag{1.51}$$

and

$$\nu_o = \frac{2r_a}{a_o} = \frac{d_o}{a_o}\ . \tag{1.52}$$

Here d_o is the minimum width of a parallel brine layer before it splits as the result of interfacial tension to produce individual brine pockets. It is commonly assumed that $d_o \approx$ constant (values of $\approx 7 \times 10^{-3}$ cm have been measured in natural sea ice).

In the elliptical cylinder model

$$\varepsilon = \frac{r_b}{r_a} \tag{1.53}$$

and

$$F = \pi r_b r_a = \frac{\pi r_b^2}{\varepsilon} \tag{1.54}$$

and eq. 1.49 becomes

$$\frac{\sigma_f}{\sigma_o} = 1 - 2\sqrt{\frac{\varepsilon\gamma}{\pi\beta_o}} \sqrt{\nu_b} \quad . \tag{1.55}$$

Additional details can be obtained from Weeks and Assur (1967, 1969).

Plots of σ_f versus ν_b, $\nu_b^{1/2}$, and $\nu_b^{2/3}$ (this last case is where all brine pockets remain of a similar shape during changes in ν_b) have been made for a number of different sets of test data, and it has commonly been found that eq. (1.49) gives the best fit. Therefore, some authors have routinely plotted strength measurements of every type against $\sqrt{\nu}_b$. As is pointed out by Mellor (1982), such indiscriminate application of eq. (1.49) is to be avoided, particularly in cases where the structure of the ice is unknown, when the orientation of the pore structure to the principal stresses and to the failure stresses is uncertain, and when the fracture mechanisms are not fully understood.

For instance, it would clearly be inappropriate to apply the details of such models to marine frazil ice (a material whose mechanical properties have not, as yet, been investigated), as its structure is very different from that of congelation sea ice (see Figures 52-56). Structurally, marine frazil ice appears to be similar to compact snow, with the included brine and air distributed between the grains (as contrasted to within the grains as is the case for congelation ice). If, as a first approximation, the brine and air inclusions can be assumed to be randomly distributed throughout the ice in a fashion similar to that observed for dense snow then we would expect strength to decrease as a linear function of ν_b since in such a case the relative plane porosity in any cross section is equal to ν_b. Also the properties of the ice would not depend upon the orientation of the sample relative to the applied stresses.

In this section we have discussed the effect of brine and gas inclusions in the ice on the bulk properties of the ice. Nothing has been said concerning the effect of the presence of solid salts. There have as of now been two attempts to develop a theory for the effect of solid salts. Assur (1958) postulated two different situations based on a model of the reinforcement of the brine pocket walls by a continuous layer of intermixed ice and solid salt. In the first case the elastic modulus (E) of the salt-ice reinforcement is similar to that of pure ice, and rupture occurs within the reinforcement. As a result a relatively thin layer of salt-ice should abruptly increase the ice strength, but little further increase would be observed with the thickening of the reinforcement. In the second case a substantial difference in E between the salt-ice mixture and the pure ice is assumed, with the strength of the mixture being sufficiently high so that the initial failure occurs in the ice as opposed to in the mixture. It was suggested that the first situation might apply to the precipitation of $Na_2SO_4 \cdot 10H_2O$, where only a small amount of salt forms, with the second case applying to the precipitation of $NaCl \cdot 2H_2O$, when much larger quantities of salt form. Further work on this problem was later undertaken by Peyton (1966), who utilized a similar salt reinforcement model.

There are two obvious problems here. First, the very limited direct observations on the nature of the distribution of solid salt crystals in brine pockets reported by Sinha (1977) show a random distribution of loosely packed crystals. This clearly does not support the geometric assumptions of Assur's model. However, more important there is still no adequate set of strength tests on real sea ice against which the predictions of any model can be tested. The closest to an adequate study of this problem was by Weeks (1962), who performed ring-tensile tests on NaCl ice. He found that the strength of $NaCl \cdot 2H_2O$ ice was a) essentially independent of temperature and the volume of $NaCl \cdot 2H_2O$, and b) comparable to the strength of freshwater ice. This plus Sinha's observations suggest that solid salts may well have little effect on the failure strength of sea ice. What is clearly needed now is a set of high quality tests that are specifically focused on resolving this problem.

Elastic Modulus. It is only recently that a similar model has been suggested to explain the variation in the elastic modulus (E) of sea ice with changes in the volume of brine included in the ice (Bergdahl, 1977). First, stress in the vertical direction (G-direction, see Figure 83) is considered. Also, the value of ν_a is assumed to be negligible and the vertical interruptions in the brine cylinders are neglected. Then the deformation for the brine inclusions and for the ice matrix are set equal ($\nu = 0$) and the loading is considered to be dynamic. Then

$$\varepsilon = pK = \sigma_1/E_1 = \sigma/E \qquad (1.56)$$

where ε = strain

p = pressure in the brine pockets

σ_1 = stress in the ice

σ = stress averaged over the cross section A, $\sigma = P/A$

E = bulk elasticity of sea ice

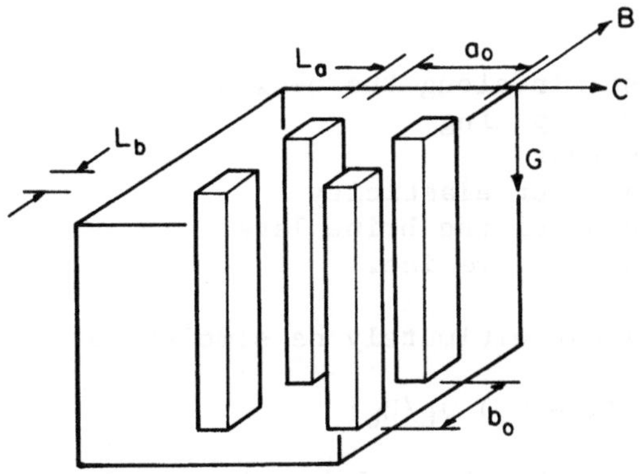

Fig. 83. An idealization of the structure of sea ice used in calculating its elastic modulus (Bergdahl 1977). The rectangular areas represent brine inclusions.

From eq. (1.56)

$$\sigma = E\varepsilon \qquad p = \varepsilon/K \qquad \sigma_1 = \varepsilon E_1 \tag{1.57}$$

and the average stress is

$$\sigma = \sigma_1 (1 - A_b) + p A_b \tag{1.58}$$

where A_b is the relative cross-sectional area of brine. These equations give

$$E = E_1 + (1/K - E_1)A_b \tag{1.59}$$

or, if $\nu_b = A_b$,

$$E = E_1 + (1/K - E_1)\,\nu_b \tag{1.60}$$

suggesting that E should prove to be a linear function of the brine volume. This, of course, has been shown to be the case experimentally (see Mellor, 1982; Weeks and Assur, 1967, or Schwarz and Weeks, 1977).

When horizontal deformations in the sea ice cover are considered, a more complex model must be utilized. In doing this Bergdahl approximates the brine pockets as rectangular tubes with a horizontal cross-section area $L_a \cdot L_b$ (see Figure 83). Deformation along the c-axis is then given by

$$a_o\varepsilon = a_o\sigma/E_h = L_a\sigma/E_b + (a_o - L_a)\,\sigma/E_1 \tag{1.61}$$

with

ε = average strain along the c-axis
a_o = brine layer spacing
σ = average stress
E_h = bulk modulus of elasticity
E_b = bulk modulus of the brine layer
E_l = elasticity of pure ice.

The bulk elasticity can ultimately be simplified to

$$E_h = E_l/(1- (1/K - E_l)\, \nu_b/E_b) \tag{1.62}$$

indicating that the elastic modulus in the horizontal direction would be expected to be a function of both the brine volume and the shape of the brine pockets. As in the equations for failure strength, to evaluate eq. (1.62) a relation must be assumed between the ratio L_b/L_a and ν_b.

There are many problems with the utilization of such models. Obviously they represent gross simplifications of the structure of real sea ice, which in actuality is very complex. One difficulty is that there is very little information on how brine pocket shapes actually change with changes in ν_b. Also, usually the total void volume is assumed to equal ν_b, the brine volume (i.e., ν_a is assumed to equal zero). Now this is not strictly true, even for young sea ice, while for low salinity multiyear ice ν_a may be much larger than ν_b. There are, at present, no good observations to support the common assumption that air bubbles are localized along the same substructure as the brine pockets. This, of course, seems reasonable, and many times the total void volume ν is taken to be the sum $\nu_a + \nu_b$. Finally, there are very few sets of measurements of the mechanical properties of sea ice available in which both ν_a and ν_b were determined. Even considering these weaknesses, the geometric models for the variations in the mechanical properties of sea ice have proven to be very useful and they undoubtedly will continue to be utilized in treating a variety of different problems.

Applications. We conclude this section with a discussion of one such application. As was discussed earlier it is known that some elements of the substructure of sea ice vary systematically with changes in the growth velocity. The growth velocity is in turn controlled by the meteorological conditions. If changes in the substructure affect the mechanical properties of the ice, it is then possible to formulate relations between the meteorology and the ensuing ice properties. For example, let us first assume that the relation between brine layer spacing a_o and growth velocity v can be expressed as

$$a_o \sqrt{v} = c \tag{1.63}$$

where c is a constant. Then if the ice growth equation can be approximated by

$$t = \frac{\rho L h}{\Delta\theta}\left[\frac{1}{e} + \frac{h}{2\kappa}\right] \tag{1.64}$$

where t is time, h is ice thickness, ρ is the density of the ice, L is the latent heat of fusion of the ice, Δθ is the difference between the ambient air temperature and the freezing temperature of seawater, κ is the thermal conductivity of sea ice, and e is the overall coefficient of surface heat transfer, by rearranging and differentiating we can obtain

$$v = \frac{dh}{dt} = \frac{\Delta\theta}{\rho L\left[\frac{1}{e} + \frac{h}{\kappa}\right]} . \tag{1.65}$$

This relation, when combined with eq. (1.63), gives

$$a_o = \sqrt{\frac{c^2 \rho L}{\Delta\theta}\left[\frac{1}{e} + \frac{h}{\kappa}\right]} . \tag{1.66}$$

For thin ice a_o is independent of ice thickness h, as $1/e >> h/\kappa$. For thick ice, on the other hand, $h/\kappa >> 1/e$, giving a_o varying proportional to $\sqrt{h}$.

The structural parameter a_o, however, appears in the ice strength equations. For instance, in the constant width model as

$$\frac{\sigma_f}{\sigma_o} = 1 - \frac{\nu_b}{\nu_o} = 1 - \frac{a_o \nu_b}{d_o} \tag{1.67}$$

and the elliptical cylinder model as

$$\frac{\sigma_f}{\sigma_o} = 1 - 2\sqrt{\frac{\gamma\, \varepsilon\, a_o}{\pi\, b_o}}\, \sqrt{\nu_b} . \tag{1.68}$$

Therefore, combining these relations with eq. (1.66) we obtain, respectively,

$$\sigma_f = \sigma_o\left[1 - \frac{c\nu}{d_o}\sqrt{\frac{\rho L}{\Delta\theta}\left[\frac{1}{e} + \frac{z}{\kappa}\right]}\right] \tag{1.69}$$

and

$$\sigma_f = \sigma_o \left[1 - 2 \sqrt{\frac{\nu_b \gamma \varepsilon c}{\pi b_o} \sqrt{\frac{\rho L}{\Delta\theta} \left[\frac{1}{e} + \frac{z}{\kappa}\right]}} \right] \tag{1.70}$$

where z is the position of a given ice layer below the surface and $\Delta\theta$ is the temperature difference when that layer formed. Combining all the constants and considering only thin ice, the two different models give, respectively,

$$\sigma_f = \sigma_o \left[1 - \text{const} \frac{\nu_b}{\sqrt{\Delta\theta}} \right] \tag{1.71}$$

and

$$\sigma_f = \sigma_o \left[1 - \text{const} \sqrt{\frac{\nu_b}{\sqrt{\Delta\theta}}} \right]. \tag{1.72}$$

These equations support the intuitive conclusions of early Soviet investigators (see Sharp, 1947), who claimed that the strength of young sea ice was in some way affected by its growth conditions. For thick ice we obtain

$$\sigma_f = \sigma_o \left[1 - \text{const}\ \nu_b \sqrt{\frac{z}{\Delta\theta}} \right] \tag{1.73}$$

and

$$\sigma_f = \sigma_o \left[1 - \text{const} \sqrt{\nu_b \sqrt{\frac{z}{\Delta\theta}}} \right]. \tag{1.74}$$

More details and a consideration of the effect of a snow cover on the sea ice can be found in Assur and Weeks (1964). The important point here is that similar studies can easily be made of the effect of vertical variations in a variety of structural characteristics of sea ice upon the profile properties of the sheet, provided that the relations between the characteristics and the ice growth conditions have been established. A related study that examines the effect of growth conditions on the salinity and brine profiles of first-year sea ice has been published by Weeks and Lofgren (1967).

11.2 Models for the Variation in the Thermal Properties of Sea Ice

By the thermal properties of sea ice we specifically refer to the parameters that compose the thermal diffusivity term in the thermal diffusion equation, i.e. the specific heat, the density and the thermal conductivity. Of these three terms the first two depend only on the temperature and the composition of the ice. The geometric distribution of the components is not a factor. Therefore they will not be discussed here. Good recent reviews and studies of these two properties can be found in Bergdahl (1977), Yen (1981) and Cox and Weeks (1983).

Thermal Conductivity. The thermal conductivity of sea ice and the thermal diffusivity which contains the thermal conductivity are dependent on the spatial arrangement of the different phases composing the sea ice: the pure ice, the brine, the entrapped gas and, if the ice is sufficiently cold, the solid salt hydrates. Figure 84 gives thermal conductivity (λ_i) of pure polycrystalline ice determined by several investigators (see Yen, 1981 for references). For practical purposes Yen suggests that the complete set be used, resulting in the relation

$$\lambda_i = 9.828 \exp(-0.0057\ T) \qquad (1.75)$$

where λ_i is in W/m K and T is in kelvin. The effect of crystal orientation on λ_i has been little studied. Landauer and Plumb (1956) suggest that λ_i parallel to the c-axis is approximately 5% higher than values measured normal to the c-axis). Note that at 273°K, λ_i is roughly 2.2 W/m °K, which is about 4 times the thermal conductivity of water at that temperature. There are, as yet, no measurements on the conductivity of seawater brine in equilibrium with ice. However, data on other salts suggest that λ_b is strongly dependent on concentration and less dependent on temperature. Based on λ_b measurements for NaCl and Na_2SO_4 solutions, Schwerdtfeger (1963) has suggested that the value for seawater brine can be approximated by

$$\lambda_b = 0.4184\ (1.25 + 0.030\ \theta + 0.00014\ \theta^2) \qquad (1.76)$$

where λ_b is in units of W/(m K) and θ is in K. As Schwerdtfeger points out, although eq. (1.76) becomes less accurate at lower temperatures, the value of λ_b is less important under such conditions because the relative volume of brine becomes negligible. The thermal conductivity of air (λ_a) is approximately 0.03 W/(m °K) at temperatures between -30°C and 0°C, or roughly 1% of the conductivity of the ice. Therefore, for most purposes λ_a is taken as nil. The conductivities of the solid hydrates at temperatures below -8.7°C are not well known and they are usually neglected inasmuch as they

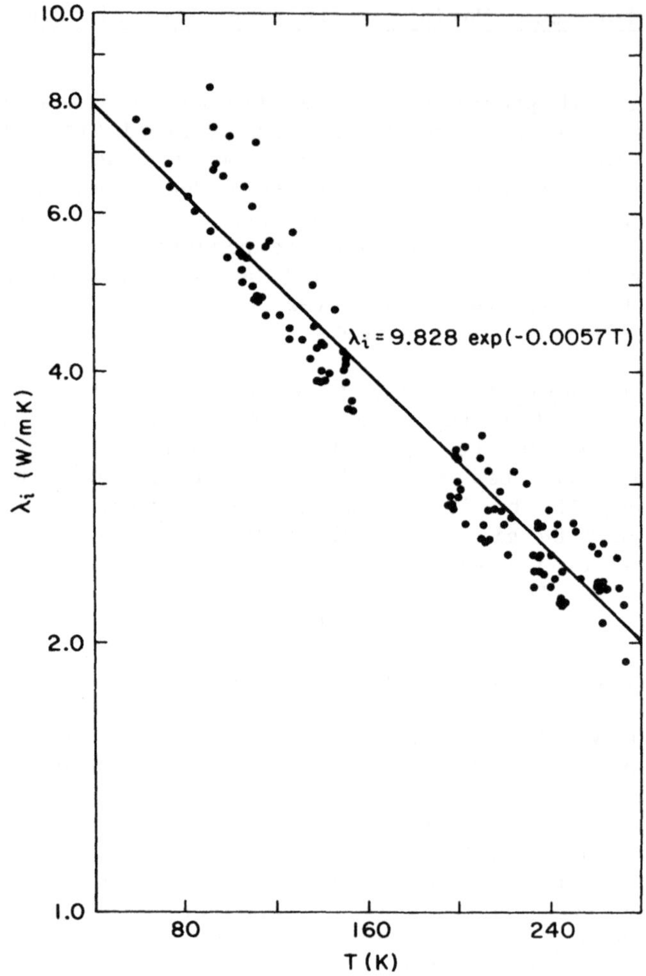

Fig. 84. Thermal conductivity of pure ice as a function of temperature. See Yen (1981) for the data sources.

would be expected to have only a very small effect on the bulk conductivity of sea ice.

Next it is necessary to consider the effect of the geometry of the brine and gas inclusions on the bulk conductivity of the sea ice. This was first done by Anderson (1958, 1960), who considered the presence of brine in several different spatial configurations. Among these were the cases with the brine distributed

a) in isolated spherical bubbles

b) in parallel cylinders or layers between ice platelets with the conductivity measured in the direction of the brine layers (analogous to electrical conduction in parallel), and

c) in parallel layers with the conductivity measured perpendicular to the layers (analogous to electrical conduction in series).

These results are shown in Figure 85. Note that the conductivity measured parallel to the c-axis (perpendicular to the brine layers) is appreciably lower than the equivalent value measured perpendicular to the c-axis. This difference in thermal conductivity with crystal orientation is undoubtedly an important contributing cause of the fact that in the transition layer in the upper part of newly formed sea ice the crystals with their c-axes horizontal are observed to grow ahead of crystals in other orientations (the c-axis-horizontal crystals have their direction of maximum thermal conductivity aligned in the vertical, parallel to the direction of heat flow (Weeks, 1958)).

Inasmuch as the heat flow in sheets of undeformed sea ice is essentially one-dimensional (in the vertical) and the great majority of congelation ice is oriented with its c-axes horizontal, this is clearly the case that should be considered in most practical problems. Schwerdtfeger (1963) expanded upon Anderson's treatment by considering the presence of air in the sea ice. This was done by assuming that the air in sea ice consists of a series of uniform, randomly distributed spherical air bubbles. Then, using a relation developed by Maxwell (1891), which incidentally is the same relation Anderson used in calculating the "porous plate" conductivity given in Figure 85,

$$\lambda_{bi} = \frac{2\lambda_i + \lambda_a - 2V_a(\lambda_i - \lambda_a)}{2\lambda_i + \lambda_a + V_a(\lambda_i - \lambda_a)}\lambda_i \tag{1.77}$$

where λ_{bi} is the thermal conductivity of pure ice containing randomly distributed bubbles as stated and V_a is the volume of air per unit volume of sea ice. Figure 86 gives the resulting values of λ_{bi} (and of the density of bubbly ice ρ_{bi}) for temperatures in the 0 to -20°C range. Here λ_i and λ_a were respectively taken to be 2.09 and 2.51×10^{-2} W/m K. Schwerdtfeger then considered sea ice as the composite material composed of layers of brine and layers of pure ice containing air bubbles, with a resulting thermal conductivity λ_{si} given by

$$\lambda_{si} = \lambda_{bi} - (\lambda_{bi} - \lambda_b)\left(\frac{\sigma\rho_{si}}{\alpha\rho_w\theta}\right) . \tag{1.78}$$

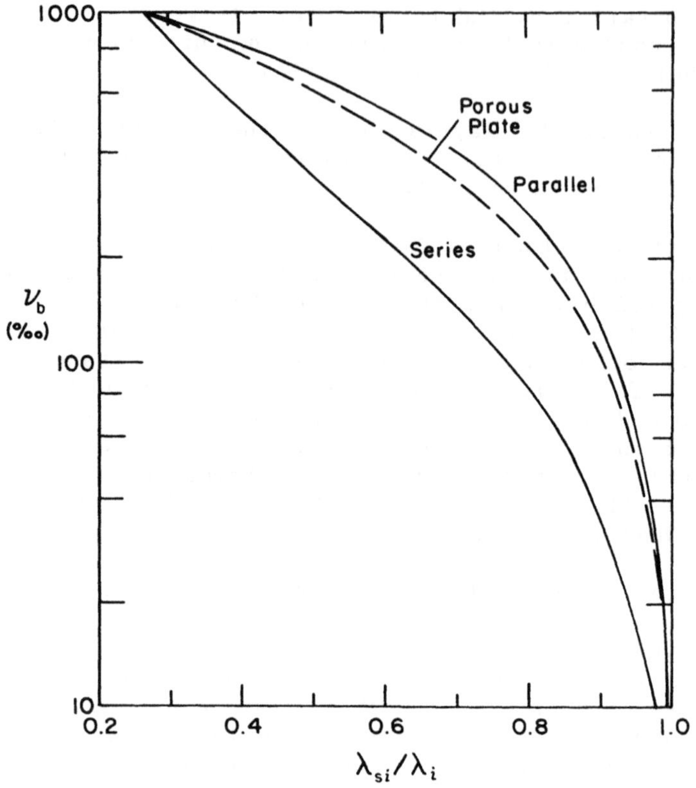

Fig. 85. Thermal conductivity of sea ice (Anderson, 1958). The different curves represent different arrangements of the brine inclusions.

Here λ_b is the thermal conductivity of the brine, σ is the salinity of the ice, α is the constant in the relation $s = \alpha\theta$ where s is the fractional salt content and θ is temperature, and ρ_w and ρ_{si} are the densities of pure water and of bulk sea ice, respectively. The results of Schwerdtfeger's calculations are shown in Figure 87, redrawn by Yen (1981). As is apparent in this figure and has been noted by Schwerdtfeger, even at temperatures of -8°C the curves in Figures 87 exhibit asymptotic behavior, with the conductivity of sea ice approaching the value for pure ice at low temperatures. Therefore at low temperatures it is the air content of the sea ice that controls its thermal conductivity, while at temperatures near the melting point it is the brine volume in the ice that is important and which in turn at a given temperature is controlled by the salinity.

The most recent work on the conductivity of sea ice is by Ono (1968) who realized through his studies of its internal structure

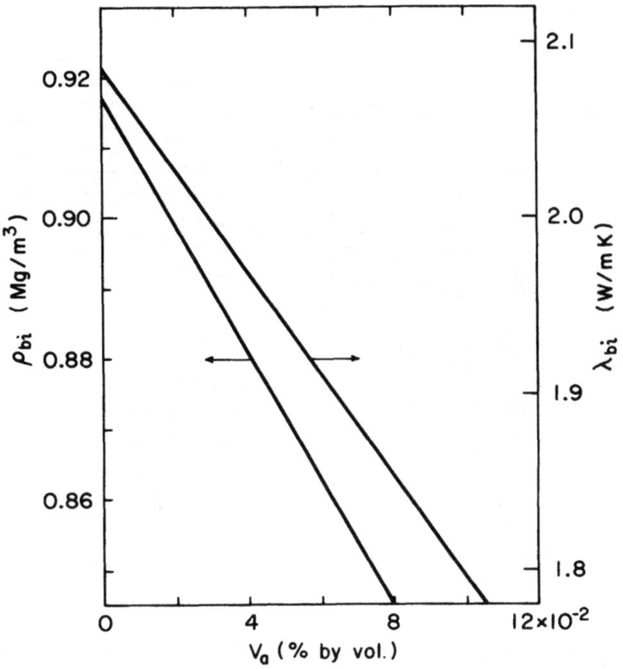

Fig. 86. Density and thermal conductivity of bubbly ice as a function of the fractional air content (Schwerdtfeger, 1963).

that it was probably not realistic to assume that air bubbles are only trapped in the ice matrix. He therefore developed a model in which the air bubbles are spherical and are uniformly dispersed through both the ice and the brine, with identical concentrations in both phases. However, as in the previous models the layers of bubbly ice and bubbly brine are assumed to be arranged in parallel. The results of Ono's calculations are shown in Figure 88. The conductivity curves are for air-free sea ice. The effect of the presence of air is included by first using the figure to determine the point specified by the ice temperature and salinity and moving vertically downward parallel to the conductivity (y) axis an amount specified by the v_a inset scale.

The requisite petrographic data necessary to quantitatively determine which model, Schwerdtfeger's or Ono's, is the most appropriate to different types of congelation ice are not available. It is our impression, however, that Ono's model is the most realistic. A further improvement would probably be to consider the case where the amounts of air bubbles in the brine and in the ice are different. However, we doubt that there is much profit in the fine-tuning of these models unless it is required by some specific

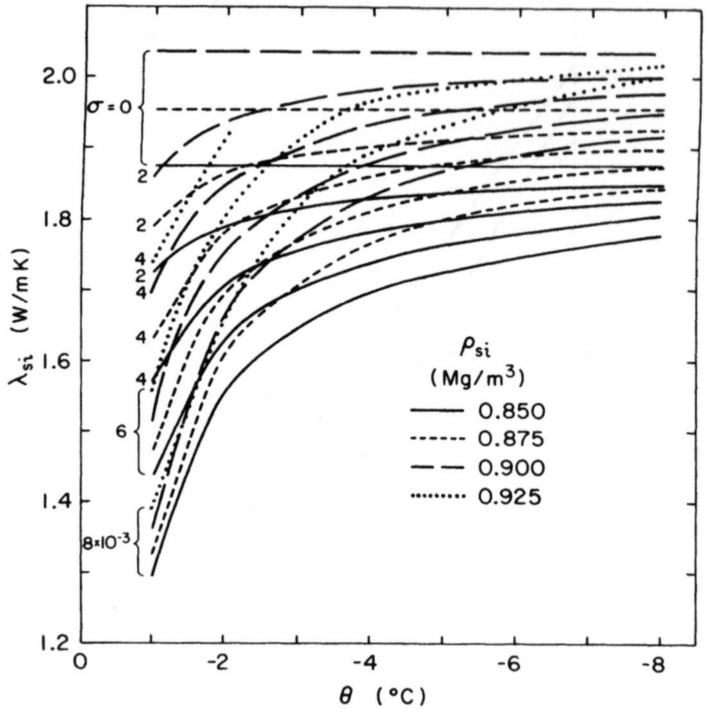

Fig. 87. Effective thermal conductivity of sea ice as a function of temperature for various salinities and densities (Schwerdtfeger, 1963).

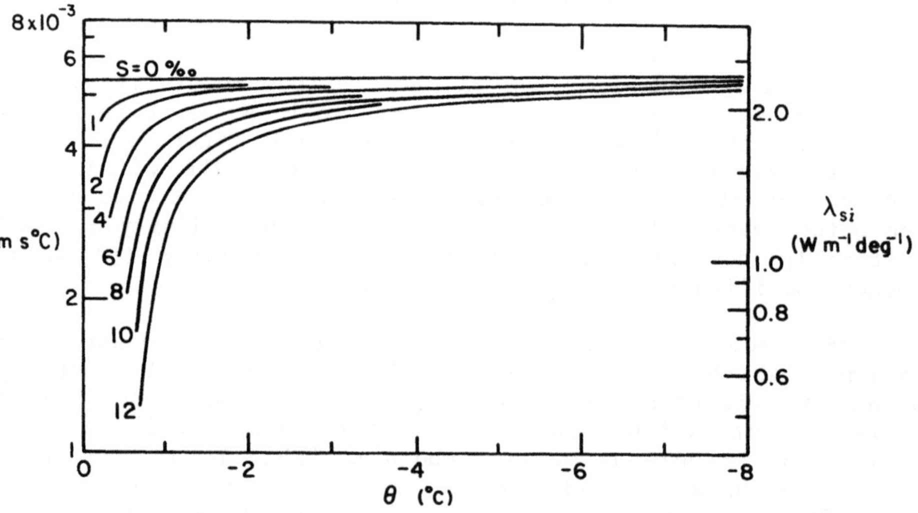

Fig. 88. Thermal conductivity of sea ice calculated using Ono's model (Ono, 1968).

problem. It should also be noted that present models do not suitably describe marine frazil ice. An adequate model of this material could undoubtedly be obtained by combining models that independently consider randomly distributed brine and air inclusions.

Thermal Diffusivity. The thermal diffusivity $\sigma_{si} = \lambda_{si}/(\rho_{si} c_{si})$ of sea ice as calculated by Ono (1968) is given in Figure 89. As can be seen σ_{si} is strongly dependent on the temperature of the ice. This is not surprising in that, as noted by Schwerdtfeger, all three terms composing σ_{si} are monotonic functions of temperature. As temperature rises λ_{si} decreases while the two terms in the denominator (ρ_{si} and c_{si}) both increase. Therefore the thermal diffusivity σ_{si} shows a decrease that is more temperature-dependent than any of its constituent parameters. However, it should be noted that the values of σ_{si} are not appreciably affected by changes in the air content of the ice. This insensitivity can clearly be seen by examining Table VI in Schwerdtfeger (1963).

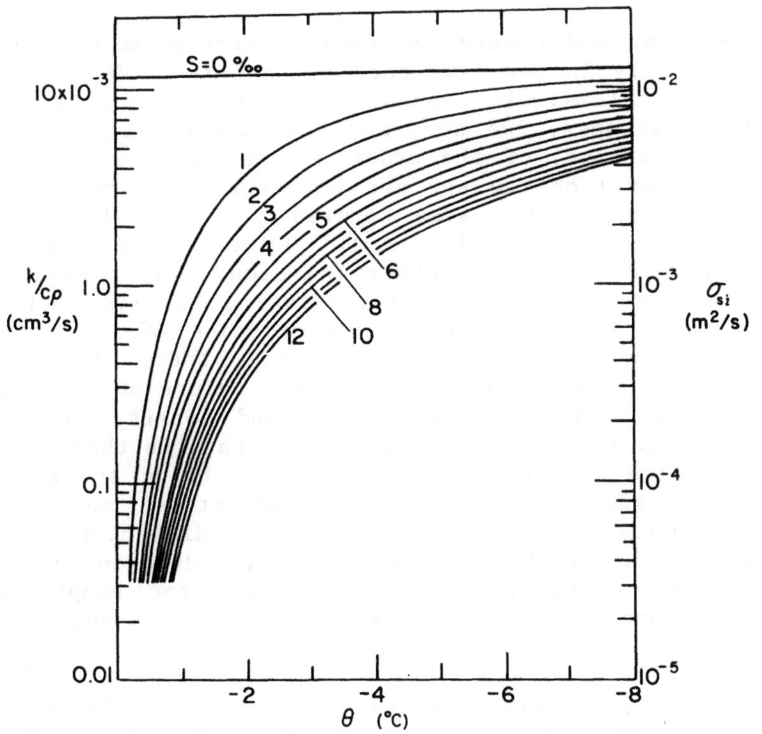

Fig. 89. Thermal diffusivity of sea ice (Ono, 1968).

The same insensitivity to changes in air content occurs in the thermal diffusivity of freshwater ice (Bergdahl, 1977, pp. 70-71).

The particular importance of σ_{si} is that it is the most directly observable thermal property, as it can be directly calculated from the rate of change of the temperature profile in an ice sheet. For a convenient method of determining σ_{si} the reader is referred to Ono (1965, 1968) and to Yen (1981). Determinations of σ_{si} from observed temperature profiles of natural sea ice have been made by Lewis (1967), Weller (1968) and Ono (1965, 1968). In all cases the observed and calculated σ_{si} and λ_{si} values proved to be in reasonable agreement. In fact we are rather surprised with the agreement in that, at least in warm first-year sea ice with large brine volumes, we would expect convective processes, that are not considered at all in current thermal conductivity models, to be very important. For instance, it is hard for us to believe that the processes occurring in brine drainage tubes that alternatively expell cold brine and take in "warm" seawater would not have an effect on the observed σ_{si} values. These matters should clearly be explored further.

11.3 Electrical Properties of Sea Ice in the 0.1 to 40 GHz Range

Active radar and passive microwave emission measurements over sea ice (generally in the frequency range between 0.1 and 40 GHz) have been used extensively over the past decade to obtain information on ice thickness, ice age, and ice concentration on vertical scales (for thickness) of the order of one to a few meters and on horizontal scales (for ice age and concentration) of the size of the Arctic Basin and the area covered by the maximum extent of Antarctic sea ice ($\approx$20 million km^2). Universal to an interpretation of this information, whether emission characteristics (passive methods) or reflection characteristics (active methods), is some model for the dielectric characteristics of the various ice types that we have discussed previously. These characteristics are manifested in the physical-structural nature of the ice cover, for example its salinity, temperature, and density (air content) profiles, and to some extent crystal structure in that it controls the configuration of the other phases (air, brine and solid salt) within the ice structure. Since these properties control the dielectric properties, variations in measured dielectric properties as observed by satellite, aircraft, or ground-based systems hold the promise of providing useful information for geophysical and engineering purposes on the variability of these physical parameters.

For sea ice, different formulations of the dielectric properties are necessary, depending on 1) the frequency range of the remote sensing device and, in combination with this, 2) the description of the ice as "high loss" (generally correlated with first-year and younger ice types) or "low-loss" (which is almost exclusively associated wth multiyear ice). The distinction arises relative to the level of electromagnetic theory needed to model the various phenomena observed. Since sea ice consists of a mixture of constituents with variable dielectric properties, in some cases (primarily at low frequencies or longer wavelengths) the sizes of the individual dielectric inhomogeneities (brine pockets, air bubbles, and solid salt particles) only influence the reflection/emission characteristics as they contribute to the average properties of the medium. This constraint, that the individual particle sizes can be neglected, and only the total volume be considered, applies as long as the particle sizes are less than approx 0.1 wavelength of the incident (or outgoing) radiation.

If, on the other hand, the particle sizes are commensurate with the wavelength, then scattering by individual particles becomes a factor. More complex theories that account for the geometry and interaction of the particles, as well as the average dielectric properties of the medium, are then necessary. The description of these materials is by Rayleigh scattering methods, and a body of theory pertaining to aerosols and atmospheric scattering problems is available. The necessity to apply scattering theory only applies to "low loss" materials (in our case multiyear ice) since "high loss" materials (first-year and younger ice) either attenuate active signals or confine thermal microwave emissions to such a thin layer that the scattering is relatively low. As well, in the younger ice types, the dielectric inhomogeneities are primarily brine pockets with size scales of a millimeter of less, while in old ice the air bubbles (with sizes approaching several millimeters) are the inhomogeneities. The air bubbles therefore affect properties at more frequencies of interest between 0.1 and 40 GHz than do brine pockets. The cutoff where scattering from air bubbles is relatively important is for frequencies above 2-3 GHz, while for brine pockets scattering may be important only above about 20-30 GHz.

We first discuss the basic translation of physical properties into electrical parameters, i.e. the dielectric constants of sea ice as functions of its constituents (and their configurations), then the effects of structural and physical properties on active radar sounding, and finally on the passive microwave emission characteristics of sea ice.

Dielectric Properties of Sea Ice. Since sea ice consists, to a greater or lesser degree, of four phases, ice, brine, air and solid salts, each with different dielectric properties, we therefore need a formulation of the dielectric properties as functions of the relative proportions of these constituents. In reality, however, the dielectric properties are dominated either by the relative amounts of ice and brine or, when the amount of brine is small, the amounts of ice and air. Solid salts do not have a strong effect since their dielectric properties (between 0.1 and 40 GHz) are generally close to those of ice so the presence of a small amount of solid salts is generally indistinguishable at these frequencies. In formulating the dielectric properties, therefore, two-phase formulations accounting for ice and brine or ice and air have generally been adopted. The dielectric properties of both brine and air present sufficient contrast with those of ice to give a distinct contribution to the dielectric properties relative to that of pure ice.

Dielectric Properties of Ice and Air. The complex dielectric constant of most materials is defined (in the pure phase) as

$$\varepsilon = \varepsilon' - j\varepsilon'' \tag{1.79}$$

where ε' is the real part (sometimes also called the dielectric constant), ε'' is the dielectric loss or imaginary part (and can also be related to the conductivity of the media) and $j = \sqrt{-1}$.

As noted earlier, the ability of ice to reject salts is large, that is the solubility of salts in the pure ice phase is invariably down by several orders of magnitude relative to the liquid phase. The ice phase in sea ice is, therefore, relatively pure, and its dielectric properties correspond closely to those observed for pure ice as reviewed, for example, in Evans (1965). The dielectric loss is of the order 10^{-2} to 10^{-4} which, in the sea ice context, is relatively insignificant as we will show later. The real part of the dielectric constant ($\varepsilon_i = 3.2$) is constant over the frequency range from 0.1 to 40 GHz. Similarly, the air component is relatively loss-less and has the value $\varepsilon_a = 1$ in its real part.

Dielectric Properties of Brine. Of the four components in sea ice, brine shows the greatest variability in dielectric properties over the frequency range of our interest (0.1 to 40 GHz). An equation of the Debye form is necessary to describe its behavior due to the relaxation brine undergoes in this frequency range. After Stogryn (1971), this is

$$\varepsilon_2 = \varepsilon_\infty + \frac{\Sigma_o - \varepsilon_\infty}{1 - j2\pi\tau f} + j\frac{\sigma}{2\pi f\varepsilon_o} \tag{1.80}$$

where Σ_o, ε_∞, ε_o, f, τ, σ are the static and high frequency dielectric constants of the brine, the free space permittivity, the frequency, the relaxation time and the ionic conductivity of the dissolved salts. As frequency increases, the contribution to dielectric loss from ionic conductivity (third term in eq. 1.80) drops off but the second term on the right-hand side contains the water dipole relaxation at ≈f = 1-25 GHz so the dielectric loss which is related to attenuation also increases for frequencies between 1 and 25 GHz. If we reformulate eq. (1.80) so that $\varepsilon_b = \varepsilon'_b - j\varepsilon''_b$ as in eq. 1.79, then Figure 90 compares the real and imaginary parts of the dielectric constant for pure water and brine at the same temperature (-10°C). Unlike air and pure ice, the dielectric properties of brine undergo substantial changes between 0.1 and 40 GHz. As also implied in this figure, the values of the dielectric constants at any given frequency are dependent on the concentration of salts in the brine. To compute dielectric properties of the brine in sea ice, it is therefore necessary to know the temperature and salinity of the mixture and then compute the brine concentration based on the phase diagram for sea ice.

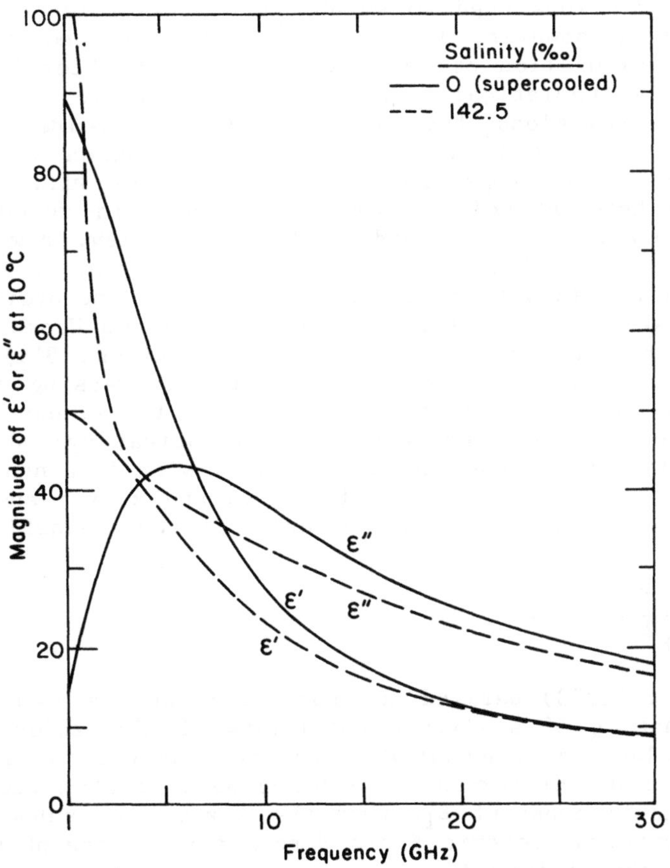

Fig. 90. Real (ε') and imaginary (ε'') parts of the dielectric constant for brine and pure water (Poe et al., 1972).

From Figure 90, we note that compared to ice, the dielectric loss of either brine or pure water is several orders of magnitude higher than that of pure ice (e.g., 10 to 100 at 1 GHz compared to 10^{-4} for ice). The change in the dielectric constant is also substantial but, relative to ice, only increases by about a factor of 10 to 30 (one order of magnitude). Therefore, since the amounts of brine in sea ice are small, the contribution of brine to the real part of the dielectric constant is small, but the electromagnetic loss (imaginary part of the dielectric constant) is almost completely accounted for by the brine. Ice, therefore, controls the behavior of the real part while brine dominates the loss behavior.

Averaged or Composite Dielectric Properties of Sea Ice. As mentioned earlier, both the low frequency and the high frequency dielectric behavior require at least a formulation of the composite dielectric behavior of ice. Poe et al. (1972) reviewed several empirical formulations used to establish a bulk dielectric constant. More recently, however, Vant et al. (1975) showed that a more analytical formulation of the mixture constant after Tinga et al. (1973) gave a theoretically justifiable dielectric constant based on analytic solutions from electrostatics. Golden and Ackley (1981) have also used Tinga et al.'s formulation in explaining some active radar properties of sea ice so we will pursue that formulation here as well. Some differences in formulation are noted between Vant et al. and Golden and Ackley, however, since the possibility of dielectric horizontal anisotropy of sea ice was not allowed for in Vant et al.'s formulation. The overall sea ice structure taken is that discussed by Anderson and Weeks (1958) and here earlier. Essentially, brine layers (Figure 91a) formed at the dendritic growth interface neck with decreasing temperature (Figure 91b) further up in the ice sheet and freeze out into cylinders (Figure 91c) and eventually into elliptical cylinders (Figure 91d). An idealized model of a brine inclusion is a non-degenerate ellipsoid (a > c > b, where a, b are in the x, y plane and c in the z plane or in the depth direction) with surface defined by:

$$\frac{x^2}{a^2} + \frac{y^2}{b^2} + \frac{z^2}{c^2} = 1. \qquad (1.81)$$

Tinga et al. (1973) derived an expression for the complex dielectric constant ε_k of a mixture consisting of ellipsoids (brine) of complex dielectric constant ε_2, relative volume V_b and uniform axial alignment, dispersed in a homogeneous medium (ice) of complex dielectric constant ε_1, when the mixture is subjected to an initially uniform electric field directed along one of the principal axes of the ellipsoid:

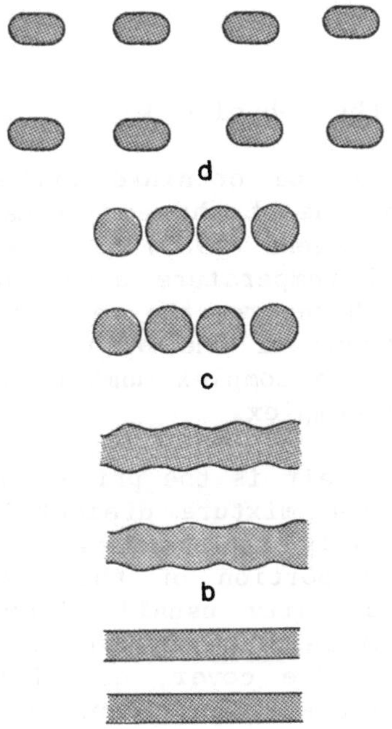

Fig. 91. Brine layers at the bottom (a) of the ice sheet and necking off as the temperature is lowered (Anderson and Weeks in Golden and Ackley, 1981).

$$\varepsilon_k = \varepsilon_1 + \left[\frac{V_b\ \varepsilon_1\ (\varepsilon_2 - \varepsilon_1)}{n_k\ (1-V_b)\ (\varepsilon_2 - \varepsilon_1) + \varepsilon_1}\right] \tag{1.82}$$

where n_k is the depolarization factor of the ellipsoid:

$$n_k = \frac{abc}{2} \int_o^\infty \frac{ds}{(k^2+s)\ \sqrt{(a^2+s)(b^2+s)(c^2+s)}}\ . \tag{1.83}$$

This solution allows an accounting for cases when the field is aligned with one of the ellipsoid axes as sometimes occurs in active sounding. If the field is not aligned with one of the axes or if the axes are randomly aligned with respect to any direction of the field, then in the case of random alignments, Ackley and Keliher (1979) show the dielectric constant is

$$\varepsilon \equiv \frac{1}{3}\,\varepsilon_a + \frac{2}{3}\,\varepsilon_b \qquad (1.84)$$

(when the E-field is in the a-b plane).

The brine volume can be obtained either directly from the phase diagram or by using one of the equations of Frankenstein and Garner (1967) or Cox and Weeks (1983) for the appropriate temperature range. Salinity and temperature also control the concentration in the brine and, therefore, the real and imaginary values of ε_2, the dielectric constant of the brine. Since the dielectric constant of the brine is a complex number, the composite dielectric constant ε_k is also complex.

For ice covers where air is the primary second component, the effects of geometry on the mixture dielectric constants are not usually included. Normally the primary electromagnetic effects are confined to the top portion of the ice cover down to the freeboard level where salinity usually increases, leading to a dominance of ice-brine dielectric mixtures below these depths. In the top portions of the ice cover, especially in multiyear ice where the air content is most significant, several factors appear to discount other geometries of the air bubbles besides spherical as being very important for dielectric modeling purposes. The ice above freeboard typically arises from several sources: a) Brine-drained ice of usually frazil structure from the previous year or years of growth. The inclusions are probably at least random if not spherical, leading to little difference than treating them as spherical. b) Recrystallized ice. Here again the tendency is for the air bubbles to trend toward more spherical shapes because of the annealing at the melting point that has taken place. c) Snow-ice or snow infiltrated by meltwater and refrozen. Again, spherical air inclusions would dominate. Even if the inclusions are non-spherical, the dielectric contrast between ice and air is small (3.2 to 1) so the ice component dominates the real part of the dielectric constant at the low percentages of air (<20%) typically found in sea ice covers. This behavior is in contrast to the imaginary part of ice-brine mixtures where the brine dominates. Since the pure ice dielectric properties are homogeneous and isotropic (at these frequencies) the air-ice dielectric properties are also mostly isotropic, even if some anisotropy exists in the inclusions. The changes in dielectric constant can be related to the density of the ice and, after Gudmandsen (1971), this is:

$$\varepsilon = (1 + 0.85\,\rho_i)^2. \qquad (1.85)$$

As the radar sounding of sea ice via the use of active methods depends directly on the dielectric properties as shown in the mixture constants we will discuss these next.

Active Radar Sounding of Sea Ice. To emphasize this point again, the mixture dielectric constants given in eq. (1.82) have an apparent dependence on the inclusion geometry. However, this geometric dependence appears only in the depolarization factor n_k which can be solved for in terms of axial ratios of the major and minor dimensions of the inclusions. Therefore the same dielectric effect would be attributed to large inclusions as to small inclusions if 1) the total volume occupied was similar and 2) the axial ratios were the same. The geometry is, therefore, decoupled from the wavelength as long as a, b, c, $\ll \lambda$. The apparent effect arises from the boundary conditions on Maxwell's equation at the surface of the inclusions, in that much more surface area of the included medium is present to particular views of the electric field if the inclusions are anisotropic.

In active sounding, an electromagnetic wave polarized in one direction travels into the sea ice and is reflected by any dielectric discontinuities it encounters. The major discontinuities are the top (air/ice) and bottom (ice/water) interfaces of the ice sheet. Recently radar measurements in the frequency range from 0.1 to 0.5 GHz (100 to 500 MHz) have been used to obtain thickness information on sea ice (when dielectric properties are either known or assumed), primarily from ground-based transceivers but also by instruments mounted in helicopters. We have formulated the dielectric constant of sea ice in a form that includes anisotropy because one of the effects seen recently (Campbell and Orange, 1974; Kovacs and Morey, 1978) is anisotropy in the strength of the return from impulse radar soundings in these thickness soundings of sea ice.

The effect seen is marked dependence of returned radar signal strength, depending on the orientation of the linearly polarizing antenna of the radar system on the surface. The strongest return also coincided with the azimuthal orientation of the c-axes in the ice cover, while the weakest return was found to be at 90° from the direction of the c-axis. This implied a structural control on the radar signal return primarily through the major elements affecting radar attenuation in sea ice, the brine inclusions. We have previously shown how brine platelet spacing aligns well with c-axis direction within any given crystal and also that large areas of sea ice have pronounced azimuthal alignments of the c-axes in response to currents under the ice cover. These alignments also control the brine inclusion geometry and affect the strength of the radar returns. To examine the observed radar anisotropy, Golden and Ackley (1981) used the mixture dielectric constants with some assumptions regarding brine pocket geometry to model the radar behavior of sea ice. Figure 92 shows a typical salinity temperature profile from the Alaskan Arctic and the computed brine volume profile. To depict the radar behavior, the profile was divided into 10-cm lengths with the dielectric constant at each

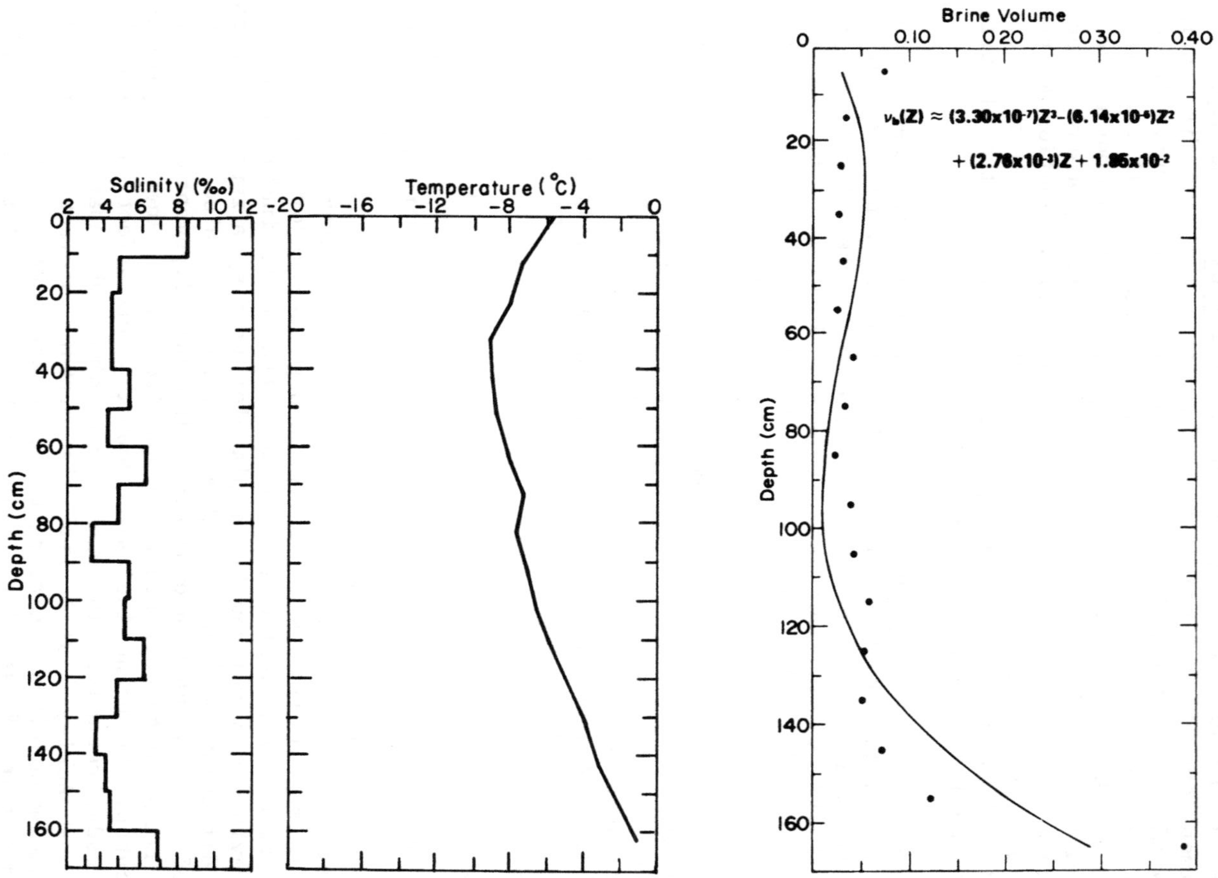

Fig. 92. Ice salinity-temperature and brine volume profile from the Alaskan Arctic (Harrison Bay) (Golden and Ackley, 1981).

level computed by eq. (1.82) from the computed brine volume at that level. Direct information on the brine inclusion geometry was not available for this data set (nor are such observations generally available), so an assumed profile was taken to compute the depolarization factor as a function of depth. Since the depolarization factor is dependent only on the axial ratio, these were varied from the top to the bottom of the ice. At the bottom, previous measurements (Anderson and Weeks, 1958; Kovacs and Morey, 1978; Weeks and Assur, 1967) indicated reasonable ratios of the ellipsoidal axes were of the order of a:b:c = 30:1:5 (here the a-axis direction is in the vertical plane and normal to the direction of the crystallographic c-axis). Near the top of the ice sheet a frazil-like initial layer probably formed, with more isotropic characteristics of the brine inclusions. These were assumed to be in the ratio a:b:c = 1:1:0.5. The brine inclusion geometries (a and c dimensions) were then taken to vary linearly between these values. Later data (Langhorne et al., 1980) on brine inclusion geometry in oriented ice showed the variation in axial ratios to follow similar patterns to the brine volumes so the most extreme excursions were concentrated in the bottom few centimeters.

As we show later, this can only enhance the behavior derivable from the assumed linear changes. The dielectric constant profiles for the two directions are shown as a function of depth in Figure 93. The normal polarization indicates the E-field parallel to the preferred c-axis (normal to the long axes of the ellipsoids) while the tangential polarization refers to the direction perpendicular to the c-axis. We see the strong anisotropy in the dielectric constants (particularly in $\varepsilon''_{n,t}$), especially near the bottom of the ice where the brine volume and axial ratios dramatically increase. It is easy to imagine that if the inclusion geometry were allowed to vary directly with the brine volume then this behavior would be even more pronounced.

Figure 94 shows the power returned as a function of depth when a multilevel reflection, attenuation, and beam spreading model is used to calculate the returned power. The returned power for the two polarizations is down by more than one order of magnitude (6% compared to 0.3%), compatible with the radar attenuation observations. Large electric field penetration into the brine layers for the highly attenuated direction allows conduction effects to attenuate the wave. This in turn reduces the power returned from the ice/water interface relative to the other direction.

The radar properties of sea ice, therefore, exhibit a response to the structural controls given by the ice sheet and are dependent specifically on the salinity-temperature profile (determining the brine volume) and, where strong anisotropy is observed, on the macroscopic manifestations of the brine inclusion

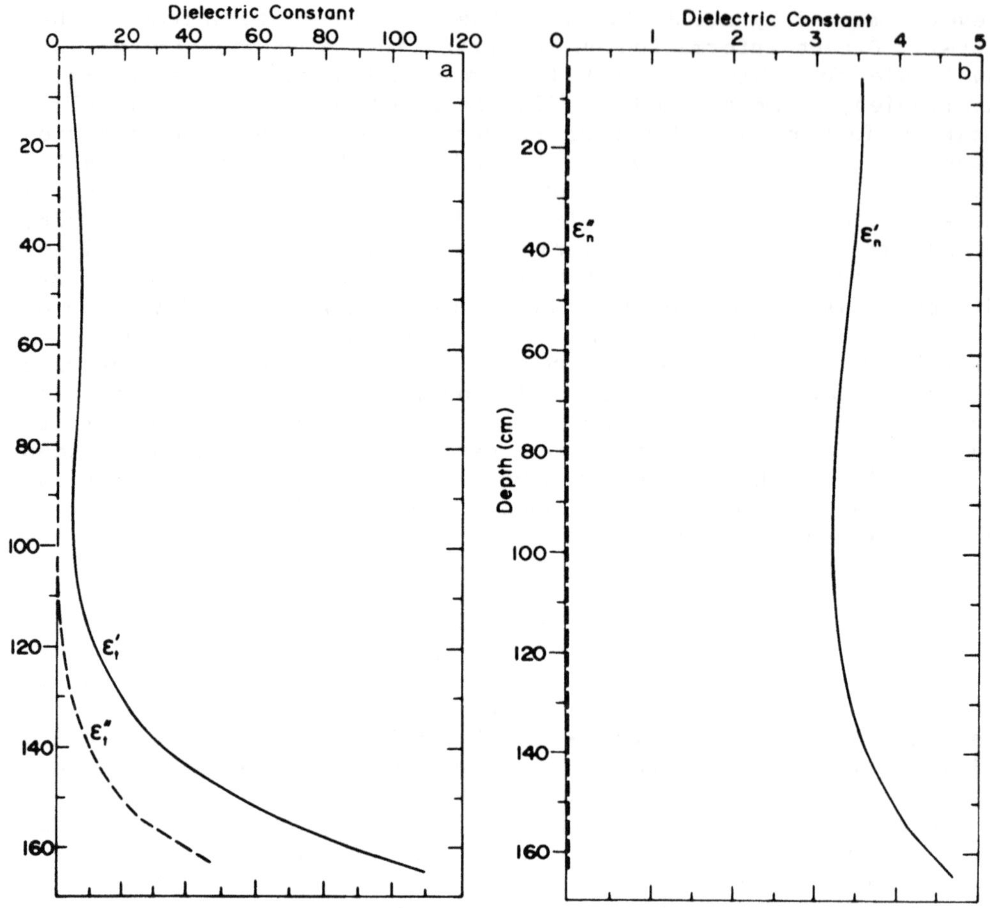

Fig. 93. Computed depth-dielectric constant profiles using the brine volume profile from Figure 92 and assumed axial ratios of the inclusions at the top a:b:c = 1:1:0.5 and at the bottom a:b:c = 30:1:5 (Golden and Ackley, 1981).

geometries as given by the depth variation of the axial ratios of the inclusions.

In the future by suitable radar modeling and sounding of sea ice, active radar measurements could provide a nondestructive means of determining sea ice structure by inverting the return radar signals and calculating the brine volume and brine inclusion geometry profiles that lead to the observed power levels.

Passive Microwave Emission Characteristics of Sea Ice. As we noted in the last section, active radar returns reflect the

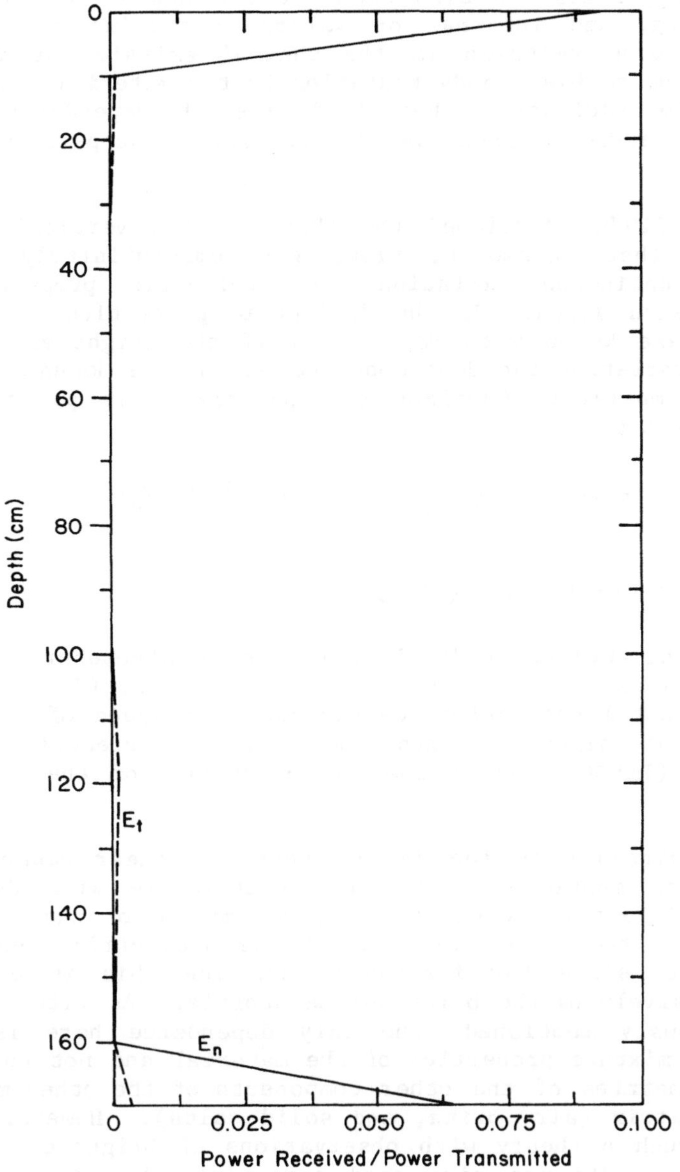

Fig. 94. Simulated returned power/incident power vs. depth for profiles shown in Figures 92 and 93 (Golden and Ackley, 1981).

physical properties of the medium. Similarly, passive emissions are another technique used to acquire information on the intrinsic dielectric properties and also upon factors such as surface roughness, layering, and internal or volume scattering of the ice. Passive microwave emission is the thermal emission at microwave frequencies due to black body radiation by the material. It therefore has the additional characteristics of dependence on the thermal temperature as well as the dielectric properties of the material.

Stogryn (1970) developed the theory for a vertically structured medium that is smooth, flat, and semi-infinitely extended and shows continuous variations in dielectric properties and temperature with depth. If the dielectric properties and thermal temperature are known with depth, and if the brightness temperature of the radiation incident upon the surface is known, then the brightness temperature (horizontally polarized) at polar angle θ and frequency ν is

$$T(\Theta,\nu) = |R_h(\Theta,\nu)|^2 T_{sky}(\Theta,\nu) + 4\left(\frac{2\pi\nu}{c}\right)^3 \cos\Theta \int_{-\infty}^{0} T(z') A(z')^2 \, \mathrm{Im}\,[\varepsilon(z',\nu)] \, dz' \qquad (1.86)$$

where T_{sky}, the incident radiation, is due to atmospheric attenuation and reradiation of cosmic noise, and R_h, c, $(T(z'),\ \varepsilon(z',f)$ are the horizontal reflection coefficient, the speed of light, the temperature at depth z′, and the complex dielectric constant respectively ($\mathrm{Im}[\varepsilon(z',\nu)]$ is the imaginary part of the dielectric constant).

For application to sea ice we note the clear dependence of the parameters mentioned, first the temperature with depth and second the dielectric constant. The dependence of the brightness temperature on the imaginary part of the dielectric constant is also of note, since for first-year sea ice this is controlled almost exclusively by the brine volume profile. As with the radar model previously mentioned, the only dependence here is on the aggregate or mixture properties of the material and not on the individual geometries of the other components of the other materials in the ice matrix (air, brine, and solid salts). However, a comparison of such a theory with observations of brightness temperature of sea ice did not give good agreement (Poe et al., 1972). Similarly, theoretical results from a surface roughness theory using Kirchoff approximations also did not provide good agreement, although parameterization of the surface backscatter coefficients appears to be a necessary component in any of these models.

A discussion of surface backscatter is both beyond our scope and somewhat independent of our theme, which is to understand some of the relationships between internal sea ice properties and observed emission characteristics. We therefore only mention it as one of the necessities in the complete description of the emission characteristics. England (1975) notes that surface roughness, while not easily treated theoretically, is visually apparent, so cautious interpretations of its effect can be made.

For the case of sea ice at microwave frequencies, a more important effect which must be included to improve the agreement between theory and observation is a consideration of volume scatterers within the ice. The field intensity calculations from scattering phenomena are somewhat burdened with notation, so the more interested reader is advised to consult the references for details. Scattering can be conveniently divided into three tasks (England, 1975). These are 1) the derivation of the direct field intensity resulting from self emission of the medium and from sky radiance, as we discussed earlier based on the work of Stogryn (1970), 2) the derivation of the diffuse field resulting from radiation singly or multiply scattered from the direct field (this calculation involves the third task), and 3) the derivation of the relative scattering strength or scattering albedo in terms of wavelength, scatterer size and concentration, and dielectric properties. The derivation of the diffuse field is the notation-ridden exercise, while task 3, the scattering albedo, gives conditions more directly comparable to our purpose since it is related most directly to the physical properties under discussion. A brief derivation of the scattering albedo, therefore, follows after England (1975). Simplifying assumptions are that the embedded particles are spherical and effectively separated so that interaction effects between scatterers are not considered. The Rayleigh scattering cross section σ is related to the size parameter (the ratio of scatterer diameter to free space wavelength) and to the dielectric constant [the ratio of the dielectric constant of the scatterer (brine or air) to that of the layer (ice)]. Spheres of diameter d and dielectric constants ε_2 embedded in a medium whose dielectric constant is ε_1 have an average dipole moment (Stratton, 1941)

$$\bar{P} = 4\pi\varepsilon_1 \varepsilon_o \left[\frac{\varepsilon_2/\varepsilon_1 - 1}{\varepsilon_2/\varepsilon_1 + 2}\right] \frac{\bar{d}^3}{8} E_o \qquad (1.87)$$

where ε_o is the free space permittivity, E_o is the incident electric field strength (or emitted radiation strength from below), and $\bar{d}$ is the mean diameter of the scatterers. The average outward energy flow P from the scatterer is

$$\bar{P} = \left(\frac{\mu}{\varepsilon_1 \varepsilon_o}\right)^{1/2} \frac{\omega^4 \bar{p}^2}{12\pi} \left(\frac{\varepsilon_1}{c^2}\right) \quad (1.88)$$

where μ is the permeability ($\mu = \mu_o$ for the dielectrics of interest), ω (= $2\pi f$) is the angular frequency of the incident wave, and c is the free space velocity of light. Poynting's relation yields the energy flow per unit area of the source wave (the energy flow prior to scattering encounter),

$$\bar{S} = \frac{1}{2} (E_o)^2 (\varepsilon_1 \varepsilon_o / \mu_o)^{1/2} \quad (1.89)$$

so that the average scattering cross section $\sigma = \bar{P}/\bar{S}$ is

$$\bar{\sigma} = \frac{2}{3} \pi^5 \left(\frac{\bar{d}}{\lambda_o}\right)^3 \frac{\varepsilon_1^2}{\lambda_o} \bar{d}^3 \left[\frac{\varepsilon_2/\varepsilon_1 - 1}{\varepsilon_2/\varepsilon_1 + 2}\right]^2 \quad (1.90)$$

where $\lambda_o = \frac{2\pi c}{\omega}$ and $\mu_o = \frac{1}{c^2 \varepsilon_o}$ are used.

If f* is the volume fraction of scatterers in the layer (related to the ice density for air inclusions) then

$$f^* = \bar{N} \frac{4}{3} \pi \left(\frac{d}{2}\right)^3 = \bar{N} \pi \frac{d^3}{6} . \quad (1.91)$$

The scattering loss is the number of scatterers times the scattering cross section per scatterer or

$$\bar{N} \bar{\sigma} = \frac{2\pi (\varepsilon_1)^{1/2} \tan\delta_1}{\lambda_o} S_c f^* \quad (1.92)$$

where $\tan \delta_1 = \varepsilon''_1/\varepsilon'_1$ is the loss tangent and

$$S_c = (2\pi)^3 \left(\frac{\bar{d}}{\lambda_o}\right) \left[\frac{\varepsilon_1}{\tan\delta_1}\right]^{3/2} \left[\frac{\varepsilon_2/\varepsilon_1 - 1}{\varepsilon_2/\varepsilon_1 + 2}\right]^2 . \quad (1.93)$$

The total loss α_1 in the layer is the sum of the scattering loss $\bar{N}\bar{\sigma}$ and the dielectric loss $2\beta_1$ where

$$\alpha_1 = \bar{N} \bar{\sigma} + 2\beta_1 \quad (1.94)$$

and

$$2\beta_1 = [2\pi(\varepsilon_1)^{1/2}/\lambda_o] \tan\delta_1 . \quad (1.95)$$

(Here the appropriate values for ε_1, $\tan\delta_1$ could be the mixture dielectric constant and loss tangent instead of the ice constants if scattering is small, in analogy to the active radar parameterization shown earlier.)

A useful parameter to estimate the influence of scattering is the scattering albedo ω_o which is defined as the ratio of scattering loss to total loss in the layer or

$$\omega_o = \frac{\bar{N}\,\bar{\sigma}}{\bar{N}\,\bar{\sigma} + 2\beta_1} \,. \tag{1.96}$$

Naturally, if ω_o is small then the contribution to total loss from scattering is small so we can define combinations of frequency ranges, concentrations of scatterers and dielectric properties where scattering has to be taken into account. We will now show a few simulations of these parameters of interest.

Figure 95 shows the scattering albedo versus d/λ_o (ratio of scatterer size to wave length) with the volume fraction as a parameter for ice overlying fresh water. We note the middle three curves f = (.05, .1, .2) are typical air content values seen in multiyear sea ice (Meeks et al., 1974). These curves indicate that "significant" contributions to the total loss from scattering occur at a minimum fractional diameter to wavelength ratio of about 0.1. From the limited data on multiyear sea ice, air bubble sizes varying from 0.5 mm to several millimeters (Poe et al., 1974) are observed, indicating a wavelength cutoff of about 0.5 to 1 cm where scattering losses become important and indeed dominate the emission character of multiyear ice. This corresponds to frequencies greater than 1-3 GHz.

The effects of brightness temperature as would be sensed by an above-surface radiometer are shown in Figure 96 where the scatter-induced brightness temperature change is plotted versus scattering albedo as a function of the layer thickness (D) to wavelength ratio. Generally for thin layers (low D/λ_o) the brightness temperature induced by scattering increases, while for thick layers (high D/λ_o) the brightness temperature drops for almost all values of scattering albedo. The conclusion is that where we have "significant" values of scattering albedo (>0.5) the major effect seen will be for thicker layers of scattering material and that change will be a drop in brightness temperature. This behavior is what is typically observed for multiyear ice in the Arctic, with lower brightness temperatures implying significantly more scattering and layer thickness at the higher frequencies than for first-year ice. As the dielectric loss (non-scattering part, $2\beta_1$) increases, the effect of layer thickness drops off. This behavior also is indicated in the equation for scattering albedo ω_o, where increasing

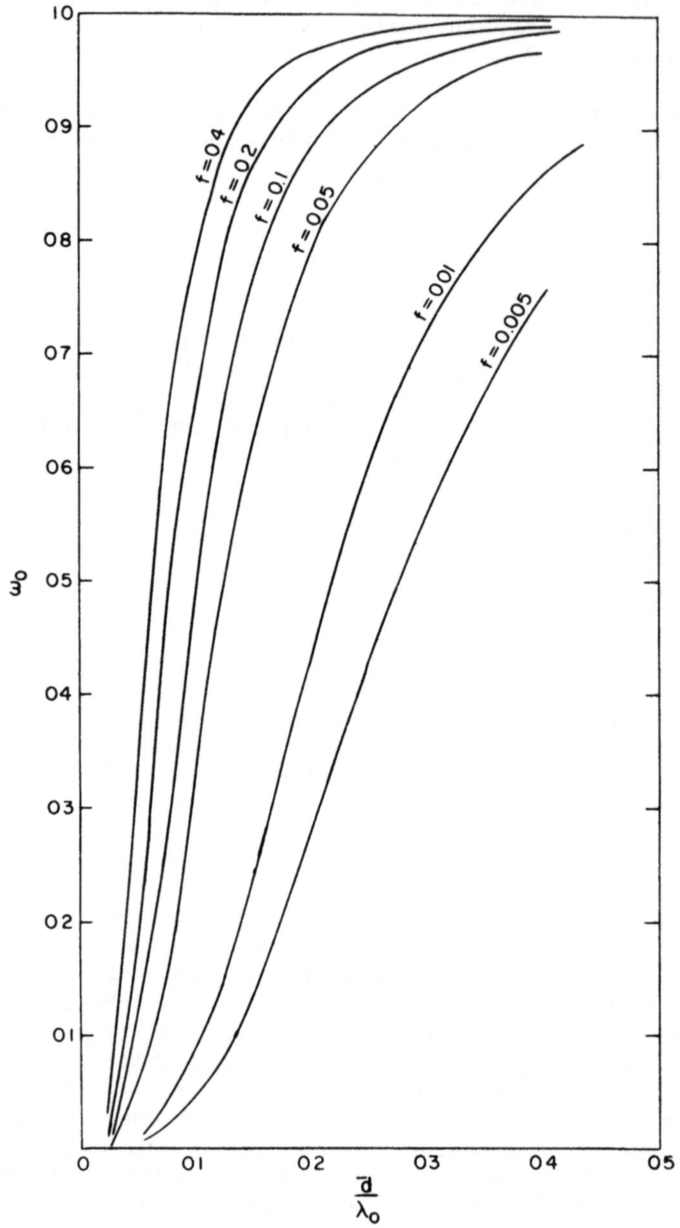

Fig. 95. Scattering albedo (ω_0) vs. scatterer size to wavelength ratio (d/λ_0) for ice overlying fresh water. The volume fraction of scatterers (f) is the parameter on the curves (England, 1975).

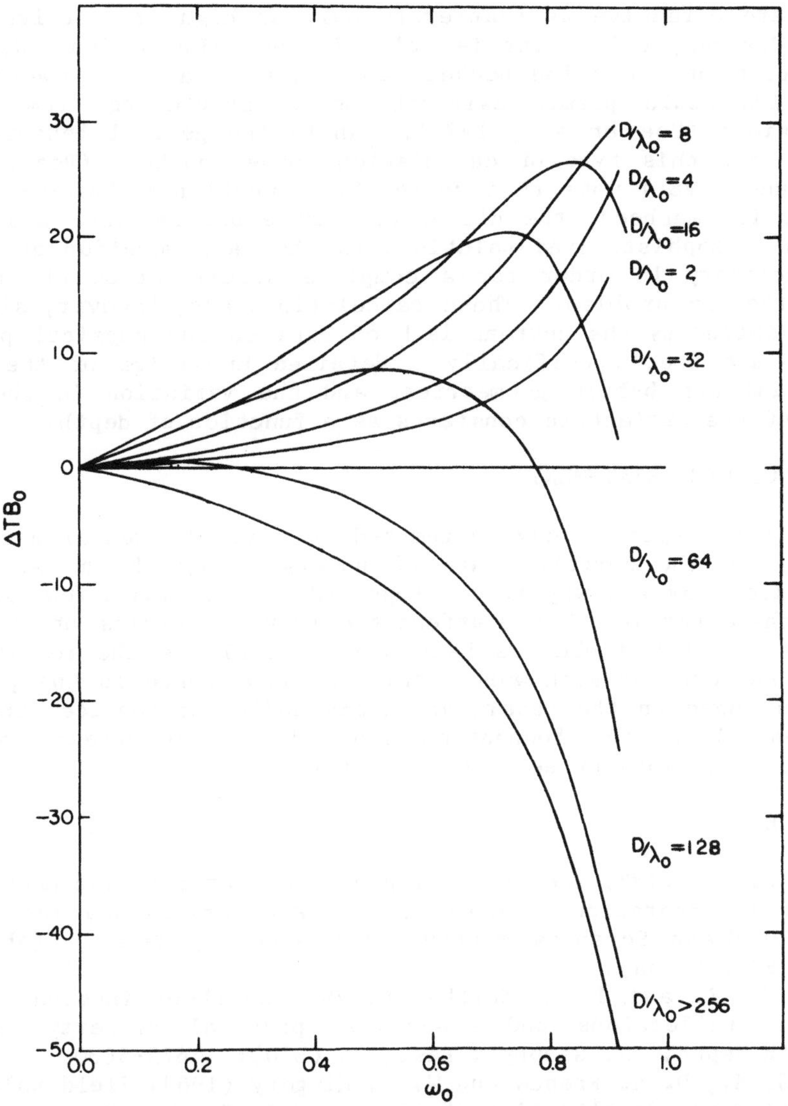

Fig. 96. Changes in brightness temperature (ΔTB_o) plotted against scattering albedo (ω_o) with the layer thickness to wavelength ratio (D/λ_o) as a parameter (England, 1975).

numbers of scatterers would be required for the scattering to dominate the ordinary dielectric loss, $2\beta_1$. For first-year ice, which is both lossy and of lower air content, the influence of scatter-induced changes on brightness temperature are commensurately less than for multiyear ice. In first-year ice, the scatterers are primarily the brine pockets, which have a dimension of about 0.1

mm and are effective as scatterers only at high frequencies 20-40 GHz). However, this point is still inconclusive without detailed information on the brine pocket geometries or a solid theoretical base which would permit calculations of scattering from strong fluctuations (Poe et al., 1974). While the general behavior can be seen from this type of calculation, other authors (Poe et al., 1972; Fung, 1981) note that correlation function techniques (small perturbation method), the distorted Born approximation, and generally more sophisticated solutions to the wave equation are probably necessary in order for a complete specification of the sea ice scattering problem. These calculations are, however, significantly limited by the current lack of data on the physical properties of sea ice, specifically a detailed knowledge of the brine pocket and air bubble geometries, and the variation in the mean values of the dielectric constants as a function of depth.

12. CONCLUDING STATEMENT

In this paper we have attempted to give the reader a general knowledge of the current state of understanding of the structure of sea ice, how and why it develops and varies, and a few examples of how these variations can affect the bulk properties of the ice. Progress in this field has been slow, largely as the result of a lack of support for such work. This is unfortunate in that, as we hope was shown in the paper, an understanding of the ice structure is essential as the foundation upon which the science of the property variations of sea ice must be built.

REFERENCES

Ackley, S. F. (1982) Ice scavenging and nucleation: Two mechanisms for incorporation of algae into newly forming sea ice. AGU-ASLO Ocean Sciences Meeting, San Antonio, Texas. (Abstract in EOS, 63:54).

Ackley, S. F. and T. E. Keliher (1979) Ice sheet internal radio-echo reflections and associated physical property changes with depth. J. Geophys. Res., 84 (B10): 5675-5680.

Adams, C. M., D. M. French and W. D. Kingery (1963) Field solidification and desalination of sea ice. In Ice and Snow: Properties, Processes and Applications (W. D. Kingery, ed.), MIT Press, Cambridge, Mass., p. 277-288.

Addison, J. R. (1977) Impurity concentrations in sea ice. J. Glaciol., 18, (78): 117-127.

Anderson, D. L. (1958) A model for determining sea ice properties. In Arctic Sea Ice, U.S. National Academy of Sciences - National Research Council, Pub. 598, p. 148-152.

Anderson, D. L. (1960) The physical constants of sea ice. Research, 13 (8): 310-318.

Anderson, D. M. and W. F. Weeks (1958) A theoretical analysis of sea ice strength. Trans. Am. Geophys. U., 39 (4): 632-640.

Arakawa, K. (1958) Studies on the freezing of water, II. J. Faculty Sci. Hokkaido Univ., Ser. II, 4: 311-339.

Arakawa, K. and K. Higuchi (1954) On the freezing process of aqueous solutions. Low Temp. Sci., A12: 73-86.

Armstrong, T., B. Roberts and C. Swithinbank (1966) Illustrated Glossary of Snow and Ice. Scott Polar Research Institute, Special Pub. 4, 60 pp.

Assur, A. (1958) Composition of sea ice and its tensile strength. In Arctic Sea Ice, U.S. National Academy of Sciences -- National Research Council, Pub. 598, p. 106-138.

Assur, A. and W. F. Weeks (1964) Growth, structure, and strength of sea ice. USA Cold Regions Research and Engineering Laboratory, Research Report 135, 19 p.

Bari, S. A. and J. Hallett (1974) Nucleation and growth of bubbles at an ice-water interface. J. Glaciol., 13 (69): 489-520.

Barrett, C. S. (1952) Structure of Metals. McGraw-Hill. New York.

Bennington, K. O. (1959) Preliminary report on sea ice crystal fabrics on Station Charlie. In Semi-Annual Report, 1 December 1959, Drifting Station Charlie, Project Husky, ONR 477(24) T.O. 307-252.

Bennington, K. O. (1963a) Some crystal growth features of sea ice. J. Glaciol., 4 (36): 669-688.

Bennington, K. O. (1963b) Some chemical composition studies on arctic sea ice. In Ice and Snow: Properties, Processes and Applications (W. D. Kingery, Ed.), MIT Press, Cambridge, Mass., p. 248-257.

Bergdahl, L. (1977) Physics of ice and snow as affects thermal Pressure. Department of Hydraulics, Goteborg, Sweden, Chalmers University of Technology, Rept. Series A:1, 158 p.

Bilello, M. A. (1961) Ice thickness observations in the North American Arctic and Subarctic for 1958-59, 1959-60. USA Cold Regions Research and Engineering Laboratory, Special Report 43, Pt. 1, 43 p.

Blinov, L. K. (1965) The salt content of sea water and sea ice. Trudy Gos. Okeanograf. Inst., 81: 5-55.

Bolling, G. F. and W. A. Tiller (1960) Growth from the melt. I: Influence of surface intersections in pure metals. J. Appl. Phys., 31: 1345-1350.

Bragg, W. H. (1922) The crystal structure of ice. Proc. Phys. Soc., 34: 98-103.

Buck, K. R. and D. Garrison (1982) Sea ice algal communities in the Weddell Sea. II: Population comparisons between the water column and sea ice. AGU-ASLO Ocean Sciences Meeting, San Antonio. (Abstract in EOS, 63.)

Burton, J. A., R. C. Prim and W. P. Slichter (1953) The distribution of solute in crystals grown from the melt. I: Theoretical. J. Chem. Phys., 21: 1987-1991.

Campbell, K. J. and A. S. Orange (1974) The electrical anisotropy of sea ice in the horizontal plane. J. Geophys. Res., 79 (33): 5059-5063.

Carslaw, H. S. and J. C. Jaeger (1959) Conduction of Heat in Solids. Oxford University Press, 510 p.

Carte, A. E. (1961) Air bubbles in ice. Proc. Phys. Soc. (London), 77 (495): 757-768.

Cherepanov, N. V. (1957) Using the methods of crystal optics for determining the age of drift ice. Problemy Arktiki, 2: 179-184.

Cherepanov, N. V. (1964) Structure of sea ice of great thickness. Trudy Arkt. Antarkt. Nauch. Issled. Inst., 367: 13-18.

Cherepanov, N. V. (1971) Spatial arrangement of sea ice crystal structure. Prob. Arkt. Antarkt., 38: 176-181.

Colbeck, S. (1979) Grain clusters in wet snow. J. Coll. Inter. Sci., 72 (3).

Coriell, S. R., M. R. Cordes, W. J. Boettinger and R. F. Sekerka (1980) Convective and interfacial instabilities during unidirectional solidification of a binary alloy. J. Cryst. Growth, 49 (1): 13-28.

Cox, G. F. N. and W. F. Weeks (1974) Salinity variations in sea ice. J. Glaciol., 13 (67): 109-120.

Cox, G. F. N. and W. F. Weeks (1975) Brine drainage and initial salt entrapment in sodium chloride ice. USA Cold Regions 85 p.

Cox, G. F. N. and W. F. Weeks (1983) Equations for determining the gas and brine volumes in sea ice samples. J. Glaciol., 29 (102): 306-316.

Dayton, P. K. and S. Martin (1971) Observations of ice stalactites in McMurdo Sound, Antarctica. J. Geophys. Res., 76: 1595-1599.

Dunbar, M. and W. F. Weeks (1975) Interpretation of young ice forms in the Gulf of St. Lawrence using side-looking airborne radar and infrared imagery. USA Cold Regions Research and Engineering Laboratory, Research Report 337, 41 p.

Edie, D. D. and D. J. Kirwan (1973) Impurity trapping during crystallization from melts. Ind. Eng. Chem. Fundam., 12: 100-106.

Eide, L. and S. Martin (1975) The formation of brine drainage features in young sea ice. J. Glaciol., 14 (70): 137-154.

Elbaum, C. (1959) Substructures in crystals grown from the melt. Prog. Met. Phys., 8: 203-253.

England, A. W. (1975) Thermal microwave emission from a scattering layer. J. Geophys. Res., 80 (32): 4484-4496.

Evans, S. (1965) Dielectric properties of snow and ice: A review. J. Glaciol., 5: 773.

Farhadieh, R. and R. S. Tankin (1972) Interferometric study of freezing of sea water. J. Geophys. Res., 77: 1647-1657.

Foldvik, A. and T. Kvinge, 1974: Conditional instability of sea water at the freezing point. Deep Sea Res., 21: 169-174.

Frankenstein, G. and R. Garner (1967) Equations for determining the brine volume of sea ice from -0.5^{o} to $-22.9^{o}C$. J. Glaciol., 6 (48): 943-944.

Fujino, K. and Y. Suzuki (1959) Observations on the process of ice rind formation on the surface of still water. Low Temp. Sci., A18: 149-155.

Fujioka, T. and R. F. Sekerka (1974) Morphological stability of disc crystals. J. Cryst. Growth, 24/25: 84-93.

Fukutomi, T., K. Kusunoki and T. Nagashima (1949) On the formation of crystal ice and the structure of ice crust. Low Temp. Sci., 2: 73-76.

Fukutomi, T., M. Saito and Y. Kudo (1953) Study of sea ice (the 16th report): On the structure of ice rind, especially on the structure of thin ice sheet and ice-sheet block. Low Temp. Sci., 9: 113-123.

Fung, A. (1981) Microwave scattering and emission from sea ice. In Proceedings of Second Workshop on Microwave Remote Sensing of Sea Ice and Icebergs, NASA Langley Research Center.

Garrison, D. and K. R. Buck (1982) Sea ice algal communities in the Weddell Sea. I: Biomass distribution and the physical environment. AGU-ASLO Ocean Sciences Meeting, San Antonio. (Abstract in EOS, 63).

Garrison, D. L., S. F. Ackley and K. R. Buck (1983) A physical mechanism for establishing algal populations in frazil ice. Nature, 306: 363-365.

Gitterman, K. E. (1937) Thermal analysis of sea water. Trudy Solyanoy Lab. Akad. Nauk SSSR, 15(1).

Glen, J. W. (1955) Comments on the paper of Professor Arakawa on the growth of ice crystals in water. J. Glaciol., 2: 483.

Golden, K. M. and S. F. Ackley (1981) Modeling of anisotropic electromagnetic reflections from sea ice. J. Geophys. Res., 86 (C9): 8107-8116.

Golovkov, M. P. (1936) K petrografii l'da Karskogo moria (The petrography of Kara sea ice). Trudy Vses. Arkt. Inst. Leningrad, 60: 7-40.

Gow, A. J. and S. Epstein (1972) On the use of stable isotopes to trace the origins of ice in a floating ice tongue. J. Geophys. Res., 77 (33): 6552-6557.

Gow, A. J. and D. Langston (1977) Growth history of lake ice in relation to its stratigraphic, crystalline and mechanical structure. USA Cold Regions Research and Engineering Laboratory, CRREL Report 77-1, 24 p.

Gow, A. J. and W. F. Weeks (1977) The internal structure of fast ice near Narwhal Island, Beaufort Sea, Alaska. USA Cold Regions Research and Engineering Laboratory, CRREL Report 77-29, 8 p.

Gow, A. J., W. F. Weeks, J. W. Govoni and S. F. Ackley (1981) Physical and structural characteristics of sea ice in McMurdo Sound. Ant. J. U.S., 16 (5): 94-95.

Gow, A. J., S. F. Ackley, W. F. Weeks and J. W. Govoni (1982) Physical and structural characteristics of Antarctic sea ice. In Third International Symposium on Antarctic Glaciology, Ohio State University, Ann. Glaciol., 3: 113-117.

Gudmandsen, P. (1971) Electromagnetic probing of ice. In Electromagnetic Probing in Geophysics (J. R. Wait, Ed.), Golem Press, Boulder, Colorado, p. 321-338.

Hallett, J. (1960) Crystal growth and the formation of spikes in the surface of supercooled water. J. Glaciol., 3: 698-702.

Hardy, S. C. and S. R. Coriell (1973) Surface tension and interface kinetics of ice crystals freezing and melting in sodium chloride solutions. J. Cryst. Growth, 20: 292-300.

Harrison, J. D. (1965) Measurement of brine droplet migration in ice. J. Appl. Phys., 36 (12): 3811-3815.

Harrison, J. D. and W. A. Tiller (1963) Controlled freezing of water. In Ice and Snow: Properties, Processes, and Applications (W. D. Kingery, Ed.), M.I.T. Press, Cambridge, Mass., p. 215-225.

Hillig, W. B. (1958) The kinetics of freezing of ice in the direction perpendicular to the basal plane. In Growth and Perfection of Crystals (R. H. Doremus, Ed.), Wiley, New York, p. 350-359.

Hillig, W. B. (1959) Kinetics of solidification from nonmetallic liquids. In Kinetics of High Temperature Processes (W. D. Kingery, Ed.), Wiley, New York, p. 127-135.

Hillig, W. B. and D. Turnbull (1956) Theory of crystal growth in undercooled pure liquids. J. Chem. Phys., 24: 914.

Hobbs, P. (1974) Ice Physics. Oxford University Press, 837 p.

Hoekstra, P., T. E. Osterkamp and W. F. Weeks (1965) The migration of liquid inclusions in single ice crystals. J. Geophys. Res., 70 (20): 5035-5041.

Johnson, N. G. (1943) Studies av isen i Gullmarfjorden. Svenska Hydrog. -- Biol. Komm. Skr., Ny Serie: Hydrografi, 18: 1-21.

Jones, D. R. H. (1973) The temperature-gradient migration of liquid droplets through ice. J. Cryst. Growth, 20: 145-151.

Jones, D. R. H. (1974) Determination of the kinetics of ice-brine interfaces from the shapes of migrating droplets. J. Cryst. Growth, 26: 177-179.

Katsaros, K. B. (1973) Supercooling at the surface of an arctic lead. J. Phys. Ocean., 3 (4): 482-486.

Kawamura, T. and N. Ono (1980) Freezing phenomena at seawater surface opening in polar winter. III: Measurement of crystallographic orientation of newly grown sea ice. Low Temp. Sci., A39: 175-180.

Kingery, W. D. and W. H. Goodnow (1963) Brine migration in salt ice. In Ice and Snow: Properties, Processes and Applications (W. D. Kingery, Ed.), MIT Press, Cambridge, Mass., p. 35-47.

Knight, C. A. (1962a) Curved growth of ice on surfaces. J. Appl. Phys., 33: 1808-1815.

Knight, C. A. (1962b) Migration of liquid inclusions parallel to the c-axis in single ice crystals: An origin for some striations. Can. J. Phys., 40: 1681-1682.

Knight, C. A. (1962c) Polygonization of aged sea ice. J. Geol., 70: 240-246.

Knight, C. A. (1962d) Studies of arctic lake ice. J. Glaciol., 4: 319-335.

Koerner, R. M. (1963) The Devon Island expedition, 1960-64. Arctic, 16: 57-62.

Kohnen, H. (1976) On the dc resistivity of sea ice. Z. Gletsch. und Glaz., 11 (2): 143-154.

Kovacs, A. and R. Morey (1978) Radar anisotropy of sea ice due to preferred azimuthal orientation of the horizontal c-axes of ice crystals. J. Geophys. Res., 83 (C12): 6037-6046.

Kovacs, A. and R. M. Morey (1980) Investigations of sea ice anisotropy, electromagnetic properties, strength, and under-ice current orientation. USA Cold Regions Research and Engineering Laboratory, CRREL Report 80-20, 18 p.

Kumai, M. and K. Itagaki (1953) Cinematographic study of ice crystal formation in water. J. Faculty Sci., Hokkaido Univ., Ser. II, 4: 235-246.

Lake, R. A. and E. L. Lewis (1970) Salt rejection by sea ice during growth. J. Geophys. Res., 75 (3): 583-597.

Landauer, J. K. and H. Plumb (1956) Measurements of anisotropy of thermal conductivity of ice. USA Snow, Ice and Permafrost Research Establishment, Research Report 16.

Langhorne, P. (1980) Crystal anisotropy in sea ice in the Beaufort Sea. In Proceedings, Workshop on Remote Estimation of Sea Ice Thickness, Memorial University, St. Johns, Newfoundland, Canada, C-CORE Pub. 80-5, p. 189-224.

Langhorne, P. J., J. R. Rossiter and T. E. Keliher (1980) Remote estimation of the properties of sea ice, ice core analysis, Beaufort Sea, March 1979. Memorial University, St. Johns, Newfoundland, Canada, C-CORE Pub. 80-7, 172 p.

Langleben, M. P. (1959) Some physical properties of sea ice. II. Can. J. Phys., 37: 1438-1454.

Lewis, E. L. (1967) Heat flow through winter ice. In Physics of Snow and Ice (H. Oura, Ed.), Institute of Low Temperature Science, Hokkaido University, Sapporo, Japan, Vol. 1, No. 1, p. 611-631.

Lewis, E. L. and R. A. Lake (1971) Sea ice and supercooled water. J. Geophys. Res., 76 (24): 5836-5841.

Lofgren, G. and W. F. Weeks (1969) Effect of growth parameters on the substructure spacing in NaCl ice crystals. J. Glaciol., 8, (52): 153-164.

Lonsdale, K. (1958) The structure of ice. Proc. Roy. Soc., 247: 424-434.

Lyons, J. B., S. M. Savin, and A. J. Tamburi (1971) Basement ice, Ward Hunt Ice Shelf, Ellesmere Island, Canada. J. Glaciol., 10, (58): 93-100.

Macklin, W. C. and B. F. Ryan (1966) Habits of ice grown in supercooled water and aqueous solutions. Phil. Mag., 14: 847-860.

Malmgren, F. (1927) On the properties of sea ice. In Scientific Results of the Norwegian North Pole Expedition "Maud," 1918-1925, Vol. 1, No. 5, 67 p.

Martin, S. (1972) Ice stalactites and the desalination of sea ice. Naval Res. Rev., XXV (9): 17-26.

Martin, S. (1974) Ice stalactites: Comparison of a laminar flow theory with experiment. J. Fluid Mech., 63 (1): 51-79.

Martin, S. (1979) A field study of brine drainage and oil entrapment in sea ice. J. Glaciol., 22 (88): 473-502.

Martin, S. (1981) Frazil ice in rivers and oceans. Ann. Rev. Fluid Mech., 13: 379-397.

Martin, S. and P. Kauffman (1974) The evolution of under-ice melt ponds, or double diffusion at the freezing point. J. Fluid Mech., 64 (3): 507-527.

Martin, S. and P. Kauffman (1981) A field and laboratory study of wave damping by grease ice. J. Glaciol., 27 (96): 283-314.

Maxwell, J. C. (1891) Electricity and Magnetism, Dover, New York, Third Ed., Vol. I.

Maykut, G. and N. Untersteiner (1971) Some results from a time-dependent thermodynamic model of sea ice. J. Geophys. Res., 76: 1550-1576.

Meeks, D. C., G. A. Poe and R. O. Ramseier (1974) A study of microwave emission properties of sea ice -- AIDJEX 1972. Aerojet Electrosystems Company, Azusa, California, Final Report No. 1786FR-1.

Mellor, M., (1983) Mechanical behavior of sea ice. USA Cold Regions Research and Engineering Laboratory, Monograph 83-1, 102 p.

Myerson, A. S. and D. J. Kirwan (1977a) Impurity trapping during dendritic crystal growth. 1: Computer simulation. Ind. Eng. Chem., Fund., 16 (4): 414-420.

Myerson, A. S. and D. J. Kirwan (1977b) Impurity trapping during dendritic crystal growth. 2: Experimental results and correlation. Ind. Eng. Chem. Fund., 16 (4): 420-425.

Nagle, J. F. (1966) Lattice statistics of hydrogen bonded crystals. I: The residual entropy of ice. J. Math. Phys., 7: 1484-1491.

Nakawo, M. and N. K. Sinha (1981) Growth rate and salinity profile of first-year sea ice in the high Arctic. J. Glaciol., 27 (96): 315-330.

Nakawo, M. and N. K. Sinha (1984) A note on brine layer spacing of first-year sea ice. Atmos.-Ocean, 22 (2): 193-206.

Nelson, K. H. and T. G. Thompson (1954) Deposition of salts from sea water by frigid concentration. J. Marine Res., 13 (2): 166-182.

Niedrauer, T. M. and S. Martin (1979) An experimental study of brine drainage and convection in young sea ice. J. Geophys. Res., 84 (C3): 1176-1186.

Ono, N. (1965) Thermal properties of sea ice. II. A method for determining the K/c value of a non-homogeneous ice sheet. Low Temp. Sci., A23: 177-183.

Ono, N. (1968) Thermal properties of sea ice. IV. Thermal constants of sea ice. Low Temp. Sci., A26: 329-349.

Owston, P. G. (1958) The structures of ice I, as determined by X-ray and neutron diffraction analysis. Adv. Phys., 7: 171-188.

Ozum, B. and D. J. Kirwan (1976) Impurities in ice crystals grown from stirred solutions. A.I.Ch.E. Sympos. Series, 72 (153): 1.

Paige, R. A. (1966) Crystallographic studies of sea ice in McMurdo Sound, Antarctica. Naval Civil Engineering Laboratory, Technical Report R494, 31 pp.

Paige, R. A. (1970) Stalactite growth beneath sea ice. Science, 167: 171-172.

Paige, R. A. and R. A. Kennedy (1967) Strength studies of sea ice: Effect of load rate on ring tensile strength. Naval Civil Engineering Laboratory, Technical Report R545, 25 p.

Pauling, L. (1935) Structure and entropy of ice and of other crystals with randomness of atomic arrangements. J. Am. Chem. Soc., 57: 2608-2684.

Perey, F. G. J. and E. R. Pounder (1958) Crystal orientation in ice sheets. Can. J. Phys., 36: 494-502.

Peterson, S. W. and H. A. Levy (1957) A single-crystal neutron diffraction study of heavy ice. Acta Crystallographica, 10: 70-76.

Petrov, I. G. (1954-55) Fiziko-mekhanicheskiye svoystva i tol'shchina ledyanogo pokrova (Physical and mechanical properties and thickness of ice cover). In Materialy Nablyudeniy Nauchno-Issledovatel'skoy Dreyfuyushchey Stantsii 1950-51 Goda (Observations of the Drifting Research Station of 1950-51) (M. M. Somov, Ed.), Arkticheskiy Nauchno-Issledovatel'skiy Institut 2, Leningrad p. 103-165.

Peyton, H. R. (1963) Some mechanical properties of sea ice. In Ice and Snow: Properties, Processes and Applications (W. D. Kingery, Ed.), MIT Press, Cambridge, Mass., p. 107-113.

Peyton, H. R. (1966) Sea ice strength. Geophysical Institute, University of Alaska Rept. UAG-182, 187 p.

Peyton, H. R. (1968) Sea ice strength: Effects of load rates and salt reinforcement. In Arctic Drifting Stations (J. E. Sater, Ed.), Arctic Institute of North America, p. 197-217.

Pfann, W. G. (1958) Zone Melting. Wiley, New York, 230 p.

Poe, G., A. Stogryn and A. T. Edgerton (1972) Microwave emission characteristics of sea ice. Aerojet General Corp., Report 1749R-2.

Poe, G. A., A. Stogryn, A. T. Edgerton and R. O. Ramseier (1974) Study of microwave emission properties of sea ice. Aerojet General Corp., Report 1804FR-1.

Pounder, E. R. and E. M. Little (1959) Some physical properties of sea ice. I. Can. J. Phys., 37: 443-473.

Ragle, R. H. (1962) The formation of lake ice in a temperate climate. USA Army Cold Regions Research and Engineering Laboratory, Research Report 107, 22 p.

Reeburgh, W. S. and M. S. Young (1982) New measurements of sulfate and chlorinity in natural sea ice. J. Geophys. Res., 88(C5): 2559-2566.

Richardson, C. and E. F. Keller (1966) The brine content of sea ice measured with a nuclear magnetic resonance spectrometer. J. Glaciol., 6 (43): 89-100.

Ringer, W. E. (1906) De Veranderinger in Samenstelling van Zeewater by Het Bevriezen. Chemisch Weekblad, 3: 223-249.

Rodhe, B. (1959) The Baltic ice code. Sveriges Meteorologiska och Hydrologiska Institut, Stockholm, Ser. E, No. 10, 59 p.

Rohatgi, P. K. and C. M. Adams (1967a) Freezing rate distributions during unidirectional solidification of solutions. Trans. Metall. Soc. AIME, 239 (6): 850-857.

Rohatgi, P. K. and C. M. Adams (1967b) Ice-brine dendritic aggregate formed on freezing of aqueous solutions. J. Glaciol., 6 (47): 663-679.

Rosenberg, A. and W. A. Tiller (1957) The relationship between growth forms and the preferred direction of growth. Acta Metallurgica, 5: 565-573.

Rutter, J. W. and B. Chalmers (1953) A prismatic substructure formed during solidification of metals. Can. J. Phys., 1: 15-39.

Ryan, B. F. (1969) The growth of ice parallel to the basal plane in supercooled water and supercooled metal fluoride solutions. J. Cryst. Growth, 5: 284-288.

Ryan, B. F. and W. C. Macklin (1968) The growth of ice in supercooled aqueous solutions. J. Cryst. Growth, 2: 337-340.

Saito, T. and N. Ono (1980) Percolation of sea ice. II: Brine drainage channels in young sea ice. Low Temp. Sci., A39: 127-132.

Savel'ev, B. A. (1958) Izucheniyu l'dov v rayone dreyfa stantsii SP-4 v period tayaniya i razrushenya ikh v 1955 g (Study of ice in the region of the drift of station SP-4 during melting and break-up in 1955). Problemy Severa, 2: 47-79.

Savel'ev, B. A. (1963) Structure, composition, and properties of the ice cover of sea and fresh waters. Izd. Moskovskogo Univ., 541 p.

Schwarz, J. and W. F. Weeks (1977) Engineering properties of sea ice. J. Glaciol., 19 (81): 499-530.

Schwarzacher, W. (1959) Pack-ice studies in the Arctic Ocean. J. Geophys. Res., 64: 2357-2367.

Schwerdtfeger, P. (1963) The thermal properties of sea ice. J. Glaciol., 4 (36): 789-807.

Seidensticker, R. G. (1965) Comment on paper by P. Hoekstra, T. E. Osterkamp and W. F. Weeks, "The migration of liquid inclusions in single ice crystals." J. Geophys. Res., 71 (8): 2180-2181.

Sekerka, R. F. (1968) Morphological stability. J. Cryst. Growth, 3/4: 71-81.

Sekerka, R. F., R. G. Seidensticker, D. R. Hamilton and J. D. Harrison (1967) Investigation of desalination by freezing. Office of Saline Water, Westinghouse Research Laboratories, Pittsburgh, Contract No. 14-01-0001-605, Final Report.

Seligman, G. (1949) Growth of glacier crystal. J. Glaciol., 1: 254-268.

Serikov, M. I. (1963) Structure of Antarctic sea ice. Infor. Bull. Sov. Antarct. Exped., 4 (5): 265-266.

Serson, H. V. (1972) Investigation of a plug of multi-year old sea ice in the mouth of Nansen Sound. Defence Research Establishment, Ottawa, Technical Note 72-6, 4 p.

Sharp, R. P. (1947) Suitability of ice for aircraft landings. Trans. Am. Geophys. Union, 28: 111-119.

Shumskii, P. A. (1955) K izucheniiu l'dov severnogo ledovitogo okeana (A study of ice in the Arctic Ocean). Vestnik Akad. Nauk SSSR, 25 (2): 33-38.

Sinha, N. K. (1977) Technique for studying structure of sea ice. J. Glaciol., 18 (79): 315-323.

Smith, D. D. (1964) Ice lithologies and structure of ice island Arlis II. J. Glaciol., 5 (37): 17-38.

Smith, V. G., W. A. Tiller and J. W. Rutter (1955) A mathematical analysis of solute redistribution during solidification. Can. J. Phys., 33: 723-745.

Stander, E. and G. A. Gidney (1980) The measurement of finite strain in sea ice by impulse radar techniques. In Proceedings Workshop on Sea Ice Field Measurement, Memorial University, St. Johns, Newfoundland, Canada, C-CORE Pub. 80-21, p. 127-164.

Stogryn, A. (1970) The brightness temperature of a vertically structured medium. Radio Science, 5 (12): 1397.

Stogryn, A. (1971) Equations for calculating the dielectric constant of saline water at GHz frequencies. IEEE Trans. Microw. Theory Tech., 19 (8): 273.

Stratton, J. (1941) Electromagnetic Theory. McGraw-Hill, New York, p. 563-573.

Suzuki, Y. (1955) Observations of ice crystals formed on sea surface. J. Ocean. Soc. Japan, 11: 123-126.

Suzuki, Y. (1967) On disorder entropy of ice. In Physics of Snow and Ice (H. Oura, Ed.), Institute of Low Temperature Science, Hokkaido University, Sapporo, Japan, Vol. 1, No. 1, p. 21-41.

Swinzow, G. K. (1966) Ice cover of an arctic proglacial lake. USA Cold Regions Research and Engineering Laboratory, Research Report 155, 43 p.

Tabata, T. (1960) Studies on the mechanical properties of sea ice. V: Measurement of flexural strength. Low Temp. Sci., A19: 187-201.

Tabata, T. and N. Ono (1957) On the structure of sea ice. Low Temp. Sci., A16: 197-210.

Tabata, T. and N. Ono (1962) On the crystallographic study of several kinds of ice. Low Temp. Sci., A20: 199-214.

Taylor, L. D. and J. B. Lyons (1959) Ice structures, Angiussaq Lake, northwest Greenland. Geophysics Research Directorate, Air Force Cambridge Research Center, TN-59-461, 33 p.

Tiller, W. A. (1962) Effect of grain boundaries on solute partitioning during progressive solidification. Acta Metallurgica, 1 (4): 428-437.

Tiller, W. A. (1963) Migration of a liquid zone through a solid. J. Appl. Phys., 34 (9): 2757-2762.

Tiller, W. A., K. A. Jackson, J. W. Rutter and B. Chalmers (1953) The redistribution of solute atoms during the solidification of metals. Acta Metallurgica, 1: 428-437.

Tinga, W. R., W. A. Voss and D. F. Blossey (1973) Generalized approach to multiphase dielectric mixture theory. J. Appl. Phys., 44 (9): 3897-3903.

Tsurikov, V. L. (1974) Statistics of salt composition in sea ice. Oceanology, 14 (3): 360-367.

Tsurikov, V. L. (1976) Liquid Phase in Sea Ice. Nauka, Moscow, 210 p.

Tsurikov, V. L. and Tsurikova, A. P. (1972) The brine content of sea ice (statement of the problem). Oceanology, 12 (5): 663-672.

Turner, J. S. (1973) Buoyancy Effects in Fluids. Cambridge University Press, 367 p.

U.S. Navy Hydrographic Office (1952) A functional glossary of ice terminology. Washington, D.C., 88 p.

Untersteiner, N. (1961) On the mass and heat budget of arctic sea ice. Arch. Meteorol. Geophys. Biok., 12: 151-182.

Untersteiner, N. (1968) Natural desalination and equilibrium salinity profile of perennial sea ice. J. Geophys. Res., 73 (4): 1251-1257.

Untersteiner, N. and F. Badgley (1958) Preliminary results of thermal budget studies on arctic pack ice during summer and autumn. In Arctic Sea Ice, U.S. National Academy of Sciences - National Research Council, Pub. 598, p. 85-98.

Vant, M. R., R. O. Ramseier and V. Makros, 1975: The complex dielectric constant of sea ice at frequencies in the range 0.1-40 GHz. J. Appl. Phys., 49 (3): 1264-1280.

Vittoratos, E. S. (1979) Existence of oriented sea ice by the McKenzie Delta. In POAC 79: Proceedings, Fifth International Conference on Port and Ocean Engineering Under Arctic Conditions, University of Trondheim, Trondheim, Norway, p. 643-650.

Wakatsuchi, M. (1977) Experiments on haline convection induced by freezing of sea water. Low Temp. Sci., A35: 249-258.

Walker, E. R. and P. Wadhams (1979) Thick sea-ice floes. Arctic, 32 (2): 140-147.

Walton, D. and B. Chalmers (1959) The origin of the preferred orientation in the columnar zone of ingots. Trans. Metal. Soc. Am. Inst. Min. Met. Eng., 215: 447-457.

Walton, D., W. A. Tiller, J. W. Rutter and W. C. Winegard (1955) Instability of a smooth solid-liquid interface during solidification. J. Metals 7, Eng., 215: 447-457.
Weeks, W. F. (1958) The structure of sea ice: A progress report. In Arctic Sea Ice, U.S. National Academy of Sciences - National Research Council, Pub. 598, p. 96-98.
Weeks, W. F. (1962) Tensile strength of NaCl ice. J. Glaciol., 4 (31): 25-52.
Weeks, W. F. and O. S. Lee (1958) Observation on the physical properties of sea ice at Hopedale, Labrador. Arctic, 11: 134-155.
Weeks, W. F. and W. L. Hamilton (1962) Petrographic characteristics of young sea ice, Point Barrow, Alaska. Am. Mineral., 47: 945-961.
Weeks, W. F. and O. S. Lee (1962) The salinity distribution in young sea ice. Arctic, 15: 92-108.
Weeks, W. F. and A. Assur (1963) Structural control of the vertical variation of the strength of sea and salt ice. In Ice and Snow: Properties, Processes and Applications (W. D. Kingery, Ed.), MIT Press, Cambridge, Mass., p. 258-276.
Weeks, W. F. and A. Assur (1967) The mechanical properties of sea ice. USA Cold Regions Research and Engineering Laboratory, Monograph II-C3, 80 p.
Weeks, W. F. and G. Lofgren (1967) The effective solute distribution coefficient during the freezing of NaCl solutions. In Physics of Snow and Ice (H. Oura, Ed.), Institute of Low Temperature Science, Hokkaido University, Sapporo, Japan, Vol. 1, No. 1, p. 579-597.
Weeks, W. F. and A. Assur (1969) Fracture of lake and sea ice. USA Cold Regions Research and Engineering Laboratory, Research Report 269, 77 p.
Weeks, W. F. and A. J. Gow (1978) Preferred crystal orientations along the margins of the Arctic Ocean. J. Geophys. Res., 84 (C10): 5105-5121.
Weeks, W. F. and A. J. Gow (1980) Crystal alignments in the fast ice of arctic Alaska. J. Geophys. Res., 84 (C10): 1137-1146.
Weller, G. E. (1968) The heat budget and heat transfer processes in Antarctic plateau ice and sea ice. ANARE Scientific Reports, Series A (IV), Glaciology, Pub. 102, 155 p.
Wernick, J. H. (1956) Determination of diffusivities in liquid metals by means of temperature-gradient zone melting. J. Chem. Phys., 25 (1): 47-49.
Whitman, W. G. (1926) Elimination of salt from sea water ice. Am. J. Sci., Ser. 5, 11 (62): 126 p.
Wilson, J. T., J. H. Zumberge, and E. W. Marshall (1954) A study of ice on an inland lake. USA Snow, Ice and Permafrost Research Establishment, Technical Report 5, 78 p.
Yen, Y. C. (1981) Review of thermal properties of snow, ice and sea ice. USA Cold Regions Research and Engineering Laboratory, CRREL Report 81-10, 27 p.

Zotikov, I. A., V. S. Zagorodnov and J. V. Raikovski (1980) Core drilling through the Ross Ice Shelf (Antarctica) confirmed basal freezing. Science, 207 (4438): 1463-1465.

CHAPTER 2

MECHANICAL BEHAVIOR OF SEA ICE

Malcolm Mellor

U.S. Army Cold Regions Research and
Engineering Laboratory
Hanover, New Hampshire 03755-1290 U.S.A.

INTRODUCTION

This review deals with properties of sea ice that are relevant to mechanical behavior on a fairly small scale. The main concern is with polycrystalline ice that is more or less intact. This sets the lower limit of scale not far below 0.1 m, and the upper limit around 10 m. However, the discussion is extended to cover the behavior of fragmented ice over broader areas. The latter discussion is intended to bridge the gap between this review, which relates to small-scale natural processes and engineering problems, and companion reviews, which deal with the mechanics of sea ice over very large areas.

Part I is a selective summary of the basic theory which is applicable to ice mechanics. This theory defines the various mechanical properties that have to be measured, and it provides a framework for the solution of practical (boundary value) problems. Since ice displays a wide range of behavioral characteristics, from elasticity and brittle fracture to ductility and plastic yielding, it is important to have a good appreciation for the various theories that might apply in different ranges of behavior.

Part II summarizes measured values for common mechanical properties. Most of the data are derived from field tests or laboratory tests on fairly small specimens (minimum linear dimension typically < 0.1 m for lab specimens, or < 2 m for in-situ field tests). The summary is selective, and it does not include some older data which are judged to be erroneous or misleading. A great deal of the information refers to ice that is non-saline, and this calls for a word of explanation.

By convention, the term "sea ice" covers all types of ice formed by direct freezing of seawater, but it does not cover icebergs (i.e. glacier ice), or freshwater ice flushed into the sea by rivers. First-year sea ice is a very complicated substance -- an anisotropic crystalline solid containing bubbles of air, pockets and films of brine, and sometimes solid salts. However, not all sea ice is significantly saline, or even anisotropic. Multi-year ice, which is of great concern in ice navigation and offshore structural engineering, may be of very low salinity and without significant anisotropy where large masses are concerned. It is therefore useful to regard sea ice not as a unique and exotic material, but simply as a variant of freshwater ice. The complications introduced by salinity and structure are important, but not to the extent that sea ice need be regarded as something wholly different from freshwater ice. In Part II of this review, the mechanical properties of sea ice are discussed in relation to the properties of freshwater ice. This is largely a matter of practicality, since non-saline has been studied in more detail than sea ice. However, there is another justification, in that the effects of salinity have to be referred to a zero salinity state, the effects of porosity have to refer to a zero porosity datum, and anisotropy has to be compared with isotropy.

Part III discusses the mechanical behavior of ice that has been fragmented, either by natural processes or by human activities. The discussion covers brash ice and ice rubble, but it does not deal with mush ice (which is somewhat like saturated snow), or with pack ice (broad expanses of ice floes).

Ideally, a review such as this ought to conclude with a major section on the application of the theory and the measured properties to the solution of boundary value problems. However, during this assignment there was insufficient time to complete such a task.

PART I. GENERAL BACKGROUND TO THE MECHANICS OF DEFORMABLE SOLIDS

Ice mechanics is concerned largely with the interactions of forces and deformations. When a force acts on a material it deforms, and the deformation may culminate in rupture of the material. We have to consider: 1) force, 2) deformation, and 3) the property of the material which controls the relation between force and deformation.

1. STRESSES, STRAINS AND THEIR TIME DERIVATIVES

1.1 Force and Stress

Force (P) is the product of mass times acceleration, and it has the dimensions MLT^{-2}, where these symbols represent respectively mass, length and time. Traditional engineering units include pounds force and kilograms force, which should properly be abbreviated as "lbf" and "kgf" as a reminder that they are units of force, not mass. The distinction is important for non-gravitational inertial forces, where pounds and kilograms have to be divided by the gravitational acceleration g. In physics, and in engineering research, dynes or newtons are more likely to be used as units of force.

Stress (σ) is force per unit area, so that it has dimensions $ML^{-1}T^{-2}$. Traditional engineering units include pounds per square inch (lbf/in.^{2} or, in archaic notation, psi), and kilograms per square centimetre (kgf/cm^2). The metric units in common use for ice mechanics research are dynes/cm^2, newtons/m^2 (or pascals), and bars, where 10^6 dynes/cm^2 = 1 bar = 0.1 MN/m^2 = 0.1 MPa = 1.02 kgf/cm^2 = 14.5 lbf/in^2. Stresses may be either: 1) normal stresses, produced by tensile or compressive forces acting perpendicular to the faces of cubic elements, or 2) shear stresses, produced by tangential forces acting parallel to the surfaces of cubic elements. In a three-dimensional stress field, the stress system can also be resolved into: 1) an isotropic component, or bulk stress, which tends only to change the volume of an element, and 2) a deviatoric component, which tends to change the shape of an element.

1.2 Displacement and Strain

Displacement (u) is the movement of one point relative to another, resulting in increase or decrease of a linear dimension. The physical dimension of displacement is L, and typical units are inches, feet, miles, millimetres, metres, etc.

Strain (ε) is displacement per unit length, and it is therefore dimensionless. Although there are no units for strain, engineers sometimes use implied multiples, e.g. "microinches per inch," a number which is strain × 10^6, or "percent strain," a number which is strain × 10^2. Where strain is displacement per unit of length in the direction of displacement, it is referred to as a normal strain. Where the strain is displacement per unit of length in a direction perpendicular to the direction of displacement, it is referred to as a shear strain. In a three-dimensional strain field the strain can be resolved into: 1) an isotropic component, or volumetric strain, representing change in volume (and density) of an element, and 2) a deviatoric component, representing change in the shape of an element.

1.3 Variation of Force, Stress, Displacement and Strain with Respect to Time

Loading rate is the rate of change of force with time, dP/dt. It has dimensions MLT^{-3}, and units such as lbf/min or N/s.

Stress rate is the rate of change of stress with time, $d\sigma/dt$ or $\dot{\sigma}$. The dimensions are $ML^{-1}T^{-3}$, and typical units are lbf/in.^{2}-s, bar/s, or Pa/s.

Velocity is the rate of change of displacement with time, du/dt or $\dot{u}$. Dimensions are LT^{-1} and typical units are ft/s, knots, m/s, etc.

Strain rate is the rate of change of strain with time, $d\varepsilon/dt$ or $\dot{\varepsilon}$. Dimensions are T^{-1} and typical units are s^{-1}, yr^{-1}.

Acceleration (f) is the rate of change of velocity with time, or the second derivative of displacement with respect to time, d^2u/dt^2 or $\ddot{u}$. Dimensions are LT^{-2} and typical units are ft/s^2 or m/s^2.

Strain acceleration is the rate of change of strain rate with time, or the second derivative of strain with respect to time, $d^2\varepsilon/dt^2$ or $\ddot{\varepsilon}$. Dimensions are T^{-2}, and typical units s^{-2}.

2. RHEOLOGICAL PROPERTIES OF MATERIALS

The relationships between stress, strain and time for typical engineering materials can seem fairly simple, for example if stress is directly proportional to strain. However, this is true only for very restricted ranges of stress, strain and temperature, and over a sufficiently broad range of conditions virtually all materials experience time-dependent inelastic deformations. To introduce some of the more complicated aspects of rheology, it is convenient to first consider simple mechanisms, or models, which have stress/strain/time characteristics similar to those displayed by deformable solids.

2.1 Elasticity

Linear elasticity means that stress is directly proportional to strain (Figure 1), and the coefficient of proportionality is called the elastic modulus. For one-dimensional stress and strain, the ratio of stress to strain, σ/ε, is called Young's modulus E, and the relationship is known as Hooke's Law. The elastic modulus has the dimensions of a stress, i.e. $ML^{-1}T^{-2}$. In a rheological model, linear elasticity is represented by a spring of modulus E, so that $\sigma = E\varepsilon$.

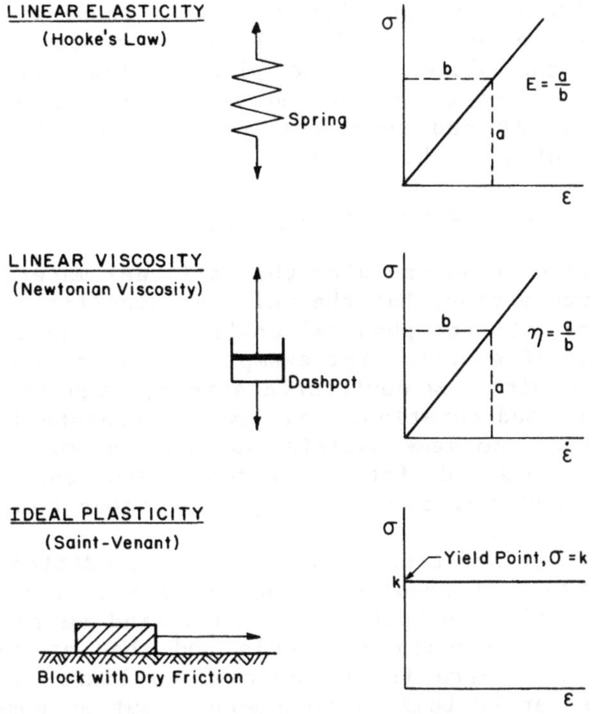

Fig. 1. Simple rheological responses and models.

2.2 Viscosity

Linear viscosity, which is also referred to as Newtonian viscosity, means that stress σ is proportional to strain rate $\dot{\varepsilon}$ (Figure 1). The coefficient of proportionality η is the viscosity, or viscosity coefficient, and it has dimensions $ML^{-1}T^{-1}$. In a rheological model, viscosity is represented by a dashpot. In the dashpot, force and velocity are proportional and we can cut the corner a little by letting the dashpot represent proportionality between stress σ and strain rate $\dot{\varepsilon}$, i.e. $\sigma = \eta \, d\varepsilon/dt = \eta \, \dot{\varepsilon}$. It is worth pointing out at this stage that viscosity is frequently non-linear, i.e. stress is a function of strain rate, but is not directly proportional. However, introduction of nonlinear viscosity into simple rheological models destroys their value as learning tools.

2.3 Plasticity

Ideal plasticity, or rigid-plastic behavior, means that there is no strain in the material until a critical yield stress k is reached, after which the material strains indefinitely at indeterminate speed (Figure 1). The name Saint-Venant is sometimes associated with ideal plastic behavior. In a rheological model, plasticity is represented by a block lying on a plane, with dry friction between the block and the plane. The upper limit of the static friction force is assumed equal to the kinetic friction force, so that it can represent a simple yield stress k. With $\sigma < k$, $\varepsilon = 0$; with $\sigma = k$, $\varepsilon \to \infty$.

2.4 Combinations of Rheological Properties

We can take it as an axiom that all real materials possess all rheological properties, but the relative significance of any component will vary with the physical environment, the stress state, and the time scale of events. For example, a material will be predominantly elastic with low deviatoric stress, high rates of stress or strain, short load-duration, and low temperature (relative to the melting point). The same material may be predominantly viscous, or plastic, with high deviatoric stress, low rates of stress or strain, long load duration, and high homologous temperature.

The real difficulties arise when the conditions are such that two or more types of rheological behavior are of comparable significance. To explore situations of this kind we can look first at some two-element rheological models and then at three- and four-element models. There is, of course, no limit to the number of elements that can be built into a model, but at some stage refinements become pointless. A model consisting of linear elements is not truly representative of real materials, and its chief merit is as a simple device for visualizing rheologic responses.

3. RHEOLOGICAL MODELS

3.1 Two-Element Models for Linear Viscoelasticity

3.1.1 The Maxwell model. This model consists of a spring and a dashpot in series, and it is easy to see how it behaves when constant force, or constant stress, is applied (Figure 2). For constant stress σ across the system, the stress/strain relation for the spring is

$$\sigma = E\varepsilon_s$$

where ε_s is the strain in the spring. The stress/strain-rate relation for the dashpot is

$$\sigma = \eta\, \dot{\varepsilon}_D$$

where $\dot{\varepsilon}_D$ is the strain rate of the dashpot. The total strain of the system at time t is the sum of the strain in the spring and the accumulated strain in the dashpot:

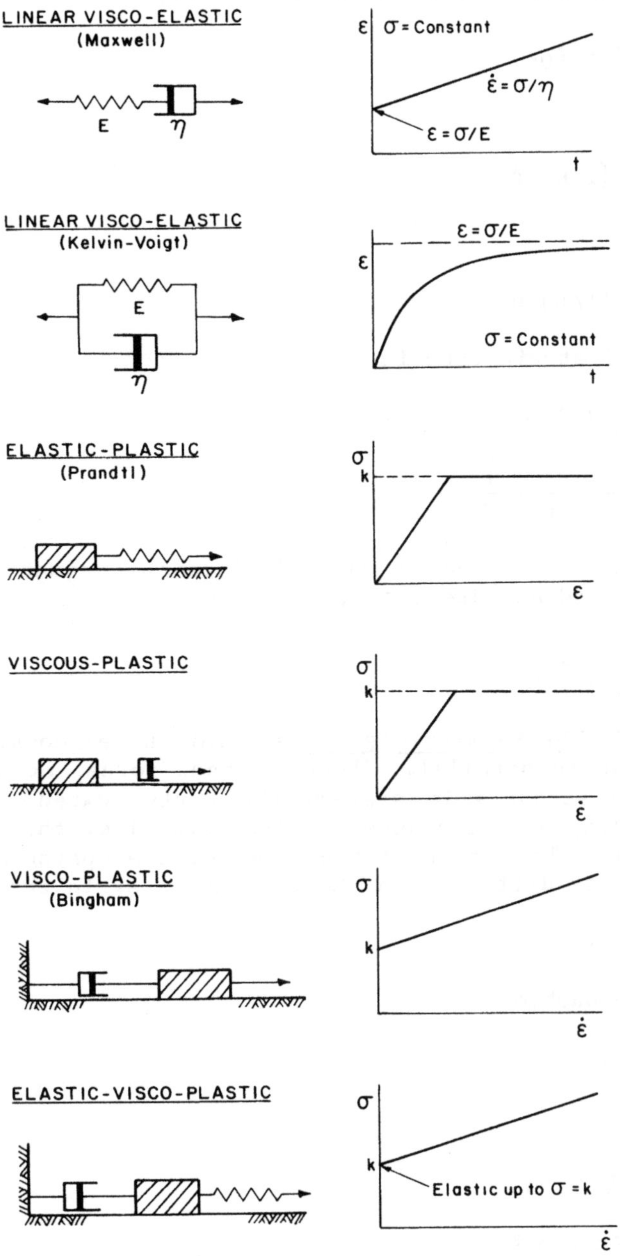

Fig. 2. Two-element rheological models.

$$\varepsilon = \varepsilon_S + \varepsilon_D$$

or, differentiating with respect to time,

$$\dot{\varepsilon} = \dot{\varepsilon}_S + \dot{\varepsilon}_D.$$

Since

$$\dot{\varepsilon}_S = (1/E)\ \dot{\sigma}$$

and

$$\dot{\varepsilon}_D = (1/\eta)\ \sigma$$

the overall strain rate is

$$\dot{\varepsilon} = (1/E)\ \dot{\sigma} + (1/\eta)\ \sigma$$

$$= \left(\frac{1}{E}\ \frac{d}{dt} + \frac{1}{\eta}\right)\ \sigma.$$

The required solution of this differential equation for constant stress applied at time $t = 0$, $\varepsilon = 0$ is

$$\varepsilon = \sigma\left(\frac{1}{E} + \frac{1}{\eta}\ t\right).$$

3.1.2 <u>The Kelvin-Voigt model</u>. This model consists of a spring and dashpot in parallel. When constant force, or constant stress, is applied across this system, the spring extends to its equilibrium length, but at a decelerating rate, i.e. the spring is damped (Figure 2). The strain is the same for the spring and the dashpot, but stress is different in the two elements. In the spring

$$\sigma_S = E\varepsilon$$

and in the dashpot

$$\sigma_D = \eta\ \dot{\varepsilon}.$$

Overall,

$$\sigma = \sigma_S + \sigma_D$$

$$= E\varepsilon + \eta\ \dot{\varepsilon}$$

$$= \left(E + \eta\ \frac{d}{dt}\right)\varepsilon.$$

The required solution for stress applied at $t = 0$, $\varepsilon = 0$ is

$$\varepsilon = \frac{\sigma}{E}\left(1 - e^{-(E/\eta)t}\right).$$

3.2 Two-Element Models for Elastoplastic and Viscoplastic Behavior

3.2.1 Elastic-plastic model. This model, sometimes associated with the name of Prandtl, consists of a spring and a friction block in series (Figure 2). For $\sigma < k$, $\sigma = E\varepsilon$, but when $\sigma = k$, $\varepsilon = \sigma/E + \varepsilon_{plas}$, where ε_{plas} is the indeterminate plastic strain.

3.2.2 Viscous-plastic model. The simplest viscous-plastic model is a series combination of a dashpot and a friction block. For $\sigma < k$, $\sigma = \eta\dot{\varepsilon}$, while at $\sigma = k$, $\dot{\varepsilon} \geq (1/\eta)k$.

3.2.3 Viscoplastic model. This model, sometimes associated with the name Bingham, consists of a dashpot and a friction block connected in such a way that the dashpot cannot move until the block slips (Figure 2). If $\sigma \geq k$,

$$\sigma = k + \eta\,\dot{\varepsilon}.$$

3.3 Multi-Element Models

Perhaps the most useful multi-element model for introducing general viscoelastic response is one known as the Burgers' model. It consists of a Maxwell model in series with a Kelvin-Voigt model (Figure 3). The general differential equation can be derived by considering constant stress across the model. Using subscripts M and K to denote respectively the Maxwell and Kelvin-Voigt units, and using additional subscripts S and D for spring and dashpot respectively, the overall strain is

$$\varepsilon = \varepsilon_{MS} + \varepsilon_{MD} + \varepsilon_{K}.$$

In the Maxwell spring

$$\sigma = E_M\,\varepsilon_{MS}$$

and in the Maxwell dashpot

$$\sigma = \eta_M\,\dot{\varepsilon}_{MD}.$$

In the Kelvin-Voigt unit

$$\sigma = E_K\,\varepsilon_K + \eta_K\,\dot{\varepsilon}_K.$$

After suitable differentiation, these relations give the general differential equation

Model		Maxwell	Kelvin–Voigt	3–Element	3–Element	Burgers
Diagram		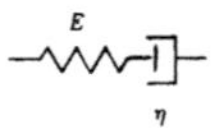	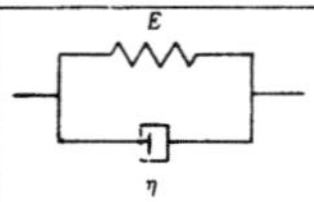	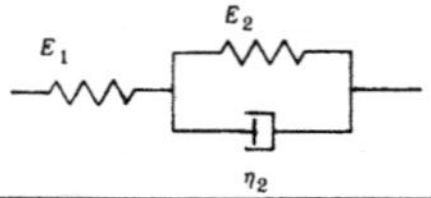	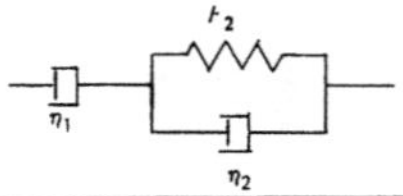	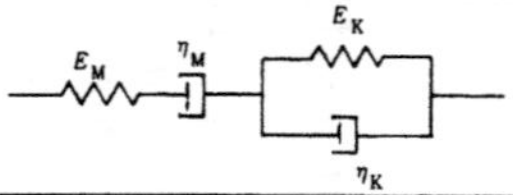
General equation		$\dot{\epsilon} = \frac{1}{E}\dot{\sigma} + \frac{1}{\eta}\sigma$	$\sigma = E\epsilon + \eta\dot{\epsilon}$	$\dot{\sigma} + \left(\frac{E_1 + E_2}{\eta_2}\right)\sigma = E_1\dot{\epsilon} + \left(\frac{E_1 E_2}{\eta_2}\right)\epsilon$	$\epsilon + \frac{E_2}{\eta_2}\epsilon = \left(\frac{1}{\eta_1} + \frac{1}{\eta_2}\right)\dot{\sigma} + \frac{E_2}{\eta_1\eta_2}\sigma$	$\sigma + \left(\frac{E_M}{\eta_M} + \frac{E_M}{\eta_K} + \frac{E_K}{\eta_K}\right)\dot{\sigma} + \frac{E_M E_K}{\eta_M \eta_K}\sigma = E_M\epsilon + \frac{E_M E_K}{\eta_K}\epsilon$
Constant stress sol'n (σ_0 applied at $t = 0$)		$\epsilon = \sigma_0\left(\frac{1}{E} + \frac{t}{\eta}\right)$	$\epsilon = \frac{\sigma_0}{E}\left[1 - \exp\left(-\frac{E}{\eta}t\right)\right]$	$\epsilon = \sigma_0\left\{\frac{1}{E_1} + \frac{1}{E_2}\left[1 - \exp\left(-\frac{E_2}{\eta_2}t\right)\right]\right\}$	$\epsilon = \sigma_0\left\{\frac{t}{\eta_1} + \frac{1}{E_2}\left[1 - \exp\left(-\frac{E_2}{\eta_2}t\right)\right]\right\}$	$\epsilon = \sigma_0\left\{\frac{1}{E_M} + \frac{t}{\eta_M} + \frac{1}{E_K}\left[1 - \exp\left(-\frac{E_K}{\eta_K}t\right)\right]\right\}$
Constant strain rate sol'n ($\dot{\epsilon} = k_1$ applied at $t \geq 0$, $\sigma = 0$)		$\sigma = k_1\eta\left[1 - \exp\left(-\frac{E}{\eta}t\right)\right]$	$\sigma = k_1(\eta + Et)$	$\sigma = k_1\left\{\eta_2\left(\frac{E_1}{E_1 + E_2}\right)^2\left[1 - \exp\left(-\frac{E_1 + E_2}{\eta_2}t\right)\right] + \frac{E_1 E_2}{E_1 + E_2}t\right\}$	$\sigma = k_1\eta_1\left[1 - \exp\left(-\frac{E_2}{\eta_1 + \eta_2}t\right)\right]$	$\sigma = k_1\left\{\frac{E_M}{r_1 - r_2}(e^{r_1 t} - e^{r_2 t}) + \frac{\eta_M}{r_1 - r_2}(r_2 e^{r_1 t} - r_1 e^{r_2 t}) + \eta_M\right\}$ where $r_1, r_2 = -\frac{1}{2}\left(\frac{E_M}{\eta_M} + \frac{E_M}{\eta_K} + \frac{E_K}{\eta_K}\right) \pm \frac{1}{2}\left[\left(\frac{E_M}{\eta_M} + \frac{E_M}{\eta_K} + \frac{E_K}{\eta_K}\right)^2 - 4\left(\frac{E_M E_K}{\eta_M \eta_K}\right)\right]^{1/2}$
Constant stress rate sol'n ($\dot{\sigma} = k_2$ initiated at $t = 0$, $\epsilon = 0$)		$\epsilon = k_2\left(\frac{t}{E} + \frac{t^2}{2\eta}\right)$	$\epsilon = k_2\left\{\frac{t}{E} - \frac{\eta}{E^2}\left[1 - \exp\left(-\frac{E}{\eta}t\right)\right]\right\}$	$\epsilon = k_2\left\{\left(\frac{1}{E_1} + \frac{1}{E_2}\right)t - \frac{\eta_2}{E_2^2}\left[1 - \exp\left(-\frac{E_2}{\eta_2}t\right)\right]\right\}$	$\epsilon = k_2\left\{\frac{t}{E_2} + \frac{t^2}{2\eta_1} - \frac{\eta_2}{E_2^2}\left(1 - \frac{\eta_2}{\eta_1}\right)\left[1 - \exp\left(-\frac{E_2}{\eta_2}t\right)\right]\right\}$	$\epsilon = k_2\left\{\left(\frac{1}{E_M} + \frac{1}{E_K}\right)t + \frac{1}{2\eta_M}t^2 - \frac{\eta_K}{E_K^2}\left(1 - \frac{\eta_K}{\eta_M}\right)\left[1 - \exp\left(-\frac{E_K}{\eta_K}t\right)\right]\right\}$
Complex modulus $Y(i\omega) = Y_1 + iY_2$	Y_1	$\frac{\omega^2\eta^2 E}{E^2 + \omega^2\eta^2}$	E	–	–	–
	Y_2	$\frac{\omega\eta E^2}{E^2 + \omega^2\eta^2}$	$\omega\eta$	–	–	–
Complex compliance $J(i\omega) = J_1 + iJ_2$	J_1	$\frac{1}{E}$	$\frac{E}{E^2 + \omega^2\eta^2}$	$\frac{1}{E_1} + \frac{E_2}{E_2^2 + \omega^2\eta_2^2}$	$\frac{E_2}{E_2^2 + \omega^2\eta_2^2}$	$\frac{1}{E_M} + \frac{E_K}{E_K^2 + \omega^2\eta_K^2}$
	J_2	$\frac{1}{\omega\eta}$	$\frac{\omega\eta}{E^2 + \omega^2\eta^2}$	$\frac{1}{\eta_2} - \frac{E_2^2}{\omega\eta_2(E_2^2 + \omega^2\eta_2^2)}$	$\frac{1}{\omega\eta_1} + \frac{1}{\omega\eta_2} - \frac{E_2^2}{\omega\eta_2(E_2^2 + \omega^2\eta_2^2)}$	$\frac{1}{\omega\eta_M} + \frac{1}{\omega\eta_K} - \frac{E_K^2}{\omega\eta_K(E_K^2 + \omega^2\eta_K^2)}$

Fig. 3. Summary of basic rheological relations.

$$\ddot{\sigma} + \left(\frac{E_M}{\eta_M} + \frac{E_M}{\eta_K} + \frac{E_K}{\eta_K}\right) \dot{\sigma} + \frac{E_M E_K}{\eta_M \eta_K} \sigma = E_M \ddot{\varepsilon} + \frac{E_M E_K}{\eta_M} \dot{\varepsilon}.$$

This equation can be solved for various sets of initial conditions and boundary conditions to give relations of special interest. In Figure 3, solutions are given for constant stress, for constant strain rate, and for constant stress rate.

4. THREE-DIMENSIONAL STRESS AND STRAIN

The preceding ideas are easy to understand in one-dimensional or simple two-dimensional situations, but three-dimensional systems are a little more complicated.

4.1 External Forces Acting on a Body

The external forces acting on a body may include surface forces and body forces. Surface forces, or surface tractions, are distributed over area. Body forces are distributed over volume. They are typically gravity, inertia, or magnetic forces.

In considering the forces on an element, it may be permissible to neglect body forces, since they diminish with the cube of a linear dimension, whereas surface forces diminish with the square of a linear dimension.

4.2 Components of Stress

Any system of stress can be resolved into a set of orthogonal components. If we define an element with respect to the axes of a cartesian coordinate system, the stress acting on that element can be resolved into nine components, which reduce by symmetry to six components.

We can denote all stress components by σ, differentiating between normal and shear stresses only by the subscripts, as in Figure 4. Normal stresses which act in a tensile sense are considered positive.

The first subscript indicates the direction of the normal to the plane on which the stress acts; the second subscript gives the direction in which the stress acts. For a normal stress, first and second subscripts are alike. For a shear stress, the two subscripts are different.

We can set out these stress components in a matrix, which has nine components:

$$\begin{vmatrix} \sigma_{xx} & \sigma_{xy} & \sigma_{xz} \\ \sigma_{yx} & \sigma_{yy} & \sigma_{yz} \\ \sigma_{zx} & \sigma_{zy} & \sigma_{zz} \end{vmatrix}$$

Note that, from consideration of the equilibrium of the element, $\sigma_{xy} = \sigma_{yx}$, $\sigma_{xz} = \sigma_{zx}$, $\sigma_{yz} = \sigma_{zy}$. This reduces the number of independent components to six, and the matrix is now symmetrical about the diagonal terms.

Now, we can make the above notation more compact in the following way. If, instead of x, y, z, we use the labels x_1, x_2, x_3 for the coordinate axes, we can then replace the subscripts x, y, z with the numerals 1, 2, 3, as in Figure 4b.

Our stress matrix becomes

$$\begin{vmatrix} \sigma_{11} & \sigma_{12} & \sigma_{13} \\ \sigma_{21} & \sigma_{22} & \sigma_{23} \\ \sigma_{31} & \sigma_{32} & \sigma_{33} \end{vmatrix} .$$

This we call the <u>stress tensor</u>. It can be written more compactly as

$$\sigma_{ij}$$

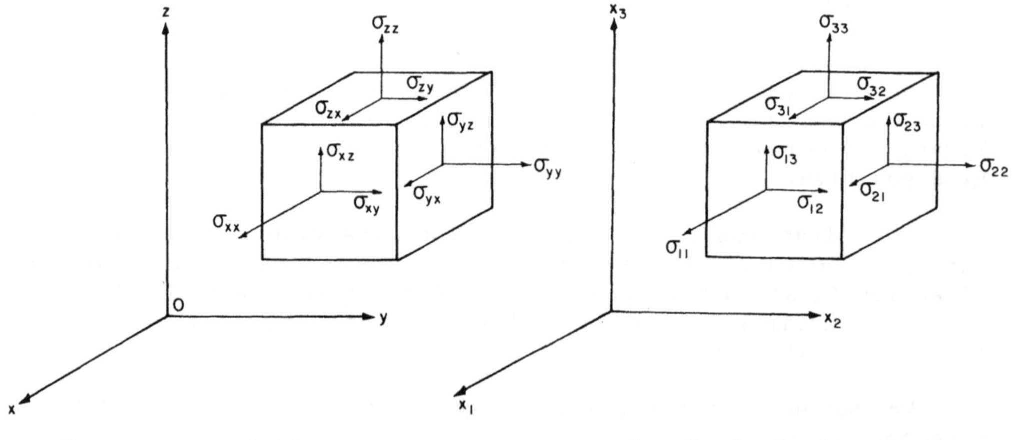

a. In x, y, z notation. b. In 1, 2, 3 notation.

Fig. 4. Components of stress.

where i,j = 1,2,3, i.e. if we write down in order all the possible combinations (permutations) of the values 1, 2 and 3 for i and j, we obtain all the terms of the stress tensor.

The stress tensor is a symmetric tensor of the second rank, or second order. As a matter of interest, vectors and tensors have the following numbers of components:

Scalar, S	1 component	
Vector V_i	3 components	(first rank tensor)
Tensor II T_{ij}	9 components	
Tensor III T_{ijk}	27 components	
Tensor IV T_{ijkf}	81 components	

The number of subscripts gives the order or rank, while 3 raised to a power equal to rank gives number of components.

4.3 Strain

In considering strain, it is convenient to assume that strains will be small and homogeneous.

4.3.1 Infinitesimal strain. These are strains sufficiently small that powers higher than the first can be neglected, e.g. squares and products of strains are negligible.

4.3.2 Homogeneous strain. Strain is homogeneous if planes remain plane, and parallel planes remain parallel.

4.4 Components of Strain

Any system of strain can be resolved into nine orthogonal components, with normal and shear strains corresponding to the components of stress.

We shall assume for a start that there is no rigid-body rotation relative to the reference axes, i.e. only deformation of the body.

When a body is deformed, the displacement of each point P can be expressed in terms of three components of displacement, u, v, w, in the directions x, y, z, respectively (Figure 5).

The gradients of u, v, w with respect to x, y, z are strains.

For simplicity we can derive the strain components in two dimensions only (Figure 6):

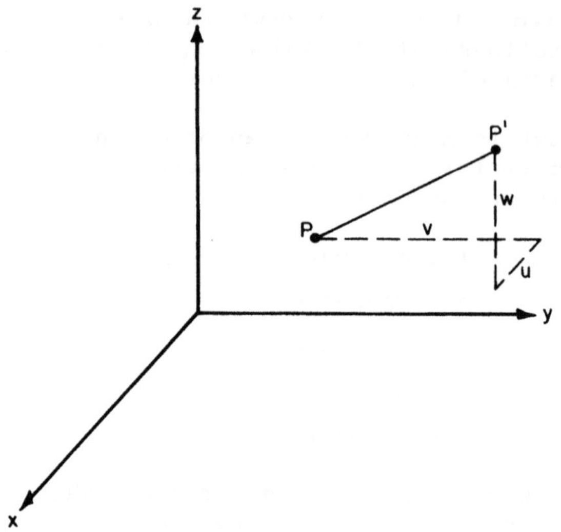

Fig. 5. Displacement of a point.

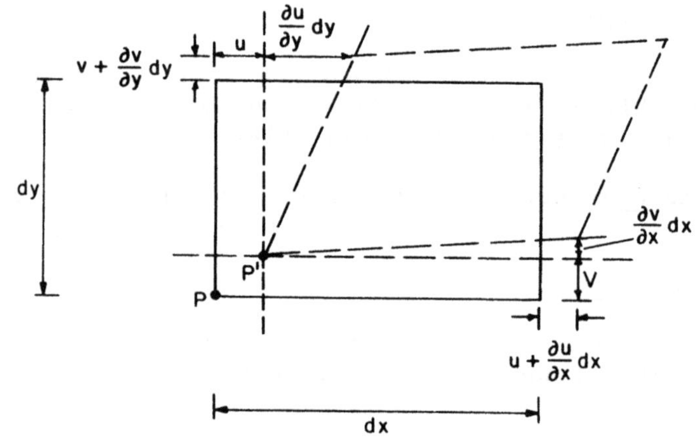

Fig. 6. Displacement gradients and strains.

$$\frac{1}{2}\gamma_{xy} = \varepsilon_{xy} = \varepsilon_{12} = \frac{1}{2}\left(\frac{\partial u}{\partial y} + \frac{\partial v}{\partial x}\right)$$

$$\frac{1}{2}\gamma_{yz} = \varepsilon_{yz} = \varepsilon_{23} = \frac{1}{2}\left(\frac{\partial v}{\partial z} + \frac{\partial w}{\partial y}\right)$$

$$\frac{1}{2}\gamma_{zx} = \varepsilon_{zx} = \varepsilon_{31} = \frac{1}{2}\left(\frac{\partial w}{\partial x} + \frac{\partial u}{\partial z}\right).$$

Now we can write the strain tensor:

$$\varepsilon_{ij} = \begin{vmatrix} \varepsilon_{11} & \varepsilon_{12} & \varepsilon_{13} \\ \varepsilon_{21} & \varepsilon_{22} & \varepsilon_{23} \\ \varepsilon_{31} & \varepsilon_{32} & \varepsilon_{33} \end{vmatrix} .$$

Or, in terms of the displacement gradients,

$$\varepsilon_{ij} = \frac{1}{2}\left(\frac{\partial u_i}{\partial x_j} + \frac{\partial u_j}{\partial x_i}\right).$$

That is,

$$\begin{vmatrix} \varepsilon_{11} & \varepsilon_{12} & \varepsilon_{13} \\ \varepsilon_{21} & \varepsilon_{22} & \varepsilon_{23} \\ \varepsilon_{31} & \varepsilon_{32} & \varepsilon_{33} \end{vmatrix} = \begin{vmatrix} \frac{\partial u_1}{\partial x_1} & \frac{1}{2}\left(\frac{\partial u_1}{\partial x_2} + \frac{\partial u_2}{\partial x_1}\right) & \frac{1}{2}\left(\frac{\partial u_1}{\partial x_3} + \frac{\partial u_3}{\partial x_1}\right) \\ \frac{1}{2}\left(\frac{\partial u_2}{\partial x_1} + \frac{\partial u_1}{\partial x_2}\right) & \frac{\partial u_2}{\partial x_2} & \frac{1}{2}\left(\frac{\partial u_2}{\partial x_3} + \frac{\partial u_3}{\partial x_2}\right) \\ \frac{1}{2}\left(\frac{\partial u_3}{\partial x_1} + \frac{\partial u_1}{\partial x_3}\right) & \frac{1}{2}\left(\frac{\partial u_3}{\partial x_2} + \frac{\partial u_2}{\partial x_3}\right) & \frac{\partial u_3}{\partial x_3} \end{vmatrix}$$

Rotation. Referring back to the diagram for strain derivation, it will be seen that if $(\partial u/\partial y)dy \neq (\partial v/\partial x)dx$, the element undergoes rotation as well as pure deformation.

Change in length in x-direction $= \left(u + \frac{\partial u}{\partial x}\,dx\right) - u = \frac{\partial u}{\partial x}\,dx$.

Unit change in x-direction $= \varepsilon_{xx} = \frac{\partial u}{\partial x}$.

Similarly, $\varepsilon_{yy} = \frac{\partial v}{\partial y}$.

Change in angle in the x-y plane at P, or γ_{xy}, is

$$\gamma_{xy} = \frac{\partial u}{\partial y} + \frac{\partial v}{\partial x} .$$

By identical considerations for the yz and zx planes we obtain the remaining components:

$$\varepsilon_{xx} = \frac{\partial u}{\partial x}, \qquad \varepsilon_{yy} = \frac{\partial v}{\partial y}, \qquad \varepsilon_{zz} = \frac{\partial w}{\partial z}$$

$$\gamma_{xy} = \gamma_{yx} = \frac{\partial u}{\partial y} + \frac{\partial v}{\partial x}$$

$$\gamma_{yz} = \gamma_{zy} = \frac{\partial v}{\partial z} + \frac{\partial w}{\partial y}$$

$$\gamma_{zx} = \gamma_{xz} = \frac{\partial w}{\partial x} + \frac{\partial u}{\partial z} .$$

γ_{xy}, γ_{yx}, γ_{zx} are the components of shear strain according to the classical, or engineering, definitions. In order to utilize the tensor system we adopt a slightly different form for the shear strains. We denote all strain components by ε, and replace subscripts x, y, z by 1, 2, 3, as in the case of stress, getting:

$$\varepsilon_{xx} = \varepsilon_{11} = \frac{\partial u}{\partial x}$$

$$\varepsilon_{yy} = \varepsilon_{22} = \frac{\partial v}{\partial y}$$

$$\varepsilon_{zz} = \varepsilon_{33} = \frac{\partial w}{\partial z}$$

$$\frac{1}{2}\gamma_{xy} = \varepsilon_{xy} = \varepsilon_{12} = \frac{1}{2}\left(\frac{\partial u}{\partial y} + \frac{\partial v}{\partial x}\right)$$

$$\frac{1}{2}\gamma_{yz} = \varepsilon_{yz} = \varepsilon_{23} = \frac{1}{2}\left(\frac{\partial v}{\partial z} + \frac{\partial w}{\partial y}\right)$$

$$\frac{1}{2}\gamma_{zx} = \varepsilon_{xx} = \varepsilon_{31} = \frac{1}{2}\left(\frac{\partial w}{\partial x} + \frac{\partial u}{\partial z}\right) .$$

Now we can write the strain tensor:

$$\varepsilon_{ij} = \begin{vmatrix} \varepsilon_{11} & \varepsilon_{12} & \varepsilon_{13} \\ \varepsilon_{21} & \varepsilon_{22} & \varepsilon_{23} \\ \varepsilon_{31} & \varepsilon_{32} & \varepsilon_{33} \end{vmatrix} .$$

Defining a displacement gradient tensor $e_{ij} = \frac{\partial u_i}{\partial x_j}$

$$e_{ij} = \frac{\partial u_i}{\partial x_j} = \frac{1}{2}\left(\frac{\partial u_i}{\partial x_j} + \frac{\partial u_j}{\partial x_i}\right) + \frac{1}{2}\left(\frac{\partial u_i}{\partial x_j} - \frac{\partial u_j}{\partial x_i}\right)$$

$\frac{1}{2}\left(\frac{\partial u_i}{\partial x_j} + \frac{\partial u_j}{\partial x_i}\right) = \varepsilon_{ij}$ is a symmetric tensor, with $\varepsilon_{ij} = \varepsilon_{ji}$

$\frac{1}{2}\left(\frac{\partial u_i}{\partial x_j} - \frac{\partial u_j}{\partial x_i}\right) = \omega_{ij}$ is an antisymmetric, or skew symmetric, tensor, with $\omega_{ij} = -\omega_{ji}$; ω_{ij} is the rotation tensor. Diagonal terms = 0.

5. TRANSFORMATION EQUATIONS

5.1 Rotation of Reference Directions for Stress Components

Having defined the components of stress and strain with respect to a given set of coordinate axes, we have to consider now the resolution, or transformation, of stress and strain into a different set of directions (Figure 7).

Denoting the direction cosine by a, the transformation matrix is:

		x_1	x_2	x_3	old directions
	x_1'	a_{11}	a_{12}	a_{13}	
new directions	x_2'	a_{21}	a_{22}	a_{23}	
	x_3'	a_{31}	a_{32}	a_{33}	

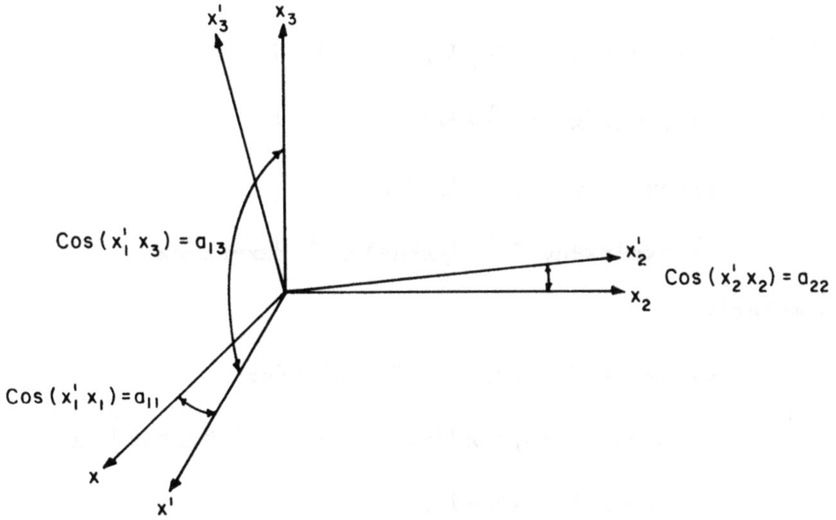

Fig. 7. Rotation of axes.

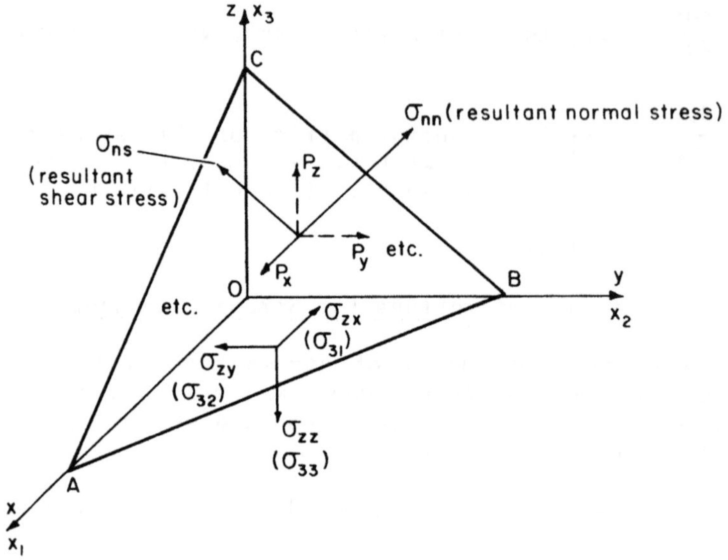

Fig. 8. Resultant stresses on a tetrahedral plane.

From a consideration of the equilibrium of the tetrahedron (Figure 8), with p_x, p_y, p_z as cartesian components of stress on the oblique plane, it can be shown that:

$$\begin{aligned}\sigma_{nn} &= a_{nx}p_x + a_{ny}p_y + a_{nz}p_z \\ &= a_{nx}(a_{nx}\sigma_{xx} + a_{ny}\sigma_{yx} + a_{nz}\sigma_{zx}) \\ &\quad + a_{ny}(a_{nx}\sigma_{xy} + a_{ny}\sigma_{yy} + a_{nz}\sigma_{zy}) \\ &\quad + a_{nz}(a_{nx}\sigma_{xz} + a_{ny}\sigma_{yz} + a_{nz}\sigma_{zz}) \\ &= \sigma_{xx}a_{nx}^2 + \sigma_{yy}a_{ny}^2 + \sigma_{zz}a_{nz}^2 \\ &\quad + 2\ \sigma_{xy}a_{nx}a_{ny} + 2\ \sigma_{yz}a_{ny}a_{nz} + \sigma_{zx}a_{nz}a_{nx}\end{aligned}$$

and similarly

$$\begin{aligned}\sigma_{ns} &= \sigma_{xx}a_{nx}a_{sx} + \sigma_{yy}a_{ny}a_{sy} + \sigma_{zz}a_{nz}a_{sz} \\ &\quad + (a_{nx}a_{sy} + a_{ny}a_{sx})\sigma_{xy} + (a_{nx}a_{sz} + a_{nz}a_{sx})\sigma_{xz} \\ &\quad + (a_{ny}a_{sz} + a_{nz}a_{sy})\sigma_{yz}.\end{aligned}$$

If we now introduce into our tensor notation a summation convention, such that all terms containing a repeated suffix are summed, we can write the above equations in compact form:

$$\sigma_{nn} = a_{ni}\, a_{nj}\, \sigma_{ij}$$

$$\sigma_{ns} = a_{ni}\, a_{sj}\, \sigma_{ij}$$

If n and s are the x' and y' axes of the transformed system, the complete transformation is given by

$$\sigma_{ns} = a_{ni}\, a_{sj}\, \sigma_{ij}$$

5.2 Principal Directions

By a suitable rotation of the axes x, y, z to a new set of directions x', y', z', the shear components of the transformed system can be made to disappear while the normal components reach extreme values. These new normal components are called principal stresses, and the planes on which they act are principal planes.

Suppose the plane ABC in Figure 8 has been rotated so as to coincide with a principal plane. Thus $\sigma_{ns} = 0$, and σ_{nn} is a minimum, which we shall just denote as σ.

Then

$$\sigma\, a_{nx} = p_x = a_{nx}\sigma_{xx} + a_{ny}\sigma_{yx} + a_{nz}\sigma_{zx}$$

$$\sigma\, a_{ny} = p_y = a_{nx}\sigma_{xy} + a_{ny}\sigma_{yy} + a_{nz}\sigma_{zy}$$

$$\sigma\, a_{nz} = p_z = a_{nx}\sigma_{xz} + a_{ny}\sigma_{yz} + a_{nz}\sigma_{zz}.$$

That is,

$$(\sigma_{xx} - \sigma)a_{nx} + \sigma_{yx}\, a_{ny} + \sigma_{zx}\, a_{nz} = 0$$

$$\sigma_{xy}\, a_{nx} + (\sigma_{yy} - \sigma)a_{ny} + \sigma_{zy}\, a_{nz} = 0$$

$$\sigma_{xz}\, a_{nz} + \sigma_{tz}\, a_{ny} + (\sigma_{zz} - \sigma)a_{nz} = 0.$$

Since the direction cosines cannot all be zero (the condition for orthogonality is $a^2_{nx} + a^2_{ny} + a^2_{nz} = 1$), the determinant of their coefficients must equal zero, i.e.

$$\sigma^3 - (\sigma_{xx}+\sigma_{yy}+\sigma_{zz})\sigma^2 + (\sigma_{xx}\sigma_{yy}+\sigma_{yy}\sigma_{zz}+\sigma_{zz}\sigma_{xx}-\sigma_{xy}^2-\sigma_{yz}^2-\sigma_{zx}^2)\sigma$$

$$- (\sigma_{xx}\sigma_{yy}\sigma_{zz}-\sigma_{xx}\sigma_{yz}^2-\sigma_{yy}\sigma_{xz}^2-\sigma_{zz}\sigma_{xy}^2+2\sigma_{xy}\sigma_{yz}\sigma_{xz}) = 0.$$

There are three real roots of this cubic equation in σ: σ_1, σ_2, σ_3. These are the principal stresses. They depend only on the state of stress at a point, and not on the system of coordinates. Thus the coefficients of the cubic equation are the same for any coordinate system, i.e. they are invariant with rotation of the reference. The first, second and third invariants of stress are:

$$I_1 = \sigma_{xx} + \sigma_{yy} + \sigma_{zz} = \sigma_1 + \sigma_2 + \sigma_3$$

$$I_2 = -(\sigma_{xx}\sigma_{yy} + \sigma_{yy}\sigma_{zz} + \sigma_{zz}\sigma_{xx}) + \sigma_{xy}^2 + \sigma_{yz}^2 + \sigma_{zx}^2$$

$$= -(\sigma_1\sigma_2 + \sigma_2\sigma_3 + \sigma_3\sigma_1)$$

$$I_3 = \sigma_{xx}\sigma_{yy}\sigma_{zz} - \sigma_{xx}\sigma_{yz}^2 - \sigma_{yy}\sigma_{zx}^2 - \sigma_{zz}\sigma_{xy}^2 + 2\sigma_{xy}\sigma_{yz}\sigma_{xz}$$

$$= \sigma_1\sigma_2\sigma_3 .$$

5.3 Strain Transformations

There are corresponding transformation equations for strain. They are completely analogous to the stress transformations if we use the tensor components of shear strain ε_{xy}, etc., rather than γ_{xy}, etc. Principal strains are analogous to principal stresses, and we also have invariants of strain analogous to the stress invariants.

6. ISOTROPIC AND DEVIATORIC COMPONENTS OF STRESS AND STRAIN

6.1 Bulk Stress and Deviatoric Stress

There is a particular scheme of stress resolution which is used a good deal in the consideration of flow laws and failure criteria. It has obvious physical significance.

Bulk stress, or volumetric stress, is that combination of stress components which tends to change the volume of an element. Bulk stress, σ, is

$$\bar{\sigma} = \frac{1}{3}(\sigma_{xx} + \sigma_{yy} + \sigma_{zz}) = \frac{1}{3}(\sigma_1 + \sigma_2 + \sigma_3)$$

i.e the average normal pressure. In the case of a fluid the bulk stress is the hydrostatic pressure ($\sigma_{xx} = \sigma_{yy} = \sigma_{zz}$).

Using the tensor rotation with the summation convention (adding terms formed from repeated suffixes):

$$\bar{\sigma} = \frac{1}{3} \sigma_{ii}$$

i.e

$$\bar{\sigma} = \frac{1}{3}(\sigma_{11} + \sigma_{22} + \sigma_{33}).$$

We note that the bulk stress is equal to one-third of the first invariant of stress:

$$\bar{\sigma} = \frac{1}{3} I_1.$$

Deviatoric stress (shear stress) tends to change the shape of an element. Deviators are denoted by primes. In tensor notation:

$$\sigma'_{ij} = \sigma_{ij} - \frac{1}{3} \delta_{ij} \sigma_{kk}.$$

Here we have introduced two new additions to the scheme of tensor notation. The symbol δ_{ij} is known as the Kronecker delta and it can take one of two values:

when $i = j$, $\delta_{ij} = 1$

when $i \neq j$, $\delta_{ij} = 0$.

The subscripts k are "dummy suffixes."

Examples of the application of this scheme are:

$$\sigma'_{11} = \sigma_{11} - \frac{1}{3}(\sigma_{11} + \sigma_{22} + \sigma_{33})$$

$$\sigma'_{12} = \sigma_{12} - 0 = \sigma_{12}.$$

A deviatoric stress is sometimes called a "reduced stress."

6.2 Bulk Strain and Deviatoric Strain

Bulk strain and deviatoric strain are analogous to bulk stress and deviatoric stress respectively. Bulk strain, which is also called volumetric strain, or dilatation, is

$$\bar{\varepsilon} = \varepsilon_{xx} + \varepsilon_{yy} + \varepsilon_{zz} = \varepsilon_1 + \varepsilon_2 + \varepsilon_3$$

or

$$\bar{\varepsilon} = \varepsilon_{ii}.$$

Note that there is a factor of 3 difference between the expressions for bulk stress and bulk strain.

Deviatoric strain (shear strain) is

$$\varepsilon'_{ij} = \varepsilon_{ij} - \frac{1}{3} \delta_{ij} \varepsilon_{kk}$$

i.e., it corresponds to deviator stress in form.

6.3 Octahedral Stresses

If we take axes parallel to the directions of principal stress, the directions of principal shear stress form a regular octahedron which has its corners on the principal axes (Figure 9).

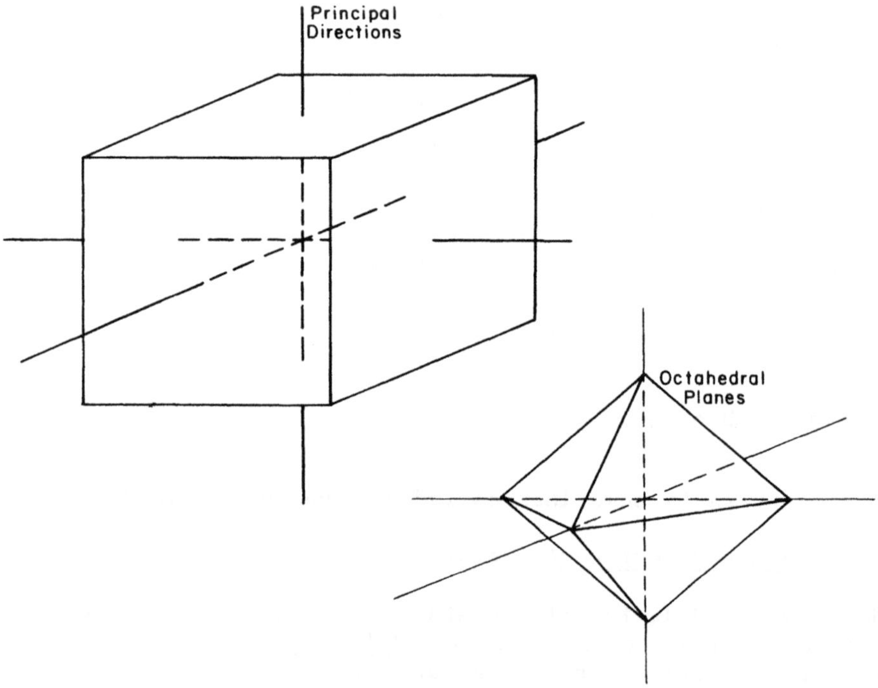

Fig. 9. Octahedral planes.

The direction cosines of the normals to the octahedral planes are $\pm 1/\sqrt{3}$.

The octahedral normal stress $\sigma_{n\ oct}$ is

$$\sigma_{n\ oct} = \frac{1}{3}(\sigma_1 + \sigma_2 + \sigma_3)$$

$$= \frac{1}{3} I_1.$$

The octahedral shear stress $\sigma_{s\ oct}$ is

$$\sigma_{s\ oct} = \frac{1}{3}[(\sigma_1 - \sigma_2)^2 + (\sigma_2 - \sigma_3)^2 + (\sigma_3 - \sigma_1)^2]^{1/2}$$

$$= \frac{\sqrt{2}}{3} [I_2^2 + 3I_2]^{1/2}.$$

If the principal directions are not immediately known, the octahedral stresses are

$$\sigma_{n\ oct} = \frac{1}{3}(\sigma_{xx} + \sigma_{yy} + \sigma_{zz})$$

$$\sigma_{s\ oct} = \frac{\sqrt{2}}{3}[\sigma_{xx}^2+\sigma_{yy}^2+\sigma_{zz}^2+3(\sigma_{xy}^2+\sigma_{yz}^2+\sigma_{zx}^2)-(\sigma_{xx}\sigma_{yy}+\sigma_{yy}\sigma_{zz}+\sigma_{zz}\sigma_{xx})]^{1/2}$$

6.4 Invariants of Stress and Strain Deviation

The components of stress deviation σ'_{ij} are formed by subtracting the mean stress from the normal components of stress, but not from the shear components. Thus the cartesian components of stress deviation are

$$\sigma'_{xx} = \sigma_{xx} - \bar{\sigma}, \qquad \sigma'_{yy} = \sigma_{yy} - \bar{\sigma}, \qquad \sigma'_{zz} = \sigma_{zz} - \bar{\sigma}$$

$$\sigma'_{xy} = \sigma_{xy}, \qquad \sigma'_{yz} = \sigma_{yz}, \qquad \sigma'_{zx} = \sigma_{zx}.$$

The principal stress deviations are

$$\sigma'_1 = \sigma_1 - \bar{\sigma} = \frac{1}{3}(2\sigma_1 - \sigma_2 - \sigma_3)$$

$$\sigma'_2 = \sigma_2 - \bar{\sigma} = \frac{1}{3}(2\sigma_2 - \sigma_1 - \sigma_3)$$

$$\sigma'_3 = \sigma_3 - \bar{\sigma} = \frac{1}{3}(2\sigma_3 - \sigma_1 - \sigma_2).$$

The invariants of stress deviation are:

$$J_1 = \sigma'_{xx} + \sigma'_{yy} + \sigma'_{zz} = \sigma'_1 + \sigma'_2 + \sigma'_3 = 0$$

$$J_2 = -(\sigma'_{xx}\sigma'_{yy} + \sigma'_{yy}\sigma'_{zz} + \sigma'_{zz}\sigma'_{xx}) + \sigma^2_{xy} + \sigma^2_{yz} + \sigma^2_{zx}$$

$$= \frac{1}{6}[(\sigma_1 - \sigma_2)^2 + (\sigma_2 - \sigma_3)^2 + (\sigma_3 - \sigma_1)^2]$$

$$= \frac{1}{2}[\sigma'^2_1 + \sigma'^2_2 + \sigma'^2_3]$$

$$J_3 = \sigma'_{xx}\sigma'_{yy}\sigma'_{zz} + 2\ \sigma_{xy}\sigma_{yz}\sigma_{zx} - (\sigma'_{xx}\sigma^2_{yz} + \sigma'_{yy}\sigma^2_{zx} + \sigma'_{zz}\sigma^2_{xy})$$

$$= \sigma'_1\ \sigma'_2\ \sigma'_3.$$

The octahedral shear stress can be expressed in terms of stress deviation as

$$\sigma_{s\ oct} = [2J_2/3]^{1/2}.$$

Components of strain deviation, denoted by the symbol e, are formed in exactly the same way as the components of stress deviation, i.e.

$$e_{xx} = \varepsilon_{xx} - \bar{e}, \qquad e_{yy} = \varepsilon_{yy} - \bar{e}, \qquad e_{zz} = \varepsilon_{zz} - \bar{e}$$

$$e_{xy} = \varepsilon_{xy}, \qquad e_{yz} = \varepsilon_{yz}, \qquad e_{zx} = \varepsilon_{zx}$$

$$e_1 = \varepsilon_1 - \bar{e}, \qquad e_2 = \varepsilon_2 - \bar{e}, \qquad e_3 = \varepsilon_3 - \bar{e}$$

where $\bar{e}$ is the mean normal strain:

$$\bar{e} = (\varepsilon_{xx} + \varepsilon_{yy} + \varepsilon_{zz})/3 = (\varepsilon_1 + \varepsilon_2 + \varepsilon_3)/3.$$

The invariants of strain deviation are exactly analogous to the invariants of stress deviation.

7. ELASTICITY AND LINEAR VISCOSITY

7.1 Elastic Constants

The basis for study of deformable solids is the theory of elasticity, which is usually introduced through consideration of an axial push or pull on a prismatic bar (Figure 10). The material is assumed to be homogeneous, isotropic and linearly elastic, with axial stress proportional to axial strain

$$\sigma_{yy} = E\ \varepsilon_{yy}$$

and transverse strain proportional to axial strain

$$\varepsilon_{xx} = \varepsilon_{zz} = -\ \nu\ \varepsilon_{yy} = -\frac{\nu}{E}\sigma_{yy}$$

where E is Young's modulus and ν is Poisson's ratio. Because the material is linearly elastic, the principle of superposition can be applied in order to obtain the strains when stresses σ_{xx}, σ_{yy}, σ_{zz} are acting, e.g.

$$\varepsilon_{xx} = \frac{\sigma_{xx}}{E} - \frac{\nu\sigma_{yy}}{E} - \frac{\nu\sigma_{zz}}{E} = \frac{1}{E}[\sigma_{xx} - \nu(\sigma_{yy} + \sigma_{zz})]\ .$$

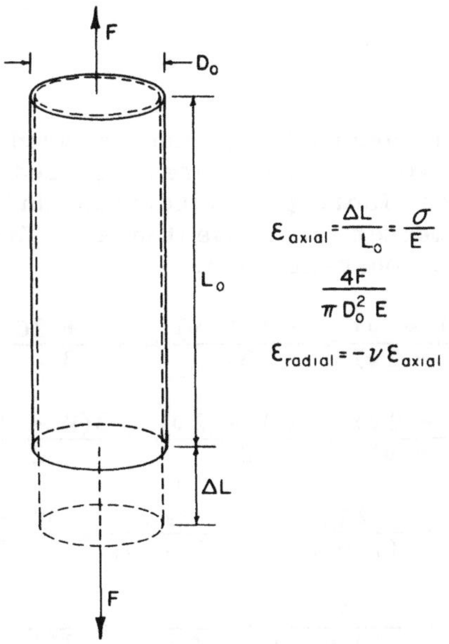

Fig. 10. Axial elastic deformation of a bar.

Shear stress and shear strain can be related more directly by the shear modulus, or modulus of rigidity, G:

$$\sigma_{xy} = 2G\ \varepsilon_{xy}.$$

It can easily be shown, e.g. by expressing ε_{xy} and σ_{xy} in terms of principal stresses, that

$$G = \frac{E}{2(1+\nu)}.$$

Young's modulus and Poisson's ratio are to some extent accidents of history, and for some purposes it is more convenient to have direct relations between bulk stress and volumetric strain, and between deviatoric stress and deviatoric strain. For isotropic volumetric deformations the proportionality constant is the bulk modulus K, where

$$K = \frac{\text{bulk stress}}{\text{volumetric strain}} = \frac{\bar{\sigma}}{\bar{\varepsilon}}$$

$$= \frac{\sigma_{ii}}{3\ \varepsilon_{ii}}.$$

Deviatoric components are related by the shear modulus G, with the general form of the relation expressable as

$$G = \frac{\sigma'_{ij}}{2\varepsilon'_{ij}}.$$

Another constant, Lamé's constant λ, can be used in association with G. However, only two constants are required to specify the elastic properties of an isotropic material, and the choice of which two is largely a matter of convenience. The relations between the various elastic constants are:

$$K = \frac{E}{3(1-2\nu)} = \frac{2G(1+\nu)}{3(1-2\nu)} = \frac{\lambda(1+\nu)}{3\nu} = \frac{3\lambda + 2G}{3}$$

$$G = \frac{E}{2(1+\nu)} = \frac{3K(1-2\nu)}{2(1+\nu)} = \frac{\lambda(1-2\nu)}{2\nu} = \frac{3(K-\lambda)}{2}$$

$$\lambda = \frac{\nu E}{(1+\nu)(1-2\nu)} = \frac{2G\nu}{(1-2\nu)} = \frac{3K\nu}{(1+\nu)} = \frac{3K-2G}{3}$$

$$\nu = \frac{E-2G}{2G} = \frac{3K-E}{6K} = \frac{\lambda}{2(\lambda+G)} = \frac{\lambda}{3K-\lambda} = \frac{3K-2G}{2(3K+G)}.$$

7.2 Solution of Problems by Theory of Elasticity

Practical problems in elasticity are solved by solving certain sets of equations for appropriate boundary conditions. The sets of equations used are: 1) equilibrium equations, 2) strain/displacement relations, 3) compatibility equations, 4) stress/strain relations.

7.2.1 Equilibrium equations. The equilibrium equations are derived from Newton's laws. For static equilibrium the stress components in any direction sum to zero. In many engineering problems, body forces can be ignored, but in ice mechanics it is often necessary to include gravity body forces. The stress equilibrium equations with body forces included in the form of gravity forces are:

$$\frac{\partial \sigma_{ij}}{\partial x_j} + \rho g_i = 0$$

where ρ is the density of the material and g_i are components of the gravitational acceleration g.

In rectangular coordinates these equilibrium equations expand as:

$$\frac{\partial \sigma_{xx}}{\partial x} + \frac{\partial \sigma_{xy}}{\partial y} + \frac{\partial \sigma_{xz}}{\partial z} + \rho g_x = 0$$

$$\frac{\partial \sigma_{xy}}{\partial x} + \frac{\partial \sigma_{yy}}{\partial y} + \frac{\partial \sigma_{yz}}{\partial z} + \rho g_y = 0$$

$$\frac{\partial \sigma_{xz}}{\partial x} + \frac{\partial \sigma_{yz}}{\partial y} + \frac{\partial \sigma_{zz}}{\partial z} + \rho g_z = 0.$$

The corresponding equations in cylindrical coordinates are:

$$\frac{\partial \sigma_{rr}}{\partial r} + \frac{1}{r}\frac{\partial \sigma_{r\theta}}{\partial \theta} + \frac{\partial \sigma_{rz}}{\partial z} + \frac{1}{r}\left(\sigma_{rr} - \sigma_{\theta\theta}\right) + \rho g_r = 0$$

$$\frac{\partial \sigma_{r\theta}}{\partial r} + \frac{1}{r}\frac{\partial \sigma_{\theta\theta}}{\partial \theta} + \frac{\partial \sigma_{\theta z}}{\partial z} + \frac{2}{r}\sigma_{r\theta} + \rho g_\theta = 0$$

$$\frac{\partial \sigma_{rz}}{\partial r} + \frac{1}{r}\frac{\partial \sigma_{\theta z}}{\partial \theta} + \frac{\partial \sigma_{zz}}{\partial z} + \frac{1}{r}\sigma_{rz} + \rho g_z = 0.$$

7.2.2 Strain/displacement relations. The strain/displacement relations have already been derived and given as

$$\varepsilon_{ij} = \frac{1}{2} \left(\frac{\partial u_i}{\partial x_j} + \frac{\partial u_j}{\partial x_i}\right).$$

In rectangular coordinates these expand as:

$$\varepsilon_{xx} = \frac{\partial u_x}{\partial x} \qquad \varepsilon_{xy} = \frac{1}{2} \left(\frac{\partial u_x}{\partial y} + \frac{\partial u_y}{\partial x}\right)$$

$$\varepsilon_{yy} = \frac{\partial u_y}{\partial y} \qquad \varepsilon_{yz} = \frac{1}{2} \left(\frac{\partial u_y}{\partial z} + \frac{\partial u_z}{\partial y}\right)$$

$$\varepsilon_{zz} = \frac{\partial u_z}{\partial z} \qquad \varepsilon_{zx} = \frac{1}{2} \left(\frac{\partial u_z}{\partial x} + \frac{\partial u_x}{\partial z}\right).$$

In cylindrical coordinates they expand as:

$$\varepsilon_{rr} = \frac{\partial u_r}{\partial r} \qquad \varepsilon_{r\theta} = \frac{1}{2} \left\{\frac{1}{r}\frac{\partial u_r}{\partial \theta} + \frac{\partial u_\theta}{\partial r} - \frac{u_\theta}{r}\right\}$$

$$\varepsilon_{\theta\theta} = \frac{u_r}{r} + \frac{1}{r}\frac{\partial u_\theta}{\partial \theta} \qquad \varepsilon_{\theta z} = \frac{1}{2} \left\{\frac{\partial u_\theta}{\partial z} + \frac{1}{r}\frac{\partial u_z}{\partial \theta}\right\}$$

$$\varepsilon_{zz} = \frac{\partial u_z}{\partial z} \qquad \varepsilon_{zr} = \frac{1}{2} \left\{\frac{\partial u_z}{\partial r} + \frac{\partial u_r}{\partial z}\right\}.$$

7.2.3 Compatibility equations for strain. If the components of displacement are given, then the components of strain can be found by appropriate partial differentiation, as indicated above. If, however, it is the components of strain which are given, then certain restrictions must be placed upon the strains in order to ensure that there are only three values of displacement, and that the displacement field is continuous. These restrictions are called the compatibility equations, or compatibility conditions.

The compatibility conditions for strain can be written in general form as

$$\frac{\partial^2 \varepsilon_{ij}}{\partial x_k \partial x_\ell} + \frac{\partial^2 \varepsilon_{k\ell}}{\partial x_i \partial x_j} = \frac{\partial^2 \varepsilon_{ik}}{\partial x_j \partial x_\ell} + \frac{\partial^2 \varepsilon_{j\ell}}{\partial x_i \partial x_k}.$$

This expression represents 81 equations, only 6 of which are independent. The independent equations are:

$$\frac{\partial^2 \varepsilon_{xx}}{\partial y^2} + \frac{\partial^2 \varepsilon_{yy}}{\partial x^2} = 2 \frac{\partial^2 \varepsilon_{xy}}{\partial x \partial y}$$

$$\frac{\partial^2 \varepsilon_{yy}}{\partial z^2} + \frac{\partial^2 \varepsilon_{zz}}{\partial y^2} = 2\,\frac{\partial^2 \varepsilon_{yz}}{\partial y\,\partial z}$$

$$\frac{\partial^2 \varepsilon_{zz}}{\partial x^2} + \frac{\partial^2 \varepsilon_{xx}}{\partial z^2} = 2\,\frac{\partial^2 \varepsilon_{zx}}{\partial z\,\partial x}$$

$$\frac{\partial^2 \varepsilon_{xx}}{\partial y\,\partial z} = -\frac{\partial^2 \varepsilon_{yz}}{\partial x^2} + \frac{\partial^2 \varepsilon_{xz}}{\partial x\,\partial y} + \frac{\partial^2 \varepsilon_{xy}}{\partial x\,\partial z}$$

$$\frac{\partial^2 \varepsilon_{yy}}{\partial z\,\partial x} = -\frac{\partial^2 \varepsilon_{zx}}{\partial y^2} + \frac{\partial^2 \varepsilon_{yx}}{\partial y\,\partial z} + \frac{\partial^2 \varepsilon_{yz}}{\partial y\,\partial x}$$

$$\frac{\partial^2 \varepsilon_{zz}}{\partial x\,\partial y} = -\frac{\partial^2 \varepsilon_{xy}}{\partial z^2} + \frac{\partial^2 \varepsilon_{zy}}{\partial z\,\partial x} + \frac{\partial^2 \varepsilon_{zx}}{\partial z\,\partial y}.$$

7.2.4 Elastic stress/strain relations. The stress/strain relations introduce the properties of the material into the consideration. For an isotropic elastic material the general relations can be written in compact form as:

$$\sigma_{ii} = 3K\,\varepsilon_{ii}$$

$$\sigma'_{ij} = 2G\,\varepsilon'_{ij}.$$

Alternatively, using E and ν and expanding for rectangular coordinates:

$$\varepsilon_{xx} = \frac{1}{E}\,\{\sigma_{xx} - \nu(\sigma_{yy} + \sigma_{zz})\}$$

$$\varepsilon_{yy} = \frac{1}{E}\,\{\sigma_{yy} - \nu(\sigma_{xx} + \sigma_{zz})\}$$

$$\varepsilon_{zz} = \frac{1}{E}\,\{\sigma_{zz} - \nu(\sigma_{xx} + \sigma_{yy})\}$$

$$\varepsilon_{xy} = \frac{1}{2G}\,\sigma_{xy}\,, \qquad \varepsilon_{xz} = \frac{1}{2G}\,\sigma_{xz}\,, \qquad \varepsilon_{yz} = \frac{1}{2G}\,\sigma_{yz}.$$

Or, putting stress in terms of strain:

$$\sigma_{xx} = \frac{E}{1+\nu}\left\{\frac{\nu}{(1-2\nu)}\,(\varepsilon_{xx} + \varepsilon_{yy} + \varepsilon_{zz}) + \varepsilon_{xx}\right\}$$

$$\sigma_{yy} = \frac{E}{1+\nu}\left\{\frac{\nu}{(1-2\nu)}\,(\varepsilon_{xx} + \varepsilon_{yy} + \varepsilon_{zz}) + \varepsilon_{yy}\right\}$$

$$\sigma_{zz} = \frac{E}{1+\nu}\left\{\frac{\nu}{(1-2\nu)}(\varepsilon_{xx}+\varepsilon_{yy}+\varepsilon_{zz})+\varepsilon_{zz}\right\}$$

$$\sigma_{xy} = 2G\ \varepsilon_{xy}, \qquad \sigma_{xz} = 2G\ \varepsilon_{xz}, \qquad \sigma_{yz} = 2G\ \varepsilon_{yz}.$$

In cylindrical coordinates the relations can be written as:

$$\varepsilon_{rr} = \frac{1}{E}\{\sigma_{rr} - \nu(\sigma_{\theta\theta} + \sigma_{zz})\}$$

$$\varepsilon_{\theta\theta} = \frac{1}{E}\{\sigma_{\theta\theta} - \nu(\sigma_{rr} + \sigma_{zz})\}$$

$$\varepsilon_{zz} = \frac{1}{E}\{\sigma_{zz} - \nu(\sigma_{rr} + \sigma_{\theta\theta})\}$$

$$\varepsilon_{r\theta} = \frac{1}{2G}\sigma_{r\theta}, \qquad \varepsilon_{rz} = \frac{1}{2G}\sigma_{rz}, \quad \varepsilon_{\theta z} = \frac{1}{2G}\sigma_{\theta z}.$$

The preceding sets of equations, together with the boundary conditions of the problem, provide everything needed for a solution. However, the various sets of equations can be transformed. For example, the equilibrium equations can be written for strain instead of stress by applying the stress/strain relations, and similarly the compatibility conditions can be given in terms of stress.

7.3 Linear Viscosity

Linear viscosity is analogous to linear elasticity, and for any solution to an elastic problem there is, in principle, a corresponding solution to a viscous problem, provided certain conditions are met.

In developing the analogy, velocity of viscous deformation corresponds to elastic displacement, and viscous strain rate corresponds to elastic strain. The equilibrium equations remain unchanged, provided that there is slow flow with zero acceleration. The strain/displacement equations are replaced by strain-rate/velocity relations of identical form, i.e. $\dot{u}$ replacing u, and $\dot{\varepsilon}$ replacing ε. Again, there is a requirement for slow, non-accelerating flow. The elastic stress/strain equations are replaced by viscous stress/strain-rate equations. If the material is homogeneous and isotropic, the viscous properties can be described by two constants, and it is convenient to choose these as a "bulk viscosity" η and a shear viscosity μ. Thus

$$\sigma_{ii} = 3\ \eta\ \dot{\varepsilon}_{ii}$$

$$\sigma'_{ij} = 2 \mu \dot{\varepsilon}_{ij}.$$

For ice of low porosity, it may be permissible to assume incompressibility in a flow problem, i.e. η is infinite, and the viscous analogue of Poisson's ratio is 1/2.

7.4 Plane Stress and Plane Strain

In this summary, the general relations have been given for three-dimensional stress and strain, but many problems can be simplified as two-dimensional problems. There are two ways of reducing to two dimensions. In plane stress, stresses in the z-direction are assumed to be zero, usually because the x-y planes at the z boundaries are free from normal stress. In plane strain, displacement gradients, or strains, in the z-direction are assumed to be zero, usually because the material is perfectly confined at the limits of z, or because the width of the body in the z-direction is effectively infinite.

8. FAILURE CRITERIA

As regards the behavior of a material, the meaning of the term "failure" is arbitrary. In other words, a material may be said to have "failed" when it ceases to meet certain arbitrary requirements. For example, when a typical engineering material fractures it clearly has failed. However, a material can fail without fracturing, and the definition of failure might be based on a transition from elastic deformation to plastic yielding, or on irreversible acceleration of strain rate, or on development of permanent strain, or yet again on the development of internal cracking. Whatever definition of failure is accepted, formal analysis requires specification of the stress or strain conditions which control the failure.

Plastic yielding of an isotropic, incompressible solid gives a fairly simple introduction to failure criteria.

8.1 Maximum Shear Stress Criteria (Tresca/von Mises)

When typical metals are deformed at moderate rates, it is found that the bulk stress σ does not influence the onset of plastic yield to any significant extent. One reasonable assumption is that yielding occurs when the maximum shear stress, $(\sigma_1 - \sigma_3)/2$, reaches a critical value. This assumption, or postulate, is usually called the Tresca criterion, although it was put forward by Coulomb almost a hundred years before Tresca thought of it.

$$\sigma_1 - \sigma_3 = \text{constant}$$

Another reasonable postulate is that yielding will occur when the second invariant of the deviator stress, J_2, reaches a critical value. This criterion, attributed to von Mises, can be written

$$J_2 = \frac{1}{2}(\sigma_1'^2 + \sigma_2'^2 + \sigma_3'^2)$$

$$= \frac{1}{6}[(\sigma_1 - \sigma_2)^2 + (\sigma_2 - \sigma_1)^2 + (\sigma_3 - \sigma_1)^2]$$

$$= \text{constant} .$$

Another interpretation of the von Mises criterion, credited to Hencky or Huber, is that yielding occurs when the elastic strain energy of distortion reaches a critical value. Since J_2 and the octahedral shear stress $\sigma_{s\ oct}$ have a simple relationship to each other, the von Mises criterion can be written in terms of octahedral shear stress:

$$J_2 = \left(\frac{3}{2}\right) \sigma_{s\ oct}^2 = \text{constant}$$

i.e. yielding occurs at a critical value of $\sigma_{s\ oct}$. In uniaxial tension or compression, where $\sigma_2 = \sigma_3 = 0$, both the Tresca and Mises criteria require that plastic yielding sets in when the uniaxial stress σ_1 reaches a critical value Y, in either tension or compression. The corresponding value of the octahedral shear stress is $\sqrt{2}$ Y/3. The Tresca and von Mises criteria differ somewhat when they are compared in detail. For example: 1) the inter-

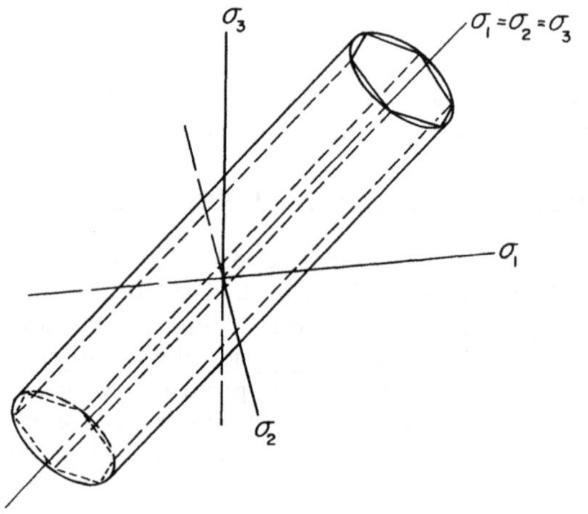

Fig. 11. Failure surfaces in three-dimensional principal stress space for the Tresca and von Mises criteria.

mediate principal stress σ_2 influences failure according to von Mises, but not Tresca, 2) the yield stress in pure shear is Y/2 according to Tresca, but $Y/\sqrt{3}$ according to von Mises, 3) the yield surface in principal stress space is a circular cylinder according to von Mises but a hexagonal prism according to Tresca, the longitudinal axis being the hydrostat line $\sigma_1 = \sigma_2 = \sigma_3$ in both cases (Figure 11).

8.2 Coulomb/Navier/Mohr Criteria

The Tresca and von Mises criteria assume that hydrostatic pressure has no effect on the resistance to shear deformation. While this may be a good approximation for metals at moderate rates of loading, it is not true for materials generally. For many materials, especially brittle non-metallic solids, the resistance to shear deformation increases as the hydrostatic pressure increases. A yield criterion representing this type of behavior evolved from the work of Coulomb, with subsequent modifications that are associated with the names Rankine, Navier and Mohr. The Coulomb-Navier, or Coulomb-Mohr, criterion assumes that failure in an isotropic material is determined by the extreme principal stresses σ_1 and σ_3, with the intermediate principal stress σ_2 exerting no influence. It is usually developed for the two-dimensional case, making the physical assumption that shear resistance on a plane is proportional to the "friction" developed on that plane by normal stress. If the constant "friction coefficient" is denoted by μ or by $\tan\phi$, the criterion becomes

$$\sigma_1[(\mu^2 + 1)^{1/2} - \mu] - \sigma_3[(\mu^2 + 1)^{1/2} + \mu] = 2c$$

or,

$$\left(\frac{\sigma_1 - \sigma_3}{2}\right) - \left(\frac{\sigma_1 + \sigma_3}{2}\right) \sin\phi = c \cos\phi$$

where c is a constant identified as the inherent shear strength of the material under zero normal stress. Since $(\sigma_1 - \sigma_3)/2$ is the maximum shear stress and $(\sigma_1 + \sigma_3)/2$ is the mean stress, the criterion says that the maximum shear stress at yielding is equal to a constant plus a multiple of the mean stress (Figure 12). It also appears to imply that the uniaxial compressive strength σ_c will differ from the uniaxial tensile strength σ_T by the factor:

$$\frac{\sigma_c}{\sigma_T} = \frac{(\mu^2 + 1)^{1/2} + \mu}{(\mu^2 + 1)^{1/2} - \mu}$$

but the implicit physical assumption that confining stress should be compressive limits the applicability of the criterion in the tension-compression quadrants, as shown in Figure 12b, where there is a cutoff at $\sigma_c/2$.

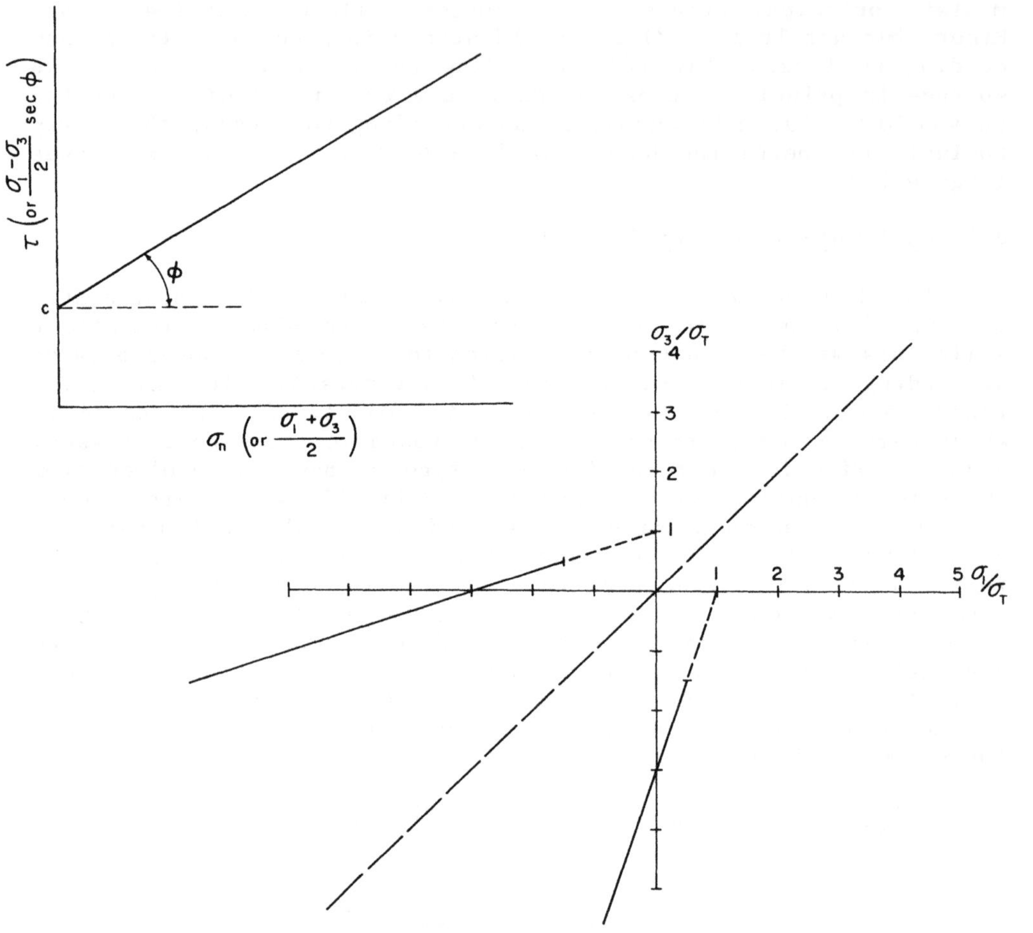

Fig. 12. Linear Coulomb-Navier-Mohr criterion for ϕ = 30°.

It might be noted that the Tresca criterion can now be regarded as a special case of the Coulomb-Navier, or Coulomb-Mohr, criterion with ϕ = o. Taking ϕ = o,

$$(\sigma_1 - \sigma_3) = 2c = \sigma_T = \sigma_C = Y.$$

The assumption that the increase of shear resistance is directly proportional to compressive bulk stress is, of course, an approximation. Experimental results show that the usual trend is for increases of confining stress to become less significant as the magnitude of the confining stress increases. In other words, the effective value of ϕ decreases as σ increases. A simple way to accommodate this effect is to assume that the relation between shear resistance and bulk stress is parabolic or, in other words, to assume a parabolic Mohr envelope. Assuming that failure is not influenced by the intermediate principal stress, the form of the relation is thus

$$(\sigma_1 - \sigma_3)^2 = A(\sigma_1 + \sigma_3) + B$$

where A and B are constants.

In terms of geometric representation, the two-dimensional linear Coulomb-Navier criterion can probably be generalized to define a failure surface in three-dimensional principal stress space. This surface would take the form of a pyramid or cone, depending on whether the intermediate principal stress is assumed to have an effect. The pyramid or cone has to be truncated in the absence of knowledge (or speculation) about behavior in the tension-tension quadrants of the principal stress planes. For a parabolic Coulomb-Navier criterion, the simplest generalization of the failure surface is a paraboloid, with the hydrostat line $\sigma_1 = \sigma_2 = \sigma_3$ as the axis of revolution and truncation at limits of the tension-tension quadrants where $\sigma_1 = \sigma_T$, $\sigma_2 = \sigma_T$, $\sigma_3 = \sigma_T$. Within the domain where all stresses are tensile, the simplest assumption is that the failure surface is a pyramid with its apex at $\sigma_1 = \sigma_2 = \sigma_3 = \sigma_T$.

8.3 The Griffith Criterion and its Derivatives

The Tresca and von Mises criteria are empirical relations, while the Coulomb-Navier-Mohr criterion is based partly on physical reasoning. There are other criteria which derive more directly from physical reasoning and associated mathematical models, the most notable being the Griffith criterion.

Starting from the observation that the bulk strength of brittle solids is, in general, orders of magnitude lower than the theoretical strength deduced from consideration of interatomic forces, Griffith postulated the existence of minute cracks and associated stress concentrations. Drawing upon the stress analysis given by Inglis for a two-dimensional elliptic crack in an elastic plate, Griffith equated the change of potential energy in the plate to the change of surface energy in the crack as the crack grew in length. For a thin elliptic crack of length 2c:

$$\frac{\partial}{\partial c}\left(\frac{\pi\sigma^2 c^2}{E}\right) = \frac{\partial}{\partial c}(4c\gamma)$$

where E is Young's modulus for the plate, γ is the specific surface energy of the material and σ is the applied stress (tensile and perpendicular to the long axis of the crack) at which crack growth occurs. Thus

$$\sigma = \left(\frac{2}{\pi}\right)^{1/2} \left(\frac{E\gamma}{c}\right)^{1/2}$$

for plane stress. For plane strain the corresponding relation is

$$\sigma = \left(\frac{2}{\pi(1 - \nu^2)}\right)^{1/2} \left(\frac{E\gamma}{c}\right)^{1/2}.$$

Where ν is Poisson's ratio. Numerically, the two versions are not much different.* Much the same result is obtained by direct consideration of theoretical material strength and stress concentration at the end of an elliptic crack. From consideration of interatomic forces as a function of separation, the theoretical tensile strength of the material σ_* is

$$\sigma_* = \left(\frac{E\gamma}{a}\right)^{1/2}$$

where a is the atomic spacing in the unstrained state. From the Inglis stress analysis for an elliptic crack with tip radius ρ, the stress at the crack tip σ_{ct} is

$$\sigma_{ct} = 2\sigma \left(\frac{c}{\rho}\right)^{1/2}$$

where σ is the applied stress in the plate. Equating σ_{ct} to σ_* for crack growth:

$$\sigma = \left(\frac{\rho}{4a}\right)^{1/2} \left(\frac{E\gamma}{c}\right)^{1/2}$$

in which ρ is considered to be of the same order of magnitude as a for a sharp crack. This version is identical to Griffith's plane stress relation if

$$\rho = (8/\pi)\ a = 2.55\ a.$$

It might be mentioned in passing that γ can be reduced by adsorption of certain surface active chemicals; this is known as the Rehbinder Effect or the Joffe Effect.

Having derived a condition for crack growth under simple tension normal to the crack axis, Griffith considered a biaxial stress

---

* For a "penny-shaped" crack, $\sigma = (\pi/4(1-\nu^2)^{1/2}(E\gamma/c)^{1/2}$, which is also not much different numerically.

field and variable orientation of the crack relative to the principal directions of stress, thus determining the critical crack orientation for initiation of crack growth. This gave a failure criterion for material which contains numerous cracks with random orientation:

$$\left(\frac{\sigma_1 - \sigma_3}{2}\right)^2 + 4\sigma_T\left(\frac{\sigma_1 + \sigma_3}{2}\right) = 0$$

if $3\sigma_1 + \sigma_3 < 0$

and

$$\sigma_1 = \sigma_T$$

if $3\sigma_1 + \sigma_3 > 0$

where σ_T is the uniaxial tensile strength, and $\sigma_1 > \sigma_3$ (tension positive. One of the implications of this criterion is that

$$\left|\frac{\sigma_c}{\sigma_T}\right| = 8.$$

Another point worth noting is that the Griffith criterion for $3\sigma_1 + \sigma_3 < 0$ is identical in mathematical form to the parabolic variant of the Coulomb-Navier-Mohr criterion. It can be generalized in three dimensions as a paraboloid, as already described for the parabolic Coulomb-Navier criterion.

Analysis of the three-dimensional Griffith criterion (Murrell, 1963; Jaeger and Cook, 1976) gives the relation for $3\sigma_1 + \sigma_3 < 0$ as:

$$\left(\frac{\sigma_1 - \sigma_2}{2}\right)^2 + \left(\frac{\sigma_2 - \sigma_3}{2}\right)^2 + \left(\frac{\sigma_3 - \sigma_1}{2}\right)^2 + 18\sigma_T\left(\frac{\sigma_1 + \sigma_2 + \sigma_3}{3}\right) = 0.$$

This gives a ratio of uniaxial compressive strength σ_c to uniaxial tensile strength σ_T that is different from the ratio for the two-dimensional criterion:

$$\left|\frac{\sigma_c}{\sigma_T}\right| = 12.$$

It also allows some different predictions to be made for multiaxial stress states, i.e. for the biaxial stress state when $\sigma_2 = 0$, for the conventional triaxial test condition when $\sigma_2 = \sigma_3$, and for the triaxial state where $\sigma_1 \neq \sigma_2 \neq \sigma_3$.

In recent times investigators have proposed various modifications to the older failure criteria in order to meet special requirements for rocks, polymers, and other materials. For example, the McClintock-Walsh criterion in rock mechanics comes from a modification of the Griffith analysis. It is assumed that when flat cracks are closed by pressure, friction develops between their faces. The resulting limit equation is identical to the Coulomb-Navier criterion if $2\sigma_T$ is substituted for the cohesion c.

A more interesting criterion for present purposes is a generalization of the two-dimensional Griffith criterion made by Babel and Sines (1968), who allowed the ellipticity of the idealized Inglis crack to vary so as to represent straight cracks, circular holes, and all elliptic shapes in between these limits. If σ_* is the theoretical strength of the material, a and b are the major and minor semiaxes of the elliptic crack respectively, and $a/b = r$, then consideration of stresses and critical ellipse orientations gives the following relations:

$$(1 + 2r)\ \sigma_1 - \sigma_3 = \sigma_*$$

when the major axis of the ellipse is normal to the direction of σ_1, and

$$\left(\frac{\sigma_1 - \sigma_3}{2}\right)^2 + \frac{2r}{(r + 1)^2}\ \sigma_* \left(\frac{\sigma_1 + \sigma_3}{2}\right) = 0$$

for other orientations. The latter is obviously a parabolic relation of the same mathematical form as the Griffith criterion and the parabolic Coulomb-Navier-Mohr criterion. The transition between the regions of applicability for the first and second equations is given by

$$\sigma_1 + \left(\frac{r - 1}{3r + 1}\right) \sigma_3 = 0.$$

Application of the first equation to the uniaxial tensile test ($\sigma_1 = \sigma_T$, $\sigma_3 = 0$) gives

$$\frac{\sigma_*}{1 + 2r} = \sigma_T$$

and application of the second equation to the uniaxial compression test gives

$$\frac{2r}{(r + 1)^2}\ \sigma_* = \frac{\sigma_c}{2}$$

and the ratio of σ_c to σ_T is thus

$$\left|\frac{\sigma_C}{\sigma_T}\right| = \frac{4r(1 + 2r)}{(r + 1)^2}.$$

A case that is of special interest is the one where the elliptic crack becomes a circular (or cylindrical) hole, with $r = 1$. The ratio σ_C/σ_T then becomes 3, as can be shown more directly by analysis of a circular hole in an elastic plate.

An important, and often overlooked, point about Griffith theory and its derivatives is that these theories treat the onset of crack growth, implicitly defining the start of crack growth as failure. In reality, while the beginning of internal crack growth may represent the start of the failure process, the applied stress needed to reach this stage may be far below the short-term structural strength of a test specimen.

8.4 Orowan/Irwin Modifications of Griffith Theory

Griffith developed his theory primarily to explain the properties of glass, and it was believed to be generally applicable to brittle solids. However, if the strength equations which contain the surface energy γ are applied to metals or polymers the predicted strength is very much lower than the actual strength of the real material. This can be explained by plastic yielding at critical stress concentrations, which has the effect of blunting the cracks.

In the late forties, Orowan and Irwin independently modified the Griffith equation for strength by taking into account the energy dissipated in localized plastic yielding, while at the same time retaining the elastic analysis for the overall effect of a crack because the plastic yield zones were considered small relative to the crack length. Orowan substituted for the surface energy γ a term which included a specific energy for plastic working γ_p:

$$\sigma = \left(\frac{2}{\pi}\right)^{1/2} \left(\frac{E(\gamma + \gamma_p)}{c}\right)^{1/2} \approx \left(\frac{2}{\pi}\right)^{1/2} \left(\frac{E\gamma_p}{c}\right)^{1/2}$$

for plane stress. The approximation follows from the fact that $\gamma_p \gg \gamma$. Irwin expressed the same idea by denoting the critical rate of change of energy with crack length by a parameter G_c. Being an energy per unit area, G_c has the dimensions of force per unit length, and it is referred to as a crack extension force. In the Irwin formulation

$$\sigma = \left(\frac{1}{\pi}\right)^{1/2} \left(\frac{EG_c}{c}\right)^{1/2}$$

for plane stress. Thus the Orowan and Irwin expressions are identical with

$$G_c = 2(\gamma + \gamma_p).$$

At this stage in the development of theoretical ideas it may seem that the elegant simplicity of Griffith's original ideas has been clouded by black magic, but as far as the failure criteria are concerned, the general approach probably remains valid for fracture which is not perfectly brittle.

9. FRACTURE MECHANICS AND FRACTURE TOUGHNESS

The name "fracture mechanics" has come to be used, somewhat restrictively, for study of the effect of cracks on the bulk strength of solid materials. It derives from the theory of Griffith, and from the later modifications of that theory by Irwin and Orowan. These theoretical ideas have already been introduced as part of the discussion of failure criteria.

9.1 Analysis of Crack Extension

Griffith's original idea was that fracture occurred when a crack extended without limit because an increment of crack extension involved a gain of surface energy U_s less than the drop of potential energy of the surrounding elastic material U_p:

$$|\delta U_p| > |\delta U_s|.$$

Irwin and Orowan introduced the idea of energy dissipation by plastic yielding at a crack tip (δW_p) and the possibility of external work input to the system (δW_e), making the critical energy balance

$$\delta U_p + \delta W_e \geq \delta U_s + \delta W_p.$$

Since $\delta W_p >> \delta U_s$ and δW_e is, for all practical purposes, zero, the condition simplifies to

$$\delta U_p \geq \delta W_p.$$

The change of potential energy δU_p as the crack extends by an increment of length δx can be equated to a unit force G multiplied by δx:

$$\delta U_p = G \ \delta x$$

or

$$G = \frac{\delta U_p}{\delta x}.$$

This is Irwin's crack extension force, which was mentioned earlier. From elastic analysis, the critical value G_c for unstable crack extension is

$$G_c = \frac{\pi c}{E} \sigma^2$$

for plane stress, and

$$G_c = \frac{\pi(1 - \nu^2)c}{E} \sigma^2$$

for plane strain, where σ is the applied stress at failure.

Analysis of the stress distribution around an idealized crack in an elastic plate gives stress fields that are geometrically similar for geometrically similar "cracks." The absolute magnitude of a given stress component is proportional to the stress applied to the plate, σ, and it is also proportional to the square root of a characteristic linear dimension of the crack, such as the half-length of the major axis c. Thus the effects of geometric scale and stress level can be expressed by a stress intensity factor K which contains the product $\sigma \times \sqrt{c}$. For convenience, K is defined as

$$K = \sigma(\pi c)^{1/2}.$$

This is obviously another way of expressing the crack extension force G. In terms of the critical values for failure, K_c and G_c:

$$G_c = \frac{K_c^2}{E}$$

for plane stress, and

$$G_c = \frac{K_c^2 (1 - \nu^2)}{E}$$

for plane strain.

9.2 Modes of Displacement and Crack Propagation

In this summary of crack analyses the basic ideas have been developed with reference to the opening or closing of a two-dimensional crack in a plate that is under uniaxial tension or compression. However, in fracture mechanics three distinct types of crack motion are recognized (Figure 13). Mode I is the simple separation considered for the foregoing discussion. Mode II is in-plane

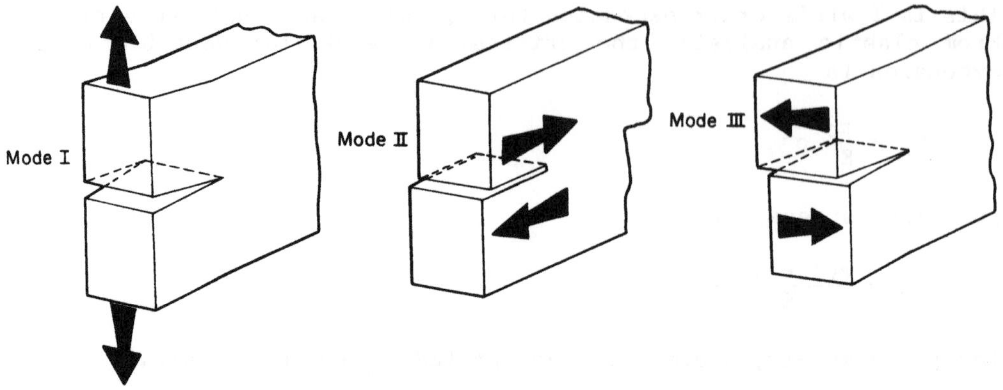

Fig. 13. Displacement and crack propagation modes.

shearing displacement, with opposite faces of a flat crack sliding across each other in the direction of the crack's major axis. Mode III involves twisting, and sliding of opposing crack faces in a direction normal to both axes of the two-dimensional crack. As far as materials testing is concerned, interest centers on Mode I, and virtually all test methods are designed to extend cracks according to Mode I. The critical value of the stress intensity factor for Mode I is denoted by K_{Ic}.

9.3 Toughness

Toughness is a poorly defined concept in engineering, but traditionally it has been associated with the capacity of a material to absorb energy before fracturing. Clearly energy alone, as represented by the area under a stress/strain curve, is not an adequate measure of toughness, since high strength and small failure strain could indicate a large energy for a very brittle material. Perhaps the best way to define and measure toughness is in terms of the ability to dissipate energy before fracturing. In other words, toughness can be associated with the integral of stress multiplied by inelastic strain, or with total strain energy minus the recoverable strain energy.

In fracture mechanics toughness, or fracture toughness, is defined as the critical value of the Irwin parameter G, which is also known as a "crack extension force." G, or G_c, has the dimensions of energy per unit area and it is, in fact, equal to the effective specific surface energy for fracture, γ, as pointed out earlier. However, with fine disregard for language and logic, many practitioners of fracture mechanics refer to the critical stress intensity factor K_c as the fracture toughness of a material, even though K_c has dimensions which have no direct relation to any reasonable definition of toughness.

PART II. THE MECHANICAL PROPERTIES OF ICE

Ice tends to display rather complicated mechanical properties, largely because it exists in nature at high homologous temperatures, commonly above 0.95, and almost always above 0.9. It behaves elastically when deformed at high strain rates, or when loaded for very brief periods, but under sustained loadings which induce deviatoric stresses and strains, ice behaves inelastically, experiencing large irreversible strains. Under constant stress ice creeps, the strain rate varying with time and strain. At any given stage of the creep process, the relationship between strain-rate and stress is strongly nonlinear, which means that the ice is viscoplastic rather than linearly viscous.

Because ice properties are so sensitive to strain rate and temperature, strength varies greatly, depending on the prevailing conditions. The general effects of multiaxial stress states, as represented by failure criteria, also change drastically as strain-rate and temperature vary.

For non-saline ice, mechanical properties have to be specified as functions of temperature, strain rate (or stress rate), porosity, grain size, and grain structure. All of these variables remain important for sea ice, and in addition there are the complications of salinity, variations of brine volume with temperature, and geometry of the pore structure.

It is not yet possible to give a clear and complete description of the mechanical properties of non-saline ice, since more and better experimental data are needed. For sea ice the situation is worse, because the material is inherently more complicated than freshwater ice, and because experimental investigations still have a long way to go. In the following notes an attempt is made to piece together a coherent story, not only by summarizing experimental data, but also by deliberately ignoring some experimental results which seem to be incorrect or misleading.

1. DEFORMATION AND FAILURE UNDER CONSTANT STRESS OR CONSTANT STRAIN-RATE

Much of what is known about the rheology and strength of ice has been derived from laboratory experiments. The two most common tests involve uniaxial compression of a right circular cylinder, with either load or displacement rate held constant. With constant load, which for small strains approximates constant stress, the test is the classic creep test, which has been favored by glaciologists interested in long-term flow under low deviatoric stress. With constant displacement rate, which approximates constant strain rate, the test is the traditional strength test, which is important in ice engineering problems involving rapid loading to failure. For comparable ranges of stress and strain-rate, the two tests ought to provide essentially the same information, as discussed later.

1.1 Constant Stress Tests

At high temperatures and relatively low stress (< 1 MPa) the creep of non-saline fine-grained ice appears to be purely ductile, with no apparent formation of internal microcracks. The general form of the creep curve, plotted according to long-established conventions, is as shown in Figure 14. If good test data are available, more informative curves can be obtained by plotting strain rate $\dot{\varepsilon}$ against either time t or strain ε, as indicated in Figure 15 for strains greater than 0.1%. A very interesting fact then emerges: the failure strain for ductile yield, as indicated by the point of minimum strain rate, is more or less constant at approximately 1% axial strain.

If the applied stress is relatively high, say greater than 1 MPa, deformation occurs both by plastic straining of the ice and by internal cracking. In fine-grained ice, cracks begin to form at

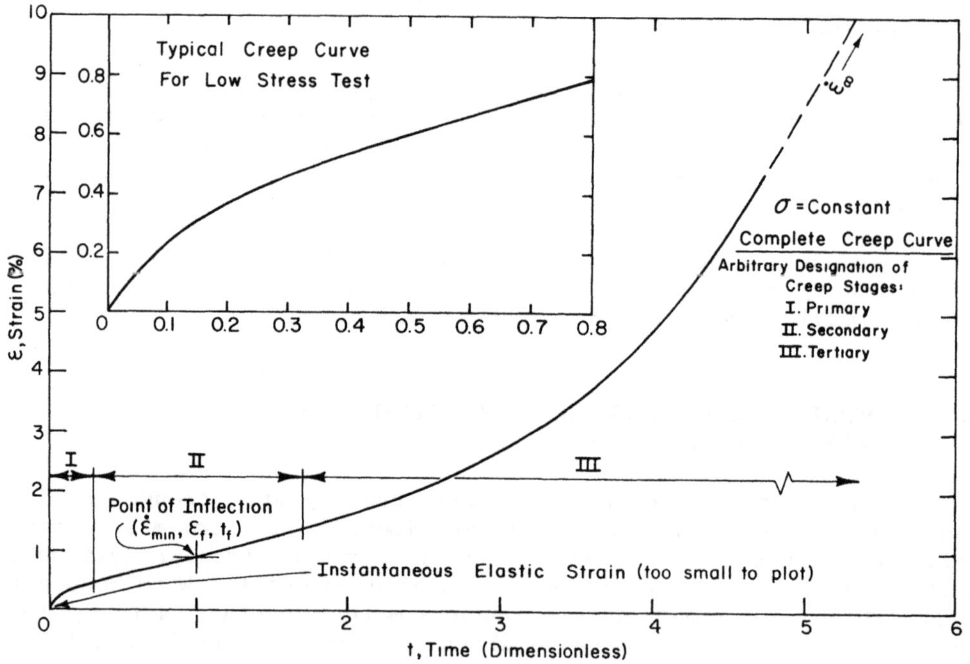

Fig. 14. Idealized creep curve for constant stress test on polycrystalline ice.

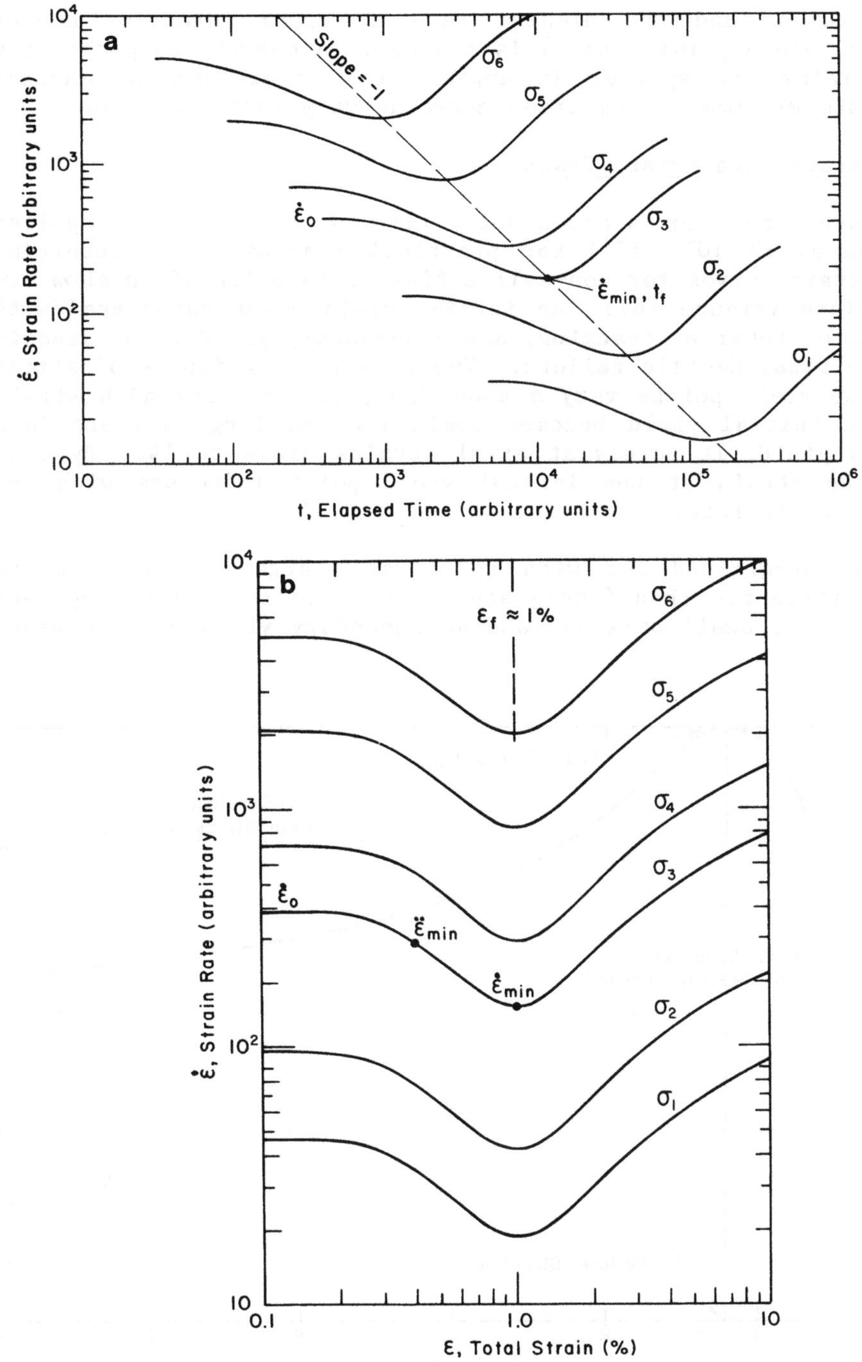

Fig. 15. Idealized creep curves for polycrystalline ice that is initially isotropic. a. Log strain rate is plotted against log time. b. Log strain rate is plotted against log strain, for strains greater than 0.1%.

small strains, and the highest rate of crack formation occurs before the yield point near 1% is reached. Although creep data for small strains are sparse, it appears that there may be another strain rate minimum in the creep curve at very small strains.

1.2 Constant Strain-rate Tests

Constant rate tests are usually limited to fairly high applied strain rates (> 10^{-7} s^{-1}) for practical reasons. The resulting stress/strain curves for non-saline fine-grained ice often show two yield points (Figure 16): an initial yield point associated with the onset of internal cracking, and a secondary yield point associated with final ductile failure. The relative magnitudes of stress at the two yield points vary a good deal, and at very high strain rates the initial yield becomes dominant. As long as there is a secondary yield, it occurs at axial strains close to 1%. By contrast, the strain at the initial yield point increases with increasing strain rate.

Columnar-grained ice with relatively large grains seems to behave differently than fine-grained isotropic ice. Yielding usually occurs at small strains, and no secondary yield point appears

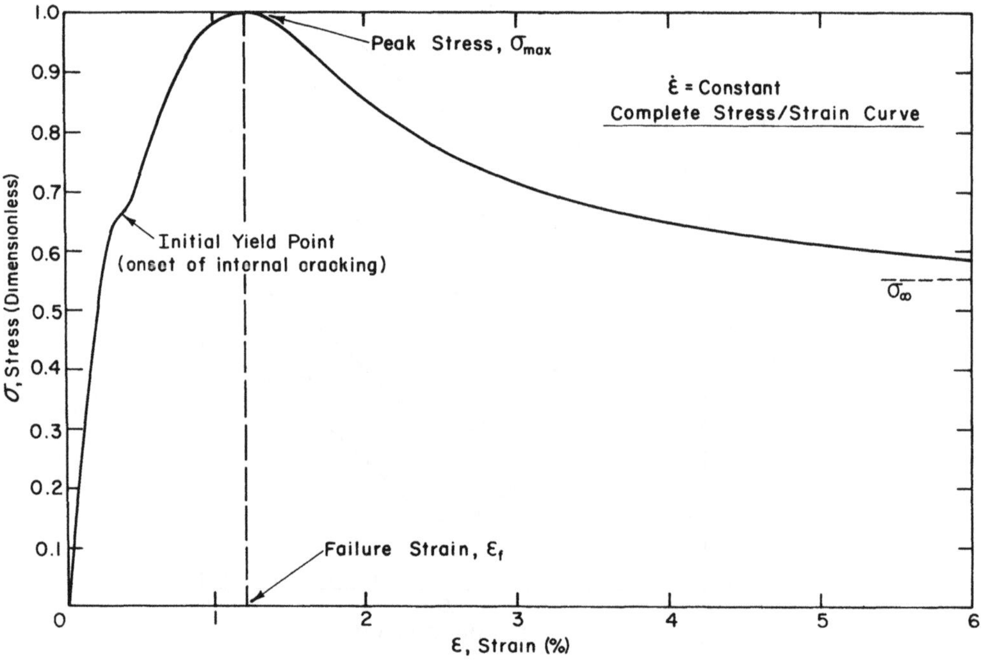

Fig. 16. Idealized stress/strain curve for constant strain-rate test on fine-grained ice.

on the stress/strain curves. In metals, a stress/strain curve with two yield points is considered to represent ductile failure, and for given conditions it is regarded as characteristic of fine-grained material. The same polycrystalline metal with larger grain size may, under the same conditions, suffer brittle fracture, giving a stress/strain curve which terminates at the first yield point.

1.3 Comparison of Data from Constant Stress Tests and Constant Strain-rate Tests

The correspondence between creep curves and stress/strain curves has been discussed in detail elsewhere (Mellor and Cole, 1982, 1983). For present purposes it is sufficient to note that a strain-rate minimum on a creep curve corresponds to a peak on a stress/strain curve. To a first approximation, the stress/strain-rate relations given by strain-rate minima in creep are the same as stress/strain-rate relations given by peaks on stress/strain curves. The significance of this for sea ice studies is that the very sparse data from creep tests can be supplemented by "creep data" derived from "strength tests."

2. MEASUREMENT OF THE MECHANICAL PROPERTIES OF ICE

Mechanical tests on ice are intended to provide information on the strength of the material (failure criteria), and on the stress/strain/time relations (constitutive relations). In principle and in practice, it is very difficult to design and execute truly valid tests. The full ranges of ideal tests required by a theoretical mechanician are not practically feasible, especially for anisotropic material, but simple tests which suffice for some other engineering materials are not always adequate for ice. Thus the real aim is to devise tests which are both valid and manageable, and this demands a thorough understanding of both the behavior of ice, and the detailed mechanics of the test. The following notes do not describe tests in detail, nor do they give recommendations on test procedures. The intention is to indicate where caution may be called for, and where detailed information can be found. Some test guidelines have been drawn up and published by an international group (Schwarz et al., 1981), but these may change as techniques and equipment improve.

2.1 Uniaxial Compression

Uniaxial compression is a much-maligned test which remains the most useful one for engineering purposes. Although apparently very simple, it is difficult to do well at high strain rates because of stress concentrations and stress field perturbations at and near the end planes, where load is applied. When a specimen in the form of a right circular cylinder is used, direct contact between the ice and steel platens is usually unsatisfactory for strain rates approaching and exceeding 10^{-3} s^{-1}, even when the specimen end planes are machined flat and square to close tolerances. Interface cushions of thin "crushable" material (such as paper or thin cardboard) may help, but cushions of "extrudable" material (such as rubber or soft plastic) are undesirable because they can introduce tensile radial stresses.

For simple field tests, special end caps filled with radially confined compliant material are useful (Haynes and Mellor, 1977). For high-quality laboratory tests, bonded end caps are satisfactory (Mellor and Cole, 1982), and if tests are to be run to large strains (several percent axial) a carefully designed dumbbell specimen with bonded end caps gives good results (details to be published later).

Measurements of axial strain are best made within the central section of the specimen itself for small strains (< 1%), but for large strains demountable displacement transducers have to be attached to the end caps to avoid undue disturbance of the transducers' attachments. So far, bonded strain gages have not been used successfully, but with some development effort this situation could change, especially on large specimens where some surface disturbance can be tolerated.

Radial or circumferential strain should be measured at the mid-section of the specimen, noting that the mid-section moves relative to the loading platens and the end caps as the specimen strains axially. If the radial strain transducer is in contact with the ice, its contact points must be free to travel axially with the ice. If non-contact proximity gauges are used, some axial displacement between the pickup and the targets can be tolerated.

2.2 Uniaxial Tension

The only reliable way to measure the uniaxial tensile strength of ice is by direct application of uniaxial stress in the classic test. Tests depending on flexure, or on diametral compression of discs, cylinders or annuli, are indirect, in that the stress field has to be calculated from elastic theory, and a failure criterion has to be assumed. In other words, in an indirect test it is necessary to assume mechanical properties in order to measure mechanical properties.

In making a uniaxial tensile test on ice, the main difficulty is gripping the specimen and straining it while maintaining true uniaxial stress in the test section. The specimen is attached to end caps by mechanical grips or by adfreeze bonding. Whatever method is used, it will certainly create a triaxial stress field near the end caps, and if the specimen is a simple right circular cylinder it is likely to fracture near the end caps, in a stress field which is not uniaxial. Thus it is necessary to use dumbbell specimens for high quality laboratory tests. The fillet radius of a dumbbell specimen controls the stress concentrations in the transition zone. If D is the diameter of the specimen "neck," a fillet radius of 8D gives a stress concentration factor of 1.02, and a fillet radius of 1.4D gives a factor of 1.05. Photoelastic model tests and practical experience indicate that a fillet radius of 2D is satisfactory for ice (Hawkes and Mellor, 1972). Precise dimensional tolerances for the specimen and the pulling device are important, and avoidance of eccentricity or twist in the pulling systems is essential.

The considerations relating to measurement of axial and radial strain are essentially the same as those for uniaxial compression.

2.3 Shear Tests

Shearing deformation is obviously very important. However, direct shear tests tend to be poorly understood, and in many cases simplistic and ill-conceived.

In plane stress or plane strain, pure shear exists when the in-plane principal stresses are equal in magnitude and opposite in sign. It is not easy to make a test which gives these conditions.

The usual experimental approach is to impose so-called "simple shear," using a prismatic specimen. A slab or block of the test material is deformed in such a way that its longitudinal cross section is changed from a square to a rhombus, or from a rectangle to a parallelogram, with zero normal stress on the surfaces which are subject to traction. This can be done with a suitably designed parallelogram frame, but there will be some perturbation of the stress field near the surfaces of the shearing plates. One research group has sheared ice by means of parallel plates which are maintained at constant separation by steel balls running in vee-grooves. This is suitable for shearing a deck of cards, but not an isotropic deformable solid.

Direct shear tests are often intended to measure the effect of normal stress on shear resistance. The general idea is to distort a specimen as for "simple shear," while applying normal stress to the traction surfaces. The traditional shear box used in soil mechanics is intended for this purpose, but it perturbs the stress field and forces failure along a particular plane. Similarly, tests which use a segmented steel cylinder to fail a cylindrical specimen in double shear are unacceptable.

A more appealing method for applying shear deformation directly is torsion of a hollow cylinder, with or without axial forces or surface pressures. In practice, torsion tests on ice are very difficult.

It is not necessary to apply surface tractions or to induce direct shear in order to define deviatoric stress/strain relations and failure criteria. Tests which induce deviatoric strains by controlling principal stresses can provide the required information.

2.4 Beam Flexure Tests

Beam flexure is used to measure the strength of ice, and the elastic modulus. Tests are made in the laboratory on small beams, and field tests are made on large beams (often in situ), using either simple beams (in 3-point or 4-point bending) or cantilevers. Since it is usually assumed that the ice is linearly elastic, with equal moduli in tension and compression, the tests should be of very short duration (of the order of 1 s or less). For beams tested in situ, there are often steep temperature gradients within the beam, and buoyancy effects may be important. The ends of in situ beams sawn in an ice sheet are not rigidly encastré; flexure of the beam can flex the adjacent ice to some extent. Thus the proportions of the beam are important, and stress-concentrating corners at the beam root have to be considered. The IAHR recommendations suggest that beam length should be 7 to 10 times the ice thickness, and beam width should be 1 to 2 times the ice thickness (Schwarz et al., 1981).

A beam breaks when the ice fails in tension at the convex surface, and the resulting crack propagates through the thickness of the beam. The critical tensile stress is first reached at the surface, and the crack propagates through ice which has a stress gradient. The calculated value of this tensile stress is usually significantly different from the uniaxial tensile strength of bulk material, not only for ice but for most non-metallic materials. Thus it is termed the "flexural group" or "modulus of rupture" to avoid confusion.

The elastic assumption is justifiable for high strain rates and low temperatures, but it is completely unrealistic for low strain rates and high temperatures. Thus beam flexure tests cannot really be used to determine variations of tensile strength with strain rate or temperature.

2.5 Diametral Compression Tests

When a solid cylinder or a disc of elastic material is squeezed from opposite ends of a diameter, the center section of the diametral plane experiences a tensile principal stress at right angles to the loading direction, and a compressive principal stress parallel to the loading direction. The magnitude of the compressive stress is three times that of the tensile stress. If contact stresses at the loading platens are properly controlled, a disc or cylinder of brittle material will fail by splitting, with the crack initiating at the center of the specimen.

For a "Griffith type" material, in which the ratio of uniaxial compressive strength σ_C to uniaxial tensile strength σ_T is 8 or greater, the tensile stress is equal to the uniaxial tensile strength, even though failure occurs in a biaxial stress field (see section on Griffith Theory). However, in ice $\sigma_C/\sigma_T < 8$, and the critical stress induced by elastic compression of a disc or cylinder is not equal to the uniaxial tensile strength. The calculated tensile failure stress is $2P/\pi dt$, where P is applied force at failure, d is disc diameter, and t is disc thickness. This stress is much smaller than σ_T, and the test result probably has to be interpreted in terms of the failure envelope for a biaxial stress field, with $\sigma_1 = -\sigma_3/3$, and $\sigma_2 = 0$.

The other diametral compression test which has been applied to ice involves diametral compression of an annulus. This is a rather complicated test, and as far as is now known it does not give a useful measure of tensile strength for ice or any other material (Mellor and Hawkes, 1971).

2.6 Conventional Triaxial Tests

The test which is usually known as a triaxial test is similar to a uniaxial compression test, but with radial pressure applied to the specimen by means of gas or liquid in a confining cylinder. The stress state is triaxial, but $\sigma_2 = \sigma_3$. The test is usually performed as a "strength test," with σ_1 increasing at approximately constant rate, and frequently with some confusion about the time relations for σ_2 and σ_3. The test can also be carried out as a constant stress creep test.

The triaxial test has not been used to any great extent in ice testing, but is potentially an important test. New equipment and techniques are being developed at CRREL.

2.7 Mixed Tension/Compression Triaxial Tests

An interesting variant of the conventional triaxial test is a test in which cell pressure alone is used, but the test specimen has a dumbbell shape. Pressure acting axially on the flared ends of the specimen induces axial tension, while at the same time the pressure exerts a radial compressive stress. Thus the specimen fails with σ_1 tensile and σ_2, σ_3 compressive, giving a point on the failure envelope for the tension-compression quadrants of principal stress space. The ratio of σ_1 to σ_2, σ_3 can be varied by changing either the neck diameter of the specimen, or the diameter of the end caps. A pressure cell for making tests of this kind was developed by I. Hawkes and D. Garfield at CRREL, and tests were made by F.D. Haynes (1973).

2.8 Polyaxial Compression Tests

In a true triaxial, or polyaxial, test $\sigma_1 \neq \sigma_2 \neq \sigma_3 \neq 0$. The test specimen has to be a cube, and all faces have to be free of surface tractions. For tests on a "creepy" material such as ice, the loading actuators which apply σ_1, σ_2 and σ_3 all have to be programmable with respect to time.

The only known polyaxial tests on ice are those done by Häusler (1981). Load is applied to each face of the specimen by a "brush" platen, which consists of a cluster of slender metal columns, each free to deflect sideways independently of its neighbors. As there is no access to any of the surfaces, strains are obtained directly from platen displacements.

2.9 Fracture Toughness Tests

Although there are three distinct modes of displacement for crack propagation (see section on Fracture Mechanics), only the tensile Mode I is used in testing. The general procedure is to propagate a crack in a tensile stress field, and sometimes to arrest the crack.

A form of test that has been used widely for ice involves four-point bending of a beam into which a "crack" has been cut on the tension side. Another popular test is one based on the so-called Compact Tension Specimen. This is a block with a sawcut, and it is pulled, or torn, apart. For crack arrest measurements a slender wedge is forced into a sawcut, propagating a crack which eventually stops as the stress drops back, to a low level. All of these tests assume linear elastic response in the test material, and it seems likely that they could give misleading results for low strain rates and high temperatures.

3. THE ELASTIC MODULI OF ICE

Since ice is markedly viscoelastic at the temperatures and strain rates which have often been used in conventional strength tests (say -5° to -10°C and 10^{-4} to 10^{-3} s^{-1}), a conventional stress/strain curve does not give an accurate measure of the elastic modulus. For example, a uniaxial test under constant strain rate, either in compression or tension, is likely to yield a stress/strain curve which is "convex up," i.e. $d\sigma/d\varepsilon$ decreases as σ

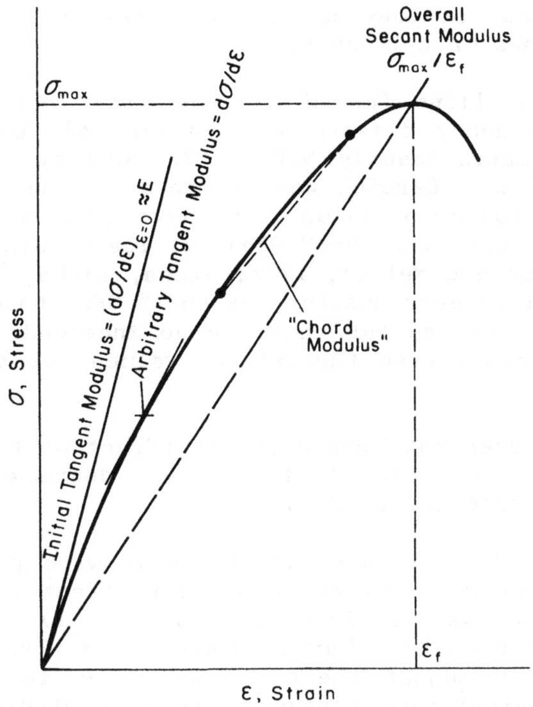

Fig. 17. Modulus definitions for nonlinear stress/strain characteristics.

and ε increase (Figure 17). Thus the "tangent modulus," given by $d\sigma/d\varepsilon$ at some arbitrary point on the curve, may have little relation to the true Young's modulus. Similarly the overall secant modulus, representing the average slope of the stress/strain curve from the origin to the peak stress, may be very different from the true Young's modulus. In principle, the initial tangent modulus, $(d\sigma/d\varepsilon)_{\varepsilon=0}$, should be equal to Young's modulus, but with typical test data the initial tangent modulus cannot be read accurately from a stress/strain curve. In fact, unless strain measurements are made on the test specimen itself (instead of across the platens of the testing machine), the initial tangent modulus cannot be measured at all. However, with good test data, $d\sigma/d\varepsilon$ can be calculated as a function of σ or ε, and when $d\sigma/d\varepsilon$ is plotted against σ or ε the zero intercept gives a value which ought to approximate Young's modulus.

A much simpler way to study Young's modulus and its variation with temperature, salinity, air content and so forth is to use high frequency vibrational methods. These may involve the propagation of small-amplitude waves or pulses in laboratory specimens or large natural ice masses, or they may involve flexural or torsional oscillations of beams or cylinders.

For polycrystalline ice of low porosity (density $\rho \rightarrow 0.917$ Mg/m^3), high frequency dynamic measurements of Young's modulus E give values of approximately 9.0 to 9.5 GPa in the temperature range -5° to -10°C. Careful measurements of the initial tangent modulus for uniaxial compression tests have given values very close to the dynamic values of E (9-10 GPa) for ice of similar type (e.g. Gold, 1958; Hawkes and Mellor, 1972; Sinha, 1978a,b). Much higher values of initial tangent modulus (up to 25 GPa) have been reported by Haynes (1978), but so far there is no independent confirmation (the highest values exceed theoretical values for a perfect monocrystal).

Temperature does not have a strong effect on the true Young's modulus of non-saline ice. As temperature decreases, E increases nonlinearly, as indicated in Figure 18.

The value of E does vary significantly with porosity in non-saline ice, and we would expect a similar effect in sea ice. The porosity of sea ice derives from two sources: 1) the air bubbles, which do not vary much in volume as temperature changes, and 2) the brine cells, which adjust their volume so as to preserve phase equilibrium as temperature changes. In most studies on sea ice, mechanical properties have been related to brine volume (as calculated from the salinity and the temperature), but air volume has not been measured in many cases. Thus we have to use "brine porosity" instead of total porosity in examining most data sets.

Figure 19 is a plot of Young's modulus E against porosity n. A data band is given for cold, non-saline ice, in which the voids contain only air. Over this band are located data envelopes for tests on small saline specimens. The results of Langleben and Pounder (1963), and also of Abele and Frankenstein (1967), were obtained by propagation of high frequency pulses; those of Tabata (1958) were obtained by flexural vibration of small beams. These results for sea ice agree quite closely with the results for non-saline ice, in spite of the fact that some of the pore volume is occupied by water. The tendency for the sea ice values to decrease more rapidly than the fresh ice ones can probably be explained by the fact that temperature is an implicit variable for the sea ice. The range of n could represent a temperature change of 20°C, and this could change E for the ice matrix by 10% or so.

The results of Kohnen (1972) were obtained from seismic wave data for broad expanses of sea ice. Those of Brown (1963), and of Anderson (1958), were also obtained from field measurements, but they involved flexural waves in the ice sheet. The data of Brown and Anderson, and perhaps those of Kohnen, appear to represent

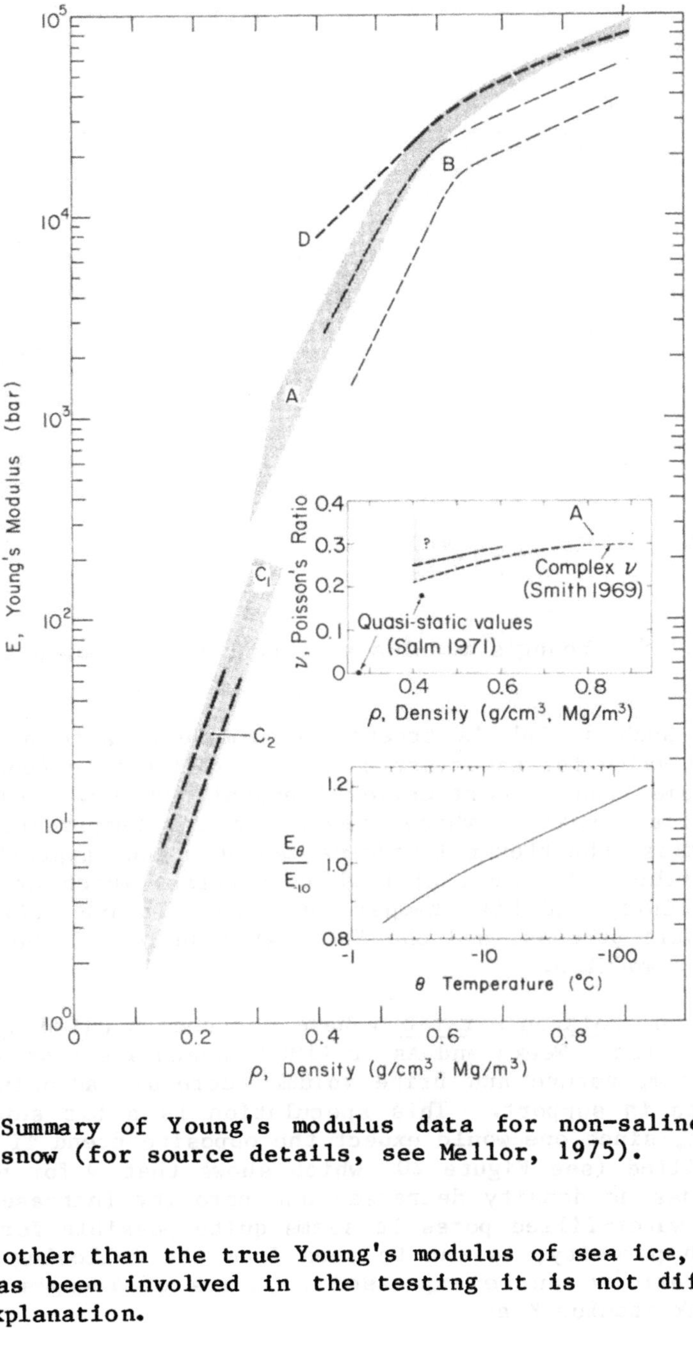

Fig. 18. Summary of Young's modulus data for non-saline ice and snow (for source details, see Mellor, 1975).

something other than the true Young's modulus of sea ice, and where flexure has been involved in the testing it is not difficult to find an explanation.

A sheet of sea ice is not a homogeneous elastic slab. It is often subject to steep temperature gradients, with the top cold and of low porosity, and the bottom warm and of high porosity. If

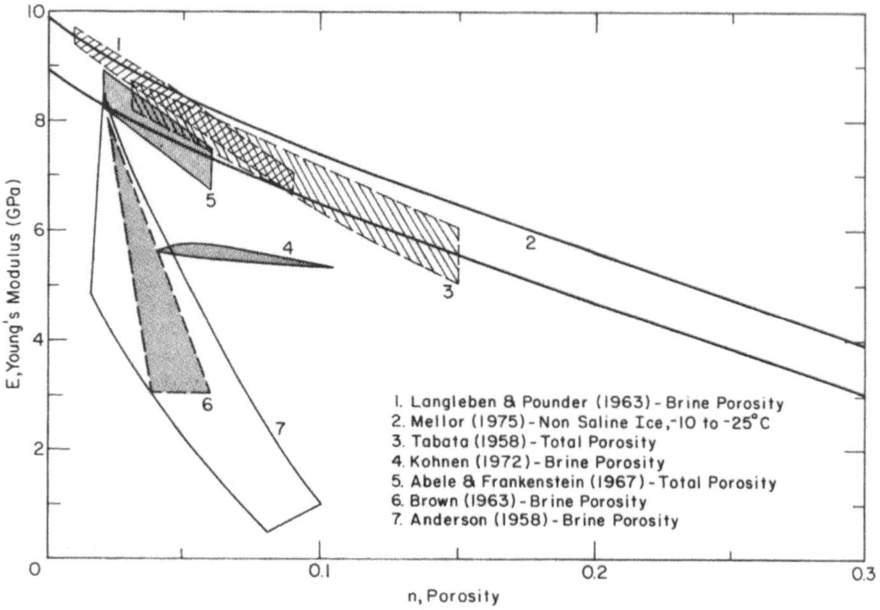

Fig. 19. Young's modulus as a function of porosity.

flexure of such a slab is treated by simple elastic beam theory, the deduced value for the elastic modulus will not be Young's modulus, but some kind of "effective" flexural modulus. For very low values of ice porosity, which prevail at low temperatures and/or low salinities, the flexural modulus may be a good approximation to Young's modulus. However, at high porosities, which are produced by high salinity and high temperature, the flexure will not be a purely elastic process, and the flexural modulus will be far lower than Young's modulus.

Values of Poisson's ratio ν have not been studied systematically for sea ice. Weeks and Assur (1967) speculate that ν will decrease as temperature and brine volume decrease, adducing Russian seismic data in support. This speculation is a bit suprising at first sight, since one would expect the opposite trend if the pores were air-filled (see Figure 20, which shows that ν for non-saline ice decreases as density decreases and porosity increases). However, with water-filled pores it seems quite possible for ν to increase with porosity, as can be seen from the following argument. Poisson's ratio ν can be expressed in terms of Young's modulus E and the bulk modulus K as

$$\nu = \frac{3K - E}{6K} = \frac{1}{6}\left(3 - E/K\right).$$

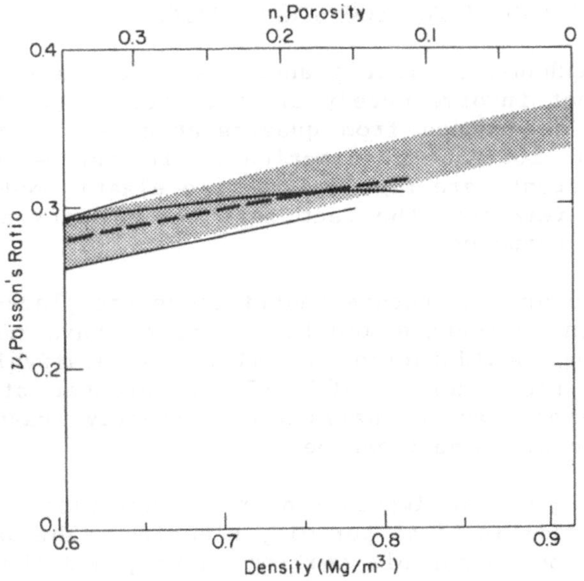

Fig. 20. Summary of data for Poisson's ratio of non-saline ice as a function of porosity (see Mellor, 1975, for details of data sources).

Since the bulk modulus of water is not greatly lower than that of ice, a small increase in the volume of water-filled pores will not have much effect on the overall value of K. Thus the variation of ν with porosity n will be controlled largely by the corresponding variation of E. Since E decreases as n increases, ν would thus increase as n increases. However, one would not expect much variation of ν. The value of ν for non-saline ice of very low porosity is about 0.33 ± 0.03, and variation with porosity in sea ice is likely to be within the limits of uncertainty for the pure ice value.

The bulk modulus K and the shear modulus G have not received much explicit attention in sea ice studies, but if it is accepted that ν will not vary much, then these constants can be given as simple multiples of E. Taking $\nu = 1/3$:

$$K = \frac{E}{3(1-2\nu)} = 1.0\ E$$

$$G = \frac{E}{2(1+\nu)} = 0.375\ E.$$

4. "EFFECTIVE" MODULI FOR LOW STRAIN RATES

In ice mechanics, elastic analysis is often applied in problems that do not involve purely elastic processes. Pseudo-elastic constants are determined from quasi-static tests, and from field observations of low-rate deformations. It can be very confusing when these constants are regarded as true elastic moduli, but there is no getting away from the fact that the numbers are useful for many practical purposes.

If no rate or temperature restrictions are placed on tests and processes, then "effective moduli" cease to have much relevance. However, if the consideration is limited to relatively high strain rates, say of the order of 10^{-4} s^{-1} or greater at temperatures below -10°C, then elastic analysis is perfectly reasonable and appropriate effective moduli can be used.

Figure 21 gives an impression of the approximate magnitude of the effective modulus E' as porosity n varies. The data by Vaudrey (1977) seem to be reasonable working values, and they are of comparable magnitude to the values obtained in a different way by Anderson and Brown (see Figure 19).

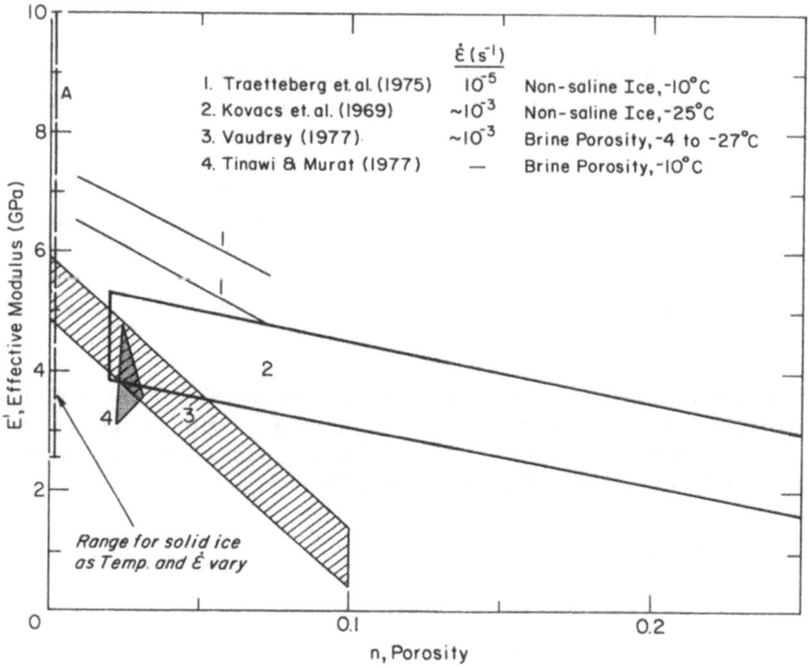

Fig. 21. Summary of data for effective modulus E' plotted against porosity.

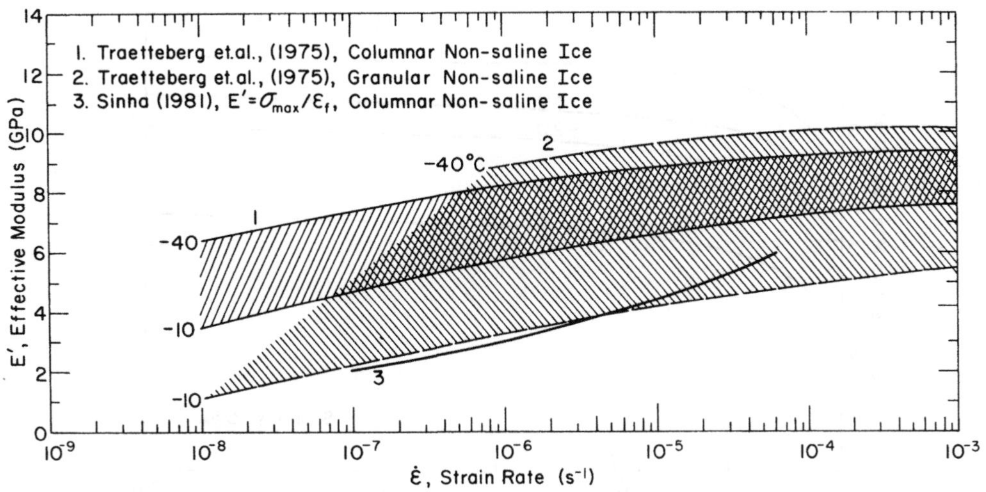

Fig. 22. Summary of data for effective modulus E' plotted against strain rate.

Effective moduli should always be used with great caution, recognizing that E' for any given specimen can vary considerably as temperature and/or strain rate vary, and even the way it is defined can vary at the discretion of the investigator. Figure 22 shows how E' varies with strain rate at various temperatures, using data for freshwater ice in the absence of information for saline ice. At low temperatures and high strain rates, E' is approximately equal to the true Young's modulus E, but at low strain rates and/or relatively high temperatures, E' may be only 25% to 30% of E. Lainey and Tinawi (1981) showed similar behavior in ice of 5 $^0/_{00}$ salinity, presenting their values of E' for beam flexure as functions of stress rate (Figure 23).

Sinha (1981c) has shown that the apparent value of the effective modulus can vary with the characteristics of the loading and measuring system. Of particular concern is the discrepancy between the apparent strain rate, as sensed across the specimen and its end contacts, and the true strain rate within the specimen. For rapid tests, the average stress rate may give a more reliable estimate of true strain rate, assuming linear elastic response.

The variation of E' with strain rate $\dot{\varepsilon}$ means that there will also be variation of E' with frequency f in the case of oscillation or vibration. Figure 24 indicates the rate of variation deduced by Sinha (1978a) from the strain-rate data for freshwater ice summarized in Figure 22.

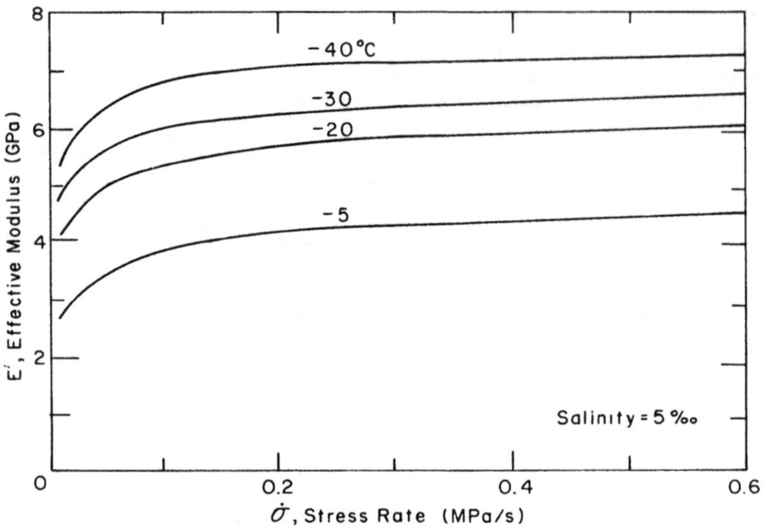

Fig. 23. Variation of E' with stress rate and temperature (Lainey and Tinawi, 1981).

It was mentioned earlier that the true Young's modulus E for freshwater ice is not very sensitive to temperature, but the same general statement cannot be made for the effective modulus E', even in freshwater ice. Figure 25, from Sinha (1978a), shows how E' for columnar-grained ice varies with temperature for various load durations and two stress levels. The values of E' in this figure are normalized with respect to the modulus E for very rapid loading. Absolute values of E', obtained from beam tests on ice of $5\,^0/_{00}$ salinity, are indicated in Figure 26, which is based on data by Lainey and Tinawi (1981). In the case of saline ice, it has to be kept in mind that a change of temperature implies a change of porosity if the ice is of fixed salinity.

When "effective" values of Young's modulus are employed, we need also "effective" values of Poisson's ratio, ν'. Effective values of ν' are only just beginning to enter the literature, but some deductions can be made. As material response becomes more ductile, the usual trend is for ν' to increase up to a limit of 1/2, which represents incompressible flow, with E'/K tending to zero. When all the pores of the ice are filled with water, there is no reason for the bulk modulus K to vary much with porosity, temperature, or strain rate, and for a first approximation it can be assumed equal to the true Young's modulus for zero porosity, E_0. Thus the effective Poisson's ratio for isotropic material can be expressed as:

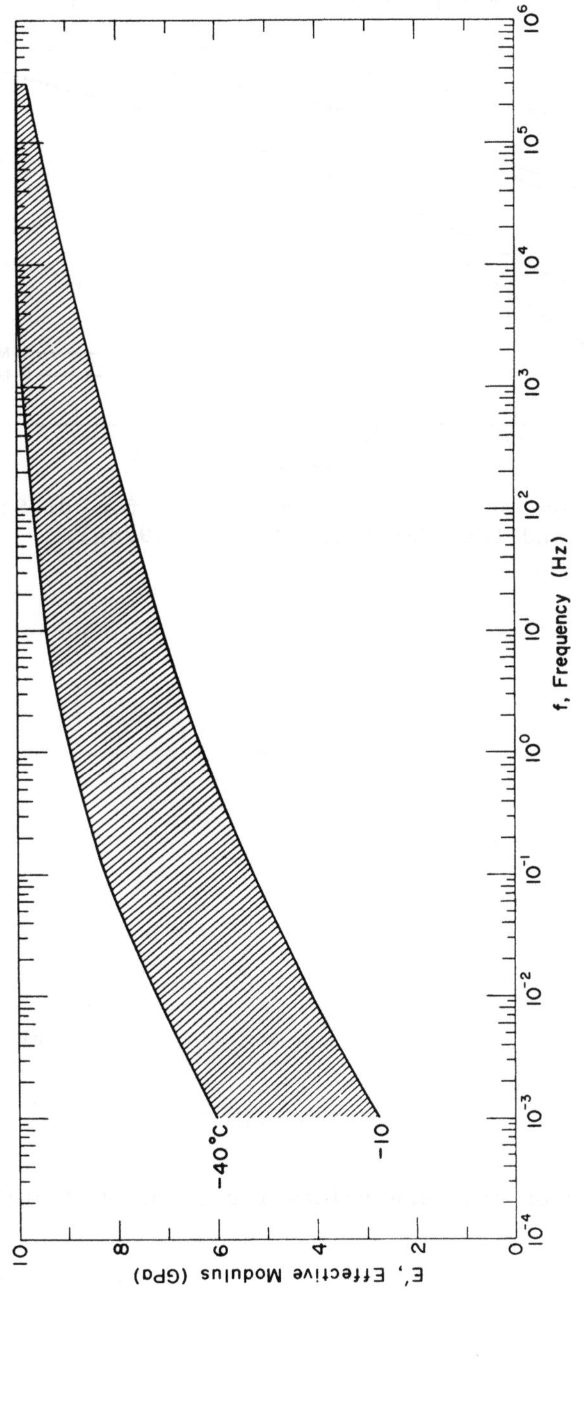

Fig. 24. Variation of E' with frequency, as deduced from strain rate data (after Sinha, 1978a).

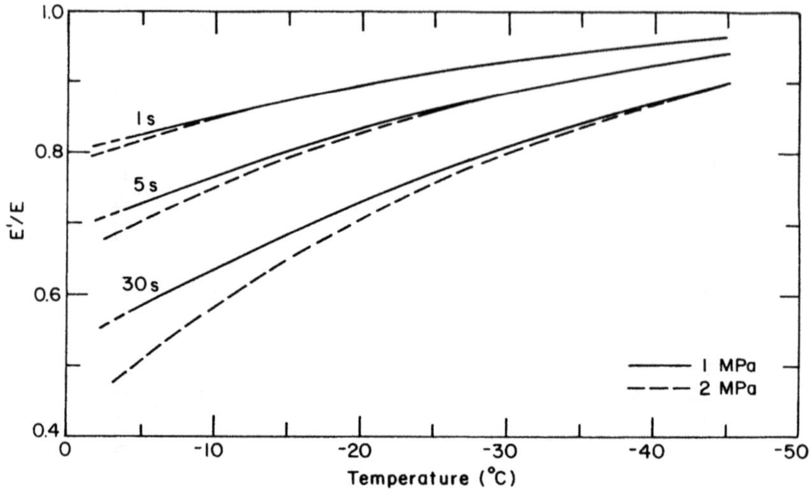

Fig. 25. Variation of E' with temperature for different stress levels and load durations (Sinha, 1978a).

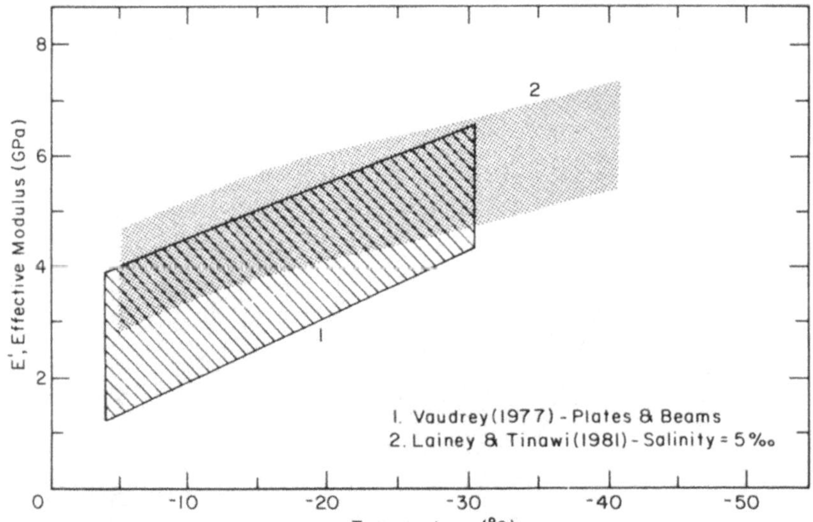

Fig. 26. Summary of absolute values for E' plotted against temperature.

$$\nu' = \frac{1}{2} - \frac{1}{6}\frac{E'}{E_o}$$

which gives a systematic variation between the limits 1/3 and 1/2. If the ice is anisotropic, more extreme values of ν' for particular directions are possible.

The expected general trend is confirmed by data from beam flexure tests made on saline ice by Murat and Lainey (1982). Values of ν' decreased from a value of approximately 0.5 as strain rate and stress rate increased. At the highest test rates (0.6 MPa/s, and up to $1.6 \times 10^{-4}\ s^{-1}$), ν' was between 0.35 and 0.4. Mean values for high rate testing decreased with decreasing temperature, from about 0.4 at -5°C to 0.37 at -30° and -40°C. This agrees with the prediction of Weeks and Assur (1967) that was discussed earlier in the context of "true" Poisson's ratio.

5. UNIAXIAL COMPRESSIVE STRENGTH

Uniaxial compressive strength σ_c is a very important parameter which appears easy to measure. In fact, accurate measurement of σ_c is very difficult for highly brittle materials. When a simple right circular cylinder is pressed between platens, the stress field is triaxial rather than uniaxial near the end planes, and stress concentrations develop at the platen/specimen interface. This means that many of the reported values of σ_c for high loading rates and/or low temperatures are probably wrong.

In assessing the uniaxial compressive strength of ice, the main variables are:

(1) Ice type (grain structure and texture, porosity, salinity, solid inclusions)

(2) Strain rate (or, with reservations, stress rate)

(3) Temperature (which automatically varies with porosity in saline ice).

For a start, we consider the strength of non-saline ice.

In the present context, uniaxial compressive strength will be regarded as the maximum stress that can be developed at a specified strain rate. Studies on fine-grained isotropic freshwater ice (Mellor and Cole, 1982; 1983) have shown that the minimum strain rate developed in a constant stress creep test gives essentially the same limiting relationship between stress and strain rate as the constant strain rate test. We can therefore combine data from

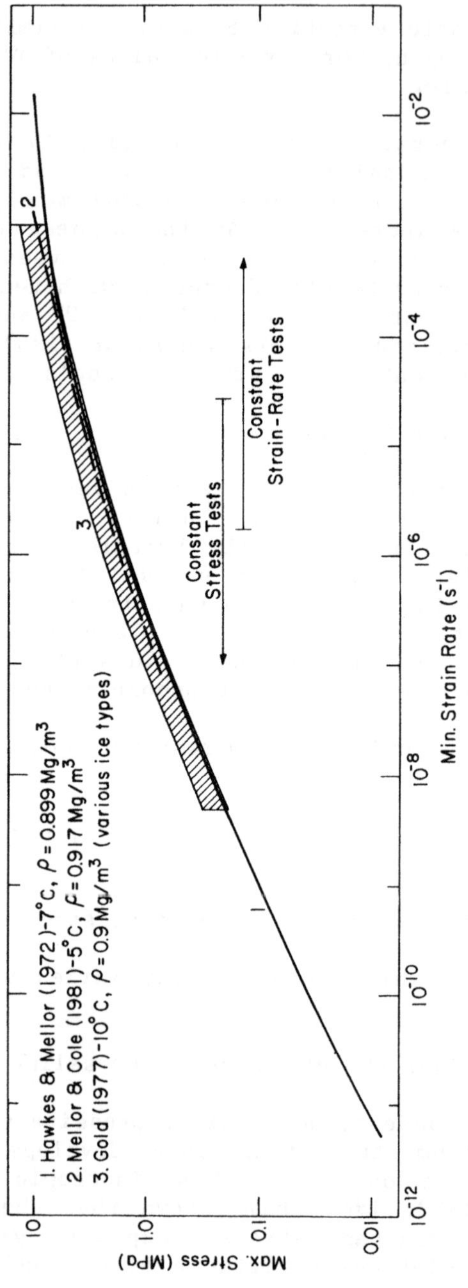

Fig. 27. Maximum yield stress as a function of strain rate for non-saline ice.

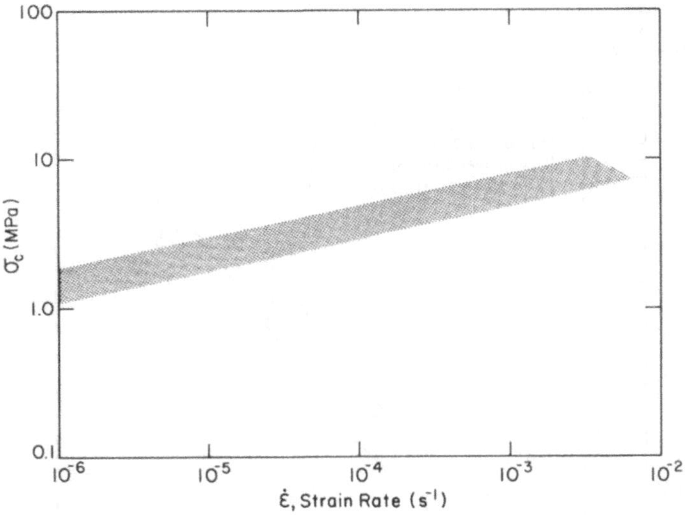

Fig. 28. Uniaxial compressive strength of sea ice as a function of strain rate (after Wang, 1979a).

creep tests and strength tests to define the relationship between strength and strain-rate over a wide range. Figure 27 shows that at low strain rates, say 10^{-7} to 10^{-5} s^{-1}, strength increases with about the one-third power of strain rate, while at higher rates, say 10^{-5} to 10^{-3} s^{-1}, strength appears to increase with a one-fourth, or smaller, power of strain rate. This relation is the inverse of the glaciologists' "flow law," which requires minimum creep rate to be proportional to the stress raised to a power of approximately 3.

Figure 28 gives some data for saline ice by Wang (1979a). These results were selected from Wang's paper because they fit the pattern of an approximate fourth-power relation between strain rate and stress. More detailed results by Wang (1979b) provide ample confirmation of the general trend, although the author fits rather complicated trend lines to the data (Figures 29, 30, 31). Some data by Schwarz (1971) are shown in Figure 32. Data for multi-year sea ice of low salinity are just beginning to appear in the open literature (Cox et al., 1984). Values of σ_c scatter widely because of great variability in the ice structure, but variations of σ_c with strain rate are not inconsistent with the power relation just mentioned (Figure 33).

For strain rates greater than 10^{-3} s^{-1}, many investigators have found that strength decreases as strain rate increases, and this is widely regarded as characteristic of brittle fracture. Such a trend would be in keeping with the ideas of fracture mechan-

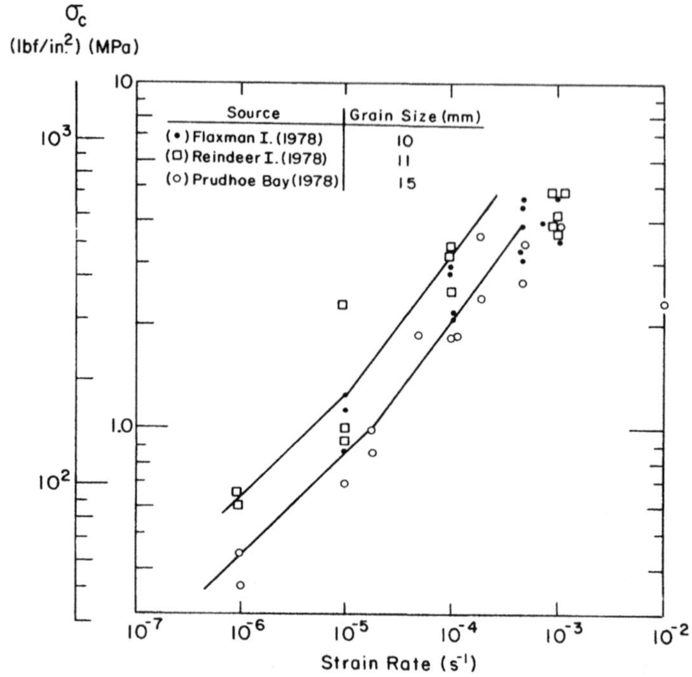

Fig. 29. Variation of σ_c with strain rate in granular sea ice at −10°C. Data are given for two ranges of grain size (after Wang, 1979b).

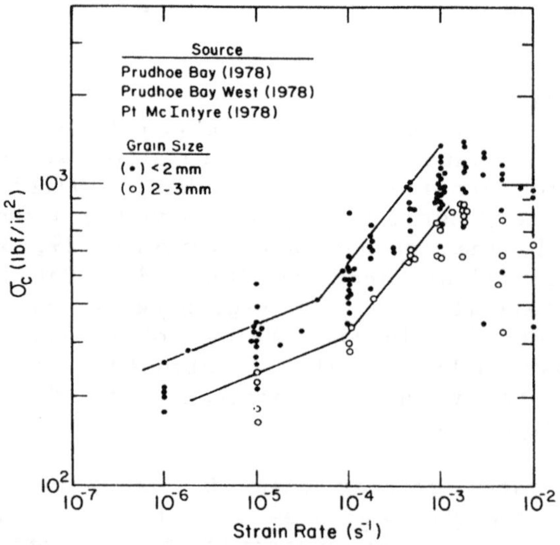

Fig. 30. Variation of σ_c with strain rate in unoriented columnar sea ice at −10°C. Samples were taken from the upper part of the ice sheet (after Wang, 1979b).

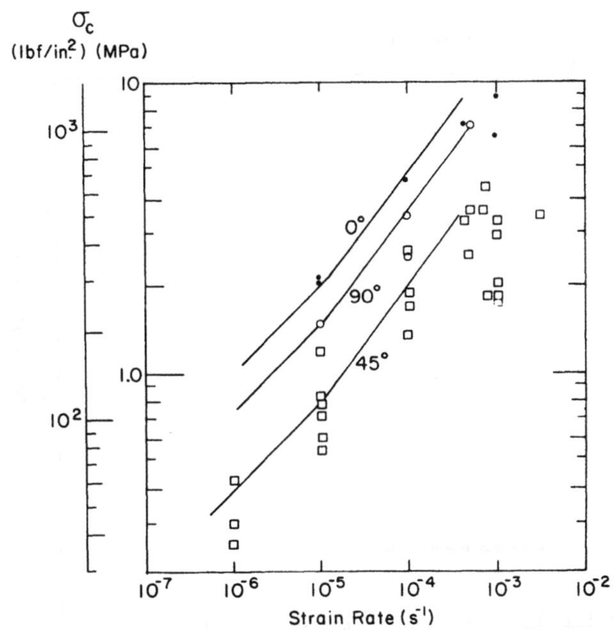

Fig. 31. Variation of σ_c with strain rate in columnar sea ice at −10°C. The three data sets show the effect of crystal orientation. The lowest values of σ_c are measured when the c-axes are at 45° to the direction of the major principal stress. Grain size for these specimens was 11 mm (after Wang, 1979b).

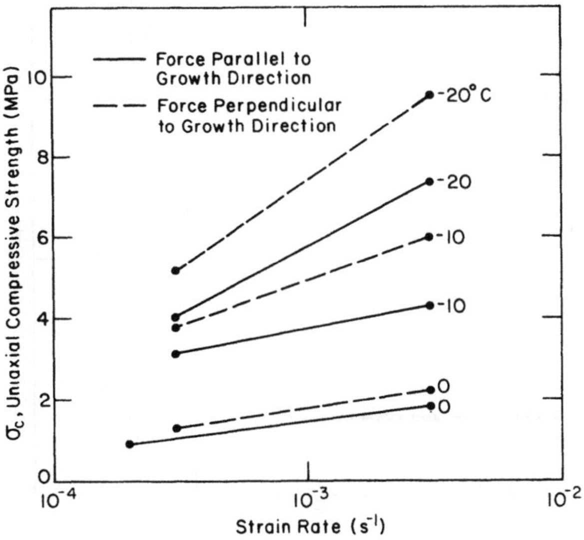

Fig. 32. Variation of σ_c with strain rate, temperature and grain orientation (data selected from Schwarz, 1971).

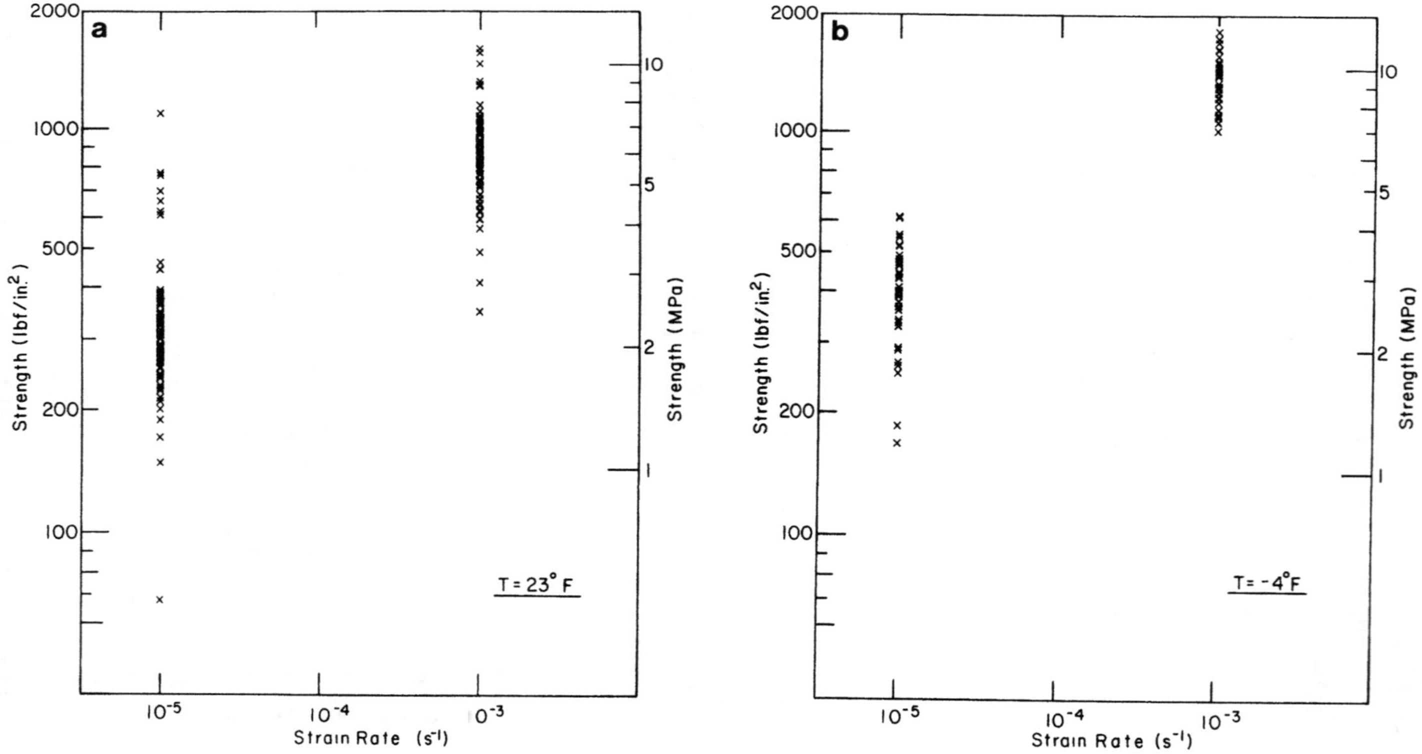

Fig. 33. Values of σ_c for multi-year sea ice at strain rates of 10^{-5} and 10^{-3} s^{-1}, and at temperatures of -5°C and -20°C. The wide scatter is produced by great variability in grain and pore structures (Cox et al., 1984).

ics, but in some cases the effect may be caused by inadequacies in specimen preparation and imperfections in test technique. Present indications are that it will be necessary to use carefully designed, and precisely machined, dumbbell specimens in order to get valid test results at very high strain rates. For these reasons, the high rate data recorded in the literature will be deliberately disregarded in this survey.

The next variable for consideration is temperature, and we can first consider how non-saline ice responds.

At very low strain rates, where ice fails by creep rupture, it has long been the practice among glaciologists to characterize the effect of temperature on stress/strain-rate relations by an Arrhenius equation, with an activation energy of approximately 70 kJ/

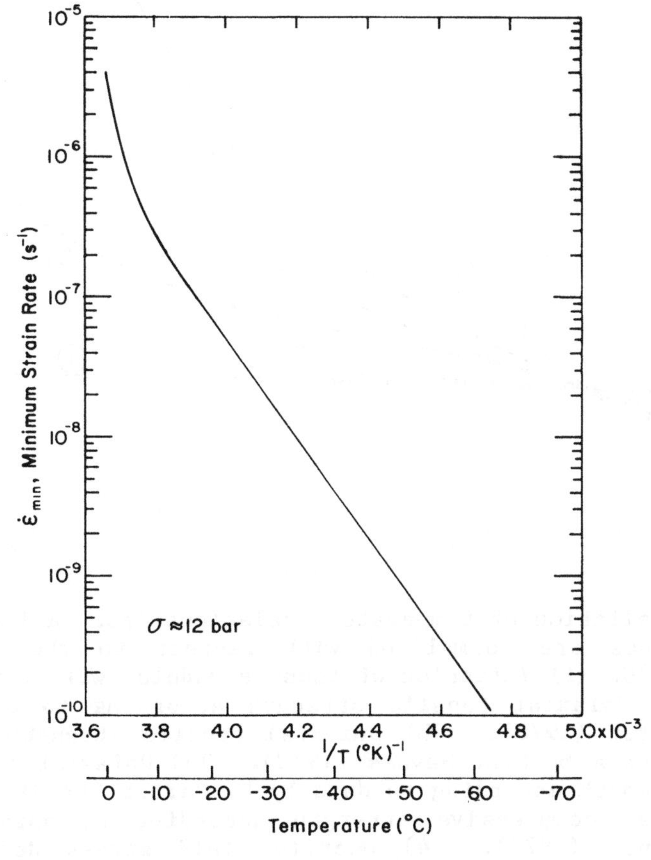

Fig. 34. Empirical relation between minimum strain rate and temperature for high stress creep (after Mellor and Testa, 1969).

mole. Such a relation is definitely not valid above −10°C, and the complete empirical relation derived from experiments (e.g. Figure 34) should be used to deduce how rupture stress varies with temperature. If creep tests give a relationship between applied stress and minimum strain rate in the form

$$\dot{\varepsilon}_{min} = A\ \sigma^n\ f(\theta)$$

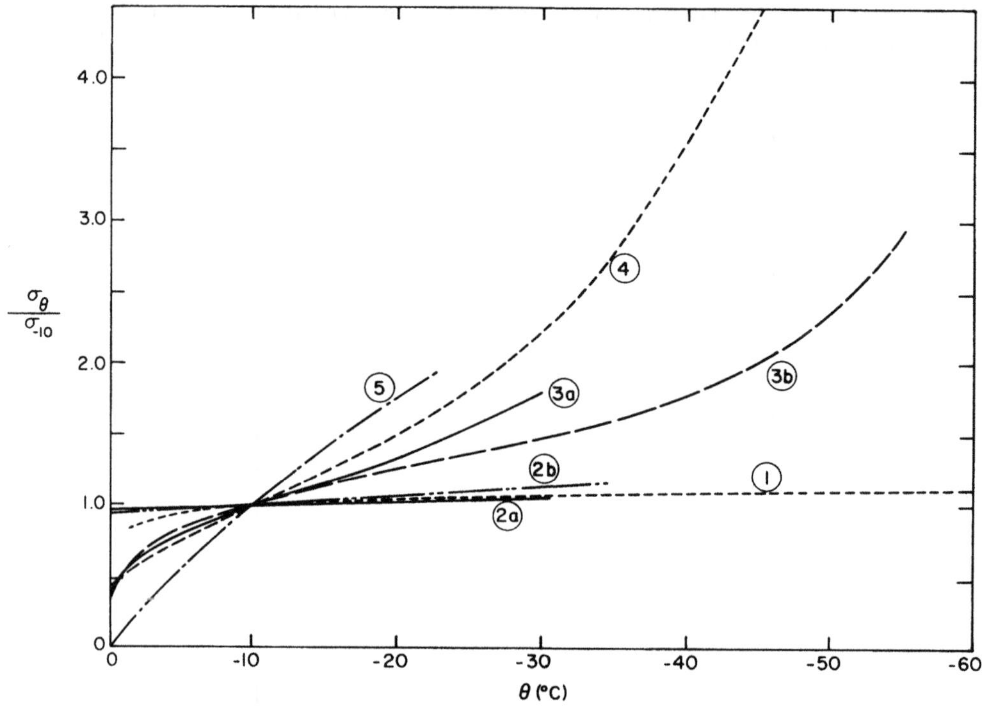

Fig. 35. Compilation of temperature relationships. All the stress values are normalized with respect to the value for −10°C. 1) Variation of Young's modulus with temperature. 2a) Uniaxial tensile strength according to data by D. Carter (1970). 2b) Uniaxial tensile strength according to data by F.D. Haynes (1978). 3a) Uniaxial compressive strength according to data by D. Carter (1970). 3b) Uniaxial compressive strength according to data by F.D. Haynes (1978). 4) Ductile yield stress deduced from creep data by M. Mellor and R. Testa (1969). 5) Pressure for phase transition from ice Ih to water under isothermal hydrostatic compression.

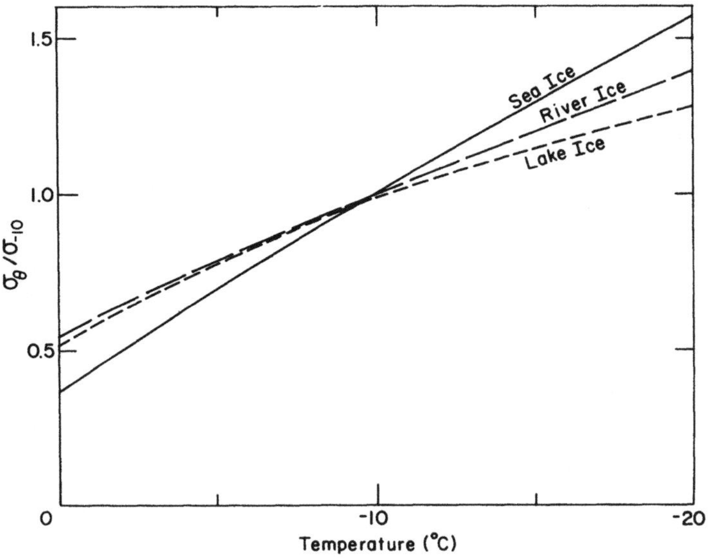

Fig. 36. Variation of uniaxial compressive strength with temperature (data from Schwarz, 1971).

where θ is temperature, then the yield stress ought to vary with temperature according to a relation of the form

$$\sigma_{max} = \{\dot{\varepsilon}/A\ f(\theta)\}^{1/n}.$$

In Figure 35, creep data have been used in this way to show how strength varies with temperature at low strain rates.

At very high strain rates, we would not expect much variation of strength with temperature, since the loading process is almost purely elastic. In the limit, strength might vary with temperature in much the same way as the true Young's modulus (Figure 35).

In Figure 35 the variation of strength is examined by normalizing all strength values with respect to the value for -10°C. Taking curves 1 and 4 as the limits for high and low strain rates separately, the curves for typical testing rates might be expected to fall in between, which does seem to be the case.

Before leaving the subject of freshwater ice, it might be mentioned that the temperature study by Haynes (1978) shows extremely high compressive strength values for high strain rates ($>10^{-3}\ s^{-1}$) and low temperatures (-10° to -55°C). Virtually all values for

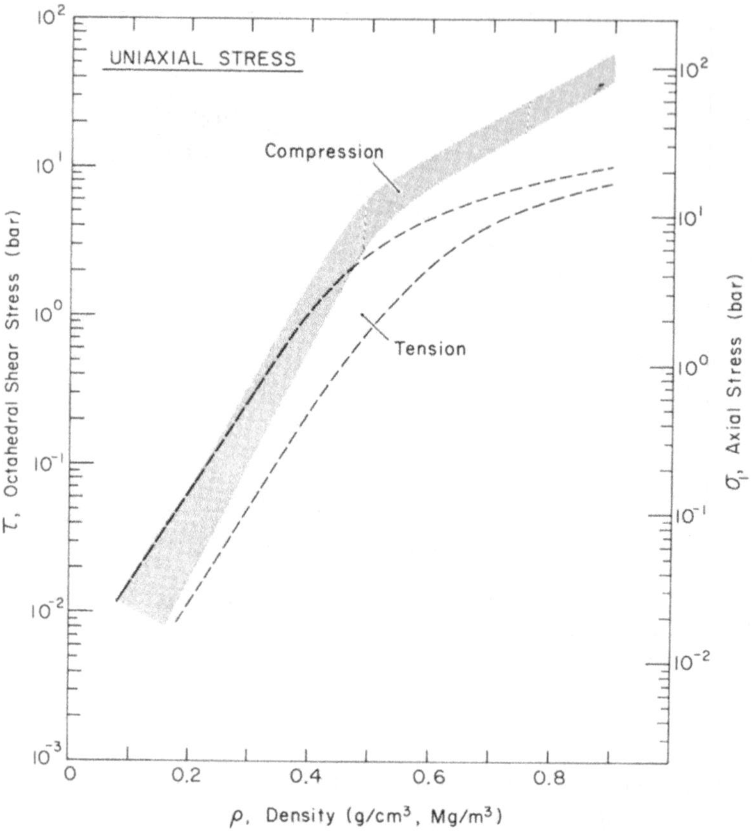

Fig. 37. Summary of strength data for non-saline ice and snow (see Mellor, 1975, for details of data sources).

this low temperature range exceeded 10 MPa, and some approached 60 MPa.

Temperature effects for saline ice include the effect of temperature on the solid ice component, as just discussed, but they also include the effect of porosity variations. All other things being equal, one might expect the strength of saline ice to increase more rapidly with decreasing temperature for the range where porosity is decreasing with temperature. Some data by Schwarz and Weeks (1971, 1977) seem to support this idea (Figure 36).

Actually, temperature has not received a great deal of attention as a primary variable for saline ice. The major emphasis has been on brine volume, which will be discussed next.

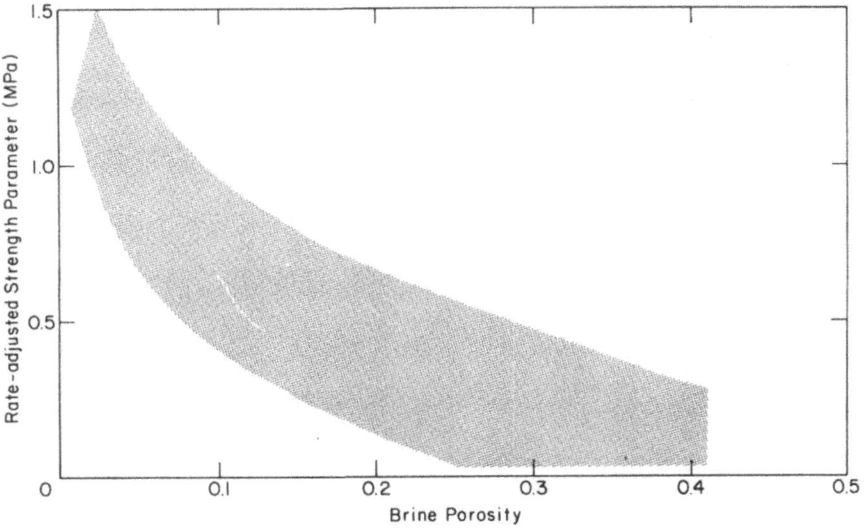

Fig. 38. Strength parameter plotted against brine porosity (data from Peyton, 1966).

The effect of porosity on the compressive strength of non-saline ice can be seen by examining the upper limits of the appropriate data band in Figure 37. As density decreases from 0.9 to 0.8 Mg/m^3, i.e. as porosity increases from 1.9% to 12.8%, the uniaxial compressive strength decreases by a factor of approximately 0.6.

In saline ice, where much of the pore volume is occupied by brine, porosity varies as temperature varies, and thus there is no easy way to separate the effects of temperature and porosity. Porosity is usually expressed as "brine volume," the volume of air rarely being known. There is a further complication in that most experimenters plot strength not against brine volume, but against the square root of brine volume. There was a rational basis for this practice in some studies, since the planar porosity of ice can be related to brine volume by assuming appropriate geometric models of pore structure. However, when the structure of the ice is uncertain, when the orientation of the pore structure to the principal stresses and the failure surfaces is variable, and when fracture mechanisms are not fully understood, there is no good reason for using the square root of brine volume as a primary variable. In this respect, the present review breaks with established tradition.

Figure 38 gives a data band representing what are usually thought to be Peyton's (1966) values of rate-adjusted compressive strength as a function of brine volume (see Schwarz and Weeks, 1977; Weeks

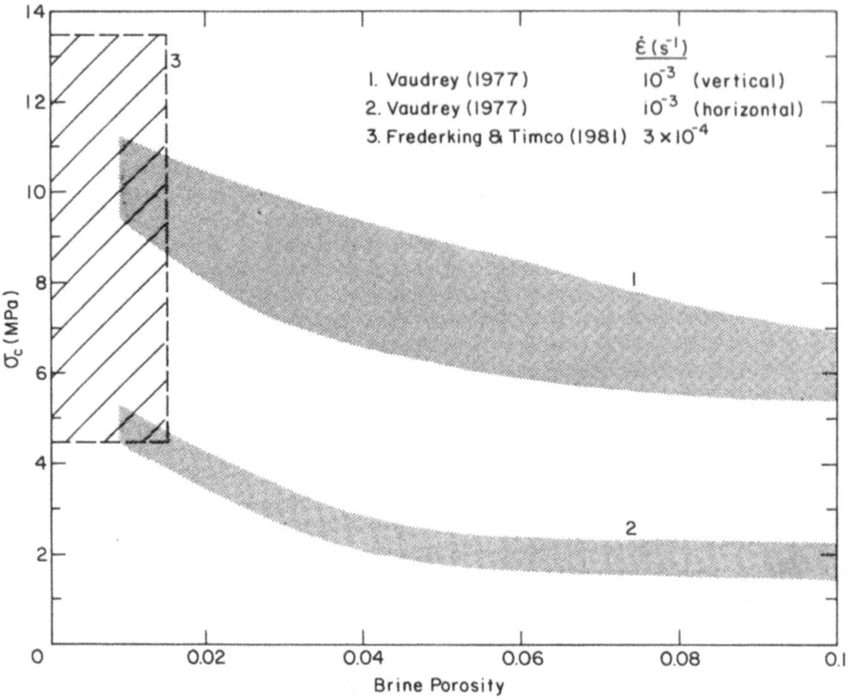

Fig. 39. Summary of data for σ_c as a function of brine porosity.

and Assur, 1967; Weeks and Assur, 1969). Actually there is real doubt as to what these numbers mean. Peyton used the strange term "strate," giving it the units of a stress and the interpretation of a rate-adjusted strength. However, "strate" does not appear to have the dimensions of a stress according to his definition; the absolute values are extremely low, and they are inconsistent with "measured strength" values shown in another graph. The brine volume values are also surprising, reaching porosities of about 40%, which is approximately the porosity of dry sand, crushed stone, or snow of density 0.55 Mg/m^3.

Less ambiguous data for the uniaxial compressive strength of saline ice have been given by Vaudrey (1977). Figure 39 gives data bands representing his results replotted against brine porosity instead of the square root of brine volume. The general magnitude of strength values measured by Frederking and Timco (1981) is also indicated in Figure 39.

6. UNIAXIAL TENSILE STRENGTH

The uniaxial tensile strength of ice, σ_T, is of both practical and theoretical importance. There is only one way to measure uniaxial tensile strength as a bulk property of the material, and that is by inducing failure in a fully determined uniaxial tensile stress field. The most reliable way to achieve this is to apply uniaxial tension to a dumbbell specimen which has been carefully designed, precisely machined, and mounted to close dimensional tolerances.

Indirect tension tests, which have to be interpreted by assuming a constitutive relation and a failure criterion, do not measure the uniaxial tensile strength of ice. In particular, exhaustive studies have shown that the ring tensile test is invalid for most real materials, and certainly for ice. The Brazil test, which involves diametral compression of a disc or cylinder, gives a direct measure of σ_T for "Griffith materials" when contact stresses are properly controlled, but for ice at typical loading rates it does not measure the uniaxial tensile strength (failure occurs in a biaxial stress field, with $\sigma_1 = -3\sigma_3$). Beam tests do not measure the uniaxial tensile strength of ice, but they are widely used and do constitute a rough analogue for typical plate fracture situations.

We can begin by looking at the behavior of non-saline ice under tension.

At very low strain rates, where deformation occurs mainly by flow and recrystallization, there is no reason to expect a significant difference between the compressive strength σ_c and the tensile strength σ_T. However, at high rates of loading, where brittle fracture occurs, σ_T is expected to be much lower than σ_c. Clearly there has to be a transition somewhere, and in Figure 40 it appears that the bifurcation between the σ_c and σ_T curves occurs at $\dot{\varepsilon} \approx 10^{-6}\ s^{-1}$ for non-saline ice at -7°C. Once the conditions for brittle, or quasi-brittle, fracture have been established, there appears to be little change of σ_T as strain rate increases, although σ_c continues to increase with $\dot{\varepsilon}$.

If σ_T is insensitive to strain rate in the brittle range, we would expect a corresponding lack of sensitivity to variations in temperature. This expectation is borne out by the limited data for the tensile strength of non-saline ice (see Figure 41).

The variation of σ_T with porosity in non-saline ice is indicated in Figure 37. As density decreases from 0.9 to 0.8 Mg/m^3, i.e. as porosity increases from 1.9% to 12.8%, σ_T decreases by a factor of approximately 0.8. Thus it appears that σ_T is less sensitive to porosity than is σ_c within the range that is of interest here.

It seems that there have not been many uniaxial tensile tests on saline ice. Peyton (1966) made tension tests, but the form of

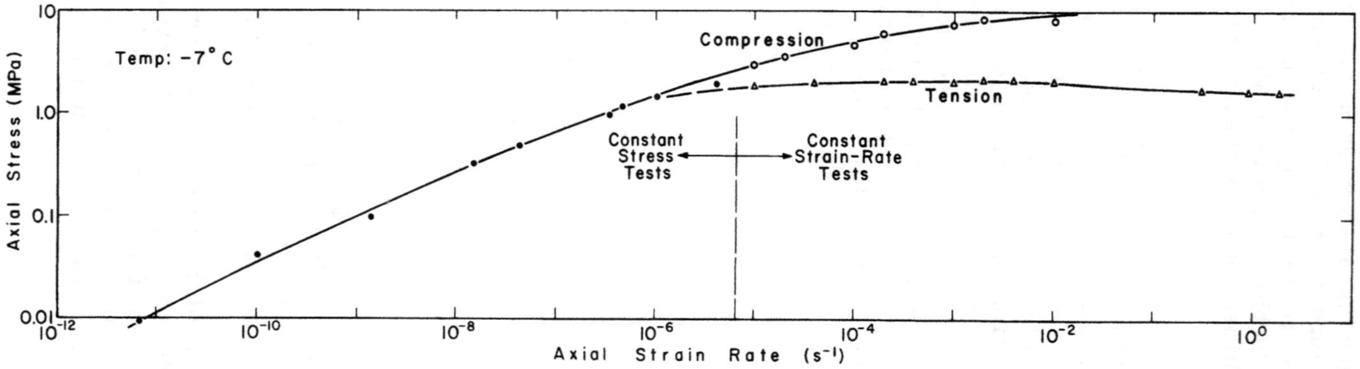

Fig. 40. Effect of strain rate on σ_c and σ_T for non-saline ice (Hawkes and Mellor, 1972).

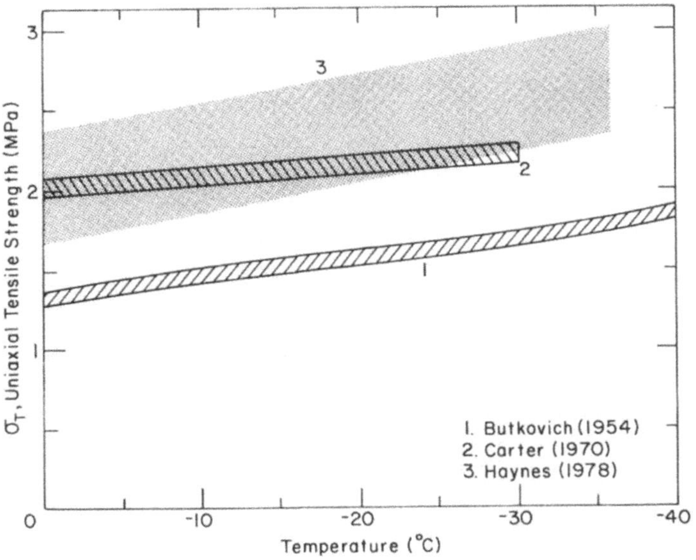

Fig. 41. Variation of uniaxial tensile strength with temperature for non-saline ice.

the report is such that the results are virtually unusable. This leaves only the work of Dykins (1970).

From consideration of the tensile strength of non-saline ice, which is relatively insensitive to strain rate, temperature, and porosity in the brittle range, we might guess that sea ice would also have relatively constant tensile strength. Just to establish perspective, if we assume that σ_T for freshwater ice varies by a factor 0.8 over the applicable range of porosity, and also by a factor of 0.8 (or 1.25) over the temperature range -30° to 0°C, then a combined effect when temperature and porosity increase together might cause σ_T to vary over the range by a factor of 0.64 (or 1.56). Figure 42 gives Dykins' (1970) data for σ_T as a function of temperature, and Figure 43 is a re-plot of his data for σ_T as a function of brine porosity. These graphs show that σ_T actually varies considerably as temperature (and therefore brine volume) varies, certainly more than might be expected from the behavior of fine-grained freshwater ice. The values of σ_T for tension applied in the vertical direction of the ice sheet are comparable to, but lower than, the values of σ_T for freshwater ice of similar temperature and porosity. The values of σ_T for tension applied in the horizontal direction of the ice sheet are much lower than values for fine-grained freshwater ice.

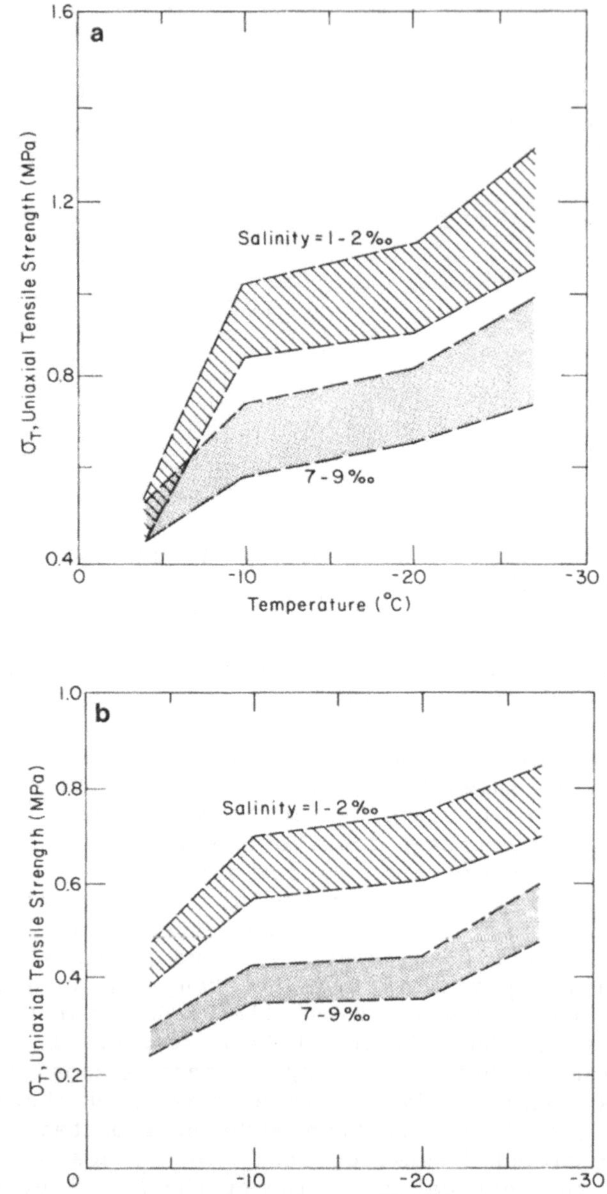

Fig. 42. Variation of uniaxial tensile strength with temperature and salinity for sea ice (from Dykins, 1970). a) Vertical specimens. b) Horizontal specimens.

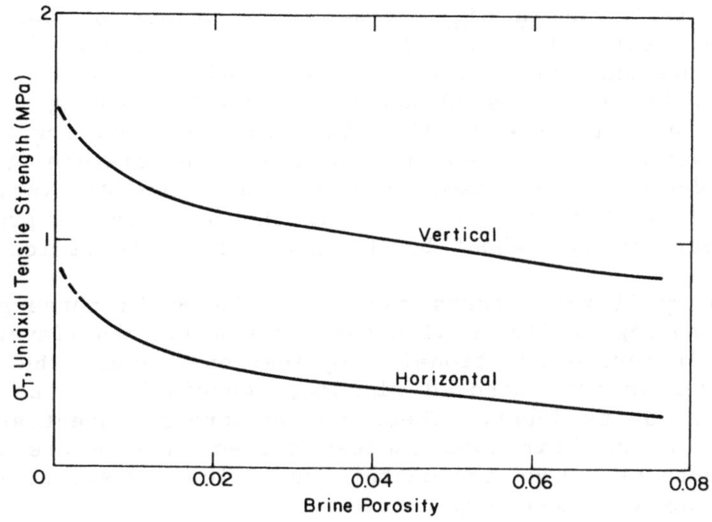

Fig. 43. σ_T as a function of brine porosity for sea ice (after Dykins, 1970).

7. FLEXURAL STRENGTH

Flexure of an elastic beam induces longitudinal stresses, compressive and tensile, which increase linearly with perpendicular distance from the neutral axis. If $\sigma_T < \sigma_c$, the beam will break when the stress at the outer surface of the convex side reaches a value equal to σ_T. For a linearly elastic material which fails by brittle fracture, a measure of σ_T can be obtained by calculation from the bending moment, the moment of inertia, and the beam depth. The simple formula is based on the following assumptions: 1) linearly elastic homogeneous material, 2) equal moduli in tension and compression, 3) small strains, with cross-section planes remaining plane and mutually parallel. It is important to note that fracture initiates at the surface, the zone of critical stress is very thin, and the fracture propagates in a stress gradient. For ice beams loaded at typical rates the behavior is not purely elastic, the effective moduli for tension and compression are not equal, and the material is often inhomogeneous. Recognizing that beam flexure does not measure the uniaxial tensile strength of the bulk material, the value of σ_T calculated from beam experiments probably ought to be called the modulus of rupture, or the flexural strength. The latter term will be used here with the symbol σ_{FT}.

One special feature of flexural strength measurements is that they include the results of both small-scale lab tests and large-scale field tests on in situ beams. The values of σ_{FT} for the two types of tests are very different for a variety of reasons. One is the inherent scale effect for the bulk strength of materials, which will be discussed under a separate heading. Another reason for the difference is that in situ beams sitting in water are often subject to steep and variable temperature gradients across the depth of the beam, to the extent that σ_{FT} varies with the time of day. Yet another reason is the variation of properties with ice depth, brought about by changes in the rate of ice growth.

Laboratory flexural tests may employ beams in three-point or four-point bending, while in situ tests often use cantilever beams. Cantilevers introduce additional complications, since the root of the cantilever is not anchored rigidly, especially if the beam is wide relative to its depth. There may be stress concentrations at the corners of the beam root (unless curved fillets are formed), and the effects of stress concentrations in ice are strongly sensitive to loading rate and temperature.

Measured flexural strength is sensitive to rate of loading, but the nature of the relationship is not well defined. There seem to be no systematic data for freshwater ice, and the results for saline ice are somewhat inconsistent. For small beams, Tabata (1967) showed σ_{FT} with a minimum at 0.01 MPa/s, and an increase with $\dot{\sigma}$ up to 0.2 MPa/s at a temperature of -9°C. In more recent work on small beams at -10° and -20°C, Tabata et al. (1976) found no correlation between σ_{FT} and $\dot{\sigma}$ for stress rates in the range 0.01 to 1.0 MPa/s. Lainey and Tinawi (1981) showed σ_{FT} dropping from a maximum at rates just below 0.1 MPa/s, with a slow but steady decrease of σ_{FT} as $\dot{\sigma}$ increased from 0.1 to 0.6 MPa/s. These data covered the temperature range -5° to -40°C. Saeki and Ozaki (1981) found a slight increase of σ_{FT} between 0.005 and 0.5 MPa/s, with a drop to lower values for the range 0.7 to 20 MPa/s (test temperature -7°C). Data given by Nippon Kokan (n.d.) for small beams at -2°C show σ_{FT} decreasing very considerably (factor of 5 or more) as deflection rate increases within the range 0.1 to 100 mm/s.

For large in situ beams, Tabata et al. (1967) showed σ_{FT} increasing greatly (factor of 5 or so) as $\dot{\sigma}$ increased from about 0.05 to 5 MPa/s at -2°C. However, Määttänen (1976) found that $\dot{\sigma}$ had no obvious influence in the range 0.01 to 10 MPa/s, provided that inertial effects in the water and the ice were eliminated.

The information on rate effects is obviously very confusing, and it is not easy to decide to what extent the various trends are influenced by real physical changes, and to what extent by imperfections of test technique.

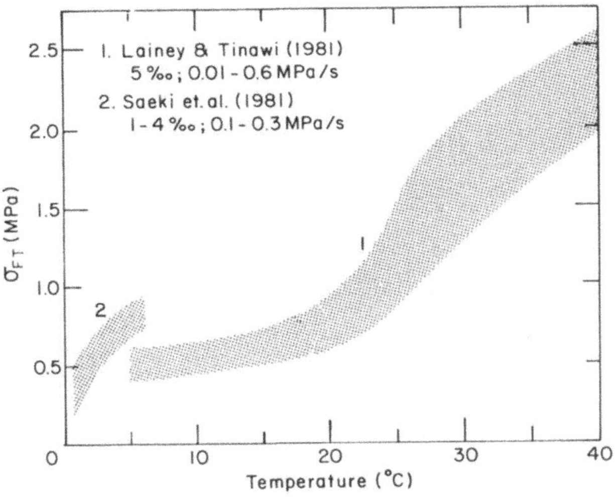

Fig. 44. Variation of flexural strength with temperature.

The effects of temperature on σ_{FT} appear at first sight to be less controversial, even though the validity of the elastic assumptions must vary with temperature. For small beams of freshwater ice, Weeks and Assur (1969) give a summary graph which shows σ_{FT} increasing linearly as temperature decreases, but for at least one of the data sources represented in the graph, the original data are perhaps less convincing than the interpreted trend. For saline ice, Lainey and Tinawi (1981) show a small increase in σ_{FT} as temperature drops from -5° to -20°C, and a much more rapid increase of σ_{FT} between -20° and -40°C (Figure 44). This perhaps says more about the relative validity of the test interpretation as a function of temperature, but as the result of an analogue test it is interesting. Saeki and Ozaki (1981) give beam data for less saline ice at temperatures between -0.5° and -6°C, showing, not surprisingly, a strong increase of σ_{FT} in this range. Butkovich (1959) tested small beams cut vertically and horizontally from sea ice, with temperatures in the range -2° to -26°C. His regression lines show a small increase of σ_{FT} over the range, but the data have a great deal of scatter. In earlier work, Butkovich (1956) carried out 80 tests on small beams, but his results show no significant correlation between σ_{FT} and temperature for the range -2° to -17°C (σ_{FT} was mainly in the range 0.4 to 1.2 MPa).

The effect of brine volume on σ_{FT} was described by Weeks and Assur (1967, 1969), who compiled test data for in situ cantilever beams. The results were represented by a linear decrease of σ_{FT} as the square root of brine volume increased from zero to 0.32 (brine porosity 0 to 0.1); for greater porosities, σ_{FT} was taken as constant. An updated compilation by Schwarz and Weeks (1977)

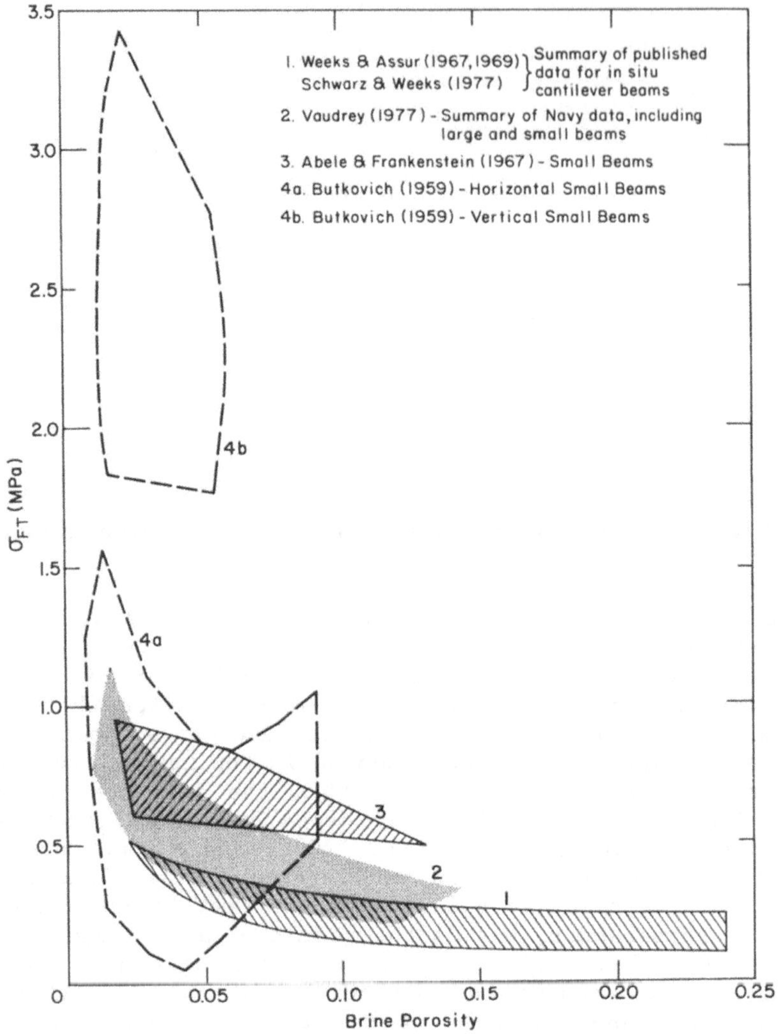

Fig. 45. Summary of data for σ_{FT} as a function of brine porosity.

weakened the case for the latter assumption, and the argument for using the square root of brine volume as the porosity variable no longer seems greatly persuasive.

When the general body of data for small beam tests is added to the plot (Figure 45), it is hard to perceive any orderly trend. Clearly, strength ought to decrease as porosity increases, but because increased porosity often reflects higher temperatures, the ductility of the material tends to increase with porosity.

8. FRACTURE TOUGHNESS

Virtually all fracture toughness measurements on ice depend on tests which flex or pry open a crack in "Mode I." Test data are thus presented in terms of the critical stress intensity factor, K_{Ic}. Because the measured values for K_{Ic} vary greatly, and because we need some intuitive "feel" when considering these values, it is worth recalling what K_{Ic} means.

Toughness is measured by the specific energy dissipation at failure G_c, which is also known as the critical crack extension force. K_{Ic} is related to G_c by

$$K_{Ic} = (EG_c)^{1/2}$$

in plane stress, and

$$K_{Ic} = \left(\frac{EG_c}{1-\nu^2}\right)^{1/2}$$

in plane strain. Thus there is a simple direct relation between K_{Ic} and G_c if E is a constant. K_{Ic} is also related to the overall tensile failure stress of the material σ:

$$K_{Ic} = \sigma\,(\pi c)^{1/2}$$

where c is the half-length of the controlling cracks. This relation implies that, if c is constant and the stress state does not change, K_{Ic} is directly proportional to the bulk strength of the material.

For ice straining at low rates, say less than $10^{-6}\ s^{-1}$, we would not expect K_{Ic} to have any relevance, since the ice is inelastic and it flows without cracking. At extremely high rates and low temperatures, ice could conceivably become perfectly elastic and perfectly brittle, so that the original Griffith theory might apply. For such a limit, with γ_p tending to the specific surface energy γ, K_{Ic} would tend to a low value:

$$K_{Ic} \rightarrow (2E\gamma)^{1/2}$$

in plane stress. Taking E = 10 GPa and the grain boundary specific energy γ = 0.1 J/m^2 for freshwater ice, the lower limit of K_{Ic} might be about 45 kN m$^{-3/2}$ for plane stress.

Measured values of K_{Ic} for ice are typically of order 100 kN m$^{-3/2}$. This is not much higher than the "Griffith" value, and it implies that $\gamma_p \approx 5\gamma$, assuming that E is more or less constant.

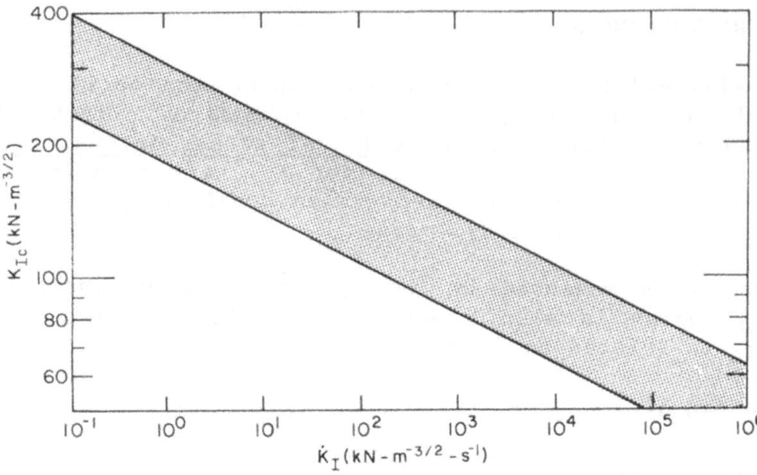

Fig. 46. Effect of loading rate on K_{Ic} for non-saline ice (from data summarized by Urabe and Yoshitake, 1981b).

When strain rate, or loading rate, is varied in a fracture toughness test for a given type of ice, we might expect K_{Ic} to decrease as $\dot{\varepsilon}$ or $\dot{\sigma}$ increases, at least for non-saline ice. While at least one set of experiments shows a trend opposite to this, the overall trend shown by compilation of published data is in the expected sense (Figures 46, 47). Rate effects were originally expressed in terms of the speed of the testing machine, which is clearly of limited interest, but now the accepted rate variable seems to be $\dot{K}_I$, which is really the inverse of the time to failure. Strain rate has been used as a variable, but there are some problems of interpretation.

In sea ice, K_{Ic} has been found to decrease with increase of loading rate for $\dot{K}_I > 10^{-2}$ kN $m^{-3/2}$ s^{-1}, or effective $\dot{\varepsilon} > 10^{-3}$ s^{-1} (Urabe et al., 1980; Urabe and Yoshitake, 1981a,b). However, for lower rates K_{Ic} appears to be insensitive to rate (Figure 48). The lowest measured values for sea ice are lower than the expected "Griffith value" for pure ice.

In discussing rate effects, it has been assumed that K_{Ic} will decrease as the material becomes more elastic and more brittle due to higher loading rates. Extending this single line of argument to <u>temperature effects</u>, it might seem reasonable to expect K_{Ic} to decrease as temperature decreases, since lower temperature undoubtedly makes ice more elastic and more brittle. However, experimental data (Figure 49) show exactly the opposite trend, with K_{Ic} increasing as temperature decreases. This observed trend is consistent with the fact that tensile strength σ_T increases as temperature decreases, since K_{Ic} is proportional to strength if

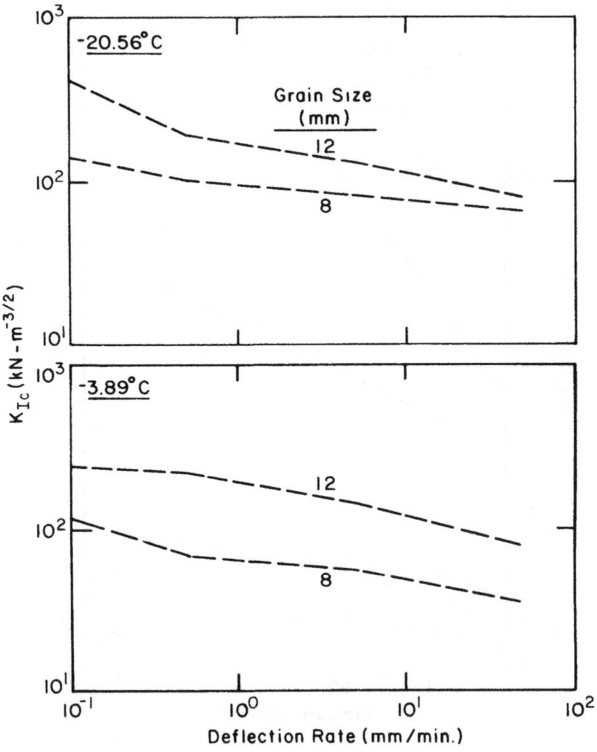

Fig. 47. Variation of K_{Ic} with loading rate, grain size and temperature for columnar freshwater ice (Hamza and Muggeridge, 1979).

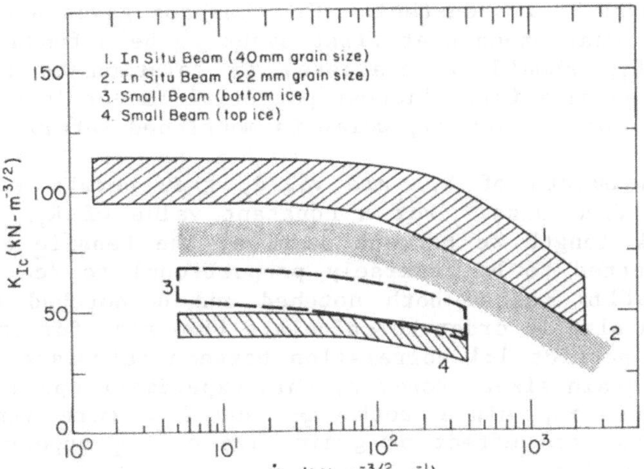

Fig. 48. Effect of loading rate on K_{Ic} for sea ice (data from Urabe and Yoshitake, 1981a,b).

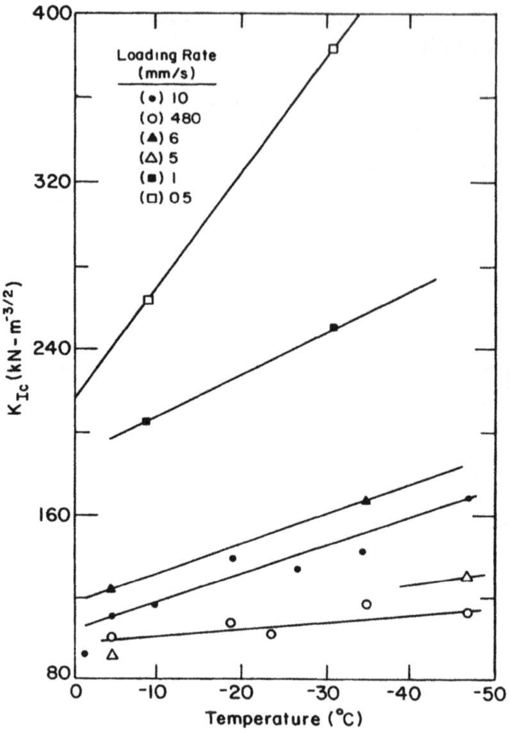

Fig. 49. Variation of K_{Ic} with temperature and loading rate (from Miller, 1980).

the crack length 2c is constant. This perhaps gives a clue to the resolution of what appears at first sight to be a fundamental contradiction: K_{Ic} should be evaluated in relation to the overall strength of the material. Another potential factor is temperature-dependence of surface energy, which is mentioned later.

If measurements of K_{Ic} are valid, they permit a systematic treatment of flaw size. For a constant value of K_{Ic} and variation of crack length 2c between samples, the tensile strength σ_T might be expected to be inversely proportional to $\sqrt{c}$. Urabe and Yoshitake (1981b) tested both notched and un-notched beams with varying grain size in order to calculate flaw size for the ice, and they found a perfect 1:1 correlation between calculated flaw size and observed grain size. However, this experiment appears to merit further discussion, since both σ_{FT} and K_{Ic} were functions of grain size, and the effect of grain size on σ_{FT} appears to be in the wrong direction.

If we refer back to the basic Griffith equation for strength as a function of modulus, surface energy, and crack length, then theoretical values of tensile strength can be calculated by taking the flaw size 2c equal to the grain size of the ice. For ice of very low porosity ($\leq$ 0.01), the true Young's modulus E is 9 to 10 GPa at typical temperatures. For non-saline ice, the vapor/solid specific energy γ is approximately 0.1 J/m^2, and the liquid/solid surface energy is about 30% of the vapor/solid value (Fletcher, 1970; Hobbs, 1974)*. The vapor/solid value is probably the appropriate one for consideration of brittle fracture in "cold" ice, but the lower liquid/solid value might be applicable in relatively warm ice which has a "liquid-like layer" or liquid-filled flaws. The latter condition might give something equivalent to the Rehbinder, or Joffe, effect, whereby γ is reduced by adsorption of certain surface-active chemicals, and strength decreases in consequence.

If we substitute into the Griffith equation E = 10 GPa and γ = 0.1 J/m^2,

$$\sigma_T = \left(\frac{2}{\pi}\right)^{1/2} \left(\frac{10^{10} \times 10^{-1}}{c}\right)^{1/2}$$

$$= \frac{2.52 \times 10^4}{\sqrt{c}} \quad \text{Pa}$$

where the half-length of the controlling flaw, c, is in metres. In Figure 50, the resulting calculated values of σ_T are given as a function of the flaw size 2c, which is assumed equal to grain size. The theoretical relation is compared with actual measured values of σ_T, and the agreement is surprisingly close. Since the controlling flaws could be substantially smaller than the grain diameter, the agreement might even be improved. The significance of this exercise is in demonstrating that simple Griffith theory cannot necessarily be rejected for ice.

Vaudrey (1977) measured K_{Ic} for sea ice at -10° and -20°C, and plotted the results against the square root of brine volume for a very narrow range. Vaudrey drew a line on the graph to indicate linear decrease of K_{Ic} with increase in the root of brine volume but, in fact, there was no significant correlation between the variables (K_{Ic} values scattered by a factor of 5). Shapiro et al. (1981) made measurements in the same range (brine porosity 0.16 to 0.38), and showed a more convincing decrease of K_{Ic} with increase of porosity, although there was still large scatter.

* Liu and Miller (1979) use values that are off by two orders of magnitude due to incorrect conversion of Fletcher's values.

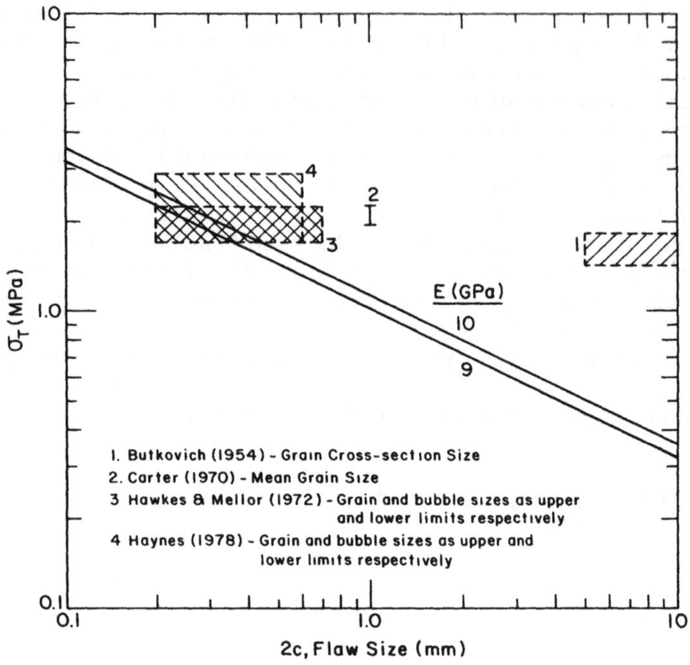

Fig. 50. Comparison of theoretical tensile strength with measured values.

9. EFFECT OF GRAIN SIZE ON STRENGTH AND DEFORMATION RESISTANCE

Early creep studies on non-saline ice suggested a strong increase of strain rate with grain size, but the experiments themselves were not fully acceptable. More recent published work by Baker (1978) showed a complicated relation between creep rate and grain size, with the direction of the trend different for different size ranges, but again the experiments seemed questionable. Subsequent studies by Duval and LeGac (1980) showed no significant variation of final creep rate with grain size.

For the strength of ice in the brittle range, the expectation is that strength of isotropic material would be inversely proportional to the square root of grain size. This is in keeping with the ideas of fracture mechanics if grain size is identified with flaw size. Limited data for σ_c tend to bear out this expectation (see Figures 29, 30, 31), but systematic verification has not yet been published. Some recent values of σ_{FT} for sea ice of varying grain size (Urabe and Yoshitake, 1981b) appear to be in direct contradiction to the expected trend, but Currier and Schulson (1982) measured σ_T at 10^{-6} s^{-1} and found the results could be described by the Hall-Petch relation, i.e. $\sigma_T = a + b\ d^{-1/2}$. In the ductile range of behavior, it seems that the maximum yield stress does not change with grain size (Jones and Chew, 1981).

To sum up, there is no convincing evidence that strength or deformation rate vary with grain size for the ductile (creep) range of behavior. For the brittle range, one might expect strength to decrease as grain size increases, but substantial experimental proof has not yet appeared.

10. SIZE EFFECTS FOR STRENGTH AND DEFORMATION RESISTANCE

In general, the measured strength of a brittle material decreases as the size of the specimen increases, or as the critically stressed volume of material increases. The accepted explanation is that bigger samples contain bigger flaws and, as shown by fracture mechanics, strength decreases as flaw size increases. Sophisticated statistical theory has been developed for consideration of the distribution of flaw sizes within a given specimen, for estimation of the probability of encountering larger flaws in larger volumes, and for the effects on strength through a "weakest link" theory.

A simple relation can be derived from rather involved theoretical arguments (see Hawkes and Mellor, 1970). If the most probable (modal) strength is S_{V_1} for a specimen of volume V_1, and S_{V_2} and V_2 are modal strength and volume for a geometrically similar specimen subjected to the same strain rate, then

$$\frac{S_{V_2}}{S_{V_1}} = \left(\frac{V_1}{V_2}\right)^{1/m}$$

where m is a characteristic constant.

Experiments on nonmetallic brittle solids indicate that m is of order 10, so that the volume effect is a weak one. In terms of a linear dimension L the effect is stronger, since $V \propto L^3$ and therefore $S \propto L^{-3/m}$. If $m \approx 10$, S is approximately inversely proportional to the cube root of L. Thus an increase in L by a factor of 10 would roughly halve the strength.

The creep resistance of homogeneous ice at low strain rates ($\dot{\varepsilon} < 10^{-6}\ s^{-1}$) is probably not much dependent on the volume of material involved, provided that the minimum linear dimension of any test sample is greater than 10 grain diameters.

11. FAILURE OF ICE IN MULTIAXIAL STRESS STATES

When isotropic ice is subjected to triaxial compression at low strain rates, it is usually said that the bulk stress $\bar{\sigma}$ has no effect on the induced deviatoric strain rate $\dot{\varepsilon}'_{ij}$, provided that

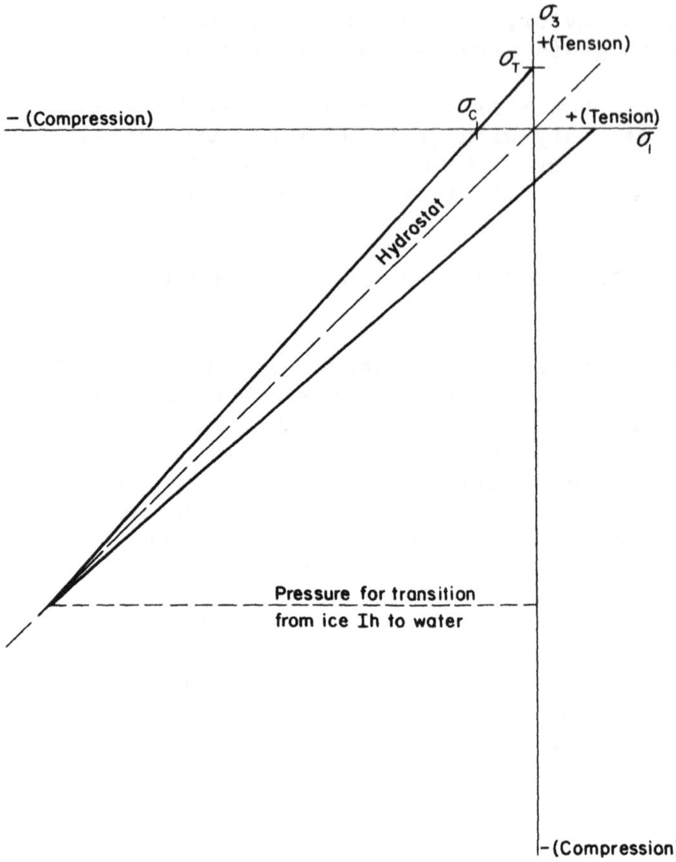

Fig. 51. Two-dimensional representation of a simple yield criterion for low-rate ductile yield of isotropic freshwater ice at temperatures above −20°C.

"allowance is made for depression of the melting point." A more straightforward interpretation is that creep rate increases slightly, and the ductile yield stress decreases slightly, when the confining pressure increases at constant temperature. If the pressure is increased sufficiently, for example to about 0.1 GPa at temperatures near −10°C, ice Ih transforms to water and its shear resistance drops to virtually zero. Thus the yield surface in principal stress space is a cone with its apex on the hydrostat line ($\sigma_1 = \sigma_2 = \sigma_3$) at the pressure for the phase transition to water (Figure 51), at least for temperatures above −20°C.

The same ice under triaxial compression at high strain rates suffers internal cracking and dilatation when unconfined, but confining pressure subdues the cracking and the deviatoric yield

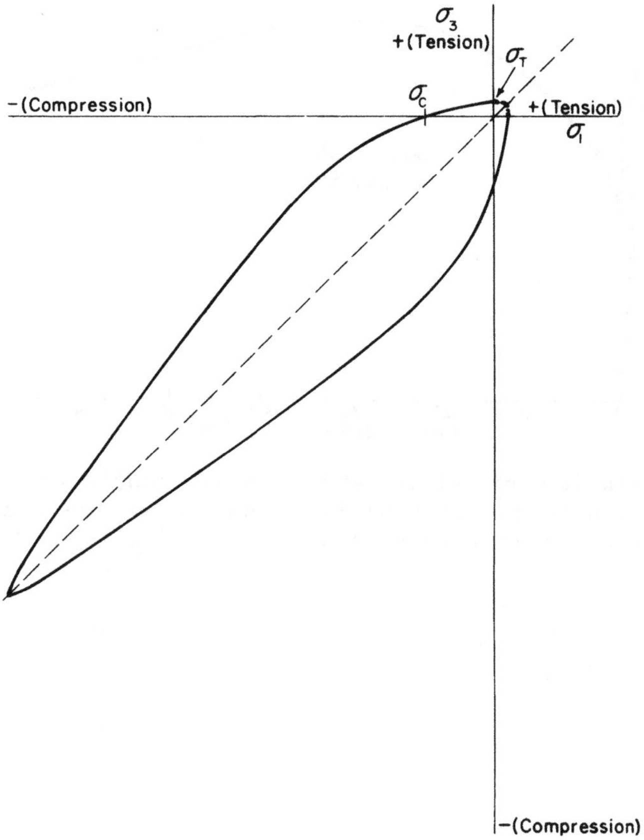

Fig. 52. Two-dimensional representation of a yield criterion which accounts for crack suppression and melting point depression in isotropic freshwater ice under moderately high strain rates at temperatures in the vicinity of -10°C.

stress is increased. A first approximation for the failure criterion under these conditions would be a linear Coulomb-Navier-Mohr criterion. In the compression quadrants of principal stress space, the yield surface corresponding to such a criterion would be a cone disposed symmetrically about the hydrostat line, with its diameter increasing indefinitely as pressure increases. Clearly such a surface contradicts the pressure melting effect which was just discussed in connection with yield for low strain rates, and the actual yield surface must account for both effects of pressure. Such an envelope is sketched in Figure 52, which shows "strength" increasing with confining pressure at first, and then decreasing as the pressure becomes very high. Triaxial tests by Jones (1978), at -11.5°C and strain rates of order 10^{-4} s^{-1}, show the effects of confining pressure reversing at about 30 to 40 MPa (Figure 53).

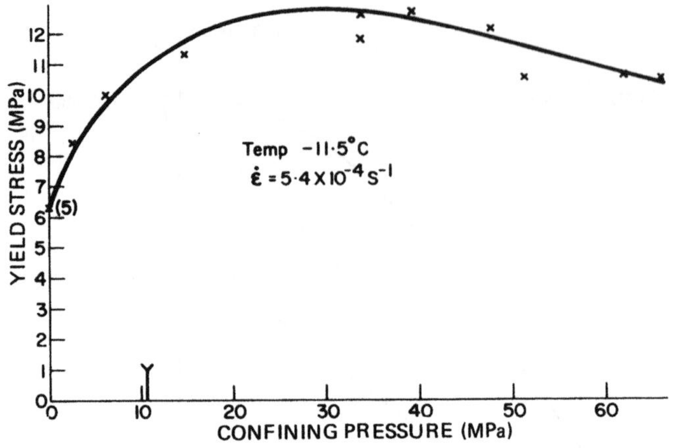

Fig. 53. Variation of yield stress with confining pressure for conventional triaxial tests on non-saline ice at -11.5°C and a strain rate of 5.4×10^{-4} s^{-1} (Jones, 1978).

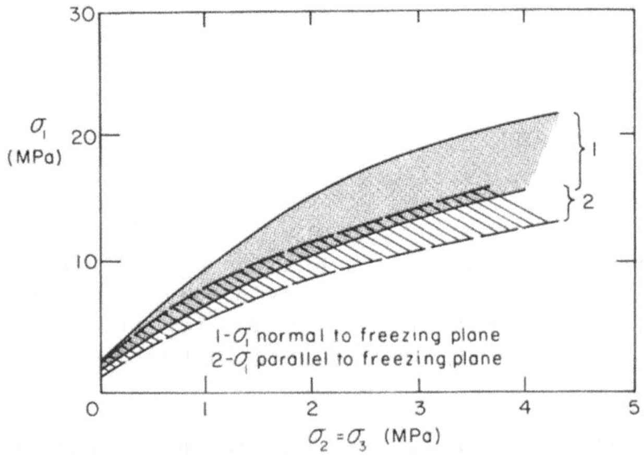

Fig. 54. Variation of axial yield stress with confining pressure for sea ice at temperatures of -2° to -3°C and salinities of 0.16 to 3.54 $^0/_{00}$ (after Panov and Fokeev, 1977).

Conventional compressive triaxial tests ($\sigma_1 \neq \sigma_2 = \sigma_3$) were made on natural and artificial sea ice by Soviet investigators (Panov and Fokeev, 1977). The axial stress σ_1 increased with σ_2, σ_3 (Figure 54), and the maximum shear stress $(\sigma_1 - \sigma_2)/2$ increased nonlinearly with the normal stress on the plane of maximum shear. When σ_2 was approximately 4 MPa, the yield values of σ_1 were an order of magnitude higher than σ_c. Under confining pressures σ_2, the failure stress σ_1 decreased as salinity and temperature increased, i.e. salinity and temperature had effects similar to their effects on σ_c. Isotropic fine-grained ice behaved differently than anisotropic ice, and strength varied with the loading direction in anisotropic material. Values of ϕ for the Mohr envelope were mostly in the range 30° to 50°, with extreme values of 14° and 55°.

Triaxial tests to very high pressures (up to 350 MPa) and very low temperatures (-78° to -196°C) were made recently on non-saline polycrystalline ice (Durham et al., 1982). Very high yield stresses were recorded, and phase transitions to higher density polymorphs were reached.

For the tension-compression quadrants of triaxial principal stress space there is little experimental information, and we have to rely largely on intelligent guesswork.

Under very low strain rates, σ_c is expected to equal σ_T, and experimental data support this idea. With $\sigma_c = \sigma_T$, the Tresca or von Mises envelope can be extended across the tension-compression quadrants without conflict (Figure 55). The pressure-melt version of these criteria (Figure 51) can similarly be extended without significant conflict for low temperatures (-10°C), but for higher temperatures an extrapolation of such an envelope would predict $\sigma_c < \sigma_T$, and there is no experimental evidence for such an effect.

At high strain rates, where $\sigma_c \neq \sigma_T$, a straight line connection of the σ_c and σ_T points is not likely to be far from the truth, especially if an adjustment is made to assure continuity of slope as the envelope passes into compression-compression where the linear Coulomb-Mohr envelope is applicable for low pressures. Experimental data by Haynes (1978) are not in conflict with this suggested trend.

The observed ratio of σ_c to σ_T gives useful insight into the failure criterion for low confining pressures and for the tension-compression quadrant. At low strain rates in ice, $\sigma_c/\sigma_T = 1$, as already discussed, which means that a von Mises or Tresca envelope is a good approximation for low pressures and for tension-compression. At higher strain rates, σ_c/σ_T increases from 1 to about 4 according to the data of Figure 40. If the linear Coulomb-Navier-Mohr criterion is extended into the tension-compression quadrant (where it is not strictly applicable), the predicted value of σ_c/σ_T is 3 for $\phi = 30°$, and 4 for $\phi = 37°$. By contrast, the simple Griffith criterion predicts $\sigma_c/\sigma_T = 8$, while the three-dimensional version predicts $\sigma_c/\sigma_T = 12$. The Babel and Sines (1968) generalization of the Griffith criterion predicts $\sigma_c/\sigma_T = 3$ if the controlling flaws are circular holes, with the value in-

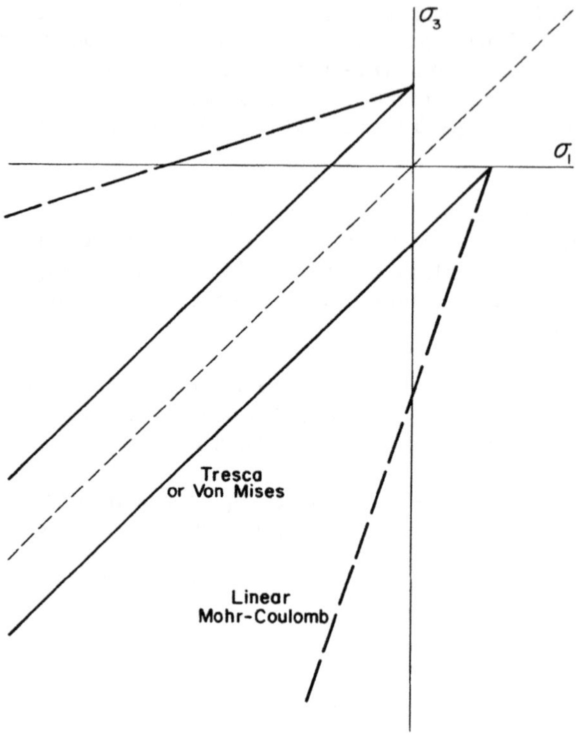

Fig. 55. Two-dimensional representation of the Tresca and von Mises criteria, and the linear Coulomb-Navier-Mohr criterion.

creasing progressively to 8 as the flaws change progressively from a circular cross section to a completely flat ellipse.

For the tension-tension stress state there are no data for ice, and in view of formidable experimental difficulties it may be a long time before there are reliable data. Perhaps the most reasonable guess is that, for isotropic ice in biaxial or triaxial tension, failure will occur when one principal stress reaches the uniaxial strength σ_T. This gives the envelope as lines parallel to the principal stress axes, with intersection on the hydrostat line at $\sigma_1 = \sigma_2 = \sigma_3$.

In Figure 52 the preceding deductions about the behavior of isotropic ice are combined to form a complete failure envelope, slightly modifying the lines deduced for each individual quadrant so as to give continuity of slope across the axes.

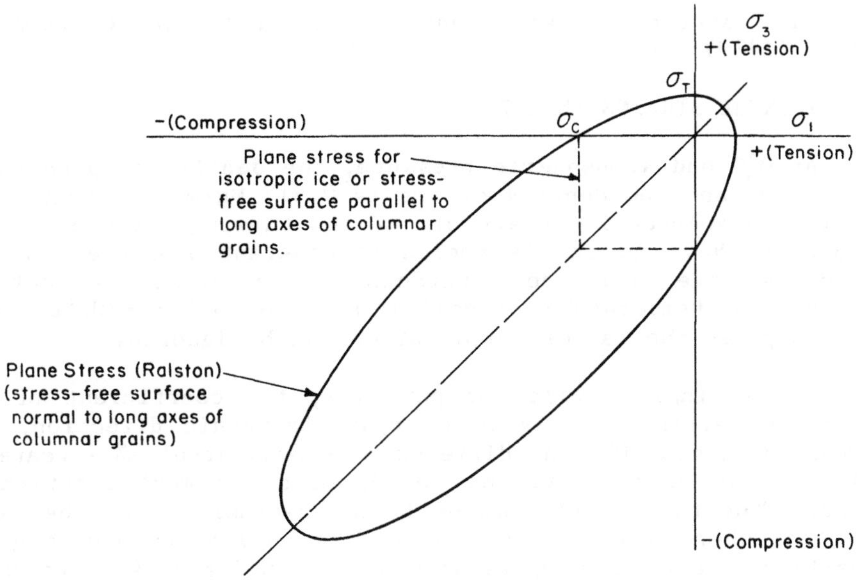

Fig. 56. Failure envelopes for biaxial stress states ($\sigma_2 = 0$)

The three diagrams discussed above are drawn in two dimensions because they are difficult to draw in three dimensions. They should not be confused with two-dimensional diagrams for biaxial stress states, i.e. for plane stress.

Biaxial tests were made by Frederking (1977), who pressed prismatic specimens between a pair of platens in one direction while rigidly restraining the ice in one of the other orthogonal directions, and leaving the remaining pair of faces stress-free. For isotropic ice the confinement had very little effect on the crushing strength, which remained approximately equal to the uniaxial compressive strength for strain rates of about 3×10^{-7} to $3 \times 10^{-4}\ s^{-1}$. This implies that, to a first approximation, the failure envelope in the compression-compression quadrant could be represented by lines parallel to the axes (Figure 56). A similar effect was found when the lateral restraint was applied to columnar ice in a direction parallel to the long axes of the columns. However, if the restraint was applied perpendicular to the long axes of the columns, the failure stress in the loading direction was roughly double the uniaxial strength.

Ralston (1980) used the data of Frederking (1977), together with a parabolic Coulomb-Navier-Mohr criterion, in order to deduce the plane stress failure envelope shown in Figure 56. It should be noted that this plane stress envelope crosses the 45° line (σ_1 =

σ_3) for a different reason than the triaxial envelope proposed by Mellor (1979), and shown in Figure 52.

12. THERMAL STRAINS IN ICE

At 0°C and atmospheric pressure, the density of pure water is 1.00 Mg/m^3 and the density of ice is 0.917 Mg/m^3, so that freezing involves a volumetric strain of -8% and thawing gives a volumetric strain of +9%. As ice is cooled at constant pressure its density increases, i.e. the ice contracts. The expansion coefficient varies with temperature, becoming negative below -200°C, but for most purposes the latter complication can be ignored.

For a single crystal of pure ice, the coefficient of linear expansion varies slightly with crystallographic direction, but at temperatures near 0°C the difference is only about 2% (greater parallel to the c-axis) and can be ignored for most practical purposes. Thus the coefficient of linear expansion α can be taken as $\alpha = \gamma/3$, where γ is the coefficient of cubical expansion given by density changes. At temperatures between 0°C and -40°C we can take $\alpha = +5 \times 10^{-5}\ C^{-1}$ (± 4%) for freshwater polycrystalline ice, noting that α decreases by approximately 10% as temperature drops from 0° to -40°C.

In sea ice, or other ice containing significant amounts of dissolved impurities, the situation is very different. At temperatures near the melting point, saline ice consists of solid ice, plus liquid inclusions which change in volume and salinity as the temperature changes. Although the ice crystals have a positive expansion coefficient, phase changes in closed brine cells could create freezing strains in a negative sense, i.e. volume increases as temperature drops. Above about -5°C, the freezing and thawing of brine inclusions is usually thought to be the dominant effect, implying that sea ice contracts as temperature increases in this range. At very low temperatures, when the brine is frozen and solid salts are precipitated, the behavior is almost identical to that of pure ice.

In the past, effective values of γ and α have been calculated as functions of temperature and salinity, using density values from the sea ice phase diagram. The usual source is a table and graph derived by Anderson (1960) on the assumption that air volume is zero and all brines remain trapped in place. Air volume was taken into account by Schwerdtfeger (1963) and by Cox and Weeks (1982), but of itself included air does little to change the variation of calculated density with temperature. While these calculations are sound in principle, they may not be very realistic, since at high temperatures brine can be expelled into air bubbles and even out through the ice boundaries. A reappraisal of the problem will be published soon by G.F.N. Cox of CRREL.

If ice is cooled or warmed very slowly, thermal strains can be accommodated by creep processes without any cracking, especially if the ice temperature is above -10°C. However, if very rapid cooling occurs at low temperatures, high strain rates are induced in ice and it cracks. For strain in one direction, the strain rate $\dot{\varepsilon}$ produced by cooling of the ice at a rate $d\theta/dt$ is

$$\dot{\varepsilon} = \alpha(d\theta/dt)$$

assuming that α is invariant with temperature θ. Strain rates developed this way are likely to be much lower than the strain rates which are associated with brittle fracture in laboratory strength testing of ice. For example, extremely rapid cooling at 10°C/hr would give $\dot{\varepsilon} = 1.4 \times 10^{-7}\ s^{-1}$ in pure ice. However, fairly large tensile strains can develop if the temperature change is great enough. For example, in non-saline ice a temperature drop of 20°C gives a tensile strain of 10^{-3} (0.1%), which is at the top end of the range of tensile failure strains measured by Hawkes and Mellor (1972).

The most extreme cooling rates and temperature changes occur at the upper surface of an ice layer. Both the amplitudes and rates of temperature change decrease with increasing depth in the ice. Thus cracking initiates at the surface, and the depth of crack penetration is limited by the attenuation of temperature waves with depth.

The effect of cooling at the surface of sea ice has been analyzed in terms of elastic plate flexure, but because strain rates are likely to be low ($< 10^{-8}\ s^{-1}$) and the bulk of the ice is likely to be "warm," the assumption of linear elasticity seems dubious.

PART III. SOME CHARACTERISTICS OF FRAGMENTED ICE

1. MECHANICAL PROPERTIES OF BRASH ICE AND ICE RUBBLE

When floating ice is fragmented, it can be regarded as a granular continuum if the dimensions of the volume under consideration are significantly greater than the block size of the broken ice. This interpretation is easiest to justify for brash ice, which can be regarded as a floating accumulation of more or less equant fragments in the size range 0.02 to 2.0 m. However, such an interpretation can perhaps be extended to heavily fragmented ice sheets when there has been rafting and ridging.

A number of research groups have treated fragmented floating ice as a granular material conforming to a linear Coulomb-Navier-Mohr yield criterion, and characterized largely by a friction angle

ϕ. Measured values of ϕ have been summarized, together with additional values of ϕ deduced from crushing experiments (Mellor, 1980). In this compilation, ϕ ranges from 42° to 58°, but some recent test results (Weiss et al., 1981) give values in the range 11° to 34° for ice of lower porosity.

Other important properties are the block density, and the porosity of the fragmented mass.

For bubble-free freshwater ice the block density is 0.917 Mg/m^3, but for bubbly ice or snow-ice the density may be as low as 0.89 Mg/m^3. For sea ice, block density varies with salinity, temperature and air content. The range for intact sea ice is approximately 0.89 to 0.93 Mg/m^3, but for fragments equilibrated with seawater the density is likely to be at the lower end of this range, say 0.9 Mg/m^3.

The porosity n of the ice mass varies with the degree of packing and the size gradation of the fragments. "Ice rubble" used in experiments has had porosity less than the 40% or so which is typical for uniform granular solids (e.g. sand, crushed stone). Keinonen and Nyman (1978) used ice with n from 0.32 to 0.37, Prodanovic (1979) had n = 0.38, and Weiss et al. (1981) used material with n ranging from 0.16 to 0.33. Sandkvist (1981) gives estimates of the water content of brash as a function of depth from actual field observations. His data suggest an almost linear decrease of porosity with depth, from virtually zero at the water surface level to 1.0 at the base of the brash layer. For fresh pressure ridges in sea ice there have been numerous inspired guesses at the overall porosity, but virtually no measurements. One might guess that porosity in a pressure ridge would vary from small values near the waterline (say about 0.05) to fairly high values (say about 0.3) near the crest and the keel, actual values varying to some extent with the mode of ridge formation and the amount of freeze-back.

When a wide layer of unbonded fragmented ice is floating in static equilibrium, the internal stresses are determined by gravity body forces, i.e. by the weight of the ice above water level and by the buoyancy below water level. If the total ice thickness t is significantly greater than the characteristic block size d, the layer can be treated as a continuum. Making the simplifying assumption that the porosity n and the friction angle ϕ were invariant with depth, Mellor (1980) gave the vertical (σ_z) and horizontal (σ_x, σ_y) components of normal stress:

$$\sigma_z = (1-n)\ \rho_i g\ \left(\frac{\rho_w}{\rho_i} - 1\right)(t-z)$$

$$= (1-n)\ \rho_i g\ \left[t_1 - \left(\frac{\rho_i}{\rho_w} - 1\right)(z-t_1)\right]$$

where ρ_i is the ice block density, ρ_w is the water density ($\approx$ 1.026 Mg/m^3 for seawater), z is the depth below the top surface of the ice, t is the total ice thickness, t_1 is the thickness of ice above water level. The values of σ_x and σ_y depend on whether the ice layer has recently been compressed or extended horizontally. The limiting states of stress according to the Rankine theory of plastic yield give:

(a) for the passive state (compression) $\sigma_x = \left(\frac{1+\sin\phi}{1-\sin\phi}\right)\sigma_z$

(b) for the active state (extension) $\sigma_x = \left(\frac{1-\sin\phi}{1+\sin\phi}\right)\sigma_z$

(c) for the neutral state $\sigma_x = [\nu/(1-\nu)]\sigma_z$.

For a loose fragmented layer of thickness t in which n and ϕ are invariant with depth, the crushing resistance per unit width, R, is

$$R = \frac{1}{2}\left(\frac{1+\sin\phi}{1-\sin\phi}\right)(1-n)\rho_i g\left(1 - \frac{\rho_i}{\rho_w}\right)t^2.$$

Although this expression is oversimplified by the assumption of constant values for n and ϕ, it serves to illustrate that the average crushing pressure R/t is proportional to t. If there is some constant cohesion c between the ice fragments, the average crushing pressure includes a term which is independent of thickness:

$$\frac{R}{t} = \frac{1}{2}\left(\frac{1+\sin\phi}{1-\sin\phi}\right)(1-n)\rho_i g\left(1 - \frac{\rho_i}{\rho_w}\right)t + 2c\left(\frac{1+\sin\phi}{1-\sin\phi}\right)^{1/2}.$$

When c is large, the second term of the previous equation dominates, and the resistance is then similar to that of an intact ice sheet with a crushing stress of $2c\sqrt{(1+\sin\phi)/(1-\sin\phi)}$. With a two-dimensional Coulomb-Navier-Mohr failure criterion, uniaxial compressive strength is usually expressed as $2c[\cos\phi/(1-\sin\phi)]$, which is identical to the crushing stress just mentioned.

The way in which a uniform homogeneous brash layer responds locally to horizontal compression against a fixed barrier ought to depend on the displacement velocity (u) relative to the velocity of propagation for the front of a plastic disturbance (U). If u/U is very low a uniform thickening of the whole layer will occur, but if u/U > 1 there can be localized thickening, "plastic wave propagation," and pressure ridge formation. If there is a discontinuity where the layer thickness increases from h_1 to h_2, conservation of mass and momentum requires that

$$u/U = (h_2/h_1) - 1 \ .$$

For an increase from h_1 to h_2, an upper limit of h_2/h_1 can be estimated in various ways. From simple energy considerations the maximum value of h_2/h_1 is

$$\left(\frac{h_2}{h_1}\right)_{max} = \left(\frac{1+\sin\phi_1}{1-\sin\phi_1}\right)$$

where ϕ_1 is the friction angle for the original layer of thickness h_1. Different values ϕ_1 and ϕ_2 for the layer sections of thickness h_1 and h_2 can be accounted for by considering the thin section to develop a state of passive pressure, and the thick section a state of active pressure. Thus

$$\left[\left(\frac{1+\sin\phi_1}{1-\sin\phi_1}\right)\Big/\left(\frac{1+\sin\phi_2}{1-\sin\phi_2}\right)\right]^{1/2} \leq \frac{h_2}{h_1} \leq \left[\left(\frac{1+\sin\phi_1}{1-\sin\phi_1}\right)\left(\frac{1+\sin\phi_2}{1-\sin\phi_2}\right)\right]^{1/2} .$$

Another estimate can be made by applying the analogy of a retaining wall at both ends of the transition slope between the two sections of the ice layer (see Mellor, 1980).

The foregoing ideas have been applied to the analysis of ship resistance in brash ice (Mellor, 1980). While the assumption of a Coulomb-Navier-Mohr type of failure criterion seems reasonable for estimating stresses in the ice layer, forces against ship hulls and structures, and pressure ridge dimensions, it may not be appropriate for consideration of in-plane indentation. From comparison of the theoretical slip surfaces in front of an indenter (i.e. ship bow) with observed slip surfaces, it appears that a von Mises type of criterion might be a better assumption for problems of this type, in which the plane stress condition prohibits indefinite increase of shear strength with increase of confining pressure.

2. PRESSURE RIDGES AND LEADS

Pressure ridges and leads in sea ice are linear features representing compressive and tensile failures respectively. In the idealized forms, pressure ridges and leads are assumed to form in a band of uniform width, running at right angles to the direction of the major principal stress in the ice sheet.

The best known and most intricate model for pressure ridge formation is one by Parmerter and Coon (1972). An existing lead is assumed to close as compression develops, and ice rubble is assumed to disturb the isostatic equilibrium of the ice sheet, leading to flexure and breakage. The ice fragments are pushed together and piled up, with horizontal thrust from the unbroken ice sheet supplying the potential energy for piling up the ice against gravity.

Vertical dimensions of the ridge are limited by the available horizontal thrust, and ridge geometry is controlled partly by the angle of repose of the ice fragments. Development of a ridge is simulated by iterative calculations for the kinematics of the process.

The actual mechanical processes of pressure ridge formation are still not firmly established, and the following notes are intended to give an idea of the forces and stresses developed by some of the mechanical processes which might be involved in forming pressure ridges and leads. In general, they suggest that processes involving quasi-elastic deformation and quasi-brittle fracture call for horizontal thrust forces which are very high compared to the thrust likely to be developed by wind shear or current action.

2.1 Wind and Current Action

Sea ice is displaced and strained in the horizontal plane by wind shear on the upper surface or by water shear on the lower surface. In either case, the shear stress τ can be expressed as

$$\tau = C_D \rho u^2$$

where ρ is the fluid density, u is the "free stream" fluid velocity (relative to the ice), and C_D is a drag coefficient. For simple calculations, air density can be taken as 1.3 kg/m^3 and water density can be taken as 1000 kg/m^3. A recent determination of C_D for water drag under first-year sea ice gives $C_D = 1.32 \times 10^{-3}$ (Langleben, 1982), but for multiyear pack ice higher values have been measured, e.g. 2.2×10^{-3}, 4.14×10^{-3}, 6.6×10^{-3}, 8.3×10^{-3} (see Langleben's paper for details). The value of C_D for air varies from 1.3×10^{-3} for strong winds over flat snowfields, to the range 1.6×10^{-3} - 3.2×10^{-3} for open sea ice, and perhaps to higher values for very rough pressured sea ice. For present purposes we can consider smooth first-year sea ice with $C_D = 1.3 \times 10^{-3}$ both on top and underneath, and calculate the shear stress as a function of fluid speed (Figure 57). For rougher ice, with $C_D = 3 \times 10^{-3}$, values of τ are also shown in Figure 57. It might be noted that water shear and wind shear are of comparable magnitude for current velocities in the range 0.3 to 2 knots and wind speeds in the range 10 to 60 knots.

The shear stresses in Figure 57 are very low, so that they have to be integrated over large areas to produce significant forces. If F' is the force per unit width which is developed by integrating τ over an upstream distance L

$$F' = L\tau .$$

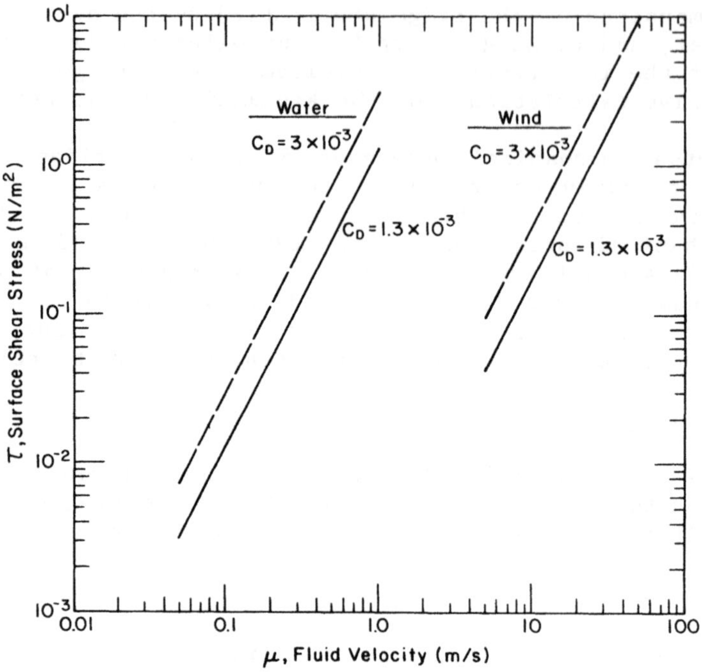

Fig. 57. Surface shear stress as a function of fluid velocity for wind and water drag against sea ice.

If $\tau = 1\ N/m^2$ is taken as a representative value for fairly fast currents (0.2 - 0.5 knot) or rather strong winds (40 - 45 knots), and it is then integrated over fetches of 10 km and 100 km, the resulting values of F' are:

L = 10 km	F' = 10 kN/m
L = 100 km	F' = 100 kN/m

Parmerter and Coon (1972) suggest that typical wind stresses on sea ice are of the order of 0.05 N/m^2. Such stresses lead to very small values of F', even with very large values of L, but they deduce that F' is only of order 1 kN/m during ridge formation (i.e. $\tau = 0.05\ N/m^2$, L = 20 km).

If the ice thickness t is of the order of 1 m, the average normal stress in the ice, F'/t, is very low, even with strong winds or strong currents acting over great distances. For example, a normal stress of 100 kN/m^2 (i.e. 1 bar) will cause ice to creep at a rate which is barely detectable over a period of days.

By contrast, if a fixed object of finite width projects through the ice sheet, or if a rigid body is embedded in the ice sheet, then wind and current shear is quite sufficient to produce very high local stresses.

2.2 Tensile and Compressive Failure of a Wide Ice Sheet

When a wide sheet of ice is subject to simple unidirectional shear from wind or current, the ice is in a state of uniaxial strain in the horizontal plane, but the upper and lower surfaces of the ice are essentially free from normal stress or significant shear stress. In the ice, the mean normal stress in the direction of the wind or current, σ_{xx}, is

$$\sigma_{xx} = F'/t$$

where F' will rarely exceed 0.1 MN/m, and t is typically between 0.1 and 2 m in first-year ice. This suggests that, while σ_{xx} might reach 1 MPa in very thin ice, it is likely to be less than 0.1 MPa in ice of appreciable thickness.

The tensile fracture strength of unflawed ice is up to 2 MPa when measured in the laboratory, but large masses of natural ice without obvious cracks or gross flaws might have a tensile fracture strength closer to 1 MPa. The compressive strength of intact ice is up to 10 MPa when measured in the laboratory, but for bulk sea ice the compressive strength might be more like 3 MPa.

The fracture strength measured in tests of short duration is not necessarily of direct relevance to the formation of leads and pressure ridges, since the sea ice may be straining for days before it fails. Nevertheless, taking into account probable magnitudes of strain rate and failure strain, it is hard to imagine average failure stresses as low as 0.1 MPa, either for compression or tension.

If the ice sheet is in tension, the stress state is such that failure can be expected when the tensile principal stress reaches the uniaxial tensile strength ($\sigma_{xx} = \sigma_T$), assuming that the sheet is uncracked to start with. However, since σ_{xx} is only likely to approach 1 MPa in very thin ice, spontaneous tensile fracture must be a rare occurrence. Because there are always cracks of some kind in the ice, especially thermal cracks or cracks caused by wave flexure, it seems more likely that leads will form by the growth of existing cracks. Referring back to the calculated values of effective tensile strength σ_T as a function of flaw size 2c, and using the same values of Young's modulus and surface energy (p. 251),

$$\sigma_T = \frac{2.52 \times 10^4}{\sqrt{c}} = \text{Pa}$$

where c is in metres. Thus, if there are existing cracks only 2 m long at right angles to the principal stress, the effective value of σ_T for instantaneous tensile fracture is only 0.025 MPa.

If the ice sheet is in compression, the stress state is such that failure could occur when the major principal stress reaches the uniaxial compressive strength, taking due account of anisotropy in the ice (crushing occurs by displacement to the stress-free boundaries). However, it is unlikely that a wide ice sheet could undergo simultaneous crushing failure along a broad front, even in very thin ice, since the maximum compression values of σ_{xx} are expected to be well below σ_c.

2.3 Buckling of a Wide Ice Sheet

For an ice sheet in compression, one possible mode of failure is crushing, as discussed above. Another possibility is buckling. From the usual analysis for an axially loaded infinite* elastic beam on an elastic foundation (Hétenyi, 1946), the critical force per unit width for buckling F' is

$$F_b' = \left[\frac{kEt^3}{3(1-\nu^2)}\right]^{1/2}$$

where k is the foundation modulus (in this case the unit weight of water), E is Young's modulus, ν is Poisson's ratio, and t is ice thickness. Sea ice is unlikely to buckle under truly elastic conditions, so that the value substituted for E should be less than the true Young's modulus. If we take E = 3 GPa, ν = 0.3, and k ≈ 1 Mgf/m^3 (or ≈ 10 kN/m^3)

$$F_b' = 3.3\ t^{3/2}\ \text{MN/m}$$

when t is in metres.

If t = 1 m, F_b' is 3.3 MN/m, i.e. well above any average value likely to be attained by F' under wind or water shear. If t = 0.1 m, F_b' is 0.1 MN/m, which is about the force level developed by strong wind or current acting over 100 km.

* The equation for the semiinfinite beam, used by some writers, gives half this value. The assumption of a semiinfinite beam seems to require that there be a hinge line where bending moments cannot be carried, and where there is no vertical force.

These estimates suggest that buckling by elastic instability across a broad front is unlikely to occur in any but the thinnest of ice covers, provided that the ice has uniform properties. However, an alternative analysis by Kerr (1980) introduces a new idea, and indicates that a semi-infinite plate of floating ice could buckle under a force smaller than F'_b as it is calculated from conventional theory. The magnitude of this effect cannot be assessed at present, as there is an undetermined coefficient in Kerr's result.

Another possibility is that the ice could buckle by creep, especially as the lower layers are likely to be warmer and softer than the top surface of the ice. If part of the ice sheet is depressed to the point where local flooding of the surface occurs, then there is no longer an elastic foundation in that area, and flexure is likely to accelerate.

2.4 Flexure of a Wide Ice Sheet

Suppose that a state of compression develops, and that the ice sheet has a discontinuity, such as an old lead, at right angles to the thrust direction. If overthrusting can develop at the discontinuity, the ice will be flexed by the wedge action of the overthrust. An analysis for a semiinfinite elastic beam on an elastic foundation is given by Hétenyi (1946), and for the case which is of particular interest the bending moment M is

$$M = -\frac{P_1}{\lambda} B_{\lambda x}$$

where P_1 is the vertical force applied to the free end of the beam. The other symbols are defined by

$$\lambda = \left(\frac{k}{4EI}\right)^{1/4}$$

$$B_{\lambda x} = e^{-\lambda x} \sin \lambda x$$

where k is the foundation modulus, E is Young's modulus, and I is the moment of inertia of the beam section. Differentiation of M with respect to horizontal distance from the free end x gives

$$\frac{\partial M}{\partial x} = P_1 e^{-\lambda x}(\sin \lambda x - \cos \lambda x).$$

M is a maximum or minimum with $\partial M/\partial x = 0$, i.e. when $x \to \infty$ or $x = \pi/2\lambda$. The required value of x for maximum bending moment in a wide plate is thus

$$x_* = \frac{\pi}{2}\left(\frac{4EI}{k}\right)^{1/4} = \frac{\pi}{2}\left(\frac{E\,t^3}{3k(1-\nu^2)}\right)^{1/4}.$$

This implies that, when the ice is flexed, it will tend to break first at a distance x_* from the free edge. Taking E = 3 GPa, k = 10 kN/m^3 and ν = 0.3, values of x_* are 5.1 m, 28.6 m, and 48.1 m for ice thicknesses of 0.1 m, 1.0 m and 2.0 m, respectively.

If there is no axial force being applied in the x-direction, the vertical force P_1 reaches a maximum value of

$$P_1 = -\frac{M_{max}\ \lambda}{B_{\lambda x*}}$$

where the maximum "failure" value of the bending moment at distance x_* is

$$M_{max} = \frac{\sigma_{FT}\ I}{t/2}$$

in which σ_{FT} is the apparent tensile strength of the ice derived from beam flexure tests. Thus, the value of P_1 at failure is

$$P_1 = -\frac{2\ \sigma_{FT}\ I}{t}\left(\frac{k}{4EI}\right)^{1/4}\frac{e^{\lambda x*}}{\sin\ \lambda x*} = -\frac{\sigma_{FT}}{6}\left[\frac{3kt^5}{E(1-\nu^2)^3}\right]^{1/4}\frac{e^{\pi/2}}{\sin\pi/2}\ .$$

From the discussion of beam flexure tests on ice, it can be seen that measured values of σ_{FT} vary widely, but for present purposes we assume σ_{FT} = 0.4 MPa. Taking E = 3 GPa, ν = 0.3, k = 10 kN/m^3,

$$P_1 = -\ 0.706\ t^{5/4} \quad \text{MN/m}$$

when t is in metres. For t = 1 m, $P_1 \approx$ 700 kN/m. For t = 0.1 m, $P_1 \approx$ 40 kN/m.

If P_1 is exerted by a wedge action as overthrusting occurs, the horizontal force F_f' that is needed to develop P_1 can be estimated in terms of an effective friction coefficient μ. For a small wedge angle,

$$F_f' \approx \mu\ P_1.$$

Taking $\mu \approx$ 0.3, F_f is about 200 kN/m for 1-m-thick ice, but only about 12 kN/m for ice that is 0.1 m thick. Thus, if conditions are favorable for overthrusting, flexural breakage seems likely to occur at thrust forces lower than those needed for buckling according to standard theory.

2.5 The Energetics of Ridge Formation

The energy needed to form a pressure ridge is assumed to be derived solely from the adjacent ice sheet, which thrusts with force F' per unit width in the x-direction. In order to build a ridge, potential energy has to be supplied to lift ice against gravity in the sail, and to push ice down against buoyancy in the keel. Energy is dissipated in the ice fracture process, and in overcoming frictional resistance between the ice fragments. For a lower limit estimate of the energy needed to build ridges, the energy dissipated in fracture and friction is neglected.

To illustrate the method, a pressure ridge can be idealized as shown in Figure 58, and the ridge and the formation process can be taken as symmetrical about the center line of the ridge, with horizontal displacements dx referred to that center line.

Work done by the ice sheet can be written as

$$dW = F' \, dx \ .$$

The mass per unit width M for half of the ridge is

$$M = M_s + M_k = \frac{1}{2} \bar{\rho}_s \, g \, h_s^2 \cot \alpha_s + \frac{1}{2} \bar{\rho}_k \, h_k^2 \cot \alpha_k$$

where M_s and M_k are masses for sail and keel respectively, $\bar{\rho}_s$ and $\bar{\rho}_k$ are mean bulk densities, and α_s and α_k are the angles of repose for sail and keel.

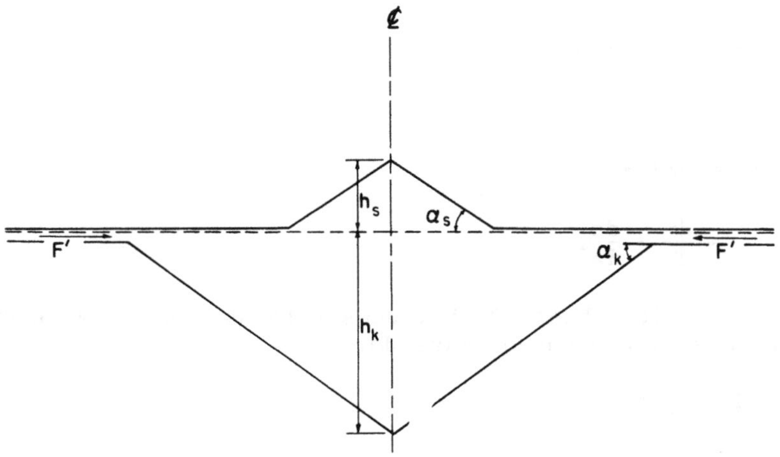

Fig. 58. Idealized cross sec'ion of new pressure ridge.

The potential energy of the sail E_s is

$$E_s = M_s\ g\ h_s/3 = \frac{1}{6}\ \bar{\rho}_s\ g\ h_s^3 \cot \alpha_s$$

and the potential energy of the keel E_k is

$$E_k = M_k\ g_{eff}\ h_k/3$$

where g_{eff} is the effective value of g for submerged ice, i.e.

$$g_{eff} = (1 - \frac{\rho_w}{\rho_i})g$$

in which ρ_i is the density of unbroken ice and ρ_w is water density. Thus

$$E_k = M_k\ (\frac{\rho_w}{\rho_i} - 1)\ g\ \frac{h_k}{3}$$

$$= \frac{1}{6}\ \bar{\rho}_k (\frac{\rho_w}{\rho_i} - 1)g\ h_k^3 \cot \alpha_k.$$

The total potential energy of the ridge (E) is therefore

$$E = E_s + E_k = \frac{g}{6}[\bar{\rho}_s\ h_s^3 \cot\alpha_s + \bar{\rho}_k(\frac{\rho_w}{\rho_i} - 1)h_k^3 \cot\alpha_k].$$

Equating work done by the ice sheet to energy gained by the ridge

$$dW/dx = dE/dx$$

or,

$$F' = dE/dx$$

$$= \frac{dE}{dh} \cdot \frac{dh}{dx}.$$

As the ice sheet of thickness t thrusts a distance dx, it contributes an increment of mass dM:

$$dM = \rho_i\ t\ dx$$

or,

$$\frac{dM}{dx} = \rho_i\ t.$$

Using this relation, dh/dx can be expressed as

$$\frac{dh}{dx} = \frac{dM}{dx} \cdot \frac{dh}{dM}.$$

In order to take the analysis further, additional assumptions have to be made. It is commonly assumed that the porosity of the ice rubble n is uniform, through both the sail and the keel. Thus

$$\bar{\rho}_s = \bar{\rho}_k = (1-n)\rho_i.$$

Some dubious assumptions have also been made about isostatic equilibrium. It has been assumed that the center of the ridge is in isostatic equilibrium, so that $h_s/h_k = (\rho_w/\rho_i)-1$. Taking $\rho_w = 1.026\ Mg/m^3$, and $\rho_i = 0.9\ Mg/m^3$ for sea ice fragments equilibrated with seawater, h_k/h_s would be about 7 according to this assumption. However, this is unrealistic, and it would be a bit closer to the mark to assume isostatic equilibrium for the complete sail against the complete keel, i.e.

$$h_s^2 \cot\alpha_s = h_k^2\left(\frac{\rho_w}{\rho_i} - 1\right)\cot\alpha_k$$

or,

$$\frac{h_s}{h_k} = \left[\frac{\cot\alpha_k}{\cot\alpha_s}\left(\frac{\rho_w}{\rho_i} - 1\right)\right]^{1/2}.$$

If geometric similarity is somehow maintained during the ridge-building process, h_s/h_k is constant, i.e.

$$\frac{h_s}{h_k} = C.$$

Making these assumptions,

$$\frac{dM}{dh} = (1-n)\rho_i\ h_s\left(\cot\alpha_s + C^2 \cot\alpha_k\right)$$

$$\frac{dE}{dh} = \frac{(1-n)\rho_i g}{2}\ h_s^2\left[\cot\alpha_s + \left(\frac{\rho_w}{\rho_i} - 1\right)C^3 \cot\alpha_k\right].$$

Finally,

$$F' = \frac{dE}{dh} \cdot \frac{dh}{dx} = \frac{dE}{dh} \cdot \frac{dM}{dx} \cdot \frac{dh}{dM}$$

$$= \frac{\rho_i\ g\ t\ h_s}{2} \cdot \frac{\frac{\cot\alpha_s}{\cot\alpha_k} + (\frac{\rho_w}{\rho_i} - 1)C^3}{\frac{\cot\alpha_s}{\cot\alpha_k} + C^2}.$$

To avoid algebraic nightmares, we can take numerical values for the dimensionless constants, e.g.

$$(\frac{\rho_w}{\rho_i} - 1) = 0.14.$$

α_s or α_k might vary between about 20° and 33°, so that $\cot\alpha_s/\cot\alpha_k$ (or the inverse ratio) might be in the range 0.56 to 1.78. Corresponding values of C would then be in the range 0.5 to 0.28 (i.e. h_k/h_s in the range 2 to 3.6).

The expression for the thrust force F' can now be written

$$F' = \frac{A}{2}\ \rho_i\ g\ t\ h_s$$

where A might have values in the range 0.71 to 0.96 (A = 1 for a rubble field, for thickened brash ice, or for a "kite-shaped" pressure ridge).

For ice which is 1 m thick, this gives F' in the range 3.14 h_s kN/m to 4.22 h_s kN/m when h_s is in metres. Taking realistic values of h_s, say 3 to 5 m, F' for t = 1 m is about 10 to 20 kN/m. This is the force level developed by a strong wind blowing over 10 to 20 km (see preceding notes on wind action). Parmerter and Coon (1972) considered that ridge-building forces were an order of magnitude lower than this, i.e. 1 kN/m.

Similar methods can be applied to brash ice using fewer assumptions. The limiting thickness of pressure ridges, including ridges formed against fixed structures and ship hulls, can be deduced by these methods, and by other methods derived from soil mechanics and plasticity theory (Mellor, 1980). Some results were outlined in the foregoing section on mechanical properties of brash ice.

2.6 The Mechanics of Pressure Ridge Formation

Where floating ice thrusts against a fixed object of limited width, or where a rigid object is embedded in straining ice, wind or current forces can cause the ice to fail by flexure, buckling or crushing. However, the forces likely to be produced by moderate winds or currents seem insufficient to crush thick ice across a broad front, or to buckle it according to the generally accepted model. Flexural breaks seem more likely, provided that local conditions favor the initiation of flexure. It is probably unrealistic to consider pressure ridge formation as a process of elastic deformation and brittle fracture. Sustained wind loads are likely to induce creep deformation and creep rupture, processes which can develop at stress levels well below those involved in elastic straining to failure. At present, the processes of pressure ridge formation are not well understood.

REFERENCES

Abele, G. and G. Frankenstein (1967) Snow and ice properties as related to runways in Antarctica. USA Cold Regions Research and Engineering Laboratory, Technical Report 176, 37 p.

Anderson, D.L. (1958) Preliminary results and review of sea ice elasticity and related studies. Trans. Eng. Inst. Can., 2(3): 116-122.

Anderson, D.L. (1960) The physical constants of sea ice. Research, 13(8): 310-318.

Babel, H.W. and G. Sines (1968) A triaxial fracture interior for porous brittle materials. J. Basic Eng., Trans. ASME, June: 285-291.

Baker, R.W. (1978) The influence of ice crystal size on creep. J. Glaciol., 21(85): 485-500.

Brown, J.H. (1963) Elasticity and strength of sea ice. In Ice and Snow: Properties, Processes, and Applications (W.D. Kingery, Ed.), MIT Press, Cambridge, Mass., p. 79-106.

Butkovich, T.R. (1954) Ultimate strength of ice. USA Snow, Ice and Permafrost Research Establishment, Research Report 11, 12 p.

Butkovich, T.R. (1956) Strength studies of sea ice. USA Snow, Ice and Permafrost Research Establishment, Research Report 20, 15 p.

Butkovich, T.R. (1959) On the mechanical properties of sea ice, Thule, Greenland, 1957. USA Snow, Ice and Permafrost Research Establishment, Research Report 54, 20 p.

Carter, D. (1970) Brittle fracture of snow ice. In Proceedings, IAHR Symposium, Ice and Its Action on Hydraulic Structures, Reykjavik, Iceland, Paper 5.2.

Cox, G.F.N. and W.F. Weeks (1982) Equations for determining the gas and brine volumes in sea ice samples. USA Cold Regions Research and Engineering Laboratory, CRREL Report 82-30.

Cox, G.F.N., J. Richter, W.F. Weeks, M. Mellor and H. Bosworth (1984) The mechanical properties of multi-year sea ice, Phase I: Test results. USA Cold Regions Research and Engineering Laboratory, CRREL Report 84-9.

Currier, J.H. and E.M. Schulson (1982) The tensile strength of ice as a function of grain size. Acta Metallurgica, 30: 1511-1514.

Durham, W.B., H.C. Heard and S.H. Kirby (1982) Deformation of ice at pressures to 350 MPa at 77 to 195 K. EOS, 63(45): 1094, T51B-01.

Duval, P. and H. LeGac (1980) Does the permanent creep-rate of polycrystalline ice increase with crystal size? J. Glaciol., 25(91): 151-157.

Dykins, J.E. (1970) Ice engineering-tensile properties of sea ice grown in a confined system. U.S. Naval Civil Engineering Laboratory, Technical Report R689.

Fletcher, N.H. (1970) The Chemical Physics of Ice. Cambridge University Press, 271 p.

Frederking, R. (1977) Plane-strain compressive strength of columnar-grained and granular-snow ice. J. Glaciol., 18(80): 505-516.

Frederking, R.M.W. and G.W. Timco (1980) NRC ice property measurements during the CANMAR KIGORIAK trials in the Beaufort Sea, winter 1979-80. National Research Council of Canada, Division of Building Research, NRCC 18722, 63 p.

Frederking, R.W. and G.W. Timco (1981) Mid-winter mechanical properties of ice in the southern Beaufort Sea. In POAC 81: Proceedings, Sixth International Conference on Port and Ocean Engineering Under Arctic Conditions, Laval University, Quebec City, Quebec, Canada, Vol. 1, p. 225-234.

Gold, L.W. (1958) Some observations of the dependence of strain on stress for ice. Can. J. Phys., 36(10): 1265-1275.

Gold, L.W. (1977) Engineering properties of fresh-water ice. J. Glaciol., 19(81): 197-212.

Goodman, D.J. (1979) Critical stress intensity (K_{IC}) measurements at high loading rates for polycrystalline ice. In Physics and Mechanicsof Ice (P. Tryde, Ed.), Springer-Verlag, Berlin, p. 129-146.

Goodman, D.J. and D. Tabor (1978) Fracture toughness of ice: A preliminary account of some new experiments. J. Glaciol., 21: 651-660.

Hamza, H. and D.B. Muggeridge (1979) Plain strain fracture toughness (K_{IC}) of fresh water ice. In POAC 79: Proceedings, Fifth International Conference on Port and Ocean Engineering Under Arctic Conditions, University of Trondheim, Trondheim, Norway, Vol. 1, p. 697-707.

Häusler, F.U. (1981) Multiaxial compressive strength tests on saline ice with brush-type loading platens. In Proceedings, IAHR International Symposium on Ice, Quebec, Canada, Vol. II, p. 526-536.

Hawkes, I. and M. Mellor (1970) Uniaxial testing in rock mechanics laboratories. Engineering Geology, 4(3): 177-285.

Hawkes, I. and M. Mellor (1972) Deformation and fracture of ice under uniaxial stress. J. Glaciol., 11(61): 103-131.

Haynes, F.D. (1973) Tensile strength of ice under triaxial stresses. USA Cold Regions Research and Engineering Laboratory, Research Report 312.

Haynes, F.D. (1978) Effect of temperature on the strength of snow-ice. USA Cold Regions Research and Engineering Laboratory, CRREL Report 78-27, 18 p.

Haynes, F.D. and M. Mellor (1977) Measuring the uniaxial compressive strength of ice. J. Glaciol., 19(81): 213-223.

Hétenyi, M. (1946) Beams on Elastic Foundation. University of Michigan Press, Ann Arbor, Michigan, 255 p.

Hobbs, P.V. (1974) Ice Physics. Clarendon Press, Oxford, 837 p.

Jaeger, J.C. and N.G.W. Cook (1976) Fundamentals of Rock Mechanics. Chapman and Hall/John Wiley, New York, 585 p.

Jones, S.J. (1978) Triaxial testing of polycrystalline ice. In Proceedings, Third International Conference on Permafrost, Edmonton, Alberta, Canada, National Research Council of Canada, p. 670-674.

Jones, S.J. and H.A.M. Chew (1981) On the grain size dependence of secondary creep. J. Glaciol., 27(97): 517-518.

Keinonen, A. and T. Nyman (1978) An experimental model-scale study of the compressible, frictional and cohesive behavior of broken ice mass. In IAHR Symposium on Ice Problems, Lulea, Sweden, Part 2, p. 335-353.

Kerr, A.D. (1980) On the buckling force of floating ice plates. In Physics and Mechanics of Ice (P. Tryde, Ed.), Springer-Verlag, Berlin, p. 163-178.

Kohnen, H. (1972) Seismic and ultrasonic measurements on the sea ice of Eclipse Sound near Pond Inlet, N.W.T. in northern Baffin Island. Polarforschung, 44(2): 66-74.

Kovacs, A., W.F. Weeks and F. Michitti (1969) Variation of some mechanical properties of polar snow, Camp Century, Greenland. USA Cold Regions Research and Engineering Laboratory, Research Report 276, 33 p.

Kovacs, A. and D.S. Sodhi (1980) Shore ice pile-up and ride-up: Field observations, models, theoretical analyses. Cold Reg. Sci. Tech., 2: 209-288.

Lainey, L. and R. Tinawi (1981) Parametric studies of sea-ice beams under short and long term loading. In Proceedings, IAHR International Symposium on Ice, Quebec, Canada, p. 440-452.

Langleben, M.P. (1982) Water drag coefficient of first-year sea ice. J. Geophys. Res., 87(61): 573-578.

Langleben, M.P. and E.R. Pounder (1963) Elastic parameters of sea ice. In Ice and Snow: Properties, Processes and Applications (W.D. Kingery, Ed.), MIT Press, Cambridge, Mass., p. 69-78.

Liu, H.W. and K.J. Miller (1979) Fracture toughness of fresh water ice. J. Glaciol., 22(68): 135-143.

Määttänen, M. (1976) On the flexural strength of brackish water ice by in-situ tests. In Proceedings, Third International Conference on Port and Ocean Engineering Under Arctic Conditions, University of Alaska, Fairbanks, Alaska, USA, Vol. 1, p. 349-359.

Mellor, M. (1975) A review of basic snow mechanics. In Symposium on Snow Mechanics, Grindelwald, Switzerland, International Commission on Snow and Ice, IAHR Publication 114, p. 251-291.

Mellor, M. (1979) Mechanical properties of polycrystalline ice. In Physics and Mechanics of Ice (P. Tryde, Ed.), Springer-Verlag, Berlin, p. 217-245.

Mellor, M. (1980) Ship resistance in thick brash ice. Cold Reg. Sci. Tech., 3(4): 305-321.

Mellor, M. and R. Testa (1969) Effect of temperature on the creep of ice. J. Glaciol., 8(52): 131-145.

Mellor, M. and I. Hawkes (1971) Measurement of tensile strength by diametral compression of discs and annuli. Engineering Geology, 5: 173-225.

Mellor, M. and D.M. Cole (1982) Deformation and failure of ice under constant stress or constant strain-rate. Cold Reg. Sci. Tech., 5(3): 201-219.

Mellor, M. and D.M. Cole (1983) Stress/strain/time relations for ice under uniaxial compression. Cold Reg. Sci. Tech., 6(3): 207.

Miller, K.J. (1979) The application of fracture mechanics to ice problems. In Physics and Mechanics of Ice (P. Tryde, Ed.), Springer-Verlag, Berlin, p. 265-277.

Murat, J.-R. and L.M. Lainey (1982) Some experimental observations on the Poisson's ratio of sea ice. Cold Reg. Sci. Tech., 6(2): 105-113.

Murrell, S.A.F. (1963) A criterion for brittle fracture of rocks and concrete under triaxial stress and the effect of pore pressure on the criterion. In Rock Mechanics (T. Fairhurst, Ed.), Proceedings, Fifth Symposium on Rock Mechanics, Pergamon Press, p. 563-577.

Nippon Kokan (NKK) (n.d.) Ice engineering, Part I. Study on ice force on offshore structure.

Nippon Kokan (NKK) (n.d.) Ice engineering. Printed technical document produced by NKK, Tokyo.

Paige, R.A. and C.W. Lee (1967) Preliminary studies on sea ice in McMurdo Sound, Antarctica, during Deep Freeze 65. J. Glaciol., 6(46): 515-528.

Panov, V.V. and N.V. Fokeev (1977) Compression strength of sea ice specimens under complex loading. Probl. Arkt. Antarkt., 49: 81-86 (English translation, p. 97-104).

Parmerter, R.R. and M.D. Coon (1972) Model of pressure ridge formation in sea ice. J. Geophys. Res., 77(33): 6565-6575.

Peyton, H.R. (1966) Sea ice strength. Geophysical Institute, University of Alaska, Report UAG R-182.

Prodanovic, A. (1979) Model tests of ice rubble strength. In POAC 79: Proceedings, Fifth International Conference on Port and Ocean Engineering Under Arctic Conditions, University of Trondheim, Trondheim, Norway, Vol. I, p. 89-105.

Ralston, T.D. (1980) Yield and plastic deformation in ice crushing failure. In Sea Ice Processes and Models (R.S. Pritchard, Ed.), University of Washington Press, Seattle, Washington, p. 234-245.

Saeki, H. and A. Ozaki (1981) Experimental study on flexural strength and elastic modulus of sea ice. In POAC 81: Proceedings, Sixth International Conference on Port and Ocean Engineering Under Arctic Conditions, Laval University, Quebec City, Quebec, Canada, Vol. 1, p. 536-547.

Sandkvist, J. (1981) Conditions in brash ice covered channels with repeated passages. In POAC 81: Proceedings, Sixth International Conference on Port and Ocean Engineering Under Arctic Conditions, Laval University, Quebec City, Quebec, Canada, Vol. 1, p. 244-252.

Schwarz, J. (1971) The pressure of floating ice-fields on piles. In Proceedings, IAHR Symposium, Ice and Its Action on Hydraulic Structures, Reykjavik, Iceland, Paper 6.3.

Schwarz, J. (1975) On the flexural strength and elasticity of saline ice. In Proceedings, Third International Symposium on Ice Problems (G. Frankenstein, Ed.), Hanover, New Hampshire, IAHR, Committee on Ice Problems, p. 373-385.

Schwarz, J. and W.F. Weeks (1977) Engineering properties of sea ice. J. Glaciol., 19(81): 499-530.

Schwarz, J. et al. (1981) Standardized testing methods for measuring mechanical properties of ice. Cold Reg. Sci. Tech., 4(3): 245-253.

Schwerdtfeger, P. (1963) The thermal properties of sea ice. J. Glaciol., 4(36): 789-807.

Shapiro, L.H., R.C. Metzner and J.B. Johnson (1981) Fracture toughness of sea ice. Geophysical Institute, University of Alaska, for Shell Development Company.

Sinha, N.K. (1978a) Short-term rheology of polycrystalline ice. J. Glaciol., 21(85): 457-473.

Sinha, N.K. (1978b) Rheology of columnar-grained ice. Experimental Mechanics, 18(12): 464-470.

Sinha, N.K. (1981a) Rate sensitivity of compressive strength of columnar-grained ice. Experimental Mechanics: 209-218.

Sinha, N.K. (1981b) Constant stress rate deformation modulus of ice. In POAC 81: Proceedings, Sixth International Conference on Port and Ocean Engineering Under Arctic Conditions, Laval University, Quebec City, Quebec, Canada, Vol. 1, p. 216-224.

Sinha, N.K. (1981c) Comparative study of ice strength data. In Proceedings, IAHR International Symposium on Ice, Quebec, Canada, p. 421-432.

Sinha, N.K. and R. Frederking (1979) Effect of test system stiffness on strength of ice. In Proceedings, Fifth International Conference on Port and Ocean Engineering Under Arctic Conditions, University of Trondheim, Trondheim, Norway, p. 708-717.

Tabata, T. (1958) Studies on the visco-elastic properties of sea ice. In Arctic Sea Ice, U.S. National Academy of Sciences - National Research Council, Pub. 598, p. 139-147.

Tabata, T. (1960) Studies of the mechanical properties of sea ice. V: Measurement of flexural strength. Low Temp. Sci., A(19): 187-201.

Tabata, T. (1966) Studies of the mechanical properties of sea-ice. X: The flexural strength of small sea-ice beams. Institute of Low Temperature Science, Hokkaido University, Sapporo, Japan, Contribution 793.

Tabata, T. (1967) Studies of the mechanical properties of sea ice: The flexural strength of small sea ice beams. In Physics of Snow and Ice (H. Oura, Ed.), Institute of Low Temperature Science, Hokkaido University, Sapporo, Japan, p. 481-497.

Tabata, T., K. Fujino and M. Aota (1967) Studies of the mechanical properties of sea ice. XI: The flexural strength of sea ice in situ. In Physics of Snow and Ice (H. Oura, Ed.), Institute of Low Temperature Science, Hokkaido University, Sapporo, Japan, Vol. I, Pt. 1, p. 539-550.

Tabata, T., Y. Suzuki and M. Aota (1975) Ice study in the Gulf of Bothnia. II: Measurements of flexural strength. Low Temp. Sci., A(33): 199-206.

Tabata, T., T. Kawamura and T. Takizawa (1976) Arctic sea ice research. I: Measurements of flexural strength of small sea ice beams. Low Temp. Sci., A(34): 201-207.

Tinawi, R. and J.-R. Murat (1977) Sea-ice testing in flexure. In POAC 77: Proceedings, Fourth International Conference on Port and Ocean Engineering Under Arctic Conditions (D.B. Muggeridge, Ed.), Memorial University of Newfoundland, St. John's, Newfoundland, Canada p. 638-653.

Traetteberg, A., L.W. Gold and R. Frederking (1975) The strain rate and temperature dependence of Young's modulus of ice. In Proceedings, Third International Symposium on Ice Problems (G. Frankenstein, Ed.), Hanover, New Hampshire, IAHR Committee on Ice Problems, p. 479-486.

Urabe, N., T. Iwasaki and A. Yoshitake (1980) Fracture toughness of sea ice. Cold Reg. Sci. Tech., 3(1): 29-37.

Urabe, N. and A. Yoshitake (1981a) Fracture toughness of sea ice: In-situ measurement and its application. In POAC 81: Proceedings, Sixth International Conference on Port and Ocean Engineering Under Arctic Conditions, Laval University, Quebec City, Quebec, Canada, Vol. 1, p. 356-365.

Urabe, N. and A. Yoshitake (1981b) Strain rate dependent fracture toughness (K_{IC}) of pure ice and sea ice. In Proceedings, IAHR International Symposium on Ice, Quebec, Canada, p. 410-420.

Vaudrey, K.D. (1977) Ice engineering: Study of related properties of floating sea-ice sheets and summary of elastic and visco-elastic analyses. U.S. Naval Civil Engineering Laboratory, Technical Report R860, 79 p.

Wang, Y.S. (1979a) Crystallographic studies and strength tests of field ice in the Alaskan Beaufort Sea. In POAC 79: Proceedings, Fifth International Conference on Port and Ocean Engineering Under Arctic Conditions, University of Trondheim, Trondheim, Norway, Vol. 1, p. 651-665.

Wang, Y.S. (1979b) Sea ice properties. In Technical Seminar on Alaskan Beaufort Gravel Island Design, Exxon, USA.

Wang, Y.S. (1981) Uniaxial compression testing of arctic sea ice. In POAC 81: Proceedings, Sixth International Conference on Port and Ocean Engineering Under Arctic Conditions, Laval University, Quebec City, Quebec, Canada, Vol. 1, p. 346-355.

Weeks, W.F. and A. Assur (1967) The mechanical properties of sea ice. USA Cold Regions Research and Engineering Laboratory, Monograph II-C3.

Weeks, W.F. and A. Assur (1969) Fracture of lake and sea ice. USA Cold Regions Research and Engineering Laboratory, Research Report 269.

Weiss, R.T., A. Prodanovic and K.W. Wood (1981) Determination of the shear properties of ice rubble. In Proceedings, IAHR International Symposium on Ice, Quebec, Canada, Vol. II, p. 648-658.

CHAPTER 3

THE ATMOSPHERIC BOUNDARY LAYER

Gordon McBean

Atmospheric Environment Service
Institute of Ocean Sciences
Sidney, B.C., Canada

1. INTRODUCTION

In consideration of any problem in air-ice-water interaction the atmospheric boundary layer (ABL) must, of course, be of central concern. In this chapter the ABL will be discussed, first, in a general way to reflect upon the general characteristics of the ABL and, then, in specific ways to explore what is known or not known about the ABL over snow, ice and water surfaces. In the general discussion an attempt is made to convey a sense of the problem and give references where the details may be found. It is appropriate to start the discussion by attempting to first define what is meant by the ABL. The ABL will be considered to be that part of the atmosphere to which the presence and characteristics of the earth's surface (land, water, ice, vegetation, etc.) are directly important. The "directness" of this importance has to be somewhat subjective in that all the atmosphere is affected in one sense or other by the characteristics of the earth's surface. Stewart (1979) defined the ABL as "that portion of the atmosphere which contains some fluid which has recently been close to the surface" and then decided that "recently" meant "within one day." In some circumstances the boundary layer might then extend to the top of a towering thunderstorm whereas for a windless polar night the depth may be only tens of meters. Most typically, the ABL is of the order of hundreds of meters to a few kilometers in depth.

The importance of the ABL can be viewed from two perspectives. First the ABL and the processes within it are significant in their own right, if for no other reason than that most human activity takes place within it. It is important to know the variations of wind, temperature, humidity, turbulence, etc., in the vertical and horizontal directions with time. The second perspective is to look at the ABL as the linkage between the free atmosphere above the ABL and the surface. The ice physicist may be only interested in the energy flux to or from the ice and the wind drag on it. The boundary-layer meteorologist is usually asked to relate these to the larger-scale meteorological parameters. The large-scale meteorological modeler is interested in the vertical fluxes and motions at the top of the boundary layer to feed into the model. Although the end result of research may be a linkage equation it is necessary to study the processes of the linkage and important for the ice physicist, for example, to have some appreciation of the limitations and assumptions inherent in the linkage relationship.

The ABL or parts of it have been the subject of several recent reviews. Stewart (1979) has given an elegant non-mathematical discussion while Wyngaard (1980), McBean (1979) and Haugen (1973) have all edited volumes covering a range of ABL topics. DeMoor (1978) and Coantic (1978) have both discussed the turbulence in the ABL while the two volumes of Monin and Yaglom (1971, 1975) are a comprehensive reference for the field. Earlier NATO ASI's have dealt with related topics; of particular relevance is the Modeling and Prediction of the Upper Layers of the Ocean (Kraus, ed., 1977; in particular Busch; pp. 72-91). Reference should be made to these works (and their extensive bibliographies) for the details relevant to the following discussion.

What are some of the characteristics of the ABL? Most important, the ABL is turbulent, continuously in space and time. This turbulent nature is responsible for the complexity of the ABL processes and makes them much more difficult to observe and understand. A turbulent boundary layer is much more efficient in transferring momentum, heat and mass in the vertical than a laminar boundary layer. It is observed that the turbulence is very dependent on the surface conditions, being more turbulent over rougher surfaces. Hence, the turbulence will be different over ice, snow, water, etc. Also the nature of turbulence is very different over a warm surface than over a cold surface (relative to the air above). In the former case, buoyant motions enchance the turbulent motions leading to rapid vertical mixing while, in the latter, the turbulence is suppressed and vertical transfers become very small. These differences will be evident over a patchy ice-water surface. Surface friction leads to an increase in horizontal wind speed with height and the ABL characteristics will determine how this low-level wind profile will match up with the free atmosphere wind above. The ABL does interact in important ways with the free atmosphere above. Sometimes the transition is abrupt, as with a capping inversion, and other times the transition is diffuse with no clear distinction between the two regions. The nature of the turbulence (Is it continuous or not in time and space?) may be the criteria for making the distinction.

The surface conditions and the free atmosphere aloft serve to impose general states on the ABL (Figure 1). If the surface is relatively warm, parcels of air originating near the surface will accelerate upward and continue to do so as long as they remain warmer than the environment. This is classed as an unstable ABL. If the vertical motion of parcels is suppressed by the thermal structure, then it is classed as a stable ABL. Generally the criteria can be made in terms of potential temperature, defined as the temperature a parcel would attain if brought adiabatically to a reference pressure (usually 1000 mb). An atmosphere of uniform

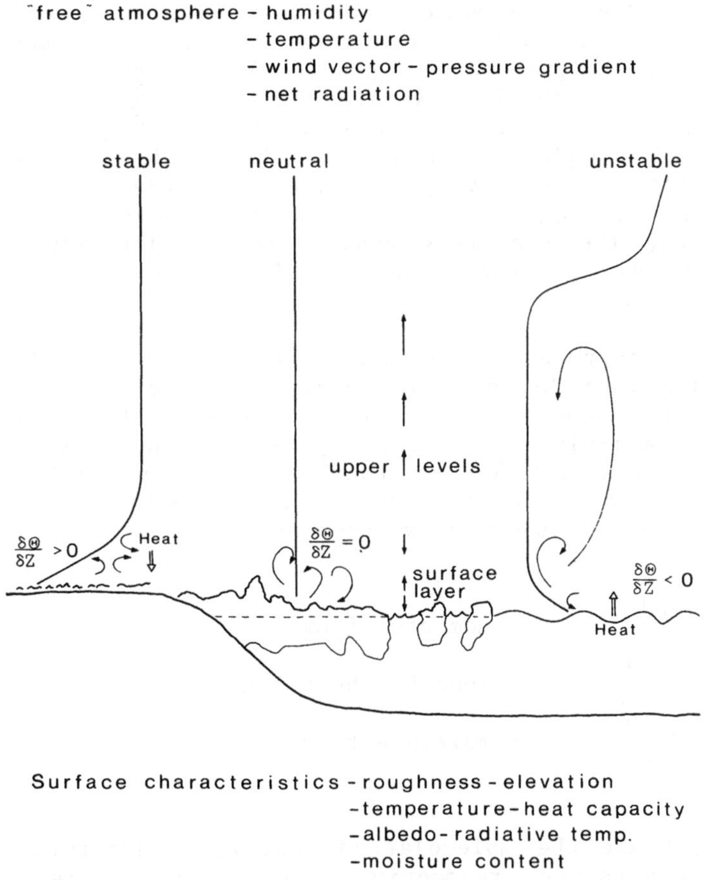

Fig. 1. Diagrammatic sketch of boundary layer over snow-ice-water surface. Three possible potential temperature profiles are indicated.

potential temperature, Θ, would have a vertical temperature gradient of:

$$\frac{\partial T}{\partial z} = -g/c_p \sim 9.8^{o}C/km; \ \frac{\partial \Theta}{\partial z} = 0 \quad (3.1)$$

where: g = acceleration due to gravity
c_p = specific heat at constant pressure

If $\partial\Theta/\partial z < 0$, then the atmosphere is unstable; if $\partial\Theta/\partial z > 0$, it is stable; and if $\partial\Theta/\partial z = 0$, it is neutrally stratified.

The vertical gradient of potential temperature will not be uniform throughout the ABL. In the case of convective conditions under sunny skies, the gradient will be very negative near the surface due to surface solar heating. Generally, however, the potential temperature will become more uniform or perhaps slightly increasing with height (stable gradient) in the middle and upper parts of the ABL. In the inversion layer at the top of the ABL the gradient will be strongly stable. Despite the near neutral or slightly stable conditions over more than half the depth of the ABL, it is still convectively dominated, and the heat flux upward (except near the top), independent of the local gradient. For these reasons, the stability designation of the ABL is usually based on the potential temperature gradient near the surface or on the surface flux to which it is related.

For many purposes it is useful to think about the ABL as being divided into layers. It should be noted that these layers do not have barriers or abrupt changes between them; rather there are smooth transitions. First, adjacent to the surface and very thin (typically about 0.1 mm thick) is the molecular sublayer where the vertical transfers or fluxes are dominated by molecular processes. The fluxes may be written as:

$$\begin{aligned} \tau &= -\rho\nu \frac{\partial U}{\partial z} && = \text{momentum flux} \\ H &= -\rho c_p K \frac{\partial \Theta}{\partial z} && = \text{sensible heat flux} \\ M &= -\rho\, D \frac{\partial Q}{\partial z} && = \text{moisture flux} \end{aligned} \quad (3.2)$$

where ν, K, D are the molecular viscosity, conductivity and diffusivity for moisture, respectively; U, Θ, Q are the mean wind, temperature and humidity (mass of water vapour per unit mass of dry air); and ρ is the density. These diffusivities are properties of the fluid (e.g., water, air, mercury). The molecular sublayer is difficult to observe directly and it tends to breakdown over rough surfaces.

About the molecular sublayer and extending upward a few tens of meters is the surface boundary layer. This is the most-studied and best-understood part of the ABL. The processes are dominated by turbulence. If u, v, w, θ, and q are the turbulent fluctuations of horizontal (along and transverse to the mean wind direction) wind, vertical wind, temperature and humidity with respect to their means and if they are correlated with w then there will be vertical transfers of momentum, heat and moisture by the turbulence. Hence the total fluxes will be:

$$
\begin{aligned}
\tau &= -\rho\ (\nu\frac{\partial U}{\partial z} + \overline{uw}) \\
H &= \rho c_p\ (-K\ \frac{\partial \Theta}{\partial z} + \overline{w\theta}) \qquad (3.3) \\
M &= \rho(-D\ \frac{\partial Q}{\partial z} + \overline{wg})
\end{aligned}
$$

where the overbar represents the average. In the surface layer, the mean wind direction is observed to be approximately constant with respect to z. Hence, there is no preferred transverse direction and $\overline{vw}$ should be zero. At higher levels where there is significant wind direction shear this term will be nonzero.

In analogy with the molecular processes, the turbulent fluxes have frequently been represented by, e.g.,

$$\overline{uw} = -K_M\ \frac{\partial U}{\partial z} \qquad (3.4)$$

where K_M is the turbulent viscosity and depends on the turbulence characteristics, i.e., a property of the flow, not the fluid. It would be expected that K_M would be of order of: a turbulent velocity scale (say 1 ms^{-1}) times a turbulent velocity length (say 100 m) so that:

$$\frac{K_M}{\nu} \sim \frac{(1\ ms^{-1})\ (100m)}{(10^{-5}\ m^2 s^{-1})} \sim 10^7 \qquad (3.5)$$

Since $\nu \sim K \sim D$ and it is expected that the turbulent diffusivities will be of similar magnitude, it can be seen that the turbulent fluxes are very much important than the molecular fluxes can be neglected. It will be shown later that the turbulent fluxes, under assumptions of horizontal homogeneity and time stationarity, are approximately constant with height in the surface layer.

Between the surface layer and the top of the ABL is a region without a well-accepted name, it is variously called: the Ekman layer, because the wind profile in certain circumstances approximates an Ekman spiral; the transition layer; the outer layer (mainly in the Soviet literature) to distinguish it from the inner or surface layer; the planetary boundary layer (which is also used to describe the whole ABL); and the upper levels. The term upper levels will be used here. The upper levels are characterized by a changing wind direction with height and a more complicated structure than the surface layer.

The governing equations for the ABL are the Navier-Stokes equations with the Boussinesq approximations (the density variations affect the weight but not otherwise the mass). After a Reynolds decomposition into mean and fluctuating components the equations for the mean flow can be written:

$$\frac{\partial U_i}{\partial t} + U_j \frac{\partial U_i}{\partial x_j} = - \frac{1}{\rho} \frac{\partial P}{\partial x_i} + \frac{g\Theta_v}{T_{v_o}} \delta_{13} - 2\,\Omega\varepsilon_{ijk}\,\eta_j U_k - \frac{\partial}{\partial x_j}\left[\overline{U_i U_j} - \nu \frac{\partial U_i}{\partial x_j}\right] \tag{3.6}$$

where: U_i is the mean velocity component in the x_i direction ($i = 1,2,3$ for the longitudinal, transverse and vertical);
P is the mean pressure deviation from a hydrostatic, dry adiabatic, motionless reference state;
$\Theta_v = \Theta(1+0.61Q)$ is the mean virtual potential temperature deviation from the same reference state. The 0.61Q term adds the effect of moisture on density and hence buoyancy;
Ω is the earth's angular velocity, and
η_j is a unit vector parallel to the earth's rotation vector. This term represents the Coriolis force due to a rotating coordinate system.

This equation represents a balance between the accelerations and advective changes on the left-hand side and the sum of the pressure, buoyancy, Coriolis and stress (turbulent and viscous forces) on the right-hand side. Note the effect of turbulence is to add an additional stress to the balance of forces. When coupled with the equations of state, thermodynamic energy, continuity of mass and of moisture the number of equations is 6 and the unknown means also 6 (U_1, U_2, U_3, Θ, P, Q). However, there are also the turbulence terms $\overline{u_i u_j}$ (6 independent ones), $\overline{u_i \theta}$(3) and $\overline{u_i q}$(3) for 12 additional unknowns.

Turbulence certainly compounds the problem. Equations for $\overline{u_i u_j}$, etc., can be derived but they unfortunately introduce a new set of unknown such as $\overline{u_i u_j u_k}$, $\overline{u_i p}$ and $\overline{u_i u_j \theta}$ (where p is the pressure fluctation). This is the closure problem of turbulence. At some point in the chain the sequence to higher order combinations of turbulent variables must be closed by, hopefully, rational assumptions relating the higher order variables to known quantities of lower order.

2. IDEALIZED BOUNDARY LAYERS

Because of the complexity of the atmospheric boundary layer under general conditions, both modelers and experimentalists have looked for and investigated idealized ABL's. The most common idealization, and one with quite a few areas of application, is that of horizontally homogeneous flow over a uniform surface. The equations reduce to:

$$\begin{aligned} \frac{\partial U}{\partial t} &= -\frac{1}{\rho}\frac{\partial P}{\partial x} + f\,V - \frac{\partial}{\partial z}\,\overline{uw} \\ \frac{\partial V}{\partial t} &= -\frac{1}{\rho}\frac{\partial P}{\partial y} - f\,U - \frac{\partial}{\partial z}\,\overline{vw} \\ \frac{\partial \Theta}{\partial t} &= -\frac{\partial}{\partial z}\,\overline{w\theta} \\ \frac{\partial Q}{\partial t} &= -\frac{\partial}{\partial z}\,\overline{wq} \end{aligned} \tag{3.7}$$

where $U = U_1$, $V = U_2$ and $(x,y,z) = (x_1,x_2,x_3)$.

Only the mean pressure field, P, is allowed to vary with x and y, as imposed by large-scale synoptic conditions. The Coriolis parameter, f, is $2\Omega\sin\phi$, where ϕ is the latitude. Phase changes, radiative heating and molecular terms have been neglected. It can be seen that local changes in Θ and Q can arise only from vertical divergences of the turbulent fluxes of heat and moisture.

The geostrophic wind $\vec{G} = \{U_g, V_g\}$ is defined as:

$$U_g = -\frac{1}{\rho f}\frac{\partial P}{\partial y} \; ; \quad V_g = \frac{1}{\rho f}\frac{\partial P}{\partial x} \tag{3.8}$$

For time-steady, frictional-less flow the actual wind must therefore be geostrophic. For just time-steady flow the equations reduce to:

$$\frac{\partial \overline{uw}}{\partial z} = f(V-V_g); \quad \overline{uw}\,\Big|_o = f \int_0^h (V-V_g)dz$$

$$\frac{\partial \overline{vw}}{\partial z} = -f(U-U_g) \; ; \quad \overline{vw}\,\Big|_o = f \int_0^h (U_g-U)dz \tag{3.9}$$

where the integration is between $z = 0$ and $z = h$, where the turbulent stresses are assumed to be zero (above the ABL). The x-axis (Figure 2) has been oriented to be along the surface wind direction. Since there is no preferred lateral direction in the surface layer, the correlation between v and w will be zero. This second equation means that $U > U_g$ for some range of heights since $U = U_g$ at the top of the ABL. The first equation can be used to deduce $(\overline{uw})_o$, the surface stress. This is called the geostrophic departure method.

2.1 The Constant-Flux Layer

At small z, the transverse wind V goes to zero because of the choice of coordinates, so (for a time-stationary case from (3.7))

$$\frac{\partial \overline{uw}}{\partial z} = \frac{\partial}{\partial z}(\tau/\rho) = -\frac{1}{\rho}\frac{\partial P}{\partial x} = -fV_g = -f\ G \sin\alpha \tag{3.10}$$

where τ is the local stress, $G = |\vec{G}|$ and α is the angle between the surface wind direction and the geostrophic wind. The relative change of the stress with respect to the surface stress u_{*o}^2 is:

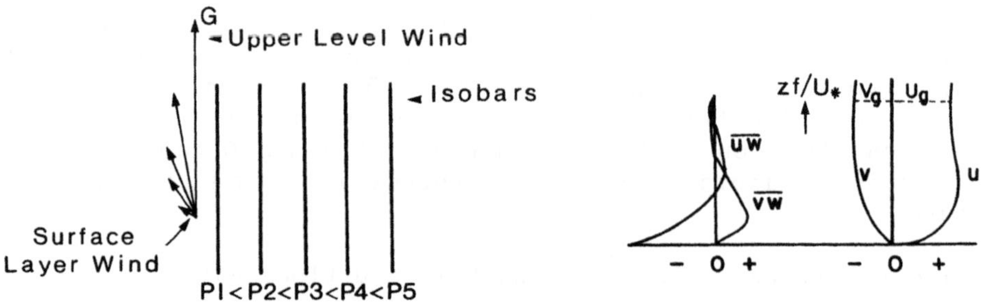

Fig. 2. On the left is shown schematically the relationship between the pressure field and the geostrophic wind G (for the northern hemisphere). On the right is shown schematically the variations of the stresses and the wind components within the ABL.

$$\frac{1}{u_{*o}^2}\frac{\partial \overline{uw}}{\partial z} \simeq \frac{f\ G\ \sin\alpha}{u_{*o}^2} \qquad (3.11)$$

Let h_c = height at which u_*^2 decreases by b percent. Then (3.11) gives:

$$h_c \sim \frac{bu_{*o}\ (u_{*o}/G)}{f\ \sin\alpha} \sim 100\ u_{*o}$$

for: $f = 1.4 \times 10^{-4}s^{-1}$; $u_{*_o}/G \sim 0.025$; $\alpha = 20^o$; and $b = 20\%$

For $G = 5ms^{-1}$, $u_* \sim .125\ ms^{-1}$ and $h_c \sim 12m$.

Therefore the stress is within 10% of its surface value for the lowest 10-20 m of the ABL. For a time-stationary regime (3.7) shows that $\overline{w\theta}$ and $\overline{wq}$ are also constant with respect to z. Therefore, this region is called the constant-flux layer.

2.2 Monin-Obukhov Similarity Theory

Monin and Obukhov (1954) postulated for a time-stationary, horizontally-homogenous surface layer that the structure of the layer and characteristics of turbulence should depend on relatively few scaling parameters, namely:

a friction velocity $u_* = (\tau/\rho)^{1/2}$

a temperature scale $\theta_* = -H/\rho c_p u_*$

a buoyancy parameter g/T_o to represent the fluid; and the height z

A second length scale can be formed from the first three parameters. This is the Monin-Obukhov length,

$$L = -\frac{u_*^3}{\frac{kg}{T_o}\ \frac{H}{\rho c_p}} = \frac{T_o\ u_*^2}{k \cdot g\ \theta_*} \qquad (3.12)$$

where k, von Karman's constant, has been inserted for historical reasons. The basis of the Monin-Obukhov similarity theory can then be stated as:

For any given characteristic of the turbulence or structure of the flow, say B, a nondimensional grouping of u_*, θ_*, g/T_o and z can be found, say B_*, so that:

$$\frac{B}{B_*} = F(\frac{z}{L})$$

This is strictly a postulate and must be verified by observations. The main advantage of the Monin-Obukhov approach is the insight it gives into how data should be interpreted.

What are the characteristics of z/L? For conditions of neutral stability, no vertical heat flux would be expected so H = 0, $L = \infty$ and $z/L = 0$. For unstable conditions the heat flux would be away from the surface (positive) so $H > 0$, $L < 0$ and $z/L < 0$. For stable conditions, the opposite would hold. Therefore z/L is referred to as the Monin-Obukhov stability parameter.

It should be noted that the choice of u_*, θ_*, L and z was somewhat arbitrary. Two other useful scales are:

$$u_f = \left(\frac{g}{T_o}\frac{Hz}{\rho c_p}\right)^{1/3} \quad \text{the free-convection velocity scale; and}$$

$$\theta_f = \left(\frac{T_o}{gz}\left(\frac{H}{\rho c_p}\right)^2\right)^{1/3} \quad \text{the free-convection temperature scale}$$

This provides two velocity scales (u_*, u_f), two temperature scales (θ_*, θ_f) and two length scales (z, L). However, only four are independent since:

$$\frac{u_f}{u_*} \propto \left(-\frac{z}{L}\right)^{1/3} \quad ; \quad \frac{\theta}{\theta_*} \propto \left(-\frac{z}{L}\right)^{-1/3}$$

The choice of which four to use will depend on the situation.

The interpretation of z/L and the other similarity parameters is best viewed by considering the turbulent energy budget:

$$\frac{\partial}{\partial t}\frac{\overline{u^2+v^2+w^2}}{2} = \frac{\partial}{\partial t}\frac{\overline{e^2}}{2} = \overline{uw}\frac{\partial U}{\partial z} + \frac{g}{T_{vo}}\overline{w\theta_v}$$

(1) (2)

$$- \varepsilon + \frac{\partial}{\partial z}\left[\frac{\overline{we^2}}{2} + \frac{\overline{wp}}{\rho}\right] \qquad (3.13)$$

(3) (4) (5)

for a time-stationary situation. The terms are:

(1) production of mechanical energy by the interaction of the turbulence and the mean shear;
(2) production of buoyant energy due to the correlation of density variations (represented by θ_v, the virtual temperature fluctuation) and vertical motion;
(3) dissipation of turbulent kinetic energy to heat by viscous forces;
(4) divergence of vertical flux of turbulent kinetic energy, e^2; and
(5) divergence of the transfer of energy through pressure-velocity interactions.

The latter two terms, which are the most difficult to measure, have usually been neglected. Recent studies of the surface layer indicate that they are significant and near-equal in magnitude but of opposite sign so that they almost cancel one another.

When multiplied by kz/u_*^3, the turbulent energy budget equation in nondimensional form, becomes:

$$\Phi_m\left(\frac{z}{L}\right) - \frac{z}{L} + \Phi_\varepsilon\left(\frac{z}{L}\right) - \frac{kz}{u_*^3}\frac{\partial}{\partial z}\left[\frac{\overline{we^2}}{2} + \frac{\overline{wp}}{\rho}\right] = 0 \qquad (3.14)$$

where the first term is nondimensional wind shear, to be discussed below. The relative magnitudes of the first three terms are shown in Figure 3. Note that the buoyant production is equal to the mechanical production for $z/L \sim -0.6$.

The production terms (1) and (2) can also be written as:

$$\text{Production} = (1) + (2) = -\overline{uw}\,\frac{\partial U}{\partial z}\,(1-Rf)$$

where Rf is the flux Richardson number which is the ratio of buoyant to mechanical (or shear) energy production. Since $\overline{uw}$ is negative in the surface layer and $\partial U/\partial z$ positive, Rf will be negative for a buoyant situation, zero for neutral conditions, and positive for stable conditions.

For Rf = 1, the turbulent energy production due to shear is balanced by the energy lost due to working against the stable gradient (negative buoyancy). However, some turbulent energy will also be lost to viscous dissipation, so the criteria for the maintenance of turbulence is Rf <1. Theoretical and observational studies put the critical value at about 0.25.

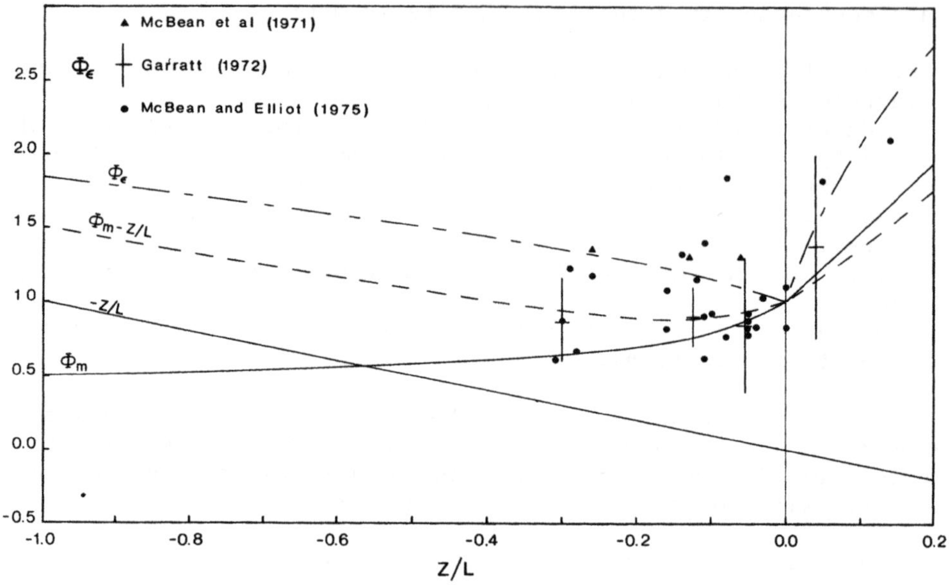

Fig. 3. Relative magnitudes of terms in the nondimensional kinetic energy budget. The curves for Φ_m and Φ_ε are from Wyngaard and Coté (1971) and the data points for Φ_ε from three experiments, as indicated. It can be seen that there is approximate balance between the local production and dissipation of turbulence.

Since Rf depends on the fluxes, and gradients were, at least originally, thought to be easier to measure, a gradient Richardson number Ri was developed. In analogy with the molecular processes the fluxes are related to the gradients by:

$$\overline{uw} = -K_M \frac{\partial U}{\partial z}$$
$$\overline{w\theta_v} = -K_H \frac{\partial \Theta_v}{\partial z} \tag{3.15}$$

where K_M and K_H are the eddy viscosity and conductivity, respectively. Then:

$$Rf = \frac{(g/T_{vo})\overline{w\theta_v}}{\overline{uw}\frac{\partial U}{\partial z}} = \frac{K_H (g/T_o)\frac{\partial\Theta_v}{\partial z}}{K_M\left(\frac{\partial U}{\partial z}\right)^2} = \frac{K_H}{K_M} Ri \tag{3.16}$$

Originally K_H/K_M was assumed to be unity, but it is actually a function of stability (e.g., a function of Rf, Ri or z/L).

The Monin-Obukhov stability parameter can be related to Rf and Ri. First consider the wind shear $\partial U/\partial z$. According to the Monin-Obukhov similarity,

$$\frac{z}{u_*}\frac{\partial U}{\partial z} = F\left(\frac{z}{L}\right) \tag{3.17}$$

For neutral conditions,

$$F(0) = \text{constant} = 1/k$$

$$\therefore \frac{z}{u_*}\frac{\partial U}{\partial z} = \frac{1}{k} \tag{3.18}$$

which can be integrated to give:

$$U(z) - U(z_o) = \frac{u_*}{K}\ln\frac{z}{z_o}\text{, the log-profile law} \tag{3.19}$$

For non-neutral conditions,

$$\frac{kz}{u_*}\frac{\partial U}{\partial z} = \Phi_m\left(\frac{z}{L}\right) \quad \text{where } \Phi_m(0) = 1 \tag{3.20}$$

Therefore,

$$Rf = \frac{(g/T_{vo})\,\overline{w\theta_v}}{\overline{uw}\;\partial U/\partial z} = \frac{z}{L}\frac{1}{\Phi_m} = \frac{K_H}{K_M}\,Ri \tag{3.21}$$

z/L can be thought of as the ratio of buoyant production to mechanical production that would occur for the same stress and for the wind profile corresponding to that stress but with neutral stability.

2.3 Applications of Monin-Obukhov (M-O) Theory

2.3.1 Turbulent Statistics. The simplest turbulent statistics of interest are the standard deviation σ_u, σ_v, σ_w, σ_θ, σ_q. Monin-Obukhov scaling suggests that, in general,

$$\frac{\sigma_{u,v,w}}{u_*} = F_{u,v,w}\left(\frac{z}{L}\right)\;;\;\frac{\sigma_\theta}{\theta_*} = F_\theta\left(\frac{z}{L}\right)\;;\;\frac{q}{q_*} = F_q\left(\frac{z}{L}\right) \tag{3.22}$$

Experimental evidence supports these relationship for w and (Figure 4) and to some extent for q. The horizontal velocity fluctuations u,v appear to be significantly influenced by large scale eddies which depend on the depth of the boundary layer and, in some circumstances, on the synoptic-scale weather patterns. These effects are not accounted for in the Monin-Obukhov theory and the theory, hence, breaks down. The statistics of humidity fluctuations are complicated by the distinction between active and passive scalars. An active scalar is one which, through its variations, significantly affects the buoyancy of the air parcel it is in; a passive scalar has a negligible effect on buoyancy. In the arctic, temperature is almost always an active scalar and humidity a passive one, although over tropical oceans this can be reversed. The active scalar has the dominant role in determining L and hence the statistics of a passive scalar, such as q, appear to conform to M-O theory only if the active scalar and passive scalar are highly correlated. Otherwise it is necessary to define an L_q in terms of $\overline{wq}$, for example, to reflect its small contribution to buoyancy and look at $\sigma_q/q^* = F(z/L_q)$ (McBean, 1971). As yet, this is not well established experimentally.

It is useful to consider σ_w and σ_θ variations further. For neutral conditions (z/L = 0), $\sigma_w/u_* = F_w(0)$ = constant, which experimentally is 1.2 - 1.4. As neutral conditions are approached, σ_θ and $\theta_* = -\overline{w\theta}/u_*$ both should go to zero. However, because of experimental noise (horizontal inhomogeneities, electrical, etc.) $\overline{w\theta}$ will likely go to zero while σ_θ remains non-zero. Hence the ratio σ_θ/θ_* will become very large and experimental data is unlikely to be able to determine the ratio in the near-neutral case.

Consider now the case where most turbulent motion arises due to buoyancy. This corresponds to $z/L \ll 0$. The vertical wind shear will be small and u_*, although not necessarily zero, does not seem to be a relevant scaling parameter. In these circumstances it seems appropriate to choose u_f, θ_f, z as the scaling parameters since u_*, θ_* and L are all dependent on u_*. This reduces the number of scaling parameters and leads to:

$$\frac{\sigma_w}{u_f} = \text{constant; and } \frac{\sigma_\theta}{\theta_f} = \text{constant} \tag{3.23}$$

Since the preferred method of presentation is in terms of z/L, these can be written as:

$$\frac{\sigma_w}{u_*} = a_w \left(-\frac{z}{L}\right)^{1/3}; \text{ and } \frac{\sigma_\theta}{\theta_*} = a_\theta \left(-\frac{z}{L}\right)^{-1/3} \tag{3.24}$$

which is reasonably well confirmed by observations (Fig. 4).

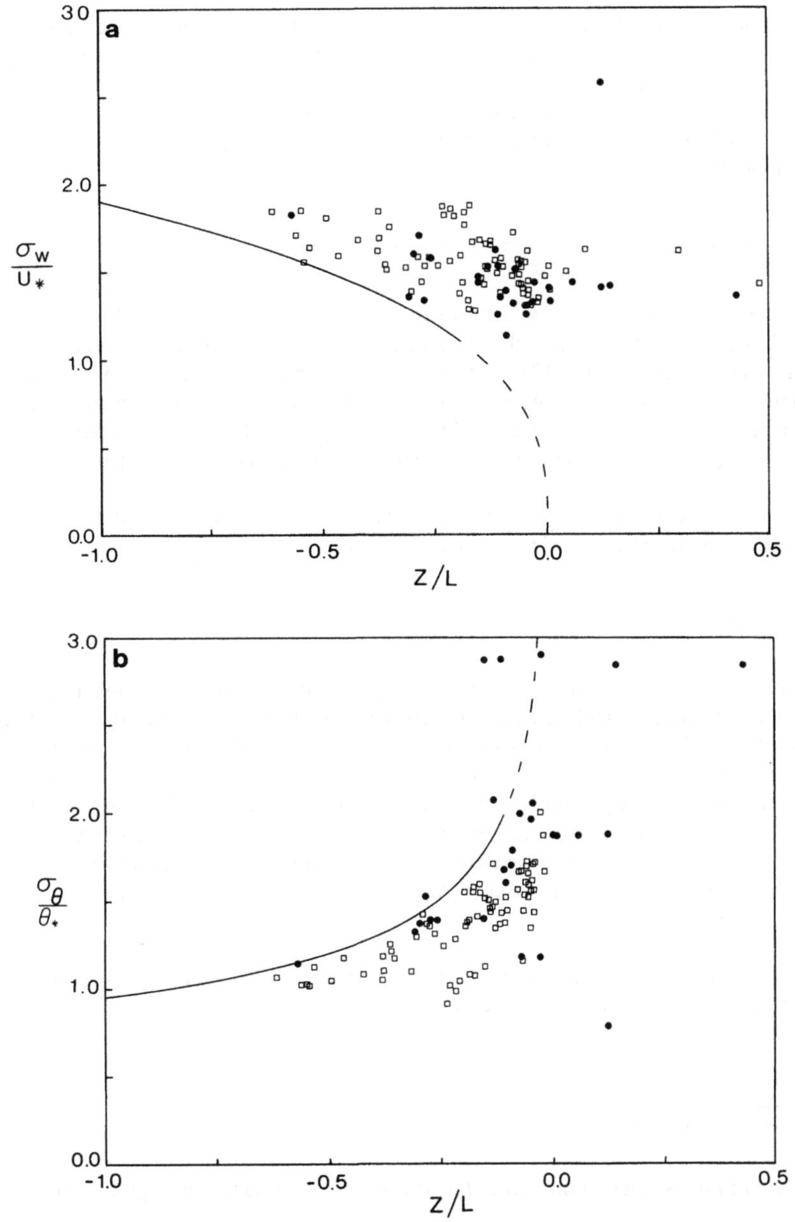

Fig. 4. The nondimensional standard deviations of vertical velocity fluctuations (4a) and temperature fluctuations (4b) plotted against the Monin-Obukhov stability parameter. The curves are from Wyngaard et al. (1971). The data points are from McBean (1971; open squares) and the Suffield experiment (dots; see McBean and Elliott 1975). The data sets generally agree with the 1/3 power law predictions but there is a variation in the constants and scatter between experiments.

The other end of the stability range is for very stable conditions, $z/L >> 0$. The vertical motions are very suppressed by the density gradient such that the turbulent eddies will not "feel" the ground. In that sense the exact value for z is immaterial and it can be excluded from the list of scaling parameters. The predictions are then

$$\frac{\sigma_w}{u_*} = \text{constant; and } \frac{\sigma_\theta}{\theta_*} = \text{constant} \tag{3.25}$$

These results have some experimental confirmation, although it is very difficult to make good measurements in stable conditions because the winds are usually light and variable and the turbulent fluctuations and fluxes very small. The arctic or antarctic surface layers of strong stability with moderate, steady winds provide a unique opportunity to test these relationships.

2.3.2. Flux-Gradient Relationships. The equation derived earlier for the neutral wind profile was:

$$U(z) - U(z_o) = \frac{u_*}{k} \ln \frac{z}{z_o} \tag{3.26}$$

k is von Karman's constant, which has experimentally been shown to reside between 0.35 and 0.42, with generally 0.35 or 0.4 being the values used. If z_o is chosen as the value at which $U(z_o)$ 0 as the log profile is extended toward $z = 0$, then z_o is called the roughness height (typically centimeters or less in height). It should be noted that the log-profile, because it neglects molecular effects and the interaction with surface roughness elements, is not valid at z_o so the wind speed is actually greater than zero there.

For the non-neutral case, the nondimensional wind shear will be a function of stability, i.e.,

$$\frac{kz}{u_*}\frac{\partial U}{\partial z} = \Phi_m\left(\frac{z}{L}\right) \tag{3.27}$$

A Taylor series expansion can be done to obtain insight into Φ_m

$$\Phi_m\left(\frac{z}{L}\right) = 1 + \Phi_m'(0)\,\frac{z}{L} + \frac{1}{2}\,\Phi_m''(0)\left(\frac{z}{L}\right)^2 + \ldots \tag{3.28}$$

where the prime denotes differentiation with respect to z/L. For small values of z/L this can be approximated by

$$\Phi_m\left(\frac{z}{L}\right) \simeq 1+\beta\frac{z}{L} \quad \text{where } \beta = \Phi_m'(o) \tag{3.29}$$

which leads to:

$$U(z) = \frac{u_*}{k}\left(\ln\frac{z}{z_o} + \beta\frac{z-z_o}{L}\right) \text{ the log-linear profile.} \tag{3.30}$$

Experiments have shown that this is a good approximation for $z/L > 0$ (stable conditions) for $\beta\sim 5$. It is interesting to note that the asymptotic state of very stable conditions ($z/L \gg 0$) leads to:

$$\frac{\partial U}{\partial z} \propto \frac{u_*}{L} \quad \text{or} \quad \frac{kz}{u_*}\frac{\partial U}{\partial z} \propto \frac{z}{L} \tag{3.31}$$

or $\Phi_m \propto \frac{z}{L}$, in agreement with the above.

Hence the log-linear law appears to be valid through a wider range of positive z/L than the Taylor series expansion would indicate.

There is not yet complete agreement on the functional dependence of Φ_m on z/L. Part of the uncertainty is due to the different values for k, von Karman's constant. Since k enters into both Φ_m and L, it is useful to consider parameters independent of k, say $S = (z/u_*)\,\partial U/\partial z$ and $L_* = kL$. A plot of 1/S versus z/L_* provides a better comparison of profile relationship (Figure 5). As can be seen, the three published relationships diverge considerably for unstable stratification and data from one other experiment effectively blankets all three curves. This reflects on the unsatisfactory state of our current knowledge of Φ_m.

In analogy to the nondimensional wind shear, the nondimensional temperature gradient is defined as:

$$\Phi_H\left(\frac{z}{L}\right) = \frac{kz}{\theta_*}\frac{\partial\Theta}{\partial z} \tag{3.32}$$

Similar formulations to those for Φ_m are available for Φ_H (see Businger et al., 1971; Dyer and Hicks, 1970).

The free convection limit of similarity theory, which works well for vertical velocity and temperature fluctuation, appears to break down for temperature gradient. The free convection expression would be:

$$\frac{z}{\theta_*}\frac{\partial\Theta}{\partial z} = \text{constant or } \Phi_m\left(\frac{z}{L}\right) \propto \left(-\frac{z}{L}\right)^{-1/3} \tag{3.33}$$

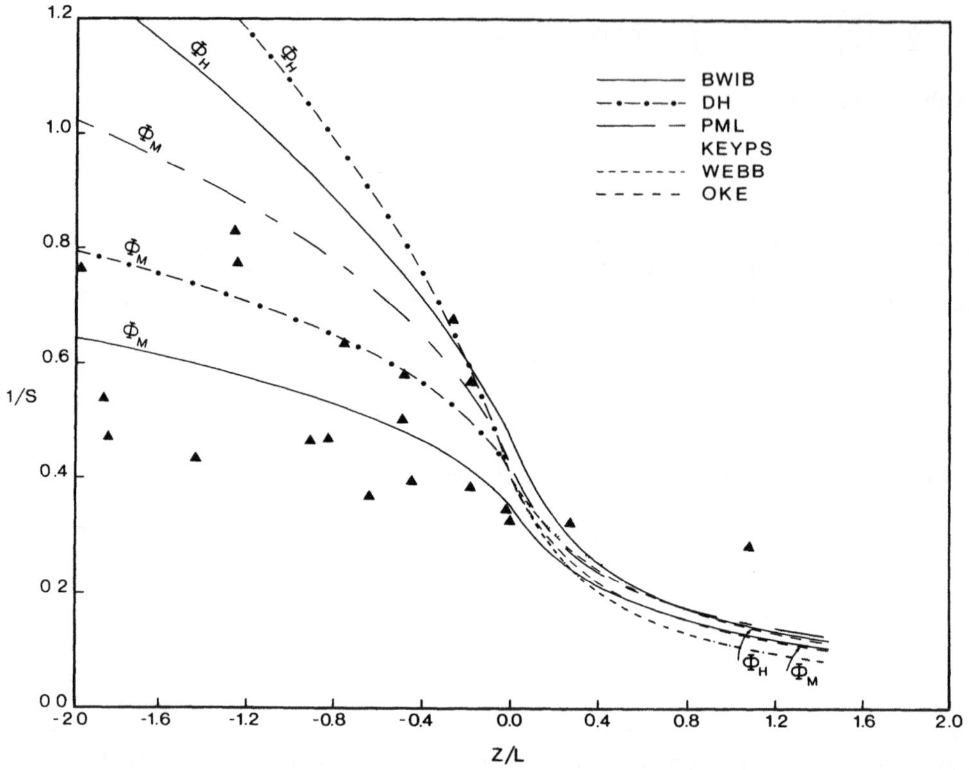

Fig. 5. Plots of nondimensional wind shear and temperature gradient against a modified Monin-Obukhov stability (excluding k dependence in both). The curves are best fits from Businger et al. (1971: BWIB); Dyer and Hicks (1970: DH); Pruitt et al. (1973: PML); Panofsky et al. (1960: KEYPS); Webb (1970) and Oke (1970). The data points for Φ_m are a selection (whenever 10 levels of DEM anemometers available) from the ITCE (Garratt et al., 1979).

while the data show

$$\Phi_H \left(\frac{z}{L}\right) \propto \left(-\frac{z}{L}\right)^{-1/2} \quad \text{for } z/L \ll 0 \tag{3.34}$$

Each of these problems will be greater for an air-ice-water situation than for over-land or warm-water surfaces for which the relations were developed and are generally used. For example, the icing of cup anemometers and thermocouples or thermistors due to freezing spray may ruin all the data. Humidity sensors usually have more difficulties at sub-zero temperatures. For stable conditions over air, ice and snow, there will be larger percentage errors due to small fluxes.

In the absence of open water, polar boundary layers are usually stably stratified due to the large radiative cooling of the surface. Data collected in Antarctica have shown extreme inversion conditions and Lettau (1979) has extended existing models to conform with these cases. He provides an interesting discussion on the interactions between profile "tilting" and "curving", which is used to explain several paradoxes.

The flux-gradient relations are usually applied in their integrated form for the momentum flux (and analogously for the other fluxes)

$$\frac{kz}{u_*}\frac{\partial U}{\partial z} = \bar{\Phi}_m\left(\frac{z}{L}\right) \tag{3.35}$$

when integrated between z_1 and z_2 gives

$$U(z_2) - U(z_1) = U_2 - U_1 = u_*/k\ [\ln z_2/z_1 - \Psi_m\ (z/L)] \tag{3.36}$$

where

$$\Psi_m\left(\frac{z}{L}\right) = \int_{z_1/L}^{z_2/L} \frac{1-\bar{\Phi}_m(\eta)}{\eta}\, d\eta \tag{3.37}$$

Since z/L depends on the heat flux as well as momentum flux it is necessary to measure temperature profiles and use:

$$\Theta_2 - \Theta_1 = \theta_*/k\ [\ln\frac{z_2}{z_1} - \Psi_H\left(\frac{z}{L}\right)] \tag{3.38}$$

and integrate toward a solution of θ_* and u_* that best fits the profile data $\Theta(z)$ and $U(z)$, consistent with the knowledge of Ψ_m and Ψ_H (see Paulson, 1970).

It is interesting to continue on to write:

$$\tau = u_*^2 = k^2[U_2 - U_1]^2/[\ln z_2/z_1 + \Psi_m(\tfrac{z}{L})]^2 \tag{3.39}$$

For z_2 = 10m, $z_1 = z_o$, $U_1 = 0$

$$\tau = \rho[k/(\ln \frac{10}{z_o} + \Psi_m)]^2 U^2(10m) = \rho C_D U_{10}^2 \tag{3.40}$$

where C_D is the drag coefficient, which therefore depends on the surface roughness, the stability, and the choice of reference height.

The corresponding equation for the heat flux is:

$$H = -u_* \theta_* = \left[\frac{k^2}{(\ln \frac{10}{z_o} + \Psi_m)\ (p \frac{10}{z_T} + \Psi_H)}\right] U_{10} (\Theta_{z_T} - \Theta_{10})$$

$$= C_H U_{10} (\Theta_{z_T} - \Theta) \tag{3.41}$$

where C_H is the heat transfer coefficient, which therefore depends on z_o, z_T, the thermal roughness height, z/L and the reference height. In sections 4 and 5 the present knowledge about C_D, C_H will be reviewed.

2.4 Upper Levels

Our knowledge and understanding of the upper levels of the atmospheric boundary layer, although still substantially less than for the surface layer, increased dramatically through the 1970's due to numerical modeling (e.g., Deardorff, 1972; Wyngaard et al., 1974b; Andre et al., 1978), field studies (Minnesota, GATE, AMTEX) and theoretical developments (Lumley, 1980; Deardorff, 1980). Although the neutral ABL is the traditional starting place it is evident that strictly neutral ABL's do not occur very frequently and the convective ABL can be considered as the best understood. Hence we will proceed from convective to neutral to stable ABL's in our consideration.

2.4.1 The Convective ABL. Convective ABL's occur when the surface is warmer than the air above, either due to solar heating of the surface or having water exposed to polar air masses. Convective elements will rise from the surface to a height at which the stratification prevents further penetration. This usually occurs at a layer of pronounced stability at a height z_i, typically hundreds of meters to a few kilometers in height. This inversion lid is eroded from below by turbulence and also moves due to large-scale influences, e.g., subsidence. Hence, it will vary diurnally, usually moving up through the day. For intense convection, it is possible for the inversion lid to be penetrated with resulting deep convection to well up into the troposphere. In these circumstances it can be claimed that the boundary extends to the top of these convective elements.

It is interesting to look at the order of magnitude of forces on a turbulent parcel. The spatial velocity gradients will generate inertia forces of the form $\vec{V}\cdot\nabla\vec{V}$ while the buoyant force per unit mass will be $(g/T_o)\theta$ where θ is the local deviation from the temperature reference state. For eddies of size, $\ell \sim 10^3$ m, and velocity, $q \sim 1 ms^{-1}$, the inertial term is of order $q^2/\ell \sim 10^{-3}$ ms^{-2}. Since g/T_o is $\sim .03$, a θ value of only 0.03^oC is required to have the forces equal. Hence, only a very small and frequently observed temperature excess is required to create a convective situation.

Because of the strong vertical mixing due to convection, most of the depth of the convective ABL will have small gradients of wind, temperature and humidity. These will be capped by sharp changes in the inversion layer where the adjustment to free-atmosphere conditions takes place (Figure 6).

In analogy with surface-layer M-O scaling, a convective mixed layer velocity and temperature scale can be defined as (see e.g., Kaimal et al., 1976):

$$w_* = \left(\frac{g}{T_o} \frac{H_o}{C_p} z_i\right)^{1/3} \quad ; \quad T_* = \left(\left(\frac{H_o}{C_p}\right)^2 \frac{T_o}{gz_i}\right)^{1/3} \tag{3.42}$$

where z_i has replaced z as the scaling height. The subscript o on H_o is to denote the surface flux value. Hence, these scales are independent of height. An example of this type of scaling is shown in Figure 7. The ratio $-z_i/L$ for convective conditions to apply apparently (based mainly on models) doesn't have to be very large ($\sim O(1)$) (Deardorff, 1972).

It is important to recognize that a convective boundary layer is driven by the buoyancy generated at the surface. Hence, although the inversion layer may be 1 km away, the rate of entrainment is still related to the surface heat flux. Using this approach several models have been developed that adequately predict the depth and mixed-layer temperature of an evolving convective boundary layer (e.g., Tennekes, 1973; Stull, 1973 and 1976). The processes of entrainment at the inversion layer are quite complicated, particularly as clouds are formed. Deardorff (1980) has provided recent reviews.

Wyngaard et al. (1974a) have derived a convective drag law

$$\frac{1}{C_g} = \frac{G}{u_*} = \frac{1}{k}[\ln - \frac{L}{z_o}] \tag{3.43}$$

where G is the geostrophic wind speed and C_g the geostrophic drag coefficient (Figure 8).

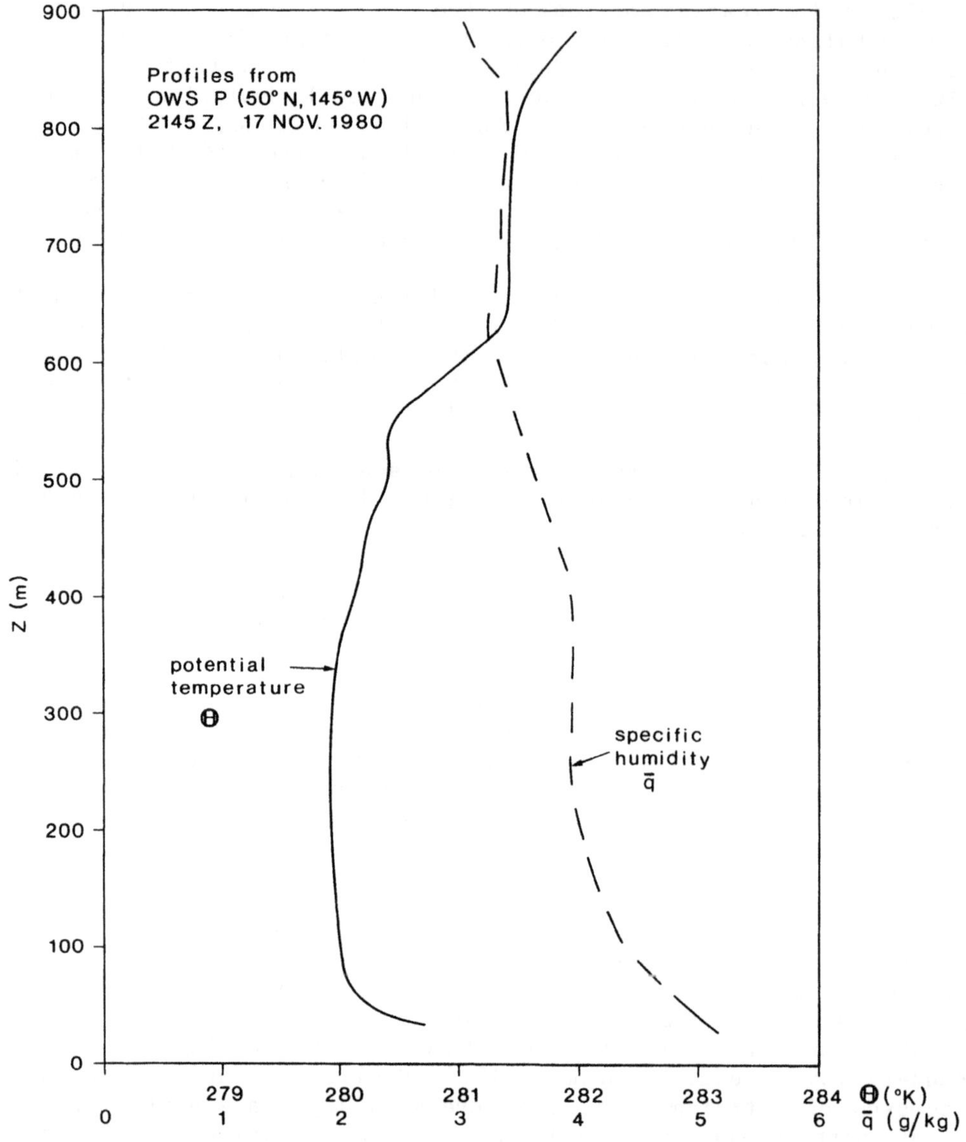

Fig. 6. An example of an inversion-capped, well-mixed, convective boundary layer over the ocean.

2.4.2 The Neutral ABL. The neutral ABL has a constant potential temperature throughout and turbulence is entirely due to wind shear. In view of the small temperature difference required for large buoyancy effects, it is unlikely that a truly neutral ABL will exist. However, because of its simplicity, in the sense of

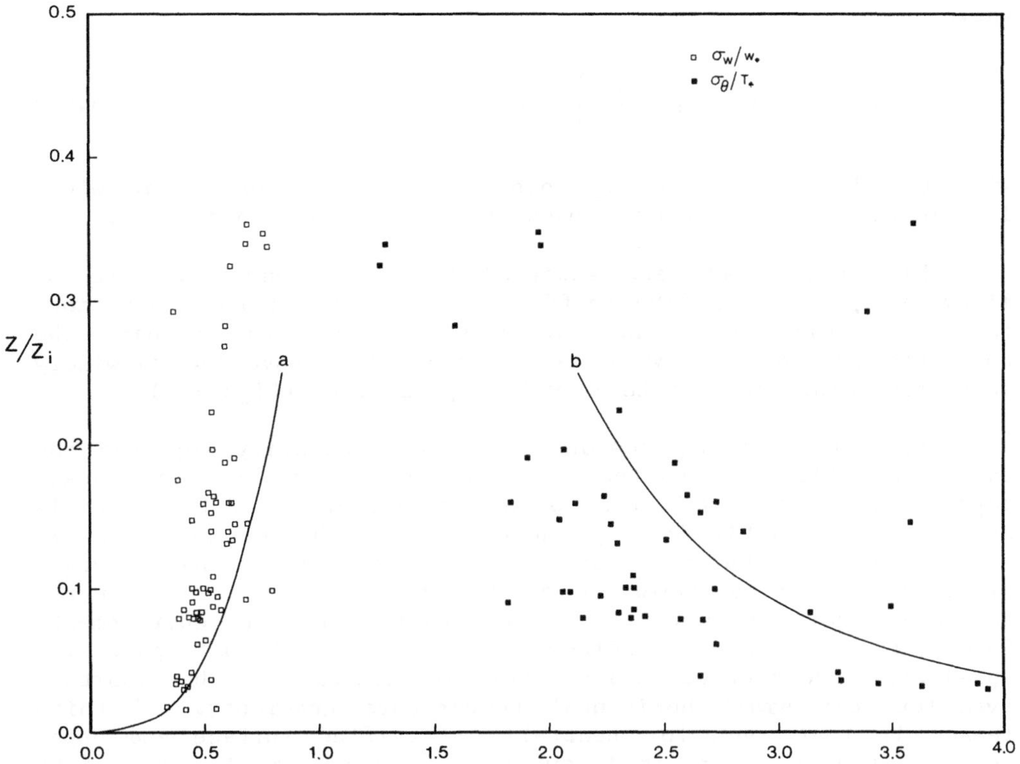

Fig. 7. The variation of nondimensional standard deviations of vertical velocity and temperature fluctuations with nondimensional height. The data are from McBean (1976) and the curves from Kaimal et al. (1976).

no buoyancy effects, it has been extensively considered theoretically. Generally, it is assumed on the basis of dimensional analysis that the depth of the ABL depends on u_*/f where f is the Coriolis parameter. Because the neutral layer blends into the free atmosphere it is somewhat difficult to say exactly where it stops. Typically values of h, the boundary layer depth, ~0.25 u_*/f are quoted. With a superimposed geostrophic wind the boundary-layer wind will exhibit an Ekman spiral. Kazanski and Monin (1961) have applied the concepts of asymptotic matching between the upper levels and the surface layer to obtain the neutral geostrophic drag law

$$\ln \frac{G}{f z_o} = B + \ln \frac{G}{u_*} + \left[\frac{k^2 G^2}{u_*^2} - A^2\right]^{1/2} \qquad (3.44)$$

where

$$A = -\frac{kV_g}{u_*} \; ; \; B = \ln\left(\frac{u_*}{fz_o}\right) = k\frac{U_g}{u_*} \tag{3.45}$$

with U_g and V_g being the x,y components of the geostrophic wind. In some papers, A and B are reversed from that presented here.

This theory has been expanded to the non-neutral case by allowing A, B and C (the coefficient for a temperature equation) to be functions of u_*/fL. To date, attempts to evaluate the constants by comparing with experimental data have led to widely scattered points subject to many interpretations (Figure 9).

There are several reasons why this similarity approach has failed in application to the upper levels. First, the time scale appropriate to the evaluation of an Ekman layer is 2 f^{-1} which is about 12 hrs at latitude 80°, and larger towards midlatitudes. It is very unlikely that the boundary conditions will not have changed either through time evolution or advection, so that the neutral boundary layer will never be in equilibrium with its environment. Second, the atmosphere is seldom truly barotropic (geostrophic wind speed independent of height) and the baroclinicity can be important even for very small horizontal temperature gradients. A third reason is that even for neutrally stratified surface and upper level of the ABL there is frequently an inversion at some height $z_i < au_*/f(a \sim .025)$ which will distinctly alter the structure. Csanady (1974) and Zeman and Tennekes (1977) discuss this problem.

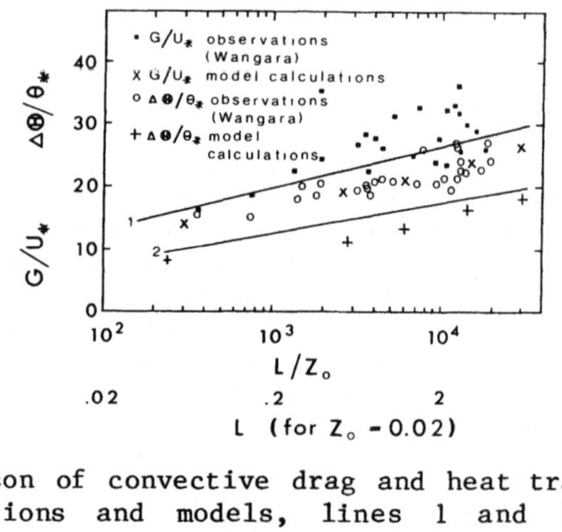

Fig. 8. Comparison of convective drag and heat transfer laws from observations and models, lines 1 and 2, respectively (adapted from Wyngaard et al., 1974a).

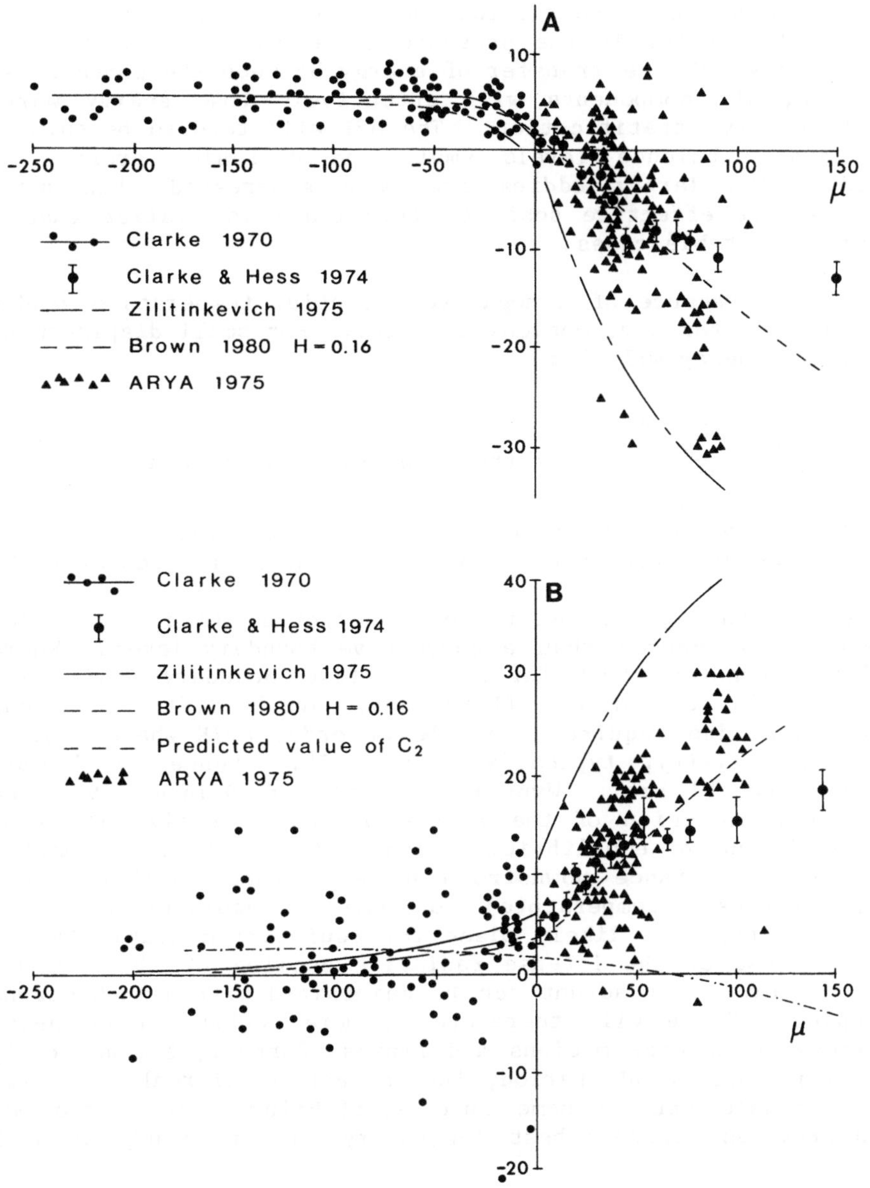

Fig. 9. Comparison of theories and observations for the similarity coefficients A and B (adapted from Brown, 1980). The stratification parameter, $\mu = ku_*/fL$.

2.4.3 The Stably-Stratified ABL. For the surface air cooler than the air aloft a stably-stratified ABL of quite different structure is set up. Energy must be expended to maintain vertical motions and transfer in the presence of a damping density structure. Because of the transfer of energy between the components of turbulence, all components will effectively lose energy working against density stratification. The ABL will tend to be thin, the fluxes and turbulence levels small and diffusion greatly suppressed. The larger eddies are more suppressed than smaller eddies, so the effective scale of turbulence is smaller than for neutral or unstable cases.

In the presence of a negative potential temperature gradient it is possible for wave motions to exist. For small displacements, the wave frequency will be:

$$N = \left(\frac{g}{T_o} \frac{\partial \Theta_v}{\partial z} \right)^{1/2} \quad = \text{the Brunt-Vaisala frequency} \qquad (3.46)$$

which typically corresponds to wave periods of 30-300 sec. These type of waves are often seen on acoustic sounder records at night.

Because the stable ABL is less energetic, it is much slower to respond to changes than a convective boundary layer. Whereas the "eddy turnover time" for a convective boundary layer is of order 5-10 min, the stable ABL responds more by diffusing a change upward. The time required would be of order h^2/K where K is the average diffusivity, typically $\sim 0.3\ m^2s^{-1}$. Hence for $h \sim 100$ m, time is about 10 hrs. Wyngaard (1975) and Delage (1974) have studied the evolution of the stable ABL and note that it is most likely to be out of equilibrium. Also a stable ABL will require a much greater distance downstream of a surface condition change (e.g., roughness, temperature) to come to equilibrium. Polar regions are the most likely place for equilibrium stable ABL's to exit. Because of the intense radiative cooling of the surface a stable temperature gradient can be maintained with a much stronger wind speed. There will therefore be more mechanically-generated turbulence to enhance motions and fluxes. Further, the polar night provides a long, steady period, independent of diurnal variations, for the stable ABL to come into equilibrium. Thus, the polar region provides nature's best laboratory for the study of stable ABL's.

The stable ABL is also sensitive to another complicating factor. Because of the small fluxes and large vertical temperature gradients, the flow is very sensitive to drainage flows due to sloping terrain. For $u_* \sim 0.1\ ms^{-1}$, $h \sim 100$ m and a lapse rate of 3°C cooler than adiabatic, a slope of only 1:1000 will give a drainage acceleration as large as the turbulent friction term (McBean, 1979, pp. 23-24).

3. FLUX ESTIMATES

There are a number of ways to obtain a measure of the surface fluxes of momentum, heat and moisture. The most direct measurements, in principle, are made by drag plates (e.g., Bradley, 1968a) for surface stress and lysimeters or evaporation pans (e.g., Pruitt et al., 1973) for evaporation. However, these techniques are fraught with difficulties in operation and interpretation, and it is unlikely that they can be used in air-ice-water situations. A related approach has been used by Den Hartog et al. (1983), where total heat loss from the surface of a small pan was estimated through rate of ice formation, with some success. The eddy correlation method, by measuring the turbulent fluctuations (w, ,u,q,) and computing covariances, gives the best practical surface flux estimate. Banke et al. (1976) and others have successfully measured momentum flux at between 3 and 10 m. Fast response equipment (better than 0.1 s time constant) and careful data processing are needed. For relatively uniform conditions the flux at 3 or 10 m will equal the surface flux.

In many instances it is desired to deduce the surface fluxes of momentum, heat and moisture from the profiles or gradients of wind, temperature and humidity near the surface. Although this approach has a long tradition, it has a number of difficulties. The accuracy of the flux determinations will depend on:

a) the accuracy of the profile data;
b) the applicability of the constant-flux layer theories to the situation (including horizontal homogeneity, time stationarity); and
c) the limitations of our knowledge of the flux-gradient relations (see Figure 5).

In addition to eddy-correlation flux measurement techniques, there are other ways to use turbulence measurements to estimate the fluxes. Some of these depend on M-O theory. For example, it is usually easier to measure temperature fluctuations than vertical wind fluctuations. For quite unstable stratifications, when the heat flux is the largest, the relationship

$$\frac{\sigma_\theta}{\theta_f} = \text{constant} \tag{3.47}$$

leads to

$$H = \rho C_p \, a_\theta \left(\frac{gz}{T_o}\right)^{1/2} \sigma_\theta^{3/2} \tag{3.48}$$

thus relating the surface heat flux to the standard deviation of temperature fluctuations. Other similar relationships have been developed by Tillman (1972), Merry and Panofsky (1976), and others.

Another set of techniques is based on the turbulent energy and temperature variance budgets and the Kolmogoroff-Obukhov hypotheses (see Champagne et al., 1977; Hicks and Dyer, 1972). The scale sizes of turbulent eddies that contribute to the various terms in the turbulent energy budget (equations 3.13 or 3.14) are split between those a meter and larger that produce turbulent kinetic energy and those centimeters and smaller that dissipate it (Figure 10). The nonlinear turbulent interactions transfer energy from production scales to dissipation scales, while the pressure-velocity interactions tend to make the eddies isotropic. In recognition of these features, Kolmogoroff postulated that the spectrum of turbulent velocity fluctuation for a range of energy

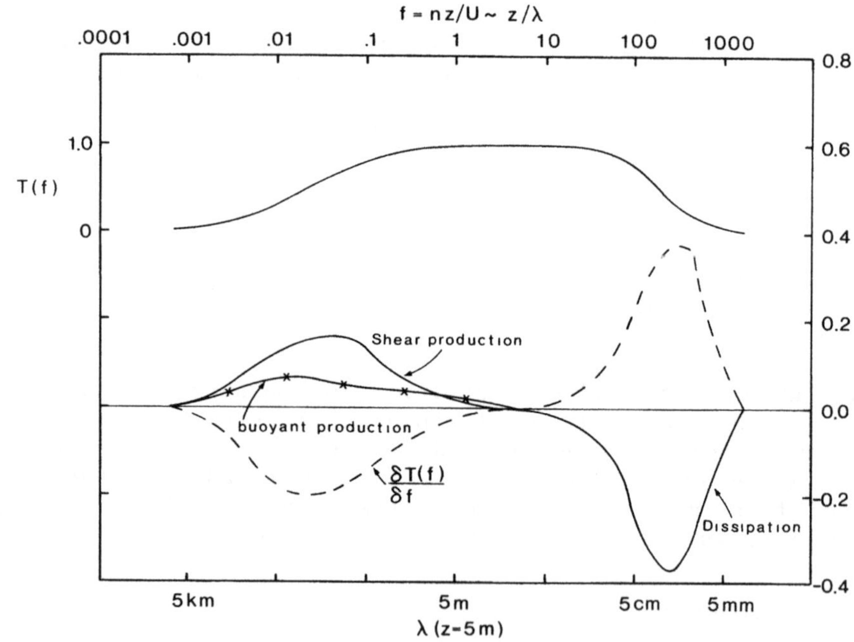

Fig. 10. Spectral energy budget terms. Adapted from McBean and Elliott (1975). The T(f) shows the transfer of energy from large scales to small scales required to balance production and dissipation.

transfer (between the production and dissipation scales, and called the inertial subrange) could be represented as:

$$S_u(k_1) = A_u \, \varepsilon^{2/3} \, k_1^{-5/3} \tag{3.49}$$

where k_1 is the wave number in the wind direction,
ε is the energy dissipation rate (which also equals the rate of energy transfer through the inertial subrange),
A_u is a constant (~0.55 experimentally).

Similarly, for temperature fluctuations, the spectrum can be represented as:

$$S\ (k_1) = A_\theta \varepsilon^{-1/3} \varepsilon_\theta \, k_1^{-5/3} \tag{3.50}$$

where ε_θ is the rate of destruction by molecular conductivity of $\overline{\theta^2/2}$;
A_θ is a constant (0.8–1.0, experimentally).

The turbulent kinetic energy and temperature variance equations can be written as:

$$\frac{kz\varepsilon}{u_*^3} = \Phi_m - \frac{z}{L} - \frac{z}{L} \frac{\partial}{\partial(z/L)} \left[\frac{\overline{we^2}}{2u_*^3} + \frac{\overline{wp}}{\rho u_*^3} \right]$$

$$\frac{kz\varepsilon_\theta}{u_*^2 \theta_*} = \Phi_H - \frac{z}{L} \frac{\partial}{\partial(z/L)} \left[\frac{\overline{w\theta^2}}{2u_* \theta_*^2} \right] \tag{3.51}$$

If both S_u and S_θ are measured, then u_* and θ_* can be determined either by neglecting the flux divergences or representing them as functions of z/L. If the S_u is measured, u_* can be estimated either by assuming neutral conditions (z/L = 0) or estimating z/L. The advantage of this technique is that only u and θ fluctuations in a limited, but high frequency ($n \geq \frac{5U}{Z}$) range need be measured accurately. Hence, the effects of platform motion (ships, aircraft, floating towers) will not usually interfere. The NISSI technique (see Section 4) is based on this principle.

A related approach is the estimation of the fluxes from structure parameters (see Wyngaard and Clifford, 1978). The advantage of the technique is that structure parameters can be obtained, in principle at least, from path-averaged wave propagation measurements. For example, the temperature structure parameter is readily obtained from optical scintillation measurements. In these cases, the path-averaging can compensate for time averaging, so Wyngaard and Clifford estimate that the path-averaged heat flux can be obtained in 1% of the averaging time of an eddy correlation measurement. The techniques also provide line averages which are useful for inhomogeneous surface conditions.

The final group of flux estimation techniques involves measurement of mean winds, temperatures and/or humidities at more than one location. In cases where inhomogeneous surface conditions are present, integral techniques (see Andreas et al., 1979) can be used. These involve, in general, the measurement of mean transports in and out of a box, which is frequently not easy. Several measurement points are needed, unless a two-dimensional situation can be assumed; then two sets of profiles may suffice.

4. ABL OVER SNOW AND ICE-COVERED SURFACES

The purpose of this chapter is to consider the ABL over snow-ice-water surfaces. Because of their complexities and the relative paucity of information on them, it was first necessary to review what is known about ABL's in general, particularly as might be applied to polar problems. In this and the next section the results from specific polar studies will be reviewed and put into perspective. Initially, the ABL's over snow and ice-covered surfaces, without significant open water at markedly different temperature, will be discussed. In the following section, the problem of open water in polar winters, such that large thermal contrasts are evident, will be discussed. This latter situation is much more difficult to address, both experimentally and theoretically.

From the point of view of ice dynamics, the most important atmospheric effect is the wind drag on the surface. Through the AIDJEX and other experiments a considerable body of information on the 10 m drag coefficient, $C_D(10\ m)$, has been obtained. In Banke et al. (1980), the eddy correlation measurements of drag coefficients from their earlier work (Banke and Smith, 1973; Banke et al., 1976; Smith, 1972) have been summarized (Table 1). They find, for near-neutral ($z/L < 0.1$) relatively ice-ridge-free areas that $C_{DN} = 1.55 \pm (0.37) \times 10^{-3}$.

The data indicate a decrease in C_D with increasing stability, approximately following $C_D = (C_{DN}^{-1/2} + 13\ z/L)^{-2}$. Banke et al. also investigated the effects of surface characteristics by running survey lines 256 m upwind of their measurement sites. The ice-surface characteristics were represented either as:

δ - the rms surface elevation at wavelengths less than 13 m
S - rms slope between adjacent points (1 m)
E - rms elevation departure from linear fit to ice surface.

Table 1. Drag coefficients for various ice surfaces.

Location	$C_D(10m) \times 10^3$	Surface conditions z_o(cm)	Surface conditions S(cm)	Comments	Reference
Drifting Station Alpha	1.37	0.02			Untersteiner and Badgley (1965)
Gulf of St. Lawrence	1.42	0.025			Smith (1972)
Beaufort Sea AIDJEX	1.68	0.058	5.6	1971 msmts.	
" " "	1.51	0.034	5.5	" "	Banke &
" " "	1.82	0.085	9.8	1972 msmts.	Smith (1973)
" " "	1.56	0.040	6.3	" "	
Robeson Channel	2.08	0.155	13.2	" "	
Robeson Channel	1.65	0.053	6.5	1974 msmts.	
Beaufort Sea AIDJEX	1.38	0.021	4.6	1975 "	Banke et al.
Flat	1.57	0.041	7.0	" "	(1976)
Hummocked	1.62	0.048	9.7	" "	
Beaufort Sea AIDJEX	1.44	0.026	4.2	1976 msmts.	Banke et al.
" " "	1.19	0.009	3.4	" "	(1980)
AIDJEX Big Bear	1.19	(nearest pressure		10 m msmts.	
" "	1.63	ridge 400 m)		22 m "	
Caribou	1.13	(pressure ridge		10 m "	Leavitt
"	1.48	in a/c 200-400 m)		25 m "	(1980)
Caribou	2.2	(pressure ridge		10 m "	
"	1.9	1.5 m, 80-110 m away)		25 m "	

The 13 m cutoff was used because the shorter wavelengths should contribute more to wind drag than do longer indulations over which the wind can flow more smoothly. The 1 m sampling interval also approximated the resolution of airborne laser altimeter estimates (Hibler, 1972). The results of linear regressions were:

$$10^3\ C_D = 1.10 + 0.072\ \delta\ (\text{cm}) \qquad r = 0.90$$
$$10^3\ C_D = 1.17 + 7.09\ \ S\ (\text{m/m}) \qquad r = 0.82$$
$$10^3\ C_D = 1.14 + 0.028\ E\ (\text{cm}) \qquad r = 0.78.$$

The first equation implies z_o = −0.03 cm + 0.012 δ. It also implies a smooth snow-covered surface drag coefficient of C_D = 1.10 x 10^{-3}. By way of comparison Smith (1980) has determined the neutral drag coefficient over the ocean as $10^{+3}C_D$ = 0.61 + 0.063U (10 m) which corresponds to C_D(5 ms^{-1}) = 0.93 x 10^{-3} and C_D(10 ms^{-1}) = 1.24 x 10^{-3}.

Thus smooth snow and the ocean appear to have much the same drag, with their relative roughness possibly depending on wind speed. For more typically AIDJEX-type snow surfaces it appears that the snow is somewhat rougher than water surface (except at high wind speeds).

Other measurements of stress and drag coefficients over snow and ice have been made by profile and NISSI (near inertial - subrange stress instrument) techniques. Langleben and Pounder (1975) found C_D x 10^3 = 1.58 ± 0.19 and 1.74 ± 0.25 at two towers 250 m apart. They also found that the horizontal wind variability over the distance was less than measurement uncertainties. Their towers were on a gently hummocked ice floe at the 1972 AIDJEX main camp. Leavitt (1980; also Leavitt et al., 1977) (see Table 1) reports average 10 m measurements of C_D's of 1.19 x 10^{-3} for the 1975 AIDJEX Big Bear Camp and 1.13 x 10^{-3} for the 1976 AIDJEX Caribou Camp before the development of a major pressure ridge close (80-100 m) to the tower. For both sites the drag coefficients showed consistent variation with wind direction reflecting the effects of different upwind roughness. NISSI measurements were made at 3 m, 10 m, and 22 or 25 m. The average value of C_D at both sites at 22-25 m was 1.6 x 10^{-3}, exceeding the lower level values by about 60%. This was interpreted as the instrument at 22-25 m sampling a longer (and rougher) upwind fetch. Further, because the stress will decrease with height, Leavitt deduces that this is an underestimate of the surface stress and that a C_D~2 x 10^{-3} is more appropriate.

The wind drag on sea ice in the Bering Sea has recently been investigated by several authors. Macklin (1983) made wind profile measurements on the ice and deduced an average near-neutral 10 m drag coefficient of 0.0031 ± 0.0005. The main variance in the drag coefficient was related to variations in orientation of the ice floe to the wind. This ice was characterized by heavy fractured 1-m thick floes with numerous small rubble piles and hence was much rougher than the situations investigated by Banke et al. (1980) and Leavitt (1980). Pease et al. (1983) and Martin et al. (1983) investigated the net drag on ice floes and used a slab inversion technique to determine both the water and air drag coefficients. Pease et al. (1983) found air-ice drag coefficients of about 0.004 for a 3-m reference level. The water drag coefficients were about 5 times larger.

Following from the above discussion, it is apparent that the presence of pressure ridges and other roughness elements will significantly alter the effective drag of the wind on the ice. Arya (1975) has developed a drag partition relation for τ_t, the total drag as:

$$\tau_t = F_D + S_D \tag{3.52}$$

where F_D - form drag of roughness elements
S_D - skin drag of surface between roughness elements.

Based on a simple analytic model and laboratory studies, it can be shown that

$$S_D = (1 - m\lambda)\ \tau_o \tag{3.53}$$

where λ is the ratio of the frontal or silhouette area of the roughness elements to the horizontal surface area (which for two dimensional elements is $\lambda = h/s$, with h being the height of and s the spacing between elements). m is a parameter depending on flow and obstacle dimensions, and is about 20. τ_o is the value of the surface stress without obstacles. The skin drag may be less than τ_o due to flow separation and probable flow reversals.

Arya derives equations for the ratio F_D/τ_t and $\ln Z_o/z_o$ where Z_o is the large-scale roughness parameter in terms of R, the ridging intensity parameter; $\langle h \rangle$ mean ridge height; and C_{Dh}, the form drag coefficient (~0.4). These allow, for example, one to compute $\langle h \rangle/z_o$ for a given R. Whittmann and Schule (1966) report values of $\langle h \rangle$ of about 1-3 m and R of 0.01-0.03. For R = 0.01 and $\langle h \rangle/z_o = 10^3$, the large-scale effective roughness parameter is about four times larger than the roughness length associated with the surface between the ridges.

Banke et al. (1976, 1980) have also investigated the form drag on ice ridges. The force per unit length of ridge was measured as the height integrated, pressure difference across the ridge. The form drag coefficient was define as:

$$C_{Dh} = 2\ F_D/\ HD^2 10 \quad (3.54)$$

and the net ridge drag coefficient

$$C' = C_{Dh} \langle h \rangle R/2 \quad (3.55)$$

They found that the form drag coefficient was dependent on α, the slope of the ridge, and fitted the regression equation

$$C_{Dh} = 0.012 + 0.012\ \alpha \qquad r = 0.91 \quad (3.56)$$

Values of ridge intensity R⟨h⟩ have been found to vary between ~4 x 10^{-3} in the AIDJEX area to ~0.03 north of Greenland. In the latter case, the ridge form drag exceeded the surface skin drag. Leavitt (1980) has shown that the Banke et al. (1976) approach gives somewhat better agreement to his measurements at the AIDJEX Caribou site (after a ridge formed) than did Arya's method.

The problem of boundary layer flow over variable surface roughness and low topography has been of considerable interest to the general boundary-layer meteorology community for some time. The idealized case of an abrupt change in surface roughness has been numerically modeled (e.g., Taylor, 1969; Peterson, 1969; Rao et al., 1974a) and experimentally studied (e.g., Bradley, 1968a; Peterson et al., 1976; Antonia et al., 1977). The most sophisticated model appears to be the simplified second-order closure model of Rao et al. which was applied to a neutral flow situation over a step change of surface roughness. The results agreed reasonably with Bradley's field data.

The internal boundary layer, IBL, that develops downwind of the step change, is considered to be the region affected by the downstream surface conditions. Defining δ, the IBL thickness, as the height at which the downstream velocity is within 1% of the corresponding upstream velocity, Rao et al. find the growth of is approximately $z\delta \propto x^{4/5}$, where x is the distance downstream of the change. The equilibrium layer, where the boundary layer is in equilibrium with the new surface conditions, is much shallower. The depth for 90% adjustment, h, is approximately h/x ~1/100 for a smooth-to-rough transition and h/x ~~1/200 for a rough-to-smooth transition. Only within this equilibrium layer would the flux-gradient relations apply.

The model indicates very rapid adjustment of the surface stress. For a step change from z_o = 0.002 cm to z_o = 0.25 cm (which are smoother and rougher, respectively, than the surfaces over which Smith, Banke, and colleagues made arctic measurements (Table 1), the surface stress initially jumps to 6-8 times its upwind value but then rapidly decreases to within 20% of the downwind equilibrium value in a 2 m fetch. For a transition in the opposite direction (rough to smooth) the surface stress decreases abruptly to about one tenth of its upwind value, but again rapidly adjusts within a few meters.

More recently, attention has been given to the problem of flow around and over topography (e.g., Hunt, 1980; Taylor, 1980; Mason and Sykes, 1979; Taylor and Gent, 1980). As yet, the models have considered simple cases of low topography, single hills, and generally neutral stratification. Even with these limitations there is not complete agreement on, for example, the net drag of a hill and the downstream area. Is the enhanced drag due to the hill compensated completely by the drag deficit downstream? Due to the increased interest in wind-energy applications, this is a rapidly expanding field and new results are to be expected regularly.

The structure of the boundary layer above the surface layer has been investigated with balloons, aircraft and remote sensors (e.g., acoustic sounders). The tethered balloon measurements of Carsey et al. (1978) can be used to demonstrate several features (Figure 11). For the low wind speed case the temperature

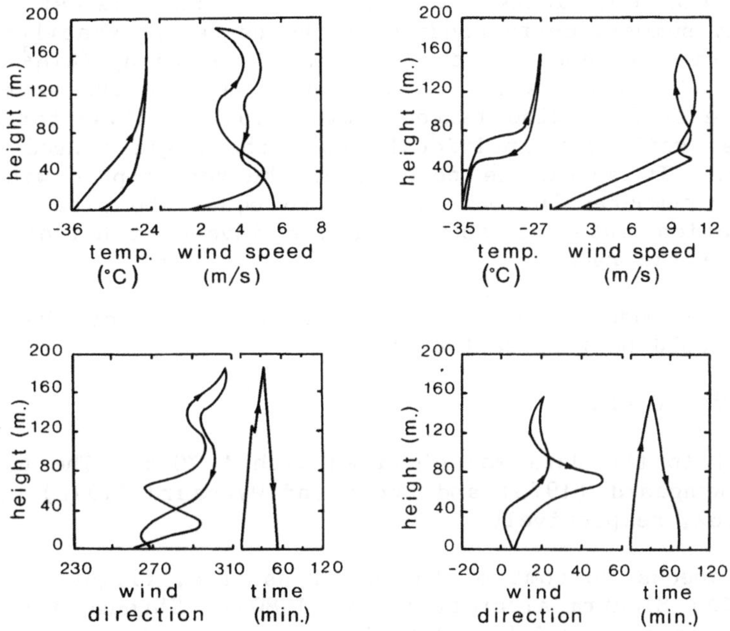

Fig. 11. Examples of tethered balloon profiles of wind speed, direction and temperature. The arrows indicate direction of sonde motion for profile (adapted from Carsey et al., 1978).

stratification is very stable up to about 100 m. Also to be noted is the variability of the wind speed and direction both with height and with time (compare the ascent and descent profiles). For the high wind speed case the temperature stratification is almost isothermal up to 50-70 m, at which point there is a very strong inversion. This almost isothermal layer results from the wind-shear-induced turbulent mixing. The wind speed has a maximum near the inversion level and the wind direction, although showing more variation between ascent and descent, has an abrupt change at the inversion level. Data sets such as this contain a wealth of interesting case studies that need to be explored.

One way of obtaining more representative stress values is to examine the wind profiles and/or turbulence over the full depth of the ABL. The geostrophic departure method (eq. 3.9), by integrating the difference between actual and geostrophic wind components over the ABL, can give such estimates. Carsey and Leavitt (1977) and Carsey (1980) report on pibal wind profile measurements and acoustic radar measurements of boundary-layer height. There is a lot of scatter and uncertainty to their results but they are able, by averaging large numbers of profiles and eliminating ones with large apparent baroclinic effects, to obtain geostrophic drag coefficients, u_*/G, of between 0.019 to 0.027 with significant seasonal variation, lowest in the winter and highest in late spring/early summer, corresponding to the degree of stability. The drag coefficients based on the 10 m wind C_D (= u_*^2/U_{10}^2) ranged from 1.3×10^{-3} to 2.1×10^{-3}. The angle between the geostrophic wind and the surface wind ranged between 19.2 to 13.4° for seasonal averages, while for individual cases the angle ranged between 0° and 80°. It should be noted that the geostrophic winds were not directly computed but were assumed to equal the pibal winds in the observation interval above z_i, the inversion height deduced from the acoustic radar.

Based on model results, it was postulated that the surface stress, τ, could be represented as:

$$\tau = A\rho f z_i \, G \sin\alpha \qquad (3.57)$$

The best fit to the data was obtained with A ~0.5. The numerical models of Wyngaard (1975) and Brost and Wyngaard (1978) give A = 0.91 and 0.62, respectively.

Simultaneous to some of the pibal ascents, flights were made with the NCAR Electra aircraft (Katz, 1980). Through the use of gust probes and inertial navigation systems, aircraft can make spatial averages of direct flux measurements. For the flight of 19 July with a boundary layer height of 220 m, the geostrophic drag coefficient was found to be 0.022 and the surface drag coefficient 1.6×10^{-3}. The stress decreased from the surface to just less than half its value at 100 m and then remained almost constant up to 400 m. It can be noted that some authors (e.g., Brost and Wyngaard, 1978) define the height of the stable boundary layer as the level at which the stress has reduced to 5% of the surface value. By that definition, the depth of the 19 July boundary layer is in excess of 400 m. Further aircraft measurements, coupled with ground based systems, are needed for these types of studies.

In comparison to the number of measurements of surface stress, there is a marked paucity of sensible heat flux and evaporation measurements. Thorpe et al. (1973) reported measurements from the 1972 AIDJEX pilot study and over Robeson Channel in July 1972. The heat transfer coefficients C_H were found to be 1.2×10^{-3} and 1.0×10^{-3}, respectively. That for moisture, C_E, was found for AIDJEX only to be 0.55×10^{-3}. The C_H values are within the range typically measured over the open ocean. Smith (1980) reports that $C_H = 1.1 \times 10^{-3}$ in unstable conditions and 0.8×10^{-3} in stable conditions, with no strong wind speed dependence. Anderson and Smith (1981) have found that, again for the open ocean, C_E had a stability variation in accordance with its empirical profile laws. For conditions of wet ice, use of oceanic values for C_E and C_H will be a reasonable approximation, particularly in view of the similarity of drag coefficients.

The ABL over snow and ice covered surfaces is usually neutrally or stably stratified. As demonstrated, there is still a large amount of scatter to the experimental results. In order to gain further insight into the ABL, it is desirable to develop models which can be tested against the limited available data. If verified, the models can then be extended to look at other cases or at detailed parameters that haven't been measured. Businger and Arya (1974), Wyngaard (1975) and Brost and Wyngaard (1978) have developed such models; the latter two emphasized the evolution of the nocturnal ABL, while the first dealt with the steady-state situation. Businger and Arya solved the ABL equation by specifying an eddy viscosity distribution of the form

$$K_* = K_M f/u^2 = k_* \; \xi \frac{\exp\,(-V_g/u_*/\xi)}{1 + \beta\xi\mu_*} \qquad (3.58)$$

where ξ = nondimensional height = fz/u_* and μ_* = stability parameter = $u_*/L\,f$. These nondimensionalizations assume that μ_*/f is the appropriate height scaling length for the neutral to stable boundary layer. All three models predict that the depth of the boundary layer h can be represented as

$$\frac{hf}{u_*} = a\left(\frac{u_*}{fL}\right)^{-1/2} \qquad (3.59)$$

but Wyngaard found a = 0.22 and Businger and Arya, and Brost and Wyngaard about a = 0.4. It should be noted that this is using the criterion that h is the height at which the stress is 5% of the surface value. For 1% of surface value Businger and Arya report a = 0.72. It appears that the depth of a steady-state stable boundary layer is strongly model dependent, even for the same definition of depth, because the turbulence depends on a balance between production and dissipation, and small differences in technique become important when dealing with small, almost vanishing turbulence levels. The models do better in predicting stress profiles, wind shears, and their time evolution, for example.

Brost and Wyngaard (1978) used a simplified version of Wyngaard's (1975) model to look at the evolution of the stable boundary layer with constant cooling rates (at 1 m) and with the surface energy budget. The constant cooling rate leads to a steady state within a few hours, while, with the surface energy budget, the boundary layer typically did not reach a steady state. The surface temperature drop behaves as $t^{1/2}$ for the first few hours and then more slowly. This lack of steady state is a reflection of the slowness of a stable boundary layer to adjust to changing surface conditions.

With the constant cooling rate, they found diffusivity profiles and geostrophic drag coefficients in good agreement with arctic and antarctic data. Of particular importance to flow over ice sheets on land was the finding that the effects of slope are very large. The depth of the boundary layer varied by a factor of 4 for a slope of 0.002 and a cooling rate of 2 K h^{-1}. To get geostrophic drag coefficients, it was found that if the depth-averaged effective pressure gradient was used to determine G, then u_*/G agreed well with barotropic, flat-terrain results. The use of layer averaged geostrophic winds has also been suggested for baroclinic convective boundary layers.

More recently Neff (1981) has completed a study of the Antarctic boundary layer, coupling a version of the Brost-Wyngaard model and acoustic sounder observations of scattering and Doppler wind profiles. The model was adapted to predict acoustic scattering cross-section (due to microscale temperature and wind fluctuations) and these results were very similar to observations. The model also produced acceptable wind profiles and boundary-layer depths and growth rates. It was found that the boundary-layer depth could be reasonably well represented by:

$$h = 1.2u_*/(fN_B)^{1/2} \text{ (m)} \qquad (3.60)$$

where N_B is the bulk bouyancy frequency. The advantage of this relationship over ones involving L or the surface heat flux is that it appears to be insensitive to baroclinicity and variations in the surface cooling rate.

Neff investigated the effects of slope-induced pressure gradients and found that additive slope-induced and synoptic pressure gradients led to a growing, nearly well-mixed boundary layer. When the gradients were opposed, there was less turbulent mixing leading to intermittent turbulent layers and patches. The acoustic sounder was especially useful in investigating these features. The origin and evolution of these layers and patches depended on the initial conditions and accelerations due to horizontal variations in temperature.

The omission of radiative transfer was identified by Neff as a limitation of his model. Garratt and Brost (1981) added radiation processes to the Brost-Wyngaard model and found that the "bulk" boundary-layer parameters (Ri_B, u_*/G, h/L, etc.), and the profile of velocity, stress and diffusivity were little affected. However, the temperature and heat flux profiles were changed with the radiative processes dominating the turbulence ones below 0.1 h and above 0.8 h, and vice versa in between. The radiative effects were greatest above the boundary layer. Although this study was aimed at nonpolar boundary layers, some of the results are transferable.

Brown (1980a, b) has specifically addressed the problem of determining surface drag from an array of pressure measurements, such as were made during AIDJEX. Brown's model consists of matching two layers: a surface layer of logarithmic velocity profile with a diabatic correction, and an outer layer with velocity turning to match the free-stream or geostrophic flow. This latter Ekman-Taylor layer is corrected for stratification effects through the variable secondary flows and for thermal winds (baroclinicity). The matching leads to relations between the geostrophic wind G, u_*, the Ekman depth $\delta = (2K/f)^{1/2}$, the patching height h_p, and z_o. The model requires z_o and the surface heat flux, and the ratio $\lambda = h_p/\delta$ must be empirically determined. The depth of the boundary layer is defined as $H = \pi\delta$ and δ is evaluated by matching the diffusivities h_p. For the surface layer $K_M = k\, u_*\, h_p \cdot (1 - \lambda\psi_s)$ and for the Ekman layer $K_M = \delta^2 f/2$, therefore $\delta = 2ku_*\lambda/f(1-\lambda\psi_s)$ where $(1-\lambda\psi_s)$ is the stability correction. For $\lambda = 0.17$, and neutral conditions, the geostrophic drag coefficient $u_*/G \sim 0.03$ and $\alpha \sim 20^o$, the angle between surface wind/stress and the geostrophic wind. The model results showed some agreement with AIDJEX data (see Figure 9); however, the AIDJEX data was generally so scattered as not to be a good test.

It is evident that the situation of boundary layer flow over snow and ice surfaces is not well understood. At the same time the situation is probably clearer than for flows over mixed ice-water surfaces to be dealt with next.

5. ABL OVER MIXED SNOW-ICE-WATER SURFACES

In the previous section the discussion was focussed on the ABL over snow and ice-covered surfaces. Although the surface roughness may have changed markedly, it was assumed that the surface thermal characteristics were relatively uniform. This would be the case for large pack-ice areas and also for summertime conditions where the surface melt-water and snow is not much different in temperature from the water below that are occasionally (or frequently) exposed. However, in the winter the situation is drastically different. The air near the surface over the ice is -20 to -30°C and similar to the ice temperature. The temperature of exposed water, which does exist even in the coldest time of winter, will be at the freezing point of water for that salinity, about -2°C. Hence, at these areas open water, or thin ice, at channels, leads, polynyas, etc., there will be intense turbulent interaction due to the air-water temperature difference. It is expected (Maykut, 1978) that the sensible heat transfer from these regions will be a significant term in the arctic heat budget, even though the percent of the total area that is exposed water is small (1-2%).

Generally air over thick-ice packs is stably stratified, the depth of the boundary layer is only a few hundred meters and the surface fluxes are small (Figure 12). Over an open water area, the stratification will be very unstable and the surface fluxes very large. The altitude to which the effects of the open water extend depends on a variety of parameters, e.g., strength of stability of the boundary layer and free atmosphere aloft, size of the area of open water, wind speed. As air moves from over pack ice to over open water it will undergo drastic changes near the

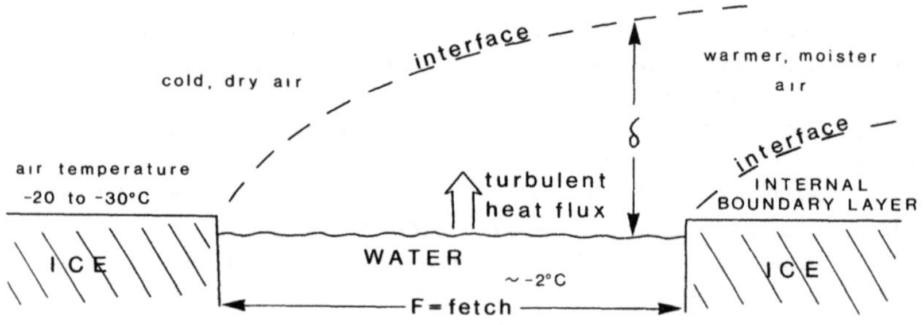

Fig. 12. Schematic of lead in the ice showing the development of internal boundary layers over the water and the downwind ice surface.

surface. A thermal internal boundary layer (IBL) will develop as the air is heated. The heating of the air will reduce the air-water temperature contrasts, but this effect will be small over a lead or a polynya. At the downwind edge of the open water a new IBL will develop, this time in response to the cold ice surface. Hence, the heated air will continue downstream aloft as a plume, gradually disappearing as it mixes with the surrounding air. The final distribution of the heat given off by the open water is not totally clear but its effect will be important.

There have been three studies in the Arctic to measure directly the heat loss from open water leads. Miyake (1965) measured wind and temperature profiles in the air flowing over an artificial pond, created by pumping water on the surface of a sheet of sea ice near Barrow. The AIDJEX Lead Experiment, ALEX (Andreas et al., 1979), was conducted in 1974 to measure the heat flux from Arctic leads in winter. Some measurements were made over natural leads of 33 m to 85 m width and over artificial pond of 6 m to 20 m. A study of a polynya several hundred meters across was conducted by Topham et al. (1983).

Miyake (1965) considered the development of an internal boundary layer in analogy to a plume from a continuous point source. Assuming that the vertical displacement of the internal boundary layer is proportional to σ_{w_2} (the subscript 2 denotes 'in the IBL') and the lateral displacement to U_1, the wind on the upwind surface, the equation for boundary propagation is:

$$\frac{dx}{dz} = \frac{U_1}{D\sigma_{w_2}} \tag{3.61}$$

where D is a constant to be determined.

For neutral, near-neutral, and extremely unstable stratification, solutions were obtained using similarity relationships for U and σ_{w_2}. In the present context, the case of extremely unstable stratification with a constant surface temperature, i.e., that of the open water, is appropriate. In this case the vertical velocity fluctuations are dominated by buoyancy forces, and free-convection formulations are appropriate. It is assumed that the upwind profile was that of a neutral ABL. It is then possible to derive for δ, the height of the internal boundary layer,

$$\frac{\delta}{z_{o_1}} = (6.41)\left(\frac{\Delta\bar{T}}{u_*\bar{T}}\right)^{5/8} z_{o_1}^{\ 5/8}\left(\frac{x}{z_{o_1}}\right)^{5/4} \tag{3.62}$$

$\Delta\bar{T}$ is the temperature difference from the surface to the top of the internal boundary layer. Note that this includes essentially a stability ($\Delta\bar{T}/u_*$) and roughness dependence. The heat flux H can be obtained as

$$H = \rho C p h \left(\frac{g}{T}\right)^{1/2} \left(\frac{\Delta T}{3}\right)^{3/2} z_{o_1}^{1/2} \left[1 + \frac{3}{2} (z_{o_2})^{1/3} \delta^{-1/3}\right] \quad (3.63)$$

This can be integrated across the width of an open-water body to give the total heat flux. Miyake applied the theory to measurements made over an artificial pond on the ice and found qualitative but not quantitative agreement. This study pointed out the importance of diabatic effects in the development of the IBL. Andreas et al. (1979) reported that z_o/L, a parameter suggested by Miyake, has a greater influence on than has the fetch. They found empirically that:

$$\frac{\delta}{z_{o_2}} = \beta\left(-\frac{z_{o_2}}{L}\right)^{0.8} \left(\frac{x}{z_{o_2}}\right)^{0.4} \quad (3.64)$$

and that β depends on upwind stability.

The main aim of the ALEX was to evaluate the heat flux from Arctic leads in winter. Profiles of wind and temperature were made both upwind and downwind of the leads (artificial and natural) and eddy flux measurements were made downwind. However, because of instrumental difficulties, the eddy heat flux measurements were not satisfactory. The main technique for obtaining the heat flux was the integral method. It is assumed that feature was two-dimensional and that the average surface flux from the lead was:

$$\bar{H}_o \simeq H_i + \frac{\rho C p}{F} \int_o^{\delta} U(T-T_i)\, dz$$

where H_i is the upwind heat flux and T and T_i are the temperature profiles at the downwind and upwind edges respectively (Figure 13). The fetch over the water is F. This technique effectively relates the surface flux to the height-integrated differences in advection of heat through the upwind and downwind boundaries. For the fetches under consideration (8-20 m) the contributions to the integral are mostly from within 1.5 m of the surface.

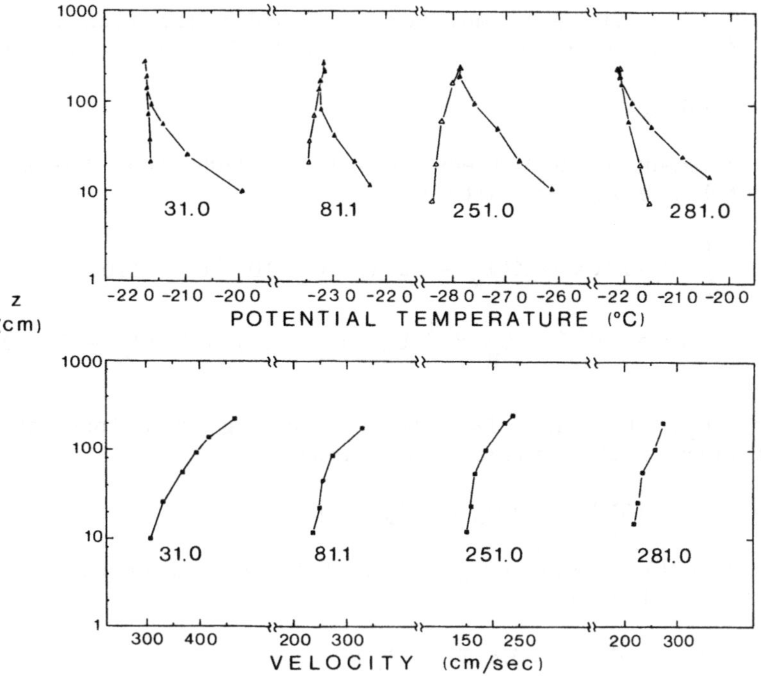

Fig. 13. Examples of temperature profiles upwind (open triangles) and wind and temperature profiles downwind (solid symbols) of an arctic lead. Numbers underneath are run numbers (from Andreas et al. 1979).

Andreas et al. present their results in terms of fetch Reynolds number, R_x, and Nusselt number, N, where

$$R_x = \frac{U(2m)F}{\nu} \tag{3.65}$$

$$N = \frac{\bar{H}_o F}{k(T_w - T(2m))} \tag{3.66}$$

U(2m) is the downwind velocity which differed little from the upwind velocity and T(2m) is the upwind temperature. ν and k are the kinematic viscosity and thermal conductivity of air, evaluated at the water temperature T_w. Andreas (1980) added data on condensate fluxes (see Andreas et al. 1981) and moisture fluxes to obtain

$$N = 0.08\ R_x^{0.76}$$

or $$N = 1.8 \times 10^{-3}\ R_x + 1100$$

where N can be for heat, moisture or condensate fluxes. Coantic and Favre (1974) reported, for latent heat flux in a wind tunnel,

$$N_E = 0.08\ R_x^{0.75}$$

in good agreement with Andreas' results. Note that these results are for fetch averaged results so

$$\bar{H}_o F = \int_o^F H_o(x)\,dx = K(T_w - T_2)(1.8 \times 10^{-3} R_x + 1100)$$

$$= (1.8 \times 10^{-3})\ C_p\ P^{-1}\ U_2\ (T_w - T_2) + k\ (T_w - T_2)\ 1100$$

Since U_2, T_w, and T_2 are not strong functions of fetch, differentiating with respect to x gives:

$$H_o = 1.8 \times 10^{-3}\ {}^{-1} \rho\, C_p\, U_2(T_w - T_2)$$

P is the Prandtl number, $\rho C_p \nu/k$ and for typical Arctic conditions is about 0.7. Therefore,

$$H_o \simeq 2.5 \times 10^{-3} \rho\, C_p\, U_2(T_w - T_2)$$

or $C_H = 2.5 \times 10^{-3}$ for 2 m reference level. Vugts and Businger (1977) found for flow from a cool ocean (T_w) onto a warm beach (T_b) that,

$$H_o \simeq 2.09 \times 10^{-3}\ \rho C_p\ U_2(T_b - T_w),$$

in reasonable agreement with the Arctic results. The corresponding equation for latent heat flux is:

$$LE = 1.8 \times 10^{-3}\ \rho L\ S_c^{-1}\ U_2(Q_o - Q_2)$$

where S_c is the Schmidt number, ν/D_w, where D_w is the molecular diffusivity of water vapor. For typical Arctic conditions S_c is 0.6 and hence

$$C_E = 3.0 \times 10^{-3}$$

How do these results compare with results from long fetches over relatively uniform bodies of water? First, it is necessary to correct the data to the normal 10 m observation height. Andreas et al. report z_o values upwind ranging from 0.002 cm to 0.177 cm and generally within a factor of two of 0.005 cm. The downwind values ranged from 0.008 cm–0.04 cm, with 0.02 cm being typical. Note that the water was generally rougher than ice, a somewhat surprising feature. The upwind profiles were near neutral and the effect of the lead did not usually extend above 2 m, so a logarithmic correction to the velocity is appropriate. Further, for a neutral condition the temperature at 10 m is the same as at 2 m. Hence, to a first approximation only the wind need be corrected. For z_o~0.005 cm the ratio $U_{10}/U_2 = 1.27$, which reduces C_H to 2.0×10^{-3} and C_E to 2.4×10^{-3}. For z_o~0.02 cm, $U_{10}/U_2 = 1.35$, reducing C_H to 1.8×10^{-3} and C_E to 2.2×10^{-3}. Both these values are larger than the typically reported values (Anderson and Smith, 1981) of $C_H = 1.1 \times 10^{-3}$ and $C_E = 1.3 \times 10^{-3}$, which are, however, for near-neutral conditions. Deardorff (1968) has derived the stability dependence of the air-sea transfer coefficients and for a bulk Richardson number of about -0.15, typical of the ALEX data, $C_{H,E}/(C_{H,E})$ neutral ~1.5 giving C_H ~1.7×10^{-3}, and C_E~2.0×10^{-3}, for 10 m values. Thus the ALEX heat and evaporation coefficients are similar to open water measurements, when corrected for stability.

From the ALEX data it is possible to infer the Bowen ratio, $B = H_o/LE_o$, and a value of 3 was typical, i.e., the sensible heat flux is typically three times larger than the latent heat flux. Over the open ocean, the ratio is usually less than 1 and frequently 0.2 - 0.3 (e.g., Anderson and Smith, 1981).

Andreas' formulae in terms of N and R_x contain a fetch dependence that is more evident when written as, for sensible heat flux,

$$\bar{H}_o = \frac{1}{F}\int_o^F H_o(x)dx = 2.5 \times 10^{-3}\rho Cp\ \bar{U}_2(T_w - T_2) + k(\dot{T}_w - T_2)1100/F$$

Hence, the fetch effect results in a correction term such that as F increases, the correction becomes smaller. For a typical ALEX case of $(T_w - T_2) = 19°C$, $U_2 = 300$ cms^{-1}, the first term corresponds to 18.6 mW cm^{-1}. For F = 8 m, the second term adds an additional 6.3 mW cm^{-2}; for F = 33 m, the addition is only 1.5 mW cm^{-2}. Hence, the average heat flux depends on fetch only for the first 30 m or less; beyond it is independent of fetch.

The modeling results of Rao et al. (1974b) are appropriate here, although they consider the case of air flowing from a smooth dry (warm) area onto a grassy wet area. The computed fluxes showed rapid changes in the first 1-2 meters and then only a gradual variation through to 18 m from the transition. The profiles adjusted more gradually.

Before leaving the ALEX flux results it is desirable to comment on their interpretation in terms of N and R_x. It should be noted that although N and R_x are well established nondimensional quantities, it is frequently useful to examine dimensional groupings as well. When plotting N versus R_x, one is in fact plotting:

$$\frac{H_o F}{k(T_w - T_2)} \quad \text{versus} \quad \frac{U_2 F}{\nu}$$

First note that F appears in both, and hence an apparent correlation will be generated if it has the largest variation of the parameters (for ALEX, F varied by about 10; H_o by 3; U_2 by 5). Also k and ν don't vary at all for these data. Hence, it is instructive to plot - $H_o/\Delta T$ vs U_2 for cases of similar F (Figure 14). For F ~8m̄, $H_o/\Delta T$~0.0035 U_2 + 0.3 (eyeball fit). For the limited data the additive constant decreases as F increases but the slope does not change significantly. The additive constant may be thought of as the heat impulse due to the abrupt temperature change, which is an important feature for short fetches but is less so for long fetches.

Thus the basic physics is similar despite the analysis. However, it is necessary to caution the extension of such analytical results beyond the range of parameters over which they were measured.

Andreas and Paulson (1979) reported on the ALEX turbulence results. They found that the u and w spectra had a more rounded shape than universal spectra reported by, e.g., Kaimal et al. (1972) and did not display an inertial subrange. The stress cospectrum had larger contributions from high nondimensional

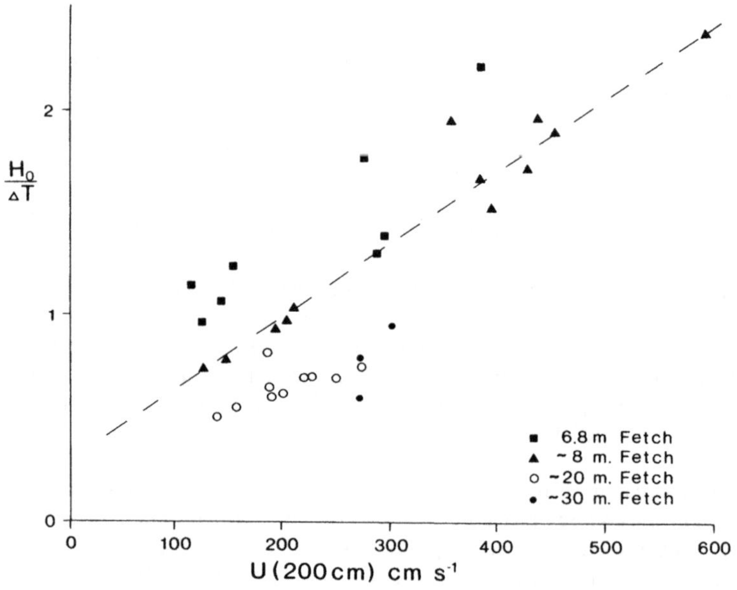

Fig. 14. Plot of $H_o/\Delta T$ against U_2 for different fetch groups.

frequencies than is usually observed. They attributed these differences to the intense mixing in the IBL and the nearness of the measurements to the surface (less than 50 cm). When grouped by Reynolds number (based on height above ground, not fetch), the spectra and cospectra grouped well. The integral statistics σ_w/U_1 and σ_u/U_2 had values smaller than usually reported, e.g., $\sigma_w/U_* = 0.99$ near neutral results compared to 1.4 for McBean's (1971) results.

As part of a comprehensive study of an Arctic polynya, Smith et al. (1983) made detailed measurements of the turbulence and the resulting heat and momentum fluxes. They found that concepts of local free convection or use of the temperature spectrum can be used to deduce the heat flux with reasonable accuracy. They also compared the turbulence statistics with results from less extreme environments. Generally similar results were found.

The studies of Miyake and ALEX dealt with quite small leads (less than 100 m) while we usually consider weather and climate on scales of tens of kilometers and more. The important question is, how do the cumulative effects of these small features affect the larger scale motions. It is not clear that the heat release from a lead is added to the large scale atmospheric heat content above the boundary layer. As discussed earlier the heat will plume downwind, gradually mixing with its environment. The warm plume will create or intensify an existing inversion such that the heat flux down to the ice surface will probably increase. The ice itself could lose the heat to the ocean below, to the atmosphere through radiation, and/or subsequently back to the atmosphere when the warm plume is no longer present. In order to investigate the processes on this scale, it is apparent that larger scale experiments, involving aircraft and arrays of profilers (radiosondes, dropsondes, remote sensors) will be required.

One of the most important air-ice ocean areas is the marginal ice zone along the seaward edge of sea-ice areas. Here there is a strong interplay between the dynamics and thermodynamics of ocean waves and currents, the oceanic boundary layer, and atmospheric forcings, resulting in the continual movement, growth or decay, break-breaking up or consolidating of the ice pack. The fluctuations in the position of the marginal ice zone (MIZ) are known to be important for climate. A good review of the problems associated with the MIZ is the proceedings of the International Workshop of the Seasonal Sea Ice Zone (Anderson et al., 1980). As yet not too much is known about the mechanisms of the MIZ, but the ABL must play an important role. It seems likely that the wind stress is the most important parameter, but the heat exchanges cannot be neglected. The Marginal Ice Zone Experiment (MIZEX) first field phase took place during the summer of 1984 in the Fram Strait between Svalbard and Greenland. The initial results from this study are now becoming available through the MIZEX Bulletin Series (published by the U.S. Cold Regions Research and Engineering Laboratory, Hanover, New Hampshire). Modelling the MIZ is reviewed in Bulletin III and preliminary results in Bulletin V.

Some studies have already been conducted in the MIZ, in this case in the Bering Sea. One interesting phenomenon that occurs when the wind blows off the ice over the warmer water is the development of roll clouds (Walter, 1980). For distances of up to a few hundred kilometers from the ice edge longitudinal roll clouds with wavelengths of 5-10 km occur. The rolls are aligned closely with the wind direction and their wavelength is related to boundary layer depth (typically, wavelength depth ratio is about 6.1). The theory of these longitudinal instabilities has been reviewed by Brown (1980c). He noted that instabilities can be due to either convective or dynamic instabilities or, most likely, a combination of both. The classical Rayleigh-Bernard thermal instability will produce cell type structure in free convection but longitudinal rolls when a moderate mean flow is present. For wind direction turning in response to Coriolis force, an inflection point instability will result in waves with crest axes approximately longitudinal and wavelengths two to four times the characteristic height. When confined to the boundary layer, the result is again longitudinal rolls.

Overland et al. (1983) have modelled the atmospheric boundary layer over the marginal ice zone. Different values of the drag coefficient were used for each ice zone and the ocean. The relative importance of each term in the momentum balance was investigated. The results suggested an atmospheric mechanism for rafting at the windward side of the marginal ice zone, divergence of the ice at the edge and ice-band formation seaward of the edge. The water movements near the ice edge were investigated by Roed and O'Brien (1983), using a coupled ice-ocean model, who found that there was weak downwelling near the ice edge due to ice motion.

A MIZ is an area of marked thermal, and probably roughness, contrast. On the small scale of leads, the effect of open water is to add heat and indirectly influence the flow. On the scale of the MIZ in the North Atlantic Ocean or around Antarctica the position of the MIZ influences the synoptic-scale flow patterns. On a somewhat smaller scale the MIZ probably generates its own circulation system in analogy to the land-sea breezes. In this case there is no diurnal cycle, so a steady-state situation may be established. The circulation would be a near-surface flow outward from the ice to the water, rising motions over the water, and a return flow aloft. Wind flows of several meters/second are observed over land stations, so appreciable flows would be possible. However, it should be noted that this case is more analogous to the nighttime land breeze than the daytime lake breeze. This would exert a drag on the ice to stretch out and separate the ice near the edge. At the same time the wind drag on the surface of the ocean would cause a transport to the right (in the northern hemisphere). As the atmospheric winds increased there would be an increased coriolis force that would cause the wind to turn to the right. The oceanic transport would than have a component back toward the ice. At the same time differential heating in the ocean due to the removal of the ice lid may generate an oceanic equivalent of this sea breeze. How these processes interact and their net effects are unknown at present. They do, however, serve to illustrate the type of interesting and important problems that may occur in the regions of an ice-ocean interaction.

6. PROBLEMS AND FUTURE DIRECTION

Throughout the preceding text an attempt was made to elucidate what is known about the ABL in general and the ABL in polar regions in particular. The impression that was intended to be left was a balanced one--that considerable is known, yet many problems remain. First, consider the cases of horizontal homogeneity. Surface, constant-flux, layer turbulence is reasonably understood. The convective ABL is also reasonably well understood, particularly from a polar perspective, because it doesn't happen as frequently as in lower latitudes. The stable boundary layer (SBL) is a problem and that means polar areas have a real problem. The problem here may be that the polar ABL hasn't had the same attention as, say, tropical boundary layers. Modelers of the SBL tend to look at diurnal variations and relatively light winds. The polar SBL can be steady-state and have high winds. These qualities make it ideal for experimental studies of the SBL and these are recommended. The WMO/CAS Working Group on ABL Problems has identified these studies as the most cost effective at the present time.

The really big problems are with moving into the three-dimensional era. Boundary-layer meteorologists are now facing the difficulties of a world that is not a flat, infinite plane and is not time-stationary. The focus is also moving to somewhat larger scales, to the microscale-mesoscale overlap and the mesoscale itself. The polar boundary layers offer lots of challenge. The major problems are dealing with the mixed ice-water surface that responds to and interacts with the ABL. The following specific problems have been identified (Anderson et al., 1980: SCAR Group of Specialists on Antarctic Climate Research; Action Plan, 1981; WMO/CAS Working Group on ABL Problems; and others).

a) development of a data base

- on the meso-to-synoptic scale through standard and specialized observations systems

- on the micro-to-meso scales through special boundary-layer measurements (remote sensors should be used to particular advantage).

b) determination and parameterization of heat, moisture, and momentum fluxes from sea ice areas for use in climate models

- particular attention should be paid to the marginal sea ice zones through special experiments (MIZEX) and modeling studies.

c) understanding and modeling the processes of wind stress on ice with irregular surfaces and distribution for the prediction of ice movement. This is related to b) but has special interests on a smaller scale for ice prediction for transportation, resource development, etc.

REFERENCES

Anderson, B. G., W. F. Weeks and J. L. Newton (Eds.) (1980) The Seasonal Sea Ice Zone: Proceedings of an International Workshop. Cold Reg. Sci. Tech. 2: 335 p.

Anderson, R. J. and S. D. Smith (1981) Evaporation coefficient for the sea surface from eddy flux measurements. J. Geophys. Res., 86: 449-456.

Andre, J. C., G. DeMoor, P. Lacarrere, G. Therry and R. du Vachat (1978) Modelling the 24-hr evolution of the mean and turbulent structures of the planetary boundary layer. J. Atmos. Sci., 35: 1861-1883.

Andreas, E. L. (1980) Estimation of heat and mass fluxes over arctic leads. Mon. Wea. Rev., 108: 2057-2063.

Andreas, E. L. and C. A. Paulson (1979) Velocity spectra and cospectra and integral statistics over Arctic leads. Quart. J. Roy. Met. Soc., 105: 1053-1070.

Andreas, E. L., C. A. Paulson, R. M. Williams, L. W. Lindsay and J. A. Businger (1979) The turbulent heat flux from arctic leads. Bound.-Layer Met., 17: 57-91.

Andreas, E. L., R. M. Williams and C. A. Paulson (1981) Observations of condensate profiles over arctic leads with a hot-film anemometer. Quart. J. Roy. Meteor. Soc., 107: 437-460.

Antonia, R. A., H. Q. Danh and A. Prabhu (1977) The response of a turbulent boundary layer to a step charge in surface heat flux. J. Fluid Mech., 80: 153-177.

Arya, S. P. S. (1975) A drag partition theory for determining the large-scale roughness parameter and wind stress on the arctic pack ice. J. Geophys. Res., 80: 3447-3454.

Banke, E. L. and S. D. Smith (1973) Wind stress on arctic sea ice. J. Geophys. Res., 78: 7871-7883.

Banke, E. G., S. D. Smith and R. J. Anderson (1976) Recent measurements of wind stress on arctic sea ice. J. Fish. Res. Bd. Can., 33: 2307-17.

Banke, E. G., S. D. Smith and R. J. Anderson (1980) Drag coefficients at AIDJEX from sonic anemometer measurements. In Sea Ice Processes and Models (R. S. Pritchard, Ed.), University of Washington Press, Seattle, Washington, p. 430-442.

Bradley, E. F. (1968a) A micrometeorological study of velocity profiles and surface drag in the region modified by a change in surface roughness. Quart. J. Roy. Met. Soc., 94: 361-379.

Bradley, E. F. (1968b) A shearing stress meter for micrometeorological studies. Quart. J. Roy. Met. Soc., 94: 380-387.

Brost, R. A. and J. C. Wyngaard (1978) A model study of the stably stratified planetary boundary layer. J. Atmos. Sci., 35: 1427-1440.

Brown, R. A. (1980a) Modeling the geostrophic drag coefficient for AIDJEX. J. Geophys. Res., 86: 1989-1994.

Brown, R. A. (1980b) Planetary boundary layer modeling for AIDJEX. In Sea Ice Processes and Models (R. S. Pritchard, Ed.), University of Washington Press, Seattle, Washington, p. 387-401.

Brown, R. A. (1980c) Longitudinal instabilities and secondary flows in the planetary boundary layer: A review. Rev. Geophys. Space Phys., 18: 683-697.

Businger, J. A., J. C. Wyngaard, Y. Izumi and E. F. Bradley (1971) Flux-profile relationships in the atmospheric surface levels. J. Atmos. Sci., 28: 181-189.

Businger, J. A. and S. P. S. Ayra (1974) The height of the mixed layer in the stably stratified planetary boundary layer. In Advances in Geophysics, 18A, Academic Press, New York, p. 73-91.

Carsey, F. D. (1980) The boundary layer height in air stress measurement. In Sea Ice Processes and Models (R. S. Pritchard, Ed.), University of Washington Press, Seattle, Washington, p. 443-451.

Carsey, F. and E. Leavitt (1977) Pibal/acoustic radar data in measurements and computation of air stress over pack ice. AIDJEX Bulletin, 36: 87-128.

Carsey, F., D. Short, R. Anderson, D. Bell, R. Brown, M. Clarke, A. Johnson, E. Leavitt and C. A. Paulson (1978) Analysis of AIDJEX data from a boundary layer profile, spring 1976. AIDJEX Bulletin, 38: 117-131.

Champagne, F. H., C. A. Friehe, J. C. LaRue and J. C. Wyngaard (1977) Flux measurements, flux estimation techniques, and fine-scale turbulence measurement in the unstable surface layer over land. J. Atmos. Sci., 34: 515-530.

Coantic, M. F. (1978) An Introduction to Turbulence in Geophysics and Air-Sea Interactions. NATO, AGARDograph No. 232, 254 p.

Csanady, G. T. (1974) Equilibrium theory of the planetary boundary layer with an inversion lid. Bound.-Layer Met., 6: 63-79.

Deardorff, J. W. (1968) Dependence of air-sea transfer coefficients on bulk stability. J. Geophys. Res., 73: 2549-2557.
Deardorff, J. W. (1972) Numerical investigation of neutral and unstable planetary boundary layers. J. Atmos. Sci., 29: 91-115.
Deardorff, J. W. (1980) Progress in understanding entrainment at the top of the mixed layer. In Workshop on Planetary Boundary Layer (J. C. Wyngaard, Ed.), American Meteorological Society, Boston, Mass., p. 36-65.
Delage, Y. (1974) A numerical study of the nocturnal atmospheric boundary layer. Quart. J. Roy. Met. Soc., 100: 351-364.
DeMoor, G. (1978) Les Theories de la turbulence dans la Couche Limite Atmospherique. Etablissement d'Etudes et de Recherches Météorologiques, Direction de la Météorologie, Paris, 288 p.
Den Hartog, G., S. D. Smith, R. J. Anderson, D. R. Topham and R. G. Perkin (1983) An investigation of a polynya in the Canadian archipelago, 3, surface heat flux. J. Geophys. Res., 88: 2911-2916.
Dyer, A. J. and B. B. Hicks (1970) Flux-gradient relationship in the constant flux layer. Quart. J. Roy. Met. Soc., 96: 715-721.
Garratt, J. R. and R. A. Brost (1981) Radiative cooling effects within and above the nocturnal boundary layer. J. Atmos. Sci., 38: 2730-2746.
Haugen, D. A. (Ed.) (1973) Workshop on Micrometeorology. American Meteorological Society, Boston, Mass., 392 p.
Hibler, W. D. (1972) Removal of aircraft altitude variation from laser profiles of the arctic ice pack. J. Geophys. Res., 77: 5954-5970.
Hicks, B. B. and A. J. Dyer (1972) The spectral density technique for the determination of eddy fluxes. Quart. J. Roy. Met. Soc., 98: 838-844.
Hunt, J. C. R. (1980) Wind over hills. In Workshop on the Planetary Boundary Layer (J. C. Wyngaard, Ed.), American Meteorological Society, Boston, Mass., p. 107-144.
Kaimal, J. C., J. C. Wyngaard, Y. Izumi and O. R. Cote (1972) Spectral characteristics of surface-layer turbulence. Quart. J. Roy. Met. Sci., 98: 563-584.
Katz, D. I. (1980) Air stress measurements from an aircraft. In Sea Ice Processes and Models (R. S. Pritchard, Ed.), University of Washington Press, Seattle, Washington, p. 452-463.
Kazanski, A. B. and A. S. Monin (1961) Dynamic interaction between the atmosphere and the earth's surface. Izv. Akad. Nank SSSR, Ser. Geofiz., 5: 786-799.
Kraus, E. B. (Ed.) (1977) Modelling and Prediction of the Upper Layers of the Ocean. Pergamon, 325 p.
Langleben, M. P. and E. R. Pounder (1975) On the air drag of an arctic ice floe. Geophys. Res. Lett., 2: 15-18.

Leavitt, E. (1980) Surface-based air stress measurements made during AIDJEX. In Sea Ice Processes and Models (R. S. Pritchard, Ed.), University of Washington Press, Seattle, Washington, p. 419-429.

Leavitt, E., D. Bell, M. Clarke, R. Andersen and C. Paulson (1977) Computation of air stress and sensible heat fluxes from surface layer profile data, AIDJEX, 1975. AIDJEX Bulletin, 36: 157-174.

Lettau, H. H. (1979) Wind and temperature profile predictions for diabatic surface layers including strong inversion cases. Bound.-Layer Met., 197: 443-464.

Lumley, J. L. (1980) A new kind of realizability in turbulence modelling of the planetary boundary layer. In Workshop on the Planetary Boundary Layer (J. C. Wyngaard, Ed.), American Meteorological Society, Boston, Mass., p. 158-175.

Macklin, S. A. (1983) Wind drag coefficient over first-year sea ice in the Bering Sea. J. Geophys. Res., 88: 2845-2852.

Martin, S., P. Kauffman and C. Parkinson (1983) The movement and decay of ice edge bands in the winter Bering Sea. J. Geophys. Res., 88:2803-2812.

Mason, P. J. and R. I. Sykes (1979) Flow over an isolated hill of moderate slope. Quart J. Roy. Met. Soc., 105: 393-395.

Maykut, G. A. (1978) Energy exchange over young sea ice in the central Arctic. J. Geophys. Res., 83: 3646-3658.

McBean, G. A. (1971) The variations of the statistics of wind, temperature and humidity fluctuations with stability. Bound.-Layer Met., 1: 438-457.

McBean, G. A. (1976) Scaling turbulence in the planetary boundary layer. Atmosphere, 14: 254-267.

McBean, G. A. (Ed.) (1979) The planetary boundary layer. World Meteorological Organization, Geneva, Technical Note 16, 201 p.

McBean, G. A. and J. A. Elliott (1975) The vertical transport of kinetic energy by turbulence and pressure in the boundary layer. J. Atmos. Sci., 32: 753-766.

Merry, M. and H. A. Panofsky (1976) Statistics of vertical motion over land and water. Quart. J. Roy. Met. Soc., 102: 255-259.

Miyake, M. (1965) Transformation of the atmospheric boundary layer over inhomogeneous surfaces. Scientific Report, University of Washington, Department of Atmospheric Sciences, 63 p.

Monin, A. S. and A. M. Obukhov (1954) Dimensionless characteristics of turbulence in the surface layer. Trudy Geofiz. Inst., Akad. Nauk SSR, 24: 163-187.

Monin, A. S. and A. M. Yaglom (1971) Statistical Fluid Mechanics: Mechanics of Turbulence. MIT Press, Cambridge, Mass., Vol. 1.

Monin, A. S. and A. M. Yaglom (1975) Statistical Fluid Mechanics: Mechanics of Turbulence. MIT Press, Cambridge, Vol. 2, 769 p.

Neff, W. D. (1981) An observational and numerical study of the atmospheric boundary layer overlying the East Antarctic ice sheet. NOAA Technical Memorandum, ERL WPL-67, 272 p.

Overland, J. E., R. M. Reynolds and C. H. Pease (1983) A model of the atmospheric boundary layer over the marginal ice zone. J. Geophys. Res., 88: 2836-2840.

Paulson, C. A. (1970) The mathematical representation of wind speed and temperature profiles in the unstable atmospheric surface layer. J. Appl. Met., 9: 857-861.

Pease, C. H., S. A. Salo and J. E. Overland (1983) Drag measurements for first-year sea ice over a shallow sea. J. Geophys. Res., 88: 2853-2863.

Peterson, E. W. (1969) Modification of mean flow and turbulent energy by a change in surface roughness under conditions of natural stability. Quart. J. Roy. Met. Soc., 95: 561-575.

Peterson, E. Q., L. Kristensen and C. C. Su (1976) Some observations and analysis of wind over non-uniform terrain. Quart. J. Roy. Met. Soc., 102: 857-869.

Pritchard, R. S. (Ed.) (1980) Sea Ice Processes and Models. University of Washington Press, Seattle, Washington, 474 p.

Pruitt, W. O., D. L. Morgan and F. L. Laurence (1973) Momentum and mass transfers in the surface boundary layer. Quart. J. Roy. Met. Soc., 99: 370-386.

Rao, K. S., J. C. Wyngaard and Q. R. Cote (1974a) The structure of the two-dimensional internal boundary layer over a sudden change of surface roughness. J. Atmos. Sci., 31: 738-746.

Rao, K. S., J. C. Wyngaard and Q. R. Cote (1974b) Local advection of momentum, heat and moisture in micrometeorology. Bound.-Layer Met., 7: 331-348.

Roed, L. P. and J. J. O'Brien (1983) A coupled ice-ocean model of upwelling in the marginal ice zone. J. Geophys. Res., 88: 2863-2872.

Smith, S. D. (1972) Wind stress and turbulence over a flat ice floe. J. Geophys. Res., 77: 3886-3901.

Smith, S. D. (1980) Wind stress and heat flux over the ocean in gale force winds. J. Phys. Oceanogr., 10: 709-726.

Smith, S. D., R. J. Anderson, G. den Hartog, D. R. Topham and R. G. Perkin (1983) An investigation of a polynya in the Canadian archipelago, 2, structure of turbulence and sensible heat flux. J. Geophys. Res., 88: 2900-2910.

Stewart, R. W. (1979) The Atmospheric Boundary Layer. World Meteorological Organization, Geneva, No. 523, 44 p.

Stull, R. B. (1973) Inversion rise model based on penetrative convection. J. Atmos. Sci., 30: 1092-1099.

Stull, R. B. (1976) The energetics of entrainment across a density interface. J. Atmos. Sci., 33: 1260-1267.

Taylor, P. A. (1969) On wind and shear stress profile above a change in surface roughness. Quart. J. Roy. Met. Soc., 95: 77-91.

Taylor, P. A. (1980) Some recent results from a numerical model of surface boundary layer flow over hills. In Workshop on the Planetary Boundary Layer (J. C. Wyngaard, Ed.), American Meteorological Society, Boston, Mass., p. 150-157.

Taylor, P. A. and P. R. Gent (1980) Modification of the boundary layer by orography. In Orographic Effects in Planetary Flows, World Meteorological Organization, Geneva, GARP Pub. 23, p. 145-165.

Tennekes, H. (1973) A model for the dynamics of the inversion above a convective boundary layer. J. Atmos. Sci., 30: 558-567.

Thorpe, M. R., E. G. Banke and S. D. Smith (1973) Eddy correlation measurements of evaporation and sensible heat flux over arctic sea ice. J. Geophys. Res., 78: 3573-3584.

Tillman, J. E. (1972) The indirect determination of stability, heat and momentum fluxes in the atmospheric boundary layer from simple scalar variable during dry unstable conditions. J. Appl. Met. 11: 783-792.

Topham, D. R., R. G. Perkin, S. D. Smith, R. J. Anderson and G. Den Hartog (1983) An investigation of a polynya in the Canadian archipelago, 1, introduction and oceanography. J. Geophsy. Res., 88: 2888-2899.

Vugts, H. F. and J. A. Businger (1977) Air modification due to a step change in surface temperature. Bound.-Layer Met., 11: 295-306.

Walter, B. A. (1980) Wintertime observations of roll clouds over the Bering Sea. Mon. Wea. Rev., 108: 2024-2031.

Wittmann, W. I. and J. J. Schule (1966) Comments on the mass budget of arctic pack ice. In Proceedings, Symposium on the Arctic Heat Budget and Atmospheric Circulation (J. O. Fletcher, Ed.), Rand Corp., Santa Monica, California, RM-5233-NSF, p. 215-246.

Wyngaard, J. C. (1975) Modelling the planetary boundary-layer extension to the stable case. Bound.-Layer Met., 9: 441-460.

Wyngaard, J. C. (Ed.) (1980) Workshop on the Planetary Boundary Layer, American Meterological Society, Boston, Mass., 322 p.

Wyngaard, J. C., S. P. S. Arya and O. R. Cote (1974a) Some aspects of the structure of convective planetary boundary layers. J. Atmos. Sci., 31: 747-754.

Wyngaard, J. C., O. R. Cote and K. S. Rao (1974b) Modelling the atmospheric boundary layer. In Advances in Geophysics, 18A, Academic Press, New York, p. 193-211.

Wyngaard, J. C. and S. F. Clifford (1978) Estimating momentum, heat and moisture fluxes from structure parameters. J. Atmos. Sci., 35: 1204-1211.

Zeman, O. and H. Tennekes (1977) Parameterization of the turbulent energy budget at the top of the daytime atmospheric boundary layer. J. Atmos. Soc., 34: 111-123.

CHAPTER 4

THE UPPER OCEAN

Miles G. McPhee

McPhee Research Company
Yakima, Washington

1. INTRODUCTION

Since the drift of the Fram nearly ninety years ago, it has been recognized that turbulent exchange in the upper ocean is a major factor in the dynamic and thermodynamic regime of high-latitude, ice-covered seas. Ekman (1905) credited the inspiration for his theory of wind drift currents, which was to become a cornerstone of physical oceanography and meteorology, to Nansen's qualitative description of the effect of the earth's rotation on the upper layers of the ocean, which he deduced from his observations that Fram drifted consistently 20-40 degrees right of the surface wind. They were among the first to appreciate the unique observational platform provided by pack ice.

Experiments performed in polar regions over the past decade or so have provided significant additions to our knowledge of how the upper ocean works, and in many ways it is an exciting time to be studying the ice-ocean boundary. Our aim in this work is to review these findings, and to place them in the framework of a boundary-layer theory which, despite its relative simplicity, turns out to be surprisingly comprehensive. We start by pointing out some basic features, illustrating them with experimental data taken mainly from the AIDJEX studies. We then review and develop theoretical descriptions for momentum transfer in the boundary layer, which leads to consideration of density structure in the upper ocean, and inclusion of the conservation equation for buoyancy (governed mainly by salinity). This provides at least a theoretical staging point for addressing the extremely complex fluid-dynamical regime of the marginal ice zone.

2. BASICS

When a steady wind blows across the surface of an ice-covered ocean, there often develops in the ice cover a nearly steady-state balance among four forces: the surface wind traction, the gradient of internal ice stress, the Coriolis force on the ice column (assuming a fixed-to-earth reference frame), and a traction exerted by the ocean on the ice undersurface. The last force is the surface manifestation of a turbulent oceanic boundary layer (OBL) which provides the mechanism for momentum exchange between the ice and ocean. The OBL is affected by rotational (Coriolis) forces and therefore falls, along with the atmospheric boundary layer (ABL), into the general category of planetary boundary layers. We shall show that the two boundary layers are closely related in many respects, but that there are also some important and interesting differences.

Consider a force balance like that described above but idealized so that first, the internal ice force is negligible, and second, the undisturbed ocean is at rest. While the latter may seem artificial, it can approximate a real state in which the entire wind-drift system is advected by the mean geostrophic flow. Since normal geostrophic currents are at most a few percent of the surface wind, this is often a useful approximation and we shall use it repeatedly below. The force balance is diagrammed in Figure 1a; its relative proportions represent average observed conditions during summer at the AIDJEX camps, when internal stress gradients were small.

The sketch demonstrates that the OBL surface stress (which is the negative of the water stress shown) acts in a direction about 25° to the left of the surface velocity, $\hat{V}_o$. From this observational fact alone, we can infer much about the structure of the OBL. It helps to consider the problem from the viewpoint of an observer drifting on the ice floe, whom we imagine to be dropping a probe that measures velocity relative to the drifting ice and turbulent stress. Near the surface the kinematic stress would be about the same as $\hat{\tau}_o$, and by analogy with other constant-stress, wall-bounded turbulent flows, we would expect a region of shear with relative velocity varying logarithmically in the same direction as turbulent stress. As the probe drops, the stress would diminish and veering of both the horizontal stress vector and the mean current would be noted. At some depth stress would vanish and the observer would measure a "free-stream" current, $-\hat{V}_o$, at this and greater depths. The currents are sketched in plan view (hodograph) in Figure 1b. That part of the profile close to the ice, where the relative velocity is logarithmic and unidirectional

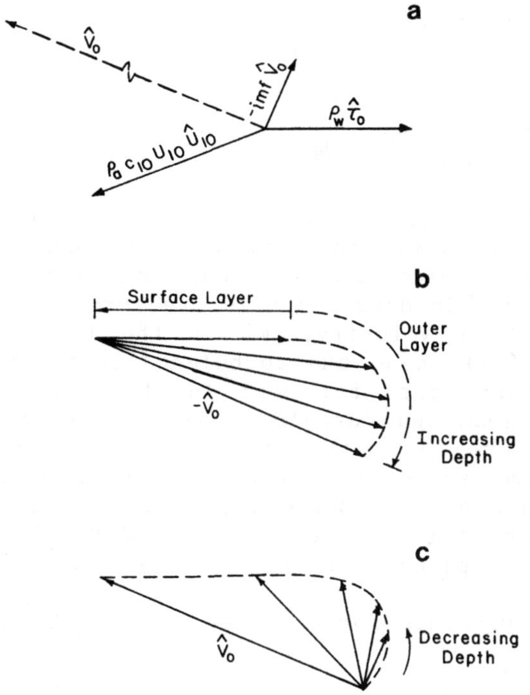

Fig. 1. Schematic of ice force balance (a), plan view of OBL velocity as viewed from drifting ice (b), velocity as viewed in frame fixed to earth (c).

is called the surface layer. The region farther from the interface, where the relative velocity veers noticeably, is called the outer, or Ekman layer. The basic structure diagrammed in Figure 1 should be studied closely, particularly the difference in currents measured in a fixed-to-earth reference frame (Figure 1c) vis-a-vis as measured in a reference frame drifting with the ice. These two ways of viewing the OBL have an exact analog in the way that the ABL is described, either in terms of the actual wind or by the "velocity defect" (see Tennekes and Lumley, 1972, Chapter 5).

2.1 Equations

Conservation equations for turbulent flow in a horizontally homogeneous fluid may be written:

$$\text{Momentum:} \quad \frac{\partial \underset{\sim}{V}}{\partial t} + f\underset{\sim}{k} \times \underset{\sim}{V} = -\frac{1}{\rho}\nabla p - g\underset{\sim}{k} - \frac{\partial}{\partial x}(\overline{w'\underset{\sim}{v'}}) \qquad (4.1)$$

$$\text{Heat:} \quad \frac{\partial \Theta}{\partial t} + \frac{\partial}{\partial z}\overline{w'\theta'} = Q_\theta/\rho c_p \tag{4.2}$$

$$\text{Salinity:} \quad \frac{\partial S}{\partial t} + \frac{\partial}{\partial z}\overline{w'S'} = Q_s \tag{4.3}$$

Where $\underset{\sim}{V}$ is the "mean" horizontal velocity (in the sense that its variation is slow relative to turbulent fluctuations); f is the Coriolis parameter (twice the vertical component of the earth's angular frequency); ρ is density; p is pressure; w' is the vertical turbulent fluctuation velocity; $\underset{\sim}{v}'$ is the horizontal turbulent velocity; Θ and θ' are mean and turbulent fluctuation of temperature; Q_θ is a source term for heat in fluid with heat capacity, c_p; S and S' are mean and turbulent fluctuation of salinity; and Q_s is a source term for salinity. Reynolds averaging has been performed over time scales commensurate with a presumed spectral gap; i.e., for an arbitrary property, a, $a = A + a'$ where A is the average over a time long compared to the scale of turbulent fluctuations, but short compared to inertial and synoptic time scales.

The equation of state for seawater is approximated by

$$\frac{\rho - \rho_o}{\rho_o} = -\alpha_\theta(\theta - \theta_o) + \alpha_s(S - S_o) \tag{4.4}$$

where ρ_o, θ_o, and S_o are reference density, temperature, and salinity.

Although the fluid is assumed horizontally homogeneous, a large-scale horizontal pressure gradient is allowed, which, through the hydrostatic relation, is identified with the gradient of the sea surface elevation, η,

$$\frac{1}{\rho}\nabla p = g\nabla\eta .$$

When no turbulence is present, eq. (4.1) defines the <u>geostrophic velocity</u> due to sea surface tilt

$$f\underset{\sim}{k} \times \underset{\sim}{V}_g = -g\nabla\eta. \tag{4.5}$$

For later convenience we introduce complex notation for horizontal vectors

$$\underset{\sim}{V} = (V_x, V_y) = \hat{V} = V_x + iV_y = Ve^{i\theta_V}, \quad i = \sqrt{-1} \ .$$

When a symbol is written with a caret (^), a complex, two-dimensional vector is indicated; when the symbol is written without a caret, magnitude only is meant: i.e., $V = |\hat{V}|$. The kinematic Reynolds stress is

$$\hat{\tau} = -\left(\overline{u'w'} + i\,\overline{v'w'}\right) \quad .$$

Substituting $\hat{V}_g$ from (4.5) into (4.1) results in the horizontal momentum equation

$$\frac{\partial \hat{V}}{\partial t} + if(\hat{V} - \hat{V}_g) = \frac{\partial \hat{\tau}}{\partial z} \quad . \tag{4.6}$$

In the Coriolis term, multiplication by i, which represents a 90° counterclockwise rotation, replaces the $\underset{\sim}{k}$ x operator. If $\hat{V}_g$ is constant over the time scale being considered, the transformation $\hat{V}' = \hat{V} - \hat{V}_g$ results in a particularly useful form of the momentum equation

$$\frac{\partial \hat{V}}{\partial t} + if\hat{V} = \frac{\partial \hat{\tau}}{\partial z}$$

where the prime is dropped and it is understood that all velocities are relative to the undisturbed, steady geostrophic flow. In what follows, this convention is observed unless otherwise stated.

2.2 Turbulence

A simple but powerful tool for describing the basic transfer mechanism in high-Reynolds-number flows is the mixing length hypothesis developed originally by Prandtl (see, e.g., Hinze, 1975). The concept is sketched in Figure 2, depicting an overturning eddy in fluid characterized by a vertical gradient of an arbitrary property, ε. To be truly turbulent, the flow must be highly diffusive, with fluid parcels caught in a turbulent eddy readily transferring their concentration of ε to the surrounding fluid in a cascade of smaller and smaller eddies. In Figure 2, Parcel 1 distributes a deficit of ε above h, Parcel 2 delivers an excess

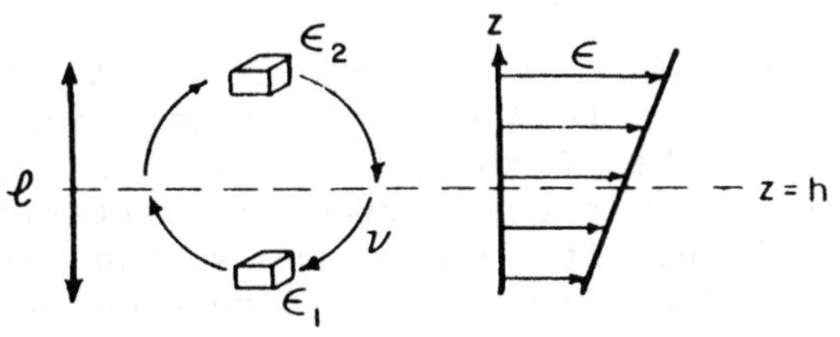

Fig. 2. Turbulent exchange in a flow with a gradient of an arbitrary property, ε.

The idea of a surface layer arises naturally from numerous observations in the atmosphere that show under typical conditions the mean wind profile varying dramatically in the first few meters, but turbulent stress changing by only a few percent over the lower tens of meters. The wind shear in the neutrally buoyant (no density variations) surface layer has been shown to satisfy very closely the relationship

$$\frac{kz}{u_*}\frac{\partial V}{\partial z} = 1 \tag{4.9}$$

where u_* is the magnitude of the surface friction velocity, defined in vector form by $\hat{\tau}_o = u_* \hat{u}_*$, where $\hat{\tau}_o$ is the kinematic interfacial stress. With stress nearly constant, eqs. (4.7) and (4.8) when combined with the empirical result eq. (4.9), imply that the turbulence scales in the surface layer are $v_m = u_*$ and $\ell_m = kz$. Integration of eq. (4.9) results in a logarithmic velocity profile relative to the surface

$$|\hat{V}(z) - \hat{V}_o| = \frac{u_*}{k} \ln \frac{|z|}{z_o} \quad . \tag{4.10}$$

The integration constant, z_o, is called the roughness length and on a surface of uniform roughness elements it corresponds in some way to the physical size of the elements. (With a surface like sea ice, punctuated by hummocks and pressure ridge keels, its meaning is more abstract as will be discussed later.) From the "law of the wall" eq. (4.10), the scales $\hat{u}_*$ for velocity and z_o for length are obvious choices for the surface layer.

Consider next the outer region of free turbulence. The steady-state form of eq. (4.6) is a balance between the Coriolis acceleration and the turbulent stress gradient.

$$if\hat{V} = \frac{\partial \hat{\tau}}{\partial z} \tag{4.11}$$

Through the outer layer the stress varies from near its surface value to zero, thus an obvious scale for stress is again the interfacial stress, $\hat{\tau}_o = u_* \hat{u}_*$.

It might be argued that the choice of $\hat{V}_o$ as the velocity scale for the outer layer is just as obvious; however, it is well to examine this in light of the assertion that the turbulence of the outer layer does not sense surface characteristics directly. The total change of velocity across the surface layer depends on z_o, and since most of the shear in the direction of surface stress occurs within the surface layer, z_o has an obvious impact on $\hat{V}_o$. But if turbulence below the surface layer does not sense z_o, then mean velocity there cannot either, and by this reasoning $\hat{V}_o$ is a below h, so that there is a net downward flux of ε. The magnitude of the flux apparently depends on the gradient, $\partial\varepsilon/\partial z$; the speed with which the eddy overturns, v ; and the size of the most energetic eddies found in the vicinity, ℓ_ε. The last two quantities represent scales at which the smaller eddies can keep pace with the large eddies at diffusing away variations in ε. The simplest expression for the flux, F_ε, is the product

$$F_\varepsilon = -v_\varepsilon \ell_\varepsilon \frac{\partial\varepsilon}{\partial z} \tag{4.7}$$

and by analogy with molecular transfer processes, an "eddy diffusivity" may be defined, $K_\varepsilon = v_\varepsilon \ell_\varepsilon$, which depends on the turbulent flow rather than the molecular properties of the particular fluid. This is a crucial aspect of high-Reynolds-number turbulence, and provides a basis for comparing, e.g., atmospheric and oceanic boundary layers.

Let ε stand for vector momentum. The momentum flux across a horizontal plane is the negative of the horizontal part of the turbulent stress tensor, and is given by

$$\hat{F}_m = -\rho_w \hat{\tau} = -\rho_w K_m \frac{\partial \hat{V}}{\partial z} \quad . \tag{4.8}$$

ε may stand for other properties, like temperature or salinity, and the same basic process is thought to apply; this is essentially Reynolds analogy. Note, however, that different properties may have different v_ε and ℓ_ε scales in the same flow, so that eddy diffusivities are not necessarily the same. Indeed, it is likely that turbulent diffusion of heat and salt occurs at different rates in the arctic thermocline (Neshyba et al., 1971).

There are obvious examples where the physics expressed by eq. (4.7) are not sophisticated enough to describe the particular flow; e.g., when flow separation occurs near a pressure ridge keel, or where there is double diffusion when a stabilizing mean salinity gradient coincides with a destabilizing temperature gradient. But despite its limitations, we shall show that the mixing length hypothesis can be very useful for interpreting the OBL, provided the scales are chosen carefully.

2.3 Scales

The planetary boundary layer (PBL) is often broken conceptually into a surface layer, where turbulence is affected directly by the hydraulic roughness of the solid surface, and an outer layer where the turbulence is "free" in the sense that it does not depend on the actual physical characteristics of the interface.

poor choice for its scale. With no other information, a good working hypothesis is that $\hat{u}_*$ is a better scale, because certainly turbulence in the outer layer must respond in some way to surface stress. In the absence of an externally imposed length scale (such as h_{BL}, the depth of the mixed layer above a strong density gradient), these choices of stress and velocity scales force the length scale to be u_*/f, found by nondimensionalizing eq. (4.11).

A consistent scaling of PBL velocity and stress by $\hat{u}_*$ and $\hat{\tau}_o$ ($= u_*\hat{u}_*$), with z_o and u_*/f the respective length scales for the surface and outer layer, has become known as Rossby-similarity scaling after pioneering work of Rossby (1932) and Rossby and Montgomery (1935), although it was later investigators (see, e.g., Blackadar and Tennekes, 1968) who recognized the significance of using u_* rather than V_o for the outer layer velocity scale.

Rossby-similarity scaling provides a convenient rule of thumb for comparing oceanic and atmospheric scales. Assuming that the stress at the ice-air interface is comparable to the ice-water stress (which is likely if the ice is thin), then $(\rho u_*^2)_{atm} \simeq (\rho u_*^2)_{ocean}$; thus velocities and lengths in the OBL will be roughly 1/30 (square root of the density ratio) of those in the ABL. An interesting consequence of this is that a pressure ridge with, say, a 1 m sail compensated by a 5 m keel, has very different aspects when viewed from the different boundary layers. To the ABL, the feature is almost part of the surface roughness, being 2 to 3 orders of magnitude less than the ABL thickness. To the OBL, the ridge is a definite topographic feature, comparable to a 150 m high hill in the ABL. The ice undersurface is often pictured either as flat, or as a series of jagged ice blocks which provide nearly all the drag and stirring. Probably neither view is appropriate, and perhaps a better model would be to think of the under-ice OBL as comparable to atmospheric flow over rolling hills.

2.4 Boundary-Layer Transport

If the steady-state momentum equation (eq. 4.11) is integrated to the surface from some level below which no turbulent stress penetrates, a remarkable result ensues.

$$\text{if} \int_{-h}^{0} \hat{V} dz = \hat{\tau}_o \quad . \tag{4.12}$$

We define a transport vector, $\hat{M} = \int_{-h}^{0} \hat{V} dz$, which is the total volume transport per unit time in the boundary layer, and from eq. (4.12)

$$\hat{M} = -\frac{i}{f}\hat{\tau}_o \quad . \tag{4.13}$$

Without any consideration of how the turbulent mechanism works, eq. (4.13) implies that the steady-state volume flux in the OBL is oriented 90° clockwise (Northern Hemisphere) from the surface stress and equal in magnitude to τ_o/f. This effect, sometimes called geostrophic departure, was point out by Ekman (1905) and the concept is used extensively in calculating the effect of wind on deep ocean circulation (see, e.g., Stern, 1975).

2.5 Inertial Oscillations

The transport vector is also useful for demonstrating another flow phenomenon associated with the rotating reference frame: inertial oscillation. If eq. (4.6) is integrated in the same manner as eq. (4.12), we have

$$\frac{\partial \hat{M}}{\partial t} + if\hat{M} = \hat{\tau}_o \quad .$$

Now suppose a constant surface stress is applied at t = 0 to a system at rest. The solution to eq. (4.13) in this case is

$$\hat{M} = \hat{M}_s(1 - e^{-ift}) \tag{4.14}$$

where $\hat{M}_s = - i\hat{\tau}_o/f$ is the stationary solution. In the complex plane, the locus of solutions eq. (4.13) traces a circle in one inertial period, $1/2\pi f$ (~12.4 hr at 76°N), centered at $\hat{M}_s$. Note that the idealized inertial amplitude is equal to the steady-state transport. M is the average velocity of the OBL multiplied by its depth, so even the surface velocity may have a significant inertial component, with an associated drift trajectory that is cycloidal.

3. OBSERVATIONS

Over the past decade a number of experimental studies have helped focus our understanding of the upper ocean under pack ice. Two of special importance regarding turbulent transfer were the 1972 AIDJEX Pilot Study and the 1975-1976 AIDJEX Main Experiment.

The most complete measurements of mean flow and turbulence in the OBL were gathered during the 1972 Pilot Study by a group from the University of Washington under the direction of J. D. Smith. The measuring apparatus included small mechanical current meters mounted along three orthogonal axes, capable of measuring the mean flow and Reynolds stress tensor at eleven levels ranging from 2 to 54 m below the drifting ice. McPhee and Smith (1976) describe the experiment and results, many of which are used below to illustrate various PBL concepts. During the April wind storm from which the data were obtained, the ice drift was comparatively straight and there was little indication of inertial oscillation in the ice/upper ocean system. On two different days of the same storm, abnormally steady conditions persisted: for about 5 hours in one case, and for 8 hours in a second. The data gathered constitute probably the most complete measurement ever made of mean and turbulent velocity structure in a fully developed, rotating boundary layer.

The AIDJEX Main Experiment, which maintained stations on the perennial pack over the Canada Basin for more than a year, lacked sophisticated equipment for precise current measurement in the OBL, but this was offset to some degree by the sheer volume of data collected. Two very important measurements were made at each of the manned stations without dropping an instrument in the water: highly accurate resolution of camp position normally at least once per hour, which provided surface velocity across the 100 km scale of the manned camp array, and continuous measurements of 10 m wind. During the summer when internal ice stress gradients were small, these data provided estimates of OBL surface velocity and stress that, as far as we know, are unparalleled in oceanographic studies. In addition, fixed current meters and daily STD and profiling current meter casts operated throughout the experiment.

3.1 Structure of the Steady OBL

Figure 3, drawn from data of McPhee and Smith (1976), shows mean flow measured at 9 levels through the mixed layer during the 1972 AIDJEX Pilot Study in both hodograph and profile form. Currents were averaged for 5 hours under very steady conditions. The pycnocline (steep density gradient) started at about 36 m below the ice. Actual currents were, of course, measured from the drifting ice, but are drawn relative to the measured current at 32 m, because concurrent measurements of the relative velocity of a fixed point on the sea floor showed that the actual current at the 32 m level was less than 2 cm s^{-1}, and was probably representative of $\hat{V}_g$.

Comparison of Figures 3 and 1 shows that, at least in its gross features, the observed flow is like the heuristic model outlined above. Near the ice, there is a region of high shear in the same direction as stress, but at greater depths the Coriolis effect is readily evident. The profiles provide an excellent example of geostrophic departure in that the integrated transport in the boundary layer is nearly all in the integral of the velocity component perpendicular to interfacial stress, despite a large surface component in the stress direction.

As detailed by McPhee and Smith (1976), turbulent stress was also measured during the 1972 experiment, and a reasonably exact

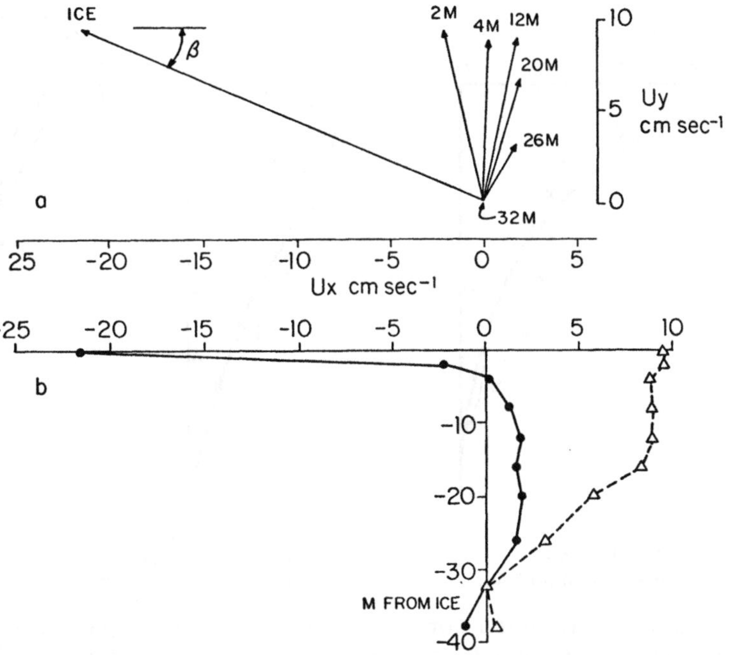

Fig. 3. Currents averaged for 5 hours, measured at AIDJEX Pilot Study station Jumpsuit, 12 April 1972. Data are referenced to the 32 m level, taken to be the geostrophic current due to sea surface ships.

estimate of interfacial stress was obtained. During the time of measurements shown in Figure 3, it was about 1 cm^2 s^{-2}. At latitude 76°, f = 1.4 x 10^{-4} s^{-1}, therefore the PBL scale depth, u_*/f, was about 70 m. Figures 4 and 5 show the six independent components of the Reynolds stress tensor measured at 9 levels, compared with results from numerical models of turbulent structure for the steady horizontally homogeneous ABL. Note that the comparison with atmospheric models is made using the Rossby-similarity nondimensionalization -- the correspondence between the ABL models and the oceanic data lends credence to that approach.

In the 1972 data, it was clear that the simple momentum equation, eq. (4.11), was not satisfied, and that some other factor was affecting the mean force balance. The explanation for this other force probably lies in the fact that the ice undersurface was not homogeneous, and horizontal variations in the topography of the ice underside would cause local accelerations. It was possible to argue consistently that such a "form drag" force could account, at least in part, for differences in both the mean flow profile and the turbulent stress components from what would be expected in a horizontally homogeneous flow (ibid.). This is an important point

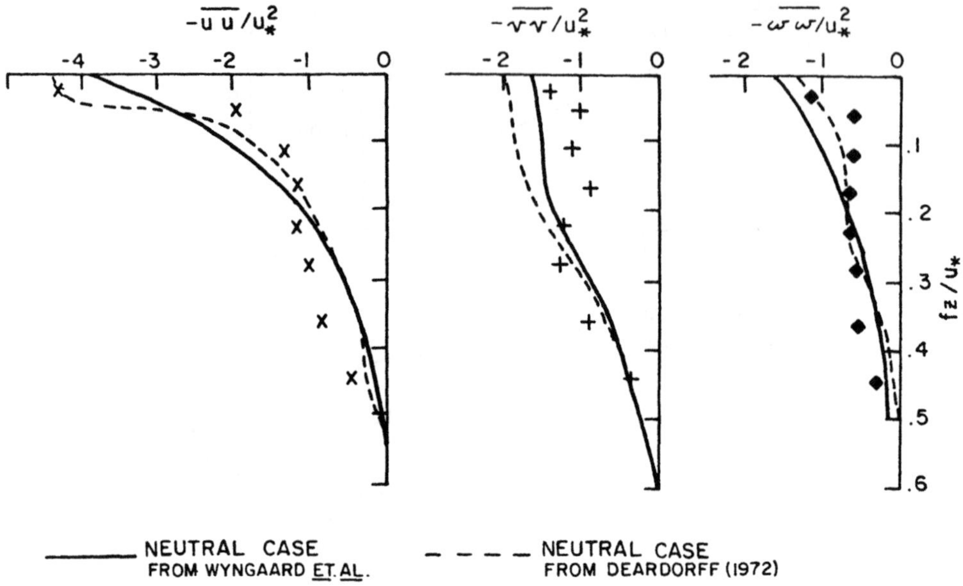

Fig. 4. Diagonal elements of the nondimensional Reynolds stress tensor corresponding to the mean currents shown in Figure 3.

regarding surface velocity. Since the ice often presents a more or less rigid surface over distances a couple of orders of magnitude greater than the OBL depth, it acts to integrate the drag contributions from many different ice types. Thus, the surface velocity that would correspond to the local turbulence structure and surface roughness if the surface were uniform, is not necessarily the one measured. From the 1972 data, we estimated that the "skin friction" drag measured under relatively smooth, undeformed ice was about half of the total drag contribution. Presumably the remainder was made up of contributions from rougher ice, and from form drag across pressure ridge keels.

Because of the form drag effect, there was no clear-cut relation between the mean velocity shear and the Reynolds stress in the 1972 measurements. Despite this, peaks in the spectra of vertical velocity fluctuations indicated length scales for the most energetic eddies, from which an effective eddy viscosity could be calculated. Results of those calculations are shown in Figure 6, again compared to ABL models. Note that past a distance of about 0.1 u_*/f from the interface, the eddy viscosity no longer increases. This is evidence that there is some limiting size to

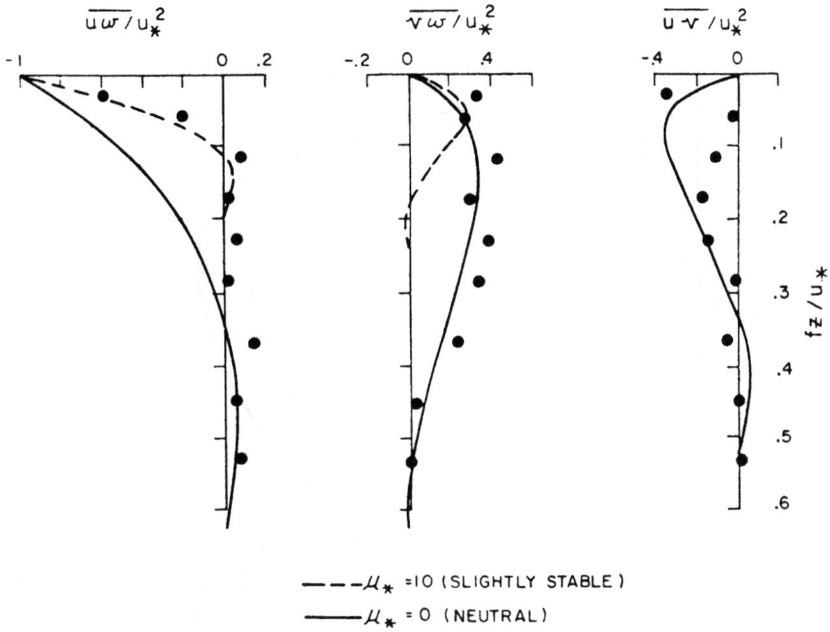

Fig. 5. Same as Figure 4, except shear stress elements. Curves are from Businger and Arya (1974).

the turbulent eddies that is considerably smaller than the scale of the boundary layer itself.

3.2 Mean Drift Statistics

One of the most basic but elusive measurements for testing steady-state PBL theory is simultaneous monitoring of surface stress and free-stream velocity. For the ABL, local surface-layer measurements of stress must be related to the mean drag over an area of at least several kilometers squared (which is straightforward only if the surface is exceptionally uniform), and this in turn must be related to the geostrophic wind, including any thermal wind effects due to horizontal temperature gradients. Adding to the difficulty in most temperate-zone experiments are time-dependent changes as the ABL goes through the diurnal cycle of heating and cooling.

In the open ocean the situation is worse. Even given that surface air stress can be measured precisely (no simple matter), determining surface or near surface currents accurately is a real challenge, particularly with well developed waves. A proper oceanic theory for near surface currents should consider gravity wave

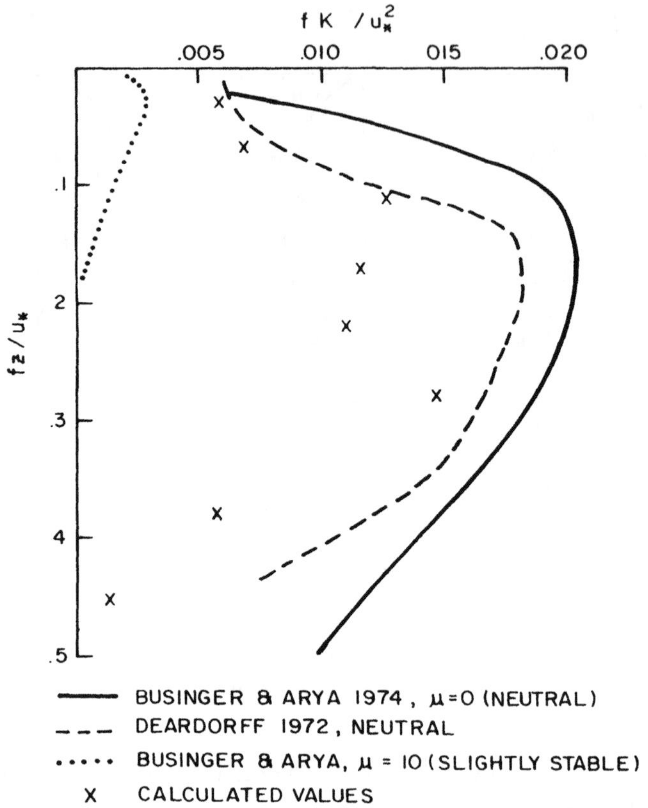

Fig. 6. Nondimensional eddy viscosity calculated from peaks in spectra of vertical velocity compared with ABL models (from McPhee and Smith, 1976).

effects like Stokes drift, and comparison with data must include an estimate of oceanic geostrophic flow.

Viewed in this light, an ice drift experiment offers exciting possibilities, provided one can identify periods of "free drift", when the internal stress gradient in the ice is negligible in the momentum balance as sketched in Figure 1. Under such conditions, the air stress can be estimated closely from surface wind measurements; surface velocity can be ascertained accurately from satellite navigation; reasonable estimates for the mean ice mass and undisturbed geostrophic flow can be made; and from these data, $\hat{V}_0$ and $\hat{\tau}_0$ calculated. In the central Arctic, dynamic height contours (e.g., Newton, 1973) are used to calculate surface geostrophic flow, which is found to be small compared with the usual ice drift. Because of the relative uniformity of the upper ice surface when compared with the ABL depth scale (~1 km), an estimate of air stress at one location is representative of a large area, provided allowance for pressure ridge sails is made. (The same cannot be said of point measurements in the ocean, because of the disparity in scales.) Therefore, the major problem is identifying free drift periods.

Although somewhat paradoxical, the best time for testing the steady-state relationship between surface velocity and stress during the AIDJEX Main Experiment occurred during the melt season of 1975 when inertial oscillation of the ice cover was most pronounced. The presence of inertial motions indicates that internal ice forces are small, and there is evidence that the mean drift of the ice is not much affected by the oscillations (McPhee, 1980b). Analysis and results of a study of drift, wind, and currents over the 60 day melt season are described by McPhee (1979). The aim of that work was to determine statistically the form of the stress to surface velocity relationship and to examine it in terms of PBL theory. Figure 7 shows the main result, which is a scatter diagram of τ_o calculated from the momentum balance of Figure 1a plotted against relative ice speed. The solid line is a best fit exponential of the form $\tau_o = aV^b$, where a and b are determined from logarithmic regression of smoothed data taken from all camps over the 60 day period. The exponent, b, was found to be 1.78 $\pm$ 0.12 at the 90% confidence level. Over a whole range of plausible values for ice mass and 10 m wind drag coefficient, this result varied by only a few percent. The relationship is obviously not linear (b = 1), but also shows a slight but significant difference from quadratic (b = 2). The dashed curve is described in the next section, where the shape of the curve is shown to support the Rossby-similarity hypothesis.

3.3 Inertial Oscillations

Most measurements of mixed layer currents in the open ocean show a strong periodic signal at the inertial frequency. Hunkins (1967) observed that summer speed records from deep current meters suspended near Fletcher's Ice Island (T-3) also contained an inertial signal, which he interpreted as evidence that T-3 and the surrounding ice pack were drifting with cycloidal loops at the inertial period. In the early AIDJEX pilot experiments, the inertial component in ice drift was detectable but of small amplitude. During the 1975 melt season of the AIDJEX Main Experiment, however, there were large inertial waves in both relative velocity records from current meters suspended below the mixed layer and in the surface drift velocity. Figure 8 shows an example of the latter from a 12-day period during August, 1975. The oscillations were initiated by the passage of two atmospheric fronts, and are described in detail by McPhee (1980a).

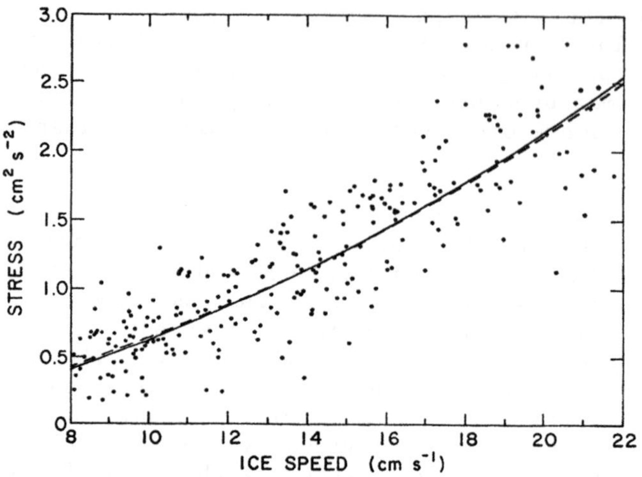

Fig. 7. Kinematic interfacial stress versus relative ice speed from smoothed AIDJEX data, summer 1975. Solid line is best fit exponential, $\tau_0 = 0.0104V^{1.78}$; dashed line is theoretical result described in text (from McPhee, 1979).

In analyzing inertial oscillations, a shorthand introduced by H. Perkins (1970), which he terms complex demodulation, is helpful. A complex inertial wave, $\hat{A}e^{-ift}$, is fit in a least-squares sense to the velocity record over a specified period, usually about twice the inertial period. The complex coefficient, $\hat{A}$, (which may be called a phasor) describes the phase and amplitude of the inertial wave. A time series of phasors may be analyzed to determine amplitude decay and phase shift of a particular wave train. Figure 9 shows such a study for surface velocity, one component of which is shown in Figure 8. The three manned stations analyzed spanned about 190 km west to east in a fairly narrow latitude band. There is a remarkable similarity in their response to the first frontal system. This is shown more clearly in Figure 10, obtained by adjusting the phase at the west and central stations by constant amounts, calculated assuming the fronts to propagate eastward at 60 km hr^{-1}. The striking uniformity in response to the first storm probably means that an oceanic coupling accompanies rapidly moving atmospheric systems. The second storm was slower moving, and the inertial response does not show the same horizontal coherence.

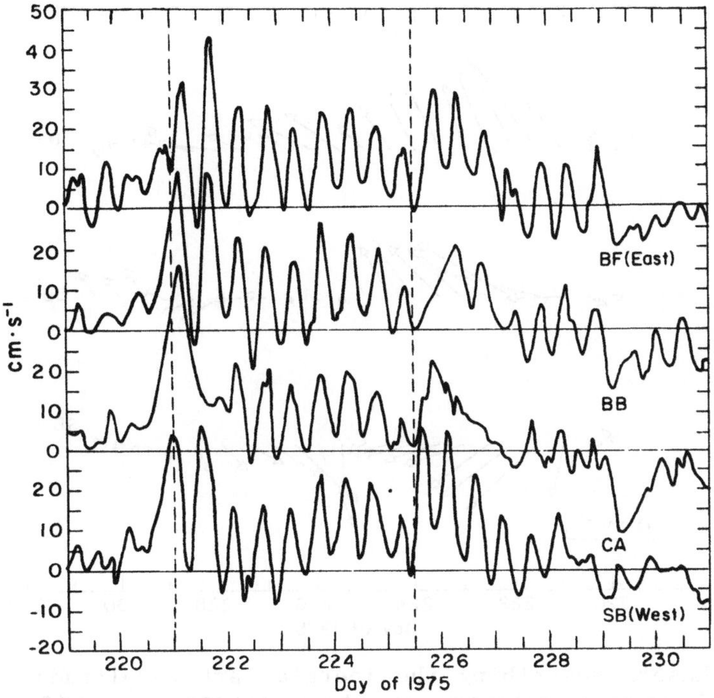

Fig. 8. Zonal (eastward) component of surface velocity determined from satellite positioning at four AIDJEX stations, showing strong inertial period motion.

4. THEORIES

4.1 OBL Momentum Transfer

Aside from intrinsic interest, there are two rather different motives for studying turbulent transfer in the upper ocean. In the first case, one is interested in how turbulence influences mean currents: a prime example is predicting surface (ice) velocity given the interfacial stress field. In the second, interest is directed more toward scalar properties of the mixed layer: e.g., predicting changes in mixed-layer temperature and salinity as a function of surface fluxes. In principle, the distinction is artificial, but in practice sometimes useful. For example, the impact on surface velocity of turbulent salinity flux at the base of the mixed layer appears to be small and often negligible unless the mixed layer is abnormally shallow (McPhee, 1979); on the other hand, it is crucial for understanding how mixed layer salinity and temperature respond to surface mixing. Viewing the same problem from these different perspectives has led to quite different theoretical formulations: e.g., second-order turbulent closure as

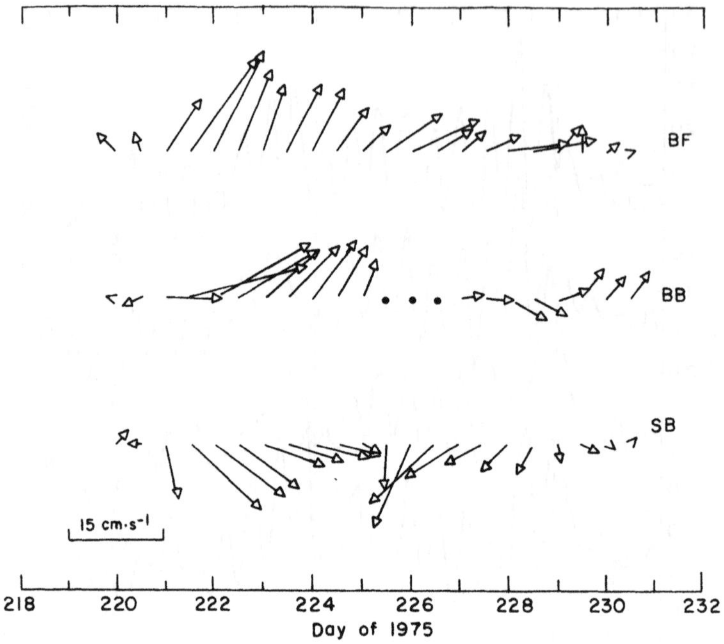

Fig. 9. Phasors describing the inertial art (amplitude and phase) of surface velocity at three AIDJEX camps (from McPhee, 1980a).

opposed to "slab" mixed-layer models. [The interested reader may wish to compare the approach of Mellor and Durbin (1975) with that of Niiler (1975).]

This section addresses primarily the question of momentum flux, particularly the relationship between surface stress and velocity. We shall review briefly the theoretical background of PBL momentum flux and outline a simple turbulence model that explains many observed features. The framework includes modifications to turbulent drag caused by rapid surface melting. In a later section on the evolution of the summer pycnocline, the model will be extended to include buoyancy flux in density gradients at lower levels.

4.2 "Geostrophic" Drag

Experience with laboratory studies of pipe and flat plate flow has shown that if the flow is laminar (exchange dominated by molecular processes), the drag relationship is linear, $\tau_o \propto V_o$; whereas in turbulent flow the drag is more nearly quadratic, $\tau_o \propto V_o^2$.

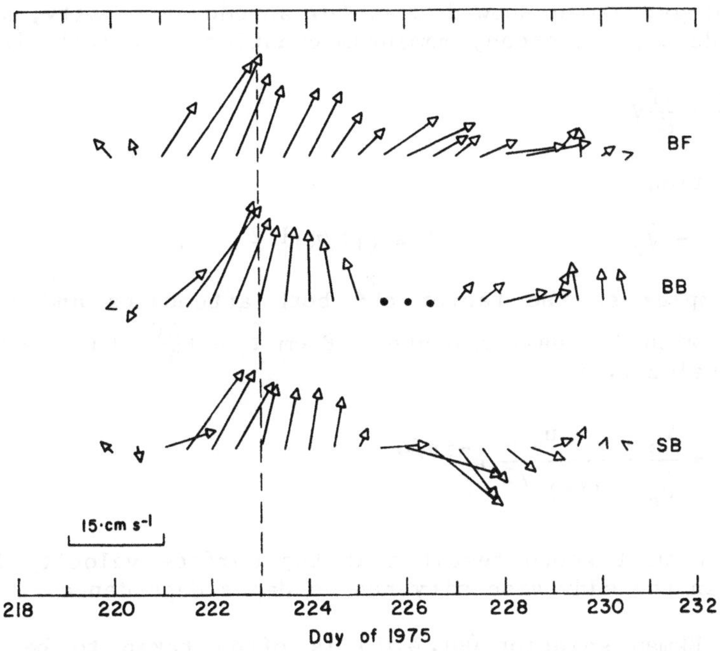

Fig. 10. Same as Figure 9, except phase adjusted by arrival time difference assuming that the generating disturbance moved eastward at 60 km hr^{-1}. Note the remarkable uniformity during the first half of the record despite zonal span of 190 km.

With quadratic drag, $V_o \propto u_*$, and it is convenient to consider the nondimensional surface velocity, $\hat{v}_o = \hat{V}_o/\hat{u}_*$. $\hat{v}_o$ is complex: its real part lies in the direction of surface stress; its imaginary part 90° counterclockwise. In polar form $\hat{v}_o = v_o e^{-i\alpha}$ where α is the boundary layer turning angle. The magnitude v_o is the inverse of what is called geostrophic drag for the atmosphere, i.e.,

$$u_* = (1/v_o)V_o, \qquad \tau_o = (\frac{1}{v_o})^2 V_o^2$$

Note that if drag is quadratic, the magnitude of the nondimensional surface velocity is constant.

4.3 Ekman Theory

Ekman's original formulation of the PBL problem (Ekman, 1905) has served as a paradigm for most subsequent PBL theory. Among other things, Ekman showed that for an eddy viscosity, K, independent of depth, the steady momentum equation (eq. 4.11) is

$$\frac{\partial^2 \hat{V}}{\partial z^2} = \frac{if}{K}\hat{V} \tag{4.15}$$

with solution

$$\hat{V}(z) = \hat{V}_o e^{\hat{\delta} z} \quad , \qquad \hat{\delta} = (if/K)^{1/2} \quad .$$

$\hat{\delta} z$ is complex so the factor $e^{\hat{\delta} z}$ both attenuates and rotates the velocity with increasing depth. From $\hat{\tau} = K\frac{\partial \hat{V}}{\partial z}$, the nondimensional surface velocity is

$$\hat{v}_o = \frac{\hat{V}_o}{\hat{u}_*} = \frac{u_*}{(fK)^{1/2}} e^{-i\pi/4} \tag{4.16}$$

whence the well known result that the surface velocity is 45° <u>cum sole</u>, when the eddy viscosity has no depth dependence.

The Ekman solution (eq. 4.16) is often taken to be synonymous with linear, or Newtonian, drag because u_* appears on the right-hand side. However, Ekman realized that K would vary depending on surface stress conditions, and in fact, he derived an empirical expression relating the Ekman depth, $D = \pi(2K/f)^{1/2}$, to the surface wind speed, u_s, and latitude, λ:

$$D \propto u_s/(\sin\lambda)^{1/2} \quad . \tag{4.17}$$

The friction velocity is proportional to the near surface wind, so Ekman's expression results in $K = K_E u_*^2$ (where K_E is constant), which makes eq. (4.16) a quadratic drag law consistent with other turbulent flows.

It is interesting that Ekman came so close to outlining what has become known as Rossby similarity. Had he chosen instead of eq. (4.17) the closely related expression

$$D \propto u_s/\sin\lambda \propto u_*/f \quad ,$$

his theory would have provided an eddy viscosity

$$K = K_* u_*^2/f \quad , \tag{4.18}$$

which implies a scaling velocity, u_*, and mixing length, $K_* u_*/f$ in the basic mixing length formulation.

For future reference, consider the Ekman problem using eq. (4.18) for eddy viscosity with nondimensional length, stress, and velocity: ξ, $\hat{T}$, and $\hat{v}$ defined by $z = u_*\xi/f$, $\hat{\tau} = \hat{T}u_*u_*$, and $\hat{V} = \hat{v}u_*$. Rather than solving directly for $\hat{v}$, we write the nondimensional form of eq. (4.11):

$$i\hat{v} = \partial\hat{T}/\partial\xi \quad , \tag{4.19}$$

differentiate and substitute $\partial\hat{v}/\partial\xi = \hat{T}/K_*$ to obtain

$$i\hat{T}/K_* = \partial^2\hat{T}/\partial\xi^2 \quad . \tag{4.20}$$

With $\hat{\gamma} = (i/K_*)^{1/2}$ and boundary conditions $\hat{T}(-\infty) = 0$, $\hat{T}(0) = 1$, and $\hat{v}(-\infty) = 0$, solutions are

$$\hat{T}(\xi) = e^{\hat{\gamma}\xi} \tag{4.21}$$

$$\hat{v}(\xi) = (iK_*)^{-1/2}\, e^{\hat{\gamma}\xi} = K_*^{-1/2}\, \hat{T}e^{-i\pi/4} \quad . \tag{4.22}$$

Hodographs of the solutions are shown in Figure 11 for $K_* = 0.021$.

While the difference may at first seem trivial, there is a subtle but important shift in perspective when the stress equation (eq. 4.20) is solved rather than the momentum equation. The heuristic view of the PBL developed in Section 2.1 requires rapid velocity change in the direction of surface stress (real axis) across the surface layer, but little change in the transverse (imaginary) component, so that the drift angle is about half of 45°. The Ekman solution for velocity in the surface layer does not predict this behavior. On the other hand, stress is presumed to change little across the surface layer, and since the change in $\mathrm{Im}(\hat{v}_o)$ is small, we can approximate the real component of nondimensional stress from eq. (4.19) by

$$\mathrm{Re}(\hat{T}) = 1 + \mathrm{Im}(\hat{v}_o)\,\xi \quad .$$

A Taylor series expansion for the stress solution eq. (4.21) is

$$\mathrm{Re}(\hat{T}) = 1 + \mathrm{Re}(\hat{\gamma}\xi) = 1 - (2K_*)^{1/2}\,\xi \quad .$$

But from eq. (4.22), this is also $1 + \mathrm{Im}(\hat{v}_o)\xi$. Thus, while the real component of surface layer velocity is not well described by depth independent eddy viscosity, there are indications that stress is. This idea can be explored further with the help of numerical models.

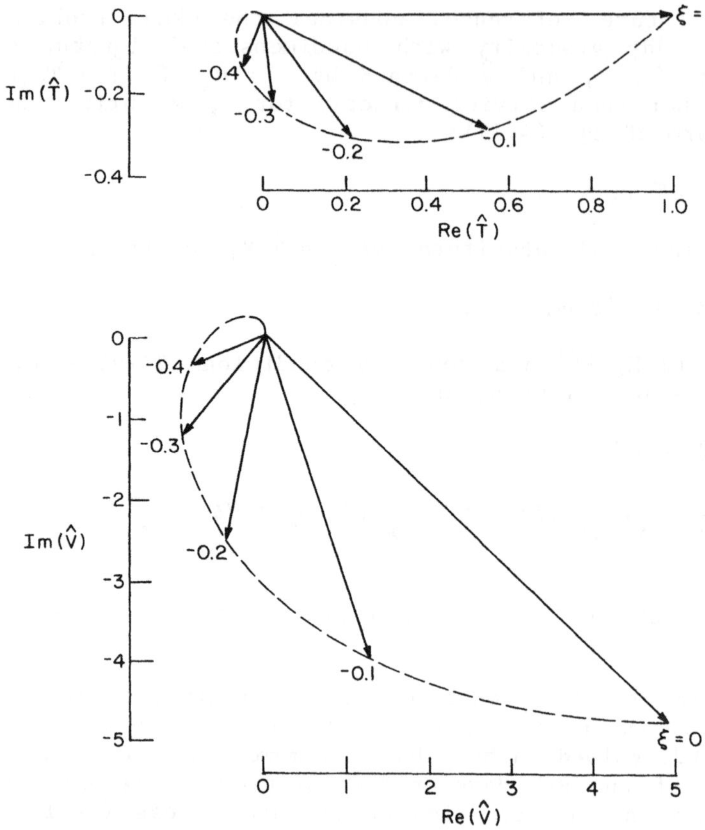

Fig. 11. The Ekman solution for nondimensional turbulent stress and mean velocity. Numbers beside hodographs indicate nondimensional depth.

4.4 Numerical PBL Models

Numerical methods for solving the boundary layer equations have elucidated much about the structure of PBL turbulence. The models range in complexity from "brute force" three-dimensional integration of the complete equations over a relatively fine mesh, where the only parameterization comes at the sub-grid scale (Deardorff, 1972), to solving the mean momentum equation given an analytic expression for the eddy viscosity. An intermediate class, referred to as "second-order-closure" models, solves simultaneous equations for mean and first-moment variables, the latter equations using some parameterization for terms involving second and higher moments. A summary of turbulent closure schemes, with hierarchical ranking, is given by Mellor and Yamada (1974). Figures 4, 5 and 6 compare data from the underice OBL with models of Deardorff (1972, average of ten realizations of time-dependent, three-dimensional numerical integration); Wyngaard et al. (1974, second-closure, simultaneous solution of 13 partial differential equations); and Businger and Arya (1974, steady state, mean momentum equation with prescribed analytic K_* distribution).

The Businger and Arya (1974) model, which for the neutrally buoyant PBL produces profiles of mean velocity and turbulent stress similar to the steady-state solutions of Wyngaard et al. (1974), is an instructive extension of the nondimensional Ekman theory of the previous section. Starting from the nondimensional stress equation (eq. 4.20), an expression for K_* is prescribed (the example here is for the neutral case only, although their treatment includes stable surface buoyancy)

$$K_* = k\xi e^{-|v_s|\,\xi} \quad . \tag{4.23}$$

K_* is an exponentially attenuated, linear function of ξ, which goes to $K_* = k\xi$ for ξ small; and v_s is the transverse (imaginary) component of surface velocity. Using an initial estimate for v_s, eq.(4.20) is solved; the calculated value of v_s is used to obtain a new solution; and the procedure is repeated as long as the change in v_s exceeds a given tolerance. Figure 12 from McPhee (1979) shows results of this model adapted to the OBL. The surface velocity is found by integrating the stress, solved by setting $\hat{T} = 1$ at $\xi = \xi_o$, where $\xi_o = fz_o/u_*$. The inverse of ξ_o is called the "surface-friction Rossby number" and is the ratio of the outer layer scale, u_*/f, to the surface layer scale, z_o. The dashed curve in Figure 7 (ibid.), which shows τ_o vs. V_o, is the locus of solutions for V_o given u_* in the range shown, with $z_o = 10$ cm. In terms of least-squares logarithmic regression of $\tau_o = aV_o^b$, the exponent, b is 1.70 $\pm$ 0.00 for the solution points. To within experimental uncertainty, this is direct confirmation that Rossby-similarity scaling governs the underice OBL.

4.5 Two Layer Theories

Before the PBL equations could be routinely solved with computers, the outer and surface layers were often reconciled by noting that since stress was nearly constant through the logarithmic surface layer, an analytic, Ekman-like solution for the outer layer, which used surface stress as it boundary condition, could be matched to a unidirectional surface layer profile (e.g., see Rossby and Montgomery, 1935). This basic approach has been used extensively to calculate oceanic drag on sea ice, under a number of different assumptions about the two layers. Several such

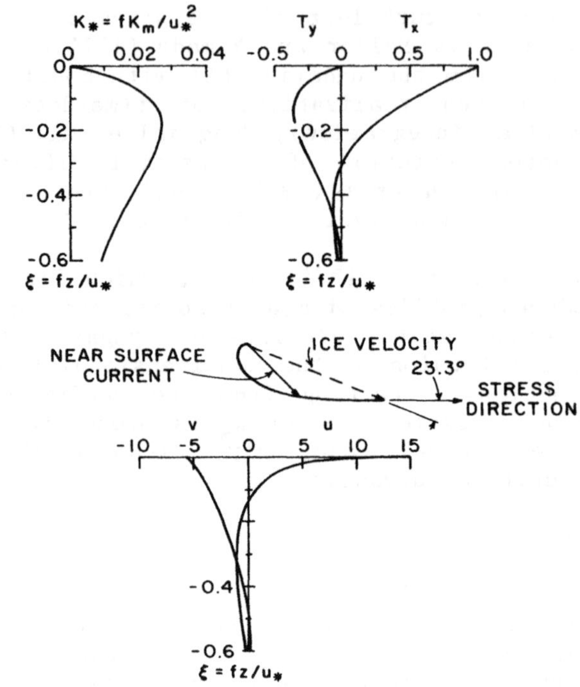

Fig. 12. Adaptation of Businger and Arya (1974) model to the neutral OBL. K_* is an exponentially attenuated, linear function of $\xi = fz/u_*$.

models (among others, Shuleikin, 1938; Reed and Campbell, 1962; Rothrock, 1975) are discussed by McPhee (1982) and compared with the AIDJEX drag relationship shown in Figure 7. A conclusion of that work was that only those simple models that somehow incorporated Rossby-similarity scaling for the outer and surface layers agreed with the shape of the stress-surface speed curve.

The simplest two-layer approach with u_* and f scaling is diagrammed in Figure 13 from McPhee (1982). We assume, first, that nondimensional eddy viscosity increases linearly with depth until it reaches the value, $K_* = k\xi_N$ (where ξ_N is the nondimensional thickness of the surface layer), which it takes for all greater depths; and second, that stress is constant ($\hat{T} = 1$) in the layer where K_* varies. The total nondimensional surface velocity is then the sum of the Ekman surface velocity and the logarithmic velocity through the surface layer, i.e.:

$$\hat{v}_o = \hat{v}_{oE} + \frac{1}{k}\ln(\xi_N/\xi_o) = (k\xi_N)^{-1/2}\, e^{-i\pi/4} + \frac{1}{k}\ln(Ro_*\xi_N) \quad (24)$$

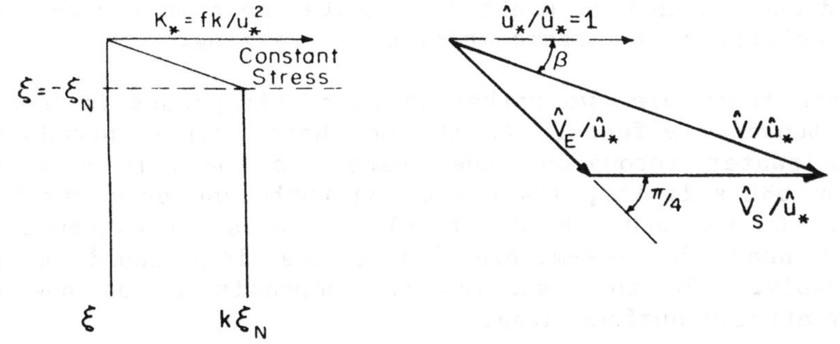

Fig. 13. Simple two-layer approach in which stress is constant and K_* varies linearly through a surface layer of thickness ξ_N. The surface velocity is vector sum of Ekman velocity, $\hat{V}_E$, and logarithmic shear across the surface layer, $\hat{V}_S$.

where ξ_0 is again the inverse of the surface-friction Rossby number, $R_{0_*} = u_*/fz_0$. With a proper empirical choice of ξ_N and z_0 (for the AIDJEX data, $\xi_N = 0.045$, $z_0 = 23$ cm), the shape of the τ_0 vs. V_0 curve and the boundary layer turning angle implied by eq. (4.24) are almost identical to the results of the Businger and Arya model. Therefore eq. (4.24) can be considered an acceptable drag law for sea ice, although conceptually it has some unsatisfying features which we shall address below.

Equation (4.24) provides a simple explanation of why the drag law is somewhat less than quadratic, despite the fact that the turbulence is fully developed. Recall that quadratic drag requires the nondimensional surface velocity to be constant. For a given surface roughness, all the terms on the right-hand side of eq. (4.24) are constant except $(\ln u_*)/k$, which contributes to the real component. As u_* increases, the nondimensional real component also increases; v_o increases slightly and the turning angle decreases. Therefore, to double the surface speed requires a little less than doubling u_*, and the drag is less than quadratic. This is the essence of Rossby-similarity theory for geostrophic drag, although it is often presented in a different way (e.g., Blackadar and Tennekes, 1968).

4.6 Surface Buoyancy

To this point we have tacitly assumed the OBL to be neutrally buoyant, meaning that surface freezing or melting has no effect on the turbulence. It is well known, however, that buoyancy (i.e., the gravitational acceleration encountered by turbulent eddies when the fluid is not of uniform density) plays an important role in ABL dynamics, and we might anticipate the same for the underice OBL, especially with extreme melting or freezing.

Again there are two rather distinct viewpoints in considering surface buoyancy effects. On the one hand, while introduction of fresh meltwater throughout the summer is bound to have a major impact on OBL salinity, its effect on turbulent drag may be unimportant. On the other hand, if melting rates are extreme, as they might be near the ice-margin front, the drag could be reduced considerably. In this section the emphasis is on how surface buoyancy affects surface drag.

So far we have considered two mixing length scales for turbulence in the OBL: kz in the surface layer, and a maximum mixing length which is some small fraction of u_*/f in the outer layer. When density variations enter the problem in the form of salinity changes at the interface with the ice, a third length scale, the Obukhov length, L, may be used to quantify how the turbulent flux of density variations, $\overline{\rho' w'}$, modifies the momentum exchange.

The surface layer turbulent kinetic energy (TKE) equation consists of a balance between shear production and buoyant production on one side against flux divergence-pressure transport and viscous dissipation on the other. (Tennekes and Lumley, 1972, Chapter 3, present a clear derivation and discussion of the TKE budget.) If the TKE equation is nondimensionalized by u_*/kz, the shear production term is

$$\phi m = \frac{kz}{u_*}\left(-\frac{\overline{\hat{V}' w'}}{u_*^2}\right)\frac{\partial \hat{V}}{\partial z} \qquad (4.25)$$

and the buoyant production term is z/L where

$$L = \rho_0 u_*^3/(gk\overline{\rho' w'}) \quad . \qquad (4.26$$

The TKE equation may be written

$$\phi_m - z/L = \phi_D + \phi_E \qquad (4.27)$$

where ϕ_D represents the nondimensional pressure-transport-flux divergence and ϕ_E is the nondimensional viscous dissipation.

The nondimensional TKE (actually, rate of energy transfer) eq. (4.27) provides a physical interpretation of L. In the surface layer the shear production, ϕ_m is of order one (see eq. (4.9) and in the absence of buoyancy, the size of the energy-containing eddies is of order z. If L is positive (stable) and also of order z, then the buoyancy term is comparable to ϕ_m, but of opposite sign, so that the total TKE production rate is much reduced. For $|L|$ large compared to z, the buoyancy term is small and shear production dominates. For L small and positive, turbulence is quelled by the gravitational force; for L small and negative (unstable), the turbulence derives most of its energy from buoyant acceleration rather than shear.

In the simplified view of the OBL developed above, we inferred from data that the maximum mixing length ($\ell_m = u_* k\xi_N/f$) is some small fraction of u_*/f. It seem reasonable that if L is several times as large as ℓ_m, turbulence through the entire boundary layer will not be much affected by buoyancy. This provides a way of estimating how important surface freezing and melting may be: If $f|L|/u_* \gg k\xi_N$ (about 0.02), we are safe in treating the OBL as neutral. L is related to ice growth rate, d, as follows. From eq. (4)

$$\frac{\overline{\rho' w'}}{\rho_o} \simeq -\alpha_\theta \overline{w'\theta'} + \alpha_s \overline{w'S'} \ .$$

Sea ice is typically only 15-20% as saline as the water from which it grows, so that when freezing (melting) occurs, there is a flux of saltier (fresher) water across the interface. The surface salinity flux is given by

$$\overline{w'S'}\Big|_o = -\frac{\rho_{ice}}{\rho_o}\left(\frac{S_w - S_{ice}}{1000}\right) d = -\Delta_s d$$

where $S_w - S_{ice}$ is the salinity difference expressed in per mille.

Unlike fresh water, the coefficient of thermal expansion for seawater, α_o, increases monotonically from the freezing point. Since α_o is small for temperatures near freezing, thermal expansion has little effect on density and can often be neglected (see, e.g., Neumann and Pierson, 1966), in which case

$$\overline{\rho' w'}\Big|_o \simeq \alpha_s \overline{w'S'}\Big|_o = -\alpha_s \Delta_s d \ .$$

The nondimensional Obukhov length is then

$$L_* = -fu_*^2/(gk\alpha_s \Delta_s d); \qquad [\theta < 0^\circ C] \ .$$

From changes in upper ocean salinity observed during the AIDJEX melt season, the maximum melting rate was estimated to be about 2.4 cm d^{-1}, or 2.8 x 10^{-5} cm s^{-1}. Under typical drift conditions,

u_* was about 1 cm s^{-1}, and with $\Delta_s = 0.0225$, $\alpha_s = 0.8$, $f = 1.4 \times 10^{-4}$ s^{-1}; from (28), $L_* = 0.71$. This is considerably larger than $k\xi_N$, thus for practical purposes we may normally neglect changes in drag due to melting, at least well within the limits of the perennial pack. In a pack with much multiyear ice, it is hard to imagine spatially average freezing rates of much more than a few cm per day, so the same arguments should apply for unstable conditions.

Near the ice margins, or in an unconsolidated pack, the situation may be quite different. In winter, cold winds blowing off the pack can expose large expanses of open water to rapid freezing, creating a widespread region of unstable boundary flux in the ocean as well as the atmosphere. Often a sharp front in mixed-layer temperature exists near the ice margin; if floes are blown into the warmer water, they will melt much more rapidly than the pack of the central Arctic, and may have a quite significant effect on oceanic drag (McPhee, 1982).

4.7 A Two-Layer System Stabilized by Surface Buoyancy

In a recent paper (McPhee, 1981) we developed analytic expressions, based on the mixing length arguments set forth above, for stress and mean velocity in the steady PBL stabilized by surface buoyancy. The theory, which borrows from scaling concepts developed by Zilitinkevich (1975) and from heuristic arguments of Businger and Arya (1974), is here briefly outlined; more complete details may be found in the paper.

From the two-layer premise, we make the following assumptions:

1. Scales formed from the parameters u_*, f, and L exist that, when used to nondimensionalize the mean velocity and turbulent stress structure in the outer layer of any flow, result in "universal" profiles of those quantities. This implies immediately that there is only one nondimensional eddy viscosity governing turbulence in the outer layer, at least for the rather restricted class of boundary layers being considered (i.e., steady-state, stabilizing buoyancy flux).

2. Turbulent stress is not affected by changes of eddy viscosity in the surface layer. This is not really a very severe restriction: In most surface layer studies stress is assumed constant through the surface layer -- here we require only that its variation across the layer is not sensitive to the magnitude of the eddy viscosity, which is given the universal value of the outer layer.

3. Finally, we assume that the maximum mixing length varies between two limits: In the neutral case it is $\ell_m = k\xi_N u_*/f$ as

before; and under highly stable conditions it is proportional to L. The proportionality constant is the critical flux Richardson number, R_c, following the reasoning of Zilitinkevich (1975). The simplest expression including these two limits is

$$\ell_m = \frac{u_* \xi_N}{f} \left(1 + \frac{\xi_N}{R_c} \cdot \frac{u_*}{fL}\right)^{-1} \tag{4.29}$$

Let $\mu_* = u_*/fL$ (the ratio of the neutral PBL scale to the Obukhov length). Substitution of eq. (4.29) into the expression for K_* (which is also equal to $k\xi_N$) yields the following nondimensional parameters:

$$\text{Depth:}\quad \xi = fz/u_*\eta_* \qquad \text{Velocity:}\quad \hat{u} = \eta_* \hat{V}/\hat{u}_*$$

$$\text{Stress:}\quad \hat{T} = \hat{\tau}/u_*\hat{u}_* \qquad \text{Eddy viscosity:}\quad K_* = fK/u_*^2\eta_*^2 \tag{4.30}$$

where $\eta_* = (1 + \xi_N \mu_*/R_c)^{-1/2}$. η_* is a stability parameter less than or equal to one (the neutral limit is one).

The velocity structure within the surface layer is governed by a nondimensional eddy viscosity which derives from the log-linear profile observed in the atmosphere:

$$K_* = -\frac{k\zeta}{\eta_*}(1 - \beta\mu_*\eta_*\zeta)^{-1}$$

where

$$\beta = \left(\frac{1}{R_c} + \frac{1}{\mu_*\xi_N}\right)(1 - \eta_*) \quad .$$

In the surface layer ($|\zeta| < \xi_N$) velocity is obtained by integration of stress divided by K_*, resulting in the following equations for stress and velocity in the entire PBL.

$$\hat{T} = e^{\hat{\delta}\zeta}$$

$$\hat{u} = \begin{cases} -i\hat{\delta}e^{\hat{\delta}\zeta} & ;\zeta \le -\xi_N \\ -i\hat{\delta}e^{-\hat{\delta}\xi_N} - \frac{\eta^*}{k}\left[\ln\frac{|\zeta|}{\xi_N} + (\hat{\delta} - a)(\zeta + \xi_N) - \frac{a}{2}\hat{\delta}(\zeta^2 - \xi_N^2)\right]; & \zeta_o \ge \zeta > -\xi_N \end{cases}$$

where $a = \eta_*(1/\xi_N + u_*/fLR_c)(1 - \eta_*)$. The "geostrophic drag law" (i.e., the nondimensional surface velocity) is

$$\hat{u}_o = \eta_* \frac{\hat{V}_o}{\hat{u}_*} = -i\hat{\delta} e^{-\delta\xi_N} - \frac{\eta_*}{k}\left[\ln\frac{|\zeta_o|}{\xi_N} + (\hat{\delta} - a)\xi_N + \frac{a}{2}\hat{\delta}\xi_N^2\right] . \tag{4.29}$$

The average ice drift and stress during the summer part of the AIDJEX Main Experiment were used to determine the one free constant in the theory, $\xi_N = 0.052$. Figure 14 shows the similarity model for stress and velocity in the outer layer compared with the neutral ($\mu_* = 0$) results of Businger and Arya (1974). Note the close correspondence in stress, even with constant K_*. Note also that stress varies appreciably through the surface layer; one feature that separates this approach from most previous two layer formulations.

From the similarity perspective provided by the scales of eq. (4.30), given a fixed surface roughness, z_o, the effect of increasing stability is analogous to increasing roughness in the Rossby-similarity treatment of the neutral PBL, i.e., the nondimensional surface speed decreases and the turning angle increases. The effect is demonstrated by Figure 15 and the explanation is straightforward: The surface roughness, which is independent of turbulence, becomes relatively larger as the physical size of turbulent eddies is inhibited by increasing buoyancy. The actual surface speed, of course, increases with more rapid melting because of reduced friction, as shown in Figure 16, which also demonstrates the reduction of PBL depth. The role of the stability parameter, η_*, becomes apparent when Figure 16 is compared with Figure 15.

For $u_* = 1$ cm s^{-1}, $\mu_* = 25$ represents a melting rate of roughly 50 cm per day, which would be considered extreme for the

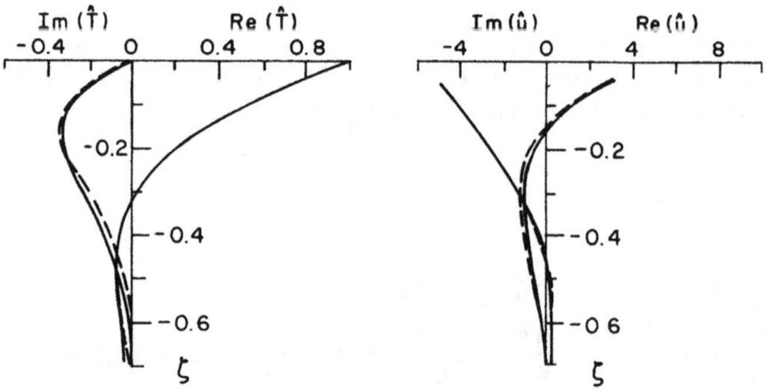

Fig. 14. Comparison of profiles from analytic similarity theory of McPhee (1981a) (solid curves) with neutral results of numerical model of Businger and Arya (1974). The profiles correspond closely even though the analytic theory uses constant K_* near the surface to calculate stress.

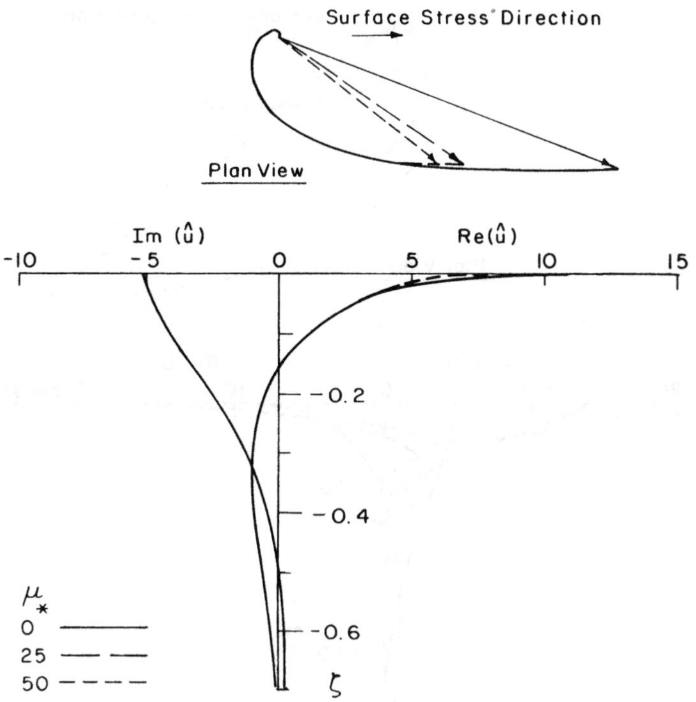

Fig. 15. Similarity velocity solutions for three values of $_*$. Other external parameters are $u_* = 1$ cm s^{-1}, $z_o = 5$ cm, $f = 1.4 \times 10^{-4}$.

interior perennial pack, but which may occasionally be encountered near the ice margin, based on observations in the Bering Sea (C. Pease, personal communication).

There are few quantitative data comparing ice drift with stress under rapid melting conditions against which to test the theory; however, the analog with the ABL can be exploited to see if the basic approach provides reasonable predictions of geostrophic drag. In the ABL geostrophic drag is often expressed in terms of the Rossby-similarity parameters A and B, thus removing explicit dependence on z_o (see, e.g., Clarke and Hess, 1974). In terms of the complex representation of surface velocity, A and B are defined by

$$B = k\,\mathrm{Im}(\hat{V}_o/\hat{u}_*) = \frac{k}{\eta_*}\,\mathrm{Im}(\hat{u}_o)$$

$$A = \ln Ro_* - \frac{k}{\eta_*}\,\mathrm{Re}(\hat{u}_o)$$

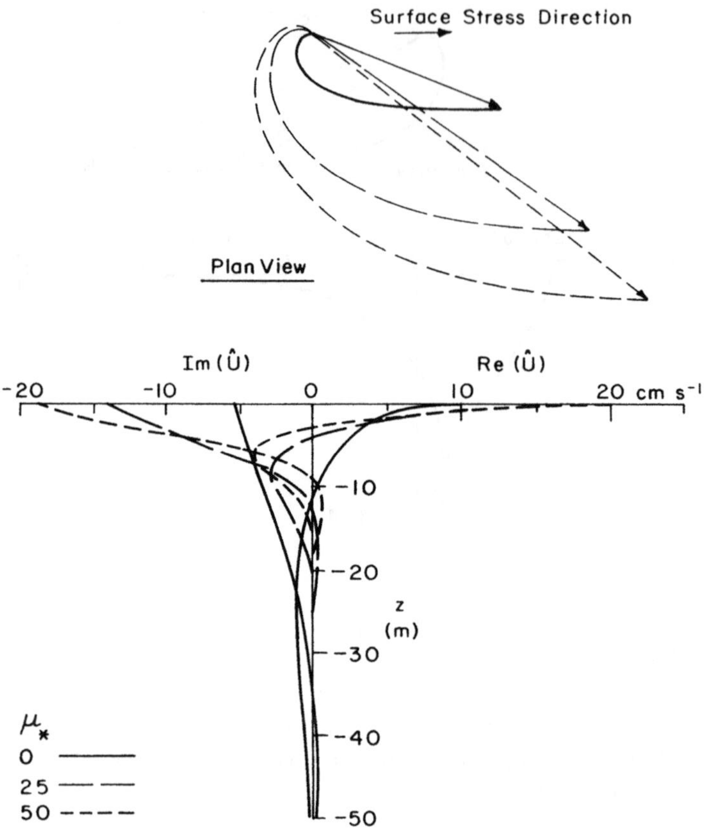

Fig. 16. Dimensional velocity components corresponding Fig. 15 where real axis is aligned with u_*.

Figure 17 shows A and B plotted as functions of $\mu_* = u_*/fL$ (the inverse of the dimensionless Obukhov length) compared with data of Clarke and Hess (1974), and with values from the models of Businger and Arya (1974), and Wyngaard (1975). The parameter curves are analytic functions; they depend only on μ_*, R_c (the critical flux Richardson number) and ξ_N, which was determined from ice drift statistics.

4.8 The Time-Dependent OBL: Inertial Oscillations

Theories of PBL momentum transfer considered so far have assumed that the inertia of the ice and upper ocean can be ignored in the force balance. In Section 2.2 it was shown, however, that drift records often contain sizable waves at the inertial period which might conceivably change the turbulent structure of the boundary layer.

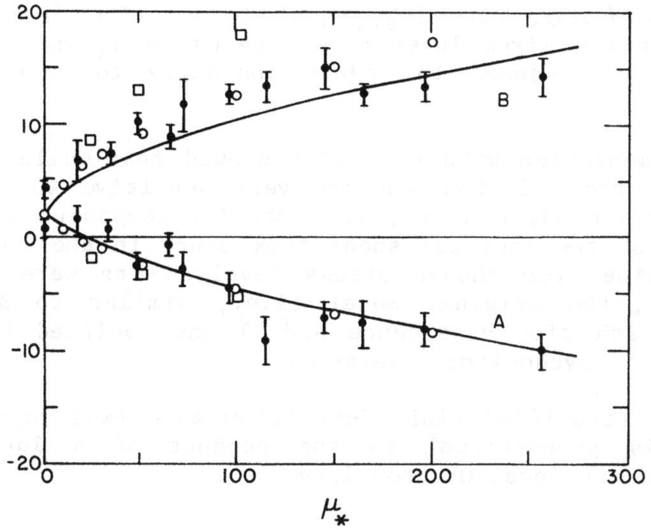

Fig. 17. The Rossby-similarity parameters A and B as functions of u_*. Solid curves are from analytic theory of McPhee (1981a); data points with error bars are from Clarke ahd Hess (1974); open circles and squares are respectively from numerical models of Businger and Arya (1974) and Wyngaard (1975).

The time-dependent nature of the PBL momentum equation was addressed by Ekman (1905, with credit to Fredholm), who presented a time varying solution for currents at various levels assuming constant eddy viscosity. Kheisin and Ivchenko (1975) extended the Ekman analysis to an ice-covered OBL.

Most simulations of observed inertial currents in the ocean have used a simpler approach (outlined in Section 2.1, see Figure 3) where mixed layer currents are related directly to transport integral, M. In most momentum integral schemes, velocity is considered uniform throughout the mixed layer, thus they are called "slab" models (e.g., Pollard and Millard, 1970). In spite of their unrealistic specification of infinite shear at the base of the mixed layer, they sometimes provide quite believable simulations of mixed layer currents. In an early analysis of AIDJEX current meter data, we used the same basic concept to simulate currents measured relative to the ice, although the mixed layer was not treated as a slab (McPhee, 1978).

In a later paper (McPhee, 1980a) a different technique was used to model observed surface velocity and upper ocean currents: The time-dependent partial differential equations for stress were solved numerically across a vertical grid using an eddy viscosity given by eq. (4.23), with $|v_s|$ equal to 6. The dimensional eddy viscosity in the mixed layer thus depends on u_* and f as well as depth, and is assumed to adjust instantly to changing surface conditions.

Experimentation with the model showed that while the stationary part of the solution was not very sensitive to details near the base of the mixed layer, the inertial component was; in fact, nearly all of the inertial shear took place in the upper part of the pycnocline even though stress levels there were small. For this reason, the original model (i.e., similar to Businger and Arya, but with time dependence added) was modified to include a more realistic pycnocline treatment.

In the stratified fluid just below the mixed layer, the eddy viscosity is proportional to the product of a local friction velocity and the local Obukhov length:

$$K_p \propto \tau_p{}^{1/2} L_p = \frac{\rho_o \tau_p^2}{gk\overline{\rho' w'}|_p}$$

where the subscript, p, refers to the pycnocline near the mixed layer interface. By Reynolds analogy, we assume that the density flux is proportional to the eddy viscosity times the mean density gradient, arriving at

$$K_p = c_{pyc} \tau_p / N \tag{4.32}$$

where N is the buoyancy (Brunt-Vaisalla) frequency defined by

$$N^2 = -\frac{g}{\rho_o}\frac{\partial \rho}{\partial z}$$

and c_{pyc} is a proportionality constant that may depend on the stratifying agent (salt in the arctic pycnocline, heat in many temperate pycnoclines).

Figure 18 from McPhee (1980a) compares in the zonal component of observed and simulated surface velocity at three AIDJEX camps for the 12 day period shown in Figure 8, using c_{pyc} = 0.20 with a timestep of 15 minutes, and a constant mixed layer depth of 25 m. Figure 19 shows one component of velocity as simulated at three levels in the mixed layer and at one 5 m below the mixed layer, along with the average current. The inertial part of the velocity signal, which obviously dominates the low frequency part at mid-depths in the mixed layer, is fairly uniform throughout the layer. Most of the inertial shear occurs near the top of the pycnocline, which helps explain the apparent success of slab models. This suggests that the inertial and low frequency motions might be modeled independently, then combined to obtain the total velocity.

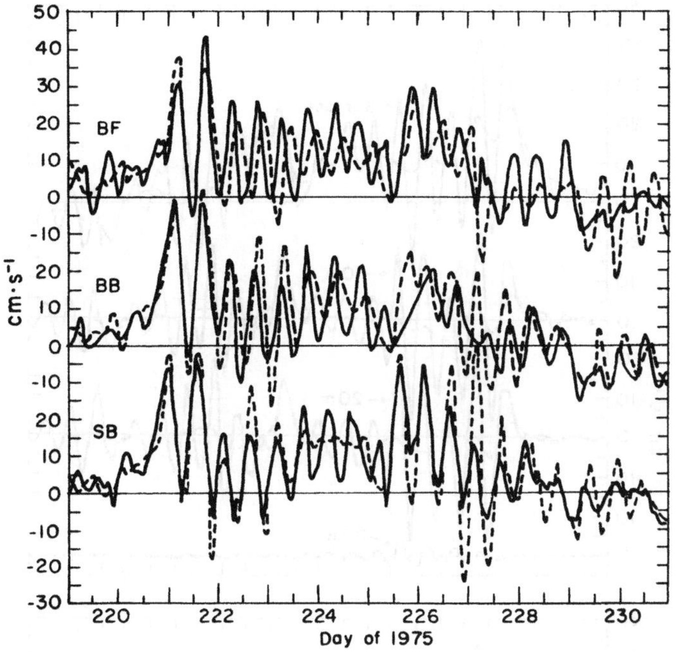

Fig. 18. Zonal component of surface velocity including inertial oscillation as observed at three AIDJEX stations (solid curves) and as simulated with model of McPhee (1980a).

The role of inertial motion in the formation and ablation of pack ice is not well understood, but could be important. During the year-long AIDJEX drift, the inertial component of ice velocity was large relative to mean drift only during a melt season that lasted about 60 days. There are probably two reasons for this: first, the mixed layer shoals because of the introduction of fresh meltwater, trapping the inertia over a smaller vertical distance; and, second, the ice pack is less capable of supporting internal stress gradients that would otherwise arrest the short period excursions. If the amplitude of inertial motion depended only on mixed layer depth, there would still be oscillations in winter about half as large as in summer; also, the transition from summer to winter conditions would be fairly gradual. Instead, the AIDJEX data showed the transition to be abrupt (it coincided with the breakup of the main station, Big Bear) and inertial amplitudes in winter to be quite small. One can thus assume that internal ice stress is mainly responsible for quelling inertial waves. This being the case, ice production could be significantly increased during the freezeup period by a mechanical pumping effect associated with the inertial motion. Ice production would be enhanced by divergences that expose areas of open water, causing more rapid cooling by sensible, latent, and radiant heat transports.

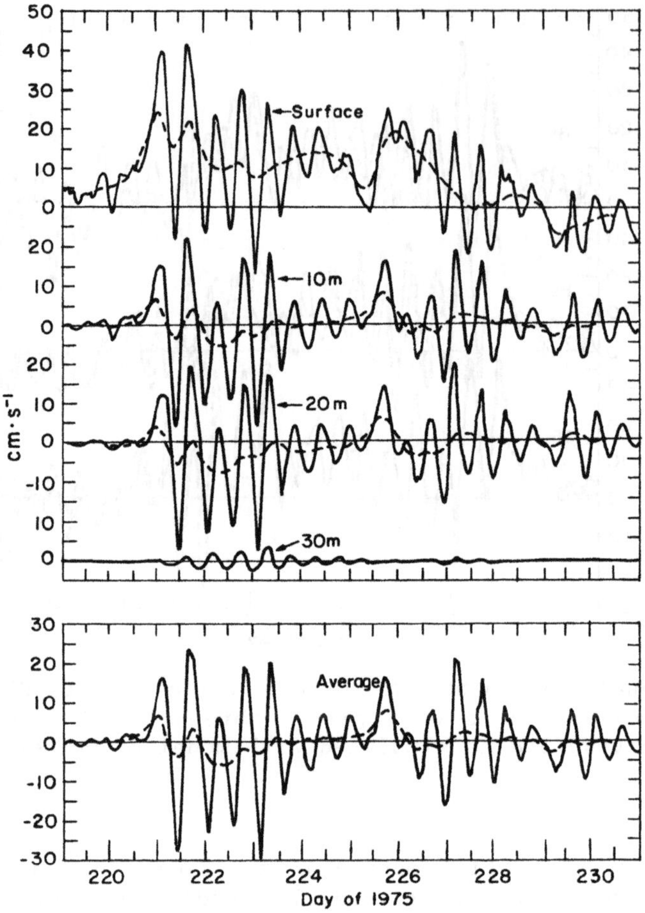

Fig. 19. Simulated velocity at three levels in mixed layer and one below, along with average velocity in the mixed layer. In the model, there is little shear in the inertial part of the velocity signal within the mixed layer.

Nansen, writing in his diary of the drift of Fram in October, 1893, described dramatic ice pressures that occurred twice daily, in between which the Fram lay in open water (see, e.g., excerpts in Manley and Lewis, 1968). Nansen thought the periodic pressures to be caused by tides, but in retrospect they seem more likely to have been the result of inertial effects as the ice tightened into its winter regime. Note that in summer, the Fram probably did drift in inertial loops, but this would have been difficult to detect without sophisticated navigation gear or continuous current measurements below the mixed layer.

In early summer, inertial waves might have the opposite effect by periodically increasing the amount of opening, which would decrease the mean albedo. In either case, the ability of the wind systems to induce inertial oscillations (e.g., sharp fronts rather than steady zonal flow) at the time of breakup or freezeup might have a decided impact on the subsequent ice characteristics.

5. DENSITY STRUCTURE OF THE UPPER OCEAN

Heat exchange across the air-ice-sea boundary is central to understanding how sea ice affects climate, and the means by which heat transfer occurs is inextricably tied to the turbulent structure of the upper ocean. The problem, of course, is not confined to high latitudes: The seasonal evolution of the upper ocean is one of the foremost problems in physical oceanography, because, among other reasons, the upper ocean is a tremendously large heat reservoir.

The polar regions differ from most of the world ocean in that heat exchange is mostly latent. If one thinks in terms of a one-dimensional system in which all the heat stored during summer is released the following winter with no net gain or loss, then for most of the perennial ice pack, the "mixed layer" problem is essentially the ice thickness distribution problem, because the upper ocean remains near its freezing point and stores little sensible heat compared with the energy that is absorbed by melting and released by freezing. In other words, if sea ice growth and decay over the seasonal cycle is well understood, a mixed layer model will add only minor corrections to the total heat budget. In this idealized view, the most important aspect of the OBL would be how the mechanical forces it exerts affect ice growth.

In reality, our knowledge of the annual ice cycle cannot be comprehensive without knowledge of the upper ocean: e.g., the temperature at which ice forms depends on mixed layer salinity. Furthermore, when low ice concentrations occur in the marginal ice zone (which includes much of the Southern Ocean about Antarctica), the upper ocean does store considerable heat, thus there are compelling reasons for studying its temperature and salinity structure.

In this section we shall briefly review some different approaches to upper ocean modeling used in the open ocean, discuss surface buoyancy and momentum flux, and outline the development of a model for predicting changes in upper ocean density during the melt season using concepts introduced in the previous section. Finally, we shall point out that one of the really interesting problems regarding oceanic heat -- namely, the fate of heat advected into the Arctic Basin from the North Atlantic -- is not amenable to simple one-dimensional modeling, but appears instead to be an advective phenomenon involving complex exchanges in the large marginal seas of the Arctic.

5.1 Upper Ocean Models

Aside from integration of the three-dimensional, primitive flow equations (Deardorff, 1972), there are essentially two classes of models for simulating changes in salinity and temperature near the surface. In one, the differential form of the horizontally homogeneous PBL equations (eqs. 4.1-4.3) is used, with some means of closing the equations in terms of known quantities. Profiles of velocity and density are calculated at each time step. Within this category, there are numerous closure techniques including specification of a mixing length dependent on stratification (e.g., Mellor and Durbin, 1975); and "second-order" closure models in which explicit equations for the second-moment quantities (e.g., $\overline{w'T'}$) are solved along with the mean quantity equations (e.g., Wyngaard, 1975). Such models have been used to successfully simulate some oceanic conditions and provide valuable research tools, but tend to be rather complex and time consuming for use, say, in climate simulation.

In the second class of upper ocean model, properties are assumed to be uniform throughout the mixed layer and integrated forms of eqs. (4.1-4.3) are used to predict its characteristics. In these models the mixed layer depth (or its time derivative, usually called entrainment velocity) becomes a calculated variable. With many variations, mixed layer models can be broken into two subclasses. In one, exemplified by Kraus and Turner (1967), the change in mixed layer depth is calculated from an energy balance in which potential energy changes associated with surface buoyancy and entrainment of denser water from below is equal to the surface energy input plus the rate of working by surface stress less the integrated turbulent dissipation. Consider a special case of a mixed layer with no surface buoyancy and constant surface wind stress. The rate of working at the surface is proportional to the cube of the surface wind (assuming no ice), and although most of the energy derived from the wind is dissipated as turbulence, what is not goes into entraining denser water upward, increasing the potential energy of the water column and thus conserving energy. Earlier versions set aside a fixed proportion of the surface work for entrainment, thus a deep mixed layer would have as much energy available for entrainment as a shallow one, which is probably unrealistic. P. Lemke (personal communication) has applied an energy balance approach (with a linear falloff of available entrainment energy from the surface to 50 m) to the arctic mixed layer with reasonable results for the seasonal variation of mixed layer salinity and depth.

The other subclass of mixed layer models places emphasis on the integrated mean momentum equation rather than the energy equation. In the prototypic model of Pollard, Rhines, and Thompson (1973), integrating the momentum through the mixed layer leaves a stress at the base which is equated to the impulse needed to boost the momentum of the entrained fluid to the value of the mixed layer slab, i.e.,

$$\tau(z = -h) = u \frac{\partial h}{\partial t}$$

where u is the average velocity of the slab and h is its thickness. Closure is accomplished by setting a bulk Richardson number (i.e., g times the ratio of the bulk density gradient to the velocity shear squared) to unity. One obvious problem with the Pollard et al., approach is that in the steady wind, no-buoyancy-flux case, the layer deepens only during the first half-inertial period. A more fundamental criticism is that mean momentum is not uniformly distributed through the layer, particularly when it is deep, which casts doubt on many of the assumptions regarding the critical Richardson number and the entrainment mechanism. Niiler (1975) has attempted to rationalize the two different mixed layer approaches in a combined model, but the conundrums posed by ignoring mean quantity gradients within the bulk of the layer make his physical arguments for energy partition obscure, and in some ways contradictory to much recent ABL theory.

A comprehensive review of upper ocean modeling is found in the proceedings of the 1975 NATO Advanced Study Institute, "Modelling and Prediction of the Upper Layers of the Ocean" (Kraus, 1977). It is perhaps obvious from the previous sections that (in the author's view) turbulence in the OBL under pack ice should be treated explicitly using eddy diffusivity concepts. Below we shall show how the analytic similarity theory can be extended to include buoyancy flux at the base of the mixed layer. The method requires relatively little computation but retains many of the physical concepts suggested by the AIDJEX OBL measurements.

5.2 Buoyancy at the Ice/Ocean Interface

Just as the summer AIDJEX measurements provided a rare opportunity to test the relationship between surface stress and velocity, they also represent a rather special data set for testing upper ocean density models, since they include daily STD (salinity-temperature-depth) profiles. By monitoring the change in salinity of the upper layers, the surface buoyancy flux can be estimated. When combined with accurate stress estimates, this provides a set of forcing functions hard to duplicate with measurements from the open ocean.

Figure 20 shows salinity and temperature of the upper 200 m of the ocean measured at AIDJEX station Snowbird on days 169, 198, and 227 of 1975. The first case is representative of late winter conditions: The mixed layer is nearly uniform to about 52 m and close to its freezing point. The second shows development of a step-like structure as fresh meltwater is introduced and mixed to varying depths. The third cast was taken near the time of maximum accumulated meltwater. Note that the mixed layer is only about half as deep, is much fresher, and has warmed slightly.

Over the course of the melt season the stations drifted about 200 km to the southeast, so that advective change in the water column must be considered. From looking at spatial variation in the profiles, it was clear that horizontal gradients in density were present, especially in one rather narrow zone through which all the camps passed. However, by looking at the shapes of the profiles, and by comparing data from drift tracks that traversed the same regions at different times, we found that at depths of about 40 m, the salinity profiles were not affected by surface conditions, even though the absolute salinity value might change with position. Thus, by assuming the value of salinity measured at 40 m at each location represented the mixed layer value there when the melt season began, we were able to calculate the cumulative salinity flux for that cast. Since the source of the negative salinity flux was melting ice (neglecting precipitation), an ablation curve for the pack can be constructed from the salinity

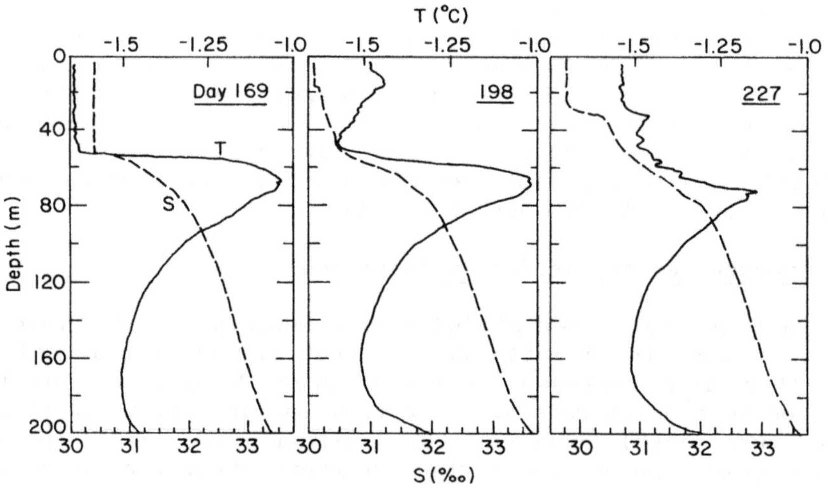

Fig. 20. Salinity and temperature measured at AIDJEX station Snowbird, Summer 1975.

measurements; i.e., for each cast the change in ice thickness from the start of the melt season is given by

$$\delta H = \frac{\rho_w}{\rho_{ice}} \cdot \frac{(\bar{S} - S_{40})h}{\Delta S}$$

where h is the mixed layer depth, $\bar{S}$ is the average salinity above 40 m, S_{40} is the salinity at 40 m, and S is the salinity difference between the mixed layer and sea ice (~25°/oo). Figure 21 shows ice ablation determined this way from the STD data. The solid curve is a simple sinusoid

$$\delta H = 0.70 \quad [1 - \cos(\frac{t - 171}{90})\pi] \text{ m} \quad . \qquad (4.33)$$

The dashed curve is the basin average ablation due to thermodynamic forcing from the sea-ice model of Hibler (1980). The amplitudes are quite similar. A plausible explanation for at least part of the time lag (about 20 days) is that it takes that long for melt-water collecting at the surface to drain through and enter the main oceanic water column.

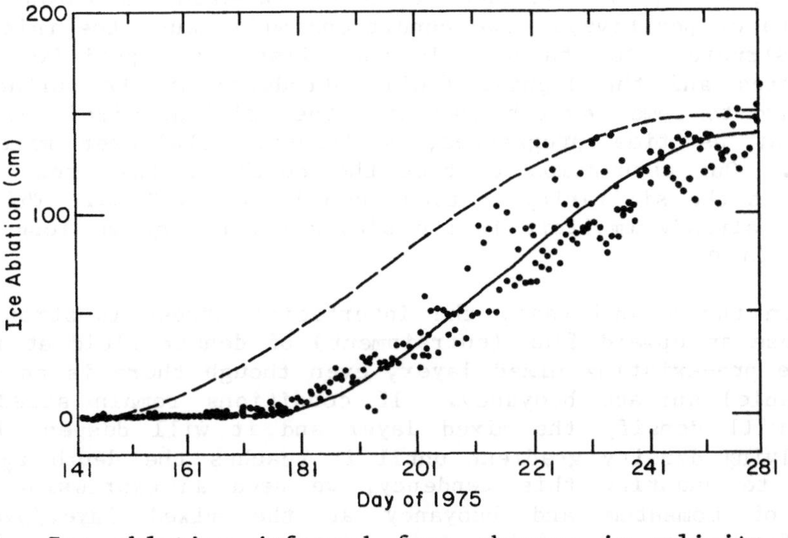

Fig. 21. Ice ablation inferred from changes in salinity of the upper 40 m at AIDJEX stations, Summer 1975 (solid line). Basin-wide, thermodynamic ablation rate from Hibler's (1980) sea ice model (dashed line).

The ablation curve eq. (4.33) can be differentiated to provide a growth rate curve which is then used to estimate the buoyancy flux as a function of time over the region in which the camps drifted.

5.3 An Upper Ocean Model Based on the Stress Equation

With time series of stress and buoyancy at the surface it seems reasonable to try a simulation of upper ocean density evolution during the melt season using most of the conceptual tools developed for the underice OBL from previous work. Here we present a previously unpublished model which represents a preliminary attempt based on an extension of the similarity theory of McPhee (1981a). It is predicated on the following assumptions.

1. Time-dependent (inertial) terms may be ignored in the turbulent stress equation, so that the stress profile is always in equilibrium with the interfacial stress at the surface.

2. Mixing of scalar properties (specifically, Θ and S) occurs to a limiting depth: $z_{dyn} = \zeta_{max}\eta_* u_*/f$, unless the depth of the pycnocline is less than z_{dyn}. This is a restatement of the familiar idea that there is a limiting depth to which turbulence penetrates in a neutrally stratified, rotating boundary layer.

Consider a stratified ocean topped by a well mixed (neutrally buoyant) layer, with surface fluxes of momentum and buoyancy (which is zero or positive). Two conditions will cause the initial density structure to change. In the first case, positive buoyancy dominates and the lighter fluid introduced at the surface mixes downward to some level higher than the original mixed layer depth so that as time progresses, a lighter, shallower mixed layer forms. Our assumption is that the depth of this new layer is given in the similarity vertical coordinate by ζ_{max}. This behavior is already implicit in the similarity theory so long as ζ_{max} is specified.

In the second case, the interfacial stress is strong enough to cause an upward flux (entrainment) of denser fluid at the base of the pre-existing mixed layer, even though there is no negative (unstable) surface buoyancy. If conditions remain steady, this flux will densify the mixed layer and it will deepen along the underlying density gradient until it reaches the depth z_{dyn}. In order to quantify this tendency, we need an expression for the flux of momentum and buoyancy at the mixed layer/pycnocline interface.

When the nondimensional pycnocline depth, ζ_p, is shallower than ζ_{max}, we have a situation like that sketched in Figure 22 where there is a finite stress and turbulent mixing at the level,

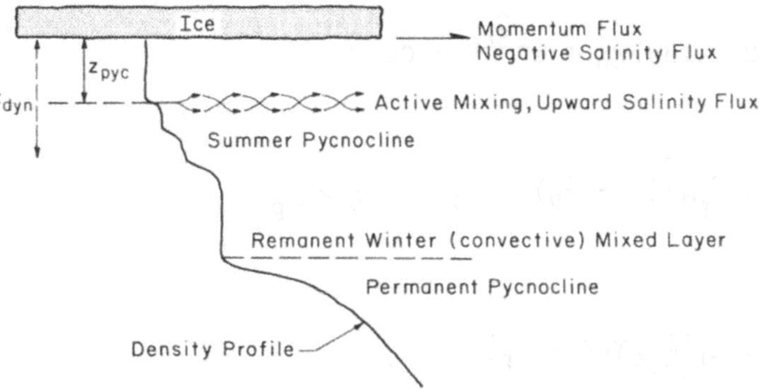

Fig. 22. Schematic of the upper ocean regime during the summer melt season.

ζ_p. In the upper (mixed) layer, the eddy viscosity is given by K_{*m} and below ζ_p we assumed that it has a small, depth-independent value, K_{*p}. The nondimensional stress equation is

$$\frac{i\hat{T}}{K_*} = \frac{\partial^2\hat{T}}{\partial\zeta^2} \tag{4.34}$$

with boundary conditions $\hat{T}(0) = 1$, $\hat{T}(-\infty) = 0$. In the mixed layer the solution is

$$\hat{T} = \hat{A}e^{\hat{\delta}\zeta} + \hat{B}e^{-\hat{\delta}\zeta}$$

where $\hat{\delta} = (i/K_{*m})^{1/2}$. Since $\hat{T}(0) = 1$, we have

$$\hat{T}(\zeta) = 2\hat{A}\sinh\hat{\delta}\zeta + e^{-\delta\zeta}\ ; \qquad \zeta \geq \zeta_p$$

and

$$\hat{v}(\zeta) = -i\frac{\partial\hat{T}}{\partial\zeta} = -i\hat{\delta}(2\hat{A}\cosh\hat{\delta}\,\zeta - e^{-\hat{\delta}\zeta})$$

away from the thin surface layer.

In the lower stratified layer the solution to eq. (4.32) is

$$\hat{T} = \hat{C}e^{\hat{\gamma}\zeta}$$

where $\hat{\gamma} = (i/K_{*p})^{1/2}$ since $\hat{T}(-\infty) = 0$. At $\zeta = \zeta_p$,

$$\hat{T}_p = 2\hat{A}\sinh\hat{\delta}\zeta_p + e^{-\hat{\delta}\zeta_p} = \hat{C}e^{\hat{\gamma}\zeta_p}$$

whence

$$\hat{T}(\zeta) = \hat{T}_p e^{\hat{\gamma}(\zeta - \zeta_p)} \quad ; \qquad \zeta < \zeta_p$$

and

$$\hat{V}(\zeta) = -i\hat{\gamma}\hat{T}_p e^{\hat{\gamma}(\zeta - \zeta_p)} \quad .$$

By matching velocities at ζ_p, $\hat{A}$ can be eliminated. In the mixed layer $K_{*m} = k\xi_N$, which is known, but in order to close the problem, an expression for K_{*p} is required. It is obtained in the same way as the expression for eddy viscosity in the time-dependent model, eq. (4.3). At the outset, the stress at the pycnocline interface, $\hat{T}_p u_* \hat{u}_*$, is unknown so the algebraic equation is implicit and is solved iteratively. Once $\hat{T}_p$ is known, the stress distribution is known, and by Reynolds analogy the flux is determined given the ratio K_ρ/K. In this way the buoyancy flux at the base of the mixed layer is calculated from surface conditions and is used along with the surface buoyancy in determining the evolution of the density structure.

Figure 23 compares the density structure predicted by the model outlined above every 5 days for a 70-day period at AIDJEX Station Snowbird. In this run the ratio K_s/K was 0.1 and the nondimensional maximum mixing depth was $\zeta_{max} = 0.4$. In the model, surface forcing is never strong enough to change the density profile below about 35 m, so for comparison, the observed profiles are forced to match the model at the 40 m level. The numbers there indicate the observed change in density (σ_t units) from the initial value on day 170, and probably represent the advective change as the ice drifted southeast.

The series of steps produced by the model result when the maximum depth of mixing ($\zeta_{max}\eta_* u_*/f$) is less than the existing mixed-layer depth, so that fresh water introduced at the surface is trapped within that layer. The effect of turbulent stirring set up by rapid ice motion is clearly demonstrated by the deepening of the mixed layer from days 210 to 225. During this time the stabilizing surface buoyancy flux is near its seasonal maximum, so that the deepening must result from shear-generated turbulence.

It is not surprising that the steps in the model profiles are more abrupt than in the observations. First, other processes like energy leakage from internal waves and molecular diffusion act to destroy sharp gradients. Perhaps more important is that turbulent mixing is intermittent and would tend to feather out more gradually

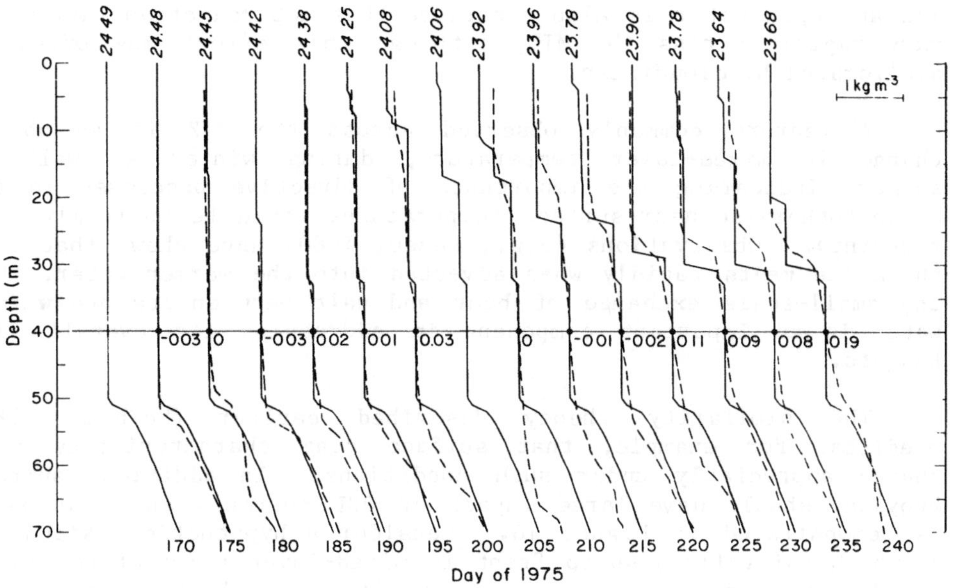

Fig. 23. Comparison of modeled (solid) and observed (dashed) density profiles at AIDJEX station Snowbird, Summer 1975.

than the abrupt limit specified in the model. Nevertheless, it is worth noting that of several values tried, $\zeta_{max} = 0.4$ does the best job of describing the formation of the lighter layers. In the limit $\eta_* \to 1$ (no surface buoyancy) this is consistent with the notion that the limiting depth of the neutral PBL is about ku_*/f.

The model is still in the developmental stage, but results so far are encouraging. Unlike the "slab" mixed-layer models, it also furnishes reasonable estimates of currents at all levels, including the surface (ice), as long as z_0 is known.

6. THE UPPER OCEAN IN THE MARGINAL ICE ZONE

The Marginal Ice Zone (MIZ), where the ice concentration falls rapidly at the edge of the ice pack, may contain some of the most complex upper-ocean structure found anywhere; mainly because the ice can induce extreme gradients in surface heat and salt flux, and can change surface drag characteristics. Since ice is an efficient insulator, the atmosphere over the interior pack can reach temperatures far below those normally encountered in maritime regimes. When this cold air is advected over open water the temperature contrast is large, with correspondingly large latent and sensible heat flux. In summer, the albedo contrast between ice and open water is also great, so that net radiation influx can vary rapidly across the MIZ, although this effect may often be ameliorated by cloudiness.

A feature commonly observed across the MIZ is an abrupt change in mixed-layer temperature, during winter as well as summer, indicating the importance of advective processes in the ocean (otherwise near-surface temperatures would be uniformly near freezing). Observations (e.g., Pease, 1980) have shown that even thick ice melts rapidly when advected into the warmer water, thus the small-scale exchange of heat and salt between ice and warmer water is an important component in addressing the overall flux budgets.

The similarity theory described earlier (McPhee, 1981) predicts, for example, that surface drag characteristics will change appreciably under such conditions. In addition, surface buoyancy should have large impact on OBL currents and transport, as demonstrated in Figure 16. Consider a hypothetical situation in which initially a sharp front in mixed-layer temperature exists at the ice margin. Now impose an off-ice wind stress that produces a boundary layer under the interior pack (where there is no melting) like that depicted in Figure 16 for $\mu_* = 0$; and for the sake of argument assume that the temperature difference across the front is sufficient to cause the melting required for $\mu_* = 25$ on the seaward side of the front. According to the theory, the whole pack should move down-wind with some rightward deflection, but ice at the leading edge will encounter less drag and will thus separate from the main pack with a slight additional rightward deflection. Note also that water in the top few meters of the boundary layer will follow the ice with a downwind (real) component, tending to insulate ice following the leading edge from intense melting. Without buoyancy flux, the trailing ice will not experience a drag reduction; therefore, fairly rapid development of a polynya behind the leading ice edge can be expected.

In the water column, mean shear will rapidly smear the sharp front that existed initially. An interesting convergence is implied by the PBL dynamics: back under the interior ice, there is a net transport toward open water in the layer from about 3 to 12 m which is compensated by opposite transport below (geostrophic departure requires the real component of velocity to integrate to zero over the whole column). Under ice that is rapidly melting, however, there is an appreciable current opposite to the wind stress throughout the same layer (in the atmosphere, the analog is called the nocturnal jet). This current would advect warm water back toward the interior, at some point meeting cold water advected by the neutral OBL; the resulting convergence could then force warm water back toward the surface where it would again interact with the ice. Effects like this might account, at least in part, for extensive finestructure in temperature observations from the MIZ (Paquette and Bourke, 1979); and for the tendency of the ice to form periodic bands parallel to the ice edge under the influence of off-ice winds. Wadhams (Chapter 15) has developed an alternative theory for band formation which involves wave-radiation stress imparting additional momentum to floes at the margin. His approach would obviate the requirement of rapid melting for band formation.

An interesting consequence of large horizontal flux gradients at the MIZ is the potential for oceanic circulations with scales larger than those of the boundary-layer processes, but still much smaller than permanent oceanic features. An obvious example occurs if the ice edge is stationary, like a glacial shelf or fast-ice boundary, in which case the abrupt change in transmission of wind stress to the water column is analogous to a coastal upwelling regime. Details of circulation that might ensue have been investigated by Gammelsrod, Mork, and Roed (1975) and by Clark (1977). The main effect is that in situations where OBL transport is away from the ice (e.g., a westerly wind with ice poleward), an appreciable upwelling can occur at the ice edge. If warmer water is to be found at depth, this represents a mechanism by which heat exchange between ocean and atmosphere might be enhanced. If the ice cover is drifting freely, this mechanism is much less dramatic because the ice, as it moves, transmits a large proportion of the wind stress to the ocean.

Buckley et al. (1979) reported evidence for upwelling near a well-defined, but apparently freely drifting, ice edge north of Spitzbergen, which they attributed to a sharp wind stress gradient during an intense storm that had traversed the region a few days prior to the time of their observations. McPhee (1980c) speculated that a different mechanism, more closely akin to a sea-breeze cell than coastal upwelling, could have induced the structure observed by Buckley et al. (1979). The basics of the argument, which is at best a crude approximation of a real situation, are presented below. It is fashioned closely on a similar exposition of sea-breeze circulation in the atmosphere by Holton (1972, Chapter 5).

Starting from the fundamental circulation theorem, which states that the time rate of change of absolute circulation about a closed loop of fluid particles is equal to the line integral of the so-called solenoidal term; we have:

---

[1]It was pointed out during discussion at Maratea that the description pertaining to Figure 13 of McPhee (1980c) wrongly labels isosteric (constant specific volume) surfaces as isopycnal. The inferred circulation remains as shown.

al. (1979). In terms of a mechanism for extensive upwelling at the ice edge, this approach suffers the same drawback as the stress-gradient proposal: namely, that the ice is mobile, and would in most circumstances move in response to the same forcing as the water, thus tending to smear out the sharp gradients in surface forcing. Røed (personal communication) has begun development of a more sophisticated combined ice-ocean model for the MIZ that shows promise of providing better understanding of these complex phenomena.

On a geophysical scale, freezing is one of the most effective distillation processes. The occurrence of sharp horizontal density gradients in the upper ocean that may result from differential freezing leads naturally to a classic problem in rotational fluid dynamics: geostrophic adjustment. Rossby (1938) was the first to explore theoretically how a region of anomalous density adjusts with surrounding fluid in a rotating system. Stern (1975, Chapter 3) treats the collapse of a lens of intermediate density water released suddenly between two stratified layers, as sketched in Figure 25. In brief outline, Stern's argument is as follows. If the fluid is assumed inviscid, Euler's equation for the intermediate lens can be written in terms of the "reduced gravity" as

$$\frac{d\underset{\sim}{V}}{dt} + f\,\hat{k} \times \underset{\sim}{V} = -g_* \nabla h$$

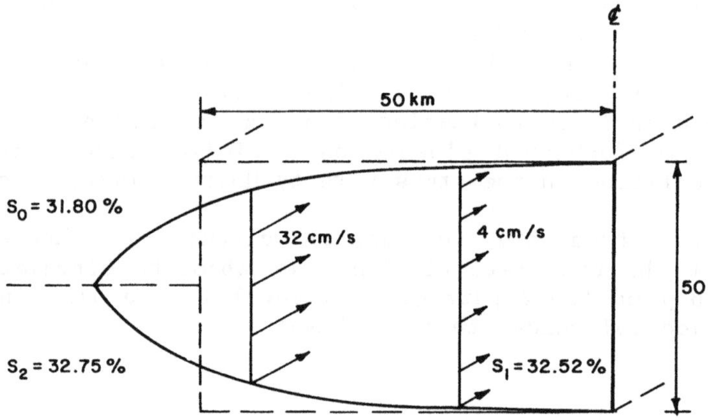

Fig. 25. Limited collapse of lens of intermediate density fluid with initial thickness 50 m, half-width 50 km, between layers in a rotating, inviscid fluid. Adapted from Stern (1975).

$$\frac{d\bar{C}_a}{dt} = - \oint \frac{dp}{\rho}$$

This is applied to the system sketched in Figure 24. Consider a 50 m deep mixed layer that is densified by about 0.1 kg m^{-3} relative to the mixed layer under the ice (this represents combined ice growth and evaporation of roughly 20 cm). Proceeding clockwise about the circuit drawn, the integral is approximated by

$$\oint \frac{dp}{\rho} \simeq \Delta P \left(\frac{1}{\bar{\rho}} + \frac{1}{\bar{\rho} + \delta\rho} \right) \simeq \frac{\Delta P \, \delta\rho}{\bar{\rho}^2}$$

where a Taylor's series expansion is used to estimate the second factor. With $\Delta P = 5 \times 10^5$ Pa, and $\delta\rho = 0.1$ kg m^{-3}, the time rate of change of mean absolute circulation is

$$\frac{d\bar{C}_a}{dt} = 0.05 \ m^2 s^{-2}$$

This implies an appreciable acceleration in the mean velocity about the circuit if no other factors affect the flow, i.e.,

$$\frac{d\bar{V}}{dt} = \frac{1}{2(H+L)} \frac{d\bar{C}_a}{dt} \quad 1.25 \times 10^{-6} ms^{-2}$$

which would amount to an increase in the clockwise velocity component of roughly 10 cm s^{-1} over a day. It is clear, however, that rotational and frictional dissipation effects would soon come into play that would completely distort this oversimplified view -- it is in fact the frictional forces that could give rise to upwelling, although probably not of the magnitude reported by Buckley et

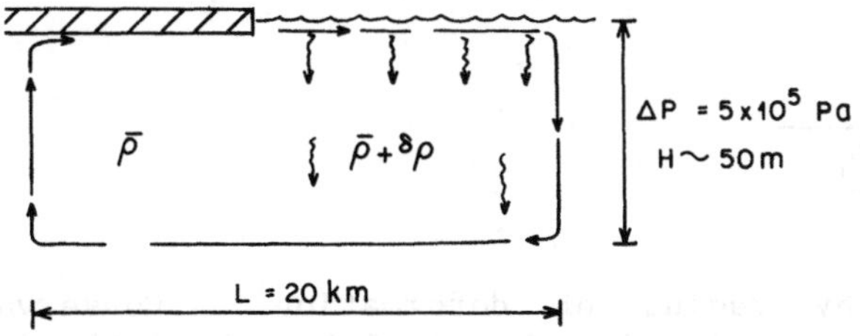

Fig. 24. Schematic of circulation cell resulting from sharp gradient in surface buoyancy flux at the ice margin. The line integral of the solenoidal term is evaluated about the circuit shown.

where h is the thickness of the lens and g_* is the reduced gravity:

$$g_* = g\left(\frac{\rho_1 - \rho_0}{\rho_1}\right)\left(\frac{\rho_2 - \rho_1}{\rho_2 - \rho_0}\right)$$

when ρ_0, ρ_1, and ρ_2 are respectively the densities of the upper, intermediate, and lower layers. Let ζ be the vertical component of local vorticity. If it is assumed that in the intermediate layer of depth, h, the horizontal velocity is independent of depth, then

$$\frac{d}{dt}\left(\frac{\zeta+f}{h}\right) = 0$$

which states that potential vorticity $(\zeta+f)/h$, is conserved. In the initial state, the fluid is at rest with potential vorticity, f/h_0. In the final state, it will achieve geostrophic balance, so that the pressure gradient within the slumping lens is in equilibrium with the Coriolis force, i.e.

$$fv = g_* \frac{\partial h}{\partial x}$$

where v is the velocity component along the lens axis. Since potential vorticity is conserved, the final lens thickness is described by the equation

$$f + \frac{g_*}{f}\frac{d^2h}{dx^2} = \frac{fh}{h_0}$$

with solution

$$h = h_0\left(1 - \frac{\cosh\ (x/\lambda_R)}{\cosh\ (L/\lambda_R)}\right)$$

where

$$\lambda_R = \sqrt{\frac{g_* h_0}{f}}$$

is the Rossby radius of deformation. Conservation of mass provides an expression for the final length of the lens:

$$\frac{L}{\lambda_R} - \tanh\frac{L}{\lambda_R} = \frac{L_0}{\lambda_R}$$

If λ_R is small compared to L_0, the lens spreads a comparatively small distance before reaching equilibrium; and, in the absence of any dissipation would remain in motion parallel to its long axis indefinitely. In reality, turbulent dissipation would slow currents near the nose, and the feature would migrate farther into the two-layer regime as it lost potential energy.

Figure 25 shows the ideal solution for intrusion of a lens of water with salinity 32.52°/oo between layers with salinity 31.80°/oo and 32.75°/oo. For these values the Rossby radius is about 12 km; the idealized solution shows that the final state departs noticeably from its initial configuration only within about two deformation radii from the nose. Note also the current parallel to the lens axis, increasing toward the nose.

The specific parameters of Figure 25 were chosen to coincide with measurements made during a helicopter survey of upper ocean CTD structure near the drift station FRAM I in April, 1979, described in more detail by McPhee, 1980d. During the course of the survey, which spanned about 6 weeks, we found a dramatic change in mixed layer characteristics throughout the survey region despite generally mild winds. The mixed layer to the east was deeper and markedly more saline than it was in the area directly west of camp; the change occurred across a narrow frontal region which meandered roughly north-south as far as we could sample. Fortuitously, the camp itself drifted over the frontal zone. Near the center, profiling current meter casts showed little or no current above about 15 m; then a strong shear between 15 and 20 m marking the upper boundary of a mostly unidirectional (current toward the northwest) jet with speeds of about 10 cm s^{-1} down to another shear layer from 50 to 70 m. Below this, currents were small and variable. Reasoning that the feature was in approximate geostrophic balance, we ran a survey line of helicopter casts perpendicular to the current direction. Salinity and temperature profiles for 7 casts in the linear transect are shown in Figures 26 and 27. Qualitative agreement with the inviscid geostrophic adjustment model is apparent; turbulent dissipation would obviously slow the currents and blunt the nose. The distance across which the two different mixed-layer regimes interact is about two Rossby deformation radii.

The origin of the denser water in Figure 26 is not known. We speculated that it had migrated (ageostrophically) from the shelves north of Svalbard where it had formed by intense freezing, maintained by continual offshore advection of ice. Whatever the origin, understanding the physics and thermodynamics of features like this is probably an important key to understanding the Marginal Ice Zone.

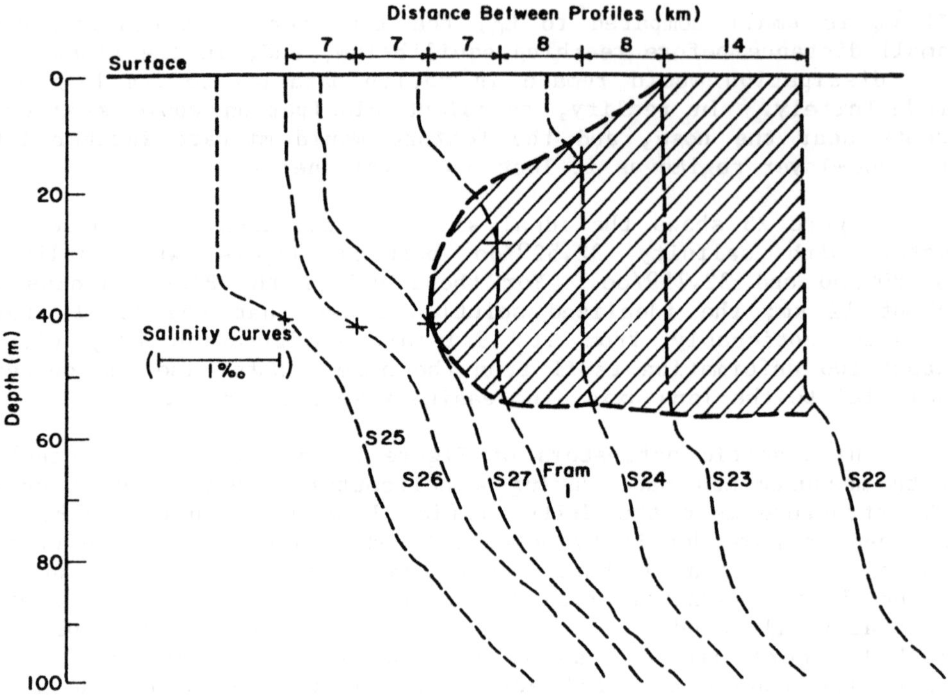

Fig. 26. Salinity profiles centered at ice station FRAM I showing intrusive layer of water with salinity 32.52‰. Pluses mark salinity 32.5‰. Samples were made by helicopter in a line perpendicular to current jet observed in a layer from 20-50 m deep at the manned station.

7. SUMMARY

Understanding of the upper ocean in contact with sea ice has increased dramatically with analysis of data from major experiments performed in the Arctic, and as general knowledge of turbulent boundary-layer processes has increased. The main thrust of the present chapter has been to show that a comparatively simply approach to parameterizing PBL turbulence, one developed primarily for the atmosphere, is an effective tool for describing and predicting characteristics of the oceanic PBL, even for phenomena such as inertial oscillations, which are not a prominent feature in the atmosphere.

Prediction of sea-ice motion requires knowledge of momentum flux at the ice undersurface. AIDJEX results have shown convincingly that drag in the interior ice pack can be described by laws analogous to atmospheric geostrophic drag laws. Equation (4.31)

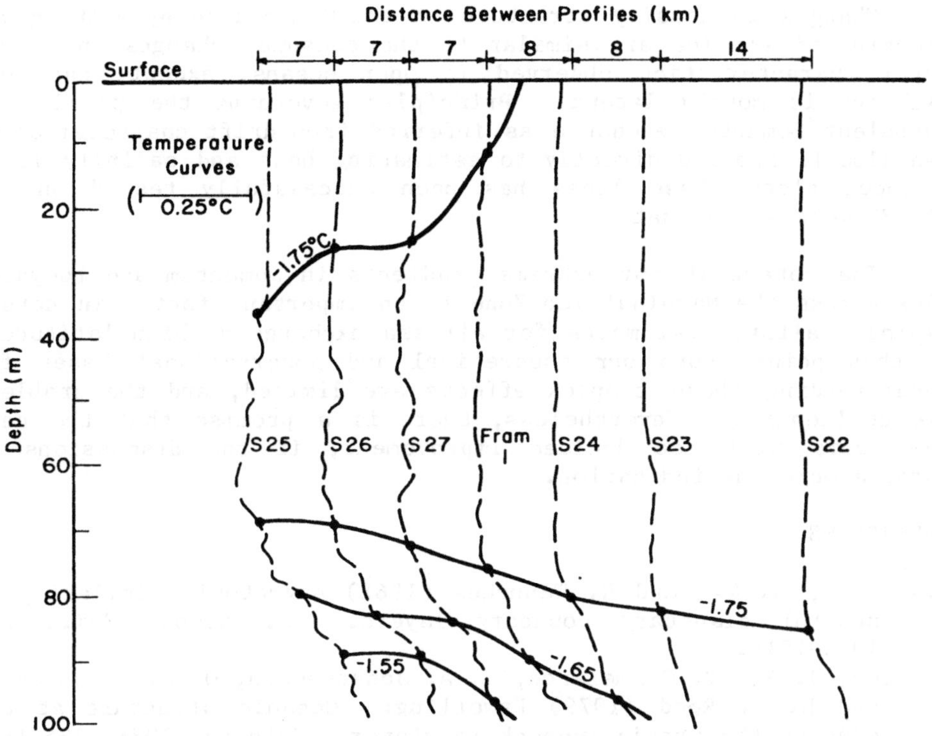

Fig. 27. Temperature profiles as in Figure 26.

represents a general statement of the drag law for sea ice. Under all but extreme circumstances, the OBL can be treated as neutrally buoyant so that η_* = 1 and a = 0. The main uncertainty then is how z_0 varies spatially and seasonally. Near the ice margins where rapid melting and freezing can modify turbulent structure, there is still much theoretical and observational work needed. In shallow waters, the effect of the bottom on OBL turbulence and mean velocity needs more investigation, especially when the water-column stratification is weak.

Inertial oscillations in the sea ice and underlying boundary layer are often present, especially during the melt season, partly because then the pycnocline is shallower, thus trapping the inertia in a smaller layer; but mostly because the ice is too weak to provide much damping. Their main source is rapid change in wind stress, but AIDJEX measurements indicated a coupling in the oscillations observed across the entire array (190 km) beyond that provided by the local surface winds.

Changes in density structure in the OBL caused by melting and freezing of sea ice are similar to the seasonal changes in mixed-layer characteristics observed in open oceans, except that heat exchange is mostly latent. Principles governing the physics of turbulent momentum exchange as inferred from drift characteristics can also be applied directly to estimating heat and salinity flux; a model along these lines has been successfully tested on the AIDJEX melt season data.

The potential for extreme gradients in momentum and buoyancy flux across the Marginal Ice Zone is an important factor in determining realistic estimates for air-sea exchange at high latitudes. At this point, both our theoretical and observational bases for understanding these complex effects are limited, and the problems are challenging. Nevertheless, there is a promise that the next few years will see decided improvement, if the discussions at Maratea were any indication.

REFERENCES

Blackadar, A. K., and H. Tennekes (1968) Asymptotic similarity in neutral planetary boundary layers. J. Atmos. Sci., 25: 1015-1019.

Buckley, J. R., T. Gammelsrod, J. A. Johannessen, O. M. Johannessen and L. P. Røed (1979) Upwelling: Oceanic structure at the edge of the arctic icepack in winter. Science, 203: 165-167.

Businger, J. A. and S. P. S. Arya (1974) The height of the mixed layer in the stably stratified planetary boundary layer. In Advances in Geophysics, 18A, Academic Press, New York, p. 73-91.

Clarke, A. J. (1977) On wind-driven quasi-geostrophic water movements near fast-ice edges. Unpublished manuscript. Ocean Modelling, 8: 9-11.

Clark, R. H. and G. D. Hess (1974) Geostrophic departure and the functions A and B of Rossby-number similarity theory. Bound.-Layer Met., 7: 267-287.

Deardorff, J. W. (1972) Numerical investigation of neutral and unstable planetary boundary layers. J. Atmos. Sci., 29: 91-115.

Ekman, V. W. (1905) On the influence of the earth's rotation on ocean currents. Ark. Mat. Astron. Fys., 2: 1-52.

Gammelsrod, R., M. Mork and L. P. Røed (1975) Upwelling possibilities at an ice edge: Homogeneous model. Mar. Sci. Comm., 1: 115-145.

Hibler, W. D., III (1979) A dynamic thermodynamic sea ice model. J. Phys. Oceanogr., 9(4): 815-846.

Hinze, J. O. (1975) Turbulence. McGraw-Hill, New York, Second Edition, 790 p.

Holton, J. R. (1972) An Introduction to Dynamic Meteorology. Academic Press, New York, 319 p.

Hunkins, K. (1967) Inertial oscillations of Fletcher's Ice Island (T-3). J. Geophys. Res., 72: 1165-1174.

Kheisin, D. Ye. and V. O. Ivchenko (1975) Water stress determination in the purely wind-induced drift. Arkt. Antarkt. Nauch. Issled. Inst., 332.

Kraus, E. (Ed.) (1977) Modelling and Prediction of the Upper Layers of the Ocean. Pergamon Press, Oxford, 325 p.

Kraus, E. G. and J. S. Turner (1967) A one-dimensional model of the seasonal thermocline. II: The general theory and its consequences. Tellus, 19: 98-106.

Manley, S. and G. Lewis (Eds.) (1968) Polar Secrets: A Treasury of the Arctic and Antarctic. Doubleday, New York, 222 p.

McPhee, M. G. (1978) A simulation of inertial oscillation in drifting pack ice. Dyn. Atmos. Oceans, 2: 107-122.

McPhee, M. G. (1979) The effect of the oceanic boundary layer on the mean drift of pack ice: Application of a simple model. J. Phys. Oceanogr., 9: 388-400.

McPhee, M. G. (1980a) A study of oceanic boundary-layer characteristics including inertial oscillation at three drifting stations in the Arctic Ocean. J. Phys. Oceanogr., 10: 870-884.

McPhee, M. G. (1980b) An analysis of pack ice drift in summer. In Sea Ice Processes and Models (R. S. Pritchard, Ed.), University of Washington Press, Seattle, Washington, p 62-75.

McPhee, M. G. (1980c) Physical oceanography of the seasonal sea ice zone. Cold Reg. Sci. Tech., 2: 93-118.

McPhee, M. G. (1980d) Heat transfer across the salinity-stabilized pycnocline of the Arctic Ocean. In Second International Symposium on Stratified Flows (T. Carstens and T. McClimans, Eds.), Tapir Press, Trondheim, Norway, p. 527-537.

McPhee, M. G. (1981) An analytic similarity theory for the planetary boundary layer stabilized by surface buoyancy. Bound.-Layer Met., 21: 325-339.

McPhee, M. G. (1982) Sea ice drag laws and simple boundary layer concepts including application to rapid melting. USA Cold Regions Research and Engineering Laboratory, CRREL Report 82-4.

McPhee, M. G. and J. D. Smith (1976) Measurements of the turbulent boundary layer under pack ice. J. Phys. Oceanogr., 6: 696-711.

Mellor, G. L. and T. Yamada (1974) A hierarchy of turbulence closure models for planetary boundary layers. J. Atmos. Sci., 31: 1791-1806.

Mellor, G. L. and P. A. Durbin (1975) The structure and dynamics of the ocean surface mixed layer. J. Phys. Oceanogr., 5: 718-728.

Neshyba, S., V. T. Neal and W. Denner (1971) Temperature and conductivity measurements under Ice Island T-3. J. Geophys. Res., 76: 8107-8120.

Neumann, G. and W. J. Pierson, Jr. (1966) Principles and Physical Oceanography. Prentice-Hall, Englewood Cliffs, N.J., 545 p.

Newton, J. L. (1973) The Canada Basin: Mean circulation and intermediate scale flow features. University of Washington, Ph.D. thesis, 158 p.

Niiler, P. P. (1975) Deepening of the wind-mixed layer. J. Mar. Res., 33: 405-433.

Paquette, R. G. and R. H. Bourke (1979) Finestructure in the vicinity of the arctic sea-ice margin. J. Geophys. Res., 84: 1155-1164.

Perkins, H. (1970) Inertial oscillations in the Mediterranean. Massachusetts Institute of Technology - Woods Hole Oceanographic Institute, Ph.D. thesis, 155 p.

Pollard, R. T. and R. C. Millard, Jr. (1970) Comparison between observed and simulated wind-generated inertial oscillations. Deep-Sea Res., 17: 813-821.

Pollard, R. T., P. B. Rhines, and R. O. R. Y. Thompson (1973) The deepening of the wind mixed layer. Geophys. Fluid Dyn., 3: 381-404.

Reed, R. J. and W. J. Campbell (1962) The equilibrium drift of the Ice Station Alpha. J. Geophys. Res., 67: 281-297.

Rossby, C.-G. (1932) A generalization of the theory of the mixing length with application to atmospheric and oceanic turbulence, Mass. Inst. Tech. Met. Pap. (4).

Rossby, C. G. (1938) On the mutual adjustment of pressure and velocity distribution in certain simple current systems, II. J. Mar. Res., 1: 239-263.

Rossby, C. G. and R. B. Montgomery (1935) The layer of frictional influence in wind and water current. Pap. Phys. Oceanogr. Met. Mass. Inst. Tech. Woods Hole Oceanogr. Inst., 3: 1-100.

Rothrock, D. A. (1975) The steady drift of an incompressible arctic ice cover. J. Geophys. Res., 80: 387-397.

Shuleikin, V. V. (1938) The drift of ice fields. Compt. Rend. (Doklady) Acad. Sci., URS, 19: 589-594.

Stern, M. E. (1975) Ocean Circulation Physics. Academic Press, New York, 246 p.

Tennekes, H. and J. L. Lumley (1972) A First Course in Turbulence. MIT Press, Cambridge, Mass., 300 p.

Wyngaard, J. C., O. R. Coté and K. S. Rao (1974) Modeling the atmospheric boundary layer. In Advances in Geophysics, 18A, Academic Press, New York, p. 193-212.

Zilitinkevich, S. S. (1975) Resistance laws and prediction equations for the depth of the planetary boundary layer. J. Atmos. Sci., 32: 741-752.

CHAPTER 5

THE SURFACE HEAT AND MASS BALANCE

Gary A. Maykut

Department of Atmospheric Sciences/Geophysics
University of Washington
Seattle, Washington 98195

5.1 INTRODUCTION

Formation of a sea ice cover produces profound changes in the state of the atmosphere and ocean, primarily by altering the surface heat exchange. The high albedo of the ice and its insulation of the atmosphere from the underlying water give rise to a climate over the polar oceans that is more characteristic of the continental ice sheets than of a marine environment. Unlike the ice sheets, however, sea ice is only a thin veneer whose thickness and areal extent are sensitive to small changes in heat input. Variations in sea ice extent have the potential to amplify small changes in climate through a variety of positive feedback mechanisms (Kellogg, 1975), leading to their central role in several ice age theories (Brooks, 1949; Ewing and Donn, 1956, 1958; Budyko, 1966).

While an ice cover reduces the transfer of momentum from the air to the water, thermal processes associated with the growth and movement of the ice have the greatest impact on the ocean. In particular, evaporative heat losses and the absorption of solar radiation in the upper ocean are drastically reduced -- lack of light is the principal factor limiting biological activity beneath the ice in the Arctic Ocean. The ice also affects vertical motions in the mixed layer. Salt rejected by the growing ice forms cold dense brine which sinks, promoting convective overturning, and causing heat entering the mixed layer to be transported upward where it is rapidly lost to the ice. In contrast to the open ocean, no significant storage of heat occurs in the mixed layer beneath the ice pack.

Little is known about the heat and mass balance of ice-covered portions of the Southern Ocean due to inaccessibility and difficulties involved in making surface observations on the dynamic seasonal ice. In the north, however, measurements have routinely been carried out on most of the U.S. and Soviet drifting stations during the past few decades, producing substantial data on incident radiation, air and water temperatures, snowfall, winds, currents, and ocean structure in many parts of the Arctic Basin (Fletcher, 1965; Aagaard and Greisman, 1975). During the past few years data buoys have been deployed across much of the Arctic Basin and have begun to produce information on the large scale fields of atmospheric pressure, ice velocity, and surface temperature (Thorndike and Colony, 1980, 1981; Thorndike et al., 1982, 1983). A few similar buoys have also been deployed in the Antarctic (Ackley, 1981), but these have been generally limited to the somewhat atypical Weddell Sea and not to the seasonal ice which makes up most of the Antarctic ice pack. Despite recent attention to the marginal ice zone (MIZ) and areas of seasonal ice, our general understanding of interactions between the ice, the ocean, and the atmosphere is still largely based on what has been seen in the perennially ice-covered portions of the Arctic Basin. The applicability of such information to more dynamically active areas such as the MIZ and the antarctic ice pack is uncertain.

In the first part of this chapter we will review the various components of the surface heat balance and examine the relative magnitude of heat fluxes observed over the arctic ice pack. We will then describe the growth pattern and mass balance of a sea ice cover, consider different methods of predicting ice growth, and look at the sensitivity of sea ice to changes in climate. Finally, we will discuss the coupling between dynamic and thermodynamic processes, then estimate how thickness variations affect the regional heat exchange and ice production.

5.2 SURFACE HEAT EXCHANGE

5.2.1 Radiation

All matter radiates energy in proportion to the fourth power of its absolute temperature

$$F^* = \varepsilon\sigma T_o^4 \tag{5.1}$$

where F^* is the emitted radiation, ε = the emissivity of the material, σ = the Stefan-Boltzman constant = $8.13 \cdot 10^{-11}$ cal cm^{-2} min^{-1} $^oK^{-4}$ ($5.67 \cdot 10^{-8}$ W m^{-2} $^oK^{-4}$), and T_o = radiative temperature of the material. Most of the radiative energy reaching the surface of the Earth lies in two spectral bands: shortwave radiation which originates from the sun and has a maximum intensity near a wavelength of 0.5 μm, and longwave radiation which originates from the earth and atmosphere and has a maximum near 12 μm. As is shown in Figure 1, there is little overlap between the radiation derived from these two sources, giving rise to a natural separation in their measurement and treatment.

Longwave radiation. Longwave radiation dominates heat exchange in the polar regions during much of the year. Incoming shortwave radiation (F_r) is negligible during most of the fall and winter, and even at the height of the summer melt season F_r is only a few percent larger than the incoming longwave radiation (F_L). The annual total of F_L is more than double that of F_r (Table 1). The magnitude of F_L depends on the vertical distribution of temperature and water vapor in the atmosphere. The 30-40°C increase in air temperature from winter to summer produces a corresponding increase in F_L on the order of 80-90%. In spite of the magnitude of F_L, the net effect of longwave radiation is to remove heat from the ice. This is because the radiative temperature of the ice usually exceeds that of the atmosphere, so that the amount of radiation emitted by the ice ($F\uparrow$) almost always exceeds F_L. Net longwave losses are at a minimum during the summer and early fall due to the presence of extensive low clouds and substantial amounts of water vapor close to the surface. These result in an effective radiative temperature for the atmosphere close to that of the surface. Fewer and higher clouds during the winter and a large decrease in the amount of water vapor combine to increase the effective temperature differential between the ground and the atmosphere, causing an increase in the net longwave losses at the surface.

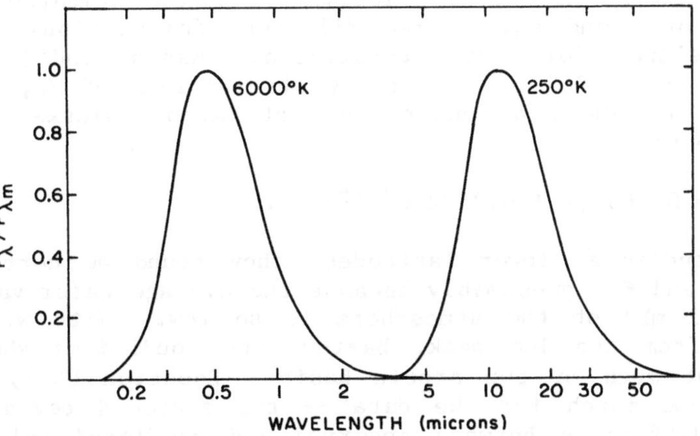

Fig. 1. Blackbody spectra for temperatures roughly corresponding to the sun (6000° K) and the earth (250° K), normalized at the wavelenth of maximum emission, λ_m. (After Fleagle and Businger, 1963).

Accurate measurements of F_L and $F\uparrow$ are particularly difficult in the polar regions where dew or frost formation on the transparent dome shielding the radiation sensor is a constant problem throughout the year. Because of the need for continual clearing and maintenance of the dome and problems with moisture inside the instrument, observations of longwave radiation are seldom carried out on a routine basis. However both F_L and F can be estimated from other information. Surface temperature data are frequently used with eq. (5.1) to calculate $F\uparrow$. Longwave emissivities (ε_L) of the ice, leads and melt ponds are all close to 0.97; $\varepsilon_L = 0.99$ when there is a snow cover.

If the vertical structure of temperature and humidity in the lower troposphere are known, F_L can be calculated by various techniques, e.g., Elsasser's method (Haltiner and Martin, 1957). Such data are generally not available in the polar regions, so these techniques are seldom used there. Because a large fraction of F_L is emitted by clouds and water vapor located relatively close to the ground, attempts to parameterize F_L on the basis of surface observations have been fairly successful. Air temperature (T_a) at screen height (2 m), the partial pressure of water vapor near the surface (e_a), and the fractional cloud cover (C) are all correlated with F_L, but the degree of correlation depends on geographic location. For example, a formula which produces good results in the tropics would not necessarily perform well in the polar regions. Longwave parameterizations have the form

$$F_L = \varepsilon^* \sigma T_a^4$$

where ε^* is an effective emissivity for the atmosphere and is a function of C and e_a. Three different formulations for ε^* have been developed for arctic conditions. Maykut and Church (1973) analyzed some 3000 sets of hourly observations of F_L, T_a, e_a, and C taken throughout an entire year at Barrow, Alaska and obtained the expression

$$\varepsilon^* = 0.7855(1 + 0.2232\, C^{2.75}) \quad . \tag{5.2}$$

Unlike studies at lower latitudes, they found no correlation between e_a and F_L, presumably because the average water vapor content ($\bar{e}_a$ = 1-2 mb) of the atmosphere is so low. With winds usually blowing from the ice pack, Barrow type conditions should obtain throughout much of the Arctic Basin. Equation (5.2) provided a fairly good match for the data -- the standard deviation of the percent difference between observed and predicted values was 12% under clear skies and 7% under overcast skies. The reason for the nonlinear dependence on cloudiness is uncertain, but a similar dependence has been reported by Laevastu (1960) for shortwave radiation and by Kirilova (1952) for longwave radiation.

Table 5.1. Radiation balance (MJ m^{-2} mo^{-1}) of perennial ice in the Central Arctic (Marshunova 1961; Maykut and Untersteiner 1971). Numbers in parentheses take into account the effects of summer melt ponds.

	Jan	Feb	Mar	Apr	May	Jun	Jul	Aug	Sep	Oct	Nov	Dec	Year
Incoming Longwave Radiation [F_L]	435	430	430	495	630	750	800	780	690	580	475	450	6945
Outgoing Longwave Radiation [$F\uparrow$]	-520	-500	-505	-600	-725	-830	-835	-820	-720	-610	-535	-520	-7720
Net Longwave Radiation [$F_L + F\uparrow$]	-85	-70	-75	-105	-95	-80	-35	-40	-30	-30	-60	-70	-775
Incoming Shortwave Radiation [F_r]			100	420	730	785	575	370	155	20			3155
Absorbed Shortwave Radiation [$(1-\alpha)F_r$]			15	80	145	170	245 (295)	115 (165)	35	5			810 (910)
Net Radiation	-85	-70	-60	-25	50	90	210 (255)	75 (120)	5	-25	-60	-70	35 (125)

An alternate expression for ε^* in the Arctic, based on monthly averaged data, was suggested by Marshunova (1961)

$$\varepsilon^* = (0.67 + 0.05\sqrt{e_a})(1 + nC) \qquad (5.3)$$

where e_a is in mb and n is an empirical parameter which varies with season, being 0.22 in mid-summer and increasing to 0.30 in the winter. The linear dependence on cloudiness was justified because conditions in the Arctic are "usually either overcast or clear," so that conditions in between are not very important to the monthly averages. The effect of e_a in eq. (5.3) is to cause a 10-15% increase in * between winter and summer, although this is partially compensated for by the decrease in n. The reason for such a dependence to show up in the monthly data and not in the hourly data of Maykut and Church is unknown.

Idso and Jackson (1969) derived a complex formula for ε^* under clear skies

$$\varepsilon_o^* = 1 - 0.261 \exp[-7.77 \cdot 10^{-4}(T_a - 273)^2] \quad . \qquad (5.4)$$

Using clear sky data from Barrow, we compared predictions made by eq. (5.4) with those of eq. (5.2) ($\varepsilon_o^* = 0.7855$) and found no discernible difference in the accuracy of the two methods. When eq. (5.4) was used in conjunction with Marshunova's (1+nC) cloud factor, the correlation between observed and predicted values was 0.78 as compared to the 0.92 obtained with eq. (5.2).

The dependence of F_L on T_a and C for the three expressions is shown in Figure 2. Equations (5.2) and (5.3) are fairly close for clear and overcast conditions, and both exhibit a stronger temperature dependence than eq. (5.4). Maximum differences in predicted values of F_L are 10-15%. Choice of the "best" parameterization is not really possible from this evidence. Equation (5.2) is supported by the most data, particularly when $0<C<1$, but was developed for use with short term averages; eq. (5.3) requires input on e_a and n(t), but appears to provide good long term averages; Idso and Jackson claim that eq. (5.4) fits their experimental data with a correlation coefficient of 0.992, but its application to the Barrow data leaves in doubt the value of the more complex expression. While the differences in F_L predictions seem small, they can lead to differences in the net longwave radiation totals of as much as 200-300%. Additional studies are needed to develop improved methods of estimating F_L -- particularly useful would be parameterizations using buoy and satellite observations.

Incoming Shortwave Radiation. Incoming shortwave radiation undergoes large seasonal variations in the polar regions, the sun being continuously above or below the horizon for 2-6 months at a

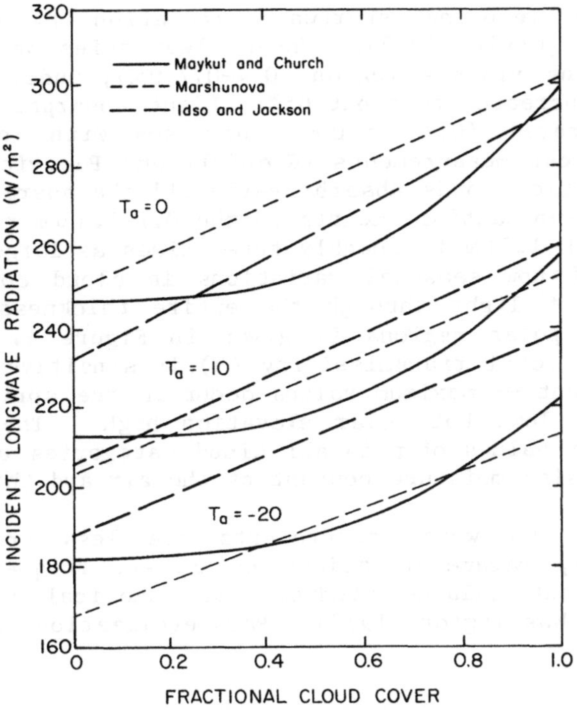

Fig. 2. Comparison of the effects of cloudiness and air temperature (°C) on incoming longwave radiation as predicted by parameterizations developed for polar conditions.

time. Because of the long days, summer values of F_r are surprisingly large. In spite of the low sun angle and extensive cloud cover, monthly totals for May and June in the Central Arctic (Table 1) equal or exceed those measured at many mid-latitude stations such as New York and Chicago. According to Marshunova (1961), there is little spatial variation in monthly F_r totals across the Arctic Basin during the summer months. Most of the shortwave radiation reaching the surface of the ice is diffuse rather than direct due to the extreme cloudiness ($\bar{C}$>0.9 from June through September). Table 1 clearly shows that F_r is asymmetric about the summer solstice -- much more shortwave energy arrives at the surface in May and early June than in late June and July. This skewness results from (i) increasing optical thickness of the atmosphere throughout the summer warming, and (ii) a decrease in the diffuse radiation caused by decreasing surface albedo.

At low and mid-latitudes, reflection and absorption by clouds commonly reduce F_r at the surface by as much as 80-90%. Because of their thinness and low water content, arctic clouds rarely decrease the incoming shortwave radiation by more than 50% (Vowinckel and Orvig, 1962). Under clear skies only about 45% of F_r lies in the visible region (0.4-0.7 μm); under overcast skies this figure increases to about 65%. Light absorption by clouds in the near-infrared (0.7-2.5 μm) increases with increasing wavelength. Optical measurements (Grenfell and Perovich, 1984) indicate that arctic clouds absorb nearly all the energy above 1.4 μm, but transmit substantial amounts in the 0.7-1.4 μm region (Fig. 3). Attenuation at 1.2 μm is roughly three times as large as at 0.7 μm. An example of how seasonal variations in cloud cover affect the transmission of light through the entire thickness of the atmosphere in the polar regions is given in Figure 4. As is to be expected, the total transmissivity (τ) is sensitive to atmospheric moisture content -- maximum values occur in the spring when T_a and e_a are still low, but solar elevation high. The general trend toward smaller values of τ in all cloud categories after April reflects increasing moisture content of the air and thicker clouds.

Although shortwave measurements are less troublesome than longwave ones, parameterizations of F_r are frequently useful in large-scale heat balance studies and numerical modeling (e.g., Parkinson and Washington, 1979). Parameterizations of F_r generally

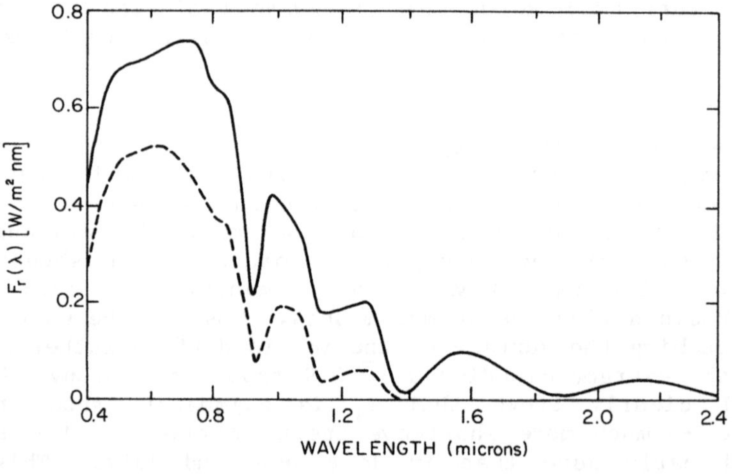

Fig. 3. The effect of a heavy overcast on the spectral distribution of incident shortwave radiation during the early summer at Point Barrow, Alaska. The decrease in the total amount of incident shortwave radiation from clear sky conditions (solid curve) to overcast conditions (dashed curve) is about 45%. (After Grenfell and Perovich, 1984).

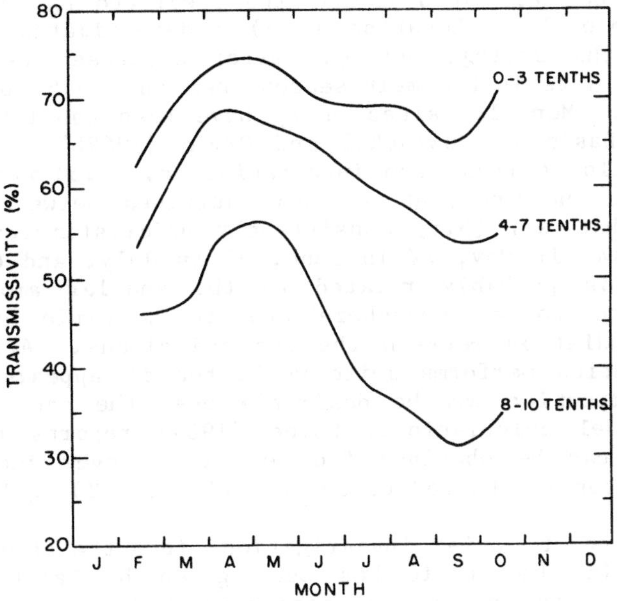

Fig. 4. The relation of total atmospheric transmissivity to cloud cover and season at Point Barrow, Alaska. (After Maykut and Church, 1973).

have the form

$$F_r = a^* F_o$$

where F_o is the downwelling shortwave radiation at the surface in the absence of clouds, and a* represents the fractional attenuation by clouds. F_o depends on hour, day and latitude. Both F_o and F_r can be calculated theoretically from various radiative transfer models of the atmosphere (e.g., Leckner, 1978; Wiscombe, 1975), but the procedures are generally too involved to be used in routine applications. Cruder estimates of F_o can be obtained from empirical relationships between air mass, water vapor content, and radiation depletion (e.g., Houghton, 1954). A particularly convenient form for F_o was given by Zillman (1972)

$$F_o = \frac{S_o \cos^2 Z}{1.085 \cos Z + (2.7 + \cos Z) e_a \cdot 10^{-3} + 0.1} \tag{5.5}$$

where S_o is the solar constant, e_a is in mb, and Z is the solar zenith angle, a function of latitude, day, and hour.

To judge the performance of eq. (5.5) in the Arctic, we looked first at daily totals of F_r from Point Barrow under clear skies and calculated the ratio (R) of the predicted F_o to the observed F_o (see Table 2). Equation (5.5) underestimated daily totals throughout the spring, but the error decreased as temperatures warmed until late in the melt season when eq. (5.5) began to overestimate F_o. Monthly values of F_o have been tabulated throughout the Arctic Basin by Vowinckel and Orvig (1962b). Comparison of monthly totals derived from integrating eq. (5.5) with those given by Vowinckel and Orvig at various latitudes between 70 and 90°N indicates that eq. (5.5) consistently underestimates F_o – by an average of 14% in May, 7% in June, 3% in July, and 6% in August. The reason is probably related to the smaller amount of water vapor in the polar atmosphere and to multiple scattering of reflected radiation between the ice and clouds. All in all, the Zillman equation performs quite well, but it appears that a high latitude correction may be desirable over the pack ice. On the basis of model calculations, Shine (1984) reports that improved performance can be obtained from eq. (5.5) over ice and snow if the denominator is changed to $\cos Z + (1 + \cos Z)\, e_a \cdot 10^{-3} + 0.046$.

A summary of Soviet investigations into the effect of cloudiness on F_r in the Arctic has been given by Marshunova (1961). They were able to obtain good agreement with observations with an expression of the form

$$a^* = 1 - kC \qquad (5.6)$$

where k depends on season (i.e., cloud thickness), increasing from 0.15 in March to 0.5 in July and August. When compared with daily values of C and F_r observed at Point Barrow, eqs. (5.5) and (5.6) generally produced agreement to within 10-15%. Equation (5.6), however, provided no consistent agreement between monthly values when used with the data of Vowinckel and Orvig (1962b, 1962c). Laevastu (1960) suggested that

$$a^* = 1 - 0.6C^3 \quad . \qquad (5.7)$$

When tested with the Point Barrow data under overcast conditions, eq. (5.7) always underestimated F_r, in some cases by as much as 40-50%. Equation (5.7) thus appears to be more suitable for describing the effects of lower latitude clouds. Surprisingly, when used with the monthly data of Vowinckel and Orvig, eq. (5.7)

Table 5.2. Ratio of predicted F_o to observed F_r for clear days at Point Barrow.

Date	1 March	22 March	7 April	30 May	12 August
R	0.76	0.80	0.92	0.95	1.08

produced monthly totals of F_r between May and September which usually agreed with their values to within 10%. The reason for such close agreement is uncertain, but the range in cloudiness was quite limited (0.75-0.90).

Optical Properties of the Ice Pack. The optical properties of the ice determine how it interacts with shortwave radiation. We shall be concerned here primarily with the albedo (α) and the extinction coefficient (κ). The albedo is defined as the ratio of reflected to incident shortwave radiation. During the winter, when the ice is covered by fairly dense ($\bar{\rho}_s$ = 0.35 Mg m^{-3}) wind blown snow, albedos are typically in the range 0.80-0.85. As the snow warms and becomes more dense, its albedo decreases to about 0.70-0.75 at the onset of melting. The albedo of bare sea ice is quite variable, depending on temperature, age, and state of the surface. Measurements on thick perennial ice give albedos between 0.70 and 0.75 when the ice is cold, and between 0.55 and 0.70 when the ice is melting; albedos of thick (1-2 m) first-year ice are 0.5-0.6 when it is cold and 0.3-0.5 during melting. The larger albedos of multiyear ice are due to a higher vapor bubble density and a thicker scattering layer at the surface. The presence of meltwater puddles during the summer melt season increase the amount of shortwave radiation absorbed by the ice pack. Pond albedos vary over a wide range, from 0.4 for shallow ponds early in the melt season to 0.15 for deep ones at the end of the summer.

Spectral albedos (α_λ) have been measured by Grenfell and Maykut (1977) over many types of surfaces in the Arctic. Figure 5 shows albedos of several kinds of first-year and multiyear ice. Albedos are fairly constant across most of the visible (400-700 nm) spectrum, but begin to decrease upon reaching the red wave lengths. There is little difference between the albedos of frozen and melting multiyear ice in the visible region, but the increased liquid water content of the melting ice increases absorption in the infrared producing a substantial decrease in $\alpha(\lambda > 700$ nm). There are two main surface types on melting first-year ice: (i) water saturated blue ice and (ii) white ice which consists of a drained, granular surface layer 5-10 cm in thickness underlain by the blue ice. Spectral albedos of first-year white ice are very close to those of the melting multiyear ice at long wavelengths, but are much smaller at shorter wavelengths. The reason for this is that light penetration is small at long wavelengths so α_λ is determined by the properties of the surface layer; at shorter wavelengths, backscatter from the ice beneath the surface layer becomes important. In the case of melting first-year white ice, there is less backscatter from the underlying blue ice in the 400-600 nm range, hence the observed decrease in α_λ. The further decrease in the albedo of the blue ice (curve d) at short wavelengths is due to the absence of the white layer, and at longer wavelengths to increased liquid water in the surface layer.

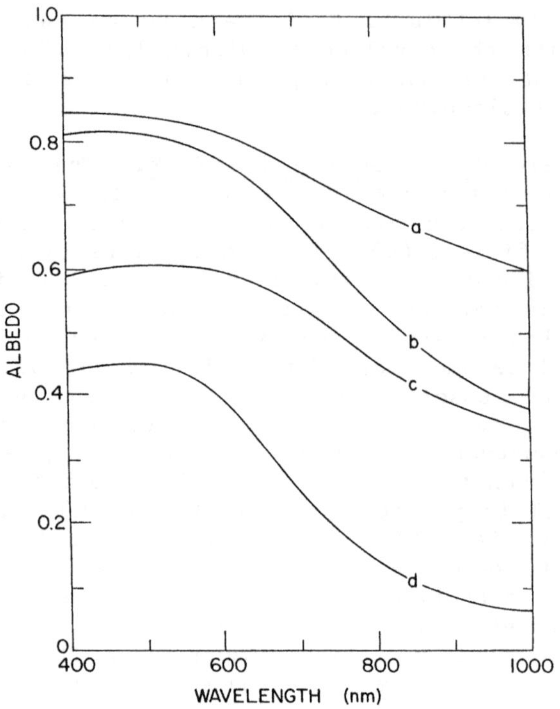

Fig. 5. Spectral albedos observed over bare sea ice: (a) frozen multiyear ice, (b) melting multiyear ice, (c) melting first-year white ice, and (d) melting first-year blue ice (after Grenfell and Maykut, 1977).

Figure 6 shows spectral albedos observed over snow and melt ponds. The albedo of dry compact snow is high and has only a weak dependence on λ. Increasing density and liquid water content decrease α_λ, particularly in the infrared. The albedos of both snow and ice decrease strongly above 1000 nm, reaching values of 0.1 or less at 1400 nm (Grenfell and Perovich, 1984). The albedo of pond covered ice is characterized by a maximum at short wavelengths and a dramatic decrease between 500 and 800 nm. Because water in the ponds is relatively transparent at short wavelengths, values below 500 nm are determined primarily by the scattering properties of the underlying ice. The transition zone (500–800 nm) represents a region where α_λ becomes increasingly insensitive to the underlying ice as absorption by the water becomes the dominant factor. Above 800 nm absorption by the water is so large that α_λ is determined only by Fresnel reflection from the pond surface. Thus most of the albedo difference between individual ponds occurs in the visible where it is readily apparent to the naked eye.

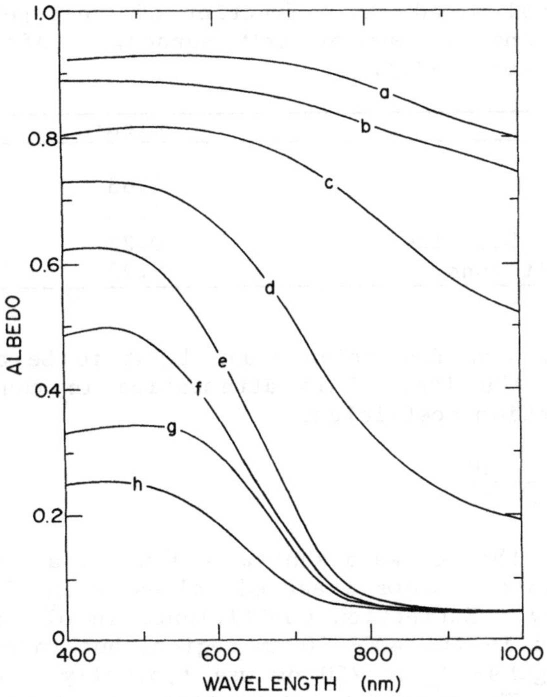

Fig. 6. Spectral albedos observed over snow and melt ponds: (a) dry snow, (b) wet new snow, (c) melting old snow, (d) frozen melt pond with 3 cm of ice, (e) early season melt pond, (f) mature melt pond, (g) melt pond on first-year ice, and (h) old melt pond (after Grenfell and Maykut, 1977).

The total albedo can be expressed as

$$\alpha = \frac{\int \alpha_\lambda F_r(\lambda)\,d\lambda}{\int F_r(\lambda)\,d\lambda} .$$

Although α_λ is independent of cloudiness, clouds tend to increase the total albedo, the size of the increase being dependent on both α_λ and the spectral distribution of the incident radiation. This is because clouds remove proportionately more energy from $F_r(\lambda)$ at long wavelengths than short ones (Figure 3), so that contributions from the more reflective shorter wavelength region are weighted more heavily in the integral when the sky is overcast. Examples of summer albedos under clear and overcast skies are presented in Table 3.

Table 5.3. Total albedo as a function of ice type and cloud cover during the summer melt season. (After Grenfell and Maykut, 1977).

Ice Type	Clear	Overcast
Old snow	0.63	0.77
White ice	0.56	0.70
First year blue ice	0.25	0.32
Mature melt pond	0.22	0.29

Scattering and absorption cause light to be attenuated as it passes through the ice. This attenuation is characterized by a spectral extinction coefficient

$$\kappa_\lambda = \frac{1}{-F(z,\lambda)} \frac{dF}{dz} \tag{5.8}$$

where $F(z,\lambda)$ is the downward shortwave flux at a depth z and wavelength λ. Figure 7 shows measured values of κ_λ for various types of ice and snow. Extinction coefficients in all cases were relatively constant in the 400-500 nm region, but increased sharply at longer wavelengths--κ_λ at 750 nm was typically 4-5 times as large as at 500 nm. The greatest attenuation was observed in cold snow where κ_λ's were roughly 4 times as large as those in the surface layer and about 20 times those of the interior ice. With the onset of melting, κ_λ's in the snow decreased by a factor of 2. The transition zone in Figure 7 describes a 10-15 cm thick region where the decomposed surface layer grades into the more uniform interior ice. The difference between κ_λ's for first-year and multiyear interior ice were caused by differences in brine volume and bubble density. The low values observed in curve (f) were the result of extremely large brine volumes (20-55%) in the ice beneath the pond.

Experimental evidence (Grenfell and Maykut, 1977) indicates that κ_λ is independent of depth as long as the ice is reasonably uniform. The flux at any depth within this homogeneous layer can be found by integrating eq. (5.8) to obtain Beer's law

$$F(z,\lambda) = F(0,\lambda)e^{-\kappa_\lambda z} \tag{8}$$

where $F(0,\lambda)$ is the net shortwave radiation at the surface at wavelength λ. For practical applications, however, it is usually the total flux in the ice, $F_z = \int_0^\infty F(z,\lambda)d\lambda$, which is of interest. If we define the bulk extinction coefficient as the ratio of the total energy absorbed at depth z to the total flux at depth z, then

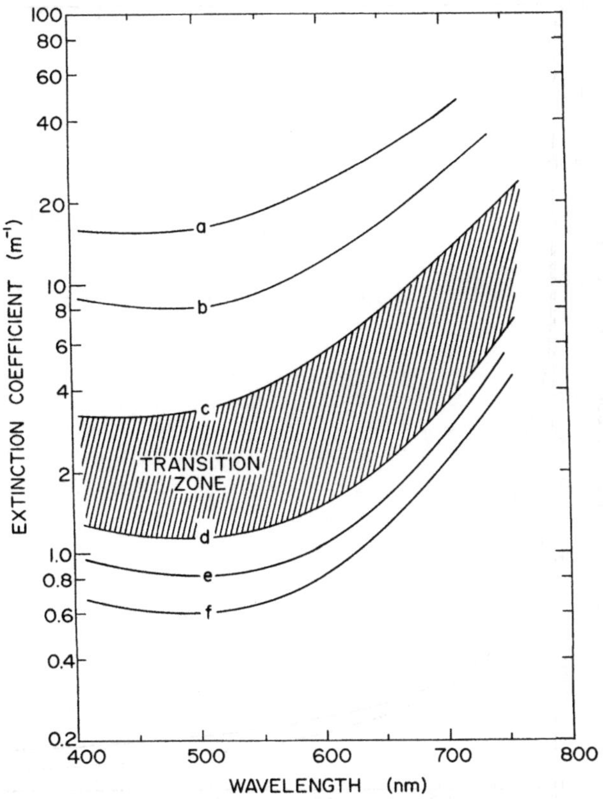

Fig. 7. Spectral extinction coefficients for various types of ice and snow: (a) dry compact snow, (b) melting snow, (c) surface layer on melting multiyear ice, (d) interior of multiyear ice, (e) interior of first-year ice, and (f) ice beneath an old melt pond (after Grenfell and Maykut, 1977).

$$\bar{\kappa}_z = \frac{\int \kappa_\lambda F(0,\lambda) e^{-\kappa_\lambda z} d\lambda}{\int F(0,\lambda) e^{-\kappa_\lambda z} d\lambda} ,$$

and it follows from Beer's law that

$$F_z = F_o \, e^{-\int \bar{\kappa}_z dz} \qquad (5.9)$$

where F_o is the net total shortwave radiation at the surface, $\int_o^\infty (1-\alpha_\lambda) F(0,\lambda) d\lambda$. Grenfell and Maykut have calculated $\bar{\kappa}_z$'s for both first-year and multiyear ice (Figure 8). They found that

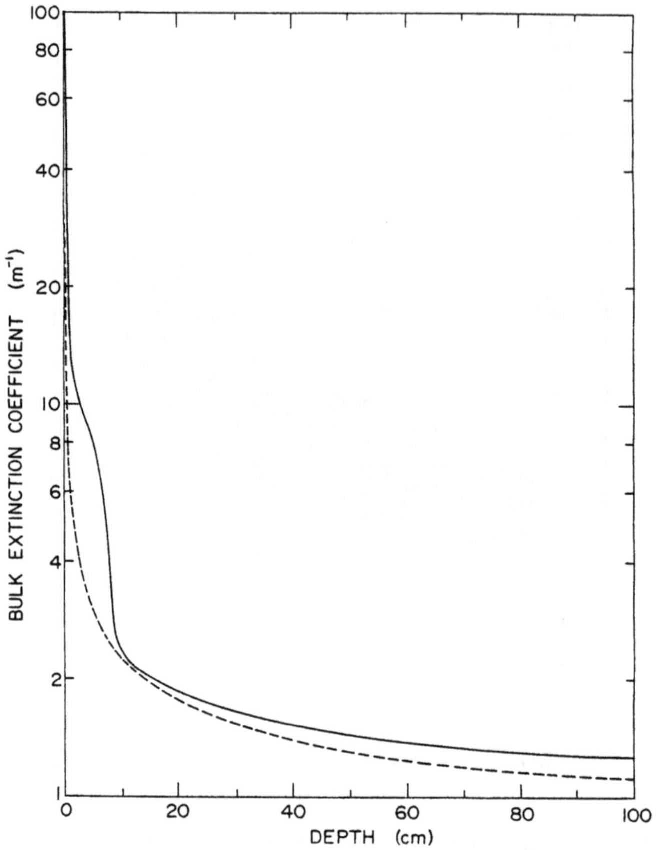

Fig. 8. Bulk extinction coefficients in multiyear ice (solid curve) and first-year blue ice (dashed curve). (After Grenfell and Maykut, 1977).

κ_z's depend strongly on z in the upper 10-20 cm, but have only a weak dependence below this. Bulk extinction coefficients below 5 cm were not affected by cloudiness. Bulk values in the lower part of the ice were in good agreement with Untersteiner (1961) and Chernigovskii (1963) who measured average extinction coefficients of 1.5 m^{-1} in perennial ice.

Absorbed and Penetrating Shortwave Radiation. The net input of shortwave radiation at the surface of the ice is simply $(1-\alpha)F_r$. Monthly averages over perennial arctic ice are shown in Table 1. About 85% of the annual total is absorbed between the first of May and the end of August. Although F_r is largest in June, the high albedo of the snow cover reduces the June total below that of July. Complicating shortwave absorption in the summer Arctic are low albedo melt ponds whose areal coverage varies greatly in both time and space. Coverage on multiyear ice may briefly reach 50-60% during the final stages of snow melt, but it quickly decreases to about 30% once the ice becomes snow-free. Deepening and consolidation reduce the areal extent of these ponds throughout the summer, producing values of about 10% by late August. Due to low surface relief, initial melting on first-year ice often results in nearly complete pond coverage. Even late in the melt cycle, ponds can account for 60% or more of the surface area on seasonal arctic ice. This pattern does not occur in the Antarctic where ponds appear to be uncommon. For this reason it is generally believed tht ponds are not a significant factor in the heat and mass balance of the Antarctic ice pack.

A rough estimate of the effects of melt ponds on shortwave absorption in the Central Arctic is presented in Table 1, assuming an average pond coverage of 30% during July and 10% during August. According to Langleben (1971), this would result in area averaged albedos of 0.49 in July and 0.56 in August, increasing shortwave input to the ice pack by 20% in July, 40% in August, and 12% for the year. Pond effects are somewhat greater in regions of seasonal ice where average albedos may drop as low as 35-40%. In coastal areas, melting is enhanced not only by the ponds, but also by the presence of wind-borne dust from the nearby land. While ponds play an important role in the decay, breakup and retreat of seasonal arctic ice, their impact on multiyear ice appears to be more limited. In the latter case much of the absorbed shortwave energy is stored as latent heat in ponds and brine pockets where it is eventually released to the atmosphere during the fall freeze-up. Theoretical calculations (Maykut and Untersteiner, 1971) indicate that this temporary heat storage has little effect on the average thickness of perennial ice.

Because ice is a translucent material, part of the net shortwave radiation (I_o) is transmitted through the surface layer and does not immediately contribute to temperature or mass changes at the surface. Accurate prediction of surface changes therefore requires some method of estimating the magnitude of I_o. Measurements made in multiyear white ice during the summer melt season (Untersteiner, 1961) indicate that about 30% of the net shortwave radiation is stored as latent heat within the ice, rather than being used directly at the surface. Theoretical calculations (Grenfell and Maykut, 1977) show that this is roughly equivalent to the amount of energy transmitted through the upper 10 cm of white ice. Thus, it appears reasonable to assume that $I_o \simeq F_z$ (z = 10 cm). If we define i_o to be the fraction of the net shortwave radiation transmitted below 10 cm, then

$$I_o = i_o(1 - \alpha)F_r \quad . \tag{5.10}$$

Grenfell and Maykut show that i_o depends on α_λ, κ_λ, and the spectral distribution of the incident radiation, then calculate the effects of cloudiness on i_o for various types of melting sea ice (Table 4). The fact that κ_z does not depend strongly on ice type, cloudiness, or depth in the interior of the ice suggests a simple way to estimate F_z below 10 cm. From eqs. (5.9) and (5.10) it follows that

$$F_z = i_o(1 - \alpha)F_r e^{-\bar{\kappa}(z - 0.1)} \tag{5.11}$$

where $\bar{\kappa}$ is the average bulk extinction coefficient in the lower part of the ice. Equation 5.11 provides the best fit to observations when $\bar{\kappa} = 1.5\ m^{-1}$

5.2.2 Turbulent Heat Exchange

Sensible heat is transferred across the lower part of the boundary layer as a result of temperature differences between the ice and atmosphere. Energy is likewise exchanged by the release of latent heat associated with evaporation, condensation, and sublimation. Apart from emitted longwave radiation, the sensible (F_s) and latent (F_e) heat fluxes are the only means of transporting energy from the surface into the atmosphere. Discussions of the surface energy balance in polar regions frequently dismiss F_s and F_e as being one to two orders of magnitude smaller than the radiation fluxes. However, while the individual radiation components are indeed large, the net radiation balance is small (Table 1) and contributions made by F_s and F_e become significant when compared with the net radiation. Theoretical calculations (Maykut and Untersteiner, 1969) have, for example, shown that the thickness of multiyear arctic ice is very sensitive to changes in F_s and F_e during the summer, indicating that the turbulent fluxes are an integral part of both the heat and the mass balance of an ice cover.

The rate at which heat is transferred by the turbulent fluxes depends on the roughness of the surface, the speed of the wind, the stability of the boundary layer, and the size of the temperature and water vapor gradients. Unfortunately the turbulent fluxes do not lend themselves readily to direct measurements. Eddy

Table 5.4. Fraction of the net shortwave radiation reaching a depth of 10 cm (i_o) in melting sea ice according to Grenfell and Maykut (1977).

Ice Type	Clear	Overcast
White ice	.18	.35
Blue ice/melt ponds	.43	.63

correlation methods are difficult under arctic conditions and have been applied for only short periods of time. Most determinations of F_s and F_e have been derived by profile methods which contain various assumptions about the structure of turbulence. The most commonly used parameterizations are

$$F_s = \rho c_p C_s u(T_a - T_o) \quad (5.12)$$

and

$$F_e = \rho L C_e u(q_a - q_o) \quad (5.13)$$

where ρ is the density of the air, c_p is the specific heat of the air, T_a is the air temperature at some reference height (usually 10 m), T_o is the temperature of the surface, u is the wind speed at the reference height, L is the latent heat of vaporization, q_o and q_a are specific humidities at the surface and reference level, and C_s and C_e are bulk transfer coefficients for sensible and latent heat (Deardorff, 1968). Specific humidity can be expressed in terms of the partial pressure (mb) of water vapor (e) and the total atmospheric pressure (P)

$$q = \frac{0.622e}{P - 0.378e} \simeq \frac{0.622e}{P}$$

since P >> e. Air at the surface of the ice is assumed to be saturated so that eq. (5.13) is frequently written

$$F_e = 0.622 \rho L C_e u(r e_{sa} - e_{so})/P \quad (5.14)$$

where r is the relative humidity at the reference level and e_s is the saturation vapor pressure, a function of temperature. Numerous expressions have been obtained for $e_s(T)$ either by integrating the Clausius-Clapeyron equation or by fitting empirical data (e.g., Richards, 1971; Maykut, 1977).

Turbulent heat flux values obtained by profile techniques over multiyear arctic sea ice are compared in Table 5. The estimates of Badgley (1966) were based on half-hourly averages taken at U.S. Drifting Station Alpha, while the results of Doronin (1963) were obtained from data gathered at several Soviet stations. The most comprehensive calculations are those of Leavitt et al. (1978) who reported 6 hour totals of F_s and F_e throughout the year using observations taken during the Arctic Ice Dynamics Joint Experiment (AIDJEX). These results show that, unlike the situation over temperate and tropical oceans, F_e is negligible during much of the polar year (November-April in the Arctic). Even during the summer, F_e fails to dominate the turbulent exchange. In July and August, for example, the average daily magnitudes of F_s and F_e are nearly equal. Values given by Doronin for

Table 5.5. Comparison of turbulent heat fluxes ($MJ/m^{-2}\ mo^{-1}$) observed over perennial ice in the Central Arctic.

		Jan	Feb	Mar	Apr	May	Jun	Jul	Aug	Sep	Oct	Nov	Dec	Year
	Doronin	49	32	30	12	-19	-12	-13	-17	-7	4	23	33	115
F_s	Badgley	29	23	25	0	-23	-28	0	-7	-13	4	11	16	37
	Leavitt et al.	15	13	11	15	7	-5	2	0	0	5	10	13	86
	Doronin	0	-1	-1	-4	-19	-29	-27	-28	-16	-8	0	0	-133
F_e	Badgley	0	0	0	0	-18	-23	-13	-2	9	11	7	2	-27
	Leavitt et al.	1	0	-2	-1	-9	-6	-7	-8	-14	-8	-1	-1	-56

both F_s and F_e tend to be larger than those of either Badgley or Leavitt et al. Doronin's winter totals of F_s are consistently 2-3 times as large as those of Leavitt et al.; in the summer Doronin shows substantial heat losses due to both F_s and F_e, while Leavitt et al. find relatively small losses due almost entirely to F_e. Badgley's results are intermediate between those of the other two. Leavitt et al. took into account the effects of stability, hence the smaller winter values for F_s could be the result of a more realistic description of conditions in the highly stable winter boundary layer. The small summer totals of Leavitt et al., however, obscure the vigor of turbulent transfer during the melt season. Daily totals obtained from Leavitt et al. indicate that F_s has about the same magnitude during the summer as during the winter. The difference is that during the summer positive and negative values alternate frequently, resulting in small monthly totals. Surprisingly, the variance of the daily totals is somewhat higher during the summer than during the winter.

It is not clear what portion of the differences in reported values for F_s and F_e are real and what portion are due to errors in measurement or analysis techniques. Probably the greatest uncertainity in the profile method is in the choice of suitable transfer coefficients. Eddy correlation measurements by Thorpe et al. (1973) yielded average values of $C_s = 1.2 \cdot 10^{-3}$ and $C_e = 0.55 \cdot 10^{-3}$. Observations in the unstable boundary layer over a refreezing lead (Lindsay, 1976; Andreas, 1980) gave average values close to $3 \cdot 10^{-3}$ for both C_s and C_e. However, roughness length measurements carried out by Untersteiner and Badgley (1965) on relatively uniform multiyear sea ice suggest temporal variations of an order of magnitude in the transfer coefficients. Differences in surface geometry (hummocks, pressure ridges, etc.) can change the drag coefficient by as much as a factor of two (Banke et al., 1980). At this point we lack sufficient information about spatial and temporal variations to say with any degree of confidence what constitutes representative values for the turbulent fluxes over the ice pack. Low level aircraft measurements can provide data on spatial changes, but ground-based eddy correlation measurements are needed to monitor changes over longer periods of time.

While a few measurements of F_s and F_e have been made near coastal sites in the Antarctic, little is known about their magnitude elsewhere in the Southern Ocean. Other evidence, however, indicates that turbulent heat transfer is generally more vigorous than in the Arctic and that the turbulent fluxes assume a correspondingly more important role in the heat and mass balance of Antarctic sea ice. An analysis of surface energy fluxes over the ice in the two hemispheres (Andreas and Ackley, 1982) suggests that lower relative humidities and higher average wind speeds in the Antarctic lead to turbulent heat losses which are sufficiently large to inhibit surface melting during the summer. Given a similar net radiation balance, Andreas and Ackley conclude that air temperatures in the Antarctic must be 2-4°C warmer than in the Arctic before surface melting can be initiated. Differences in surface roughness and cloudiness may also affect the relative importance of F_s and F_e in the two regions, but considerably more data must be acquired in the Antarctic before a final assessment will be possible.

5.2.3 Conductive Heat Flux

Except for the summer melt season, temperatures in the upper part of the ice are colder than those at the underside, causing an upward conduction of heat. The amount of heat reaching the surface depends on the temperature gradient at the surface and the thermal conductivity of the ice (k_i)

$$F_c = k_i(\frac{\partial T}{\partial z})_o \quad .$$

In pure ice, the conductivity (k_o) depends on temperature (T) and can be approximated by the relation $k_o = 9.828 \exp(-0.0057T)$, where T is in °K and k_o in W m^{-1} $°K^{-1}$ (Yen, 1981). A typical value for k_o in warm ice is 2.1 W m^{-1} $°K^{-1}$. Heat transport in sea ice is complicated by the presence of brine pockets which decrease its thermal conductivity relative to that of fresh ice. Detailed theories, taking into account the effects of brine pocket distribution, air bubbles, and precipitation of solid salts on k_i, have been worked out (Anderson, 1958; Schwedtfeger, 1963). Untersteiner (1961) found that k_i was directly related to the brine volume and introduced the approximate formula

$$k_i = k_o + \frac{\beta S_i}{T_i} \tag{5.15}$$

where S_i is the salinity of the ice in ppt, T_i is in °C, and $\beta = 0.13$ W m^{-1}. Although somewhat crude, eq. (5.15) is particularly convenient in applications requiring a simple functional relationship between R_i, T_i and S_i. Under natural conditions k_i varies by 10-20% in perennial ice and by up to 50% in young ice. Monthly totals of F_c in perennial ice are remarkably constant between September and April, varying between 30 and 40 MJ m^{-2} mo^{-1} (Maykut and Untersteiner, 1971). This is due to the increasing thickness of the snow cover as temperatures cool during the fall, and to the relatively constant snow depth and temperature averages during the winter.

Except during the melt season there is an upward flux of heat in the ice which supplies much of the energy needed to balance the longwave loss at the surface. Over an annual cycle the amount of heat conducted to the surface of multiyear ice is the sum of the sensible heat input from the ocean (F_w), the latent heat released by bottom accretion, and the shortwave energy temporarily stored in brine pockets and surface melt ponds. For perennial arctic ice, F_c averages about 8.8 W m^{-2}, of which 20% is supplied by the ocean, 40% by bottom accretion, 30% by release of heat stored in the brine pockets and 10% by the freezing of surface meltwater. In young ice, F_c can be more than two orders of magnitude larger than in multiyear ice. Over the course of the entire growth season the amount of heat conducted to the surface of seasonal ice is roughly 2 to 3 times the annual total for perennial ice.

Because of its low thermal conductivity (k_s = .31 W m^{-1} $^oK^{-1}$), snow insulates the ice from the air, decreasing the rate at which the ice loses heat during the winter. Figure 9 shows the dependence of F_c on snow depth (h_s) in perennial ice. There is roughly a factor of three decrease in F_c as h_s increases from 0 to 100 cm. The rate of decrease in F_c is most rapid when h_s is small. As h_s increases, temperatures in the underlying ice become less sensitive to T_a and more sensitive to the temperature at the bottom of the ice, producing the decreasing effect on F_c with increasing h_s.

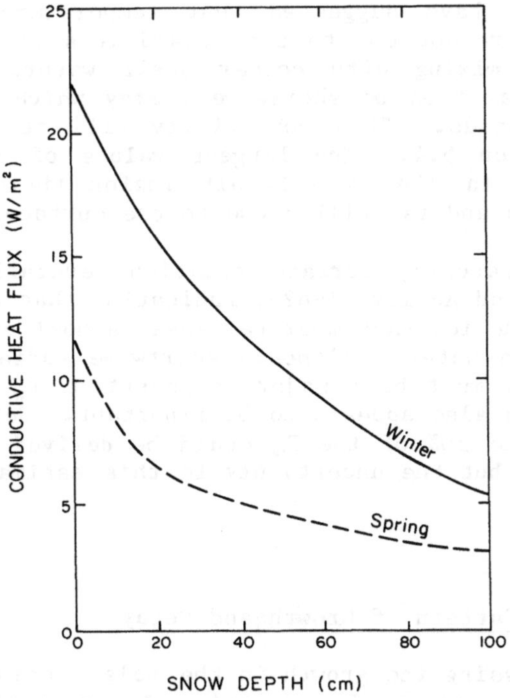

Fig. 9. Effects of snow depth on heat conducted to the surface of perennial ice during mid-winter (solid curve, $\overline{T}_a$ = -34°C) and early spring (dashed curve, $\overline{T}_a$ = -22°C).

Heat conduction through young sea ice (Figure 10) is much more sensitive than the thicker ice to changes in h_s when h_s is small, but is less sensitive when h_s is large. This is because the thermal conductance $[k_i\ k_s/(k_s H + k_i\ h_s)]$ of the combined ice-snow slab is dominated by the properties of te snow cover when the ice is thin (Maykut, 1978). For the cases shown in Figure 10, F_c is essentially constant once h_s exceeds 15-20 cm.

5.2.4 Oceanic Heat Flux

Observations of ablation at the underside of the ice (Badgley, 1966) and theoretical calculations (Untersteiner, 1964; Maykut and Untersteiner, 1971) indicate that there is a net flux of heat from the ocean (F_w) to the underside of the ice on the order of 2 W m^{-2} in the Central Arctic; little is known directly about F_w in areas of seasonal ice. The source of this energy has usually been ascribed to heat advected into the Arctic Ocean from the Atlantic. A relatively warm layer of Atlantic water can be identified throughout much of the central basin. Estimates of the rate of heat loss from the Atlantic layer in the Central Arctic (Panov and Shpaikher, 1964) indicate average rates of 2.0-2.6 W m^{-2}, consistent with the other evidence. Other investigators (Aagaard et al. 1981), however, have suggested that temperature changes in the Atlantic layer are not due to the upward loss of heat to the ice, but rather to mixing with colder shelf water. A more likely source for F_w may thus be shortwave energy which enters the upper ocean through leads. This possibility will be explored in more detail in Section 5.4. The largest values of F_w in the Arctic probably occur in the Fram Strait region where Atlantic water enters the basin and is still close to the surface.

In the Antarctic, surface ablation generally appears to be small (Andreas and Ackley, 1982), indicating that the seasonal disappearance of the ice pack must be caused almost entirely by energy contained in the water. Although shortwave radiation absorbed in the upper ocean must be a major component of F_w there, heat from the deeper ocean also appears to be important. Gordon (1981) estimates that up to 50% of the F_w could be derived from water below the pycnocline, but the uncertainty in this estimate appears to be large.

5.3 ICE GROWTH

5.3.1 General Pattern of Growth and Decay

When discussing ice growth in the polar oceans, it is convenient to consider areas of seasonal and perennial ice separately. Ice in the seasonal sea ice zone (SSIZ) forms in the fall or winter and melts away completely during the following spring and summer. Seasonal sea ice may reach a maximum thickness anywhere from a few

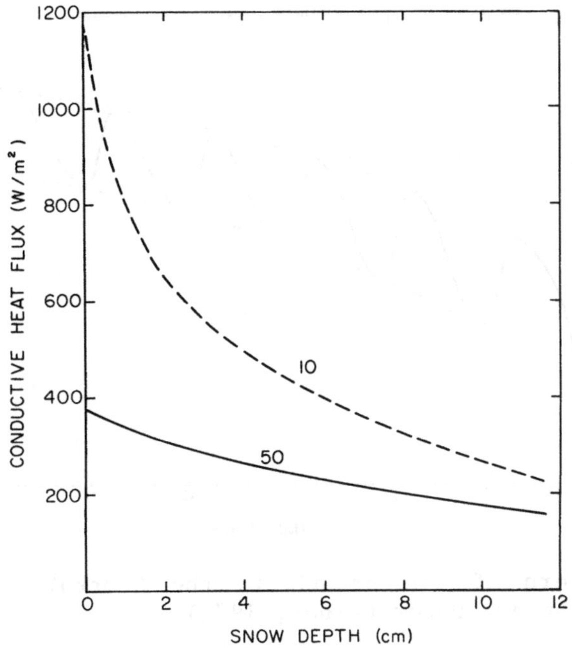

Fig. 10. Effects of snow depth on heat conduction in young ice during mid-winter (T_a = -34^oC). The dashed curve shows 10 cm thick ice and the solid curve 50 cm thick ice.

centimeters to a couple of meters. In areas of perennial ice, the climate is such that ice formed during the fall does not melt completely during succeeding summers. Figure 11 follows the growth of ice in the Central Arctic under climatic conditions which are assumed to be uniform from year to year. Note that the net increase in ice thickness slowly decreases as the average thickness becomes larger. Eventually year to year thickness changes become negligible and the ice is said to be in equilibrium with its environment. Figure 12 shows in nomographic form how, under steady climatic conditions and no dynamic motions, thickness dependent changes in ablation and accretion cause the ice to approach an equilibrium thickness (H_e). While changes in climate and advection of ice into different climatic regions prevent this ideal from ever being attained exactly, the concept is useful in numerous heat and mass balance problems. Equilibrium thickness in the Central Arctic is on the order of 3 m, but the average thickness tends to be somewhat larger because of dynamic effects discussed in Section 5.4.

Sea ice extent in the Northern Hemisphere is sketched in Figure 15.1 where we see that most of the Central Arctic Ocean is

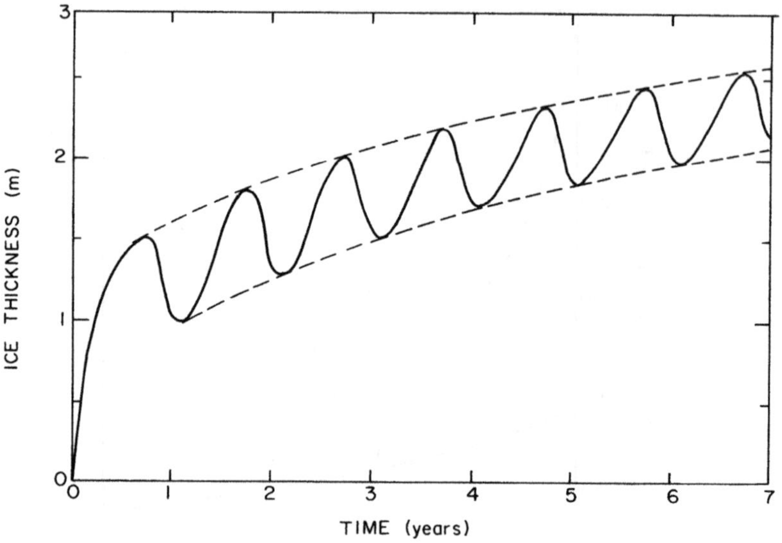

Fig. 11. Pattern of ice growth in the Central Arctic. (After Maykut and Untersteiner, 1971).

covered by perennial ice and most of the peripheral seas by seasonal ice. There have been numerous studies of the formation and growth of young ice in the SSIZ, but most of the work in the Central Arctic has focused on thick multiyear ice. How well ice growth equations derived from observations in the SSIZ describe the growth of young ice in the Central Arctic is not known. Decay of ice in the northern hemisphere SSIZ generally proceeds rapidly following the onset of snow melt. Usually within a week or so shallow meltwater puddles begin to cover the ice and a network of surface drainage canals and vertical melt holes develop. Melt ponds decrease the average surface albedo to as low as 0.4 (Langleben, 1971), greatly increasing the shortwave absorption. In addition, warm winds from nearby land masses often accelerate the surface melting of coastal ice where ablation rates can reach .04 m day^{-1}. At the same time, shortwave radiation absorbed within the salty seasonal ice causes internal melting and structural weakening. Cracking of the weakened ice produces leads which absorb several times as much solar energy as the ice. Heat absorbed in the water causes bottom ablation and lateral melting on the floe edges, increasing the area of open water and forming a positive feedback loop. These processes can result in the complete disappearance of the annual ice cover from sheltered bays and fjords along the arctic coast in a month or less after breakup begins (Langleben, 1972).

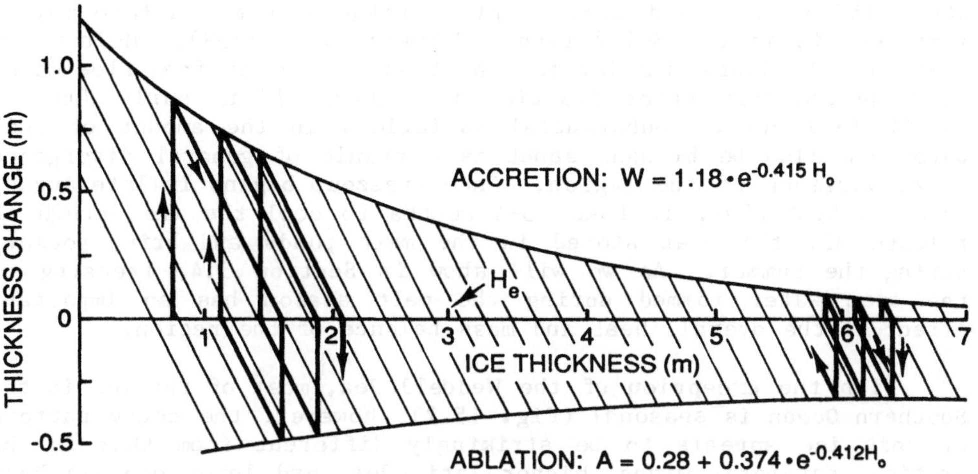

Fig. 12. Nomogram showing the theoretical dependence of accretion (September-May) and ablation (June-August) on ice thickness under present climatic conditions in the Central Arctic (after Untersteiner, 1962; Maykut and Untersteiner, 1969, 1971). Seasonal mass changes for any initial thickness (H_o) are obtained by first moving upward or downward until the accretion (W) or ablation (A) curve is encountered; following the slanted lines back to the horizontal axis then gives the new ice thickness. This procedure is illustrated by the heavy lines which show examples of thickness changes over several annual cycles. It can be seen that ice thickness approaches H_e regardless of whether $H_o < H_e$ or $H_o > H_e$.

Similar melt processes take place in areas of perennial ice, but with less dramatic results. Snow begins melting in June, with most of the ice melt taking place in July and August -- total melting at the upper surface is some 500-600 kg m^{-2}, about .4 m of snow and .4-.5 m of ice. While some melting (.1-.2 m) has been observed on the underside of the ice (Untersteiner, 1961; Hanson, 1965), the situation there is complicated by freshwater runoff which brings about the formation of an ice layer ("false bottom") at the interface between the fresh and salty water (Hanson, 1965; Martin and Kauffman, 1974). Field studies of the overall effects of this freshwater on the mass balance of the ice pack have not been made. Enhanced melting in ponded areas contributes another .3-.4 m to the total mass loss, but part of this is returned when the ponds remaining in the fall refreeze. Strong shortwave absorption in the melt ponds can result in brine volumes in the underlying ice of as high as 50%, compared to maximum values of about 16% in unponded ice. Typical brine volumes in bare multi-year ice lie in the 8-12% range (Schwarzacher, 1959). Melting of young ice in leads and lateral ablation or erosion from floe edges increase the open water fraction from about 2% in early June to 10-15% in August. Substantial variations in the amount of open water can also be brought about as a result of general divergence or convergence of the region. Once freezeup begins in late August or early September, it takes 3-4 months to cool the ice enough to release all the heat stored in the melt ponds and brine pockets during the summer. As we will show in Section 5.4, freezing of the open water formed during the melt season has an important effect on the overall heat and mass balance of the region.

With the exception of the Weddell Sea, most of the ice in the Southern Ocean is seasonal (Fig. 15.2), however, the decay pattern of this ice appears to be strikingly different from that in the Arctic. Katabatic winds, higher latitudes, and lower oceanic heat fluxes cause melt rates near the continental boundaries to be small compared to those near the free boundary of the Antarctic ice pack. Moreover, the melt ponds which are so characteristic of ice decay in the Arctic have rarely been observed in the Antarctic, and surface melting appears to be minimal. The reason is related to differences in the meteorological variables that control the surface heat balance. Surface melting in the Arctic often takes place with air temperatures significantly below freezing. In the Antarctic, Andreas and Ackley (1982) argue that lower relative humidities and stronger winds enhance turbulent heat losses from the ice and allow surface melting to occur only when air temperatures are above 0 °C, a condition rarely encountered over the ice cover in the Southern Ocean. Thus, decay and retreat of the Antarctic ice pack are controlled primarily by the rate of heat transfer at the ice-water interface. At this point, there are few details regarding this process.

5.3.2 Young Ice Growth

Empirical Relations. Field studies have shown that the thickness (H) of young sea ice is closely related to the cumulative number of freezing-degree days θ, where

$$\theta \equiv \int_0^t (T_f - T_a)dt \quad ,$$

T_f is the freezing point of the water and T_a is the air temperature at screen height (usually 2 m). Typical results are those obtained by Anderson (1961) near Thule, Greenland (Figure 13). As the ice becomes thicker, the relationship between θ and H becomes weaker, e.g., Anderson observed ice thicknesses varying between 90 and 140 cm for θ = 3000°C day. Although Anderson's data cannot be

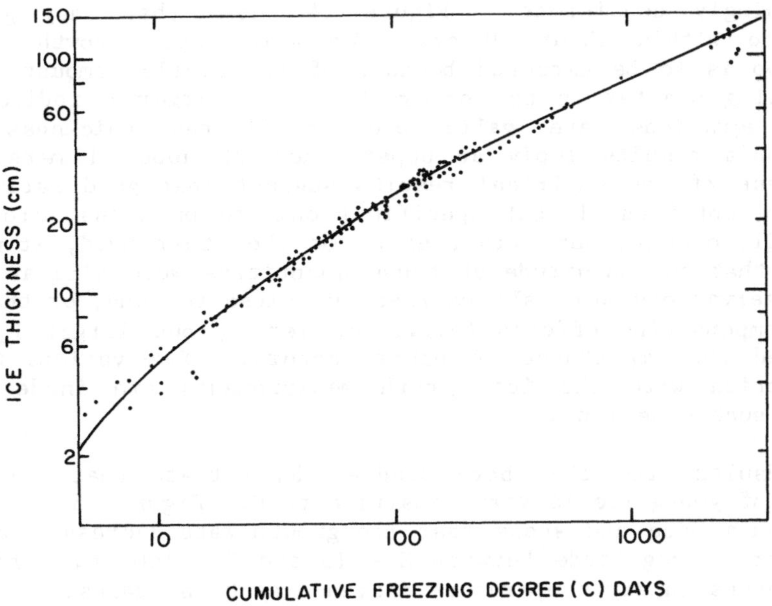

Fig. 13. Relationship between the thickness of young sea ice and the cumulative number of freezing-degree days. (After Anderson, 1961.)

fitted with a simple power law across its entire range, he found that a good approximation was obtained with

$$H^2 + 5.1H = 6.7\theta \tag{5.16}$$

where H is in centimeters (10<H<80 cm) and θ has units of °C day. Other relationships have been derived by Zubov (1943) from observations near Cape Schmidt

$$H^2 + 50H = 8\theta \tag{5.17}$$

and by Lebedev (1938) using 24 stations years of observations from various locations in the Soviet Arctic

$$H = 1.33\theta^{.58} \quad . \tag{5.18}$$

More complex, semi-empirical methods have also been developed by Lee and Simpson (1954) and Bilello (1961).

According to Anderson, eq. (5.16) describes ice with little or no snow cover, while Lebedev states that eq. (5.18) describes growth under "average" snow conditions. Despite the differences in location, climate, and snowfall, eqs. (5.16)-(5.18) are surprisingly consistent. Figure 14 shows that they generally agree to within about 10 cm. The more rapid growth given by Anderson is to be expected because of the smaller amount of snow. Increasing scatter in the observations at larger H indicate that these equations are valid over a limited thickness range; Anderson's results imply an upper bound of about 1 meter. The closeness of the empirical results suggest that good estimates of H can be obtained without specific knowledge of winds, cloudiness, snowfall, oceanic heat flux, etc. On the other hand, it might be argued that the magnitude of these quantities were similar because the observations were all carried out close to land, or that there were compensating effects (e.g., smaller h_s but larger F_W) which obscured the importance of other factors. Observations taken in conjunction with the ice growth measurements are inadequate to settle such questions.

Results from the above studies demonstrate that the rate of growth of young ice is very sensitive to H. Figure 15 is based on Anderson's data and shows that the growth rate decreases by almost an order of magnitude between H = 10 and H = 100 cm. Effects of differences in air temperature decrease as H increases.

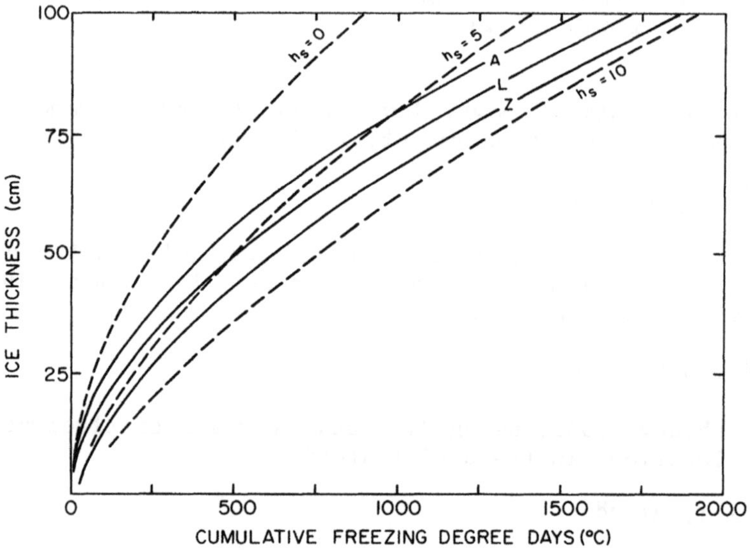

Fig. 14. The solid curves show ice growth predictions by the empirical formulas of Anderson (A), Lebedev (L), and Zubov (Z). Dashed curves show theoretical relationships between ice thickness and degree-days obtained from eq. (5.23) for different thicknesses (h_s) of snow.

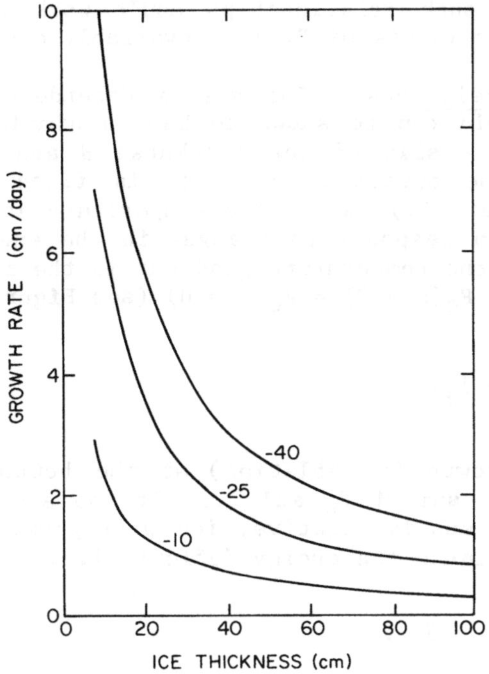

Fig. 15. Dependence of growth rates in young sea ice on thickness for air temperatures of -10, -25, and -40°C.

It does not seem likely that analogous attempts to relate the rate at which sea ice decreases in thickness during the summer would be successful. Variations in cloudiness, area of open water, oceanic heat flux, surface albedo and the like would seem to be equally important. However, Bilello (1961) found a good fit to near-shore data with the equation

$$\Delta H = 0.55\theta'$$

where ΔH = the total decrease in ice thickness and θ' = the accumulated degree-days above T_f. A least squares computation yielded a correlation coefficient of 0.93 and a standard deviation of 16.4 cm. Bilello also mentions another relationship developed by Karelin for the Soviet Arctic

$$\Delta H = 0.51(\theta'' - 32)$$

where θ'' = the accumulated degree days above -5°C. The reason for the correlations found by Bilello and Karelin must be due to frequent incursions of warm air from nearby land masses. It is doubtful whether such relationships could work away from coastal areas where summer values of T_a are invariably close to 0°C.

Analytical Relations. The strong dependence of H on θ observed in the field can be shown to have a simple physical basis. Let us consider a slab of ice thickness H and conductivity k_i. Temperature at the bottom of the ice is fixed at the freezing point of the water (T_f), while the temperature at the surface (T_o) is free to vary in response to changes in the energy balance. If the ice is thin, the temperature gradient in the ice will be essentially linear and $F_c(z = 0) = F_c(z = H)$ (see Figure 16), hence

$$F_c = \frac{k_i}{H}(T_o - T_f) \ . \tag{5.19}$$

The amount of growth (or ablation) at the bottom of the ice is determined by the sum of F_w and F_c. If the sum is positive, ice will melt; if the sun is negative, ice will grow, releasing enough latent heat to balance the energy deficit, i.e.

$$-\rho_i L\frac{dH}{dt} = F_c + F_w \ . \tag{5.20}$$

If we take the simplest possible case where $F_w = 0$ and $T_o = T_a$, eq. (5.20) becomes

$$\rho_i L\frac{dH}{dt} = \frac{k_i}{H}(T_f - T_a) \ .$$

Assuming that H(t=0)=0, integration of this equation gives

$$H^2 = \frac{2k_i}{\rho_i L}\int_o^t (T_f - T_a)dt = \frac{2k_i}{\rho_i L}\,\theta \ .$$

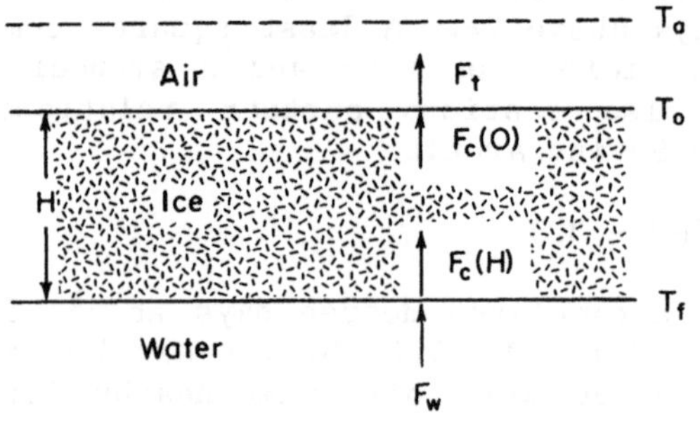

Fig. 16. Idealized picture of the ice slab and heat fluxes.

This equation is unrealistic since T_o can be much warmer than T_a when H is small. To obtain the dependence on H noted by Anderson and Zubov, we must relax the assumption that $T_o = T_a$. Instead we will assume that the net rate of heat exchange between the ice and atmosphere (F_t) is proportional to the difference between T_o and T_a, i.e.,

$$F_t = C_t(T_a - T_o)$$

where C_t is some sort of an average transfer coefficient which describes both sensible and latent heat exchange. Heat gain at the upper surface must equal heat loss, so $F_c = F_t$ and

$$\frac{k_i}{H}(T_o - T_f) = C_t(T_a - T_o) \quad .$$

Solving this equation for T_o gives

$$T_o = \frac{k_i T_f + C_t H T_a}{k_i + C_t H} \tag{5.21}$$

Substituting eq. (5.21) into eq. (5.19) gives

$$F_c = \frac{k_i C_t}{k_i + C_t H}(T_f - T_a) \quad .$$

Again assuming that $F_w \ll F_c$ and H(t=0)=0, integration of eq. (5.20) now yields

$$H^2 + \frac{2k_i}{C_t} H = \frac{2k_i}{\rho_i L} \theta \ . \tag{5.22}$$

How well does this equation fit the observations? Letting C_t = 50 cal cm^{-1} day^{-1} $^oC^{-1}$ (slightly higher than the nominal value for sensible heat flux alone), k_i = 419.9 cal cm^{-1} day^{-1} $^oC^{-1}$, and $\rho_i L$ = 65 cal cm^{-3} (corresponding to an average brine volume of 10%, eq. (5.22) becomes

$$H^2 + 16.8H = 12.9\theta$$

which is graphed in Figure 14 as the h_s = 0 curve. Equation (5.22) clearly predicts much more rapid growth than observed in nature. One possible reason for this is neglect of the snow cover. If we include a snow layer of depth h_s and specify that the conductive flux in the snow must be equal to that in the ice, it is straightforward to show that

$$H^2 + \left[\frac{2k_i}{k_s} h_s + \frac{2k_i}{C_t}\right] H = \frac{2k_i}{\rho_i L} \theta \tag{5.23}$$

or

$$H^2 + (13.1\ h_s + 16.8)H = 12.9\theta\ .$$

When $h_s = 0$, the linear term in eq. (5.23) rapidly decreases in importance once H exceeds about 20 cm, so that H becomes increasingly a function of $\theta^{1/2}$. A small amount of snow, however, magnifies the importance of the linear term. For example, when $h_s = 5$ cm, the linear term dominates the expression until H reaches about 75 cm. A few cm of snow also acts to decrease the importance of temperature differences between the air and the surface. Equation (5.23) is graphed in Figure 14 for snow depths of 5 and 10 cm. At first glance it appears that choosing h_s in the 5-10 cm range produces reasonable agreement with the empirical results. This apparent agreement is somewhat deceptive, however, because the good fit at larger H can only be obtained by seriously underestimating growth rates in the thin ice. The problem is with the assumption of a constant h_s. By imposing a layer of snow on the very thin ice, we cause an unrealistic decrease in its rate of growth. It is possible to take into account changes in h_s by stepwise integration of eq. (5.20). This was the approach taken by Nakawo and Sinha (1981) in analyzing ice growth data from Pond Inlet. They found, however, that growth predictions made by including observed variations in h_s were less accurate than those obtained assuming a large constant snow depth. They attributed the apparent failure of the more rigorous method to changes in the physical properties of the growing ice and to uncertainties in the snowfall when the ice was very thin. A more likely explanation is that heat input from the ocean was responsible for the slower than expected growth. Calculations by Lee and Simpson (1954) indicate that at least 40-80 MJ m^{-2} are released from the upper ocean in the SSIZ during the fall and winter, following the initial formation of the ice cover. Allison (1981) infers F_w values beneath seasonal ice near Mawson, Antarctica of 40-120 MJ m^{-2} mo^{-1} (15-50 W m^{-2}) during the first month of growth, dropping to 20-40 MJ m^{-2} mo^{-1} (7-15 W m^{-2}) in succeeding months. Analysis of the Anderson data indicate that an eventual snow accumulation of 15-20 cm is needed to explain differences between observations and theoretical predictions. Since there was actually very little snow, the most likely explanation for the slower growth is again a substantial F_w. This implies that, to be generally applicable, our analytical ice growth formula needs a term describing the effect of F_w on H. Although eq. (5.20) can be solved in some special cases for $F_w \neq 0$, more general solutions require numerical treatment.

Numerical Modeling. Empirical relationships such as those given by Anderson and Zubov contain implicit assumptions regarding $F_w(t)$ and $h_s(t)$. As we have seen above, variations in these quantities can have a large effect on the response of H to changes in θ. Application of such relationships to areas with different ocean heat flux or snowfall patterns (e.g., the Central Arctic and Southern Ocean) is thus questionable. Suitable forecasting equations for these areas should be developed independently, either directly from ice growth measurements or indirectly from heat balance data and theoretical models. It is the latter approach we shall discuss below.

Let us consider a slab of young ice of thickness H, overlain by a layer of snow of thickness h_s. We shall assume that the surface is in thermal equilibrium, i.e., heat gain is exactly balanced by heat loss. We can then write a heat balance for the upper boundary

$$(1 - \alpha)F_r - I_o + F_L - F\uparrow + F_s + F_e + F_c + F_m = 0 \qquad (5.24)$$

where F_m is the heat loss due to melting of ice or snow, and

$$F_m = [\rho L \frac{d(H + h_s)}{dt}]_o \,. \qquad (5.25)$$

When the surface temperature is below freezing, $F_m = 0$. If T_o is at the melting point, any surplus of energy flux toward the surface will be balanced by melting and a change in $H + h_s$. Since we are considering the case of young ice, we will assume that temperature gradients in the ice and snow are linear, so that

$$F_c = \frac{k_i k_s}{k_i h_s + k_s H}(T_f - T_o) = \gamma(T_f - T_o) \qquad (5.26)$$

where γ is the thermal conductance of the combined ice/snow slab. Substituting eqs. (5.1), (5.10), (5.12), (5.14), (5.25) and (5.26) into eq. (5.24), the upper boundary condition can be written

$$(1 - \alpha)(1 - i_o)F_r + F_L - \varepsilon_L \sigma T_o^4 + K_s(T_a - T_o)$$

$$+ K_e(re_{sa} - e_{so}) + \gamma(T_f - T_o) + [\rho L \frac{d(H + h_s)}{dt}]_o = 0 \qquad (5.27)$$

where $K_s = \rho_a C_p C_s u$ and $K_e = .622\, \rho_a L C_e u/P$. Because the albedo of cold, thin ice increases rapidly with thickness, problems involving spring ice formation may need to take this change into account. An approximate relationship between α and H can be derived from the observations of Weller (1972).

$$\alpha(H) = .44\ H^{.28} + .08$$

where H is in meters and $0<H<.8$ m.

The lower boundary condition is obtained from eqs. (5.20) and (5.26)

$$-\rho_i L \left(\frac{dH}{dt}\right)_H = \frac{k_i k_s}{k_i h_s + k_s H}(T_f - T_o) + F_w \quad . \tag{5.28}$$

Equations (5.27) and (5.28) form the young ice model. The rate of growth for any given values of H, h_s, and F_w can be calculated from eq. (5.28) if T_o is known. T_o can be obtained from eq. (5.27) by specifying F_r, F_L, T_a, h_s, u, and r. Growth of an ice cover from arbitrary initial conditions can be calculated in this way for any time dependent, thermal forcing.

The advantage of this approach over empirical formulas is generality. Providing that sufficient information is available regarding the thermal forcing, eqs. (5.27) and (5.28) can be used not only to describe heat exchange in areas such as the Central Arctic, Southern Ocean, and SSIZ, but also to calculate ice decay or growth when the radiation balance is not close to zero. It can be argued that there are too many unknown variables for this approach to be useful in a routine forecasting mode. Data on F_L, for example, will almost never be available. In practice, however, we can reduce the number of variables significantly. By using the parameterizations for F_r and F_L discussed in Section 5.2 (eqs. 5.2, 5.5 and 5.6), incident radiation at a particular location can be expressed in terms of cloudiness and air temperature. During most of the year F_e is small relative to F_s and r can be taken as 0.9 or 0.95 with little error. Wind speed tends to be relatively constant over much of the Arctic Basin and an average value of 5 m sec^{-1} is frequently used in calculating the turbulent fluxes. The remaining quantities which must be specified are T_a, C, h_s, and F_w. The first three can be measured directly, estimated from climatology, or, in the case of cloudiness, determined from satellite data. F_w cannot be measured directly and must instead be inferred from salinity and temperature changes in the upper ocean or from its effect on ice thickness.

We have described here a practical method of estimating young ice growth which includes an explicit dependence on those factors which are most likely to undergo large changes from one region to another. Known differences in climatic or oceanographic conditions can therefore be taken directly into account. The model can be used to generate simple predictive equations like those of

Zubov and Anderson and to learn how uncertainties in the magnitudes of the input parameters affect the predicted growth of the ice. Probably the greatest strength of the theoretical approach is that it provides a framework for utilizing whatever information is available.

5.3.3 Multiyear and Thick First Year Ice

A fundamental step in the development of the young ice model was the assumption of a linear temperature gradient in the ice which caused ice thickness to respond immediately to any changes in thermal forcing at the surface. Although reasonable in thin ice, this assumption is inadequate to describe heat transport through multiyear and thick first year ice where changes in surface temperature do not necessarily result in immediate changes in ice growth. The growth of thicker ice depends more on its thermal history than on the surface heat balance at any particular instant of time. In the early summer, for example, accretion can be observed at the underside of the ice while melting is taking place at the surface. Similarly, in October when young ice is growing vigorously in leads, thick ice is generally decreasing in thickness due to bottom melting. Figure 17 shows the annual temperature field observed in perennial ice in the Central Arctic. It is evident from this that the fall cooling which begins in September is not felt at the bottom of the ice until mid-December. Changes in surface temperature like the January warming are largely damped out by the time they reach the bottom. Even during the

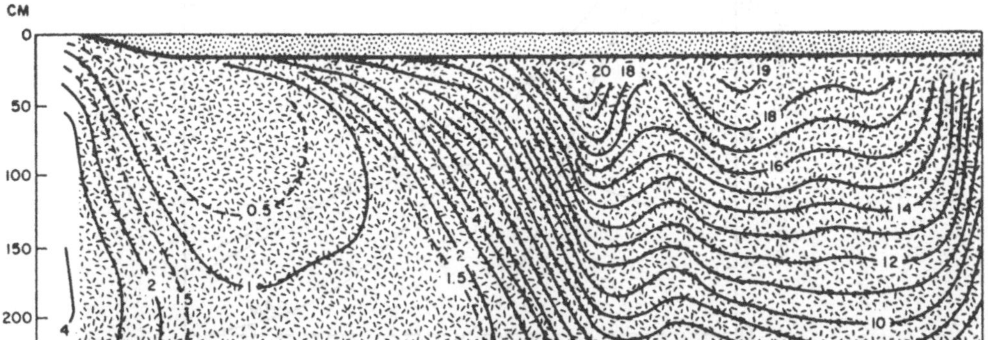

Fig. 17. Temperature and thickness variations observed in perennial ice in the Central Arctic. (After Untersteiner, 1961.)

middle of the winter there appear to be significant departures from linearity within the ice.

Figure 18 shows theoretical predictions of how growth rates (f) in thicker arctic ice vary with season and thickness. The curves demonstrate that differences in thermal mass strongly affect how the ice responds to changes at the upper surface. In November, over 2 months after freeze-up, ice thicker than about 3 m continues to ablate because the fall cooling has not yet begun to affect the lowest part of the ice. Above 4 m there is a slight decrease in ablation rate with increasing thickness, suggesting that the effects of summer warming have not yet reached a maximum. By January it is evident that the fall cooling has penetrated all but the thickest ice. During April there is typically a 5°C increase in T_a which is reflected in decreased growth rates in ice thinner than 3 m. Growth rates in thicker ice, however, are unaffected by this warming and increase in response to continued cooling near the bottom. A 10-15°C increase in T_a during May causes a sharp drop in f for the thinner ice, while f(H>5 m) again increases. The magnitude of the spring warming effect decreases with increasing thickness and explains the predicted increase in f between 1-3 m. Clearly the assumption of a linear temperature profile is invalid for thicker ice.

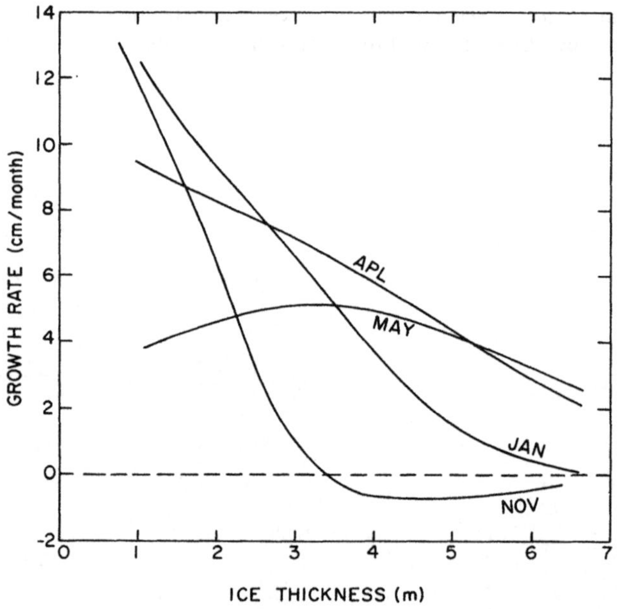

Fig. 18. Growth rates in multiyear arctic sea ice as a function of thickness and season according to the theoretical model of Maykut and Untersteiner (1971).

Brine pockets are an important factor contributing to non-linearities in the temperature profile of thick ice. Cooling of the ice causes some of the brine to freeze, releasing latent heat and slowing the overall rate of cooling; increasing temperatures cause melting of ice surrounding the brine pockets and a decrease in the rate of warming. Brine pockets thus act as thermal buffers which retard temperature changes in either direction. Because of these temperature dependent changes in brine volume, the specific heat of sea ice (c_i) must include not only the heat required to alter the temperature of the ice and brine, but also the latent heat associated with melting and freezing in the brine pockets. Ono (1967) derived a theoretical expression for c_i applicable to temperatures above -8 °C,

$$c_i = c_o + aT_i + \frac{b\,S_i}{T_i^2}$$

where T_i is in °C, $a = 7.53$ J kg^{-1} °C^{-2}, $b = 0.018$ MJ °C kg^{-1}, and $c_o = 2113$ J kg^{-1} °C^{-1} is the specific heat of pure ice. A simpler form which gives similar results is the empirical formula of Untersteiner (1961): $c_i = c_o + \zeta S_i/T_i^2$, where $\zeta = 0.0172$ MJ °C kg^{-1}. Tables and analytical expressions are also available for colder ice (Anderson, 1958; Schwedtfeger, 1963). As with k_i, the largest changes in c_i occur at temperatures above -4 to -6°C, but the magnitude of these changes are much larger. Under natural conditions, c_i varies by a factor of 10-20 as the ice warms toward the melting point. Both k_i and c_i are dependent on depth because of the vertical variations in ice salinity.

The dramatic increase in $c_i(T_i,S_i)$ at temperatures near 0°C means that mass changes within the brine pockets are consuming an increasing share of the available energy, making it increasingly difficult to change T_i. At first glance it would appear that the equations for c_i and k_i predict infinite values as $T_i \rightarrow 0$°C. This is not the case, however, because T_i cannot approach 0°C unless $S_i \rightarrow 0$. Whereas pure ice has a well-defined temperature (0°C) at which phase transitions occur, sea ice does not. Any change in T_i produces internal phase changes. It can be shown that for ice of a given salinity there is a temperature (T_m) above which the ice cannot exist without completely melting away. According to Ono (1967), $T_m \simeq -0.05411\, S_i$. This is consistent with the claim that the temperature of sea ice can reach 0 °C only if $S_i = 0$.

General Thermodynamic Model. To avoid the limitations imposed on the treatment of thin ice, Maykut and Untersteiner developed a detailed sea ice model which takes into account effects of salinity and internal heating on ice growth. To describe heat transport and temperatures within the ice, they used a modified form of the heat conduction equation

$$\rho_i c_i \frac{\partial T_i}{\partial t} = \frac{\partial}{\partial z}\left(k_i \frac{\partial T_i}{\partial z}\right) + I_o \kappa_i e^{-\kappa_i z} \tag{5.29}$$

where the last term describes the depth dependent internal heating which occurs in the ice as a result of absorption of penetrating shortwave radiation (see Section 5.2.1). A similar equation was used to describe temperatures in the snow

$$\rho_s c_s \frac{\partial T_s}{\partial t} = k_s \frac{\partial^2 T_s}{\partial z^2} \tag{5.30}$$

where k_s was assumed to be constant with depth and the I_o term was taken to be negligible because light attenuation in the snow is large and most of the solar energy is absorbed in the upper 0.1 m. This means that I_o in the ice is zero when $h_s>0$.

Heat conduction at the ice-snow interface was assumed to be equal in both media

$$k_i \left(\frac{\partial T_i}{\partial z}\right)_{h_s} = k_s \left(\frac{\partial T_s}{\partial z}\right)_{h_s} .$$

Equations (5.24) and (5.20) were used for the boundary conditions at the top and bottom of the ice. An important difference between this model and the young ice model described previously is that the conductive heat fluxes at the boundaries depend on the local temperature gradients, hence are generally different at the top and bottom of the slab. A schematic illustration of this model is shown in Figure 19. When the model was driven with estimates of incident energy fluxes and snowfall in the Central Arctic, it provided temperatures and mass changes in good agreement with observations (Figure 20).

A serious drawback to this approach is that the solution to eqs. (5.29) and (5.30) must be obtained by finite difference techniques which involve considerable computer time if good temperature resolution is desired. This makes it impractical to use the full model in many applications which require simultaneous predictions at numerous locations or thicknesses. In an effort to develop a simpler version suitable for numerical climate experiments, Semtner (1976) devised a streamlined numerical method for solving the above equations which required an order of magnitude fewer grid points. He showed that this method produced equilibrium thicknesses which agreed with results from the more elaborate procedures of Maykut and Untersteiner to within about 25 cm over a wide range of environmental conditions. As part of a study of transient

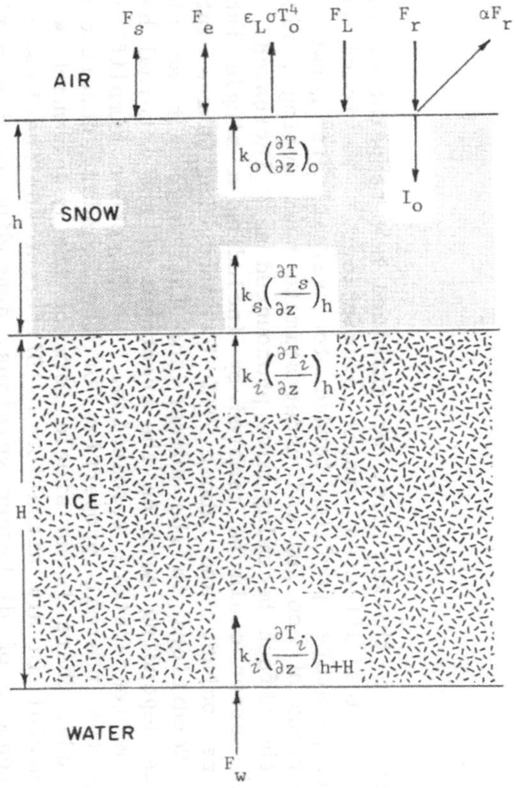

Balance of fluxes

$$(1-\alpha)F_r - I_o + F_L - \epsilon_L \sigma T_o^4 + F_s + F_e + k_o\left(\frac{\partial T}{\partial z}\right)_o + \left[\rho L \frac{d}{dt}(h+H)\right]_o = 0$$

$$(\rho c)_s \frac{\partial T_s}{\partial t} = k_s \frac{\partial^2 T_s}{\partial z^2} \quad \text{Heat conduction}$$

$$k_s\left(\frac{\partial T_s}{\partial z}\right)_h = k_i\left(\frac{\partial T_i}{\partial z}\right)_h \quad \text{Balance of fluxes}$$

$$(\rho c)_i \frac{\partial T_i}{\partial t} = k_i \frac{\partial^2 T_i}{\partial z^2} + \kappa_i I_o \exp(-\kappa_i z)$$

Heat conduction with internal heat source

$$k_i\left(\frac{\partial T_i}{\partial z}\right)_{h+H} - F_w = \left[\rho L \frac{d}{dt}(h+H)\right]_{h+H} \quad \text{Balance of fluxes}$$

$$\text{where} \quad (\rho c)_i = \rho c_o + \frac{\gamma S(z)}{(T_i - 273)^2}, \quad k_i = k_o + \frac{\beta S(z)}{T_i - 273}$$

Fig. 19. Schematic illustration of the sea ice model developed by Maykut and Untersteiner (1971).

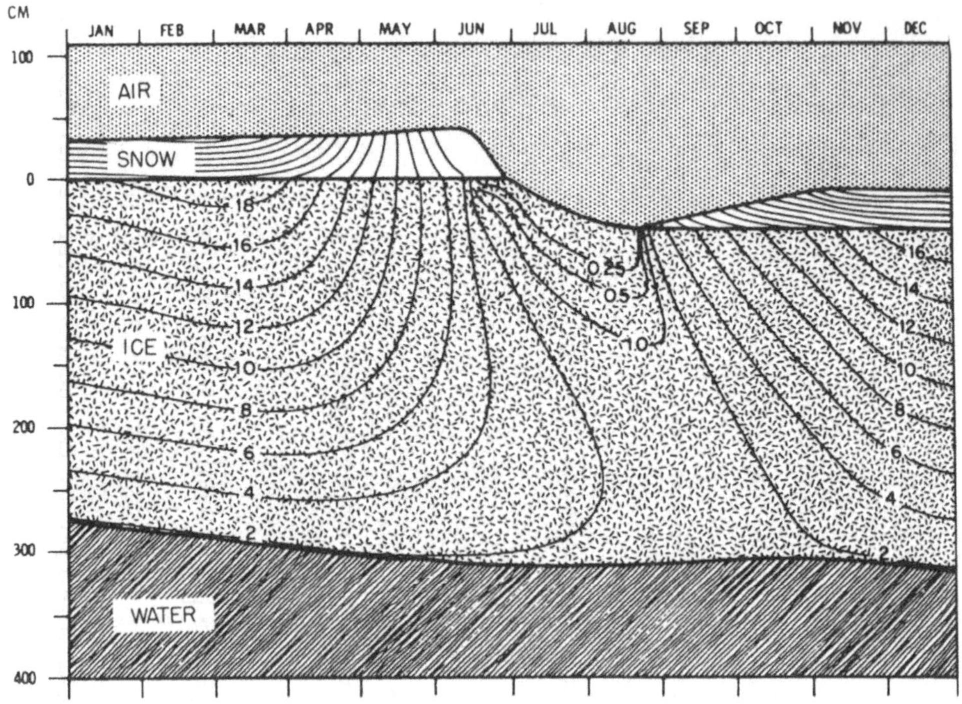

Fig. 20. Predicted values of equilibrium temperature and thickness in the Central Arctic (after Maykut and Untersteiner, 1971).

growth rates in thicker ice, we examined how temperature resolution in the ice affected winter predictions. We found that a linear temperature profile tended to overestimate growth rates during periods of cooling and underestimate them during periods of warming, the effect becoming more pronounced as H increased. About 75% of the error could be removed by including a single internal temperature point, and about 95% by including three interior points. Conditions in the summer and fall were not so simple, with rates depending more on the treatment of the internal heating than on the temperature resolution. While other simplifications involving the thermal properties or the boundary conditions can also be made, there is no single form of the model which will be equally useful in all situations. Different applications require the development of different versions, depending on the variables of interest, the desired temporal and spatial resolution, and the data available.

5.3.4 Response of Sea Ice to Changes in Thermal Forcing

Because of its probable role in climate, there has been a great deal of interest in the long term response of the ice to changes in thermal forcing and in techniques whereby ice extent or thickness might be artificially modified. The complete ice model described in the preceding section provides a convenient way to quantitatively evaluate the relative importance of various climatic factors and the efficiency of different modification techniques. The effects of a particular change can be studied by integrating the model through many annual cycles until an equilibrium temperature and thickness pattern is obtained. Equilibrium is reached when the net annual growth at the bottom of the ice equals the net annual ice ablation at the surface. An increase in the amount of surface ablation, for example, would force the ice to thin until the larger bottom accretion produced by the steeper temperature gradient exactly balances the increased mass loss at the surface; a decrease in surface ablation would provoke the opposite response. The time required for the ice to reach equilibrium depends on the final equilibrium thickness (H_e), being only a few years when H_e is small (1 m) and tens of years when H_e is large (> 6 m).

The turbulent transfer of heat from the ocean to the ice has been subject to numerous estimates and speculations. Wide variations in F_w are an integral part of some ice age theories, and numerous schemes for artificially influencing the extent of the ice in the Arctic depend on changing F_w. The feasibility of such proposals is difficult to establish without quantitative knowledge of the sensitivity of the ice to F_w. Maykut and Untersteiner (1971) carried out a series of simulations in which F_w was allowed to vary between 0 and 8 W m^{-2} (Figure 21). Under present surface conditions, perennial ice can only grow to a maximum thickness of 5-1/2 to 6 meters. The ice cover melts completely away when F_w exceeds about 7 W m^{-2}. The results shown in Figure 21 were obtained assuming that F_w was constant throughout the year, but another series of unpublished calculations indicate that temporal variations have little net effect on the ice. For example, when all the oceanic heat flux was added to the ice during the summer and fall, H_e was somewhat thinner during the fall than in the constant F_w case, but showed an overall thickness increase by the following summer because of enhanced winter growth; H_e was essentially unchanged. In agreement with other independent estimates, F_w = 2 W m^{-2} provides the best agreement between seasonal variations in H_e and field observations in the Central Arctic. As we shall discuss in more detail later, it now appears that F_w in areas of perennial ice is more closely related to the ice concentration than to the amount of heat carried into the Arctic Basin from the Atlantic Ocean. If this is the case, then artificial modification methods and ice age theories which depend on variations in this heat exchange should be reexamined.

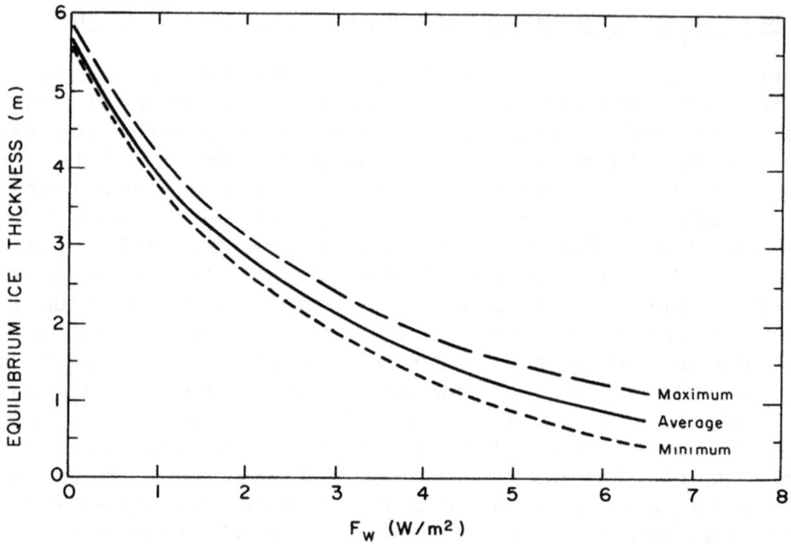

Fig. 21. Equilibrium thickness of arctic sea ice as a function of of oceanic heat flux. Average annual thickness, as well as the absolute annual maximum and minimum, are shown (after Maykut and Untersteiner, 1971).

Because of the magnitude of the incoming shortwave radiation, a small reduction in the surface albedo would provide a large increase in available energy at the surface. It has been frequently suggested that the ice pack might be removed by lowering its albedo through the dispersal of some dark substance like coal dust on its surface. Although some small scale albedo modification experiments have been performed on sea ice (Arnold, 1961), it is not known how feasible such efforts would prove to be on a large scale. Numerical experiments indicate that a 10% reduction in albedo (from 0.64 to 0.54) during the melt season would reduce H_e to about 1 m, while a 20% decrease would result in the complete melting of the ice cover after about three years. Thus, as expected, we see that relatively small changes in α have a large effect on H_e. However, due to the presence of melt ponds, area-averaged summer albedos on perennial ice probably already lie in the .50-.55 range and we must ask why H_e appears to be closer to 3 m than to 1 m over much of the Arctic Basin. The answer is that melt ponds seem to have little net effect on the mass balance of perennial ice. Most of the increase in absorbed shortwave radiation goes into deepening of the ponds and increasing the brine volume in the underlying ice. Fall and winter cooling cause the ponds to refreeze, increasing the ice mass and releasing their stored heat to the atmosphere. This cycle of melting and refreezing in the ponded areas does not have a large effect on the surrounding ice, hence H_e remains about the same.

Modifying the surface of the ice also alters the amount of penetrating shortwave radiation. A series of calculations were therefore carried out to determine the sensitivity of the ice to i_o (Figure 22). The results showed that increasing the amount of internal heating increased H_e, primarily because of a decrease in the energy available for melting at the upper surface. In direct analogy with the melt ponds, shortwave energy absorbed and stored in the ice as latent heat had little impact on the mass balance over an annual cycle.

Winter temperatures in the polar regions appear to have increased by as much as 10-15°C from the ice age minimums. The direction of future changes is of course unknown, but there is concern regarding the possible effects of increasing CO_2 in the atmosphere. Theoretical studies (e.g., Manabe and Wetherald, 1980) suggest that the expected doubling in CO_2 concentrations during the next century could lead to temperature increases of 7-8°C in the Arctic. Relatively simple heat balance models (Budyko, 1974; Parkinson and Kellogg, 1979) predict that summer temperatures only 4-5°C warmer than present would result in an ice-free Arctic Ocean, at least during the summer. The ice cover thus appears to be extremely sensitive to T_a. While this statement is certainly true, one must be cautious about using predictions such as those quoted above. The rapid ice decay relies on the assumption that it is possible to maintain temperature gradients of several degrees next to the melting ice, thereby pumping large amounts of sensible and latent heat to the surface for months at a time. This is clearly unrealistic. What is needed to evaluate the consequences of atmospheric warming during the summer is a coupled atmospheric

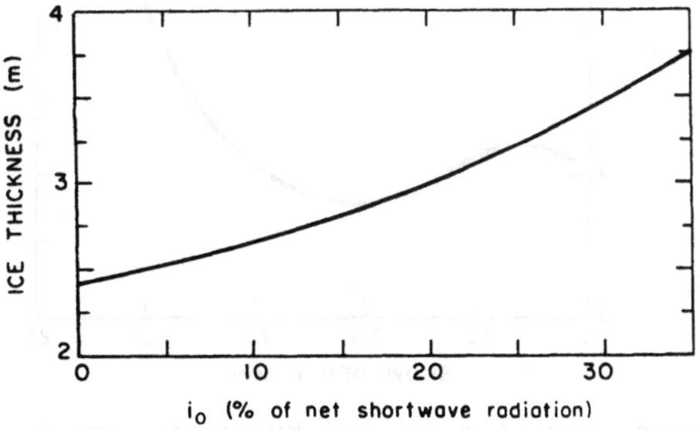

Fig. 22. Average equilibrium thickness of arctic sea ice as a function of i_o (after Maykut and Untersteiner, 1971).

model that will more realistically describe the downward transport of heat from the atmosphere to the ice.

Accompanying changes in temperature are likely to be changes in cloudiness, incident radiation, and precipitation. While we don't know exactly how large such changes might be, it is possible to get some idea of what could happen by simple parameter studies. Figure 23, for example, shows theoretical predictions of the relationship between X_e and h_s. Snow accumulations from 0 to twice the present value of 40 cm have a minimal effect on H_e. This is surprising because variations in h_s alter surface ablation and ice temperatures, suggesting the potential for substantial changes in H_e. The reason this doesn't happen is that the two major effects oppose one another. With less snow, colder ice temperatures promote bottom accretion, but at the same time the summer snow cover melts away sooner, decreasing the average albedo and extending the duration of ice melting; the result is an increase in both bottom accretion and surface ablation which tends to balance out over the range $0 < h_s < 80$ cm. Above 80 cm, decreasing summer ice ablation becomes the dominant factor and H_e increases rapidly with h_s.

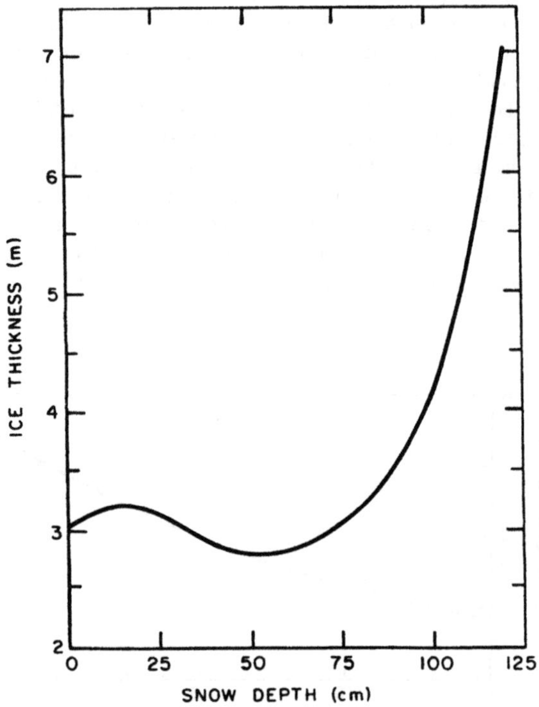

Fig. 23. Average equilibrium thickness of arctic sea ice as a function of maximum annual snow depth; the present day value is 40 cm. (After Maykut and Untersteiner, 1971).

For any climatic region there will be some critical value of snowfall (about 1.2 m in the central Arctic) which is just able to be removed during the melt season. If h_s exceeds this value, there will be a net annual accumulation of snow and the ice becomes a "floating glacier," growing from above and ablating from below. The combination of a small F_w and high snowfall can produce ice of almost any thickness and has been used to explain the origin of the very thick ice found in Nansen Sound and along the north coast of Ellesmere Island (Walker and Wadhams, 1979).

Although we have not carried out a systematic investigation of how variations in cloudiness, winter temperatures, or incident radiation affect the ice cover, a few miscellaneous calculations from the Central Arctic can be mentioned. In one case we looked at what could happen if there were an increase in the average cloudiness during the winter (October-April) from 0.55 to 0.85. This is roughly equivalent to increasing T_a by 5-6°C. There was a 10% increase in the incoming longwave radiation, but no change in the net longwave balance since $F\uparrow$ increased in response to a warmer T_o. The increase in T_o decreased temperature gradients in the ice forcing H_e to decrease in order to maintain large enough growth rates to match the surface ablation. The increased winter temperatures produced almost a 1 m decrease in H_e. In another case we looked at the ice response to a 10% increase in F_r, corresponding to a decrease in the summer value of $\bar{C}$ of about 0.1. Surface ablation increased by 50%, decreasing H_e by 1.2 m. Small variations in summer cloudiness can thus produce large changes in H_e. As with summer air temperatures, however, it is not clear whether changes in $\bar{C}$ of even 0.1 are reasonable in the presence of a melting ice cover. While these cases attempted to look at situations where variations in C affect only one of the incident radiation fluxes, changes in cloudiness are also likely to be accompanied by changes in snowfall, air temperature and turbulent heat exchange near the surface. The above results therefore do not describe the total response of the system to changes in C. The most important thing to keep in mind is that no matter how well we can individually model the ice, ocean or atmosphere, we will not be able to fully understand the relationship between ice and climate until the various interactions and feedback processes occurring within the system have been included in coupled, large scale simulations.

5.4 LARGE SCALE HEAT AND MASS BALANCE

In Section 5.2 we described the local heat balance observed over perennial ice. We know, however, that the ice pack is in almost continual motion and that strains within the ice cause it to fracture and expose the ocean. Temperature gradients over winter leads can reach 30 to 40°C over a vertical distance of only a few meters, causing large losses of sensible and latent heat from the ocean. The source of energy for these fluxes is latent heat generated by freezing of the water. As a result, ice production in winter leads is correspondingly large. Continual opening and closing of the ice, combined with continual ice formation and new growth, give rise to a horizontally nonuniform ice cover composed of ice of many different thicknesses. Although subject to similar thermal forcing, the heat and mass balance of each of these different thicknesses is distinct. For example, winter rates of ice growth, turbulent heat exchange with the atmosphere, and salt rejection to the ocean can be up to two orders of magnitude larger over a refreezing lead than over perennial ice. Although decreasing rapidly with increasing ice thickness, these rates can still be an order of magnitude larger over 50 cm ice than over 3 m ice. During the summer there is little thermal distinction between ice of different thicknesses, but leads admit large quantities of shortwave radiation to the upper ocean which ultimately affect the mass balance of the ice pack. Because current estimates of heat exchange and ice production in the Arctic are based largely on measurements made over perennial ice, they do not directly take into account contributions made by thin ice and open water. For some components such as the incident radiation this omission is of little consequence, but for quantities which are sensitive to ice thickness the result may be a serious distortion of the large scale picture. For example, if the ice pack were composed of 99% perennial ice and 1% refreezing leads, winter ice production in the small area covered by leads would be roughly equal to that in the area covered by perennial ice; total ice production for the region would be twice as large as that predicted on the basis of local measurements made only in the thicker ice.

5.4.1 Regional Averaging

Clearly there is a strong coupling between dynamic and thermodynamic processes in the ice. Large changes in the amount of thin ice and open water can occur on time scales of hours to days, producing corresponding changes in the regional heat and mass balance independent of any changes in the incident energy fluxes. Because they do not take into account contributions made by areas of thin ice and open water, data gathered over thick ice do not necessarily provide good estimates of the area-averaged heat and mass fluxes. In most cases such large scale fluxes cannot be measured directly. Instead it is necessary to determine, first, how each component of the heat and mass balance varies with ice thickness and, second, the area covered by ice of any given thickness. The contribution made by a particular thickness category is then found by multiplying the fractional area covered by that category times the magnitude of the flux over that thickness. Summing over all categories yields the large scale total.

To quantify thickness variations within a particular region, Thorndike et al. (1975) introduced a probability density function g(H), called the ice thickness distribution, where $\int_{H_1}^{H_2} g(H)dH$ is the the fractional area covered by ice in the thickness range $H_1 \leq H < H_2$ and g(H) is normalized such that $\int_0^{\infty} g(H)dH = 1$. If we let ϕ be some property which depends on H, then the average of that property within the region for which g(H) is defined is given by

$$\bar{\phi} = \int_0^{\infty} \phi(H)g(H)dH \quad . \tag{5.31}$$

We shall refer to $\bar{\phi}$ as the regional or large scale average. In practice we usually consider discrete thickness categories. If g(H) is partitioned into n categories, eq. (5.31) can be approximated as

$$\bar{\phi} = \sum_{i=1}^{n} \Delta G_i \bar{\phi}_i \tag{5.32}$$

where $\bar{\phi}_i$ is the average value of ϕ in the ith category and $\Delta G_i = g_i \Delta H_i$. Since both ϕ and g are in general functions of time, we use the notation

$$<\phi> = \int_0^t \bar{\phi}(t)dt$$

to denote quantities integrated over time.

5.4.2 Energy Exchange Over Young Sea Ice

To obtain regional estimates of the heat and mass balance, we must first know how each of the components vary with thickness and time. A detailed study of this problem has been carried out for conditions in the Central Arctic (Maykut, 1978). It was assumed that the ice pack is predominantly composed of ice thicker than 1 m and that the young ice is found mostly in narrow refrozen leads tens to hundreds of meters in width. It was also assumed that air temperatures measured over the thick ice could be used to estimate the turbulent heat fluxes over adjacent areas of thin ice. While not strictly true, the air mass modification which does occur close to the lead surface does not appear to have a large effect on the turbulent heat exchange (Vowinckel and Orvig, 1972). Thus the results obtained below describe conditions in regions of perennial ice but will not apply to the initial formation and growth of ice in the SSIZ where the boundary layer evolves with the growing ice.

With radiation data from Marshunova (1961; Table 1) and air temperatures derived from the data of Doronin (1963), a thin ice model similar to eqs. (5.27) and (5.28) was used to infer ice growth and energy fluxes over various thicknesses of ice throughout the year. Figure 24 shows predicted seasonal variations in surface temperature for different values of H. When the air was cold, T_0 decreased rapidly with increasing H, reaching 50% of the perennial ice value at H = 10 cm and 90% of this value at H = 80 cm. The dependence of T_0 on H became more subdued in the late spring and early fall when the ice was near the melting point.

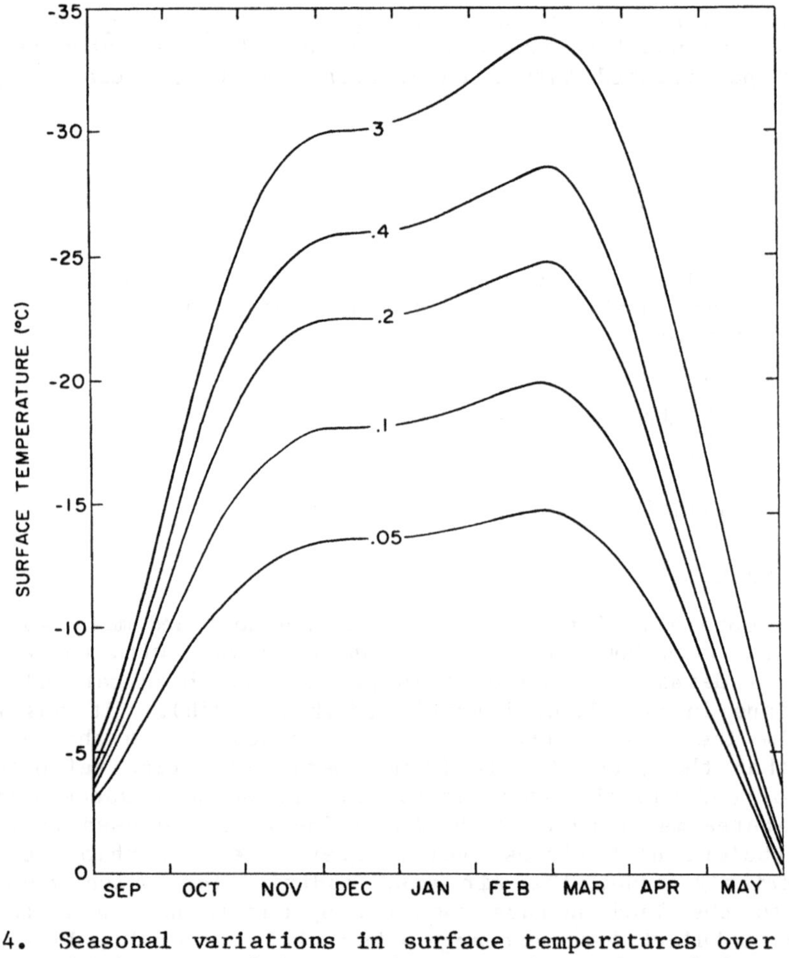

Fig. 24. Seasonal variations in surface temperatures over various thicknesses of ice in the Central Arctic. (After Maykut, 1978). Lines of constant ice thickness are given in meters.

Seasonal fluctuations in T_o increased as H increased. When the ice was thin, F_c dominated the heat balance so that temperatures in the young ice were only mildly sensitive to changes in atmospheric forcing. Growth rates (Figure 25) were obtained from T_o using eq. (5.28). F_W was assumed to be made up of two contributions: (i) a constant term (F_{WO} = 2 W m^{-2}) that includes shortwave energy stored in the upper ocean and any heat loss from the Atlantic layer, and (ii) a variable term that was taken to be 35% of the shortwave radiation transmitted through the ice to the ocean. This is roughly the amount of solar energy absorbed between the bottom of thin ice and the bottom of the surrounding thick ice.

Seasonal variations in the turbulent fluxes are shown in Figures 26 and 27, with a summary of all component presented in Table 6. The results show that F_s dominates the heat exchange

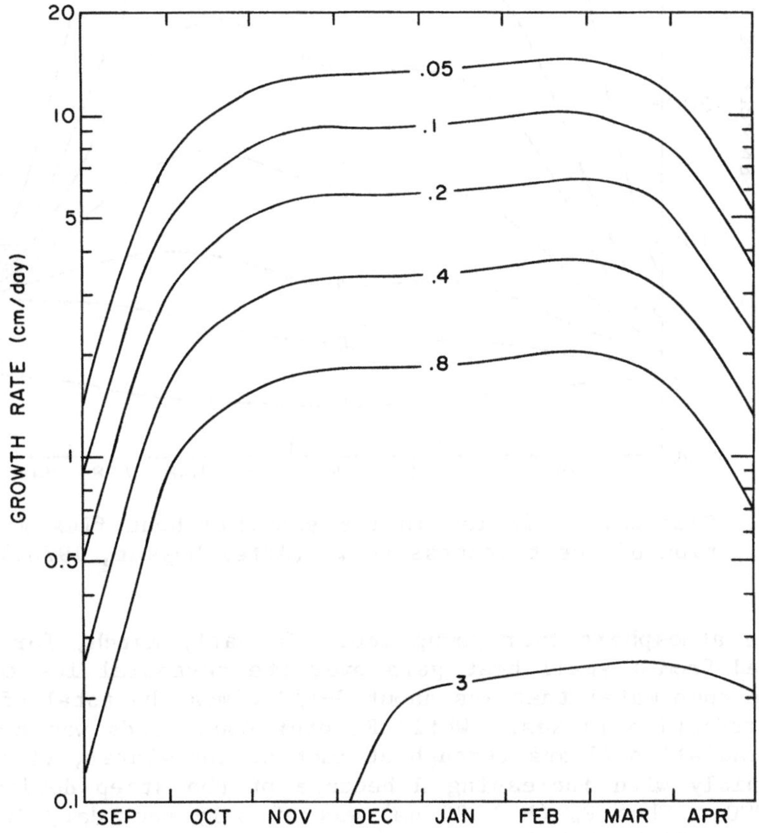

Fig. 25. Seasonal variations in growth rates as a function of ice thickness (cm). Note the logarithmic scale.

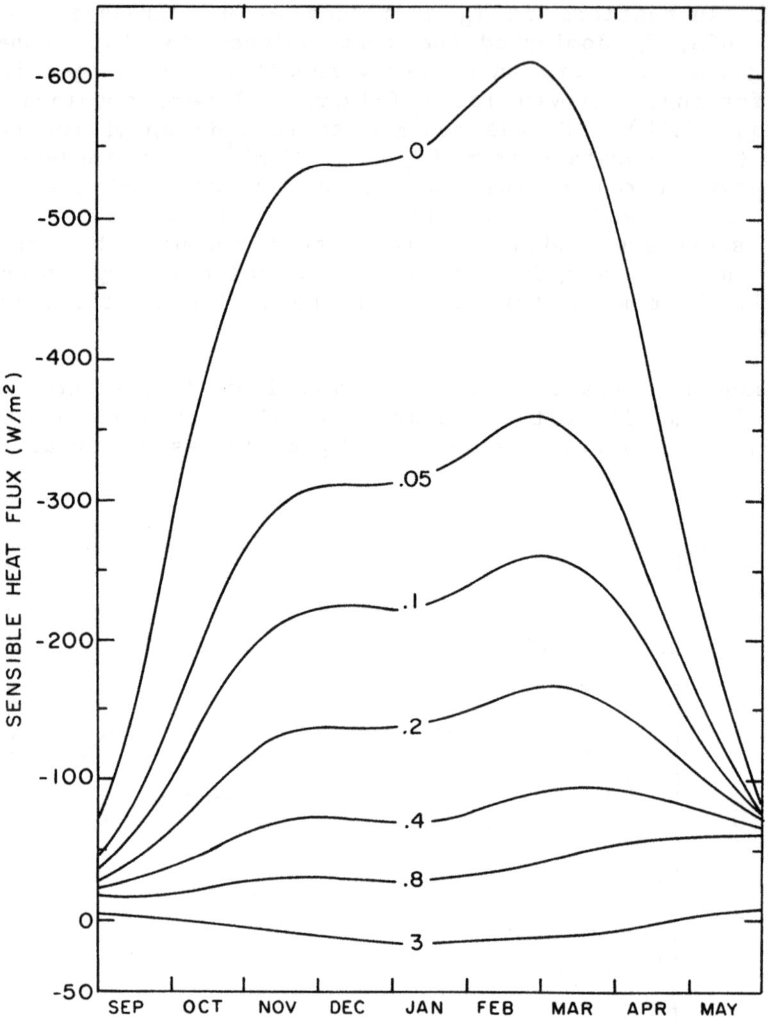

Fig. 26. Seasonal variation in the sensible heat flux as a function of ice thickness (m). (After Maykut, 1978.)

with the atmosphere over young ice. In early March, for example, F_s ranged from a small heat gain over the perennial ice to a loss over the open water that was about 3-1/2 times the total of the incident radiation fluxes. While F_e over open leads was comparable to the radiation fluxes throughout much of the winter, it decreased very rapidly with increasing H because of the steep decline in e_s below 0^oC. Thus F_e ($H > 20$ cm) was of only secondary importance in the overall heat balance.

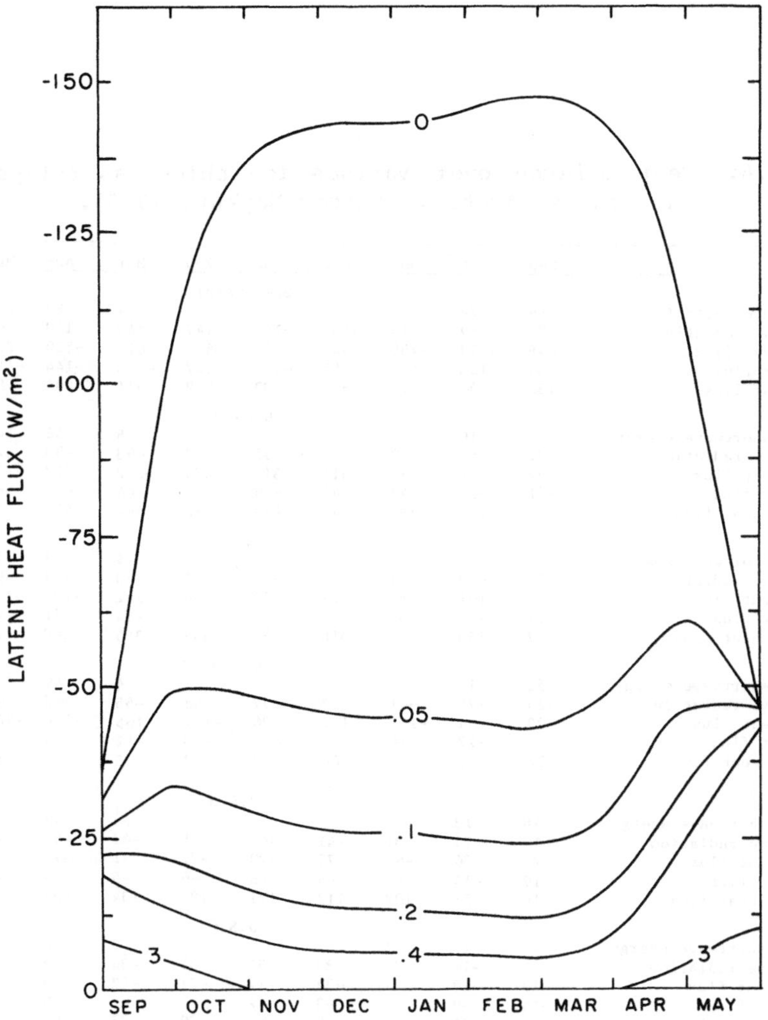

Fig. 27. Seasonal variation in the latent heat flux as a function of ice thickness (m). (After Maykut, 1978.)

The results make it clear that the net heat input to the atmosphere (F_n) over very thin ice is controlled primarily by the rate at which turbulent heat is transferred. Over thicker first year ice F_s is comparable to the net longwave radiation and both make an equal contribution to F_n. Over thick ice F_s and F_c must both supply heat to the surface to compensate for longwave losses. The calculations also demonstrate that the surface heat balance becomes insensitive to thickness once H exceeds 80-100 cm. As we saw in Figure 18, however, this is the range where the growth rates

Table 5.6. Heat balance over various ice thickness categories in the central Arctic. (After Maykut, 1978).

	Sep	Oct	Nov	Dec	Jan	Feb	Mar	Apr	May	June
					Open water					
Net shortwave radiation	89	24					7	83	209	281
Net longwave radiation	-28	-69	-114	-138	-142	-147	-149	-140	-98	-43
Sensible heat flux	-68	-259	-458	-540	-543	-575	-615	-520	-276	-33
Latent heat flux	-34	-108	-139	-145	-145	-147	-150	-144	-112	-31
Oceanic heat flux	130	436	711	823	830	869	914	804	486	107
					0.05 m					
Available shortwave energy*	60	16					5	56	141	189
Net longwave radiation	-22	-42	-70	-87	-91	-93	-93	-93	-76	-51
Sensible heat flux	-44	-142	-261	-311	-312	-334	-362	-313	-179	-74
Latent heat flux	-31	-50	-49	-47	-46	-45	-44	-51	-62	-47
Conductive heat flux	37	218	380	445	449	472	494	401	176	-17
					0.1 m					
Available shortwave energy	56	15					4	52	131	175
Net longwave radiation	-20	-33	-55	-69	-73	-75	-73	-77	-68	-50
Sensible heat flux	-36	-101	-186	-222	-222	-238	-262	-232	-143	-73
Latent heat flux	-27	-34	-30	-27	-26	-25	-24	-31	-47	-47
Conductive heat flux	27	153	271	318	321	338	355	288	127	-5
					0.2 m					
Available shortwave energy	53	14					4	49	124	166
Net longwave radiation	-19	-25	-41	-53	-57	-58	-55	-62	-61	-49
Sensible heat flux	-29	-64	-115	-137	-136	-147	-165	-154	-109	-68
Latent heat flux	-23	-22	-16	-14	-14	-13	-12	-17	-34	-46
Conductive heat flux	18	97	172	204	207	218	228	184	80	-3
					0.4 m					
Available shortwave energy	48	13					4	46	114	153
Net longwave radiation	-17	-19	-31	-41	-45	-45	-42	-51	-55	-47
Sensible heat flux	-22	-36	-62	-72	-70	-77	-91	-93	-82	-60
Latent heat flux	-19	-14	-9	-6	-6	-6	-5	-10	-26	-44
Conductive heat flux	10	56	102	119	121	128	134	108	49	-2
					0.8 m					
Available shortwave energy	45	12					3	42	104	140
Net longwave radiation	-16	-16	-24	-33	-37	-37	-34	-43	-51	-46
Sensible heat flux	-17	-18	-27	-30	-27	-31	-42	-53	-61	-54
Latent heat flux	-16	-9	-4	-3	-2	-2	-2	-5	-19	-39
Conductive heat flux	4	31	55	66	66	70	75	59	27	-1
					3.0 m					
Available shortwave energy	16	4					1	17	42	59
Net longwave radiation	-14	-12	-18	-26	-31	-29	-25	-32	-38	-35
Sensible heat flux	-5	0	6	11	17	16	12	9	-2	-8
Latent heat flux	-9	-4	0	0	0	0	0	0	-4	-10
Conductive heat flux	12	12	12	15	14	13	12	6	2	-6

Energy fluxes are given in watts per square meter and describe conditions on the first day of the month.

*$(1 - i_o)(1 - \alpha)F_r$.

become increasingly dependent on the thermal history of the ice rather than just on H.

5.4.3 Ice Thickness Variations

The second step in obtaining regional estimates of the heat and mass balance of the ice pack is to find out the area covered by each thickness category and how this area changes with time. Spatial variations on H can be measured with submarine sonar (Wadhams and Horne, 1978), but the coverage is too limited in time to provide much data on temporal variations in the distribution of thin ice and open water, information of particular concern in the heat and mass balance calculations. Satellite imagery, particularly microwave, can provide frequent coverage but the thickness resolution is still poor.

In the absence of suitable observational methods, theoretical calculations offer a possible way to estimate changes in the ice thickness distribution on a routine basis from strain and ice growth data. The governing equation for g(H,t) can be written (Thorndike et al., 1975)

$$\frac{\partial g}{\partial t} = -\frac{\partial}{\partial H}(fg) - g \operatorname{div} \vec{V} - \vec{V} \cdot \overline{\nabla} g + \psi(H,\vec{V}) \tag{5.33}$$

where f(H,t) is the growth rate of the ice, $\vec{V}$ is the velocity field of the ice, and ψ is the redistribution function which describes how thin ice is ridged into pressure ice. The four terms on the right-hand side of eq. (5.33) describe the effects on g(H,t) of: (i) thermodynamic changes in H, (ii) expansion or contraction of the region for which g is defined, (iii) advection of ice into or out of the region, and (iv) mechanical changes in H caused by convergence or shear. Information on $\vec{V}$ is available from the motion of drifting stations and unmanned buoys, while f(H,t) can be calculated from the heat balance data using a thermodynamic ice growth model. Although eq. (5.33) is quite general, there are some difficulties with its application. In particular, the model parameterizes a number of complex and poorly understood processes, and we have no real way to test these parameterizations or predictions made by the model. Another problem involves the large scale strain field which must be calculated from the movement of points several hundreds of kilometers apart. We have assumed that there exists a functional relationship between these large scale strains and the local opening and closing of leads, but this relationship is not based on measurements. Local opening is probably not only a function of the large scale strain but also of small scale random motions unrelated to this strain. How large a role these random motions play in the local deformation is yet to be established.

In spite of these uncertainities, numerical experiments with the thickness distribution model (Thorndike et al., 1975) showed that predictions were consistent with what is known about g(H,t) in the Central Arctic. It was therefore decided to utilize the model for estimating regional energy fluxes and rates of ice production. Ice thickness distributions were calculated for three different sets of strain data taken from drifting stations in the Central Arctic. The first two sets were obtained during the Arctic Ice Dynamics Joint Experiment (AIDJEX) in the Beaufort Sea and include year-long strain histories from the four manned camps and from the extensive data buoy network (Thorndike and Cheung, 1977; Colony, 1978). The third set was derived from the motions of Ice Stations T-3, ARLIS II and NP-10 during a 2-year period between 1962 and 1964. These data differ markedly in terms of space scales, time scales, and measurement accuracy. Because they contain the least smoothing and were constructed from the most accurate positions, strains from the AIDJEX manned camp array probably provide the best measure of deformational activity within the ice pack. The thickness distribution based on these strains is shown in Figure 28. The amount of ice in the thinnest category averaged only .2-.3% during the winter, essentially reflecting strains which occurred during the preceding day. The total amount of young (H<80 cm) ice varied from 4-6% during this period. Satellite data (Ahlnas and Wendler, 1977) indicate that "young ice" coverage in the Beaufort Sea averaged at least 15% in March 1973 and 6% in March 1974. Submarine data (Wittmann and Schule, 1966) indicate that "young ice" comprises 8-12% of the ice pack between January and May. The maximum thickness of the young ice included in these estimates is unknown, but seems unlikely to exceed 80 cm. The young ice estimates in Figure 28 may thus be somewhat conservative during winter and early spring. The substantial increase in the amount of young ice in May appears to reflect persistent divergence occurring during April and May. The large amount of open water in August results from the melting of young ice and from divergence of the array during July and August. While the 17% figure seems excessive, an aerial mosaic of the array made during mid-August revealed an average of 12-15% open water (R. T. Hall, personal communication). In light of this evidence, the calculated summer values may not be unreasonable.

5.4.4 Regional Fluxes

Using the incident radiation fluxes of Marshunova (1961) and the turbulent fluxes of Doronin (1963), area weighted averages of all thickness dependent quantities were calculated at 6 hour intervals throughout the year for the AIDJEX manned camp array, then integrated over time to obtain the monthly flux totals shown in Table 7. Because of their weak dependence on H, heat flux contributions from the thicker ice were collected into a single category ($0.8-\infty$ m). The results show that effects of g(H) on the individual

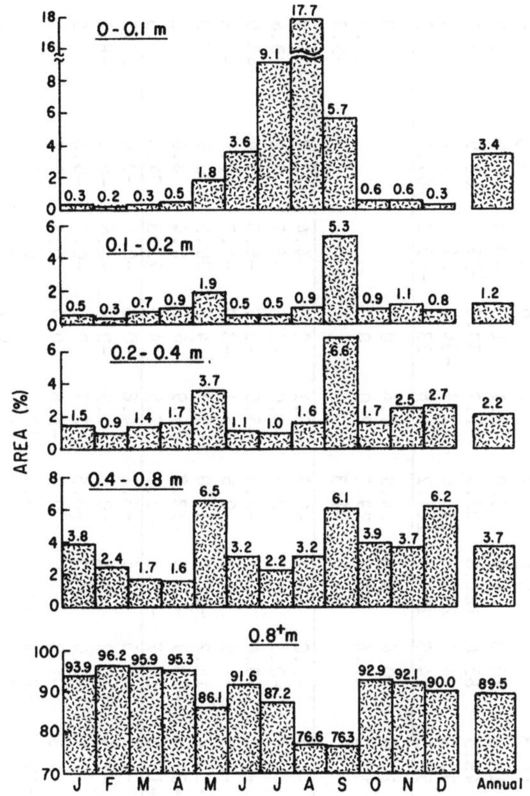

Fig. 28. Ice thickness distribution in the region defined by the AIDJEX manned camps. (After Maykut, 1982.)

radiative fluxes are small on a percentage basis, but significant in relation to the net heat balance. Greater emission of longwave radiation by young ice during the winter increased net longwave losses 3-6% above those over thick ice; summer differences were negligible. The low albedo of leads and thin ice increased the summer absorption of shortwave radiation by 10-20%. As a result the annual net radiation balance for the region was over twice as large as the value over perennial ice. Most of this additional energy was absorbed in the water. Part of the solar radiation absorbed by the water is needed to maintain the water temperature (T_w) at or above the freezing point. The remainder, I_{ow}, is stored in the water, i.e.,

$$I_{ow} = (1 - \alpha_w)F_r + F_L - \sigma T_w^4 + F_s + F_e \quad .$$

Table 5.7. Area weighted heat flux (MJ m^{-2} mo^{-1}) over the AIDJEX manned array. (After Maykut, 1982.)

	Thickness category	Jan	Feb	Mar	Apr	May	Jun	Jul	Aug	Sep	Oct	Nov	Dec	Annual
	0-0.1	-0.6	-0.4	-0.6	0.6	7.4	21.6	44.4	46.6	5.5	-0.7	-1.2	-0.8	121.8
	0.1-0.2	-0.8	-0.5	-0.8	1.4	6.7	2.1	1.5	1.7	3.5	-0.7	-1.6	-1.3	11.2
	0.2-0.4	-2.0	-1.1	-1.1	2.7	11.7	4.4	3.1	2.8	2.9	-0.9	-2.7	-3.3	16.5
Net	0.4-0.8	-4.1	-2.4	-0.9	1.9	18.0	10.8	5.9	5.0	2.1	-1.3	-3.3	-5.9	25.8
radiation	0.8- ∞	-74.2	-65.2	-50.2	-10.9	48.5	85.6	181.8	86.5	-1.7	-30.8	-55.4	-60.9	53.1
	TOTAL	-81.7	-69.6	-53.6	-4.3	92.3	124.5	236.7	142.6	12.3	-34.4	-64.2	-72.2	228.4
	3 m	-79.0	-67.8	-52.4	-10.8	56.5	94.3	207.9	111.2	-1.8	-33.1	-60.2	-67.6	97.2
Energy excess in leads $\langle I_\ell \rangle$				0.3	2.4	10.6	10.6	18.8	24.9	6.4	0.3			74.3
Energy absorbed below ice $\langle I_w \rangle$				0.5	4.1	17.2	17.4	29.9	29.3	8.0	0.3			106.7
Radiation stored in ice $\langle I_i \rangle$								64.0	31.3					95.3
	0-0.1	-2.1	-1.4	-2.9	-2.9	-4.7	-1.0	-1.2	-0.7	-8.1	-2.9	-4.4	-2.7	-35.0
	0.1-0.2	-2.3	-1.6	-3.4	-3.2	-3.5	-0.2	-0.1	-0.2	-5.0	-2.6	-4.7	-3.8	-30.6
Sensible	0.2-0.4	-3.9	-2.6	-4.4	-4.0	-5.7	-0.6	-0.1	-0.3	-5.0	-2.7	-6.3	-7.4	-43.0
heat flux	0.4-0.8	-4.7	-3.6	-2.8	-2.6	-9.3	-1.9	-0.3	-0.6	-3.1	-3.0	-4.7	-8.2	-44.8
$\langle F_s \rangle$	0.8- ∞	43.7	32.8	27.8	11.1	-13.4	-15.2	-11.8	-12.0	-5.7	4.6	20.5	30.9	113.3
	TOTAL	30.7	23.6	14.3	-1.6	-36.6	-18.9	-13.5	-13.8	-26.9	-6.6	0.4	8.8	-40.1
	3 m	46.5	34.1	29.0	11.3	-15.6	-16.7	-13.6	-15.7	-7.5	4.9	22.3	34.3	113.3
	0-0.1	-0.3	-0.2	-0.4	-0.7	-2.2	-1.3	-2.5	-3.1	-4.8	-0.8	-0.8	-0.4	-17.5
	0.1-0.2	-0.3	-0.2	-0.3	-0.7	-1.5	-0.2	-0.1	-0.3	-2.9	-0.6	-0.6	-0.4	-8.1
Latent	0.2-0.4	-0.4	-0.2	-0.4	-0.8	-2.5	-0.6	-0.3	-0.5	-2.7	-0.6	-0.7	-0.8	-10.5
heat flux	0.4-0.8	-0.5	-0.3	-0.2	-0.5	-4.3	-1.8	-0.6	-0.9	-1.8	-0.9	-0.5	-0.8	-13.1
$\langle F_e \rangle$	0.8- ∞	0.0	0.0	0.0	-3.2	-16.2	-26.1	-24.0	-20.5	-12.7	-7.3	-0.1	0.0	-110.1
	TOTAL	-1.5	-0.9	-1.3	-5.9	-26.7	-30.0	-27.5	-25.3	-24.9	-10.2	-2.7	-2.4	-159.3
	3 m	0.0	0.0	0.0	-3.5	-18.8	-28.4	-27.5	-26.7	-16.8	-7.9	-0.1	0.0	-129.7
	0-0.1	3.1	2.0	4.1	4.1	6.1	0.1	0.0	3.3	13.7	4.6	6.4	3.9	51.4
	0.1-0.2	3.4	2.3	4.7	4.2	3.6	0.0	0.0	0.1	8.1	3.9	6.7	5.5	42.5
Conductive	0.2-0.4	6.3	3.9	6.2	4.6	4.1	0.0	0.0	0.0	7.8	4.4	9.8	11.4	58.5
heat flux	0.4-0.8	9.3	6.2	4.2	2.3	3.6	0.0	0.0	0.0	4.4	5.5	8.5	14.9	58.9
$\langle F_c \rangle$	0.8- ∞	30.6	32.4	22.4	3.1	-19.0	-4.5	0.0	0.0	20.1	33.5	35.0	29.9	183.5
	TOTAL	52.7	46.8	41.6	18.3	-1.6	-4.4	0.0	3.4	54.1	51.9	66.4	65.6	394.8
	3 m	32.5	33.7	23.4	3.0	-22.1	-5.0	0.0	0.0	26.1	36.1	38.0	33.3	199.0

Let us now resolve I_{ow} into two components: I_w = the amount of shortwave radiation absorbed beneath the bottom (3 m) of the surrounding ice, and I_ℓ = the difference between the shortwave energy absorbed in the upper 3 m of the leads and the net heat loss to the atmosphere at the surface of the leads. Relative motions between the ice and water tend to distribute L_w uniformly in the horizontal, while stability of the water in summer leads inhibits downward mixing of I_ℓ. Regional values of I_o can thus be expressed as the sum of three terms

$$\langle I_o \rangle = \langle I_i \rangle + \langle I_w \rangle + \langle I_\ell \rangle$$

where $\langle I_i \rangle$ = the amount of solar energy stored within the ice. I_i slowly returns to the atmosphere as the brine pockets freezing during the fall, but I_w and I_ℓ primarily affect the mass balance of the ice pack through bottom ablation and lateral melting on floe edges. The calculated increase in net radiation went almost entirely into I_w and I_ℓ.

The amount of heat conducted to the surface doubled when variations in ice thickness were taken into account. Most of this additional heat was lost to the atmosphere via the turbulent fluxes. Figure 29a shows total turbulent heat losses from each of the thickness categories. It is notable that the ope water categorty (0-0.1 m) did not dominate the balance, rather it was the intermediate thicknesses (0.2-0.8 m) of young ice that were responsible for over half the total loss. The greatest differences between regional and 3 m values occurred during the fall (Figure 29b) in conjunction with the rapid ice production.

Regional ice production $\langle f \rangle$ was dramatically larger when contributions from young ice were considered (Figure 30a). The mass of ice produced in the 0-0.1 m category was equivalent to a layer .154 m in thickness covering the entire array. Annual ice production for the entire region totaled .886 m, in contrast to the zero mass balance expected for 3 m ice. Growth in areas of first year ice accounted for essentially all the ice production. Despite the extremely large growth rates in new leads, ice growth in the 0-0.1 m category made up only about 16% of the total first year ice production, reflecting the small area occupied by open water during the winter. Seasonal variations in $\langle f \rangle$ are shown in Figure 30b, and are compared with the corresponding values for 3 m ice. Although $\langle f \rangle$ was 2-3 times the 3 m values during the winter, the largest differences occurred in the fall. In thick ice heat stored during the summer retarded conduction at the bottom, allowing most of the oceanic heat flux to go directly into bottom ablation; in thin ice conductive fluxes in the ice were much larger than F_w and ice growth was rapid. Ice formation in areas of open water created during the summer was responsible for the large rates of ice production during the fall.

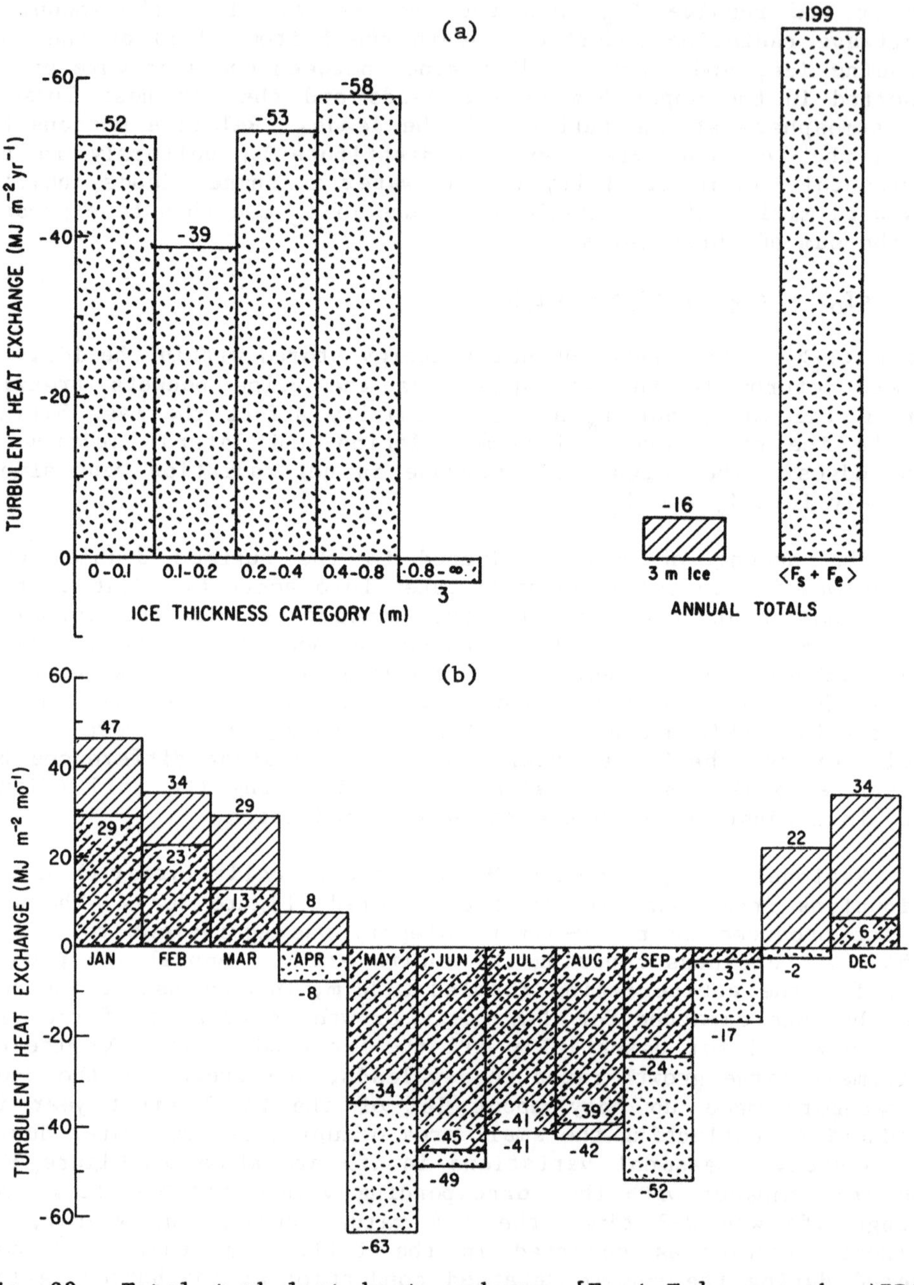

Fig. 29. Total turbulent heat exchange [$F_s + F_e$] over the AIDJEX manned array. (a) Annual totals in each thickness category. (b) Monthly totals obtained by taking into account ice thickness variations (stippled) and by assuming a uniform 3 m ice cover (cross-hatched). (After Maykut, 1982.)

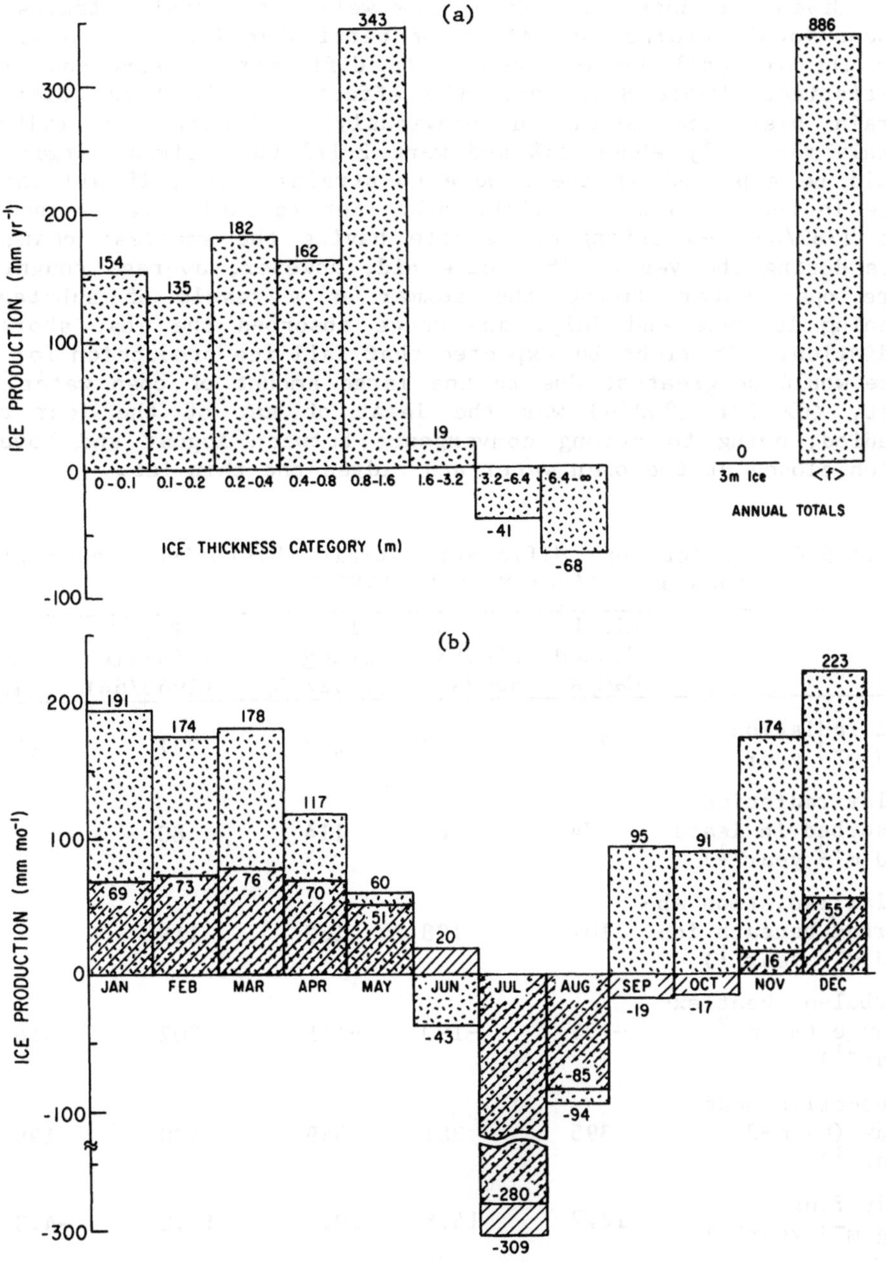

Fig. 30. Ice production in the AIDJEX manned array. (a) Annual totals in each thickness category. (b) Monthly totals obtained by taking into account ice thickness variations (stippled) and by assuming a uniform 3 m ice cover (cross-hatched). (After Maykut, 1982).

Given the uncertainties in how well large scale strains represent local deformation, it is of particular interest to examine how the regional fluxes respond to different strains and strain statistics. Table 8 compares the effects of the three different strain histories mentioned previously. Annual net radiation totals varied by about 25% and were 2-1/2 to 3 times larger than would be expected if there were no strains (i.e., if all the ice were in thermodynamic equilibrium). Net radiation was largest in the 1962/63 T-3 triangle, despite having the smallest change in area during the year. This case did, however, average roughly 5% more open water during the summer as a result of substantial opening in June and July, and hence absorbed the most shortwave radiation. It might be expected that fall ice production for this case would be greatest due to the large amount of open water. In fact, $\langle f \rangle$ for 1962/63 was the least of any of the four cases studied, owing to strong convergence between August and November which closed up the open water and ridged the thin ice.

Table 5.8. Effect of different strain histories on regional totals. (After Maykut, 1982.)

	AIDJEX Manned Camps	AIDJEX Buoys	T-3 Triangle (1962/63)	T-3 Triangle (1963/64)	No Strains
Net radiation (MJ m^{-2} $year^{-1}$)	228	267	292	237	97
Solar radiation absorbed in leads (MJ m^{-2} $year^{-1}$)	74	86	114	80	
Solar radiation absorbed beneath ice (MJ m^{-2} $year^{-1}$)	107	128	160	119	
Turbulent heat exchange (MJ m^{-2} $year^{-1}$)	-199	-190	-155	-202	-16
Conductive heat flux (MJ m-2 $year^{-1}$)	395	381	349	400	199
Salt flux (kg m^{-2} $year^{-1}$)	12.7	14.8	10.2	19.2	-4.3
Ice production (m $year^{-1}$)	.89	.95	.71	1.16	0
Change in area of strain array (% per year)	18	17	9	50	

The most anomalous case was that of the 1963/64 T-3 triangle which underwent a 50% increase in area during the year. The persistent divergence observed arose because the array spanned two different flow regimes -- NP-10 and ARLIS II were in the Transpolar Drift Stream, while T-3 was in the Beaufort Gyre. Although such strains are not typical of the arctic ice pack, they offer an opportunity to study the response of the heat and mass balance to an extreme situation. Surprisingly, the regional heat exchange was not greatly affected by the large total divergence. Neither net radiation, $\langle F_s \rangle$, $\langle F_e \rangle$, nor $\langle F_c \rangle$ exhibited any consistent departures from the other cases. Only $\langle f \rangle$ was correlated with the overall level of divergence. However, in spite of ice production in excess of 1.1 m $year^{-1}$, $\bar{H}$ decreased by 0.3 m as the thermodynamics could not quite keep pace with the divergence. The tendency for increased divergence to cause increased ice production suggests that the amount of ice in the Arctic Basin is probably not very sensitive to variations in ice export. Net convergence, on the other hand, produces an immediate increase in $\bar{H}$ because there is no quick way for the thermodynamics to balance out the increase in mass.

Ice strains typically alternate between divergence and convergence over periods of days, so there are continual changes in the production of open water, the growth of new ice, and the deformation of existing ice. The weak correlation between the average divergence and the heat flux totals (both monthly and annual) suggest that the short term variations in strain are more important to the regional thermodynamics than the long term averages. In fact, other numerical experiments indicate that it is possible to reproduce monthly heat flux totals fairly well with different strain histories possessing similar variance. Effects on the mass balance, however, appear more closely related to the time average of the strain components.

One of the least understood aspects of the large scale heat and mass balance of the ice pack is the role played by heat stored in the ocean. Our results show that a great deal of shortwave energy enters the upper ocean through summer leads. To estimate how this energy might interact with the ice cover, we kept track of I_w and I_ℓ separately. Although part of I_w can be conducted upward into leads and occasional lead closures can transport I_ℓ beneath the ice, we assume as a first approximation that I_w goes primarily into F_w and that I_ℓ represents the amount of energy potentially available for lateral melting. In the four years studied, I_ℓ ranged between 75 and 115 MJ m^{-2} $year^{-1}$. This amount of energy is sufficient to decrease the ice concentration by 6-10%. However, the situation is complicated by ice movement, wind mixing, changing stratification of the surface waters, and uncertainities regarding the relative importance of various heat transfer processes in summer leads. At this point we have no reliable way to estimate what fraction of $\langle I_\ell \rangle$ actually goes into lateral melting. While it seems evident that lateral melting is an important element in the summer mass balance, particularly in the marginal ice zone, systematic studies to define and quantify the dominant melt processes are needed before adequate models can be developed.

Shortwave radiation absorbed beneath the ice (I_w) was about 1.5-2.0 times as large as that needed to maintain a uniform 3 m thick ice cover. Some of this energy may escape to the atmosphere through leads, but most must ultimately interact directly with the ice. If F_w included both this energy and the heat lost from the Atlantic layer, theoretical calculations predict an equilibrium thickness of less than 1 m. Unless there are substantial energy sinks associated with the nonuniform nature of the ice cover, this suggests that little heat from the Atlantic layer reaches the ice, at least in the Central Arctic. Even if we neglect possible contributions from the Atlantic water, there appears to be a surplus of energy beneath the ice. A probable sink for much of this heat is the melting of ice eroded from pressure ridge keels. Measurements made on a pair of 9-12 m ridges over a three month period in the summer of 1975 (Rigby and Hanson, 1976) showed thickness changes of 1-4 m, apparently due to both melting and mechanical erosion. Presumably much of the ice eroded from the keels would spread out under the ice and be eventually melted by heat contained in the water. If we assume an effective F_w of 2.0-2.6 W m^{-2} beneath level ice, then the 3.4 W m^{-2} calculated for $\langle I_w \rangle$ in the AIDJEX manned triangle requires a net annual bottom ablation of 0.85-1.25 m from ice over 6 m in thickness, an amount consistent with the 1.5 m average observed by Rigby and Hanson.

These calculations show that thin ice and open water resulting from differential ice movement cause large-scale interactions between the ocean and atmosphere to be much more vigorous than would be expected on the basis of drifting station data. Dynamic effects during the winter are confined largely to increased ice production, salt flux to the ocean and sensible heat input to the atmosphere; during the summer dynamics strongly influence the interaction of shortwave radiation with the ice and ocean. Despite the numerous uncertainties, the results described above appear to provide a reasonable picture of how dynamic processes and thickness variations affect the heat and mass balance in regions of perennial ice.

ACKNOWLEDGMENTS

Special thanks go to Steve Warren and Norbert Untersteiner for many helpful comments and suggestions during the preparation of the manuscript. This review was made possible by support from the Office of Naval Research, Arctic Program, under Contract N00014-76-C-0234.

REFERENCES

Aagaard, K. and P. Greisman (1975) Toward new mass and heat budgets for the Arctic Ocean. J. Geophys. Res., 80: 3821-3827.

Aagaard, K., L. K. Coachman and E. Carmack (1981) On the halocline of the Arctic Ocean. Deep Sea Res., 28: 529-545.

Ackley, S. F. (1981) Sea-ice-atmosphere interactions in the Weddell Sea using drifting buoys. In Sea Level Ice and Climatic Change (I. Allison, Ed.), IAHS, Washington, D.C., Publication 131, p. 177-191.

Ahlnas, K. and G. Wendler (1977) Arctic sea ice conditions in early spring viewed by satellite. Arct. Alp. Res., 9: 61-72.

Allison, I. (1981) Antarctica sea ice growth and oceanic heat flux. In Sea Level Ice and Climatic Change (I. Allison, Ed.), IAHS, Washington, D.C., Publication 131, p. 161-170.

Anderson, D. L. (1961) Growth rate of sea ice. J. Glaciol., 3: 1170-1172.

Anderson, D. L. (1958) A model for determining sea ice properties. In Arctic Sea Ice, National Academy of Sciences, Washington, D.C., National Research Council, Publication 598: 198-152.

Andreas, E. L. (1980) Estimation of heat and mass fluxes over arctic leads. Mon. Wea. Rev., 108: 2057-2063.

Andreas, E. L. and S. F. Ackley (1981) On the differences in ablation seasons of the Arctic and Antarctic sea ice. J. Atmos. Sci., 39: 440-447.

Arnold, K. C. (1961) An investigation into methods of accelerating the melting of ice and snow by artificial dusting. In Geology of the Arctic, University of Toronto Press, Toronto, Ontario, Canada, Vol. 2, p. 989-1013.

Badgley, F. I. (1966) Heat balance at the surface of the Arctic Ocean. In Proceedings, Symposium on the Arctic Heat Budget and Atmospheric Circulation (J. O. Fletcher, Ed.), Rand Corp., Santa Monica, California, RM-5233-NSF, p. 215-246.

Banke, E. G., S. D. Smith and R. J. Anderson (1980) Drag coefficients at AIDJEX. In Sea Ice Processes and Models (R. S. Pritchard, Ed.), University of Washington Press, Seattle, Washington, p. 430-442.

Bilello, M. A. (1961) Formation, growth and decay of sea ice in the Canadian Arctic Archipelago. Arctic, 14: 3-24.

Brooks, C. E. P. (1949) Climate Through the Ages. McGraw-Hill, New York, 395 p.

Budyko, M. I. (1966) Polar ice and climate. In Proceedings, Symposium on the Arctic Heat Budget and Atmospheric Circulation (J. O. Fletcher, Ed.), Rand Corp., Santa Monica, California, RM-5233-NSF, p. 3-22.

Budyko, M. I. (1974) Climate and Life. International Geophysical Series 18. Academic Press, New York, 508 p.

Chernigovskii, N. T. (1966) Radiational properties of the central Arctic ice coat. Trudy Arkt. Antarkt. Nauch. Issle. Inst., 253: 249-260. (Translated by the Rand Corp., Santa Monica, California, RM-5003-PR.)

Colony, R. (1978) Daily rate of strain of the AIDJEX manned triangle. AIDJEX Bulletin, 39: 85-110.

Deardorff, J. W. (1968) Dependence of air-sea transfer coefficients on bulk stability. J. Geophys. Res., 73: 2549-2557.

Doronin, Yu. P. (1963) On the heat balance of the central Arctic. Trudy Arkt. Antarkt. Nauch. Issle. Inst., 253: 178-184.

Ewing, M. and W. L. Donn (1956) A theory of ice ages: 1. Science, 123: 1061-1066.

Ewing, M. and W. L. Donn (1958) A theory of ice ages: 2. Science, 127: 1159-1162.

Fleagle, R. G. and J. A. Businger (1963) An Introduction to Atmospheric Physics. Academic Press, New York, 346 pp.

Fletcher, J. O. (1965) The Heat Budget of the Arctic Basin and Its Relation to Climate. Rand Corp., Santa Monica, California, R-444-PR, 179 p.

Gordon, A. L. (1981) Seasonality of Southern Ocean sea ice. J. Geophys. Res., 86: 4193-4197.

Grenfell, T. C. and G. A. Maykut (1977) The optical properties of ice and snow in the Arctic Basin. J. Glaciol., 18: 445-463.

Grenfell, T. C. and D. K. Perovich (1984) Spectral albedos of sea ice and incident solar irradiance in the southern Beaufort Sea. J. Geophys. Res., 89(C3): 3573-3580.

Haltiner, G. J. and F. L. Martin (1957) Dynamical and Physical Meteorology. McGraw-Hill, New York, 470 p.

Hanson, A. M. (1965) Studies of the mass budget of arctic pack ice floes. J. Glaciol., 5: 701-709.

Houghton, H. G. (1954) On the annual heat balance of the Northern Hemisphere. J. Met., 11: 1-9.

Idso, S. B. and R. D. Jackson (1969) Thermal radiation from the atmosphere. J. Geophys. Res., 74: 5397-5403.

Kellogg, W. W. (1975) Climatic feedback mechanisms involving the polar regions. In Climate of the Arctic (G. Weller and S. A. Bowling, Ed.), Geophysical Institute, University of Alaska, p. 111-116.

Kirilova, T. V. (1952) On the dependence of the counter-radiation of the atmosphere on the degree of cloudiness. Proceedings, Main Geophysical Observatory, 37.

Laevastu, T. (1960) Factors affecting the temperature of the surface layer of the sea. Comment. Phys.-Math., 25(1): 8-134.

Langleben, M. P. (1971) Albedo of melting sea ice in the southern Beaufort Sea. J. Glaciol., 10: 101-104.

Langleben, M. P. (1972) The decay of an annual cover of sea ice. J. Glaciol., 11: 337-344.

Leavitt, E., M. Albright and F. Carsey (1978) Report on the AIDJEX meteorological experiment. AIDJEX Bulletin, 39: 121-148.

Lebedev, V. V. (1938) Rost l'do v arkticheskikh rekakh i moriakh v zavisimosti ot otritsatel' nykh temperatur vozdukha. Problemy Arktiki, 5: 9-25.

Leckner, B. (1978) The spectral distribution of solar radiation at the earth's surface: Elements of a model. Solar Energy, 20: 143-150.

Lee, O. S. and L. S. Simpson (1954) A practical method of predicting sea ice formation and growth. U.S. Naval Hydrographic Office, Washington, D.C., Technical Report 4, 27 p.

Lindsay, R. W. (1976) Wind and temperature profiles taken during the arctic lead experiment. University of Washington, M.S. thesis, 89 p.

Manabe, S. and R. T. Wetherald (1980) On the distribution of climate change resulting from an increase in CO_2 content of the atmosphere. J. Atmos. Sci., 37: 99-118.

Marshunova, M. S. (1961) Principal characteristics of the radiation balance of the underlying surface and of the atmosphere in the Arctic. Trudy Arkt. Antarkt. Nauch. Issle. Inst., 229. (Translated by the Rand Corp., Santa Monica, California, RM-5003-PR, 1966.)

Martin, S. and P. Kauffman (1974) The evolution of underice melt ponds, or double diffusion at the freezing point. J. Fluid Mech., 64: 507-527.

Maykut, G. A. (1978) Energy exchange over young sea ice in the Central Arctic. J. Geophys. Res., 83: 3646-3658.

Maykut, G. A. (1982) Large scale heat exchange and ice production in the Central Arctic. J. Geophys. Res., 87(C10): 7971-7984.

Maykut, G. A. and P. E. Church (1973) Radiation climate of Barrow, Alaska, 1962-66. J. Appl. Met., 12: 620-628.

Maykut, G. A. and N. Untersteiner (1969) Numerical prediction of the thermodynamic response of arctic sea ice to environmental changes. The Rand Corporation, Santa Monica, California, RM-6093-PR, 173 p.

Maykut, G. A. and N. Untersteiner (1971) Some results from a time dependent, thermodynamic model of sea ice. J. Geophys. Res., 76: 1550-1575.

Nakawo, M. and N. K. Sinha (1981) Growth rate and salinity profile of first year sea ice in the high Arctic. J. Glaciol., 27: 315-330.

Ono, N. (1967) Specific heat and heat of fusion of sea ice. In Physics of Snow and Ice, 1 (H. Oura, Ed.), Institute of Low Temperature Science, Hokkaido University, Sapporo, Japan, p. 599-610.

Panov, V. V. and A. O. Shpaikher (1964) Influence of Atlantic waters on some features of the hydrology of the Arctic Basin and adjacent seas. Deep-Sea Res., 11: 275-285.

Parkinson, C. L. and W. M. Kellogg (1979) Arctic sea ice decay simulated for a CO_2-induced temperature rise. Climatic Change, 2: 149-162.

Parkinson, C. L. and W. M. Washington (1979) A large scale model of sea ice. J. Geophys. Res., 84: 311-337.

Perovich, D. K. and T. C. Grenfell (1981) Laboratory studies of the optical properties of young sea ice. J. Glaciol., 27: 331-346.

Richards, J. M. (1971) A simple expression for the saturation vapour pressure of water in the range -50° to $140^{o}C$. Brit. J. Appl. Phys., 4: L15-L18.

Rigby, F. A. and A. Hanson (1976) Evolution of a large arctic pressure ridge. AIDJEX Bulletin, 34: 43-71.

Schwarzacher, W. (1959) Pack ice studies in the Arctic Ocean. J. Geophys. Res., 64: 2357-2367.

Schwerdtfeger, P. (1963) The thermal properties of sea ice. J. Glaciol., 4: 789-807.

Semtner, A. J. (1976) A model for the thermodynamic growth of sea ice in numerical investigations of climate. J. Phys. Oceanogr., 6: 379-389.

Shine, K. P. (1984) Parameterization of shortwave flux over high albedo surfaces as a function of cloud thickness and surface albedo. Q. J. R. Meteorol. Soc., 110: 747-760.

Thorndike, A. S., D. A. Rothrock, G. A. Maykut and R. Colony (1975) The thickness distribution of sea ice. J. Geophys. Res., 80: 4501-4513.

Thorndike, A. S. and J. Y. Cheung (1977) AIDJEX measurements of sea ice motion. AIDJEX Bulletin, 35: 1-149.

Thorndike, A. S. and R. Colony (1980) Arctic Ocean buoy program, 19 January 1979 - 31 December 1979. Polar Science Center, University of Washington, 131 p.

Thorndike, A. S. and R. Colony (1981) Arctic Ocean buoy program, 1 January 1980 - 31 December 1980. Polar Science Center, University of Washington, 132 p.

Thorndike, A. S., R. Colony and E. A. Munoz (1983) Arctic Ocean buoy program, 1 January 1981 - 31 December 1981. Polar Science Center, University of Washington, 137 p.

Thorndike, A. S., R. Colony and E. A. Munoz (1983) Arctic Ocean buoy program, 1 January 1982 - 31 December 1982. Polar Science Center, University of Washington, 132 p.

Thorpe, M. R., E. G. Banke and S. D. Smith (1973) Eddy correlation measurements of evaporation and sensible heat flux over arctic sea ice. J. Geophys. Res., 78: 3573-3584.

Untersteiner, N. (1961) On the mass and heat budget of arctic sea ice. Arch. Met. Geophys. Bioklim., A(12): 151-182.

Untersteiner, N. (1962) The ice budget of the Arctic Ocean. Proceedings of the Arctic Basin Symposium, Arctic Institute of North America, p. 219-226.

Untersteiner, N. (1964) Calculations of temperature regime and heat budget of sea ice in the Central Arctic. J. Geophys. Res., 69: 4755-4766.

Untersteiner, N. and F. Badgley (1965) The roughness parameters of sea ice. J. Geophys. Res., 70: 4573-4577.

Vowinckel, E. and S. Orvig (1962a) The relation between solar radiation income and cloud type in the Arctic. J. Appl. Met., 1: 552-559.

Vowinckel, E. and S. Orvig (1962b) Insolation and absorbed solar radiation at the ground in the Arctic. McGill University, Montreal, Publication in Meteorology No. 53, 50 p.

Vowinckel, E. and S. Orvig (1962c) Cloud amount and type over the Arctic. McGill University, Montreal, Publication in Meteorology No. 51, 27 pp.

Vowinckel, E. and S. Orvig (1972) Synoptic energy budgets from the Beaufort Sea. In Energy Fluxes over Polar Surfaces. World Meteorological Organization, Geneva, p. 143-166.

Wadhams, P. and R. J. Horne (1978) An analysis of ice profiles obtained by submarine sonar in the AIDJEX area of the Beaufort Sea. Scott Polar Research Institute, Technical Report 78-1, 65 p.

Walker, E. R. and P. Wadhams (1979) Thick sea ice floes. Arctic, 32: 140-147.

Weller, G. (1972) Radiation flux investigation. AIDJEX Bulletin, 14: 28-30. (Available as PB 220/859, National Technical Information Service, Springfield, Virginia.)

Wiscombe, W. J. (1975) Solar radiation calculations for arctic summer stratus conditions. In Climate of the Arctic (G. Weller and S. A. Bowling, Ed.), Geophysical Institute, University of Alaska, p. 245-254.

Wittmann, W. I. and J. J. Schule (1966) Comments on the mass budget of arctic pack ice. In Proceedings, Symposium on the Arctic Heat Budget and Atmospheric Circulation (J. O. Fletcher, Ed.), Rand Corp., Santa Monica, California, RM-5233-NSF, p. 215-246.

Yen, Y. C. (1981) Review of thermal properties of snow, ice and sea ice. USA-CRREL Res. Rept. 81-10, Hanover, NH, 27 p.

Zillman, J. W. (1972) A study of some aspects of the radiation and heat budgets of the Southern Hemisphere oceans. In Meteorological Studies, 26, Bureau of Meteorology, Department of the Interior, Canberra, Australia, 562 p.

Zubov, N. N. (1945) L'dy Arktiki. (Arctic Ice.) (Izdatel'stvo Glavsevmorputi), Moscow, 360 p. (U.S. Navy Hydrographic Office, Translation 217, 1963; available as AD426972 from NTIS, Springfield, VA.)

CHAPTER 6

ARCTIC STRATUS CLOUDS

Gerald F. Herman

Department of Meteorology
University of Wisconsin,
Madison, Wisconsin 53706 U.S.A.

1. INTRODUCTION

Arctic stratus clouds are extremely relevant to the problem of air-sea-ice interaction in high latitudes because: 1) due to their persistence and extent they are a dominant climatological feature of the central Arctic during the summer; 2) they strongly influence the amount of solar and infrared radiation that is incident on the surface in the Arctic, thus affecting the energy balance of the pack ice and the heat available to the ocean mixed layer in ice-free regions; 3) they are an example of a cloud type whose structure is strongly dependent on the disposition of solar and infrared radiation within the cloud and its interaction with boundary layer turbulence; 4) at present they are not realistically simulated in any major general circulation or weather forecasting model, and thus pose an obstacle to the proper simulation of the coupled atmosphere-ice-ocean system.

A detailed review of research conducted prior to 1974 on Arctic stratus appears in the report by Herman (1975). Virtually all of the cloud climatological data reviewed in that report were based on surface or aircraft observations, since prior to 1970 there was very little use made of satellite data in Arctic cloud studies. Since that time a great deal of interest has developed in obtaining global cloud climatologies from satellites (see NASA, 1981), and work is now underway at several institutions to construct Arctic cloud climatologies based on visible and infrared satellite imagery.

There additionally has been a great deal of interest in the general problem of layered cloud formation particularly with regard to the formation of sub-tropical marine stratocumulus. The original mixed layer theories of Ball (1960) and Lilly (1968) have been elaborated upon to include the effects of advection, and more details of the radiative energy exchanges near the cloud top. The role of radiation in the cloud top entrainment process has been studied in the series of papers by Deardorff (1976; 1980; 1981), Randall (1980a,b) and Lilly and Schubert (1980). While these theories have been developed primarily for situations in which active convective layers underlie regions of strong subsidence, as off the coast of California, they have some applicability to Arctic situations, especially for explaining the nature of the convection generated by the cloud top energy loss.

Prior to 1974 very few in situ measurements of stratus clouds properties were available, either for the Arctic or elsewhere. Some radiative characteristics are cited by Marshunova (1961) and by Koptev and Voskresenskii (1962), but very little information is available about the details of the experiments. Until recently the principal source of data on the radiative properties of mid-latitude stratus was the early balloon flights of Neiburger (1949), and other cloud studies reported by Robinson (1958), Platt (1972, 1976), and Paltridge (1971). Since 1974 there have been aircraft flights made in Arctic stratus as part of AIDJEX (see Herman, 1977; 1980) and as part of the June, 1980, Arctic stratus experiment (see Sect. 5). Subsequently a large number of flights have been conducted in California stratus, and in stratus occurring over the English Channel, in the North Atlantic, off the coast of Australia, and in the North Pacific. An excellent summary of all stratus missions completed prior to 1981, together with the details of some planned programs, has been prepared by Platt (1981).

2. STRATUS CLOUD MORPHOLOGY

2.1 Seasonal Behavior

The seasonal variations of Arctic cloud cover have been best illustrated by the studies of surface-based cloud observations by Huschke (1969) and Vowinckel and Orvig (1970). Figure 1 shows the annual march of total cloud amount in various regions of the Arctic. The variability of the total cloud amount is dominated by the variability of low or stratiform clouds, as shown in Figure 2. Both the frequency of occurrence and the total cloud amount increase to their maximum summertime values over a very short transitional period, and this behavior is evident in each of the five years of the Huschke study (Figure 3). The lower values for the Canadian Arctic during summer presumably are due to the larger number of reports which are included from the continental stations in that sector, which ordinarily have much lower total cloudiness.

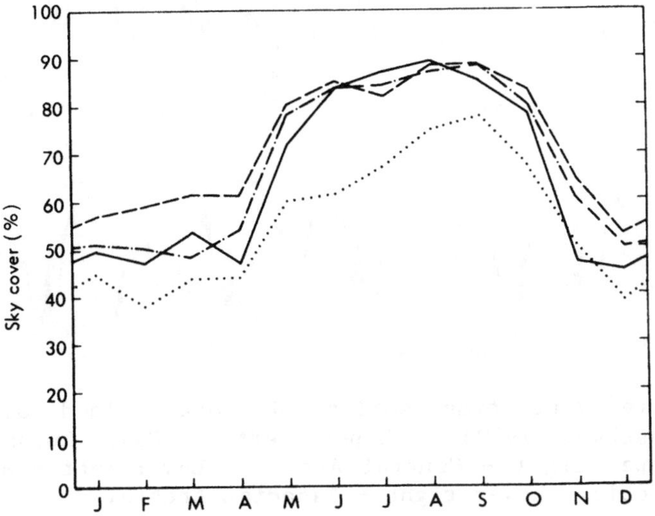

Fig. 1. Annual march of total cloud amount in the Arctic (after Huschke, 1969). The lines correspond to the following regions: West Eurasian Arctic -----; East Eurasian Arctic -•-•-; Canadian Arctic .....; Central Arctic _____.

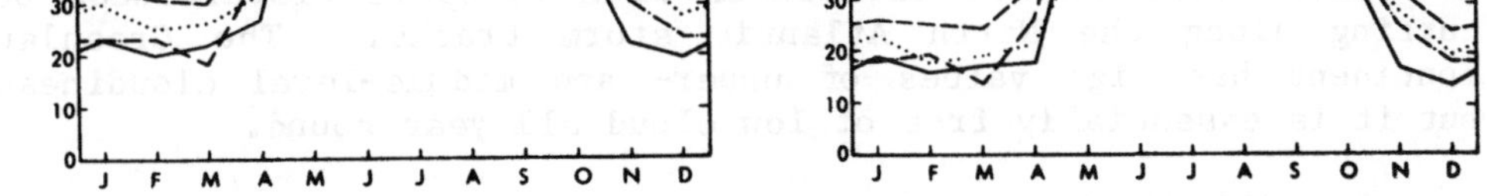

Fig. 2. As in Figure 1, except for low cloud amount and low cloud frequency.

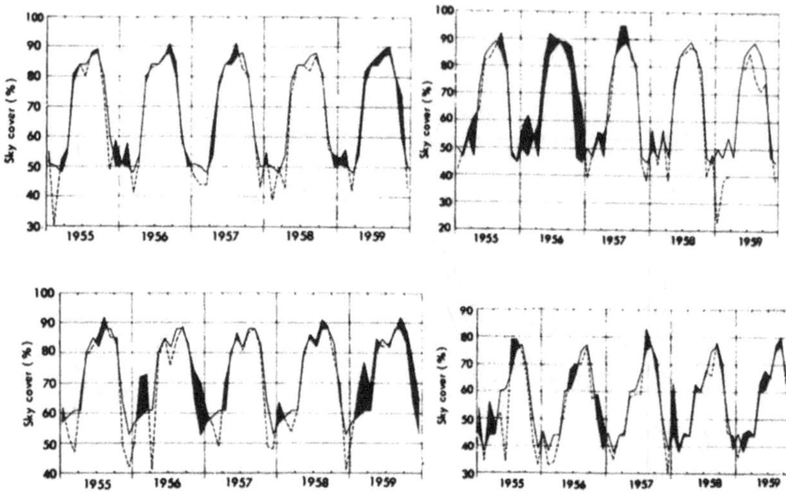

Fig. 3. Five year time series of total cloud amount (after Huschke, 1969). Upper left - East Eurasian Arctic. Upper right - Central Arctic. Lower left - West Eurasian Arctic. Lower right - Canadian Arctic.

2.2 Geographic Distribution

The isopleths of total cloud amount in the Arctic Basin show that values generally increase toward the pole (Figure 4). The low cloud frequencies (Figure 5) also show the same feature.

Well-developed convective clouds occur infrequently in the Arctic Basin during the summer, and when they do they are associated with migratory depressions which traverse the region. The convective cloud cover tends to be higher over the continents than in the central Arctic since some convection occurs when the ground is heated during the summer. The Norwegian and Barents Seas are extremely cloudy during all seasons, but this cloudiness is of a convective type and is associated with synoptic disturbances occurring along the North Atlantic storm tracks. The Greenland continent has high values of upper- and middle-level cloudiness, but it is essentially free of low cloud all year round.

It will be instructive to learn if our view of Arctic cloud conditions will change with the advent of satellite cloud climatologies. In general there are substantial differences between remote and surface-based measures of cloudiness. For example, ground based observations are normally given in terms of the fraction of the celestial dome obscured to the eyes of an observer (i.e., in terms of observed visible radiation), and thus could

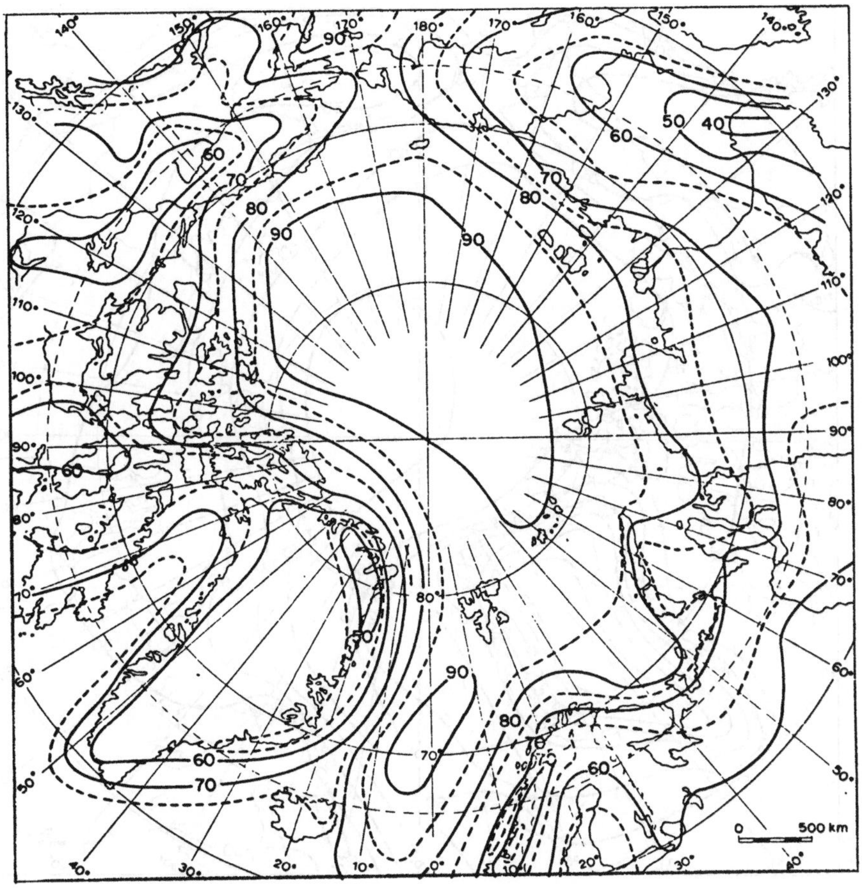

Fig. 4. Isopleths of total cloud amount in July (after Vowinckel and Orvig, 1970).

give a completely different measure of cloudiness than a satellite observation viewing at near-normal incidence. An ambiguity in the definition of "cloudiness" exists in the Arctic as it does elsewhere. The numerous measures of cloudiness that are cited in the literature (cloud cover, cloud amount, cloud frequency, cloud density, effective cloudiness, apparent cloudiness, relative cloudiness, etc.) clearly need to be reconciled or related to one another.

2.3 Vertical Structure

Arctic stratus cloud formation is not limited to the planetary boundary layer as are other types of marine stratus. This is evident if one looks at the various profiles of cloudiness that have appeared in the literature (see Herman, 1975, p. 19; Herman,

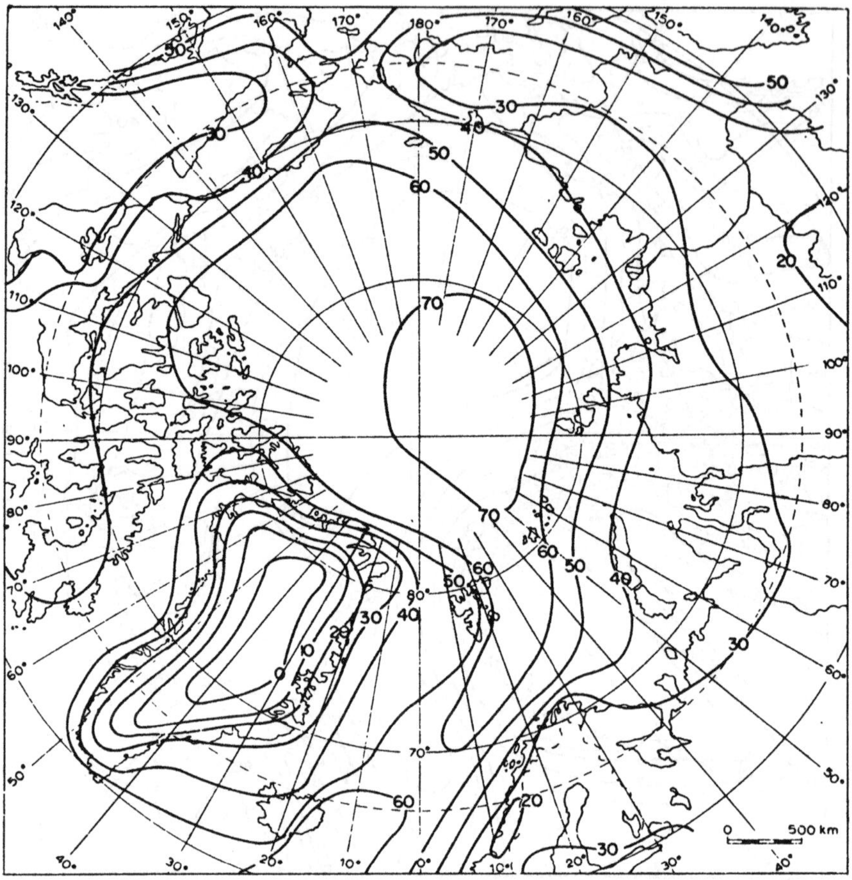

Fig. 5. As in Figure 4, except for low cloud frequency.

1977; also Figures 8 of this chapter). Though sometimes occurring as a single cloud layer under a PBL inversion, sheets of liquid water stratus are found throughout the lower troposphere in the summertime Arctic. Thus, Arctic stratus has come to be termed "multilayered." Typically the cloud lamina are 300-1000 m thick, although some are so thin (~50 m) as to make direct measurement with an aircraft impossible. Sometimes the stratus is located directly at the surface, and is indistinguishable from fog.

3. RADIATIVE PROPERTIES AND RADIATIVE EFFECTS

The role that stratus clouds play in the Arctic energy balance is illustrated in Table 1, which shows the observed monthly radiation balances at the surface in the Central Arctic based on the tabulations of Vowinckel and Orvig (1964). The values shown

Table 1. Average monthly radiation fluxes at the surface in the Arctic (W m^{-2}) (from Herman, 1980).

	Jan	Feb	Mar	Apr	May	Jun	Jul	Aug	Sep	Oct	Nov	Dec
					Actual Cloud Conditions							
Net thermal	-32	-26	-29	-40	-36	-31	-15	-16	-11	-13	-23	-26
Absorbed solar	0	0	2	26	52	68	81	42	7	2	0	0
Total	-32	-26	-27	-14	16	37	66	26	-4	-11	-23	-26
					No Clouds							
Net thermal	-60	-61	-59	-65	-71	-71	-69	-71	-74	-68	-61	-58
Absorbed solar	0	0	5	34	64	91	145	80	14	2	0	0
Total	-60	-61	-54	-31	-7	20	76	9	-60	-66	-61	-58

include measurements taken only under clear sky conditions. The role of the clouds in the solar and thermal regions of the spectrum is evident: The clouds reduce the outgoing longwave radiation by 28-55 W m^{-2}, and thus increase the net radiation at the surface ("the greenhouse effect"). At the same time they reduce the solar radiation at the surface by reflecting incident radiation back to space, thus decreasing the net radiation balance ("the albedo effect"). Except during the month of July the cloud effect on the net radiation is dominated by its effect on the longwave component. The conditions at the surface of the pack ice would clearly be very different if no cloud were present, and all other terms in the energy balance remained the same.

3.1 Longwave Radiative Characteristics

In addition to its role in influencing the surface energy budget, the infrared absorption of stratus clouds is important for several other reasons. First, since their liquid water content is so relatively low, clouds of average thickness are capable of transmitting a portion of the longwave radiation that is incident upon them. This property complicates the study of the Arctic using infrared satellite measurements, since it is difficult to distinguish upwelling radiation originating near the cloud top from that which is emitted at the surface and transmitted through the cloud. Secondly, the infrared radiative energy loss from the cloud top is a source of buoyancy for the cloud layer, and this has important consequences for planetary boundary layer structure and cloud development.

The upward flux of radiation, F^{+}, at a depth, h, below the top of an isothermal cloud of total thickness, H, is

$$F^{+}(h) = \pi B(h) + F^{+}(H)T(h,H) - \pi B(H)T(h,H) \qquad (6.1)$$

where T(h,H) is the spectrally-averaged transmission of the cloud between the base and the level h over the spectral region of interest, which is usually the water vapor window region, and B is the Planck function corresponding to the local temperature, θ.

Similarly, the downward flux, F^-, is

$$F^-(h) = \pi B(h) + F^-(0)T(h,0) - \pi B(0)T(0,h) \qquad (6.2)$$

Either eq. (6.1) or (6.2) may be used to solve for the transmission of the cloud, T. When scattering is assumed to be negligible in the infrared, the cloud emissivity is related to the transmission simply by $\varepsilon(h) = 1 - T(h)$.

In practice eq. (6.2) is used more commonly than eq. (6.1) to determine the cloud emissivity from measurements of the downward flux with an airborne radiometer, since there is so little variation of the upward flux near cloud top.

The results of a series of aircraft measurements made during the AIDJEX field program are shown in Figure 6. The emissivity is unity for clouds with thicknesses greater than about 320 m (corresponding to an extinction depth of 59 m), and thus according to eq. (6.1) the upward flux is given simply by the Planck function, πB. However, for thicknesses less than 320 m a large fraction of thermal radiation is transmitted through the cloud, and thus arises the problem of distinguishing cloud from surface using infrared measurements.

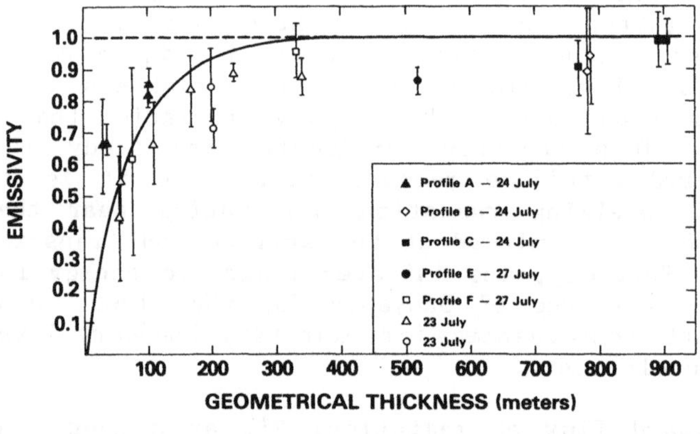

Fig. 6. Emissivity of Arctic stratus as a function of h, the distance below cloud top. Solid line is fit to data for an extinction length, λ, of 59 m. (Reprinted from Herman, 1980).

The energy loss from the cloud top plays an important dynamical role. The total cooling, Q, of a cloud layer is $\Delta F/(\rho C_p H)$, where ΔF is the divergence of the net infrared flux in the layer. For a black cloud of temperature θ in an isothermal atmosphere of the same temperature, $Q = (\rho\ C_p)^{-1}\sigma(\theta)^4\ (1-\varepsilon_a)H^{-1}$ where ε_a is the emissivity of the slab of atmosphere overlaying the cloud. For θ = 273K, H = 500 m and ε_a = .3, this would correspond to a total cooling of the layer of about 24°C per day, as compared with typical atmospheric cooling rates of 2°C per day.

However, this cooling is generally not distributed uniformly through the depth of the cloud, and in fact the relation between the location of the cooling and the generation of convective turbulence is a disputed question in theories of stratus formation. The early mixed layer model of Lilly (1968) assumed (somewhat unrealistically) that all of the cooling took place in a very thin layer near the inversion, while Kahn and Businger (1979) distribute it throughout the mixed layer. Most recently Deardorff (1981) related the fraction of the radiative energy loss that was available for buoyant energy production to the depth of the mixed layer and to the radiative extinction length. The main conclusion to his work was that in deep mixed layers with a thick cloud cover, the radiatively produced buoyancy was confined to a thin layer near cloud top. In more tenuous clouds, it was distributed more uniformly through the mixed layer.

3.2 Solar Radiation

The task of making measurements of solar radiation in Arctic stratus suffers from many of the same difficulties that plague *in situ* measurements in clouds elsewhere. These include:

Cloud Non-homogeneity. The picture of stratus clouds as infinite, plane-parallel homogeneous layers is seldom realized, even in the relatively stable Arctic situations. In general the cloud tops have turret elements, which are caused by the convective cells within the cloud, and various forms of scud protrude from the base. Thus there is an ambiguity in defining the true "top" and "base" of a cloud layer. A variety of holes and gaps in the cloud layer also occur in real cloud situations, and substantial horizontal variations of cloud thickness and liquid water content frequently are observed.

Non-stationarity. Clouds seldom remain fixed in space and time for the benefit of an experimenter. Clouds continually form, grow and dissipate, and advect with the mean flow, which due to vertical shear, may vary with height. Thus it is seldom possible to assemble a true "vertical" profile, since the successive legs of an aircraft flight generally are not flown over precisely the same part of the same cloud. This is a problem since the principal goal of aircraft radiation missions is to relate the vertical profiles of radiation to the vertical distribution of temperature, liquid water, and water vapor.

Cloud Contamination by Cirrus. It is difficult to locate stratus cloud decks in areas that are completely free of other cloud forms, especially cirrus. These clouds contaminate the measurements of the downward flux of both solar and infrared radiation, and are frequently difficult to observe with the naked eye, except at very oblique viewing angles.

i. Cloud reflectance and transmittance. The details of the solar reflectance and transmittance measurements obtained during the AIDJEX flights are found in Herman (1977). In general, the reflectance of the cloud layers was nearly constant over a wide range of cloud thicknesses, indicating the importance of the high surface reflectivity in determining the albedo of the cloud-plus-surface system, which is the quantity measured by downward-looking radiometers. The following range of reflectances were obtained: Over the total solar spectrum, 60-70%, visible spectrum, 70-85%, near infrared, 50-65%.

ii. Cloud absorptance. The true absorptance of stratus clouds is a particularly difficult quantity to measure directly because of the observational difficulties mentioned above, and there presently is some difficulty in validating experimentally the predictions of very small visible absorption made on the basis of laboratory measurements and theoretical calculations. The bulk absorption of a cloud layer, A, is given by

$$A = \frac{F^-(0) + F^+(H) - F^-(H) - F^+(0)}{F^-(0)} \qquad (6.3)$$

where F^- and F^+ denote downward and upward fluxes, and h=0, h=H denote the top and base of the cloud respectively. Note that the absorption computed according to eq. (6.3) is the residual of the net flux at the top of the cloud and base of the cloud, and thus contains the errors of each of the four flux measurements. Values of cloud absorption obtained during AIDJEX are shown in Table 2. The bulk absorptance averaged over all cloud decks that were sampled was 7% over the total solar spectrum, and 5% and 9% in the visible and near infrared, respectively. For comparison, we show in Table 2 other measurements of cloud absorption that have appeared in the literature. (The large range of absorptances shown here is associated with the wide variety of liquid water paths encountered.)

The problem with interpreting these measurements arises with reference to the visible (0.29 = 0.68 μm) absorption. Based on the current data for the real and imaginary parts of the index of

Table 2a. Average absorptance in stratus cloud layers from AIDJEX flights (percent). (After Herman, 1977).

Flight leg	Total solar (0.29-2.8μm)	Visible (0.29-0.68μm)	Near infrared (0.68-2.8μm)
A	11 ± 3	6 ± 5	5 ± 3
B	6 ± 6	3 ± 6	9 ± 6
C	12 ± 9	6 ± 9	18 ±10
D	3 ± 2	4 ± 3	2 ± 1
E	6 ± 8	3 ± 7	9 ±10
F	9 ± 6	0 ± 6	11 ± 5
All flights	7 ± 8%	5 ± 8%	9 ± 5%

Table 2b. Observations of solar absorption in other stratus situations.

Investigator	Absorption (%)	Comment
Neiburger	7	Average of all observations
Robinson	19	Average of many stratiform clouds
Koptev and Voskresenskii	4 - 10	200-500 m Arctic stratus
Griggs	4	Average of 3 sc decks
Reynolds et al.	23	Average of 3 sc decks

refraction for liquid water, there should be essentially no absorption in this part of the spectrum. A measure of the importance of droplet absorption is the single scatter albedo, $\tilde{\omega}_0$, which is the ratio of the scattering extinction coefficient to the total extinction coefficient. Thus, $1 - \tilde{\omega}_0$ is the fractional extinction due to absorption. The wavelength dependence of $1 - \tilde{\omega}_0$ is illustrated in Figure 7, and shows that this parameter is extremely small in the visible region of the spectrum, but increases by several orders of magnitude in the near infrared.

Note, however, that the values in the visible obtained from the aircraft flights (Table 2a) are about 5%, but with a standard error of $\pm$ 8%. Such values would be consistent with small visible absorption, but unfortunately it is not possible to state precisely how small, since eq. (6.3) provides only an upper bound to the true absorption because of the noisiness of each individual flux measurement. Absorptance of less than a few percent simply cannot be resolved. If some slight visible absorption could be resolved, this would pose some discrepancy with theoretical predictions. However, absorption by liquid water droplets is only one way that visible radiation may be lost in a cloud. Another possibility that has frequently been mentioned, especially with reference to

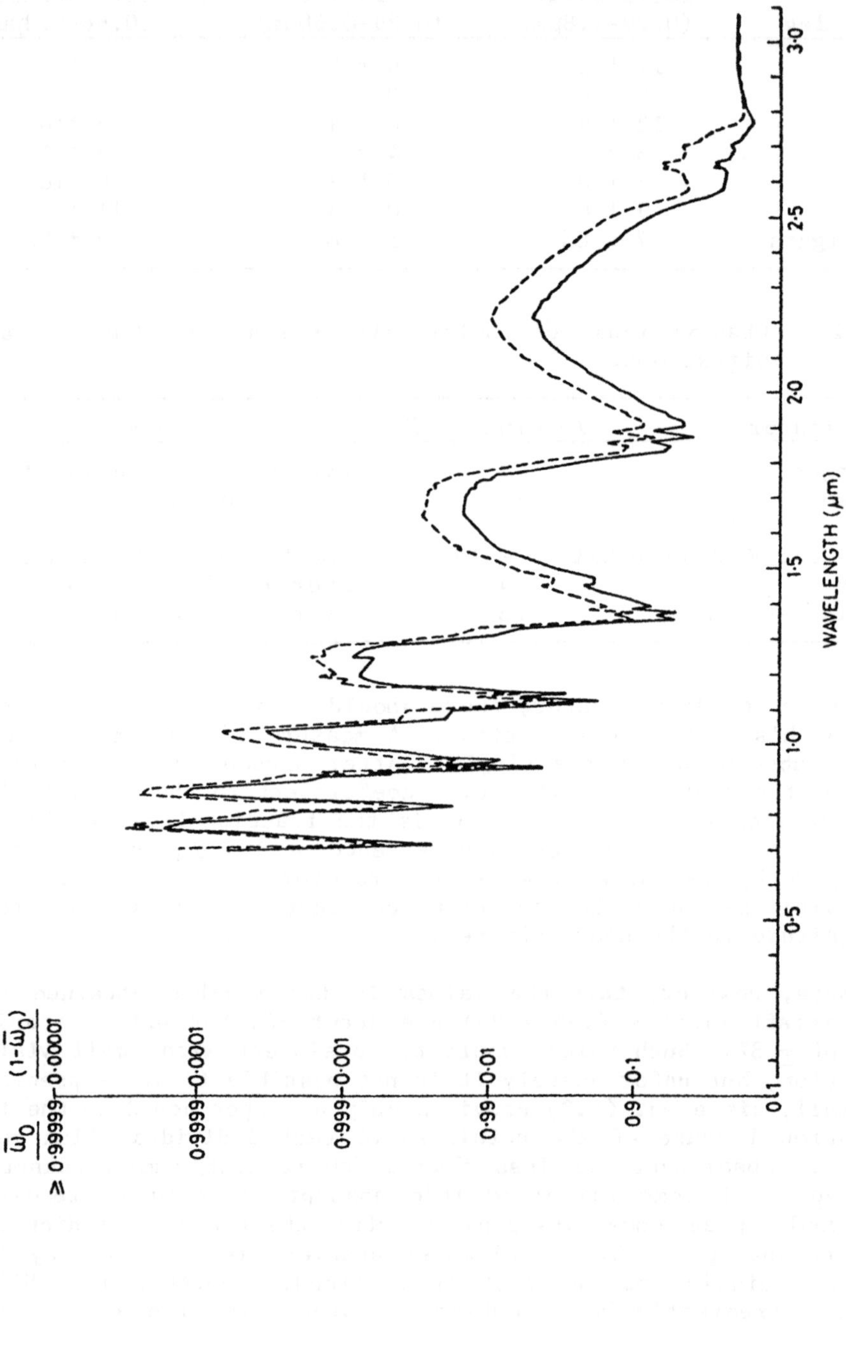

Fig. 7. The parameter $1-\bar{\omega}_0$ for liquid water. (After Twomey, 1976.)

anomalously high near infrared absorption (see Twomey, 1972) is the absorption by aerosol particles embedded in the cloud. Newiger and Bähnke (1981) have computed solar absorptances in clouds with haze assumed to have typical properties, and found that the visible single scatter albedo was reduced to levels comparable with that in the near infrared when aerosol was present. Aerosol layers are known to be present throughout the Arctic, (e.g., Heintzenberg, 1980) and some information about their optical properties is available. Another sound possibly is the loss of energy out of the lateral boundaries of the clouds, the so-called three dimensional effects. These occur when the height-to-width ratio of a cloud is sufficiently large to permit photons to be scattered out of the side of a cloud, and thus be mistakenly included in the absorption when using a budget equation such as eq. (6.3). As an example, Newiger and Bähnke estimate that 14% of the flux incident on a conservatively scattering stratus cloud is scattered from the sides when the cloud is 500 m thick and 4 km wide, although this figure is somewhat reduced when cloud, gas and aerosol absorption are included.

It is not clear whether the amount of visible radiation absorbed in stratus clouds will ever be resolved by in situ measurements because of the observational problems mentioned previously. It is possible that improved measurements of cloud geometry might be obtained with fast-response lidar systems, and that some sampling problems might be eliminated by the use of several aircraft simultaneously. However, it is likely that the problems imposed by the non-homogeneity of the clouds will prove insurmountable.

3.3 Mean Structure and Turbulent Fluxes From Aircraft Studies

The vertical distributions of temperature and water vapor mixing ratio in stratus situations encountered during the AIDJEX radiation missions are shown in Figures 8a through 8c. The lapse rate of temperature in the sub-cloud layer was sometimes adiabatic, although there were exceptions. A small temperature inversion (~3°C) was located at the cloud top, but no such inversion was to be found in the moisture profiles. No general statement can be made about the temperature or moisture profiles in the clear interstices between the stratus layer. For comparison, we note that in typical California stratus situations the temperature inversion at cloud top may be as large as 15-20°C (Lilly, 1968), and as a rule the sub-cloud layer is well-mixed.

Turbulent fluxes in Arctic stratus situations were calculated from the NCAR Electra's gust probe data by Chen (1981). The fluxes calculated in the Chen analysis are extremely small as compared with those obtained in other marine stratus situations: Typical values of both heat and moisture fluxes were less than 10 W m^{-2}, and showed no systematic organization with height.

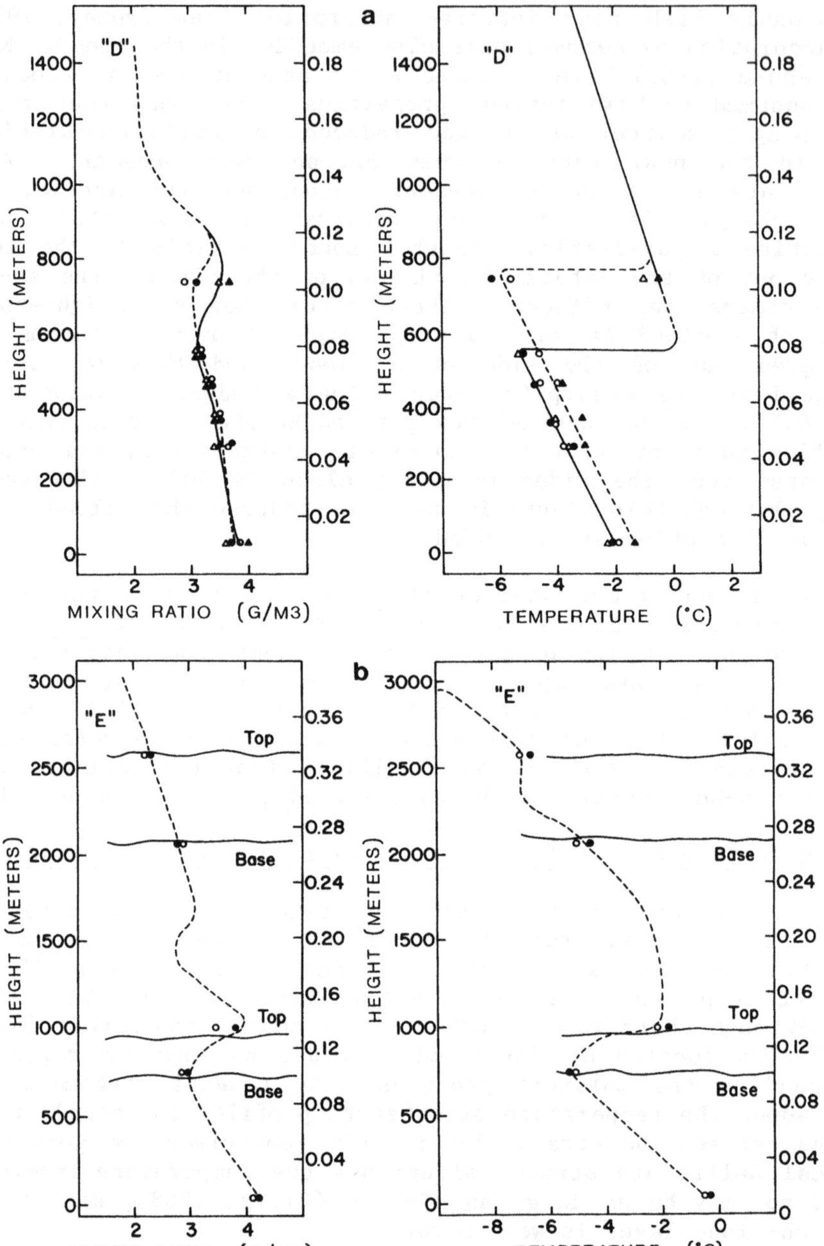

Fig. 8. Vertical profiles of temperature (^{O}C) and mixing ratio (g/kg) for (a) flight "D", 27 July 1975, (b) flight "E", 27 July 1975, and (c) flight "F", 27 July 1975.

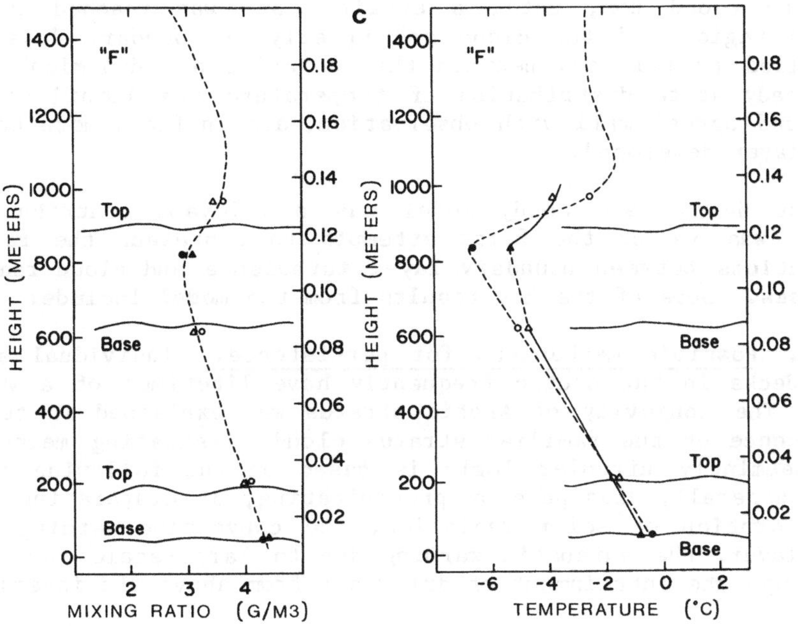

Fig. 8c.

4. FORMATION AND PERSISTENCE OF ARCTIC STRATUS CLOUDS

A theory for Arctic stratus cloud formation was developed by Herman and Goody (1976). We review here the key features of that theory, and note the various aspects that could be subject to some modification in view of the data that subsequently have become available from the AIDJEX and 1980 Arctic Stratus Experiment Field Programs.

Herman and Goody viewed Arctic stratus formation as a problem in air mass modification:* As moist air was advected into the Arctic basin it was cooled to condensation both radiatively and diffusively until a liquid water cloud formed. (These processes are indicated schematically in Fig. 9.) Once liquid water was present, cloud growth was accelerated by longwave radiative loss

---

*The statement by Loomis (1868, p. 97) is of interest here: "During an Arctic summer the temperature of the earth rises much above that of the sea, portions of which are covered with immense fields of ice. The air resting upon the earth partakes of its temperature, and when this warmer air is brought in contact with the colder ocean, a portion of its vapor is condensed, and a heavy mist is formed."

from the cloud top, until a steady state was reached in which various regions of the cloud and underlying boundary layer were controlled by balances between the radiative and dynamical terms. The steady state distribution of temperature and liquid water in the cloud agreed well with observation, and in fact, more than one cloud layer developed.

The Herman and Goody model was a relevant contribution in that it was one of the first attempts to represent the range of interactions between boundary layer turbulence and cloud radiative processes. Some of the key results from the model include:

i. Possible mechanisms for persistence. Individual stratus cloud decks in the Arctic frequently have lifetimes of a week or more. The longevity of Arctic stratus was explained in terms of the absence of the familiar stratus cloud dissipating mechanisms. This seemingly circular logic is based on the following result: Clouds generally dissipate by precipitating, or evaporating due to the absorption of solar radiation, the convective warming of the mixed layer, the adiabatic warming due to large-scale subsidence, or through the entrainment of drier air from above the inversion.

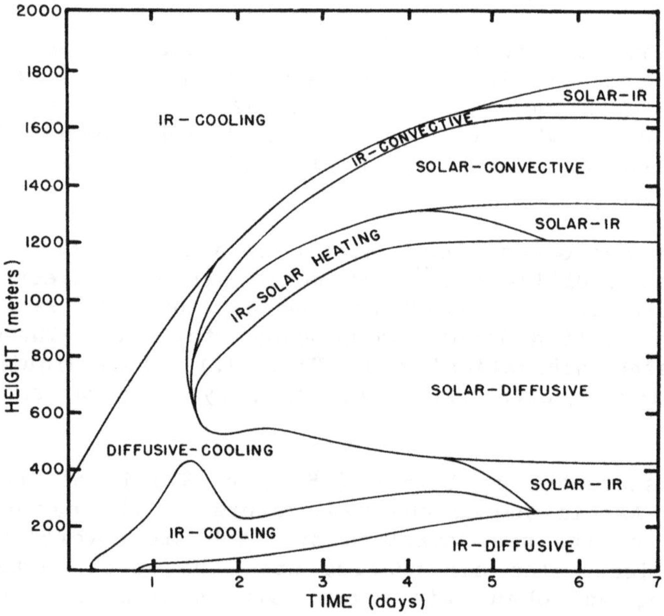

Fig. 9. Components of the radiative and turbulent regimes computed for case of warm air flowing over a colder ice surface. (From Herman and Goody, 1976).

In the summertime Arctic, these processes are either weak or absent. Radiative heating of the cloud is weaker than in middle latitudes because of lower solar elevations, and convection due to heating of the surface is essentially absent since any excess in the surface energy budget is expended in the melting of ice rather than in warming the surface. Large-scale subsidence is also weaker than in the sub-tropics or middle latitudes, except in transient synoptic disturbances. Hence once a stratus deck forms in the Arctic, it has the opportunity to meander about for a relatively longer period of time that it would if it formed under typical mid-latitude conditions.

ii. Dependence on radiative parameters. A discouraging aspect of the Herman and Goody model was that the results were extremely sensitive to the cloud droplet absorption coefficient, k_c. Relatively small variations of k_c determined large variations of thickness, cloud top temperature, initiation of condensation, and the occurrence of layering. This is a troublesome result because, following the discussion in Section 6b, the cloud's absorption parameters are extremely difficult to assess accurately.

It is conceivable that our ability to predict accurately the radiative energy balance at the surface, which enters into ice thickness calculations, may be limited by the uncertainity in the determination of this key empirical parameter.

5. RADIATIVE EQUILIBRIUM IN THE CLOUD INTERIOR

Because the turbulent fluxes in Arctic stratus clouds are so small, it is instructive to consider a description of the cloud based only on radiative considerations.

For the sake of argument, consider a homogeneous cloud layer with a total optical thicknesses in the infrared of τ^*_T, and in the solar spectrum of τ^*_S. Let the cloud rest on a surface whose temperature is determined by radiation and other processes (such as the melting of sea ice).

In radiative equilibrium the temperature change in the cloudy atmosphere due to radiative processes is zero, i.e., $\partial F/\partial z = 0$, where F is the net flux consisting of a solar component, F_S, and a thermal component, F_T,

$$F = F_S + F_T \tag{6.4}$$

The fluxes are easy to compute using a two-stream approximation. The solar flux is simply

$$F_S = C_1 e^{\beta\tau_S} + C_2 e^{-\beta\tau_S} \tag{6.5}$$

where C_1 and C_2 are constants that are determined by boundary conditions at the top of the cloud (downward flux = solar irradiance) and at the surface (upward flux = reflected flux). β is the increase in the absorption path length due to multiple scatterings.

In the infrared region of the spectrum, the two stream equations (1.24) may be written as a differential equation,

$$\frac{d^2F_T}{d\tau_T^2} = 3F_T - 2\pi\sqrt{3}\ \frac{dB}{d\tau_T} \tag{6.6}$$

where B is the Planck function.

We may combine eqs. (6.4) and (6.6) to yield

$$-3F = \left(\frac{d^2}{d^2\tau_T} - 3\right)F_T - 2\pi\sqrt{3}\ \frac{dB}{d\tau_T} \tag{6.7}$$

In a homogeneous layer the solar and thermal optical depths may be considered proportional to one another, i.e., $\tau_S = \gamma\tau_T$, and eq. (6.7) can be integrated once to obtain the Planck function, B, from which the temperature can be determined. We obtain,

$$2\ B = \left(\frac{\beta\gamma}{\sqrt{3}} - \frac{\sqrt{3}}{\beta\gamma}\right)\left[C_1\ e^{\beta\gamma\tau_T} - C_2\ e^{-\beta\gamma\tau_T}\right] + \sqrt{3}\ F\ \tau_T + \alpha_o \tag{6.8}$$

where F and α_o are determined by the infrared boundary conditions (downward flux = 0 at top of atmosphere; upward flux = $\sigma(T^*)^4$ at bottom).

The solution to eq. (6.8) depends only on the total opacity of the cloud, τ_T^*, the temperature T* of the lower boundary (which appear in the coefficients C_1 and C_2), and the parameter $\beta\gamma/\sqrt{3}$, which includes the cloud's two-stream absorption and scattering parameters.

The radiative equilibrium temperature profiles corresponding to the solutions to eq. (6.8) are plotted in Figure 10 for T* = 273K (corresponding approximately to melting ice conditions), and for $\beta\gamma/\sqrt{3}$ = 0.07, (corresponding to the best estimates of Arctic stratus optical properties).

The temperature in the cloud is controlled by the apportionment of solar heating and infrared cooling. For small optical depths (τ_T = 3) thermal radiation escapes to space from all levels in the cloud, and temperatures decrease monotonically from base to top. As the cloud becomes more opaque to thermal radiation,

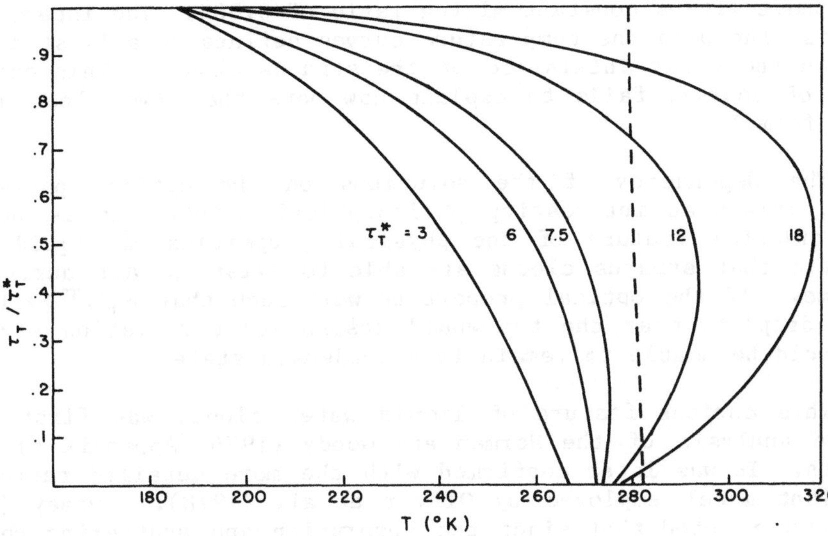

Fig. 10. Radiative equilibrium temperature profiles. (From Herman and Goody, 1976.)

maximum thermal emission to space takes place at levels that are increasingly closer to the cloud top, while the interior continues to be heated by absorption of solar radiation. For τ_T greater than about 7.5, temperatures actually increase with height. For these cases the solar radiation is trapped in the interior to cause heating, but cannot penetrate far enough into the cloud to counteract the cooling at the boundary. Consequently, a cold-warm-cold structure develops, cooled at the top by cooling-to-space, at the bottom by the cold boundary, and warmed in the middle by solar radiation.

The solutions for $\beta\gamma/\sqrt{3} \ll 1$ are referred to as "greenhouse solutions," and occur when the absorption of solar radiation takes place deep within the cloud, while thermal radiation is emitted near cloud top. The solutions occurring when the optical properties are such that $\beta\gamma/\sqrt{3} \gg 1$ (not shown here) are referred to as "thermospheric solutions," and occur when intense solar heating occurs at the top of the atmosphere, with infrared cooling taking place from deeper layers.

The structure of the greenhouse-type temperature profiles were suggested by Herman and Goody as a possible explanation for the layered structure of Arctic stratus clouds: When the interior of the cloud becomes so warm that the dew point temperature is exceeded, the cloud must evaporate. The broken line in Figure 10 runs parallel to the lapse rate of dew point temperature in an atmosphere with a constant mixing ratio of 0.003. The intersection of this line with the temperature curves defines to a first approximation the clear interstice of the stratus cloud. (This explanation, of course, fails to explain how more than two cloud layers would form.)

The dependency of the solutions on the optical parameter, $\beta\gamma/\sqrt{3}$, raises an interesting philosophical point. It is perhaps an accidental feature of the physical properties of liquid water droplets that stratus clouds are able to exist at all during the daytime. If the optical properties were such that $\beta\gamma/\sqrt{3} >> 1$, the cloud droplets near the top would absorb solar radiation strongly and would be unable to remain in a condensed state.

This curious feature of liquid water clouds was first noted in the analysis of the Herman and Goody (1976, Appendix A) model results. It was later confirmed with the more detailed radiative-turbulent model employed by Oliver et al. (1978). Twomey (1978) has further noted that since the absorption and scattering characteristics of the cloud depend on drop size, cloud occurrence depends indirectly on the aerosol concentration of the atmosphere as well.

6. RESULTS FROM THE 1980 ARCTIC STRATUS EXPERIMENT

In June of 1980 a comprehensive aircraft investigation of Arctic stratus was conducted over the Beaufort Sea for the purposes of making simultaneous measurements of the clouds' radiative and microphysical properties, and boundary layer turbulence, and relating these to synoptic-scale meteorological conditions. (The experiment was conducted jointly by K. Jayaweera of Univ. of Alaska, W. Barchet of Battelle, V. Saxena of N. Carolina State Univ., and this author.) The experiment provided one of the first sets of cloud radiation measurements to be obtained with simultaneous measurements of cloud droplet number density and size distribution. The main results are summarized below.

Solar Radiation. Solar radiation was measured with two systems, namely, the Eppley flux pyranometers, and a silicon detector which provided the additional measurement of the diffuse component of the radiation. The Knollenberg FSSP probe was used to determine the number of droplets in 15 size bins from 1.6 to 23.5 μm. The radiation measurements were compared to calculated values obtained with a detailed radiation model developed by Slingo and Schrecker (1982). A pervasive feature of the droplet measurements was the large variability of droplet radii, size distribution, and number density from cloud to cloud, and within a single cloud. This result casts doubt on the likelihood of representing successfully the transfer of radiation in the cloud in terms of a single parameter, such as the equivalent radius. Instead, it appears that the complete drop size spectrum and its height variation will need to be represented explicitly in the radiation calculations.

The ratio of diffuse-to-total radiation above the cloud varied from 0.12 to 0.30, with a general tendency for the ratio to increase toward longer wavelengths due to the stronger absorption of diffuse radiation at longer near infrared wavelengths. This ratio was very close to unity in the interior of the cloud since there almost all of the radiation had been scattered from the incident beam. Interestingly, the Slingo-Schrecker radiation model was unable to produce the observed diffuse-to-total ratio unless the model included a background distribution of aerosol. However, the inclusion of an aerosol did not affect the absorption within the cloud.

The transmittance, reflectance, and absorptance obtained with the two radiometer systems were in good agreement with each other, and with the Slingo-Schrecker model, except for very thin clouds. Values of these radiative parameters were tabulated as a function of optical depth. A complete description of the solar radiation analysis appears in the paper by Herman and Curry (1984).

Infrared Radiation. The infrared, or longwave, properties of the clouds were studied with Eppley pyranometers in conjunction with the microphysical properties obtained with the Knollenberg FSSP. The mass absorption coefficient, which is a crucial parameter that affects the cloud transmissivity, was found to be a sensitive function of the droplet size distribution. Typically, the larger drops were associated with the smaller absorption coefficients. A similar sensitivity was obtained for the cloud reflectance. Infrared cooling rates near cloud top were inferred to be as large as 6°C per hour. As with the solar radiation results, the sensitivity to the droplet sizes again makes it difficult to characterize the clouds' infrared radiative properties in terms of a single descriptive parameter, such as the cloud liquid water path. The infrared properties of Arctic stratus are fully summarized in the paper by Curry and Herman (1985).

Relationship to Synoptic Conditions. Curry and Herman (1985) investigated the large-scale conditions that support Arctic stratus formation in a combined analysis of satellite cloud data and the operational analyses from the European Center for Medium-Range Weather Forecasts (ECMWF). Large amounts of low cloud cover during the June experiment were attributed primarily to the low level advection of moisture, and cooling due to radiation and boundary layer turbulence. This interpretation of low-level stratus formation is consistent with the formation mechanisms implied in the numerical model results of Herman and Goody (1976). The occurrence of clouds at higher levels was associated with the large-scale transport of both moisture and cooler temperatures. At upper levels a mean large-scale relative humidity convergence was observed which was due to rising motion and advection, which in turn was balanced by a residual moisture sink resulting from condensation. Curry and Herman also examined the daily amounts of mid-level cloudiness in terms of the prevailing synoptic situation, and found that in contrast to the low cloud cover, the mid-level cloud cover varied primarily in response to variations in the large scale transports of heat and moisture associated with the synoptic flow. Curry and Herman noted that while their study illustrated plausible mechanisms for summertime Arctic stratus formation, it was not possible to quantify the implied relationships into a form that would prove useful for their parameterization in global GCMs. The authors conclude their study with the philosophical observation that, although it is not yet known precisely how accurately Arctic clouds will need to be calculated, it is likely that the accuracy will need to be high, given the importance of the clouds for the energy balance of the pack ice. They suggest that the current diagnostic techniques for modeling clouds in GCMs is probably inadequate, and will need to be replaced with physical models, perhaps also including a prognostic equation for liquid water.

ACKNOWLEDGMENTS

These lectures were written while the author was supported at the University of Oxford under a joint NASA/NOAA/NSF research program supported under NSF Grant ATM-8014177.

REFERENCES

Ball, F. K. (1960) Control of inversion height by surface heating. Quart. J. Roy. Met. Soc., 86: 483-494.

Chen, C. F. (1981) Turbulent flux measurements in the arctic planetary boundary layer. University of Wisconsin, M.S. thesis, 96 p.

Curry, J. A. and G. F. Herman (1985) Infrared radiative properties of summertime Arctic stratus clouds. J. Cli. App. Met., 24: 525-538.

Curry, J. A. and G. F. Herman (1985) Relationships between large scale heat and moisture budgets and the occurrence of Arctic stratus clouds. Mon. Wea. Rev., 113.

Deardorff, J. W. (1976) On the entrainment rate of a stratocumulus-topped mixed layer. Quart. J. Roy. Met. Soc., 102: 563-582.

Deardorff, J. W. (1980) Cloud top entrainment instability. J. Atmos. Sci., 37: 131-147.

Deardorff, J. W. (1981) On the distribution of mean radiative cooling at the top of a stratocumulus-capped mixed layer. Quart. J. Roy. Met. Soc., 107: 191-202.

Griggs, M. (1968) Aircraft measurements of albedo and absorption of stratus clouds, and surface albedos. J. Appl. Met., 7: 1012-1017.

Heintzenberg, J. (1980) Particle size distribution and optical properties of arctic haze. Tellus, 32: 251-260.

Herman, G. F., 1975: Radiative-diffusive models of the arctic boundary layer. Mass. Inst. Tech., Department of Meteorology Report COO-2195-21, 170 p.

Herman, G. F. (1977) Solar radiation in summertime arctic stratus clouds. J. Atmos. Sci., 34: 1425-1432.

Herman, G. F. (1980) Thermal radiation in arctic stratus clouds. Quart. J. Roy. Met. Soc., 106: 771-780.

Herman, G. F. and R. M. Goody (1976) Formation and persistence of summertime Arctic stratus clouds. J. Atmos. Sci., 33: 1537-1553.

Herman, G. F. and J. A. Curry (1984) Observational and theoretical studies of solar radiation in Arctic stratus clouds. J. Cli. Appl. Met., 23: 5-24.

Huschke, R. E. (1969) Arctic cloud statistics from air calibrated surface weather observations. RAND Corp. Santa Monica, California, RM-6173-PR, 79 p.

Kahn, P. H. and J. A. Businger (1979) The effect of radiative flux divergence on entrainment of a saturated convective boundary layer. Quart. J. Roy. Met. Soc., 105: 303-306.

Koptev, A. P. and A. I. Voskresenskii (1962) On the radiation properties of clouds. Trudy Ark. Antarkt. Nauch. Issled. Inst., 239: 39-47. (Translated by the Rand Corp., Santa Monica, California, RM-5003-PR, 1966.)

Lilly, D. K., 1968: Models of cloud-topped mixed layers under a strong inversion. Quart. J. Roy. Met. Soc., 94: 292-309.

Lilly, D. K. and W. H. Schubert (1980) The effects of radiative cooling in a cloud-topped mixed layer. J. Atmos. Sci.: 482-487.

Liou, K. N. (1976) On the absorption, reflection, and transmission of solar radiation in cloudy atmospheres. J. Atmos. Sci., 33: 798-805.

Loomis, E. (1868) A Treatise on Meteorology. Harper Brothers.

Marshunova, M. S. (1961) Principal characteristics of the radiation because of the underlying surface and of the atmosphere in the Arctic. Trudy Arkt. Antarkt. Nauch. Issled. Inst., 229: 5-53. (Translated by the Rand Corp., Santa Monica, California, RM-5003-PR, 1966.)

National Aeronautics and Space Administration (1981) Clouds in Climate: Modelling and Satellite Observational Studies. Report of Workshop at NASA Goddard Institute for Space Studies, New York, 1980.

Neiburger, M. (1949) Reflection, absorption and transmission of insolation by stratus cloud. J. Met., 6: 98-104.

Newiger, M. and K. Bahnke (1981) Influence of cloud composition and cloud geometry on the absorption of solar radiation. Beitr. Phys. Atm., 54: 370-382.
Oliver, D. A., W. S. Lewellen and G. G. Williamson (1978) The interaction between turbulent and radiative transport in the development of fog and low level stratus. J. Atmos. Sci., 35: 301-316.
Paltridge, G. W. (1971) Solar and thermal radiation flux measurements over the east coast of Australia. J. Geophys. Res., 76: 2857-2865.
Platt, C. M. R. (1972) Airborne infrared radiance measurements (10- to 12-micron wavelength) off the tropical east coast of Australia. J. Geophys. Res., 77: 1597-1609.
Platt, C. M. R. (1976) Infrared absorption and liquid water content in stratocumulus clouds. Quart. J. R. Met. Soc., 102: 553-561.
Platt, C. M. R. (1981) Extended clouds and radiation: A report compiled for the International Radiation Commission and presented at IAMAP, Hamburg. International Council of Scientific Unions, World Meteorological Organization, Geneva, 35 p.
Randall, D. A. (1980a) Entrainment into a stratocumulus layer with distributed radiative cooling. J. Atmos. Sci., 37: 148-159.
Randall, D. A. (1980b) Conditional instability of the first kind, upside down. J. Atmos. Sci., 37: 125-130.
Robinson, G. D. (1958) Some observations from aircraft of surface albedo and the albedo and absorption of cloud. Arch. Met. Geophys. Bioklim., B(9): 28-41.
Schubert, W. H. (1976) Experiments with Lilly's cloud-topped mixed layer model. J. Atmos. Sci., 33: 436-446.
Slingo, A. and H. M. Schrecker (1982) On the shortwave properties of stratiform water clouds. Quant. J. R. Met. Soc., 108: 407-426.
Twomey, S. (1972) The effect of cloud scattering on the absorption of solar radiation by atmospheric dust. J. Atmos. Sci., 29: 1156-1159.
Twomey, S. (1976) Computations of the absorption of solar radiation by clouds. J. Atmos. Sci., 33: 1087-1091.
Twomey, S. (1978) Influence of the aerosol on optical properties of clouds. In Report of the JOC Study Conference on Parameterization of Extended Cloudiness and Radiation for Climate Models, International Council of Scientific Unions, Geneva.
Vowinckel, E. and S. Orvig (1964) Energy balance of the Arctic. II: Longwave radiation and total radiation balance at the surface in the Arctic. Arch. Met. Geophys. Bioklim., B(13): 451-479.
Vowinckel, E. and S. Orvig (1970) The climate of the North Polar Basin. In World Survey of Climatology, Climates of the Polar Regions (S. Orvig, Ed.), Elsevier, Amsterdam, Vol. 14, 370 p.

CHAPTER 7

KINEMATICS OF SEA ICE

Alan S. Thorndike

Department of Physics
University of Puget Sound
Tacoma, Washington 98416

7. INTRODUCTION

Kinematics is that branch of physical science dealing with motion itself and not with the causes of the motion. Attention is focused, in this chapter, on how sea ice moves. The deeper question of why sea ice moves as it does is addressed in the chapter on sea ice dynamics. For some scientific and practical applications, knowledge of the ice kinematics (as opposed to the dynamics) is sufficient. As examples, consider gouging of the sea floor in shallow water, the transport of natural debris or pollutants, the impact loads on structures, navigability in ice regions, and the movement of the ice edge. Aside from these applications, the study of kinematics is a basic step toward understanding the dynamics.

Figure 1 illustrates the idea that the motion of sea ice is the response of the ice pack to external forces. To understand the response of the ice one needs to monitor both the driving forces and the motion. Since we will discuss only the motion here, there will be the underlying ambiguity of whether the results

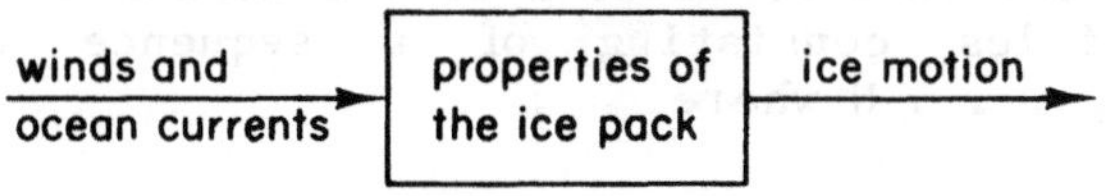

Fig. 1. Ice motion represents the response of a particular material to certain external forces.

inform about properties of the ice itself or about the driving forces. We will see for example that the ice motion is nearly nondivergent. But we should not conclude on this basis alone that the ice is an incompressible material. The correct explanation could be that the net driving force is itself nearly non-divergent. Because of this ambiguity in the interpretation of kinematic data, it is useful to think of the study of kinematics as part and parcel of the broader study of sea ice dynamics.

The fundamental kinematic notion is that all pieces of ice have an identity which is preserved in time. If we identify a piece of ice by making a small mark on the ice surface at position X at time o, the assumption is that at some later time t, the piece of ice as identified by that mark will be at a position x. This defines the position function x(t,X) with x(o,X) = X. We do not mean to imply by this that ice floes retain their identity indefinitely, only that it is possible to track individual points. Two points originating on the same ice floe may very well wander apart in time, but we assume we could keep track of them.

There are difficulties with the notion of a position function. It is natural to consider the space domain to be the two dimensional surface of the Arctic Ocean, say $\underline{X}$. At t=o there will generally be some points in $\underline{X}$ which are not covered by ice, so x(t,X) is not defined for some $X \in \underline{X}$. Further, ice is always being created and destroyed thermodynamically, so the position function x(t,X) is only defined for an interval of time.

Finally, the idea that the small mark retains its identity is questionable. This is a common problem in the definition of the motion of turbulent fluid. After a time the marked fluid has mixed with the unmarked and it is no longer possible to assign a position to the mark. Something of the same sort happens with sea ice. Despite these shortcomings, the notion of a position function is the best we have. We will use the term trajectory for the function x(t,X) where X is held constant. The initial position X serves to label ice particles and we refer to the trajectory x(t,X) of the particle X.

5.1 Observations

The most common observations of ice motion are of trajectories of ice particles consisting of a sequence of measurements (z_i, t_i), i=1, . . . N where

$$z_i = x(t_i,X) + \varepsilon_i$$

The measurement error is represented by ε_i.

In passing we mention that other kinds of measurements of ice motion have been made. Hunkins (1967) and McPhee (1978) for instance inferred the ice velocity using current meters suspended from the ice into what they assumed to be essentially a static ocean. These data were used to study motion on the time scales of hours. An attempt to study motions on much shorter time scales using accelerometers was made by Craig (1972).

Many techniques have been used to measure ice trajectories. These are summarized in Table 1. Each technique has its good and bad features. Over the past decade most data have been obtained by satellite positioning which works somewhat as follows. Suppose a stable frequency f_o is transmitted by a device on the ice. The signal received at the rapidly moving satellite will have a frequency f which has been shifted by the Doppler effect: $f = f_o + \Delta f$. The Doppler shift Δf is related to the rate of change of distance between the device on the ice and the satellite:

$$\Delta f \sim \frac{d}{dt}||x_{sat} - x_{ice}||.$$

The received frequency f is measured at several times. If the satellite coordinates are known at these times, then each measurement produces an equation with unknowns f_o and x_{ice}. Generally several measurements are made during the 10 to 20 minutes it takes for the satellite to pass by. The several equations are solved simultaneously for f_o and x_{ice}. Precise positioning requires a stable transmit frequency, precise measurement of the received frequency, and precise knowledge of the satellite coordinates. With care, errors can be controlled to the order of tens of meters, as in the best use of the Navy Navigation Satellite System.

Satellite systems which serve primarily to relay data from automatic data platforms (or buoys) to data processing centers also determine the location of the transmitting data platform, using the same Doppler positioning principle. The ARGOS system currently on the NOAA-B satellite can relay data from up to 30 sensors and determine platform locations to an accuracy of a few hundred meters about ten times per day. Fully automatic platforms cost in the neighborhood of $6,000 with additional costs depending on the desired sensors.

Satellite imaging systems can also be used to measure ice motions, provided that some features on the ice can be identified in a sequence of images. Because of their all season and all weather capability and good resolution, imaging radar systems will probably be best. The basic angular resolution of these systems is approximately the ratio of the wavelength of the radar signal to the diameter of the antenna. To achieve an angular resolution of 10^{-5} (10 meters at a range of 1,000 kilometers) with a wave-

Table 1. Techniques for measuring sea ice motion.

Technique	Basic measurement	Sampling rate	Accuracy	Comments
Hunkins, et al., 1971 celestial navigation	Azimuth and elevation of heavenly bodies	1 per day	10^3 m	Weather dependent
Martin et al., 1978 satellite navigation	Doppler shift of stable transmitted signal	30 per day	30-500 m	
Thorndike, 1973 acoustic tracking	Travel time from ice to fixed reference on the ocean floor	up to 1 per minute	5 m	Limited range, elaborate instrumentation, relative position only
Tucker et al., 1980 Tabata et al., 1980 radar	Travel time from source on shore to to target on ice	no limit	1 m	Limited to near shore
Hall, 1980 Landsat imagery	Location of identifiable ice features	erratic	80 m	Weather dependent, good space resolution, poor time resolution
Hall & Rothrock, 1981 synthetic aperature radar imagery (SEASAT)	Location of identifiable ice features	1 per day	80 m	No system available at present, properties of future systems unclear

length of 25 cm (1.2 GHz) requires an antenna 25 km in diameter. Although such large antennas cannot be constructed in space, it is possible to synthesize large antennas by using data from several points along the satellite's orbit. The determinations of the geographical position of an ice feature with the SEASAT data contained errors of up to 3 kilometers (see Hall and Rothrock, 1981). Unless these errors can be reduced the data are not particularly valuable for measuring the displacements over intervals of a few days or less. However, the errors are highly correlated in space and are essentially eliminated in estimates of the spatial variability of the ice motion.

The attractive feature of imaging radar systems is their potential to sample densely in space. Hall and Rothrock's work suggests that it will be possible to track roughly one feature per square kilometer, which will resolve most of the spatial structure of the field of motion. Techniques for extracting data from the images or from the raw data are still rather primitive. No doubt satisfactory automated techniques for identifying and tracking features will be developed when the need arises. At present there is no imaging radar system in space. Planning is underway for a system to be in operation perhaps by 1985.

5.1.1 Sources of Data. The earliest ice motion data are the trajectories of ships beset in the ice. These are followed by the trajectories of numerous Soviet and U.S. drifting research stations, and more recently by the trajectories of automatic data buoys. The following list of data sources is by no means complete.

Experiment	Reference	Description
T-3	Hunkins et al. 1971	Monthly positions May 1962-December 1970
Beset ships and drifting stations before 1970	Hastings, 1971	A chart showing most prior trajectories, one point per month; includes: Fram, Jeanette, Maud, Sadko, Sedov, Tegetthoff, British Trans-arctic Exp., Alpha, Arlis I, Arlis II, Charlie, North Pole 1-20, T-3.
AIDJEX 1972	AIDJEX Staff, 1972	100 km triangle, Beaufort Sea, March and April 1972.
AIDJEX 1975-76	Thorndike and Cheung, 1977	Manned camp positions and velocities tabulated at 6 hr intervals, daily buoy positions.

Experiment	Reference	Description
1976-77 buoys	Thorndike and Cheung, 1977	Tabulated daily positions Beaufort Sea buoys.
1979 buoys	Thorndike and Colony, 1980	Tabulated daily positions, 25 buoys, analyses of surface pressure plotted daily.
1980 buoys	Thorndike and Colony, 1981	As above.
1981 Fram I, 1979	Hunkins et al. 1979	March-May 1979; 84°N, 9°W.
LOREX 79	Popelar et al. 1981	Three stations, 100 km spacing April, May 1979; 88°-90°N.

5.2 The General Circulation

The main features of the long term circulation, Figure 2, are the clockwise circulation of the Beaufort Sea -- and the motion of ice from the Siberian coasts across the North Pole and through the Greenland-Spitsbergen passage. Time honored nomenclature for these

Fig. 2. The general circulation of pack ice in the Arctic Ocean.

features are the Beaufort Gyre and the Transpolar Drift Stream. Some handy numbers for these long term features are:

Center of Beaufort Gyre: 80°N, 155°W, half way between Point Barrow, Alaska and the North Pole

Time to make 1 circuit: 5 years

Time to traverse Transpolar Drift Stream: 3 years

Area flux through Greenland-Spitsbergen Passage: 300 km^2/day or about 20% of the area of the basin per year.

This pattern of motion exists only as an average over several years. On shorter time scales there are departures from the long term pattern. Consider for example the trajectories plotted in Figure 3 for the year 1979 and Figure 4 for 1980. The trajectories are characteristically meandering and convoluted showing that on monthly time scales the ice motion differs markedly from the long term mean. On shorter time scales, not resolved in these figures, the motion is even more irregular. Notice the major anomaly which occurred in the summer of 1980 when the motion of several buoys for several months was in the opposite sense from the long term mean. These departures of the actual motion from the long term pattern in most cases represent the response of the ice to the passage of atmospheric systems.

If we ignore the forces and examine only the ice motion, the departures from the mean circulation appear aperiodic and chaotic. The departures can be thought of as random but they are not without structure. Our objective now is to clarify this structure.

5.3 Ice Velocity

The ice velocity field can be defined by the relationship

$$u(x,t) = \lim_{h \to 0} \frac{x(t+h,X) - x(t,X)}{h} \tag{7.1}$$

Here the particle label X plays no role.

This definition is meaningful only if the limit exists. From a practical point of view it is useful only if the limit exists <u>and</u> is approached when the interval h decreases to the time interval τ between observations.

In Figure 5 several sets of observations of ice motion are plotted, showing the variation of one coordinate of position versus time. Successive data sets divide the sampling interval τ

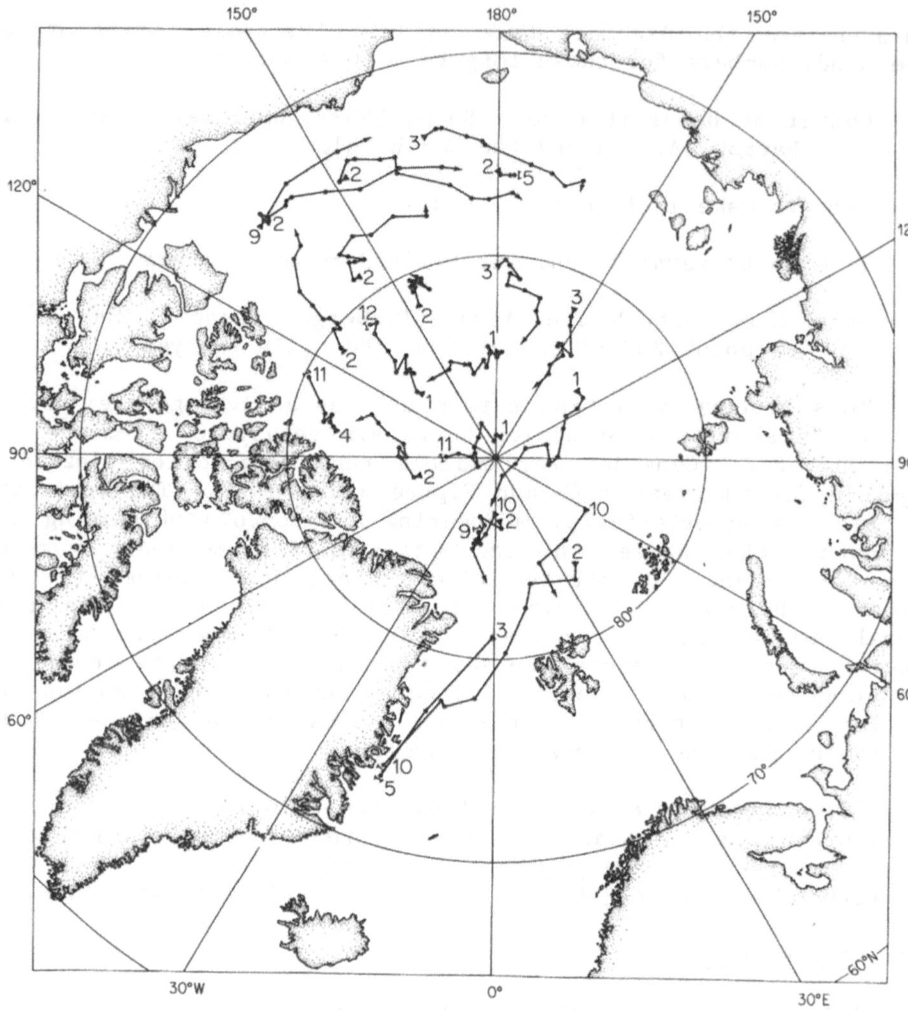

Fig. 3. Trajectories of drifting buoys during 1979.

by 15 and improve space resolution by the same factor, giving a sequence of closer and closer perspectives on the motion. By constructing velocity estimates, $u_k(t) = [x(t+k\tau,X) - x(t,X)]/k\tau$ for k decreasing to 1, one can examine the limiting process. In the first two figures of the sequence, u_k continues to change appreciably for small k. In the last two figures, $u_k(t)$ at most times t becomes almost independent of k for small k, implying that one these time scales, $\tau \lesssim 15$ minutes, the ice indeed possesses a velocity.

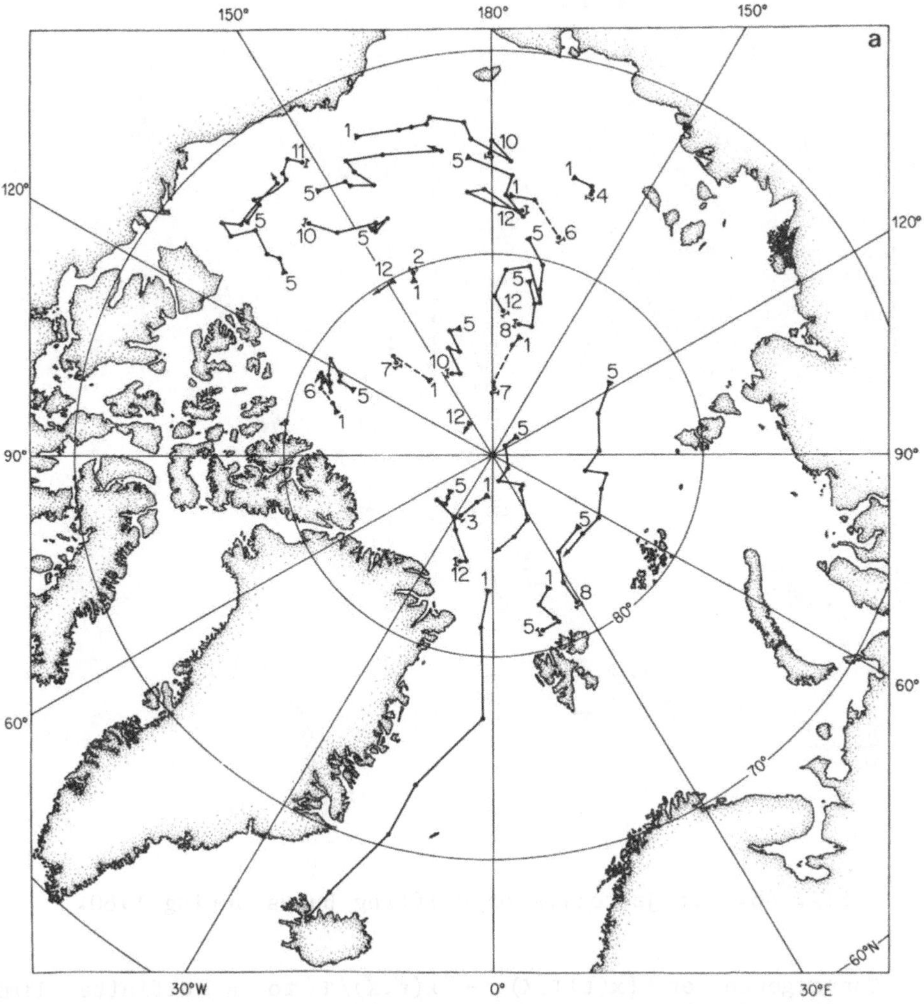

Fig. 4a. Trajectories of drifting buoys during 1980.

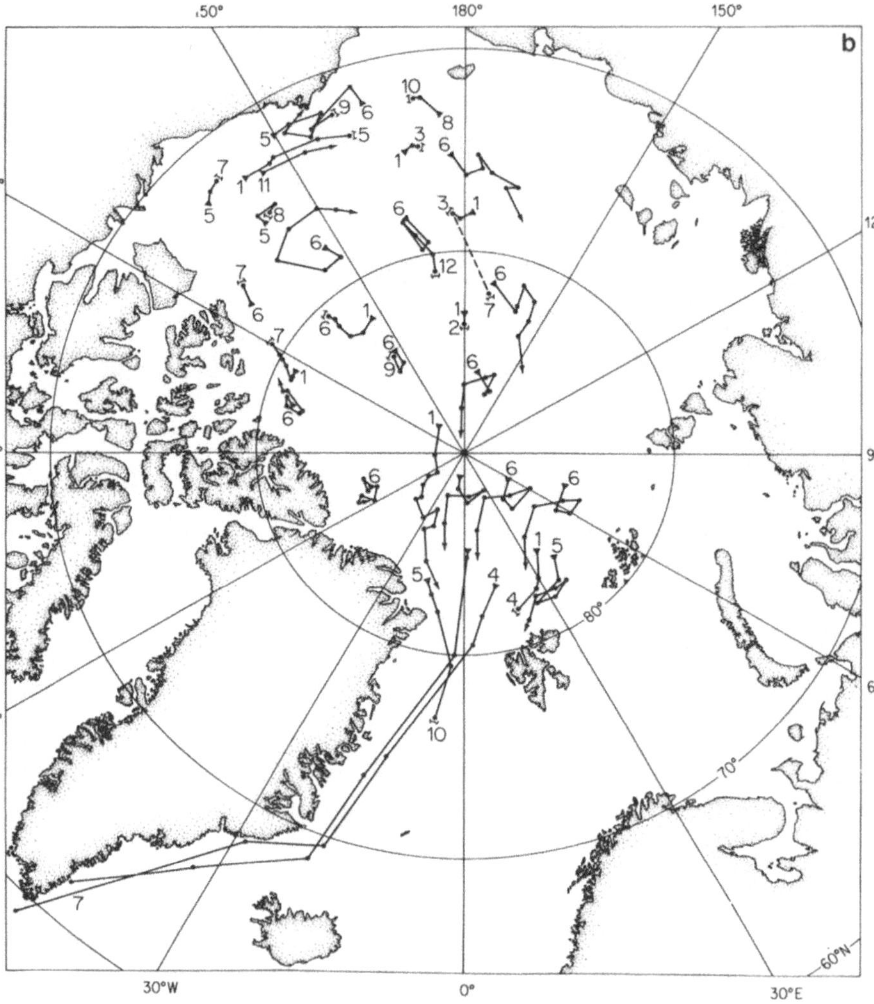

Fig. 4b. Trajectories of drifting buoys during 1980.

Convergence of $[x(t+\tau,X) - x(t,X)/\tau$ to a definite limit implies that the increment $x(t+\tau,X) - x(t,X)$ is proportional to τ for small t. For random processes, a useful condition for convergence is mean squared convergence in which the variance of the increment becomes proportional to τ^2 for small τ. This can be readily tested for ice trajectories by plotting $E(x(t+\tau,X) - x(t,X))^2$ versus τ on a log-log plot, as is done in Figure 6. The process is differentiable if the graph has slope of 2 for small τ.

From a practical point of view, the process ceases to be differentiable at measurement intervals τ for which the slope of

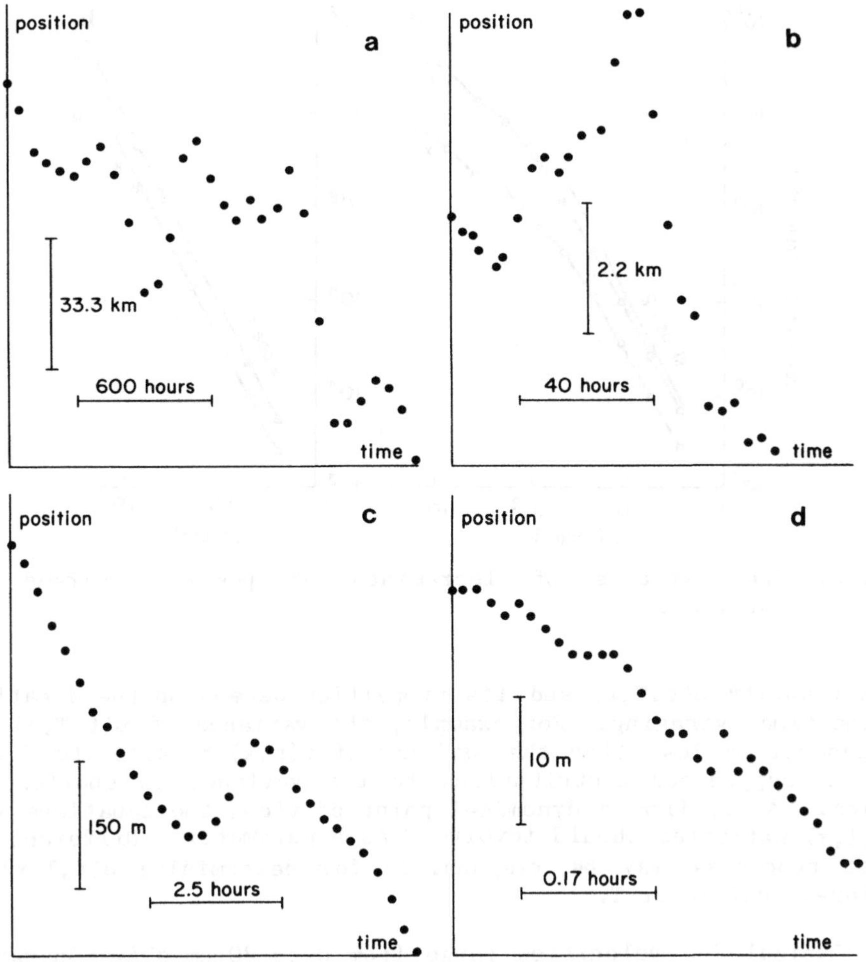

Fig. 5. x vs. τ on different time and space scales.

the graph departs appreciably from 2. The evidence in Figures 5 and 6 implies that the ice does have a velocity and that it can be resolved with a sampling interval of about three hours.

Time averaged velocities can be defined without reference to the limiting process. Let u(t,T,x) be the time averaged velocity at t.

$$u(t,T,x) = \frac{1}{T}\left[x(t+T/2,X) - x(t-T/2,X)\right] = \frac{1}{T}\int_{t-T/2}^{t+T/2} u(s,x)ds. \tag{7.2}$$

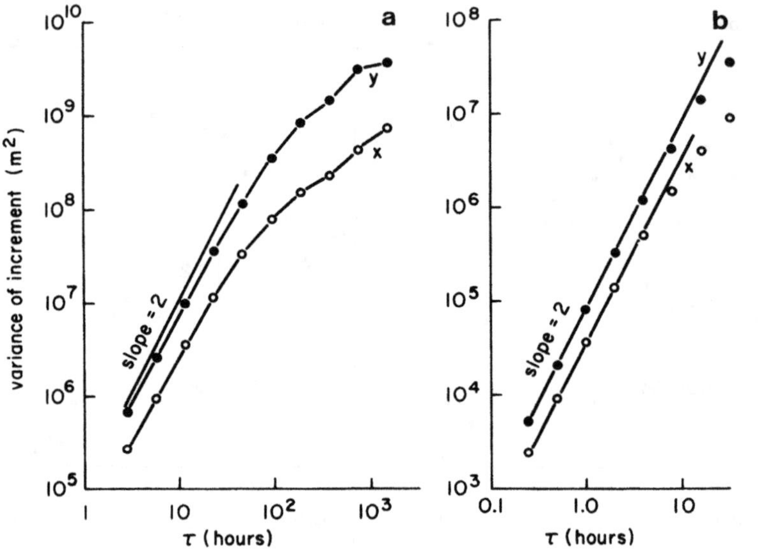

Fig. 6. The variance of increments of position versus time interval.

This quantity u(t,T,x) and its properties depend on the duration T of the time averaging. For example, the variance of u(t,T,x) will in general be less than the variance of u(t,x) because the T average has suppressed contributions to the variance on shorter time scales. Also, from a dynamical point of view, the equations which u(t,T,x) satisfies should involve T as a parameter. Different physical processes may be responsible for determining u(t,T,x) for different values of T.

Typical ice velocities range from 0 to 20 cm s^{-1}. An extreme velocity of 140 cm s^{-1} has been observed. Two histograms of ice speed are shown in Figure 7, one corresponding to a full year of observations, the other restricted to summer observations. The winter data contained several periods of essentially zero motion. During the summer the ice was never observed to stop.

In the following pages particular importance is attached to the time and space variability of the ice motion. We will often refer to the variance of velocity, $E((u-\bar{u})^2 + (v-\bar{v})^2) = q^2$. This quantity has been evaluated from many observations; it varies appreciably with season and with location. Table 2 gives estimates of the velocity variance for each of the buoys shown in Figure 3.

5.3.1 Velocity Time Series. The ice velocities deduced from trajectories should be interpreted from the Lagrangian point of

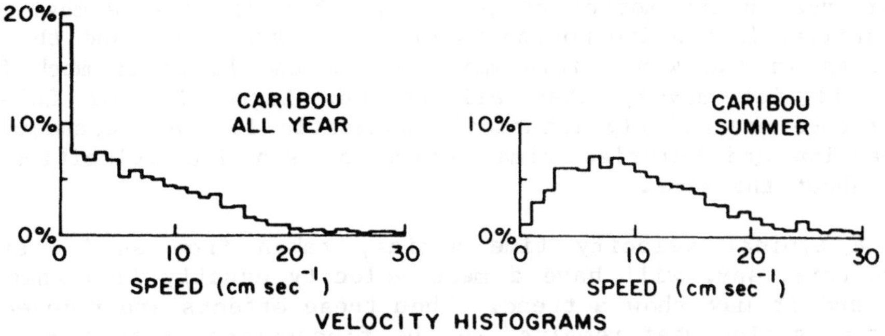

Fig. 7. Histograms of ice speed.

view since the measurements are made following the material particle. If many trajectories are measured simultaneously it is possible to obtain by interpolation velocity time series at fixed Eulerian points. Although the Lagrangian description is more directly measurable, it has the drawback of sampling both the time and space variations in velocity. The differences between the Lagrangian and Eulerian points of view is not of great practical

Table 2. Techniques for measuring sea ice motion.

Buoy ID	Velocity variance	Number of data points
1901	59 $cm^2\ s^{-2}$	342
1902	48	295
1903	50	284
1905	210	231
1906	81	299
1907	37	229
1908	174	61
1909	38	267
1911	19	62
1913	84	314
1914	83	282
1915	172	31
1916	57	172
1918	35	283
1917	29	219
1920	54	293
1923	63	302
1924	167	72
1925	67	280
1926	168	59
1927	62 (median)	311

importance in the motion of sea ice. This is because most of the variability in the ice motion is driven by the wind, and the space patterns in the wind field move across the basin so much faster than the ice moves, that all points, Lagrangian or Eulerian, experience essentially identical stochastic forcing. Consequently, Lagrangian and Eulerian time series of sea ice velocities look just about the same.

A typical velocity time series, taken from an ice station trajectory, say, will have a mean velocity usually less than 3 cm s^{-1}, and it may show a trend. When these effects are removed from the time series what remains are the fluctuations with time scales longer than the sampling interval and shorter than the length of record T. The AIDJEX position data, for instance, serve to resolve fluctuations on time scales from a few hours up to a few months, ($\tau \approx 1$ hour, $T \approx 1$ year). Over this range of time scales, the ice velocity has a power spectral density as sketched in Figure 8.

The power spectral density is plotted for positive and negative frequencies corresponding to counter-clockwise and clockwise rotations of the velocity vector. At the end of this section an algorithm is given for calculating the spectra of vector time series. We interpret the velocity vector time series u(t,x) as a complex time series with the Fourier decomposition

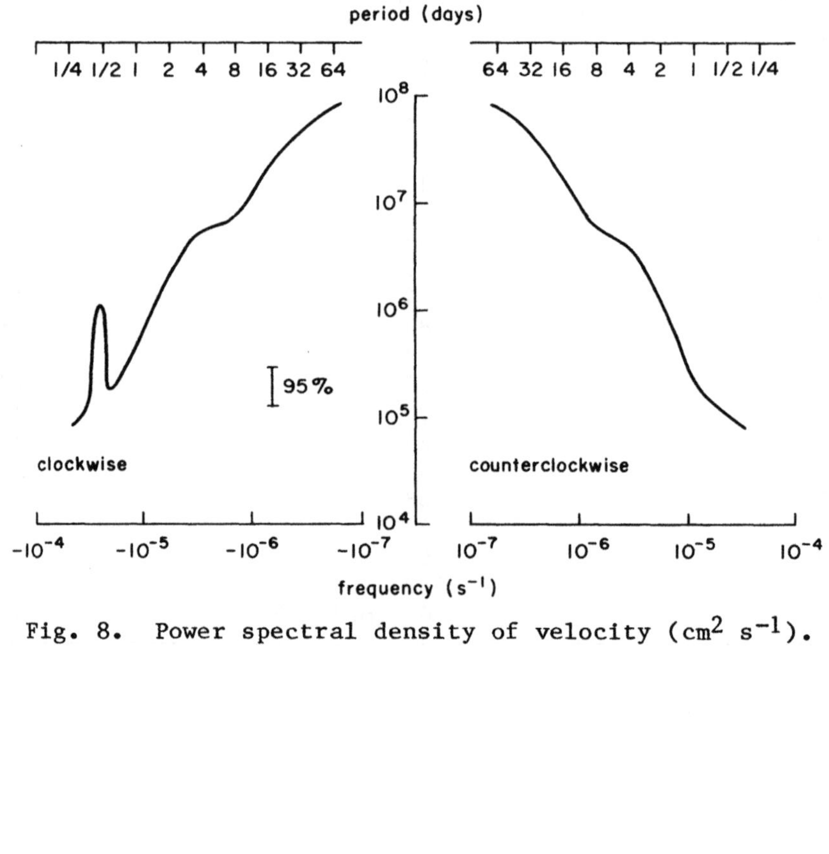

Fig. 8. Power spectral density of velocity (cm^2 s^{-1}).

$$u(t,x) = \int_{-\infty}^{\infty} a(\omega)e^{i\omega t}d\omega. \tag{7.3}$$

The power spectral density is the real function $S(\omega) = a(\omega)a^*(\omega)$, defined in the frequency range $\frac{1}{2\tau} < \omega < \frac{1}{2\tau}$.

The integral of the spectral density over this frequency range is the variance q^2, and the integral over any frequency band is the part of the total variance contributed by fluctuations with frequencies in that band. A useful way to summarize the information contained in the spectrum is to state the fraction of the total variance coming from frequencies greater than a certain value (Table 3).

The equation of motion for sea ice balances the ice acceleration against air stress, water stress, pressure gradient forces due to the sloping sea surface, and internal ice stress gradients. While it is not the purpose here to examine the ice dynamics, it is useful to relate features of the ice velocity spectrum to these driving forces. In the central basin about 75% of the variance of the ice velocity can be explained by the local geostrophic wind (Thorndike and Colony, 1982). In fact, the ice velocity fluctuations are roughly proportional to the local wind fluctuations. This implies that the spectra of the ice velocity and of the wind should have approximately the same shapes (see Figure 9).

The water stress depends on the difference in velocity between the ice and the upper ocean. If the ocean is at rest the water stress is simply a drag opposing the ice velocity and its only effect is to reduce the ice response near the inertial frequency. If the ocean is in motion, the ice will be carried along with it, and should acquire spectral traits similar to those of the ocean. Unfortunately the spectral signature of motion in the Arctic Ocean is not known. The long term circulation of the upper ocean appears to be similar to that of the ice, with a clockwise circulation in the Beaufort Sea and a transpolar current flowing from Siberia through the Greenland-Spitsbergen passage. It probably is not productive to ask whether the ice drives the long term

Table 3. Fraction of total variance exceeding certain frequencies.

Frequency: 1 cycle per month	2	4	15	30	60
Period: 1 month	2 weeks	1 week	2 days	1 day	12 hours
58%	45%	34%	12%	7%	3%

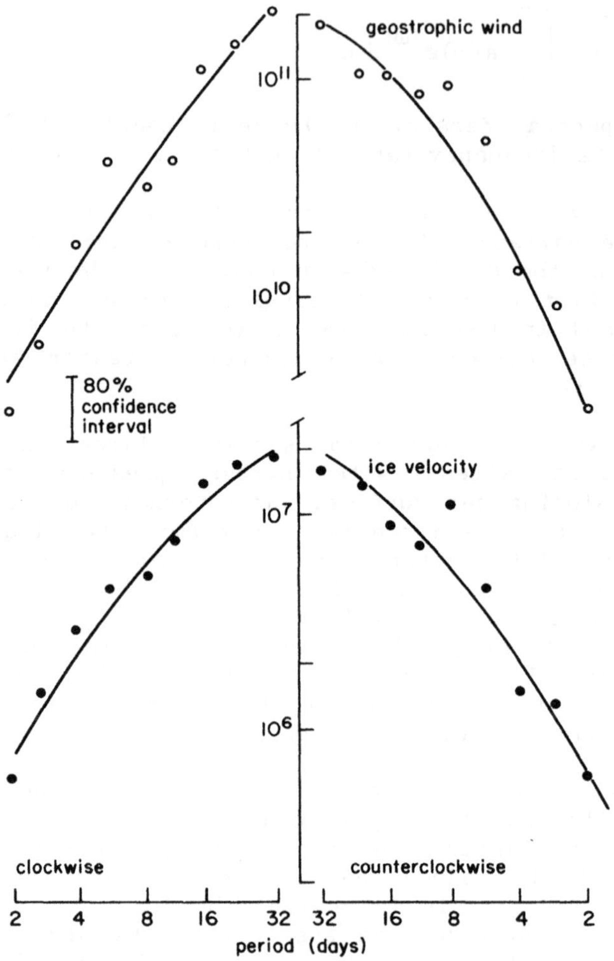

Fig. 9. Wind and ice spectra ($cm^2\ s^{-1}$).

ocean circulation or vice versa because, in the long term, the ice should be thought of as part of the upper ocean. In any case the long term ocean behavior does not affect the spectrum we are considering since we have subtracted out the long term mean velocity.

Fluctuations in ocean currents on shorter time scales have been observed but it is not yet possible to say how great their effect on the ice velocity may be. Some evidence for these variable currents is listed here.

1. Monthly variations are apparent in the currents deduced from wind and ice motion observations by Thorndike and Colony, 1982.

2. Subsurface eddies. Manley's (1981) study of subsurface eddies during AIDJEX described numerous features with velocity of about 50 cm s^{-1}, a length scale of 10 kilometers, and a depth of 50-200 meters. He found no expression of these structures in the ice motion.

3. Inertial oscillations. The balance between the ice acceleration and the Coriolis force leads to oscillatory ice motion with a period which decreases from 12.77 hours at 70°N to 12 hours at 90°N. Inertial motions are always clockwise -- hence the negative value for the frequency -- in the Northern Hemisphere because the Coriolis force always accelerates the velocity to the right. Inertial motions in sea ice were first described by Hunkins (1967) and have received subsequent study by McPhee (1979) and Colony and Thorndike (1980). The amplitude can reach 0.20 m s^{-1} during summer when the ice pack is comparatively loose. Their effect on the summer ice velocity spectrum is striking (see Figure 10).

4. Tides. Tidal currents in the central basin are small because of the great depth. Theoretical estimates are in the range of 1-2 cm s^{-1}. Over the shallow continental shelves the amplitudes are predicted to be at least an order of magnitude greater (Kowalik and Untersteiner, 1978). These theoretical estimates are for the lunar semi-diurnal tide. Evidence from tide gauges around the basin summarized by Sverdrup (1926) implies that the lunar semi-diurnal (period 12.47 hr) and the solar semi-diurnal (12 hr) tides are the dominant tidal constituents in the Arctic Ocean.

The prediction of a large amplitude of the tidal current over the shelf is confirmed by Sverdrup's observations from the Maud; Nansen may also have observed tidal motion in the ice pack surrounding the Fram. Because the tidal and inertial periods are so close it may be difficult to diagnose observed motions correctly. Nevertheless there are several differences between the two kinds

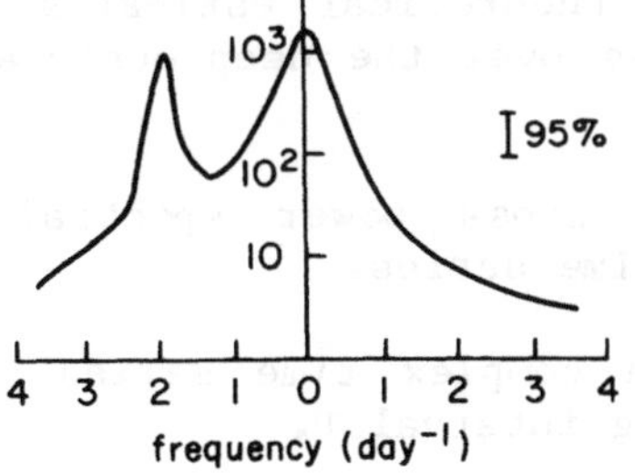

Fig. 10. Velocity spectrum, showing inertial oscillations, from AIDJEX ice camp Big Bear, summer 1975, units: m^2 s^{-1}.

of motions which can sometimes be used to distinguish them. First the tidal vector traces out an ellipse during one period. An ellipse can be viewed as the sum of a clockwise circle and counter-clockwise circle. Thus, unless the tidal ellipse should happen to be exactly a clockwise circle, it should have some expression on the counter-clockwise side of the spectrum. Inertial motion is strictly clockwise. Second, at a fixed point, tidal motions should have a fixed phase. Inertial motion on the other hand, acquires a new phase every time the ice receives a sudden impulse of momentum. Third, the phase of tidal motion should vary smoothly and slowly in space. Inertial motions at different points may have no fixed phase relationships.

The motion described by Sverdrup can hardly be mistaken for inertial motion. The smooth variations in phase as the Maud moved from Wrangel Island to the New Siberian Islands, and the clearly elliptical cycle described by the measured velocity vector are not compatible with inertial motions.

Most ice trajectories which have been analyzed in the western literature have been over the deep basin where the tidal motion is small. Careful observations and analysis would be required to detect a tidal component of order 1 cm s^{-1} since it would be mixed with an inertial component which is often much larger and with a rich spectrum of other types of motions.

The Soviet literature contains many references to tidal motion in sea ice. Doronin and Kheisin (1977) and Zubov (1943) each devote several pages to the subject.

A consequence of the tidal motion is the associated cycle of convergence and divergence caused by the difference in phase of the tidal cycle at different points. Periodic opening and closing of the ice in the shallow seas has an effect on the heat exchange between the atmosphere and the ocean and on the rate of ice production. The theoretical calculations of Kowalik and Untersteiner indicate maximum divergence rates in the shallow seas exceeding 10^{-6} s^{-1} which is enough to produce one percent opening during the tidal cycle. Their theoretical estimates of the divergence rate associated with tides over the deep ocean are three orders of magnitude smaller.

Calculation of cross power spectral density for two two-dimensional vector time series.

Given: two discrete complex time series u_i, v_i, $i = 1, \ldots N$ with sampling interval τ.

Step 1. Select M, the number of lags; a good rule is $M < N/3$.
Step 2. Remove the mean and trend from each time series.
Step 3. Compute cross covariances

$$r_\ell = \begin{cases} \frac{1}{N} \sum_{j=1}^{N-\ell} u_{j+\ell} V_j^* & \ell = 0, 1, \ldots, M, \\ \frac{1}{N} \sum_{j=1-\ell}^{N} u_{j+\ell} V_j^* & \ell = -M, -M+1, \ldots, -2, -1. \end{cases} \quad (7.4)$$

Step 4. Define spectral window

$$w_\ell = \frac{1}{2} (1 + \cos \frac{\pi\ell}{M+1}) \qquad \ell = -M, \ldots, M. \quad (7.5)$$

Step 5. Calculate spectrum S and frequency

$$S_k = \tau \sum_{\ell=-M}^{M} w_\ell \, r_\ell \, e^{-\pi\ell k\sqrt{-1}/M} \qquad k = -M, \ldots, M$$

$$\omega_k = \frac{k}{2M\tau} \, . \quad (7.6)$$

5.3.2 Velocity Correlations in Time. An alternative characterization of a random function is its autocorrelation function. The discussion in the previous section of the power spectral density of the time function u(t) at a fixed or moving point could have been given in terms of the autocorrelation function

$$R(\tau) = \frac{E\, u(t+\tau)\, u^*(t)}{E\, u(t)\, u^*(t)} = \frac{1}{q^2} \, E\, u(t+\tau)\, u^*(t). \quad (7.7)$$

The two functions $S(\omega)$ and $R(\tau)$ are Fourier transforms of each other and therefore contain equivalent information. Which description is the more useful depends on the application. The spectrum is useful for distinguishing physical processes with distinct characteristic frequencies; for example separating the free inertial oscillations from the wind forced motion. On the other hand the autocorrelation function is more useful for questions related to prediction or experiment design. For example the question: "How well can tomorrow's ice velocity be predicted on the basis of today's velocity," has an answer involving the autocorrelation function $R(\tau)$ evaluated at $\tau = 1$ day.

The autocorrelation function is complex. Its real part contains information about the lagged correlations of the x component of velocity with itself and the y component with itself. The imaginary part has information about the lagged correlations between the x and y velocity components. If the velocity of a piece of ice undergoing an inertial oscillation has a positive x component at a certain time, it will have a negative y component three hours later. The imaginary part of the correlation should have negative peaks at 3 hours, 15 hours, etc.

The autocorrelation function for the velocity time series at the AIDJEX ice station Caribou is shown in Figure 11. The autocorrelation function estimated by the 1979 buoy data for the central basin is shown in Figure 12. The real part of the correlation falls to about 0.7 after one day, 0.4 after two days, and decreases slowly at longer lags. The large correlation at lag one day indicates that persistence (the forecast strategy which predicts that the future will be the same as the present) will have some success for one day forecasts.

As expected the autocorrelation from Caribou has negative peaks in the imaginary part at 3 hours, 15 hours, etc. Generally though the imaginary part is small. If the inertial motions are not of interest in a particular application, the imaginary part of

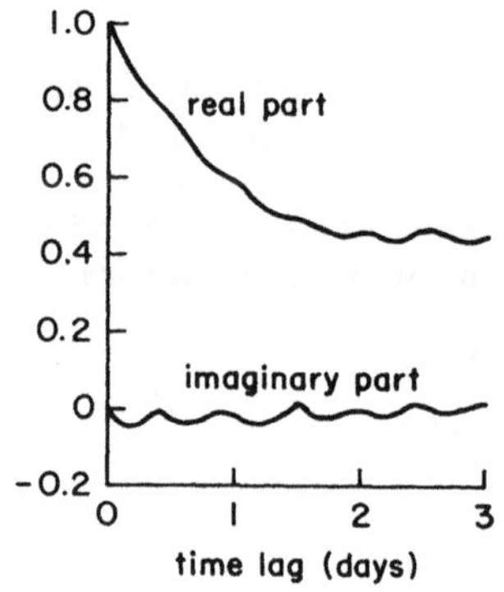

Fig. 11. Time correlation from Caribou.

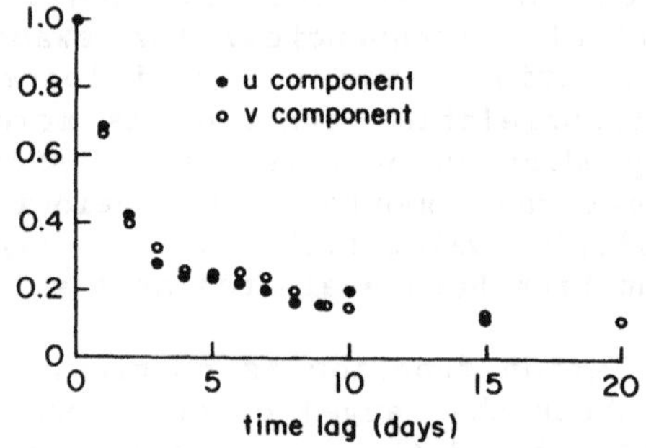

Fig. 12. Time correlation from 1979 buoys.

the autocorrelation function can safely be ignored. This is equivalent to treating the two velocity components as independent time series.

5.3.2.1 <u>Application of the time autocorrelation function.</u> It is desired to estimate the ice velocity u at time t given observations z_j at times t_j, $j = 1,\ldots, N$. Suppose the observations have zero mean random errors $\varepsilon_j = z_j - u(t_j)$ which have covariance

$$E\varepsilon_j\varepsilon_k^* = \sigma^2\delta_{jk}. \tag{7.8}$$

The delta function expresses the independence of errors at different times. Finally suppose the errors are independent of the actual velocities. Choose an estimator $\hat{u}$ which is a linear sum of the observations.

$$u = \sum_{j=1}^{N} \alpha_j z_j \; . \tag{7.9}$$

It is desired to find the complex constants α_j which give the best estimate $\hat{u}$ in the sense that $F(\alpha_1,\ldots, {}_N) = E(\hat{u} - u)(\hat{u} - u)^*$ is minimized. This is accomplished by differentiating F with respect to the real and imaginary parts of each α_j and equating to zero. The result is the system of complex equations for the unknowns $\alpha_1,\ldots, \alpha_N$:

$$\sum_j \alpha_j \; E\, z_j z_\ell^* = (Ez_\ell \; u^*)^* \quad . \tag{7.10}$$

A further reduction is achieved by noting that

$$Ez_j z_\ell^* = Eu_j u_\ell^* + \sigma^2\delta_{j\ell} \text{ and } Ez_\ell \hat{u}^* = Eu_\ell u^* \; . \tag{7.11}$$

Use of the matrix notation

$$\underset{\sim}{z} = \{z_j\},\; A = \{\alpha_j\},\; M = \{Eu_j u_\ell^* + \sigma^2\delta_{j\ell}\}, \text{ and } P = \{Eu_\ell u_j^*\} \tag{7.12}$$

gives the compact expressions

$$\hat{u} = A^T z \quad \text{and}$$
$$M^* A = P, \tag{7.13}$$

which involve he autocorrelation function R and the measurement error variance σ^2. The estimation error is

$$F = q^2 - P^T M^{-1^T} P^*. \tag{7.14}$$

With these two expressions we can answer a number of practical problems.

1. What sampling rate is required to insure good interpolation to the mid point between two observations? For given sampling interval τ and measurement error variance σ^2, we write eq. 13 as

$$\begin{pmatrix} q^2 + \sigma^2 & q^2 R(\tau)^* \\ q^2 R(-\tau)^* & q^2 + \sigma^2 \end{pmatrix} \begin{pmatrix} \alpha_1 \\ \alpha_2 \end{pmatrix} = \begin{pmatrix} q^2 R(\frac{\tau}{2})^* \\ q^2 R(-\frac{\tau}{2})^* \end{pmatrix} \tag{7.15}$$

The solutions for the α_j, which in this case are complex conjugates for reasons of symmetry, and for F are sketched in Figure 13. The

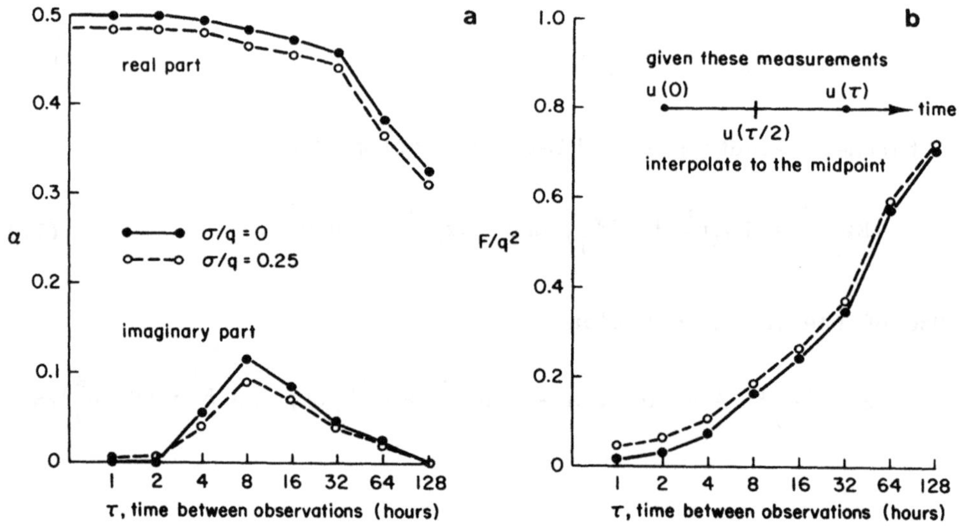

Fig. 13. Two point interpolation in time.

solution technique extends trivially to interpolation involving more points.

The optimal choice of the weights α_1 and α_2 is not $\alpha_1 = \alpha_2 = 1/2$. This linear interpolation between the two data points is the most natural scheme, and it is quite good for small τ, and small σ^2/q^2, but it is not the optimal interpolation. Especially at large τ, a smaller error variance is achieved by giving less weight to the observations. For very large τ the optimal estimate is simply $\hat{u} = 0$, ($\alpha_1 = \alpha_2 = 0$) since, in this case the two observations are so removed in time as to have no correlation with the desired velocity.

The measurement error σ affects α_j and F slightly. The dependence of F on the sampling interval τ is quite strong. It appears that if one needs to know each velocity component to $\pm$ 3 cm s^{-1}, $F = E\varepsilon_u^2 + \varepsilon_v^2 = 18$ cm^2 s^{-1}, a sampling interval of about six hours is required even with perfect measurements. Here we have used $q^2 = 50$ cm^2 s^{-2}, a representative value from Table 2.

2. What errors are expected in forecasts based only on the present velocity? Here we obtain

$$\alpha = \frac{R(\tau)^*}{1 + \sigma^2/q^2}$$
$$F/q^2 = 1 - \frac{R(\tau)R(\tau)^*}{1 + \sigma^2/q^2} \tag{7.16}$$

The forecast error is plotted for two values of initial error in Figure 14.

The prediction error grows with time, approaching the velocity variance q^2 for large τ. From the sketch we see that an optimal 3 hour forecast captures about 75% of the velocity variance, an optimal one day forecast captures about 35% of the variance.

The velocity was assumed to have zero mean in this analysis and, in constructing the autocorrelation function (Figures 11 and 12), the long term mean was first removed from the data. In an actual application to a region where the mean is known, the best forecast would be

$$\hat{u} = \bar{u} + \alpha(z - \bar{u}) . \tag{7.17}$$

Note that the best forecast is not simple persistence ($\alpha = 1$). For long forecasts, the best estimate is the mean ($\alpha = 0$), as equation 16 correctly indicates.

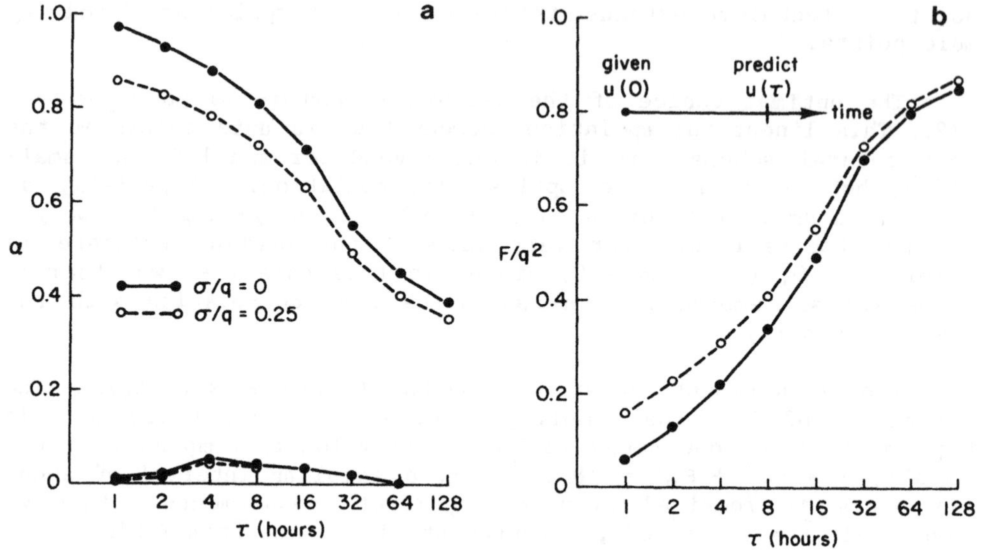

Fig. 14. Predictions of ice velocity based on a single observation.

5.3.2.1 <u>Acceleration</u>. The time autocorrelation function for velocity, Figures 11 and 12, behaves like $1-\tau/c$ for small τ. It follows from this that the variance of the increment in velocity $u(t+\tau) - u(t)$ is proportional to τ (not τ^2). This implies that the velocity is not time differentiable in the mean square sense. Attempts to measure accelerations precisely by sampling over shorter and shorter time intervals may merely result in larger and larger estimates of acceleration. Of course, time averaged accelerations exist and have finite variance.

$$\sigma_a^2 = \text{var}\ \{\frac{1}{T}\ [u(t+T) - u(t)]\} = \frac{2q^2}{T^2}\ (1 - R(T)) \qquad (7.18)$$

which for small T is $2q^2/cT$ with $c \approx 10^5$ s.

It is frequently claimed that the inertial forces due to acceleration of the ice are small compared to other forces acting on the ice. To support this claim a typical average acceleration σ_a is compared, for instance, to a typical value of the Coriolis acceleration, $f\sigma_u \simeq 10^{-4}\ s^{-1} \cdot 10\ cm\ s^{-1}$. Using the above estimate for σ_a, and taking $q^2 = 100\ cm^2\ s^{-2}$, we find $\sigma_a < f_u$ provided $T > 2 \times 10^3$ s. Thus, roughly a half hour time average is required to bring the acceleration down to the level of other terms in the momentum balance.

5.3.3 Velocity Correlations in Space. In earlier sections, the emphasis has been on the structure of the complex function u of the real argument t, at a fixed or moving point. In this section, the emphasis is on u as a function of position for fixed time. Certain results are more easily expressed if u and x are regarded as two dimensional vectors rather than as complex numbers. The autocorrelation function

$$q^2 R(\underset{\sim}{x}_1, \underset{\sim}{x}_2) = E\, \underset{\sim}{u}(\underset{\sim}{x}_1)\underset{\sim}{u}^T(\underset{\sim}{x}_2) = \begin{pmatrix} E\, u(\underset{\sim}{x}_1)u(\underset{\sim}{x}_2) & E\, u(\underset{\sim}{x}_1)v(\underset{\sim}{x}_2) \\ E\, u(\underset{\sim}{x}_2)v(\underset{\sim}{x}_1) & E\, v(\underset{\sim}{x}_1)v(\underset{\sim}{x}_2) \end{pmatrix} \tag{7.19}$$

is a 2 x 2 matrix involving possibly four different functions, whereas the time autocorrelation function involves just two.

The structure of the spatial autocorrelation matrix for sea ice velocities reveals some of the properties of the velocity field. To establish the basic ideas we first determine the structure of the autocorrelation matrix for a homogeneous, isotropic, two dimensional velocity field, borrowing from Kolmogoroff's (1941) classic discussion of three dimensional turbulence.

A random field is said to be homogeneous if its statistical properties are invariant under a translation of coordinates, and isotropic if they are invariant under rotations and reflections of coordinates. In particular, the autocorrelation matrix for a homogeneous isotropic field must satisfy

$$R(\underset{\sim}{x}_1,\underset{\sim}{x}_2) = R(\underset{\sim}{x}_1+\underset{\sim}{a},\ \underset{\sim}{x}_2+\underset{\sim}{a}) \qquad \text{for any } \underset{\sim}{a}, \text{ and}$$

$$q^2R(\underset{\sim}{x}_1,\underset{\sim}{x}_2) = E\underset{\sim}{u}(\underset{\sim}{x}_1)\underset{\sim}{u}^T(\underset{\sim}{x}_2) = E\ M\underset{\sim}{u}(M\underset{\sim}{x}_1)\ [M\underset{\sim}{u}(M\underset{\sim}{x}_2)]^T \tag{7.20}$$

for any orthonormal matrix M. Multiplication by an orthonormal matrix accomplishes an arbitrary rotation or reflection of coordinates. Note that the change of coordinates is applied to both the position and the velocity vectors. These definitions make precise the notions that the statistics do not depend on position or direction. By choosing the translation vector $\underset{\sim}{a}$ or the rotation and reflection matrix M suitably, we can see how these definitions of homogeneity and isotropy constrain the form of the autocorrelation matrix.

First, $R(\underset{\sim}{x}_1, \underset{\sim}{x}_2)$ depends only on the vector difference $\underset{\sim}{x}_2 - \underset{\sim}{x}_1$. To see this, take $\underset{\sim}{a} = -\underset{\sim}{x}_1$. Note that this does not imply that R depends only on the distance $||\underset{\sim}{x}_2 - \underset{\sim}{x}_1||$.

Second, R is symmetric. This requires using $\underset{\sim}{\alpha} = -\underset{\sim}{x}_1$, M = -I and $\underset{\sim}{\alpha} = \underset{\sim}{x}_2$, to obtain this sequence of equalities:

$$Eu(\underset{\sim}{x}_1)v((\underset{\sim}{x}_2) = Eu(\underset{\sim}{o})v(\underset{\sim}{x}_2 - \underset{\sim}{x}_1) = E(-u(-\underset{\sim}{o}))(-v(\underset{\sim}{x}_1 - \underset{\sim}{x}_2)) =$$
$$Eu(\underset{\sim}{x}_2)v(\underset{\sim}{x}_1).$$

Third, $Eu(o,o)v(x,o) = o$. Here use the reflection invariance in the definition of isotropy by taking

$$M = \begin{pmatrix} 1 & o \\ o & -1 \end{pmatrix}$$

which changes the sign of the y axis. Then

$$Eu(o,o)v(x,o) = Eu(o,-o)[-v(x,-o)] = -Eu(o,o)v(x,o)$$

which is only true when $Eu(o,o)v(x,o) = o$.

Consequently, the autocorrelation function evaluated at $x_1 = (o,o)$, $x_2 = (r,o)$ has the form

$$R\left[(o,o),\ (r,o)\right] = \frac{1}{q^2}\begin{pmatrix} E\ u(o,o)u(r,o) & o \\ o & E\ v(o,o)v(r,o) \end{pmatrix}$$

$$= \begin{pmatrix} B_{||}(r) & o \\ o & B_{\perp}(r) \end{pmatrix} = B(r) \qquad (7.21)$$

involving only two functions of distance. $B_{||}(r)$ is the correlation between the components of velocity parallel to the line joining two points separated by a distance r. $B_{\perp}(r)$ is the correlation between the components of velocity perpendicular to that line.

More generally if $\underset{\sim}{x}_1$ and $\underset{\sim}{x}_2$ are arbitrary points with $r^2 = (x_1 - x_2)^2 + (y_1 - y_2)^2$, $\cos\theta = (x_2 - x_1)/r, \sin\theta = (y_2 - y_1)/r$, then

$$\underset{\sim}{u} = \begin{pmatrix} u \\ v \end{pmatrix} = A\begin{pmatrix} u_{||} \\ u_{\perp} \end{pmatrix} = A\ \underset{\sim}{\tilde{u}} \text{ where } A = \begin{pmatrix} \cos\theta & -\sin\theta \\ \sin\theta & \cos\theta \end{pmatrix}. \qquad (7.22)$$

Then

$$R(\underset{\sim}{x}_1,\underset{\sim}{x}_2) = \frac{1}{q^2} E\, \underset{\sim}{u}(\underset{\sim}{x}_1)\underset{\sim}{u}^T(\underset{\sim}{x}_2) = \frac{1}{q^2} E\, A\, \tilde{\underset{\sim}{u}}(\underset{\sim}{x}_1)\;\; A\, \tilde{\underset{\sim}{u}}(\underset{\sim}{x}_2)^T = A\, B(r) A^T \text{ or}$$

$$R(\underset{\sim}{x}_1,\underset{\sim}{x}_2) = \begin{pmatrix} \cos^2\theta\, B_{||}(r) + \sin^2\theta B_{\perp}(r) & \cos\theta\, \sin\theta\, (B_{||}(r) - B_{\perp}(r)) \\ \cos\theta\, \sin\theta (B_{||}(r) - B_{\perp}(r)) & \sin^2\theta\, B_{||}(r) + \cos^2\theta\, B_{\perp}(r) \end{pmatrix} \quad (7.23)$$

which is the general from for the autocorrelation matrix for a homogeneous, isotropic two dimensional velocity field. It involves two functions, $B_{||}$ and $B_{\perp}$, which we will estimate below from observations. Further it involves the separation r and the orientation θ of the two points $\underset{\sim}{x}_1$ and $\underset{\sim}{x}_2$ at which the autocorrelation is being evaluated. It may seem strange that θ enters the statistics for an isotropic field. The condition of isotropy does not imply that Eu(o,o)u(r,o) equals Eu(o,o)u(o,r) however. A rotation of coordinates affects both the velocity and the position vectors. A valid statement is Eu(o,o)u(r,o) = Ev(o,o)v(o,r) (see Figure 15).

There is another constraint on the spatial correlation function. Let $\underset{\sim}{U}_i = (u_i, v_i)$ be the velocities at any set of points x_i, i=1,..., N and let δ_i, η_i be arbitrary real numbers. Then the linear sum

$$s = \sum_{i=1}^{N} \delta_i u_i + \eta_i v_i \quad (7.24)$$

has a variance which can be expressed in terms of R. The added constraint is that var(s) $\geq$ o for all choices of x_i, δ_i, and η_i. This requires that the matrix consisting of the correlations of the velocity component at any set of points $\{x_i\}$ must be positive definite (all positive eigenvalues), and is equivalent to the condition that the two dimensional spatial power spectrum of velocity be positive for all pairs of wave numbers (n_x, n_y).

Contradictions can arise if correlation functions are used which do not satisfy this constraint. For example, suppose we wish to consider a velocity field for which $B_{||}(o) = 1$, $B_{||}(L) = 0.9$ and $B_{||}(2L) = 0.5$. The attempt to evaluate the variance of s

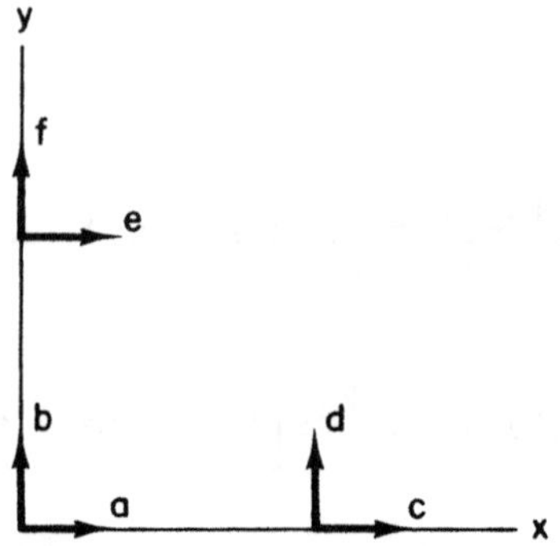

Fig. 15. Correlations between velocity components.

= u(o,o) - 1.3 u(L,o) + 0.4 u(2L,o) in terms of these correlations produces a negative result. Since variance is intrinsically positive, we must conclude that no velocity field could have the presumed correlations.

As an example of an analytical form for R which satisfies the positive definite constraint and bears some resemblance to reality, define

$$B_{||}(r) = e^{-(r/A)^2}, \quad B_{\perp}(r) = (1-2(\tfrac{r}{A})^2)\, e^{-(r/A)^2}, \quad A \sim 900 \text{ km} . \tag{7.25}$$

This example was constructed by taking a stream function Φ with the positive definite correlation function $R(r) = \exp(-r^2/A^2)$. The corresponding power spectrum is the Hankel transform of $\exp(-r^2/A^2)$. This is $\pi A^2 \exp(-\omega^2 A^2/4)$, $(\omega^2 = n_x^2 + n_y^2)$, which is always positive. Now define velocity components $u = -\partial\Phi/\partial y$, $v = \partial\Phi/\partial x$ and deduce $B_{||}(r) = \frac{1}{r}\frac{\partial R(r)}{\partial r}$ and $B_{\perp}(r) = \frac{\partial^2 R(r)}{\partial r^2}$.

5.3.3.1 Estimates of the autocorrelation functions. In order to estimate the autocorrelation matrix, simultaneous velocity observations are required at pairs of points separated by distances ranging from roughly 1 to 10^3 kilometers. We present data from the 1979 buoy program and from AIDJEX which define the functions

$B_{||}$ and $B_{\perp}$ only for distances exceeding 100 kilometers. The 1979 data were first interpolated to give velocity estimates at a uniform grid in space and time.

$$\underset{\sim}{u}_{ijk} = u(x_i, y_j, t_k),\ \Delta x = \Delta y = 400 \text{ km},\ \Delta t = 1 \text{ day},$$

$$i = 1,\ldots N_x,\ j = 1,\ldots,\ N_y,\ k = 1,\ldots,\ N_t.$$

The mean velocity over all points i,j,k was removed. Then for lags l,m,n the lagged correlations were found using

$$R_{\ell,m,n} = \frac{1}{N_x N_y N_t} \sum_{i=1}^{N_x-\ell} \sum_{j=1}^{N_y-m} \sum_{k=1}^{N_t-n} u_{i+\ell,\ j+m,\ k+n}\, u^T_{ijk}\,. \tag{7.26}$$

For these calculations $N_x = 7$, $N_y = 4$, and $N_t = 200$. The results are sketched in Figure 16 for the lags $\ell = 1,\ldots,7$, $m = n = 0$, and $\ell = n = 0$, $m = 1,\ldots,4$. The results show that Eu(o,o)u(r,o) and Ev(o,o)v(o,r) are similar functions of r, and that Ev(o,o)u(r,o), Eu(o,o)v(o,r), Eu(o,o)v(r,o), and Ev(o,o)u(o,r) are all small. These observations are roughly consistent with the assumption that the field is homogeneous and isotropic. We will proceed then on the assumption, supported by these data, that the ice has a homogeneous, isotropic velocity field with the functions $B_{||}$ and $B_{\perp}$ empirically determined from Figure 16.

Correlations for the AIDJEX data were constructed by choosing two of the ice camps, resolving their velocities into components parallel and perpendicular to the line joining the camps, and correlating these components. These correlations are also plotted in Figure 16.

5.3.3.2 Applications of the space correlation. The spatial autocorrelation function is the basis for sound experiment design and data processing. For instance, suppose observations of velocity $\underset{\sim}{u}_i$ are made at N points $\underset{\sim}{x}_i$ with independent zero mean errors having variance σ^2. Estimates of velocity are to be made at arbitrary points $\underset{\sim}{x}$ using these observations. What is the maximum estimation error? We use the tools developed earlier. To estimate the u component of velocity at $\underset{\sim}{x}$, we use the observations $\underset{\sim}{u}_1, \ldots, \underset{\sim}{u}_N$ and construct the matrices

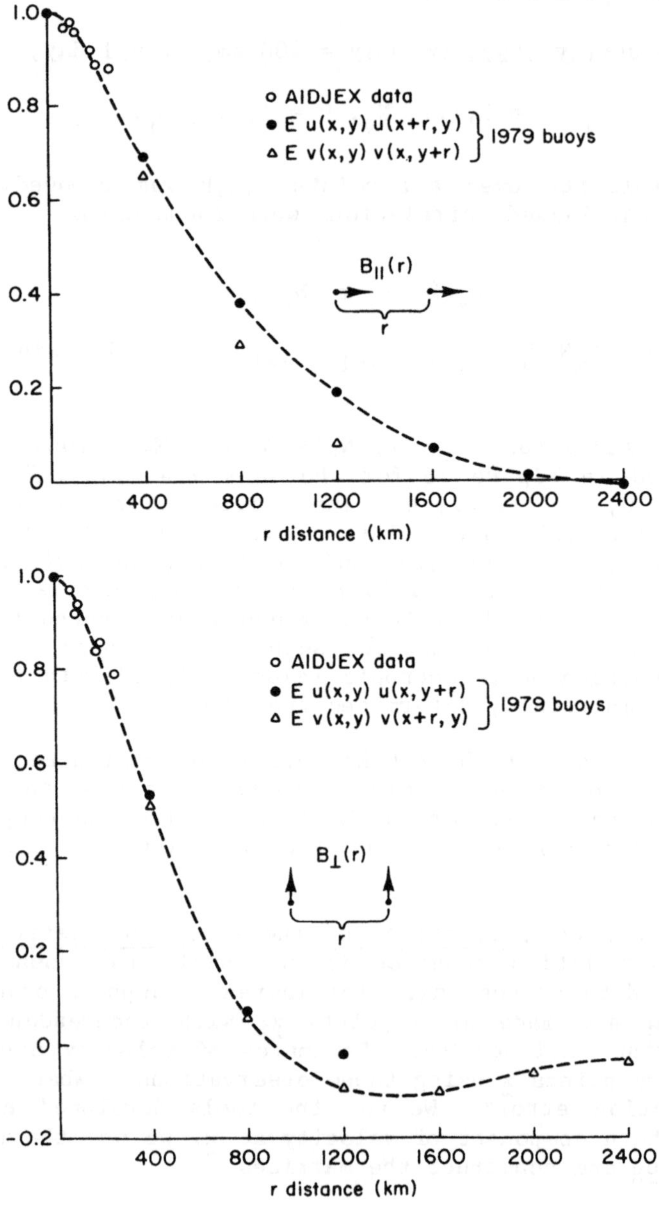

Fig. 16. Correlations for velocity components; AIDJEX and FGGE data.

$$\begin{pmatrix} Ru_1u_1+\sigma^2 & Ru_1v_1 & \vdots & Ru_1u_2 & \cdots\cdots & Ru_1v_n \\ Rv_1u_1 & Rv_1v_1+\sigma_2 & \vdots & Rv_1u_2 & & \vdots \\ \cdots\cdots & \cdots\cdots & & & & \vdots \\ Ru_2u_1 & Ru_2v_1 & & & & \vdots \\ \vdots & & & & & \vdots \\ Rv_nu_1 & \cdots & \cdots & \cdots & \cdots & Rv_nv_n+\sigma^2 \end{pmatrix} \begin{pmatrix} \alpha_1 \\ \beta_1 \\ \vdots \\ \alpha_n \\ \beta_n \end{pmatrix} = \begin{pmatrix} Ru\,u_1 \\ Ru\,v_1 \\ Ru\,u_2 \\ \vdots \\ Ru\,u_n \\ Ru\,v_n \end{pmatrix} \qquad (7.27)$$

where $Ru_iv_j = Eu(\underset{\sim}{x}_i)v(\underset{\sim}{x}_j)$ etc. as given by eq. 7.23. The estimation error variance is given by eq. 7.14. Figure 17 shows how the maximum estimation error depends on the separation between measurement stations and on the measurement accuracy. The curve labelled $\sigma^2/q^2 = 0$ refers to the idealized condition in which the measurement errors are zero. Thus it represents the interpolation error which is due to the intrinsic spatial variability of the velocity field.

The standard deviation of the ice velocity itself is about q ~ 10 cm s^{-1}. A reasonable observational goal is to keep the interpolation errors below 2 or 3 cm s^{-1}. This can be achieved if the raw measurements of velocity are good to about 3 cm s^{-1} and the grid spacing is about 400 km. At larger grid spacings the interpolation accuracy deteriorates rapidly.

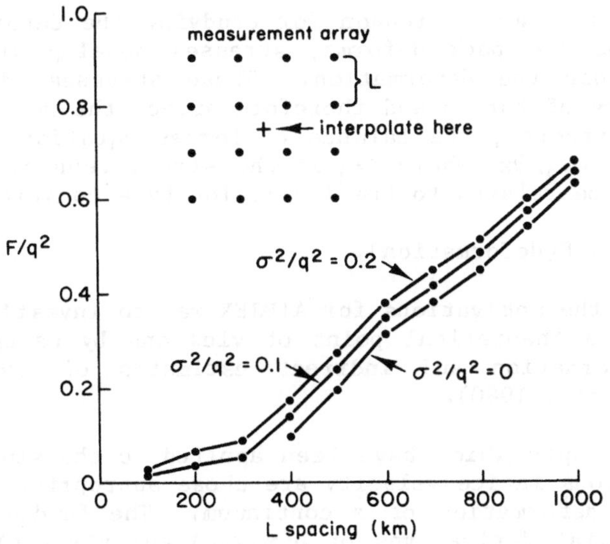

Fig. 17. Interpolation error for any array of measurements.

5.4 Deformation

The differences in velocity from place to place are responsible for the characteristic morphology of the ice. When the velocity difference between neighboring pieces of ice is such that they tend to move apart, a lead forms and widens exposing the ocean surface to the atmosphere. During the winter, ice growth is rapid over open leads. If the motion changes -- perhaps because of a change in the winds -- so that these pieces of ice move toward each other, the open lead will close, and any new ice which formed there will need to rearrange itself to occupy a smaller area. Typical mechanisms for this rearrangement are rafting where one part of the new ice sheet overrides the other, and ridging where the ice is crushed into pieces which pile into ridges rising a meter or two above and sometimes many meters below the surrounding ice. Ridging and rafting are not restricted to thin ice. If the closing motion continues, the original pieces of ice come in contact and one may override the other or, by grinding together, pieces may break off and pile up and down to form a ridge. The essential ideas here are that the ice accommodates divergent motion by increasing the area of open water rather than by stretching and thinning. It accommodates convergent motion by reducing the area of open water and by ridging and rafting. These processes link the morphology of the ice, characterized by such features as floes, ridges, and leads, to the spatial differences in the ice velocity. By studying the spatial variations in velocity we may be able to understand better why the ice pack has the form it does.

There is a second reason for studying the deformation of sea ice. As the ice pack deforms, stresses develop within it which tend to oppose the deformation. These stresses figure into the local balance of forces and therefore affect the motion of the ice. To be more precise, the balance of forces equation contains terms of the form $\partial\sigma_{ij}/\partial x_i$ where σ_{ij} is the stress tensor. The stresses may in turn be related to the ice motion by a constitutive law

stress = F(deformation).

One of the motivations for AIDJEX was to investigate the function F from a theoretical point of view and by using observations of the deformation and indirect estimates of the stress (see Rothrock et al., 1980).

The concepts which have been applied to the study of the spatial variations in ice velocity are those appropriate to the analysis of the deformation of a continuum. The fundamental concepts are the partial derivatives of u(x,y,t) and v(x,y,t) with respect to x and y. The line of thought is that a description of the large scale deformation in terms of large scale average derivatives should give some idea of how much opening and closing is occurring on smaller scales and some idea of the state of stress. For example, under a purely divergent motion only leads are formed. On a small scale, shearing along a crack is of no morphological consequence. But large scale shearing is often expressed locally by opening at some cracks and closing at others.

Mathematical formulation of these ideas requires two steps. Required first is an appropriate quantitative description of the large scale deformation in terms of observable parameters. The second requirement is for a functional relationship giving the opening, closing, or state of stress in terms of observable parameters.

Current practice is to describe the motion in terms of the large scale strain rate invariants, which themselves are constructed from the large scale velocity derivatives. The first invariant

$$E_I = \frac{\partial u}{\partial x} + \frac{\partial v}{\partial y} \tag{7.28}$$

quantifies the large scale divergence and convergence, and the second invariant

$$E_{II} = [(\frac{\partial u}{\partial x} - \frac{\partial v}{\partial y})^2 + (\frac{\partial u}{\partial y} + \frac{\partial v}{\partial x})^2]^{1/2} \tag{7.29}$$

quantifies the rate of shearing. A useful alternative pair of invariants is $(|E|,\theta)$, where

$$|E| = [E_I^{\,2} + E_{II}^{\,2}]^{1/2} \tag{7.30}$$

expresses the total rate of deformation and

$$\theta = \tan^{-1}\frac{E_{II}}{E_I}, \quad o \leq \theta < \pi \tag{7.31}$$

indicates whether the motion is predominantly divergence $\theta \sim 0$, shear $\theta \sim \pi/2$ or convergence $\theta \sim \pi$.

In the theory developed by Rothrock and presented by Thorndike et al. (1975), opening and closing are assumed to be known functions of θ and proportional to $|E|$.

A = total opening = $|E|\alpha_0(\theta)$,

C = total closing = $|E|\alpha_r(\theta)$.

The functions α_o and α_r which give the total opening and closing in terms of the strain rate invariants may have a form somewhat as sketched in Figure 18. Some constraints on the shapes of these functions have been deduced by Nye, 1975 and by Rothrock and Hall, 1975. From pairs of Landsat images of the ice these investigators were able to measure the strain invariants and the total opening and closing.

The functional relationships for A and C in terms of $|E|$ and θ can only be regarded as approximate since they do not take into account the actual geometry of the floes in the region being considered. Different regions with different geometries will respond differently to similar large scale strain rates.

In practical applications of these ideas the procedure has been to take measured velocities at several points in a region of interest -- perhaps the three corners of a 100 km triangle. Velocity derivatives are estimated from the measured velocities. The estimated velocity derivatives are combined to form the strain rate invariants $|E|$ and θ, and these are used to find the rates of opening and closing A and C, and the state of stress σ.

In the following sections we examine some of these ideas, particularly the notion of velocity derivatives, the influence of the ice pack geometry on opening and closing, and the difficulties in basing inferences about these quantities on small numbers of measurements.

5.4.1 Velocity Derivatives. Since the ice pack is made up of discrete pieces moving relative to each other, the variations in velocity have two types:

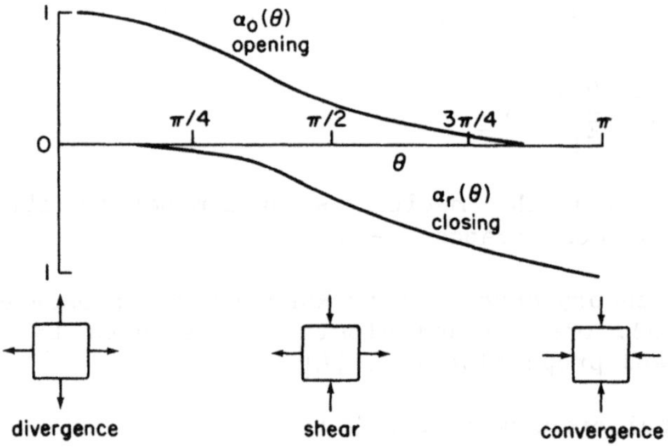

Fig. 18. $\alpha_o(\theta)$, $\alpha_r(\theta)$.

1) for any $\underset{\sim}{x}_1$ and $\underset{\sim}{x}_2$, on the same rigid piece, the linear relation

$$\underset{\sim}{u}(\underset{\sim}{x}_2) - \underset{\sim}{u}(\underset{\sim}{x}_1) = \begin{pmatrix} \cos\omega & -\sin\omega \\ \sin\omega & \cos\omega \end{pmatrix} (\underset{\sim}{x}_2 - \underset{\sim}{x}_1 \qquad (7.32)$$

must hold, where ω is a constant related to the rate of rotation of the rigid piece.

2) for $\underset{\sim}{x}_1$ and $\underset{\sim}{x}_2$ on separate pieces,

$\underset{\sim}{u}(\underset{\sim}{x}_2) - \underset{\sim}{u}(\underset{\sim}{x}_1)$ is arbitrary.

If we select a random point (x,y) on an ice floe, the partial derivative of velocity,

$$\frac{\partial u}{\partial x}(x,y,t) = \lim_{h \to o} \frac{u(x+h,y,t) - u(x,y,t)}{h} \qquad (7.33)$$

is well defined, because for small enough h, (x,y) and (x+h,y) almost always lie on the same floe and the linearity of the first type of motion mentioned above implies the existence of the limit. Thus the partial derivative exists, but it describes only the rigid body rotation of a floe. It says nothing about the relative motion between floes which is after all the motion causing the opening and closing and the exchange of momentum between floes.

To include the effect of velocity differences between floes we cannot let h become arbitrarily small. Consider the velocity increment u(x+h,y,t) - u(x,y,t) where (x,y) is fixed and h begins at 100 km and decreases to the smallest value h_o such that $(x+h_o,y)$ and (x,y) lie on separate floes. This increment captures the variation in velocity from floe to floe and has nothing to say about the rigid rotation of the floe (x,y) lies on. The velocity increment will decrease as h approaches h_o, and we can examine the characteristic rate of decrease by evaluating the variance $E(u(x+h,y,t) - u(x,y,t)^2$. If this quantity is proportional to h^2 for small h, the velocity at least has a derivative in the mean square sense.

The best data available for examining the variance of velocity increments for a range of spacings were extracted from SEASAT synthetic aperture imaging radar by Hall and Rothrock, 1981. By comparing two images separated by three days, they were able to measure the displacement of enough ice features to resolve considerable spatial detail in one dimension of the ice motion. After interpolating to evenly spaced points x_i their data have the form

$$\{u(x_i),\ v(x_i)\} \qquad i = 1,\dots,N,\ \Delta x = 2\ \text{km}.$$

For h = 1,..., the h-increments were defined as

$$I_u(h,i) = u(x_{i+h}) - u(x_i)$$
$$I_v(h,i) = v(x_{i+h}) - v(x_i).$$

The variance of the increments was estimated from the $\{I_u(h,i), I_v(h,i)\}$, $i = 1,\dots, N-h$ with the results plotted in Figure 19. The linearity of the log-log plot for small h supports a power law relationship with $\mathrm{var}(I(h)) = ch^{\alpha}$ with α in the neighborhood of 1.1 to 1.3. It appears then that the increments do not decrease as fast as h^2, and therefore the velocity does not have a derivative in the mean square sense.

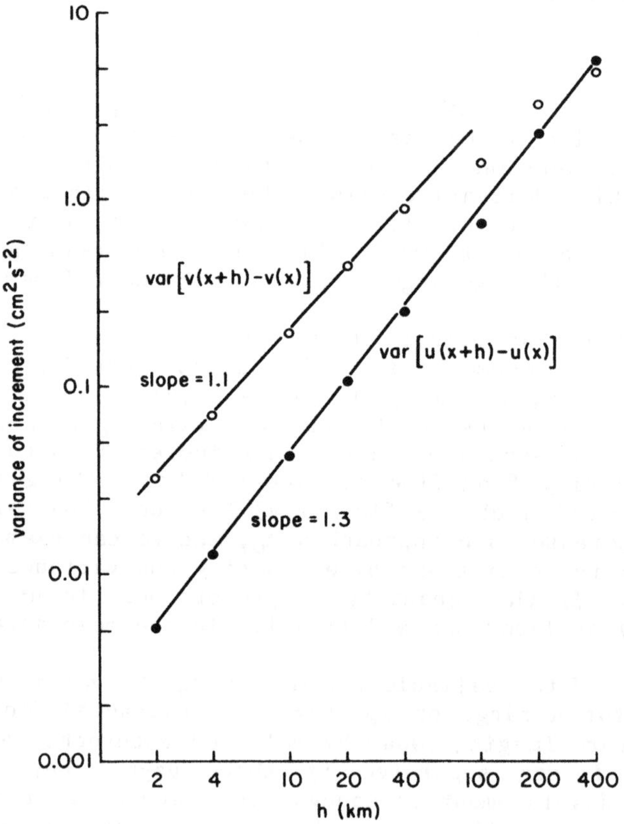

Fig. 19. The variance of the increments of velocity versus interval length.

The statement just made for the variance of increments can be recast in terms of the autocorrelation function. It implies that for small r

$$B_{||}(r),\ B_{\perp}(r) \sim 1 - \frac{c}{2} r^{\alpha} \tag{7.34}$$

thus helping to define the shape of $B_{||}$ and $B_{\perp}$ at r = 0. At r = 0, $B_{||}$ and $B_{\perp}$ have the value unity, are continuous, have a continuous first derivative of zero, but fail to have a second derivative.

5.4.2 Velocity Differences. The definition of the derivative as a limit could be discarded in favor of a velocity difference over a finite distance,

$$u_{x,L}(x,y,t) = \frac{1}{L}[u(x+L,y,t) - u(x,y,t)] \tag{7.35}$$

and similar definitions for $u_{y,L}$, $v_{x,L}$, and $v_{y,L}$.

There is no question about the existence of these differences. Definitions for strain rate invariants follow naturally by associating $u_{x,L}$ with $\partial u/\partial x$ etc., in the earlier definitions. When L is chosen to be of order 100 km, these definitions can be used to describe the large scale deformation.

There are several problems with this approach. First, the L-differences carry no information about smaller scales. Second, no basis has been given for choosing a particular value for L. Third, measurements of velocity are rarely available at uniformly spaced points. Still most analyses of sea ice strain have been done with a length scale implicitly fixed by the scale of the observations. To interpret these analyses we must determine the role played by L.

Consider the covariances between the various L-differences. These follow from the covariance structure of the velocity itself. For example

$$\begin{aligned} E\, u_{x,L}(o,o,t)u_{x,L}(r,o,t) &= \frac{1}{L^2}\ E[\{u(L,o,t) - u(o,o,t)\} \\ &\quad \{u(r+L,o,t) - u(r,o,t)\}] \\ &= -\frac{q^2}{L^2}\ [B_{||}(r+L) - 2\,B_{||}(r) + B_{||}(r-L)]\ . \end{aligned} \tag{7.36}$$

To interpret this equation, suppose the L-difference $u_{x,L}$ is measured at two points separated by a distance r along the x axis. The right hand side of eq. 7.36 approximates the second derivative of

$B_{||}$ at r. For small r, $B_{||}$ is concave down so the two L-differences are positively correlated. At some value of r, $B_{||}(r)$ has an inflection point. Over such a distance the L-differences become uncorrelated; at longer distances the L-differences are negatively correlated. This example has been worked for the $u_{x,L}$ component at points separated along the x-axis, but the same procedure can be used to correlate any two first velocity differences at any two points.

It is apparent from this and Figure 16 that quantities related to the L-differences are highly correlated in space only over distances of less than about 400 km. Furthermore this statement is not sensitive to the choice of L, since the expression 5 will change sign at approximately the same r, independent of L. The correlation length for velocity by comparison is roughly twice as large (see Figure 16).

The ice velocities are well correlated at distances of up to 1000 km, as evidenced by the patterns in the long term ice motion (Figures 2-4) and in shorter term motions (see Thorndike and Colony, 1980 for instance). It may seem surprising not to find similar patterns in the strain fields. The reason why such patterns are not found is that the strain field has, by eq. 7.36, and Figure 16, a correlation length scale of only a few hundred kilometers. Measurement arrays like the 1979 and 1980 buoy arrays with a buoy spacing of roughly 500 kilometers are too coarse to resolve patterns in the strain field.

For the example just worked we can find the variance of the L-difference by setting r = o.

$$E\,u^2_{x,L} = \frac{2q^2}{L^2}\,(1 - B_{||}(L)). \tag{7.37}$$

The symmetric matrix of covariances between pairs of L-differences at the same point is

$$\begin{array}{c|cccc}
 & u_{x,L} & v_{y,L} & u_{y,L} & v_{x,L} \\
\hline
u_{x,L} & \frac{2q^2}{L^2}\left[1-B_{||}(L)\right] & \frac{-q^2}{2L^2}\left[B_{||}(\sqrt{2}L)-B\ (\sqrt{2}L)\right] & \frac{q^2}{L^2}\left[1+\bar{B}(\sqrt{2}L)-2\bar{B}(L)\right] & \\
v_{y,L} & & \frac{2q^2}{L^2}\left[1-B_{||}(L)\right] & 0 & \frac{q^2}{L^2}\left[1+\bar{B}(\sqrt{2}L)-2\bar{B}(L)\right] \\
u_{y,L} & & & \frac{2q^2}{L^2}\left[1-B_{\perp}(L)\right] & \frac{-q^2}{2L^2}\left[B_{||}(\sqrt{2}L)-B_{\perp}(\sqrt{2}L)\right] \\
v_{x,L} & & & & \frac{2q^2}{L^2}\left[1-B_{\perp}(L)\right]
\end{array}$$

where $\bar{B} = \frac{1}{2}(B_{||} + B_{\perp})$.

This matrix can be evaluated at L = 400 km for example by reading points from the plotted functions $B_{||}$ and $B_{\perp}$ in Figure 16.

	$u_{x,L}$	$v_{y,L}$	$u_{y,L}$	$v_{x,L}$	
$u_{x,L}$	$3.9 \times 10^{-14} s^{-2}$	$-0.8 \times 10^{-14} s^{-2}$	$1.4 \times 10^{-14} s^{-2}$	0	
$v_{y,L}$		$3.9 \times 10^{-14} s^{-2}$	0	$1.4 \times 10^{-14} s^{-2}$	L = 400 km
$u_{y,L}$			$6.1 \times 10^{-14} s^{-2}$	$-0.8 \times 10^{-14} s^{-2}$	$q^2 = 10^2 cm^2 s$
$v_{x,L}$				$6.1 \times 10^{-14} s^{-2}$	

The covariance matrix for the L-differences illustrates several properties of the way the ice deforms. Notice that the variances of the parallel differences $u_{x,L}$ and and $v_{y,L}$ are smaller than the perpendicular differences $u_{y,L}$ and $v_{x,L}$. The parallel differences relate to the linear stretching and contracting of the ice pack. The perpendicular differences relate to shear and rotation. Apparently the second type of motion is greater, but recall the word of caution given earlier that the interpretation of this kinematic evidence is ambiguous. We cannot say whether the observed motion reflects some property of the ice pack or some characteristics of the driving forces.

With the L-differences one could define L-strains in a natural way. The statistics in Table 4 of the L-strains follow from the above matrix. These values predict, on the basis of the spatial correlations above, that the large scale divergence will typically be smaller than the vorticity and shear, a prediction verified by many different sets of observations.

5.4.3 Deformation Measurements. Several studies have been made over the past decade or so using observations of velocity at an array of stations, with the objective of monitoring the large scale deformation of the region covered by the array. While it is never stated just this way, the implicit definition of large scale deformation appears to involve spatially averaged derivatives over the region,

$$\overline{\frac{\partial u}{\partial x}} = \frac{1}{A} \int_R \frac{\partial u}{\partial x} \, da \qquad (7.38)$$

where A is the area of the region R sampled by the array. Similar expressions define the other large scale average velocity gradients. Application of the Green-Gauss theorem implies

Table 4. Typical strain rates as a function of gauge length L, in percent per day, using $q^2 = 10^2\ cm^2\ s^{-2}$.

L-strain		Variance	L: 100 km	200 km	400 km	800 km
Divergences	$u_{x,L}+v_{y,L}$	$\frac{q^2}{L^2}\left[4-4B_{\|\|}(L)-B_{\|\|}(\ 2L)+B_{\|}(\ 2L)\right]$	1.6	2.2	2.2	1.6
Vorticity	$u_{y,L}-v_{x,L}$	$\frac{q^2}{L^2}\left[4-4B_{\|}(L)+B_{\|\|}(\ 2L)-B_{\|}(\ 2L)\right]$	4.3	3.7	3.2	2.2
Shear	$u_{x,L}-v_{y,L}$	$\frac{q^2}{L^2}\left[4-4B_{\|\|}(L)+B_{\|\|}(\ 2L)-B_{\|}(\ 2L)\right]$	3.1	3.0	2.6	1.8
	$u_{y,L}+v_{x,L}$	$\frac{q^2}{L^2}\left[4-4B_{\|}(L)-B_{\|\|}(\ 2L)+B_{\|}(\ 2L)\right]$	3.4	3.1	2.8	2.0

Table 5. Statistics of sea ice deformation. All deformations are expressed in percent per day. (1%/d = $1.16 \times 10^{-7}\ s^{-1}$)

Length scale diameter (km)	Season	Divergence		Shear		Vorticity		Strain magnitude mean	Experiment
		Mean	St.dev.	Mean	St.dev.	Mean	St.dev.		
20	spring	--	0.65	--	0.86	--	1.1	--	AIDJEX 1972
100	spring	--	0.43	--	0.54	--	0.61	--	AIDJEX 1972
200	winter	0.07	1.0	1.6	1.6	-0.52	2.0	2.1	1975 AIDJEX manned array
	summer	-0.03	1.6	3.5	2.2	--	--	--	
800	winter	0.02	0.56	1.0	0.86	--	--	1.3	1975 AIDJEX buoy array
	summer	0.16	1.0	1.6	0.86	--	--	2.2	
800	winter	0.07	0.58	0.77	0.54	--	--	1.1	1962-64 T3 NP-10, Arlis II
	summer	0.06	0.80	1.0	0.68	--	--	1.5	
1000	winter	0.02	0.45	0.95	0.82	-0.78	1.5	1.1	1979 buoys
	summer	-0.11	0.76	1.5	0.95			1.9	

Sources: Hibler et al., 1974; Colony and Thorndike, 1981; Maykut, 1982.

$$\overline{\frac{\partial u}{\partial x}} = \frac{1}{A} \oint_C u \, \underset{\sim}{n} \cdot \hat{\underset{\sim}{i}} \, d\ell \tag{7.39}$$

where $\underset{\sim}{n}$ is the outward normal to the perimeter C of R, and $\hat{\underset{\sim}{i}}$ a unit vector in the x direction.

Note the difference between the L-difference and the large scale average derivative. The former can be measured exactly as u(x+L,y) - u(x,y); the latter requires measurements at every point around a closed curve. In the next section we will discuss how many measurements are required to achieve a desired accuracy in the large scale deformations. But first we review some of the results of the studies just alluded to.

In these studies the deformation estimates were made by finding the linearly varying velocity field which most closely fit the observed velocities at an array of points within R.

$$\underset{\sim}{u}(\underset{\sim}{x}) = \underset{\sim}{u}_o + M\underset{\sim}{x}$$

where the matrix M contains the four large scale average velocity partial derivatives. M and $\underset{\sim}{u}_o$ are determined using a least squares fit of the observed velocities $\underset{\sim}{u}$ and positions $\underset{\sim}{x}$. The strain rate invariants were then computed from the elements in M.

The values, given in Table 5, confirm some of the results deduced above from the observed spatial correlation functions. For example, the standard deviations of vorticity and shear are generally larger than the standard deviation of divergence. Note also the mean divergence is small compared to the mean vorticity and shear.

The distribution of the invariant θ is plotted in Figure 20. Note that nearly always $\pi/4 < \theta < 3\pi/4$. In this range, the two principal values of the strain rate tensor have opposite signs. The figure indicates that it is almost always the case that when the ice pack is extending in one direction, it is contracting in the perpendicular direction. Hibler et al. 1974 also comments on this for strain measurements on a 20 kilometer scale.

It is also evident from Table 5 that the summer values are usually somewhat greater than the winter and spring values. Presumably the ice pack is weaker and offers less resistance to deformation in the summer. (The alternative hypothesis, that the driving forces are larger in the summer, is not true.)

The mean quantities in Table 5 produce large strains over the course of a year. The year long deformations at the AIDJEX arrays are shown in Figure 21. The region experienced a net clockwise

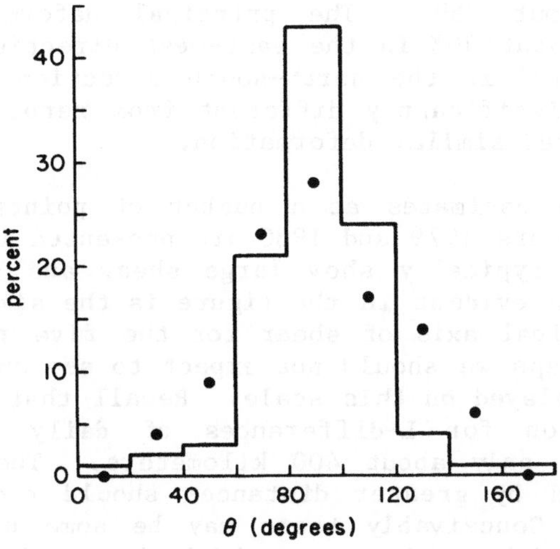

Fig. 20. The distribution of θ.

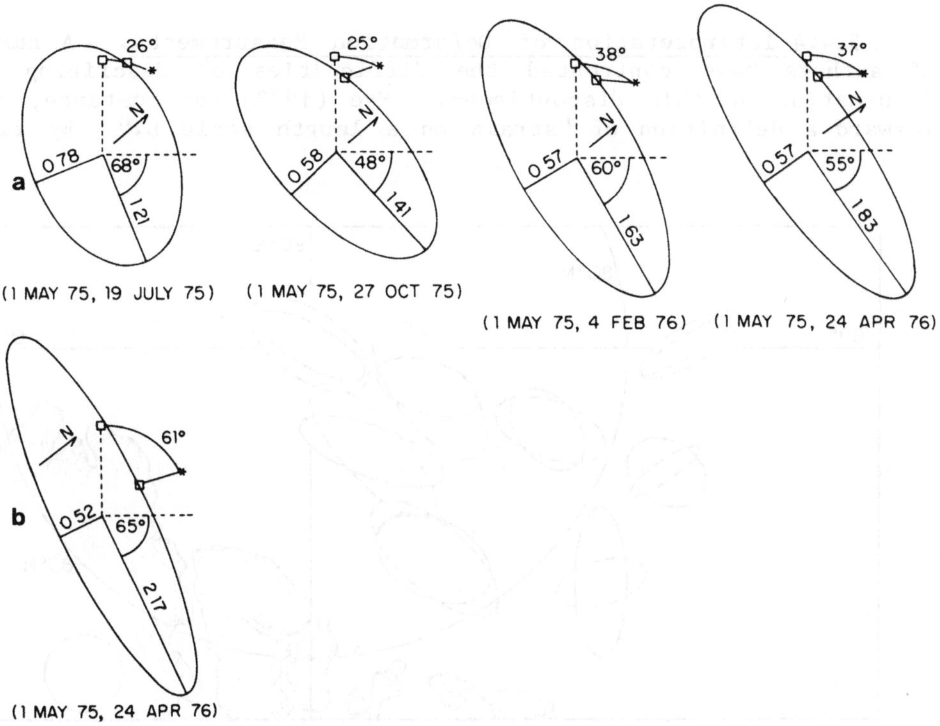

Fig. 21. AIDJEX long term strains.

rotation of about 35°. The principal deformation involved a stretching of about 90% in the east-west direction and a contraction of about 40% in the north-south direction. The net divergence was not significantly different from zero. The two nested arrays experienced similar deformation.

Deformation estimates at a number of points in the central basin for the years 1979 and 1980 are presented in Figure 22. The strain ellipses typically show large shear and small divergence. The only pattern evident in the figure is the similar alignment of the major principal axis of shear for the five points closest to the pole. Perhaps we should not expect to see any patterns in the deformation displayed on this scale. Recall that the spatial correlation function for L-differences of daily velocities has a length scale of only about 400 kilometers. The deformations at points separated by greater distances should evolve more or less independently. Conceivably there may be some underlying spatial pattern to the deformation rate which is small on a daily basis but which accumulates to a large spatially organized strain over the course of a year. The strain ellipses for 1979 and 1980 do not support this idea. On the other hand the long term average vorticity is clockwise throughout the central basin. It amounts to about 40° per year.

5.4.4 Interpretation of Deformation Measurements. A number of authors have confronted the difficulties of describing the deformation of this discontinuum. Nye (1973) for instance, puts forward a definition of "strain on a length scale L." By first

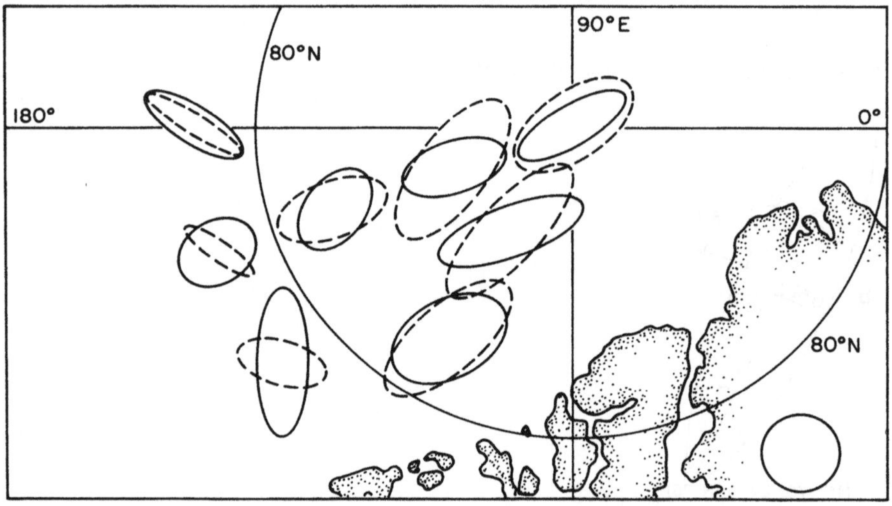

Fig. 22. Deformations from 1979 and 1980.

smoothing the velocity field using a kernel scale L, he obtains a new velocity field which is differentiable and for which the usual notions of deformation based on partial derivatives are valid. Papers by Hibler et al. (1974) and Thorndike and Colony (1977) take a similar point of view by attempting to partition the ice deformation into two parts. The first part is assumed to be associated with the spatial variation of the atmospheric and oceanic forcing fields and with the geometry of the basin, to have a characteristic length of 10^3 km, and to be differentiable. The second part is associated with the irregular, discrete geometry of the ice pack. It has length scales less than 10 km and is characterized by discontinuous variations in velocity. In these papers, the first kind of variation is referred to as the large scale, underlying, continuum, or linear part of the velocity field, while the second is referred to as the small scale, local, perturbation, fluctuating, or non-linear part. Velocity variations of the first type are regarded as signal and those of the second type as noise. This nomenclature reflects the hope that any physical process of interest can be parameterized in terms of the large scale signal, with small scale noise only making it difficult to measure the signal well.

This convention obscures the true nature of the velocity field which is that it has variations on all length scales, with a smooth decrease in amplitude for decreasing length scales. There is no clear division between large and small scale. Furthermore there is no clear reason to associate large scale with signal and small scale with noise. For studies involving the actual opening and closing of leads, the small scale phenomena may indeed be the signal. Further we must be prepared for the possibility that the small scale phenomena cannot be parameterized in terms of the large scale motion.

Perhaps a better conceptual model of the spatial structure of the ice velocity is a system which accepts a smooth input and produces a discontinuous output. (A simple example of such a system is the greatest integer function: [x] = greatest integer less than or equal to x.) The shift in emphasis from the earlier model is this: the earlier model viewed the velocity as the sum of separate contributions, one smooth, one discontinuous. The alternative views the velocity as a discontinuous response to a smooth input. This point of view may lead more naturally to phenomenological descriptions of the properties of the ice pack, through a comparison of the input and output fields. To my knowledge this has not been attempted because data with adequate spatial resolution of the velocity field are still too scarce. Such a study would not explain why the velocity field had certain properties. That explanation must be based on rather deeper understanding of the geometry of the ice pack and the forces which act between floes that we have at present. Still it would be useful to compare some of the properties of the ice velocity with properties of the external driving forces, i.e., the geostrophic wind and the ocean currents.

5.4.5 Errors in Estimating the Large Scale Deformation. The large scale deformation estimates of the previous section are subject to sampling variations since they depend on the particular array of measurements points. There have been several attempts to quantify the sampling errors. Hibler et al. (1974), Nye (1975) and Thorndike and Colony (1977) examined the departures of the measured velocities from the best fit linear relationships. These departures, called residuals, homogeneity variations, or nonlinear fluctuations, represent the variability of the actual velocity field on scales smaller than the diameter of the region of interest.

From observations over a 20 km scale taken during the spring of 1972, Hibler et al. found rms velocity residuals of 0.06 cm s^{-1}, compared to typical linear changes in velocity ($L\partial u/\partial x$) over 20 km of 0.14 cm s^{-1}. Thorndike and Colony used observations from a 100 km scale taken during the spring of 1975 and found rms residuals 0.4 cm s^{-1} and typical linear changes over 100 km of 1.1 cm s^{-1}. Their summer values were somewhat larger: 1.1 cm s^{-1} and 1.8 cm s^{-1} for the nonlinear and linear contributions. These residuals can be regarded as errors with respect to the large scale average derivatives for the region. When the number of observations is small, the estimated large scale deformations are strongly contaminated by these errors.

Using what we know about the correlation functions for velocity, we can profitably address the sampling question from a different angle. Taking the line integral definition for the large scale velocity derivatives, we ask: how many points around the perimeter of the region must be sampled to resolve the integral to some desired accuracy? Intuition suggests that the measurements should be spaced closely enough to permit good interpolation but not so closely as to be highly redundant. A correlation between velocities at neighboring measurement points of 0.5 might be a fair guess; this would indicate a spacing of about 400 km.

To get a better answer, we estimate the line integral in equation 7.39 around a circular region of radius r as

$$D_M = \frac{1}{\pi r^2} \sum_{i=1}^{M} u_i \cos\theta_i \frac{2\pi r}{M}, \quad \theta_i = (i-1)\frac{2\pi}{M}, \quad i=1, \ldots, M. \tag{7.40}$$

Then as the number of measurements M increases, D_M approaches $\partial u/\partial x$. Since D_M is a linear combination of velocity components, we find its variance in terms of the variances and covariances of the velocity components at all M points, and these can be evaluated in terms of the correlation functions $B_{||}$ and $B_{\perp}$. Let $\underset{\sim}{U}$ be the vector involving all 2M velocity components, $\underset{\sim}{A}$ be the vector of coefficients, and R be EUU^T. Then $D_M = A^T U$ and $ED_M^2 = A^T RA$. This calculation has been done for M = 60 which was large enough to resolve $\overline{\partial u/\partial x}$. The dependence of $(ED_{60}^2)^{1/2}$ on the radius r is shown in Figure 23a. This curve depends on $B_{||}$ and $B_{\perp}$ which for this purpose were taken to be $(1-r/a) \exp - (r/b)^2$ with a = 2000 km, b = 1300 km for $B_{||}$ and a = 900 km, b = 1000 km for $B_{\perp}$. These functions incorporate the features of the observed correlations in Figure 16. The leading factor (1 − r/a) makes the assumed functions vary linearly near r = 0. The behavior of the correlation functions near r = 0 is related to the variance of increments by

$$\text{var}\ \{u(r+\Delta) - u(r)\} = 2q^2\ (1-B_{||}(\Delta)) \approx 2q^2\ \Delta/a \quad \text{for small } \Delta \tag{7.41}$$

which is a rough approximation to Figure 19 where the variance of increments is seen to be proportional to $\Delta^{1.3}$. The results in Figure 23 are quite sensitive to the behavior of the correlation function. They should not be taken too seriously until more is known about $B_{||}$ and $B_{\perp}$ near $\Delta = 0$.

With some caution then we interpret Figure 23a as follows. Typical values for the large scale average derivatives are 1% per day decreasing somewhat with the size of the region over which the average is taken. The derivatives $\overline{\partial u/\partial y}$ and $\overline{\partial v/\partial x}$ are generally larger than $\overline{\partial u/\partial x}$ and $\overline{\partial v/\partial y}$ as a consequence of $B_{\perp}$ falling off more rapidly than $B_{||}$.

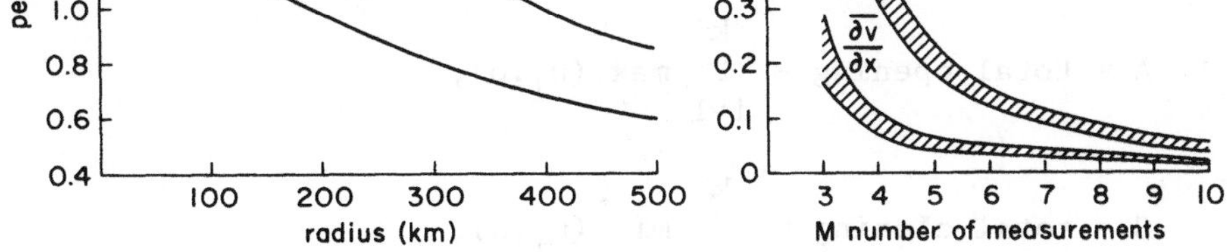

Fig. 23. a) Standard deviation of large scale average strain rates versus radius. b) Estimation errors for large scale average strain rates versus number of measurements.

The variance of the error in estimating $\overline{\partial u/\partial x}$ using only M points, $E(D_M - D_{60})^2$ can be calculated in the same manner simply by redefining the coefficients in the vector $\underset{\sim}{A}$. These results are presented as a fraction of the signal variance $E(D_{60}^2)$ in Figure 23b. When M is 3, as was the case for AIDJEX 1972, the 1975-76 manned AIDJEX array, and LOREX 79, the ratio of estimation error variance to signal variance was about 0.7.

With six measurements around the circumference of the region, the average derivatives are resolved fairly well, and only small improvement is realized by adding more measurements. The results are insensitive to the size of the region. The broad curves in Figure 23b contain results for discs ranging in radius from 100 to 500 km.

In this section we have asked how many measurement stations are needed to provide good estimates of the large scale average velocity derivatives. This is a different issue than how densely must one sample in space to resolve the spatial patterns in the deformation. Observations from a 500 km grid can be expected to give fairly good estimates of the average deformation, but will not resolve the spatial pattern very well.

5.4.6 <u>The Relationship Between Measured Large Scale Deformation and Total Opening and Closing.</u> Suppose a region of interest is intersected by a number of cracks, each of which is opening or closing at some rate. If the rates were known, the total rates of opening and closing for the region could be found. In practice we are not able to measure the motion at each crack, but only motion of a few points in the region. How are we to use these few measurements to estimate the opening and closing?

In an analogous situation in one dimension, we have k cracks each with opening or closing rate u_i $i=1,\ldots, k$. We can imagine the velocity u to be a random function of x having random discontinuities at random points and being constant between the points. Suppose we have measured only the motion at the end points of the region, an interval of length L. Then

$$A = \text{total opening} = \sum_{i=1}^{k} \max (u_i, o),$$

$$C = \text{total closing} = \sum_{i=1}^{k} \min (u_i, o),$$

$$U = \text{net opening or closing} = A + C = \sum_{i=1}^{k} u_i, \qquad (7.42)$$

and the problem is to estimate A and C given U. At first sight the situation seems hopeless. U clearly contains information about the net opening or closing but not about the total opening and closing. However, knowledge about the random variables k, u_1, $u_2, \ldots, u_k$ can be used to make probability statements about the opening and closing. Suppose for instance k has the Poisson distribution with parameter λ. This means that λL is the average number of cracks in an interval of length L. Then the probability of finding k cracks in a random interval of length L is

$$p(k;\lambda,L) = (\lambda L)^k e^{-\lambda L}/k! \quad . \qquad (7.43)$$

If $\lambda L = 0.1$, for instance, the probability of getting $k = 0$ is 0.9051, prob(k=1) = 0.0905, and prob(k>1) = 0.0045. Therefore, with high probability, there is either no crack or just one in the interval and the observed value of the velocity difference U would itself indicate the total opening and total closing. Of course for larger λL it becomes more likely that several cracks intersect the test interval, in which case the observed U cannot separate the opening from the closing.

A similar approach for the two dimensional problem is to imagine the ice pack to be crisscrossed by a family of random cracks, defined by the random straight lines

$$x \cos \phi_i + y \sin \phi_i = r_i \qquad (7.44)$$

where each ϕ_i has the uniform distribution on $(0,2\pi)$ and the normal distances r_i from the origin to each line form a sequence of Poisson points. Imagine that associated with each crack is a velocity discontinuity $\underset{\sim}{u_i}$ having the Gaussian distribution. This describes a random vector field having structure akin to the velocity field of sea ice. The observations of Hall and Rothrock can be used to estimate the parameters in the Poisson and Gaussian distributions. These random fields have the following properties. They consist of discrete rigid floes. The floes have a distribution of sizes determined by the Poisson field of lines. The Poisson lines are isotropic and homogeneous. The velocity difference between any two points is the vector sum of the velocity discontinuities encountered getting from one point to the other.

At each crack the opening and closing is determined by the projection of the velocity difference vector onto the normal to the crack.

$$\text{opening} = \max(o,\ u_i \cos\phi_i + v_i \sin\phi_i)$$

$$\text{closing} = \min(o,\ u_i \cos\phi_i + v_i \sin\phi_i)$$

Thus it is a simple matter to evaluate the total opening and closing for any realization of the random field.

Of course one can also imagine measuring the velocity at a few points and constructing the L-strain rate invariants from the observed velocities. In this way one can test for a relationship between the L-strain rate invariants calculated from a few sampled velocities and the total opening and total closing found by tallying up the activity at every crack.

In an attempt to carry out this program, parameter values were taken to be $\lambda = 0.08\ km^{-1}$, $\mu = 0$, $\sigma^2 = 1$ (arbitrary units). An L = 100 km triangle was used to simulate the sampling procedure using during AIDJEX. A large number of random fields were generated.

Each realization of the random field is defined by the sequences of random numbers k, ϕ_i, u_i, v_i, r_i for $i = 1, \ldots, k$. Here the ϕ_i are drawn independently from the uniform distribution on $(0, 2\pi)$. The values for r_i form a Poisson process with parameter λ. This is achieved by drawing the increments $\Delta = r_{i+1} - r_i$ independently from the exponential distribution with density $\lambda \exp(-\lambda\Delta)$. The process is terminated as soon as r_k exceeds 100 kilometers since none of the subsequent lines would intersect the 100 kilometer region. Finally the u_i and v_i were drawn independently from the normal distribution with zero mean and unit variance.

For each field the velocity was measured at the three specified points and the L-strain rate invariants $|E|$ and θ were evaluated. Also for each field the total opening and closing were evaluated using eq. 7.42. These were normalized by $|E|$ and displayed versus θ in Figure 24. For each realization of this random field two points are plotted, $(\theta, A/|E|)$ and $(\theta, C/|E|)$. From the plot it is clear that there are not unique values of $A/|E|$ and $C/|E|$ corresponding to a given θ. Instead there is a distribution of values for $A/|E|$ and for $C/|E|$, and this distribution changes with θ. The distributions sketched in Figure 25 are broad in the sense that probable departures from the mean are at least as large as the mean.

The interpretation of this exercise is that the total opening and closing are only weakly determined by the L-strain rate invariants based on three measurements 100 kilometers apart. Had the relationship been a strong one, the distributions in Figure 25 would have been narrower, or to say it differently, the points in Figure 24 would have clustered more closely around curves like those in Figure 18. It may still be useful to imagine these smooth curves but only with the recognition that the actual opening and closing scatter widely about the imagined curves:

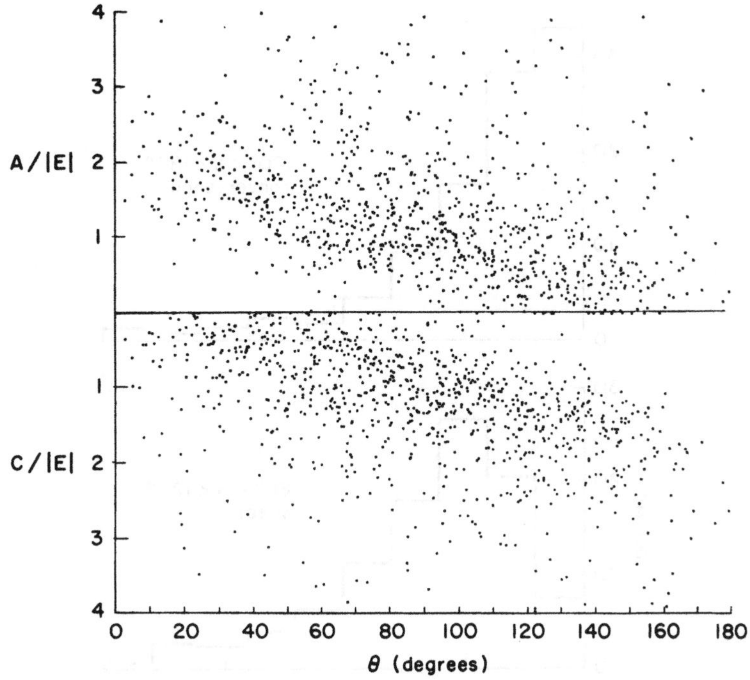

Fig. 24. Total opening and closing for a random model of the velocity field.

$$A/|E| = \alpha_o(\theta) \quad + \text{ random error}$$
$$C/|E| = \alpha_r(\theta) \quad + \text{ random error} \qquad (7.45)$$

where the random terms and the θ-dependent terms make roughly equal contributions to the total opening and closing.

Some of the scatter in these figures arises from having used only three points to estimate the deformation invariants. With more than three points, the errors in the invariants could be reduced, somewhat as shown in Figure 23. But even if the invariants are determined without error, scatter will remain in Figures 24 and 25. This is because many different fields of motion with different total opening and closing could have exactly the same large scale average deformation.

The total opening and closing are an essential part of the theory of the ice thickness distribution. The above results suggest that it may not be satisfactory to represent the total opening and closing as functions of the large scale average velocity derivatives. When many velocity measurements are available in

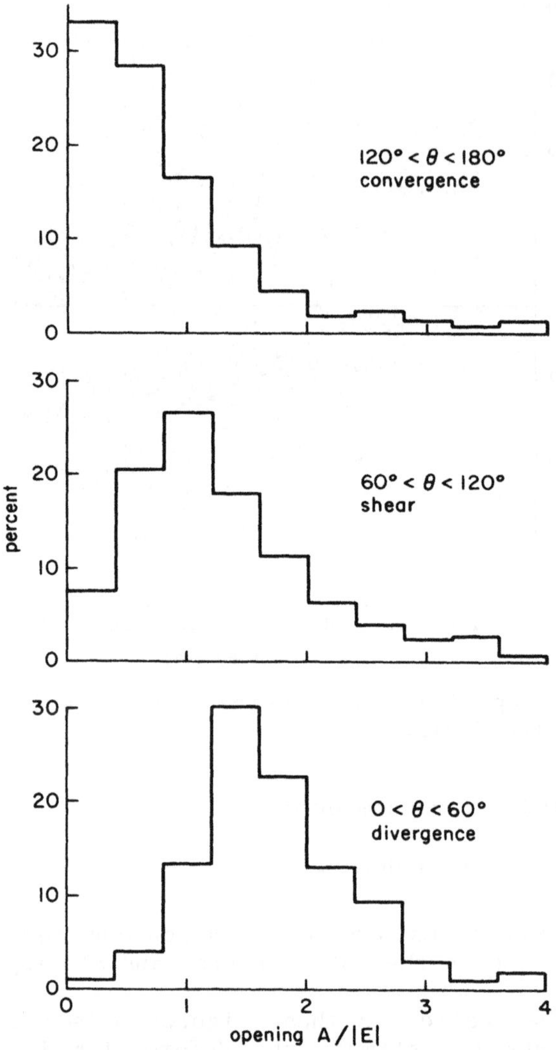

Fig. 25. The distribution of opening for given θ for a random model of the velocity field.

a region, it is possible to estimate more of the spatial structure than simply the average velocity derivative. An attempt to estimate the variability of velocity within the region may help to reduce the scatter which is anticipated on the basis of Figure 24. Or perhaps it will be enough to assume that the variability is constant in time. One might then drive the thickness distribution calculations with opening and closing time series which have the right statistical properties even if they may have large errors on a day to day basis. The statistical properties can be inferred from Figure 24. Of course this figure is the consequence of a particular conceptual model motivated by a limited data set -- the SEASAT SAR data from early October 1978. More data need to be collected and studied before these ideas can be extended to other times and places with confidence.

5.4.7 The Relationship Between Kinematics and Stress. Kinematic data can be used for studying the relationship between stress and strain for sea ice. Suppose for instance that the state of stress tensor σ suitably averaged over some region, is related to kinematic quantities ε and the ice geometry s by some expression of the form

$$\sigma = F(\varepsilon, s). \qquad (7.46)$$

Although σ cannot be measured directly, its divergence $\nabla \cdot \sigma$ can be inferred indirectly from the observed momentum balance when all the other terms in that balance are known. If a particular function F is hypothesized, it can be tested by evaluating both sides of the equation $\nabla \cdot \sigma = \nabla \cdot F(\varepsilon, s)$, the left hand side as a residual from the momentum balance and the right hand side in terms of observed kinematic and ice geometry quantities. This was one of the objectives of AIDJEX (Maykut et al. 1972). In practice the test is difficult to make since neither $\nabla \cdot \sigma$ nor $\nabla \cdot F$ can be determined very accurately from observations. (See Rothrock et al., 1980).

Under special circumstances, meaningful tests may be possible, however. When the ice deformation is strongly divergent, ice floes tend to move apart. With no floe-to-floe contacts there can be no $\nabla \cdot \sigma$ forces. The left hand side should differ from zero only by measurement error. These errors are probably small enough to permit a useful test.

Another special situation of interest is when the ice is being forced up against the coasts by the wind. As the ice moves toward the coast it converges, becomes stronger, and eventually becomes strong enough to resist further deformation. If the on shore wind persists, a zone of motionless ice can widen to several hundred kilometers (Pritchard, 1977). In this situation, intuition suggests that the $\nabla \cdot \sigma$ vector should point off shore and should increase in time until it balances the on shore wind stress. With simultaneous ice trajectories at 100, 200, 300 and 400 km from the coast, it should be possible to observe the amount of convergence required to produce the required resistance to the wind.

At greater distances from the coasts, there is evidence that the ice stresses embodied in $\nabla \cdot \sigma$ are usually small. Little success is anticipated in trying to observe them. Any stress-strain law which provides adequate resistance to deformation near the shore is probably adequate for full basin dynamic modeling. The observations may never be adequate to discriminate very selectively between candidate stress-strain laws.

5.5 A Statistical Model of Ice Kinematics

If the velocity of the ice was known exactly for all locations and times, it would be possible to deduce the future path of an ice floe and its past history as well. Precise statements could be made about where a given piece of ice would be two years hence, or where it had been two years ago. In reality the velocity is known only approximately. Therefore, any predictions will contain uncertainty.

The ice velocity can be decomposed into a time independent or slowly varying field, and a time dependent, rapidly varying field. This distinction may not be clear cut, but the historical record does give the impression of a general circulation which is the same in our day as it was in Nansen's, and a varying field with major fluctuations on time scales of days to months. If the general circulation is regarded as known, and the time varying field as an unknown, random process, then predictions of the future or past location of a piece of ice can be made. The predictions will be statistical in nature, expressing the probability that a floe in location X at time t will be found in a region R at some future or past time.

Colony realized in 1983 that a statistical description of sea ice trajectories could be developed using the concepts of Markov processes. The idea was to consider the motion u at any location x and any time t to be a random vector. The probability distribution for this random vector was to be specified by defining, from observations, the mean field $\bar{u}$, a function of location but not time, and a variance σ^2 characteristic of the departures from the mean.

After a time interval τ, the actual location of the particle is not known, but the probability that it has moved to any specified region can be calculated based on the probability distribution of the random process. Repeated applications of this idea permit predictions over arbitrarily long periods.

To construct their Markov model, Colony and Thorndike (1984, 1985) made a new estimate of the general circulation using about 100 station years of drifting station data. Figure 26, reprinted from the 1984 paper, is similar to Gordienko's field in Figure 2. The difference to note is that this figure is based on more data, and can be relied on for quantitative estimates of the velocity

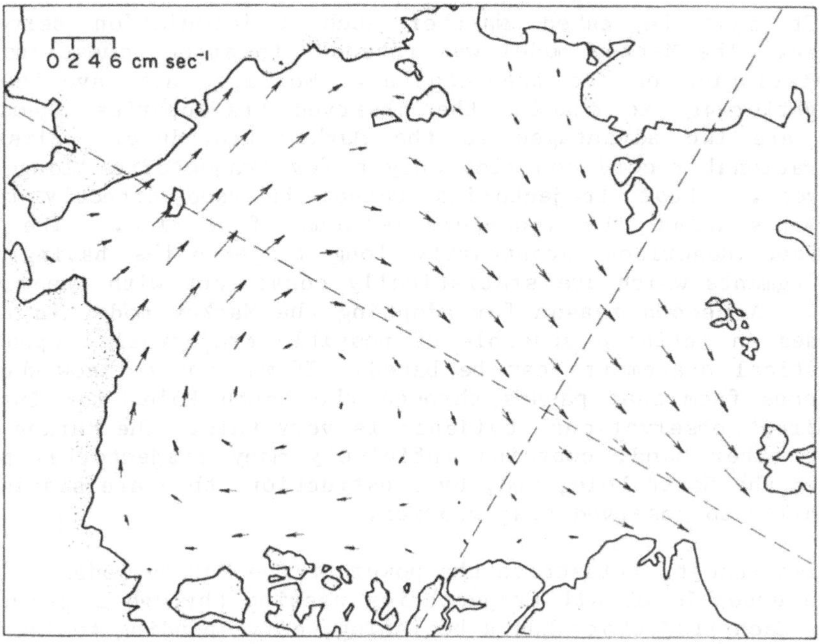

Fig. 26. Field of interpolated mean ice motion. The vectors originate at the center of the boxes.

field. Except for the row of vectors closest to Siberia, the data base is large enough to give confidence in the field shown.

The statistical model of the random motions was also based on observations. Suppose two stations pass over the same location at different times t_1 and t_2. The difference between the subsequent positions at times $t_1+\tau$ and $t_2+\tau$ provides a measure of the random motion. About 100 cases of drift stations occupying the same location were identified by looking for intersecting trajectories. From these, the mean squared distance was found between the two positions after the time interval τ. In calculations, τ was set equal to 90 days. The corresponding mean squared separation was 10^5 km^2, which should be thought of as twice the variance for a single station. Rough numbers for 90 days are mean displacements of 0-250 km depending on location, and random displacements of about 200 km.

In formulating the ice motion as a Markov process, some assumptions were made which were not justified by the data -- the isotropy of the random displacements, for example. These were made to keep the formulation as uncluttered as possible.

It must be asked whether such a formulation serves any purpose. The Markov model was intended to answer questions about the statistics of ice trajectories. Would it not have been more straightforward to examine the observed trajectories themselves? There are two advantages to the Markov procedure. First, the observational record contains only a few trajectories longer than one year. Most trajectories cannot be used directly to make statements about the long-term motions of a floe. The Markov technique describes arbitrarily long trajectories having short-term segments which are statistically consistent with the observations. A second reason for adopting the Markov model is that it provides an infinite ensemble of possible trajectories upon which statistical statements can be based. If we want to know where the ice comes from that passes through the North Pole, for instance, the direct observational evidence is very thin. The Markov model, on the other hand, contains infinitely many trajectories passing through the North Pole, and, by construction, they are statistically similar to observed trajectories.

Two results illustrate the power of the Markov model. Consider the ensemble of all trajectories passing through a given location. Each trajectory has a beginning, corresponding to ice formation in one of the ice-producing regions, and an end, corresponding to ice melt or to passage through one of the exits from the basin. (Consult the original paper for details about the ice formation processes.) Therefore for each trajectory it is possible to determine the time elapsed since formation, and the time remaining before melting or exiting from the basin. These two times are random variables. Their probability distributions can be calculated directly from the Markov model. Results from the Colony and Thorndike calculation are reproduced in Figure 27. The power of this model resides in its ability to predict quantities, like the mean age of the ice in a given region, which cannot be deduced more directly from the kinematic observations.

5.6 Discussion

I have tried in this paper to bring together some of what is known about the motion of sea ice. The emphasis has been on the departures of the true motion from the long term mean circulation. This is a comparatively new topic, the investigation of which has only recently been made possible, mainly by the developments of satellite positioning techniques. Perhaps because of its newness, the field lacks a clear agenda of questions to be answered. Instead we are still trying to characterize the motion, to determine the magnitude of the motion on different scales and to identify motions which have some intrinsic interest or are related to other phenomena of interest.

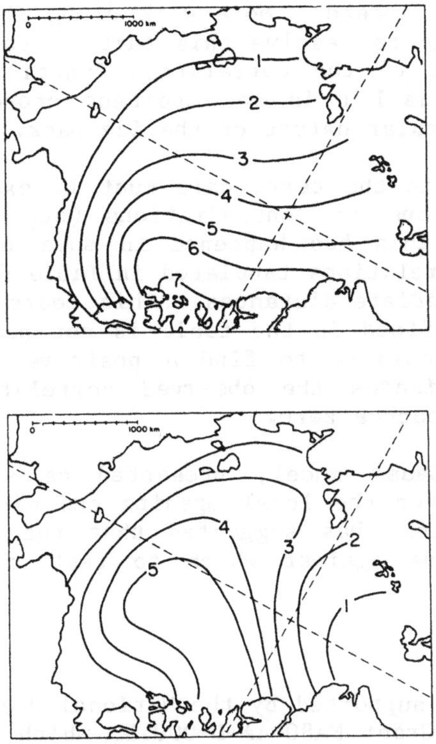

Fig. 27. Contours of the mean age of the ice (a) and mean time until melting or leaving the basin (b); in years.

The time and space correlations have been used extensively here for several reasons. First they give a compact description of the motion. In principle, the correlation function involves one time and two space variables, but in practice a great deal of information can be extracted from three functions, each of a single variable: R(r), $B_{||}(r)$ and $B_{\perp}(r)$. Second, these three functions are accessible to observation. Third, properties of many kinematic quantities can be deduced from these functions, as illustrated in the text. Fourth, these functions form the rigorous basis for answering questions related to interpolation, prediction, and experiment design.

Many of the results presented here are based on sample autocorrelation functions deduced from limited data. The data available for estimating correlations at small space lags (less than

100 km) are meager indeed. More work along the lines of Hall and Rothrock would help to resolve this part of the correlation function. The behavior of the correlation functions $B_{||}(r)$ and $B_{\perp}(r)$ in the limit of small r is an important property of the motion related to the granular nature of the ice pack.

As mentioned in the text, care must be exercised in choosing correlation functions or contradictions (negative variances) can occur. In fact, this has happened in some of the calculations done using the correlations tabulated in Table 6 with linear interpolation to intermediate distances. This means that the piecewise linear function defined in the table is not positive definite. A useful objective would be to find a positive definite analytical form which approximates the observed correlations including the behavior as r approaches zero.

The Poisson-Gauss model, presented as a way to study the relationships between the local opening and closing and the large scale ice deformation, has suggested that the relationship may be weak. Observations against which to test this suggestion are needed.

ACKNOWLEDGMENT

This work was supported by the National Oceanic and Atmospheric Administration Grant NA80-AA-D-00015, which was funded in part by the Global Atmospheric Research Program and the Office of Climate Dynamics, Division of Atmospheric Sciences and the Meteorology Program, Division of Polar Programs, of the National Science Foundation, and the Office of Naval Research, Arctic Programs. The paper also reports on work supported by the Office of Naval Research Contract N00014-79-C-0418.

Some of the ideas discussed here grew out of informal discussions with Drew Rothrock and with Roger Colony. I take this opportunity to acknowledge their contributions and to thank them.

Table 6. Spatial correlation functions for sea ice velocity.

Distance (km)	$B_{\|\|}$	$B_{\perp}$
0	1.00	1.00
100	.98	.95
200	.91	.84
400	.68	.51
800	.37	.06
1200	.19	-.09
1600	.10	-.10
2000	.01	-.06
2400	.00	.00

REFERENCES

AIDJEX (1972) Station positions, azimuths, weather, 1972 AIDJEX pilot study, preliminary data. AIDJEX Bulletin, 14: 63-208.

Colony, R. and A. S. Thorndike (1980) The horizontal coherency of the motion of summer arctic sea ice. J. Phys. Oceanogr., 10(8): 1281-1289.

Colony, R. and A. S. Thorndike (1981) Sea ice strans during 1979. In POAC 81: Proceedings, Sixth International Conference on Port and Ocean Engineering Under Arctic Conditions, Laval University, Quebec City, Quebec, Canada, p. 619-628.

Craig, L. W. (1972) High frequency accelerations of arctic pack ice. University of Washington, Master's thesis.

Doronin, Yu. P. and D. E. Kheisin (1975) Morskoi Led (Sea Ice). English translation, Amerind Publishing Co., New Delhi, India, 1977, 323 p.

Hall, R. T. (1980) AIDJEX modeling group studies involving remote sensing data. In Sea Ice Processes and Models (R. S. Pritchard, Ed.), University of Washington Press, Seattle, Washington, 474 p.

Hall, R. T. and D. A. Rothrock (1981) Sea ice displacement from SEASAT synthetic aperture radar. J. Geophys. Res., 86(C11): 11078-11082.

Hastings, A. D., Jr. (1971) Surface climate of the Arctic Basin. Geographic Sciences Laboratory, U.S. Army Engineer Topographic Laboratories, Report ETL-TR-71-5.

Hibler, W. D. III, W. F. Weeks, A. Kovacs and S. F. Ackley (1974) Differential sea-ice drift. I: Spatial and temporal variations in sea ice deformation. J. Glaciol., 13(69): 437-455.

Hunkins, K. (1967) Inertial oscillations of Fletcher's Ice Island (T-3). J. Geophys. Res., 74(4): 1165-1173.

Hunkins, K. and J. Hall (1971) The navigation of Fletcher's Ice Island (T-3). AIDJEX Bulletin, 7: 47-58.

Hunkins, K., T. Manley and W. Tiemann (1979) Observations on position, ocean depth, ice rotation, magnetic declination and gravity taken at the Fram I drifting ice station. Office of Naval Research, CU-1-79, Technical Report 1, 60 p.

Kolmogoroff, A. (1941) The local structure of turbulence in incompressible viscous fluid for very large Reynolds numbers. Compt. Rend. (Doklady) Acad. Sci. URSS, 30(4): 301-305.

Kowalik, Z. and N. Untersteiner (1978) A numerical study of the M_2 tide in the Arctic Ocean. Deut. Hydrogr. Zeit., 31: 216-229.

Manley, T. (1981) Eddies of the western Arctic Ocean: Their characteristics and importance to the energy, heat and salt balance. Columbia University, Ph.D. thesis, 212 p

Martin, P., C. R. Gillespie, A. S. Thorndike and D. Wells (1978) Position measurements of AIDJEX manned camps using the Navy Navigation Satellite System. AIDJEX Bulletin, 40: 83-102.

Maykut, G. A. (1982) Large scale heat exchange and ice production in the central Arctic. J. Geophys. Res., 87(C10): 7971-7984.

Maykut, G. A., A. S. Thorndike and N. Untersteiner (1972) AIDJEX scientific plan. AIDJEX Bulletin, 15: 67 p.
McPhee, M. G. (1978) A simulation of inertial oscillations in drifting pack ice. Dyn. Atmos. Oceans, 2: 107-122.
Nye, J. F. (1973) The meaning of two-dimensional strain-rate in a floating ice cover. AIDJEX Bulletin, 21: 9-17.
Nye, J. F. (1975a) The use of ERTS photographs to measure the movement and deformation of sea ice. J. Glaciology, 73(15): 429-436.
Nye, J. F. (1975b) A test of the ice thickness distribution equations by measurements of ERTS pictures. AIDJEX Bulletin, 28: 141-150.
Nye, J. F. (1976) A coordinate system for two dimensional stress and strain rate and its application to the deformation of sea ice. AIDJEX Bulletin, 33: 131-143.
Popelar, J., J. Kouba and D. Wells (1981) LOREX satellite positioning, Earth Physics Branch, Department of Energy, Mines and Resources, Ottawa, Ontario, Canada, Draft Report.
Pritchard, R. S. (1977) An estimate of the strength of arctic pack ice. AIDJEX Bulletin No. 34: 94-113.
Rothrock, D. A. (1979) Modeling sea ice features and processes. J. Glaciol., 24(90): 359-375.
Rothrock, D. A., R. Colony and A. S. Thorndike (1980) Testing pack ice constitutive laws with stress divergence measurements. In Sea Ice Processes and Models (R. S. Pritchard, Ed.), University of Washington Press, Seattle, Washington, p. 102-112.
Rothrock, D. A. and R. T. Hall (1975) Testing the redistribution of sea ice thickness from ERTS photographs. AIDJEX Bulletin, 29: 1-20.
Sverdrup, H. U. (1926) Dynamics of tides on the north Siberian shelf: Results from the Maud expedition. Geofysiske Publikajoner, IV(5): 75 p.
Tabata, T., T. Kawamura and M. Aota (1980) Divergence and rotation of an ice field off Okhotsk Sea Coast of Hokkaido. In Sea Ice Processes and Models (R. S. Pritchard, Ed.), University of Washington Press, Seattle, Washington, p. 273-282.
Thorndike, A. S. (1973) An integrated system for measuring sea ice motions. In Ocean 73: IEEE International Conference on Engineering in the Ocean Environment, IEEE Publ. 73 CHO 774-0, p. 490-499.
Thorndike, A. S. and J. Y. Cheung (1977a) AIDJEX measurements of sea ice motion, 11 April 1975 to May 1976. AIDJEX Bulletin, 35: 1-149.
Thorndike, A. S. and J. Y. Cheung (1977b) Measurements of sea ice motion, 14 May 1976 to 30 November 1976, October 1975 to December 1976, and January 1977 to September 1977. Unpublished data reports, available from the authors.
Thorndike, A. S. and R. Colony (1977) Estimating the deformation of sea ice. In POAC 77: Proceedings, Fourth International Conference on Port and Ocean Engineering Under Arctic Condi-

tions (D. B. Muggeridge, Ed.), Memorial University of Newfoundland, St. John's, Newfoundland, Canada, Vol. 1, p. 506-517.

Thorndike, A. S. and R. Colony (1980a) Large scale ice motion in the Beaufort Sea during AIDJEX, April 1975 to April 1976. In Sea Ice Processes and Models (R. S. Pritchard, Ed.), University of Washington Press, Seattle, Washington, p. 249-260.

Thorndike A. S. and R. Colony (1980b) Arctic Ocean buoy program: Data report, 19 January 1979 - 31 December 1979. University of Washington, 131 p.

Thorndike, A. S. and R. Colony (1981) Arctic Ocean buoy program: Data Report, 1 January 1980 - 31 December 1980. University of Washington, 127 p.

Thorndike, A. S. and R. Colony (1982) Sea ice motion in response to geostrophic winds. J. Geophys. Res., 87(C8): 5845-5852.

Thorndike, A. S., D. A. Rothrock, G. A. Maykut and R. Colony (1975) The thickness distribution of sea ice. J. Geophys. Res., 80(33): 4501-4513.

Tucker, W. B., W. F. Weeks, A. Kovacs and A. J. Gow (1980) Nearshore ice motion at Prudhoe Bay, Alaska. In Sea Ice Processes and Models (R. S. Pritchard, Ed.), University of Washington Press, Seattle, Washington p. 261-272.

Zubov, N. N. (1943) L'dy Arktiki (Arctic Ice). Izdatel'stvo Glavsevmorputi, Moscow, 360 p. (U.S. Navy Hydrographic Office, Translation 217, 1963; available as AD426972 from NTIS, Springfield, Va.)

CHAPTER 8

ICE THICKNESS DISTRIBUTION - MEASUREMENT AND THEORY

D. A. Rothrock

Polar Science Center, Applied Physics Laboratory
University of Washington
Seattle, Washington 98195

1. INTRODUCTION

In considering the historical literature relevant to the subject of ice thickness distribution, I was struck by the paper by Wittmann and Schule (1966) at the Lake Arrowhead symposium on the Arctic heat budget. They presented a synthesis of visual observations from aircraft showing ice concentration, lead spacings, percent coverage of several ice types, and ridging indices in subregions of the Arctic Ocean. Their format emphasizes the basic elements of present thinking--first, that we want properties averaged over space scales which include many ice features, and second, that these average properties vary on yet longer scales over the basin and from one season to the next, in ways of interest to the climatologist.

Wittmann and Schule showed what data were available from operational observations just as Soviet scientists were attempting to forecast ice concentration (Nikiforov et al., 1967; Doronin, 1970). In the intervening fifteen years, effort has been directed to defining what data we want for climatological and forecast modeling of sea ice, and why we want them. One data set we want is ice thickness distribution. A major reason is that it tells us the mass of ice in the ice cover. Other reasons are the subject of other chapters. Suffice it to say that ice thickness and its distribution influence components of the surface heat and momentum balance, the salt balance of the upper ocean, light penetration and biological productivity, and trafficability on top of, through, and beneath the ice cover. It is unfortunate that thickness alone does not determine these phenomena; they are affected by snow cover, temperature, ice formation history and surface relief, so that writing some flux as a function only of thickness sweeps a lot under the rug.

2. THICKNESS AND ITS DISTRIBUTION DEFINED

Although one can draw a slab of ice or a cross section of a ridge schematically with several straight lines, when one sets about measuring thickness, it is no longer such a clear-cut notion. There are cavities in ridged ice--some connected to the air or water. Ponds of fresh or brackish water covered with a skim of ice can lie in the troughs of both the upper and lower surfaces. How should a slush of ice crystals be treated? We suppose that one has satisfactory conventions for dealing with these phenomena and that there is a top surface of the ice (not the snow) at a height h_t measured up from the sea surface, and a bottom surface at draft h_b measured down from the sea surface. The thickness is $h = h_t + h_b$.

From thickness measurements which resolve ridges and leads, we want to know average properties over regions hundreds of kilometers in diameter. For instance, to evaluate mean thickness H for a region R centered on (x,y), we write

$$\hat{H}(x,y) = \frac{1}{A} \iint_{x' \varepsilon R} h(x',y')dx'dy', \quad (\text{where } A = \iint_R dx'dy')$$

as an integral, over thickness, of the area covered by each thickness

$$= \iint_R h \frac{da(h)}{A} \equiv \iint_{\text{all } h} h\, g(h)dh$$

The ice thickness distribution g(h)dh is defined as the fraction of R with thickness between h and h + dh. If we want, say, an average heat flux Q and Q depends on h, we can again integrate over all thicknesses

$$\overline{Q} = \frac{1}{A} \iint_R Q[h(x',y')]dx'dy' \equiv \int_{\text{all } h} Q(h)\, g(h)\, dh$$

Since many fluxes at the air-sea-ice interface are strongly related to thickness, measurements of thickness distribution would have many applications.

Two observed distributions of draft (which is nearly the same as thickness) are illustrated in Figure 1. The Beaufort Sea distribution shows 5% coverage of 0-to-1 meter ice and 40% coverage of 2-to-3 meter ice. On the other side of the ocean, the Fram Strait ice is much heavier, with roughly five times as much thick ice as in the Beaufort Sea ice. The 2-to-3 meter category is comparatively empty with only 3.5%. Thus the Fram Strait ice has a greater mass, is frictionally rougher by virtue of its deformed ice, and yet with its 10% of 0-to-1 meter ice is producing ice faster than the ice in the Beaufort Sea sample.

Although the distributions in Figure 1 look strikingly different, the question must be answered: are these real geophysical differences or could these two distributions be different samples from the "same" ice cover? An empirical answer is obtained by taking several samples from each area, and noting whether all samples from one region are similar to each other and different from the samples from another region.

An analytical technique for answering questions about sampling errors is rooted in probability theory. We adopt the notion that $h(\underset{\sim}{x})$ is a stochastic process; the thickness h at a fixed position $\underset{\sim}{x}$ is a random variable. In this framework, the thickness distribution is a probability density function.

$$p = \int_a^b g(h)dh = \text{probability that } a<h<b$$

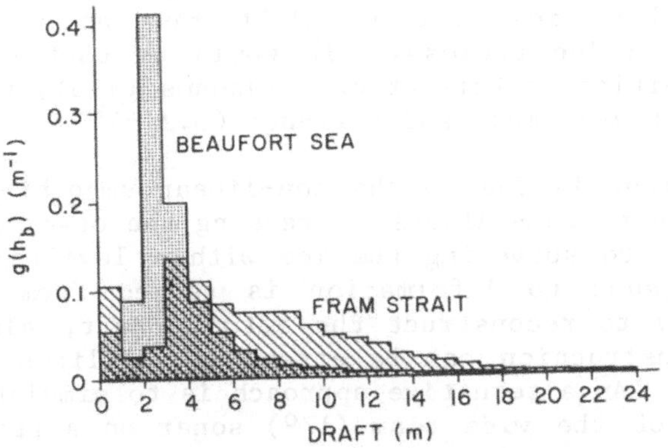

Fig. 1. Distributions of ice draft for two regions. The bin size is one meter. The Beaufort Sea data are from Gurnard profiles (the 60 km offshore profile of Rothrock and Thorndike, 1980), and the Fram Strait data are from Sovereign profiles (a 97 km profile, section no. 2 in Wadhams, 1980).

The expected value of thickness is the mean of the distribution $H = \int_0^\infty hg\,dh$. By taking the process to be nearly stationary (varying only on scales of hundreds of kilometers) and ergodic, we can estimate statistics of the process, such as the mean, or distribution, or spectrum, from spatial averages of data. In addition, we can find confidence intervals for these estimates.

3. MEASURING THICKNESS AND ESTIMATING ITS STATISTICS

There are two aspects to measurements of thickness statistics: measuring thickness accurately at a point and sampling it at enough points to estimate a statistic satisfactorily. All methods are seriously limited either by measurement errors or sampling errors or both.

3.1 Submarine Sonar

The most satisfactory measurements of thickness distribution are obtained from upward-looking submarine sonar. Table 1 lists several cruises from which data have been published. The sonar does not directly measure thickness, but the distance from the transducer mounted on the submarine to the closest point of the bottom surface within the cone of the sonar beam. A pressure sensor measures the transducer depth. The difference between these distances is some approximation of the ice draft.

There are several sources of measurement error. First, the depth of the transducer is not precisely known; in the more modern data the accuracy is about ± 0.2 m or 0.3 m. The error is not so critical for thick ice, but it might make an estimate of the coverage of thin ice useless. It would be useful to know more about the properties of this error. Wadhams (1981) takes it to be white noise with zero mean and variance $(0.2\ \text{m})^2$.

Another error is due to the non-linear smoothing of the surface by the sonar beam--always increasing the observed draft. The problem is akin to surveying the ice with a leveling rod 15 m in diameter. Because no information is sensed from some troughs, there is no way to reconstruct the actual draft, although an approximate reconstruction can be attempted (Williams et al., 1975; Wadhams, 1981). An alternative approach is to simulate numerically the smoothing of the wide beam (17°) sonar on a profile measured with a narrower beam. The wide beam causes the draft distribution to be shifted to the right, underestimating thin ice and overestimating deep drafts and the mean draft by about 10% (Williams et al., 1975; Rothrock and Thorndike, 1980; Wadhams, 1981).

Table 1. Submarine sonar data.

Submarine	Period	Location	Beam width along track	Submarine depth uncertainty	Length	Reference
Nautilus, Skate, Sargo, Seadragon	1957-62	Whole Arctic		2.0 m	10^4 km	Lyon (1963, 1966) LeShack (1980)
Dreadnought	March 71	86°-90°N 6°- 7°E	20°	0.3 m	1000 km	Swithinbank (1972) Williams et al. (1975)
Sovereign	October 76	80°-90°N 25°E-70°W	17°	0.2 m	4000 km	Wadhams (1981)
Gurnard	April 76	70°-77°N 138°-155°W	3°	0.1 m	1400 km	Wadhams and Horne (1980)

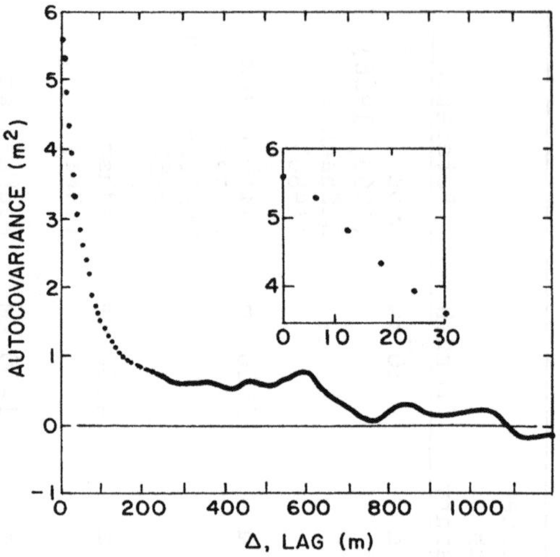

Fig. 2. Autocovariance of the Beaufort offshore profile (from Rothrock and Thorndike, 1980).

Table 2. The sampling error of an estimate of mean thickness from a line record of length L.

L, km	$(\mathrm{var}\ \hat{H})^{1/2}$, m
10	0.38
50	0.17
100	0.12
500	0.05
1000	0.04

Although no error model has been developed to treat these several measurement errors and the sampling error as a whole, it is likely that mean thickness estimates from 100 km sonar profiles have an uncertainty of 0.2 to 0.4 m.

How large is the sampling error for distributions shown in Figure 1? In particular, how accurate is an estimate of p, for the bin $a<h<b$? To answer this question, we need to characterize the spatial arrangement of the regions contributing to this bin. These regions are identified by the indicator function

$$\pi(\underset{\sim}{x},t) = \begin{cases} 1, & a<h(\underset{\sim}{x},t)<b \\ 0, & \text{else} \end{cases}$$

The desired quantity is thickness, not draft. So, although isostatic balance is achieved only over tens of meters, it is assumed to hold at each point in order to convert $h_b(x)$ to $h(x)$ by

$$h(x) = \frac{\text{water density}}{\text{ice density}} h_b(x) \simeq 1.11\ h_b(x)$$

The error thus introduced has zero mean and is probably not important.

To estimate a statistic of thickness, we integrate over some length L of the profile. For instance

$$\hat{H} = \frac{1}{L} \int_0^L h(x)dx$$

is an estimate of the true mean thickness H. If L is chosen too small, the estimate is poor. The estimator $\hat{H}$ is a random variable with a distribution. Its mean is H: it is an unbiased estimate. The important quantity is the variance of $\hat{H}$ which tells how close $\hat{H}$ is likely to be to the true mean. The variance is

$$\text{var}\ \hat{H} = E(\hat{H}-H)^2 = E\left\{\left[\frac{1}{L}\int_0^L h(x_1)dx_1 - H\right]\left[\frac{1}{L}\int_0^L h(x_2)dx_2 - H\right]\right\}$$

$$= \frac{1}{L^2}\int_0^L\int_0^L E[h(x_1) - H]\ [h(x_2) - H]dx_1dx_2$$

E is the expected value operator. The integrand is the autocovariance of h, C_h; assuming the thickness is homogeneous and isotropic, it is a function of separation only, $s = |x_1 - x_2|$. The double integral reduces to

$$\text{var}\ \hat{H} = \frac{2}{L}\int_0^L \left(1 - \frac{s}{L}\right) C_h(s)ds$$

by transforming to new variables s and $t = x_1 + x_2$ and integrating over t. For the Gurnard profile whose autocovariance is shown in Figure 2, this sampling error is listed in Table 2. A typical mean thickness of 3-to-4 m can be sampled by a record ten-to-twenty kilometers long with a sampling error of 10%.

The regions where $\pi = 1$ are the regions we must sample to estimate p. Given profile data (and so, ignoring the y and t dependence), our estimate of p is the fraction of the profile intersecting these regions

$$\hat{p} = \int_0^L \pi(x)dx$$

The estimate is unbiased, since

$$E\hat{p} = \frac{1}{L}\int_0^L E\pi \, dx = \text{probability } [a<h(x)<b] \cdot \frac{1}{L}\int_0^L dx = p$$

assuming the process is spatially stationary. The remaining issue is: how large is the variance of p? We have

$$\text{var } \hat{p} = E(\hat{p}-p)^2$$
$$= E\left\{\frac{1}{L}\int_0^L [\pi(x_1)-p]dx_1 \cdot \frac{1}{L}\int_0^L [\pi(x_2)-p]dx_2\right\}$$
$$= \frac{1}{L^2}\int_0^L\int_0^L C\pi(x_1, x_2)dx_1dx_2$$

where

$$C_\pi(x_1,x_2) = E\,[\pi(x_1) - p]\,[\pi(x_2) - p]$$

is the autocovariance of the indicator function. Stationarity provides that C_π depends only on the distance between points $s = |x_1 - x_2|$, and the double integral reduces, as before, to

$$\text{var } \hat{p} = \frac{2}{L}\int_0^L \left(1 - \frac{s}{L}\right) C_\pi(s)ds \tag{8.1}$$

The autocovariance C_π has magnitude $p(1-p)$ at $s = 0$ [since $E\pi(x)\pi(x) = E\pi(x) = p$], and is zero at $s = \infty$ [since $E\pi(x)\pi(x+s) \rightarrow p^2$ as $s \rightarrow \infty$]. If we define the correlation coefficient

$$r(s) = \frac{C_\pi(s)}{p(1-p)}$$

(8.1) becomes

$$\text{var } \hat{p} = p(1-p)\ell/L \tag{8.2}$$

where

$$\ell = \int_0^L 2\left(1 - \frac{s}{L}\right) r(s)ds$$

In practice we are interested in the case where r(s) drops off in a distance much smaller than L; then

$$\ell \simeq \ell_\infty = 2\int_0^\infty r(s)ds$$

(When $L/\ell_\infty > 5$, we find ℓ within 10% of ℓ_∞.) The quantity ℓ_∞ is the autocorrelation length scale; it is also a measure of the characteristic length of individual profile segments with = 1.

The first portion of Table 3 gives values of ℓ_∞ computed from the Beaufort Sea profile. Values as small as 10 m have considerable uncertainty because the horizontal resolution of the Gurnard data is several meters. The resulting sampling error for $\hat{p}$ is also shown there and in Figure 3. The length scale for thin ice is much greater than that for thick ice, because thin ice is spread out in leads and polynyas, whereas thick ice is clumped in ridges. In the table, we see that the thinnest ice and the 5-to-6 meter ice are equally plentiful, but that the error $\sigma_{\hat{p}}$ is an order of magnitude smaller for the thicker ice bin. To get an equally good estimate of the thin ice would require a record 5,600 km long (although not necessarily in a straight line).

Assuming ℓ_∞ is not very different for another mature ice cover, the values of ℓ_∞ from the Beaufort Sea profile have been used in (8.2) to evaluate the sampling error for the Fram Strait distribution (in Table 3 and in Figure 3). Within the sampling error, the percentage of thin ice (0-to-1 meter) in the two distributions is marginally distinguishable, but in all other bins, the distributions are measurably different.

Heat balance estimates are sensitive to the thinnest 5 or 10 centimeters of ice (see Chapter 5). Given the measurement errors

Table 3. Length scale and sampling error for ten bins of two draft distributions. The quantity $\sigma_p = (\text{var } \hat{p})^{1/2}$ is the standard deviation of $\hat{p}$; $\sigma_p/\hat{p}$ is the relative error.

	Beaufort Sea $L \simeq 60$ km				Fram Strait $L \simeq 100$ km		
Bin (m)	ℓ_∞ (m)	$\hat{p}$	$\sigma_{\hat{p}}$	$\sigma_{\hat{p}}/\hat{p}$ (%)	$\hat{p}$	$\sigma_{\hat{p}}$	$\sigma_{\hat{p}}/\hat{p}$ (%)
0 to 1	1300	.053	.033	62	.105	.035	33
1 to 2	230	.089	.018	20	.026	.0076	29
2 to 3	170	.42	.026	6	.036	.0077	21
3 to 4	34	.20	.0095	5	.137	.0063	5
4 to 5	20	.087	.0051	6	.110	.0044	4
5 to 6	14	.051	.0034	7	.084	.0033	4
6 to 7	11	.031	.0023	8	.073	.0027	4
7 to 8	10	.021	.0019	9	.073	.0026	4
8 to 9	12	.014	.0017	12	.073	.0028	4
9 to 10	11	.011	.0014	13	.057	.0024	4

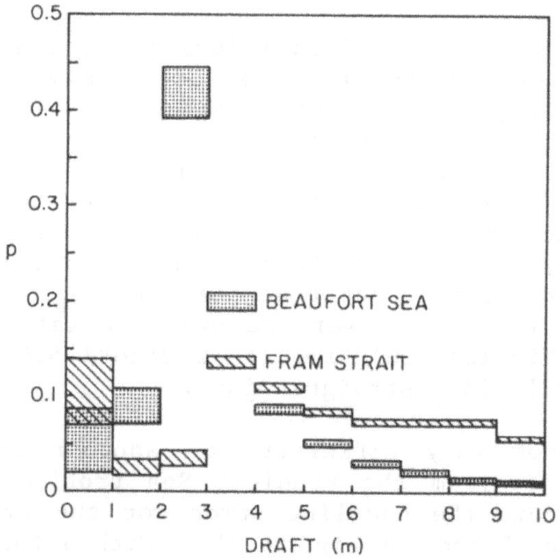

ig. 3. The range or probable values for the distributions in Figure 1. In each bin, the range $\hat{p} - \sigma_{\hat{p}}$ to $\hat{p} + \sigma_{\hat{p}}$ has been filled in--stippled for the Beaufort Sea distribution and crosshatched for Fram Strait.

and sampling difficulty, present submarine sonar records cannot provide good estimates of such narrow categories of thin ice.

3.2 Boring Holes

The most direct way to measure thickness is to bore a hole or cut a core. Thickness can be measured to whatever accuracy is appropriate to the condition of the surfaces. So there is negligible measurement error.

It can take a few minutes' work to bore a hole. A reasonable strategy for estimating thickness statistics might be to determine the thickness h_i by boring n holes. To be independent the samples must be roughly 1 km apart; this is the spacing at which the autocorrelation function of $h_b(x)$ has its first zero (Figure 2). How close is the estimate of mean thickness

$$\hat{H} = \frac{1}{n} \sum_{i=1}^{n} h_i$$

to the true mean H? The variance of $\hat{H}$ is

$$\text{var } \hat{H} = \sigma^2/n$$

where σ^2 is the variance of the thickness distribution. Table 4 shows a few values of n required to achieve a desired sampling accuracy using the variance $\sigma^2 = C_h(0) = 5.6 \text{ m}^2$ from Figure 2. So we could sample H to ± 0.5 m by coring once, flying a kilometer away, kicking out a bag of dye to mark an unbiased site, landing and boring at the dye mark and repeating this twenty times. This argument has been advanced by Untersteiner and Maykut (1969), with the result that 50 samples are needed to establish the mean with a standard deviation of 1/3 meter.

Such requirements to obtain the average ice thickness are difficult to meet in practice, but not outlandish. To obtain an

Table 4. The standard deviation of an estimate of mean thickness from n independent samples.

n	$(\text{var } \hat{H})^{1/2}$, m
23	0.5
62	0.3
560	0.1
2240	0.05

entire thickness distribution by drilling holes in the ice requires logistics bordering on the prohibitive:

From n samples at positions $\underset{\sim}{x}_i$, we estimate the fraction of area in the bin a<h<b by

$$\hat{p} = \frac{1}{n} \sum_{i=1}^{n} \pi(\underset{\sim}{x}_i)$$

which has expected value p and variance $\sigma_{\hat{p}}^2$ equal to

$$\frac{p(1-p)}{n^2} \sum_{i=1}^{n} \sum_{j=1}^{n} r\left(\left|\underset{\sim}{x}_i - \underset{\sim}{x}_j\right|\right)$$

The object is to choose the hole positions $\underset{\sim}{x}$ for the most efficient sampling. If r is never negative, the minimum variance is obtained by taking all holes so far apart that r_{ij} is zero for $i \neq j$. From the submarine data we know this separation should be several kilometers (several times ℓ_∞) if we are sampling the 0-to-1 meter bin. The situation reduces then to n independent Bernoulli trials with chance of success p; the variance is simply p(1-p)/n. Table 5 shows that to resolve the distribution into ten bins each with ten percent of the area to an accuracy of 30% (that is, p = 0.10 $\pm$ 0.03) would require one hundred holes. Boring might be practical only in applications which require a fairly coarse resolution of the distribution with generous tolerances.

3.3 Other Methods

There is a class of observing techniques that may eventually provide measurements of ice thickness from aircraft or satellite. Some are discussed in Chapter 16, but not necessarily with regard to sensing ice thickness. All are in an early stage of development

Table 5. Number of independent samples n required to estimate fraction of area p with standard deviation σ_p equal to some fraction of p.

p	n for $\sigma_{\hat{p}} = 0.3$ p	n for $\sigma_{\hat{p}} = 0.1$ p
0.3	26	233
0.2	44	400
0.1	100	900
0.05	211	1900
0.01	1100	9900

as thickness sensing instruments. Included are radio echo sounding, visual and infrared photography, various microwave and radar sensors, and laser altimeters. Only radio echo sounding senses both the upper and lower surface. Altimeters sense the height of the snow surface; to estimate the ice surface h_t and then multiply by

$$\left(1 - \frac{\text{ice density}}{\text{water density}}\right)^{-1} \simeq 10$$

to obtain thickness h introduces unsatisfactory errors. The microwave radiometers sense radiative properties at or near the top surface--properties only weakly tied to ice thickness for thick ice.

Some measurements of thin ice fraction have been obtained from LANDSAT visual images with accompanying ground truth. The area of gray ice is measured from an image; the thickness of the gray ice is determined by coring or drilling. This method has provided data for tests of thickness distribution theory (Thorndike, 1980).

4. THICKNESS DISTRIBUTION THEORY

Having observed the thickness distribution and having at least a rough idea of the thermal and ridging processes which determine it, the ice physicist has naturally been tempted to formulate a mathematical theory to describe the distribution and its evolution. Such a theory was introduced as part of AIDJEX modeling work (Thorndike et al., 1975), and is gradually being introduced into forecasting and climatological models (e.g., Hibler, 1980). Its use does involve several difficulties. First, since the distribution is difficult to observe, often no initial conditions are available, and model results cannot be verified (or refuted). Second, largely for the same reason, model parameters describing the ridging process are not well known, so that the model results can only be considered rough approximations. And third, generally one is modeling the ice cover as a function of two (horizontal) space coordinates and of time, and the addition of another independent variable, ice thickness, considerably enlarges the computational burden. Being thus forewarned that the theory is no panacea, we will nonetheless review it and see what strengths we find in it.

The equation for the time rate of change of thickness distribution $g(h,\underset{\sim}{x},t)$ is

$$\frac{\partial g}{\partial t} = \psi - \operatorname{div}(\underset{\sim}{u}g) - \frac{\partial}{\partial h}(fg) + \phi \qquad (8.3a)$$

The four terms on the right hand side, discussed one at a time below, represent opening and ridging, flux divergence, ablation and accretion at the upper or lower surface, and lateral melting. The Lagrangian form is

$$\frac{Dg}{Dt} = \psi - g \operatorname{div} \underset{\sim}{u} - \frac{\partial}{\partial h}(fg) + \phi \qquad (8.3b)$$

These equations are primarily statements relating geometry and kinematics. The real physics of the ice cover enters in the structure of ψ, f, and ϕ. All three of these functions depend on the thickness distribution itself, making (8.3) a non-linear functional equation. Early work with this equation treated only the feedback of the thickness distribution g through ridging and opening ψ. More recent work (Hibler, 1980; Maykut, 1982) has included the feedback by which summer melting (ϕ and f) is determined by the amount of radiation absorbed and hence the open water.

To consider flux divergence, we focus on an arbitrary region R and the exchange of h-ice (ice of thickness between h and h+dh) across its boundary curve C. The flux of h-ice across a line with normal $\underset{\sim}{n}$ is $\underset{\sim}{n} \cdot \underset{\sim}{u}g\,dh$. The total flux into or out of R is written first as the integral of the flux around C

$$\oint_C \underset{\sim}{n} \cdot \underset{\sim}{u}g \, dh \, dl$$

and then, by Gauss' theorem, as

$$\iint_R \operatorname{div}(\underset{\sim}{u}g \, dh) da$$

This term accounts for the net loss of h-ice from R

$$-\frac{\partial}{\partial t} \iint_R (g dh) da$$

Since R is arbitrary and fixed in space, the equation takes the differential form

$$\frac{\partial g}{\partial t} = -\operatorname{div}(\underset{\sim}{u}g)$$

This flux divergence term has two parts: $-g \operatorname{div} \underset{\sim}{u}$, denoting the influx of h-ice by convergence, and $-\underset{\sim}{u} \cdot \operatorname{grad} g$ denoting advection.

Thermal growth $f = dh/dt$ carries ice to a new thickness, in just the same way as velocity $\underset{\sim}{u} = d\underset{\sim}{x}/dt$ carries ice to a new position, giving a term completely analogous to the flux divergence term

$$\frac{\partial g}{\partial t} = - \frac{\partial}{\partial h} (fg) \tag{8.4}$$

This is a mathematical statement. For a more physical view of thermal growth, it is best to turn to the cumulative distribution $G(h,t)$. The governing principal is that ice thinner than h at time t is thinner than $h + fdt$ at time $t + dt$ giving

$$G(h,t) = G(h + fdt, t + dt)$$

Expanding the right hand side about (h,t) gives

$$G(h,t) = G(h,t) + \frac{\partial G}{\partial h} fdt + \frac{\partial G}{\partial t} dt + O(dt^2)$$

or

$$\frac{\partial G}{\partial t} = -f \frac{\partial G}{\partial h}$$

which is the integral form of (8.4).

Lateral melting ϕ eats away at the edges of floes to reduce their area. The process conserves area: an amount of open water is formed equal to the total ice-covered area melted so

$$\phi = s\big(\delta(h) - \ell(h)\big)$$

where s depends on the available heat (but not on h), δ is the Dirac delta function and $\ell(h)$ is positive with $\int_0^\infty \ell(h)dh = 1$. The volume loss of h-ice $sh\ell(h)$ is proportional to the area of floe edges which we can take to be proportional to $hg(h)$; hence, $\ell(h) \propto g(h)$, and the integral constraint determines that $\ell(h) = g(h)$. The unknown s is determined by stating that

$$Q = \rho L \int_0^\infty h\phi dh$$

where Q is the rate at which heat becomes available for lateral melting (per unit area of R), ρ is the density of ice, and L is the latent heat of fusion. Solving for s and substituting gives

$$\phi = \frac{Q}{\rho LH}\left(\delta(h) - g(h)\right)$$

where $H = \int_0^\infty hg\,dh$.

Ridging and opening processes are embodied in the redistribution function ψ, on which there are two constraints. These processes can only rearrange existing ice, ridging thin ice to produce thick, but cannot alter the mean ice thickness, so by assumption

$$\int_0^\infty h\psi\,dh = 0$$

Furthermore, the formation of new area of open water less the loss of area by ridging must exactly accommodate the area of ice imported by convergence. Hence, ignoring the thermodynamics (setting f $= \phi = 0$) and integrating (8.3) over all h gives

$$\int_0^\infty \psi\,dh = \text{div}\,\underset{\sim}{u}$$

The definition of ψ is hypothetical, relying on assumptions and reasoning about idealized situations (Thorndike et al., 1975; Hibler, 1980); there are few hard facts about redistribution. Open water formation is represented by a delta function at $h = 0$, and ridging by a function $\mu(h)$ which is a sink of thin ice and a source of thick ice. The degree to which each of these processes occurs in redistribution depends on strain rate invariants. We denote the sum of the principal strain rates by $\dot{\varepsilon}_I$ (= div $\underset{\sim}{u}$) and the difference by $\dot{\varepsilon}_{II}$. These can be expressed in polar form as a modulus $|\dot{\varepsilon}| = (\dot{\varepsilon}_I^2 + \dot{\varepsilon}_{II}^2)^{1/2}$, which tells the quantity of the deformation, and an angle $\Theta = \tan^{-1}(\dot{\varepsilon}_{II}/\dot{\varepsilon}_I)$, which tells the quality of the deformation.

The combination of ridging and opening into ψ is assumed to be proportional to strain rate in the form

$$\psi = |\dot{\varepsilon}|\ \{\alpha_o(\theta)\ \delta(h) + \alpha_r(\theta)[-\mu_L(h) + \mu_G(h)]\} \tag{8.5}$$

The delta function provides a source of open water. The quantity $-\mu_L$ is a loss of thin ice and, μ_G is a gain of thick ice; the combination represents a normalized $\left[\int_0^{\infty}\left(-\mu_L + \mu_G\right)dh = -1\right]$ reduction in the ice-covered area. The coefficients α_o and α_r need to be observed in nature--the idea is that divergence (small θ) will favor opening (larger α_o, smaller α_r) and conversely that convergence (θ getting up toward π) will favor ridging. But even in pure shear ($\theta \approx \pi/2$) the separation of floes in some locations and coming together of floes elsewhere will insure that both opening and ridging occur simultaneously.

Just how the ridging functions μ_L and μ_G vary with h depends on the present state of the thickness distribution (since we don't want to ridge ice that isn't there), and on two parameters G* and k. G* is the cumulative fraction of the thickest ice assumed to ridge. It is taken to lie in the range 0.05 to 0.15, allowing ice up to about one meter to ridge. The second parameter k is the ratio of the thickness of ridged ice to its original thickness. Of course, even the notion that this ratio is a constant is a strong contradiction of reality. The value originally assumed (k = 5) seems to have been too small, producing no ice thicker than 6 m and allowing the pack to deform too easily. (Ice strength is thought to be tied to ridging and to increase with the size of ridges being formed.) A value of k = 15 is better in both respects.

An alternative formulation of redistribution has been advanced in which ridged ice is quantified not by its thickness distribution but in terms of mean ridge height and ridge intensity (Bugden, 1979). The motivation, of course, is that ridge statistics are readily available from aircraft visual observations and laser profiles, so a model predicting them would be testable. This approach has its drawbacks, though, and has not been hotly pursued. Ridged ice is not the sheet threaded with long triangular rubble piles seen in schematic diagrams. In reality old and new ridges overlap, making the identification of a pressure ridge a matter of each observer's particular definition. So ridge statistics are not uniquely related to areas or volumes of ridged ice. Hence an equation involving areas of undeformed ice and ridged statistics for deformed ice is less clean than the formulation involving only ice areas. The more desirable path is to state the theory wholly in terms of thickness distribution and improve techniques for extracting it from data.

The first theoretical solutions for thickness distribution (Thorndike et al. 1975) were calculated for a single Lagrangian point from (8.3b); that is, space was not a variable. We will examine solutions of that problem here. Deformation rate is a function of time, known from sets of buoy tracks, often lacking frequencies greater than 1 cycle per day. Growth rate is a function of thickness and time, known from climatology, and lateral melting is neglected. Starting with an observed distribution as an initial condition, solutions change in time but hover near a realistic distribution. Thus the theory is compatible with our understanding of ice dynamics.

The theory, of course can do more than predict the distribution. It shows us which terms are important to the balance in various ranges of thickness and just how large these terms are. Table 6 shows the annual average of net thermal production (growth into the range less growth out), ridging and open water gain, and import by convergence. The thinnest category is maintained by newly formed open water growing thermally into the next category; some ice is ridged out of this category. For the rest of the thin ice (0.1 to 1.6 m), the net thermal production is nearly balanced by ridging loss. For equilibrium ice (1.6 to 6.4 m) both thermodynamics and ridging provide sources. A balance is only made possible by the divergence. Thick ice (> 6.4 m) is not in balance. It is gaining 0.7% per year because the ridging gain cannot be compensated for by this divergence and the net melt.

Table 6. An average over one year of the terms in the thickness distribution equation (8.3b) and (8.5). Each term is also integrated over a thickness range. Units of terms are percent per day, of thickness distribution, percent.

Terms in (8.3b) and (8.5)		0-0.1 m	0.1-1.6 m	1.6-6.4 m	6.4-25.6 m
Thermal flux gradient	$-\frac{\partial}{\partial h}(fg)$	-0.59	0.43	0.013	-0.003
Open water production	$\lvert\dot{\varepsilon}\rvert\alpha_o\delta$	0.73			
Ridging loss	$-\lvert\dot{\varepsilon}\rvert\alpha_r\mu_L$	-0.14	-0.44	0	0
Ridging gain	$\lvert\dot{\varepsilon}\rvert\alpha_r\mu_G$	0.00	0.01	0.015	0.014
Import	$-$ g div u	0.00	0.00	-0.028	-0.009
Rate of change of g		0.00	0.00	0.000	0.002
Thickness distribution	$\int_a^b gdh$	2.9	20.0	58.9	18.1

Whereas ice thinner than the mean adjusts fairly rapidly to a given deformation and growth regime, it is probably futile to think about an equilibrium distribution for thick ice. For one thing, the determining variables are poorly known: the long term net divergence, the infrequent large convergences capable of making very thick ice, and the melt rate and rate of erosion for thick ice. For another, the adjustment time for thick ice is slow: it takes thirty years for twenty-five meter ice to approach equilibrium thickness. Only the Beaufort Sea may contain ice that old.

The balance in Table 6 is no surprise but simply shows that the model behaves much as the ice it was designed to mimic: open water grows to thin ice and is ridged into thick ice which continually ablates or is exported. The theory elucidates some unforeseen details, and most importantly tells the rates of these various transfers and shows how nearly balanced they are.

5. A WORD ON NUMERICAL TECHNIQUES

One numerical procedure for solving the integral form of (8.3b) involves integrating the functional differential equation

$$\frac{\hat{D}G}{\hat{D}t} = \int_0^h \psi dh - G \operatorname{div} \underset{\sim}{u}, \text{ where } \frac{\hat{D}}{\hat{D}t} = \frac{D}{\hat{D}t} + f\frac{\partial}{\partial h}$$

for each time step along characteristics satisfying

$$\frac{\partial h}{\partial t} = f(h,t)$$

and then interpolating the new G(h + fdt, t + dt) to some fixed h-grid G(h, t + dt). A large area of open water freezing at the same time (late summer appears as a step in G(h) or a spike in g(h). This step cannot be resolved by any fixed grid. But to make matters worse, repeated interpolation at each time step continues to smooth out the step. The numerical solution then is only a shadow of the correct solution. An alternative procedure (Colony, personal communication) is to allow the h-grid to float, moving with the characteristics. A step can be followed by two nearly identical characteristics on which the values of G are quite distinct. Figure 4 shows solutions using Colony's characteristic grid, a coarse fixed grid (à la Thorndike et al., 1975), and a much denser fixed grid (153 h-values at spacings varying from 2 to 50 cm). Table 7 shows further comparisons. The integration starts with all thin ice, proceeding in time steps of one day for a year; the deformation rates are those computed from the AIDJEX manned camps (Colony, 1978). The values k = 15 and G* = 0.1 were

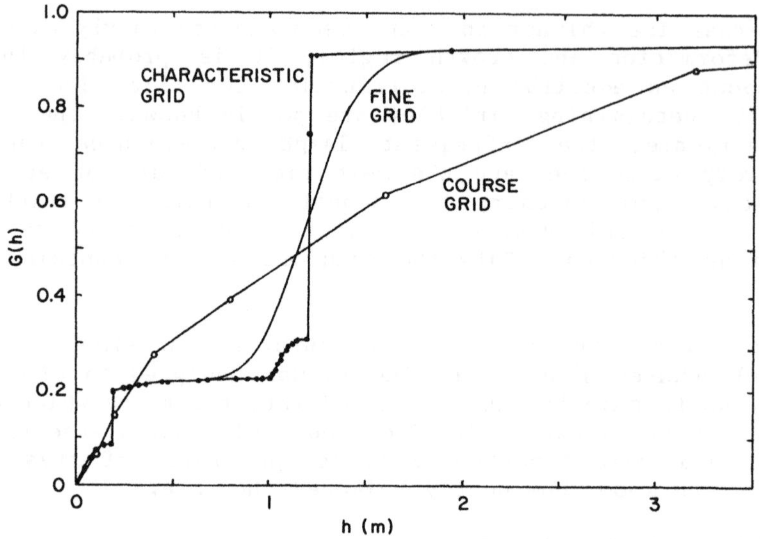

Fig. 4. Cumulative thickness distribution computed from theory using three numerical schemes: characteristic grid, coarse fixed grid, and fine fixed grid.

Table 7. Variables computed by different numerical procedures.

Numerical procedure	mean thickness m	thermal production m	coverage by ice <1 m G(1), %	coverage by ice between 1 and 2 m G(2)-G(1), %	coverage by ice >6 M 1-G(6), %
Characteristic grid	144	163	23	69	95
Dense fixed grid	144	171	34	58	98
Coarse fixed grid	176	246	45	23	96

used, and α_0 and α_r were those identified with the 30° teardrop in Rothrock and Hall (1975).

The large step at 1.2 m is the remains of the initial thin ice. The smaller step at 18 cm is from the freeze up several days before the integration stops. These are clearly not going to be captured by a coarse fixed grid. It is more surprising that the finer fixed grid smears the large step out over nearly one meter. Another curiosity is that even the coarse fixed grid represents the thick ice better than the thin ice from which it is produced. I see two lessons here. First, special numerical care is warranted for these discontinuous functions. Second, some variables are robust against numerical errors, and some are quite sensitive; it would seem prudent to know into which class one's favorite variable falls.

G can also have steps in space. If there is ice of thickness H to the left of an ice edge at x = X, G has a jump along the curve shown in Figure 5. Representing this function G(h,x) accurately will require more than single values on a fixed (h,x)-grid. And as H and X change with time, the jump in G will be lost if special care is not taken.

6. TESTING THE THEORY

Several tests of the theory (Nye, 1975; Thorndike et al., 1975; Rothrock and Hall, 1975; Coon et al., 1977; Thorndike, 1980) have focused primarily on the dynamics of thin ice and open water and have compared these quantities to satellite observations. They show that:

i) the theory adequately simulates thin ice with the chosen forms of α_o, α_r, and μ_L, and the value $G^* \sim 0.05$ to 0.10,

ii) the considerable uncertainty in deformation estimated from the motion of only several points can cause considerable discrepancy between computations and observations, in particular of summer open water,

iii) the value k = 5 produces ridged ice that is insufficiently thick.

As another test of the theory, focusing particularly on the thick ice, the following situation has been simulated. Ice forms in the Siberian marginal ice zone, and drifts across the ocean. It passes the pole in roughly three-to-five years, judging from the drift tracks of North Pole 6 and 10 and the Sedov. Starting on September 1, with an initial condition of 100% open water, the thickness distribution model just discussed (k = 15, G^* = 0.1, 30°

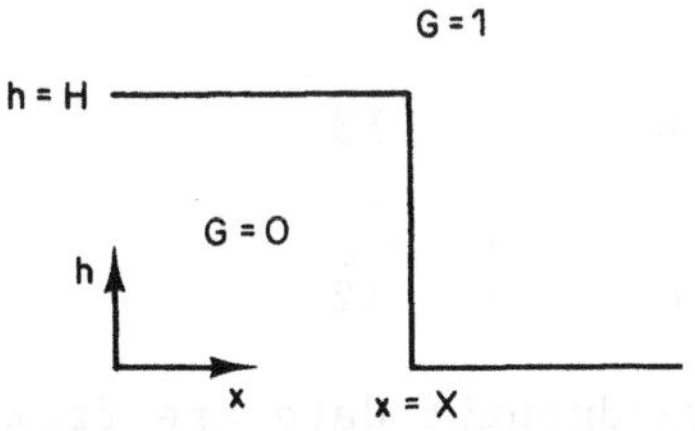

Fig. 5. A line of discontinuity in G on the (x,h) plane. Open water lies to the right of x = X.

teardrop, AIDJEX manned camp deformations, climatological growth rates, no lateral melting) was integrated for several years.

Some results are compared in Table 8 with submarine data from the vicinity of the pole. The theory has not produced enough thick ice. That was a shortcoming in the calculation by Thorndike et al. (1975), but there k was 5, and no ice over 6 m was produced. Here, k is 15, and ice is produced in the whole thick end of the distribution, up to 23 m. Furthermore, the shape of the thick end of the distribution is similar to observations. Ice of the right thickness is being made but in too small an amount. The more likely cause for the discrepancy is in the deformations driving the ridging. They were observed in a mature ice cover in the Beaufort Sea, and may be quite inappropriate for young ice in the marginal ice zone, and for ice being pushed against the north coast of Greenland.

The open water at the start of the simulation has, after five years, grown to 1.9 m and covers 30% of the area. The Dreadnought thickness distribution shows a strong peak at 2.9 meters' thickness (Williams et al., 1975, Figure 10). The Sovereign peaks occur at slightly greater thicknesses. This deficiency of the model could be due to poor growth rates or to an underestimation of rafting and ridging of the young Siberian ice.

Table 8. Comparison of theory to data in a transpolar drift.

	Mean thickness m	Coverage of ice thinner than 2.2 m %	Coverage of ice between 2.2 and 5.6 m %	Coverage of ice thicker than 5.6 m %
Theory, after 3 years, October	3.0	74	12	14
Theory, after 5 years, October	3.5	64	16	20
Sovereign, sec. 17-21 October	4.4	13	59	33
Dreadnought, March	4.6	12	55	33

Sovereign and Dreadnought data are from Wadhams (1981). Drafts are multiplied by 1.11 to obtain thickness.

The procedure of combining inputs and thickness distribution observations from different times and places is not going to provide any further improvement in the theory. As Rogers and Hammerstein put it, "the'v gone about as fer as the' kin go." A more controlled and complete experiment is necessary, in which a Lagrangian region is monitored repeatedly by submarine and satellite, and the deformation history and thermal environment is documented continuously.

7. WHITHER THICKNESS DISTRIBUTION?

Several avenues are available for advancing our knowledge of ice thickness distribution.

1) The parameters in the model are pinned down as well as available data allow. A controlled mass balance experiment would be valuable in which the model inputs and outputs were simultaneously monitored.

2) Sampling theory can determine how large a sample is required to resolve the differences we think we observe between two distributions. This should be a prerequisite for a mass balance experiment.

3) Remote sensing is unlikely to resolve the thickness of thick ice, but a program combining satellite estimates of thin ice and submarine sampling of thick ice would be valuable.

4) Substantial concentration of a single thickness creates numerical problems that need more attention.

5) The ice cover moves in pieces and not as a continuum as widely assumed in theory and in the analysis of deformation. (Some thoughts on this piece-like motion are presented in Section 4.6 of Chapter 7.) A more direct approach to redistribution could be developed in terms of the movement of pieces.

ACKNOWLEDGMENTS

The sampling ideas arose from discussions with Alan Thorndike. Roger Colony provided code for thickness distribution computations. I am grateful to both for their contributions. This work was supported by the Office of Naval Research, the National Science Foundation, and the National Aeronautics and Space Administration.

REFERENCES

Bugden, G. L. (1979) The deformation of pack ice by ridging. J. Geophys. Res., 84: 1793-1796.

Colony, R. (1978) Daily rate of strain of the AIDJEX manned triangle. AIDJEX Bulletin, 39: 85-110.

Coon, M. D., R. T. Hall and R. S. Pritchard (1977) Predictions of arctic ice conditions for operations. Proceedings, 9th Annual Offshore Technology Conference, Dallas, Texas, Vol. 4, p. 307-314.

Doronin, Yu P. (1970) K metodike rascheta splochennosti i dreifa l'dov. (On a method of calculating the compactness and drift of ice floes.) Trudy Arkt. Antarkt. Nauch.-Issled. Inst., 291: 5-17. (English translation, AIDJEX Bulletin, 3: 22-39, 1971.)

Hibler, W. D. (1980) Modeling a variable thickness sea ice cover. Mon. Wea. Rev., 108(12): 1943-1973.

LeShack, L. A. (1980) Arctic Ocean sea ice statistics derived from the upward-looking sonar data recorded during five nuclear submarine cruises. Technical Report to the Office of Naval Research, Contract N00014-76-C-0757/NR 307-374.

Lyon, W. (1963) The submarine and the Arctic Ocean. Polar Record, 11: 699-705.

Lyon, W. (1966) Under surface profiles of sea ice observed by submarine. 11th Pacific Science Congress, Tokyo.

Maykut, G. A. (1982) Large scale heat exchange and ice production in the central Arctic. J. Geophys. Res., 87(C10): 7971-7984.

Nikiforov, Ye. G., Z. M. Gudkovich, Yu. I. Yefimov, M. A. Romanov (1967) Osnovy metodiki rascheta pereraspredeleniya l'da b arkticheskikh moryakh b navigatsionnyy period pod vosdeystviem vetra, (Principles of a method for computing ice redistribution in arctic seas under the influence of wind during the navigation system). Trudy Arkt. Antarkt. Nauch.-Issled. Inst., 257: 5-25. (English translation, AIDJEX Bulletin, 3: 40-64.)

Nye, J. (1975) A test of ice thickness redistribution equations by measurements on ERTS pictures. AIDJEX Bulletin, 28: 141-49.

Rothrock, D. A. and R. T. Hall (1975) Testing the redistribution of sea ice thickness from ERTS photographs. AIDJEX Bulletin, 29: 1-19.

Rothrock, D. A. and A. S. Thorndike, 1980. Geometric properties of the underside of sea ice. J. Geophys. Res., 85(C7): 3955-3963.

Swithinbank, C. (1972) Arctic pack ice from below. In Sea Ice: Proceedings, International Conference (T. Karlsson, Ed.), National Research Council, Reykjavik, Iceland, p. 246-254.

Thorndike, A. S. (1980) Tests of the ice thickness distribution theory. In Sea Ice Processes and Models (R. S. Pritchard, Ed.), University of Washington Press, Seattle, Washington, p. 144-150.

Thorndike, A. S., D. A. Rothrock, G. A. Maykut and R. Colony (1975) The thickness distribution of sea ice. J. Geophys. Res., 80(33): 4501-4513.

Untersteiner, N. and G. A. Maykut (1969) Arctic sea ice. Naval Res. Rev., (Office of Naval Research), p. 12-13.

Wadhams, P. (1980) A comparison of sonar and laser profiles along corresponding tracks in the Arctic Ocean. In Sea Ice Processes and Models (R. S. Pritchard, Ed.), University of Washington Press, Seattle, Washington, p. 283-299.

Wadhams, P. (1981) Sea ice topography of the Arctic Ocean in the region 70°W to 25°E. Phil. Trans. Roy. Soc., 302: 45-85.

Wadhams, P. and R. J. Horne (1980) An analysis of sea ice profiles obtained by submarine in the Beaufort Sea. J. Glaciol., 25(93): 401-424.

Williams, E., C. Swithinbank and G. de Q. Robin (1975) A submarine sonar study of arctic pack ice. J. Glaciol., 15(73): 349-362.

Wittman, W. I. and J. J. Schule (1966) Comments on the mass budget of arctic pack ice. In Proceedings, Symposium on the Arctic Heat Budget and Atmospheric Circulation (J. O. Fletcher, Ed.), Rand Corp., Santa Monica, California, RM-5233-NSF, p. 215-246.

CHAPTER 9

ICE DYNAMICS

W.D. Hibler III

U.S.A. Cold Regions Research and
Engineering Laboratory
72 Lyme Road
Hanover, NH 03755

9.1 INTRODUCTION

Sea ice dynamics deals with the way momentum is transferred through the sea ice system. The essential features of this transfer may be characterized by equations representing the following elements: 1) a momentum balance describing ice drift including air and water stresses, Coriolis force, internal ice stress, inertial forces, and ocean current effects; 2) an ice rheology which relates the ice stress to the ice deformation and strength; and 3) an ice strength determined primarily as a function of the ice thickness distribution. In discussing these components it is implicitly assumed that a description of sea ice as a two-dimensional continuum is possible. Such a continuum description does not, however, rule out using these equations to represent rapidly changing ice conditions as occur, for example, near the ice margin. This can be accomplished by allowing various quantities (e.g., ice mass, ice strength, water stress) to vary rapidly in space. Also, most present-day treatments of ice dynamics employ highly nonlinear rheologies which can, as will be demonstrated later, introduce discontinuities into the ice velocity field, even though the forcing and mass fields may be rather smooth.

In the following review, these three elements will be discussed in order. In this discussion the emphasis will be on understanding the effect of ice rheology on ice drift and deformation, and both interior and ice edge dynamics will be examined. In addition, the final section will examine the role of ice dynamics in ice-atmosphere-ocean coupling.

9.2 MOMENTUM BALANCE

The momentum balance describes the forces that determine drift and deformation of sea ice. Denoting the ice mass per unit area by m, the momentum balance in a two-dimensional Cartesian coordinate system is

$$m\, D_t\, \underline{u} = -\, m\, f\, \underline{k} \times \underline{u} + \underline{\tau}_a + \underline{\tau}_w + \underline{F} - mg\nabla H \qquad (9.1)$$

where $D_t = (\partial/\partial t + \underline{u} \cdot \nabla)$ is the substantial time derivative, $\underline{k}$ is a unit vector normal to the surface, $\underline{u}$ is the ice velocity, $\overline{f}$ is the Coriolis parameter, $\underline{\tau}_a$ and $\underline{\tau}_w$ are the forces due to air and water stresses, H is the dynamic height of the sea surface, and $\underline{F}$ is the force due to variations in internal ice stress. In this formulation, $\underline{\tau}_w$ is assumed to include frictional drag due to the relative movement between the ice and the underlying ocean. Also, H can, in principle, include the variation of the sea surface height due to atmospheric pressure changes as well as to geostrophic current balance. In this momentum balance, the air and water stresses are normally determined from idealized boundary layers, assuming constant turning angles (McPhee, 1979; Brown, 1980; Leavitt, 1980):

$$\underline{\tau}_a = c'_a\, (\underline{U}_g \cos\phi + \underline{k} \times \underline{U}_g \sin\phi) \qquad (9.2)$$

$$\underline{\tau}_w = c'_w [(\underline{U}_w - \underline{u}) \cos\theta + \underline{k} \times (\underline{U}_w - \underline{u}) \sin\theta\,] \qquad (9.3)$$

where $\underline{U}_g$ is the geostrophic wind, $\underline{U}_w$ the geostrophic ocean current, c'_a and c'_w air and water drag constants, and ϕ and θ air and water turning angles. In practice both the currents and the sea surface tilt are estimated from geostrophic considerations by setting H equal to the dynamic heights and computing currents by $\underline{U}_w = gf^{-1}\underline{k} \times \nabla H$ where g is the acceleration due to gravity. In general, both c'_a and c'_w are nonlinear functions of the winds and currents. The two most commonly used formulations are 1) linear, where c'_a and c'_w are taken to be constant, and 2) quadratic, where

$$c'_a = \rho_a\, c_a\, |\underline{U}_g| \text{ and} \qquad (9.4)$$

$$c'_w = \rho_w\, c_w |\underline{U}_w - \underline{u}\, | \qquad (9.5)$$

with c_a and c_w being dimensionless drag coefficients (with typical values of 0.0012 and 0.0055 respectively; McPhee, 1980).

There are also a variety of other nonlinear oceanic boundary layer formulations, several of which have been reviewed by McPhee (1982). In some of these boundary layers, a precise drag law cannot be specified in the form of eq. 9.3 and 9.5, but rather the boundary layer characteristics are determined by the stress at the ice/water interface.

Basically, in this formulation the essential causes of the ice motion can be thought of as the geostrophic wind above the atmospheric boundary layer and the ocean current beneath the oceanic boundary layer. These forces are transmitted to the ice via simple integral boundary layers. In addition, a steady ocean current introduces a tilt (∇H) of the sea surface height (or alternatively a pressure on the ice), which also affects the ice motion. A subtle but important point with regard to the oceanic boundary layer is that there is no stress transferred from the bottom of the boundary layer to the deeper ocean. Instead, the drag upon the ice arises because of a combination of shearing and rotation in the boundary layer. This point has important ramifications for ice-ocean coupling. Specifically, while the stress transferred to the ocean is equal to the wind stress less the ice interaction, it will in general be somewhat different than the water drag on the ice. This can be seen by taking the curl of equation 1 with the internal ice force term set equal to zero. When this is done the curl of the water drag will differ from the curl of the wind stress by a term proportional to $mf\nabla \cdot \underset{\sim}{u}$ which arises because of the Coriolis force on the ice.

9.2.1 Relative Magnitude of Forces

With respect to the dominant components of the momentum balance, observations in conjunction with dimensional analysis show that the force balance is largely dominated by the Coriolis term, the air and water stresses, and the ice interaction. Specifically, measurements of water and air stresses show them to be of the order of 0.1 N m^{-2}. For 3-m-thick ice moving at 0.1 m s^{-1}, the Coriolis force is about 0.05 N m^{-2}, a value commensurate with the wind and water stresses. With respect to inertial force, 3-m-thick ice must accelerate up to a typical drift rate of 0.1 m s^{-1} in less than 15 minutes to yield an inertial force of 0.1 N m^{-2}. However, the acceleration term can be significant in inertial oscillations. In this case an acceleration is induced by changing the direction of the ice motion on the same time scale as the inverse of the Coriolis parameter (about 3 hours). For most applications the momentum advection term is very small, since velocity changes of 0.1 m s^{-1} per kilometer would be needed for a 0.1 N m^{-2} force.

With respect to geostrophic current effects and the ocean tilt terms, their importance depends on the magnitude of the current fields, which are generally smaller than the ice drift rates. In the central Arctic, for example, empirical model studies (Hibler and Tucker, 1979) indicate that steady current effects are normally quite small (several percent relative to the wind-driven components) for ice velocities averaged over a few days. However, even in the central basin, current and tilt effects are significant over a long time period since the wind effects tend to average out (see Figure 1). Also, in relatively shallow regions where barotropic effects are significant, the role of currents in ice drift may be critical.

The remaining term in the momentum balance is the ice interaction. Under typical winter conditions in the central Arctic this has been estimated to be comparable to the other terms by residual calculations. A force balance obtained by Hunkins (1975) employing such a procedure is shown in Figure 2. A somewhat more direct method is to make use of the fact that in enclosed areas, such as

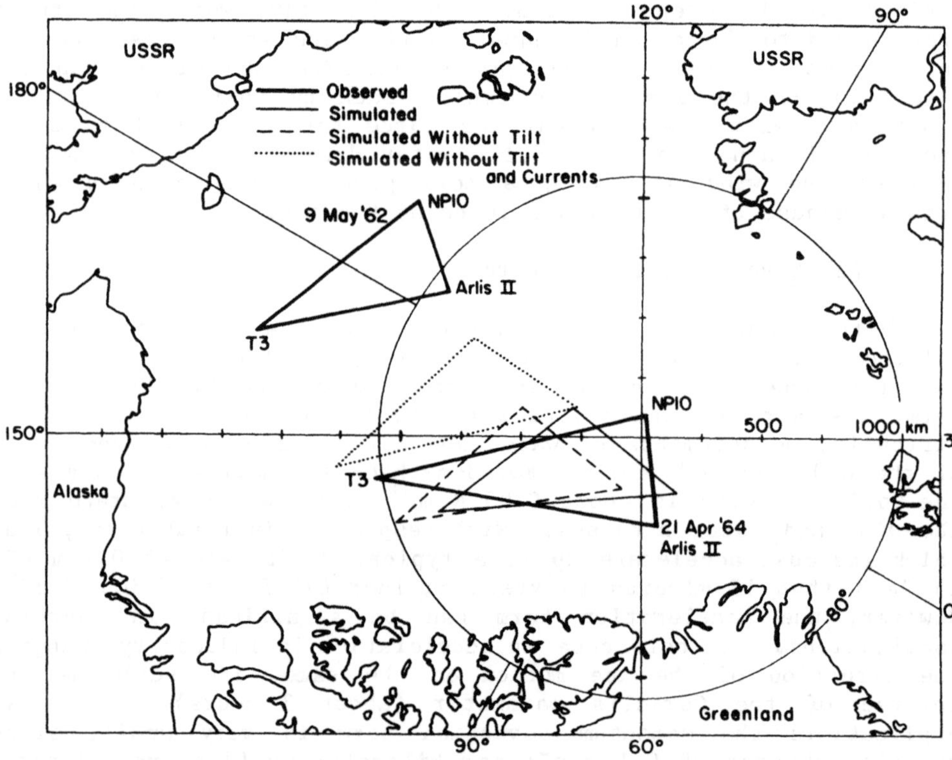

Fig. 1. Simulated and observed net drift of three ice stations over a two-year period employing an empirical linear ice drift model. Simulations are shown both with and without current effects. (From Hibler and Tucker, 1979.)

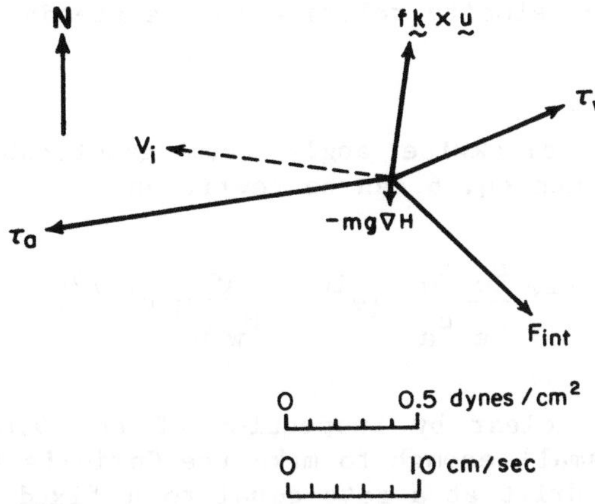

Fig. 2. An estimate of the force balance on sea ice for winter conditions based on wind and water stress measurements. (From Hunkins, 1975.) In this balance the force due to internal ice stress is determined as a residual and the dashed line shows the ice velocity.

the Bay of Bothnia, the ice does not move at all, even with significant winds (Lepparanta, 1980; Omstedt and Sahlberg, 1977). This indicates that forces due to variations in ice stresses are at least of the same order of magnitude as the wind and water drag. It is also possible to estimate ice stresses by considering the mechanics of the ridging process. More details on this approach and the relation of the ice interaction to the ridging process will be discussed in a later section.

9.2.2 Free Drift Analysis

Much of this chapter will be concerned with analyzing the effects and characteristics of the internal ice force term $\underset{\sim}{F}$. However, it is instructive to first examine a special case of the momentum balance in which there is no ice interaction. This case is usually referred to as free drift. To further simplify matters only steady free drift will be considered, so that the inertial terms can also be removed.

Following McPhee (1980) and Thorndike and Colony (1982) it is convenient to use complex notation to analyze this case. Writing all vectors in complex form, i.e., $\underset{\sim}{A} = Ae^{i\gamma}$, and noting that the force on the ice due to sea surface tilt is given by $-imf\underset{\sim}{U}_w$, the steady state momentum balance can be written in the form

$$\rho_a c_a U_g \underset{\sim}{U}_g e^{i\phi} = imf\underset{\sim}{U} + \rho_w c_w U \underset{\sim}{U} e^{i\theta} \tag{9.6}$$

where $\underset{\sim}{U}$ is the ice velocity relative to the steady currents:

$$\underset{\sim}{U} = \underset{\sim}{u} - \underset{\sim}{U}_w \tag{9.7}$$

Letting δ be the clockwise angle from geostrophic wind to ice drift direction, then eq. 6 can be rewritten

$$U_g^2 e^{i\delta} = U^2 e^{-i\phi} \frac{\rho_w c_w}{\rho_a c_a} \left(e^{i\theta} + \frac{mf}{\rho_w c_w U} e^{i\pi/2}\right) \tag{9.8}$$

Several things are clear by inspection of eq. 9.8. For one, if the ice mass is small enough to make the Coriolis term negligible, then the ice will drift at a rate equal to a fixed fraction of the geostrophic wind with a drift angle $\theta-\phi$ relative to the geostrophic wind. As the mass increases the ice will tend to drift further to the right of the geostrophic wind. Similar arguments also apply to the ice speed, with larger ice speeds causing the ice to drift more nearly parallel to the geostrophic wind. A final point of note is (referring back to eq. 9.6) that for linear boundary layers, the ratio and angle between the ice drift and the wind will be fixed for a given ice mass, with increasing ice mass causing the ice to drift further to the right of the wind.

A more quantitative illustration of the nonlinear boundary layer case is given in Figure 3, which shows solutions of eq. 9.8 in the form of ice speed and turning angle versus geostrophic wind speed for different ice masses. In this figure drag coefficients and Coriolis values similar to those used by Hibler (1979) were employed ($\rho_w = 10^3$ kg m^{-3}, $\rho_a = 1.3$ kg m^{-3}, $c_w = 0.0055$, $c_a = 0.0012$, $f = 1.46 \cdot 10^{-4}$ s^{-1}, $\theta = \phi = 25°$). One interesting result of this plot is the very nearly linear relation between the wind speed and ice drift for all but very small speeds, despite the fact that boundary layer drags are nonlinear. This situation arises because both boundary layers are quadratic, and hence, except at low speeds, largely balance one another. However, the turning angle does show somewhat greater sensitivity to wind speed. Also note that although U and U_g are linearly related, the best linear fit does not go through the origin.

9.2.3 Effects of Ice Interaction: A Qualitative Discussion

It is possible to get a qualitative idea of the effect of ice interaction by examining the free drift force balance. The most obvious effect of the ice stress is to change the ice drift direction and magnitude from the free drift case. While there can be exceptions, in many cases the net effect of the ice interaction on the local force balance is to cause a force opposing the wind stress and the Coriolis force, roughly in the manner shown in

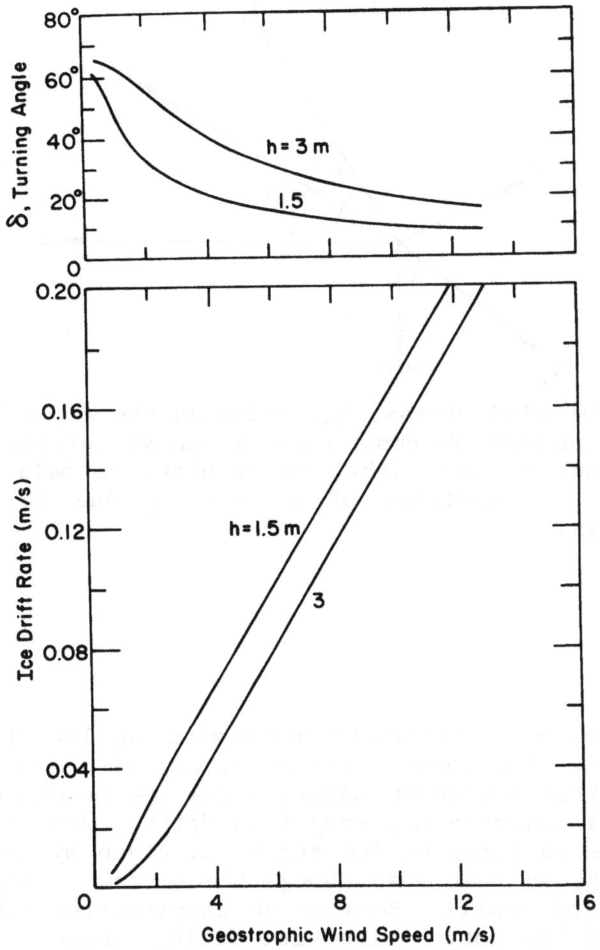

Fig. 3. The ice drift rate versus the geostrophic wind speed, and the clockwise turning angle from the geostrophic wind to the ice velocity. These curves were calculated assuming free drift, and quadratic boundary layers. Curves for different ice thicknesses are shown.

Figure 4. As a consequence, to achieve the same ice velocity under these circumstances a larger wind stress more nearly parallel to the ice velocity is required.

Less obvious effects arise from the nonlinear nature of the ice interaction. As will be discussed later, over a wide range of strain rates the ice stress is expected to change relatively little. As a consequence the force due to ice stress is less sensitive to ice speed. Because of this, the modifications to free

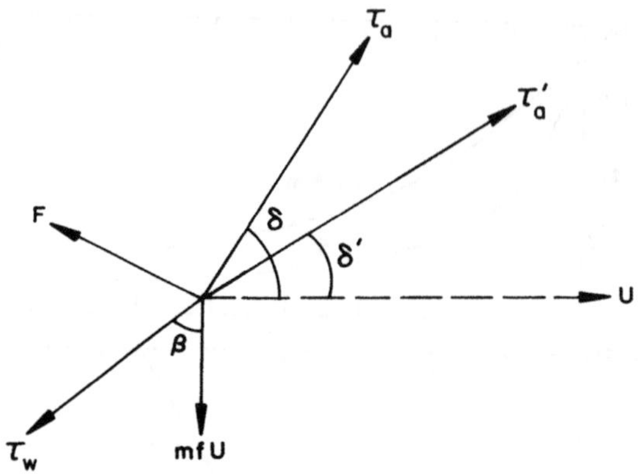

Fig. 4. If the wind stress, τ_a, balances the water drag, Coriolis and current forces, then a larger stress τ_a', turned farther to the right, is required to balance the forces with the addition of a force F due to internal ice stress.

drift will become particularly pronounced at low wind speeds and ice velocities. Nonlinear interaction can also give rise to the interesting situation where, although the ice is strongly interacting, its local velocity may obey free drift. This arises because the force due to internal ice stress is given by the gradient of the ice stress, so that even though the stress is high its gradient may be very small. Because of complexities like this, high correlation of ice motion with wind fields cannot be, by itself, taken as evidence for the absence of ice interaction.

A third subtle situation which can arise due to nonlinear ice interaction is a much more "rigid body" motion of large regions of sea ice. Within the framework of a plastic rheology this state can arise where the ice stress is not large enough to exceed the plastic yield value. Under this situation the motion of the rigid body will behave very similarly to a spatially "averaged" free drift, with the main difference being that the deformation will be much different than what would be expected from free drift consideration. This particular situation is discussed in more detail in a later section.

As these general remarks indicate, the effects of ice interaction can manifest themselves in subtle and sometimes counterintuitive ways. In the remainder of this chapter some of these subtleties will be examined in more detail.

9.3 SEA ICE RHEOLOGY

Analysis of the momentum balance has shown the importance of the internal ice stress in ice drift. Because of this, the determination of a realistic constitutive law relating the ice stress to the deformation is critical to understanding ice drift and deformation. Some of the main questions relating to ice rheologies are a) the degree of nonlinear versus linear behavior for both shearing and compressive failure, b) the relative magnitudes of shear and compressive strengths, c) the presence (if any) of an unconfined pressure term due to random bumping of floes, and d) the magnitude and dependence of ice strength on ice thickness and floe size characteristics. These questions are important for understanding the way the ice responds to external forcing and for understanding the behavior of the coupled air-ice-ocean system. In this section observational and theoretical work relating to the first three questions is discussed. The dependence and relationship of the ice strength to thickness characteristics is discussed in the next section. An attempt is made to present first a fairly unified general framework, and then to expand on the framework with various pedagogical examples employing analytical, numerical and observational work.

In discussing the rheology of sea ice it will be assumed that a description of sea ice as a two-dimensional continuum is possible. While such an approximation is reasonable, it should be remembered that pack ice does consist of individual floes, which can be as large as 20 km in the interior of the Arctic Basin and only tens of meters across near the ice margins. Because of these considerations the scale at which the continuum approximation is valid will vary. More generally, continuum motion has to be considered a statistically defined quantity for sea ice.

9.3.1 Theoretical Framework

Considering sea ice to be a two-dimensional isotropic continuum, the most general constitutive law relating stress and strain rates (see, e.g., Glen, 1970; Malvern, 1969) can be written in the following form:

$$\sigma_{ij} = 2\eta\dot{\varepsilon}_{ij} + [(\zeta-\eta)(\dot{\varepsilon}_{kk}) - P]\,\delta_{ij} \tag{9.9}$$

where σ_{ij} is the two-dimensional stress tensor and $\dot{\varepsilon}_{ij}$ the two-dimensional strain rate, $\delta_{ij} = 1$ for i equal to j or 0 for i not equal to j. In eq. 9.9, η, ζ and P are in general functions of the two invariants of the strain rate tensor and can vary spatially. These invariants can be taken as the principal values of ice strain rate tensor, or alternatively as $\dot{\varepsilon}_{kk}$ and $\dot{\varepsilon}_{ij}\,\dot{\varepsilon}_{ij}$. [In these expressions and equations throughout this chapter repeated subscripts are summed over. For example $\dot{\varepsilon}_{kk} = \dot{\varepsilon}_{xx} + \dot{\varepsilon}_{yy}$.] Within the framework of this equation, η and ζ can be thought of as nonlinear shear and bulk viscosities, while P can be viewed as a pressure term. The motivation for this nomenclature is the fact that the compressive stresses and deviatoric stresses can be written as (as easily shown by manipulation of eq. 9.9):

$$\sigma_{kk} = 2\zeta\,\dot{\varepsilon}_{kk} - 2P \tag{9.10}$$

$$\sigma'_{ij} = 2\eta\,\dot{\varepsilon}'_{ij} \tag{9.11}$$

where $\dot{\varepsilon}'_{ij} = \dot{\varepsilon}_{ij} - \dot{\varepsilon}_{kk}/2$ is the deviatoric strain rate. To obtain an internal ice force for the momentum balance, the stress tensor is differentiated:

$$F_i = \frac{\partial\sigma_{ij}}{\partial x_j} \tag{9.12}$$

Particular forms of this constitutive law include both viscous and rigid plastic rheologies. In general, a "viscous" rheology is defined as one in which stress can only be sustained through non-recoverable dissipation of energy by deformation. A "plastic" rheology, on the other hand, is one in which stress can be sustained through lack of deformation or deformation where the energy is recoverable (i.e., elastic deformation; e.g., Malvern, 1969). Special cases of these rheologies often used in sea ice are a "linear viscous" rheology, where the stress depends linearly on the strain rates, and an "ideal rigid plastic" rheology, where the stress state is either indeterminate or independent of the magnitude of the strain rates. In the linear viscous case, η , ζ and P are constant. This results in a stress state linearly dependent on the strain rates. In the rigid plastic case the stress state is taken to be fixed, independent of the magnitude of the strain rate invariants, provided the ratio of the invariants does not change. In this case, η and ζ will be nonlinear functions of the invariants. In the rigid plastic case, however, η and ζ are not well defined for zero strain rates. Under these conditions the ice is understood to move rigidly with the stress determined by external balances.

Some of the linear viscous and plastic rheologies proposed for sea ice are shown in Figure 5, where the allowable stress states for two particular linear viscous cases of eq. 9.9 are shown. These states are a viscous fluid with only a shear viscosity (η = constant, ζ = P = 0), and a fluid having resistance to dilation and compression but not to shear (ζ = constant, η = P = 0). For an ideal plastic constitutive law the stress state lies

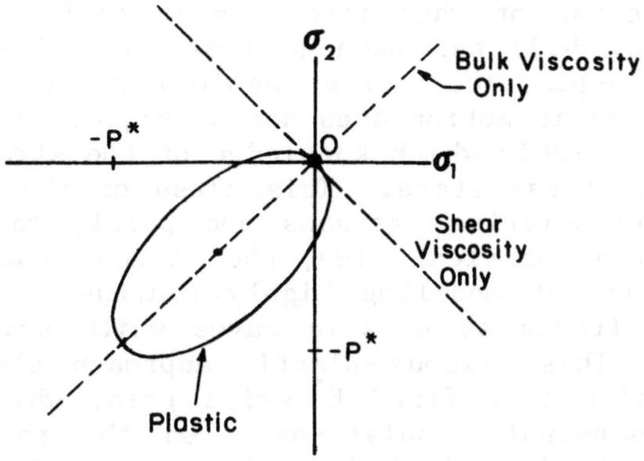

Fig. 5. Allowable stress states for a linear viscous rheology with either a bulk or shear viscosity, and for an ideal rigid plastic rheology with an elliptical yield curve. The stress states are plotted as a function of the principal components of the two-dimensional stress tensor.

on or inside some fixed yield curve. In Figure 5 an elliptical yield curve is shown. For a plastic rheology a rule is needed to uniquely relate the stress state to the strain rates. The most common rule (e.g. Malvern, 1969) is to take the ratio of the strain rates for a given stress state to be that of a vector normal to the surface (this is referred to as the normal flow rule). For the ellipse, the normal flow rule yields η, ζ and P values of (see e.g., Hibler, 1977)

$$P = P^*/2 \tag{9.13}$$

$$\zeta = P^*/2\Delta \tag{9.14}$$

$$\eta = \zeta/e^2 \tag{9.15}$$

where

$$\Delta = [(\dot{\varepsilon}_{xx}^2 + \dot{\varepsilon}_{yy}^2)[1+(1/e^2)] + \frac{4}{e^2}\dot{\varepsilon}_{xy}^2 + 2\dot{\varepsilon}_{xx}\dot{\varepsilon}_{yy}(1-1/e^2)]^{1/2}, \tag{9.16}$$

-P* is the maximum compressive stress allowable along any principal stress axis, and e is the ratio of the lengths of principal axes of the ellipse. For rigid motion, the stress state will lie within the ellipse. Note that both linear viscous laws allow large amounts of tensile strength (i.e., positive stresses), whereas this particular plastic law allows only small tensile strengths.

A general class of rheologies almost equivalent to the most general case (eq. 9.9) has been proposed by Hibler (1979) under the name "viscous-plastic." The rheology refers to the case of eq. 9.9 in which rigid motion does not occur, or, stated differently, any case of eq. 9.9 in which knowledge of the strain rate uniquely specifies the stress state. This class of rheologies was proposed partly for numerical reasons and partly based on physical arguments. On the numerical side, the viscous-plastic concept is an efficient means of modeling highly nonlinear sea ice dynamics without such artifacts as elastic waves which arise in elastic-plastic models. This viscous-plastic approach also lends itself well to formulation in a fixed Eulerian grid, which is essential for long-term seasonal simulations. On the physical side the viscous-plastic rheology includes a more general class of yield curves than can be considered employing elastic-plastic rheologies (see, e.g., Pritchard, 1975). Specifically, with the viscous-plastic approach it is not necessary to employ the associated flow rule. Some of these more general cases may be particularly applicable to the marginal sea ice zone.

9.3.2 Linear Versus Nonlinear Behavior

A particular problem in early developments in ice dynamics was the degree of linear versus nonlinear behavior of the ice rheology. Early efforts to approximate the ice interaction made use of linear viscous models. Laikhtman (1964), for example, proposed a linear Newtonian viscous model employing only a shear viscosity, which was easy to deal with numerically and was subsequently used by a number of authors (Egorov, 1970; Doronin, 1970) to carry out empirical-numerical studies. Probably the most well known application of this rheology was a calculation by Campbell (1965) of the steady circulation of the Arctic Basin using an idealized mean annual wind field. These empirical model studies demonstrated that the Newtonian viscous rheology had some ability to simulate the effects of ice interaction. However, a particular defect of this rheology is that with only a shear viscosity it has no resistance to convergence (see, e.g., Rothrock, 1975a).

Empirical studies using a more "general" linear viscous law containing both bulk and shear viscosities (and hence having compressive strength, see eqs. 9.9-9.11) have successfully simulated observed ice drift and deformation far from shore, both over short time intervals (Hibler, 1974) and over a seasonal cycle (Hibler and Tucker, 1979). In particular, Hibler (1974) showed that a linear viscous rheology could explain observations showing the ice pack to converge in a low pressure system in the winter and diverge in the summer under similar forcing. Best estimates of the viscosity magnitudes yielded viscosities of the order of 10^{12} kg s^{-1}, with the bulk viscosity (ζ) about twice as large as the shear viscosity (η). In a seasonal study Hibler and Tucker (1979) showed the "best fit" viscous parameters to vary in a regular seasonal manner. This seasonal variation was commensurate with the physical notion that sea ice has less thin ice in winter (due to the higher freezing rates) and hence has higher strength. The seasonal viscosity also tended to correctly simulate observed seasonal variations of the angle of ice drift relative to the wind.

However, while generally performing well far from shore, the linear viscous rheology has one major problem in that viscosities for reasonable simulations in the central pack are quite different from viscosities needed to simulate near-shore behavior (Rothrock, 1975a). For example, viscosities (two-dimensional) of the order of 10^{11} to 10^{12} kg s^{-1} (Hibler, 1974; Hibler and Tucker, 1979) are needed to model pack ice behavior far from shore. Near shore, however, the effective viscosities may be as small as 10^8 kg s^{-1} (Hibler et al., 1974a), and in very small channels they can be as small as 10^5 kg s^{-1} (Sodhi and Hibler, 1980). Such variations suggest that inclusion of some type of nonlinear behavior is necessary.

An additional problem with linear viscous laws is that their physical basis is questionable. In particular, the local characteristics of sea ice appear to be nonlinear and plastic in nature. However, even though sea ice may not exhibit linear viscous characterisics on a small scale, Nye (1973) has noted that such linear laws still might be appropriate for a time or spatially averaged response. A particular case of this concept has been proposed by Hibler (1977), who suggested that on appropriate time scales stochastic fluctuations in the deformation rate might cause the average stress-strain rate relation to be essentially linear. This concept may have application to the marginal ice zone and is discussed in more detail later in this chapter.

To avoid the inherent deficiencies in linear viscous rheologies and to better describe the local real-time behavior of sea ice, Coon (1974) proposed a plastic rheology for sea ice. As discussed earlier, in a plastic constitutive law the stress state of the ice is presumed to be independent of the magnitudes of the stress and strain rate. Such a law allows highly nonlinear behavior, which is helpful in explaining ice flow both near and far from shore. In addition, it contains a simple way to specify a low tensile strength concurrently with a high compressive strength. To justify such a plastic law physically, Coon and Rothrock argued that such behavior is consistent with the local physics of sea ice. The essential argument for rate independence is that the work done in ice deformation is primarily due to ridging. The energy expended in ridging, however, would be expected to be independent of the rate of ridge building. This concept can be extended into a complete explanation (in terms of ridging and frictional losses) of all energy dissipated during plastic flow. This particular approach is discussed in the next section. With regard to small tensile strengths, it was argued that the presence (over a large scale) of many cracks prevented any effective cohesion under dilation.

While the concept of an ideal rigid plastic law is useful theoretically, it is difficult to implement in practice. In particular, to examine a plastic rheology numerically, some approximation must be made to describe the "rigid" behavior, i.e., the behavior when the ice is not flowing plastically. Two approaches have been suggested: the elastic-plastic approach (Pritchard, 1975) and the viscous-plastic approach (Hibler, 1979). In the elastic-plastic method the ice is assumed to behave elastically for certain strain states. The advantage of this is that it allows rigid behavior in the sense that the ice can sustain a variety of stresses while possibly undergoing no deformation. However, inclusion of the elasticity is inherently unwieldy and has the disadvantage of adding elastic waves to the behavior of the model (Colony and Rothrock, 1980).

In the viscous-plastic approach (Hibler, 1979), sea ice is considered to be a nonlinear viscous continuum characterized by both nonlinear bulk and shear viscosities and a pressure term. These nonlinear viscosities are adjusted to give plastic flow for normal strain rates. To close the system, for very small deformation rates the ice is assumed to behave as a very stiff linear viscous fluid. Thus the rigid behavior is approximated by a state of very slow flow. This technique is computationally advantageous. Its disadvantage, however, is that ice that is actually stationary may be modeled as slowly creeping.

Simulations using plastic rheologies have reproduced many aspects of sea ice dynamics. Nearshore simulations by Pritchard et al. (1977) have shown the capability of the elastic-plastic rheology to model both relatively stationary behavior under stress and intense shearing behavior near coasts. With respect to seasonal simulations, Hibler (1979) has shown that in addition to producing nearshore shear zone effects, a plastic rheology yields geographical ice thickness buildup and ice outflow from the Arctic Basin which agree well with observations. Also, for sufficiently high strengths, Hibler (1980a) found the modeled ice cover in the Arctic Basin stopped moving, even though the wind forcing is significant. Such a state of no motion in a constricted region, even though the external forcing is significant, is an important manifestation of a nonlinear rheology. This characteristic is a salient feature of ice drift in regions with narrow land constrictions such as the Bay of Bothnia or the Great Lakes of the United States.

9.3.3 Analytic Analysis of a One-Dimensional System

To get some feeling for the relative behavior of linear versus nonlinear rheologies and to examine the effects of nonlinear ice interaction on ice drift rates it is useful to analyze a special case of the momentum balance. Consider the one-dimensional case of the momentum balance employing only a linear water drag term cu, an external constant wind stress τ, and a one-dimensional ice interaction stress σ:

$$cu - \frac{\partial}{\partial x}[\sigma] = \tau \tag{9.17}$$

Consider the region bounded by rigid walls at x = 0 and x = L. Let us now examine the linear viscous case where $\sigma = \zeta\ \partial u/\partial x$ with ζ a constant. Substituting this expression for σ into eq. 9.17 gives

$$cu - \zeta \frac{\partial^2 u}{\partial x^2} = \tau \tag{9.18}$$

A general solution of equation eq. 9.18 is

$$u = A \exp[(c/\zeta)^{1/2}x] + B \exp[-(c/\zeta)^{1/2}x] + \tau/c \tag{9.19}$$

Utilizing the boundary conditions u(0) = u(L) = 0 we obtain

$$u(x) = \frac{\tau}{2c\sinh(\lambda L)} \{[e^{-\lambda L} - 1]e^{\lambda x} + [1 - e^{\lambda L}]e^{-\lambda x} + \tau/c \tag{9.20}$$

where $\lambda = (c/\zeta)^{1/2}$.

Let us now consider a rigid plastic case with constant strength. For our plastic rheology we will assume that $\sigma = -P$ for $\partial u/\partial x < 0$ and $\sigma = 0$ for $\partial u/\partial x > 0$. These assumptions define a rigid plastic rheology with no tensile strength. A solution for this case may be constructed by noticing a) that for any convergent deformation $\sigma = -P$ while for no motion $0 \geq \sigma \geq -P$, and b) that the maximum force would be expected to take place at the right-hand boundary since the wind stress has built up at this point. With these considerations in mind we see first that if L < P no motion of any kind will occur and the system will be rigid. If, on the other hand, $\tau L > P$ then a solution of eq. 9.17 can occur for a σ satisfying the above plastic assumptions and $\sigma(L) = -P$. Integrating eq. 9.17 from x to L we have

$$\int_x^L (cu - \tau) = -P - \sigma(x) \tag{9.21}$$

However, we must also have $\partial u/\partial t = 0$ except at the boundary, otherwise $\sigma(x) = -P$ and eq. 9.17 cannot be satisfied. Therefore, u is a constant, as is τ, and we obtain

$$\sigma(x) = -P + (cu_o - \tau)(x - L) \tag{9.22}$$

The value of u_o may be obtained by noting that the total force acting on the rigid block moving is $-cu_oL + \tau L - P$, and since there is no acceleration

$$u_o = \frac{\tau L - P}{cL} \tag{9.23}$$

Another way to see this is to note that $\sigma(x = 0) = 0$ since divergence is occurring there and we have assumed no tensile strength. With a little thought, this type of simple analysis can be extended to a) the equations with inertial terms (rigid block motion occurs with the block gradually speeding up to equilibrium with a spin-up time based on the water drag) and b) a case with strong wind gradients (for sufficiently large gradients the region of convergence will occur over a finite distance rather than at a single location). In this latter case once convergence has occurred the stress will be constant and hence free drift will occur even though the ice is interacting. These more general cases will be illustrated with some numerical examples discussed below.

This analysis is also easily extended to include nonlinear water drag terms, an extension that makes it more comparable to the free drift analysis carried out earlier in the momentum balance section. Using the notation of eq. 9.6, the steady-state momentum balance for this one-dimensional system is

$$\rho_a c_a U_g^2 = \rho_w c_w U^2 - \frac{\partial \sigma}{\partial x} \tag{9.24}$$

By the same methods as employed in the linear case for a constant wind the solution for this system is

$$U = \sqrt{\frac{\rho_a c_a U_g^2 - P/L}{\rho_w c_w}} \tag{9.25}$$

if $P < \rho_a\ c_a\ U_g^2\ L$ and $U = 0$ otherwise. For a numerical comparison relevant to the Arctic Basin we take $L = 2.5 \cdot 10^3$ km which is the scale of the Arctic Basin, and $P = 8.25 \cdot 10^4\ Nm^{-1}$, an empirical best fit ice strength for 3 m thick ice in the Arctic Basin in winter (Hibler and Walsh, 1982). Using these numbers together with the drag coefficients used in Figure 3, we obtain a comparison of free drift and drift with ice interaction as shown in Figure 6. Note that for large wind speeds there is little difference between free drift and the rigid plastic result, whereas for smaller wind speeds the difference is very marked, with the rigid plastic system totally stopping. A significant difference from the linear water drag case is that the slope of the ice drift rate versus wind speed decreases asymptotically to the free drift slope for large geostrophic wind speeds. In the one-dimensional linear water drag case the free drift and ice interaction curves would be parallel but would have different x-axis intercepts.

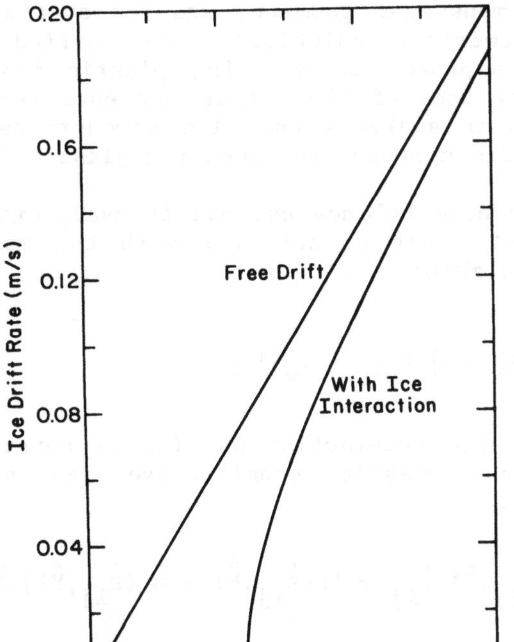

Fig. 6. Comparison of ice drift rates for an idealized one-dimensional system with and without plastic ice interaction. Quadratic boundary layers were used, and for the ice interaction case typical strength and length scales relevant to the Arctic Basin were assumed. Other constants are the same as used in Figure 3 for 3-m-thick ice.

A mirror reflection of this figure holds for negative ice velocities. Consequently, if one were to force this model with a zero mean Gaussian distribution of wind speeds and then fit the results with a linear model, the effect of the ice interaction would be to reduce the ratio of ice drift to wind speed. This ratio would, however, be dependent on the variance of the wind speeds, with smaller variances yielding more of a slope reduction. These considerations are relevant to linear correlations of ice drift rates with winds (Thorndike and Colony, 1982), which yield only slight reductions in winter. This simple analysis suggests that part of the reason for such results may be the increased intensity and hence variance of winds in winter. This analysis also indicates that although useful, such linear correlations may be masking some of the essential physics of ice drift.

9.3.4 Some Numerical Solutions for an Idealized Geometry

To examine the behavior of a plastic rheology for a more realistic two-dimensional geometry and for spatially varying wind forcing some numerical solutions were carried out using the viscous-plastic approach to modeling plastic flow. These solutions help verify some of the simple concepts mentioned above in the one-dimensional analysis and also demonstrate the utility of the viscous-plastic approach for modeling plastic flow.

For the momentum balance eq. 9.1 is used, with the geostrophic current U_w set equal to zero and with the momentum advection term set equal to zero:

$$m \frac{\partial \underline{u}}{\partial t} = - mf\underline{k} \times \underline{u} + \underline{\tau}_a + \underline{\tau}_w + \underline{F} \qquad (9.26)$$

For modeling the ice interaction the ice is considered to have a nonlinear viscous plastic constitutive law similar to that mentioned earlier.

$$\sigma_{ij} = 2\eta(\dot{\varepsilon}_{ij},\overset{*}{P})\, \dot{\varepsilon}_{ij} + [\zeta(\dot{\varepsilon}_{ij},\overset{*}{P}) - \eta\, (\dot{\varepsilon}_{ij},\overset{*}{P})]\, \dot{\varepsilon}_{kk}\, \delta_{ij} - \overset{*}{P}\delta_{ij}/2 \qquad (9.27)$$

where σ_{ij} is the two-dimensional stress tensor, $\dot{\varepsilon}_{ij}$ the strain rate tensor, $P^*/2$ a pressure term, and ζ and η are nonlinear bulk and shear viscosities. For normal strain rates the dependence of ζ and η on $\dot{\varepsilon}_{ij}$ and P^* is given by eq. 9.13-9.16 so that the stress state lies on an elliptical yield curve passing through the origin with a no-stress condition applying for pure divergence. For very small strain rates the viscosities specified by eq. 9.13-9.16

become arbitrarily large. To avoid this they are chosen to be the minimum of the plastic values and some large limiting values dependent on the ice strength P*. For the calculations performed here the standard limiting values were taken to be

$$\zeta_{max} = (5.0 \times 10^7 s)P^* \tag{9.28}$$

$$\eta_{max} = \zeta_{max}/e^2 \tag{9.29}$$

Sensitivity tests (Hibler et al., 1981) show these simulation results to be relatively insensitive to these maximum viscosities unless they are taken to be about two orders of magnitude smaller. However, the maximum viscosity values do affect computational speeds, especially at finer resolution (see Hibler et al.,1981).

To solve eq. 9.26-9.29 and 9.12-9.16 numerically, a time stepping procedure using finite difference techniques is employed. The computer code for this procedure is documented in Hibler (1980b). Briefly, the numerical procedure works as follows. The viscosities used in the momentum balance are based on the deformation field from the previous time step. Using these viscosities a new velocity field is obtained, a new set of viscosities are estimated, and another linearized equation solved. We will refer to each of these relaxation solutions as an "iterative time step." By carrying out several iterative time steps at each "physical time step," ideal plastic flow may be approached. In the simulations in this section 15 iterative time steps are carried out at each physical time step, so that plastic equilibrium is approached at each time step. The effect of smaller numbers of iterative time steps is discussed in more detail by Hibler et al. (1981).

Using these equations and numerical method a series of simulations over a region 148 km by 148 km with fixed boundaries and fixed geostrophic wind in the x-direction were carried out (see Figure 7). The boundary conditions consisted of specifying zero velocity at the boundaries. For later reference the standard simulation utilized the parameters shown in Table 1. The equilibrium velocity field for the standard simulation is shown in Figure 8. While there are some deviations due to the two-dimensional character of the rheology and boundaries this result basically represents a rigid block motion that is consistent with the simple one-dimensional arguments presented in the previous subsection. In this block motion the main converging deformation occurs at the right-hand boundary, while divergence occurs at the left-hand boundary. There is also some slight reduction of velocity at the side boundaries due to shear stresses.

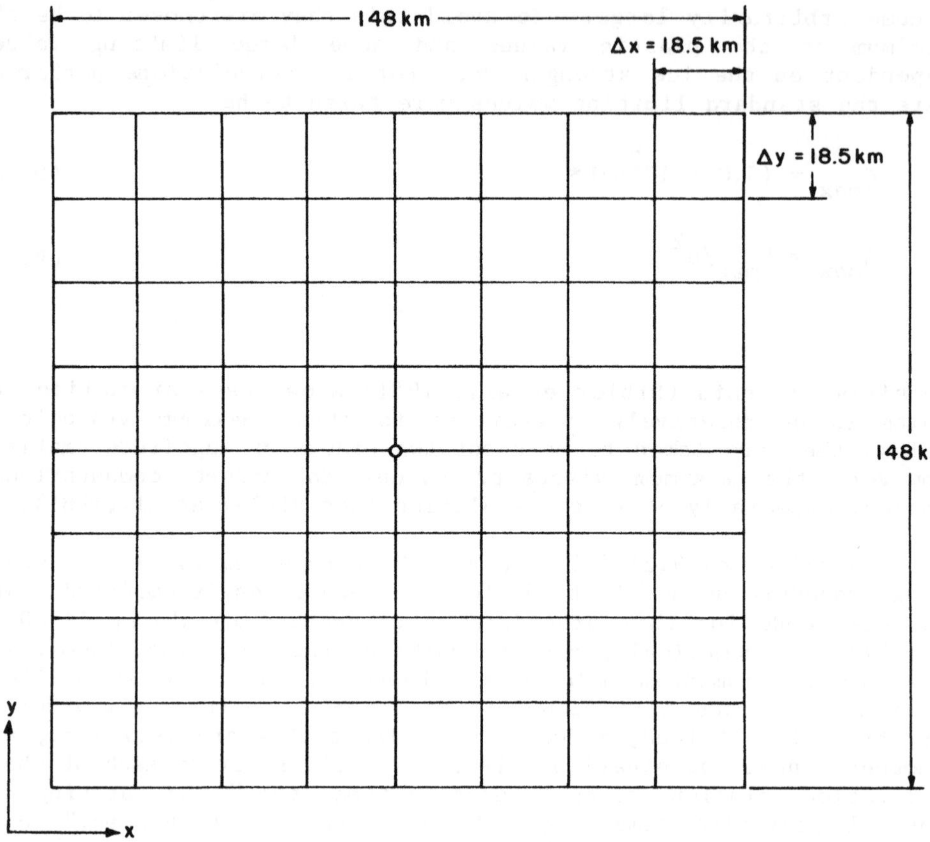

Fig. 7. Grid configuration used in idealized simulations.

Table 1. Numerical parameters used in standard case simulation.

$C_a = 0.0012$	$\rho_a = 1.3$ kg m^{-3}	$\Delta t = 3$ hours	$U_g = 8$ m s^{-1}
$C_w = 0.005$	$P^* = 5000$ N m^{-1}	$\zeta_{max} = (5\cdot 10^7 s)P^*$	$V_g = 0.0$
$e = 2$	$m = 0.91\cdot 10^3$ kg m^{-2}	$\eta_{max} = \zeta_{max}/e^2$	$U_w = 0.0$
$f = 1.46\cdot 10^{-4} s^{-1}$	$\Delta x = \Delta y = 18.5$ km	$\phi = \theta = 25°$	$V_w = 0.0$

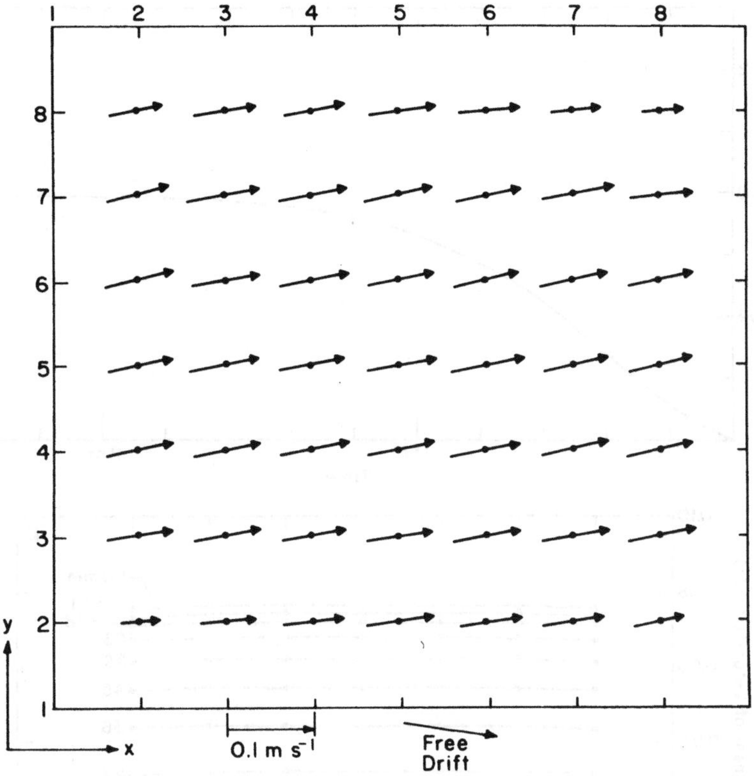

Fig. 8. Equilibrium velocity field for the standard simulation.

Note that, as discussed above, the plastic solution is qualitatively very similar to free drift (shown in Figure 8 as a single vector ouside the grid) in that the whole sea ice region moves practically at the same velocity. However, there are quantitative differences. As in the one-dimensional case (see Figure 6), a major difference is a smaller velocity magnitude in the plastic system than for free drift. A second significant difference is a substantially reduced turning angle (~15° reduction) between the plastic ice velocity vectors and the surface winds. As discussed in the earlier sections on the momentum balance (see Figure 4) this can be explained by the fact that the force due to internal ice stress approximately opposes the ice motion. Moreover, as the ice strength increases this turning angle difference becomes even more pronounced.

The spin-up results for the standard case are shown in Figure 9. In these spin-up results 3-minute physical time steps were used as compared to 3 hours used in the other tests. Figure 9a shows the x component of the center point velocity versus time,

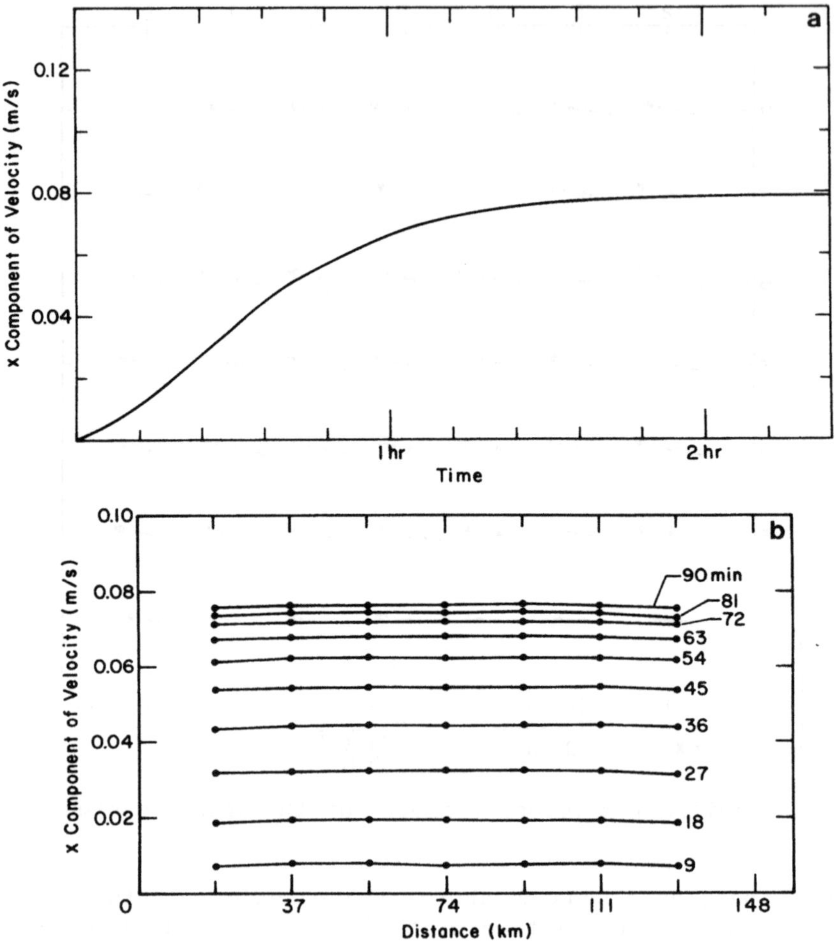

Fig. 9. Inertial spin-up characteristics of idealized viscous-plastic sea ice model. Part (a) shows the x component of the center point velocity versus time while (b) shows profiles (taken parallel to the x axis and through the center point) of the x velocity component.

while Figure 9b shows velocity profiles (along the x axis) at 9-minute intervals. As can be seen the spin-up is also a rigid motion, with the block motion gradually increasing in accordance with the magnitude of the inertial term. A notable feature of these numerical spin-up results is that they are free of elastic wave effects that result from elastic-plastic rheologies (see, e.g., Colony and Rothrock, 1980). As a result, the viscous plastic approach allows a more direct analysis of the physics of ice interaction with minimal numerical artifacts. Tests were also done with different strengths. Spin-up times for these cases were slightly different, due to the fact that different equilibrium velocities resulted in different water stress terms. However, with a linear water drag the spin-up times were essentially the same for different strengths.

The response of this simple system was also examined for a variety of forces of different magnitudes and having spatial variation. Analysis of the results shows that the equilibrium velocity characteristics depend on the physical size of the basin, external forcing, and ice strength. From dimensional analysis it can be deduced that near the "motion" to "no motion" transition a key dimensionless parameter is $P^*/(\tau_a \cdot L) = \beta$, where L is the size of the region, P* the ice strength, and τ_a the wind stress. To examine the dependence on this parameter a series of tests were done with fixed resolution (Δx), size of area (L), and wind stress (τ_a). In each test a different value of P* (fixed in time and space) was used. All other parameters were the same as the standard case. To verify the scaling, tests were also done with L and τ_a varying, with similar results as long as β was fixed. (To obtain identical results the water drag and ice mass would also have to be modified.) Figure 10 shows the component of the center point velocity versus strength from these simulations. This figure shows the strength for which the system behaves rigidly and also shows that once the strength is low enough to allow the ice to

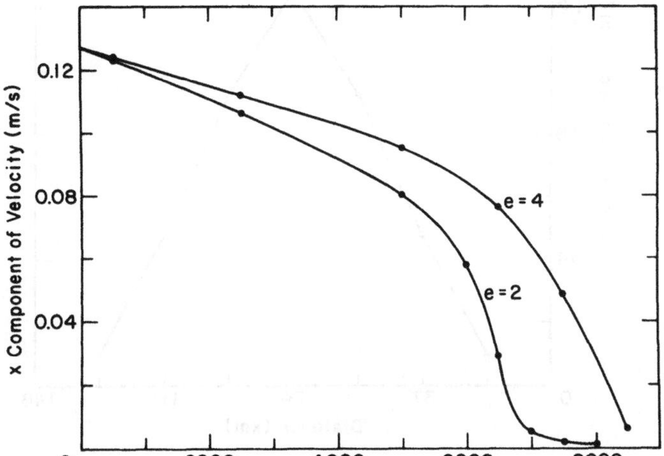

Fig. 10. Equilibrium x-velocity component of the center point as a function of ice strength and for different yield curves. Curves are shown for the standard yield curve (e = 2) and for a narrower elliptical yield curve with less shear strength (e = 4).

move, halving the strength brings the velocity field almost to free drift values. The value of β for the rigid "motion" to "no-motion" transition is about half of the value estimated for the one-dimensional case. This reduced value of β is primarily due to the effect of shear stress, which was not considered in the one-dimensional analysis. This hypothesis is verified in Figure 10, which shows equilibrium velocities versus P* for a smaller ratio of shear to compressive strength (e = 4).

Figure 11 shows the effect of a spatially varying wind field. Illustrated in part a of this figure is the equilibrium ice velocity field for a wind field varying linearly from 16 m s^{-1} at the left-hand boundary to zero at the right-hand boundary. As can be seen, this large velocity field gradient causes the region of rigid motion to recede to the left-hand side, with the convergence now occurring over a larger region at the right-hand side. This convergence has to occur over a band because the gradient of the stress in the rigid region now must change in order to match the

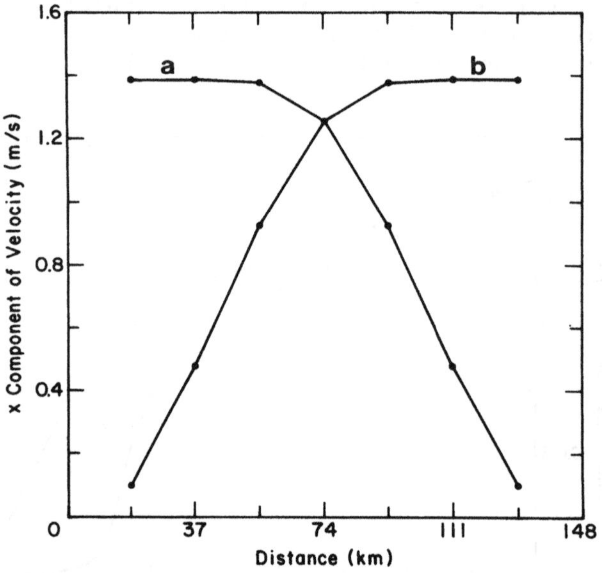

Fig. 11. Effect of gradients in the wind fields on the equilibrium velocity field for the idealized domain in Figure 7. In curve (a) the wind is in the positive x direction and varies from 16 m s^{-1} at the left-hand boundary to 0 at the right-hand boundary. In curve (b) the wind (also in the positive x direction) starts at 0 at the left hand boundary and goes to 16 m s^{-1} at the right hand boundary.

external force, and the changes available for a given ice strength are limited. Figure 11 also shows the case where there is an offshore wind gradient with wind varying from zero at the left-hand boundary to 16 m s^{-1} at the right-hand boundary. In this case the region of divergence is enlarged.

An interesting feature of Figure 11 is the symmetry between the two ice drift curves for opposite gradients in the wind fields. This occurs even though the diverging ice has effectively no stress, whereas the converging ice has a stress state at the plastic yield. However, in the converging case even though the stress is large, there is effectively no stress gradient and hence no significant force due to ice interaction. In general this figure demonstrates some of the counter-intuitive results that can occur with a nonlinear ice interaction. In this case, even though the ice is interacting strongly the local drift still obeys free drift.

These gradient results, in conjunction with the spin-up results, have application to rapidly moving fronts. Such fronts contain large wind gradients, and on the basis of these simulations would be expected to cause moving zones of deformation to propagate ahead of the front. The block motion results also give insights into why simple free ice drift ice forecasts have generally been quite useful (see, for example, Zubov, 1943). Basically, in the rigid motion case the rigid block behaves very much like freely drifting ice, except that the plastic stress tends to renormalize the wind field (or whatever external forcing exists) to a smaller value. Also, such free drift rates are usually invoked to predict the drift and not the deformation, and hence are primarily based on the smoothly varying components of the wind fields.

9.3.5 Stochastic Effects and an Unconstrained Pressure Term

Far from the ice edge, where the floes are large and closely packed, the ideal rigid approximation is probably good. However, near the edge there may be a great deal of random bumping of floes. This in turn may introduce an unconstrained pressure term which can do work on the continuum system. While such a rheology is still a special case of the viscous plastic rheology, it can cause a different type of ice edge dynamics than is yielded by rigid plastic analysis. However, random bumping is likely localized at the edge and will be superimposed on a larger scale behavior that has more typical plastic characteristics. Specifically, on the large scale it is likely that the ice interaction has a rectifying effect on the ice motion. Under off-ice winds, for example, the plastic nature of ice will cause it to diverge relatively freely, whereas under on-ice winds the ice interaction will prevent further convergence.

Preliminary simulations of the East Greenland pack (Tucker and Hibler, 1981) have demonstrated such a rectifying effect, and indicate that it is an important factor in the tendency of the ice to drift parallel to the coast. Figure 12, for example, shows simulated and observed trajectories for a buoy off East Greenland, assuming a viscous-plastic ice interaction. As can be seen, the ice interaction tends to prevent onshore motion of the buoy, even though the winds tend to drive it against the coast.

One way to examine the effect of random bumping and other stochastic fluctuations is to assume a plastic law locally and then average over stochastic effects in space and time. Such an approach was employed by Hibler (1977) in an attempt to relate plastic law parameters to those used in viscous models. To demonstrate the general idea let us consider the specific case of a rigid plastic law having a circular yield curve. Referring back to eq. 9.9-9.16 the stress strain rate relation for such a system is:

$$\sigma_{ij} = (P^*/\Delta)\ [\varepsilon_{ij}] - \delta_{ij}\ P^*/2 \qquad (9.30)$$

where

$$\Delta = [(\dot{\varepsilon}_{11}^2 + \dot{\varepsilon}_{22}^2)2 + 4\dot{\varepsilon}_{12}^2]^{1/2} \qquad (9.31)$$

To obtain a statistical average, assume that $\dot{\varepsilon}_{ij}$ fluctuates stochastically about a mean value $<\dot{\varepsilon}_{ij}>$ (the fluctuations could be in either time or space) with some Gaussian probability density

$$G(\dot{\varepsilon}_{ij}) = [\lambda(2\pi)^{1/2}]^{-1} \exp\ [-(\dot{\varepsilon}_{ij} - <\dot{\varepsilon}_{ij}>)^2/2\lambda^2] \qquad (9.32)$$

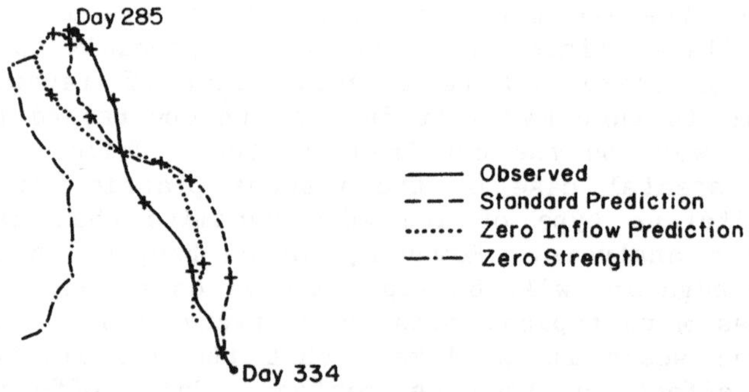

Fig. 12. Simulated and observed trajectories of a buoy off East Greenland, assuming either a viscous plastic ice interaction or no ice interaction.

where λ is the standard deviation of the strain rate fluctuations. Consequently, the statistical average of σ_{ij}, denoted by $\langle\sigma_{ij}\rangle$, is given by

$$\langle\sigma_{ij}\rangle = [\lambda^3(2\pi)^{3/2}]^{-1} \int \exp \{-[(\dot{\varepsilon}_{11} - \langle\dot{\varepsilon}_{11}\rangle)^2 + (\dot{\varepsilon}_{22} - \langle\dot{\varepsilon}_{22}\rangle)^2$$

$$+ (\dot{\varepsilon}_{12} - \langle\dot{\varepsilon}_{12}\rangle)^2]/2\lambda^2\}\sigma_{ij}(\dot{\varepsilon}_{11},\dot{\varepsilon}_{22},\dot{\varepsilon}_{12})d\dot{\varepsilon}_{11}d\dot{\varepsilon}_{22}d\dot{\varepsilon}_{12} \qquad (9.33)$$

where $\sigma_{ij}(\dot{\varepsilon}_{11},\dot{\varepsilon}_{22},\dot{\varepsilon}_{12})$ is given by eq. 9.30. From this equation, two limiting cases are clear. In particular, for very small λ, which corresponds to very small fluctuations in $\dot{\varepsilon}_{ij}$, $\dot{\varepsilon}_{ij}$ in σ_{ij} may be replaced by $\langle\dot{\varepsilon}_{ij}\rangle$, in which case the plastic law is obtained for the average values. For very large λ, which corresponds to large fluctuations compared to the mean value, the right-hand side of eq. 9.33 may be expanded in powers of $\langle\dot{\varepsilon}_{ij}\rangle/\lambda$, a procedure which would yield to first order a viscous-like law, i.e., $\langle\sigma_{ij}\rangle$ proportional to $\langle\dot{\varepsilon}_{ij}\rangle$. It is worth noting here that this expansion argument is independent of the type of distribution and constitutive law assumed. For example, considering an arbitrary constitutive law and a general probability distribution, we could still expand in powers of $\langle\dot{\varepsilon}_{ij}\rangle/\lambda$, although the actual coefficients would depend on the functional law assumed.

For the special case of a circular yield curve it is possible, by using eq. 9.33, to estimate analytically the coefficients of the linear terms in $\langle\dot{\varepsilon}_{ij}\rangle/\lambda$. Substituting eq. 9.30 and 9.31 into eq. 9.33 and changing variables, after some algebra (see Hibler,1977), one obtains

$$\langle\sigma_{ij}\rangle = P^*\{0.35 \langle\dot{\varepsilon}_{ij}\rangle/\lambda - 0.5\sigma_{ij}\} \qquad (9.34)$$

where 0.35 is an approximation to two significant figures of a combination of logarithms of integers and pi. Referring back to eq. 9.9, this expression is basically a linear viscous law with a pressure term and with $\eta = \zeta = 0.173\ P^*/\lambda$.

For the elliptical yield curve, analytic estimation of η and ζ is difficult, although their ratio can be deduced. In particular, it is clear by symmetry that in expanding eq. 9.33 in powers of $\langle\dot{\varepsilon}_{ij}\rangle/\lambda$ the lowest-order terms will yield an average stress-strain relationship of

$$\langle\sigma_{ij}\rangle = 2\eta\langle\dot{\sigma}_{ij}\rangle + (\zeta - \eta)\langle\dot{\varepsilon}_{kk}\rangle\ \delta_{ij} - \delta_{ij}/2P^* \qquad (9.35)$$

where $\zeta/\eta = e^2$. Consequently, as the ellipse becomes more elongated, the ratio ζ/η will increase.

For a more specific application of this concept one can consider the case of fluctuations in time of the deformation at a given location. Using observed and simulated deformation results, Hibler (1977) was able to show that averaging over a period of only a few days yielded a more linear-like behavior, together with a finite pressure term. Considering other effects, such as inertial oscillations that are probably prominent near the edge, it is likely that necessary averaging intervals for these stochastic arguments will be quite small, and likely shorter than typical temporal wind variations.

The ramifications of such a pressure term and nonlinear behavior near the edge are not clear. However, one interesting case has been investigated by Roed and O'Brien (1981). They considered the case of an unconstrained pressure term only, without shear strength of any kind, and without water drag. When this rheology was coupled to a Coriolis force and to a simple continuity equation for the ice compactness, geostrophic adjustment tended to cause a jet-like effect to appear at the ice edge. While this calculation is a bit unrealistic it is suggestive of some of the rheology-induced phenomena that may occur near the ice edge.

9.3.6 Relative Magnitudes of Shear and Compressive Strengths

The above stochastic arguments give some justification for using linear viscous laws in certain circumstances. One example of such a calculation is an idealized infinite boundary calculation performed by Hibler (1974). The main reason for this calculation was to compare observed deformation models. However, the calculations also gave some insight into the relative magnitudes of shear and compressive strengths. Specifically, these idealized calculations yield a best fit bulk viscosity about twice as large as the shear viscosity. This in turn suggests that although of the same order of magnitude, compressive strengths are likely larger than shear strengths.

The basic idea in the calculation was to consider a linearized momentum balance (linear air and water drag) employing a linear constitutive law with η and ζ constant,

$$\sigma_{ij} = 2\eta\dot{\varepsilon}_{ij} + (\zeta - \eta)\,\dot{\varepsilon}_{kk}\,\delta_{ij} \tag{9.36}$$

Solving these equations in the case of an infinite boundary, and taking the limit of large viscosities, yielded the result that the divergence rate over the vorticity is given by

$$\frac{\nabla \cdot \underline{u}}{k \cdot \nabla x \underline{u}} = - \tan\phi \frac{2\eta}{(\eta + \zeta)} \tag{9.37}$$

where ϕ is the turning angle of the atmospheric boundary layer. Comparison of observed ice vorticity and divergence rate over a month-long period (see Hibler, 1974) did show a negative correlation, with the vorticity being about twice as large as the divergence rate. This in turn indicated that $\zeta \approx 2\eta$ and suggests that the compressive strength is larger than the shear strength.

9.4 RELATION OF ICE STRENGTH TO ICE THICKNESS CHARACTERISTICS

In the studies above we have discussed the behavior of ice dynamics under conditions where a constant strength is specified. However, in practice the ice strength is related to ice thickness characteristics in some way. To parameterize this concept there have been two general approaches to date: a multi-level approach developed by Thorndike et al. (1975) and a two-level approach developed by Hibler (1979). In the multi-level approach the ice thickness distribution is broken down into a large number of levels and the strength determined by a hypothesized model of the ice ridge building process. In the two-level approach the ice strength is empirically related to the compactness and the mean ice thickness. In the next two subsections we discuss these approaches. In the first subsection a critique of the Thorndike et al. (1975) ridge building process is presented, and other redistributors are presented and discussed. This section also gives estimates of ice strength based on mechanical considerations and compares these strengths to best fit strengths used in numerical simulations to obtain agreement with observed drift. The two-level formulation of ice strength is discussed in the second subsection. A brief comparison of one-year simulations of the Arctic Basin ice cover employing the multi-level and two-level formulations is given in the third subsection. Finally, in the fourth subsection some of the small scale ramifications of a nonlinear coupling between ice strength and ice thickness characteristics are discussed.

9.4.1 Variable Thickness Parameterization of Ice Strength

Recent work on the strength of a variable thickness ice cover has been largely based on the concept, first suggested by Parmerter and Coon (1972) in a kinematic ridge model, that the major work done in ridge building is due to potential energy changes caused by ice pile-up. Rothrock (1975b) followed up this idea by suggesting that this concept be extended to two dimensions by insisting that two-dimensional deformational work be explicitly related to the amount of work done by ridging. However, if one accepts this hypothesis it then becomes critical to determine how ice ridges.

To formulate the ridge building process on the geophysical scale Thorndike et al. (1975) suggested that thin ice be redistributed into thicker ice categories. The precise manner in which this redistribution should be carried out is not clear. As an initial guess, Thorndike et al. (1975) suggested a redistribution which transfers ice into categories that are fixed multiples of the initial thickness. Ridge observations and theoretical considerations suggest that such a redistribution is unrealistic. In order to provide a more realistic parameterization of the ridging process, Hibler (1980a) proposed a scaling law for ridge building, and used this scaling law in a seasonal equilibrium simulation of the Arctic Basin ice cover. In the discussion below we examine in more detail the strength characteristics of different redistribution processes, and compare these processes to ridge morphological data. In addition, annual ridge buildup results from the seasonal equilibrium simulation of Hibler (1980a) are reported and compared to observations of ice roughness.

General Framework. Following Thorndike et al. (1975) the ice thickness characteristics are described by an areal ice thickness distribution $g(h,\underline{x},t)$, where $g(h,\underline{x},t)dh$ is the fraction of area (in a region centered at position $\underline{x}$ at time t) covered by ice with thickness between h and h + dh. This distribution evolves in response to deformation, advection, growth and decay. Neglecting lateral melting effects Thorndike et al. (1975) derived the following governing equation for the thickness distribution:

$$\frac{\partial g}{\partial t} + \nabla\cdot(\underline{u}g) + \frac{\partial(fg)}{\partial h} = \psi \qquad (9.38)$$

where f is the vertical growth (or decay) rate of ice of thickness h and ψ is a redistribution function (depending on h and g) that describes the creation of open water and the transfer of ice from one thickness to another by rafting and ridging. In general, f is considered to be a function of g(h), and ψ a function of both g(h) and the rate of deformation. Except for the last two terms, eq. 9.38 is a normal continuity equation for g. The last term on the left-hand side can also be considered a continuity requirement in thickness space since it represents a transfer of ice from one thickness category to another by the growth rates. An important feature of the Thorndike et al. (1975) theory is that it presents an "Eulerian" description in thickness space. In particular, growth occurs by rearranging the relative areal magnitudes of different thickness categories.

This equation can be formally generalized (Hibler, 1980a) to include lateral melt by simply adding a source and sink term $F_L(g,h)$ such that

$$\int_0^\infty F_L dh = 0. \qquad (9.39)$$

Equation 9.39 follows from the fact that, by definition, lateral melting of ice will be compensated for by a change in the extent of open water. In addition, $F_L > 0$ for $h = 0$ and $F_L < 0$ for $h > 0$. These conditions reflect the fact that lateral melting decreases the areal extent of ice while increasing open water extent. Adding this lateral growth term to eq. 9.38, the more general thickness distribution equation becomes

$$\frac{\partial g}{\partial t} + \nabla \cdot (\underline{u}g) + \frac{\partial (fg)}{\partial h} = F_L + \psi. \tag{9.40}$$

<u>Mechanical Redistributor</u>. Consider now a two-dimensional continuum with ice described by a plastic rheology such that the stress state $\sigma_{ij}(\dot{\varepsilon}_{ij}, P^*)$ is a function of the strain rate $\dot{\varepsilon}_{ij}$ and a strength P^*. Then, following Thorndike et al. (1975) and Rothrock (1975b), the mechanical redistribution function ψ can be reasonably postulated to satisfy the constraints

$$\int_0^\infty \psi dh = \nabla \cdot \underline{u} \tag{9.41}$$

$$\int_0^\infty h\psi dh = 0 \tag{9.42}$$

$$C \int_0^\infty h^2 \psi(h)\, dh = \sigma_{ij}\dot{\varepsilon}_{ij} \tag{9.43}$$

where $\underline{u}$ is the ice velocity field and C a constant. Considering only the work done on the ice by gravitational and buoyancy forces, C would be given by

$$C_b = \frac{1}{2} \rho_I \left[\frac{\rho_w - \rho_I}{\rho_w}\right] g \tag{9.44}$$

with ρ_I the density of ice, ρ_w the density of water, and g the acceleration due to gravity. Equation 9.41 follows from the constraint that ψ renormalizes the g distribution to unity due to changes in area. Equation 9.42 follows from conservation of mass and basically states that ψ does not create or destroy ice but merely changes its distribution. (An additional assumption in eq. 9.42 is that the ice mass is related in a fixed manner to the thickness.) The third constraint (Rothrock, 1975b) specifies that work done in building ridges is equal to the deformational work. An important feature of this constraint is that, for an arbitrary plastic yield curve, it forces some ridging to occur even though there may be no net convergence. In its present form (with $C = C_b$) eq. 9.43 ignores frictional losses occurring in ridging and considers only the potential energy change due to deformed ice being lifted against gravity and being forced down against buoyancy. However, Rothrock (1975b) has estimated these frictional losses to be the same order of magnitude as the potential energy changes. Consequently, as a crude approximation for frictional losses, eq. 9.43 is retained here with a constant C equal to $2C_b$.

A redistribution which satisfies these constraints is (using the convention that repeated subscripts are summed over)

$$\psi = \delta(h)[(P^*)^{-1}\sigma_{ij}\dot{\varepsilon}_{ij} + \dot{\varepsilon}_{kk}] + (P^*)^{-1}\sigma_{ij}\dot{\varepsilon}_{ij}W_r(h,g), \qquad (9.45)$$

where $\delta(h)$ is the dirac delta function and

$$W_r(h,g) = \frac{[-P(h)g(h) + \int_0^\infty \gamma(h',h)P(h')g(h')dh']}{\int_0^\infty [P(h)g(h) - \int_0^\infty \gamma(h',h)P(h')g(h')dh']dh} \, . \qquad (9.46)$$

The first term in eq. 9.45 specifies the amount of open water created, while the second term describes the transfer of thin ice to thicker categories by ridging. In this formalism $\gamma(h_1,h_2)dh_2$ can be thought of as the area of ice put into the thickness interval $[h_2,h_2 + dh_2]$ when a unit area of ice of thickness h_1 is used up. Basically the integral over γ specifies where the ice is transferred to during ridging. The function P(h), on the other hand, is some probability function specifying which categories are being destroyed by ridging. Written in this form γ automatically satisfies eq. 9.41. To satisfy eq. 9.42 and 9.43 leads to the additional constraints, respectively,

$$\int_0^\infty \gamma(h',h)hdh = h', \qquad (9.47)$$

$$C\int_0^\infty W_r h^2 dh = P^*. \qquad (9.48)$$

Equation 9.47 requires that transfer of ice from one category to another does not change its mass, and eq. 9.48 serves as a definition of ice strength P*.

The two main unknowns in the redistribution theory of ridging are P(h) and $\gamma(h_1,h_2)$. Thorndike et al. (1975) have suggested that P(h) be given by

$$P(h) = \max\left[\left(1 - \int_0^h g(h)dh/c_1\right), 0\right] \qquad (9.49)$$

where c_1 is a constant, say 0.15. The basic idea here is that under ridging conditions the closure of both thin ice and thick ice occurs. However, the thinnest ice is allowed to be removed to a greater extent. By fixing c_1 to be 0.15, only the thinnest 15% of the ice is deformed. Note that if there is, say, 25% open water, only open water will be removed and no ridging will occur.

The other unknown is $\gamma(h_1,h_2)$, which specifies how ice is transferred from one category to another by ridging. Thorndike et al. (1975) have suggested that

$$\gamma(h_1,h_2) = \delta(h_2 - Kh_1)K^{-1} \tag{9.50}$$

where K is a constant. Equation 9.50 effectively states that ice is transferred into a fixed multiple of its own thickness. This form is computationally simple and is useful for examining the behavior of the ice thickness distribution model.

However, the redistribution of eq. 9.50 ignores two important physical features of the ridging. One feature is that under deformation, ridging transfers ice to a variety of thickness categories. This, for example, is evident from field observations of ridges (see, e.g., Weeks et al., 1971; Kovacs, 1972) which show them to be roughly triangular in shape. These observations indicate that under deformation, leads containing thin ice of a given thickness typically form ridges having a triangular cross section. To satisfy eq. 9.50 such ridges would have to have vertical sides.

The second physical feature is that typical ridge heights divided by the thickness of ice being ridged appear to decrease with increasing thickness (see, e.g., Tucker and Govoni, 1981). This particular scaling is an important feature of the Parmerter and Coon (1972) mechanistic ridge model. In particular, Parmerter and Coon (1972) found that simulated ridges using their mechanistic model tended to have a maximum limiting height. This limiting height, however, tended to increase more slowly than the thickness of ice being ridged.

To avoid some of these shortcomings Hibler (1980a) proposed a redistributor which distributed ice over a variety of thickness categories and, perhaps more importantly, the limiting height was taken to scale as the square root of the thickness of ice being ridged. Hibler (1980a) further noted that such a square root dependence is consistent with one's intuition concerning ridging. Consider, for example, two equal width leads undergoing ridging. Assuming all the thin ice is deformed into piles of relatively small blocks with similar slope angles, the two ridges formed would have heights scaling with the thickness of ice being ridged. If the distribution of lead widths is independent of the ice thickness in leads, such an intuitive argument should have application.

A Comparison of Redistributors. To examine the ramification of different redistributors let us compare eq.9.50 to several other forms for $\gamma(h_1,h_2)$ that have been suggested, or are introduced here for pedagogical purposes:

Redistribution

$$\gamma(h_1,h_2) = \delta(h_2-Kh_1)\frac{1}{K} \quad \text{with } K = \text{constant} \qquad (9.51)^1$$

$$\gamma(h_1,h_2) = \frac{1}{2(H^* - h_1)} \quad \text{for } 2h_1 \le h_2 \le 2\sqrt{H^*}\sqrt{h_1} \quad \text{with } H^* = \text{constant} \qquad (9.52)^2$$

$$\gamma(h_1,h_2) = \frac{2}{h_1(K^2-4)} \quad \text{for } 2h_1 \le h_2 \le Kh_1 \quad \text{with } K = \text{constant} \qquad (9.53)^3$$

$$\gamma(h_1,h_2) = \delta(h_2-H')\frac{h_1}{H'} \qquad (9.54)^4$$

The scaling characteristics of different redistributors can be tested by field examination of ridge height and block size characteristics. A particularly useful data set for this purpose has been obtained and analyzed by Tucker and Govoni (1981). This data set was taken off the North Slope of Alaska. Figure 13 shows the salient characteristics of the data set relevant to the redistribution theory. The error bars are the standard error in the mean ridge height estimate. As can be seen from this figure the square root scaling fits these limited data better.

To approximately convert the ridge height scaling to an ice thickness scaling, a 4:1 ridge keel to sail ratio is assumed (Kovacs and Mellor, 1974). With this scaling the square root case is equivalent to an H* of 100 m in the Hibler redistributor, whereas the linear scaling represents a modified Hibler redistributor. The rubble case represents a fixed thickness H′ = 20 m.

---

[1]Thorndike et al. (1975).
[2]Hibler (1980a).
[3]Modified Hibler (1980a).
[4]Rubble redistribution.

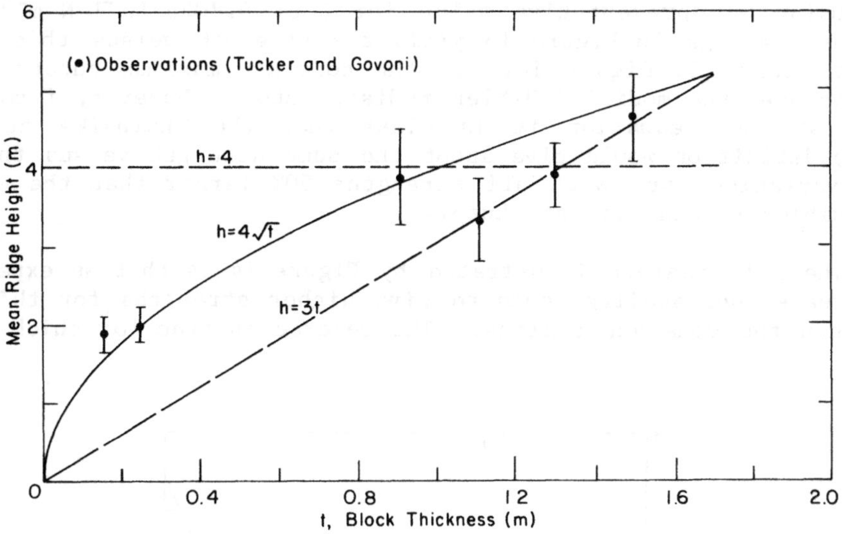

Fig. 13. Comparison of different redistribution scaling laws with observed data. The solid curve represents the Hibler (1980 a) scaling (eq. 9.52), the large dashed curve the Thorndike et al. (1975) scaling (eq. 9.51) and the small dashed curve the "rubble" scaling.

Of particular relevance to ice mechanics are the ice strengths generated by different redistributors. To get some feeling for these strengths, consider the special case of only one thickness of ice, say h_o, being ridged. Formally representing this condition by $P(h) = \delta(h-h_o)$ one can substitute eq. 9.46 in eq. 9.48 and obtain analytical results for the ice strength. After some algebra different redistributors yield the following strength equations:

$$P^* = \frac{Ch_o^2\ (4\Gamma^3 - 3\Gamma^2 - 1)}{3(\Gamma-1)\ \Gamma} \tag{9.55}$$

where $\Gamma = \sqrt{H^*/h_o}$ for the Hibler redistributor and $\Gamma = K/2$ for the modified Hibler case.

For the Thorndike et al. (1975) redistribution

$$P^* = Ch_o^2K \tag{9.56}$$

For the rubble redistributor

$$P^* = Ch_oH' \tag{9.57}$$

Using a representative value for C of $0.8026 \cdot 10^{+3}$ N m^{-3} the various scalings in Figure 13 yield the strength versus thickness results shown in Figure 14. In the constant maximum cutoff case we have used the modified Hibler redistributor. However, from the above strength equation it is clear that the Thorndike et al. (1975) definition would give about the same strength versus thickness variation, but with all strengths 50% larger than the modified Hibler (1980a) redistributor.

The main feature illustrated by Figure 14 is that as expected the square root scaling tends to give higher strengths for thinner ice than the constant scaling. The reverse is true for thick ice.

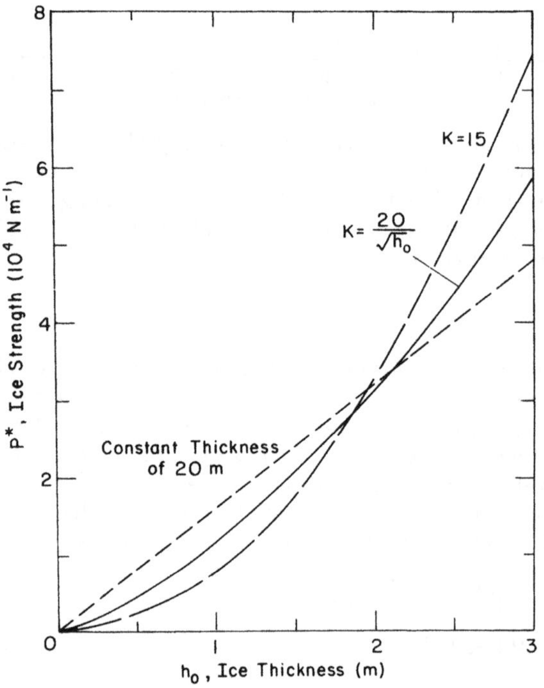

Fig. 14. Ice strength versus thickness for different redistributors. The solid and long-dashed lines are for a uniform redistribution with the maximum thickness scaling as the square root (solid line) of the thickness of ice being ridged (eq. 9.52), and linearly (long-dashed line) with the thickness of ice being ridged (eq. 9.51 and 9.53). The short-dashed line is for the "rubble" redistributor (eq. 9.54) where all ice is transferred into a 20-m thickness. In these curves, frictional losses are assumed to be equal to changes in gravitational potential energy.

However, in both these cases the three-dimensional stress will increase as the ice becomes thicker. This is in contrast to the rubble case where the three-dimensional stress will be a constant at ~$1.6 \cdot 10^4$ N m^{-2} (or ~2.3 psi).

Selected Seasonal Numerical Results. The Hibler (1980a) redistributor (eq. 9.52) has been used as one component of a coupled variable thickness dynamic thermodynamic sea ice model. This model has been used to carry out a seasonal equilibrium simulation of the Arctic Basin ice cover by integrating the model for five years with the same seasonal forcing being repeated each year. Of particular interest to this review are the ridge buildup results and the sensitivity of the model to different assumed frictional energy dissipation values in the ridging process. With regard to ridge buildup, while the model does not explicitly keep track of ridge advection it is possible to determine the volume of deformed ice created over an annual cycle at a fixed location. Figure 15 shows fifth-year contours of this quantity using the H* value of 50 m in

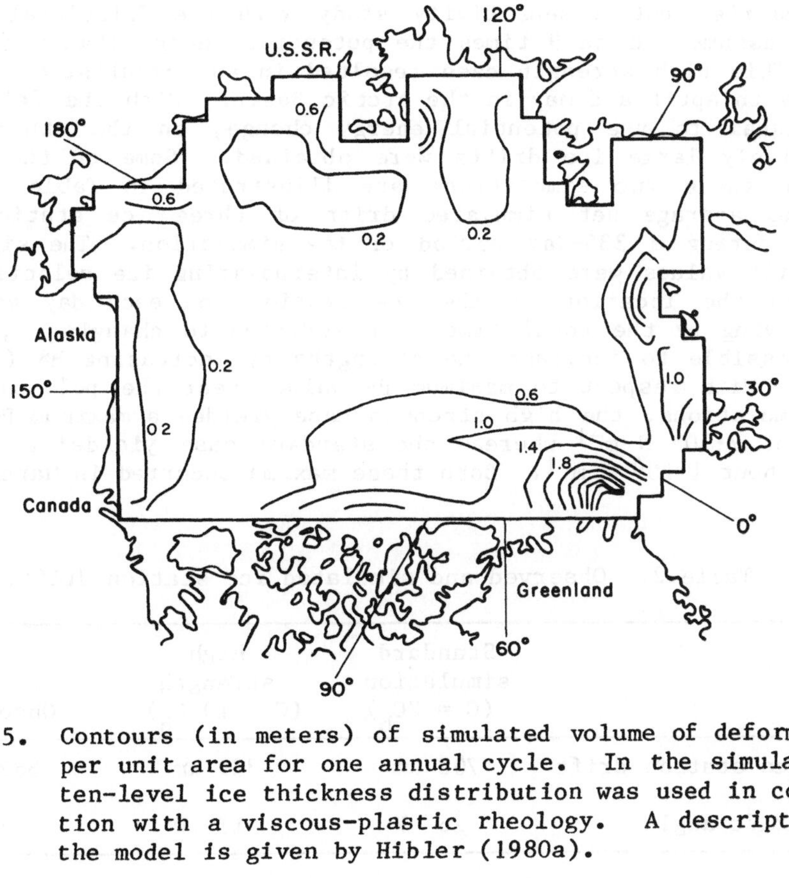

Fig. 15. Contours (in meters) of simulated volume of deformed ice per unit area for one annual cycle. In the simulation a ten-level ice thickness distribution was used in conjunction with a viscous-plastic rheology. A description of the model is given by Hibler (1980a).

eq. 9.52 and a C value in eq. 9.48 of $2C_b$ (see eq. 9.44). The general shape of the roughness contours agrees well with surface roughness observations in the western Arctic Basin (Weeks et al., 1971; Hibler et al., 1974b). The two major features are a heavy buildup of ridging off the Canadian Archipelago and less ridging in the Beaufort Sea than near the pole. Note, however, that there is a zone of heavy ridging just off the North Slope, in agreement with observations by Tucker et al. (1979). (This feature is partially masked in Figure 15 by the large contouring interval. Finer contouring intervals show a peak in the volume of deformed ice per unit area between the 0.2-m contour in Figure 15 off the North Slope.) The other major feature is a tongue of high ridging further offshore near the tip of Greenland. Roughness data obtained from submarine sonar profiles by LeShack (private communication) also show such a tongue.

With regard to ice strengths on the geophysical scale, a major unknown factor is the amount of frictional energy loss occurring during ridging. To get an estimate of how increases in this assumed frictional loss affect drift simulations, Hibler (1980a) also carried out a sensitivity study with the frictional energy losses assumed to be 9 times the potential energy change ($C = 10\ C_b$). This high strength case resulted in an unrealistic stoppage of flow in April and May in the Arctic Basin. With the frictional loss equal to the potential energy change, on the other hand, excessively large ice drifts were obtained. Some of the ranges between these two simulations are illustrated in Table 2, which compares average net simulated drift of three ice stations obtained during a 335-day period of the simulation. The simulated net drift values were obtained by interpolating ice velocity vectors to the location of the ice stations on each day and then multiplying by the total time. In addition to changing C, it is also possible to increase the strengths by increasing H* (see eq. 9.46). With respect to maximum P* values near the pole in these two simulations, the high strength case yielded a maximum P* value of about $7 \cdot 10^4$ N m^{-1} whereas the standard case yielded a maximum P* of about $1 \cdot 10^4$ N m^{-1}. Both these maxima occurred in March.

Table 2. Observed and simulated ice station drift.

	Standard simulation ($C = 2C_b$)	High strength ($C = 10\ C_b$)	Observed
Net ice station drift	702 km	339 km	562 km
Net drift angle	26° W	13° W	6° W

9.4.2 Two-Level Formulation of Ice Strength

The multi-level parameterization provides a consistent way of treating a variable thickness ice cover. However, as is apparent from the above discussion it tends to transfer much of the complexity of the problem to the mechanical redistribution function, which is rather complicated and inherently contains a number of assumptions. The idea in the two-level approach (Hibler, 1979) is to approximate the whole thickness distribution by two categories of ice: thick and thin. When this is done the treatment of the ridging process becomes much simpler. The strength is then taken to be an empirical function of these two categories, with the most important strength dependence being on the thin ice. This empirical function does contain some assumptions, but as discussed above the mechanical redistribution function also contains a large number of assumptions. Also, a two-level Arctic sea ice simulation, discussed in the next section, yields results for ice thickness and drift in the Arctic Basin very similar to those obtained from the multi-level model, which indicates that for many climatic studies the two-level approach may be quite useful.

The basic concept in the two-level approach is that many of the dynamic and thermodynamic characteristics of sea ice are dominated by the thin ice. Consequently, for many purposes the ice thickness distribution may be approximately characterized by breaking it into two parts (thick and thin). The thought here is that if the thin ice component is monitored, then uncertainties in the remaining ice thickness distribution may not be critical. Within this two-level approach the ice cover is broken down into an area A (often called the compactness), which is covered by thick ice, and a remaining area (1-A), which is covered by thin ice, which, for computational convenience, is always taken to be of zero thickness (i.e., open water). The idea here is to have the open water approximately represent the combined fraction of both open water and thin ice up to some cutoff thickness h_0. The remainder of the ice is distributed arbitrarily. However, since the thin ice mass is normally small, the mean thickness of the remaining "thick" ice is approximately equal to h/A.

For the mean thickness h and compactness A the following continuity equations are used:

$$\partial h/\partial t = -\partial(uh)/\partial x - \partial(vh)/\partial y + S_h \tag{9.58}$$

$$\partial A/\partial t = -\partial(uA)/\partial x - \partial(vA)/\partial y + S_A \tag{9.59}$$

where $A < 1$, u is the x component of the ice velocity vector, v the y component of the ice velocity vector, and S_h and S_A are thermodynamic terms. While eq. 9.58 is a simple continuity equation for ice mass (characterized by the mean thickness h), with thermodynamic source and sink terms, eq. 9.59 is somewhat more complex. By including the restriction that $A < 1$, a mechanical sink term for the areal fraction of ice has been added to a simple continuity equation for the ice concentration. This sink term turns on when $A = 1$ (i.e., no open water left) and under converging conditions removes enough ice area through ridging to prevent further increase in A. Although the sink term does not change the ice mass, it can cause the "thick" ice thickness to increase by allowing h to increase while A does not. This simple sink term is the equivalent of the Thorndike et al. (1975) mechanical redistribution. Hence, within the two-level model the whole ridging process is approximated by specifying that $A \leq 1$. Note that however this ridging problem is formally treated in this two-level model it does not affect the conservation of ice mass, which is explicitly guaranteed by eq. 9.58. This explicit treatment of the ice mass continuity equation directly is one of the main simplifications of the two-level model over the multi-level treatment.

Treatment of the thermodynamic terms S_A and S_h in eq. 9.58 and 9.59 varies with investigators, and can make a significant difference in the physics. In ice forecasting applications (e.g. Nikiforov, 1957; Doronin, 1970), the S_A term is often omitted. In longer term seasonal simulations (Hibler, 1979; Parkinson and Washington, 1979) S_A has been used to cause the thin ice fraction to disappear in proportion to the ice growth rates. Under these formulations (1-A) represents the combined fraction of thin ice and open water.

To couple the ice strength to the ice thickness characteristics the ice strength P* is taken to be a function of compactness and thickness according to

$$P^* = P_o h \exp\left[-C(1 - A)\right], \qquad (9.60)$$

where P_o and C are fixed empirical constants and h is in meters. This formulation makes the strength strongly dependent on the amount of thin ice (characterized by the compactness A), while also allowing the ice to strengthen as it becomes thicker.

While directly intuitive, the strength expression in eq. 9.60 can also be qualitatively related to the strength assumptions used in the multi-level thickness approach discussed above. There, under converging conditions, the closure of both thin and thick ice (and also open water) occurs simultaneously. The relative amount of thin ice being deformed, however, is weighted by the area of thick ice. Hence, for small amounts of thin ice, more thick ice is deformed, and thus the ice has a high strength which is dependent on the thick ice characteristics. For a large amount of thin ice, on the other hand, mostly thin ice is deformed, which yields low strength. These features are represented in eq. 9.60 by the exponential dependence on (1 - A), which characterizes the amount of thin ice being ridged, and the proportionality to h, which approximates the strength of thick ice undergoing deformation. The assumption in the h dependence is that the distribution of thick ice is reasonably well described by its mean (h/A). However, while qualitatively correct, this two-level ice thickness scheme does not allow the particular thickness of ice involved in ridging to be determined.

9.4.3 A Numerical Comparison of Thickness-Strength Coupling

To give some feeling for the role of strength-thickness coupling in the two-level and multi-level models we briefly discuss here the results of comparative simulations of the Arctic Basin ice cover. In order to compare the behavior of the two-level model to the multi-level model, one-year simulations employing the same forcing of the Arctic Basin ice cover are examined. These simulations consisted of integrating the coupled continuity and momentum balance equation for each model over one seasonal cycle for the Arctic Basin ice cover. While not to seasonal equilibrium, the results are useful for comparing ice strength formulations. Complete details on the numerical framework for these simulations and the forcing fields are described in Hibler (1979, 1980a). Since our main concern is the ice buildup we will not go into the details of the forcing fields (and heat storage in the oceanic boundary layers) which are discussed in the above references. The two main parameter sets of interest for this discussion are P_o (= $5 \cdot 10^3$ N m^{-2}), C = 20, the strength constants for the two-level case, and H* (= 50 m), the ridging specification parameter for the multi-level case (see eq. 9.52). In the two-level case the ice strength is given by eq. 9.60, while in the multi-level case the strength is specified by combining eq. 9.48 with eq. 9.44, 9.46, 9.49 and 9.52. In all cases a uniform initial thickness of 3.3 m was specified everywhere.

Figure 16 shows the December thickness contours for both models. The boundaries employed in the simulation are specified by the solid lines. Also, at the Greenland-Spitsbergen passage an "open boundary" condition (see Hibler, 1979) was used, which allowed ice inflow and outflow. As can be seen, both the multi-level and two-level models yield very similar ice thickness buildup, with thicker ice near the archipelago and thinner ice near the pole and off the Alaskan and Siberian coasts. This thickness buildup is due to the ice dynamics, which tends to pile up ice off the archipelago while removing it from other regions. In both models, as this ice gets thicker it gets stronger and tends to prevent further convergence. The main feature illustrated by this figure is, however, the similarity of the thickness contours in the two simulations, despite the differences in ice thickness parameterizations. Both

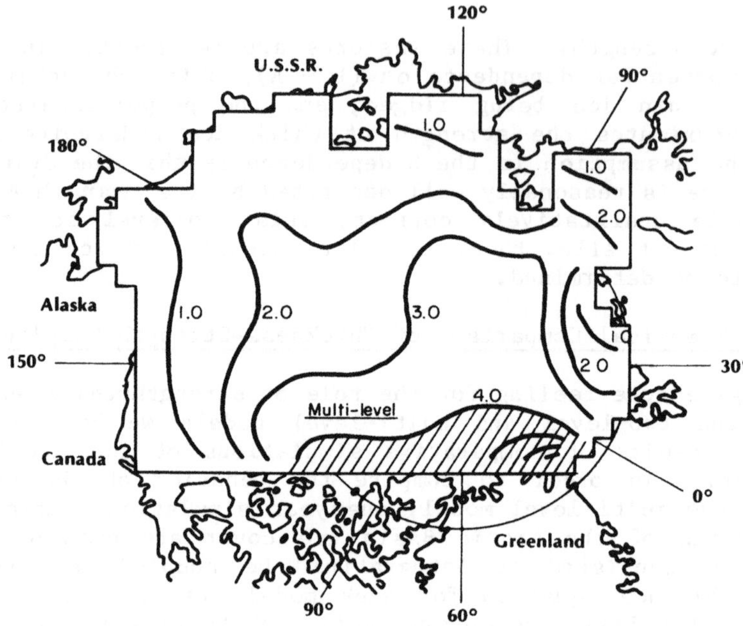

Fig. 16a. Simulated 30-day (day 331-360) averaged ice thickness contours in meters for comparative one-year simulations. Part (a) shows results from a multi-level simulation employing eight thickness levels and an ice strength based on an assumed redistribution process (see eq. 9.44, 9.46, 9.48, 9.49 and 9.52).

these thickness contours are in good agreement with observations of thickness characteristics in the Arctic Basin, which show a greater buildup along the archipelago, with typical mean thicknesses of about 5-7 m (see, e.g., Wadhams, 1981).

It should be noted that one quantitative difficulty with both these simulations is the use of eight-day averaged winds (from May 62 to May 63) in the forcing fields. More recent two-level simulations with daily time varying winds of the Arctic ice cover (Hibler and Walsh, 1982) and portions of the Antarctic ice cover (Hibler and Ackley, 1982, 1983) have yielded higher best fit strength constants, P_o, for the two-level model of $2.75 \cdot 10^4$ N m^{-2}. This value is about five times larger than the best fit P* value used by Hibler (1979) and also used above. The main reason for this increase is that due to the nonlinear wind stress, daily wind fields yield a larger average wind force. As a consequence, a larger ice interaction is needed to yield modification of the ice drift in agreement with observations.

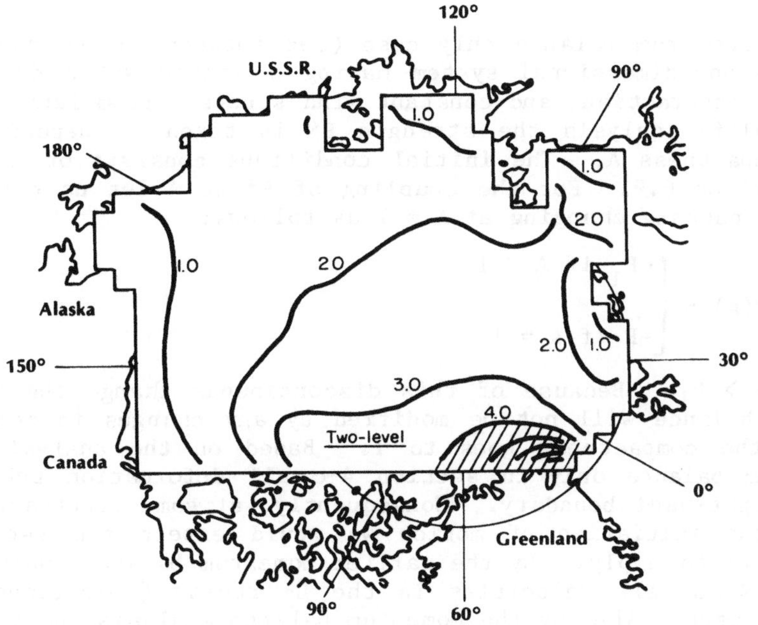

Fig. 16b. Simulated 30-day (day 331-360) averaged ice thickness contours in meters for comparative one-year simulations. Part (b) is from a two-thickness-level simulation using an empirical ice strength formulation (eq. 9.60).

9.4.4 Small Scale Ramifications of Coupling Between Ice Strength and Thickness Characteristics

As discussed above, the main effect of thickness strength coupling on the large scale is for the ice to become stronger in regions of convergence, which in turn prevents further convergence. While this same behavior also occurs on smaller scales, there can be a number of fluctuating effects, such as kinematic waves, which are masked on larger scales. Some of these effects depend on the nonlinear ice interaction and the nonlinear coupling between ice strength and thickness characteristics. To yield some insights into these small scale effects Hibler et al. (1983) analyzed some special idealized cases of the two-level model using the same grid as shown in Figure 7. The essential feature of these results is that for a nonlinear model it is possible for ice velocities at a given location to fluctuate in time, even though the wind forcing may be fixed.

Analytical Analysis of Coupled One-dimensional Systems. To demonstrate the overall nature of plastic buildup on a small scale it is useful to analyze an idealized one-dimensional system. As in the momentum balance only case (see subsection 4-c above) consider a one-dimensional system having a linear water drag, rigid plastic interaction, and constant wind stress from left to right. To simplify analysis the strength P* is taken to depend only on the compactness A. The initial conditions consists of a compactness (A) of 0.9. For the coupling of P* to A let us consider P* discontinuously changing at A = 1 as follows:

$$P^*(A) = \begin{cases} -P_1 \text{ if } A < 1 \\ -P \text{ if } A = 1 \end{cases} \tag{9.61}$$

where $P > P_1$. Because of this discontinuous change the local momentum balance will not be modified by any changes in compactness until the compactness goes to 1. Based on the analysis of the momentum balance only in section 4-c all deformation takes place at a right-hand boundary. Consequently, at some arbitrary time Δt after the initiation of motion we would expect the geometry in Figure 17 to apply. By the earlier momentum balance analysis, for $(P-P_1) < \tau L_3$ all velocities in the L_3 region (compactness of 1) will be zero. Also by the momentum balance analysis the velocities in the L_2 region will be given by

$$U_2 = (\tau L_2 - P_1)/cL_2 \tag{9.62}$$

so that as L_2 becomes smaller U_2 will slow down somewhat unless $P_1 = 0$. Since with the L_3 region motionless all the deformation is occurring at the L_2, L_3 boundary, it is clear that the L_2, L_3 boundary will propagate outward (to the left) at a speed of $U_2/0.1$. However, note that for L_3 large enough (i.e., $L_3 > (P-P_1)/\tau$) the L_3 region will start moving again with a speed

$$U_3 = (\tau L_3 - (P-P_1))/cL_3 \tag{9.63}$$

When this occurs the propagation of the ridging front will slow down somewhat to $(U_2-U_3)/0.1$. Also, once all the ice is converted

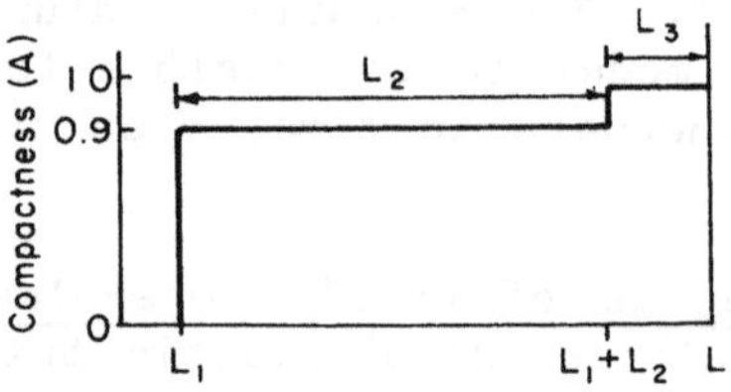

Fig. 17. Geometry of idealized one-dimensional initial compactness conditions.

to 100% compactness, the speed will then slowly decrease as the length L_3 decreases due to removal of ice by ridging at the right-hand boundary. Based on these considerations one can construct a qualitative speed versus time of two arbitrary points, say p and p′, where p′ is located further left than p (Figure 18).

Obviously, by varying strengths and/or the wind forcing the various slopes, magnitudes and spacing in time of these profiles can be modified. One of the more interesting features of the system is that due to the complexities of the thickness coupling, for appropriate strength differences, points near the right-hand coast will actually stop for a while and then speed up, even though external forcing is constant. The essential physics here is that because of the discontinuous nature of the ridging front and the nonlinear rheology, points in the L_3 region will not move at all until L_3 becomes large enough to accumulate an adequate wind fetch to overcome the differential of plastic stresses at the boundaries.

<u>Numerical Investigation of Response Characteristics of Coupled System.</u> A series of two-dimensional numerical simulations yielded results that generally agree with the one-dimensional analysis. However, there are significant differences associated with the fact that P* does not change discontinuously but is instead a smooth function of A. For the standard case study, the coupled equations were integrated for 18 hours at 15-minute time steps using the grid of Figure 7. The initial conditions consisted of 10% open water together with a mean ice thickness of 0.5 m. To simplify analysis the ice strength is taken to depend only on the compactness, not the ice thickness. As a consequence the ice strength in these particular idealized studies P* is given by

$$P^* = (0.5\ \text{m})\ P_o\ e^{-C(1-A)} \qquad (9.64)$$

To make the test more comparable to the one-dimensional analysis, turning angles and the Coriolis parameter were set equal to zero. A constant wind speed of 9.23 ms^{-1} in the positive x direction was used, together with an ice strength constant P_o of $2.0 \cdot 10^4\ \text{Nm}^{-2}$,

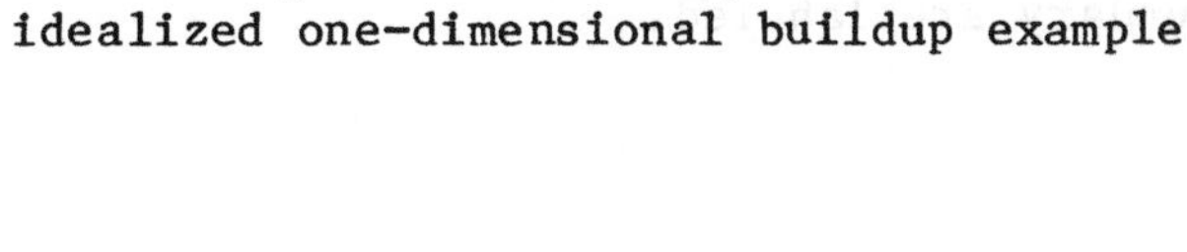

Fig. 18. Qualitative plot of velocity versus time for two points in idealized one-dimensional buildup example.

and a constant C of 30. Other parameters are the same as in Table 1.

The basic characteristics of this idealized numerical simulation are shown in Figure 19, which shows x velocity components versus time for several grid points. Both these plots were taken from grid points and grid cells centered in the y direction. The general behavior of the velocity time series is commensurate with the idealized one-dimensional analysis, with the points nearest the coast rapidly decreasing in speed, with a later increase as the region of ridging moves further out. However, there are significant differences from the analytic case associated with the fact that the motion near the coast is balanced by a gradient in compactness rather than a discontinuous change. This gradient means a) that the points do not ever totally stop (since some convergence must occur everywhere to allow the gradient to move outward) and b) that the kinematic waves are allowed to propagate upstream (or more precisely down the stress gradient).

These waves arise from the fact that the velocity of ice is balanced in part by the stress gradient, while the stress gradient is maintained by a convergence of the ice velocity field. As a consequence, a perturbation of either one of these quantities can cause a wave to propagate down the stress gradient slope. This wave can naturally develop in the evolution of the system since the initial buildup can be out of equilibrium with regard to

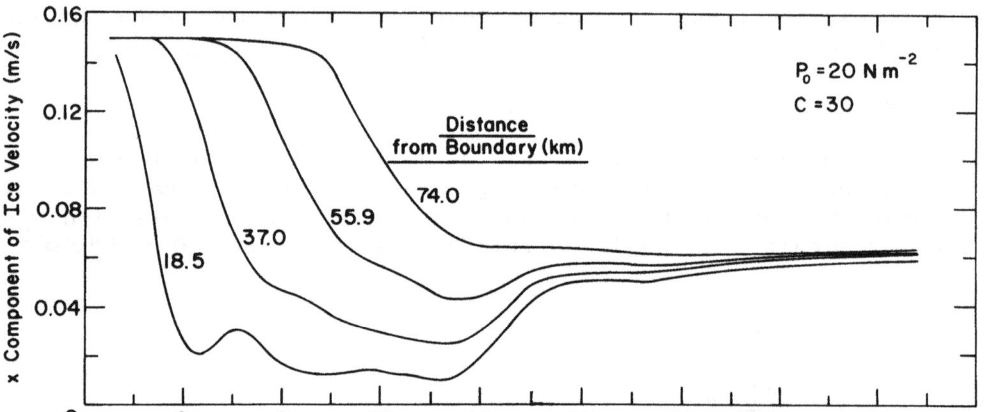

Fig. 19. Time series of x component of ice velocity progressively further from right-hand boundary of numerical buildup experiment. The grid points plotted were centered in the y direction and the distances from the right-hand boundary are labeled.

kinematic waves (when the dependence of strength on compactness is nonlinear). Such a kinematic wave is apparent in Figure 19 at hour 6 in the velocity time series of the point nearest the coast. Tests at smaller time steps verified that this wave was not a numerical artifact. Closer examinations of this wave show it to slow down somewhat and diffuse as it progresses outward. More detailed investigations of the characteristics of these waves are needed.

9.5 THE ROLE OF ICE DYNAMICS IN THE ATMOSPHERE-ICE-OCEAN SYSTEM

9.5.1 General Discussion

The fact that ice is moving and deforming rather than static has important ramifications. A feature that has been emphasized a great deal recently is the fact that dynamics creates thin ice and leads in an ice cover (Maykut, 1978; Washington et al., 1976). However, in addition to creating leads the dynamics also causes advective effects which can play a pivotal role in the atmospheric and oceanic circulation. In the presence of a free ice edge, for example, advective effects can transfer ice to the marginal regions to be rapidly melted. When viewed from a satellite photo such an ice cover may appear static, since the ice extent does not rapidly change. However, the actual physics maintaining such a "static" extent can be complex and largely dependent upon processes occurring at the ice margin.

Advection also causes net imbalances of ice growth by transferring ice from regions of high growth to regions of greater melt. Since most of the salt is taken out of the ice when it freezes, the high growth regions will be regions of high salt flux into the ocean, whereas the high melt regions will be areas of fresh water input. These features can significantly alter the oceanic circulation and vertical convection. With regard to the atmosphere, regions of high growth are regions of high heat exchange with the atmosphere due to the latent heat of freezing of the ice cover. A third, rather indirect way ice dynamics can affect the atmosphere-ocean heat exchanges has recently been noted by Hibler and Ackley (1983). They carried out several simulations invoking an ice feedback mechanism similar to that used in many simple climate models, whereby the presence of ice caused the lower atmosphere to be cooled. With this feedback mechanism, simulations without dynamics produced very unrealistic summer ice margins, whereas simulations with dynamics produced a realistic delay. Basically, the dynamics affords a mechanism for "turning off" the feedback link, which is not possible in the purely static thermodynamic models.

Finally, in addition to modifying the thermal and salinity fluxes at the air-ocean interface, sea ice can substantially modify the momentum transfer by wind to the ocean. Specifically, if we consider the ice-ocean system to be coupled, then the momentum transfer into the ocean is given by the wind stress less the force due to ice interaction. As discussed earlier, due to the highly nonlinear ice interaction the ice stress terms can be very complex.

To illustrate the relative importance of some of these features, results from two series of simulations are discussed in the following subsections. The first subsection describes simulation of the Weddell Sea pack ice, while the second subsection examines the salient characteristics of an ice-covered Arctic Ocean model. A more complete description of these simulations is given elsewhere (Hibler and Ackley, 1983; Hibler and Bryan, 1984).

9.5.2 Results from a Weddell Sea Pack Ice Model

Model Description. The model essentially consists of a momentum balance employing a plastic rheology coupled to the two-thickness evolution equations for the ice thickness and compactness discussed earlier (eq. 9.58-9.60). These equations also include ice growth and decay due to surface heat budget considerations and heat storage in a motionless fixed depth oceanic boundary layer. For the ice strength (see eq. 9.60) we take $P_o = 2.75 \cdot 10^4$ N m^{-2} and C = 20. This P_o value is about five times larger than the P_o value used by Hibler (1979). This increase was necessitated by the fact that daily wind fields were used here as opposed to the eight-day average wind fields used by Hibler (1979).

Following formulations developed by Hibler (1979), the thermodynamic terms in the continuity eq. 9.58 and 9.59 are given by

$$S_h = f(h/A)A + (1 - A)\, f(0) \qquad (9.65)$$

$$S_A = \begin{cases} (f(0)/h_o)(1 - A), & \text{if } f(0) > 0 \\ 0, & \text{if } f(0) < 0 \end{cases} + \begin{cases} 0 & \text{if } S_h > 0 \\ (A/2h)S_h & \text{if } S_h < 0 \end{cases} \qquad (9.66)$$

with f(h) the growth rate of ice of thickness h, and h_o a fixed demarcation thickness between thin and thick ice.

The S_h term specifies the net ice growth or melt. Within the two-level approximation, S_h is given by the sum of the ice grown on open water plus the additional growth over the portion of the cell covered by thick ice. To approximate the growth and decay rate of this thick ice, its mean growth rate is estimated to be that of ice of constant thickness h/A. For melting conditions the same sum over open water and thick ice is used. A critical assumption here is that the heat absorbed by open water will horizontally mix and melt additional ice until the mixed layer returns to freezing.

The S_A term characterizes the way in which growth and decay change the relative areal extents of thin and thick ice. The basic physical notion embodied by this term is that the areal fraction of thin ice will decrease rapidly under freezing conditions, and increase slowly under melting conditions. To parameterize simply the freezing effect, the fraction of open water (1 - A) is allowed to decay exponentially with a time constant of $h_o/f(0)$, which gives the first term in eq. 9.66. The constant h_o is chosen to be small compared to mean ice thicknesses but large enough so that heat fluxes through h_o-thick ice are substantially less than through open water. In the Arctic simulations performed by Hibler (1979) h_o = 0.5 m is used. However, for these Antarctic simulations h_o is set equal to 1.0. This larger value is based on field observations reported in Ackley et al. (1980) and Gow et al. (1982), which suggest that frazil ice formation may well prolong ice production in open water regions. Other than h_o and P_o, the numerical parameters used in the simulation are identical to those used by Hibler (1979) and Hibler and Walsh (1982).

The second term in eq. 9.66 accounts for melting. Its magnitude is derived by assuming that the thick ice is uniformly distributed between 0 and 2 h/A in thickness, and all melts at the same rate. Therefore, over a time Δt the ice of thickness less than S_h Δt will melt and form open water. By the assumption of uniform distribution this ice covers a fraction of area equal to S_h Δt A/2h, which for Δt small yields the second term in the equation for S_A.

The vertical growth rates (f(h)) are estimated by employing Semtner's (1976) sea ice thermodynamic model in conjunction with a surface heat budget calculation similar to that of Parkinson and Washington (1979) and Manabe et al. (1979). The basic surface energy budget calculation (for details see Hibler, 1980a, Appendix B) includes incoming shortwave and longwave radiation and sensible and latent heat fluxes, and provides for the determination of the surface temperature of the ice by numerical iteration.

Following Manabe et al. (1979), the effects of snow cover are approximated by allowing the ice surface albedo to be that of snow when the calculated surface temperature is below freezing, and that of snow-free ice when the surface temperature of the ice is at the melting point. The model also includes a motionless oceanic mixed layer of fixed depth (30 m) and assumes a constant oceanic heat flux of 2 Wm^{-2}. When open water is absorbing heat, all of the absorbed heat is used to further melt the ice. If there is no ice the absorbed heat raises the temperature of the mixed layer. Under growth conditions, no ice is allowed to form until the mixed layer reaches the freezing temperature of seawater.

Input Fields. Input fields to drive the model consist of average daily surface atmospheric air temperatures, mixing ratios, and surface pressure taken from the 1979 Australian analysis. Preliminary simulations with these analyzed atmospheric temperature fields yielded unrealistically small ice concentrations in summer. Comparison of analyzed surface temperature with buoy observations indicated that the main problem was the excessively high temperatures in the late summer and fall (February-April) in the Australian surface analysis, which did not include these buoy observations. Two procedures were examined to construct a more realistic surface analysis. In a crude reanalysis method, hereafter referred to as the standard case, air temperatures and mixing ratios were not allowed to rise above 271.2 K and 0.25 g kg^{-1} in the region designated by the square box in Figure 20 in all months except January. Slightly different regions were experimented with but had little effect on the results. For the more fully coupled case, the air temperature and the humidity were related to the presence of ice by specifying maxima of 271.2 K and 0.25 g kg^{-1} for the temperature and mixing ratios if the ice cover concentration was greater than 0.50 (or in the case of the thermodynamics-only model if the ice thickness was less than 25 cm). It is notable that such a "feedback" mechanism is consistent with field observations (S.F. Ackley, private communication) near the ice edge suggesting that the air temperature undergoes an abrupt change as the ice edge is crossed. Also, measurements of atmospheric boundary layer modification (Andreas et al., 1979) support the assertion of a feedback between atmospheric temperature and ice extent.

For the calculation of geostrophic ocean current fields, mean dynamic topography values reported by Gordon et al. (1978) were used. In the radiation calculations parameterizations similar to those employed by Parkinson and Washington (1979) were used. Specifically, daily global solar radiation under cloudless skies was obtained by integrating an empirical equation by Zillmann (1972) over solar zenith angles for any particular day. (Zenith angles at half-hour intervals for this purpose were obtained from a numerical solution of Kepler's equation (Holloway, personal communication).) Incoming longwave radiation was obtained using Idso and Jackson's (1969) formula for clear skies. For cloud cover, a constant value of 0.85 was assumed at all points. This value is commensurate with maximum estimates by Van Loon (1972).

Simulation Results. Using the modified input fields and feedback parameterization discussed above, two-year simulations were carried out using a) the full coupled model and, for comparison,

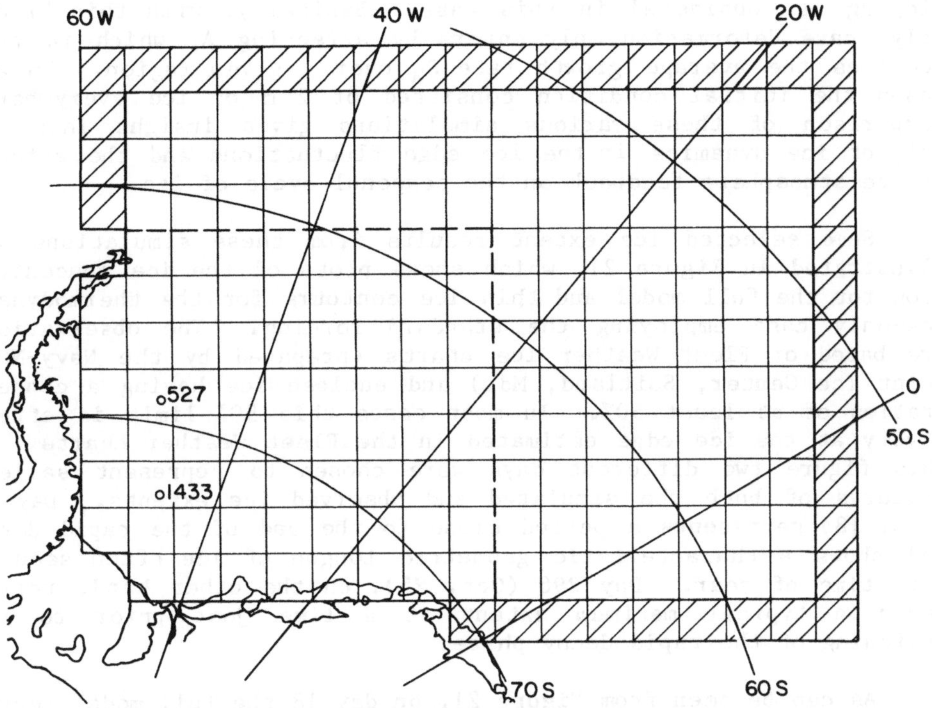

Fig. 20. Grid used for numerical simulations of the Weddell Sea pack ice. The dashed line encloses the area where the temperature and humidity fields were reanalyzed. The hatched grid cells represent open boundaries which allow inflow or outflow of ice and are taken to have zero strength for ice dynamics calculations (see Hibler (1979) for a more complete discussion of these open boundaries). The open circles denote the 1 Jan 1979 positions of two drifting buoys.

b) only the thermodynamic portion of the model, and c) a version including leads but no advection. This latter case was obtained by using the continuity equations

$$\frac{\partial A}{\partial t} = (\nabla \cdot \underline{u})A + S_a \tag{9.67}$$

$$\frac{\partial h}{\partial t} = S_h + f(h/A)A + (1-A)f(0) \tag{9.68}$$

To generate the leads, the ice velocity field from the second year of the standard fully coupled simulation was used. Note that no ridging is considered in this case. Basically, with this "leads only" case deformation only enters by affecting A, which in turn modifies the average growth rate S_h over a given region. In all cases the initial condition consisted of 2 m of ice everywhere. Comparison of these various simulations gives insight into the role of ice dynamics in the ice edge fluctuations and the effects of ice-atmosphere feedback on the seasonal cycle of ice.

Some selected ice extent results from these simulations are illustrated in Figure 21, which shows plots of the ice concentration for the full model and thin ice contours for the thermodynamics-only case employing the standard forcing. The observations are based on Fleet Weather ice charts (prepared by the Navy-NOAA Joint Ice Center, Suitland, Md.) and enclose ice having a concentration of at least 50%. In most cases this 50% limit is effectively at the ice edge estimated in the Fleet Weather charts. In this figure two different days were chosen to represent salient features of both the simulated and observed ice extents. Day 18 (Jan. 18) represents a period close to the end of the rapid decay and shows a characteristic geometric tongue of ice often seen at this time of year. Day 298 (Oct. 25), on the other hand, represents a typical maximum extent at a time just prior to the beginning of the rapid decay phase.

As can be seen from Figure 21, on day 18 the full model reproduces the tongue effect and ice extent much better than the thermodynamics-only case. Analysis of the ice velocity field (see the next section) shows that this tongue is largely due to northward and eastward advection of ice by the ice velocity field. (The tongue is, however, a bit too far south, partially due to insufficient northward ice transport as discussed in the next section.) In the thermodynamics-only case the structure of the ice edge is totally incorrect and there is too much ice.

On day 298 the results are, however, quite similar, with both the full model and the thermodynamic case showing extents slightly less than observed. This in turn indicates that the expansion to maximum is largely thermodynamic in nature and controlled (in both simulations) by the temperature field.

The effect of dynamics on the seasonal variation is illustrated more graphically in Figure 22, which shows the observed time series of the total grid area covered by ice with concentrations greater than 15%. This figure includes simulations using both the "feedback" (part b) forcing and the "standard" forcing (part a). Also shown in this figure are the time series of the thermodynamics-only case and of the "leads-only" simulation. In the thermodynamics-only case the area given is for ice of non-zero thickness. As can be seen, the coupled model, with the dynamics included, gives a bigger seasonal swing of extent and a more rapid decay of

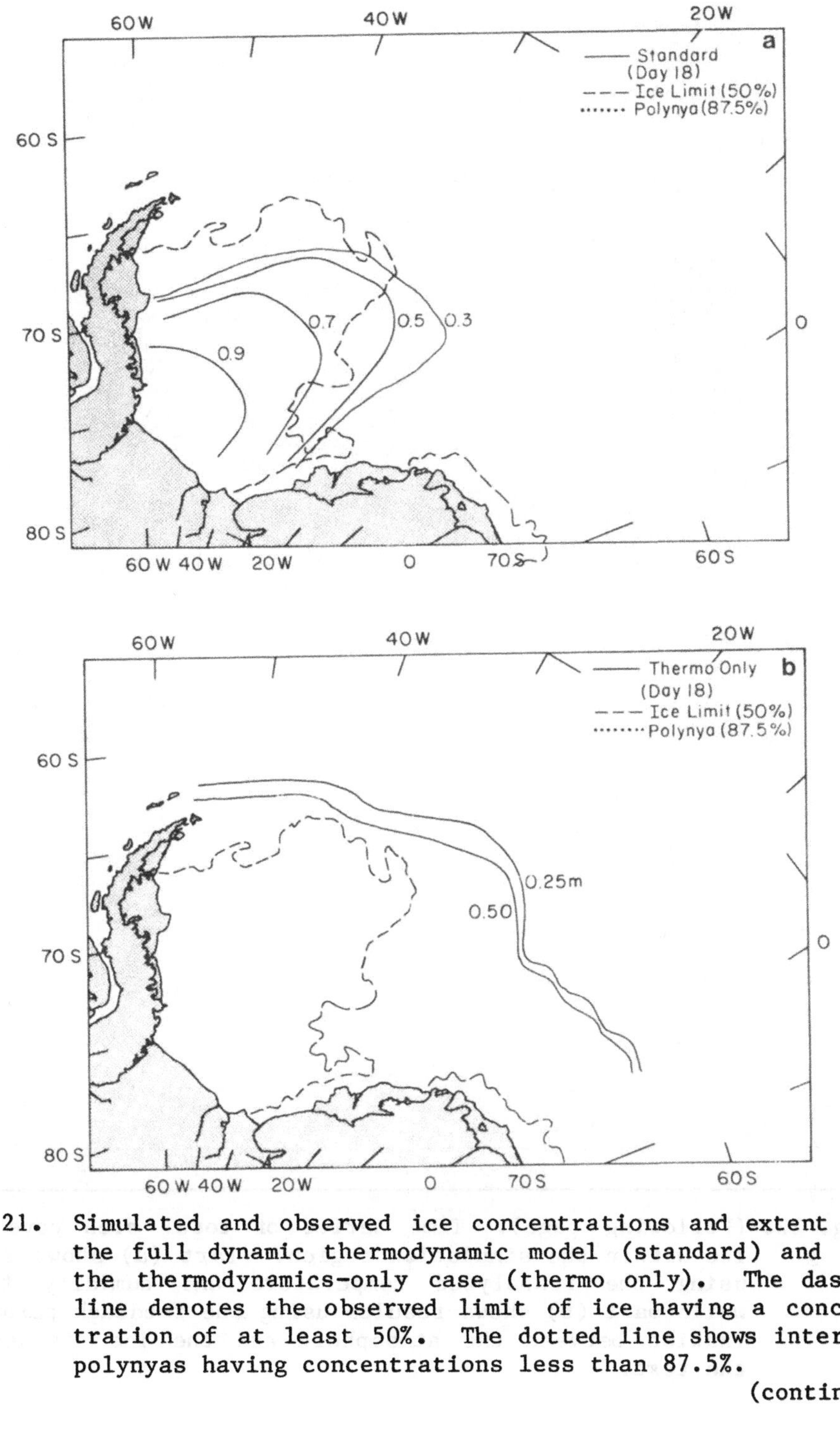

Fig. 21. Simulated and observed ice concentrations and extent for the full dynamic thermodynamic model (standard) and for the thermodynamics-only case (thermo only). The dashed line denotes the observed limit of ice having a concentration of at least 50%. The dotted line shows interior polynyas having concentrations less than 87.5%.

(continued)

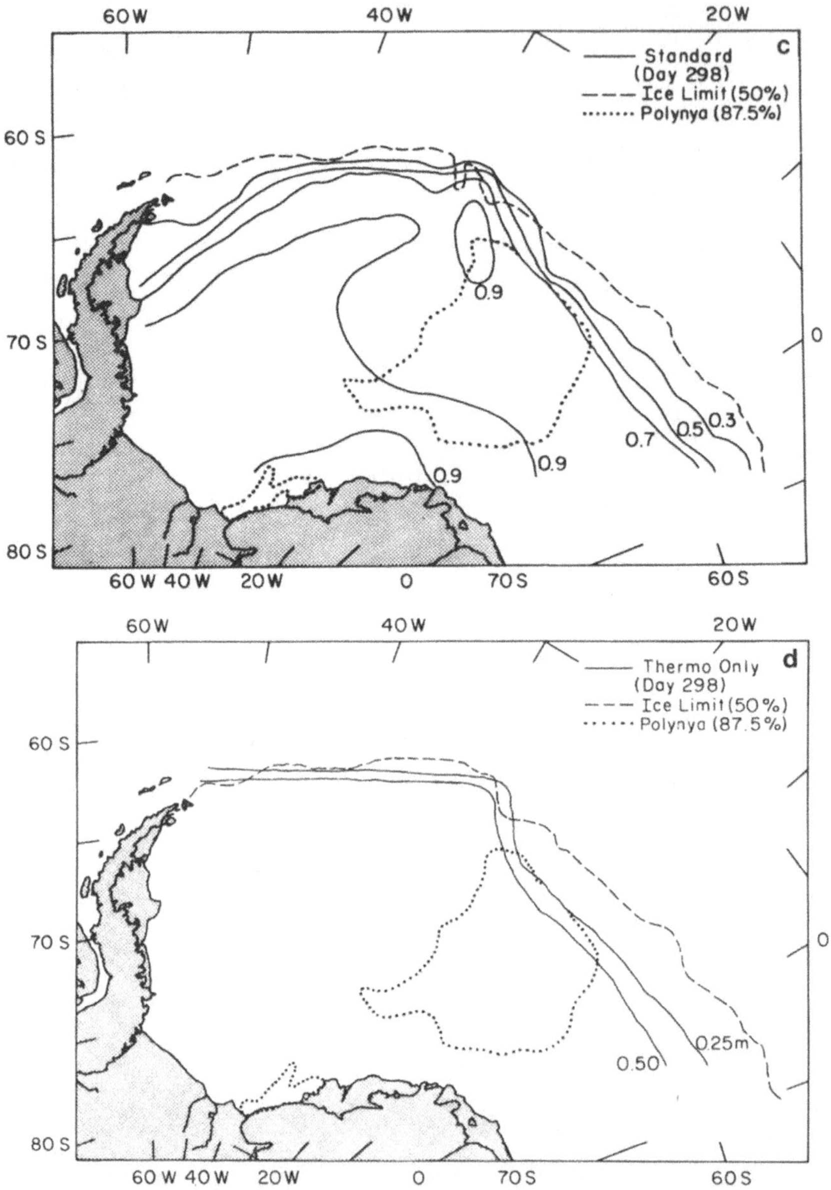

Fig. 22. (Following page). Time series of total area covered by ice within the simulation region. Part (a) shows results using the reanalyzed temperature and humidity fields, while part (b) shows results using the feedback parameterization between the atmosphere and the ice discussed in the text.

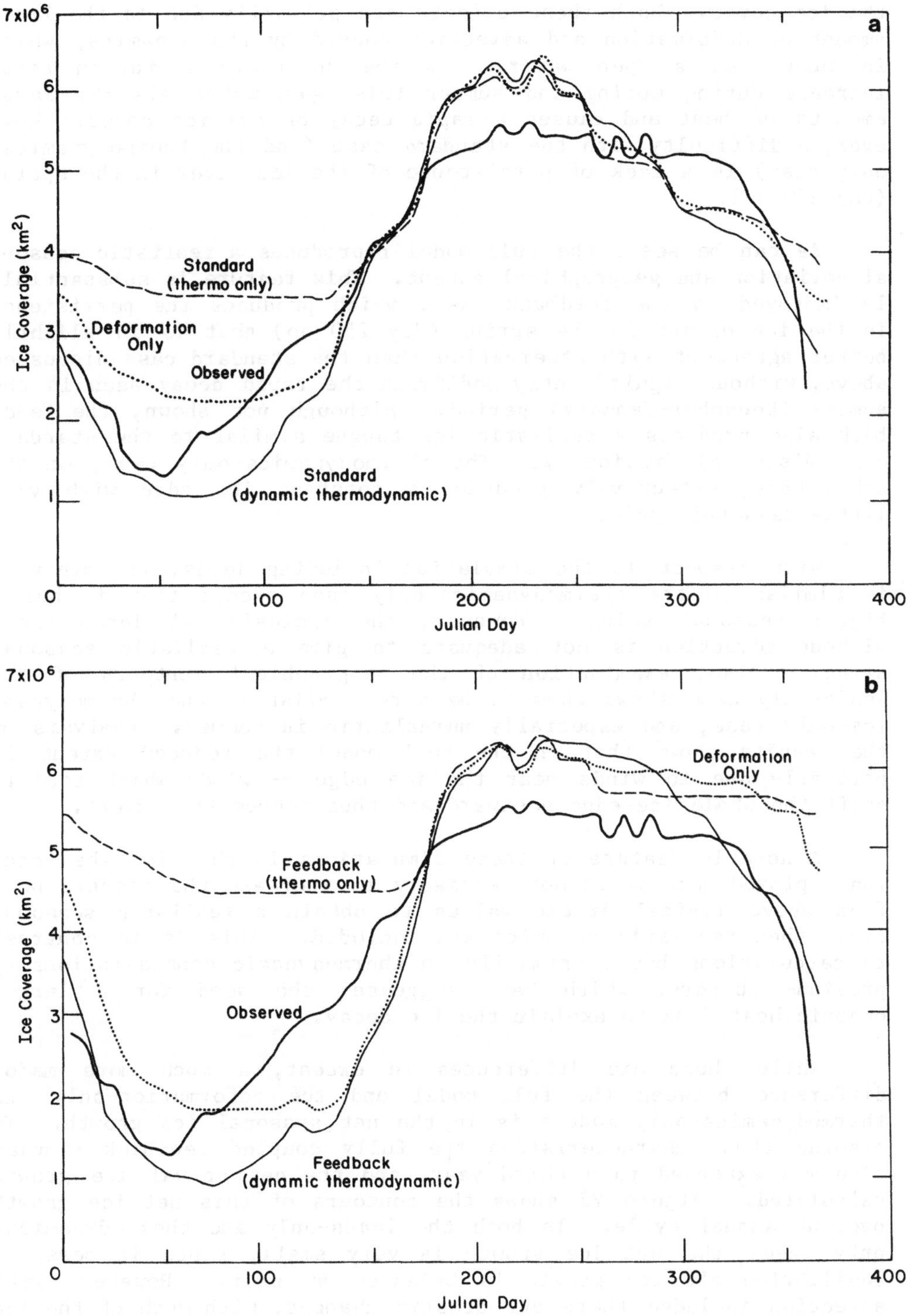
a
7x10^6
6
5
4
3
2
1
0
100
200
300
400
Julian Day
Ice Coverage (km^2)
Standard
(thermo only)
Deformation
Only
Observed
Standard
(dynamic thermodynamic)
b
7x10^6
6
5
4
3
2
1
0
100
200
300
400
Julian Day
Ice Coverage (km^2)
Deformation
Only
Feedback
(thermo only)
Observed
Feedback
(dynamic thermodynamic)

the ice cover. Both these effects are primarily due to the large amount of deformation and advection caused by the dynamics, which in turn creates open water. As the shortwave radiation terms increase during spring and summer this open water absorbs large amounts of heat and causes a rapid decay of the ice cover. However, a difficulty with the standard case (and the thermodynamics-only case) is a lack of persistence of the ice cover in the spring (day 250 on).

As can be seen, the full model reproduces a realistic seasonal variation and geographical extent. This feature is substantially improved in the feedback case, which produces the persistence in the ice extent in the spring (day 250 on) that is in slightly better agreement with observation than the standard case discussed above, without significantly modifying the rapid decay seen in the summer (December-January) period. Although not shown, the feedback also produces a realistic ice tongue similar to the standard case discussed previously. The thermodynamics-only case, on the other hand, effectively produces a constant ice edge with very little seasonal cycle.

With respect to the simulation including leads, its behavior is similar to the thermodynamics-only case except that it has a bigger seasonal swing. However, the inclusion of leads alone without advection is not adequate to give a realistic seasonal swing. Also, examination of the geographical contours of the leads-only case shows them to be very similar to the thermodynamics-only case, and especially unrealistic in summer. Analysis of the results shows that in the full model the reduced extent is partially due to winds near the ice edge -- winds which tend to shift the whole ice edge eastward and thus reduce the extent.

A notable feature of these simulations is that for the forcing employed here it is not necessary to increase the oceanic heat flux above typical Arctic values to obtain a realistic seasonal cycle when realistic dynamics are included. This is in contrast to calculations based primarily on thermodynamic considerations by previous authors, which have suggested the need for a larger oceanic heat flux to explain the ice decay.

While there are differences in extent, a much more major difference between the full model and the deformation-only and thermodynamics-only models is in the net seasonal ice growth. To examine these characteristics the fully coupled feedback simulation was extended to a third year, and the net annual ice growth calculated. Figure 23 shows the contours of this net ice growth over an annual cycle. In both the leads-only and thermodynamics-only cases the net ice growth is very small, since in seasonal equilibrium all ice growth is balanced by melt. However, with advection included there are drastic changes, with much of the ice

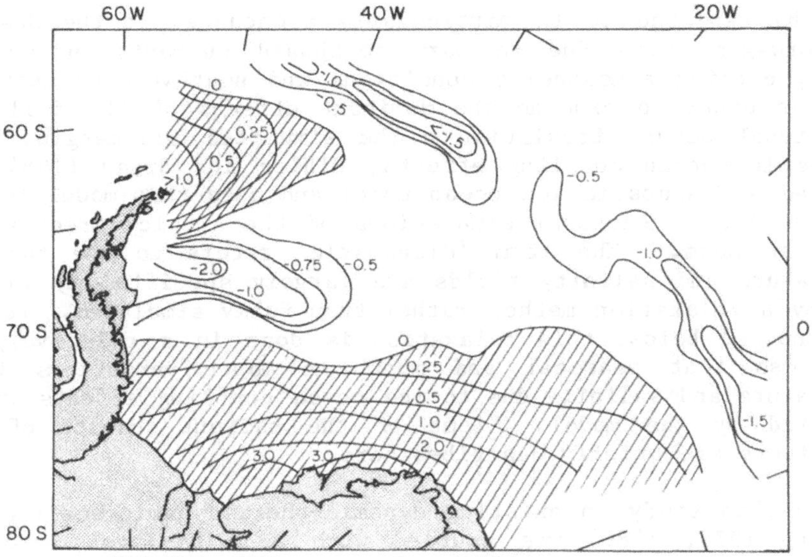

Fig. 23. Contours (in meters) of the net annual ice growth over an annual cycle for the fully coupled dynamic thermodynamic model. The contours were compiled from the third year of the feedback simulation.

growth taking place in southern parts of the grid and ice melt occurring farther north near the Equator. Off the Filchner Ice Shelf the ice growth exceeds 3 m. High growth in this region is known to be a major factor in bottom water formation, and the numbers obtained from the simulation are commensurate with maximum estimates in the literature (see, e.g., Ackley, 1980) based on oceanographic considerations. Also, the fact that ice melting takes place at different locations than growth means that large amounts of fresh water can effectively be transported to different regions of the Weddell Sea, a factor of some importance in the pycnocline structure there. Note also that much of the melt is highly concentrated at the ice edge, illustrating a conveyor-belt-like effect whereby ice is transported to the edge to be melted.

9.5.3 Ice Edge Characteristics of an Ice Covered Arctic Ocean Model

In the simulations of Antarctic sea ice discussed above, realistic ice edge results were obtained with the ocean parameterized by only a motionless fixed depth mixed layer. In the Greenland/Norwegian Seas, however, such an approximation (see, e.g., Hibler and Walsh, 1982) yields unrealistically large ice extents. This is not surprising in that large scale oceanic circulation effects would be expected to be particularly pronounced in the Greenland and Norwegian Seas due to warm northward currents encountering rapidly cooling atmospheric conditions and southward advancing sea ice. In order to examine the dominant effects of the full three dimensional ocean circulation o the simulated ice margin, and to examine ice-ocean coupling effects, Hibler and Bryan (1984) constructed a diagnostic ice ocean model and used this model to carry out a series of seasonal simulations of the Arctic, Greenland and Norwegian seas. The term 'diagnostic' refers to the fact that temperature and salinity fields are largely specified by observed data by a relaxation method rather than fully simulated. However, as discussed below, this relaxation is done in a relatively weak manner so that seasonal and month to month variations in the temperature and salinity due to ice ocean coupling effects will be simulated by the model. Some of the salient results of these simulations are briefly described here.

In this study an existing dynamic-thermodynamic sea ice model (Hibler, 1979, 1980) was coupled with a multi-level baroclinic ocean model (Bryan, 1969). The sea-ice model (essentially equivalent to the Weddell Sea pack ice model described above) supplied heat flux, salt flux, and momentum exchange boundary conditions for the top of the ocean. The ocean model, in turn, supplied current and heat exchange information to the ice model. Since the concern here was to examine the effect of ocean circulation on sea ice, observed oceanic temperature and salinity data (Levitus, 1982) were used to weakly force the ocean model so that its equilibrium time scale is similar to that of the ice model (3 to 5 years). This forcing is done by adding additional source terms to the heat and salt conservation equations of the form R $[T_o - T]$ and R $[S_o - S]$. Here R is a relaxation constant taken to be 0.333 yr^{-1}, and T_o and S_o are the mean annual observed temperature and salinity fields. This "diagnostic" method allows the ocean model to be forced to available climatological ocean data, while at the same time allowing considerable adjustment in the upper ocean due to the effects of ice/ocean interaction. In addition, the barotropic mode of the ocean is fully simulated so that temporally varying currents due to surface stress fluctuations are part of the model predictions. The basic idea here is to specify rather than model the ocean circulation over climatic time scales while modeling the shorter term fluctuations.

The most noticeable effect of the modeled ocean circulation is to greatly improve the ice margin simulation, as shown in Figure 24a. This improved ice margin is due to the large oceanic heat flux from the deeper ocean into the upper mixed layer in this region. The heat flux occurs primarily in winter and analysis shows that much of the enhancement is due to deep convection which brings up warm water and prevents ice formation in early winter.

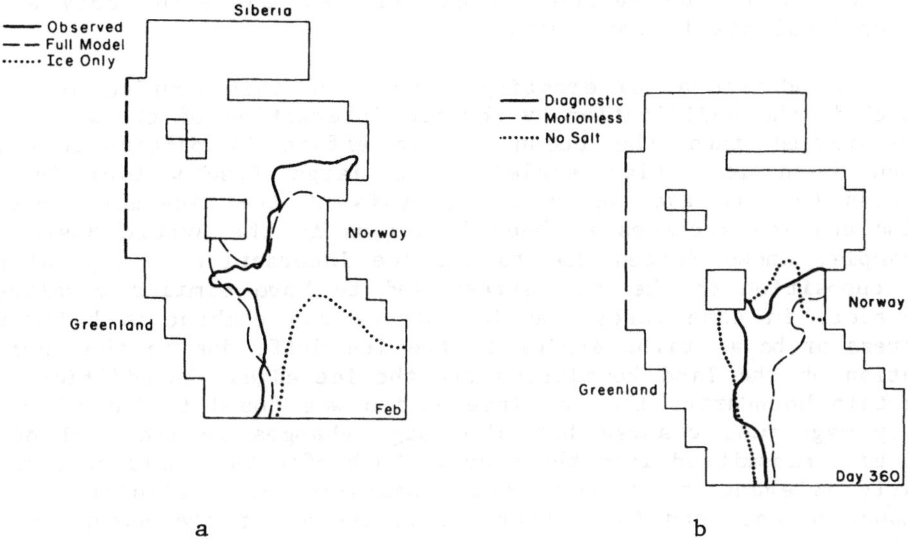

Fig. 24. Comparisons of simulations of ice concentration and ice thickness (from Hibler and Bryan, 1984). (a) February 50% concentration limits from the full coupled ice ocean model and for the ice-only model, which includes a fixed-depth mixed layer. The observed limit is taken from Fleet Weather (Suitland, Md., U.S. Navy) charts for the end of February. (b) Simulated 0.05 m thickness contours for day 360 for the full model, a "motionless" ocean model, and a coupled ice-ocean model with no salt fluxes due to ice freezing and melting. The "motionless" model and the no-salt simulations were run for only 1 year and were initialized to ice thickness and ocean fields at the end of the fourth year of the full diagnostic model simulation.

This convection explains the absence of ice in the full model simulation far from the ice margin.

Near the ice margin a similar physics applies, but the precise location of the ice edge becomes more sensitive to the surface salt balance and to lateral effects in the oceanic circulation. The sensitivity simulations shown in Figure 24b, for example, indicate that the freshwater flux from melting at the advancing ice edge tends to seal off the ice margin to upward oceanic heat flux and allows a farther advance than would otherwise occur. A sensitivity simulation without lateral motion in the ocean (Figure 24b) also tends to produce a more excessive edge at the end of one year, since the lateral transport of heat to both the deep and upper layers is missing. This latter result demonstrates the difficulty of simulating the seasonal cycle of sea ice using only a one-dimensional mixed layer model.

An additional interesting feature of this coupled ice-ocean model is the modification by the ice interaction of the wind stress transferred into the ocean. This effect is particularly pronounced on long time scales since large fluctuations in wind stress tend to average out. Analysis of the nine-month average wind and ice stresses at buoy locations in the Arctic Basin, for example, shows forces due to the ice interaction to typically be in opposition to the wind stress and to have similar magnitudes. However, in some cases the ice stress can combine with the wind stress or be at right angles to the ice drift due to the configuration of the land boundaries and the ice edge. In addition, near certain boundaries the ice interaction was found to introduce not only magnitude changes but also sign changes in the curl of the stress transmitted into the ocean. Such effects should be particularly relevant to longer term prognostic ocean simulations, and emphasize the need for better understanding of the nature of ice interaction.

9.6 CONCLUDING REMARKS

Sea ice dynamics deals with the way momentum and energy are distributed at the air-ice-ocean interface. As such, it plays a pivotal role in describing air-sea interaction in polar regions. Because sources of energy and momentum are often external to the ice, understanding sea ice dynamics entails studying the coupled air-sea-ice system. This review has shown that sea ice dynamics can cause substantial modification to the air-sea momentum, heat, and salt exchanges. Examination of the consequences of these modifications for the atmosphere and ocean is just beginning. From the work reviewed here, however, it is evident that to understand these modifications the nonlinear interaction of sea ice must be considered. Further studies are needed to improve our understanding of the interaction, especially near the ice margin.

ACKNOWLEDGMENTS

The author would like to thank A. Semtner, M. Lepparanta and P. Lemke for valuable comments on this manuscript. In addition I. Udin and A. Ullerstig provided computational aid and valuable discussions on the small scale idealized simulations. Aid in preparation of the manuscript was provided by S. Bowen and D. Harp. This work was supported by the National Aeronautics and Space Administration, and by the Office of Naval Research. In addition computer support for small scale simulations was provided by the Swedish Meteorological and Hydrological Institute.

REFERENCES

Ackley, S. F. (1980) Mass balance aspects of the Weddell Sea pack ice. J. Glaciol., 24: 391-405.

Ackley, S. F., A. J. Gow, K. R. Buck and K. M. Golden (1980) Sea ice studies in the Weddell Sea aboard USCGC Polar Sea. Ant. J. U.S., 15(5): 84-86.

Andreas, E. L., C. A. Paulson, R. M. Williams, R. W. Lindsay and J. A. Businger (1979) The turbulent heat flux from arctic leads. Bound.-Layer Met., 17: 57-91.

Brown, R. A. (1980) Planetary boundary layer modeling for AIDJEX. In Sea Ice Processes and Models (R. S. Pritchard, Ed.), University of Washington Press, Seattle, Washington, p. 387-401.

Bryan, K. (1969) A numerical method for the study of the circulation of the world oceans. J. Comput. Physics, 4: 347-376.

Campbell, W. J. (1965) The wind-driven circulation of ice and water in a polar ocean. J. Geophys. Res., 70: 3279-3301.

Colony, R. and D. A. Rothrock (1980) A perspective of the time-dependent response of the AIDJEX model. In Sea Ice Processes and Models (R. S. Pritchard, Ed.), University of Washington Press, Seattle, Washington, p. 124-133.

Coon, M. D. (1974) Mechanical behavior of compacted arctic ice floes. J. Petrol. Tech., 26: 466-470.

Doronin, Yu. P. (1970) On a method of calculating the compactness and drift of ice floes. Trudy Arkt. Antarkt. Nauch. Issled. Inst., 291: 5-17. (English translation, AIDJEX Bulletin, 3: 22-39, 1971.)

Egorov, K. L. (1970) Theory of drift of ice floes in a horizontally heterogeneous wind field. Probl. Arkt. Antarkt., 34: 71-78. (English translation, AIDJEX Bulletin, 6: 37-45, 1971.)

Glen, J. W. (1970) Thoughts on a viscous model for sea ice. AIDJEX Bulletin, 2: 18-27.

Gordon, A., E. Molinelli and T. Baker (1978) Large scale relative dynamic topography of the southern ocean. J. Geophys. Res., 83: 3023-3032.

Gow, A. J., S. F. Ackley, W. F. Weeks and J. Govoni (1982) Physical and structural characteristics of Antarctic sea ice. Ann. Glaciol., 3: 113-117.

Hibler, W. D. III (1974) Differential sea ice drift. II: Comparison of mesoscale strain measurements to linear drift theory predictions. J. Glaciol., 13: 457-471.

Hibler, W. D. III (1977) A viscous sea ice law as a stochastic average of plasticity. J. Geophys. Res., 82: 3932-3938.

Hibler, W. D. III (1979) A dynamic thermodynamic sea ice model. J. Phys. Oceanogr., 9: 815-846.

Hibler, W. D. III (1980a) Modeling a variable thickness sea ice cover. Mon. Wea. Rev., 108: 1943-1973.

Hibler, W. D. III (1980b) Documentation for a two-level dynamic thermodynamic sea ice model. U.S.A. Cold Regions Research and Engineering Laboratory Special Report 80-8, 35 p.

Hibler, W. D. III, S. F. Ackley, W. K. Crowder, H. W. McKim, and D. M. Anderson (1974a) Analysis of shear zone ice deformation in the Beaufort Sea using satellite imagery. In The Coast and Shelf of the Beaufort Sea (J. C. Reed and J. E. Sater, Ed.), Arctic Institute of North America, p. 285-296.

Hibler, W. D. III, S. J. Mock and W. B. Tucker (1974b) Classification and variation of sea ice ridging in the western Arctic Basin. J. Geophys. Res., 79: 2735-2743.

Hibler, W. D. III and W. B. Tucker (1979) Some results from a linear viscous model of the Arctic ice cover. J. Glaciol., 22: 293-304.

Hibler, W. D. III, I. Udin and A. Ullerstig (1981) On modeling mesoscale ice dynamics using a viscous plastic constitutive law. In POAC 81: Proceedings, Sixth International Conference on Port and Ocean Engineering Under Arctic Conditions, Laval University, Quebec City, Quebec, Canada, p. 1317-1328.

Hibler, W. D. III and S. F. Ackley (1982) On modeling the Weddell Sea pack ice. Ann. Glaciol., 3: 125-130.

Hibler, W. D. III and J. E. Walsh (1982) On modeling seasonal and interannual fluctuations of Arctic sea ice. J. Phys. Oceanogr., 12: 1514-1523.

Hibler, W. D. III and S. F. Ackley (1983) Numerical simulation of the Weddell Sea pack ice. J. Geophys. Res., 33: 2873-2878.

Hibler, W. D. III, I. Udin and A. Ullerstig (1983) On forecasting mesoscale ice dynamics and buildup. Ann. Glaciol., 4: 110-115.

Hibler, W. D. III and K. Bryan (1984) Ocean circulation: Its effects on seasonal sea-ice simulations. Science, 224: 489-492.

Hunkins, K. (1975) The oceanic boundary layer and stress beneath a drifting ice floe. J. Geophys. Res., 80: 3425-3433.

Idso, S. and R. D. Jackson (1969) Thermal radiation from the atmosphere. J. Geophys. Res., 74: 5397-5403.

Kovacs, A. (1972) On pressured sea ice. In Sea Ice: Proceedings of an International Conference. National Research Council, Reykjavik, Iceland, p. 276-295.

Kovacs, A. and M. Mellor (1974) Sea ice morphology and ice as a geologic agent in the southern Beaufort Sea. In The Coast and Shelf of the Beaufort Sea (J. L. Reed and J. E. Sater, Ed.), Arctic Institute of North America, p. 113-161.

Laikhtman, D. L. (1964) Physics of the boundary layer of the atmosphere. (Translated by U.S. Dept. of Commerce, Washington, D.C., 200 pp.)

Leavitt, E. (1980) Surface-based air stress measurements made during AIDJEX. In Sea Ice Processes and Models (R. S. Pritchard, Ed.), University of Washington Press, p. 419-429.

Lepparanta, M. (1980) On the drift and deformation of sea ice fields in the Bothnian Bay. Winter Navigation Research Board, Helsinki, Finland, Research Report 29.

Levitus, S. (1982) Climatological atlas of the world ocean. NOAA Professional Paper 13, 173 p.

Malvern, L. E. (1969) Introduction to the Mechanics of a Continuous Media. Prentice Hall, Englewood Cliffs, New Jersey.

Manabe, J., K. Bryan and M. J. Spelman (1979) A global ocean-atmosphere climate model with seasonal variation for future studies of climate sensitivity. Dyn. Atmos. Ocean., 3: 393-426.

Maykut, G. A. (1978) Energy exchange over young sea ice in the central Arctic. J. Geophys. Res., 83: 3646-3658.

McPhee, M. G. (1979) The effect of the oceanic boundary layer on the mean drift of pack ice: Application of a simple model. J. Phys. Oceanogr., 9: 388-400.

McPhee, M. G. (1980) An analysis of pack ice drift in summer. In Sea Ice Processes and Models (R. S. Pritchard, Ed.), University of Washington Press, Seattle, Washington, p. 62-75.

McPhee, M. G. (1982) Sea ice drag laws and simple boundary layer concepts, including application to rapid melting, U.S.A. Cold Regions Research and Engineering Laboratory Report 82-4.

Nikiforov, Y. G. (1957) A change in the concentration of the ice cover in connection with its dynamics. Problemy Arktiki, 2: 59-71. (Translated by American Meteorological Society.)

Nye, J. F. (1973) Is there any physical basis for assuming linear viscous behavior for sea ice? AIDJEX Bulletin, 21: 18-19.

Omstedt, A. and J. Sahlberg (1977) Some results from a joint Swedish-Finnish sea ice experiment, March 1977. Winter Navigation Research Board, Norrkoping, Sweden, Research Report 26.

Parkinson, C. L. and W. M. Washington (1979) A large-scale numerical model of sea ice. J. Geophys. Res., 84: 311-337.

Parmerter, R. R. and M. D. Coon (1972) Model of pressure ridge formation in sea ice. J. Geophys. Res., 77: 6565-6575.

Pritchard, R. S. (1975) An elastic-plastic constitutive law for sea ice. J. Appl. Mech, 43E: 379-384.

Pritchard, R. S., M. D. Coon and M. G. McPhee (1977) Simulation of sea ice dynamics during AIDJEX. J. Press. Vess. Tech., 99J: 491-497.

Roed, L. P., and J. J. O'Brien (1981) Geostrophic adjustment in highly dispersive media: An application to the marginal ice zone. J. Geophys. Astrophys. Fluid Dyn., 18: 263-278.

Rothrock, D. A. (1975a) The mechanical behavior of pack ice. Ann. Rev. Earth Planet. Sci., 3: 317-342.

Rothrock, D. A. (1975b) The energetics of the plastic deformation of pack ice by ridging. J. Geophys. Res., 80: 4514-4519.

Semtner, A. J. Jr. (1976) A model for the thermodynamic growth of sea ice in numerical investigations of climate. J. Phys. Oceanogr., 6: 379-389.

Sodhi, D. S. and W. D. Hibler III (1980) Nonsteady ice drift in the Strait of Belle Isle. In Sea Ice Processes and Models (R. S. Pritchard, Ed.), University of Washington Press, Seattle, Washington, p. 177-186.

Thorndike, A. S., D. A. Rothrock, G. A. Maykut and R. Colony (1975) The thickness distribution of sea ice. J. Geophys. Res., 80: 4501-4513.

Thorndike, A. S. and R. Colony (1982) Sea ice motion in response to geostrophic winds. J. Geophys. Res., 87: 5845-5852.

Tucker, W. B. III and W. D. Hibler III (1981) Preliminary results of ice modeling in the East Greenland area. In POAC 81: Proceedings, Sixth International Conference on Port and Ocean Engineering Under Arctic Conditions, Laval University, Quebec City, Quebec, Canada, p. 867-878.

Tucker, W. B. III and J. W. Govoni (1981) Morphological investigations of first-year sea ice pressure ridge sails. Cold Reg. Sci. Tech., 5: 1-12.

Tucker, W. B. III, W. F. Weeks and M. Frank (1979) Sea ice ridging over the Alaskan continental shelf. J. Geophys. Res., 84: 4885-4897.

Van Loon, H. (1972) Cloudiness and precipitation in the Southern Hemisphere. In Meteorology of the Southern Hemisphere (C. W. Newton, Ed.), Meteorological Monographs 13, p. 101-111.

Wadhams, P. (1981) Sea ice topography of the Arctic Ocean in the region 70°W to 25°E. Phil. Trans. Roy. Soc. London, A302: 1464-1504.

Washington, W. M., A. J. Semtner, C. Parkinson and L. Morrison (1976) On the development of a seasonal change sea-ice model. J. Phys. Oceanogr., 6: 679-685.

Weeks, W. F., A. Kovacs and W. D. Hibler III (1971) Pressure ridge characteristics in the arctic coastal environment. In Proceedings, First International Conference on Port and Ocean Engineering Under Arctic Conditions, Technical University of Norway, Trondheim, Norway, Vol. 1, p. 152-193.

Zillmann, J. W. (1972) Solar radiation and air-sea interaction south of Australia. In Antarctic Oceanology. II: The Australian-New Zealand Sector (D. E. Haynes, Ed.), AGU Antarctic Research Series, Vol. 19, p. 11-40.

Zubov, N. N. (1943) L'dy Arktiki (Arctic Ice). Izdatel'stvo Glavsevmorputi, Moscow, 360 p. (U.S. Navy Hydrographic Office, Translation 217, 1963; available as AD426972 from NTIS, Springfield, Va.)

CHAPTER 10

CIRCULATION AND MIXING IN ICE-COVERED WATERS

Eddy C. Carmack

National Water Research Institute
West Vancouver Laboratory
4160 Marine Drive
West Vancouver, BC V7V 1N6

1. INTRODUCTION

This chapter discusses the connections between physical oceanography and ice cover: namely, how does water circulation influence ice state, and what effects do ice growth and decay have on water movement. For this purpose a review of the hydrography of the Arctic Ocean, Greenland Sea and Southern Ocean is first given, followed by a review of heat and mass transfer mechanisms which are either caused by or affected by the presence of ice.

There are a number of timely reasons for better understanding the dynamics of ice-covered waters. The interactions among the ocean, ice cover and climate are of obvious importance (SCOR, WG58, 1979; Baker, 1979a; Goody, 1980; Polar Group, 1980). Conditioning of water masses by salt fractionation due to freezing is largely responsible for circulation in the deep ocean (cf. Reid and Lynn, 1971; Warren, 1981). An understanding of circulation and mixing in the proximity of ice is also important from the point of view of waste disposal (Aagaard, 1981a), resource development (Campbell and Martin, 1973), release of thermal discharges (Ehlin, 1976), the utilization of icebergs as a freshwater source (Weeks and Campbell 1973), and massive river diversion (Micklin, 1981). Air-sea exchange and deep convection at high latitudes may also provide a sink for anthropogenic CO_2 (Chen, 1982). In lakes and rivers the behavior of ice can cause serious problems for navigation and hydroelectric facilities, and can increase flood risk (Michel, 1971; Ashton, 1978; Martin, 1981).

This review is not complete. In particular, non-English language material, and the literature of government and/or industrial reports are largely omitted. Emphasis is placed on refereed papers from major journals, and a few university dissertations which address important topics. Although not covered in this chapter, the physical oceanography of the Bering Sea is described by Coachman, Aagaard and Tripp (1975); Sayles, Aagaard and Coachman (1979); and Hood and Calder (1980-81); the hydrography of Baffin Bay is covered by Muench (1971); and the dynamics of the Antarctic Circumpolar Current are reviewed by Baker (1979b).

Recent overviews exist emphasizing physical processes relevant to our understanding of ice-covered waters: for example, internal waves are discussed by Leblond and Mysak (1978); topographically trapped waves are discussed by Mysak (1980); finestructure and microstructure are covered by Gregg and Briscoe (1979); small-scale mixing processes are described by Turner (1981); double-diffusive convection is dealt with by Huppert and Turner (1981); mixed-layer theory is reviewed by Garwood (1979); variability in the ocean is discussed in the book by Monin, Kamenkovich and Kort (1977); and mixing processes in lakes is the subject of a review by Imberger and Hamblin (1982). The seasonal sea ice zone is the subject of a number of articles edited by Anderson, Weeks and Newton (1980), and the present volume covers a number of subject areas relating directly to ice dynamics and polar meteorology.

2. GENERAL HYDROGRAPHY

2.1 Physical Features

2.1.1 Arctic Ocean. The Arctic Ocean (Figure 1) is about 10 x 10^6 km^2 in area, and about 13.5 x 10^6 km^3 in volume. It is surrounded by a continental shelf which is interrupted only by Fram Strait, the deep passage between Greenland and Spitsbergen. The Polar Basin is divided by the Lomonosov Ridge (sill depth about 1400 m) into two deep basins, the Eurasian Basin (maximum depth about 4200 m) and the Canadian Basin (maximum depth about 3800 m). The latter is further divided by the Alpha Rise which lies parallel to the Lomonosov Ridge.

The shelf from Alaska to Greenland is narrow (50 to 100 km), but from Spitsbergen eastward (i.e., the Barents, Kara, Laptev, East Siberian and Chukchi seas) the shelf is very broad, typically about 800 km. These five seas occupy approximately 36% of the area of Arctic Ocean. The major rivers which enters via the Bering Sea, flow into the Eurasian seas. A number of submarine canyons indent the continental shelf, the largest being the Svataya Anna Canyon in the Kara Sea, over 500 km long. These canyons likely influence exchange between the shelf and deep ocean.

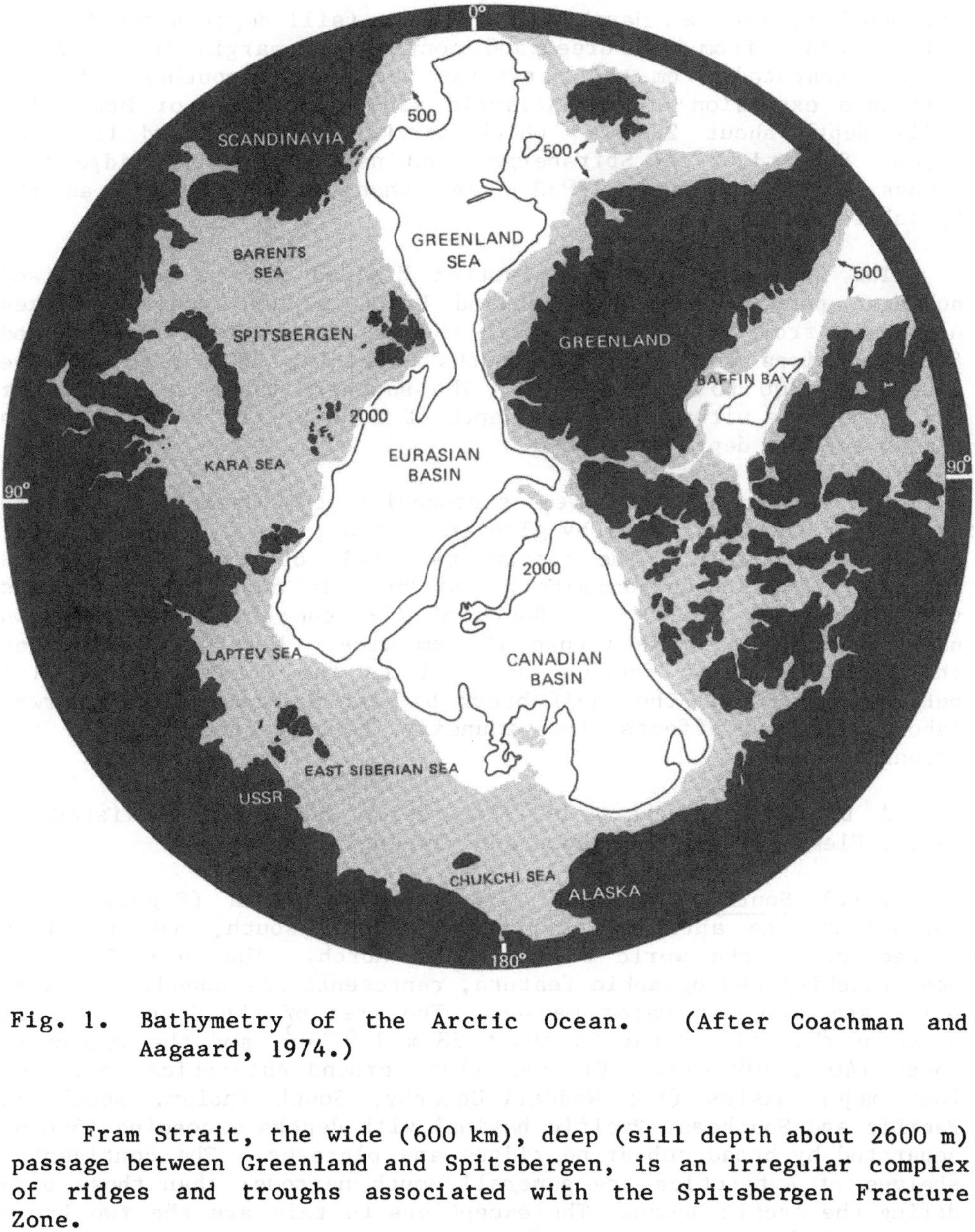

Fig. 1. Bathymetry of the Arctic Ocean. (After Coachman and Aagaard, 1974.)

Fram Strait, the wide (600 km), deep (sill depth about 2600 m) passage between Greenland and Spitsbergen, is an irregular complex of ridges and troughs associated with the Spitsbergen Fracture Zone.

The bathymetry of the Arctic Ocean is presented in the chart published by Pinther, Heezen and Tharp (1975).

2.1.2 Greenland Sea. The combined Greenland, Iceland, and Norwegian seas occupy the region between Greenland and Scandinavia. The Greenland Sea, itself, is separated from the Iceland Sea to the south by the Jan Mayen Fracture Zone (sill depth about 1400 m) which extends from the Greenland continental margin at 71°-72°N; it is separated from the Norwegian sea to the southeast by the northward extension of the Atlantic mid-ocean ridge or Mohn Rise (sill depth about 2500 m) which strikes northeastward from Jan Mayen. Seaward of the Spitsbergen Continental Rise the ridge continues north to about 77°33'N and then north-northwest as the Spitsbergen Fracture Zone.

The Greenland Basin is further divided into a southern and northern portion by the Greenland Fracture Zone, which strikes northwest from its juncture with the Mohn Rise at about 74°N and 8°E, and rises abruptly from depths exceeding 3800 m to a ridge generally 1800 to 2800 m deep. The southern basin is the larger and deeper one with an average depth of 3600 m; the northern basin has an average depth of 3100 m.

The continental shelves surrounding the Greenland Sea are typical of those bordered by glaciated areas; i.e., they are wide, deep and irregular. The continental shelf off eastern Greenland south to 77°N is approximately 300 km wide and lies at depths varying between 200-400 m. South of 77°N the shelf break narrows until at 75°N it is less than 100 km wide. West of Spitsbergen the shelf break is roughly parallel to the coastline and lies between 100-300 m. The shelf break between Spitsbergen and Norway (about 200 m) affects the boundary between the Barents and Greenland seas.

A bathymetric chart of the region has been published by Perry, Fleming, Cherkis, Feden and Massengill (1977).

2.1.3 Southern Ocean. The Southern Ocean (Figure 2) is bounded by the antarctic continent to the south, but has free connection to the world ocean to the north. The Polar Front, a deep-reaching hydrographic feature, represents the boundary between polar and subpolar water masses. The area of the Southern Ocean south of the Polar Front is about 38 x 10^6 km^2, and the volume is about 140 x 10^6 km^3. The sea floor around Antarctica comprises four major basins (the Weddell-Enderby, South Indian, Southwest Pacific and Southeast Pacific basins) with depths exceeding 4500 m, separated by broad submarine ridges and plateaus. The continental shelves of Antarctica are generally much narrower than those bordering the Arctic Ocean. The exceptions to this are the two broad (ca. 400 km) and deep (ca. 400 m) shelves of the Weddell and Ross seas. These two shelves are important in that they are believed to be the source of Antarctic Bottom Water. The shelf regions are further characterized by numerous depressions and submarine canyons, and by floating ice shelves which extend seaward from the continent.

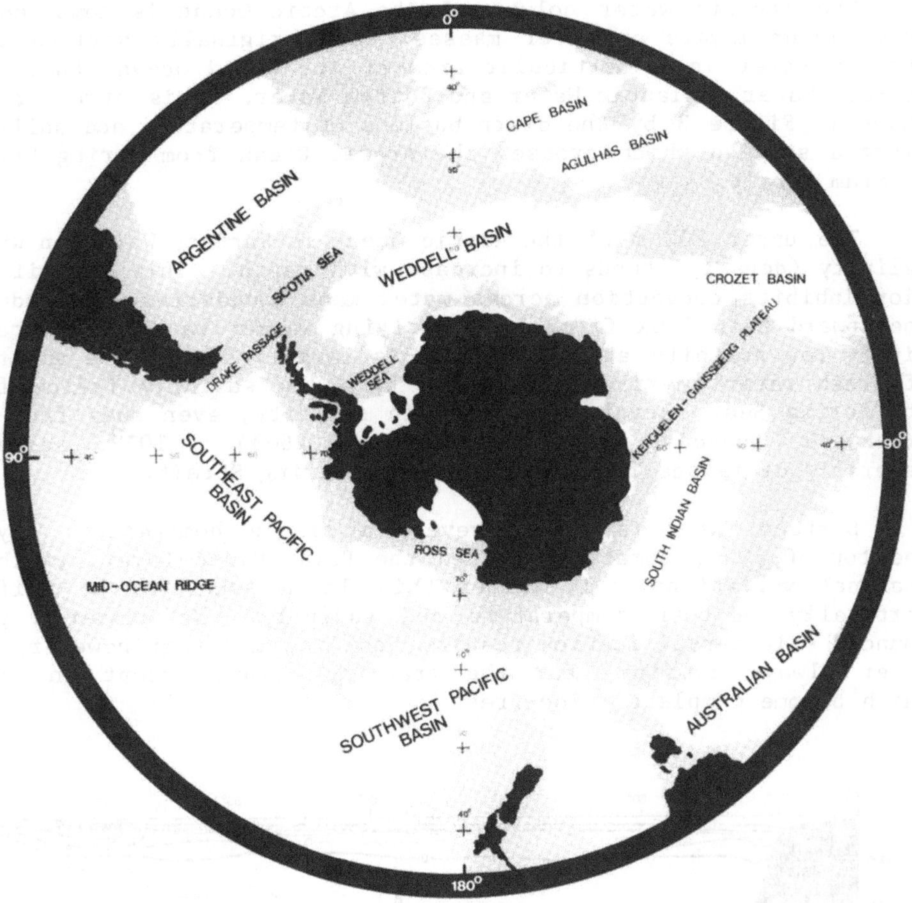

Fig. 2. Bathymetry of the Southern Ocean. (After Heezen, Tharp, and Bently, 1972.)

Bathymetric charts of the Southern Ocean have been published by Heezen, Tharp and Bently (1972) and Johnson, Vanney, Drewry and de Q. Robin (1980).

2.2 Water Masses

2.2.1 Arctic Ocean. The general hydrography of the Arctic Ocean is treated by Zubov (1945), Coachman and Aagaard (1974), Treshnikov (1977), and SCOR WG58 (1979). Mass and heat budgets are discussed by Fletcher (1965), Aagaard and Griesman (1975), and SCOR WG58 (1979). A recent overview of arctic oceanography has been given by Aagaard (1979).

The oceanic water column of the Arctic Ocean is composed of three major layers or water masses, each originally derived from surface water in a particular area of the world ocean; these are Surface Water, Atlantic Water and Bottom Water. This structure is shown in Figure 3 by the distributions of temperature and salinity along a section which crosses the Arctic Ocean from Bering Strait to Fram Strait.

The upper 200 m of the Arctic Ocean is Surface Water in which salinity (density) tends to increase with depth. This stratification inhibits convection across water mass boundaries, and reduces the upward heat flux from the underlying warmer water. The relatively low salinity at the surface is largely due to the addition of fresh water by river inflow and by a low salinity inflow from the Bering Sea. Areal variations in salinity, even away from the freshwater sources, are large, between 26-34 x 10^{-3}; salinity generally decreases from Fram Strait to Bering Strait.

Surface Water is not, however, a single homogeneous layer. The top 30-50 m, sometimes called the Polar Mixed Layer, exhibits seasonal variations. In winter this layer tends to be uniform vertically in both temperature and salinity. In summer a pronounced salt stratification results from ice melting; however, the water always remains near the freezing point, except in areas which become completely ice-free.

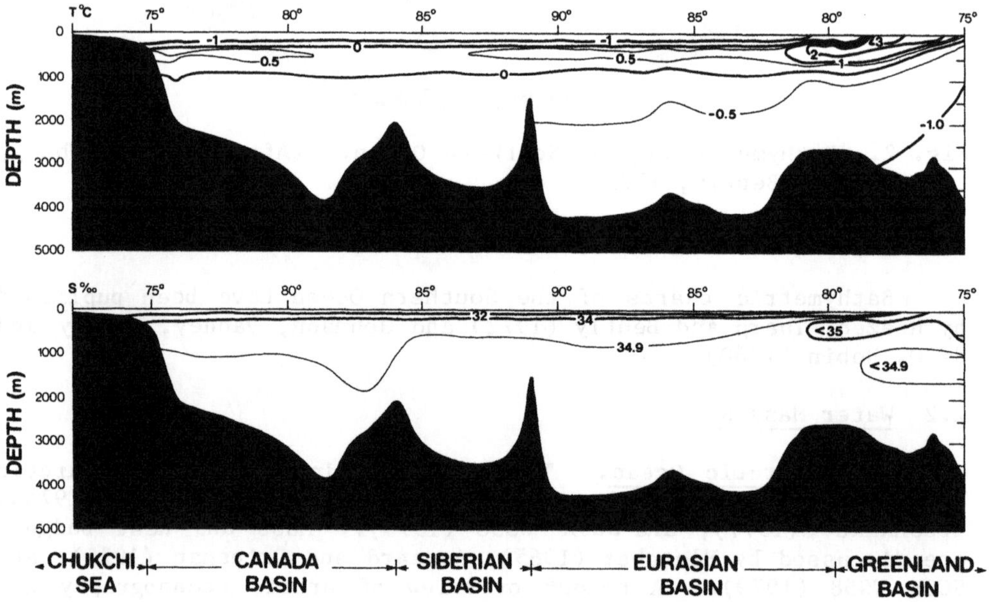

Fig. 3. Longitudinal section of temperature and salinity across the Arctic Ocean. (From R. B. Tripp, unpublished.)

The lower portion of the surface layer comprises the main halocline, one of the most important hydrographic features of the Arctic Ocean. In the Eurasian Basin the salinity increased rapidly with depth, reaching 34.9 to 35.0 x 10^{-3} at 200 m, while temperature remains between -1.9 and -1.5°C to 100 m and then also begins to increase, reaching 0°C at about 200 m. A temperature minimum frequently occurs at 75-100 m (cf. Nansen, 1902). This cold, saline water is probably formed on the continental shelves during winter (Aagaard, Coachman and Carmack, 1981). In the Canadian Basin the salinity increases less rapidly with depth. There is a relative maximum in temperature, about 0°C, between 50-100 m, followed by a minimum about -1.5°C, at 150 m. The relative maximum is maintained by warm water coming from the Bering Sea in summer, while the minimum comes from the Bering Sea in winter, and possibly from other adjacent shelf regions as well.

Below the Surface Water lies Atlantic Water. This water flows northward in the Norwegian and West Spitsbergen Currents and enters the Arctic Ocean via Fram Strait. Once into the Polar Basin the initially high temperature and salinity (over 3°C and 35 x 10^{-3} in Fram Strait) of Atlantic Water are quickly reduced as it sinks and spreads eastward along the Eurasian continental slope. This water is identifiable over the entire basin by a temperature maximum ($T > 0.5$°C) and high, nearly uniform salinities (about 34.95 x 10^{-3}). There is virtually no density stratification beneath the main halocline.

Beneath the Atlantic Water lies cold ($T < 0$°C) Bottom Water. Its origin is uncertain: the classical notion is that it is probably formed in the Greenland Sea; however, some local formation now appears to be of importance (cf. Aagaard, 1981b; Swift, 1982). Bottom Water is nearly isothermal within each of the two major basins, but below 1400-1500 m a temperature difference of 0.5°C exists between the two basins, resulting from the sill depth of the Lomonosov Ridge.

A volumetric (T/S) analysis of water masses in the Arctic Ocean has not yet been published. However, a rough estimate is that surface water comprises about 15% (ca. 2 x 10^6 km^3), Atlantic water about 25% (ca. 3.5 x 10^6 km^3), and bottom water about 60% (ca. 8 x 10^6 km^3) of the total volume of water in the Arctic Ocean.

2.2.2 Greenland Sea. The Greenland Sea is often viewed as a transition region for surface inflows from the Atlantic and Arctic Oceans; it is also a region in which local processes result in the assembly of unique water masses (Figure 4). The classic studies of the Greenland Sea are those of Helland-Hansen and Nansen (1909) and Kiilerich (1945); more contemporary descriptions of water mass distributions are given by Carmack and Aagaard (1973), and Swift

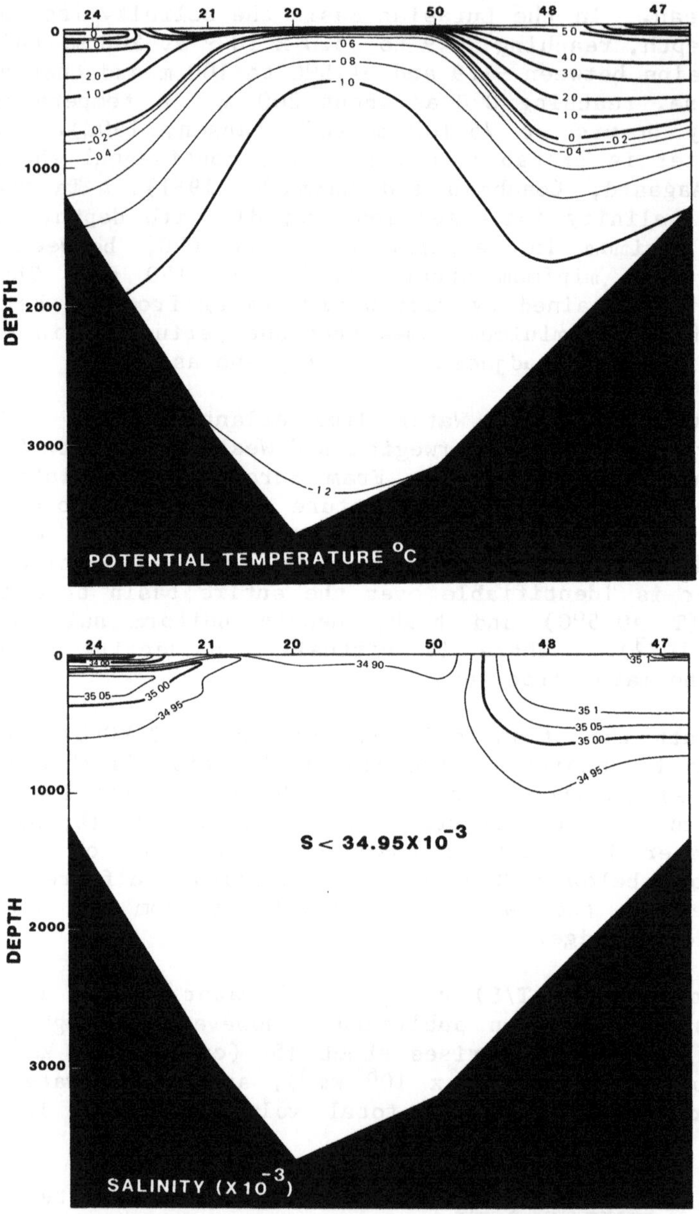

Fig. 4. Section of potential temperature and salinity extending north-northwest across the Greenland Sea. (Data from U.S.S. Atka, 1958.)

and Aagaard (1981); the water mass nomenclature used below follows from the latter work.

Atlantic Water, the warm, saline water of Atlantic origin, is carried into the Greenland Sea as a branch of the Norwegian Atlantic Current. It can be defined as any water in the Greenland Sea with salinities greater than 34.9 x 10^{-3} and temperatures above 3°C.

Polar Water carried southward from the Arctic Ocean by the East Greenland Current can be recognized by salinities below 34.4 x 10^{-3} and temperatures below 0°C. A transitional form of this water, called Polar Intermediate Water, is recognized by salinities in the range 34.4 to 34.7 x 10^{-3}.

Separating Atlantic and Polar water masses is a large, hydrographically unique region occupying the central portion of the Greenland Gyre; Swift and Aagaard (1981) refer to this region as the Arctic domain. During the summer, the upper part of the Arctic domain is comprised of Arctic Surface Water, which overlies a temperature minimum at 50-150 m and a temperature maximum at 200-400 m. Upper Arctic Intermediate Water is defined to include much of the water column between the temperature maximum and temperature minimum; i.e., temperatures less than 2°C and salinities in the range 34.7 to 34.9 x 10^{-3}. During winter upper Arctic Intermediate Water extends to the surface, an indication of seasonal formation.

Below upper Arctic Intermediate water is lower Arctic Intermediate Water. This layer contains the temperature and salinity maxima. Temperatures in this water mass range from 0 to 3°C and the salinity is above 34.9 x 10^{-3}. This water is believed to be derived from the cooling and sinking of Atlantic Water, especially in the northern portion of the Greenland Sea.

An important characteristic of the Arctic domain is that the static stability of the water column is markedly less than that of the Atlantic and Polar domains. Because of this it has long been supposed that winter cooling can produce a dense, well-mixed layer that may reach to the ocean floor. Surface water densities are highest in the central gyre region; however, winter surface water sufficiently dense to displace the existing bottom water has not been observed (Carmack and Aagaard, 1973; Killworth 1979a).

Greenland Sea Deep Water is the densest water mass in the Greenland Sea. Its temperature is below 0°C and salinities are typically 34.88 to 34.90 x 10^{-3}. This water is found only in the central part of the Greenland Sea where, presumably, it is formed by deep convection in winter. This water mass should be distinguished from the slightly warmer and more saline (34.90 to 34.94 x

10^{-3}) Norwegian Sea Deep Water, which comprises the densest water mass of the Norwegian and Iceland seas. Norwegian Sea Deep Water is also observed around the perimeter of the Greenland Sea (cf. Metcalf, 1960). While the origin of Norwegian Sea Deep Water is unclear, some association with Greenland Sea Deep Water is strongly suspected; e.g., oxygen distributions suggest Norwegian Sea Deep Water originates at mid-depth in the central gyre immediately above Greenland Sea Deep Water (Carmack 1972).

A volumetric (T/S) analysis of water masses in the Greenland Sea for the summer and winter periods has been published by Carmack and Aagaard (1973). Due to lack of adequate data, their census does not include the water masses of the surrounding shelf domains. Applying the above classification system to this census yields the following quantitative distribution: of the total volume of water in the Greenland Sea (1.4×10^{-6} km^3; excluding the shelf domains) the deep waters comprise 85.4%; Atlantic Water is 4.49; Polar Water and Polar Intermediate Water make up 0.1%; and the various forms of Arctic water are 10.1%.

Numerous authors have constructed water and heat budgets for the combined Norwegian and Greenland seas (e.g., Vowinkel and Orwig, 1961; Mosby, 1962; Fletcher, 1965; and Worthington, 1970). Estimates of surface heat exchange are given by Timofeyev, Smirnova and Nepop (1970). The important task of determining the heat budget of the Greenland Sea alone has not been carried out.

2.2.3 Southern Ocean. Our general understanding of the physical oceanography of the Southern Ocean is founded on the early work of Brennecke (1921), Wust (1933), Mosby (1934), and Deacon (1937). Recent discussions by Gordon (1971), Gordon, Taylor and Georgi (1977) and Deacon (1981) have emphasized the importance of water mass zonation. Mass and heat budgets are discussed by, among others, Gordon (1975), Toole (1981), and Georgi and Toole (1982).

The ocean water column of the Southern Ocean south of the Polar Front can also be separated into three main layers: Surface Water, Warm Deep Water (or Circumpolar Deep Water) and Antarctic Bottom Water. This general structure is illustrated by the distribution of temperature and salinity along a section which encircles Antarctica, beginning at the base of the continental slope of the Weddell Sea, proceeding eastward around the continent, and ending in the Drake Passage (Figure 5).

Surface Water includes water within and above a shallow (50-300 m) temperature minimum layer, and with salinities below 34.5×10^{-3}. This temperature minimum layer is thought to be a remnant of wintertime convection, and is commonly called Winter Water. Toole (1981) has discussed the seasonal maintenance of the

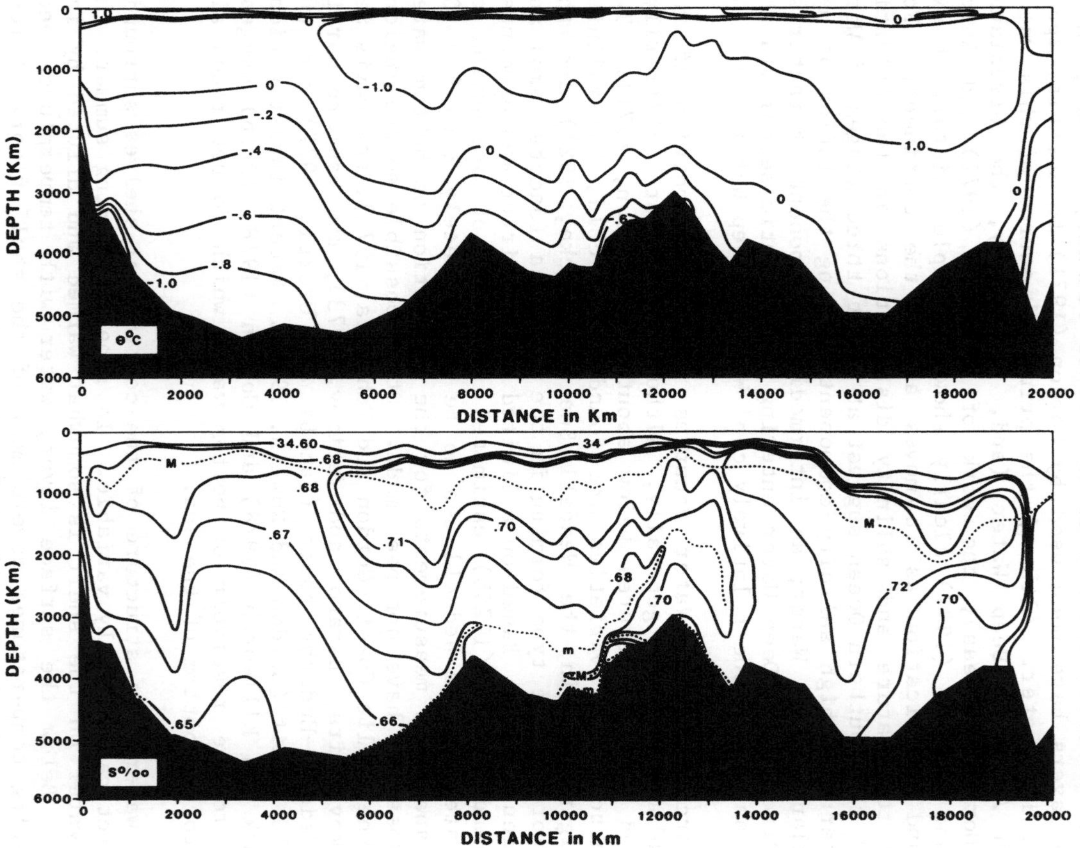

Fig. 5. Section of potential temperature and salinity encircling the Antarctic Continent, beginning at the base of the continental slope of the Weddell Sea, proceeding eastward around the continent, and ending in the Drake Passage. (From Reid and Carmack, in preparation.)

temperature minimum layer. Near the continental margin, and in the central Weddell and Ross gyres, Winter Water has temperatures near the freezing-point and salinities approaching the critical value for the onset of the cabbeling instability (Fofonoff, 1956).

Warm Deep Water is characterized by an intermediate depth temperature maximum, and a somewhat deeper salinity maximum; these two core layers have been used by Gordon (1971) to identify upper and lower deep water. Although the principal source of this water is North Atlantic Deep Water (and, ultimately, the Greenland/Iceland/Norwegian seas), the work of Callahan (1972) and Reid, Nowlin and Patzert (1977) clearly shows a complex history of exchange and modification as it moves around the continent. For example, temperature and salinity distributions in the Atlantic sector of the Southern Ocean suggest at least three stages of Warm Deep Water: a high salinity component marking the influx of new North Atlantic Deep Water; an intermediate component characteristic of Circumpolar Deep Water entering the Atlantic sector via the Drake Passage; and the relatively cool Weddell Deep Water.

Antarctic Bottom Water occupies the lowest portion of the water column and is generally, although somewhat arbitrarily, recognized by temperatures below about $0^{\circ}C$ (Gordon, 1971). Two major bottom water varieties are found: the cold, low salinity type originating in the Weddell Sea (Brennecke, 1921) and the warmer, more saline type forming in the Ross Sea (Jacobs, Amos and Bruchhausen, 1970). Based on a limited set of direct measurements, Carmack and Foster (1975b) estimated the rate of production of bottom water in the Weddell Sea to be approximately 2 to 5 x 10^6 m^3 s^{-1}; analogous measurements for the production of bottom water in the Ross Sea have not been made. Other possible source regions include the Adelie Coast (Gordon and Tchernia, 1972), the Davis Sea (Treshnikov, Girs, Baranov and Yefimov, 1973), the eastern margin of the Weddell-Enderby basin (Jacobs and Georgi, 1977) and the Bransfield Strait (Gordon and Nowlin, 1978). Recent investigations by Carmack and Killworth (1978) and Gordon (1978, 1981b) suggest loadings to the reservoir of bottom water which do not reach all the way to the bottom of the sea.

The water mass structure of the continental shelves surrounding Antarctica is highly variable. In general, during summer, the surface water on the shelf is somewhat warmed and diluted by ice melting. Below the surface layer, water with temperatures near the freezing temperatures covers most of the shelf floor. Gordon (1974) examined shelf salinity distributions and noted a close relationship between shelf width and bottom salinity, with the most saline water lying above wide continental shelves. Correspondingly, the highest shelf water salinities, up to 34.9 to 35.0 x 10^{-3}, are found on the broad shelves of the Ross and Weddell Seas. In relation to bottom water production, it is convenient to distinguish below Low Salinity Water and High Salinity Shelf Water by the 34.6 x 10^{-3} isohaline, since this value appears to represent the minimum salinity shelf water must attain to participate in bottom water formation (cf. Mosby, 1934).

Modified Warm Deep Water, formed by the upwelling of Warm Deep Water onto the continental shelf, is observed as an intermediate layer interleaving between the surface layer and High Salinity Shelf Water, particularly in the Ross and Weddell seas. The intrusion of this water onto the shelf is believed to partially compensate the offshore flow of newly formed bottom water, and to contribute to bottom water formation by mixing with High Salinity Shelf Water (Gill, 1973; Foster and Carmack, 1976a).

The final shelf water mass, apparently unique to Antarctic waters, is the body of water well below the freezing-point at one atmosphere pressure, primarily observed near the Filchner, Ronne and Ross ice shelves. Because the freezing point of water is depressed by increasing hydrostatic pressure, the extreme temperatures are likely affected below the sea surface by cooling along the undersides of floating glacial ice shelves at depths of 200 m or more. This water is thus designated as Ice Shelf Water; and potential temperatures below -2.0°C are common (cf. Jacobs, Amos and Bruchhasuen, 1970; Carmack and Foster, 1975a; Jacobs, Gordon and Ardai, 1979).

A volumetric (T/S) census of water masses in the Southern Ocean south of the Polar Front was published by Carmack (1977). Warm Deep Water is by far the most voluminous water mass comprising nearly 58% of the total volume; followed by Antarctic Bottom Water with about 28%. The remaining 14% of water is composed of surface and transitional water masses.

2.3 Primary Fronts

While the principal water masses are distinguished on the basis of vertical stratification, it is also important to recognize the significance of frontal regions as water mass boundaries and as factors influencing ice distribution.

In the Greenland Sea frontal systems are correlated with both the south-flowing Polar Water of the East Greenland Current, and the north-flowing Atlantic Water of the West Spitsbergen Current. Seasonal fluctuations in ice cover generally are constrained to the region between these two frontal systems. In the Barents Sea a front lies between the inflowing Atlantic Water and outflowing Polar Water. The location of this front, marked by strong horizontal temperature gradients, is well-correlated with topography (Johannessen and Foster, 1978). During winter this front marks the pack ice boundary. A primary front is also observed in the Baffin Bay/Labrador Sea are associated with the Labrador Current.

The dominant hydrographic feature in the Southern Ocean is the Antarctic Polar Front, which marks the boundary between the polar and subpolar surface waters, and is associated with the core of the Antarctic Circumpolar Current. The Polar Front is the largest and, probably, best understood of all primary fronts in polar waters (cf. Gordon, Taylor and Georgi, 1977; Joyce, Zenk and Toole, 1978). This feature occurs at different latitudes around the continent, reaching beyond 50°S in the Atlantic sector and as far south as 60°S in the Pacific sector; its average latitude is 53°S. The position of the Polar Front is known to vary temporally, perhaps by 2-4° latitude.

Other primary fronts in Antarctic waters include the Weddell-Scotia confluence, formed at the juncture of the Antarctic Circumpolar Current and the north arm of the Weddell Gyre, and the shelf break fronts which separate the shelf waters from those of the open ocean. The latter is apparently associated with a relatively narrow current near the shelf break and is an important aspect of bottom water production.

2.4 Circulation

Wind and topography dominate the upper ocean circulation and ice-drift in polar seas (cf. Campbell, 1965; Aagaard, 1970, 1972; Galt, 1973; Semtner, 1976).

2.4.1 Arctic Ocean. The map of dynamic height (0/1200) for the Arctic Ocean, adapted from Coachman and Aagaard (1974), is shown in Figure 6. Surface waters of the Eurasian Basin tend to move toward the North Pole. Mean speeds are slow in the central ocean, about 2 cm sec^{-1}, but increase as water exits the basin as part of the East Greenland Current (see also Wadhams, Gill and Linden, 1979). In the Beaufort Sea surface waters move in a clockwise direction having a gyre axis which coincides with that of the mean atmospheric anticyclone. Aspects of flow through the Canadian Archipelago are given by Sadler (1976), and of flow along the Alaskan shelf by Aagaard (1981a).

2.4.2 Greenland Sea. A representative map of dynamic height (0/800 db) for the Greenland Sea, taken from Carmack (1972), shows

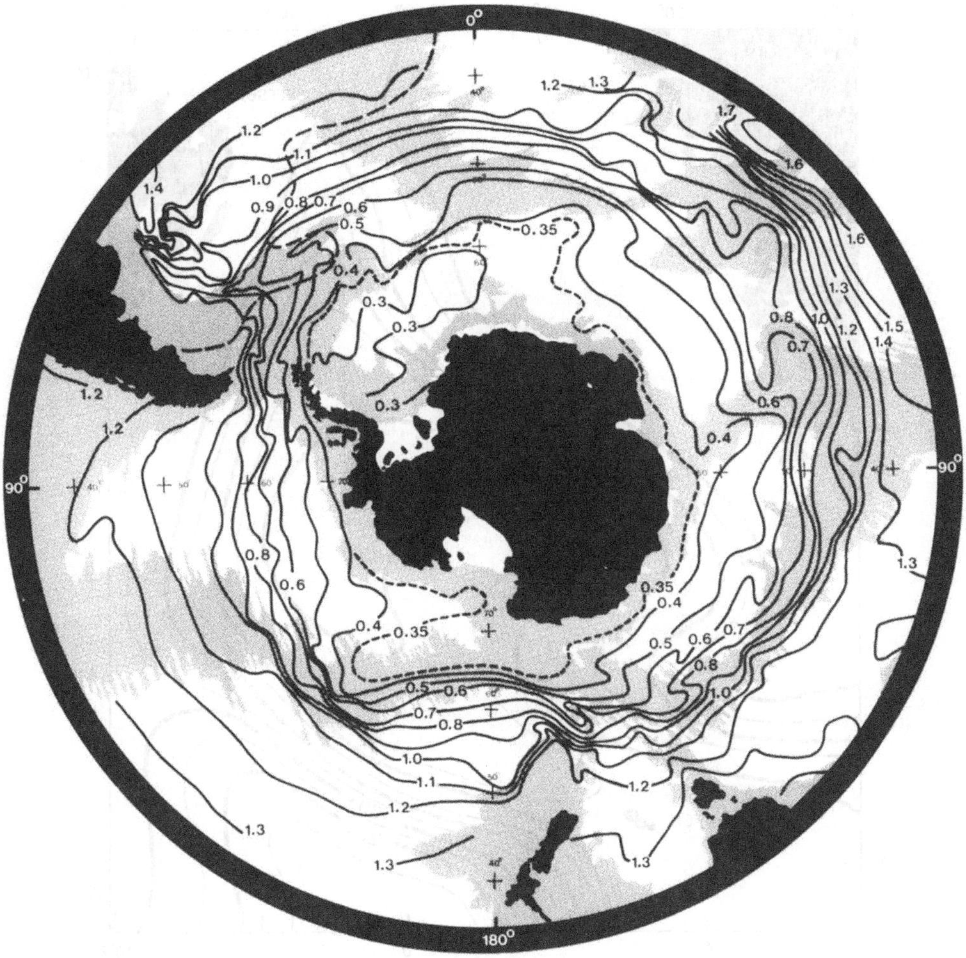

Fig. 6. Dynamic topography (0/1200 db) of the Arctic Ocean. (After Coachman and Aagaard, 1974.)

that the flow is cyclonic around the central gyre, centered on the prime meridian at about 74-75°N (Figure 7).

The northward moving Norwegian-Atlantic current enters the Greenland Sea as a relatively intense flow following the eastern flank of the Jan Mayen Fracture Zone. At about 71°N the current changes direction and flows east-northeast along the contours of the Mohn Rise. As the current moves along this ridge, some of the water it carries is diverted south, effecting an anticyclonic gyre in the northern Norwegian Basin. The remaining water continues along the length of the ridge to about 74°N, where it turns and flows northward along the mid-ocean ridge as the West Spitsbergen

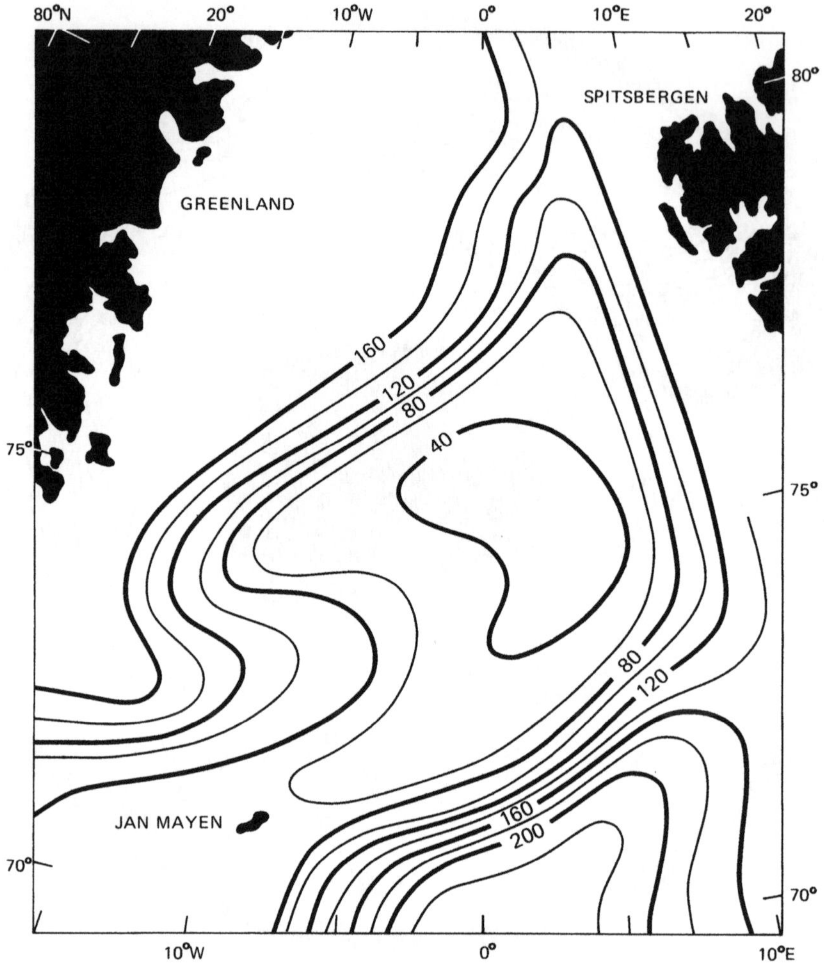

Fig. 7. Dynamic topography (0/800 m) of the Greenland Sea. (Data from U.S.S. Atka, 1958; after Carmack, 1972.)

Current. Although not shown in Figure 7, the West Spitsbergen Current is augmented by a flow of low salinity water from the East Spitsbergen Current which extends westward from the Barents Sea around the south tip of Spitsbergen.

The East Greenland Current originates in the Arctic Ocean and enters the Greenland Sea as a boundary current along the western side of the Greenland Sea. At about 73°N a branch of the current, called the Jan Mayen Polar Current, extends eastward to form the southern margin of the central gyre.

2.4.3 Southern Ocean. The map of dynamic height (0/1000 db) for the Southern Ocean, adapted from Gordon, Molinelli and Baker (1981), is shown in Figure 8. Surface water tends to flow around the continent in a clockwise direction. The zone of strong baroclinicity (average latitude, 53°S) is the Antarctic Circumpolar Current, an intense flow with little attenuation with depth. Recent measurements by Fandry and Pillsbury (1980) place the volume transport of flow through Drake Passage at $127 \pm 24 \times 10^6\ m^3\ sec^{-1}$. The Eastwind Drift, although poorly shown in Figure 8, is a narrow westward flowing current along the continental margin. This current is not completely circumpolar, but instead becomes part of the clockwise gyres in the Weddell, Ross and Bellingshausen Seas.

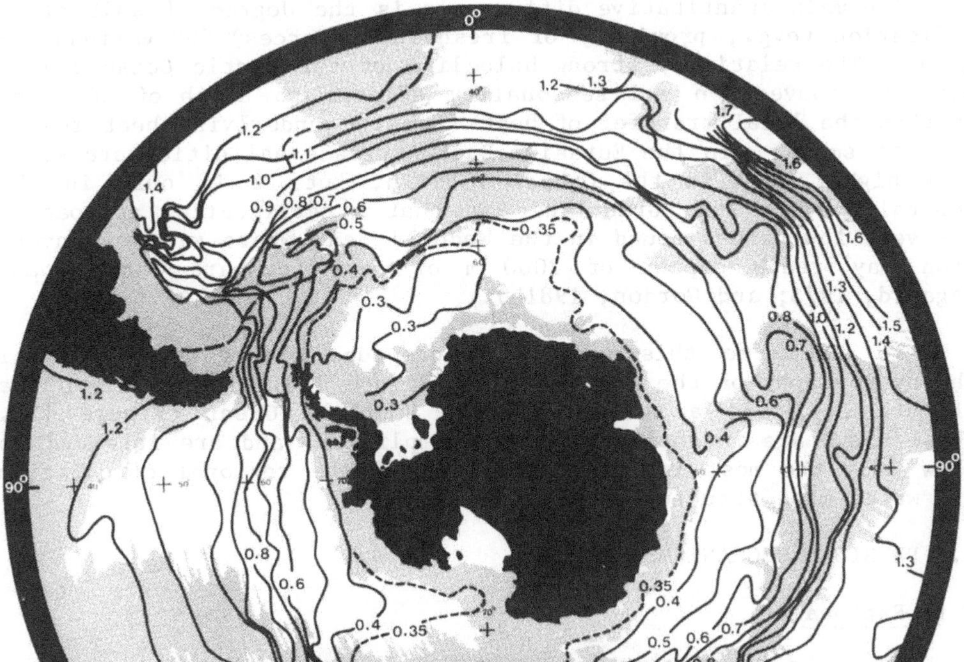

Fig. 8. Dynamic topography (0/1000 m) of the Southern Ocean. (After Gordon, Molinelli and Baker, 1981.)

The Weddell Gyre is marked by a trough in the dynamic topography. This gyre may also have a significant westward intensification along the Antarctic Peninsula (Gordon, Martinson and Taylor, 1981).

2.5 Regional Comparison and Summary

Foster (1978) gives a comparison of the Arctic and Southern Oceans. A common feature of the temperature and salinity structure is that the deep areas typically have a surface layer of cold, dilute water; a mid-depth layer of warmer, saltier water and a bottom layer of relatively cold water. This structure is shown for a number of polar regions in Figure 9.

The main quantitative differences is the degree of salt stratification (e.g., proximity of freshwater sources) for various regions. The relatively strong halocline of the Arctic Ocean limits vertical convection and seasonality effects to depth of 30-50 m, so that the local transfer of heat from the underlying heat reservoir is small. In the Eurasian Basin surface salinities are somewhat higher than in the Canadian basin, but it is only in the central gyre of the Greenland Sea that stratification approaches the very low values found in the Weddell Sea, where winter convection may reach depths of 2000 m or more (compare Carmack and Aagaard, 1973; and Gordon, 1981b).

To summarize this section, conceptual models are shown for the hydrography of the Arctic Ocean (Figure 10), the Greenland Sea (Figure 11), and a portion of the Southern Ocean (Figure 12). These views are intentionally over-simplified, and are intended to show only the most important features of the regional circulation and water mass structure.

3. PHYSICAL MECHANISMS

3.1 Rationale

The scales of motion encountered in polar oceans range from the turbulent field (10^{-3} m) up to the wind-driven and buoyancy-driven mean flows (10^{7} m), while the corresponding time scales range from milliseconds up to and beyond the seasonal time scale (cf. Killworth, 1979b). Depending upon scale, local topography and specific forcing, a wide variety of physical phenomena may be important. While dynamical scaling criteria offer, perhaps, the best way of categorizing processes associated with circulation and mixing in ice-covered waters, no such study has yet been carried out. In this section a more personal approach is taken by constructing a hypothetical "section" extending from the coast into the deep ocean, and listing processes which may be important at each location along the section. Phenomena are thus arranged

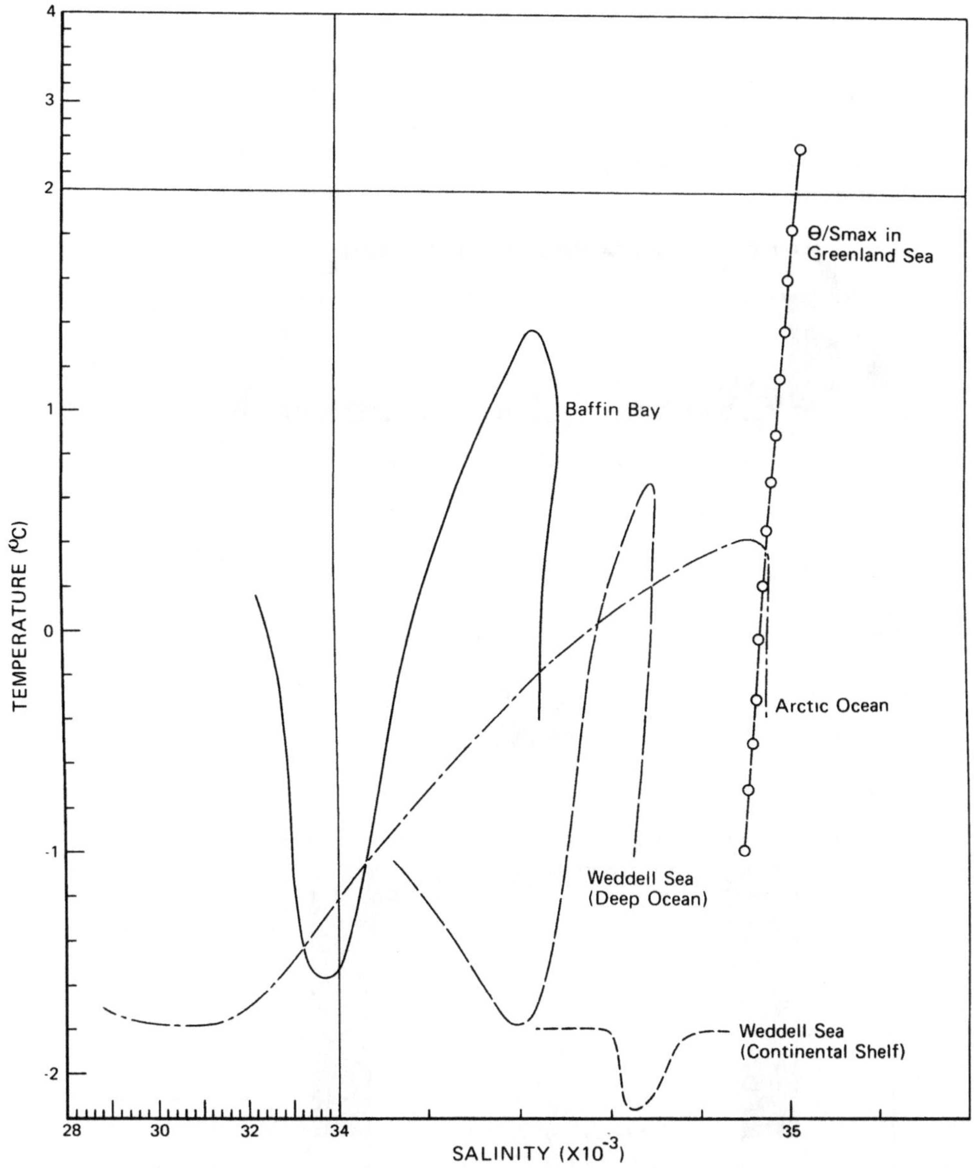

Fig. 9. Representative T/S correlation curves for various polar regions.

in relation to bathymetry, oceanographic conditions and proximity to ice. Such a presentation has the advantage of providing an intuitive feeling for subsequent process interactions.

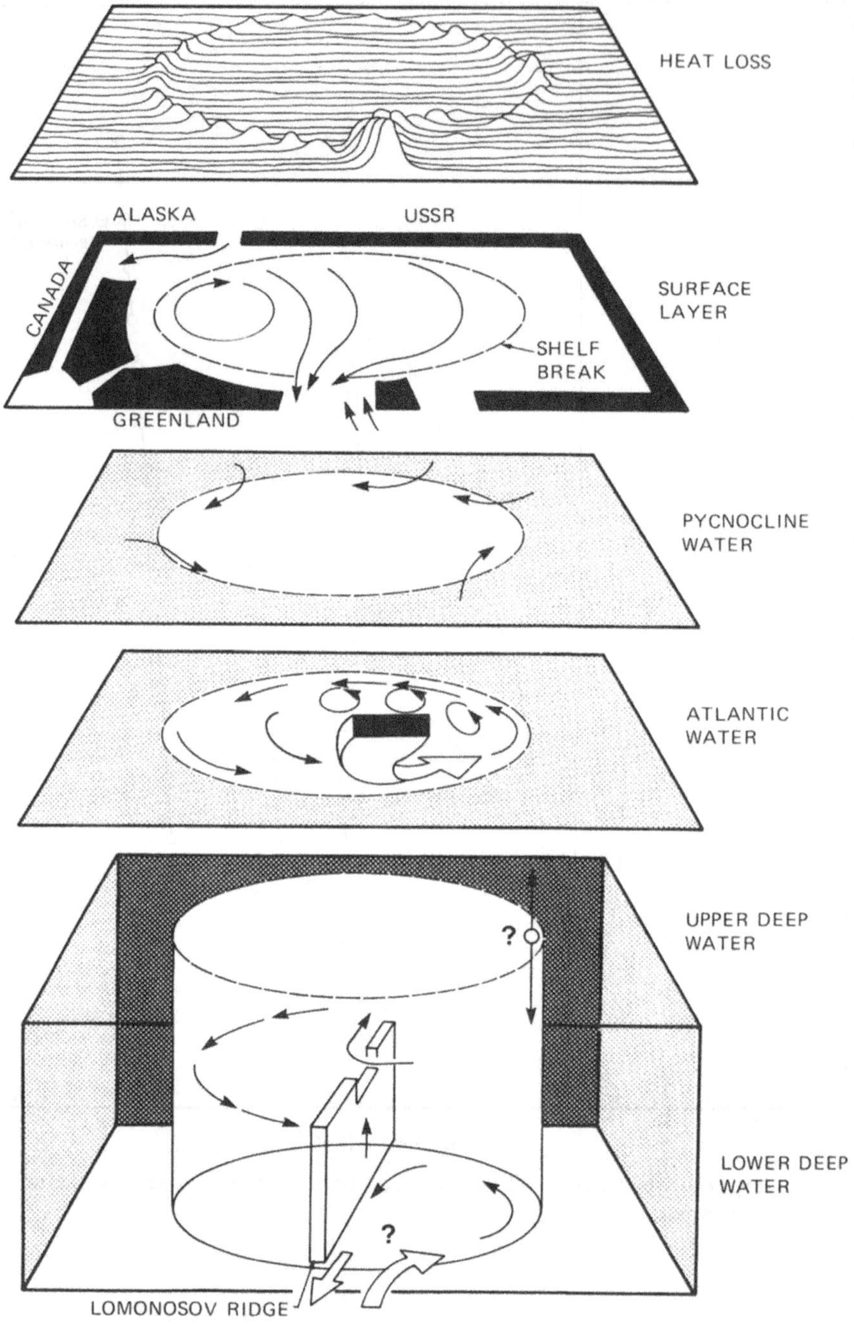

Fig. 10. Simplified schematic diagram of Arctic Ocean hydrography and circulation.

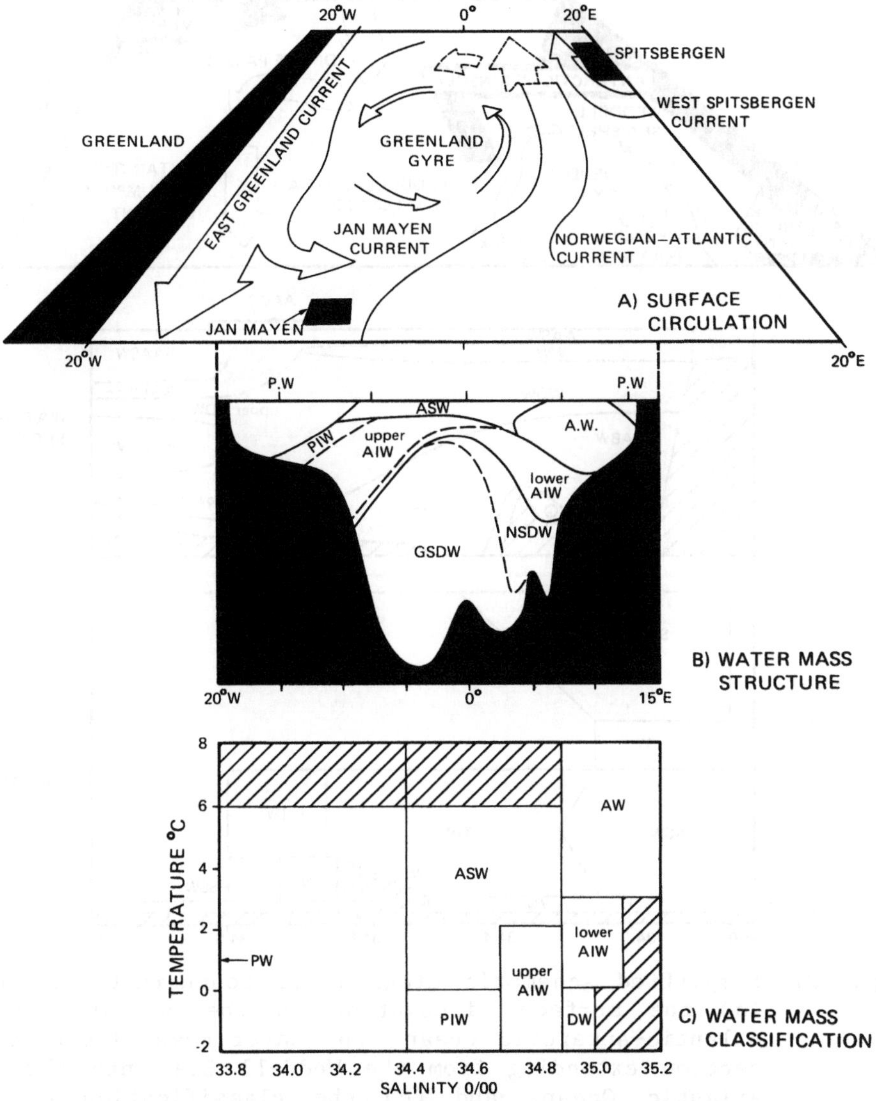

Fig. 11. Simplified schematic diagrams of Greenland Sea showing (a) the surface circulation; (b) water mass structure on a section across the central gyre; and (c) the classification of water masses on a T/S correlation diagram. The following abbreviations are used: Polar Water (PW); Polar Intermediate Water (PIW); Atlantic Water (AW); Arctic Surface Water (ASW); upper Arctic Intermediate Water (upper AIW); lower Arctic Intermediate Water (lower AIW); Greenland Sea Deep Water (GSDW); and Norwegian Sea Deep Water (NSDW).

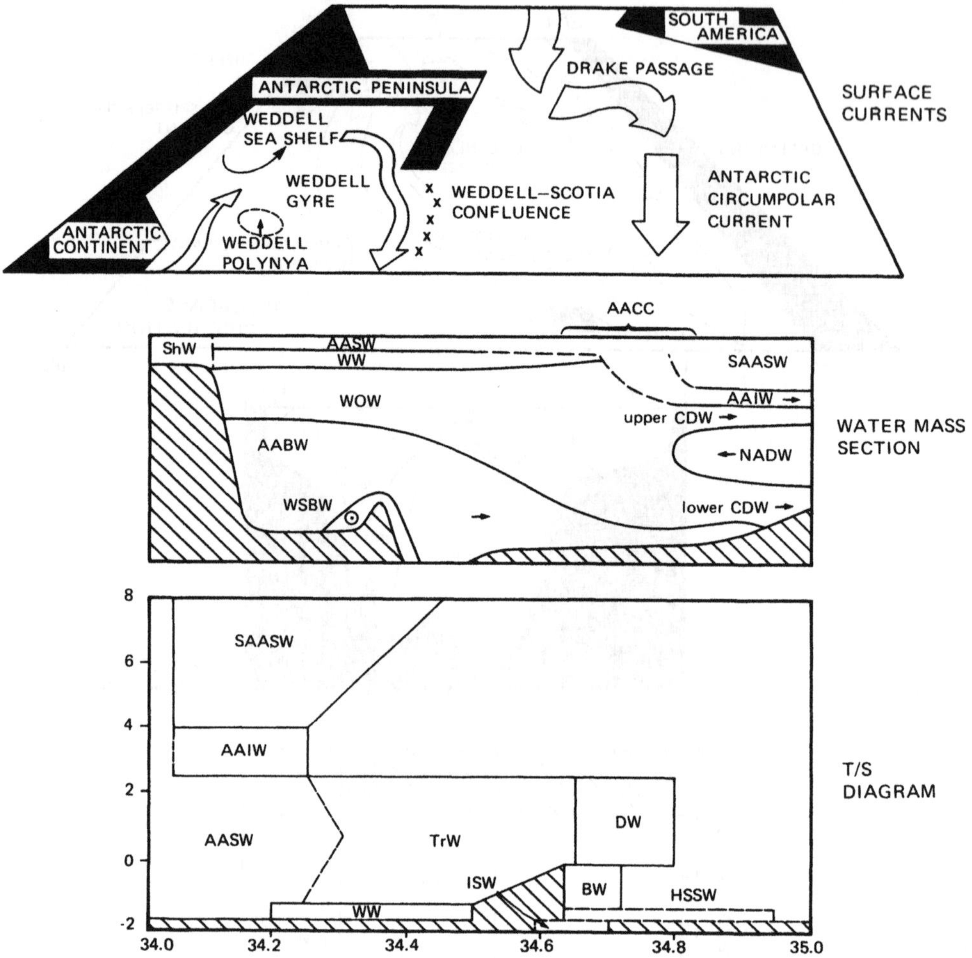

Fig. 12. Simplified schematic diagrams of Southern Ocean showing (a) the surface circulation in the vicinity of the Atlantic-Antarctic Ocean; (b) water mass structure on a section extending from the Weddell Sea into the South Atlantic Ocean; and (c) the classification of water masses on a T/S correlation diagram. The following abbreviations are used: Sub-Antarctic Surface Water (SAASW); Antarctic Surface Water (AASW); Antarctic Winter Water (AAWW); Antarctic Intermediate Water (AAIW); North Atlantic Deep Water (NADW); upper Circumpolar Deep Water (upper CDW); lower Circumpolar Deep Water (lower CDW); Warm Deep Water (WDW); Antarctic Bottom Water (AABW); Weddell Sea Bottom Water (WSBW); and Shelfwater (ShW). (Note that Warm Deep Water is taken to include Circumpolar Deep Water and Weddell Deep Water.)

The polar "section" and associated physical processes are shown in Figure 13. The section comprises a coastal zone influenced by the shoreline and continental inputs, the continental shelf and slope region, and the deep ocean. In the latter a distinction is made between areas near the ice edge and areas of the ice interior. Admittedly, not all physical processes are included in this scheme; some overlap is unavoidable and the relationship between site and phenomena is not single valued.

3.2 Coastal Regions

An overview discussion of coastal circulation in temperate oceans has been given by Csanady (1981): attention is focused

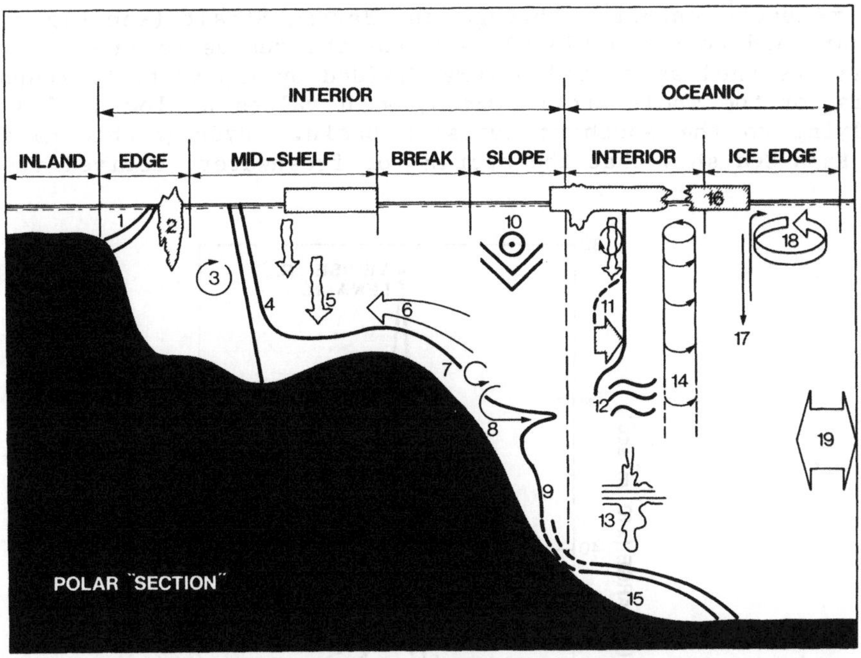

Fig. 13. Simplified schematic diagram of some physical processes associated with circulation and mixing in polar seas: (1) river discharge; (2) glacial melting; (3) coastal circulation; (4) shelf front; (5) haline convection; (6) shelf-break upwelling; (7) shelf drainage; (8) mid-depth intrusion; (9) gravity plumes; (10) shelf-break jet; (11) mixed-layer dynamics; (12) halocline formation by lateral advection; (13) double diffusion; (14) chimney; (15) thermohaline circulation; (16) ice-drift; (17) ice-edge upwelling; (18) eddies; (19) exchange with the world ocean.

below on the effects of freshwater inputs by rivers (Arctic Ocean) and glaciers (Southern Ocean) on nearshore oceanographic conditions.

3.2.1 River Inflow. In some respects the Arctic Ocean may be considered as a great estuary, with a surface layer diluted by river inflow spreading above warm, high salinity water derived from the oceans to the south; the dynamical similarity is so strong that Stigebrandt (1981a) has used what is essentially an estuarine model to describe the surface layer characteristics. The bulk of the freshwater comes from the Yenisei, Lena, Ob, Kolyma and Indigirka Rivers in the Soviet Union; from the Mackenzie River in Canada; and from the Yukon River via the Bering Strait. Various compilations of runoff data are given by SCOR WG58 (1979); while individual estimates differ, the annual discharge to the Arctic is roughly 0.10×10^6 m^3 sec^{-1}, with an additional 0.05×10^6 m^3 sec^{-1} entering through the Bering Strait (see Figure 14). Aagaard and Coachman (1975) estimate the residence time of freshwater (defined as stored volume divided by input) to be roughly 10 years for the whole Arctic Ocean, with values as low as 2-3 years applying to the southern Eurasian Basin. Hydrographic data from the Kara Sea show that the storage of freshwater, derived from the

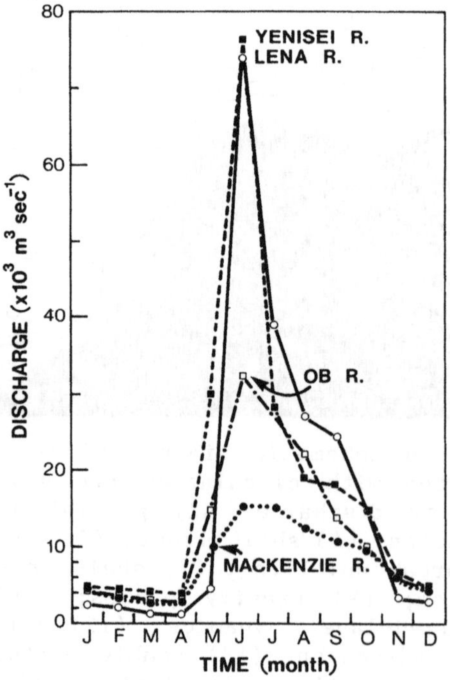

Fig. 14. Mean monthly discharge of four major rivers entering the Arctic Ocean.

Yenisei and Ob Rivers, is roughly equivalent to 2.5 years of river discharge.

Freshwater distributions are clearly important with regards to the growth and destruction of ice cover. During winter cooling the presence of a low-salinity surface layer inhibits deep convection mixing and allows ice growth. Conversely, during spring heating, the presence of a shallow, brackish layer reduces the downward mixing of heat absorbed from solar radiation and thus speeds ice melting, an effect prevalent along the advancing fronts of river plumes. Both the above processes, by altering the degree and competence of ice cover, also modify the surface albedo and the exchange of heat at the air-sea interface. Long-term considerations suggest that any reduction in river input will ultimately alter the Arctic ice cover, particularly in the Eurasian Basin (Aagaard and Coachman, 1975; Micklin, 1981).

Quantitative studies of river plume dynamics in polar seas are lacking. Along these lines the models described by Takano (1954) and Garvine (1977) for the ocean, and by Hamblin and Carmack (1978) and Svensson and Larsson (1980) for lakes may be extended for use in polar seas. Here, account must also be taken of seasonal and/or interannual variations in river discharge; at present, the length scales over which such variability occurs are largely unknown. It would be desirable, for example, to identify regions in which the seasonal advance and retreat of riverine water dominates exchange processes, as opposed to outer regions which respond to change on a much longer time scale (e.g., see Figure 15).

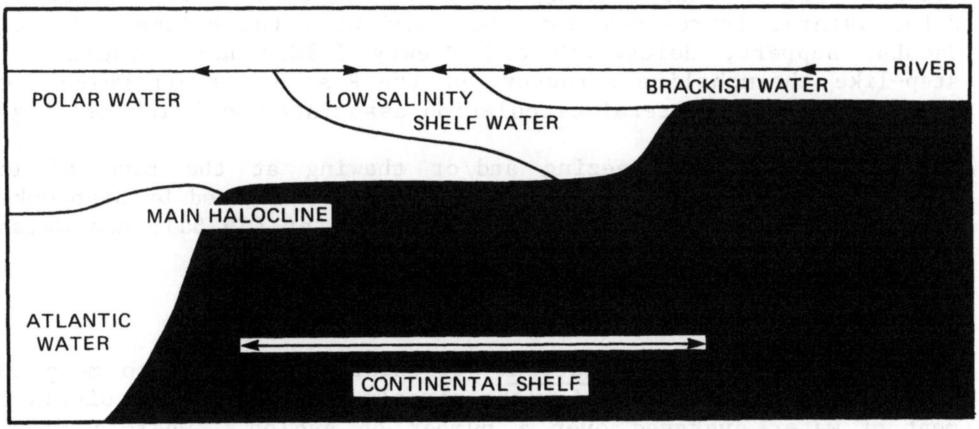

Fig. 15. Schematic diagram of "multiple plume" structure supposed to hold on broad shelves adjacent to river input.

3.2.2 Glacial Inputs. The combined effects of low temperature and high pressure at the base of a floating ice sheet can have a number of effects on the adjacent ocean. When ice melts in seawater, or when seawater freezes to the base of the ice, an altered water mass is formed at the ice/water interface, the temperature of which corresponds to the freezing-point at ambient salinity and pressure (cf. Gade, 1979). This temperature is depressed both by increasing pressure [about -7.6×10^{-4} °C $(dbar)^{-1}$] and by increasing salinity [about -5.5×10^{-2} °C per 10^{-3}]. One example of this effect is the common occurrence beneath floating ice shelves of water with temperatures below the freezing point at 1 atmosphere pressure (Figure 16; Jacobs, Amos and Bruchhausen, 1970; Carmack and Foster, 1975a; Jacobs, Gordon and Ardai, 1979).

Several circulation schemes arising from the melting of icebergs and ice shelves have been suggested (e.g., Figure 17). Foldvik and Kvinge (1974) noted the potential for a conditional instability near the base of ice shelves in which ice crystal formation, induced by any upward displacement of a parcel of melt water, would form adjacent to a melting ice face which would then entrain ambient water. They assumed that the rate of entrainment would be such that when the rising plume reached the surface its density would equal that of surface water, and subsequently calculated an entrainment ratio of 60:1 for conditions in the Weddell Sea. Greisman (1979) proposed a model of upwelling at the face of ice shelves driven by melt water, and calculated an entrainment ratio for the ascending plume based on the amount of heat required from the boundary layer to melt the ice.

A number of authors (e.g., Huppert and Turner, 1978, 1980; Huppert and Josberger, 1980) have examined the influence of a vertical salinity gradient on the behavior of the meltwater. These studies suggest that melting along the face of ice shelves may drive lateral intrusions into the ambient water column. Indeed, Jacobs, Huppert, Holdsworth and Drewry (1981) have described a step-like thermohaline structure of the Ross Ice shelf which they suggest is due to lateral circulation associated with ice melting.

The effect of freezing and/or thawing at the base of ice shelves on bottom water production has been discussed by Seabrooke, Hufford and Elder (1971), Carmack and Foster (1975a), and Weiss, Ostlund and Craig (1979).

3.3 Continental Shelf and Slope Regions

3.3.1 General Circulation. Circulation is taken to mean the longer term net movement of water parcels or the residual displacement of water averaged over a number of cycles. Whereas tidal, intertidal and shelf wave currents affect the turbulence and can

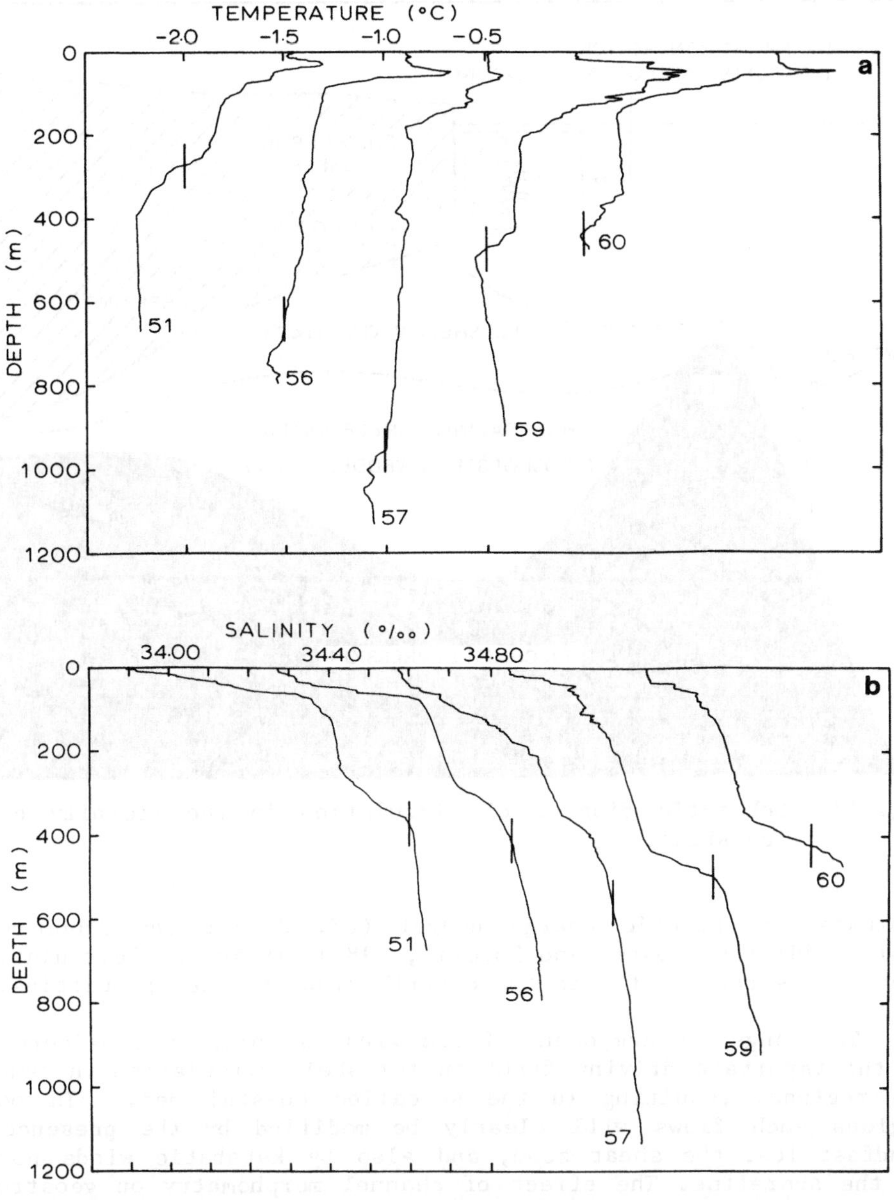

Fig. 16. Continuous vertical profiles of (a) temperature and (b) salinity for stations near the edge of the Fikhner Ice Shelf. Profiles are offset laterally for each successive station; a short vertical mark denotes a temperature of -2°C in (a) and a salinity of 34.6 x 10^{-3} in (b).

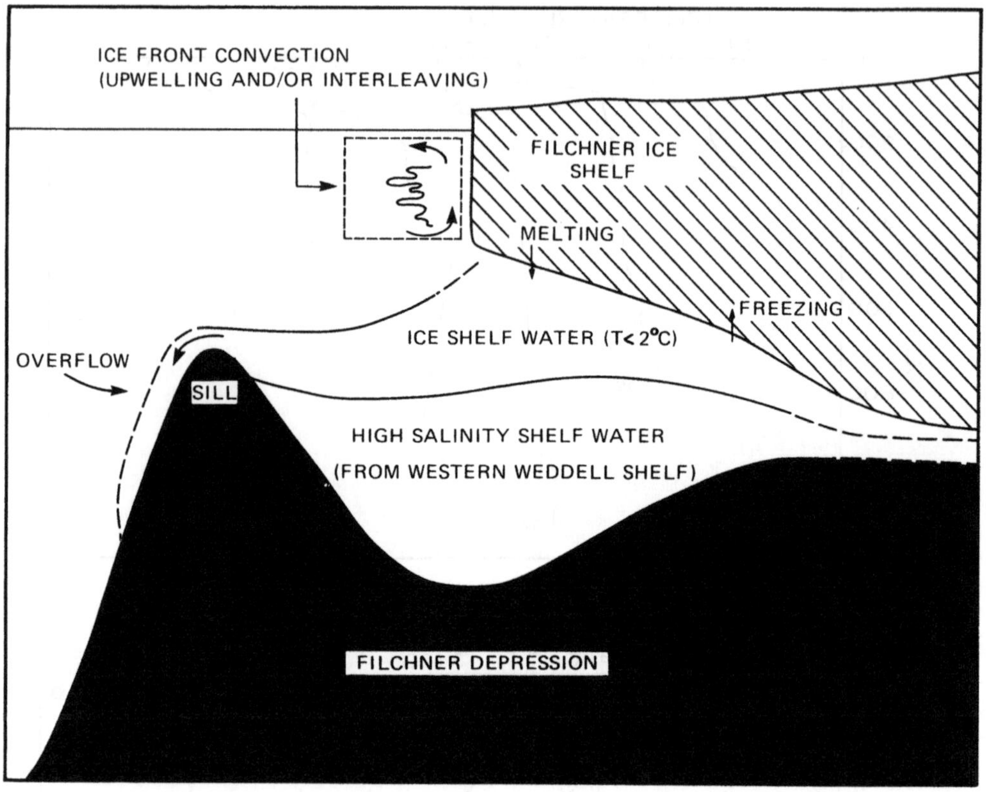

Fig. 17. Schematic diagram of circulation in the vicinity of an ice shelf.

dominate the kinetic energy budget (cf. Kinder and Schumacher, 1981; Middleton, Foster and Foldvik, 1982), it is the "circulation" that most strongly affects the distribution of water properties.

The longshore component of the wind is normally considered to be the important driving function for shelf circulation in temperate regions, resulting in the so-called coastal jet. In polar regions such flows will clearly be modified by the presence of landfast ice, the shear zone, and also by katabatic winds normal to the shoreline. The effect of channel morphometry on geostropic flow in the Canadian Arctic Archipelago is discussed by Leblond (1980).

Significant horizontal density differences can be caused by freshwater inflow, heating and cooling, and brine rejection, all of which act to form their own thermohaline circulation. Further, the superposition of wind-driven and thermohaline flows may make interpretation of hydrographic data (i.e., dynamic topography) difficult. This is because the usual assumption of velocity decreasing with depth may be violated in a region of dense water formation. A good example of this is shown by the dynamic height map of the continental shelf for the southwestern Weddell Sea (Figure 18). A traditional interpretation of this mapping indicates a clockwise circulation whereas the true flow is thought to be counterclockwise.

Another type of shelf circulation in which the Beaufort Current on the outer shelf off Alaska may be driven by the pressure difference between the Pacific and Atlantic oceans has been discussed by Aagaard (1981a).

3.3.2 Shelf Fronts. A front is a zone of enhanced horizontal gradients of some oceanographic property. In horizontal mappings they appear as closely packed isopleths; on sections they appear as close, nearly vertically oriented isopleths. Fronts on or near continental shelves in polar seas have been described, among others, by Gill (1973), Foster and Carmack (1976a), Foster and Middleton (1979, 1980), Schumacher, Kinder, Pashinski and Charnell (1979), and Paquette and Bourke (1979). Formation of T/S fine structure in the vicinity of shelf fronts is discussed by Coachman and Charnell (1977) for the case of a laterally intruding water mass, and by Paquette and Bourke (1981) for the case of a melting (retreating) ice-cover.

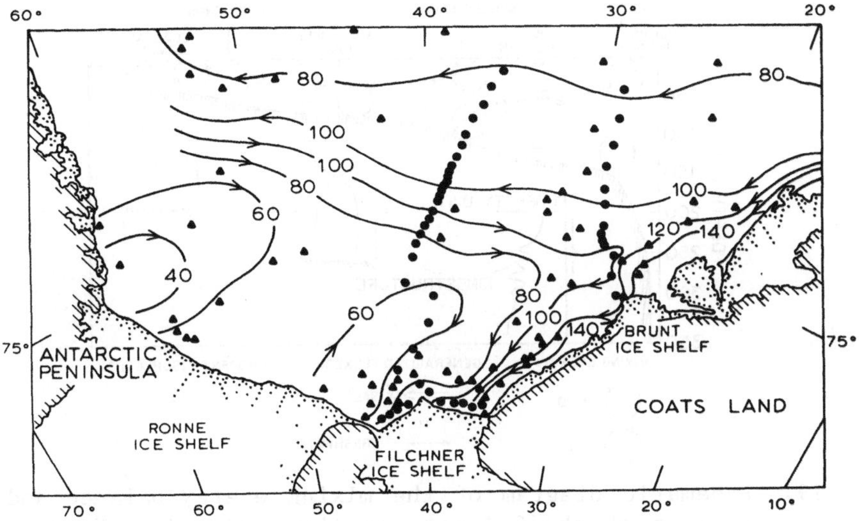

Fig. 18. Dynamic topography of the 50/300 db surface (units in dynamic mm) in the Southwestern Weddell Sea.

An excellent description of the "overall" shelf frontal system in the southeastern Bering Sea, based on over 2000 CTD stations, is given by Coachman, Kinder, Schumacher and Tripp (1980). They propose a conceptual model (Figure 19) based on the occurrence of three mixing zones. In· the shallow nearshore zone the surface and bottom mixed-layer "merge" to form a well-mixed water column. In the middle zone the two mixed-layer are just barely separated from each other resulting in a sharp interface at mid-depth. In the outer zone the two mixed-layers are well-separated and interface is weak. This model appears to agree, at least qualitatively, with the shelf front model proposed by Stigebrandt (1981b).

3.3.3 Formation of Dense Water. Through salt fractionation due to freezing dense water is formed on the Arctic and Antarctic shelves during winter. The formation of dense water is especially rapid in the vicinity of polynyas or other open water areas (Foster, 1972b; Schumacher et al. 1983). Such water may also be formed by the cooling of upwelling Atlantic Water and Warm Deep water. Drainage of this water offshore contributes to the formation of Antarctic Bottom Water, to maintenance of the Arctic Ocean halocline, and possibly to the ventilation of other layers in the adjacent ocean. Despite its importance the formation of dense water on shelves, including sites, time, mechanisms and rates, is very poorly understood.

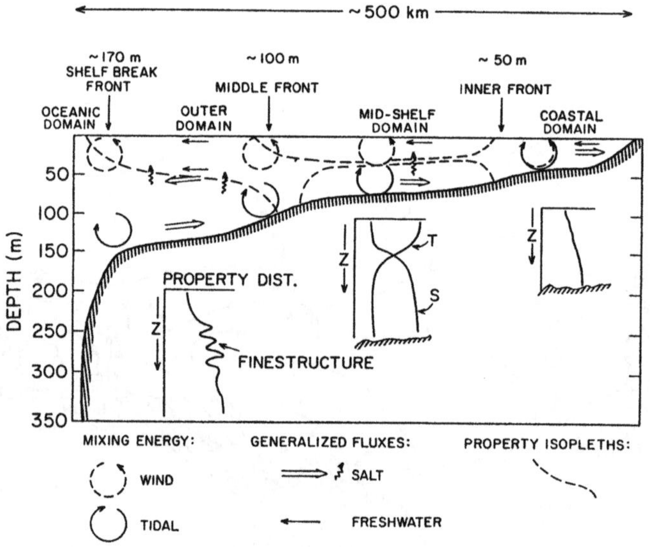

Fig. 19. Schematic diagram of the mixing energy balance and maintenance of shelf fronts on the Bering Sea Shelf. (From Coachman et al., 1980.)

Foster (1968, 1969) has examined the convective scales for haline convection and found good agreement between predictions scaled by Rayleigh criteria and laboratory results. Lake and Lewis (1970), on the other hand, have argued that growing ice platelets and drainage channels can affect convective scales in the underlying water. Weber (1977) has discussed ice-formation allowing for a coupling of heat loss and the associated changes in salinity due to salt rejection. Wakatsuchi and Ono (1983) have carried out both laboratory and field measurements of the volume flux and salinity of brine excluded from growing sea ice. However, it remains largely unknown how dense water moves from the surface, where it is produced, to the bottom, where it is observed (Figure 20).

Gill (1973) concluded that there is a net brine release over antarctic shelves because the pack ice is continually being blown offshore. Aagaard, Coachman and Carmack (1981) used simple ice-growth computations to estimate the potential for dense water production in various regions of the Arctic Ocean (Figure 21). Melling and Lewis (1982) constructed a detailed salt budget for the eastern Beaufort Sea using satellite data to estimate ice divergence. The weakness of such budget calculations, however, is the uncertainty of the longshore and cross-isobath components of advection.

3.3.4 Upwelling. Most observations of cross-isobath flow appear to involve the upwelling of warm, saline water possibly because such water is so distinct from normal shelf waters (e.g., Figure 22).

Foster and Carmack (1976a) have documented the upwelling of Warm Deep Water in the Weddell Sea, and suggested that this is an important stage in bottom water formation. More detailed measurements by Foster, Foldvik and Middleton (1981) show that the onshore intrusion is marked by intense fine-structure development. Coachman and Charnell (1977) have also described the formation of fine structure by water mass intrusion.

Evidence for strong upwelling onto Arctic shelves has been presented by Hufford (1974) and Aagaard, Coachman and Carmack (1981). Garrison and Paquette (1982) described an intrusion of warm Atlantic water into Barrow Canyon. In his thorough study of the continental shelf off Alaska, Aagaard (1981a) concluded that upwelling events were common on the outer shelf, that the associated thermal effects persisted about one week, that upwelling events involved a vertical excursion of offshore water of order 100 m, and that neither the directional effect of the wind nor bottom Ekman-layer effects could account for the observed upwelling events.

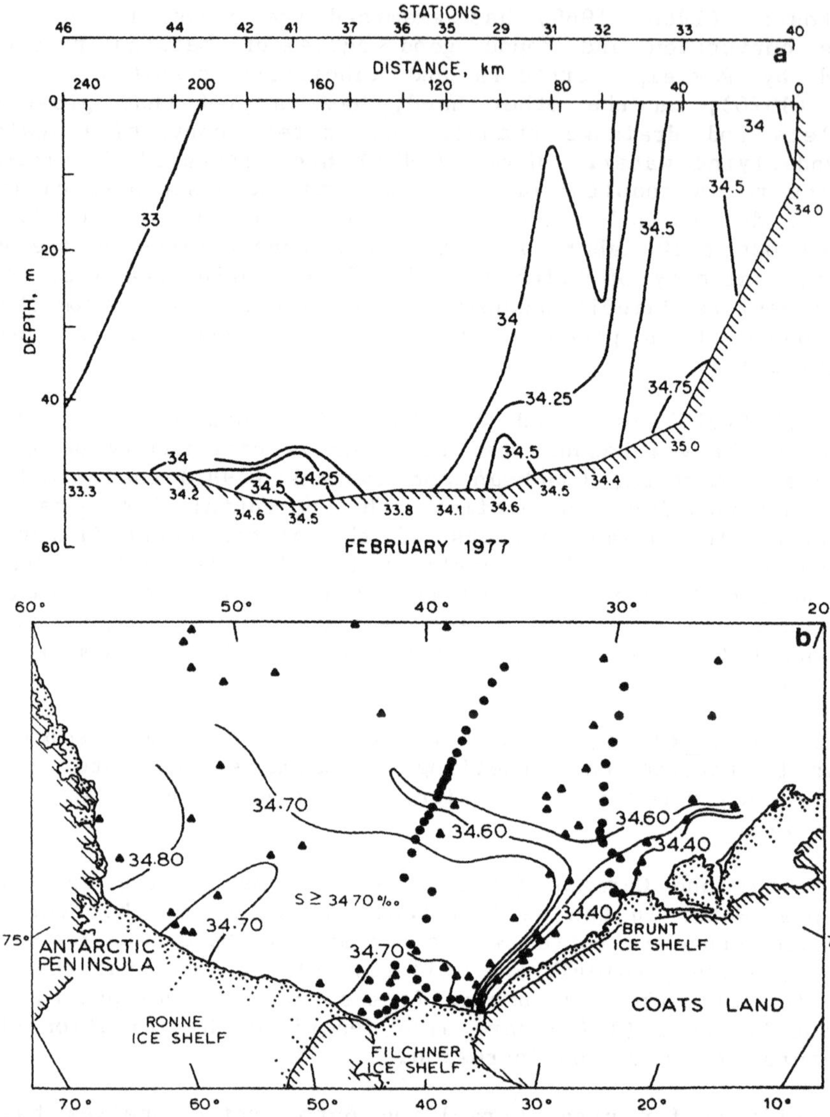

Fig. 20. Observations of high salinity shelf water in the (a) Chukchi Sea and (b) Weddell Sea. (From Aagaard et al., 1981; and Carmack and Foster, 1975a.)

There are, apparently, no theoretical models of upwelling in polar oceans. However, Aagaard (1981a) has suggested that onshore flow could be considered in terms of a geostrophic re-adjustment problem.

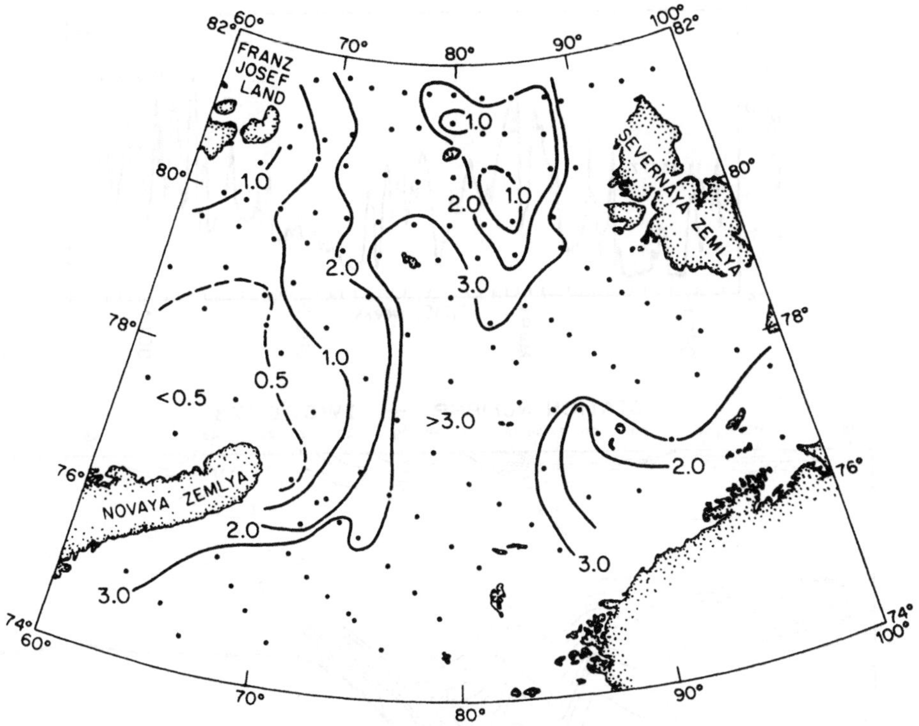

Fig. 21. Ice growth (m) required to raise the salinity of the column on the shelf to 34.1×10^{-3}. This salinity corresponds to the break in the T/S diagram at the closest off-shelf stations. (From Aagaard, Coachman and Carmack, 1981).

3.3.5 Shelf Drainage and Plume Dynamics. Closely related to the problem of dense water formation is the mechanism whereby the dense water drains from the shelf into the ocean interior. Direct evidence of shelf drainage and/or the sinking of dense water above the slope has been reported by Carmack (1973), Foster and Carmack (1976a), Carmack and Killworth (1978) and Foldvik (personal communication) for the southern Ocean, and by Garrison and Becker (1976), Mountain et al., (1976), Melling and Lewis (1982) and Aagaard et al., (1981) for the Arctic Ocean (e.g., Figures 23 to 26).

Killworth (1977) examined the sinking of dense water along the antarctic continental slope using a variety of entraining plume models (cf. Smith, 1975). His results showed that entrainment of Warm Deep Water caused a significant increase in mass flux; that bottom friction was required to force an imbalance between buoyancy and the Coriolis force and cause offshore flow and sinking; and that the variation of the coefficient of thermal

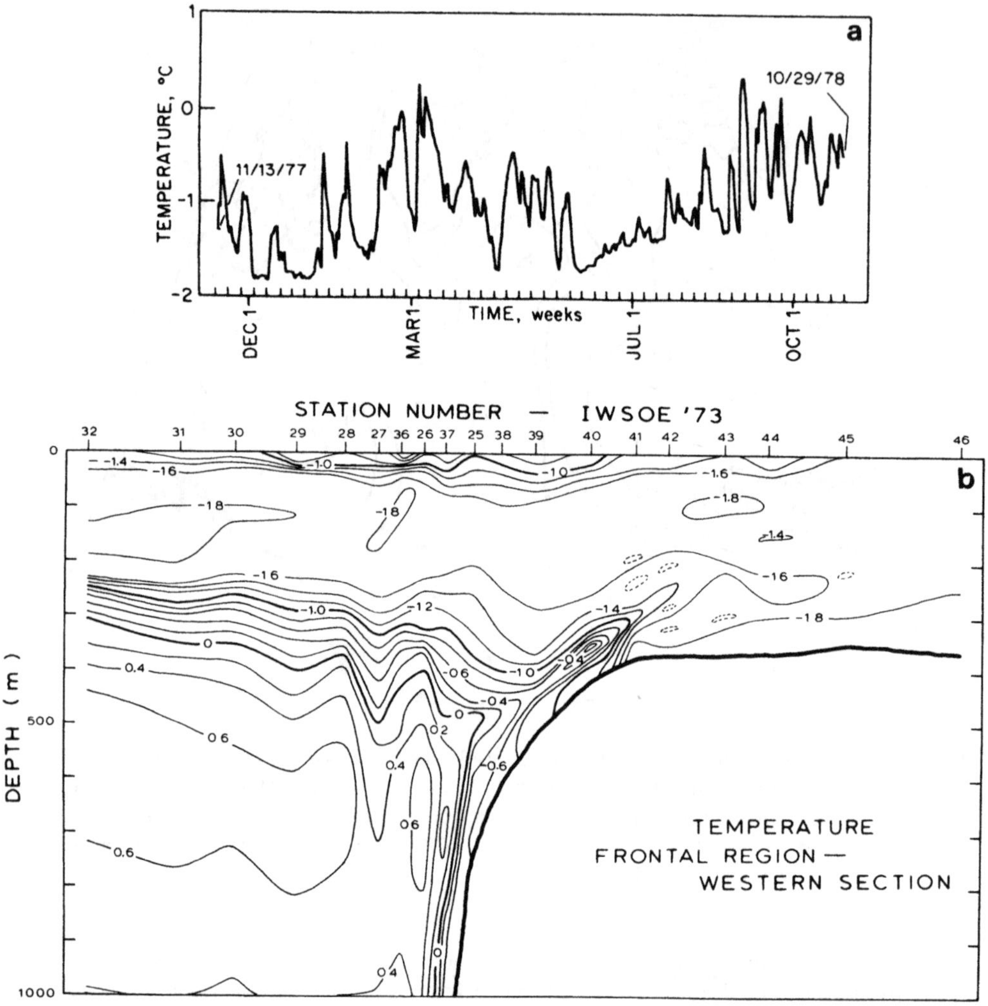

Fig. 22. Observations of a shelf-break upwelling in (a) the Beaufort Sea; and (b) the Weddell Sea. (From Aagaard, Coachman and Carmack, 1981; and Foster and Carmack, 1976a).

expansion with pressure strongly influenced the plume's bouyancy. Carmack and Killworth (1978) applied this model to the case of mid-depth convection off Wilkes Land. Melling and Lewis (1982) constructed a two-stage entraining plume model to examine the effects on plum dynamics of changes in bottom slope and ambient hydrography between the shelf and slope regions (Figure 27). They found that the greatest increase in volume flux occurred during descent of the slope, whereas the greatest change in density

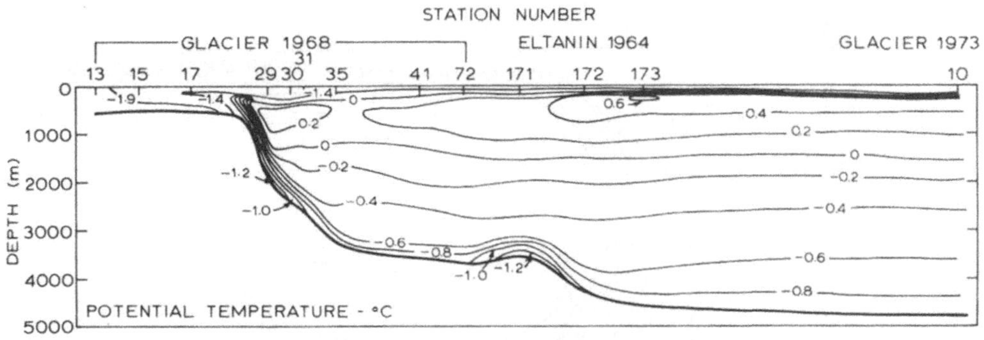

Fig. 23. Potential temperature section showing the core of the Weddell Sea Bottom Water from the southwest to the northeast Weddell Sea. (From Foster and Carmack, 1977.)

occurred as the plume crossed the shelf; this is because the small amount of water entrained on the gently sloping shelf has a relatively large difference in density.

A final aspect of sinking plumes, which has not yet received attention, is that they are sometimes observed to be stratified. For example, data published by Carmack (1973; see Figure 28) and Foster and Middleton (1980, their Figure 4) show that the abyssal water column is comprised of two layers of nearly constant vertical temperature gradient with the gradient of the lower layer being of order 100 larger, that the thickness of the lower layer decreases with depth and/or distance from the source, and that the coldest water is at the shallowest depths but involves the warmest juncture point with overlying water. It is probable that a stratified plume sinking into a stratified water column will exhibit some dynamical features not revealed by entraining plume models.

3.4 Deep Ocean Regions

3.4.1 Mixed-Layer Dynamics. The mixed-layer is generally taken to mean that layer of the ocean whose structure is due to one-dimensional surface forcing (cf. Turner, 1973). The upper layers of polar seas reflect the combined action of the wind, which may be transmitted to the water through ice motion, and the buoyancy flux arising from heat and salt exchanges.

The simplest mixed-layer model is that of Zubov (1945, page 70) in which the mixed-layer depth is assumed to extend to the depth of convective equilibrium, i.e., to precisely the depth where the densities of the mixed-layer and local water are equal. Despite its obvious limitations, this model works well in general applications (cf. Killiworth, 1979a).

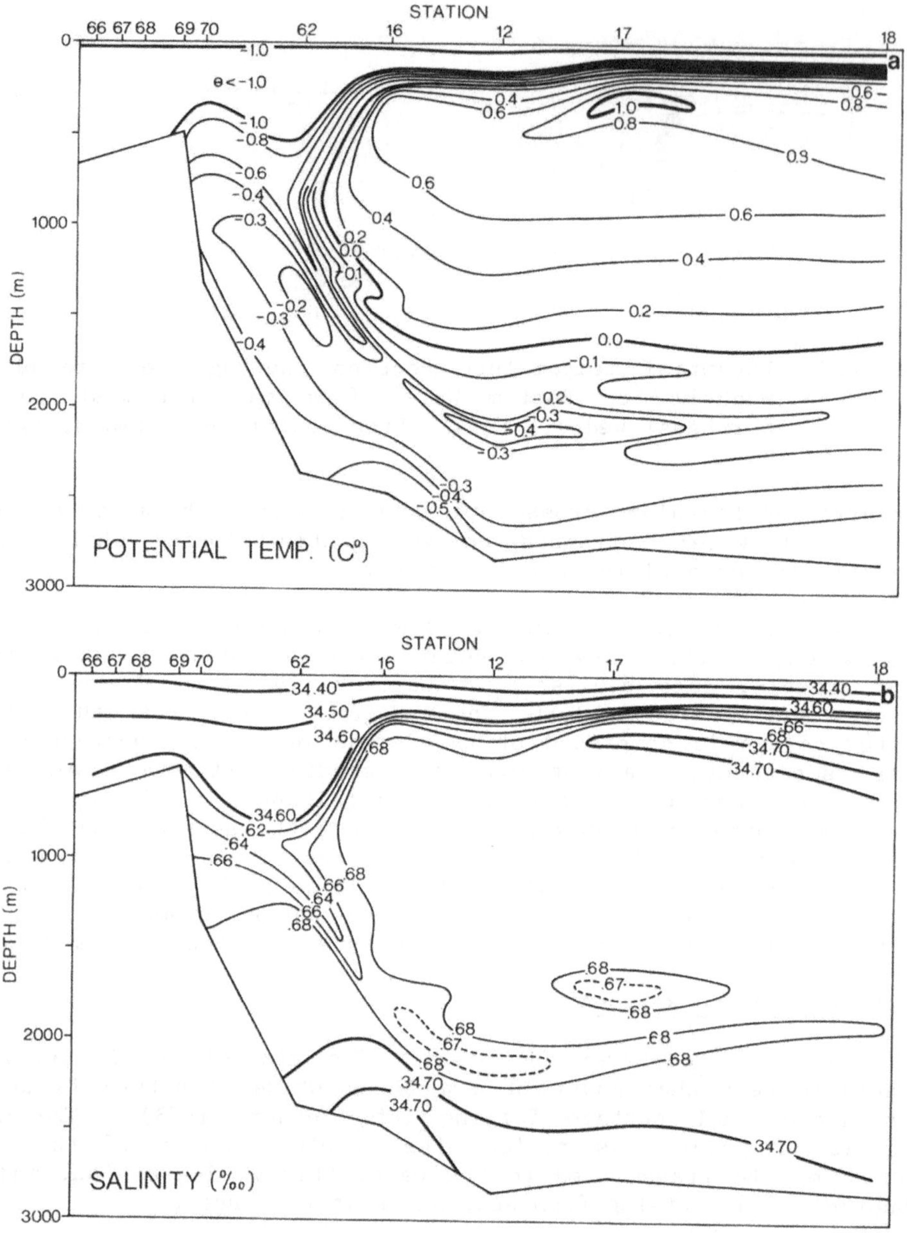

Fig. 24. Vertical sections of (a) potential temperature and (b) salinity off Wilkes Land, Antarctica. (From Carmack and Killworth, 1978.)

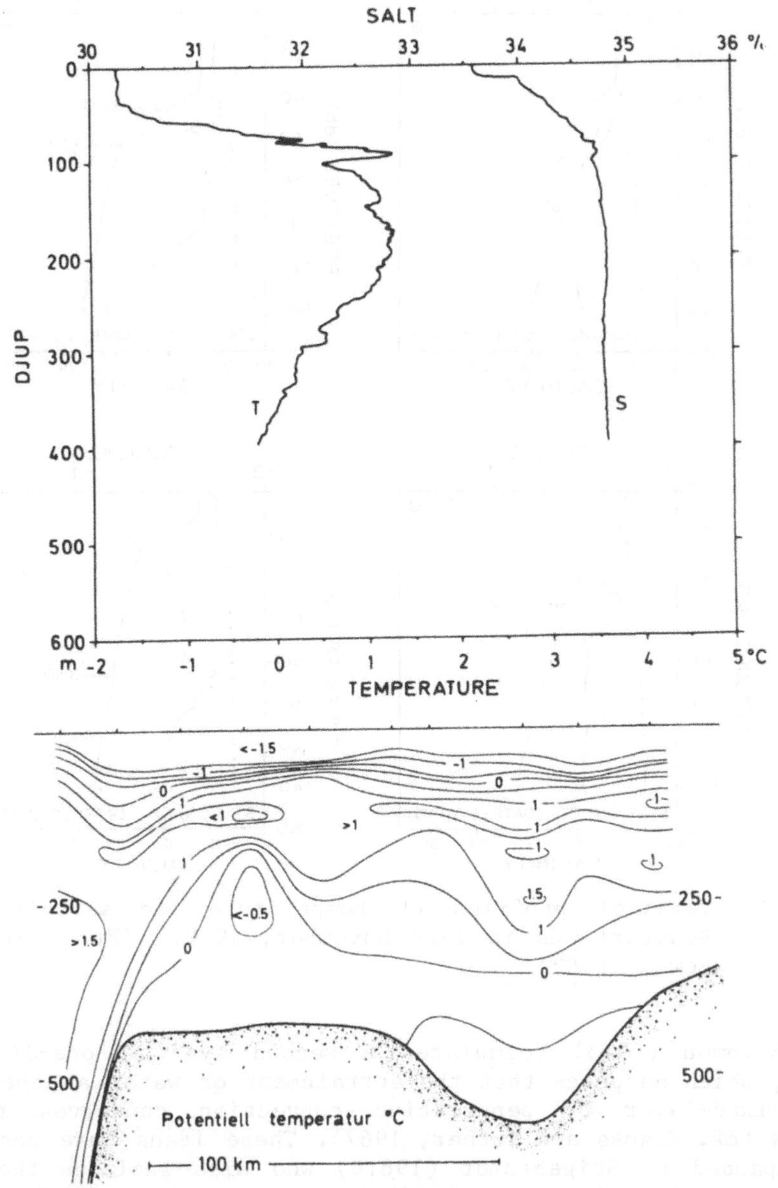

Fig. 25. Vertical profiles of temperature and salinity north of the Barents Sea at 81°35′N, 41°32′E. (From Aagaard, Foldvik and Rudels, 1981).

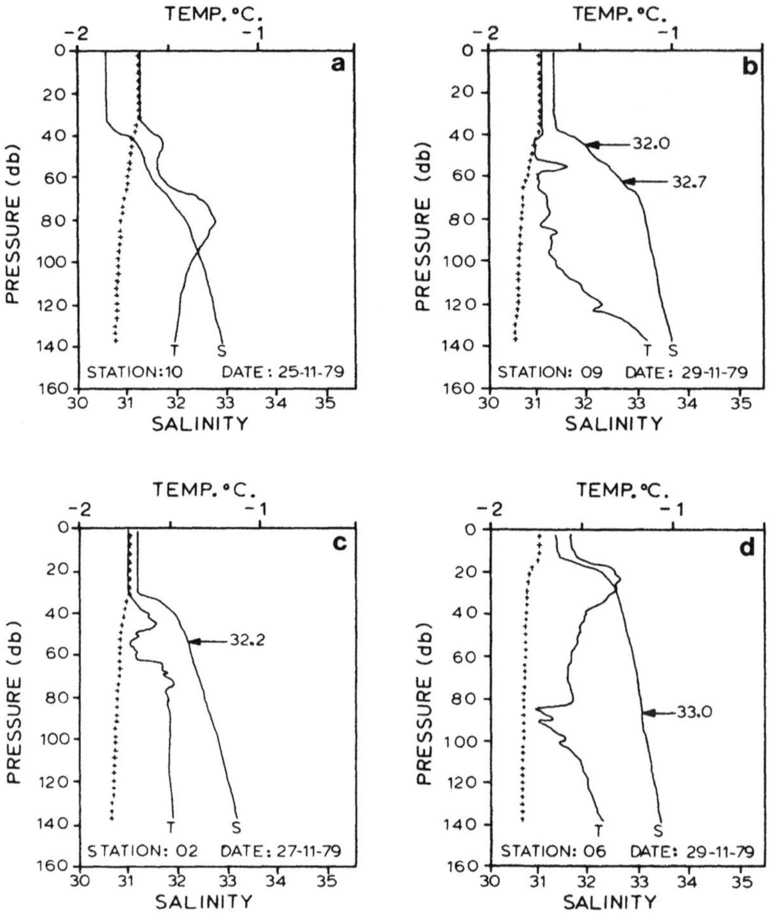

Fig. 26. Vertical profiles of temperature and salinity in the Beaufort Sea in late November, 1979. (From Melling and Lewis, 1982).

Solomon (1973) formulates a second type of one-dimensional model, which supposes that the entrainment of water at the base of the mixed-layer by penetrative convection conserves potential energy (cf. Krauss and Turner, 1967). These ideas have been greatly expanded by Stigebrandt (1981b) who also includes the effects of river inflow, ice export, and advective exchange with the world ocean.

Of major importance to mixed-layer structure and air-sea interactions is the degree to which the sea surface is covered by ice (cf. Maykut and Untersteiner, 1971). This has prompted a

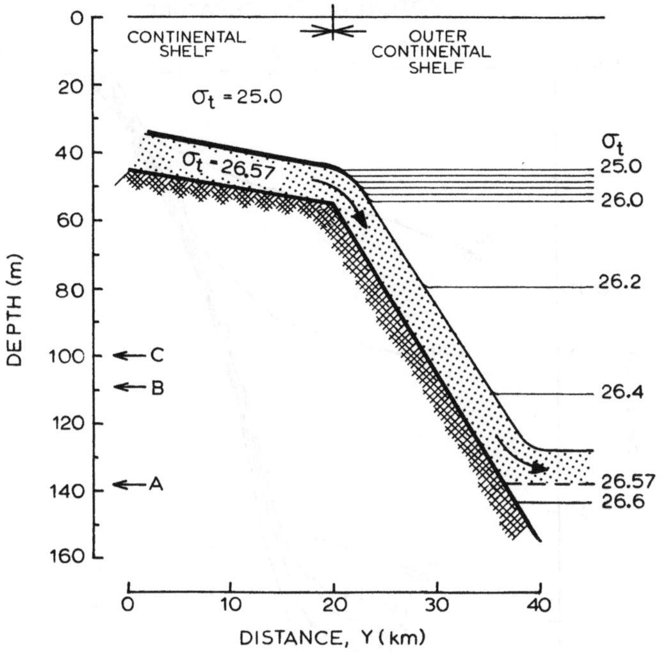

Fig. 27. Schematic section showing the application of a "stream-tube" model to the particular seafloor geometry and stratification encountered by a density current flowing off-shelf in the Beaufort Sea. (From Melling and Lewis, 1982.)

number of studies on convection in leads and polynyas (see Foster, 1972b; Schaus and Galt, 1973; and Martinson, Killworth and Gordon, 1981, for references). Recent measurements in a large semi-permanent polynya in the Canadian Archipelago set the heat flux from open water in winter at roughly 400 Wm^{-2} (Lewis, personal communication), sufficient to cool a 100 m deep column of water by 1°C in about 10 days. There appear to be no studies on convection at the seasonal ice edge; such work will likely need to include the dual effects of wind and bouyancy forcing.

It is important to recognize that in the Arctic Ocean the depth of the isothermal layer is generally much greater than that of the (wintertime) isohaline layer due to the seaward advection of cold, relatively saline water from adjacent shelf regions (Figure 29). This suggests that the one-dimensional view of mixed-layer dynamics should be modified to include the effects of upwelling from the pycnocline layer.

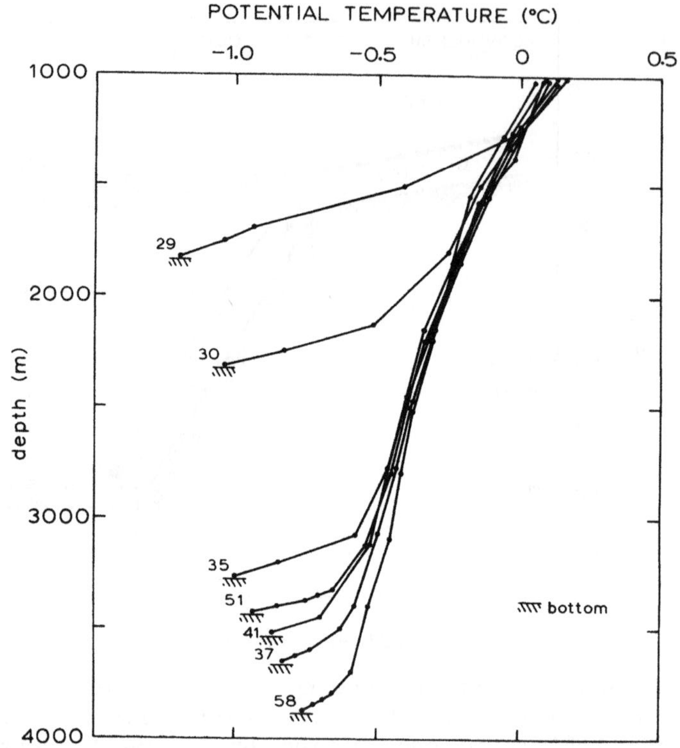

Fig. 28. Vertical profiles of potential temperature in the Weddell Sea obtained by the U.S.C.G.C. Glacier during February-March, 1968. (From Carmack, 1973.)

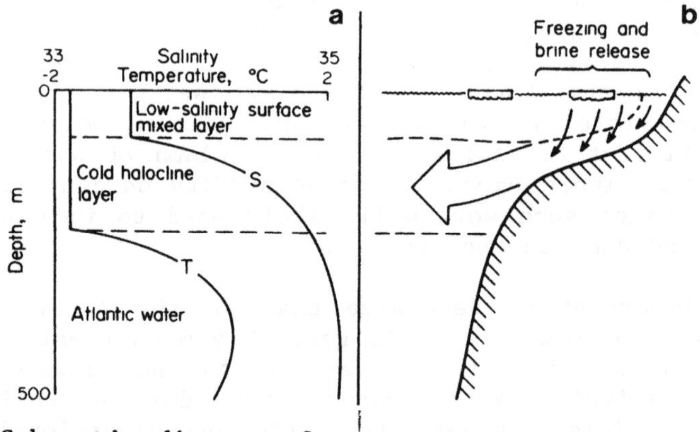

Fig. 29. Schematic diagram of temperature and salinity structure in the upper Arctic Ocean, and effects of lateral advection on mixed-layer dynamics. (From Aagaard, Coachman and Carmack, 1981.)

3.4.2 Thermohaline Convection. The unique temperature/salinity structure and frequent occurrence of regions of weak stability provide the setting for a number of thermohaline instability mechanisms.

Studies of thermohaline convection, more than any other aspect of physical oceanography, require accurate and precise equations for the calculation of water properties. The following references are suggested: the equation of state of Chen and Millero (1978); the equation for the freezing temperature of Fujino, Lewis and Perkin (1974), and the definition of salinity given by Lewis and Perkin (1981).

a) Double-diffusion -- Two requirements for double-diffusive convection are that the fluid contain two or more components of different molecular diffusivities, and that these components make opposite contributions to the vertical density gradient. Instability occurs when diffusion acts to release the potential energy of the destabilizing component. Two types of double-diffusive convection may be important in the ocean (Figure 30); the salt-finger instability in which warm saline water overlies cold, more dilute water (temperature stabilized), and the diffusive instability in which cold, fresh water overlies warm saline water (salt stratified). In polar oceans conditions suitable for the diffusive instability are found at the base of the surface layer, while conditions for salt fingering are found below the core of the warm, mid-depth water (Figure 31a). For a discussion of double-diffusive phenomena the reader is referred to Turner (1973) and Stern (1975). A simple model of the diffusive instability has recently been developed by Foldvik and Rudels (1979).

Gill and Turner (1969) proposed that the diffusive instability may lead to bottom water formation along the continental slope of Antarctica (Figure 31b). Carmack and Aagaard (1973) presented hydrographic data suggesting that Greenland Sea Deep Water is formed by the subsurface modification of Atlantic Water involving two large-scale, convective regimes coupled by a double-diffusive process (Figure 31c). Some evidence for diffusive instabilities in the Weddell Gyre is given by Foster and Carmack (1976b) and Middleton and Foster (1980).

There are other situations in polar oceanography in which the differing diffusivities of heat and salt are important including the transfer of heat in under-ice melt ponds (Martin and Kauffman, 1974) and the formation of underwater ice by heat transfer through the boundaries of vertically sinking brine plumes (Lewis, personal communication). It is also possible that the transient upwelling of Atlantic and/or Warm Deep Water along the continental slope may lead to the formation of multiple intrustions extending back into the interior of the ocean (cf. Ruddick and Turner, 1979; Figure 31d).

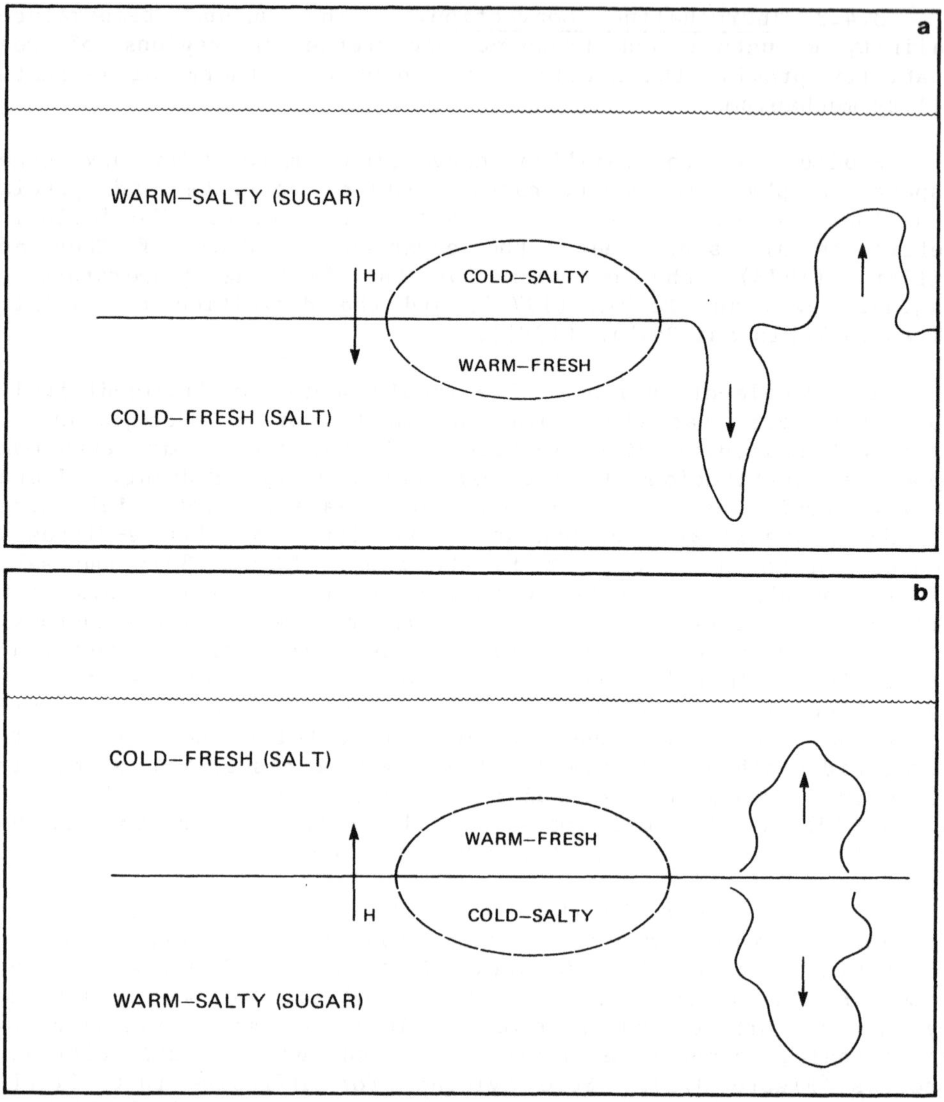

Fig. 30. Schematic diagram of double-diffusion: (a) conditions leading to the salt finger instability; and (b) conditions leading to the diffusive instability.

b) Cabbeling -- volume contraction occurs whenever two dissimilar water types are mixed; cabbeling refers to the special case in which the mixture is denser than either parent water type.

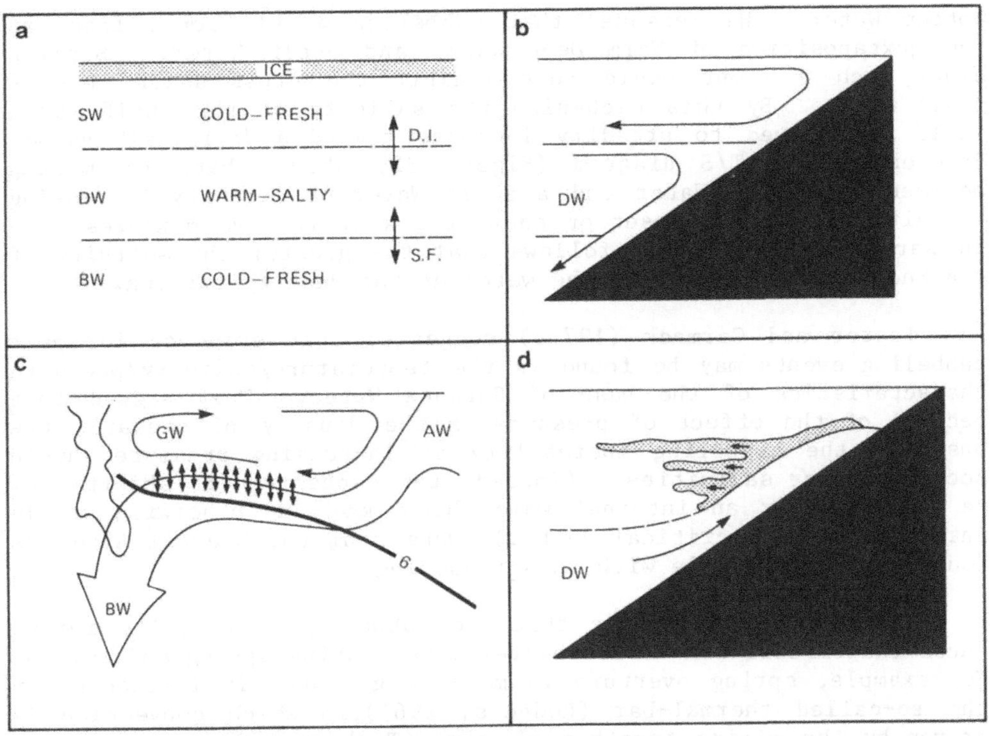

Fig. 31. Schematic diagrams of hypothetical situations involving double-diffusive convection in polar seas: (a) the case of diffusive instabilities at the interface between Surface Water and Warm Deep Water (or Atlantic Water), and salt finger instabilities below the core of Warm Deep Water; (b) the model of Antarctic Bottom Water formation originally proposed by Gill and Turner (1969); (c) the model of Greenland Sea Deep Water formation proposed by Carmack and Aagaard (1973); and (d) the potential for lateral interleaving associated with transient upwelling events.

There are two concerns associated with the cabbeling instability: the first involves the hydro-thermodynamics events leading to the production of a local instability; the second involves the response of the fluid to the presence of density extrema thus formed. For a discussion of the thermodynamic aspects see Mamayev (1975); for various treatments of the dynamic response of a fluid obtaining a local instability see Foster (1972a), Bowman and Okubo (1978), and Garrett and Horne (1978).

The ideas of cabbeling were introduced to polar oceanography by Fofonoff (1956) in reference to the formation of Antarctic

Bottom Water. He reasoned that cabbeling would form a front at the juxtaposition of Warm Deep Water and Shelf Water. Sinking along such a front would thus inhibit the cross shelf flow of shelf water. By this mechanism the salinity of the shelf water would be allowed to steadily increase toward a "critical" value. Reference to a T/S diagram (Figure 32) shows that for mixing between Warm Deep Water and a Shelf Water of steadily increasing salinity the initial onset of cabbeling will involve mixtures rich in Warm Deep Water. It follows that the greater the salinity of the Shelf Water the colder the water at the density extrema.

Foster and Carmack (1975b) suggested that evidence for past cabbeling events may be found in the temperature/salinity/pressure characteristics of the base of Surface Water. They argued that because of the effect of pressure on the density of seawater the onset of the cabbeling instability at increasing pressure should occur at lower salinities. (Indeed, the cabbeling instability may be triggered by an internal wave which moves an otherwise stable interface down to critical depth.) Data from the Weddell Gyre was found to be compatible with this reasoning.

There is also evidence that the cabbeling instability influences the circulation of freshwater lakes during spring and autumn. For example, spring overturn in many large lakes is influenced by the so-called thermal-bar (Rodgers, 1965),in which convection is driven by the mixing together of warm (T above 4°C) inshore water and cold (T below 4°C) offshore water to form a common, dense water mass. A circulation analogous to the thermal bar in which river inflow drives the cabbeling instability has been documented by Carmack (1979). An example of the riverine thermal bar is shown in Figure 33.

c) Pressure-related instabilities - Ekman (1934) pointed out that cold water is more compressible than warm water and that the vertical gradient of potential density at 1 atmosphere pressure is an inaccurate measure of stability at depth.

Fig. 32. Schematic diagram showing the author's interpretation of Fofonoff's (1956) hypothesis on the effects of cabbeling on the formation of Antarctic Bottom Water Formulation: (a) T/S correlation diagram showing the principal water masses, tangents to lines of constant density (broken lines), and mixing curves (1, 2, and 3) between Warm Deep Water of constant salinity and Shelf Water of variable salinity; (b) vertical section showing the distribution of water masses near the shelf-break; (c, d, e) circulations due to mixing between Warm Deep Water and Shelf Waters of variable (increasing) salinities.

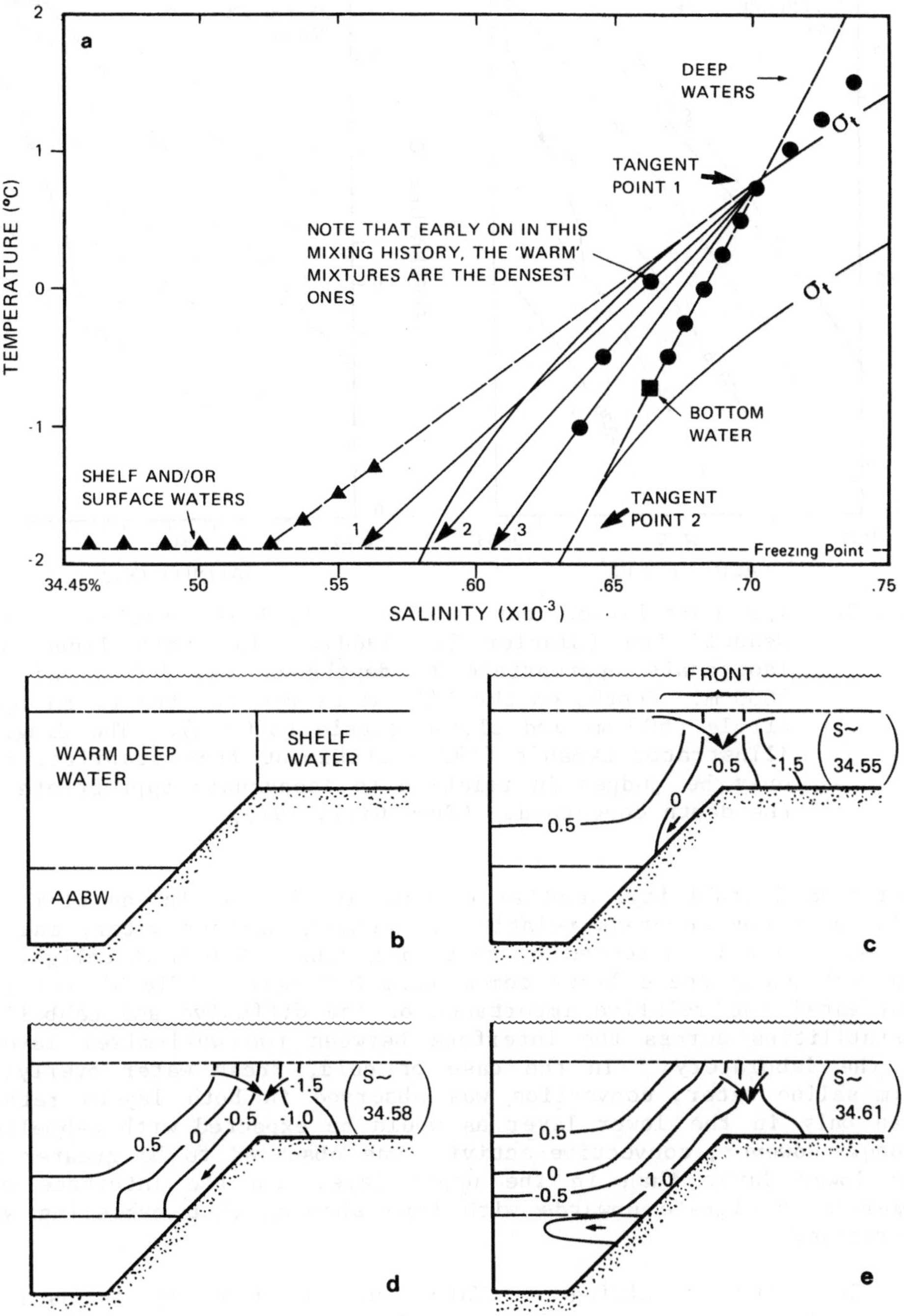
a
TEMPERATURE (°C)
SALINITY (X10^{-3})
34.45‰
.50
.55
.60
.65
.70
.75
2
1
0
-1
-2
DEEP WATERS
σ_t
TANGENT POINT 1
NOTE THAT EARLY ON IN THIS MIXING HISTORY, THE 'WARM' MIXTURES ARE THE DENSEST ONES
BOTTOM WATER
SHELF AND/OR SURFACE WATERS
TANGENT POINT 2
1
2
3
Freezing Point
WARM DEEP WATER
SHELF WATER
AABW
b
FRONT
S~ 34.55
-0.5
-1.5
0
0.5
c
S~ 34.58
-1.5
-0.5
-1.0
0.5
0
d
S~ 34.61
0.5
0
-0.5
-1.0
e

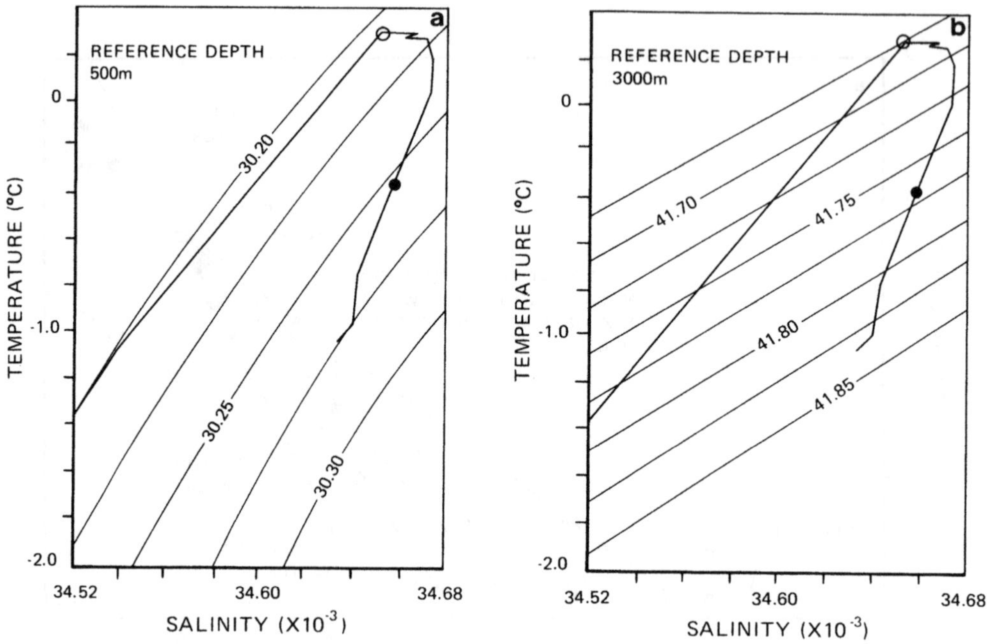

Fig. 34. T/S correlation curves for a typical station in the Weddell Sea (Glacier 21, 1968). The thin lines are isopycnals appropriate to depths of (a) 500 m and (b) 3000 m. Depths on the T/S curves are denoted by an open circle (500 m) and closed circle (3000 m). The diagram illustrates Ekman's (1932) statement that stability can only be judged in relation to isopycnals appropriate to the depth concerned. (From Gill, 1973).

diffusive instability; another example is that a sinking plume of cold water may compress relative to warmer, ambient water, and at the same time be affected by salt fingering. The most enlightening work along these lines comes from McDougal (1981a,b) who investigated the relative importance of the diffusive and cabbeling instabilities across the interface between two well-mixed layers in the laboratory. In the case of cold, fresh water overlying warm saline water, convection was observed in both layers rather than only in the lower layer as would be expected with cabbeling alone. However, convective activity was observed to be greater in the lower layer than in the upper layer and the interface was observed to migrate upwards with time showing that cabbeling was operative.

e) Stability to mixing -- Water column stability is usually expressed in terms of the buoyancy frequency, N, which gives the specific force required to displace a water parcel from its

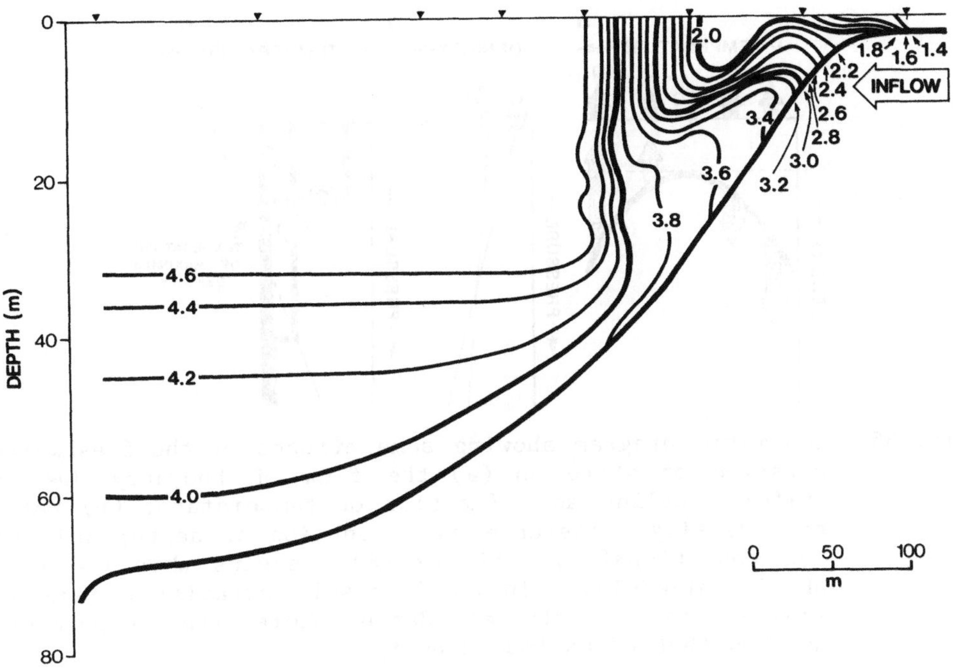

Fig. 33. An example of the Riverine Thermal Bar.

Gill (1973) noted the curious feature that if Surface Water of the Weddell Gyre was somehow moved to the sea floor compression would render it denser than ambient bottom water (Figure 34). Subsequent observational (Foster and Carmack, 1976a) and theoretical (Killworth 1977) studies have shown pressure related effects to be an important aspect of the sinking of newly formed bottom water.

Farmer and Carmack (1982) describe an additional way in which pressure affects vertical convection in fresh water lakes (Figure 35). Because the temperature of maximum density decreases with depth, at about 0.02°C $(\mathrm{bar})^{-1}$, an inversely stratified lake can be rendered conditionally unstable if the base of the mixed layer is lowered to a "critical" depth where its temperature matches that of maximum density at ambient pressure. Above this depth the wind must work against buoyancy during cooling, while below it the water is gravitationally unstable. This process explains why many temperate and high-latitude lakes have waters colder than 4°C down to depths greater than one would expect from wind mixing alone.

d) Mixed mixing mechanisms -- The internal structure of the ocean may be altered by mechanisms which work in concert with one another (cf. Joyce, 1977). For example, the temperature/salinity structure required for the onset of cabbeling is also supportive of the

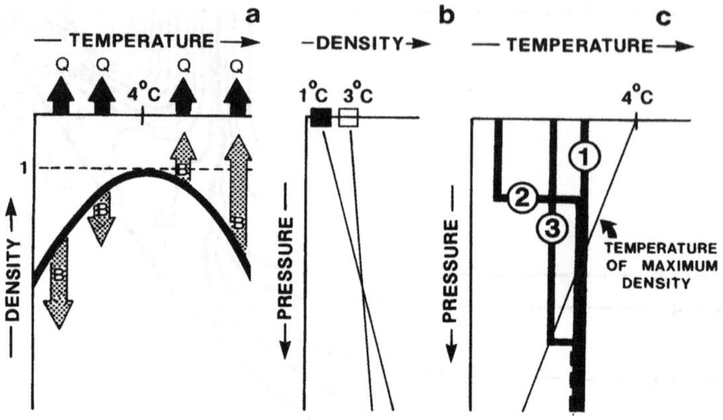

Fig. 35. Schematic diagram showing some effects of the freshwater equation of state on (a) the flux of buoyancy due to uniform cooling as a function of temperature; (b) relative density difference as a function of depth; and (c) the conditional instability that results when the base of the mixed-layer in an inversely stratified lake is lowered to a critical depth where its temperature matches that of maximum density.

equilibrium position assuming that the parcel is neither compressed nor allowed to mix during its excursion. Fofonoff (1961) considered the problem more realistically in that he accounted for the energy change subsequent to deformation within a water column (Figure 36). He derived an expression for total stability ET as given by ET = N^2 + EM + EC, where the buoyancy frequency is

$$N^2 = -\frac{g^2}{V^2}\left\{\frac{\delta V}{\delta T}\left[\frac{dT}{dP} - \left(\frac{dT}{dP}\right)_A\right] + \frac{\delta V}{\delta S}\left(\frac{dS}{dP}\right)\right\}$$

and the stability to mixing and compression is

$$EM = -\frac{Pg^2}{V^2}\left\{\frac{\delta^2 V}{\delta T^2}\left[\frac{dT}{dP} - \left(\frac{dT}{dP}\right)_A\right]^2 + 2\frac{\delta^2 V}{\delta T\delta S}\left[\frac{dT}{dP} - \left(\frac{dT}{dP}\right)_A\right]\cdot\frac{dS}{dP} + \frac{\delta^2 V}{\delta S^2}\left(\frac{dS}{dP}\right)^2\right\}$$

$$EC = -\frac{Pg^2}{V^2}\left\{\frac{\delta}{\delta T}\left(\frac{\delta V}{\delta P}\right)_A\left[\frac{dT}{dP} - \left(\frac{dT}{dP}\right)_A\right] + \frac{\delta}{\delta S}\left(\frac{\delta V}{\delta P}\right)_A\left(\frac{dS}{dP}\right)\right\}$$

This expression shows that the release of energy subsequent to a mixing event may actually exceed the amount of energy required for the initial displacement (Figure 36); Fofonoff (1961) suggested

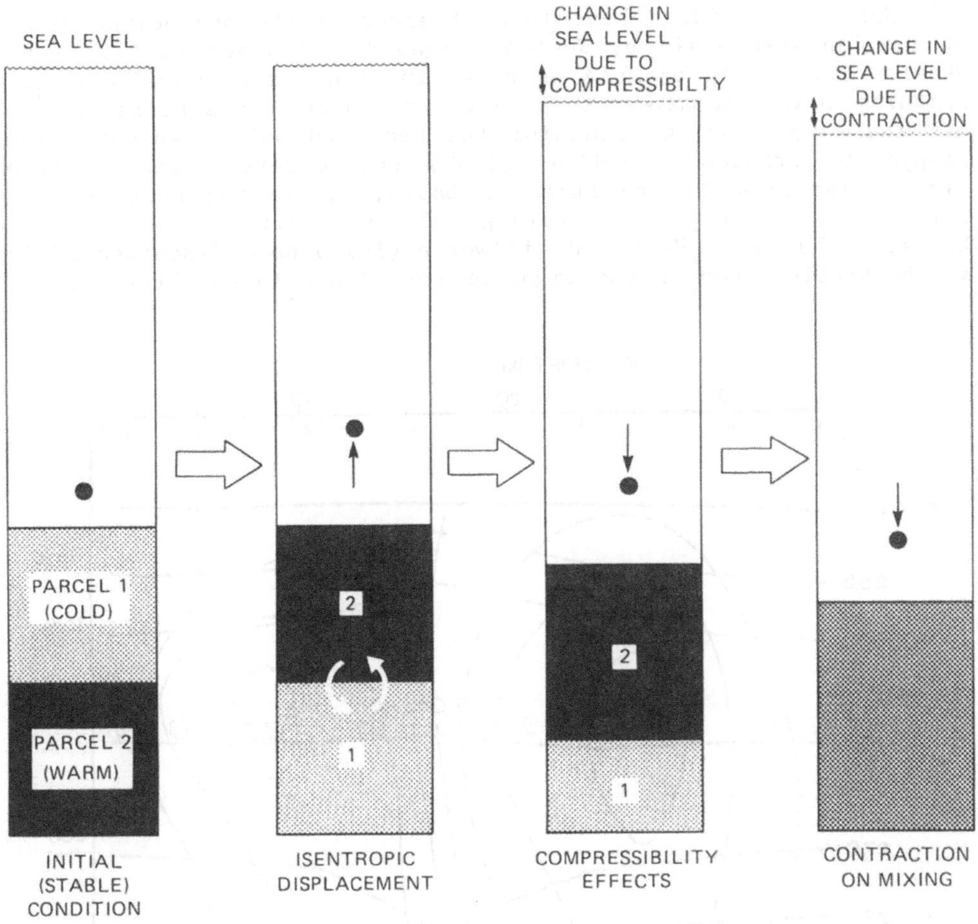

Fig. 36. Schematic diagram of energy changes associated with the displacement and deformation of two water parcels.

that such an energy conversion mechanism may actually help drive ocean currents.

3.4.3 Eddies, Chimneys, and Open-ocean Convection. Observations of transient, mesoscale gyres, which appear to scale with the Rossby radius, have been made in both the Arctic and Southern Oceans during the past decade. Eddies are usually associated with the shedding of kinetic energy by larger currents, perhaps through barocline instabilities. Chimneys are deep-reaching, convective features formed by intense buoyancy fluxes. Chimneys may be the most important vehicle for deep convection in the open-ocean (Killworth, 1981).

Subsurface eddies have been observed in the pycnocline waters of the Canadian Basin by Newton, Aagaard and Coachman (1974) and Hunkins (1974); an example is shown in Figure 37. These lens-like features have anomalous temperature/salinity characteristics and may thus represent a mechanism for heat and salt transport along isopycnal surfaces. McPhee (1980) has observed and modelled similar features in the Eurasian basin, suggesting that they are ultimately caused by the drainage of cold saline water from the Eurasian shelves. Hart and Killworth (1976) have discussed eddies in the Arctic Ocean in the light of baroclinic instability theory.

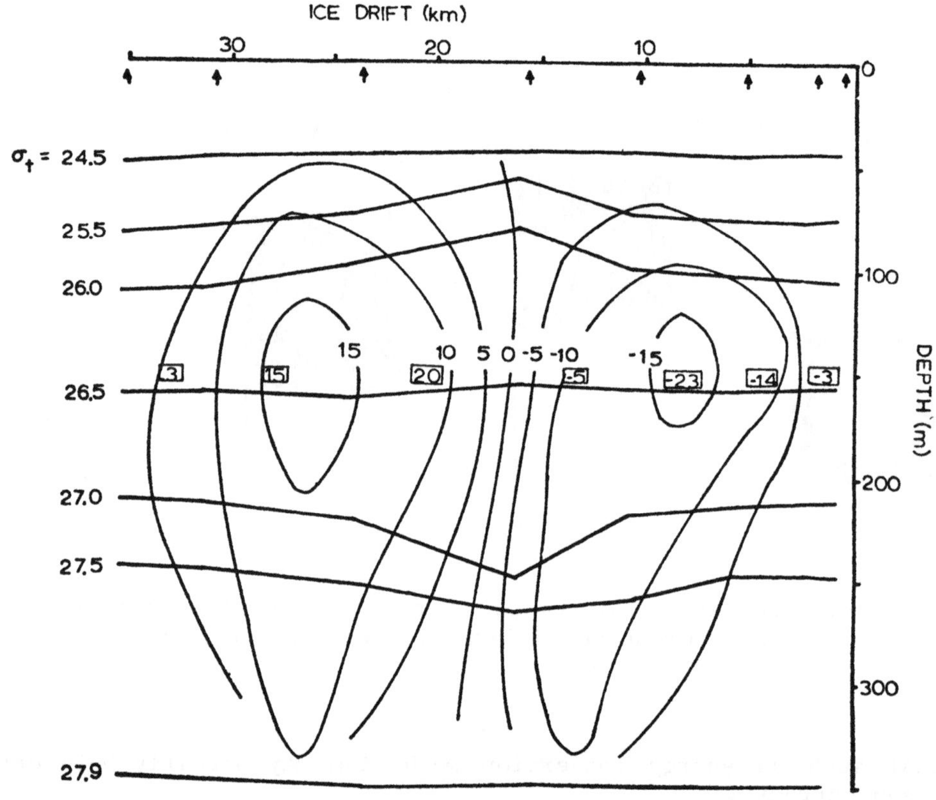

Fig. 37. Example of an eddy observed in the pycnocline of the Arctic Ocean. The vertical arrows locate the hydrographic stations. The isotachs of computed geostrophic currents are relative to 20 m (taken to represent the surface layer); the measured current differences (in boxes) are taken between the surface layer (represented by the measurements at 30 m) and the center of the pycnocline (150 m) containing the current maximum. (From Newton, Aagaard and Coachman, 1974.)

Chimneys represent a second gyral circulation (apparently cyclonic) which may strongly influence heat and salt transfer; in this case primarily in the vertical direction. One such feature, observed in the Weddell Sea, has been documented in vivid detail by Gordon (1978; Figure 38). Within this column of water, nearly 30 km in diameter and 4000 m deep, the normal stratification sequence was absent, replaced by cold, low-salinity and high-oxygen water. Gordon supposed that this cyclonic flowing water was forced by wintertime convection in a large semi-permanent polynya in the central Weddell gyre. He speculates that many such convective chimneys may be formed in the polynya region and that these cells may actively carry heat from the deep layers to the sea surface.

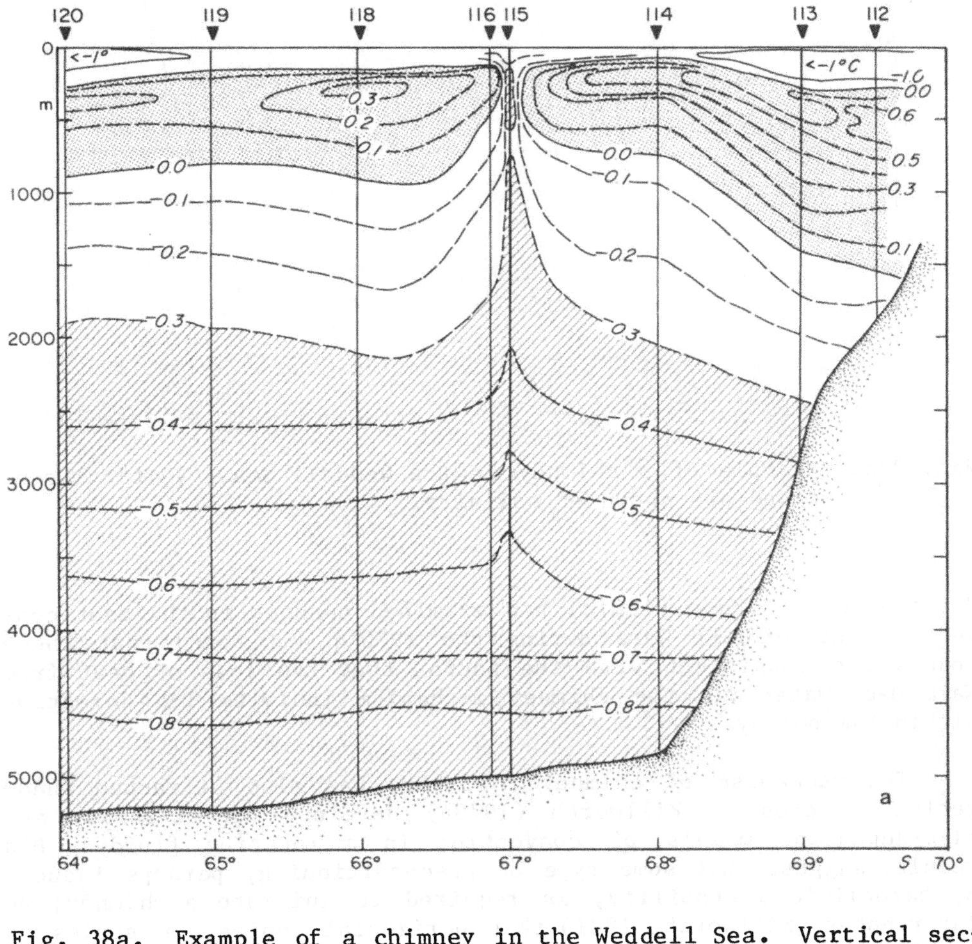

Fig. 38a. Example of a chimney in the Weddell Sea. Vertical sections are shown of temperature. (From Gordon, 1978.)

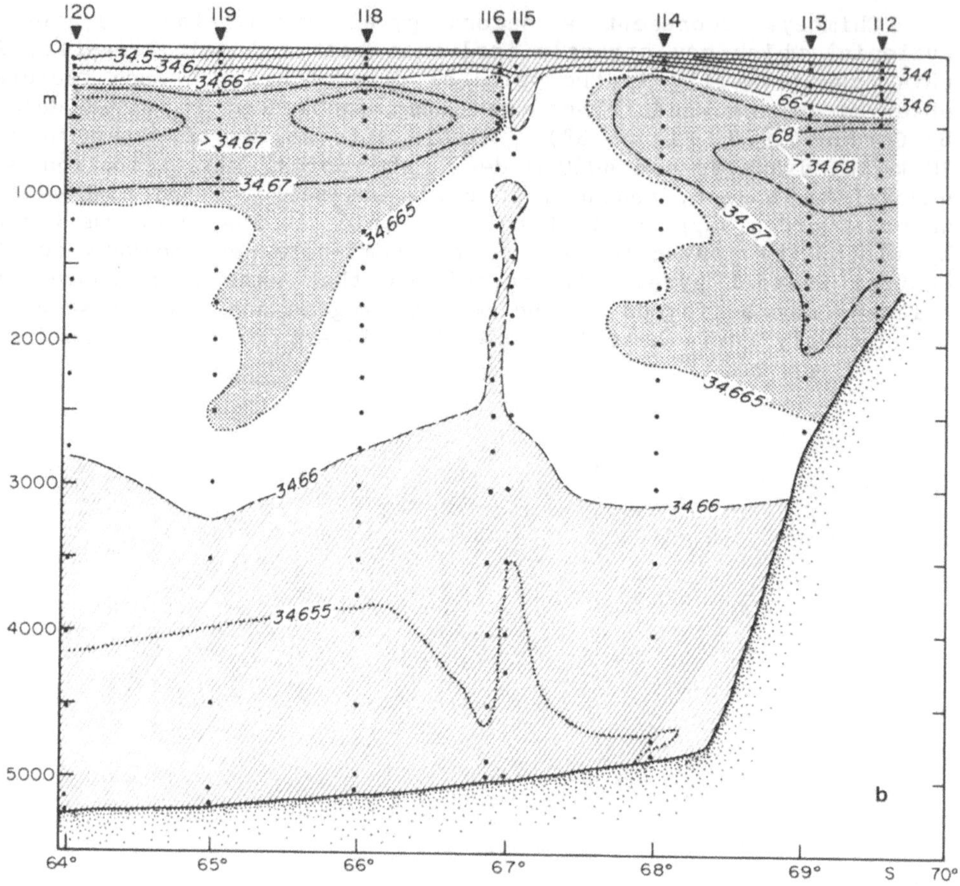

Fig. 38b. Example of a chimney in the Weddell Sea. Vertical sections are shown of salinity.

Subsequently, Gordon (1981a) has examined changes in the heat content of the Weddell Gyre during the 1970's and found that these losses are consistent with the notion that the loss of heat from Warm Deep Water via the chimney mechanism inhibits ice formation within the polynya.

The mechanism of chimney formation has also attracted theoretical interest. Killworth (1979a) developed quasi-steady and time-dependent models of convection in a rotating fluid. His results suggest that some type of preconditioning, perhaps induced by baroclinic instability, is required to initiate a chimney; he later notes (Killworth, 1981) that horizontal eddies, on scales as small as 10 km, have been observed at or near the boundary between overturning and non-overturning water.

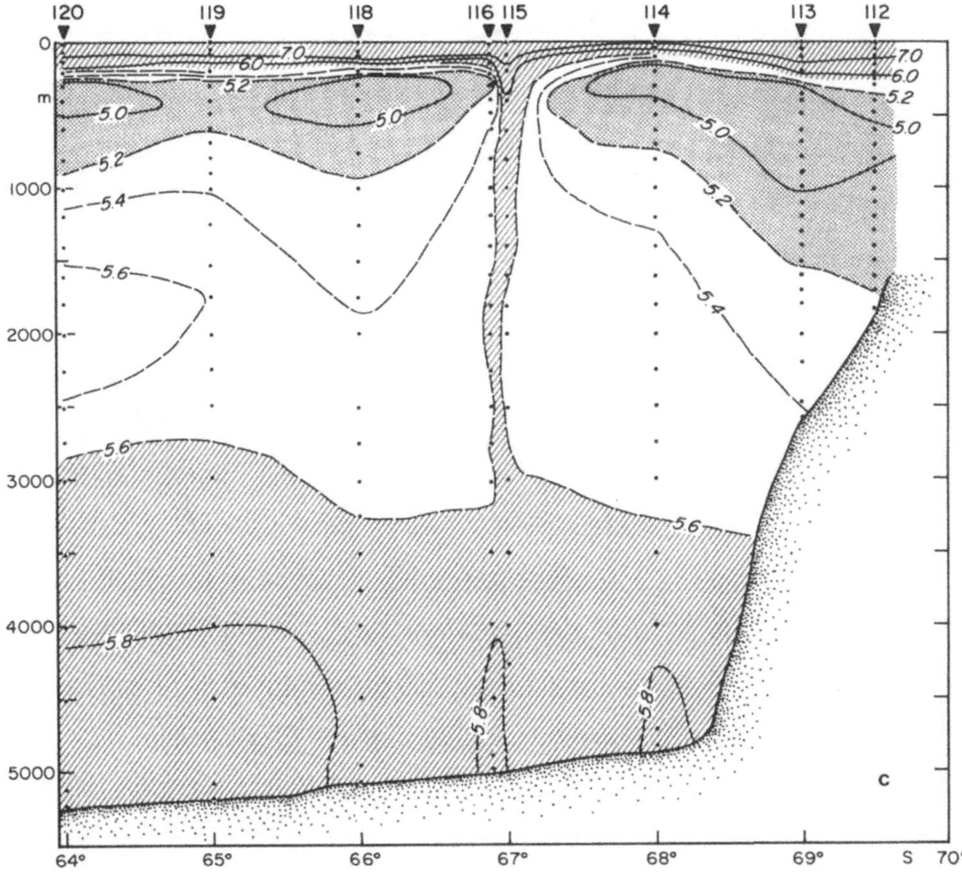

Fig. 38c. Example of a chimney in the Weddell Sea. Vertical sections are shown of oxygen.

Guided by the above findings, Martinson, Killworth and Gordon (1981) published a two layer model of the Weddell Polynya. They suppose that local, wind-induced upwelling initially shallows the pycnocline so that air-sea exchanges are more effective in driving the deep water overturn. In a closed system convection thus induced would tend to recur each subsequent winter. However, such a steady condition is likely to be disrupted by interannual variation in freshwater input and by the westward drift of the large scale circulation.

3.4.4 Internal Waves. At present very little is known about the internal wave field in polar seas; Morrison (1983; see Special Topic 2) has reviewed the available observations and theory of internal waves in the Arctic Ocean. Most descriptions of internal wave characteristics are based on following structural features in

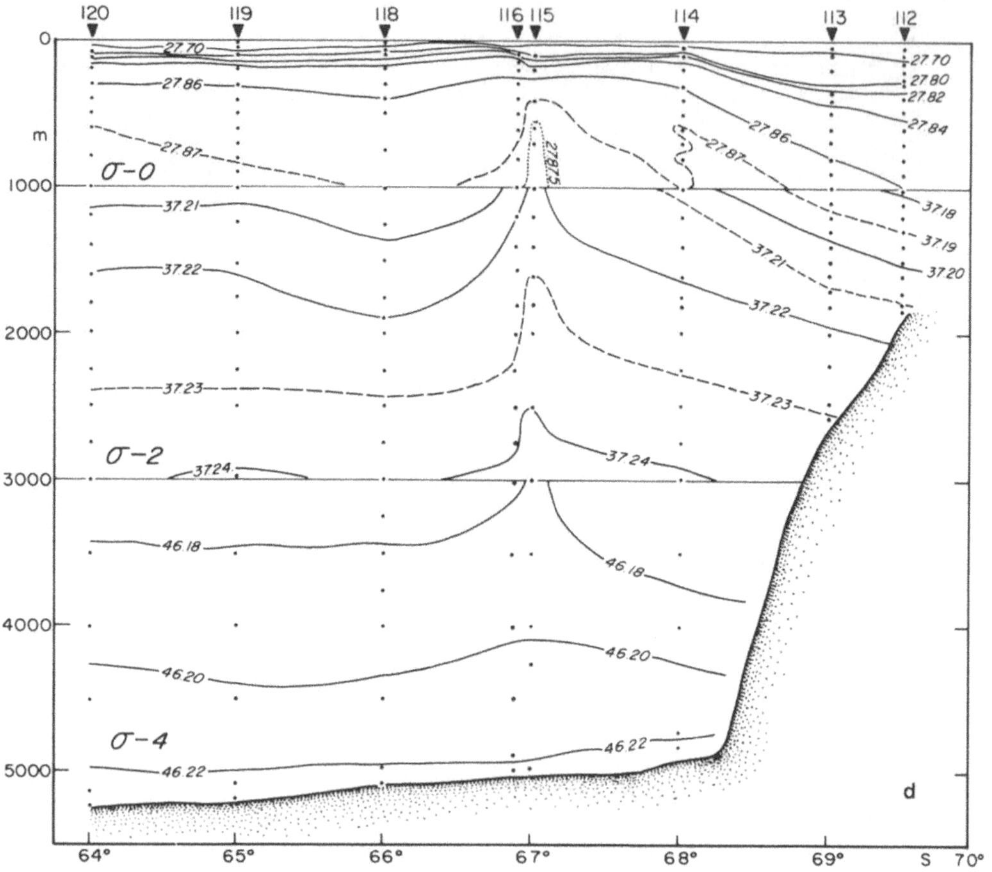

Fig. 38d. Example of a chimney in the Weddell Sea. Vertical sections are shown of potential density.

vertical profiles of temperature and salinity (see Neshyba, Neal and Denner, 1971). Georgi (1978) examined fine structure in the Drake Passage and attempted to separate variability due to internal waves from that due to interleaving on isopycnal surfaces. Middleton and Foster (1980) examined the variability in fine structure in the halocline of the Weddell Sea. They concluded that much of the variability between 90-300 m was due to internal waves with 8 to 10 hour periods and discussed their results in the view of the internal wave model of Garrett and Munk (1975).

Future studies may show internal waves to be an important component in edge mixing (see Stigebrandt, 1976; Armi, 1978 and

Ivey and Corcos, 1981, for reviews). Some support for this comes from the studies of Perkin and Lewis (1978) in Cambridge Bay of the Canadian Archipelago where the breaking of internal waves on the boundary influences the downward transport of salt by gravity currents (Figure 39).

3.4.5 The Ice Boundary. The oceanography of the seasonal ice zone is the focus of much current research. Phenomena associated with the ice margin include upwelling, convection, fronts, eddies, interleaving, and along-margin geostrophic flows. A review of this subject is given in the proceedings of a workshop on the seasonal ice zone edited by Anderson, Weeks and Newton (1980). In view of this the present report will touch only briefly on one aspect of the ice boundary, namely, upwelling.

The potential for upwelling at ice edges was noted by Gammelsrod, Mork and Roed (1975) using a homogeneous ocean model with stationary ice. In their model the presence of the ice edge causes an abrupt change in the divergence of the wind-driven circulation which in turn drives upwelling (Figure 40). Clarke (1978) extended their model to account for stratification, and Niebauer (1982) included the effects of meltwater on water column stability near the ice edge.

The above models all assume a stationary ice cover. However, Buckley, Gammelsrod, Johannessen, Johannessen and Rød (1979) point out that the ice edge moved several tens of kilometers during an upwelling episode. Because of this Rød and O'Brien (1981) have proposed an alternative explanation for upwelling at the ice edge. They note that the rheology of the ice boundary is highly dependent on local ice concentration; as the ice is stressed compression is resisted but not divergence. This results in a kind of "pressure" field normal to the ice boundary so that the leading edge of the ice moves outward from its initial position (Figure 41). When treated as a geostrophic adjustment problem, predicts the existence of an along-marine jet. Coupling of this jet to the underlying water column then drives upwelling.

3.5 Ice-covered Lakes

Brief attention is focused here on mechanisms which may affect circulation and ice distributions in lakes and their possible relationships to fjords and semi-enclosed basins.

Thermal stratification is inverse in an ice-covered lake. The formation of an ice-cover requires a period of both low

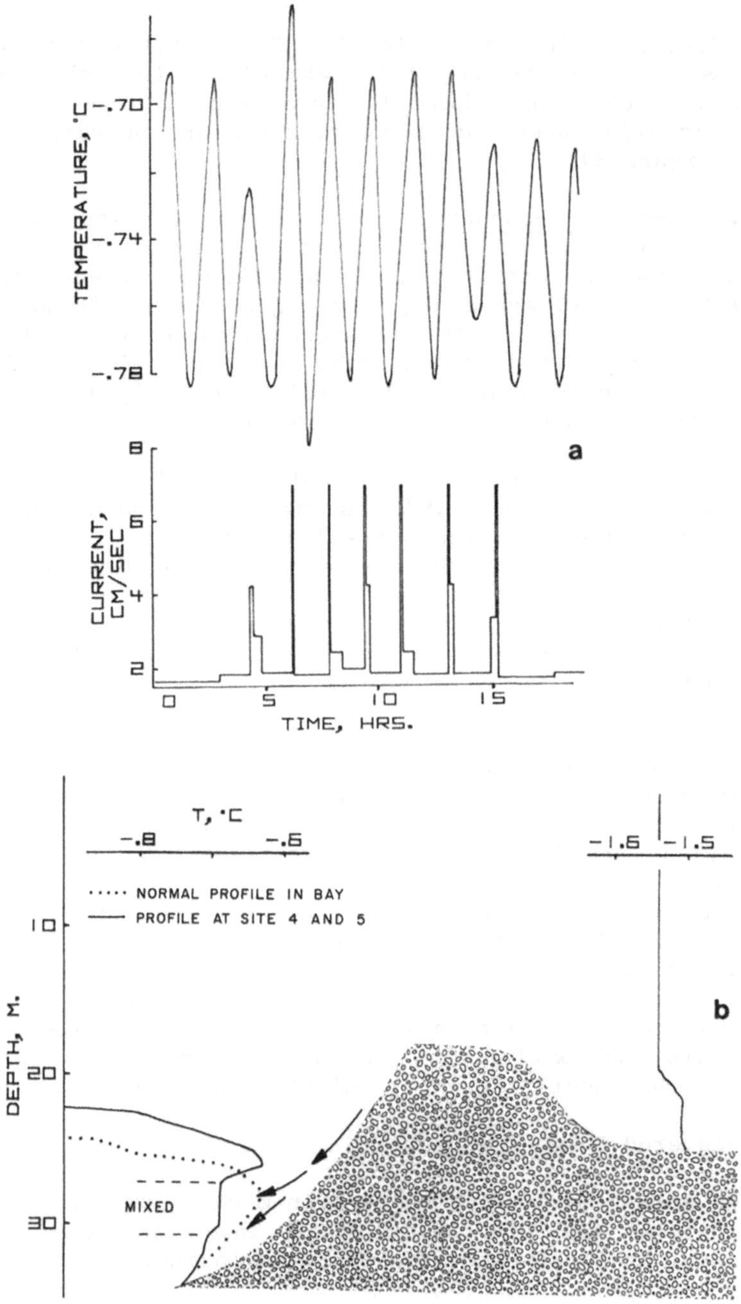

Fig. 39. Example of internal waves in an Arctic fjord: (a) temperature and current records from just below the pycnocline; (b) temperature profiles just within and without the sill compared to a typical profile in the interior of the fjord.

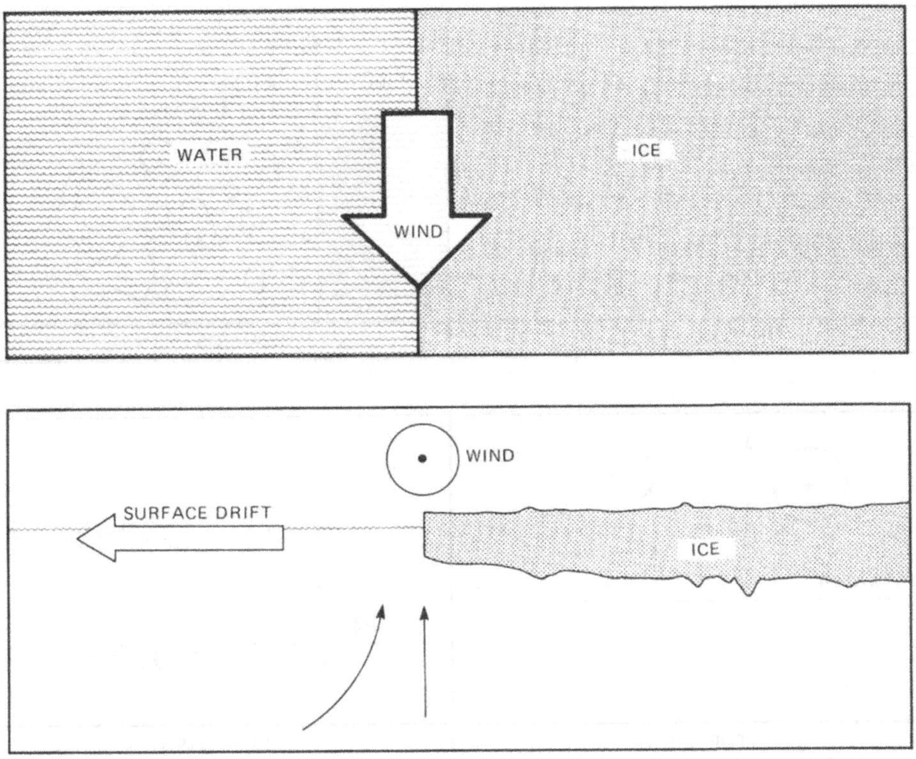

Fig. 40. Schematic diagram showing wind-induced upwelling along a stationary ice edge.

temperatures and calm wind conditions and many lakes continue to mix until deep temperatures drop well below 4°C (Yoshimura, 1936; Farmer and Carmack, 1981). Once ice is formed it serves both to shield the underlying water from wind mixing and to reduce heat loss to the atmosphere. Insulation due to ice and snow-cover is so effective that the heat content of some lakes actually increases during winter due to solar radiation and/or heat flow from sediments (cf. Hutchinson, 1957; Bengtsson, 1978). That circulation under ice may be driven by the release of heat and dissolved solids from bottom sediments into the overlying water which subsequently sink to greater depths as bottom currents was shown by Mortimer and Mackereth (1958). Spring overturn in deep lakes may be initiated by the deep convection driven by the absorption of solar radiation beneath ice-cover (e.g., Farmer, 1975). Circulation and ice distribution patterns may also be influenced by the inflow of river water as shown by Tesaker (1973) and Svenson and Larsson (1980).

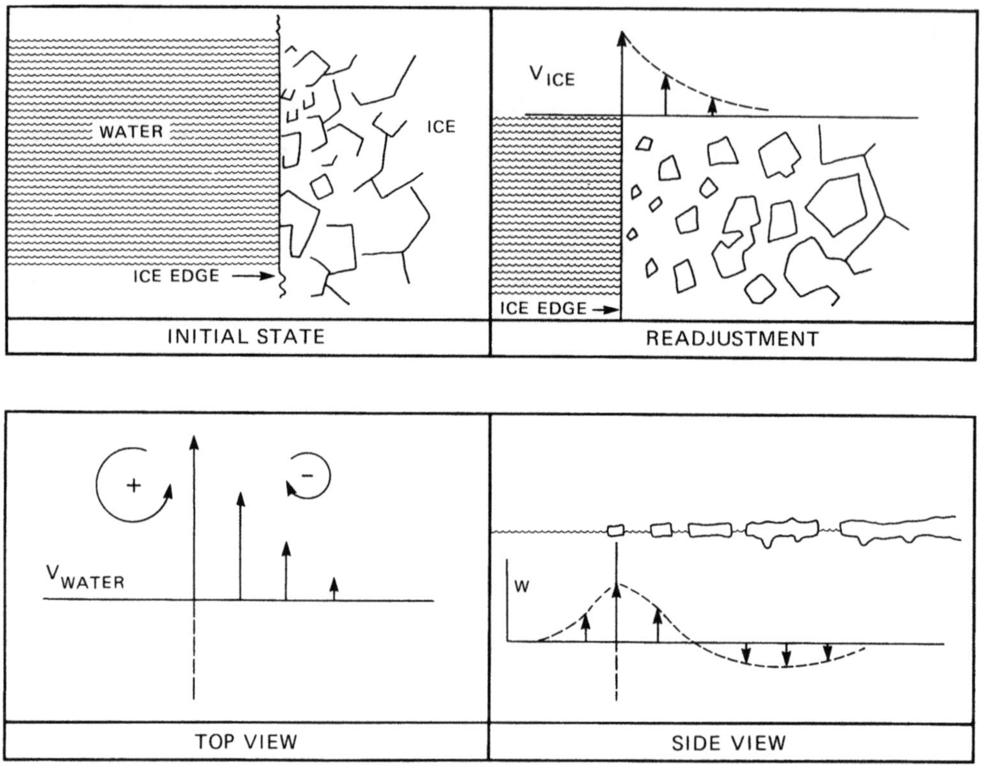

Fig. 41. Schematic diagram showing upwelling along a mobile ice edge.

Stigebrandt (1978) defined three dynamic regimes in an ice-covered lake with river through-flow (Figure 42): (a) the inflow region, where incoming river water mixes with ambient lake water to form the riverine water mass; (b) the outflow region, where the process of selective withdrawal operates; and (c) the lake proper, where baroclinic flow dominates. A common feature of ice-covered lakes with through-flow is that the outflow and/or inflow regions remain ice free throughout winter and thus act as centers of vertical mixing and heat exchange.

The behavior of river water upon entering the basin depends upon its density relative to that of lake water, the volume flux, and the geometry of the inlet, all of which may be expressed in terms of the densimetric Froude number (see Stigebrandt, 1978, for discussion). Normally, the inflow water is colder (less dense) than lake water and thus enters the lake as a surface overflow immediately below the ice-cover. An outstanding problem here is that of quantifying the mixing between lake and river water, e.g.,

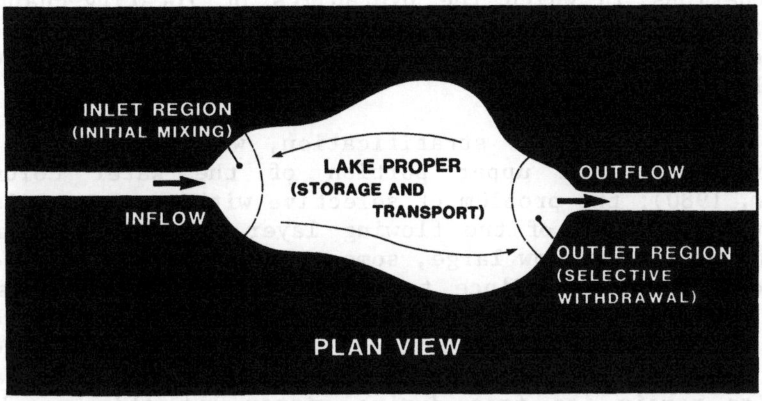

Fig. 42. Schematic diagram of a typical ice-covered lake showing the different dynamical regimes.

determining the net volume fluxes through the border between the inflow region and the lake proper.

After forming in the inflow region, the mixture leaves the mixing zone. If the lake is wider than the internal Rossby radius of deformation, generally a few hundred meters, the current may occur along the right hand shoreline in the direction of flow (northern hemisphere) due to the Coriolis force. The parameter characterizing the relative importance of rotation is the Rossby number (u_*/fD); with decreasing Rossby number the effects of rotation become more and more important. Furthermore, friction, both at the ice-water interface and below, results in Ekman pumping, a secondary circulation normal to mean flow direction, which tends to lift warmer water towards the surface along the right hand shoreline (cf. Stigebrandt, 1978; Svenson and Larsson, 1980).

Subsequent to freeze-up large lateral variations in density structure may be found in a single lake having two or more orographic and/or bathymetric regions. For example, in a lake with two basins separated by a sill, the larger, more exposed basin will tend to have deeper isopleths of a given density than the smaller, more sheltered one. With the ice-cover shielding further wind effects, the baroclinic flow will dominate interbasin exchange. However, dense water from the smaller basin will move only a small distance downslope of the sill before rotational effects become important. Whitehead (1980) has reviewed the hydraulics of rotating-channel flow and given a number of transport criteria based on specific geometry and stratification; for sill flow in a rotating channel whose width is greater than the Rossby radius the transport is given by $V_h = g'H^2/2f$.

Situations in which the hydraulics of rotating-channel flow become important also include ice-covered fjords connected to the open ocean via sills and the channels and passages of the Canadian Archipelago.

In the presence of stratification, water leaving a lake is drawn only from the upper portion of the water column (cf. Imberger, 1980); the problem of selective withdrawal entails determining the thickness of the flowing layer. If stratification is weak, and/or the outflow large, some water will be drawn from below the discharge depth. Since temperature generally increases with depth in an ice-covered lake, selective withdrawal may represent an important mechanism of vertical heat transport (Stigebrandt, 1978). Indeed, the outlet regions of many high-latitude lakes and reservoirs remain ice free during winter but this results also from the increase in flow speed.

4. FUTURE DIRECTIONS

A number of recent reports and publications are available which discuss the outstanding problems of the physical oceanography of polar regions: e.g., SCOR WG-58 (1978), Polar Group (1980), Weeks, Anderson and Newton (1980), Wadhams, Martin, Johannessen, Hibler and Campbell (1981), Deacon (1981), and Carsey, Ramseier and Weeks (1982), among others.

In this chapter an attempt has been made to give an overview of what is presently known about the regional hydrography of polar seas and of the physical processes that effect heat and mass transport. From this approach it appears that more is known, generally, about the physical oceanography of the deep Southern Ocean than that of the deep Arctic Ocean. For example, the large-scale oceanography of the southern Ocean is sufficiently well-known to allow, at least, the efficient planning of interdisciplinary and process-related studies. This does not appear to be the case for the central Arctic Ocean, where even basic problems such as the origin of bottom water or the thermal history of Atlantic Water are largely unresolved. It follows that a well-planned hydrographic survey of the Arctic basin is of the highest priority.

Additional work should be carried out in both polar regions which will allow a better understanding of mass and energy budgets. For example, more accurate estimates are required of the advective/diffusive flux of heat across the Polar Front of the Southern Ocean (cf. Gordon, 1974; Bryden, 1979), and a similar challenge exists concerning the exchange through Fram Strait (SCOR WG58, 1979). Within the general framework of heat budgets it is also crucial to understand processes governing exchanges across and along the continental margins.

Strong efforts should be made to obtain a general knowledge of seasonal variability in polar regions: e.g., the seasonality of sea-ice (cf. Gordon, 1981a), the effects of seasonal ice-cover on circulation and mixing (cf. Anderson, Weeks, and Newton, 1980); and the seasonality of continental shelf regions (cf. Aagaard, 1981b).

A comprehensive survey of theoretical work applicable to hydrographic conditions in polar seas has not been attempted in this chapter. However, owing to the scarcity of such work relative to other regions of the world ocean, one can only assume that the prospects for new discoveries in this area of study are good.

REFERENCES

Aagaard, K. (1970) The wind-driven transports in the Greenland and Norwegian Seas. Deep-Sea Res., 17: 281-291.

Aagaard, K. (1972) On the drift of the Greenland ice pack. In Sea Ice, National Research Council, p. 17-22.

Aagaard, K. (1979) Polar oceanograph I: Arctic Ocean. Rev. Geophys. Space Phys. 17: 1576-1578.

Aagaard, K. (1981a) On the deep circulation in the Arctic Ocean. Deep-Sea Res. 28: 251-268.

Aagaard, K. (1981b) Current measurements in possible disposal region of the Beaufort Sea. Unpublished manuscript.

Aagaard, K. and L. K. Coachman (1975) Toward an ice-free Arctic Ocean. EOS, 56: 484-486.

Aagaard, K. and P. Griesman (1975) Towards new mass and heat budgets for the Arctic Ocean. J. Geophys. Res., 80: 3821-3827.

Aagaard, K., L. K. Coachman and E. C. Carmack (1981) On the halocline of the Arctic Ocean. Deep-Sea Res. 28.

Aagaard, K., A. Foldvik and B. Rudels (1981) Fysdisk Oceanography. YMER Expedition Reports, p. 111-121.

Anderson, B. G., W. F. Weeks and J. L. Newton (Eds.) (1980) The Seasonal Sea Ice Zone: Proceedings of an International Workshop. Cold Reg. Sci. Tech., 2: 357 p.

Armi, L. (1978) Some evidence for boundary mixing in the deep ocean. J. Geophys. Res., 83: 1971-1979.

Ashton, G. D. (1978) River ice. Annu. Rev. Fluid Mech., 10: 369-392.

Baker, J. D. (1979a) Ocean - atmosphere interaction in high southern latitudes. Dyn. Atmos. Oceans, 3: 213-229.

Baker, J. D. (1979b) Polar oceanography. 2: Southern Ocean. Rev. Geophys. Space Phys., 17: 1578-1585.

Bengtsson, L. (1978) Winter stratification in a lake dominated by through-flow. In: Modelling, the Dynamic Phenomena in Lakes (L. Bengtsson, Ed.), University of Lulea, Sweden, p. 25-45.

Bowman, M. J. and A. Okubo (1978) Cabbeling at thermohaline fronts. J. Geophys. Res., 83: 6173-6178.

Brennecke, W. (1921) Die Ozenographischen Arbeiten der Deutschen Antarktischen Expedition, 1911-12. Aus dem Archiv de Deutschen Seewarte, 39, 214 p.

Bryden, H. (1979) Poleward heat flux and conversion of available potential energy in Drake Passage. J. Mar. Res., 37: 1-22.

Buckley, J. R., T. Gammelsrød, J. A. Johannessen, O. M. Johannessen and L. P. Røed (1979) Upwelling: Oceanic structure at the edge of the arctic ice pack in winter. Science, 203: 165-167.

Callahan, J. E. (1972) The structure and circulation of deep water in the Antarctic. Deep-Sea Res., 19: 563-575.

Campbell, W. J. (1965) The wind driven circulation of ice and water in a polar ocean. J. Geophys. Res. 70: 3279-3301.

Campbell, W. J. and S. Martin (1973) Oil and ice in the Arctic Ocean: Possible large-scale interactions. Science, 181: 56-58.

Carmack, E. C. (1972) On the hydrography of the Greenland Sea. University of Washington, Ph.D. thesis, 185 p.

Carmack, E. C. (1973) Silicate and potential temperature in the deep and bottom waters of the western Weddell Sea. Deep-Sea Res., 20: 927-932.

Carmack, E. C. (1974) A quantitative characterization of water masses in the Weddell Sea during summer. Deep-Sea Res., 21: 431-443.

Carmack, E. C. (1977) Water characteristics of the Southern Ocean south of the polar front. In: A Voyage of Discovery, George Deacon 70th Anniversary Volume (M. Angel, Ed.), Pergamon Press, Oxford, p. 15-41.

Carmack, E. C. (1979) Combined influence of inflow and lake temperatures on spring circulation in a riverine lake. J. Phys. Oceanogr., 9: 422-434.

Carmack, E. C. and K. Aagaard (1973) On the deep water of the Greenland Sea. Deep-Sea Res. 20: 687-715.

Carmack, E. C. and T. D. Foster (1975a) Circulation and distribution of oceanographic properties near the Filchner Ice Shelf. Deep-Sea Res., 22: 77-90.

Carmack, E. C. and T. D. Foster (1975b) On the of flow water out of the Weddell Sea. Deep-Sea Res., 22: 711-724.

Carmack, E. C. and P. D. Killworth (1978) Formation and interleaving of abyssal masses off Wilkes Land, Antarctica. Deep-Sea Res., 25: 357-369.

Carmack, E. C. and D. M. Farmer (1982) Cooling processes in deep temperate lakes: A review with examples from two lakes in British Columbia. J. Mar. Res., 40: Supplement 85-111.

Carsey, F. D. (1980) Microwave observations of the Weddell Polynya. Mon. Wea. Rev. 108: 2032-2044.

Carsey, F. D., R. O. Ramsier and W. F. Weeks (1982) Sea-ice mission requirements for the U.S. FIREX and Canada RADASAT programs. Jet Propulsion Laboratory, Pasadena, California, Publication 82-24, 60 p.

Chen, C.-T. (1982) On the distribution of anthropogenic CO_2 in the Atlantic and Southern Oceans. Deep-Sea Res., 29: 563-580.

Chen, C. T. and F. J. Millero (1978) The equation of state of pure water determined from sound speeds. J. Mar. Res., 36: 657-691.

Clarke, A. J. (1978) On wind-driven quasi-geostrophic water movement at fast ice edges. Deep-Sea Res., 25: 41-51.

Coachman, L. K. and K. Aagaard (1974) Physical oceanography of arctic and subarctic seas. In: Marine Geology and Oceanography of the Arctic Seas. Springer-Verlag, New York, p. 1-72.

Coachman, L. K., K. Aagaard and R. B. Tripp (1975) Bering Strait: The Regional Physical Oceanography. University of Washington Press, Seattle, Washington, 1972 p.

Coachman, L. K. and R. L. Charnell (1977) Fine structure in outer Bristol Bay, Alaska. Deep-Sea Res., 24: 869-889.

Coachman, L. K., T. H. Kinder, J. D. Schumacher and R. B. Tripp (1980) Frontal systems of the southeastern Bering Sea strait. In: Proceedings, International Symposium on Stratified Flows, 2nd (T. Carstens and T. McClimans, Eds.), Trondheim, p. 917-927.

Csanady, G. T. (1981) Circulation in the coastal ocean: Part 1. EOS, 62: 9-11.

Deacon, G. E. R. (1937) The hydrology of the Southern Ocean. 'Discovery' Reports, 15: 1-124.

Deacon, G. E. R. (1981) Zonation in the Southern Ocean. Deep-Sea Res.

Ehlin, U. (1976) Cooling-water discharge in Sweden: Studies in models and prototype. La Tribune du Cebedeau: Centre Belgie d'Etude et de Documentation des Eaux, 29: 302-208.

Ekman, V. W. (1934) Review of "Das Bodenwasser und die Gliederund der Atlantischer Tiefesse" by Georg Wust. J. Conseil, 9: 102-104.

Fandry, C. G. and R. D. Pillsbury (1979) On the estimation of absolute geostrophic volume transport applied to the Antarctic Circumpolar Current. J. Phys. Oceanogr., 9: 449-455.

Farmer, D. M. (1975) Penetrative convection in the absence of mean shear. Qart. J. Roy. Met. Soc. Society, 101: 869-891.

Farmer, D. M. and E. C. Carmack (1981) Mixing and restratification in a lake near the temperature of maximum density. J. Phys. Oceanogr. 11: 1516-1533.

Fletcher, J. O. (1965) The heat budget of the Arctic Basin and its relation to climate. Rand Corp., Santa Monica, California, R-444-PR, 108 p.

Fofonoff, N. P. (1956) Some properties of sea water influencing the formation of Antarctic bottom water. Deep-Sea Res., 4: 32-35.

Fofonoff, N. P. (1961) Energy transformations in the sea. Fisheries Research Board of Canada, Manuscript Report Series, Oceanographic and Limnological, 109, 82 p.

Foldvik, A. and T. K. Kvinge (1974) Conditional instability of sea water at the freezing point. Deep-Sea Res., 21: 169-174.

Foldvik, A. and B. Rudels (1979) Double diffusive convection into deep homogeneous layers. Geophysical Institute, Division A, University of Bergen, Norway, Report 53, 22 p.

Foster, T. D. (1968) Haline convective induced by freezing of sea water. J. Geophys. Res. 73: 1933-1938.

Foster, T. D. (1969) Experiments on haline convection induced by freezing of sea water. J. Geophys. Res., 74: 6967-6974.

Foster, T. D. (1972a) An analysis of the cabbeling instability in sea water. J. Phys. Oceanogr., 2: 294-301.

Foster, T. D. (1972b) Haline convection in leads and polynyas. J. Phys. Oceanogr., 2: 462-469.

Foster, T. D. (1978) Polar oceans: Similarities and differences in their physical oceanography. In: Polar Research: To the Present, and the Future (M. McWhinne, Ed.), Westview Press, p. 117-140.

Foster, T. D. and E. C. Carmack (1976a) Frontal zone mixing and Antarctic bottom water formation in the southern Weddell Sea. Deep-Sea Res., 23: 301-317.

Foster, T. D. and E. C. Carmack (1976b) Temperature and salinity structure in the Weddell Sea. J. Phys. Oceanogr., 6: 36-44.

Foster, T. D. and E. C. Carmack (1977) Antarctic bottom water formation. In: Polar Oceans (M. Dunbar, Ed.), Arctic Institute of North America, p. 167-177.

Foster, T. D. and J. H. Middleton (1979) Variability in the bottom water of the Weddell Sea. Deep-Sea Res., 26: 743-762.

Foster, T. D. and J. H. Middleton (1980) Bottom water formation in the Weddell Sea. Deep-Sea Res., 27: 367-382.

Foster, T. D., A. Foldvik and J. H. Middleton (1981) Recent oceanographic results from the Weddell Sea. Unpublished manuscript presented at AGU spring meeting, Baltimore.

Fujino, K., E. L. Lewis and R. G. Perkin (1974) The freezing point of seawater at pressures up to 100 bars. J. Geophys. Res., 79: 1792-1797.

Gade, H. G. (1979) Melting of ice in sea water. A primitive model with applications to the Antarctic ice shelf and icebergs. J. Phys. Oceanogr., 9: 189-198.

Galt, J. A. (1973) A numerical investigation of Arctic Ocean dynamics. J. Phys. Oceanogr., 3: 379-396.

Gammelstrød, T., M. Mork and L. P. Røed (1975) Upwelling possibilities at an ice edge: Homogeneous model. Mar. Sci. Comm., 1: 115-145.

Garrett, C. and W. Munk (1975) Space-time scales of internal waves: A progress report. J. Geophys. Res. 80: 291-297.

Garrett, C. and E. Horne (1978) Frontal circulation due to cabbeling and double diffusion. J. Geophys. Res., 83: 4651-4656.

Garrison, G. R. and P. Becker (1976) The Barrow Submarine Canyon: A drain for the Chukchi Sea. J. Geophys. Res., 81: 4445-4453.

Garrison, G. R. and R. G. Paquette (1982) Warm water interactions in Barrow Canyon. J. Geophys. Res., 87: 5853-5859.

Garwood, R. W. (1979) Air-sea interaction and dynamics of the surface mixed-layer. Rev. Geophys. Space Phys, 17: 1507-1524.

Garvine, R. W. (1977) River plumes and estuary fronts. In: Estuaries, Geophysics and the Environment (C. B. Officer, Ed.), National Academy of Sciences, Washington, D.C., p. 10-35.

Georgi, D. T. (1978) Fine structure in the Antarctic polar front zone: Its characteristics and possible relationship to internal waves. J. Geophys. Res., 18: 4579-4588.

Georgi, D. T. and J. M. Toole (1982) The Antarctic Circumpolar Current and the oceanic heat and freshwater budgets. J. Mar. Res., 40, Supplement: 183-197.

Gill, A. E. (1973) Circulation and bottom water production in the Weddell Sea. Deep-Sea Res., 20: 111-140.

Gill, A. E. and J. S. Turner (1969) Some new ideas about the formation of Antarctic bottom water. Nature, 224: 1287-1288.

Goody, R. (1980) Polar processes and world climate: a brief overview. Mon. Wea. Rev., 108: 1935-1942.

Gordon, A. L. (1971) Oceanography of Antarctic water. In: Antarctic Oceanography: I. Antarctic Research Series (J. L. Reid, Ed.), American Geophysical Union, 15, p. 169-203.

Gordon, A. L. (1974) Varieties and variability of Antarctic bottom water. Processus de formation des eaux oceaniques profondes. In: Colloques International du Centre National de la Recherche Scientifique, No. 215, p. 33-47.

Gordon, A. L. (1975) General ocean circulation. In: Numerical Models of Ocean Circulation. National Academy of Sciences, Washington, D.C.

Gordon, A. L. (1978) Deep Antarctic convection west of Maud Rise. J. Phys. Oceanogr., 8: 600-612.

Gordon, A. L. (1981a) Seasonality of Southern Ocean sea ice. J. Geophys. Res., 86: 4193-4197.

Gordon, A. L. (1981b) Weddell deep water: Source and variability. Ant. J. U.S., 16(5): 99-100.

Gordon, A. L. and P. Tchernia (1972) Waters of the continental margin off Adelie coast, Antarctica. In: Antarctic Oceanography. II: The Australian-New Zealand Sector, Antarctic Research Series (D. E. Hayes, Ed.), American Geophysical Union, 9, p. 59-69.

Gordon, A. L., H. W. Taylor and D. T. Georgi (1977) Antarctic oceanographic zonation. In Polar Oceans (M. Dunbar, Ed.), Arctic Institute of North America, p. 45-76.

Gordon, A. L., E. Molinelli and T. Baker (1978) Large-scale relative topography of the Southern Ocean. J. Geophys. Res., 83: 3023-3032.

Gordon, A. L. and W. D. Nowlin (1978) The basin waters of the Bransfield Strait. J. Phys. Oceanogr., 8: 258-264.

Gordon, A. L., D. G. Martinson and H. W. Taylor (1981) The wind-driven circulation in the Weddell-Enderby Basin. Deep-Sea Res., 28: 151-163.

Gordon, A. L., E. Molinelli and T. Baker (1981) Southern Ocean Atlas. Columbia University Press, New York.

Gregg, M. C. and M. G. Briscoe (1979) Internal waves, fine structure, microstructure and mixing in the ocean. Rev. Geophys. Space Phys., 17: 1524-1548.

Griesman, P. (1979) On upwelling driven by the melt of ice shelves and tidewater glaciers. Deep-Sea Res., 26: 1051-1066.

Hamblin, P. F. and E. C. Carmack (1978) River-induced current in a fjord lake. J. Geophys. Res., 83: 855-899.

Hanzlick, D. and K. Aagaard (1980) Freshwater and Atlantic water in the Kara Sea. J. Geophys. Res., 85: 4937-4942.

Hart, J. E. and P. H. Killworth (1976) An open ocean baroclinic instability in the Arctic. Deep-Sea Res., 23: 637-645.

Heezen, B., M. Tharp and C. Bently (1972) Morphology of the earth in the Antarctic and sub-Antarctic. In: Antarctic Map Folio Series, 16, American Geographic Society, New York.

Helland-Hansen, B. and F. Nansen (1909) The Norwegian Sea: Its physical oceanography based on the norwegian researches 1900-1904. Report on Norwegian Fishery and Marine Investigations, 2, 390 p.

Hood, D. W. and J. A. Calder (Ed.) (1980-81) The eastern Bering Sea shelf: Oceanography and resources. U.S. Department of Commerce, NOAA/OMPA, Washington, D.C., Vol. 1 (1980) and Vol. 2 (1981).

Hufford, G. L. (1974) On apparent upwelling in the southern Beaufort Sea. J. Geophys. Res., 79: 1305-1306.

Hunkins, K. L. (1974) Subsurface eddies in the Arctic Ocean. Deep-Sea Res., 21: 1017-1033.

Huppert, H. E. and J. S. Turner (1978) On melting icebergs. Nature, 271: 46-48.

Huppert, H. E. and E. G. Josberger (1980) The melting of ice in cold stratified water. J. Phys. Oceanogr., 10: 953-960.

Huppert, H. E. and J. S. Turner (1980) Ice blocks melting into a salinity gradient. J. Fluid Mech., 100: 367-384.

Huppert, H. E. and J. S. Turner (1981) Double-diffusive convection. J. Fluid Mech., 106: 299-329.

Hutchinson, G. E. (1957) A Treatise on Limnology, Volume 1. Wiley, New York, 1015 p.

Imberger, J. (1980) Selective withdrawal: A review. In: Proceedings, International Symposium on Stratified Flows, 2nd (T. Carstens and T. McClimans, Ed.), Trondheim, Norway, p. 381-400.

Imberger, J. and P. F. Hamblin (1982) Dynamics of lakes, reservoirs and cooling ponds. Annu. Rev. Fluid Mech., 14: 153-187.

Ivey, G. N. and G. M. Corcos (1981) Boundary mixing in a stratified fluid. J. Fluid Mech., 121: 1-26.

Jacobs, S. S., A. F. Amos and P. M. Bruchhausen (1970) Ross Sea oceanography and Antarctic bottom water formation. Deep-Sea Res., 17: 935-962.

Jacobs, S. S. and D. T. Georgi (1977) Observations on the southwest Indian/Antarctic Ocean. In: Voyage of Discovery (M. V. Angel, Ed.), Deep-Sea Res., supplement to Vol. 24, p. 43-84.

Jacobs, S. S., A. L. Gordon and J. L. Ardai (1979) Circulation and melting beneath the Ross Ice Shelf. Science, 203: 439-443.

Jacobs, S. S., A. L. Gordon and A. F. Amos (1979) Effect of glacial ice melting on the Antarctic surface water. Nature, 277: 469-471.

Jacobs, S. S., H. E. Huppert, G. Holdsworth, D. J. Drewry (1981) Thermohaline steps induced by melting of the Erebu Glacier tongue. J. Geophys. Res., 86: 6547-6555.

Johannessen, O. M. and L. A. Foster (1978) A note on the topographically controlled oceanic polar front in the Barents Sea. J. Geophys. Res., 83: 4567-4571.

Johnson, G. L., J. R. Vanney, D. J. Drewry and G. de Q. Robin (1980) General bathymetric chart of the ocean, No. 518. Canadian Hydrographic Service, Ottawa.

Josberger, E. G. and S. Martin (1982) A laboratory and theoretical study of the boundary layer adjacent to a vertical melting ice-wall in salt water. J. Fluid Mech., 111: 439-473.

Josberger, E. G. and S. Neshyba (1981) Iceberg melt-driven convection inferred from field measurements of temperature. Ann. Glaciol, 1: 113-117.

Joyce, T. M. (1977) A note on the lateral mixing of water masses. J. Phys. Oceanogr., 1: 626-629.

Joyce, T. M., W. Zenk and T. M. Toole (1978) The anatomy of the Antarctic polar front in the Drake Passage. J. Geophys. Res., 83: 6093-6113.

Kiilerich, A. (1945) On the hydrography of the Greenland Sea. Medd. Gron. 144, 63 p.

Killworth, P. D. (1973) A two-dimensional model for the formation of Antarctic bottom water. Deep-Sea Res., 20: 941-971.

Killworth, P. D. (1974) A baroclinic model of motions of Antarctic continental shelves. Deep-Sea Res., 21: 815-839.

Killworth, P. D. (1977) Mixing on the Weddell Sea continental slope. Deep-Sea Res., 24: 427-448.

Killworth, P. D. (1979a) On "chimney" formations in the oceans. J. Phys. Oceanogr., 9: 531-554.

Killworth, P. D. (1979b) A discussion of deep circulation processes. Dyn. Atmos. Oceans, 3: 275-281.

Killworth, P. D. (1981) Deep convection in the world ocean. JSC/CCCO Meeting on Time Series of Ocean Measurements, Tokyo.

Kinder, T. H. and J. D. Schumacher (1981) Circulation over the continental shelf of the southeastern Bering Sea. In: The Eastern Bering Sea Shelf: Oceanography and Resources, Vol. I (D. W. Hood and J. A. Calder, Eds.), U.S. Dept. Commerce/NOAA/OMPA, Washington, D.C.

Krauss, E. B. and J. S. Turner (1967) A one-dimensional model of the seasonal thermocline. II: The general theory and its consequences. Tellus, 19: 89-106.

Lake, R. A. and E. L. Lewis (1970) Salt rejection by sea ice during growth. J. Geophys. Res., 75: 583-597.
Leblond, P. H. (1980) On the surface circulation in some channels of the Canadian Arctic Archipelago. Arctic, 33: 189-197.
Leblond, P. H. and L. A. Mysak (1978) Waves in the Ocean. Elsevier, Amsterdam, 602 p.
Lewis, E. L. and R. G. Perkin (1981) The practical salinity scale 1978: conversion of existing data. Deep-Sea Res., 28: 307-328.
Lewis, E. L. and G. R. Perkin (1982) Seasonal mixing processes in an Arctic fjord system. J. Phys. Oceanogr., 12: 74-83.
Mamayev, O. I. (1975) Temperature-salinity analysis of world ocean waters. Elsevier, New York, 374 p.
Martin, S. (1981) Frazil ice in rivers and oceans. Ann. Rev. Fluid Mechanics, 13: 379-397.
Martin, S. and P. Kauffman (1974) The evolution of under-ice melt ponds, or double diffusion at the freezing point. J. Fluid Mech., 64: 507-527.
Martinson, D. G., P. D. Killworth, and A. L. Gordon (1981) A convective model of the Weddell Polynya. J. Phys. Oceanogr. 11: 466-488.
Maykut, G. A. and N. Untersteiner (1971) Some results from a time-dependent thermodynamic model of sea ice. J. Geophys. Res., 76: 1550-1575.
McDougall, T. J. (1981a) Double-diffusive convection with a non-linear equation of state. Part I: The accurate conservation of properties in a two-layer system. Progress in Oceanography, 10: 71-89.
McDougall, T. J. (1981b) Double-diffusive convection with a non-linear equation of state. Part II: The relative importance of cabbeling and double-diffusive convection. Ibid. 91-121.
McPhee, M. G. (1980) Heat transfer across the salinity-stabilized pycnocline of the Arctic Ocean. In Proceedings, International Symposium on Stratified Flows, 2nd (T. Carstens and T. McClimans, Eds.), Trondheim, Norway, p. 527-537.
Melling, H. and E. L. Lewis (1982) Shelf drainage flows in the Beaufort sea and their effect on the Arctic Ocean pycnocline. Deep-Sea Res., 29(84): 967-985.
Metcalf, W. G. (1960) A note on water movement in the Greenland-Norwegian Sea. Deep-Sea Res., 7: 190-200.
Michel, B. (1971) Winter regime of rivers and lakes. USA Cold Regions Research and Engineering Laboratory, Monograph III-Bla, 131 p.
Micklin, P. P. (1981) A preliminary systems analysis of impacts of proposed Soviet river diversion on Arctic sea ice. EOS, 62: 489-493.
Middleton, J. and T. D. Foster (1980) Fine-structure measurements in a temperature-compensated halocline. J. Geophys. Res. 85, 1107-1122.

Middleton, J. H., T. D. Foster and A. Foldvik (1982) Low-frequency currents and continental shelf waves in the southern Weddell Sea. J. Phys. Oceanogr., 12: 618-634.

Monin, A. S., V. M. Kamenkovich and V. G. Kort (1977) Variability of the oceans. Wiley-Interscience, New York.

Morison, J. (1983) Internal waves in the Arctic Ocean: a review. In: The Geophysics of Sea-Ice (N. J. Untersteiner, Ed.), NATO Advanced Study Institute, Plenum.

Mortimer, C. H. and F. J. H. Mackereth (1958) Convection and its consequence in ice-covered lakes. Verh. Intern. Verein., 13: 923-932.

Mosby, H. (1934) The waters of the Atlantic Antarctic Ocean. In: Scientific Results of the Norwegian Antarctic Expeditions 1927-1928, I(11), 131 p.

Mosby, H. (1962) Water, salt and heat balance of the North Polar Sea and of the Norwegian Sea. Geofysiske Publikasjoner, 24: 1-66.

Mountain, D. G., L. K. Coachman and K. Aagaard (1976) On the flow through Barrow Canyon. J. Phys. Oceanogr., 6: 461-470.

Muench, R. D. (1971) The physical oceanography of the Northern Baffin Bay region. Arctic Institute of North America, The Baffin Bay - North Water Project, Scientific Report, 1, 150 p.

Mysak, L. A. (1980) Topographically trapped waves. Ann. Rev. Fluid Mech., 12: 45-76.

Nansen, F. (1902) Oceanography of the North Polar Basin. The Norwegian North Polar Expedition, 1893-1896, Scientific Results, 3(9), 427 p.

Neshyba, S. (1977) Upwelling by icebergs. Nature, 267: 507-508.

Neshyba, S. and E. G. Josberger (1980) On the estimation of Antarctic iceberg melt rate. J. Phys. Oceanogr., 10: 1681-1685.

Neshyba, S., V. T. Neal and W. Denner (1971) Temperature and conductivity measurements under Ice Island T-3. J. Geophys. Res., 76: 8107-8120.

Newton, J. L., K. Aagaard and K. L. Coachman (1974) Baroclinic eddies in the Arctic Ocean. Deep-Sea Res., 20: 707-719.

Niebauer, H. J. (1982) Wind- and melt- driven circulation in a marginal sea ice edge frontal system: A numerical model. Cont. Shelf Res., 1(1): 49-98.

Paquette, R. G. and R. H. Bourke (1979) Temperature fine structure near the sea-ice margin of the Chukchi Sea. J. Geophys. Res., 84: 1155-1164.

Paquette, R. G. and R. H. Bourke (1981) Ocean circulation and fronts as related to ice melt-back in the Chukchi Sea. J. Geophys. Res., 86: 4215-4230.

Perkin, R. G. and E. L. Lewis (1978) Mixing in an arctic fjord. J. Phys. Oceanogr., 8: 873-880.

Perry, R. K., H. S. Fleming, N. Z. Cherkis, R. H. Feded and J. V. Massengill (1977) Bathymetry of the Norwegian-Greenland and Western Barents Seas. U.S. Naval Research Laboratory, Acoustics Division, Environmental Sciences Group. Williams and Heintz Map Corporation, Washington, D.C.

Pinther, M., B. C. Heezen and M. Tharp (1975) Map of the Arctic Region. American Geographical Society, New York.

Polar Group (1980) Polar atmosphere-ice-ocean processes: A review of polar problems in climate research. Rev. Geophys. Space Phys., 18: 525-543.

Reid, J. L. and R. J. Lynn (1971) On the influence of the Norwegian, Greenland and Weddell seas upon the bottom waters of the Indian and Pacific Oceans. Deep-Sea Res., 18: 1063-1088.

Reid, J. L., W. D. Nowlin and W. C. Patzert (1977) On the characteristics and circulation of the southwestern Atlantic Ocean. J. Phys. Oceanogr. 7: 62-91.

Rodgers, G. K. (1965) The thermal bar in the Laurential Great Lakes. In: Proceedings, 8th Conference on Great Lakes Research, Ann Arbor, Great Lakes Research Division, University of Michigan, p. 358-363.

Rφed, L. P. and J. J. O'Brien (1981) Geostrophic adjustment in a highly dispersive media: An application to the marginal ice zone. J. Geophys. Astrophys. Fluid Dyn., 18(3/4): 263-278.

Ruddick, B. R. and J. S. Turner (1979) The vertical length scale of double-diffusive intrusions. Deep-Sea Res., 26: 903-914.

Sadler, H. E. (1976) Water, heat and salt transports through Nares Strait, Ellesmere Island. J. Fish. Board Can., 33: 2286-2295.

Sayles, M. A., Aagaard and L. K. Coachman (1979) Oceanographic Atlas of the Bering Sea Basin. University of Washington Press, Seattle, Washington, 158 p.

Schaus, R. H. and J. A. Galt (1973) A thermodynamic model of an arctic lead. Arctic, 26: 208-221.

Schumacher, J. D., K. Aagaard, C. H. Pease and R. B. Tripp (1983) Effects of a shelf polynya on flow and water properties in the Northern Bering Sea. J. Geophys. Res., 88.

Schumacher, J. D., J. H. Kinder, D. J. Pashinski and R. L. Charnell (1979) A structured front over the continental shelf of the eastern Bering Sea. J. Phys. Oceanogr., 9: 79-87.

SCOR Working Group 58 (1979) The Arctic Ocean heat budget. Geophysical Institute Division A, University of Bergen, Report 52, 98 p.

Seabrooke, J. M., G. L. Hufford and R. B. Elder (1971) Formation of Antarctic bottom water in the Weddell Sea. J. Geophys. Res., 76: 2164-2178.

Semtner, A. J. (1976) Numerical simulation of the Arctic Ocean circulation. J. Phys. oceanogr., 6: 409-425.

Smith, P. C. (1975) A streamtube model for the bottom boundary currents in the ocean. Deep-Sea Res., 22: 853-874.

Solomon, H. (1973) Wintertime surface layer convection in the Arctic Ocean. Deep-Sea Res., 20: 269-283.

Stern, M. E. (1975) Ocean Circulation Physics. Academic Press, New York, 246 p.

Stigebrandt, A. (1976) Vertical diffusion driven by internal waves in a sill fjord. J. Phys. Oceanogr., 6: 486-495.

Stigebrandt, A. (1978) Dynamics of an ice-covered lake with through-flow. Nordic Hydrology, 9: 219-244.

Stigebrandt, A. (1981a) A model for the thickness and salinity of the upper layer in the Arctic Ocean and the relationship between the ice thickness and some external parameters. J. Phys. Oceanogr., 11(10): 1407-1422.

Stigebrandt, A. (1981b) Cross thermocline flow on continental shelves and locations of shelf fronts. In: Ecohydrodynamics (J. Nichovl, Ed.), Elsevier, Amsterdam, pp. 51-65.

Svensson, U. and R. Larsson (1980) A one-dimensional numerical model study of some basic features of the flow in ice-covered lakes. J. Hyd. Res., 18: 251-267.

Swift, J. H. (1982) The arctic waters of the Norwegian, Greenland and Iceland Seas. (In press.)

Swift, J. H. and K. Aagaard (1981) Seasonal transitions and water mass formation in the Iceland and Greenland Seas. Deep-Sea Res., 28: 1107-1129.

Takano, K. (1954) On the salinity and the velocity distribution off the mouth of a river. J. Oceanogr. Soc. Japan, 10: 92-98.

Tesaker, E. (1973) Horizontal cross-flow temperature gradients in a lake due to coriolis' force. In:Hydrology of Lakes Symposium, International Association of Hydrological Sciences, Helsinki, p. 72-80.

Timofeyev, V. V., A. I. Smirnova and K. I. Nepop (1970) Centers of ocean-atmosphere interaction in the North Atlantic. Oceanology, 10: 586-589.

Toole, J. M. (1981) Sea ice, winter convection, and the temperature minimum layer in the Southern Ocean. J. Geophys. Res., 86: 8037-8047.

Treshnikov, A. F. (1977) Water masses of the Arctic Basin. In: Polar Oceans (M. Dunbar, Ed.), Arctic Institute of North America, p. 17-31.

Treshnikov, A. F. and G. I. Baranov (1972) Structure and circulation of water in the Arctic Basin. Leningrad, 158 p.

Treshnikov, A. F., A. A. Girs, G. I. Baranov and V. A. Yefimov (1973) Preliminary programme of the polar experiment for the South Polar region. Arctic and Antarctic Research Institute, Leningrad, 55 p.

Turner, J. S. (1973) Buoyancy Effects in Fluids. Cambridge University Press, Cambridge, 367 p.

Turner, J. S. (1981) Small-scale mixing processes. In: Evolution in Physical Oceanography (B. A. Warren and C. Wunsch, Ed.), MIT Press, Cambridge, Mass., p. 236-262.

Vowinckel, E. and S. Orvig (1961) Water balance and heat flow in the Arctic Ocean. McGill University, publication in Meteorology No. 44, 35 p.

Wadhams, P., A. E. Gill and P. F. Linden (1979) Transects by submarine of the East Greenland Polar Front. Deep-Sea Res., 26: 1311-1327.

Wadhams, P., S. Martin, O. M. Johannessen, W. D. Hibler, III and W. J. Campbell (1981) MIZEX: A program for mesoscale air-ice-ocean interaction experiments in arctic marginal ice zones. USA Cold Regions Research and Engineering Laboratory, Special Report 81-19, 20 p.

Wakatsuchi, M. and N. Ono (1983) Measurements of salinity and volume of brine excluded from growing sea ice. J. Geophys. Res. 88.

Warren, B. A. (1981) Deep circulation of the world ocean. In: Evolution of Physical Oceanography (B. Warren and C. Wunsch, Eds.), MIT Press, Cambridge, Mass., p. 6-41.

Weber, J. E. (1977) Heat and salt transfer associated with formation of sea ice. Tellus, 29: 151-160.

Weeks, W. F. and W. J. Campbell (1973) Icebergs as a freshwater source: An appraisal. J. Glaciol., 12: 207-233.

Weiss, R. F., H. G. Ostlund and H. Craig (1979) Geochemical studies of the Weddell Sea. Deep-Sea Res., 26: 1093-1120.

Whitehead, J. A. (1980) Rotation critical flows in the ocean. In: Proceedings, International Symposium on Stratified Flows, 2nd (T. Carstens and T. McClimans, Eds.), Trondheim, Norway, p. 72-80.

Worthington, L. V. (1970) The Norwegian Sea as a mediterranean basin. Deep-Sea Res., 17: 77-84.

Wust, G. (1933) Das Bodenwasser und die Gliederung der atlantischen Tiefsee. In: Wissenschaftliche Ergebnisse der Deutschen Atlantischen Expedition auf dem Vermessungs- und Forschungschiff 'Meteor,' 1925-27, 9(1), 107 p.

Yoshimura, S. (1936) A contribution to the knowledge of deep water temperatures of Japanese Lakes. Part II: winter temperatures. Japan. J. Astron. Geophys., 14: 57-83.

Zubov, N. N. (1945) L'dy Arktiki (Arctic Ice). Isdatel'stvo Glavsevmorputi, Moscow, 360 p. (U.S. Navy Hydrographic Office, Translation 217, 1963; available as AD426972 from NTIS, Springfield, Va.)

9781489953544VOL01

The Geophysics of Sea Ice

NATO ASI Series

Advanced Science Institutes Series

A series presenting the results of activities sponsored by the NATO Science Committee, which aims at the dissemination of advanced scientific and technological knowledge, with a view to strengthening links between scientific communities.

The series is published by an international board of publishers in conjunction with the NATO Scientific Affairs Division

A Life Sciences **B Physics**	Plenum Publishing Corporation New York and London
C Mathematical and Physical Sciences	D. Reidel Publishing Company Dordrecht, Boston, and Lancaster
D Behavioral and Social Sciences **E Engineering and Materials Sciences**	Martinus Nijhoff Publishers The Hague, Boston, and Lancaster
F Computer and Systems Sciences **G Ecological Sciences**	Springer-Verlag Berlin, Heidelberg, New York, and Tokyo

Recent Volumes in this Series

Volume 139—New Vistas in Nuclear Dynamics
edited by P. J. Brussaard and J. H. Koch

Volume 140—Lattice Gauge Theory: A Challenge in Large-Scale Computing
edited by B. Bunk, K. H. Mütter, and K. Schilling

Volume 141—Fundamental Problems of Gauge Field Theory
edited by G. Velo and A. S. Wightman

Volume 142—New Vistas in Electro-Nuclear Physics
edited by E. L. Tomusiak, H. S. Caplan, and E. T. Dressler

Volume 143—Atoms in Unusual Situations
edited by Jean Pierre Briand

Volume 144—Fundamental Aspects of Quantum Theory
edited by Vittorio Gorini and Alberto Frigerio

Volume 145—Atomic Processes in Electron-Ion and Ion-Ion Collisions
edited by F. Brouillard

Volume 146—The Geophysics of Sea Ice
edited by Norbert Untersteiner

Series B: Physics

The Geophysics of Sea Ice

Edited by

Norbert Untersteiner
University of Washington
Seattle, Washington

Springer Science+Business Media, LLC

Based on the proceedings of the NATO Advanced Study Institute on
Air-Sea-Ice Interaction,
held September 28-October 10, 1981,
in Acquafredda di Maratea, Italy

Library of Congress Cataloging in Publication Data

The Geophysics of sea ice.

(NATO ASI series. Series B, Physics; vol. 146)
"Published in cooperation with NATO Scientific Affairs Division."
"Based on the proceedings of the NATO Advanced Study Institute on Air-Sea-Ice Interaction held September 28-October 10, 1981, in Acquafredda di Maratea, Italy"—T.p. verso.
Includes bibliographical references and index.
1. Sea ice—Congresses. I. Untersteiner, Norbert, 1926- . II. North Atlantic Treaty Organization. Scientific Affairs Division. III. NATO Advanced Study Institute on Air-Sea-Ice Interaction (1981: Acquafredda di Maratea, Italy) IV. Series: NATO ASI series. Series B, Physics; v. 146.
GB2401.2.G46 1986 551.3′43 86-22651
ISBN 978-1-4899-5354-4 ISBN 978-1-4899-5352-0 (eBook)
DOI 10.1007/978-1-4899-5352-0

Originally published by Plenum Press, New York in 1986
Softcover reprint of the hardcover 1st edition 1986

PREFACE

The intent of this book is to present the topic of sea ice in the broad and interdisciplinary context of atmospheric and oceanographic science. It is the result of a NATO Advanced Study Institute for which the authors had provided preprints of their respective chapters. The material was presented orally in several lectures by each author, followed by discussions among the approximately fifty attendees of the Institute. By prior agreement between the editor and the authors, the material is introduced on a tutorial level, and developed toward problems currently under study.

Thanks to the willing cooperation by the authors, it was possible to present the material in a more or less structured form, although it was clear from the outset that a book written by twenty authors would not be a text in the usual sense. Even so, we hope that it will help the beginning researcher to survey the subject of sea ice and its role in the global fluid system, and will provide the expert with summaries of less familiar topics.

Instruments and methods of making observations in the field receive inadequate attention. This is, in part, due to the immense variety of methods needed to approach the interdisciplinary subject of sea ice. Also, owing to the use of microprocessors, new power sources, sophisticated means of data storage and transmission, and data acquisition by remote sensing, there is a rapid development toward automation. It seemed an inopportune time to review techniques of observation that would be outdated by the time of their publication.

Seattle, July 1986 N. Untersteiner

Note concerning literature citations

It is widely known and appreciated that researchers in the Soviet Union have played a leading role in sea ice research. They demonstrated, in 1938, that ordinary ice floes can be safely used as manned research platforms. Shortly thereafter (1943) N.N. Zubov published his monumental treatise "Arctic Ice" which, to this day, remains a source of valuable information and insight. During the past three decades, researchers at the Arctic and Antarctic Research Institute, Leningrad, have produced prodigious amounts of literature on sea ice, polar oceanography, meteorology, and climatology.

Even though much of that literature is translated eventually, typical lag times are several years. When the Soviet work becomes available in English it is no longer "fresh" and, as a result, receives inadequate attention. We trust that the sparsity of Soviet citations in this book will not be interpreted as disregard, but simply as the regrettable consequence of the language barrier.

ACKNOWLEDGMENTS

My most sincere thanks to the authors and, among them, the Editorial Committee consisting of

G. Herman, Madison, Wisconsin, USA,
P. Lemke, Hamburg, Federal Republic of Germany
G. McBean, Sidney, British Columbia, Canada, and
M. McPhee, Naches, Washington, USA,

who not only wrote their own chapters but also arranged for peer reviews of all sections of the book. All authors were extraordinarily punctual in providing pre-prints of their section for use at the Institute. Their cooperative spirit in providing the material in the requested form and level of sophistication is also greatly appreciated. That it has taken five years to produce this book is entirely my fault and the result of the erroneous assumption that such a job can be done "on the side" with part-time help. In that regard, too, the authors were generous and updated their material. Most of the chapters including their final corrections were received during 1985.

After a number of false starts, preparation of the camera-ready manuscripts finally ended up in the hands of Marcus Duke, Publications Office, School of Fisheries, University of Washington, whose competence and efficiency were instrumental in finishing that large task. Generous and invaluable help was also given by Stephen Bowen, Chief Editor of the U.S. Army Cold Regions Research and Engineering Laboratory, Hanover, N.H.

The original suggestion to hold an Advanced Study Institute on the subject of this book came from the Special Program Panel on Air-Sea Interaction of the NATO Scientific Affairs Division, especially from Professor Clayton Paulson. With that encouragement, I submitted a proposal to NATO, outlining a program of structured lectures and discussions spanning ten working days, with the idea that the Institute would result in a more or less systematic treatise on air-sea-ice interaction rather than a volume of conference proceedings.

Our collective thanks are due to the NATO Scientific Affairs Division for a grant in support of travel and subsistence for instructors, subsidies for student participants, and other expenses associated with the Institute. With advice and help from Dr. Mario di Lullo and Dr. Tilo Kester, the choice of a location for the Institute fell on Hotel Villa del Mare at Acquafredda di Maratea, about 150 km south of Naples. At the time of the Institute (28 September to 10 October 1981), we had the hotel virtually to ourselves and enjoyed superb hospitality (Mr. A. Guzzardi, Hotel Director) and conference and recreational facilities, and a collection of spirited and hard working attendees. Unfortunately, all local expenses in Italy were accountable based on the U.S. dollar, which experienced a sharp increase in value during the interval between issuance of the grant and its actual expenditure, resulting in the loss of a sizable fraction of the NATO grant in Belgian francs. Thanks to the financial rescue by the Office of Naval Research (Arctic Program) and the National Oceanic and Atmospheric Administration (Research and Development) it was possible to finance the Institute as planned.

I am grateful to the Oceanographer of the Navy and to the Applied Physics Laboratory at the University of Washington for their generous help, especially to Dr. John Harlett.

CONTENTS

Selected Special Topics

CHAPTER 11

ATMOSPHERIC MODELLING AND AIR-SEA-ICE INTERACTION

Gerald F. Herman

Department of Meteorology
University of Wisconsin,
Madison, Wisconsin 53706 U.S.A.

PART I. BASIC ASPECTS OF GENERAL CIRCULATION MODELLING IN HIGH LATITUDES

1. INTRODUCTION

General circulation models have been relied upon extensively for weather and climate studies in the Arctic, and are an important tool for studying air-sea-ice interactions in high latitudes. The field of GCM modelling is a well-developed branch of the atmospheric sciences, and its evolution has greatly depended upon techniques developed in other disciplines, especially numerical analysis, geophysical fluid dynamics, turbulence, and radiative transfer.

A complete review of the subject of General Circulation modelling at this Institute is neither possible nor appropriate because of the scale of the task. Reference is made to numerous reviews and surveys [see, for example, Chang (1977), Gates (1979) and Haltiner and Williams (1980)] from which the interested reader can locate further information in any required detail. Instead, we present here a summary of the essential features of general circulation modelling, focusing on those aspects which are most relevant for high latitude weather prediction and climate studies. In part I we state the basic equations and consider special aspects of the numerical methods and parameterization schemes that are relevant in high latitudes. In part II we review some modelling studies relevant to high latitude weather and climate.

2. BASIC EQUATION SYSTEMS

The state of the atmosphere may be presented in the most general form by Newton's Second Law of Motion, The First Law of Thermodynamics, the conservation of mass, and the perfect gas law. In component forms these are:

$$\frac{du}{dt} = \frac{\tan\phi}{a} uv + fv - \frac{g}{a\cos\phi} \frac{\partial z}{\partial \lambda} + F_\lambda \tag{11.1}$$

$$\frac{dv}{dt} = - \frac{\tan\phi}{a} u^2 - fu - \frac{g}{a} \frac{\partial z}{\partial \phi} + F_\phi \tag{11.2}$$

$$\frac{\partial z}{\partial p} = - \frac{RT}{gp} \tag{11.3}$$

$$\frac{1}{a \cos\phi} \left(\frac{\partial u}{\partial \lambda} - \frac{\partial}{\partial \phi} v \cos\phi\right) + \frac{\partial \omega}{\partial p} = 0 \tag{11.4}$$

$$\frac{dT}{dt} = \frac{\kappa T \omega}{p} + \frac{Q}{C_p} \tag{11.5}$$

The time derivatives in these equations, d/dt, are individual derivatives, i.e., they refer to the rate of change at a point which moves with the flow. The variables are:

t	:	time
u,v	:	eastward, northward velocity components
f	:	coriolis force
p	:	pressure
g	:	gravitational acceleration
F	:	frictional force per unit mass
C_p	:	specific heat at constant pressure
Q	:	rate of heating per unit mass
T	:	temperature
R	:	gas constant for air
ϕ,λ	:	latitude, longitude
a	:	radius of the earth
ω	:	pressure vertical velocity

Also $\kappa = R/C_p$. An equation for the vertical acceleration does not appear, since for most classes of large-scale atmospheric motion there is hydrostatic equilibrium, i.e., an approximate balance between gravity and the vertical pressure gradient force. When this is the case, the vertical component of the vectorial prognostic equation of motion is replaced by the diagnostic hydrostatic equation (11.3). Note also that when the pressure p is the vertical coordinate, p becomes an independent variable and the height, z, a dependent variable. The remaining horizontal components of the equation of motion, together with the first law of thermodynamics, the equation of continuity and the perfect gas law are referred to as the primitive equations. All general circulation models discussed hereafter are based on some form of these equations, and are thus termed primitive equation models.

A conservation equation for water vapor concentration, q, is also included:

$$\frac{dq}{dt} = C + E \qquad (11.6)$$

where C and E are rates of condensation and evaporation, respectively.

Many general circulation models, including some discussed in these lectures, utilize the sigma-coordinate system to circumvent the most serious problem of the p-coordinate system, namely, that in the vicinity of mountains a given coordinate surface may not exist above the ground. Moreover, a surface pressure itself changes with time, further complicating the lower boundary conditions. The common procedure for dealing with this problem is to use the sigma system introduced by Philips (1957), where

$$\sigma = (p-p_t)/\pi \qquad (11.7)$$

where p_t is a constant pressure corresponding to the model top, and $\pi = (p_s-p_t)$, where p_s is the surface pressure. Thus, the earth's surface is always represented by $\sigma = 1$, and the top of the model by $\sigma = 0$.

3. MODEL DYNAMICS AND MODEL PHYSICS

The description of general circulation models can conveniently be organized along the two principal parts of the calculation, namely, model dynamics and model physics. The term model dynamics usually includes the numerical methods employed in solving the adiabatic parts of the equations of motion (i.e., those without friction or diabatic heat sources). Also relevant here for the numerical solutions are considerations of stability, to ensure that the numerical solutions remain bounded after a specified integration time as the time step becomes very small; convergence, which requires that the solution to the finite difference form of the equations approach the true solution to the differential equation as the differencing increments in space and time become vanishingly small; consistency, which requires that the finite difference form of the equation approaches the differential equation as the grid lengths are reduced to zero. Other considerations are the truncation error, which is the difference between the finite difference form of the equation and the differential equation itself; well-posedness, which requires that the continuous equations together with their boundary conditions indeed have solutions, and round-off errors, which are cumulative differences between the analytical and numerical solution that are due to the particular algorithm employed by a machine to arrive at the least significant digit in a computed number. A detailed discussion of these topics may be found in the texts by Haltiner and Williams (1980) and Richtmyer and Morton (1957).

Model physics may be viewed as everything that is not included in model dynamics. Included in this category are the procedures used to calculate radiative transfer processes; cloud formation and precipitation; ground hydrology; ocean and ice processes, including the transfer of momentum, mass and heat between the atmosphere and the ocean or sea ice; stratospheric processes and other processes which in nature would occur above the top layer of a model; i.e., in the stratosphere or mesosphere; and so-called sub-grid scale processes.

Central to the discussion of model physics is the concept of parameterization. The atmosphere is a continuous fluid, and within it occur processes ranging in scale from the microscopic (e.g., nucleation of cloud droplets) to the global (e.g., planetary scale waves). The numerical schemes employed in the GCM require and provide values of the variables at a discrete set of grid points. Consequently the model does not resolve certain processes because they occur on a space smaller than the grid resolution of the model, and may also alias them by misrepresenting their true scale because of insufficient resolution. A relevant example of the former problem concerns leads in the Arctic ice cover: the length of leads is usually on the order of tens of kilometers, while their width is of the order of tens or hundreds of meters. Yet the highest horizontal resolution used in so-called "ultrafine" versions of current GCM's is about 60 km. Thus the model simply does not "see" the lead.

In many cases it is impossible to calculate explicitly such processes, even if the resolution could be greatly increased, because they depend on variables that are not even represented in the model. The size of the lead, for example, depends on the heat balance of the ocean mixed layer and the large-scale deformation of the pack ice. These variables are seldom available in current GCM's.

The effect of such a sub-grid scale process may nonetheless be crucial to the GCM simulation, and it may therefore be necessary to represent in some way the effect of the process without explicitly calculating the process itself. This representation of a non-resolvable or non-calculable process in terms of the available model variables is termed parameterization.

In addition to the problem of leads mentioned above, there are other special parameterization problems that are relevant at high latitudes. These include sub-grid scale variations of surface albedo and emissivity due to variable ice type, ice cover, or presence of open water or meltwater; variations of heat flux and stress due to sub-grid scale homogeneities or ice characteristics; orographic influences of the Greenland and Antarctic ice sheets; stratus cloud formation and cloudiness associated with transient synoptic events; haze formation and its effect on radiative transfer and cloud formation; sub-grid scale (eddy) heat transfer, especially between the planetary boundary layer and the lower troposphere; evaporation and other hydrologic processes over variable soil and vegetation types; and sub-grid scale diffusion and dissipation, especially in the decay of mid-latitude cyclones that migrate into the Arctic.

The following is a brief summary of some current problem areas in GCM parameterization at high latitudes:

3.1 Boundary Layer Exchanges

In general terms the boundary layer is the region of the atmosphere which interacts with the earth's surface. This region is of vital importance, since there the atmosphere exchanges heat, water vapor, and momentum with ocean and ice surfaces. Boundary layers may be discussed under several headings.

(i) The planetary boundary layer. This is the region where eddy friction is of comparable importance to the coriolis force and pressure gradient force in maintaining steady flow. The actual distribution of horizontal wind components with height is dependent upon stratification and surface roughness. Satisfactory theories for its structure exist for near neutral conditions but the effects of either vertical or horizontal temperature gradients have not been satisfactorily accounted for. Similarity arguments have been used to calculate the surface stress, which is proportional to u_*^2 where u_* is the frictional velocity, based on the geostrophic wind speed, G. Under near neutral conditions, "universal" relations between u_* and G can produce satisfactory surface stress fields for ice drift calculations if gradient wind level or geostrophic winds are known. The planetary boundary layer, also referred to as the Ekman layer, has a typical scale of 300-1000 m, and thus is not resolved in most GCM's.

(ii) The surface layer. The term surface layer usually denotes the lowest 10-100 m of the planetary boundary layer. Here momentum fluxes are nearly constant with height, and depend on the surface roughness and stratification. Similarity laws for the wind, temperature and moisture give satisfactory results for a range of stability conditions. The usual modelling procedure is to base transfer coefficients for momentum and heat, C_m and C_h, respectively, on empirically determined similarity laws, then use them in conjunction with a low level wind (say 10 m) to determine momentum fluxes, u_*^2, and heat fluxes $\overline{w'T'}$ (see Businger, 1973; Deardorff, 1972) according to drag laws of the form

$$u_*^2 = C_m v_a v_a \tag{11.8}$$

$$\overline{w'T'} = - C_h v_a (T_a - T_o) \tag{11.9}$$

where v_a and T_a are wind speeds and temperature at "anemometer level" and T_o is the temperature of the lower boundary.

Variants of (11.8) and (11.9) are used to calculate fluxes in many GCMs. T_a and v_a may be obtained by introducing specifically a surface layer into the GCM calculation (see Washington and Williamson, 1977), or by interpolating from the lowest σ-level.

While this procedure is generally sound, the results obtained with it have not always proven satisfactory in terms of their agreement with observations. Some of the commonly-cited explanations include:

(a) Breakdown of similarity laws under extremely stable or extremely unstable conditions. Similarity laws generally are most applicable in near-neutral conditions. Under very stable conditions, as occur over ice during the Arctic night, the laws yield a critical Richardson number beyond which the transfer coefficients are zero, corresponding to the suppression of both shear-driven and buoyancy-driven turbulence. The nature of the atmosphere-surface exchanges under these conditions is not well understood. Extremely unstable conditions in high latitudes occur during air mass modification, as, for example, when cold air flows from eastern Canada over the warm waters of the Western North Atlantic.

(b) Dependence of surface roughness on windspeed. The similarity laws involve the surface roughness, z_o, and over the oceans this is related to the wave height, which itself is a function of the surface stress. This can be accounted for empirically. Cardone (1969) suggests the empirical relation

$$z_o = Au_*^{-1} + Bu_*^2 + C \quad , \tag{11.10}$$

where A, B, and C are empirical constants. Other dependencies are noted by Garrett (1977).

(c) Difficulties in defining v_a and T_a. While these quantities are relatively straightforward to obtain observationally, they are difficult to represent in a GCM that does not have a level close to the surface. The wind and temperature at the top of the surface layer are a function of the dynamics of the entire Ekman layer, and these are not explicitly represented in GCMs. The GLAS GCM, for example, derives v_a and T_a simply by interpolating between the surface and the lowest sigma layer.

(iii) The mixed layer. Under unstable conditions the profile of potential temperature in the boundary layer is adiabatic, and the layer is said to be well-mixed. During these conditions, the distribution of temperature and moisture conditions in the boundary layer is determined by the convective heating at the surface, entrainment, and large scale descending motion. Relatively simple descriptions (see, for example, Lilly, 1968) of the boundary layer are obtained under these conditions, which are frequently useful in high latitudes since the boundary layer over the North Atlantic and North Pacific in winter frequently fall into this mixed layer category, and frequently so do those in the central Arctic during the later spring, summer, and early fall.

3.2 Radiative Transfer

Radiative heating is a significant diabatic term entering into the heat equation (11.5), and the radiative fluxes at the surface are dominant terms in the energy balance of the ice, the ocean mixed layer, and the ground surface. The theory of radiative transfer is well-developed, and, in principle, it is possible to compute fluxes and flux divergences to a high degree of accuracy if the temperature and optical properties of the atmosphere and the surface are completely specified. The term optical properties here refers to the absorption, scattering and emission characteristics of the atmosphere's gaseous constituents, clouds, and other aerosol particles, and the emissivity and reflectivity of the surface. The radiative properties of ice and snow are discussed in Chapter 5 (Maykut).

The net radiative heating rate, Q_{rad}, may be calculated as the divergence of the vertical component of the net flux,

$$Q_{rad} = \frac{g}{c_p} \frac{\partial}{\partial p} (F_s + F_t) \tag{11.11}$$

where F_s and F_t are the fluxes of solar radiation and thermal radiation respectively. In certain complicated cloud situations involving ensembles of cumulus clouds there may be significant horizontal components of the net flux, and computational techniques are available for treating these circumstances. (See McKee and Cox, 1976.) The importance of these so-called three-dimensional effects has not yet been determined for Arctic and high latitude cloud conditions.

The calculation of F_t and F_s requires the solution of the equation of radiative transfer with and without a thermal source function, respectively. This can be accomplished efficiently and accurately for the case of a clear atmosphere in which the only absorbers and scatterers are gaseous constituents. Numerous complications arise when cloud droplets or other aerosol particles are present.

It is convenient to discuss separately the solar and thermal heating rate calculations, since the peaks of the source function emission for the sun and for terrestrial radiation are separated by a broad spectral gap.

(i) Thermal radiation. The flux obtained as the solution of the monochromatic two stream radiative transfer equation is

$$F_t = \int_{\tau}^{\tau^*} \pi\, B_\nu(t) \exp[-r(t-\tau)] r\, dt + \int_{\tau}^{\infty} \pi B_\nu^* \exp[-r(t-\tau)] r\, dt$$

$$\int_{0}^{\tau} \pi B_\nu(t) \exp[r(t-\tau)] r\, dt \qquad (11.12)$$

where B is the monochromatic Planck function, τ is optical depth, r is a diffusivity factor, and an asterisk denotes conditions at the ground. Equation (11.12) is derived under the assumption that scattering is negligible, which is almost always the case in the thermal portion of the spectrum, except in some narrow spectral regions when liquid water cloud droplets are present (see Hunt, 1973).

The integrals in (11.12) can be evaluated in a straightforward manner using numerical quadrature that is suited to the vertical resolution of the temperature and optical data. The difficulty in evaluating (11.12) is due to the presence of the optical depth, τ, which in fact contains all of the information relating to the absorption of atmospheric gases, cloud and aerosol. This monochromatic equation must be integrated with respect to frequency to obtain the net flux in a spectral region, and thus the absorption band characteristics of the atmospheric constituents are required.

The absorption of the major atmospheric gases is well understood, and the reader is referred to the treatises of Goody (1964) and Liou (1980) and the report by Rodgers (1976) for a discussion of ways of representing this absorption in calculations.

Equation (11.12) is highly sensitive to the presence of cloud through the optical depth, τ. The spectral properties of liquid water clouds in the infrared are well known (see Hunt, 1973), and clouds can be treated in the equation of transfer either by including them in the definition of τ and integrating through them, or by treating them as opaque boundaries. Although the computational techniques for dealing with cloud are straightforward, it has proven difficult to treat them in a satisfactory way in most major general circulation models, and this is due to inadequacies in the cloud prediction schemes rather than in the treatment of the radiation.

The optical properties of ice crystals and aerosols which also enter into the calculation of optical depth are not well known.

(ii) Solar radiation. Solar radiation can be treated efficiently and accurately using once of the available variants of the two-stream approximation. (A detailed comparison and discussion of a number of two-stream approximations is given by Zdunkowski et al., 1980.) As with thermal radiation, the principal difficulty is defining the optical depth when complicated cloud forms or aerosol are present. The following are some of the major difficulties in the solar radiation calculations in high latitudes:

1. Defining representative values of the surface reflectivity for a model grid square when heterogeneous surfaces are present, e.g., mixtures of open ocean, snow covered ice, and meltwater ponds. [A discussion of surface albedo in the Arctic is given in the lecture by Maykut (Chapter 5).]

2. Defining scattering and absorption coefficients for cloud. Solar radiation calculations are critically dependent on the value of the single scatter albedo, $\tilde{\omega}_0$, which is the ratio of the scattering extinction coefficient to the total (absorbing plus scattering) extinction coefficient. Present values obtained from theoretical Mie scattering results do not agree satisfactorily with those obtained from atmospheric measurements, and absorption by aerosol, three dimensional transfer effects, and data sampling problems are among the possible causes of this discrepancy.

3. Representing the effect of scattering and absorption by aerosol. Haze and aerosol are components of the radiative environment of the Arctic. Their potential effect on the radiation field has been speculated upon, but they are not represented in most major GCM's because of the uncertainties in their optical properties.

4. Representing fractional cloudiness or ensembles of mixed cloud types. The fluxes and heating rates computed for non-uniform cloud fields differ substantially from those for plane paral-

lel slabs (see Newiger and Bahnke, 1981). At present the details of cloud geometry cannot be predicted by the GCM's, nor are convenient techniques available for dealing with clouds that have other than plane parallel geometry.

4. SPECIAL MODELLING PROBLEMS IN HIGH LATITUDES

4.1 The Spherical Grid

A natural and convenient grid scheme to use on a sphere has been the latitude-longitude grid, so called because it consists of equal divisions in latitude and longitude. Unfortunately, the most serious drawbacks to this grid occur in polar regions. The pole itself is a singularity where the northward (v) and the eastward (u) velocity components cannot be defined. Also, the distance between longitude lines decreases in proportion to the cosine of the latitude as the meridians converge, and thus also the maximum time step allowed by the Courant-Friedrichs-Lewy criterion, which is a necessary condition for computational stability. Although this problem does not arise in spectral versions of GCM's (see below), a number of techniques for dealing with this problem in grid point models have been tested over the years. These include:

(i) Reducing the number of usable grid-points in high latitudes so as to maintain a nearly uniform geometric spacing between longitude lines. Two general methods have been employed to accomplish this reduction: A grid based on equal-area rectangles, the number of which remains constant around a latitude circle, and a so-called split grid where the longitudinal spacing of grid points is systematically increased in the poleward direction. The properties of this constant grid area Kurihara scheme (see Kurihara, 1965) have been extensively tested (see Kasahara, 1977, for summary), and the general conclusion is that the loss of spatial resolution at high latitudes causes a serious degradation of results both in forecasts and in climate simulations. The split-grid scheme has been employed in second-order GLAS GCM calculations (see Herman and Johnson, 1978). In the GLAS split grid scheme (see Figure I.1) the longitudinal spacing is 5° up to latitudes of 62°, then is increased to 10° between 66 and 78°, and to 20° poleward of 78°. The scheme retains the same quadratic conservation properties as original schemes for equal latitude-longitude spacing. The scheme allows the integration step to be doubled while eliminating instabilities and reducing significantly the amount of smoothing required polar regions. Model accuracy is not affected outside of polar regions.

(ii) Reducing the time step in polar regions. Spatially variable time steps were employed in early versions of the British

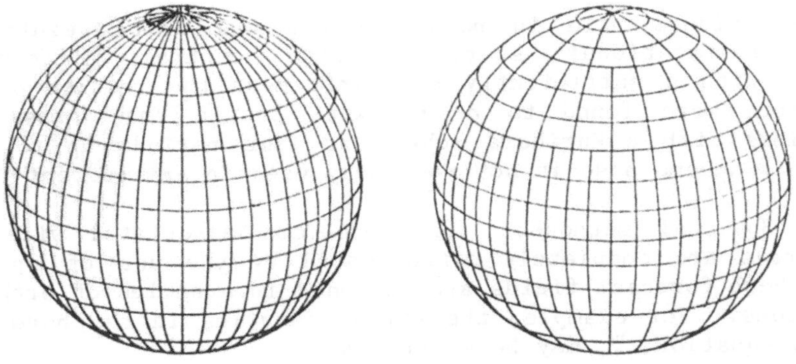

Fig. I.1. Schematic diagram of the GLAS GCM grid structure.

Meteorological Office GCM. A polar cap region where the time step systematically decreased poleward was patched to the global domain which had a uniform time step. However, the price paid for this modification was high: 50% of the computing time was spent on 2% of the hemisphere area (see Grimmer and Shaw, 1967). For obvious reasons, this approach was never adopted operationally, nor has it been further pursued.

(iii) Fourier filtering of short waves. This technique truncates the Fourier spectrum at a critical wavelength beyond which computational instability should occur. Typically, the prognostic model variables are Fourier analyzed around each latitude circle, the series is truncated at a critical wavelength (twice the zonal grid spacing), and then resynthesized with only the desired wavelengths present. This technique has frequently been used in versions of the GFDL general circulation model (see Holloway et al., 1974).

(iv) Using a grid that is nearly homogeneous over a sphere. For historical perspective we note the grid system with nearly uniform spacing that was developed by Williamson (1968) and Sadourny et al. (1968). The basis was the spherical geodesic consisting of nearly equal area equilateral spherical triangles covering the sphere. The spherical triangles are based on the 12 vertices of the icosahedron, which is the geometric solid having 20 equilateral triangular faces. Each spherical triangle is subdivided into smaller grid triangles. Williamson (1970) also developed energy conserving finite difference equations for the grid, and tested them with a barotropic model. It was found that first order errors in the mass flux calculations arose as a consequence of the slight non-uniformity of grid sizes. This error was somewhat reduced with a smaller grid size, with a concomitant reduction in time step being required. With centered time differencing, for example, a time step comparable to that of a latitude-longitude grid in polar regions was still required to maintain computational stability with central differencing schemes. With a sufficiently fine grid mesh the results of test integrations were found to be somewhat better than those from integrations with a Kurihara scheme at time periods up to eight days. This scheme also is not routinely used in any current GCM.

(v) Spectral methods. Many problems associated with grid point models are completely circumvented by the use of spectral models. Here forecast fields are expanded in a series of orthogonal functions. For example, the stream function in the barotropic vorticity equation, Ψ, may be written as

$$\Psi(x,y,t) = \sum_{i=1}^{N} A_i(t)\ \phi_i(x,y) \tag{11.13}$$

where the functions ϕ_i are orthogonal and suitably normalized such that

$$\int \phi_i\ \phi_j\ dS = \delta_{ij} \begin{cases} = 0, & i \neq j \\ = 1, & i = j \end{cases} \tag{11.14}$$

where S is the region of integration.

The functions $\phi_i(x,y)$ usually satisfy

$$\nabla^2 \phi_i + \alpha_i^2\ \phi_i = 0 \tag{11.15}$$

If (11.13) is substituted into the barotropic vorticity equation, multiplied by ϕ_j, integrated over the whole domain, and use made of the orthogonality relation (11.14) one obtains

$$\alpha_i^2 \frac{dA_i}{dt} = \sum_p \sum_q \alpha_q^2\ A_p\ A_q\ L_{pqi} \tag{11.16}$$

where

$$L_{pqi} = - \int \phi_i (k \times \nabla\phi_p \cdot \nabla\phi_q) dS \tag{11.17}$$

Equation (11.16) now represents a series of N first-order differential equations in time, which may be solved by a variety of iterative methods. The choice of N depends upon the scale of the meteorological feature whose resolution is desired.

Spectral methods have numerous advantages, especially on a sphere where the eigenfunctions ϕ_i are spherical harmonics, and the problem of coordinate singularities does not arise. The spatial derivatives of the ϕ_i's may be computed exactly, so that phase errors associated with finite differences are eliminated.

The main drawback to the spectral method is that for N very large, there is a very large number of interaction coefficients (L_{pqi}) that need to be computed, although substantial improvement has been obtained by employing fast-fourier transform techniques. The spectral method is also cumbersome when it comes to evaluating physical processes, such as convection, radiation, or boundary layer exchange, since these processes must be computed in the grid domain and transformed back to the spectral domain.

Spectral domain circulation models have been developed and extensively utilized as NCAR (Washington, 1982), GFDL (Manabe et al., 1979), The Australian Numerical Meteorology Research Center (Bourke et al., 1977) and ECMWF (Simmons, 1983).

4.2 Sea Ice Processes

Sea ice conditions enter the GCM calculation as a lower boundary condition in several places. The albedo of sea ice is required in solar radiation calculations, and a single value is generally assumed for the entire solar spectrum. Empirical corrections are sometimes made to the surface albedos to account for their dependency on the solar zenith angle (see Paltridge and Platt, 1976). Surface temperature and surface emittance are similarly required for the lower boundary condition in the longwave radiation calculations. The roughness parameter is required for the calculation of surface stress, and the heat and moisture fluxes are required in atmospheric PBL calculations, as are the surface temperature and surface specific humidity. Ice thickness, snow cover, and thermal conductivity are required to compute the conductive heat flux in the surface temperature calculation.

These properties of sea ice may be non-interactive with respect to the rest of the GCM calculation. That is, they may be fixed, or vary according to a predetermined climatology and thus are independent of the changing conditions in the GCM's lowest layers and permit no ice-atmosphere feedback. Most operational weather forecasting models, and some current GCM's use non-interactive sea ice boundaries.

Less commonly, the distribution of sea ice is fully coupled to the atmospheric model, and evolves as part of the solution to the complete system of equations governing atmospheric and sea ice processes. A wide range of highly parameterized climate models fall into this category (for review, see Saltzman, 1978), but there have been very few experiments with fully-coupled ice-atmosphere general circulation models. Further discussion of sea ice parameterizations in GCM's can be found in Parkinson and Herman (1980), and in Chapter 9 (Hibler).

5. AN EXAMPLE OF A GCM CLIMATE MODEL

The general circulation models used at NASA's Goddard Laboratory for Atmospheres (GLA), during the period 1977-81, are examples of primitive equation models that have been used in weather forecasting and climate studies. A review of their calculations serves to illustrate the typical structure of a GCM calculation.

As is the case with most major GCM's, the GLA GCM computer code was seldom "frozen", but rather evolved continually as improvements in various aspects of the model were made. The model was originally developed by Mintz and Arakawa at UCLA, and was "brought up" at the Goddard Institute for Space Studies in New York for the execution of the first observing system simulation experiments (OSSE's) in preparation for the Global Weather Experiment. Its most complete description, though somewhat dated, appears in Somerville et al. (1974). (As of the final editing of this chapter the GLA family of GCMs included a fourth-order version (Rivas et al., 1977) and a new version of a UCLA GCM developed by Randall.) The main features of the Somerville et al. version that were used in several early ice and climate experiments are described below:

5.1 Model Hydrodynamics

The model is a primitive equation model in sigma coordinates, and solves the system (11.1) through (11.6) in flux form on a latitude-longitude grid. The most commonly-used resolution is 4° in latitude, 5° in longitude and with nine equally-spaced sigma layers (see Figure I.1). "Ultrafine" versions are available with a horizontal resolution of 2.5° in latitude by 3° in longitude, and other models with variable vertical resolution are being developed.

Most simulations with this version of the GLA GCM have involved second order spatial differencing procedures whereby horizontal derivatives of a quantity, Ψ, are computed with centered differences of the form

$$\frac{\partial \Psi}{\partial \lambda} \doteq (\Psi^{\tau}_{i+1,j} - \Psi^{\tau}_{i-1,j})/2\Delta\lambda \qquad (11.18)$$

where the superscript τ denotes the time step, and the indices i and j denote longitude ($\Delta\phi$) and latitude ($\Delta\lambda$) increments, respectively.

A fourth-order version was also introduced by Rivas et al. (1977), whereby spatial derivatives are of the form

$$\frac{\partial\psi}{\partial\lambda} \doteq \frac{4}{3}\,(\psi^{\tau}_{i+1,j} - \psi^{\tau}_{i-1,j})/2\Delta\lambda - \frac{1}{3}\,(\psi^{\tau}_{i-2,j} - \psi^{\tau}_{i-2,j})/4\Delta\lambda \qquad (11.19)$$

Recent experiments with the fourth-order versions using the data from the Global Weather Experiment have demonstrated a substantial improvement in forecast accuracy as compared with second-order versions, especially in the high mid-latitudes.

Experiments have been conducted with several time differencing schemes. A commonly-used scheme is due to Matsuno whereby for some function F of the variables x and t,

$$x^{*}_{n+1} = x_n + F(x_n)\,\Delta t \qquad (11.20)$$

$$x_{n+1} = x_n + F(x^{*}_n)\,\Delta t \qquad (11.21)$$

Ten minute time steps have been employed in most climate simulations.

5.2 Model Physics

(i) Radiation. The parameterization of short wave radiation in this version of the GLA GCM follows Lacis and Hansen (1974). The solar beam is depleted by a predetermined distribution of O_3 in the stratosphere, and its absorption in the clear atmosphere by water vapor is based on well-established laboratory measurements. When clouds are present scattering is computed according to a two-stream model, although no absorption by the cloud droplets is allowed. The surface reflectivity varies according to a predetermined seasonal climatology. Cloud optical properties are specified according to cloud type.

Thermal radiation is computed according to a model developed by Wu (1980). The Wu calculation is elaborate, computing water vapor absorption in 10 spectral regions, and including ancillary absorption by overlapping CO_2 and O_3 bands. Spectral band models based on laboratory measurements of infrared radiation are employed for H_2O vapor, while CO_2 and O_3 absorption is based on theoretical calculations of the gaseous spectra. Clouds are treated simply as black surfaces with unit emissivity.

(ii) Cloud formation. Two kinds of clouds are allowed to form, namely, convective and supersaturation clouds. Convective cloud formation in this version of the GLA model (see Helfand, 1979) follows the Arakawa three-level cumulus cloud parameterization.

Convection occurs if air at a cloud base layer becomes buoyant during its moist adiabatic ascent to the cloud top layer. Convective clouds (see Figure I.2) are said to be penetrative when the lowest 6 model layers participate in the vertical mixing, while low-level convection involves σ-layers 7 and 8, and upper-level convection involves layers 5 and 6. Supersaturation clouds in the model, which represent stratiform clouds in nature, form simply when the water vapor mixing ratio at a grid point exceeds the saturation mixing ratio for the ambient temperature.

Note that the cloud formation process in the model are fully coupled to the cloud radiative processes insofar as clouds grow and dissipate in response to changes in temperature, stability and surface heating, and these depend in turn on the flux and flux divergence of solar and thermal radiation. However, the model's cloud formation schemes are incapable of predicting fractional cloudiness, or the way that cloud layers are vertically stacked. All model clouds are assumed to occur as 4° latitude by 5° longitude horizontal slabs. This has an important and potentially detrimental effect when this cloud information is passed to the longwave radiation code. Because the clouds also are totally opaque in this spectral region, the radiation balances remain too positive, and wintertime surface temperatures over continental regions, especially Siberia and Canada, are too warm. Further experiments have shown that the use of relatively simple parameterizations of fractional cloudiness alleviate this problem, but at present these are not used routinely.

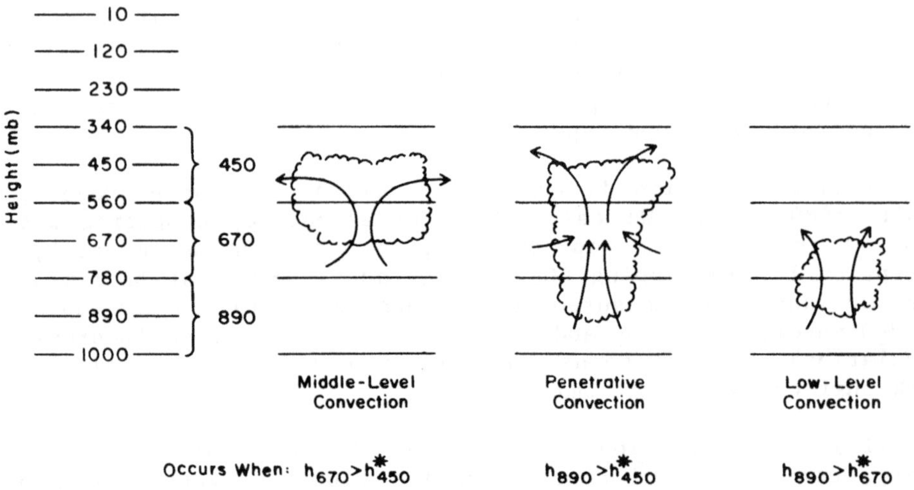

Fig. I.2. Convective cloud formation in a version of the GLA GCM.

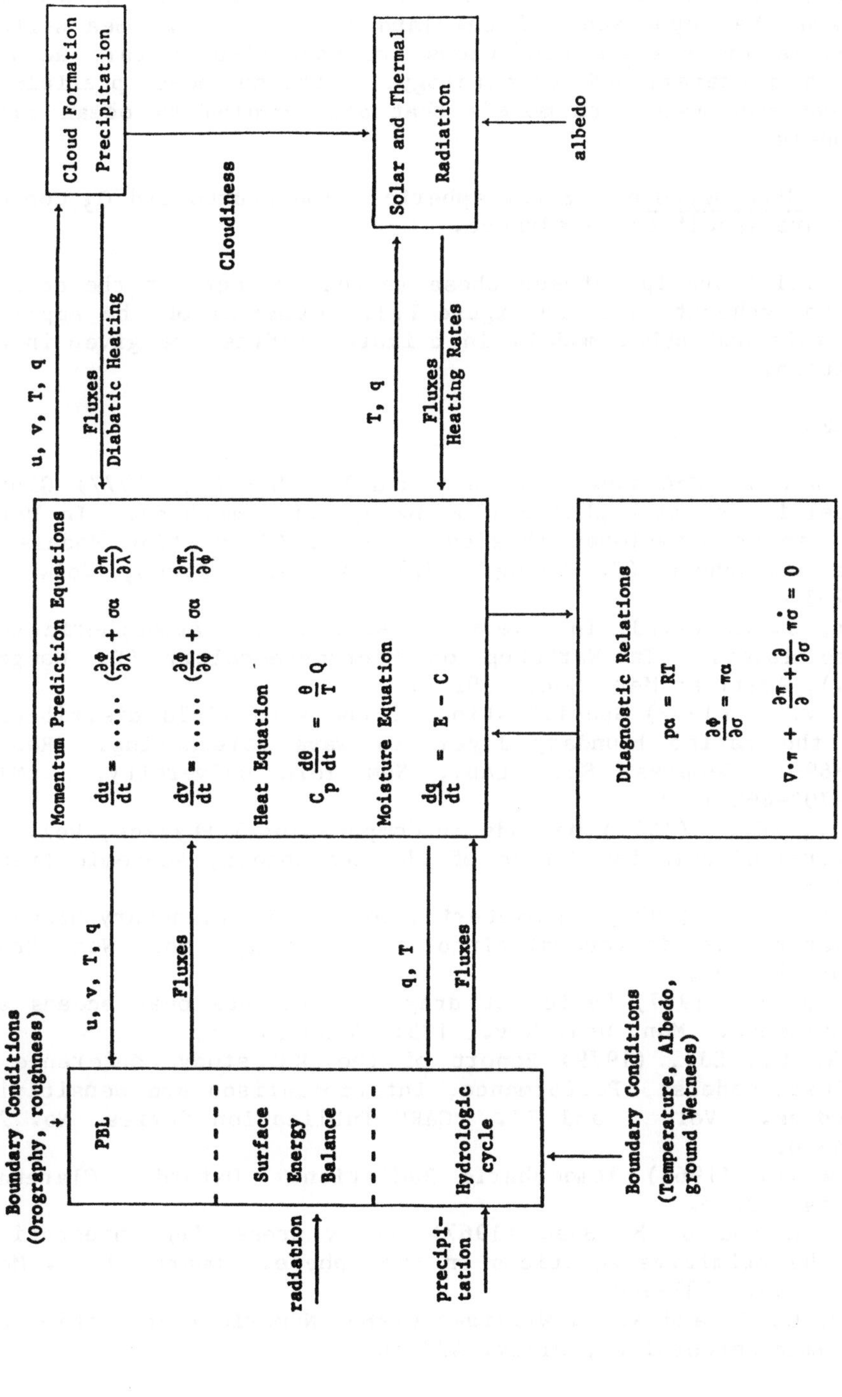

Fig. I.3. Schematic diagram of GCM calculation.

(iii) Lower boundary conditions. The lower boundary may be land, sea, snow, land ice or sea ice. Soil moisture values are fixed, and the topography of the land is specified. Sea surface temperatures and sea ice conditions are specified to vary according to a predetermined climatology, although some preliminary tests have been made with models that are coupled to ocean mixed layer models.

(iv) Stratosphere. Stratospheric temperatures and O_3 concentrations are specified in advance.

The relationship between these various aspects of the code is illustrated schematically in Figure I.3. Examples of the application of this and other models in climate studies are given in the next section.

REFERENCES

Bourke, W., B. McAvaney, K. Puri and R. Thurling (1977) Global modelling of atmospheric flow by spectral methods. In Methods in Computational Physics, General Circulation Models of the Atmosphere (J. Chang, Ed.), Academic Press, Vol. 17, 267-324.

Businger, J. A. (1973) Turbulent transfer in the atmospheric surface layer. In Workshop on Micrometeorology (D. Haugen, Ed.), American Met. Soc., 392 p.

Cardone, V. J. (1969) Specification of the wind field distribution in the marine boundary layer for wave forecasting. Report TR-69-1. Geophys. Sci. Lab., New York University. [NTIS AD-702-490.]

Chang, J., Ed., (1977) Methods in Computational Physics, Vol. 17, General Circulation Models of the Atmosphere, Academic Press, 337 p.

Deardorff, J. W. (1972) Parameterization of the planetary boundary layer for use in general circulation models. Mon. Wea. Rev., 100: 93-106.

Garrett, J. R. (1977) Review of drag coefficients over oceans and continents. Mon. Wea. Rev., 105: 915-929.

Gates, W. L., Ed., (1979) Report of the JOC study conference on climate models: Performance, intercomparison and sensitivity studies. Vol. I and II. GARP Publication Series, No. 22, 1049 p.

Goody, R. M. (1964) Atmospheric Radiation. Oxford: Clarendon Press, 436 p.

Grimmer, M. and D. B. Shaw (1967) Energy preserving integrations of the primitive equations on the sphere. Quart. J. R. Met. Soc., 93: 337-349.

Haltiner, G. J. and R. T. Williams (1980) Numerical Prediction and Dynamic Meteorology, Wiley, 477 pp.

Helfand, H. M. (1979) The effect of cumulus friction on the simulation of the January Hadley circulation by the GLAS model of the general circulation. J. Atmos. Sci., 36: 1827-1843.

Herman, G. F. and W. T. Johnson (1978) The sensitivity of the general circulation to Arctic sea ice boundaries: A numerical experiment. Mon. Wea. Rev., 106: 1649-1664.

Holloway, J. L., M. J. Spelman and S. Manabe (1973) Latitude-longitude grid suitable for numerical time integration of a global atmospheric model. Mon. Wea. Rev., 101: 69-84.

Hunt, G. E. (1973) Radiative properties of terrestrial clouds at visible and infrared thermal window wavelengths. Quart. J. Roy. Met. Soc., 99: 346-369.

Kasahara, A. (1977) Computational aspects of numerical models for weather prediction and climate simulation. In Methods in Computational Physics, (J. Chang, Ed.), Vol. 17, p. 1-66.

Kurihara, Y. (1965) Numerical integration of the primitive equations on a spherical grid. Mon. Wea. Rev., 93: 399-415.

Lacis, A. A. and J. E. Hansen (1974) A parameterization for the absorption of solar radiation in the earth's atmosphere. J. Atmos. Sci., 31: 118-133.

Lilly, D. K. (1968) Models of cloud-topped mixed layers under a strong inversion. Quart. J. Roy. Met. Soc., 94: 292-309.

Lion, K.-N. (1980) An Introduction to Atmospheric Radiation. Academic Press. 392 p.

Manabe, S., D. G. Hahn and J. L. Holloway (1979) Climate sublimation with GFDL spectral models of the atmosphere. GARP Publ. Ser. 22, World Meteorological Org., Geneva.

McKee, T. B. and S. K. Cox (1976) Simulated radiance patterns for finite cubic clouds. J. Atmos. Sci., 31: 2014-2020.

Newiger, M. and K. Bänke (1981) Influence of cloud composition and cloud geometry on the absorption of solar radiation. Beitr. Phys. Atm., 54: 370-382.

Paltridge, G. W. and C.M.R. Platt (1976) Radiative Processes in Meteorology and Climatology. Elsevier Press, 318 pp.

Parkinson, C. L. and G. F. Herman (1980) Sea ice simulations based on fields generated by the GLAS GCM. Mon. Wea. Rev. 108: 2080-2091.

Phillips, N. A. (1957) A coordinate system having some special advantages for numerical forecasting. J. Meteorology, 14: 184-185.

Richtmyer, R. D. and K. W. Morton (1957) Difference Methods for Initial-Value Problems. Wiley-Interscience, 399 p.

Rivas, E. K., A. Bayliss and J. Storch (1977) The fourth-order GISS model of the global atmosphere. Beitr. Phys. Atmos., 50: 299-311.

Rodgers, C. D. (1976) Approximate methods of calculating transmission by bands of spectral lines. NCAR Technical Note TN-116+1A, 48 p.

Sadourny, R., A. Arakawa, and Y. Mintz (1968) Integration of the nondivergent barotropic vorticity equation with an icosahedral-hexagonal grid for the sphere. Mon. Wea. Rev., 96: 351-356.

Saltzman, B. (1978) A survey of statistical-dynamical models of the terrestrial climate. Advances in Geophysics, Academic Press, Vol. 20, p. 184-295.

Simmons, A. J. (1983) Adiabatic formulation of the ECMWF forecasting system. ECMWF Workshop, 1982, Interpretation of numerical weather prediction products.

Somerville, R. C. J., P. H. Stone, M. Halem, J. E. Hansen, J. S. Hogan, L. M. Druyan, G. Russell, A. A. Lacis, W. J. Qurik, and J. Tenenbaum (1974) The GISS model of the global atmosphere. J. Atmos. Sci., 31: 84-117.

Washington, W. M. and D. L. Williamson (1977) A description of the NCAR global circulation models. In Methods in Computation Physics, Academic Press, Vol. 17, 337 p.

Washington, W. M., Ed. (1982) Documentation for the Community Climate Model (CCM) Version ϕ. Climate Section, National Center for Atmospheric Research, Boulder, CO.

Williamson, D. L. (1968) Integration of the barotropic vorticity equation on a spherical geodesic grid. Tellus, 20: 642-653.

Williamson, D. L. (1970) Integration of the primitive barotropic model over a spherical geodesic grid. Mon. Wea. Rev., 98: 512-520.

Wu, M. L. (1980) The exchange of infrared radiative energy in the troposphere. J. Geophys. Res., 85: 4084-4090.

Zdunkowski, W. G., R. M. Welch and G. Korb (1980) An investigation of the structure of typical two-stream methods for the calculation of solar fluxes and heating rates in clouds. Beitr. Phys. Atm., 53: 147-166.

PART II. APPLICATION OF GCMs TO STUDIES OF HIGH LATITUDE CLIMATE PROCESSES

1. INTRODUCTION

General circulation models have commonly been employed in experiments designated forecast experiments and climate experiments.

Forecast experiments attempt to predict, in a deterministic sense, the state of the atmosphere at specified locations and times by integrating from a set of observationally-derived initial conditions. The resulting forecast is thus dependent on the initial state, the internal physics of the model, and the boundary conditions and other external forcings.

The skill of a forecast model, which is a measure of the accuracy of its forecasts, deteriorates with time to a point where the model-generated fields are generally unrelated to the initial state, and are determined by the model's own internal adjustment to the external forcing and conditions at the boundary. This deterioration is partly due to the model's inadequate representation of the physical processes actually occurring in the atmosphere, partly due to errors associated with the definition of the initial state and partly due to the numerical errors associated with the solutions of the equations. However, even if the physical processes were perfectly represented and the initial state perfectly defined, a further error would derive from dynamical processes, most importantly convection and advection, that occur on sales not resolved by the model's discretization schemes. Through non-linear interactions the uncertainties associated with these unresolvable scales eventually contaminate successively larger scales until a limit to predictability is reached. At present this limit is thought to be about 2 weeks.

Beyond this theoretical limit to predictability the models have potential utility for providing information about certain time- and spatially-averaged features. Model integrations carried out for periods longer than this two-week interval fall into the realm of climate experiments. Lorenz (1975) distinguishes between climate predictability of the first kind, which is the prediction of the natural variations of the climate system, and climate predictability of the second kind, which involves the sensitivity or response of the average fields to perturbations in the external forcing.

2. SOME EXAMPLES OF GCM CLIMATE STUDIES

GCM climate studies can be divided into several categories: Simulation studies, which attempt to reproduce the observed time- and spatially-averaged state of the atmosphere, and its seasonal and interannual variations; Sensitivity studies, which focus on the response of the model to perturbations or changes in the boundary conditions or external forcing; and Predictability studies, which focus on the inherent variability of the model-generated fields.

We may also distinguish between atmospheric models in which the ice cover and ocean surface temperature boundary conditions are either fixed or prescribed to vary in a predetermined manner, and those in which the boundary conditions change due to an ice or ocean model's response to the forcing imposed by the atmospheric model. The latter are referred to as coupled models, and the former simply as uncoupled models.

An extensive comparison of GCM simulations recently was conducted by the Joint Organizing Committee (JOC), and details of simulations conducted with the major general circulation models are reported. (See International Council of Scientific Unions, 1979). It is beyond the scope of these lectures to review or to summarize the performance of the major GCM experiments. Instead, we will focus on the performance of the GLA GCM, and will summarize other major GCM experiments that pertain to high latitude and Arctic climate processes. The version of the model considered here was used extensively by GLA for its climate experiments during the period 1975-1983.

2.1 The Performance of a GLA (Goddard Laboratory for Atmospheres) GCM

The accuracy of simulations with this version of the GLA GCM was assessed by Herman and Johnson (1980), with special emphasis on its performance in high latitudes. A model climatology for the January-February period was constructed by averaging eight 30-day simulations, each of which spanned the period from 15 January to 14 February. A mean July climatology was similarly defined on the basis of the averages of seven 30-day simulations, each spanning the period 1-31 July. The comparisons were made with a variety of climatological observations that were available for the high latitude regions.

The model's simulation of the following features is relevant for Arctic air-sea-ice interaction studies:

Sea level pressure. (See Figures II.1 and II.2). The principal centers of action in the North Atlantic and North Pacific, namely, the Icelandic and Aleutians lows, were adequately simulated, although in winter the Icelandic low was too deep by about 4 mb, and the center of the Aleutian low located about 25-30° too far to the east. The most striking deficiency of the winter simulation occurred over Siberia, where the intensity and areal extent of the Siberian anticyclone was seriously underestimated. (At present this deficiency is common to a number of GCM's.) This shortcoming is attributed to the failure to simulate properly cloud-radiative feedback in the Siberian region as discussed by Herman (1981). The failure of the high to develop caused simulated sea level pressure gradients in the Soviet sector of the Arctic to be weaker, a result which could cause error in sea ice drift calculations. The elongated structure of the low in the Barents Sea also was not simulated.

Potentially serious discrepancies between model simulations and observations occurred during the summer as well (Figure II.3 and II.4.). Unrealistically high pressure was obtained in the Atlantic sector of the Arctic, and pressure gradients were about

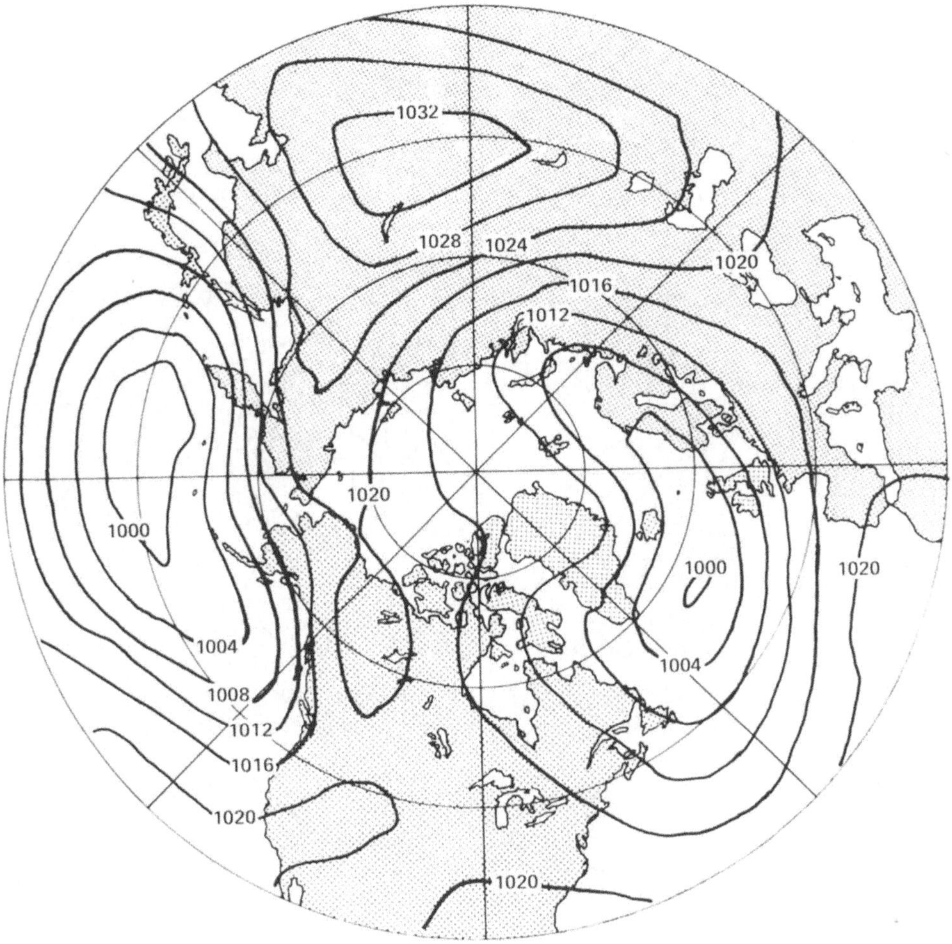

Fig. II.1. Observed sea level pressure during the Northern Hemisphere winter.

50% stronger than observed. The simulations in the Beaufort Sea sector of the Arctic were also characterized by unrealistically strong sea level pressure gradients.

500 mb circulation. The observed and model generated geopotential height fields are shown in Figures II.5 and II.6. The North American trough and strong jet over the Atlantic were simulated by the model, but the axis of the trough appears to be located about 10-15° too far to the east, and its intensity was slightly diminished. The Eurasian trough was also simulated, but it, too, was displaced toward the east. The most striking discrepancy between the model and the observations was the absence in the simulations

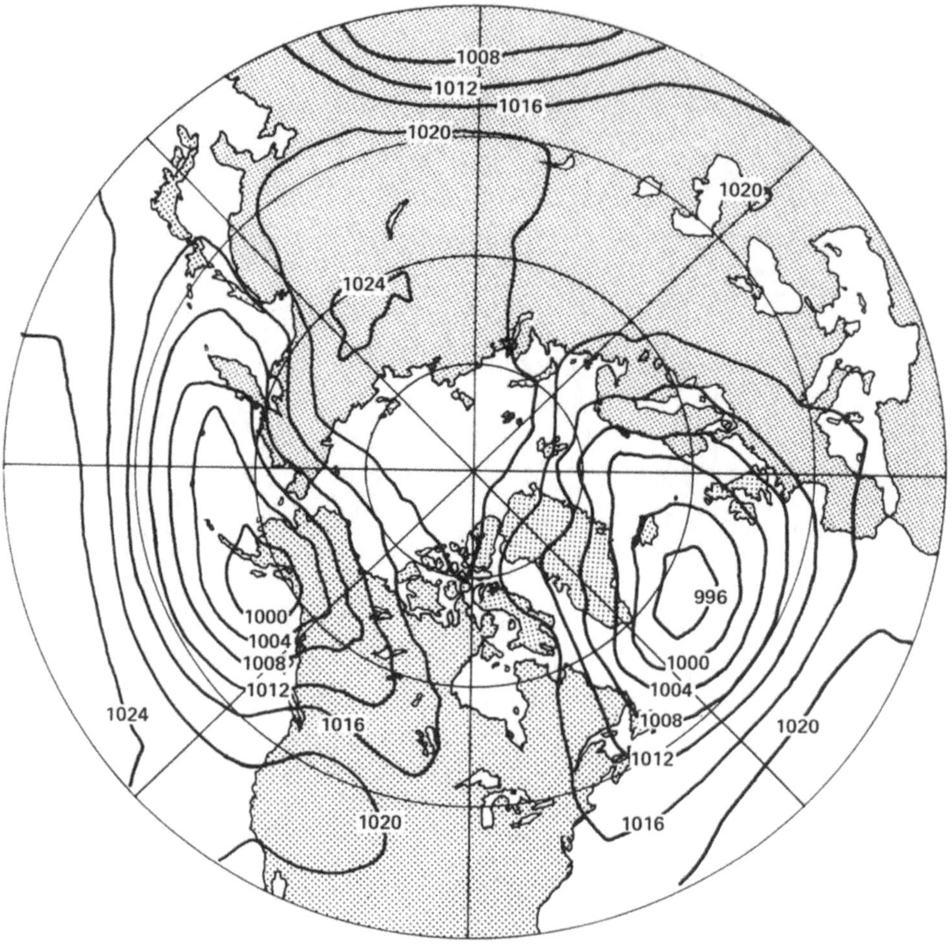

Fig. II.2. As in Fig. II.1, except simulated by the GLA GCM.

of the deep trough and associated meridional flow over the North Pacific. In contrast to the observations, the model produced a strong zonal flow over Japan and the northwestern Pacific. A slight ridge over northwestern Canada also is apparent. It is likely that the failure of the Pacific trough to develop was due to the model's difficulty in simulating adequately the interaction of the largest scale planetary waves with the topography of the Himalayan plateau.

Surface temperature. The observed mean surface air temperature in the central Arctic (Figure II.7) during January-February ranges between -35 and -40°C and is strongly controlled by the net longwave radiation at the surface, and the heat conducted through

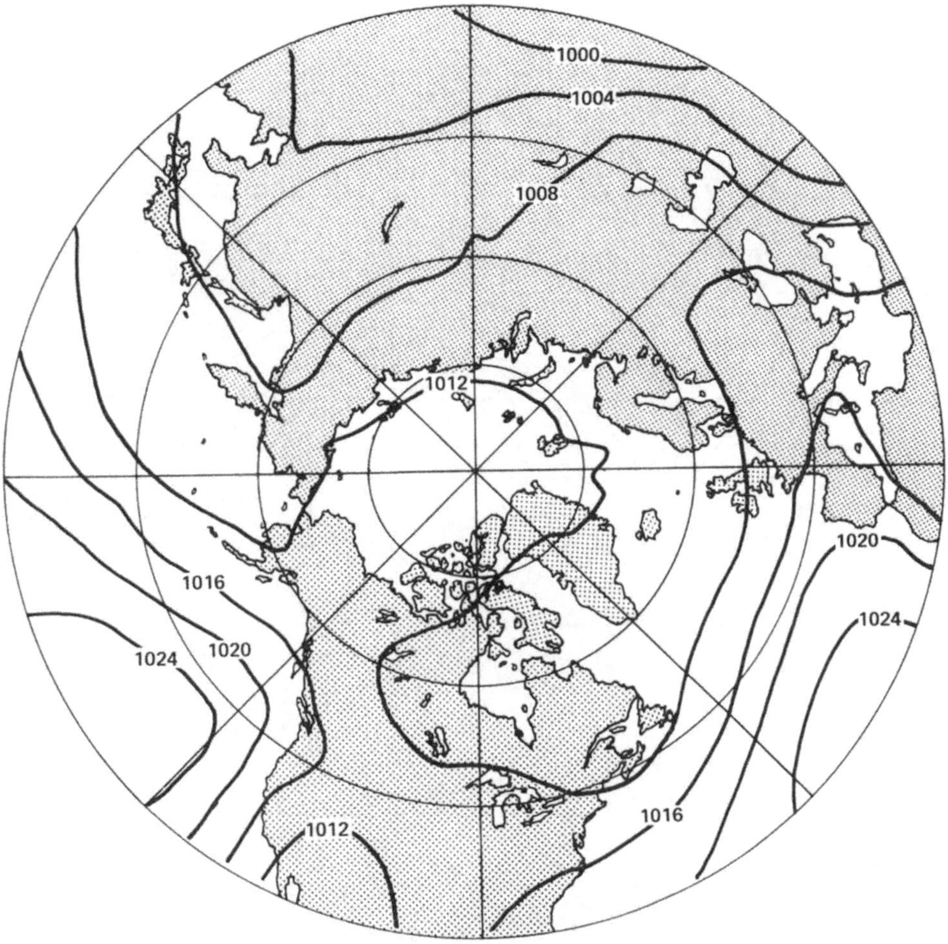

Fig. II.3. Observed sea level pressure during Northern Hemisphere summer.

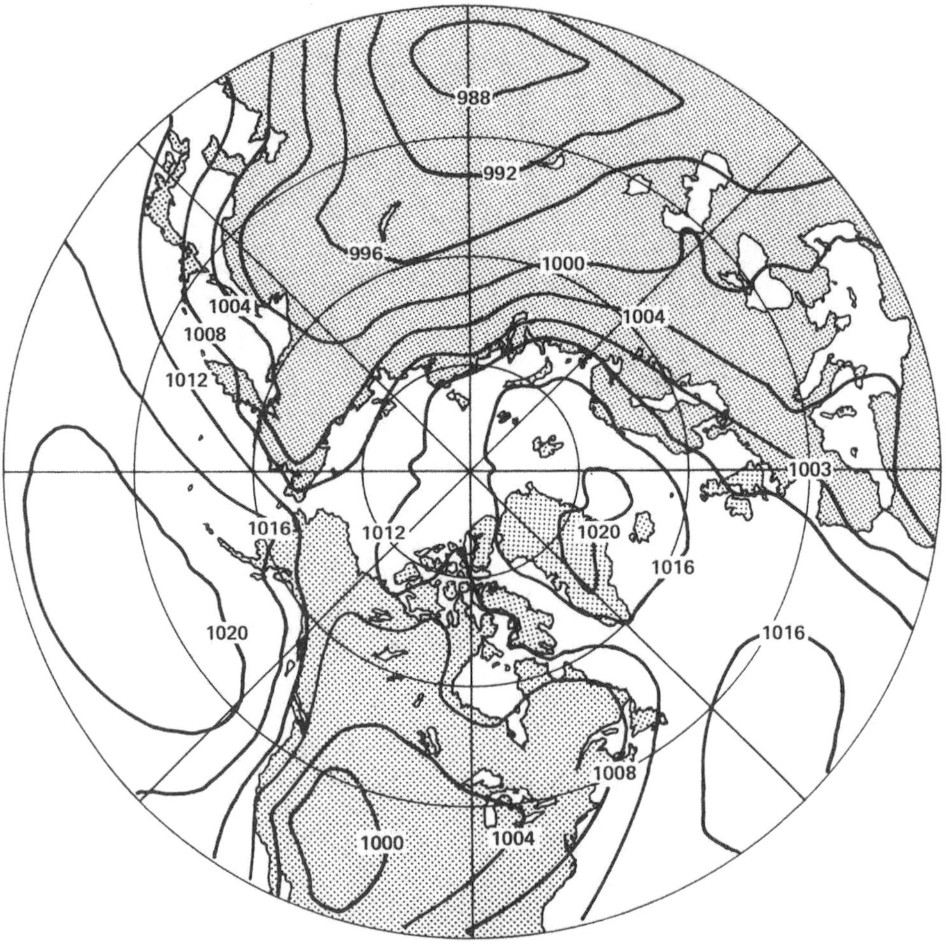

Fig. II.4. As in Fig. II.3, except simulated by the GLA GCM.

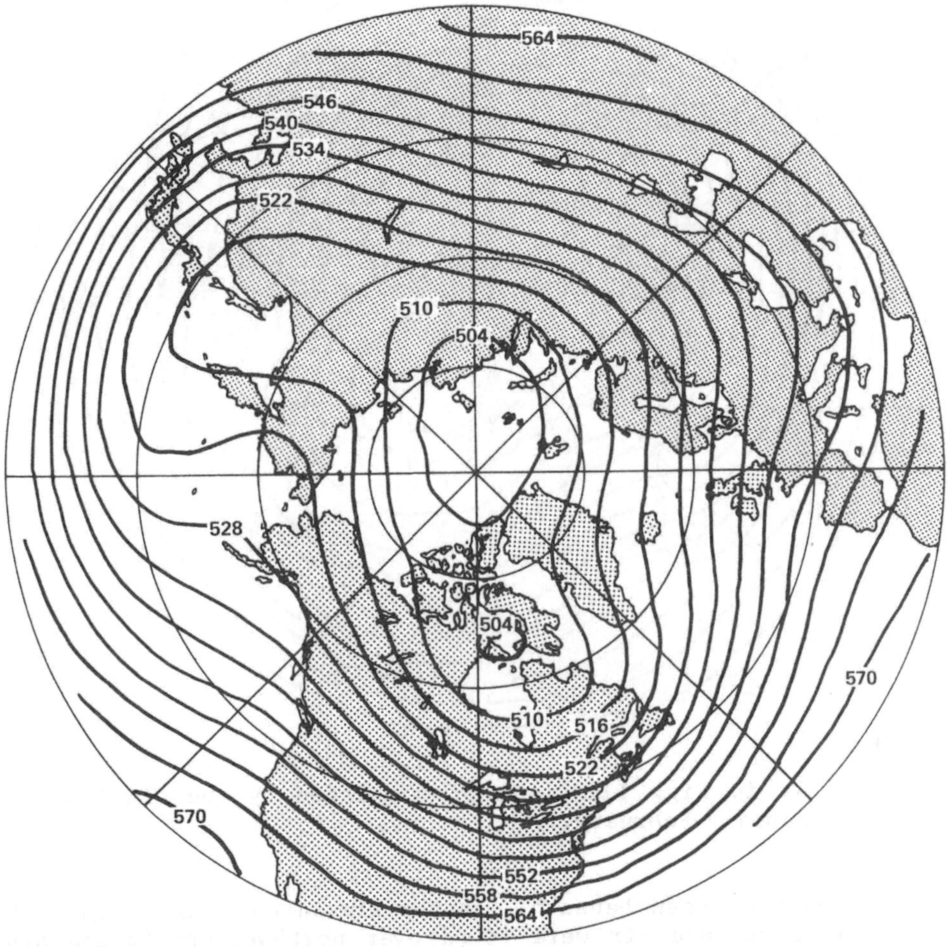

Fig. II.5. Observed 500 mb geopotential during January.

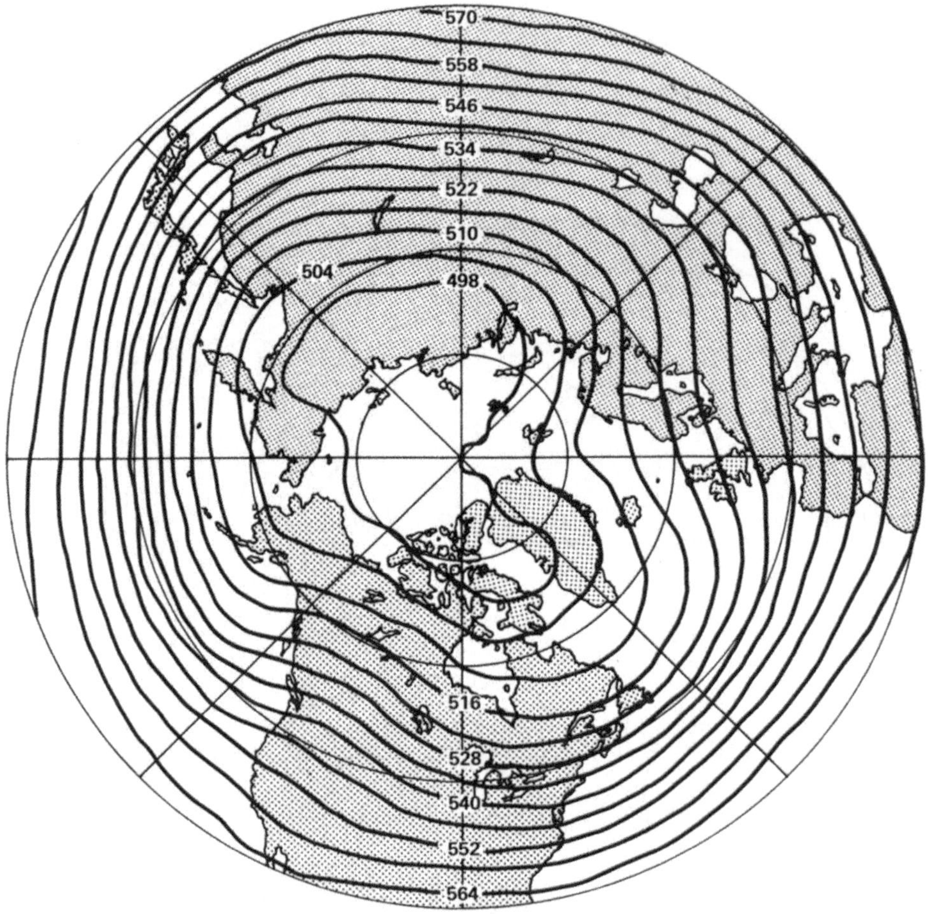

Fig. II.6. As in Fig. II.5, except simulated by the GLA GCM for January-February.

the ice from the ocean beneath. In the simulated climatology very cold pools of surface air were found over northwestern Canada and northeastern Siberia in conjunction with the surface anticyclones located in those regions. In general the surface air temperatures computed by the GLA GCM (Figure II.8) tended to be too warm throughout the high latitudes. The discrepancies with observations were most obvious over the central Arctic, where differences were 5-9°C, and over Siberia, where differences exceeded 10°C.

During the summer the surface temperature in the central Arctic is influenced strongly by the melting sea ice, and thus remains very close to 0°C, and the model results agree well with the observations over these ice covered regions.

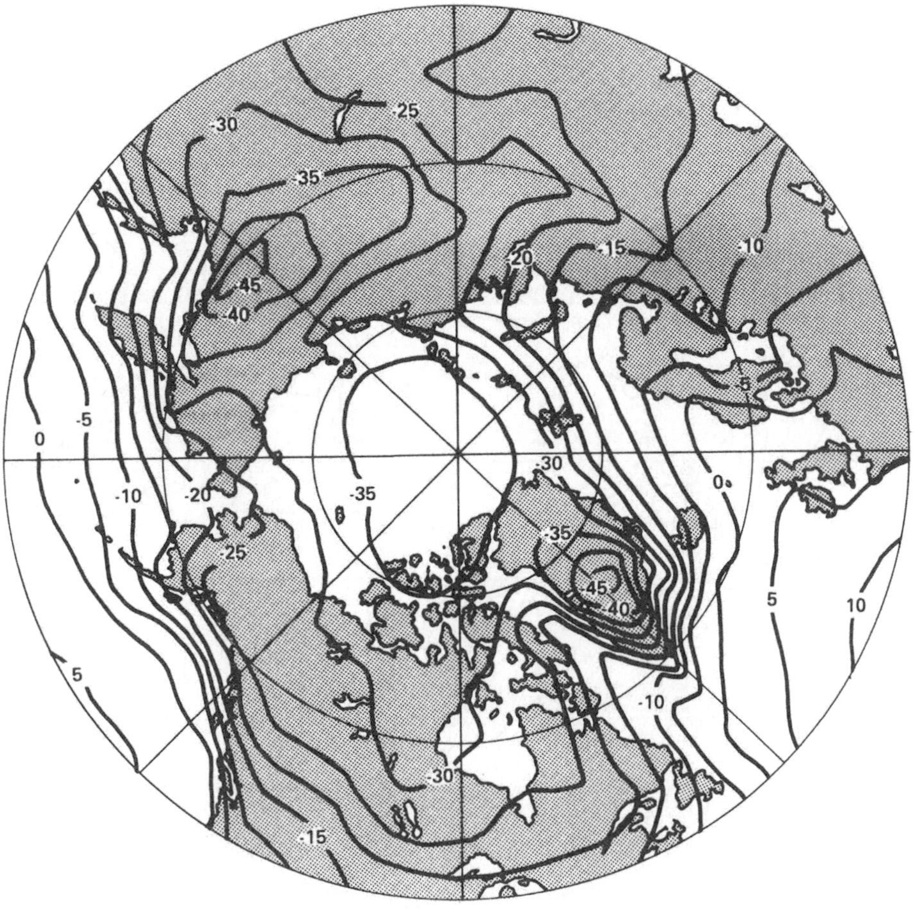

Fig. II.7. Observed surface air temperature during January.

Cloudiness. The importance of Arctic cloudiness is discussed in detail in the chapter on Arctic stratus. The observed and simulated zonally-averaged cloudiness is shown in Figures II.9 and II.10. The high latitude distribution of cloud frequency was clearly unrealistic, as the model significantly underpredicted the summertime cloud cover, and overpredicted the wintertime cover.

2.2 Sensitivity Studies

It is generally very difficult to make unambiguous associations between changes in the general circulation or in climate, and changes in surface conditions, atmospheric composition, or radiative forcing. The atmosphere seldom provides clearly-defined ensembles of anomaly/no anomaly situations since it is uncommon for one anomaly to occur independently of all other processes. It

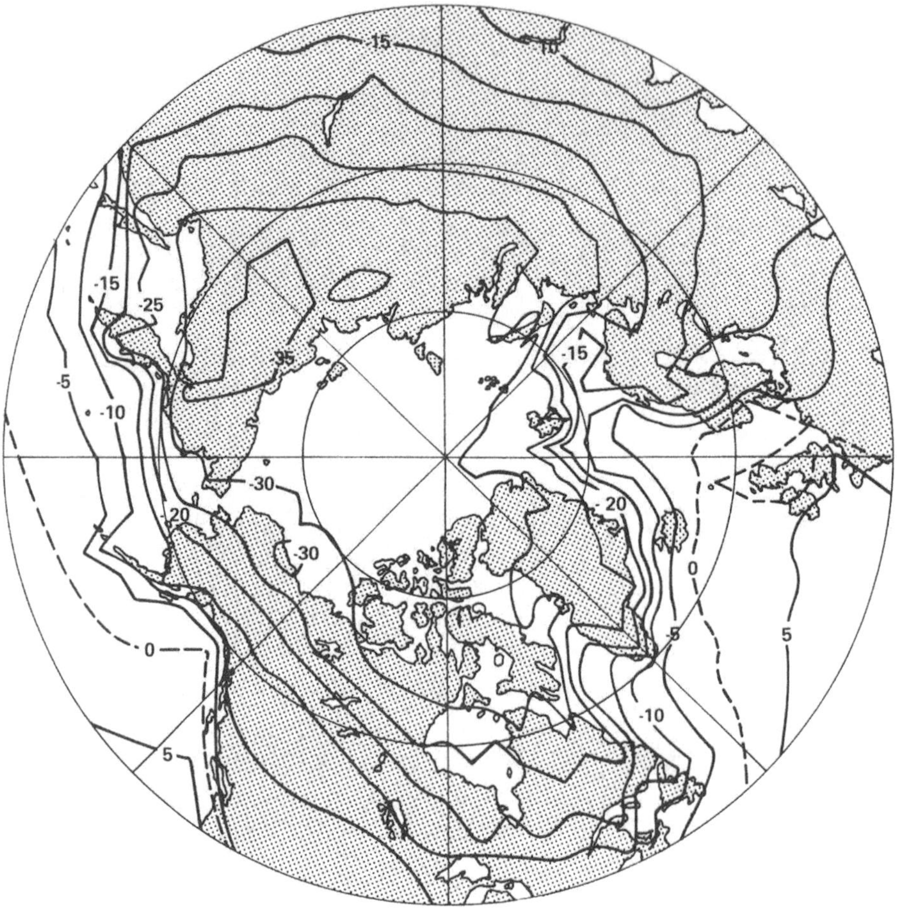

Fig. II.8. As in Fig. II.7, except simulated by the GLA GCM.

is here that the GCM can be used to advantage, since the models, unlike the atmosphere, can be run in a carefully-controlled experimental mode. Thus the model potentially provides the means of assessing the atmosphere's response to changes in sea surface temperature, sea ice conditions, surface albedo, CO_2, O_3 or concentration of minor constituents, ground wetness and solar irradiance. Moreover, the role of certain invariant features, such as mountains, can be illustrated by deleting them from the calculation, although it usually is difficult to decide which effects result directly from the change, and which have resulted from other processes that were related to the change in more complicated ways. Experiments of this nature are termed <u>sensitivity</u> studies, and have been used extensively in Arctic climate studies. The following are the principal results of the most relevant of these experiments:

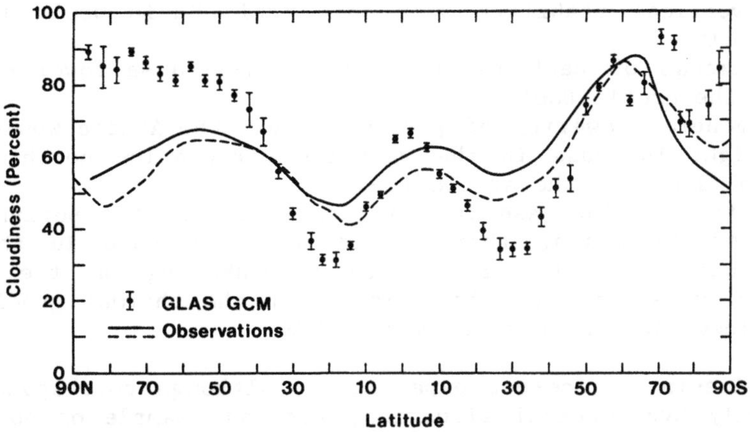

Fig. II.9. Observed and simulated zonally-averaged cloudiness (January).

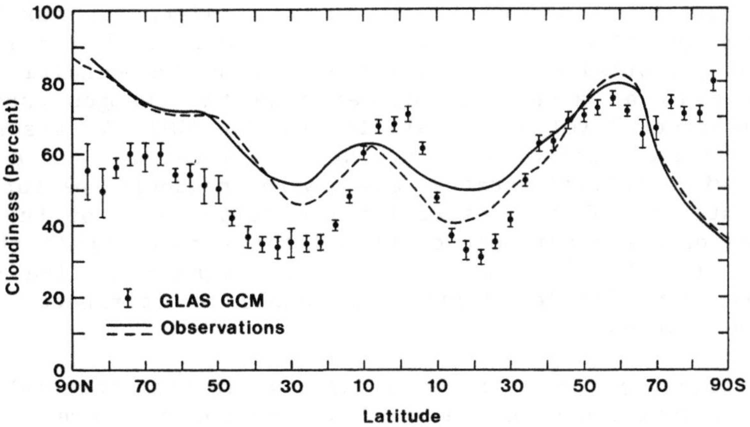

Fig. II.10. As in Fig. II.9, except for July.

(i) High latitude sensitivity experiments. The earliest sensitivity studies with GCM's were investigations into the climatic consequences of removing completely the Arctic ice cover (see Fletcher et al., 1971; Warshaw and Rapp, 1973). The Fletcher et al. experiments were conducted with early versions of the 2-level UCLA general circulation model, and consisted of month-long runs with and without an ice cover in the central Arctic. Based on the differences between the ice-in and ice-out experiments, the authors argued that in the absence of ice,

(1) The vertical stability over the Arctic Basin would decrease markedly.
(2) The excess of heat in the Arctic would cause lower pressure over the Arctic Basin.
(3) The general lowering of pressures over the Arctic would cause a slight increase in the westerly circulation north of the normal limit of the arctic ice.
(4) Despite this increase in the circulation on the periphery of the Arctic Basin, there would be a decrease in the mid-latitude westerlies reflecting a weakening of the global polar vortex due to the decrease in the thermal gradient in the lower levels north of about 45°N.

The ice-in -- ice-out experiments, although corresponding to an extremely hypothetical situation, were an example of how a GCM could be used to provide insight into the mechanisms through which sea ice might affect climate. The experiments also provided some quantitative validation of Fletcher's (1965) hypothesis relating sea ice and climate.

The Warshaw and Rapp experiments also focused on the ice-in -- ice-out question, and utilized the UCLA model. It further addressed the question of statistical significance of the results: Are the changes obtained for the two different ice conditions distinguishable to an acceptable degree from the changes associated with the inherent level of variability in the GCM? To answer this question, several additional runs were carried out for each ice condition with initial states that were randomly perturbed by small amounts in order to establish the noise level of the model. The results of the experiment confirmed the strong impact that the removal of the ice had on the lower troposphere in the central Arctic, but very little significance could be attributed to the midlatitude changes.

More recently there has been interest in the potential impact of ice cover variations in the marginal ice zones, where the largest annual and interannual variability of ice extent is found. Herman and Johnson (1978) used the GLAS GCM to investigate the sensitivity of monthly-averaged conditions to ice boundary variations in the north Atlantic and north Pacific. In their experiments the control consisted of "minimum" ice conditions, corresponding to an ice distribution that would correspond to the ice cover in all sectors being simultaneously at their climatologically minimal value. The perturbation run similarly had ice margins that corresponded to a simultaneous sea ice maximum in all sectors.

The January-February means of six runs with minimum ice cover were compared with the means of two runs with maximum ice cover. The most important differences between the two sets of runs were:

(1) Sea level pressure in the Baffin-Davis region and in the East Greenland-Barents-Norwegian Sea regions were higher when the ice was at its maximum extent. Sea level pressure was lower in the North Atlantic between Iceland and Great Britain.
(2) 700-mbar and 500-mbar geopotential height fields were lowered substantially with maximum ice extent.
(3) The zonally average poleward eddy energy transport increased with maximum ice cover.
(4) Statistically significant changes also occurred in the subtropical regions of the eastern north Atlantic and eastern north Pacific.

The first three of these results were not entirely unexpected, and possibly could have been deduced in advance using mass conservation arguments. Probably only the magnitude of the response was in doubt. The fourth result is intriguing, since it points to the possibility of relationships between events in the high latitudes of the north Atlantic and those in the subtropics ("the north Atlantic oscillation"). A satisfactory explanation for this result has not been offered, although similar patterns of positive anomalies in one region of the Atlantic occurring in conjunction with negative anomalies in other regions have been obtained in sea surface temperature anomaly experiments, and in data-based empirical orthogonal function analyses. Lau (1981) reported the results of a 15 year simulation with the GFDL model which produced a similar pattern of anomalies without perturbations of sea surface temperature or sea ice conditions.

The Herman and Johnson experiments contained many unrealistic features, and other GCM experiments are still required to answer the basic questions about the potential effect of sea ice on climate. For example, the imposed signal due to the composite maximum and minimum ice conditions used here is probably stronger than any ever realized in nature, at least in recent times. Moreover, ice cover in the model was of uniform thickness, so that the larger heat transfer through thin ice in the marginal ice zone was not represented. Furthermore, there was no ice or ocean feedback to provide damping of the original anomaly due to air-sea exchange processes.

(ii) <u>CO_2 Simulation Experiments.</u> Several potentially important aspects of ice-atmosphere interaction have emerged from recent GCM experiments designed to investigate the climatic impact of an enhanced atmospheric CO_2 concentration. The basic CO_2 problem is how the climate would respond to the modifications in the surface and atmospheric radiation budgets that would result from increasing the CO_2 opacity of the atmosphere. (The basic radiative mechanism involves increased absorption and emission in the 15 μm region of the infrared spectrum, which increases the downward flux of longwave radiation at the surface, and the upward flux of radiation from the atmosphere to space.)

A complete and systematic review of GCM studies of the CO_2 question has been prepared by Schlesinger (1983), and reference is made to that report for a detailed intercomparison of the models employed, and their results.

Virtually all CO_2 experiments predict an increase in the globally-averaged surface temperature due to a doubling or quadrupling of atmospheric CO_2, but produce widely varying results on a region-by-region basis. Some of these regional differences may be associated with the differences in the parameterizations of sea ice, ocean temperature and heat transport, cloud formation and ground hydrology. In some cases the inclusion of these parameterized processes has introduced positive or negative feedback mechanisms into the basic relationship between surface temperature and atmospheric opacity. The most important of these feedbacks that relate to the high latitudes include:

a. Ice-albedo feedback. The ice-albedo feedback is one of the basic mechanisms usually thought to be responsible for the "polar amplification" of the CO_2-induced surface warming, i.e., greater surface warmings in the polar regions than elsewhere. The largest polar amplification appears to have been obtained in the GFDL simulations of Manabe and Stouffer (1980), although varying degrees of amplification are present in most models that contain an interactive sea ice cover. The basic ice-albedo feedback process is simple, i.e., darker snow-ice surfaces absorb more energy than bright ones, and thereby melt and become even darker. Maykut and Untersteiner (1971) have demonstrated the strong sensitivity of the melting rate of the ice to its albedo, which in turn depends on the composition and optical properties of the mixed snow-ice surface, and in particular on the distribution of snow, meltwater, and open ocean. Grenfell and Maykut (1977) have also noted the dependency on the ratio of direct to diffuse radiation, which depends on the cloud cover.

An accurate simulation of the ice-albedo feedback requires a calculation of the surface reflectivity as a function of the changing surface conditions. Ice albedos in current GCMs are either fixed (e.g., at 0.7 in the GLA GCM), or change as a crude function of temperature or snowcover (e.g., GFDL). The database required to support the development of parameterizations in this area is somewhat limited. While several local studies have been conducted (e.g., Grenfell and Maykut, loc. cit.; Grenfell and Perovich, 1983), no adequate basin-wide climatology is available. An excellent summary of ice albedo data has been prepared by Barry et al. (1983). Wiscombe and Warren (1980) have developed a physical radiative transfer model for the reflectivity of snow surfaces, but it does not appear that similar calculations have been carried out for sea ice or meltwater ponds.

b. Cloud feedback. There are two important aspects to the cloud feedback problem in CO_2 experiments. First, does the CO_2 perturbation affect the distribution of cloudiness on a global or regional basis, and secondly, if it does, is there an associated change in the surface radiation budget? These questions take on special importance in the Arctic in view of the strong effect that Arctic clouds have on the surface radiation budget.

The effect of clouds on the surface radiation balance relates to the issue of whether the clouds' greenhouse or albedo effect dominates. Observational evidence (Ohring et al., 1981) and theoretical calculations (Cess, 1976) both suggest that the albedo effect of clouds dominates their greenhouse effect, although the infrared greenhouse properties may be more important in certain regional situations, or when the clouds occur at high levels.

The nature of cloud radiation feedback in the polar regions is uncertain at the present time. There is some evidence from the analyses of Vowinckel and Orvig (1964) that the clouds' effect on the infrared component of the radiation budget dominates during most months of the year. Other data sets, such as those compiled by Maykut and Church (1973) and Marshunova (1961) have the potential to provide some input into this question, but it appears that no analysis has yet been undertaken.

The experiments conducted with the GFDL sector model (Manabe and Wetherald, 1980) indicated only a very slight sensitivity to the inclusion of cloud-radiative feedback. On the other hand, experiments with the NCAR community climate model (Washington and Meehl, 1983, 1984) and with the OSU model (Schlesinger, 1983) have indicated that the amplitude of the polar warming due to CO_2 was diminished when cloud feedback was included.

c. Lapse rate feedback. A second process that has been sometimes associated with the polar amplification of the CO_2 induced surface warming (Manabe and Wetherald, 1980) is the so-called lapse rate feedback. Relative to conditions in the tropics, the planetary boundary layer and lower troposphere in the Arctic and high latitudes are stably stratified. As a consequence, it is very difficult for excess energy at the surface to affect upper layers. On the other hand, moist convective instability in the tropics produces a fast and effective means of distributing heat throughout the troposphere.

The problem of simulating accurately the vertical distribution of CO_2 heating involves parameterizations of the Arctic planetary boundary layer, and the modes through which the PBL is coupled to the lower troposphere. Note that the failure of many climate models to reproduce realistic Arctic stratus cloud conditions, and the unrealistically warm wintertime surface temperatures are both potential manifestations of deficient boundary layer treatments.

(iii) Ice-atmosphere interactions in ice-age simulations. Extremely large variations in sea ice limits have occurred in conjunction with the major glaciations of 18-20,000 years ago. Several GCM experiment (Williams et al., 1974; Gates, 1976; Manabe and Hahn, 1977) have attempted to simulate the climatic consequences of the associated changes in the surface boundary conditions. While these experiments have provided valuable insight into the possible nature of ice-age circulations, they do not quite provide a clear indication of the role of sea ice variations, since the sea ice boundary conditions were varied along with other boundary conditions an external forcings. The Williams et al. simulations of 20,000 BP conditions produced a general southerly displacement of the Aleutian and Icelandic lows and northern hemisphere cyclonic activity, but contained variations in sea surface temperature, and continental ice and topography. The Gates and the Manabe and Hahn simulations with the 18,000 BP CLIMAP boundary conditions both produced stronger anticyclonic conditions and reduced precipitation over the mid-latitude continents, but also included changes in ocean surface temperature, continental ice, and in the case of the GFDL simulation, changes in the solar forcing.

While these experiments have provided useful insight into the possible nature of an ice-age general circulation, it does not appear possible to determine the precise causes of the changes in the circulation, i.e., those ascribed to ice edge variations, ocean temperature changes, cloud feedback, etc.

(iv) Statistical reliability of results. A major problem in the interpretation of sensitivity experiments is differentiating between the response caused by changes in the boundary conditions or external forcings of the model, and the noisiness or inherent variability of the model. For example, in an experiment in which it is desired to examine the sensitivity of the model to the location of the sea ice margin, it is necessary to establish the confidence with which a quantity in the control run, $x_{control}$, is distinguishable from its value in the perturbation run, $x_{perturbed}$. For example, x could be the zonally-averaged-temperature. This noisiness or inherent variability of the model is analogous to the natural variability of the atmosphere, and is due to the range of dynamic instabilities that the model is capable of producing.

The Warshaw and Rapp (1973) analysis of the ice-free Arctic Ocean experiment was one of the first to deal with the question of statistical significance. The Warshaw and Rapp strategy for estimating noise was to redo the control run with new initial conditions derived by superimposing small, random perturbations on the initial conditions used in the control. The variance of an ensemble of runs thus initialized is a measure of the model's "natural variability" and defines the level below which differences between the control and the anomaly cannot be resolved. This procedure has also been adopted in the Herman and Johnson ice margin experiments, and has been discussed in further detail by Chervin and Schneider (1976).

An alternative approach was adopted by Shukla (1975) in his analysis of the atmosphere's response to sea surface temperature anomalies. Following an earlier analysis of Leith (1973) Shukla computed the model noise, σ_T^2, as

$$\sigma_T^2 = \frac{2\sigma^2}{T} \int_0^T \left(1 - \frac{\tau}{T}\right) R(\tau)\, d\tau \tag{11.22}$$

Here σ is the standard deviation of the time series whose mean is being estimated by averaging over period T, and $R(\tau)$ is the autocorrelation of two values in time separated by an interval τ. Statistical significance is then judged on the basis of the ratio $\Delta x/\sigma_T$, where Δx is again the "signal", or difference between the control run and anomaly run. Further tests of statistical significance are given by Laurmann and Gates (1977), and Chervin et al., 1974.

2.3 Coupled Simulations

There are numerous ways in which the atmosphere interacts with the surface of the ocean and with the pack ice. Dynamically, the momentum transfer from the atmosphere to the ice-covered surface contributes to the pack ice drift, drives the circulation of the surface layers of the ocean, and contributes to the dissipation of atmosphere kinetic energy. Thermodynamically, heat and moisture exchanges during convection simultaneously heat the atmosphere and cool the ocean. Radiatively, the reflectivity and transmissivity of the surface to solar and thermal radiation determine the amount of energy scattered or emitted to the atmosphere. Moreover, there can be a transfer of mass through the cycle of evaporation and precipitation.

These ocean-ice-atmosphere couplings are represented in GCM's with varying degrees of realism. In a large number of simulations to date the interactions have been one-way: The temperature and vapor pressure at the surface determine the exchange of heat and moisture to or from the atmosphere [e.g., (11.9)], but themselves do not change, or else change in a fashion that is predetermined.

Similarly, the roughness of the surface affects the momentum transfer through the roughness parameter z_o, but no redistribution of ice or changes in the state of the sea are allowed to occur. The albedo of the ice may remain fixed even though in nature the melting process would darken the surface by introducing meltwater ponds or open ocean.

The need to represent explicitly each interaction between the atmosphere, the ice, and the ocean is determined partly by the strength of the interaction, and partly by the characteristic time over which significant changes in the various media occur. For example, if the characteristic lifetime of an ice margin anomaly were very long as compared with the climatic period of interest, say a month, then it would be adequate to specify the ice margin as a fixed lower boundary condition. If, on the other hand, ice margin changes were strongly linked to atmospheric processes, and occurred on time scales shorter than a month, it would be necessary to calculate explicitly the mutual adjustment of ice and atmosphere.

The likely consequence of not properly simulating the ice-atmosphere coupling in the case of the ice margin experiments is to produce an atmospheric response that is probably stronger than that which occurs in nature. As an example, consider the case of air mass modification which occurs during winter when cold continental air flows off the North American continent and over the comparatively warmer and ice-free waters of the North Atlantic. The initial heat loss during such an event could be as large as 500 W m^{-2}. However, in nature a heat loss of this magnitude would probably not be sustained. Since the heat transfer from ocean to atmosphere is roughly proportional to the air-sea temperature difference, the flux would diminish as the atmosphere heated or the ocean cooled. The ocean might even cool sufficiently to allow a new layer of sea ice to begin to accrete. If the ocean were not allowed to cool, the imposed anomaly would give a stronger signal to the atmosphere since the large air-sea temperature difference would persist as long as some heat sink were available to the atmospheric boundary layer, or new cold air masses were advected over the anomaly.

However, attempts to model the coupled system have met with limited success. This is due principally to the fact that there is a strong dependence of ocean and ice surface processes on atmospheric energy and momentum fluxes, and to the fact that atmospheric models generally are incapable of providing fluxes with the accuracy required by the ice or ocean models.

An an example, the energy balance at the top surface of the ice is

$$F_s + F_\ell + F_{conv} + F_{cond} + F_{evap} = \begin{matrix} 0 \ , & T < 273 \\ Q_m, & T > 273 \end{matrix} \qquad (11.23)$$

where F_s, F_ℓ, F_{conv}, F_{cond} and F_{evap} are the surface fluxes due to solar and longwave radiation, convection, conduction and evaporation, respectively and where Q_M is the rate of melting of the ice. Thus, a systematic error of 10 W m^{-2} in the net flux would translate into an error of approximately 9 cm per month in the summer melting ice.

We note here the coupled ice ocean atmosphere simulations of Manabe et al. (1975; 1979) and Bryan et al. (1975). The early versions of the GFDL coupled model produced sea level pressure distributions in mid and high latitudes that were generally unrealistic. The Antarctic regions tended to be too warm near the surface due to the extra heat that the oceans transported polewards, and this in turn caused even more heating because some Antarctic sea ice melted, decreasing the surface albedo and allowing more solar radiation to be absorbed. Excessive ice growth in the Arctic occurred, although deficiency was later removed when the mean annual solar radiation was replaced with a realistic seasonal cycle. As previously noted, coupled ice-atmosphere-ocean simulations have been used in investigations of the CO_2 issue.

REFERENCES

Barry, R. G., A. Hednerson-Sellers and K. P. Shine (1983) Climate sensitivity and the marginal cryosphere. Fourth Biennial Ewing Symposium, Am. Geophys. Union (To appear).

Bryan, K., S. Manabe and R. C. Pacanowski (1975) A global ocean-atmosphere climate model. Part II: The oceanic circulation. J. Phys. Oceanogr., 5: 30-46.

Cess, R. D. (1976) Climatic change: An appraisal of atmospheric feedback mechanisms employing zonal climatology. J. Atmos. Sci., 33: 1831-1843.

Chervin, R. M., W. L. Gates and S. H. Schneider (1974) The effect of time averaging on the noise level of climatological statistics generated by atmospheric general circulation models. J. Atmos. Sci., 31: 2216-2219.

Chervin, R. M. and S. H. Schneider (1976) On determining the statistical significance of climate experiments with general circulation models. J. Atmos. Sci., 33: 405-412.

Fletcher, J. O. (1965) The heat budget of the Arctic Basin and its relation to climate. RAND Corp., Santa Monica, California, Report R-444-PR, 179 p.

Fletcher, J. O., Y. Mintz, A. Arakawa and T. Fox (1971) Numerical simulation of the influence of arctic sea ice on climate. In Proceedings, IAMAP/IAPSO/WMO Symposium on Energy Fluxes over Polar Surfaces, Moscow, WMO Technical Note 129.

Gates, W. L. (1976) The numerical simulation of ice-age climate with a global general circulation model. J. Atmos. Sci., 33: 1844-1873.

Grenfell, T. C. and G. A. Maykut (1977) The optical properties of ice and snow in the Arctic basin. J. Glaciol., 18: 445-464.

Grenfell, T. C. and D. K. Perovich (1983) Spectral albedos of sea ice and incident solar irradiance in the southern Beaufort Sea. J. Geophys. Res. (To appear).

Herman, G. F. (1981) Cloud-radiation experiments conducted with GLAS general circulation models. In Proceedings, Workshop on Radiation on Cloud-Radiation Interaction in Numerical Modelling, European Center for Medium Range Weather Forecasts.

Herman, G. F. and W. T. Johnson (1978) The sensitivity of the general circulation to arctic sea ice boundaries. A numerical experiment. Mon. Wea. Rev., 106: 1645-1664.

Herman, G. F. and W. T. Johnson (1980) Arctic and Antarctic climatology of a GLAS general circulation model. Mon. Wea. Rev., 108: 1974-1991.

International Council of Scientific Unions (1979) Report of the JOC Study Conference on Climate Models: Performance, Intercomparison and Sensitivity Studies. World Meteorological Organization, GARP Publication Series, (W. L. Gates, Ed.), Vol. I and II, 1040 p.

Lau, N. C. (1981) A diagnostic study of recurrent meteorological anomalies appearing in a 15-year simulation with a GFDL general circulation model. Mon. Wea. Rev., 109: 2287-2311.

Laurmann, J. A. and W. L. Gates (1977) Statistical considerations in the evaluation of climatic experiments with atmospheric general circulation models. J. Atmos. Sci., 34: 1187-1199.

Leith, C. E. (1973) The standard error of time-averaged estimates of climatic means. J. Appl. Met., 12: 1066-1069.

Lorenz, E. N. (1973) Climatic predictability. In The Physical Basis of Climate Modelling, World Meteorological Organization, GARP Publication Series, No. 16, p. 132-136.

Madden, R. A. (1976) Estimates of the natural variability of time-averaged sea level pressure. Mon. Wea. Rev., 104: 942-952.

Manabe, S., K. Bryan and M. J. Spelman (1975) A global ocean-atmosphere climate model. Part I: The atmospheric circulation. J. Phys. Oceanogr. 5: 3-29.

Manabe, S. and D. Hahn (1977) Simulation of the tropical climate of an ice age. J. Geophys. Res., 82: 3889-3911.

Manabe, S., K. Bryan and M. J. Spelman (1979) A global ocean-atmosphere climate model with seasonal variation for future studies of climate sensitivity. Dyn. Atmos. Oceans, 3: 393-426.

Manabe, S. and R. J. Stouffer (1980) Sensitivity of a global climate model to an increase of CO_2 concentration in the atmosphere. J. Geophys. Res., 85: 5529-5554.

Manabe, S. and R. T. Wetherald (1980) On the distribution of climate change resulting from an increase in CO_2-content of the atmosphere. J. Atmos. Sci., 37, pp. 99-118.

Marshunova, M. S. (1961) Principal characteristics of the radiation balance of the underlying surface and of the atmosphere in the Arctic. Tr. Ark. Nanchno-Issled. Inst., 229: 5-53.

Maykut, G. A. and N. Untersteiner (1971) Some results from a time-dependent model of sea ice. J. Geophys. Res., 76: 1550-1575.

Maykut, G. A. and P. E. Church (1973) Radiation climate of Barrow, Alaska, 1962-1966. J. Appl. Met., 12: 620-628.

Mitchell, J. F. B. (1981) The seasonal response of a general circulation model to changes in CO_2 and sea temperatures. Meteorological Office, Brackness, UK, Met. 0.20. Technical Note II/171.

National Academy of Sciences (1979) Carbon dioxide and climate: A scientific assessment. Washington, D.C.

Ohring, G., P. F. Clapp, T. R. Heddinghaus and A. F. Krueger (1981) The quasi-global distribution of the sensitivity of the earth-atmosphere radiation budget to clouds. J. Atmos. Sci., 38: 2539-2541.

Schlesinger, M. E. (1983) Climate model simulation of CO_2-induced climatic change. Unpublished manuscript, Climatic Research Institute, Oregon State Univ.

Shukla, J. (1975) Effect of Arabian Sea surface temperatures anomaly on India summer monsoon: A numerical experiment with the GFDL model. J. Atmos. Sci., 32: 503-511.

Somerville, R.C.J., P. H. Stone, M. Halem, J. E. Hansen, J. S. Hogan, L. M. Druyan, G. Russell, A. A. Lacis, W. J. Quirk and J. Tenenbaum (1974) The GISS model of the global atmosphere. J. Atmos. Sci., 31: 84-117.

Vowinckel, E. and S. Orvig (1964) Energy balance of the Arctic. II. Longwave radiation and total radiation balance of the surface in the Arctic. Arch. Met. Geophy. Biokl., B, 13: 451-479.

Warshaw, M. and R. R. Rapp (1973) An experiment on the sensitivity of a global circulation model. J. Appl. Met., 12: 43-49.

Washington, W. M. and G. A. Meehl (1982) General circulation model experiments on the climatic effects due to a doubling and quadrupling of carbon-dioxide concentration. National Center for Atmospheric Research, Boulder, Colorado, Report NCAR/0306/81-2(E), 41 p.

Washington, W. M. and G. A. Meehl (1983a) General circulation model experiments on the climatic effects due to a doubling and quadrupling of carbon dioxide concentration. J. Geophys. Res., 88: 6600-6610.

Washington, W. M. and G. A. Meehl (1984) Seasonal cycle experiments on the climate sensitivity due to a doubling of CO_2 with an atmospheric general circulation model coupled to a simple mixed layer ocean model. J. Geophys. Res., 89: 9475-9503.

Williams, J., R. G. Barry and W. M. Washington (1974) Simulation of the atmopshere using the NCAR global circulation model with ice age boundary conditions. J. Appl. Met., 13: 305-317.
Wiscombe, W. J. and S. G. Warren (1980) A model for the spectral albedo of snow, I. Pure snow. J. Atmos. Sci., 37: 2712-2733.

CHAPTER 12

DIAGNOSTIC STUDIES OF LARGE-SCALE AIR-SEA-ICE INTERACTIONS

John E. Walsh

Department of Atmospheric Sciences
University of Illinois
Urbana, IL 61801

Two approaches to the problem of large-scale air-sea-ice interaction can be distinguished: numerical modelling and diagnostic (statistical) analysis. The diagnostic approach is the subject of this chapter. Stochastic modelling, which combines the two approaches, is described in Chapter 13 (Lemke).

Diagnostic (statistical) analyses are generally based on time series of historical data sampled at (ideally) regular time intervals. We begin this chapter with a brief survey of the data series available for diagnostic studies of large-scale air-sea-ice interactions. Particular attention will be given to the features of these data sets that can be problematic in the context of statistical analysis. Some of the statistical procedures most commonly used in diagnostic studies of large-scale interactions are summarized in Section 2. After a survey of the literature of diagnostic studies of air-sea-ice interactions (Section 3.), we conclude with a discussion of the implications for the predictability of sea ice (Section 4.).

1. HIGH-LATITUDE DATA SERIES

1.1 Sea Ice

A surprisingly large number of sea ice data sets have been compiled over the past several decades. Many of these data sets have been compiled by regional and national ice centers using a variety of observations from ships, aircraft and, since the mid 1960's, satellites. The data sets include indices of regional ice severity, maps and charts for specific regions, and hemispheric or global ice analyses. Chapter 17 (Barry) lists these data sets, together with their resolution and periods of record.

The regional charts often include relatively detailed information, on features such as the ice type and ridging distribution over limited areas, whereas hemisphere or global charts are most appropriate for studies of large-scale interactions. Unfortunately, the questionable reliability and representativeness of many of the large-scale data sets from the pre-satellite era limit their suitability for statistical analysis of air-sea-ice interactions. A particularly difficult problem is the blending of diverse data sets compiled for different regions and/or time periods. Inhomogeneities resulting from integrations of data sets may introduce trends or other biases into time series analyses. The use of sea ice data in diagnostic studies is therefore characterized by a general trade-off between the record length and homogeneity.

A common practice in the analysis of hemispheric sea ice data sets is the consolidation of the digital information into the areal ice coverage within various longitudinal sectors (e.g., 36 x 10°). Because the maringal ice zone is relatively narrow, this compression of the data is achieved with a minimal loss of informational content. Several such uses of hemispheric data sets will be discussed in Section 3.

1.2 Meteorological Data

While the number of data sets containing high-latitude meteorological data is clearly too large to permit a thorough treatment here, several of the longest and most extensively used data sets will be mentioned in order to point out some of the problems inherent in their usage.

Grids of Northern Hemisphere sea level pressure are available on a daily basis from 1899 to the present (Jenne, 1975). In a recent evaluation of trends and discontinuities in this data set, Trenberth and Paolino (1980) found the high-latitude pressures to be spuriously high prior to the 1930's. Even after the 1930's, observations from drifting ice stations have generally not been incorporated into the historical sea level pressure data sets. Errors of 2-4 mb in monthly (30 day) analyses and 5-10 mb in daily analyses can occasionally be found (e.g., Albright, 1980). Historical sets of pressure grids have also been compiled by the National Meteorological Agencies of West Germany (Deutscher Wetterdienst) and the USSR; the high-latitude errors in these data sets have been found to be comparable to those in the U.S. analyses.

While records of surface air temperature extend back to varying portions of the 19th and early 20th centuries for a network of up to 200 stations north of 55°N, the distribution is obviously biased toward the more heavily populated land areas. The most extensive gridded temperature data sets are the sets of monthly grids for 1881-present compiled at the University of Anglia (see Kelly et al. 1982) and at the Main Geophysical Observatory of the USSR (see Borzenkova et al. 1976). The latter set of analyses depicts temperature anomalies derived from an irregularly defined set of normals. We also note that routinely compiled daily temperature analyses can be seriously deficient in high latitudes. For example, the operational temperatures of the U.S. National Meteorological Center tend to be 2-5°C too high during the summer in the central Arctic because of excessive smoothing in the objective analysis routines.

Since the upper air observing network was established in the 1940's, the time series of upper air data are now 30-35 years in length. The distribution of upper air data also shows a heavy bias toward the mid-latitude land areas of the Northern Hemisphere (Fig. 1a). Because the data coverage is quite poor over the ice-covered oceans, only the broad-scale features can be considered reliable in the high-latitude portions of the gridded data sets.

In comparison with the Northern Hemisphere, the meteorological data base for the Southern Hemisphere is, in many respects, inadequate for time series analysis. Although surface and upper-air data are available since the IGY for 15-20 stations on the Antarctic continent (Fig. 1b), gridded analyses are limited to the most recent 10-15 years and are highly suspect over the vast expenses of the Southern Hemisphere oceans (e.g., van Loon, 1980). A voluminous set of ship observations of surface air temperature, pressure and wind for the period 1861-1960 have recently been compiled by the multinational Historical Sea Surface Temperature Project (HSSTP). The data are stored as monthly values for latitude-longitude sectors which increase in size as the density of observations decreases, e.g., in the Southern Ocean (see Fletcher et al., 1982).

1.3 Ocean Data

Because measurements of salinity, current velocity and subsurface quantities are highly irregular in space and time, the time series of oceanic data used in studies of air-sea-ice interaction have been limited largely to data sets of sea surface temperature (SST). Monthly grids of Northern Hemisphere SST from 1854-present have been compiled on the 5° x 5° grids of the TDF11 data set by the U.S. National Weather Records Center (see Jenne, 1975); these grids were subjected to a "quality-control" procedure by the British Meteorological Office. The coverage and reliability of the data improve considerably through the period of record. SSTs for fourteen regions of the North Atlantic have been compiled on a

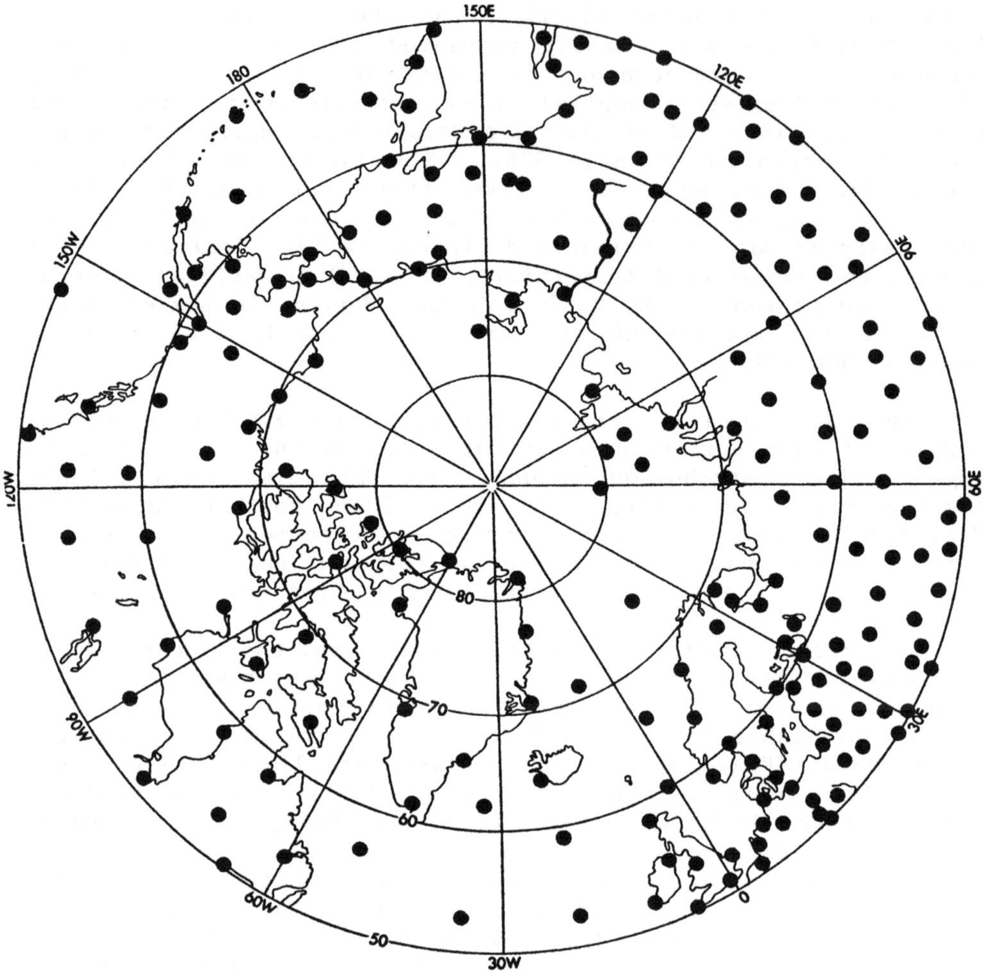

Fig. 1a. Radiosonde stations in the Northern Hemisphere. [Adapted from Oort, 1975, p. 69.]

monthly basis since 1876 by the ICES (International Council for the Exploration of the Sea) Hydrographic Service; several applications of the ICES data set will be discussed in Section 3. The most thorough compilation of the North Pacific SSTs is the set of monthly $5^{\circ} \times 5^{\circ}$ grids produced by the Scripps Institute of Oceanography for the period 1947-present. Daily grids of the northern hemisphere SST have been compiled since 1962 at the U.S. Navy Fleet Numerical Weather Center. As an illustration of the reliability of the SST data sets, a comparison of the North Atlantic portions of the monthly TDF11 and U.S. Navy grids for the years 1963-1967 showed that the variance common to the two data sets

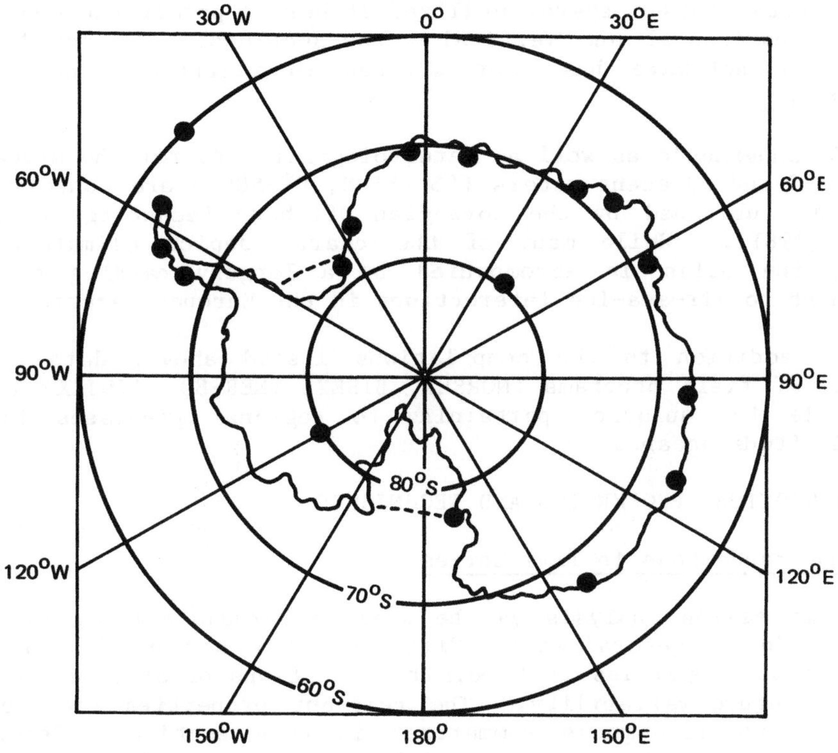

Fig. 1b. Antarctic meteorological stations in operation in 1968. [Adapted from Schwerdtfeger, 1970, p. 253.]

ranged from approximately 50% to 80% of the total variance (Walsh and Sater, 1981).

A set of monthly sea surface temperatures for the oceans of the Southern Hemisphere as well as the Northern Hemisphere is contained in the HSSTP data compilation discussed in Section 2.2. The period of record, 1861-1960, is the same as for the meteorological HSSTP data. As is the case of the Northern Hemisphere data, the Southern Hemisphere SST data are clearly more reliable in the major shipping lanes of the southern oceans (Fletcher et al., 1982).

Among the regional oceanic data sets, a few deserve special mention because of their potential value for studies of regional-scale air-sea-ice interactions. The first of these is the Bergen Ocean Data Archive, which contains a wealth of oceanographic profiles from the European Arctic. The Archive includes data from more than 200,000 Nansen casts taken during 1900-1979 in the Greenland, Norwegian and Barents seas. The spatial and temporal formats of the data storage system utilized at Bergen permit the construction of horizontal and vertical corss-sections. A considerable amount of software has been written to facilitate interactive analyses.

Oceanographic as well as meteorological data for the Norwegian Sea Shelf and adjacent waters (55°-85°N, 0°-50°E) are contained in an atlas published by the Norwegian Offshore Technology Research Group (1981). While many of the charts depict climatological means, the atlas is accompanied by a lengthy catalog of data pertinent to air-sea-ice interactions in the European Arctic.

In addition to the compilations listed above, data from a variety of field programs (NORSEX, BISEX, YMER-80, AIDJEX) remain available for anlayses pertaining to regional processes in the high-latitude oceans.

2. STATISTICAL PROPERTIES AND TECHNIQUES

2.1 Non-randomness in Time Series

Time series analysis may be used to formulate empirical predictions in two general ways. First, the time series of a variable may possess properties that permit predictions of at least some of its own future variability. The relevant properties include the types of non-randomness summarized in this section. Secondly, predictive relations for one variable may be formulated in terms of the time series of other variables. Multilinear regression, an example of this second form of empirical prediction, is outlined in Section 2.3.

The types of non-randomness which may permit the prediction of a variable from its own past are persistence, trend and periodicities.

(i) Persistence. As outlined in Section 2.2, the persistence of a variable may be evaluated as a lagged autocorrelation. The square of a lagged autocorrelation is a measure of the fraction of variance that is predictable from persistence alone. In studies of atmospheric predictability, persistence is often used to construct a "control" forecast. Because of the stronger persistence of oceanic and sea ice variables, persistence can be competitive with, or even superior to, other oceanographic and ice forecasting strategies.

(ii) Trend. In many geophysical time series, computed autocorrelations are contaminated by the presence of a trend. Cross-correlations can also be considerably exaggerated when both series contain trends. Nevertheless, a trend can provide the basis for the predictability of a (generally small) fraction of the variance of a time series. Various tests of the statistical significance of trends are available. These include the Mann-Kendall and the Spearman rank statistics (Mitchell et al., 1966, p. 64-65), both of which are valid tests of nonlinear as well as linear trends. The significance of a linear trend can be evaluated by comparing a regression-derived trend or slope (see Section 2.3) with the standard deviation of the slopes obtained from an infinite number of samples containing the same variance and record length as the series under investigation (Brownlee, 1965, Chap. 11).

We also note that persistence and trend will reduce the effective number of independent sample members to $N(1-r_1)/(1+r_1)$, where N is the actual sample size and r_1 is the autocorrelation at lag one.

(iii) <u>Periodicities (spectral analysis)</u>. Periodicities or cycles in a time series can aid in both the interpretation and the prediction of the fluctuations of a variable. Non-randomness of a quasi-periodic or cyclic nature is often represented quantitatively by the frequency spectrum, which is also known as the power spectrum. In its continuous form, the spectrum gives the distribution of the variance of a time series over all possible frequencies or periods, ranging from an infinite period to the shortest resolvable period (equal to twice the interval between successive observations). While alternative procedures (the fast Fourier transform, Bartlett's method) are available for the computation of the power spectrum, the simplest procedure is the serial convariance method summarized below. A comparative discussion of the different procedures is given by Jenkins and Watts (1968).

In case of a finite sample of size N, the set of m spectral estimates can be obtained from the serial covariances, C_τ, at lags $\tau = 0$ to $\tau = m$ ($m < N$):

$$C = \frac{1}{N-\tau} \sum_{n=1}^{N-\tau} (x_n - \bar{x})(x_{n+\tau} - \bar{x}) \quad (12.1)$$

where $\bar{x}$ is the mean of x_n for the entire series. Raw spectral estimates are computed as cosine transforms of the serial covariances (see Mitchell et al., 1966). The raw spectral estimates are generally smoothed with a 1/4, 1/2, 1/4 weighted moving average (the "Hamming" method). Significance levels of computed spectra may be estimated using the fact that each spectral estimate is distributed as chi-square divided by the degrees of freedom, $(2N-m/2)/m$ (Tukey, 1950), where the maximum lag, m, can be chosen arbitrarily subject to the constaint that $m \leq N$. Since the number of frequency intervals in the computed spectrum is m+1, the resolution of the spectrum is proportional to m. However, the stability of the spectral estimates will decrease as m increases. The maximum lag should therefore not exceed approximately one-third of the total record length (Mitchell et al., 1966, p. 4).

The spectrum of a random time series ("white noise") will tend to be relatively "flat", i.e., the spectral estimates will tend to be independent of frequency. By contrast, a time series with persistence will exhibit a spectrum containing the largest amplitudes at the longest periods ("red noise"). A quasi-periodicity in a time series will be indicated by relatively large spectral estimates in the appropriate range of frequencies.

The major limitations of spectral analysis include (i) the absence of information about the phase of the fluctuations and (ii) the problem of the "aliasing" of the high-frequency (unresolvable) fluctuations onto the lower-frequency (resolvable) variations. Phase information can be obtained, however, in "cross-spectral" analysis, which is based on the Fourier transform of the cross-covariance rather than the serial covariance. Cross-spectral analysis produces a "coherence" function characterizing the frequency-dependence of the relationship between two variables, and the "phase" function describing the lag between the two variables.

2.2 Correlation Coefficients

Quantitative anlayses of interactions between two variables, x and y, are often based on the coefficient of linear correlation, r, given by

$$r = \frac{\sum_{n=1}^{N} [(x_n-\bar{x})\ (y_n-\bar{y})]}{\left[\sum_{n=1}^{N} [(x_n-\bar{x})^2]\ [\Sigma\ (y_n-\bar{y})^2]\right]^{\frac{1}{2}}} \qquad (12.2)$$

where the overbars denote sample means. The quantities $\Sigma(x_n-\bar{x})^2$ and $\Sigma(y_n-\bar{y})^2$ are N times the sample-derived estimates of the variances of the two quantities; the standard deviations of x and y are the square roots of the respective variances. The standard error of estimate of r is $(1-r^2)\sqrt{N-1}$. If samples of size N are drawn from two uncorrelated populations, the distribution of r is approximately normal with a standard deviation of $\sigma_r \simeq (N-2)^{-1/2}$, provided that N is large (Panofsky and Brier, 1968). We therefore say that correlations differing from zero by $1.96\sigma_r$ and $2.58\sigma_r$ are statistically significant at the 95% and 99% levels, respectively. Other criteria (e.g., t-test; z-transformation) may also be used to evaluate the significance of a correlation, although, except for very small samples, the significance level will differ only negligibly from the estimate obtained using σ_r.

Lagged autocorrelations, often used as a measure of the temporal "persitence" of a departure from the mean of a variable, can be computed from (12.2) by setting y equal to the value of x at a prescribed lag.

When correlative methodologies are employed with meteorological, oceanographic and/or cryospheric data, several properties of geophysical time series deserve consideration. The first such property is the non-normality found in some data (e.g., precipitation; ice coverage in land-blocked high-latitude regions). Although the literature of mathematical statistics abounds with warnings against the use of non-normal data in statistical methodologies formulated for normally distributed data, the effect of non-normality can be shown to be generally quite small if not negligible (McDonald, 1960). Indeed, a survey of sampling studies leads to the conclusion that "the normal bivariate surface may be distorted and mutilated to a remarkable degree without seriously influencing the frequency-distribution of 'r' in the case of small parent-population correlation ..." (Haldane, 1949).

A more problematic consideration is the presence of persistence in the time series of geophysical variables. While the lack of independence of successive sample members permits the analysis of air-sea-ice interactions in terms of Markhov processes, tests of the statistical significance of cross-correlations must take into account the effective reduction of the sample size:

$$N' = N \frac{1-\rho_1\rho_2}{1+\rho_1\rho_2} \qquad (12.3)$$

where N′ and N are the effective (reduced) and actual sample sizes, and ρ_1 and ρ_2 are the lag 1 autocorrelations of the two series (Hannan, 1970, p. 368). Since correlations will be inflated not only by persistence but also by the smoothing (filtering) of time series and by the presence of trends in the data, the more stringent significance criteria implied by (12.3) are often necessary in correlative analyses of relationships within the air-sea-ice system.

We conclude this discussion with two notes of caution. First, (12.2) describes only linear correlations between two variables. If the variables are nonlinearly related, reliance on (12.2) can lead to misleading interpretations. Secondly, large and/or significant correlations do not imply cause and effect. In many geophysical systems, two variables will be related because both are affected by the same other variable(s).

2.3 Regression Analysis

Statistical studies of a diagnostic or predictive nature often use multilinear regression as a means to synthesize a collection of correlative results. Briefly, a prediction $\hat{y}$ of a predictand y is formulated as a linear combination of a set of predictors x_1, x_2 ..., x_m:

$$\hat{y} = \sum_{m=1}^{M} a_m x_m \tag{12.4}$$

The root mean square error of the prediction is minimized by taking

$$a_m = \sum_{\ell=1}^{M} \langle x_\ell x_m \rangle^{-1} \langle x_\ell y \rangle \tag{12.5}$$

where the brackets denote a sample covariance and the negative exponent denotes a matrix inverse. The a_m are often referred to as "regression coefficients". The skill, S, of a set of predictions is commonly taken as the fraction of the variance that is successfully predicted:

$$S \equiv 1 - \frac{\overline{(y-\hat{y})^2}}{\overline{y^2}} \tag{12.6}$$

S is the square of the correlation between the predicted and observed quantities. The computed values of S will include the true skill, S_o, and an artificial component, $S-S_o$, attributable to the differences between the sample and population covariances. The artificial skill is the amount by which the computed skill exceeds the skill expected in the case of an infinite sample drawn from the same statistical system. When a sample-derived regression relation is applied to an independent (second) sample, the explained variance can be expected to be less than S by an amount to the artificial skill, $S-S_o$. S and S_o satisfy the approximate relation

$$S \simeq S_o + \frac{M}{N-1} \tag{12.7}$$

The derivation of (12.7) is outlined by Lorenz (1977). If the time series of the x_m and y are autocorrelated, (12.7) must be replaced by an expression of the form

$$S \simeq S_o + \frac{M}{N}\frac{\tau}{\Delta t} \tag{12.8}$$

where Δt is the sampling interval and τ is an "autocorrelation time scale" (see Davis, 1976; Leith [1973] also describes a procedure for the estimation of τ). It is apparent from either (12.7) or (12.8) that the spurious or artificial component of the skill will increase as the number of predictors is increased and as the sample size is decreased. The statistical reliability of a predictive relation will therefore be enhanced by a maximization of the record length and by a minimization of the number of predictors. Since M is the number of predictors chosen a priori (i.e., prior to a screening procedure), techniques which effectively compress predictor data sets are clearly advantageous. One such technique will be described in Section 2.4.

2.4 Empirical Orthogonal Functions (EOFs)

The summary of regression-derived forecast skill in Section 2.3 illustrates the desirability of restricting the number of variables in predictive applications. Moreover, the spatial coherence exhibited by fluctuations of atmospheric, oceanic and sea ice variable suggests that considerable compression of the data fields should be possible. One strategy used to achieve an effective data reduction is the representation of the fluctuations in terms of empirical orthogonal functions (also known as principal components). EOFs may be interpreted as independent (orthogonal) patterns of variability of a field of variables. While temporal as well as spatial patterns may be computed, most geophysical applications of EOFs employ spatial functions.

Although EOFs represent valuable statistical tools, their usage in geophysical applications has been somewhat limited by a formulation in terms of matrix algebra and by an often bewildering assortment of jargon that has evolved among nongeophysical proponents of pattern analysis.

In nonmathematical terms, EOFs represent the dominant modes of variability in the set of data fields under investigation. Unlike Fourier series, the EOFs form a set of basis functions that are determined from the particular data set in such a way that each successive EOF is a spatial pattern representing a maximum of the variability unexplained by the previously computed EOFs. Any anomaly field can be expressed exactly as a linear combination of the entire set of spatial EOFs.

A mathematical outline of the formulation of EOFs is based on a matrix of observations $\underset{\sim}{F}$:

$$\underset{\sim}{F} = (\underset{\sim}{f}_1, \underset{\sim}{f}_2, \ldots, \underset{\sim}{f}_I) = \begin{pmatrix} f_{11} & f_{21} & & f_{I1} \\ f_{12} & f_{22} & & f_{12} \\ \vdots & \vdots & \cdots & \vdots \\ f_{1J} & f_{2J} & & f_{IJ} \end{pmatrix} \qquad (12.9)$$

In the above symbolism, the column vector $\underset{\sim}{f}_i$ represents the ith observation (array, map, grid) containing values for each of J points. The ith observation at the jth point is represented by the scalar f_{ij}. Following Kutzbach (1967, p. 792-793), we determine the column vector (map) $\underset{\sim}{e}$ having the highest resemblance to all $\underset{\sim}{f}_i$ simultaneously. This is done by maximizing the quantity $\underset{\sim}{e}'\underset{\sim}{R}\underset{\sim}{e}$, where $\underset{\sim}{e}'$ is the transpose of $\underset{\sim}{e}$ and $\underset{\sim}{R}$ is the J x J matrix of covariances between the $\underset{\sim}{f}_i$. The maximization of $\underset{\sim}{e}'\underset{\sim}{R}\underset{\sim}{e}$ leads to an eigenvector equation for which there are J solutions. These solutions are the columns $\underset{\sim}{e}_1, \underset{\sim}{e}_2, \ldots, \underset{\sim}{e}_J$ of the eigenvector matrix, $\underset{\sim}{E}$. The $\underset{\sim}{e}_j$ and the corresponding eigenvalues, λ_j, are generally arranged in order of decreasing λ_j. The transpose matrix, $\underset{\sim}{E}'$, can be used to construct an I x J matrix of coefficients, $\underset{\sim}{C} = \underset{\sim}{E}'\underset{\sim}{F}$. The original observations can then be expressed as linear combinations of the J eigenvectors:

$$\underset{\sim}{f}_i = \sum_{\ell=1}^{J} c_{\ell i}\underset{\sim}{e}_\ell \qquad (12.10)$$

The $c_{\ell i}$ play the role of amplitudes or weighting functions. The observation vectors corresponding to the largest $c_{\ell i}$ will most closely resemble the eigenvector $\underset{\sim}{e}_\ell$. (Note that, because $c_{\ell i}$ can be either positive or negative, resemblance to both $\underset{\sim}{e}_\ell$ and $-\underset{\sim}{e}_\ell$ can be distinguished.) Since the I observations will generally correspond to I different times, the elements of the ℓth row of $\underset{\sim}{C}$ constitute a time series of the coefficients associated with the ℓth eigenvector.

Because of the orthogonally constraint, the eigenvectors and eigenvalues satisfy the respective relations $\underset{\sim}{E}'\underset{\sim}{E} = \underset{\sim}{I}$ (=identity matrix) and $\underset{\sim}{C}\underset{\sim}{C}' = \underset{\sim}{L}$ N, where $\underset{\sim}{L}$ is the J x J diagonal matrix of eigenvalues. The eigenvectors are therefore orthogonal (uncorrelated) in space, while the coefficients are orthogonal (uncorrelated) in time.

The fraction V_k of the variance explained by the k largest eigenvectors is given by

$$V_k = \frac{\sum_{j=1}^{k} \lambda_j}{\sum_{j=1}^{J} \lambda_j} \tag{12.11}$$

In nearly all geophysical applications, most of the variance is described by the first few eigenvectors:

$$\underset{\sim}{f}_1 = \sum_{\ell=1}^{J} c_{\ell i} \underset{\sim}{e}_\ell \simeq \sum_{\ell=1}^{J^*} c_{\ell i} \underset{\sim}{e}_\ell \quad i = 1, 2, \ldots, I \tag{12.12}$$

where J* is considerably smaller than J. In many studies involving synoptic scale domains, only 3-4 EOFs are required to describe 60-95% of the variance. An EOF representation can therefore permit the highly desirable minimization of the number of predictors in the formulation of regression-derived predictive relations. The fact that severe truncations often produce only small errors in (12.12) is a consequence of the spatial coherence of the $\underset{\sim}{f}_i$ and permits the effective compression of the data. Since the desired percentage of retained variance will vary from one application to another, it is not possible to define precisely a "useful" percentage. However, tests are available for the identification of the subset of EOFs which are statistically significant (e.g., Preisendorfer and Barnett, 1977).

Diagnostic studies of air-sea-ice interactions can therefore be performed by representing one or more of the data fields in terms of EOFs. The time series of the coefficients of the first several EOFs can then be used to compute statistical quantities such as cross-correlations, frequency spectra, etc. The corresponding EOF patterns permit physical interpretation of the statistics of the coefficients. The results of several EOF-based analyses are summarized in Section 3.

Finally, we note that the advantages of the statistical independence (orthogonality) of EOFs may be offset by the fact that the dominant modes of actual geophysical fluctuations are not constrained to be orthogonal. A host of strategies exist for "rotating" eigenvectors into corresponding sets of oblique patterns that may more realistically represent certain types of variability (e.g., smaller-scale fluctuations). A survey of rotational techniques is given by Richman (1981).

3. APPLICATIONS TO THE ICE-ATMOSPHERE-OCEAN SYSTEM

The statistical tools of Section 2. have been used in a wide variety of diagnostic studies of the ice-atmosphere-ocean system.

While a detailed examination of all these studies is not possible here, we will summarize some of the most pertinent data-based analyses. The studies will be grouped into three categories: (3.1) evaluations of statistics of ice fluctuations, (3.2) analyses of ice-atmosphere interactions, and (3.3) diagnostic studies incorporating all three components of the ice-atmosphere-ocean system. Within each category, the studies will be highlighted in approximate chronological order.

3.1 Statistics of Sea Ice Fluctuations

In the more than 100 years since the earliest publications of time series of sea ice data, the statistics of sea ice fluctuations have been evaluated with increasing levels of sophistication. The marginal ice zone of the North Atlantic was the subject of most of the earlier analyses. For example, Meinardus (1906) used over 100 years of data describing the duration of ice near Iceland to support a more qualitative assessment by Lyell (1872) that a cycle of 4-5 years is present in Icelandic sea ice severity. Meinardus also evaluated the seasonal cycle of East Greenland Sea ice. Brooks and Quennell (1928) examined the 4-5 year periodicity by computing lagged autocorrelations of the Icelandic time series. The autocorrelations were not statistically significant, ranging from -0.16 at two years to 0.06 at 5 years. The lack of significance was attributed to the variations in the intensity of successive maxima and minima of the cycle. Brooks and Quennell also found modest correlations ($r \simeq 0.2 - 0.4$) between indices of ice extent in the East Greenland Sea and those in the Barents and Kara Seas during the same calendar year. Some year-to-year persistence of ice anomalies was also apparent in Brooks and Quennell's results.

Evaluations of trends of North Atlantic ice extent include the work of Koch (1945) and Ahlmann (1949), both of whom showed a substantial decrease of ice severity in the Icelandic region during the first half of the twentieth century. The trend is also apparent in Walker's (1949) compilation of ice anomalies based on the historical series of Danish ice charts. A reversal of this trend after 1950 was noted by Lamb (1977). One of the few attempts to assess corresponding changes in the North Pacific is the work of Dunbar (1967), who found some (albeit inconclúsive) evidence that the wintertime Bering Sea ice extremes were not much different in the late 1800's from those of the 1954-65 period. There were indications, however, of heavier ice in the western Bering Sea and of later springtime breakup in the late 1800's. Finally, Zakharov and Strokina (1978) summarize the more recent trends of summetime ice extent in the Arctic Ocean; these included an increase from the mid 1950's to the mid 1960's, followed by a decrease through the mid 1970's.

While the studies summarized above are highly regional, the relationships between ice anomlies in different regions have been systematically evaluated during the past decade. Sanderson's (1975) analysis of the British Meteorological Office ice charts for 1967-74 showed substantial year-to-year variability of ice extent in various longitudinal sectors of the Northern Hemisphere. However, it was also apparent that concurrent anomalies in different sectors tended to be of opposite sign. For example, an opposition of wintertime ice anomalies in the Eastern Canadian Waters and the Greenland/Barents Seas was noted by Sanderson as well as by Rogers and van Loon (1979). Smirnov (1980) used 30 years of Soviet ice data to deduce that the ice anomalies in the western Arctic (Chukchi and Beaufort Sea) and eastern Arctic tend to be of opposite sign. Itin's (1936) analysis of Soviet data showed a similar opposition. Walsh and Sater (1981) confirmed this opposition but found that the actual correlations between ice anomalies in the two regions are only about -0.4. Similar values are apparent in Lemke et al.'s (1980) cross-correlations based on data for 1966-76. The corresponding fractions of variance are therefore generally less than 20%.

These regional interrelationships are further illustrated by several recent EOF analysis of sea ice variability (Walsh and Johnson, 1979a, 1979b; Lemke et al. 1980). Because the majority of data points in grids or maps of arctic sea ice are either ice-free or ice-covered throughout the year, the EOF representations of sea ice fluctuations are formulated somewhat differently from the more common two-dimensional representations of meteorological and oceanographic data fields. In the EOF analyses referenced here, each grid is described by 36 values, each of which is the area covered by ice in a 10° longitudinal sector. The spatial fluctuations are thus represented in terms of 36 EOFs that are orthogonal functions of longitude only. The first EOFs computed from non-normalized data by both Walsh and Johnson (1979b) and Lemke et al. (1980) represent anomalies of opposite sign in the eastern Canadian waters and the Greenland/Barents Seas (see Fig. 2), thereby confirming the results of Sanderson (1975) and Rogers and van Loon (1979). The first two patterns of Walsh and Johnson also contain anomalies of opposite sign in the Alaskan and North Atlantic sectors, thereby confirming Smirnov's (1980) results. Since the first three patterns of each set of EOFs contain both positive and negative anomalies in large sectors of longitude, a note of caution must be added to statements about trends or changes of hemispheric ice extent. More specifically, in the decadal-scale samples used by the investigators listed above, trends and/or fluctuations in one sector of the Arctic trend to be accompanied by trends and/or fluctuations of the opposite sign in other sections of the Arctic. We note that the first 3 of the 36 EOFs of each set describe over 50% of the total ice variance.

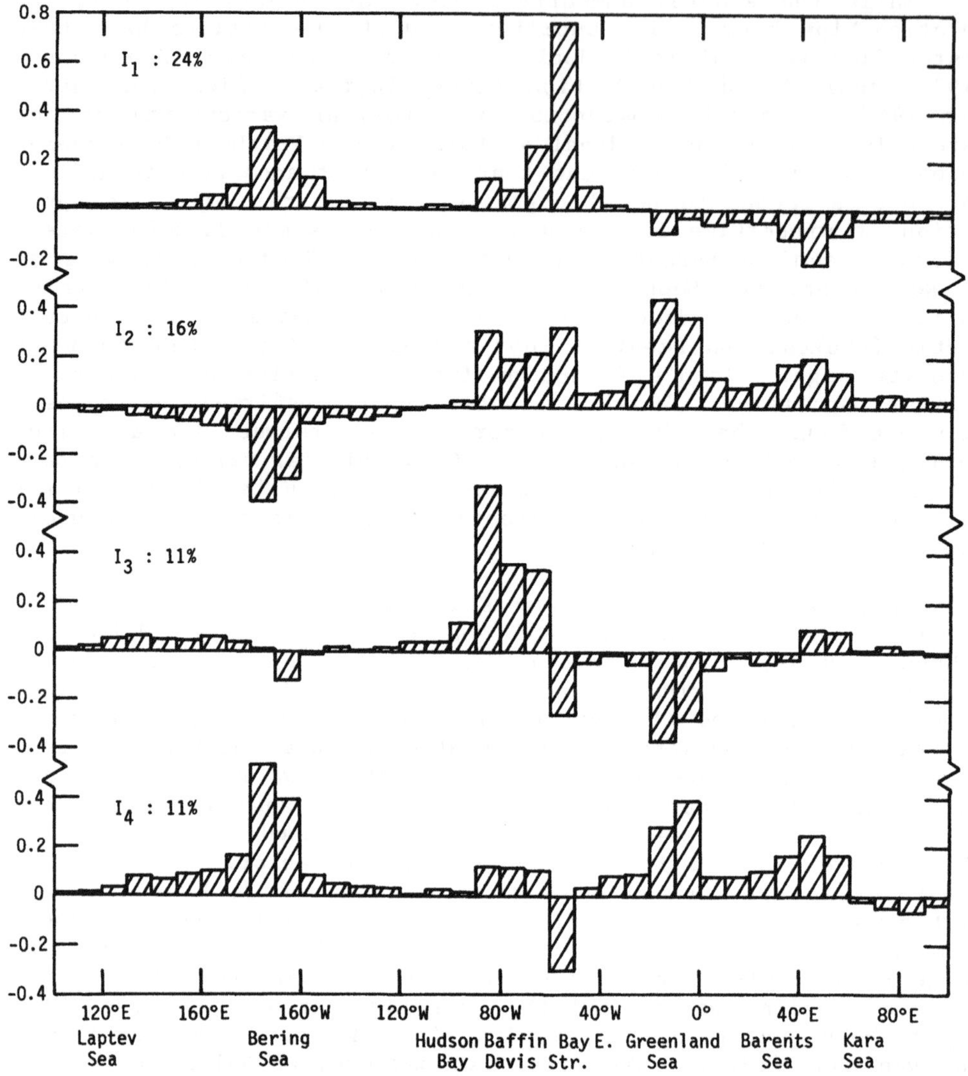

Fig. 2. The first four eigenvectors of northern hemisphere ice extent. Horizontal axis is longitude.

While Lemke et al. discuss frequency spectra of EOF amplitudes for the years 1966-76, the 11-year record length is too short for an evaluation of multiyear cycles. Volkov and Sleptsov-Shevevich (1971), in an analysis of 30-35 years of data for (presumably) the Soviet Arctic, describe in a qualtitative manner the apparent

2-year, 4-5 year and 7-year cycles revealed by a graphical analysis. The 4-5 year cycle is stated to be "more random" and "less definite" than the other cycles, although Itin (1936) also noted a 5-year cycle in ice coverage north of the Soviet Union during the years 1900-1934. More recently, Barnett (1980) has presented a rather striking illustration of a 5-year cycle in 25 years of summertime ice data for the northern Alaskan waters. Rogers (1977) detected the same periodicity in Barrow temperature data for 1948-74, but found that the signal was not statistically significant in a longer (1921-1975) time series of Barrow temperatures. Roger's results therefore appear to cast some doubt on the validity of the 5-year cycle in the Alaskan sector. Yet a 4-5 year periodicity, as well as a 7-year cycle, has also been detected by Kelly and collaborators in a preliminary analysis of North Atlantic ice fluctuations during the years 1901-1956. This analysis is based on Kelly's (1979) digitized sea ice data set. While apparent cycles such as these have been reported by various investigators, they clearly do not represent regular sinusoidal fluctuations. Moreover, the variety of purported cycles in the 2-10 year range will accomodate virtually any multiyear interannual fluctuation. In a broader context, we note that the meteorological literature is strewn with the wreckages of periodicities that were long ago abandoned after failing the test of time.

Because the corresponding Antarctic analyses have been severely limited by data availability, we will mention only several evaluations of Antarctic ice fluctuations. Iciness in the Weddell Sea, as inferred from the data for Scotia Bay, showed a general increase from 1900-1930 and a general decrease thereafter (Fletcher, 1969). Budd's (1975) analysis, however, shows considerable longitudinal dependence in the fluctuations of Antarctic ice extent. Moreover, the longitudinal EOFs of the Antarctic ice extent, while based on only six years of data, show that most of the variance is contained in patterns representing anomalies of opposite sign in different longitudes (Lemke et al., 1980). The only suggestion concerning periodicities in Antarctic ice coverage in Heap's (1965) reference to 2-4 year and 10-year variations in ice conditions between 7°W and 92°W. However, no evidence or supporting data has been presented to substantiate these periodicities.

3.2 Atmospheric Variability and Sea Ice Fluctuations

Studies of associations between large-scale atmospheric and sea ice variability have appeared rather regularly in the literature for almost a century. While the investigators and the statistical techniques have changed, the goal remains the same: an untangling of the physical mechanisms by quantitative analyses in order to improve our understanding of ice-atmosphere-ocean interactions and in order to permit more reliable forecasts.

The studies during the early twentieth century were limited to the North Atlantic largely because of the availability of sea ice data. Among the first to employ correlative methodologies were Meinardus (1906), Wiese (1922, 1925), Brooks and Quennell (1928) and Frommeyer (1928). Common findings included (i) the association between heavy ice conditions and atmospheric pressure gradients consistent with enhanced southward advection of ice, (ii) the association between heavy (light) ice, below- (above-) normal temperatures, and equatorward (poleward) shifts of cyclone trajectories, and (iii) weak associations between ice fluctuations and the distribution of meteorological variables in selected regions thousands of kilometers distant from the North Atlantic ice margin. A thorough summary and comparison of the studies of the early twentieth century is given by Schell (1939).

Perspective was added to the subject of ice-atmosphere interactions by Walker (1947), who noted the following:

(1) A temporal asymmetry is evident in ice-atmosphere interactions, as "it appears that in general the relations with pressure conditions are closer before than after the ice season of spring and summer, and that the initiative lies rather with the pressure distribution than with the ice" (Walker, 1947, p. 237).

(2) Large-scale ice-atmosphere relationships derived empirically from data for one period may not be present in data for other periods. Regarding interactions in the North Atlantic, "conditions which were so closely associated for twenty and thirty years that their connection inevitably seemed to be a physical reality have in some cases change materially and their relationships have greatly diminished or even reversed" (Walker, 1947, p. 226).

(3) Correlative results of earlier studies must be interpreted very cautiously because the ice indices were sometimes biased by ship reports of drifting ice of undetermined extent and because the correlations were often exaggerated by smoothing and/or trend effects.

Walker also presented examples of ice forecasting formulas derived from multi-linear regression onto meteorological variables.

While studies of the North Atlantic ice-atmosphere interactions have continued through the past few decades, the recent results have been based on somewhat more reliable data. Strubing (1967) notes that the amount of ice in the East Greenland Sea during April-August is a function of the antecedent wind-induced influx of ice from the Pacific gyre into the transpolar drift stream. Some predictability over the monthly to seasonal time scale is indicated. Bjornsson (1969) shows that ice conditions in the waters north of Iceland are also dependent on the intensities of the Icelandic and Barents Sea lows; intensification of the former favors light ice conditions, while intensification of the latter favors heavy ice. Bjornsson also notes that northerly winds are associated with colder temperatures, thereby obscuring the distinction between dynamic and thermodynamic influences on sea ice. Haupt and Kant (1976) use satellite data to estimate ice coverage in various sectors of the North Atlantic during 9 years; the authors argue that the physical mechanisms influencing ice fluctuations vary from one region to another.

The interannual variability of ice conditions in the Canadian Arctic has been shown by Keen (1977) and Crane (1978) to be primarily a response to the surface airflow, which in turn is closely associated with the position of the axis of the eastern North American trough. In the Pacific sector, Rogers (1978) has demonstrated the sensitivity of northern Alaska ice conditions to the distribution of surface pressure in the Alaskan area. Northward airflow, which is associated with both offshore ice drift and enhanced melt, favors light ice conditions. Overland and Pease (1982) have shown that the location of the Bering Sea storm track is correlated ($r = 0.71$) with Bering Sea ice extent, thereby supporting the viewpoint that wind-driven advection dominates the advance of the Bering Sea ice edge. Bering Sea ice variability therefore appears to be "controlled by an externally determined variation in storm-track position related to large-scale differences in the general circulation" (Overland and Pease, 1982, p. 5).

The studies summarized above are regional in scope. Yet the results consistently support the conclusion that a substantial portion of sea ice variability is attributable to atmospheric forcing. The same conclusion is found in Lamb and Morth's (1978) discussion of global teleconnections. While Lamb and Morth's (p. 19) correlative results are enhanced by data smoothing and by trend effects, the results are very consistent with the interpretation that change of ice extent are primarily responses to changes in the amplitudes and wavelengths of the planetary-scale stationary waves. However, we note here the suggestion by Barry (1980, p. 145) that, in view of the data uncertainities and the considerable regional differences in ice extent, "it seems unwise at this time to try and correlate long-term trends with circulation trends within a hemisphere or between hemispheres." A broader interpretation is that interannual of sea ice extent may indeed by dominated by interannual changes in the planetary wave pattern, while longer-term changes such as the decrease of ice extent in the early twentieth century might be associated with global temperature trends. Wallace and Gutzler (1981) suggest this type of reasoning in a recent study of interannual variability of geopotential heights in the Northern Hemisphere.

More detailed analyses of lag relationships have been made possible by the digitization of historical records of sea ice extent (Kelly, 1978; Walsh and Johnson, 1979b; Lemke et al., 1980). Walsh and Johnson correlated amplitudes of EOFs of monthly ice extent during 1953-1977 with the amplitudes of the corresponding EOFs of sea level pressure, surface temperature, and 700 mb height and temperature. The ice-atmosphere correlations were statistically significant at atmospheric lags of up to 2 months and sea ice lags of up to 4 months. (Note, however, that the interpretation of the lagged cross-correations is complicated by the persistence of the sea ice fluctations, the autocorrelations of which are approximately 0.70, 0.55 and 0.45 at 1, 2 and 3 months). Of all the meteorological fields, the surface temperature field showed the strongest relationship to the sea ice fluctuations. However, because the different meteorological fields (e.g., pressure and temperatures) are highly correlated, it was not possible to isolate the dynamic and thermodynamic portions of the atmospheric effects on sea ice. The importance of longitudinal asymmetries was shown by the stronger coupling between the amplitudes of the temperature and sea ice EOFs than between total arctic sea ice extent and the area-averaged temperature of the Arctic. Walsh and Johnson's cross-correlations also showed that, during the (autumn) season of ice growth, the response of the atmosphere to the ice was statistically comparable to the response of the ice to the atmosphere. However, the cross-correlations were very asymmetric about zero-lag during the period of ice retreat, when the ice anomalies depend on the forcing by the atmosphere over a longer antecedent period. The time scales of the significant ice-atmosphere correlations were very similar to those shown by the Lemke et al. (1980) cross-correlations between 1000-500 mb thickness and sea ice extent in the Davis Strait.

In a spectral analysis of EOF amplitudes, Kelly (1978) has found a spectral peak at 7-8 years in the sea-level pressure pattern describing the intensity of the depression track in the Kara Sea. These peaks are consistent with those found in the sea ice data, although their interpretation is subject to the same caveats noted in Section 3.1. Kelly et al. (1982) also report similar spectral peaks in the amplitudes of EOFs describing surface temperature anomalies in the subarctic seas of the North Atlantic. Moreover, the same authors find significant correlations between Siberian and Barents Sea temperatures at 1-3 year lags of the latter. Since these regions are close to the major areas of sea ice growth and decay, the authors speculate that "the time scales of these lag relationships, which are well above those of atmospheric advection, may be related to oceanic or cryospheric advection" (Kelly et al., 1982, p. 82). Similar arguments were made by Haupt and Kant (1976) in an attempt to explain the extreme ice cover in the Greenland Sea during 1969.

Owing to the short record length of the Antarctic data sets, methodologies such as EOF representations and spectral analysis have not been exploited in the study of Antarctic ice/atmosphere interactions. Nevertheless, Schwerdtfeger and Kachelhoffer (1973) present evidence of a statistically significant relationship between the position of the Antarctic ice edge and the latitudinal band of maximum frequency of cyclones. Budd (1975) shows that the annual mean temperature is highly correlated with the latitudinal extent of sea ice near the Antarctic periphery. Finally, Ackley and Keliher (1976) used two years of Antarctic data to argue that a rapid regional decrease of ice extent may be maintained by enhanced cyclonic activity attributable to increased diabatic heating of the atmosphere over the open water. Further discussion of ice/atmospheric feedbacks can be found in Chapter 14.

3.3 Three-component (Ice-Atmosphere-Ocean) Analyses

According to the results of the studies summarized in the previous section, the percentage of sea ice variability that can be specified in a hindcast sense from meteorological data varies considerably as a function of the region, the temporal resolution, the quality of the ice data and the choice of the meteorological variables. Commonly cited sources of the "unexplained" ice variance are oceanic factors such as variable ocean currents and sea surface temperature (SST) distributions. Since a host of studies (e.g., Ratcliffe, 1973; Namias, 1976; Davis, 1976) have shown the existence of statistically significant associations between oceanic variables and the atmospheric circulation, the link between the high-latitide sea ice distribution and SST fields must be viewed in the context of Figure 3. The "direct association", in which high SSTs enhance melting and retard freezing, is to be distinguished from the "indirect association", in which the sea ice distribution is influenced by SST through an SST-induced effect on the atmospheric circulation. The evaluation of the ocean-ice coupling from observational data is complicated by the possibility that both coupling mechanisms may act simultaneously.

Among the first evaluations of ocean-ice coupling was Wiese's (1924) computed correlation of $r = -0.79$ between the summertime ice area in the Barents Sea and the subsequent sea surface temperature in the same region. More detailed anlayses of air-sea-ice coupling in the North Atlantic were performed by Schell (1956, 1970), who found sea ice extent to be "a better index of the subsequent sea surface temperatures than of those contemporary with the ice" (Schell, 1956, p. 57). When extreme years of Icelandic sea ice extent were used by Schell to construct composite difference fields of SST and sea level pressure, the differences were generally small ($< 1^{o}C$ and $\lesssim 4$ mb, respectively). The statistical significance of the results, although not presented by Schell, does not appear to be high.

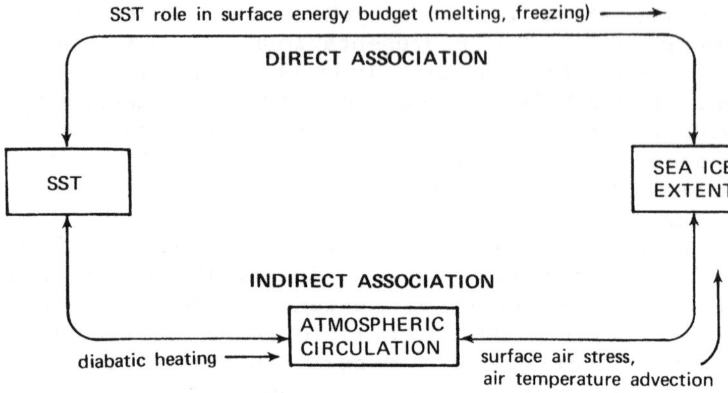

Fig. 3. Schematic representation of the coupling mechanisms between sea surface temperature (SST) and sea ice extent.

Kirillov and Khromtsova (1972) examined the predictability of Greenland Sea ice extent in terms of atmospheric and oceanic variables. Using over 40 months of ice data for the spring and summer months, the authors found the best predictors to be Barentsburg air temperature ($r \simeq -0.5$) and the atmospheric pressure gradient normal to the East Greenland current. The quantity of Atlantic water entering the Arctic through Fram Strait correlated modestly ($r = -0.2$ to -0.4) with summertime ice extent, although the authors do not describe their procedure for estimating the Atlantic inflow. Regression relations derived from the 40 years of data described over 50% of the (dependent) sample variance of Greenland sea ice. Vinje (1977) presents a similarly detailed analysis of the atmospheric and oceanic factors contributing to the interannual variability of East Greenland sea ice extent.

Rogers and van Loon (1979) have evaluated the sea ice and SST anomalies associated with the North Atlantic "seesaw" in winter temperatures between Greenland and Europe. The SST anomalies were obtained from the TDF11 dataset (Section 1.). Although less than half the winters since the mid-1900's satisfied the seesaw criteria, the two modes of the seesaw were found to be associated with statistically significant differences of sea ice in Davis Strait and the Baltic Sea (but not in the East Greenland/Barents Sea) and with statistically significant SST differences during the seesaw winter and the following spring and summer. Interestingly, the largest SST differences were found in middle and high latitudes. Antecedent SST anomalies did not correlate highly with the occurrences of the seesaw mode.

Relationships between the air, sea and ice fluctuations in the eastern Bering Sea were evaluated by Niebauer (1980). Plots of the seasonal cycles of the variables showed a tendency for sea ice and SST to lag the air temperature (freezing degree days). Reduced sea ice extent was associated with higher SSTs (r = -0.64), fewer freezing degree days (r = 0.66) and more frequent southerly winds. The correlations are somewhat enhanced by the presence of trends (less ice and higher SSTs during the late 1970's). Niebauer's data also suggested that compaction and melting can both contribute to Bering Sea ice retreat.

An EOF analysis formed the basis of a recent comparison of air-sea-ice interactions in the North Pacific and the North Atlantic (Walsh and Sater, 1981). The data consisted of four indices of regional sea variability (northern Alaska, Bering Sea, Denmark Strait, East Greenland Sea) and thirty years of monthly SST and sea level pressure (SLP) grids. Separate sets of EOFs were computed for high-latutide SLP, midlatitude SLP, and Atlantic and Pacific SST.

The time series of the EOF amplitudes computed by Walsh and Sater showed that the high-latitude atmospheric circulation is significantly correlated with the circulation fields over both the North Atlantic and the North Pacific. However, the midlatitude circulation shows no systematic tendency to lead or lag the arctic circulation on the monthly time scale. The best predictor of the interannual sea ice variability was found to be the north-south component of the high-latitude geostrophic air-flow. Integration of this component over a period of up to 9 months increased to 0.84 the correlation with northern Alaskan ice variability, while integration of the atmospheric forcing over 3-4 months was found to maximize the Bering Sea ice correlations. The analysis showed that the coupling between the SST distribution and subsequent sea ice extent is stronger in the North Pacific than in the North Atlantic. The results support the "direct association" between positive (negative) SST anomalies and light (heavy) ice in the Bering Sea. The nature of the SST/ice association north of Alaska is apparently one in which the atmosphere serves as an intermediate link, although such a conclusion is quite tentative in the absence of more detailed information on ocean current variability in the vicinity of the Bering Strait.

In agreement with earlier findings, the results of Walsh and Sater indicate that East Greenland ice extent is significantly correlated with the intensity of the geostrophic airflow parallel to the transpolar drift stream during the previous 5-7 months. However, in comparison with the North Pacific fluctuations, the East Greenland and Denmark Strait fluctuations appear to be largely decoupled from the SLP and SST fluctuations in middle latitudes. Such conclusions must be qualified by a cautionary note about the quality of the North Atlantic SST data (Section 1.3). Moreover, the limited areal coverage of the SST data sets used by Walsh and Sater prevented the inclusion of SST data from latitudes north of

62.5°N. Systematic measurements of high-latitude oceanic fluctuations are clearly needed to permit more definitive analyses of air-sea-ice interactions in the subarctic.

The recent study by Fletcher et al., (1982) of SST and wind anomaly changes in the Southern Ocean raises interesting questions pertaining to air-sea-ice interactions in the Antarctic waters. The HSSTP data (see Section 1.3) revealed two distinct modes of oceanic variability with respective time scales of several years and 30-100 years. Wind and SST changes were found to be highly correlated in the latitude belt 40°S to 50°S. The authors suggest that this region plays a leading role in the long-term variability of the ocean climate through the formation of cold water masses which spread northward to the tropics beneath the thermocline. An understanding of the role of Antarctic sea ice in this long-term variability will undoubtedly require further study incorporating numerical models in addition to the limited observational data.

4. PREDICTABILITY: AN OVERVIEW

It is apparent from the preceding section that a rather extensive literature exists on the subject of large-scale air-sea-ice interactions. We conclude our discussion with a general overview of these diagnostic studies and their implications concerning predictability within the air-sea-ice system.

Two of the most pertinent conclusions regarding atmospheric and sea ice predictability on weekly to seasonal time scales are:

(1) Concurrent fields of meteorological and sea ice variables are significantly correlated.

(2) The distribution of sea ice correlates more highly with antecedent than with subsequent meteorological fields.

For practical purposes, the predictability of ice-induced atmospheric variability is therefore largely limited to applications of zero-lag relationships in conjunction with an assumed evolution of the ice anomalies. The forecast evolution of the ice anomalies may be derived from persistence, trends, cycles, or any other forecasting aids. It is perhaps appropriate here to stress the fact that data-based studies do not support the existence of a profound effect of sea ice on the interannual variability of large-scale weather and climate. While sea ice may well be a crucial climatic variable on the longer time scales, particularly in the context of CO_2-induced changes, the atmospheric effects of its interannual variability appear to be quite localized.

Regarding the predictability of atmosphere-induced ice variability, two strategies can be distinguished: (a) evaluations of the integral effects of prior (observed) atmospheric forcing, and (b) the use of predicted meteorological fluctuations to construct forecasts of ice variability. Nearly all the diagnostic studies of Section 3. have followed the "hindcast" approach of (a) in the sense that the meteorological forcing is assumed to be known perfectly (subject ot observational/analysis errors). Ice fluctuations are then explained in terms of the antecedent atmospheric forcing. The consensus of the more recent studies (e.g., Rogers, 1978; Rogers and van Loon, 1979; Walsh and Johnson, 1979b; Barnett, 1980) is that "in hand" meteorlogical observations will permit forecasts of statistically significant fractions of sea ice variability out to one or two seasons. The fractions of ice variance predictable from meteorological data alone decay from over 50% at near-zero lag to insignificant fractions beyond the 1-2 season forecast range. However, the utility of meteorologically-derived sea ice forecasts is reduced by the fact that the simple persistence of sea ice anomalies will often permit forecasts of comparable accuracy.

The predictability of sea ice fluctuations at ranges greater than 1-2 seasons in advance can be viewed as a more controversial topic or, more vividly, as "science at its ragged edge." Since predictions at the longer ranges are generally based on cycles and/or trends in the atmospheric forcing fields, these predictions are examples of strategy (b) of the preceding paragraph. The major limitation to this strategy is that the "signals" represented by cycles and trends in meteorological time series generally correspond to small fractions of the series variance (e.g., Walker, 1947).

On the larger scale, the teleconnections between fluctuations in high latitudes and those in lower latitudes may be exploitable in predictive applications if the existence and physical bases of the teleconnections can be clearly demonstrated. The recent resurgence of interest in the Southern Oscillation (e.g., Horel and Wallace, 1981) is pertinent in the regard.

On the subject of longer-range forecasts, we note parenthetically that the value of a forecast generally increases with the range of the forecast. Individual users must determine whether this increase in value offsets the decrease of the predictable fraction of the variance.

Finally, mention must be made of the potentially major role of the oceans in the predictability of the ice-atmosphere-ocean system. Because of the large spatial scales and temporal persistence of many oceanic anomalies, predictive value of oceanic variables is not surprising. The SST-derived correlative results of

Schell (1970) and Walsh and Sater (1981) may be regarded as rather intriguing in the context of ice-atmosphere predictability. In these studies, however, the statistical computations may well have outpaced the corresponding physical interpretations required for a convincing demonstration of applicability to long-range forecasting.

ACKNOWLEDGMENTS

The preparation of this manuscript was supported by the National Science Foundation, Division of Polar Programs, through Grant DPP80-25218. The comments of R. Madden, J. Rogers and J. M. Wallace were very helpful in the revision of the manuscript.

REFERENCES

Ackley, S. F. and T. E. Keliher (1976) Antarctic sea ice dynamics and its possible climatic effects. AIDJEX Bulletin, 33: 53-76.

Ahlmann, H. W. (1949) The present climatic fluctuation. Geographical Journal, 12: 165-195.

Albright, M. (1980) Geostrophic wind calculations for AIDJEX. In Sea Ice Processes and Models (R. S. Pritchard, Ed.), University of Washington Press, Seattle, Washington, p. 402-409.

Barnett, D. G. (1980) A long-range ice forecasting method for the north coast of Alaska. In Sea Ice Processes and Models (R. S. Pritchard, Ed.), University of Washington Press, Seattle, Washington, p. 360-372.

Barry, R. G. (1980) Meteorology and climatology of the seasonal sea ice zone. Cold Reg. Sci. Tech. 2: 133-150.

Bjornsson, H. (1969) Sea ice conditions and the atmospheric circulation north of Iceland. Jokull, 19: 11-17.

Borzenkova, I. I., K. Ya Vinnikov, L. P. Spirina and D. I. Stekhovskii (1976) Changes in the air temperature of the Northern Hemisphere for the period 1881-1975. Sov. Met. Hydrol., 7: 21-26.

Brooks, C. E. P. and W. A. Quennel (1928) Influence of arctic ice on the subsequent distribution of pressure over the eastern North Atlantic and Western Europe. British Meteorological Office, Geophysical Memoirs, 4(41).

Brownlee, K. A. (1965) Statistical Theory and Methodology, Wiley, Somerset, N.J., 590 p.

Budd, W. F. (1975) Antarctic sea-ice variations from satellite remote sensing in relation to climate. J. Glaciol., 15(73): 417-427.

Crane, R. G. (1978) Seasonal variations of sea ice extent in the Davis Strait - Labrador Sea area and relationships with synoptic-scale atmospheric circulation. Arctic, 31: 434-447.

Davis, R. E. (1976) Predictability of sea surface temperature and sea level pressure anomalies over the North Pacific Ocean. J. Phys. Oceanogr., 6: 249-256.

Dunbar, M. (1967) The monthly and extreme limits of ice in the Bering Sea. In Physics of Snow and Ice (H. Ours, Ed.), Institute of Low Temperature Science, Hokkaido University, Sapporo, Japan, Vol. 1, p. 687-703.

Fletcher, J. O. (1969) Ice extent on the Southern Ocean and its relation to world climate. Rand Corp., Santa Monica, California, Memo. RM-5793-NSF, 111 pp.

Fletcher, J. O., U. Radok and R. Slutz (1982) Climatic signals of the Antarctic Ocean. J. Geophys. Res., 87: 4269-4276.

Frommeyer, M. (1928) Die Eisverhältnesse um Spitzbergen und ihre Beziehungen zu klimatischen Factoren. Ann. D. Hydr., 56: 209-214, 240-248.

Haldane, J. B. S. (1949) A note on non-normal correlation. Biometrika, 36: 467-468.

Hannan, E. J. (1970) Multiple Time Series, John Wiley, New York, 424 p.

Haupt, I. and V. Kant (1976) Satellite ice surveillance studies in the Arctic in relationship to the general circulation. In Proceedings, Symposium on Meteorological Observations from Space: Their Contributions to the First GARP Global Experiment, COSPAR, ICSU, Paris, p. 179-187.

Heap, J. A. (1965) Antarctic pack ice. In Antarctica (T. Hatherton, Ed.), F. Praeger Publ. Co., p. 187-196.

Horel, J. D. and J. M. Wallace (1981) Planetary-scale atmospheric phenomena associated with the southern oscillation. Mon. Wea. Rev., 109: 813-829.

Itan, V. (1936) The variability of ice conditions in the arctic seas of the USSR. In Severnyi Morskoi put′ i Karskie Ekspeditsii, (N. Sibertsev and V. Itin, Ed.), Zapadno-Sibirskoe Kraevoe Izdatel′stvo, Novosibirsk, p. 123-172.

Jenkins, G. M. and D. G. Watts (1968) Spectral Analysis and its Applications. Holden-Day, San Francisco, 517 p.

Jenne, R. L. (1975) Data sets for meteorological research. National Center for Atmospheric Research, Boulder, Colorado, Technical Note NCAR-TN/IA-111, 194 p.

Keen, R. A. (1977) The response of Baffin Bay ice conditions to change in atmospheric circulation patterns. In Proceedings, Fourth International Conference on Port and Ocean Engineering Under Arctic Conditions, (D. B. Muggeridge, Ed.), Memorial University of Newfoundland, St. John′s, Newfoundland, Canada, p. 962-971.

Kelly, P. M. (1978) Forecasting the arctic ice on time scales of a few months to many years. Climate Monitor, 1(3): 95-99.

Kelly, P. M. (1979) An arctic sea ice data set, 1901-1956. Glaciological Data, GD-5, World Data Center A for Glaciology, Boulder, Colorado, p. 101-106.

Kelly, P. M., P. D. Jones, C. B. Sear, B. S. G. Cherry and R. K. Tavakol (1982) Variations in surface air temperatures. Part II: Arctic regions, 1881-1980. Mon. Wea. Rev., 110: 71-83.
Kirillov, A. A. and M. S. Khromtsova (1972) Many-year variations of the ice coverage of the Greenland Sea and methods of forecasting it. AIDJEX Bulletin, 16: 46-55.
Koch, L. (1945) The East Greenland Ice. Medd. Groen. 130(3): 374 p.
Kutzbach, J. E. (1967) Empirical eigenvectors of sea-level pressure, surface temperature and precipitation complexes over North America. J. Appl. Met., 6: 791-801.
Lamb, H. H. (1977) Climate: Present, Past and Future. Methuen, Co., London, Vol. 2.
Lamb, H. H. and H. J. Morth (1978) Arctic ice, atmospheric circulation and world climate. Geographical Journal, 144: 1-22.
Leith, C. E. (1973) The standard error of time-average estimates of climatic means. J. Appl. Met., 12: 1066-1069.
Lemke, P., E. W. Trinkl and K. Hasselmann (1980) Stochastic dynamic analysis of polar sea ice variability. J. Phys. Oceanogr., 10: 2100-2120.
Lorenz, E. N. (1977) An experiment in nonlinear statistical weather forecasting. Mon. Wea. Rev., 105: 590-602.
Lyell, C. (1872) Principles of Geology. Murray Press, London.
McDonald, J. E. (1960) Remarks on correlation methods in geophysics. Tellus, 12: 176-183.
Meinardus, W. (1906) Periodic variations of drifting ice near Iceland. Annalen der Hydrogrpahie und Maritimen Meteorologie, 34: 148-162, 227-239.
Mitchell, J. M., B. Dzerdzeevskii, H. Flohn, W. L. Hofmeyr, H. H. Lamb, K. N. Rao and C. C. Wallen (1966) Climatic Change. World Meteorological Organization, Technical Note 79, WMO No. 195, TP 100, 79 p.
Namias, J. (1976) Negative ocean-air feedback systems over the North Pacific in the transition from warm to cold seasons. Mon. Wea. Rev., 104: 1107-1121.
Niebauer, J. (1980) Sea ice and temperature variability in the eastern Bering Sea and the relation to the atmospheric fluctuations. J. Geophys. Res., 85: 7507-7516.
Norwegian Offshore Technology Research Group (1981) Environmental Data Atlas and Catalog for the Norwegian Sea Shelf and Adjacent Waters. Publication N-7034, Oslo, 460 p.
Oort, A. H. (1975) Year-to-year variations in the energy balance of the arctic atmosphere. In Climate of the Arctic (G. Weller and S. A. Bowling, Ed.), Geophysical Institute, University of Alaska, p. 68-75.
Overland, J. E., and C. H. Pease (1982) Cyclone climatology of the Bering Sea and its relation to sea ice extent. Mon. Wea. Rev., 110: 5-13.
Panofsky, H. S., and G. W. Brier (1968) Some Applications of

Statistics to Meteorology. Pennsylvania State University Press, University Park, Pennsylvania, 224 p.
Preisendorfer, R. W., and T. P. Barnett (1977) Significance tests for empirical orthogonal functions. In Preprints, Fifth Conference on Probability and Statistics in Atmospheric Sciences, Las Vegas, Nevada, American Meteorological Society, p. 169-172.
Ratcliffe, R. A. S. (1973) Recent work on sea surface temperature anomalies related to long-range forecasting. Weather, 28: 106-117.
Richman, M. B. (1981) Obliquely rotated principal components: An improved meteorological map typing technique? J. Appl. Met., 20: 1145-1159.
Rogers, J. C. (1977) A meteorological basis for long-range forecasting of summer and early autumn sea ice conditions in the Beaufort Sea. In POAC 77: Proceedings, Fourth International Conference on Port and Ocean Engineering Under Arctic Conditions (D. B. Muggeridge, Ed.), Memorial University of Newfoundland, St. John's, Newfoundland, Canada.
Rogers, J. C. (1978) Meteorological factors affecting interannual variability of summertime ice extent in the Beaufort Sea. Mon. Wea. Rev., 106: 890-897.
Rogers, J. C. and H. van Loon (1979) The seesaw in winter temperatures between Greenland and northern Europe. Part II: Some oceanic and atmospheric effects in middle and high latitudes. Mon. Wea. Rev., 107: 509-519.
Sanderson, R. M. (1975) Changes in the area of Arctic sea ice, 1966 to 1974. Meteorological Magazine, 104: 313-323.
Schell, I. I. (1939) Polar ice as a factor in seasonal weather. In Reports on Critical Studies of Methods of Long-Range Weather Forecasting, Mon. Wea. Rev., Supplement No. 39, p. 27-51.
Schell, I. I. (1956) Interrelations of arctic ice with the atmosphere and the ocean in the North Atlantic: Arctic and adjacent areas. J. Met., 13: 46-58.
Schell, I. I. (1970) Arctic ice and sea temperature anomalies in the northeastern North Atlantic and their significance for seasonal foreshadowing locally and to the eastward. Mon. Wea. Rev., 98: 833-850.
Schwerdtfeger, W. (1970) The climate of Antarctic. In Climates of the Polar Regions (S. Orvig, Ed.), Elsevier, Amsterdam, p. 253-322.
Schwerdtfeger, W. and St. Kachelhoffer (1973) The frequency of cyclonic vortices over the southern ocean in relation to the extension of the pack ice belt. Ant. J. U.S., 8: 234.
Smirnov, V. I. (1980) Opposition in ice redistribution in the waters of the foreign Arctic. Meteorol. Gidrol., 3: 73-77.
Strubing, K. V. (1967) Uber Zusammenhange zurischen der Disfuhrung des Ostgronlandstroms und der atmosphareschen Zirkulation uber dem Nordpolarmeer. Deut. Hydrogr. Zeit. 20: 257-265.

Trenberth, K. E. and D. A. Paolino, Jr. (1980) The Northern Hemisphere sea level pressure data set: Trends, errors and discontinuities. Mon. Wea. Rev., 108: 855-872.

Tukey, J. W. (1950) The sampling theory of power spectrum estimates. Symposium of Applications of Autocorrleation Analysis to Physical Problems, U.S. Office of Naval Research, Washington, D.C., NAVEXOS-Pp735, p. 47-67.

van Loon, H. (1980) Transfer of sensible heat by transient eddies in the atmosphere of the Southern Hemisphere: An appraisal of the data before and during FGGE. Mon. Wea. Rev., 108: 1774-1781.

Vinje, T. E. (1977) Sea ice conditions in the European sector of the marginal seas of the Arctic, 1966-1975. Norsk Polarinstitutt Arbok 1975, p. 163-174.

Volkov, N. A. and B. A. Sleptsov-Shevlevich (1971) Cyclic variations of the ice-cover coefficient of the arctic sea. Translation in AIDJEX Bulletin, 16: 1-34 (1972).

Walker, G. (1974) Arctic conditions and world weather. Quart. J. Roy. Met. Soc., 73: 226-256.

Wallace, J. M. and D. S. Gutzler (1981) Teleconnections in the geopotential height field during the Northern Hemisphere winter. Mon. Wea. Rev., 109: 784-812.

Walsh, J. E. and C. M. Johnson (1979a) An analysis of arctic sea ice fluctuations, 1953-1977. J. Phys. Oceanogr., 9: 580-591.

Walsh, J. E. and C. M. Johnson (1979b) Interannual atmospheric variability and associated fluctuations in arctic sea ice extent. J. Geophys. Res., 84: 6915-6928.

Walsh, J. E. and J. E. Sater (1981) Monthly and seasonal variability in the ocean-ice-atmosphere systems of the North Pacific and the North Atlantic. J. Geophys. Res., 86(C8): 7425-7445.

Wiese, W. (1922) Die Einwirkung das Polareises in Gronlandischen Meere auf die nord-atlantische zyklondale Tatigkeit. Ann. D. Hydr., 50: 271-280.

Wiese, W. (1924) Polareis und atmospharische Schwankungen. Geog. Annaler., 6: 273-299.

Wiese, W. (1925) The significance of ice in spring in the Greenland Sea and of the East Iceland Polar Current for the following winter temperature distribution in Europe. Inst. Hydr. No. 14.

Zakharov, V. F. and L. A. Strokina (1978) Recent variations in extent of Arctic Ocean ice cover. Meteorol. Gidrol. 7: 35-43.

CHAPTER 13

STOCHASTIC DESCRIPTION OF ATMOSPHERE - SEA ICE - OCEAN INTERACTION

Peter Lemke

Max-Planck-Institut fur Meteorologie
2000 Hamburg 13
Federal Republic of Germany

1. INTRODUCTION

1.1 Sea Ice as a Member of the Climate System

Growth, decay and movement of sea ice are among the most important features of air-sea interaction at high latitudes. To understand, and ultimately to predict the location of the sea ice edge, the ice thickness and ice drift is of interest not only concerning the effects on regional and global climate, but has also practical applications, since the polar regions have become an area of intense human activity and resource development.

An adequate modeling of sea ice variability can only be accomplished by simulating the interaction of the complete physical system atmosphere - ocean - ice. Of special interest is the marginal sea ice zone, where this coupling is particularly intense (see Chapter 15). It includes those regions inward and outward from the ice boundary that are significantly influenced by the presence of the margin. Figure 1 demonstrates schematically the complexity of the interaction of the three sub-systems in this area. The position of the ice edge is mainly determined by two processes: sea ice drift and freezing and melting. The ice movement is dominated by air and water stress, coriolis force, ocean tilt and internal ice stress (see Chapter 9), whereas freezing and melting are affected by radiation and heat exchange (see Chapter 5), all these factors being dependent on the state of atmosphere, ocean and sea ice, which in turn is influenced by the moving sea ice boundary.

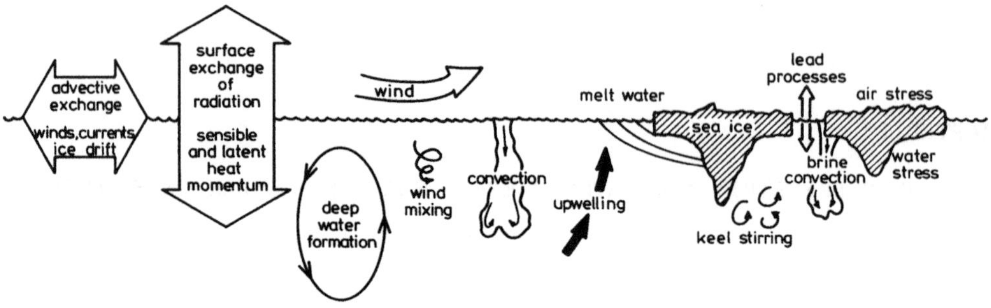

Fig. 1. Processes near the sea ice margin.

The difficulty in modelling this closed feedback loop is not only due to the variety of complex process in the marginal sea ice zone, but to a great extent also due to the fact that the three interacting components are characterized by different internal time-scales (Hasselmann, 1976, 1978): the atmosphere - hours to several days, sea ice and oceanic mixed layer - several months, deep ocean - several years to thousand years. Even if all processes governing the dynamics of the system were completely understood, the fact that different sub-systems respond with different time-scales represents a serious numerical modelling problem.

Therefore models describing the geophysical fluid system with realistic geographical resolution are all designed only for a single time-scale range (see Figure 2). The influence of sub-grid and short time-scale processes on the prognostic regime is usually introduced by parameterizing the interaction through a statistical treatment of the short time-scale variables. On the other hand the components which vary on time-scales larger than the prognostic regime (typically of the order of 10^3 - 10^4 basic time steps Δt) are simply taken as constant boundary values or external parameters.

Thus in a complete model describing the processes in the marginal sea ice zone the atmosphere is placed in the statistical, diagnostic regime of the model. The coupling of the short time-scale atmospheric system to the more slowly changing sea ice variations is not only a problem of parameterization in order to close the system of equations, but includes also stochastic forcing terms, that introduce long-period fluctuations in sea ice extent. This may well represent the origin of the observed sea ice variability, as will be shown in the following sections.

The incorporation of small-scale atmospheric processes in the prognostic equations of sea ice extent, thickness and drift is only one side of the coin. The other side is the back-interaction of

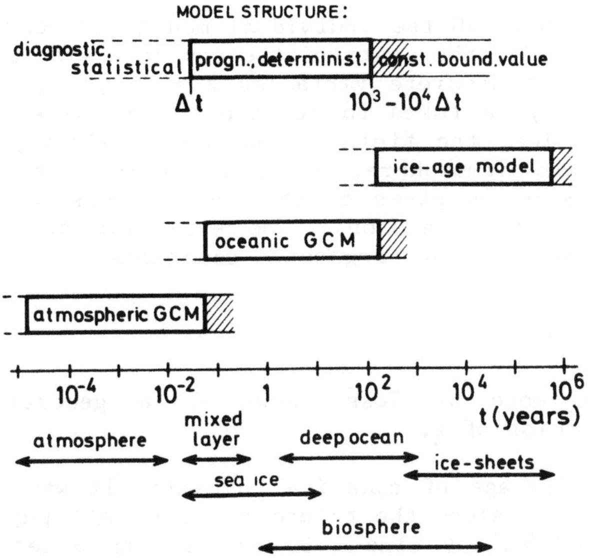

Fig. 2. Natural time scales of the climatic system and time-scale band width covered by particular models (from Hasselmann, 1978).

sea ice variations on the state of the atmosphere, the long-term behaviour of which is believed to be closely related to slow variations of the external boundary conditions of the atmospheric system. It has long been suggested, that the moving sea ice boundary may significantly affect the atmospheric circulation pattern, i.e., the trajectories and frequency of cyclones near the ice margin. However, space- and time-scales of this relationship, which presumably is dominated by heat and momentum transfer at the surface, have not been quantitatively established (see Chapter 11). Since only time-scales larger than the internal time-scale of the atmosphere are of interest here this is a problem of identifying the mean atmospheric response to external forcing in the presence of the natural variability of the atmosphere. In general, the significance of the atmospheric response to sea ice variations will probably not be established without pattern-filtering and an improvement of the signal-to-noise ratio (Hasselmann, 1979).

2. STOCHASTIC MODELING

2.1 Coupling Between Fast and Slow Components

The interaction of subsystems with different response times is a general problem often found in physical science (see next section). It is, therefore, useful to introduce a formal notation, which is independent of the individual model structure. Following Hasselmann (1976) let the state of the complete system be described by a finite set of discrete variables $\tilde{z} = (z_1, z_2, \ldots)$. The state vector $\tilde{z}$ may be taken to represent the thickness and drift velocity of sea ice, the fields of density, velocity and temperature of ocean and atmosphere, etc., as defined at discrete grid points or levels or as given by the coefficients of some suitably truncated functional expansion. The evolution of the system is then described by a set of prognostic equations

$$\frac{d z_i}{d t} = w_i (\tilde{z}) \tag{13.1}$$

where w_i is a more or less known - in general complicated nonlinear - function of $\tilde{z}$.

In the golden age of classical dynamics it was believed that with given initial values the future state of all physical systems can be predicted at any time t by integrating a set of differential equations of the form (13.1) which determine the dynamics of the system. But soon it was realized that this perfect prediction can only be achieved for very few idealized problems. In most cases - although the governing prognostic equations were often known - the classical dynamic approach was not applicable because of the large number of interacting variables, the uncertainty of the initial conditions or the multiple time-scale behaviour of the system. Nevertheless, in such many-particle or multiple time-scale systems information about the future state can be obtained, because prediction of certain macroscopic or long time-scale variables is possible, although only in a statistical sense. This is the main problem of statistical physics, which describes the evolution of the expectation (i.e., mean) value, the variance or other statistical functions of random, rather than deterministic, variables.

It is a common experience that many physical processes which exhibit reasonable variations on a large time scale are caused by accumulation of a large number of highly random short time-scale events (i.e., Brownian motion).

In analogy to classical statistical mechanics where certain macroscopic variables (like temperature or pressure in a fluid) are determined by the statistics of the microscale variables (the fluid molecules), the evolution of the slow variables in a multi-time-scale system can be expressed from the statistics of the fast variables with the help of stochastic differential equations.

The basic assumption which enables and limits the applicability of such stochastic models is that the complete state vector of the system $\tilde{z}$ can be divided into at least two subsystems, $\tilde{z} = (\tilde{x}, \tilde{y})$, which are characterized by extremely different response times.

Let $\underset{\sim}{x} = (x_1, x_2 \; . \; . \; . \;)$ denote the state vector of the fast components (e.g., the atmosphere) and $\underset{\sim}{y} = (y_1, y_2, \; . \; . \; . \;)$ the state vector of the slow system (e.g., ocean, sea ice) such that the prognostic equations (13.1) are now given in terms of the two subsystems

$$\frac{dx_i}{d\,t} = u_i\,(\underset{\sim}{x}, \underset{\sim}{y}) \tag{13.2}$$

$$\frac{d\,y_i}{d\,t} = v_i\,(\underset{\sim}{x}, \underset{\sim}{y}) \tag{13.3}$$

and the characteristic time scales τ_x and τ_y satisfy the two-scale inequality

$$0\left(x_i \Big/ \frac{dx_i}{d\,t}\right) = \tau_x << \tau_y = 0\left(y_i \Big/ \frac{dy_i}{d\,t}\right) . \tag{13.4}$$

The first step to treat the coupled equations (13.2) and (13.3) is to restrict one's attention to a limited time-scale band, as it is done, for example, in a typical atmospheric general circulation model (GCM) experiment, where the integration time τ lies in the range

$$\tau_x \quad << \quad \tau \quad << \tau_y \tag{13.5}$$

In this case the slow variables can be regarded as constant and the problem reduces to the integration of

$$\frac{d\,x_i}{d\,t} = u_i\,(\underset{\sim}{x}, \underset{\sim}{y}_o) \tag{13.6}$$

with given boundary conditions $\underset{\sim}{y} = \underset{\sim}{y}_o$ (i.e., sea ice extent, sea surface temperature, etc.). On the other hand, the integration time τ is still sufficiently long to define the statistics of the fast (atmospheric) variables $\underset{\sim}{x}$ for a given state of the slow variables $\underset{\sim}{y}_o$. In general, in numerical experiments it is not possible to extend the integration time sufficiently far to include variations of the order of τ_y. Therefore it is essential - and the only way to obtain sufficient information about the slow variables - to formulate the problem with the well-known methods of statistical physics in which both $\underset{\sim}{x}$ and $\underset{\sim}{y}$ are regarded as random variables. In order to deduce statistical statements about the evolution of the slow (macroscopic) components $\underset{\sim}{y}$, eq. (13.3) is averaged over the statistical ensemble of all realizations of the rapidly varying (microscopic) states $\underset{\sim}{x}$, given the slowly varying state $\underset{\sim}{y}$. The ensemble may be understood as a set of very many similarly prepared systems. Since the ensemble average, denoted by <•>, and the time derivative commute, we obtain

$$\frac{d}{d\,t} \langle y_i \rangle = \langle v_i(\underset{\sim}{x}, \underset{\sim}{y}) \rangle. \tag{13.7}$$

Introducing $v_i = \langle v_i \rangle + v_i'$ and $y_i = \langle y_i \rangle + y_i'$, where v_i' and y_i' denote the deviations of v_i and y_i from the ensemble - averages $\langle v_i \rangle$, and $\langle y_i \rangle$, the complete prognostic equation for $\underset{\sim}{y}$ is

$$\frac{d\,y_i}{dt} = \langle v_i\,(\underset{\sim}{x}, \underset{\sim}{y}) \quad + v_i'\,(\underset{\sim}{x}, \underset{\sim}{y}) \rangle \tag{13.8}$$

where v_i' is a random function with a correlation time of the order τ_x, i.e., the variance-spectrum of v_i' is practically white for climatic time scales ($\tau >> \tau_x$). In the following we will therefore regard v_i' as a stationary white-noise process.[1]

Statistical dynamical climate models (SDM's) (see Schneider and Dickinson, 1974) usually treat equations of the form (13.7), assuming that the ensemble average can be replaced by a time average with an averaging time τ characterized by eq. (13.5). To simplify and close the problem the expectation values for the $\underset{\sim}{x}$ terms appearing in $\langle v_i(\underset{\sim}{x},\underset{\sim}{y}) \rangle$ are parameterized in terms of the instantaneous state $\underset{\sim}{y}$, e.g., in Budyko-Sellers type climate models the meridional eddy heat fluxes are expressed in terms of the mean meridional temperature distributions. Although SDM's contain an averaging procedure and a statistical closure hypothesis, eq. (13.7) is in fact treated determinisitcally, and the solution is

---

[1]A simple observed time series of a random phenomenon is called a sample record. A stochastic process is the set of all possible sample records which the random phenomenon might have produced. An observed time series of a random phenomenon may therefore be regarded as one particular physical realization of a stochastic process. The basic properties of stochastic processes are described by statistical functions like mean, covariance, power- (or variance-) spectrum and probability density function (see Bendat and Piersol, 1971; Jenkins and Watts, 1968). A stochastic process is called stationary if these statistical functions do not vary with time.

The characterization of a variance-spectrum as "white," "red" or "blue" is adapted from optics. A red spectrum shows increasing variance for smaller frequencies, whereas a blue spectrum increases for larger frequencies, and a white spectrum is constant.

only part of the information that can be gained, i.e., only the mean, but not the variance is described.

In general the complete information about the system $\underset{\sim}{y}$ is given from the stochastic differential equation (13.8) in terms of the probability density function $p(\underset{\sim}{y},t)$, which for a Markov-process[2] of the form (13.8) is governed by a Fokker-Planck equation (see Wang and Uhlenbeck, 1975; Hasselmann, 1976).

In practice analytical integrations of the Fokker-Planck equation will not be possible, but numerical solutions may be constructed by Monte-Carlo methods.

Attention is now focused on linear feedback models, which are applicable for systems with small deviations about an equilibrium state. After introducing an appropriate closure hypothesis, that relates the expectation values of the x terms appearing in $\langle v_i \rangle$ to the instantaneous state of $\underset{\sim}{y}$, $\langle v_i(\underset{\sim}{x},\underset{\sim}{y}) \rangle = \hat{v}_i(\underset{\sim}{y})$ in eq. (13.8) may be approximated by a linear function of $\underset{\sim}{y}$. Since the feedback terms must vanish for the equilibrium state, eq. (13.8) reduces to a first order liner stochastic differential equation.

$$\frac{dy_i}{dt} = - \sum_j \lambda_{ij} \; y_j + v_i' \tag{13.9}$$

where y_i now denotes the deviation form the equilibrium state $\tilde{y}_i$. The Central Limit Theorem states that, under very general conditions, the response of a linear system driven by a statistically stationary input consisting of a continuous sequence of infinitely short, statistically independent pulses is Gaussian, independent of the detailed statistical structure of the input. Therefore, the statistical structure of the slow components $\underset{\sim}{y}$ is completely determined by the first and second moments defined by

$$[y_i(t)] \text{ and } C_{ij}(t,\tau) = [\{y_i(t+\tau) - [y_i(t+\tau)]\}\cdot\{y_j(t) - [y_j(t)]\}],$$

[2]A special class of stochastic processes are Markov processes whose relationship to the past in the discrete case do not extend beyond the immediately preceding observation. A continuous Markov process is governed by a first order differential equation of the form (13.8). Markov processes are named after the Russian mathematician A. A. Markov (1856-1922), who did the basic pioneering work on this class of stochastic processes. A special class of Markov processes are auto-regresive processes, the continuous form being defined by eq. (13.9). The discrete auto-regressive process is explained in the example in section 3.2.

where the square parenthesis [•] denote averages over the ensemble of the slow variables $\underset{\sim}{y}$.

In general, these moments are time-dependent functions that can be determined from eq. (13.9). In analogy to ordinary differential equations the general solution of the stochastic differential eq. (13.9) is in matrix notation given by (for a more rigorous mathematical derivation see Arnold, 1974)

$$\underset{\sim}{y}(t) = e^{-\underset{\sim}{\lambda}(t-t_0)} \underset{\sim}{y}(t_0) + e^{-\underset{\sim}{\lambda} t} \int_{t_0}^{t} e^{\underset{\sim}{\lambda} t'} \underset{\sim}{v}'(t')dt' \tag{13.10}$$

For convenience we will take $t_0 = 0$ and denote $\underset{\sim}{y}(t_0) = \underset{\sim}{y}^0$. It is clear that a stable solution requires a positive definite feedback matrix $\underset{\sim}{\lambda}$. The first moment (mean) is then given by

$$[\underset{\sim}{y}(t)] = e^{-\underset{\sim}{\lambda} t}[\underset{\sim}{y}^0] \tag{13.11}$$

i.e., the expectation value of $\underset{\sim}{y}(t)$ will decay from the initial value $[\underset{\sim}{y}^0]$ towards the equilibrium state $[\underset{\sim}{y}(t)] = 0$. Inserting (13.10) into the definition of the second moment and assuming that $\underset{\sim}{v}'(t)$ resembles a stationary white noise process, i.e.,

$$[\underset{\sim}{v}'(t)\ \underset{\sim}{v}'^{+}(t')] = \underset{\sim}{D}\delta(t-t') \tag{13.12}$$

with given constant covariance-matrix $\underset{\sim}{D}$, we find for the covariance-function of the slow variables $\underset{\sim}{y}(t)$

$$\underset{\sim}{C}(t,\tau) = e^{-\underset{\sim}{\lambda}(t+\tau)} \{R^0 + \int_0^{t} e^{\underset{\sim}{\lambda} t'} \underset{\sim}{D} e^{\underset{\sim}{\lambda}^{+} t'} dt'\} e^{-\underset{\sim}{\lambda}^{+} t} \tag{13.13}$$

where $z = \min(t, t+\tau)$, $\underset{\sim}{R}^0 = \underset{\sim}{C}(0,0)$ and the + denotes the adjoint (transposed and conjugate complex) matrix. The second moment is therefore given by

$$\underset{\sim}{C}(t, \tau) = \begin{cases} e^{-\underset{\sim}{\lambda}\tau} \underset{\sim}{R}(t) & \tau \geq 0 \\ \underset{\sim}{R}(t+\tau)\, e^{\underset{\sim}{\lambda}^{+}\tau} & \tau \leq 0 \end{cases} \tag{13.14}$$

where

$$\underset{\sim}{R}(t) \equiv \underset{\sim}{C}(t, \tau=0) = e^{-\underset{\sim}{\lambda} t} \left\{ \underset{\sim}{R}^0 + \int_0^t e^{\underset{\sim}{\lambda} t'} \underset{\sim}{D} e^{\underset{\sim}{\lambda}^+ t'} dt' \right\} e^{-\underset{\sim}{\lambda}^+ t} \tag{13.15}$$

The dependence of the covariance-function $\underset{\sim}{C}(t,\tau)$ on the reference time t is totally described by $\underset{\sim}{R}(t)$ which is the solution of the differential equation (found by differentiating eq. (13.15)).

$$\frac{d\underset{\sim}{R}(t)}{dt} = -\underset{\sim}{\lambda}\underset{\sim}{R} - \underset{\sim}{R}\underset{\sim}{\lambda}^+ + \underset{\sim}{D} \tag{13.16}$$

With the initial condition $\underset{\sim}{R}(t=0) = \underset{\sim}{R}^0$ this is

$$\underset{\sim}{R}(t) = \underset{\sim}{R}^\infty - e^{-\underset{\sim}{\lambda} t} (\underset{\sim}{R}^\infty - \underset{\sim}{R}^0)\, e^{-\underset{\sim}{\lambda}^+ t} , \tag{13.17}$$

where $\underset{\sim}{R}^\infty$ is the asymptotic stationary solution of eq. (13.16)

$$\underset{\sim}{\lambda}\, \underset{\sim}{R}^\infty + \underset{\sim}{R}^\infty\, \underset{\sim}{\lambda}^+ - \underset{\sim}{D} = 0 \tag{13.18}$$

These expressions become particularly simple for a diagonal feedback matrix $\underset{\sim}{\lambda}$.

$$[y_i(t)] = e^{-\lambda_i t} [y_i^0] \tag{13.19}$$

$$R_{ij}(t) = R_{ij}^\infty \left(1 - e^{-(\lambda_i + \lambda_j)t}\right) + R_{ij}^0\, e^{-(\lambda_i + \lambda_j)t} \tag{13.20}$$

$$R_{ij}^\infty = \frac{D_{ij}}{\lambda_i + \lambda_j} \tag{13.21}$$

$$C_{ij}(t,\tau) = \begin{cases} e^{-\lambda_i \tau} R_{ij}(t) & \tau \geq 0 \\ R_{ij}(t+\tau)\, e^{\lambda_j \tau} & \tau \leq 0 \end{cases} \tag{13.22}$$

The upper panel of Figure 3 shows the decay of the expectation value [y(t)] from the initial value [y(0)] towards the equilibrium state [y(∞)] = 0 for a single component system. The dashed lines demonstrate the evolution of the variance according to eq. (13.20). For larger time t ($t \gg \frac{1}{\lambda_{ij}}$ for all i,j) the covariance matrix $\underset{\sim}{R}$ and therefore the time-lagged covariance function $\underset{\sim}{C}$ become independent of t. This asymptotic stationary solution

$$\overline{\underset{\sim}{C}}(\tau) = \begin{cases} e^{-\underset{\sim}{\lambda}\tau} \underset{\sim}{R}^{\infty} & \tau \geq 0 \\ \underset{\sim}{R}^{\infty} e^{\underset{\sim}{\lambda}^{+}\tau} & \tau \leq 0 \end{cases} \tag{13.23}$$

is of particular interest for practical purposes, since it can be compared with the statistical properties of observed quasi-stationary climatic time series. $\overline{\underset{\sim}{C}}(\tau)$ is the covariance function of a multivariate first order Markov or auto-regressive process (see Jenkins and Watts, 1968). The τ-dependence of $\underset{\sim}{C}(\tau)$ for a single component system is shown as a solid line in the middle panel of Figure 3.

Another quantity of practical importance is the structure function (Tatarski, 1961), since it is a measure of the deviation of a random function from an initial value. It is easily seen, that the structure function or deviation covariance $\underset{\sim}{S}$ defined as

$$\underset{\sim}{S}(t,\tau) \equiv \left[\left(\underset{\sim}{y}(t+\tau) - \underset{\sim}{y}(t) \right) \left(\underset{\sim}{y}(t+\tau) - \underset{\sim}{y}(t) \right)^{+} \right] \tag{13.24}$$

can be rewritten in terms of the covariance function $\underset{\sim}{C}$.

$$\underset{\sim}{S}(t,\tau) = \underset{\sim}{C}(t+\tau, 0) + \underset{\sim}{C}(t, 0) - \underset{\sim}{C}(t,\tau) - \underset{\sim}{C}^{+}(t,\tau)$$
$$+ \left([\underset{\sim}{y}(t+\tau)] - [\underset{\sim}{y}(t)] \right) \cdot \left([\underset{\sim}{y}(t+\tau)] - [\underset{\sim}{y}(t)] \right)^{+} \tag{13.25}$$

The asymptotic stationary solution $\overline{\underset{\sim}{S}}(\tau)$ is given by ($t \gg \lambda_{ij}^{-1}$)

$$\overline{\underset{\sim}{S}}(\tau) = \overline{\underset{\sim}{C}}(0) - \overline{\underset{\sim}{C}}(\tau) + \overline{\underset{\sim}{C}}^{+}(0) - \overline{\underset{\sim}{C}}^{+}(\tau) \tag{13.26}$$

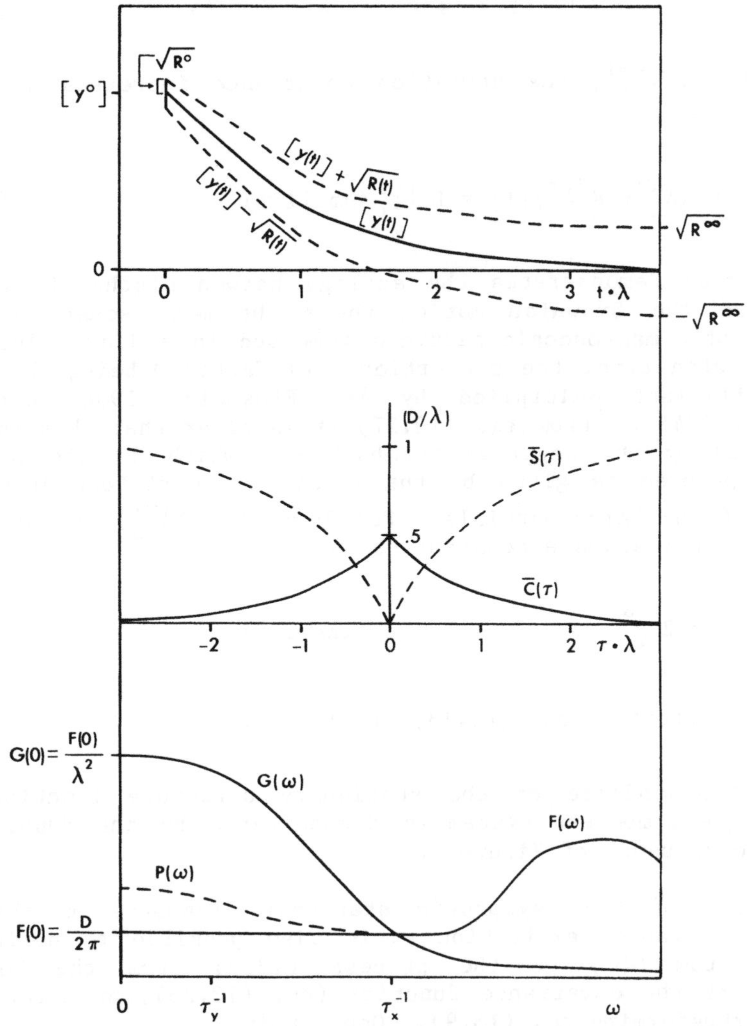

Fig. 3. Time evolution of the mean $[y(t)]$ and the variance $R(t)$ (upper panel), τ-dependence of the asymptotic stationary covariance function $\bar{C}$ and structure function $\bar{S}$ (middle), and input spectra $F(\omega)$, $P(\omega)$ and response-spectrum $G(\omega)$ (lower panel). The definition of $P(\omega)$ is given in section 5.1.

$\bar{\underset{\sim}{S}}(\tau)$ is symmetric, $\underset{\sim}{S} = \underset{\sim}{S}^+$, and an even function of τ, $\bar{\underset{\sim}{S}}(\tau) = \bar{\underset{\sim}{S}}(-\tau)$, since $\bar{\underset{\sim}{C}}(-\tau) = \bar{\underset{\sim}{C}}^+(\tau)$.

With eq. (13.23) we find

$$\bar{\underset{\sim}{S}}(\tau) = (1 - e^{-\underset{\sim}{\lambda}|\tau|})\,\underset{\sim}{R}^{\infty} + \underset{\sim}{R}^{\infty}\,(1 - e^{-\lambda^{+}|\tau|}) \,. \tag{13.27}$$

For small $\tau(\tau<<\lambda_{ij}^{-1})$ the deviation covariance increases linearily with time lag τ:

$$\bar{\underset{\sim}{S}}(\tau) = (\underset{\sim}{\lambda}\underset{\sim}{R}^{\infty} + \underset{\sim}{R}^{\infty}\underset{\sim}{\lambda}^{+})|\tau| = \underset{\sim}{D}\,|\tau|\ (\text{small } \tau) \tag{13.28}$$

This equation demonstrates the analogy betwen stochastic climate models and the Brownian motion where the mean square velocity increment of a macroscopic particle immersed in a liquid increases linearily with time, the proportionality factor D being the diffusion coefficient multiplied by λ^2 (Einstein, 1906; Wang and Uhlenbeck, 1945). From eq. (13.27) it is clear that this increase in confined by the negative feedback $-\underset{\sim}{\lambda}$, which in the Brownian movement problem is given by the friction coefficient divided by the mass of the large particle. For large $\tau(\tau>>\lambda_{ij}^{-1})$ the deviation covariance approaches a constant value,

$$\bar{\underset{\sim}{S}}\ (\tau) = 2\,\underset{\sim}{R}^{\infty} \qquad (\text{large } \tau) \tag{13.29}$$

where $\underset{\sim}{R}^{\infty}$ is given by eq. (13.18) or (13.21).

The τ-dependence of the stationary structure function $\bar{S}(\tau)$ for a single component system is demonstrated by the dashed line in the middle panel of Figure 3.

Because of the asymptotic stationary response of the slow variables to random excitation it is also possible to define the cross-spectrum $\underset{\sim}{G}(\omega)$ of the process (13.9) from the Fourier-transform of the covariance function (eq. (13.23)) or directly by Fourier-transforming eq. (13.9). One obtains

$$\underset{\sim}{G}\ (\omega) = [\underset{\sim}{y}(\omega)\ \underset{\sim}{y}^{+}(\omega)] = \underset{\sim}{H}(\omega)\ \underset{\sim}{F}(\omega)\ \underset{\sim}{H}^{+}(\omega) \tag{13.30}$$

where $\underset{\sim}{H}(\omega) = (i\omega\underset{\sim}{I} - \underset{\sim}{\lambda})^{-1}$ ($\underset{\sim}{I}$ = unit matrix), and $\underset{\sim}{F}(\omega)$ is the cross-spectrum of the fast variables $\underset{\sim}{x}$. For diagonal $\underset{\sim}{\lambda}$ eq. (13.30) becomes

$$G_{k\ell}(\omega) = \frac{F_{k\ell}(\omega)}{(\omega-i\lambda_k)\ (\omega+i\lambda_\ell)} \tag{13.31}$$

In the frequency range $\omega \ll \tau_x^{-1}$, the spectrum of the fast variables is regarded as essentially white, such that $F(\omega)$ can be replaced by $F(0) = D/2\pi$ in eqs. (13.30) and (13.31). [See Fig. 3 lower panel. For a correction due to the back interaction of the slow variables onto the fast variables, see Section 5.1.] In the intermediate range of $\tau_y^{-1} \ll \omega \ll \tau_x^{-1}$ the response spectrum of slow variables $G(\omega)$ is red (proportional to ω^{-2}) and for $\omega \ll \tau_y^{-1}$ $G(\omega)$ is approximately white due to the stabilizing negative feedback.

The general form of the idealized input spectrum $F(\omega)$ and the response spectrum $G(\omega)$ for a single component system is shown in the lower panel of Figure 3.

$\bar{C}(\tau)$, $\bar{S}(\tau)$ or $G(\omega)$ completely describe the asymptotic stationary variance of a linear system with stabilizing feedback (eq. 13.9). Once one of the functions is known, the others can be derived.

2.2 Application of Stochastic Models

The method of describing the interaction of physical systems with completely different relaxation times is well known from the treatment of the Brownian motion problem (Wang and Uhlenbeck, 1945; Reif, 1965). In that case $y(t)$ in eq. (13.9) denotes the velocity of a large particle suspended in a liquid, λ is the friction coefficient, divided by the mass of the particle and $v'(t)$ is the fluctuating random force caused by the bombardment of the particle by the molecules of the liquid. The mean square velocity increment is then given by the structure function $\bar{S}(\tau)$ (eqs. (13.28), (13.29)), where D is calculated from statistical thermodynamics of closed systems ($D = 2kT\lambda/m$, where k is the Boltzmann constant, T is temperature and m is the mass of the particle). The mean-square displacement of the particle can be determined from the correlation function of the velocity (eq. (13.23), see Reif, 1965).

This two-timing method was introduced in climate research by K. Hasselmann (1976, 1977, 1978), who pointed out that the relaxation times for the atmospheric variables are much smaller than those of the remaining part of the climate system (ocean, sea ice, ice sheets). Therefore a stochastic description of the evolution of the slow components of the climate system should be appropriate. The concept outlined above has been applied to the climate system and was verified quantitatively for sea-surface temperature variations and thermocline variability (Frankignoul and Hasselmann, 1977; Reynolds, 1978; Herterich and Hasselmann, 1985) and sea ice anomalies (Lemke et al., 1980) in the time-scale range of months to a few years. It was also applied to longer time scales ($10 - 10^4$ years) in global, stochastically forced Budyko-Sellers climate models (Lemke, 1977; Fraedrich, 1978, 1979; Robock, 1978). The main variable in these models is the zonally and yearly averaged surface temperature of the earth. It is determined from the zonally averaged, vertically integrated equation of the heat balance of the earth-atmosphere system. The linear feedback matrix $\tilde{\lambda}$ in eq. (13.9) then resembles the positive feedback from the temperature dependent albedo and the negative feedback from the infrared radiation and the mean meridional heat fluxes. The main contribution to the stochastic forcing term comes from the turbulent part of the meridional heat fluxes. The model calculations yielded red variance spectra, eq. (13.30), which are consistent with observations both in the qualitative structure of the spectrum and the order of magnitude of the energy levels.

The observations show several marked peaks superimposed on the continuous red spectrum, the most dominant one at a period of 100,000 years. Although linear stochastic models with constant coefficients are able to represent the continuous red distribution, they cannot describe the observed peaks. Several attempts to extend the stochastic modeling approach in this direction have been reported recently. The application of linear stochastic models with time-dependent parameters (Ruiz de Elvira and Lemke, 1982) is outlined in Section 5.2. Certain features of simple non-linear stochastic models are discussed in Nicolis and Nicolis (1981), Nicolis (1982), Saltzman (1982) and Benzi et al. (1982). In contrast to linear stochastic models the statistical verification of non-linear models is very complicated and often impossible, since the response of a non-linear system to stochastic forcing is no longer Gaussian, and therefore third- and higher order statistical moments have to be included in addition (Hasselmann et al., 1963; Muller, 1985).

2.3 Comparison with Deterministic Models

The relationship between deterministic models of atmosphere, ocean and ice and stochastic models may be explained in terms of the Brownian motion problem. Let the slow variables $\tilde{y}$ (ocean, ice) and the fast variables $\tilde{x}$ (atmosphere) be interpreted in the analogous particle picture as the position and momentum coordinates of large and small particles, respectively. The deterministic description of ocean and ice is then equivalent to determining the paths of the large particles by considering only the interaction between the large particles themselves and the mean interaction through pressure and stress fields exerted by the small-particle motion. In addition, the influence of variable external forces is included. For example, deterministic sea ice models (except Hibler's (1979)) describe sea ice variations through the interaction between the ice itself and the underlying oceanic mixed layer and the mean (seasonally varying) atmospheric fields (wind, air temperature, solar radiation). Deterministic models of the atmosphere (GCM's) compare in this picture to the explicit computation of all paths of the small particles for fixed positions of the large particles. Stochastic models, finally, correspond to the classical statistical treatment of the Brownian motion, in which the large particles are no longer deterministic, but the statistical properties (mean and variance) of the large particles are inferred from the statistics of the small particles with which they interact. As a consequence of the random forcing of the fast variables the prediction of the state of the slow variables must be formulated as a statistical problem. The evolution of the slow variables from a given initial state can no longer be described by a single curve in the phase space $\underset{\sim}{y}$, but has to be characterized for linear models by the evolution of the mean and the variance of $\underset{\sim}{y}$.

Despite their stochastic nature, systems with stabilizing feedback still have a certain degree of predictability. This is because the feedback terms, in contrast to the stochastic forcing, represent deterministic functions. In a statistically stationary system the deterministic feedback and the stochastic forcing terms are generally of the same order, and the resultant predictive skill (for a definition see Chapter 12) is maximally of the order of 50% (Hasselmann, 1976).

According to the discussion outlined in Section 2.1, long-period fluctuations will occur naturally in any multi-time-scale system through the integration of the short-time-scale fluctuations by the slowly varying components of the system. The theory predicts red spectra (eq. (13.30)), which are characterized by two terms, the spectrum of the fast variables $\underset{\sim}{F}(0)$ at zero frequency and the feedback matrix $\underset{\sim}{\lambda}$, describing the interaction between the slow variables itself and the mean conditions of the fast variables. The noise level $\underset{\sim}{F}(0)$, which in section 2.1 was assumed to be independent of the state of the slow variables $\underset{\sim}{y}$, may obviously depend on $\underset{\sim}{y}$. This functional relationship as well as the parameterization of the expectation values of the x-terms in $\langle v_i \rangle$ (eq. (13.8)) can be determined empirically from a series of relatively short sensitivity runs with atmospheric GCM's characterized by (eq. (13.25)) for a number of different boundary conditions (sea surface temperature, ice extent) or from real atmospheric, oceanic and ice data.

The basic characteristics of observed climatic time series, covering time-scales from a month up to tens of thousands of years, is the continuous red distribution (increase towards low frequencies) of the variance spectrum with a flattening for very small frequencies due to a stabilizing feedback. This principal feature of climate variability cannot realistically be simulated on all time-scales by deterministic models. In order to identify the basic processes which are responsible for the observed statistical properties of climate variability it is feasible to construct the simplest conceivable model which is able to reproduce the observed features. Such simple models are the stochastic models which, in contrast to deterministic models, generate this spectral shape in a very simple way as the integral response of a slow system to continuous random forcing by short-time-scale (atmospheric) disturbances.

By fitting stochastic models (in a least squares sense) to observe data it is possible to yield information on the statistics of the atmospheric forcing as well as on the deterministic feedback governing the system.

Although stochastic models simulate the general shape and magnitude of the observed variance spectra, it has to be demonstrated that random forcing by atmospheric weather disturbances is indeed one of the principal cause of observed climate fluctuations by verifying those models reliably against data. The general problem of the construction and verification of statistical models from data is discussed in the next section.

3. MODEL CONSTRUCTION AND VERIFICATION

3.1 Model Construction (Model Hierachy)

Linear models are usually tested by comparing the theoretical and observed auto-spectra and the cross-spectra of the input and output time-series (Jenkins and Watts, 1968). This technique is often inapplicable in long-term climate variability studies because the daily atmospheric input data is not available for all space- and time-scales of interest. In these cases the validity of linear stochastic models is tested by comparing the observed and theoretical cross-covariance (eq. (13.23)) or cross-spectra (eq. (13.30)) of the slow components $\underset{\sim}{y}$ for optimally fitted input statistics D_{ij} or $F_{ij}(0)$ and feedback matrix λ_{ij}.

A model containing a large number of adjustable parameters can always be tuned to reproduce the observed data reasonably well. However, the significance of the model will be small in the sense that simpler models may yield equally acceptable simulations. Therefore, a hierachy of models, starting with a simple, few parameter model and then introducing more and more coefficients, should be tested in order to systematically establish the degree of detail about the physics of the system that can be inferred from the existing limited data set. An error-function (see next section) that denotes the deviations of the model from the data provides a stopping rule within the model hierarchy. The simplest model yielding an error below a beforehand specified critical value is taken to be the most realistic representative within the model hierarchy at a certain significance level. Since the statistical conclusions drawn from such an approach are limited to the a priori specified hierarchy, it is advisable to construct the most realistic model hierarchy based on the known physics of the system.

3.2 Multivariate Least Squares Fit

A statistical test of the validity of models fitted to statistically variable data sets usually consists of the following steps:

1. Define an error function ε as a suitably weighted measure of the deviations of the optimal fitted model vector $\hat{G}(a_i)$, depending on the model parameters, a_i, from the data vector $\tilde{G}$.
2. Calculate the probability density function of ε under the hypothesis that the model would fit the error-free data perfectly.
3. Choose a critical ε_c, defining a particular probability level for $\varepsilon<\varepsilon_c$ (usually 95%). If the value ε of the fitted model exceeds ε_c, the model is then rejected at that significance level.

As additional information, confidence limits for the estimated parameters can be calculated from the diagonal elements of the covariance matrix of the parameters-errors (eq. (13.42)). And a white-noise test for the residuals between optimal fitted model and data may be applied.

The error-function for a multivariate least squares fit is given by (Linnik, 1961; Muller et al., 1978; Hasselmann, 1979b)

$$\varepsilon = \sum_{i,j=1}^{P} M_{ij}(G_i - \hat{G}_i)(G_j - \hat{G}_j)^* \qquad (13.32)$$

where $\tilde{M}$ is a weighting matrix and p is the number of fitted data points. For simplicity of notation the data set $\tilde{G}$ and the model values $\hat{\tilde{G}}$ are represented as a vector, i.e., with only one index. In case of fitting a single model auto-spectrum $\hat{G}(\omega)$ to spectral estimates $G(\omega)$ from observed data the indices i and j in (13.32) represent the discrete frequency variable. If a set of cross-spectra is fitted the indices i and j represent the time series indices as well as the frequency.

The data set $\tilde{G}$ is regarded as statistically variable. For spectral estimates the statistical errors are sampling errors introduced by the finite length of the time series. The statistical errors in the data set also induce errors in the fitted model. Since $\underset{\sim}{G}$ and $\hat{\underset{\sim}{G}}$ are statistical variables, ε has to be regarded as a statistical variable, too, and can be characterized by a probability density function. In order to determine this probability density function the model errors are expressed in terms of the data errors, and it is usually assumed that the data errors are Gaussian. In the following it is shown how this procedure is applied to a multivariate least squares fit.

The deviations $\delta\underset{\sim}{G} = \underset{\sim}{G} - \bar{\underset{\sim}{G}}$ of the data from their true value $\bar{\underset{\sim}{G}}$ induce deviations $\delta\hat{\underset{\sim}{G}} = \hat{\underset{\sim}{G}} - \bar{\hat{\underset{\sim}{G}}}$ of the fitted model from the true model $\bar{\hat{\underset{\sim}{G}}}$. With the assumption that the true model would fit the error-free data perfectly, i.e., $\bar{\hat{\underset{\sim}{G}}} = \bar{\underset{\sim}{G}}$, the error function ε is in matrix notation given by

$$\varepsilon = (\delta\underset{\sim}{G} - \delta\hat{\underset{\sim}{G}})^{+}\,\underset{\sim}{M}(\delta\underset{\sim}{G} - \delta\hat{\underset{\sim}{G}}). \tag{13.33}$$

The + denotes the adjoint (transpose and conjugate complex) matrix. The statistical errors of the parameters $\delta\underset{\sim}{a}$ and the model $\delta\hat{\underset{\sim}{G}}$, introduced by the fitting of the model to statistically variable data, can be calculated in terms of the data-errors from the least squares condition

$$\frac{\partial\,\varepsilon}{\partial\,\underset{\sim}{a}} = \underset{\sim}{K}^{+}\,\underset{\sim}{M}\,(\delta\underset{\sim}{G} - \delta\hat{\underset{\sim}{G}}) = 0 \tag{13.34}$$

where K_{ij} represents the derivative of the model $\hat{\underset{\sim}{G}}$ with respect to the model-parameters $\underset{\sim}{a}$.

$$K_{ij} = \partial\hat{G}_i/\partial a_j \tag{13.35}$$

The dimension of $\underset{\sim}{K}$ is p x q, where p is the number of data points and q is the number of model-parameters. With $\delta\underset{\sim}{G} = \underset{\sim}{K}\delta\underset{\sim}{a}$ the parameter and model errors are given by

$$\delta\underset{\sim}{a} = \underset{\sim}{B}\;\delta\underset{\sim}{G} \tag{13.36}$$

$$\delta\underset{\sim}{G} = \underset{\sim}{T}\;\delta\underset{\sim}{G} \tag{13.37}$$

where

$$\underset{\sim}{B} = (\underset{\sim}{K}^{+}\underset{\sim}{M}\underset{\sim}{K})^{-1}\underset{\sim}{K}^{+}\underset{\sim}{M} \tag{13.38}$$

and

$$\underset{\sim}{T} = \underset{\sim}{K}\underset{\sim}{B} \tag{13.39}$$

The covariance matrix $\underset{\sim}{A} = \langle \delta \underset{\sim}{a} \delta \underset{\sim}{a}^{+} \rangle$ of the parameter errors $\delta \underset{\sim}{a}$ is then calculated as

$$\underset{\sim}{A} = \underset{\sim}{B}\ \underset{\sim}{R}\ \underset{\sim}{B}^{+} \tag{13.40}$$

where $\underset{\sim}{R}$ is the covariance-matrix of the data errors.

$$\underset{\sim}{R} = \langle \delta \underset{\sim}{G}\ \delta \underset{\sim}{G}^{+} \rangle \tag{13.41}$$

If the weighting-matrix $\underset{\sim}{M}$ is taken to be the inverse of the covariance-matrix of the data-errors, i.e., $\underset{\sim}{M} = \underset{\sim}{R}^{-1}$, eq. (13.40) reduces to

$$\underset{\sim}{A} = (\underset{\sim}{K}^{+} \underset{\sim}{R}^{-1}\ \underset{\sim}{K})^{-1} \tag{13.42}$$

where we have used the fact that $\underset{\sim}{R}$ and therefore $\underset{\sim}{M}$ are hermitian, i.e, $\underset{\sim}{M} = \underset{\sim}{M}^{+}$. Accordingly the covariance matrix $\underset{\sim}{N} = \langle \delta \hat{\underset{\sim}{G}} \delta \hat{\underset{\sim}{G}}^{+} \rangle$ of the model-errors $\delta \hat{\underset{\sim}{G}}$ is given by

$$\underset{\sim}{N} = \underset{\sim}{K}\ \underset{\sim}{A}\ \underset{\sim}{K}^{+} \tag{13.43}$$

If the data errors $\delta \underset{\sim}{G}$ are Gaussian (for spectral estimates this is approximately the case) the model errors $\delta \hat{\underset{\sim}{G}}$ and the parameter errors $\delta \underset{\sim}{a}$ are also Gaussian, since they are linear transformations of $\delta \underset{\sim}{G}$. Confidence intervals for each model parameter a_i and each model value G_i may then be calculated form the diagonal elements of $\underset{\sim}{A}$ and $\underset{\sim}{N}$, respectively (see Jenkins and Watts, 1968).

Often one is not so much interested in confidence intervals of single model parameters or single model values, but in the validity of the model as a whole. In this case a model validity test is applied (see Bendat and Piersol, 1971, p. 115ff). The error-function ε, defined by eq. (13.33), being a quadratic form of Gaussian variables, is accordingly approximately χ^2 - distributed. The model validity test is then essentially a multivariate chi-square goodness-of-fit test. It can be shown (cf. Linnik, 1961; Lemke, 1980) that the number of degrees of freedom ν of the χ^2-distribution of ε is a maximum for the optimal weighting matrix $\underset{\sim}{M} = \underset{\sim}{R}^{-1}$ and is then given by $\nu = p-q$, where p is the number of data and q the number of parameters of the model.

Since the probability density function is now known, a reference value ε_c can be defined from tables such that the probability of $\varepsilon < \varepsilon_c$ is equal to a beforehand specified significance level α (usually 95%, see Fig. 4). From eqs. (13.32) and (13.41) (with $\underset{\sim}{M} = \underset{\sim}{R}^{-1}$) the actual error function ε for a particular model-fit can be calculated and compared with ε_c. If $\varepsilon > \varepsilon_c$, the model is

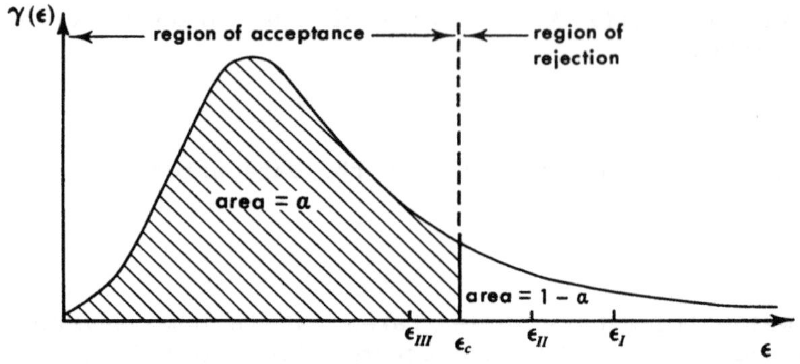

Fig. 4. Regions of acceptance and rejection for a model validity test with arbitrary probability density function $\gamma(\varepsilon)$ of the model-error ε (α = significance level).

rejected at the significance level α. If $\varepsilon < \varepsilon_c$ the model is not rejected and taken to represent the observed data at the significance level α. The shaded area in Figure 4 represents the region of acceptance, whereas the unshaded area is called the region of rejection.

Two possible errors can occur during a model validity test: A true model may be rejected (the probability of this error is $1-\alpha$, i.e., equal to the unshaded area in Figure 4) and a false model may be accepted. The probability of both errors cannot be reduced simultaneously by shifting the significance level α, i.e., ε_c. The only way to reduce both errors at the same time is to increase the number of data used in the fitting process (see Bendat and Piersol, 1971).

Suppose a hierarchy of fitted models of increasing complexity is characterized by errors ε_I, ε_{II}, ε_{III}.... In this case the critical value ε_c provides a stopping rule within the model hierarchy. The simplest model yielding an error below ε_c (model III in Figure 4) is accepted to be the most realistic representative within the model hierarchy at the significance level α.

Example: Modeling a single auto-spectrum. In our application of stochastic models to Arctic and Antarctic sea ice variability, the data consist of auto- and cross-spectral estimates. For simplicity the procedure outlined above is applied in this example to the modeling of a single auto-spectrum.

Since the original data consist of discrete time-series with a spacing of Δ (1 month for the Arctic data set and 1 week for the Antarctic data set) the spectrum $G(\omega_i)$ is estimated only at discrete frequencies ω_i. The single-component, discrete first-order

Markov process corresponding to the continuous form (13.9) is given by

$$y(t) = a_1\, y(t-\Delta) + \tilde{x}(t) \tag{13.44}$$

where

$$a_1 = 1 - \lambda\Delta \text{ and } \tilde{x}(t) = \int_{t-\Delta/2}^{t+\Delta/2} x(t')\, dt' \tag{13.45}$$

The power-spectrum is calculated as (see Jenkins and Watts, 1968)

$$\hat{G}_i = \frac{a_2}{1+a_1^{\,2} - 2a_1 \cos(\omega_1 \Delta)} \tag{13.46}$$

where a_2 represents the (constant) power-spectrum of the white-noise forcing $\tilde{x}(t)$.

The general covariance matrix of the finite sampling errors of auto-spectral estimates is given by (Jenkins and Watts, 1968)

$$R_{ij} = \langle \delta G_i \; \delta G_j \rangle = \frac{2}{\mu} \Gamma_i \, \Gamma_j \, \delta_{ij} \tag{13.47}$$

where Γ_i is the true-cross-spectrum at the frequency ω_i, μ denotes the degrees of freedom of each spectral point and δ_{ij} represents the Kronecker-symbol. Since we are testing the hypothesis of a true model $\hat{G}$, Γ is set equal to $\hat{G}$ in (13.47). For cross-spectral estimates (13.47) is more complicated, the indices i and j including the time-series indices as well as the frequency index.

After inserting the observed spectrum G_i and the model $\hat{G}_i$ (from 13.46) into eq. (13.32) and setting $\underset{\sim}{M} = \underset{\sim}{R}^{-1}$, where $\underset{\sim}{R}$ is taken from (13.47), the actual weighted error $\tilde{\varepsilon}$ can be calculated and minimized by varying the two parameters a_1 and a_2, which determine the model $\hat{G}$. The minimum value of ε is then compared with the 95% value of a χ^2_{p-2} distribution, where p is the number of fitted frequencies. This concludes the model validity test. Examples of observed and fitted autospectra and associated model parameters are given in Figures 5 and 6.

If one is also interested in confidence limits for parameter and model values, the matrix $\underset{\sim}{K}$ of the derivatives of the model $\hat{G}$ with respect to the model-parameters has to be calculated (eq. (13.35)). For this example $\underset{\sim}{K}$ is a p x 2 matrix given by

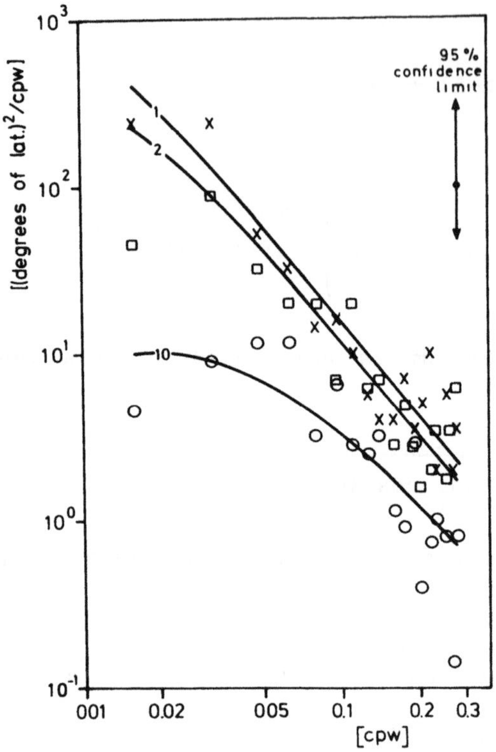

Fig. 5. Auto-spectra of the amplitudes of the first (x), second (□) and the tenth (o) EOF of Antarctic sea ice variability and fitted first-order MARKOV models (solid lines) (from Lemke, 1980).

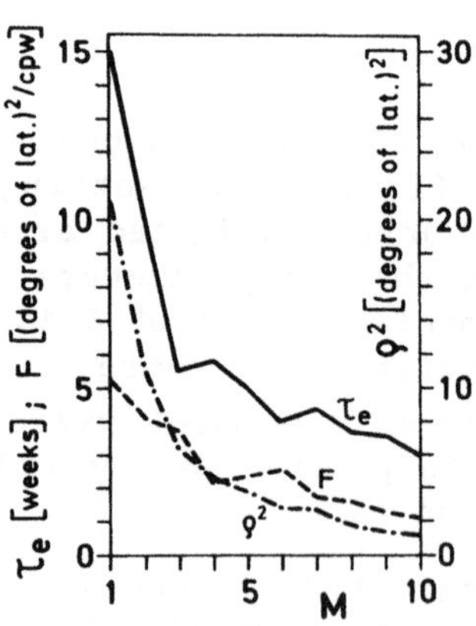

Fig. 6. White-noise excitation level F, relaxation time τ_e and variance ρ^2 of the amplitudes of the first ten EOF's in Antarctica (from Lemke, 1980).

$$K_{i1} = \frac{\partial \hat{G}_i}{\partial a_1} = -\frac{a_2(2a_1-2\cos(\omega_i\Delta))}{(1+a_1^2-2a_1\cos(\omega_i\Delta))^2}$$

$$K_{i2} = \frac{\partial \hat{G}_i}{\partial a_2} = \frac{1}{1+a_1^2-2a_1\cos(\omega_i\Delta)} \qquad (13.48)$$

With (13.47) and (13.48) the covariance-matrices of the parameter-errors A and model-errors N are calculated according to (13.42) and (13.43), respectively. Confidence intervals for the parameters a_1 and a_2 and for the model-values $\hat{G}_i$ are then given in terms of the diagonal elements of $\underset{\sim}{A}$ and $\underset{\sim}{N}$.

4. APPLICATION TO ARCTIC AND ANTARCTIC SEA ICE VARIABILITY

4.1 General Characteristics of Arctic and Antarctic Sea Ice Variability

The analysis of existing data on polar sea ice extent clearly demonstrates the pronounced variability in space and time (Sanderson, 1975; Walsh and Johnson, 1979a ,b; Rayner and Howarth, 1979; Lemke et al., 1980; see also Chapter 12). Results shown in this paper are based on an eleven-year Arctic data set (January 1966-December 1976, digitized from the monthly Northern Hemisphere sea ice charts published by the United Kingdom Meteorological Office, Bracknell) and on a six-year Antarctic data set (4 January 1973-22 February 1979), digitized from the weekly Southern Hemisphere sea ice charts produced by the U.S. Fleet Weather Facility, Suitland, Maryland). Fluctuations of sea ice extent are generally larger in the Southern Ocean than in the land-locked Arctic basin, interannual variations being of the order of 1/3 of the seasonal excursion in both regions. Decomposition of the sea ice anomaly data (defined as the residual after subtracting the mean annual cycle from the original data set) into empirical orthogonal functions (EOF's, for a definition see Chapter 12) indicates that lower-order EOF's (large eigenvalues) are associated generally with larger spatial scales and higher-order EOF's (small eigenvalues) have the appearance of spatial white noise. The autocorrelation functions of the (time-dependent) EOF-amplitudes demonstrate that the correlation time scales decrease with increasing order of the EOF's, indicating that sea ice variations on large spatial scales (lower EOF's) are associated with longer relaxation times (Lemke et al., 1980).

The pronounced asymmetry of the observed spatial correlation of the ice-edge variations at different longitudes suggests that advection of sea ice anomalies plays an important role in large scale sea ice dynamics in the Southern Ocean. Ice anomalies seem to drift eastward in most parts of East Antarctica and westward in the Weddell Sea, the Ross Sea and the Amundsen Sea. This is in qualitative agreement with straightforward advection of sea ice or with lateral heat fluxes due to prevailing winds or ocean currents (Antarctic Circumpolar Current, gyres in the Weddell Sea, the Ross Sea and the Amundsen Sea), that induce spatially correlated melting and freezing processes. In the Arctic basin zonal advection of sea ice seems to be of minor importance. The zonal correlations at different longitudes in this region exhibit only a weak asymmetry, in accordance with the dominating sea ice drift pattern in the Arctic (Transpolar Drift Stream), which is meridional rather than zonal (Lemke et al., 1980). Examination of the complete spectral matrix of sea ice anomalies indicates that strong correlations between ice anomalies at different longitudes exist essentially only for longitude lags of 10 - 20°. In general, the spatial scale of the Antarctic cross-correlations is slightly larger than in the Arctic (Lemke, 1980). A unique feature of the variance auto-spectra of the anomalies of Arctic and Antarctic sea ice extent is the continuous red distributions with a flattening for very small frequencies (Walsh and Johnson, 1979a; Lemke et al., 1980). This spectral characteristic suggests the modeling of sea ice variablity with stochastic forcing models.

In contrast, to deterministic models which are usually designed to simulate the seasonal cycle of sea ice cover, thickness and drift as the response to observed time varying forcing fields [winds, air temperature, solar radiation (see Hibler, 1979; Parkinson and Washington, 1979)], the stochastic models, discussed in section 4.2, and 4.3, are applied in the inverse modeling sense to determine certain time independent dynamical parameters (local feedback, diffusion, advection and white noise atmospheric forcing) as optimal least squares fits to the observed variations of sea ice anomalies in space and time. The mean annual cycle is subtracted from the original data set, since it represents a deterministic signal that cannot be described by a stochastic model. The main purpose of applying a hierarchy of linear stochastic models with time-independent coefficients is to estimate the characteristic dynamical parameters required to explain the observed anomalies as a guide for constructing more sophisticated dynamical models of the coupled ocean-ice-atmosphere system based on more detailed physics.

4.2 Local Response Model

The interaction between sea ice anomalies and the atmospheric circulation is considered first in the simplest form in which sea ice variability is modelled as the local response to a statistically independent, white-noise forcing field, which may be associated with short-lived weather disturbances. The stablizing feedback parameter may be attributed to the mean heat transfer from the ice to either the atmosphere or the ocean.

Consider a static ice cover in a 10°-longitudinal sector with an ice thickness h that is constant in space and ice, i.e., only lateral freezing and melting is discussed in the following example. The rate of change of the ice covered area is then given by

$$\rho \, Lh \frac{dA}{dt} = -Q \tag{13.49}$$

where ρ is the density of sea ice, L is the latent heat of fusion, A is the ice covered area and Q is the net heat supplied by ocean, atmosphere and solar radiation along the ice edge within the 10° sector. Since Q depends on oceanic and atmospheric variables, it is divided into a mean term, $\bar{Q}$, describing mainly oceanic and mean atmospheric fluxes, and a fluctuating term, Q', which is mainly due to short time-scale atmospheric fluxes being considered to represent white-noise forcing.

$$\rho \, Lh \frac{dA}{dt} = - \bar{Q} - Q' \tag{13.50}$$

Q is mainly a function of latitude. From small deviations about an equilibrium ice cover, the ice covered area A is proportional to the latitude of the ice edge θ. In this case both sides of eq. (13.50) can be expanded into a Taylor series about the equilibrium value of either A or θ, where only the linear terms have to be considered. After subtracting the solution for the equilibrium state, eq. (13.50) turns into

$$\frac{dy}{dt} = - \lambda y + n_{A,\,\theta} \tag{13.51}$$

where y denotes the deviation from equilibrium of either the ice covered area A or the latitude of the ice edge θ, n_A and n_θ are proportional to Q' and differ by the proportionality factor $dA/d\theta$ at the equilibrium ice edge and λ is in both cases the same feedback coefficient, mainly determined by the meridional gradient of $\bar{Q}$ at the equilibrium ice edge. Such a decoupled first order Markov-process is applied in a least squares sense (see example in Section 3.2) to anomalies of both Arctic sea ice cover and Antarctic sea ice edge on a 10°-longitudinal grid as well as to the amplitude of the corresponding EOF-representation.[3] The observed

[3]It should be noted here that for simplicity of notation only Markov processes which are continuous in time are discussed in this paper, although actually discrete processes are fitted to the discrete data sets as explained in the example in Section 3.2 (for further detail, see Lemke et al., 1980).

variance auto-spectra of the amplitudes of various EOF's of Antarctic sea ice are shown in Figure 5 together with the best-fit model auto-spectra determined from eq. (13.31) with $i=j$.[4] In the Arctic, the first order Markov model yielded a statistically valid description of the first 15 amplitudes at the 95% confidence level, with the exception of the 11th amplitude. In the Antarctic a Markov fit was also achieved for the first 15 amplitudes at the same confidence level, with the exception of the second and fourth amplitude, which were only acceptable at the 99% confidence level.

The dependence of the optimally fitted white-noise excitation level F and relaxation time $\tau_e = 1/\lambda$ on model number M is shown in Figure 6 for Antarctic sea ice. Also shown for comparison is the variance of each model, ρ^2, which is given by $\rho^2 = \pi F \tau_e / 2$. The figure shows clearly that the lower EOF's have larger relaxation times, and that according to the above expression the larger variance of the lower EOF's is due primarily to the large relaxation time of these modes, rather than increased excitation. The same features are also found for Arctic sea ice (Lemke, 1980).

The fitted relaxation times of the Antarctic sea ice are generally smaller than those in the Arctic (4.7 months for the first Arctic EOF). This is probably due to an enhanced ice-ocean interaction in the Southern Ocean (upwelling, divergent Ekman drift, strong ocean currents), and the fact that nearly all Antarctic sea ice is annual ice of 1-2 m thickness, whereas 80% of the Arctic sea ice is in thicker layers older than 2 years and 60% older than 5 years.

The results of a Markov-fit to sea ice anomalies on a 10° longitudinal grid are shown in Figures 7 and 8. The zonal distribution of the excitation level F for both the Arctic and Antarctic sea ice is very similar to the distribution for the annual signal and the anomaly variance (see Figures 2 and 5 in Lemke et al., 1980), whereas the zonal distribution of the relaxation time τ_e shows no clear correlation with the magnitude of the forcing or the anomaly.

In the Arctic (Fig. 7), the feedback factor, $\lambda = 1/\tau_e$, is seen to be large in the Siberian and Canadian sectors. This may be attributed to land locking by the Eurasian and American continents, which supports persistence during wintertime, and also to river runoff, which tends to stabilize the upper layers of the Arctic Ocean, thereby decreasing the thermal inertia and the relaxation time. The largest relaxation times in the Arctic, of the order of 3 to 4 months, are found in the Davis Strait, the East Greenland current and the Barents Sea, in areas in which the

---

[4]In the following figures, cpw = cycles per week and cpm = cycles per month.)

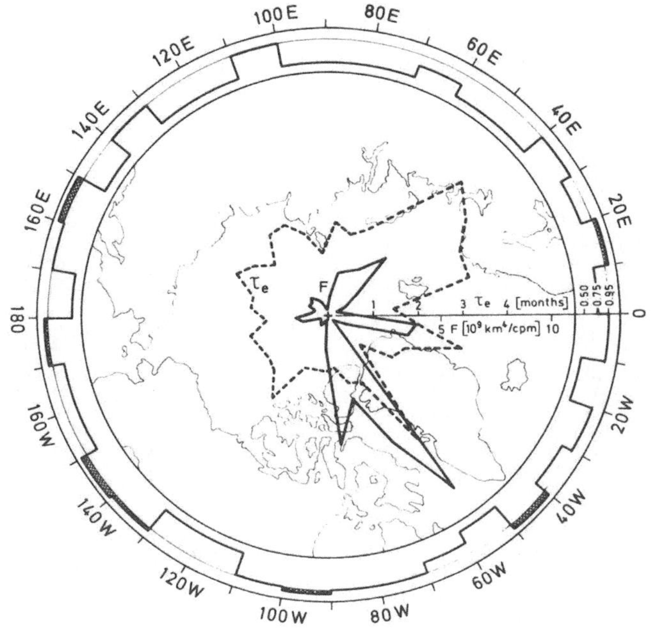

Fig. 7. Longitudinal distribution of white-noise excitation level F and relaxation time τ_e in the Arctic (from Lemke et al., 1980).

interaction between sea ice, ocean and atmosphere is not disturbed by continental influences. In the Southern Ocean (Figure 8) the relaxation times (1 to 2 months) are generally smaller than in the Arctic Basin. As pointed out above in the discussion of the EOF relaxation times, this is presumably a consequence of the enhanced thermal contact between sea ice and the underlying ocean due to upwelling and ocean currents.

4.3 Zonally Coupled Models

The local reponse model yields a satisfactory description of the auto-spectra both for the EOF representation and for most of the individual longitude sectors. However, although these models provide some useful numbers on the general magnitude of the forcing and feedback rates, they need to be generalized in order to explain the structure of the non-diagonal elements of the cross-spectra of sea ice anomalies between different longitudes. Local response models on a longitudinal grid will obviously provide a complete description only in those regions without strong zonal winds an ocean currents, where sea ice variability is dominated by variations of the vertical and meridional heat fluxes. But in most parts of the polar seas the sea ice is exposed to strong

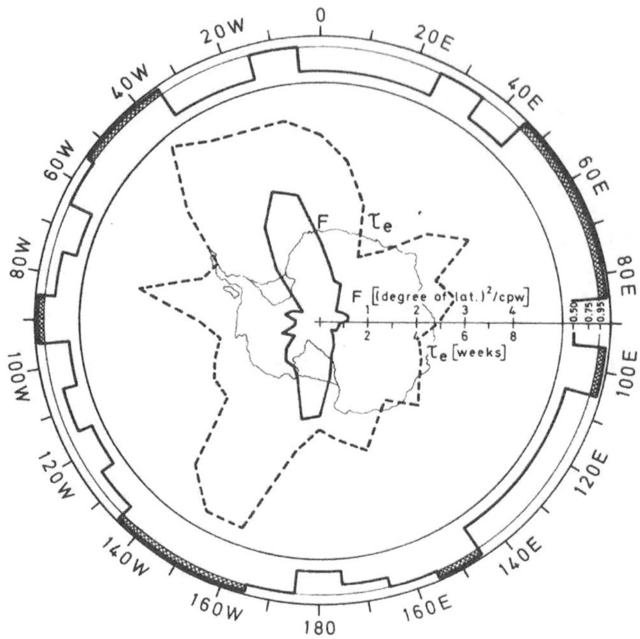

Fig. 8. Longitudinal distribution of white-noise excitation level F and relaxation time τ_e in the Antarctic (from Lemke et al., 1980).

lateral motions in ocean and atmosphere creating significant zonal cross-correlations. As discussed in section 4.1, an advective or diffusive type of direct interaction between different longitudes is indicated by the scale dependence of the relaxation times of the EOF coefficients and by the asymmetry of the cross-correlation functions. These effects can be most easily modelled by considering a vector Markov process of the form (13.9), where $\underset{\sim}{y}$ denotes the sea ice extent or ice covered area and i,j represent longitudinal indices. There are generally two mechanisms that generate cross-spectra in a coupled Markov model (see eq. (13.30)): 1) zonal interactions in the fast system (atmosphere) give rise to non-diagonal elements in the spectral matrix of the forcing $\underset{\sim}{F}$ and 2) zonal interactions in the slow system (see ice, ocean) introduce non-diagonal elements in the feedback matrix $\underset{\sim}{\lambda}$.

The relative importance of both mechanisms in the sea ice-ocean-atmosphere system can be evaluated by testing a hierarchy of models that introduce these effects separately. In the same way the longitudinal structure of feedback and forcing coefficients can be resolved. In view of the narrow width of the cross-correlation function of sea ice anomalies (strong correlations exist only for longitude lags of 10-20°) and the finite correlation length in the atmosphere at high latitudes (≈500 km, see Ghil et al., 1981) we shall limit ourselves to models where only the two closest off-diagonal elements are introduced in the feedback and forcing matrices, in addition to the local feedback and forcing coefficients, $\underset{\sim}{F}$ and $\underset{\sim}{\lambda}$ therefore representing tri-diagonal matrices.

Eq. (13.9) then reduces to

$$\dot{y}_i(t) = \sum_{j=i-1}^{i+1} \lambda_{ij}\, y_j(t) + V_i'(t) \tag{13.52}$$

and

$$F_{ij}(\omega) = \langle V_i'(\omega)\, V_j'(\omega)\rangle \approx \begin{cases} F_{ij}(o) & i-1 \leq j \leq i+1 \\ o & \text{otherwise} \end{cases} \tag{13.53}$$

From complex symmetry properties of a spectrum it follows further that $\underset{\sim}{F}(o)$ is symmetric,

$$F_{ij}(o) = \int_{-\infty}^{\infty} \bar{C}_{ij}(\tau)dt = F_{ji}(o) \tag{13.54}$$

The model (13.52) with tri-diagonal interaction matrix can be interpreted from eq. (13.49), where the ice cover is no longer at rest but is changing due to advective and diffusive fluxes in addition to melting and freezing. In this case eq. (13.49) is replaced by the general transport equation

$$\frac{\partial A}{\partial t} + \text{div}\, \underset{\sim}{J} = q \tag{13.55}$$

where q denotes the source term and $\underset{\sim}{J}$ represents the lateral sea ice flux, consisting of an advective part mainly due to the effect of prevailing winds and ocean currents and a diffusive part mainly created by winds from variable directions.

$$\underset{\sim}{J} = A\underset{\sim}{u} - D\underset{\sim}{\nabla} A \tag{13.56}$$

For small deviations about an equilibrium state and purely zonal flow $\underset{\sim}{u}$ and diffusion D -- these can only be determined from the ice cover or ice edge data sets -- the problem can be treated as one-dimensional. The transport equation as compared to eq. (13.51) then takes the specific form

$$\frac{\partial y(x, t)}{\partial t} = -\tilde{\lambda}(x)\, y(x, t) + \frac{\partial}{\partial x}\left[D(x)\, \frac{\partial y(x, t)}{\partial x}\right] - \frac{\partial}{\partial x}[y(x, t)\, U(x)] + n(x, t) \qquad (13.57)$$

where x denotes the longitude and y represents the deviation of the ice covered area or the ice edge from their equilibrium values. The local feedback factor $\tilde{\lambda}$ and the diffusion and advection coefficients D and U are taken to be time-independent quantitites.

Comparing the discretized form of (13.57) with (13.52) it is evident that the symmetric part of the next neighbor interaction in (13.52) is determined by the lateral diffusion coefficient D and the antisymmetric part by the lateral advection U, modified by a term arising from the longitudinal gradient of D:

$$\begin{aligned} \lambda_{ii} &= \tilde{\lambda}_i + 2D_i + \frac{U_{i+1} - U_{i-1}}{2} \\ \lambda_{ii+1} &= D_i - \frac{U_i}{2} + \frac{D_{i+1} - D_{i-1}}{4} \qquad (13.58) \\ \lambda_{ii-1} &= D_i + \frac{U_i}{2} - \frac{D_{i+1} - D_{i-1}}{4} \end{aligned}$$

The complete model thus contains three feedback parameters for each longitudinal grid point: the local feedback $\hat{\lambda}_i$, the diffusion D_i and the advection coefficient U_i, in addition to the two parameters that describe the atmosphere forcing: F_{ii} and F_{ii+1}. How many parameters are actually necessary to yield a statistically significant description of the data has to be determined from a model hierarchy.

Application of different types of models demonstrates that rotationally invariant models with longitude independent feedback or forcing matrices had to be rejected (Lemke, 1980), basically because they were unable to reproduce the strong longitudinal variations of forcing and feedback apparent in Figures 7 and 8. Therefore, all models considered in the following hierarchy allow full longitudinal freedom of all parameters. The simplest extension of the local response model assumes that sea ice correlations between different longitudes arise entirely through statistical correlations in the forcing functions: the feedback matrix $\tilde{\lambda}$ in (13.52) is assumed to contain only diagonal elements, i.e., the local feedback terms without advection and diffusion (Model I). In the next higher order model either an advective (Model IIa) or a diffusive flux (Model IIb) is introduced additionally into $\underset{\sim}{J}$ (eq.(13.56)). The highest order model, finally, includes both advective and diffusive fluxes, without restriction of the general form of $\underset{\sim}{J}$ (Model III). The results of the model-fitting (for details see Lemke et al., 1980) show that at several longitudes, especially in the Arctic, the simple local feedback Model I is acceptable (at a 95% significance level), but in most sectors the highest order model including both symmetric and antisymmetric next neighbour interactions has to be introduced to achieve an acceptable description of the data. In the Arctic, especially in the Atlantic sectors and the East Siberian Sea, Model III yields a distinct improvement compared to Model I. In Antarctica, Model III is clearly better than Model I in the eastern Weddell Sea, East Antarctica, the Ross Sea and the Amundsen Sea, indicating that in these areas the interaction with the next neighbour longitude (lateral diffusion and advection) plays an important role in the overall sea ice dynamics.

In both polar regions Model IIb (local feedback and lateral diffusion) is in general more realistic than Model IIa (local feedback and lateral advection). This is especially true for the Arctic sea ice variability, where the diffusion-model (IIb) is signficiantly better than the advection-model (IIa), but the highest order model (III) provides only a small improvement compared to Model IIb. Thus, lateral advection of sea ice anomalies appears to be of relatively minor importance in the Arctic Ocean. In the Antarctic, however, lateral advection is found to be of comparable importance to lateral diffusion.

In the Arctic Model III - including non-diagonal atmospheric forcing, local feedback, lateral advection and diffusion in all sectors - is acceptable at the 80% significance level. The Arctic diffusion model (IIb) is similarly acceptable at the less restrictive 95% significance level. Models I and IIa must be rejected.

Except at six sectors in the Bellinghausen Sea (90W), the South Indian Basin (50E, 70E) and the western Weddell Sea (40W - 60W), which obviously demand additional dynamical parameterization, Model III is also an acceptable description at the 95% significance level of antarctic sea ice anomalies.

The diffusion coefficient for the Arctic sea ice determined from Model III is of the order of $0.5 \cdot 10^4 m^2/sec$. This is comparable to values often assumed in large scale ocean circulation studies. The mean absolute values of the lateral advection of sea ice anomalies in the Arctic Ocean is of the order of 4 cm/sec.

The advection and diffusion coefficients inferred for the Antarctic (Figure 9) are considerably larger than in the Arctic Ocean. This is an expected because of the greater sea ice-ocean

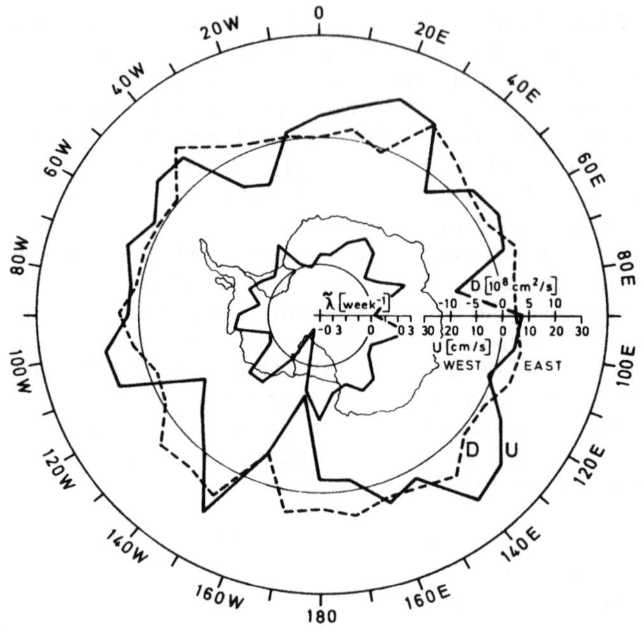

Fig. 9. Longitudinal distribution of lateral diffusion D, advection U and local feedback $\tilde{\lambda}$ for sea ice anomalies in the Antarctic (from Lemke et al., 1980).

interaction due to upwelling and strong zonal ocean currents in the Antarctic. The mean diffusion coefficient (dashed line, Figure 9) in the Southern Ocean is of the order of $3 \cdot 10^4 m^2/sec$, and the mean absolute value of the advection is 11 cm/sec, in rough agreement with typical velocities of ocean surface currents in that area. The advection (outer solid line, Figure 9) indicates a significant westward flow of sea ice anomalies in the Weddell Sea, the Ross Sea and north of the Amery Ice Shelf (80E) and the Getz Ice Shelf (120W). In all other parts of the Southern Ocean the sea ice flow is eastward, most strongly east of the Weddell Sea (20E), in the South Indian basin (140E) and in the Amundsen Sea (100W).

The general flow pattern is in good agreement with the observed ocean surface circulation around Antarctica (Weddell Sea gyre, Ross Sea gyre, Antarctic Circumpolar Current; Treshnikov, 1964; see Figure 10).

The significance of cross-sector interactions is also demonstrated by the differences in the local feedback parameters λ_i for the decoupled model (Figure 8, where $\tau_e = \lambda^{-1}$) and the complete interactive Model III (inner solid line, Figure 9). In the

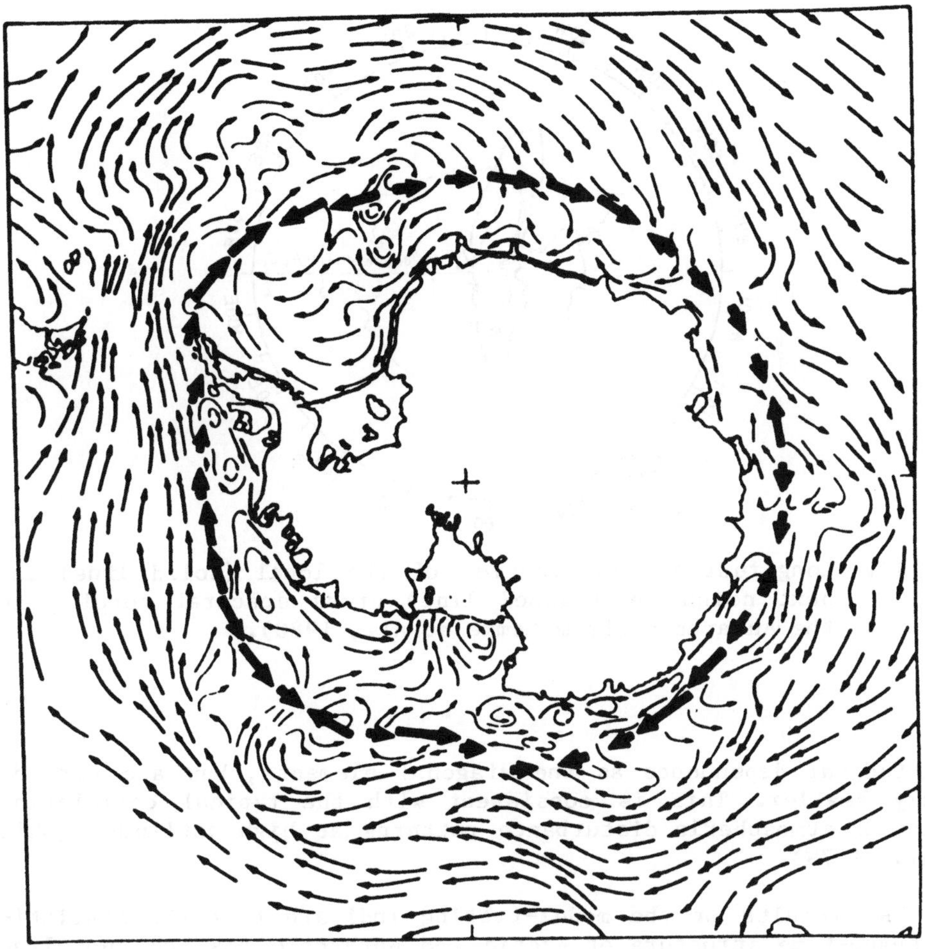

Fig. 10. Southern Ocean surface circulation (after Treshnikov, 1964) and Antarctic sea ice model-advection pattern from Figure 9 (thick arrows) (from Lemke et al. 1980).

complete model a significant fraction of the net relaxation rate of sea ice anomalies is attributed to advection and diffusion rather than local negative feedback (cf. eq. (13.58)).

The longitudinal distribution of the antarctic forcing parameters is shown in Figure 11. The diagonal elements of the forcing matrix $\tilde{F}$ (solid lines) exhibit the same longitudinal dependence, found previously for the decoupled model, Figure 8. Thus, the introduction of lateral interaction between sea ice anomalies does not affect the local forcing. The nondiagonal forcing terms (dashed lines), representing the cross-spectrum of the forcing with the eastern neighbour longitudinal gridpoint, show the same

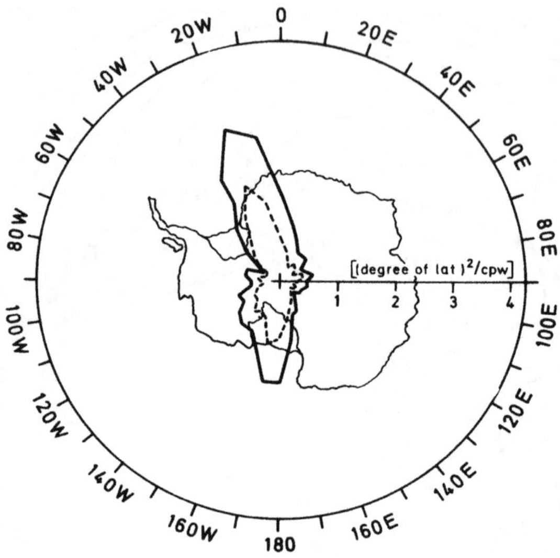

Fig. 11. Longitudinal distribution of the local (solid line) and next neighbour (dashed line) cross spectral forcing in the Antarctic (from Lemke et al., 1980).

longitudinal dependence as the diagonal elements, but are considerably smaller. This is consistent with the typical correlation scales of atmospheric disturbance patterns at high latitudes (Ghil et al., 1979).

The results of the model-fitting indicate that the statistical space-time structure of Arctic and Antarctic sea ice anomalies described by the variance cross-spectra can be satisfactorily explained by a dynamical model based on white-noise atmospheric forcing, local stabilising relaxation and lateral diffusion and advection. Obviously, the empirical parameters derived only from sea ice data cannot clarify the mechanisms of the local feedback and the lateral coupling between sea ice anomalies at adjacent longitudes. These processes could be due to either oceanic or atmospheric interactions. The resolution of this question will require a more detailed joint analysis of both ice-anomaly and atmospheric and sea-surface temperature anomaly fields.

Nevertheless, the stochastic model (eq. (13.52)) provides a basis for constructing more sophisticated dynamical models or for developing empirical sea ice prediction models for periods up to approximately six months (Johnson et al., 1985).

5. FUTURE PROBLEMS

5.1 Response of the Atmosphere to Sea Ice Anomalies

The stochastic forcing term in eq. (13.52) interpreted as arising from short-time scale weather disturbances was taken to be independent of the sea ice variability. Since the ocean-atmosphere heat transfer changes significantly from ice-free to ice-covered regions of the ocean, sea ice anomalies may nevertheless be expected to appreciably affect the structure of the atmospheric circulation. One way to incorporate this effect into the stochastic model approach is to describe the influence of sea ice anomalies on the atmosphere with a linear time-independent response function. The time-independence seems to be justified because the atmosphere immediately adjusts to slow variations in sea ice extent. The closed feedback-loop of the atmosphere-sea ice interaction is then given by

$$\dot{\underset{\sim}{y}}(t) = - \underset{\sim\sim}{\lambda y}(t) + \underset{\sim}{V}'(t) \tag{13.59}$$

$$\underset{\sim}{V}'(t) = \underset{\sim}{L}\ \underset{\sim}{y}(t) + \underset{\sim}{n}(t) \tag{13.60}$$

where $\underset{\sim}{L}$ is the time independent response function. The atmospheric variability is described by two terms, one accounting for the sea ice induced external part, $\underset{\sim\sim}{Ly}$, and the second denoting the internal white noise of the atmosphere, $\underset{\sim}{n}$.

From (13.59) and (13.60) the spectrum of sea ice, $G(\omega)$ is again given by

$$\underset{\sim}{G}(\omega) = \underset{\sim}{H}(\omega)\ \underset{\sim}{F}(o)\ \underset{\sim}{H}^{+}(\omega) \tag{13.61}$$

where $\underset{\sim}{F}$ is the spectrum of the white noise part of the atmosphere, $\underset{\sim}{F}(\omega) = \langle \underset{\sim}{n}(\omega)\underset{\sim}{n}^{+}(\omega)\rangle$, and $\underset{\sim}{H}(\omega)$ now includes also $\underset{\sim}{L}$

$$\underset{\sim}{H}(\omega) = (i\omega\underset{\sim}{I} + \underset{\sim}{\lambda}-\underset{\sim}{L})^{-1} \tag{13.62}$$

The spectrum of the atmospheric forcing, $\underset{\sim}{P}(\omega) = [\underset{\sim}{v}'(\omega)\underset{\sim}{v}'^{+}(\omega)]$, is no longer white but is calculated as

$$\underset{\sim}{P}(\omega) = (i\omega\underset{\sim}{I}+\underset{\sim}{\lambda})\ \underset{\sim}{G}(\omega)\ (i\omega\underset{\sim}{I}+\underset{\sim}{\lambda}^{+}) \tag{13.63}$$

The general form of $\underset{\sim}{P}$ for a single component system with $0 < L < \lambda$ is shown as a dashed line in Figure 3.

The important question to be answered at this stage is whether the mean atmospheric response to sea ice variabilty, $\underset{\sim}{L}\underset{\sim}{y}$, can be identified in the presence of the natural variaiblity of the atmosphere, $\underset{\sim}{n}$. Since the response represents a multi-dimensional vector, the problem is essentially one of pattern recognition, which usually will fail a multi-variate significance test without signal processing in order to enhance the signal-to-noise ratio (Hasselmann, 1979). Application of such statistical techniques to the response of the atmosphere to SST-anomalies is a topic of current research (Hannoschock and Frankignoul, 1985).

5.2 Seasonally Dependent Forcing and Feedback Processes

The ice-ocean feedback representing an important part of the feedback matrix $\underset{\sim}{\lambda}$ is obviously dependent on the variability of the mixed layer, which is shallow and fresh during summer and deep and more saline in winter time. And, also, the mean ice-atmospheric interaction and the variance of the atmospheric noise are seasonally variable.

Therefore, an obvious extension of the stochastic model which may be expected to improve the dynamical description and extend the range of prediction is the harmonic expansion of the statistics and the model parameters with respect to the annual cycle. Results presented in Chapter 4 may be regarded as the zero'th harmonic component of this expansion.

The extension of eq. (13.9) and (13.12) to a Langevin-equation with seasonally dependent feedback λ and forcing variance D is straightforward (Ruiz de Elvira and Lemke, 1982). The main difference to the model discussed in Chapter 4 is that the covariance function is no longer stationary but cyclo-stationary, therefore, depending on two variables instead of one, the first variable describing time lag and the second denoting the seasonal dependence of the response (Bendat and Piersol, 1971). The spectrum therefore also depends on two frequency coordinates a continuous frequency according to the time lag and a discrete frequency corresponding to the periodic time-dependence of the correlation function. The statistics of the seasonal model shows marked departures from the stationary model. The effect of a periodic forcing is that each line input at frequency ω is split into two side bands at $\omega+\omega_a$ and $\omega-\omega_a$ (ω_a is here the annual frequency but it can represent any other frequency the forcing depends on), i.e. the discrete frequency coordinate can take only three values ω, $\omega+\omega_a$ and $\omega-\omega_a$. The introduction of a periodic feedback results in a splitting of the response into an infinite set of side bands with decreasing energy and a spacing of ω_a. The data has to show how important the seasonal dependence of forcing and feedback is for the interaction of fast and slow components in the climate system.

ACKNOWLEDGMENTS

The author is grateful to M. Ghil, G. Herman, and G. R. North for a critical review of the original manuscript and to Esther Olsen and Paula Constantini for typing. This work was partly supported by the Geophysical Fluid Dynamics Program, Princeton University/NOAA Grant No. 04-7-022-44017 and by the Max-Planck-Institut fur Meteorologie.

REFERENCES

Arnold, L. (1974) Stochastic Differential Equation: Theory and Applications. Wiley-Interscience, New York, 228 p.

Bendat, J. S. and A. G. Piersol (1971) Random Data: Analysis and Measurement Procedures. Wiley-Interscience, New York, 407 p.

Benzi, R., G. Parisi, A. Sutera and A. Vulpiani (1982) Stochastic response in climatic change. Tellus, 34: 10-16.

Einstein, A. (1906) Zur Theorie der Brownschen Bewegung. Ann. Physik, IV(19): 371-381.

Fraedrich, K. (1978) Structural and stochastic analysis of a zero-dimensinal climate system. Quart. J. Roy. Met. Soc., 104: 416-474.

Fraedrich, K. (1979) Catastrophe and resilience of a zero-dimensional climate system with ice-albedo and greenhouse feedback. Quart. J. Roy. Met. Soc., 105: 147-168.

Frankignoul, C., and K. Hasselmann (1977) Stochastic climate models. Part II. Application to sea-surface temperature anomalies and thermocline variability. Tellus, 29: 289-305.

Ghil, M., M. Halem and R. Atlas (1979) Time continuous assimilation of remote-sounding data and its effect on weather forecasting. Mon. Wea. Rev.,107: 140-171.

Ghil, M., R. Balgovind and E. Kalnay-Rivas (1981) A stochastic-dynamic model for global atmospheric mass field statistics. NASA Technical Memorandum 82009.

Hannoschock, G. and C. Frankignoul (1985) Multivariate statistical analysis of a surface temperature anomaly experiment with the GISS general circulation model. J. Atmos. Sci., 42: 1430-1450.

Hasselmann, K. (1976) Stochastic climate models, Part I: Theory. Tellus, 28: 473-485.

Hasselmann, K. (1977) Application of the two-timing methods in statistical geophysics. J. Geophys., 43: 351-358.

Hasselmann, K. (1978) On the problems of multiple time scales in climate modeling. In Man's Impact on Climate: Proceedings of an International Conference, Berlin.

Hasselmann, K. (1979a) On the signal-to-noise problem in the analysis of the atmospheric response to external forcing. In Meteorology of Tropical Oceans, Royal Meteorological Society, London, p. 251-259.

Hasselmann, K. (1979b) Linear statistical models. Dyn. Atmos. Oceans, 3: 501-521.

Hasselmann, K., W. Munk, and G. MacDonald (1963) Bispectra of ocean waves. In Time Series Analysis (M. Rosenblatt, Ed.), John Wiley and Sons, New York, p. 125-139.

Herterich, K. and K. Hasselmann (1985) Extraction of mixed layer advection velocities, diffusion coefficients, feedback factors and atmospheric forcing parameters from the statistical analysis of North Pacific SST anomaly fields. Submitted to Phys. Oceanogr., 9: 815-846.

Jenkins, G. M. and D. G. Watts (1968) Spectral Analysis and Its Application. Holden-Day, 525 p.

Johnson, C. M., P. Lemke, and T. P. Barnett (1985) Linear prediction of sea ice anomalies. J. Geophys. Res., 90:5665-5675.

Lemke, P. (1977) Stochastic climate models. Part 3: Application to zonally-averaged energy models. Tellus, 29: 385-392.

Lemke, P. (1980) Application of the inverse modeling techniques to Arctic and Antarctic sea ice anomalies. Hamburger Geophysikalische Einzelschriften, A(49).

Lemke, P., E. W. Trinkl and K. Hasselmann (1980) Stochastic dynamic analysis of polar sea ice variability. J. Phys. Oceanogr., 10: 2100-2120.

Linnik, Yu. V. (1961) Methods of Least Squares and Principles of the Theory of Observations. Pergamon, London, 360 p.

Muller, D. (1985) Non-Gaussian properties of SST-anomalies. Ph.D. thesis, Hamburger Geophysikalische Einzelschriften, A(72).

Muller, P., D. J. Olbers and J. Willebrand (1978) The IWEX spectrum. J. Geophys. Res., 83: 479-500.

Nicolis, C. (1982) Stochastic aspects of climatic transitions: Response to a periodic forcing. Tellus, 34: 1-9.

Nicolis, C. and G. Nicolis (1981) Stochastic aspects of climate transitions: Additive fluctuations. Tellus, 33: 225-234.

Parkinson, C. L., and W. M. Washington (1979) A large-scale numerical model of sea ice. J. Geophys. Res., 84: 311-337.

Rayner, J. N. and D. A. Howarth (1979) Antarctic sea ice: 1972-1975. Geographical Review, 69: 202-223.

Reif, F. (1965) Fundamentals of Statistical and Thermal Physics. McGraw-Hill, New York.

Reynolds, R. W. (1978) Sea surface temperature anomalies in the North Pacific Ocean. Tellus, 30: 97-103.

Robock, A. (1978) Internally and externally caused climatic changes. J. Atmos. Sci., 35: 1112-1122.

Ruiz de Elvira, A. and P. Lemke (1982) A Langevin equation for stochastic climate models with periodic feedback and forcing variance. Tellus, 34: 313-320.

Saltzman, B. (1982) Stochastically-driven climatic fluctuations in the sea ice, ocean temperature, CO_2 feedback system. Tellus, 34: 97-112.

Sanderson, R. M. (1975) Changes in the area of Arctic sea ice 1966 to 1974. Meteor. Mag., 104: 313-323.

Schneider, S. H. and R. E. Dickinson (1974) Climate Modeling. Rev. Geophys. Space Phys., 12, 447-493.

Sutera, A. (1981) On stochastic perturbation and long-term climate behavior. Quart. J. Roy. Met. Soc., 107: 137-151.

Tatarski, V. I. (1961) Wave Propagation in a Turbulent Medium. McGraw-Hill, New York.

Treshnikov, A. F. (1967) Surface water circulation in the Antarctic Ocean. Soviet Antarctic Expedition, Vol. 5, p. 81-83 (English translation).

Walsh, J. E. and C. M. Johnson (1979a) An analysis of Arctic sea ice fluctuations, 1953-77. J. Phys. Oceanogr., 9: 580-591.

Walsh, J. E. and C. M. Johnson (1979b) Interannual atmospheric variability and associated fluctuations in Arctic sea ice extent. J. Geophys. Res., 84: 6915-6928.

Wang, M. C. and G. E. Uhlenbeck (1945) On the theory of the Brownian motion. Rev. Mod. Phys., 17: 323-342.

CHAPTER 14

THE SEASONAL ICE ZONE

Peter Wadhams

Scott Polar Research Institute
University of Cambridge
Cambridge, CB2 IER England

PART A. SPECIFIC ICE FORMS, REGIONAL AND SEASONAL

The seasonal sea ice zone is defined as the area between the minimum and maximum seasonal ice limits plus the region of the ice margin that is significantly affected by the ice-ocean boundary. This definition includes almost all of the Antarctic ice cover, the ice in the marginal seas around the Arctic and the ice cover overlying the shelves of the Arctic Ocean itself. Such a vast area includes a very large number of ice types. We shall consider three main divisions:

Fast Ice. This is ice which grows seaward from a coast and which stays in place throughout the winter, while breaking up, drifting away or melting in spring. Off a gently-sloping shore facing an unbounded sea, such as the Beaufort Sea coasts of Alaska and Canada, the fast ice is stabilized by the presence of grounded pressure ridges in its outermost parts, and therefore extends out to the draft limit of such ridges, about 20-30 m water depth. In constricted channels, such as the Canadian Arctic Archipelago, the fast ice fills the whole width of the channel, regardless of water depth which may be 200 m or more, and should really be described as "quasi-fast ice" since the stabilization is not complete, allowing some movement and deformation throughout the winter.

The Shear Zone. This ice forms when drifting pack ice moves against a coastal boundary which can be either fast ice or the land itself. The shearing and convergence of the pack's motion generates a band of highly deformed ice which has a higher density of ridging than the ice further out in the ocean. The most highly developed Arctic shear zone extends from the north coast of Alaska across the fringe of the Canadian Arctic Archipelago to northeast Greenland. Because the shear zone generally overlies shelf areas bearing mineral resources, a knowledge of its properties is of great practical importance; also, it represents the boundary condition for models of Arctic Ocean ice drift. Other transition conditions can occur if ice drift has an offshore trend, in which case it may leave zones of open water to seaward of the fast ice edge.

The Marginal Ice Zone (MIZ). The MIZ is that part of the ice cover which is close enough to the open ocean boundary to be affected by its presence. Therefore, all cold drift currents (e.g., East Greenland, Labrador) can be considered as MIZ, as well as ice lying within 150-200 km of the Antarctic ice edge. Ice in the MIZ has much greater freedom of movement than ice in the central polar pack, and a number of ice-ocean-atmosphere interaction phenomena occur which are peculiar to this area.

A.1 FAST ICE

A.1.1 Grounded Landfast Ice - Occurrence, Development, and Deformation

Ice which is fast both to the shore and to anchoring points on the seabed occurs in winter around the entire rim of the Arctic Ocean as well as in many sub-Arctic areas and around the coasts of Antarctica.

On steep shorelines, or in places where rapid currents or large tides prevent normal fast ice formation, there is usually an icefoot which is wide enough to facilitate sledge travel in winter. This, the first ice to form in autumn, is a belt of sea ice which adheres to the shore and which is unaffected by the tide. It is separated from the grounded landfast ice proper by one or more tidal cracks. Feyling-Hanssen (1953) gives a review of icefoot observations in Greenland and Vestspitsbergen, and McCann and Carlisle (1972) give a very thorough description of the icefoot on Devon Island. In Vestspitsbergen (Isfjord) the icefoot forms in November and breaks down in May on the open coast, but does not disappear from fjords until mid-July. On exposed and projecting headlands the icefoot does not form at all. After the icefoot breaks up, remnants still remain buried in the beach sediments, and thus the icefoot plays a major role in shore erosion (Nansen, 1922). The mechanism of initial formation, according to Bentham (1937), is that on the ebb tide the water left behind in pools and puddles on the shore freezes, and not all of it melts again on the following flood. Successive tides build up a continuous rim of ice along the shore. The icefoot cannot grow very far past the high-water mark before the bending stress at low tide causes it to fracture. Here occur the tidal crack(s) which separate the icefoot from the rest of the fast ice zone. On gently sloping shorelines, or in places where there is little tide (e.g., north coast of Alaska) the tidal crack occurs far out from the shore, and instead of speaking of the icefoot we should describe the innermost zone as bottomfast ice, ice which is completely grounded.

Beyond the icefoot, a consolidated ice sheet forms early in the winter because of:

i) the shallow water, which reduces the depth of convection necessary to cool the water column to the freezing point;
ii) the lower salinity caused by river discharge;
iii) the generally more quiescent sea surface conditions.

(i) is the major factor.

The genesis of an ice sheet growing in calm water has been described in earlier chapters. Its subsequent development on a gently sloping shoreline (Figure A.1) has been reviewed by Kovacs and Mellor (1974) and Reimnitz et al., (1977) for the case of the Beaufort Sea coast of Alaska, although the east Siberian shelf (Laptev Sea), the Mackenzie Delta area and parts of Svalbard have similar conditions. The fast ice grows seaward beyond the icefoot and tidal crack(s) until it encounters the moving pack. The events which happen at this stage largely determine the winter morphology of the fast ice zone. The wind field over the Beaufort Sea tends to drive the pack ice towards land. Each shoreward drive of the pack deforms the nascent fast ice sheet. If the drive is of short duration and limited intensity, only the outer part of the ice sheet will be affected, yielding an irregular and low-lying relief of rafted blocks. If the pressure is higher and more prolonged, the ice sheet will be deformed into ridges which may grow large enough to become grounded, producing scour marks in the sea bed in the process. The grounded ridge then provides an anchoring point for the fast ice sheet. The major grounded ridge systems form on pronounced linear shoals yearly. Other grounded ridges originate from the pack itself but become incorporated in the fast ice zone, as do grounded ice island fragments, driven ashore during autumn storms. More extensive and prolonged pressure may cause further deformation into a continuous field, of rubble-like hummocks or oriented pressure ridges. Such a hummock field may remain in the fast ice zone or may break away and drift with the pack. Eventually the fast ice sheet becomes thick enough and strong enough to resist further deformation, and takes on its permanent winter morphology.

The outer limit of the fast ice zone is roughly the inner draft limit of essentially continuous grounded pressure ridges, and off Alaska occurs at a water depth of about 18 m (Stringer, 1974), but with occasional bulges, possibly temporary, out to

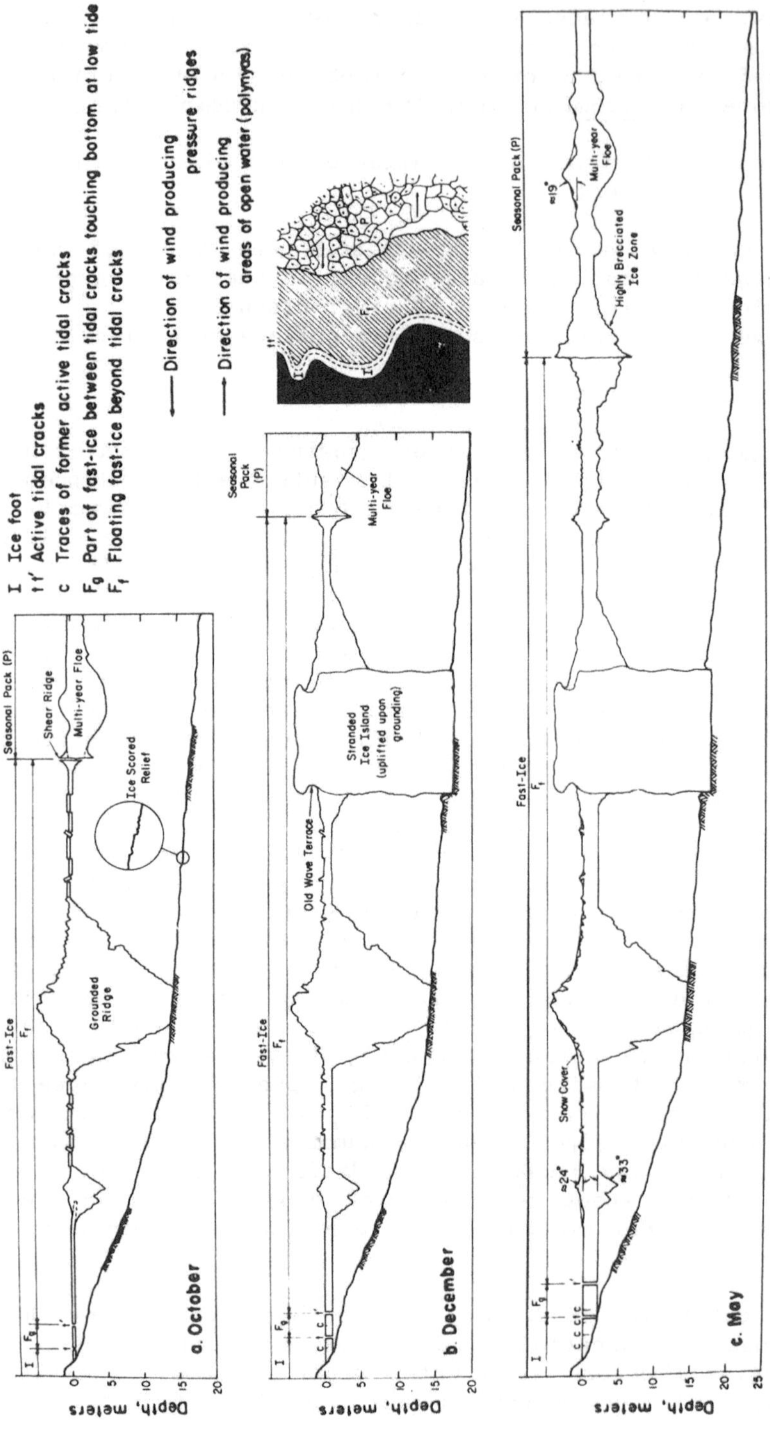

Fig. A.1. Three stages in the development of fast ice zone. (After Kovacs and Mellor, 1974.)

27 m. During winter parts of the shear zone become temporarily attached to the fast ice zone, since ridges in the shear zone are often deep enough to touch bottom and hold the surrounding ice stationary out to water depths of 35-40 m for days or weeks at a time.

Even when bound to anchoring points, the fast ice still makes small movements during the winter. Cooper (1974) measured displacements of the order of 10 m in Mackenzie Bay. Using a laser ranging system Tucker et al., (1980) observed similar movements off the Alaskan coast, but in the outer fast ice zone they measured displacements of up to 160 m. They ascribed the small progressive displacements to thermal expansion of the ice, and the larger motions (which were poorly correlated with the local wind) to the compressive effect of the offshore pack moving against the fast ice. The ice always returned to approximately its former position after a large excursion of this kind. The cause and magnitude of these motions are important because of the possible danger to drilling rigs operating through artificially thickened fast ice sheets. A further possible cause of fast ice motion is winter storm surges; by temporarily raising sea level they facilitate ice push onto beaches. Kovacs and Sodhi (1980) give a thorough review of ice pileup and rideup onto beaches, and the mechanical forces involved.

Landfast ice of this kind can sometimes exist on steeply shelving coasts - e.g., east Greenland and Antarctica - without grounded ridges to fortify it. It is, however, subject to periodic break-out at times of intense storm activity. Most of the fast ice along the east Greenland coast which survives the entire winter is fortified by outlying islands or islets, or by the geometry of the coast. Such ice is described in the next section.

A.1.2 Fast Ice In Constricted Channels

Ice which is fast only to the shore, and which is held in place through the winter by the restrictive geometry of the channels, bays, or coastline where it grows, is found throughout the Canadian Arctic Archipelago, between the Soviet Arctic islands and in the Kara Sea, in parts of Svalbard, and in much of Greenland and Antarctica.

A review of ice in the Canadian Arctic has been given by Lindsay (1969), who described the characteristics of ice in the Queen Elizabeth Islands (i.e., the islands north of Parry Channel). On average the new ice forms a consolidated sheet by 15 October, and by the end of May has reached a thickness of 2 m. It breaks up by about 15 July, after which the broken ice either partially melts and partially drifts out of the channels in a generally northwest-to-southeast direction under the prevailing wind and current, or else stays around to choke the channels until the next ice season begins. Thus, ice years can be divided into "good," "average," or "bad" from the point of view of navigation, depending on the behavior of the ice after breakup. 1962 was a good year and 1964 a bad, and the difference is shown in the percentage of sea area among the channels that remained ice-covered:

1962	June	87	July	48	August	32	September	65
1964		90		85		72		79

It is seldom that two good years occur in succession, since the removal of the winter ice in the first good summer permits Polar ice to enter the western channels, mainly via Gustav Adolf Sea and M'Clure Strait, and to move further east in the following summer to disrupt navigation in Parry Channel. Thus, there is a natural period of 2-4 years in ice conditions, on which are superimposed fluctuations due to climate and wind variations. Lindsay analyzed the natural history of good and bad years in various parts of the Queen Elizabeth Islands in a quest for an "indicator area" whose behavior in autumn could give a prediction of the ice conditions to be expected in the following summer. He suggested Jones Sound as such an indicator, since the patterns of freezeup and breakup occur in a number of distinct stages whose dates correlate well with the "goodness" or "badness" of the ice year.

Ice in this configuration has a considerably greater tendency to move than grounded landfast ice. Some measurements from the Canadian Arctic are again the best illustration. Wetzel et al., (1974) set up an acoustic transponder system in November on the ice of Belcher Channel, using a bottom reference. They found that the ice moved 100 m in the first 7 days and finally drifted bodily out of range of the transponder at a rate of 300 m in 12 minutes. While within range the ice moved in elliptical orbits at the inertial frequency, but with net translations of at least 100 m per orbit. It is probable that a very wide sheet such as this is able to suffer internal deformation, say along a line of weakness near the shore, driven by wind or current stress integrated over a large area.

In the central part of East Greenland the fast ice which fills the deep fjord systems forms in October and melts in place in June or July. The fjord ice is usually very smooth and undeformed, except when it has embedded within it old floes, brash and bergy bits that have remained in the fjord from the previous summer. The final stages of melting are accelerated by rainfall and by increased absorption of solar radiation in the meltwater ponds. The typical appearance of rotting fjord ice from the air is that of an army of giant spiders, the "legs" being meltwater channels draining into a thaw hole which is the dark "body." This pattern is also visible in spring off the mouths of rivers, where runoff flows out over the fast ice surface and drains away violently through cracks or seal breathing holes, a phenomenon known as ice strudel (Reimnitz and Bruder, 1972; Reimnitz et al., 1974).

The fast ice near the mouths of fjords and along the open coast (called "land ice" by Koch, 1945) also forms in October but often breaks up in the storms of November to January since it is more exposed than the fjord ice. It then grows again during March and April, reaching its maximum at the end of April. During May, weakened by radiation absorption, it begins to break up, but the time at which it actually drifts away is critically dependent on weather conditions. In a bad year, with easterly winds keeping the sky overcast, plugs of land ice may remain blocking the mouths of fjords as late as July, which causes considerable difficulty to navigation along the coast. Usually the mode of breakup includes a single major fracture line which is observed about a week before the ice to seaward of it breaks up and drifts away, the position of the line being a function of the geometry of the fjord. Once the land ice has broken out and the fjord ice has melted, pack ice from the East Greenland Current can enter the fjord, and icebergs calving from tidewater glaciers within the fjord can proceed more easily out to sea. As in the Canadian Arctic, the amount of ice of whatever source that remains in the fjord at the end of summer then has an important bearing on whether the following summer will be a good or bad season for navigation, since the remanent ice stays trapped in the fjord all winter.

The large latitudinal variation in fast ice thickness in east Greenland is shown in Figure A.2, drawn from data for 1964 given in Isforholdene i de Grønlandske Farvande (published annually by Danske Meteorologiske Institut, Charlottenlund). This figure shows a yearly cycle at four locations along the East Greenland coast. The ice-free season grows rapidly shorter with increasing latitude, which is not surprising given that the mean January temperature falls from -4°C at Angmagssalik to below -18°C in northeast Greenland. In Svalbard fjords are icebound from late November to early June, except on the north and northeast coasts where the ice season last from mid-October to late July. In the milder climate of Iceland fast ice is confined to the very heads of fjords in a normal year.

Some current research problems on fast ice are:

i) What is the contribution of Arctic Ocean ice to the mass balance of the Arctic Archipelago?
ii) Do we now have indicators which will predict the goodness or badness of the forthcoming navigation season from the dates of freezeup, the ice characteristics at freezeup or from meteorological indicators?

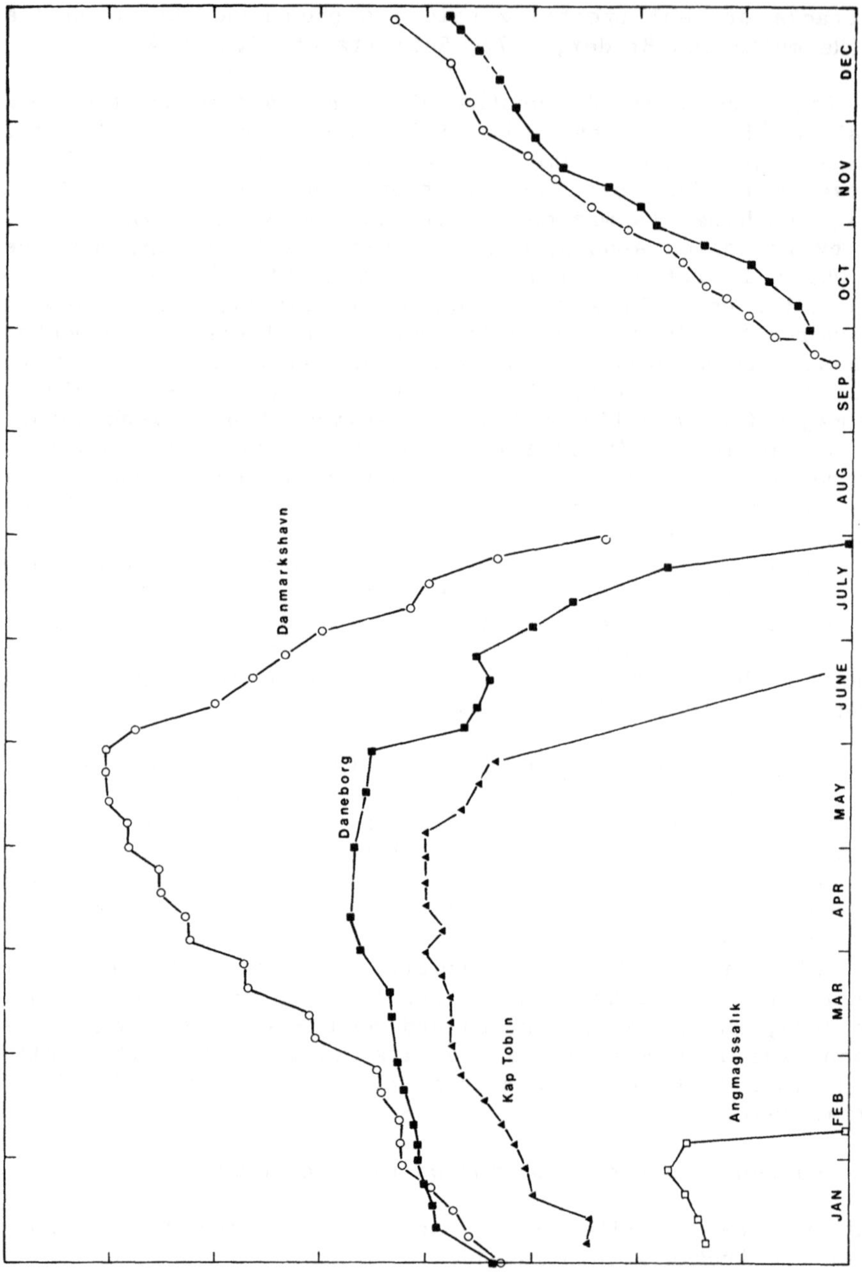

Fig. A.2. Fast ice thickness at four locations on the East Greenland coast during 1964. (After Wadhams, 1981b.)

iii) Is there a critical size of basin or channel above which fast ice never consolidates during the winter? In other words, are size and shape the major factors determining whether or not a basin grows a fast ice cover, or are the mean wind and current fields equally important? It is noteworthy that basins do not have to be completely closed to possess a fast ice cover; Lancaster Sound, for instance, is open to the east. In ice modeling terms, there is a certain length scale for a basin within which the internal ice stress is sufficient to arrest entirely the dynamics.

A.1.3 Sikussak-Fast Ice of Great Thickness

Undeformed ice floes of exceptional thickness are observed very occasionally in the Arctic Ocean. There are two well-documented observations. The first was by Cherepanov (1964), who studied the fabric of the floe on which the Soviet drifting station NP-6 was set up in 1955. The floe was of area 80 km^2, and of thickness 10-12 m with a crystal structure clearly indicative of slow congelation growth. The second observation was in October 1976 at the North Pole during a 3900 km under-ice passage by HMS SOVEREIGN. A floe of length 1 km and draft varying between 6.5 m and 12.5 m (mean 9.2 m) was profiled by upward-looking sonar (Walker and Wadhams, 1979). A.R. Milne (personal communication) has reported encountering a floe of 10-12 m thickness during an icebreaker voyage west of Prince Patrick Island in 1965 and there are a small number of other anecdotal reports, but it is clear that such floes are very rare in the drifting pack.

Thermodynamic models of sea ice growth (e.g., Maykut and Untersteiner, 1971) predict that undeformed floes in the Arctic Ocean will reach an equilibrium thickness of little more than 3 m, which is in reasonable accord with observations. Therefore we must look elsewhere for the origin of floes of exceptional thickness, and the natural conclusion is that they are formed as fast ice, and represent the result of many years of growth in place before a final break-out into the pack. This conclusion is supported by both theoretical and observational evidence.

On the theoretical side Walker and Wadhams (1979) ran the Maykut and Untersteiner model with modified energy inputs, setting the oceanic heat flux to zero and the annual snowfall to 1.0 m. This is what would be expected for the interior ramifications of a High Arctic fjord or bay. When the model was run for 65 years the ice thickness reached 12 m, and we estimated that the final equilibrium thickness would be about 20 m, to be reached in 200-300 years.

Direct observations of very thick fast ice come partly from the Canadian Arctic and partly from Greenland. Serson (1972) described a "plug" of multiyear fast ice in Nansen Sound,

Ellesmere Island, with a thickness of 10 m. From historical evidence he believed that it had existed since 1932, although it broke up and moved out of Nansen Sound some time between 1971 and present. Serson (1974) also reported a plug of sea ice in Sverdrup Channel (80°N, 98°W), about 6 m in thickness. Recent observations of Ward Hunt Ice Shelf have revealed very thick fast ice growing in bights on the edge of the shelf where ice islands have broken out (R. Pilkington, personal communication). In Greenland, according to Koch (1945), the sikussak is found chiefly in the north and northeast, i.e., the part with the most severe climate. He reports three regions of sikussak that were observed in May 1938 (Figure A.3), in Frederick E. Hyde Fjord, De Long Fjord and J.P. Koch Fjord, all in Peary Land. In 1933 Koch had observed that old sikussak held icebergs in place against the edges of the glaciers throughout North Greenland. He concluded that an exceptionally warm period between 1934 and 1937 allowed much of the sikussak to melt away or break out, leaving only three remnants. In August 1980 the present author observed ice of sikussak type in the fast

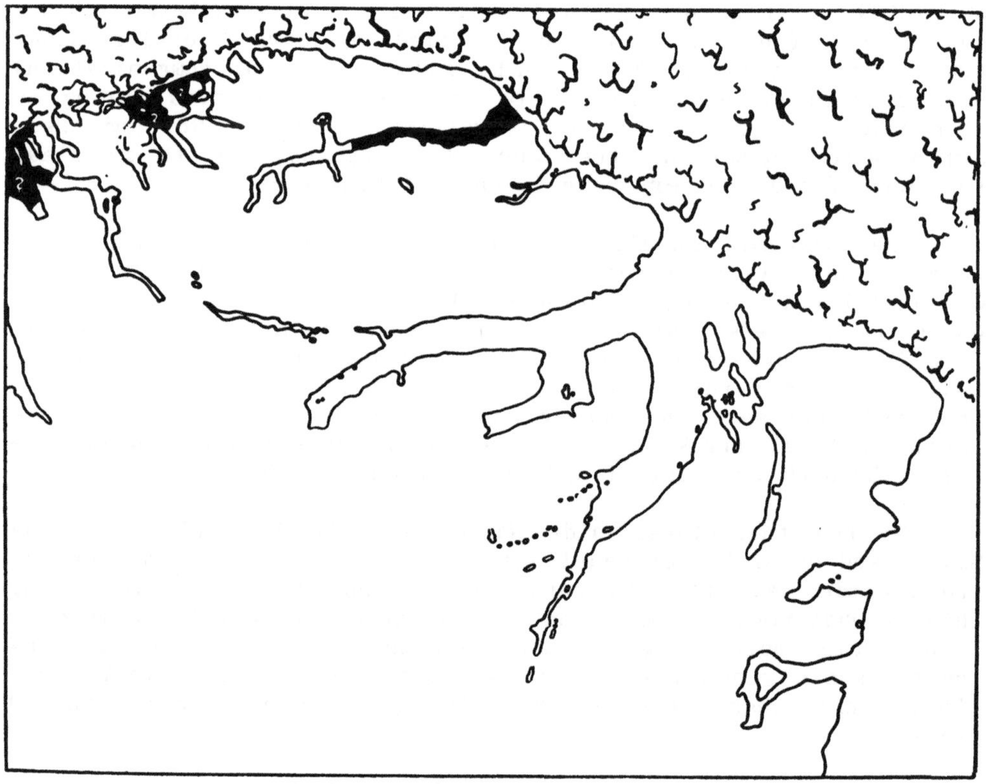

Fig. A.3. Distribution of sikussak (black) in north Greenland in May 1938. (After Koch, 1945.)

ice at the mouth of Danmarks and Independence Fjords at 82°24'N, 16°22'W (Figure A.4). The ice surface had a highly developed drainage system of streams, canyons and mesas, with at least 2 m of topographic relief, and a core drilled from one of the lowest points failed to find bottom at 6.1 m. The salinity record from the uppermost 4.6 m of the core (Figure A.5) shows evidence of many years of alternating growth and surface melting.

Sikussak probably exists in other areas of the Arctic where ice "shelves" have been reported, e.g., Severnaya Zemlya, Franz Josef Land and off the southern coast of Nordaustlandet and the east coast of Edgeøya in Svalbard. Koch's observation of sikussak fringing glacier snouts is reminiscent of observations of Koettlitz Glacier in Antarctica (Gow and Epstein, 1972) which reveal a transition to sea ice structure in its tongue.

It is clear that sikussak is the cause of very thick floes found in the pack ice, an important piece of supporting evidence being Cherepanov's (1964) observation that the ice of NP-6 showed strong horizontal c-axis orientation, characteristic of sea ice that has grown with a fixed orientation relative to a prevailing current (Weeks and Gow, 1978). The existence of sikussak in place, and the nature and causes of its occasional breakout, are

Fig. A.4. Sikussak at mouth of Danmarks Fjord, August 1980. (After Wadhams, 1981b).

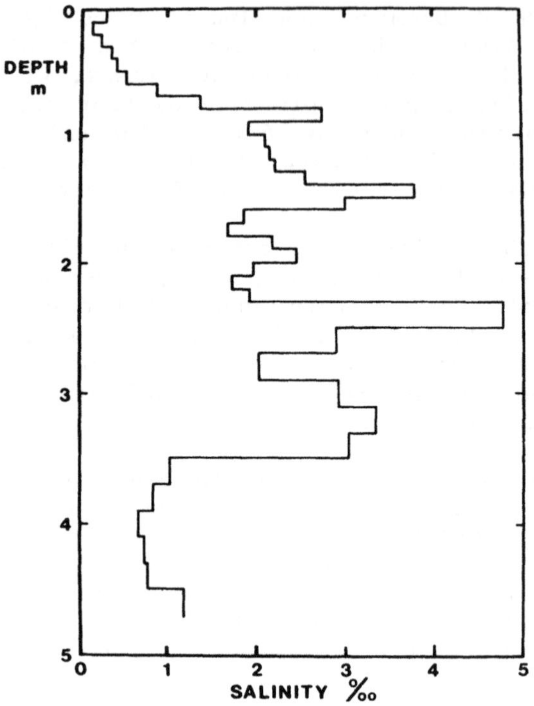

Fig. A.5. Salinity profile of uppermost 4.6 m of a core taken through sikussak in northeast Greenland. (After Wadhams, 1981b).

subjects worthy of future investigation, not least because sikussak would provide an excellent platform for studies of sea ice structure and thermodynamics under conditions of extremely slow growth.

A.2 THE SHEAR ZONE AND OTHER TRANSITIONAL PHENOMENA

A.2.1 The Shear Zone Off Alaska

The shear zone off the Beaufort Sea coasts of Alaska and northern Canada is a narrow region where in winter the polar pack, driven by the wind pattern over the Arctic Ocean, moves against the landfast ice and becomes heavily deformed by shear and convergence, generating a zone of pressure ridging which is heavier than any occurring in the central Arctic. In fact we shall define the shear zone somewhat arbitrarily as that region within which the ice characteristics differ significantly from those of the central Arctic. Harsh weather conditions limit detailed studies of winter shear zone processes. Three major questions in studying the zone are:

i) How wide is the zone, i.e., for what distance from the shore does the ice have properties distinctly different from ice in the central Arctic?
ii) How does the distribution of ice thickness and ridging compare with other parts of the Arctic, and why is it different?
iii) What is the field of motion of the shear zone, on short and long scales of time and distance? Does it approximate that of a boundary layer or is it more complex?

The shear zone begins at the edge of the fast ice, i.e., at about 20 m water depth, and extends out into the Arctic Ocean, possibly reaching across the continental slope into deep water before its properties merge with those of the central Arctic ice. The most accurate way of measuring ice thickness characteristics in the shear zone is from submarine sonar profiles, but this can be done only in the outer shear zone where the water depth exceeds 100 m. A profile was obtained in April 1976 by USS GURNARD beginning at the 100 m isobath north of Barter Island (Figure A.6). The track was analysed in 50 km sections (Figure A.7) by Wadhams and Horne (1980). We found that Sections 1 and 2 of the profile differed significantly from Sections 3-25, which formed a homogeneous whole. Sections 26 and 27 also differed slightly, presumably because the shear zone was again being approached. We shall look at Sections 1 and 2 in relation to 3-25, since these show the clearest progressive transformation of ice conditions across the shear zone:

		Section 1	Section 2	Central Beaufort Sea (3-25)
Position	N	70°36'-71°04'	71°04'-71°31'	71°31'-75°30'
	W	144°13'	144°13'	137°-151°
Ice draft	Mean	5.09 m	4.22 m	3.67 m
	Percent 0-1 m	4.5	0.8	3.5
	Percent > 5 m	41	25	17
Percent undeformed ice		35	44	56
Pressure	Number per km	5.1	3.1	1.4
ridges	Mean draft m	12.2	12.2	11.8
(> 9 m)	Maximum draft	23.1 m	28.8 m	31.1 m

The mean ice draft and percentage of thick ice (>5 m) both decline roughly linearly with increasing distance from the coast. The percentage of undeformed ice (defined as ice with a local gradient of less than 1 in 40) increases, but the most drastic change is in the frequency of pressure ridges, which is greater by a factor of 3-4 in Section 1 than in the central Arctic. We note, however,

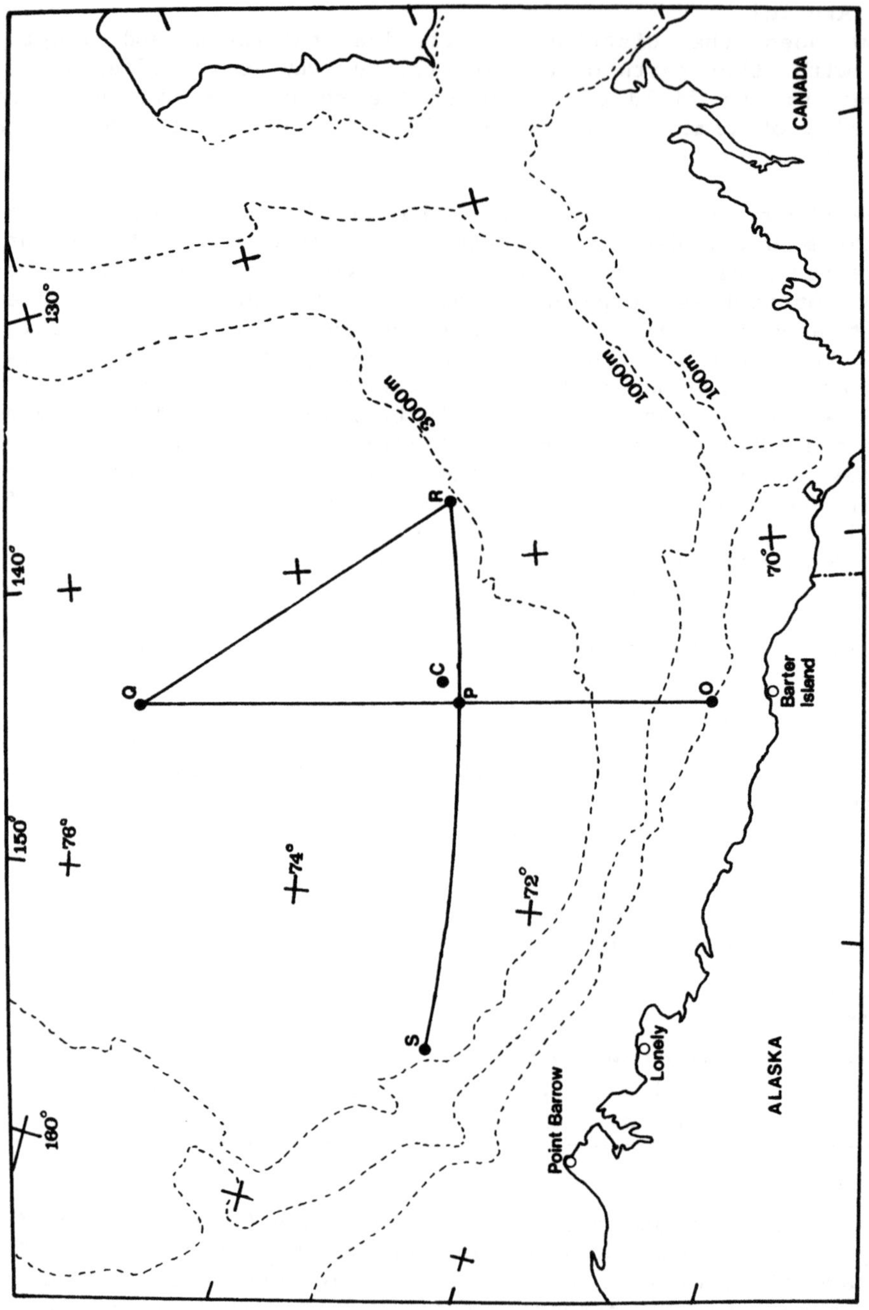

Fig. A.6. Route of U.S.S. Gurnard, 7–10 April 1976. (After Wadhams and Horne, 1980.)

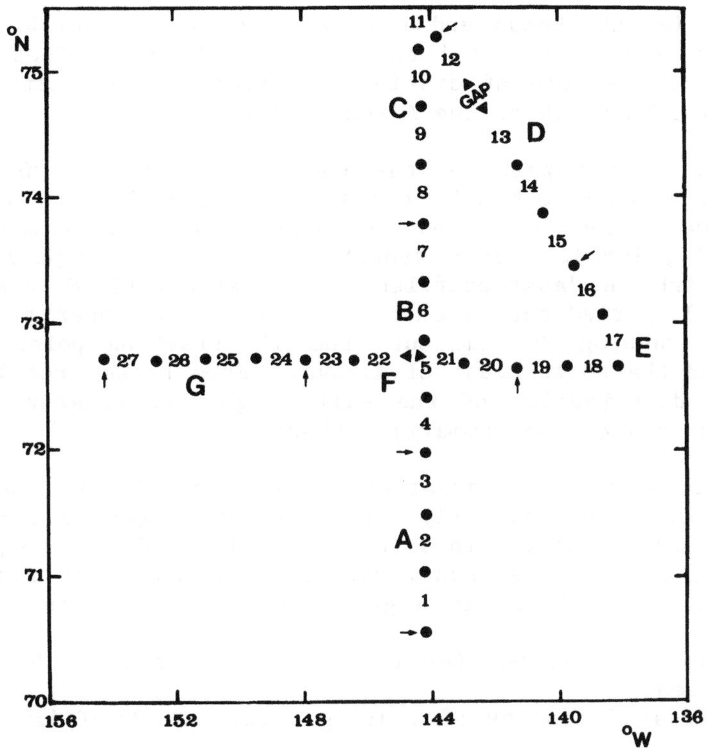

Fig. A.7. Division of track in 50 km sections. (After Wadhams and Horne, 1980).

that the mean draft of pressure ridges is not significantly higher in the shear zone. Similarly, the deepest keel in the track was found in the central Beaufort Sea, although this is explained by the greater track length. Another interesting phenomenon is the occurrence of thin ice, which is generally rare throughout the shear zone except in one small part of Section 1 where it reached 13%. Elsewhere the percentage is less than the average for the central Beaufort Sea, which raises the question of how the intense ridging in the shear zone can occur if there is little thin ice to provide the raw material. One could postulate that occasional intervals of divergence occur, which allow thin ice to grow for a short while between the longer intervals of compression, or else that the ridging takes place progressively as the shear zone ice moves downstream from the Arctic Archipelago towards the Beaufort Sea (i.e., that shear zone ice remains in the shear zone rather than being advected into the central Arctic), or finally that the forces of compression and shear are so great that ridges can be created out of ice sheets of normal thickness as well as thin ice. The question is unresolved. However the results clearly show that first, mean ice thickness and ridging intensity diminish steadily in the outer shear zone until they merge with Arctic Ocean conditions; second, the zone of distinctly different ice conditions extends some 160 km north of the Alaskan coast.

Over the inner part of the shear zone, in 20–100 m water depth, we have only a single dataset from the ice underside, a one-year time series from a bottom-mounted sonar in 30 m of water (Hoare et al., 1980). For a synoptic view we must depend on the results of airborne laser profiles. Weeks et al. (1980) and Tucker et al. (1979) carried out an extensive program of laser profiling, with tracks running 200 km out from six starting points spaced evenly around the north coast of Alaska. From these profiles they derived the distribution of the sail heights of pressure ridges. Their conclusions can be summarized thus:

i) There is a seasonal variation in mean sail height, being low (1.1-1.2 m) in summer and early winter, and high (up to 1.8 m, all relative to 0.98 m cutoff) in late winter, i.e., February-April.
ii) There is a similar seasonal variation in number of ridges per km, varying from 1.4–2.7 in August-December to 4.4 in February-April.
iii) There is slightly heavier ridging in the Beaufort Sea than in the Chukchi Sea.
iv) The largest frequency of ridging occurs at 20–60 km from the coast.

The 20-60 km distance range corresponds almost exactly to a 20–100 m depth range. Obviously in winter little ridging is seen inside the 20 m contour, since this is the edge of the fast ice zone within which the only ridges are relics of pack ice pressures during freezeup. The 20–100 m range corresponds to the zone of greatest shear and presumably greatest deformation. This can be seen when we consider the motion of ice in the shear zone.

The accepted picture of shear zone dynamics is that of a region in which the clockwise motion of the Beaufort Gyre is gradually dissipated by a number of mechanisms until the net longshore velocity at the boundary between the shear zone and the fast ice is quite small. Many of these mechanisms generate extensive ice deformation. The velocity shear across the shear zone is therefore similar to the velocity profile in a boundary layer and can be represented by some smooth function, possibly logarithmic. Unfortunately, observations have shown this simple picture to be not entirely true.

The long-term average ice drift in winter within the Beaufort Gyre has been given as 2 to 2.5 km/day (Coachman and Barnes, 1961), and an analysis of satellite images during a single month (April) gave 3 km/day (Campbell et al., 1975). Hibler et al.,

(1974) carried out a satellite image analysis of how this drift rate diminishes across the shear zone, and found that, rather than being continuous, the shear all took place in a narrow zone about 50 km wide near the fast ice boundary. Beyond this the Gyre behaves as a relatively cohesive mass. These results referred only to a short period in March 1973 and may not be representative. The authors' conclusion was that the shear zone cannot be modeled by the assumption of a single ice viscosity and a no-slip boundary condition. Instead one should either use two viscosities (a lower viscosity applying to the region nearest the fast ice boundary) or else a single viscosity with slip allowed at the boundary. The shear zone is thus seen as something approaching a slip plane rather than being a continuous zone, in terms of dynamics if not of ice characteristics.

Tucker et al. (1980) measured shear zone drift velocities directly using radar transponders mounted on floes and tracked from stations on Narwhal and Cross Islands. The furthest station was 37 km seaward of the islands, i.e. 60 km from the coast, and gave the surprising result that between November 1976 and March 1977 the net westward displacement was only 5 km. Stations closer to the coast showed even lower net drifts, although cyclic displacements of 0.5-1.5 km amplitude were frequent. The displacements showed only a slow, weak response to local winds and were therefore determined by stresses from the distant pack. The authors' conclusion was that throughout the winter the Gyre was effectively stationary, exerting a net compressive force on the shear zone which prevented motion. Only during early to midsummer is the Gyre free to move. Again, however, the conclusions refer to a single winter, which may not be representative of normal conditions. The inference that we draw from comparing Tucker et al. (1979) with Hibler et al. (1974) is that general conclusions about shear zone motion cannot be drawn from short-term observations, and that what is required is a long-term analysis of satellite imagery.

The nearest approach to such a study has been a review by Reimnitz et al. (1978), who carried out an intensive investigation of the shear zone off the north coast of Alaska which combined ground observations with the analysis of airborne and satellite imagery. Their conclusions were as follows:

i) There is a distinct development from the shear zone which they named the stamukhi zone (after Zubov, 1945), consisting of a belt of lineated ridges and hummocks, the deepest points of which are aground on shoals. This zone is initiated during early winter as the pack grinds against the fast ice zone, and is distinct from the ungrounded outer shear zone. The stamukhi zone develops further throughout the winter, constantly increasing its width as Gyre ice grinds and shears against its seaward edge, generating ridges which are deep enough to ground. Within the stamukhi zone further deformation continues to occur, with active bottom gouging taking place, and thus the stamukhi zone represents a vast energy sink.

ii) Shear occurs intermittently either within or to seaward of the stamukhi zone, and takes place within the small compass of a few hundred m, far less than even the 50 km stated by Hibler et al. (1974). The whole shear between the "stationary" ice, (i.e., the stamukhi zone, or grounded part of the inner "shear" zone) and the normal Gyre ice takes place in this very narrow slip plane, so there is no such thing as the continuous velocity shear normally postulated. The stamukhi zone extends out to about 40 m water depth. The existence of stamukhi beyond a fast ice zone that extends to 20 m is also characteristic of the East Siberian Sea (Zubov, 1945).

The picture of Reimnitz et al. (1978) is not entirely incompatible with the observations of Hibler et al., since Hibler's drift net employed a small number of identifiable floes on a 15 km grid. A large shear occurring in a narrow zone between two grid points would be interpreted as a shear over 15 km; this still does not explain the 50 km figure, however. The picture also fits the observations of Tucker et al. if we suppose that their transponder stations were in the stamukhi zone. The picture is an original one because of its implications for modeling. It suggests that the winter shear zone can be modeled in the following way:

(a) Intermittently there is perfect slip along a single line between the Beaufort Gyre ice and a grounded segment of the inner shear zone known as the stamukhi zone, in this case behaving as if it were fast ice.
(b) Between these periods of perfect slip there is no slip at all; the Gyre ice grinds against the stamukhi, and its energy is dissipated in bottom gouging and in the creation of new expanses of linear shear ridges and hummocks (probably by the grinding up of thick ice blocks rather than thin ice from leads).

The nearshore ice oscillates between these two modes which stand out in stark contrast to one another. In (a) there is no shear zone and perfect slip. In (b) these is no slip but heavy energy loss. In the case of (b), is there some kind of continuous velocity shear further out in the outer "shear zone"? In other words, do the times when the Gyre ice is grinding against the stamukhi always coincide with the times when the whole Gyre itself almost stops moving (as it did during the AIDJEX experiment)? If so, then there is never a shear zone in the accepted sense of the word. If not, then there presumably is a shear zone, i.e., a continuous velocity shear, but it lies further out to sea, in the outer part of what we have labeled as the shear zone in terms of its ice characteristics but which was originally named the offshore province by Weeks et al., (1971). This would allow ice deformation and ridging to occur out to the 160 km distance found from submarine observations, while still leaving the stamukhi-grinding as a significant energy-dissipating effect which should be included in modeling studies.

Two other interesting features of the Alaskan shear zone are the intermittent appearance of a wide flaw lead between the moving ice and the fast ice, which will be treated in Section A.2.4, and the phenomenon of bottom scouring. Linear scour marks have been reported throughout the Alaska and Mackenzie Delta offshore zones, as well as on the East Siberian Shelf, in the Great Lakes and off the Labrador coast, although the marks on the Labrador shelf and Grand Banks are due to icebergs. The scours are generally 0.5 to 1 m deep, although they can be up to 5.5 m. They are found in water as deep as 100 m, although the scour density is greatest at 10-30 m. Clearly the scouring is produced by grounding pressure ridges and ice islands, although at depths beyond 60 m the scours are probably relics of a period when sea level was lower. Repeated sidescan sonar surveys over identical areas show new annual scour marks at all water depths up to those (30-40 m) at which pressure ridges are commonly found.

A.2.2 The Shear Zone Off North Greenland and the Archipelago

North Greenland and north Ellesmere Island possess a shear zone which is wider and more heavily ridged even than the Alaskan shear zone. A single submarine profile has been obtained across it, obtained by HMS SOVEREIGN in 1976 (Wadhams, 1978b, 1980, 1981a). The profile (Figure A.8) ran almost parallel to the north Greenland coast for some 100 km, the closest approach being 80 km in Sections 5 and 6; point B lay 180 km off the northern tip of Ellesmere Island. Very heavily deformed ice was observed throughout Sections 1 to 10 and for some distance further into the Arctic Ocean, but with a change in characteristics after Section 5. Table A.1 shows the ice draft distribution for the 100 km Sections identified in Figure A.8 (after Wadhams, 1981a). Sections 1-5 have mean ice drafts of 5.1-5.8 m with large percentages of ice less than 0.5 m thick. Sections 6-10 have distinctly higher mean drafts of 6.0-7.5 m with much lower percentages of thin ice. This suggests different origins for the two ice types, and the inference is that Sections 1-5 are a shear zone for the Trans Polar Drift Stream (coming from the vicinity of the North Pole) while 6-10 are a shear zone for Beaufort Sea ice.

Numerical models of Arctic ice drift (e.g., Hibler, 1979) support this idea. For instance, in Hibler's simulation the entire offshore zone as far west as Ellesmere Island has a mean annual

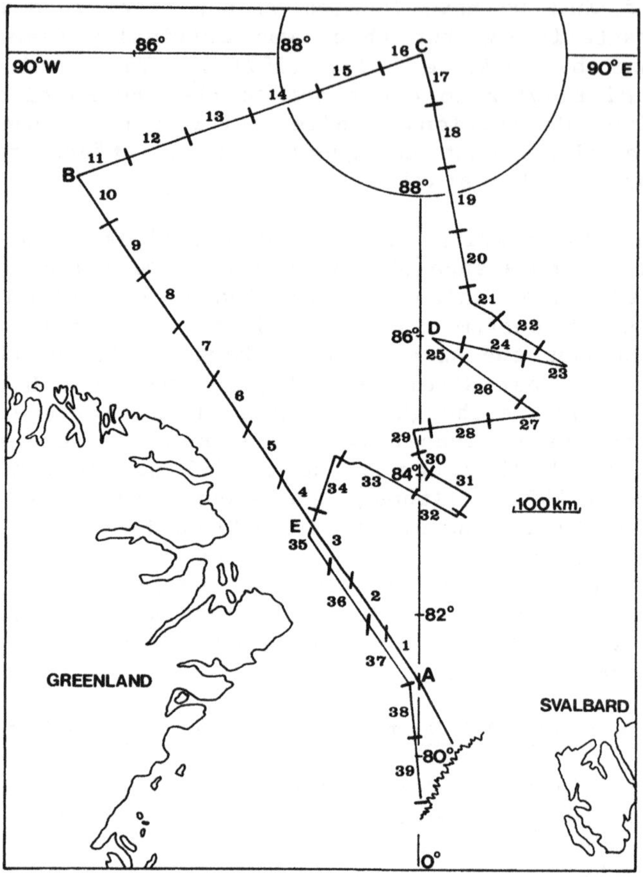

Fig. A.8. Track of HMS SOVEREIGN in the Arctic Ocean, with division into 100-km sections. (After Wadhams, 1981a).

ice flow out towards Fram Strait, with a velocity which increases monotonically from almost zero off Ellesmere Island to a large value at the Fram Strait entrance. This would cause a divergence which accounts for the large amounts of thin ice in Sections 1-5. Hibler's simulation also shows, however, that in some of the monthly averages the ice drift splits, with the western portion flowing westward towards the Beaufort Sea coasts. This would account for the differing characteristics of Sections 6-10. The occurrence of heavy ridging associated with little or no thin ice in Sections 6-10 is reminiscent of conditions in the Beaufort Sea offshore zone in April described in Section A.2.1, and has to be explained.

When we examine the track that runs northward along 70° in Figure A.8 we see that the transition from highly deformed ice to a mildly ridged central Arctic regime occurs quite suddenly in

Table A.1. Distribution of ice draft in 100-km sections of the under-ice profile shown in A.8. (After Wadhams, 1981a).

Section	Percentage of ice				Mean draft
	<0.5 m	0.5-2 m	2-5 m	>5 m	m
1	3.2	9	42	45	5.84
2	10.0	10	40	40	5.77
3	9.4	10	39	42	5.78
4	13.4	13	38	35	5.07
5	10.1	11	44	36	5.38
6	3.3	6	43	47	6.66
7	3.6	4	47	46	6.32
8	2.1	6	46	46	6.31
9	3.7	5	48	44	6.05
10	2.5	4	38	55	7.49
11	2.6	5	51	42	6.14
12	5.1	5	49	40	6.41
13	8.3	9	46	37	5.72
14	4.3	9	62	25	4.38
15-32 (average)	2.7	11	58	28	4.53
33	9.6	17	44	30	4.80
34	2.6	11	43	43	5.97
35	2.1	14	41	43	6.35
36	11.2	20	37	32	5.02
37	4.7	17	49	29	4.63
38	16.3	16	45	22	3.77
39	10.1	21	55	14	3.07

Section 13, which is over 400 km from the nearest coast. From Section 14 onwards to 33 the ice conditions are statistically homogeneous, and have been summarized by a single average in Table A.1; again this central Arctic regime resembles a homogeneous central Beaufort Sea regime found by Wadhams and Horne, although the mean ice draft is some 0.8 m greater. Sections 34-36 show the way in which the offshore zone is again encountered further to the southeast, while 38-39 show the onset of marginal ice zone conditions in Fram Strait. The sudden transition at Section 13 (87°N) was also seen in an earlier analysis of the data (Wadhams and Lowry, 1977), in which a sharp decline in pressure ridge frequency and mean draft occurs at 87°N. Wadhams (1978b) found that of 45 pressure ridge keels of draft greater than 30 m occurring in the entire 3750 km of the SOVEREIGN profile, no less than 39 were found in the 1200 km comprising the leg across the Greenland offshore zone from 81°N and the section at 70°W extending to 87°N (see Figure A.9). In fact the greatest concentration of all was in the western part of the offshore zone in the region 60°-70°W, 85°-86°N

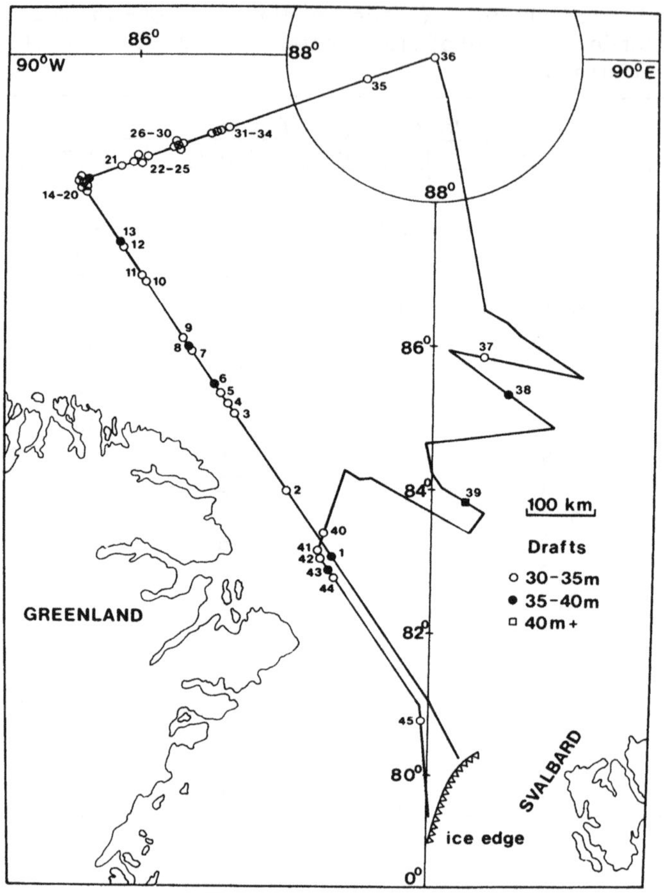

Fig. A.9. Keels exceeding 30 m in draft plotted on track of SOVEREIGN. (After Wadhams, 1978b).

(Figure A.10). In terms of total pressure ridge concentration, the offshore zone had an average of 2.8-4.3 keels per km deeper than 10 m while the central Arctic regime had concentrations of only 1.8-2.9 per km (Wadhams and Lowry, 1977). Corresponding rms keel drafts were 14.6-16.3 m and 13.5-14.3 m respectively.

The great width of the offshore zone around north Greenland (400 km) compared to that off Alaska (160 km) requires explanation, as does the sparseness of thin ice in the west and its higher occurrence in the east. A possible phenomenological scheme is as follows. In the vicinity of the North Pole, ice originating from both the Drift Stream and the Beaufort Gyre moves toward the north coast of Greenland and Ellesmere Island. On its way, somewhere in the region 87°-88°N, 60°-90°W, it begins to feel

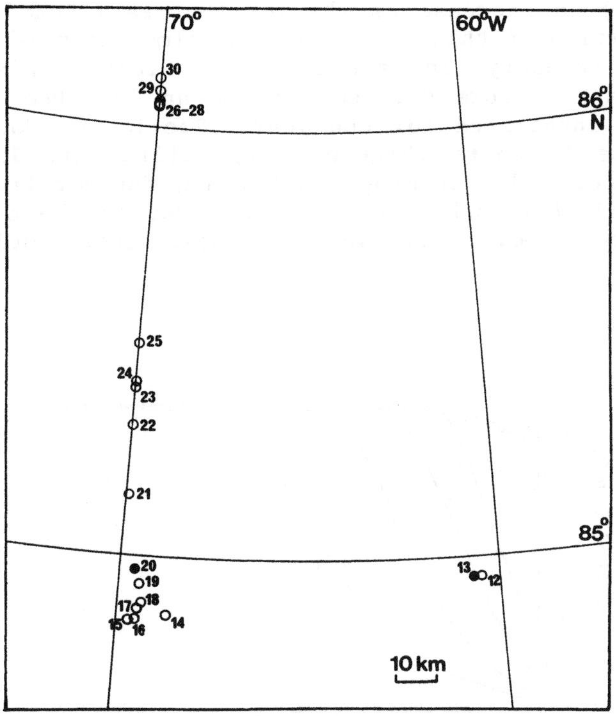

Fig. A.10. Deep keels in vicinity of 70°W. (After Wadhams, 1978b).

the influence of the coast, transmitted as internal stress through the ice from the downstream land boundary. At this time the icefield is fairly open, with a net divergence in the east-west direction. This allows extensive ridging to take place as soon as the distant influence of the land is felt, so that even at 87°N the ice is taking on the new character of "offshore zone ice." The heavy pressure ridging found in the shear zone off Ellesmere Island is thus mainly created far out in the Arctic Ocean. Near to shore there is little possibility of further ridge building, and here (Section 6-10) heavy ridging and the absence of thin ice can coexist. Fresh divergence takes place in the eastern part of the offshore zone because of the acceleration towards Fram Strait. To put it another way, the offshore zone begins 400 km from land off Ellesmere Island because the ice is moving towards land, and only 160 km from land off Alaska because it is moving parallel to land.

A quantitative version of this concept is seen in a simulation by Hibler (1979, Appendix C). Hibler's model for Arctic ice dynamics and thermodynamics couples the dynamics to the ice thickness distribution by allowing the ice interaction (which is assumed to be plastic) to increase as the ice becomes thicker or contains a lower proportion of thin ice. In this simulation of ice buildup at a coastal boundary (in fact the North Slope of Alaska) he imposed a constant onshore surface wind on an ice sheet of constant thickness and integrated for one year. The ice thickness steadily built up close to shore (Figure A.11) falling off linearly with onshore distance and extending further and further from the coast with time, while the onshore drift rate close to the coast rapidly decreased with time. The ice thickness distribution seen in

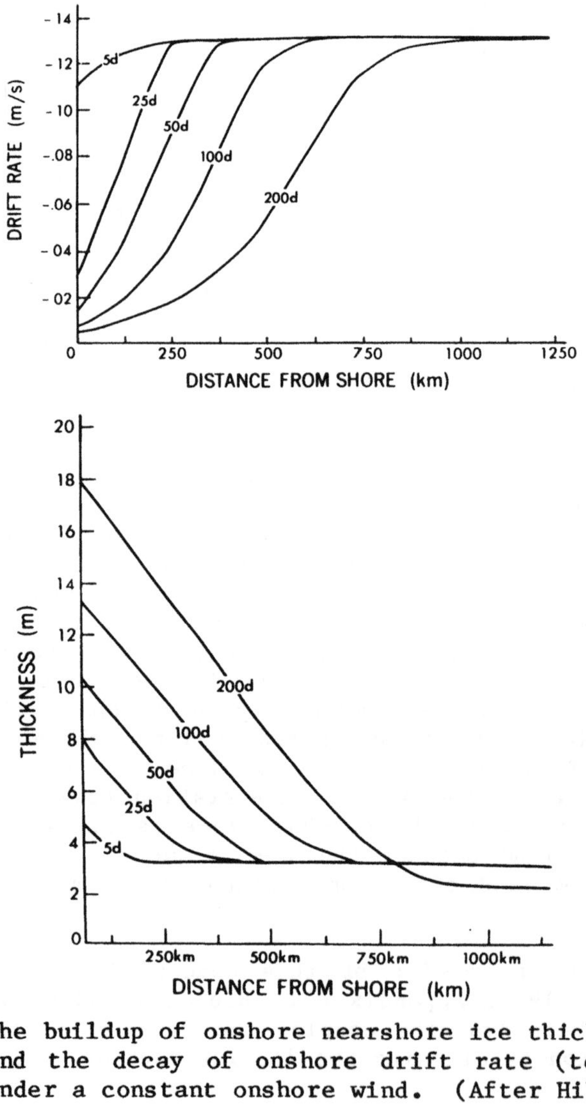

Fig. A.11. The buildup of onshore nearshore ice thickness (bottom) and the decay of onshore drift rate (top) with time under a constant onshore wind. (After Hibler, 1979).

Sections 10-14 of Table A.1 roughly corresponds with the situation 25-50 days into Hibler's simulation, i.e., the ridging is equivalent to an onshore wind of 8.5 m s^{-1} blowing steadily over an initially uniform 3.3 m ice sheet for 25-50 days. Clearly the comparison is not really valid because of the difference of location and the presence of an ice outflow, but the result is suggestive of the processes that are occurring.

Data from the shear zone of the northwest coasts of the Canadian Arctic Archipelago are more sparse, but two submarine transects have been made across it, in February 1960 by USS SARGO and in August 1960 by USS SEADRAGON. An analysis of the data (McLaren et al., 1984) reveals extremely heavy ridging off the southwest tip of Prince Patrick Island near the entrance to M'Clure Strait, where mean ice drafts of 6.4-7.8 m (averaged over 50 km of track) occur. It is noteworthy that the nearshore zone off Prince Patrick Island is thought to be the source of "multiyear hummock fields" (R.D. Hudson, personal communication), extensive coherent ice masses, several km in diameter and intensively ridged, which drift down into the southern Beaufort Sea.

A.2.3 Winter and Spring Polynyas

A feature of the pack ice along the East Greenland coast noted by Koch (1945), and also of the Antarctic, is the presence of open water in winter and spring at certain well-defined localities. The polynyas lie just beyond the edge of the fast ice; they may persist for days or weeks, and can open and close repeatedly. The areas of occurrence in Greenland are:

A. From just south of Nordostrundingen (81°30′N), extending southward as far as about 80°30′N.

B. Off the southern part of Store Koldeway, sometimes extending northward to Ile de France (76°-78°N). This is not found in winter, but often forms in early spring and persists until autumn.

C. From Bass Rock southward to Jackson Ø (73°30′-75°N).

D. In the mouth of Scoresby Sund.

Koch proposed a different explanation for each of these polynyas. Polynya A is caused by strong northwesterly winds, the sheltering effect of Nordostrundingen on the main ice stream coming from the Arctic Ocean via the Greenland offshore zone, and the presence of an eddy or local gyre in this lee. Polynya B is entirely due to the presence of an eddy or a branch of the East Greenland Current entering Dove Bugt and running up the west side of Store Koldeway. Polynya C is due purely to winds (northerly gales), and Polynya D to powerful tidal currents and an outflow of fresh water.

These explanations require closer attention. With the advent of earth sensing satellites it has become possible for the first time to observe the polar oceans synoptically. The first definitely identifiable winter polynyas seen in this way lie in the Antarctic. Knapp (1972), using Nimbus imagery, reported polynyas with areas up to 4500 km^2 around the edge of the fast ice. These open quickly to their maximum extent--within about 2 days--then persist for a much longer period, slowly getting smaller and/or acquiring an increased ice concentration, until they either disappear or else reopen in another rapid growth episode. Knapp adopted a mechanism proposed by Zubov (1945) for the creation of such polynyas. Zubov considered the ice drift and deformation caused by the passage of a cyclone, on the simple assumption that the ice drifts in the direction of the isobars (since the turning between the geostrophic wind and the surface wind is opposite and approximately equal to the turning between the surface wind and the ice drift). Depending on the relationship between the isobar spacing and the speed of translation of the depression center, a variety of complex ice patterns can result. If a coastline or fast ice edge with a pattern of promontories and inlets lies within the influence of the passing cyclone, ice will be piled up against some portions of the boundary and drawn away from others. A polynya created in this way persists (in the absence of a prevailing strong current) because once the cyclone passed the floes ceased their motion. Zubov applied his explanation to the polynya often found in Vilkitski Straits and to the Kara Sea, while Knapp found that the explanation fitted one of the Antarctic polynyas for which concurrent satellite imagery and meteorological data were available.

A simpler case of the Zubov mechanism is the straightforward effect of persistent offshore winds, which just blow the pack away from the coast. This tends to create a long narrow strip of open water along the coast, the so-called "land water" which is considered in Section 2.4. However, a purely local offshore wind will create a polynya rather than a narrow lead, and in Greenland these occur in the form of föhn or katabatic winds which blow off the ice cap and funnel down certain fjords with especial force. Koch (1945) himself names three fjords most susceptible to föhn winds as Ardencaple Fjord (near Shannon Ø), Bessels Fjord and Mørkefjord (just north of Shannon Ø). These winds, it seems, could account at least in part for Polynyas B and C. In a similar way northerly winds in the Bering Sea in winter account for the persistent coastal polynyas in Norton Sound and off St. Lawrence Island.

A more recent satellite study in the Antarctic, by Zwally and Gloersen (1977), has drawn attention to a class of larger polynya. A prominent feature shown up by passive microwave imagery is a 200 by 1000 km area of reduced sea ice concentration in the eastern Weddell Sea. This was explained as being an upwelling of warmer deep water caused by the Antarctic divergence with its associated transition in the prevailing wind direction from easterly in the south to westerly in the north. The feature existed from 1973-6 but has not been seen since (Zwally et al., 1983). No such large persistent area exists in the central Arctic, although a film compiled from a year's successive frames of microwave imagery (Ramseier et al., 1979) revealed what appeared to be a very large polynya which developed deep in the Arctic pack north of Franz Josef Land; recent analysis suggests that the polynya may be only a zone of somewhat reduced ice concentration. Oceanographic factors can also produce smaller-scale polynyas. One mechanism is that of a rapid surface current--for instance through a strait or past a cape--which advects growing ice away from an area with a fixed upstream boundary (such as a fast ice sheet) as quickly as it forms, leaving an open water area near the boundary. These are sometimes called "current holes" when they are purely local, and the mechanism on a larger scale has been suggested as a cause of the North Water in Smith Sound. This is probably the explanation for Polynya D, at the mouth of Scoresby Sund, where tidal currents are the scouring agent together with the branch of the East Greenland Current which enters Scoresby Sund and forms a small gyre within it.

Polynya A, as the largest and most northerly in east Greenland, is of especial interest, and its mechanism of formation must owe something to the geometry of its position in the lee of Nordostrundingen as well as to the local anticyclonic gyre which is known to exist close to shore. The precise mechanism by which this gyre keeps the area free of ice, and the degree to which offshore wind stress contributes, are unknown. It is possible that the offshore wind in such a semi-enclosed bay generates upwelling which contributes to maintaining the polynya. Polynya A is seen quite frequently in satellite imagery. It appears mainly in late spring (May-July) or, less frequently, in early winter (October-December) but never in mid-winter. It may persist for several months. Measurements in summer 1979 (Newton and Piper, 1981) have shown the presence of 0.5°C water at the bottom of a trough separating the Belgica from the Ob' Bank, extending west up to the coast just south of Nordostrundingen. This indicates one route for the water which keeps the polynya open.

Figure A.12, a Tiros-N image (very high resolution radiometer, VHRR, infra-red) obtained on 13 March 1979, shows small winter polynyas of diameter about 10-15 km in positions approximating to B and C of Koch. It is clear that, in terms of the prevailing surface current, each polynya is on the lee side of a protruding cape of fast ice around an island or headland. The four polynyas correspond to the following geographical features: (1) Kap Broer Ruys; (2) Clavering Ø; (3) Sabine Ø and Lille Pendulum Ø; (4) Shannon Ø.

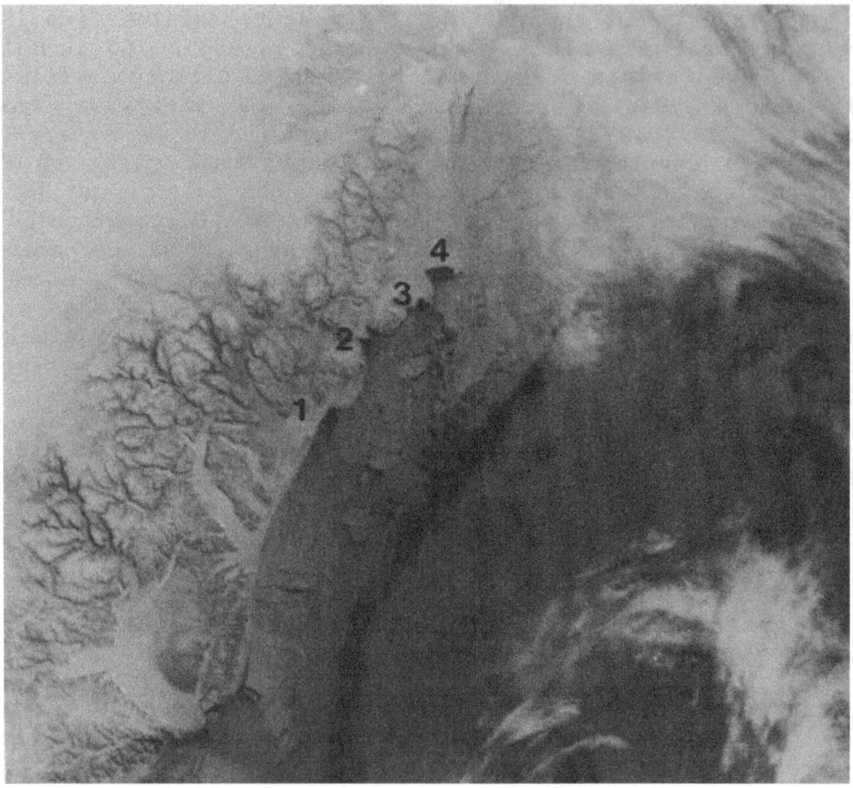

Fig. A.12. Tiros-N imagery of four winter polynyas off the East Greenland coast, March 13, 1979. (After Wadhams, 1981b).

A.2.4. The Landwater Phenomenon

Land water is the effect described by Koch (1945) whereby a continuous narrow strip of open water is formed against the fast ice edge by offshore winds. Sensible heat brought by wind-induced upwelling may add to the purely mechanical effect of the wind in keeping the landwater channel open, as may the effect of thaw water runoff from land in spring.

In east Greenland the land water typically opens in May, and the presence of such a channel of open water behind the offshore pack can be of great benefit to navigation. Koch reports that in July 1933 land water extended from Scoresby Sund to Hold With Hope, and on 1 August it reached from Bass Rock to north of Danmarkshavn. He mentions three cases of sealing ships which employed the land water to progress far north early in the season: the first reached Germania Havn from south of Scoresby Sund in August; the second reached Clavering Ø starting from 69°N in July; the third reached Young Sund from south of Scoresby Sund in July.

Modern data on land water are available from analyses of satellite images, and Table A.2 shows the extent and width of land water detected from 1968 to 1978. These results ignore the general breakup of ice in fjords from Scoresby Sund northwards, which occurs in July before the retreat of the offshore pack. It can be seen that land water is most often detected in May (once in April) and that the latest occurrence is in July. None is detected in mid-winter, indicating that the effect is due to the offshore winds which prevail once the summer high has established itself over the Greenland ice cap. The only data on duration come from 1968 (June 12-19) and 1969 (May 23-June 6); these are simply dates on which clear satellite images were received and do not necessarily reflect the lifetime of the land water. According to Koch it is of the order of days, however.

Table A.2. Reported occurrence of land water along the east Greenland coast, 1968-1978. (After Wadhams, 1981b, summarizing data from T. E. Vinje.)

Date of detection	Latitudinal extent	Width km	Nature	Within lead oktas
June 12 and 19, 1968	68.6°-75°N	30	Continuous except for Kap Brewster area	<1
May 23, 1969	69.1°-75.2°N	50	Continuous	1-3
June 6, 1969	69.1°-75.2°N	75-100	Expanding latterally towards pack edge	1-3
July 2, 1969	75.6°-79°N	15-25	Continuous	1-3
May 10, 1970	74.4°-82°N	15	Continuous except for break at Shannon Ø	
May, 1971	74.1°-75°N 76.6°-81.4°N	15	In two segments	<1
May, 1972	71.2°-72.9°N	50	Continuous	1-3
May, 1973	74.4°-78.6°N	15	Interrupted by Shannon Ø and Ile de France	<1
June, 1973	77.3°-79°N	25	Lies inside outer line of islands	<1
April, 1974	69.2°-74.2°N	10-15	Continuous	<1

Figure A.13(a) is a Tiros-N image (VHRR visual) showing land water which has opened up outside the fast ice along the entire length of the Liverpool Land coast from Scoresby Sund to Davy Sund to an average width of 10 km. The photograph was obtained on 17 July 1979; five days later (Figure A.13(b)) the land water has disappeared and the pack, although still containing some large polynyas, lies close up against the fast ice edge.

Land water, or long narrow leads similar to it, are by no means confined to east Greenland. Zubov (1945) calls them "wind leads" and reports their presence in spring across the entire coast of the East Siberian Sea (the "great Siberian lead" and the "Kolyma lead"). In the Beaufort Sea the opening of a wide lead (the flaw lead) in April-May across the edge of the fast ice from Amundsen Gulf to Point Barrow is a well-known phenomenon which precedes the breakup of the fast ice, and north of Ellesmere Island in spring is the "Big Lane" reported by Peary (1910). In each case the lead opens in the position of the winter shear zone, and is due to the greater mobility of the Arctic pack in spring which permits bodily offshore motion of the entire pack under wind stress.

The Beaufort Sea flaw lead is associated with a kindred phenomenon. This is the occasional observation of spatially periodic lead patterns on Landsat imagery. The flaw lead forms the southern limit of the pattern. The pattern is an intersecting mesh of parallel leads spaced some 100 km apart. Explanations include planetary waves and semibrittle failure (Marko and Thomson, 1975, 1976) but it is likely that the phenomenon is analogous to an effect seen in the Chukchi Sea where, if southward ice drift occurs, the land boundaries form a converging throttle. This produces a mechanical effect on the ice similar to the pattern of failure of wet sand in a hopper and Sodhi (1977) has presented satellite pictures of this effect in the Chukchi Sea and Amundsen Gulf.

Land water is also observed around Svalbard and Franz Josef Land. Vinje (1969) by interpreting satellite pictures discovered that in the 1967 season a lead of width 20-30 km opened in April around the entire north coast of Svalbard, curving round to enclose the east and south coasts of Nordaustlandet (Figure A.14). It closed in May, but a lead of similar width opened around the entire south coast of Vestspitsbergen and Edgeøya, stretching up the east coast to reach Nordaustlandet. By June both leads were open simultaneously and had expanded to participate in the general retreat of the ice edge. Studies of Svalbard ice in earlier years (e.g., Lunde, 1963) showed a more orderly retreat of the ice, but the opening of a narrow coastal lead around the uninhabited parts of Svalbard could easily have been missed because of the unavailability of satellite pictures.

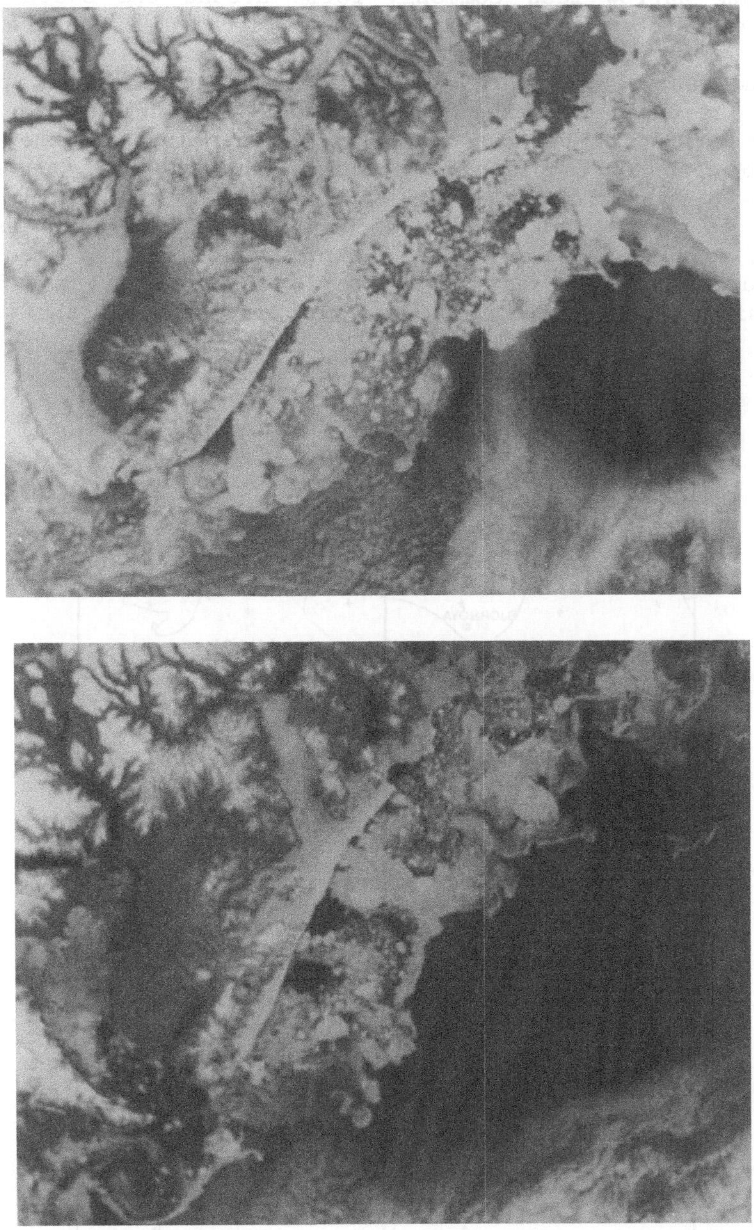

Fig. A.13. Tiros-N imagery of Greenland coast north of Scoresby Sund, showing (a) land water off the fast ice edge, 17 July 1979, and (b) closure of the land water, 22 July 1979. (After Wadhams, 1981b).

A long narrow lead commonly forms around the south side of Franz Josef Land as early as the end of March. As the ice in the northern Barents Sea disintegrates in spring, this lead gradually widens and reaches south to join the disintegration that proceeds from the ice edge, breaking through into the open water in July or August so that this becomes the first archipelago of the eastern Arctic to open up completely in summer. Sometimes another long narrow lead opens to the north of Franz Josef Land and reaches westward to join open water which is spreading northeast from Svalbard, so that in June-August a completely isolated area of sea ice is sometimes left east of Svalbard. It is clear that for such rapid disintegration to occur there can be little in the way of a

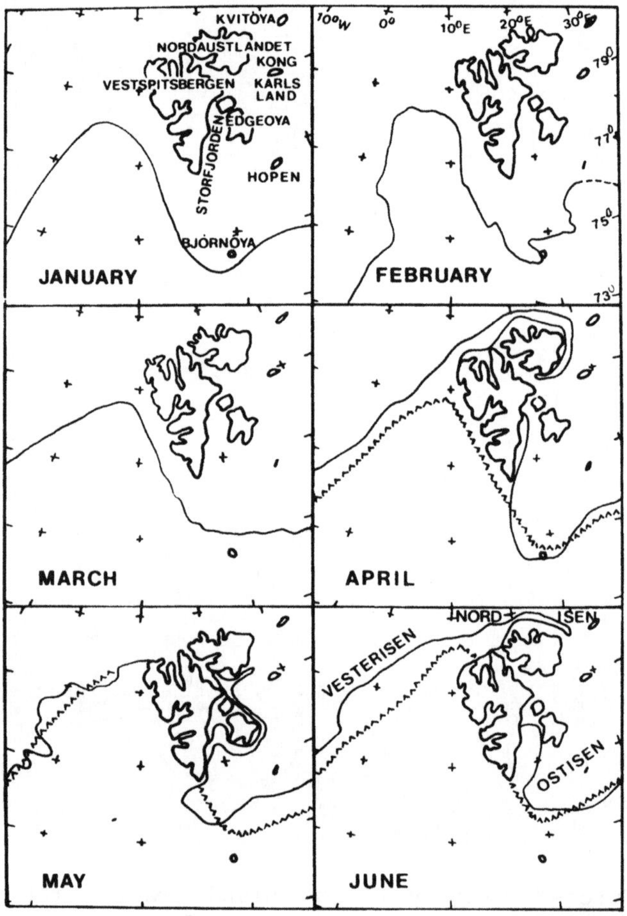

Fig. A.14. The development of land water around Svalbard in 1967. (After Vinje, 1969). The boundaries refer to the 3/10 pack ice limit at the end of the month. The monthly mean limit (1919-43) is shown hatched.

dominant drift stream of polar ice, and indeed Vinje and Steinbakke (1977) found that Nimbus-6 buoys placed in the area in August 1975 circulated for a long time with no strongly preferred vector. The earliest reported opening of a lead is 12 February; this exceptionally early date demonstrates how thin, mobile, and divergent the ice cover near Franz Josef Land must be in relation to the rest of the Arctic Ocean.

A.3 ICE FORMS IN THE SEASONAL PACK

This is a brief survey of ice types found in various seasonal drift ice zones of the world. Further material on the physical and mechanical properties of each type of ice will be found elsewhere in this volume.

A.3.1 The Antarctic

Almost all of the Antarctic is a seasonal ice zone. It has been estimated (Treshnikov, 1966) that the yearly variation in the area of the ice cover is 75% of the maximum (from 4 to 20 x 10^6 km^2) compared with only 25% for the Arctic. The main areas in which pack ice remain during the summer are the Weddell Sea and parts of the Ross and Bellingshausen Seas (Atlas Antarktiki, 1966; Mackintosh, 1972), and so it is only in these areas that significant amounts of multiyear ice can be found. The vast perimeter of the Antarctic winter ice cover means that a large proportion of the Antarctic ice can be considered as MIZ -- in fact, with the enormous fetch and strong winds of the Southern Ocean and the velocity shear across the Antarctic Polar Front, the Antarctic ice is the classic MIZ. Modeling of the Antarctic ice cover is only in its early stages (Kulakov et al., 1979; Parkinson and Washington, 1979; Hibler and Ackley, 1983) yet is of great climatic importance since during the southern winter the Antarctic sea ice covers a greater area than the Arctic ice does during the northern winter (Fletcher, 1969).

Studies of the climatic effects of Antarctic ice have been greatly assisted by the data provided by the electronically scanning microwave radiometer (ESMR) aboard the Nimbus-V satellite (Zwally and Gloersen, 1977). Not only does this provide data on the overall extent of the ice cover, but it also gives a measure of the percentage ice cover within the ice margin; using ESMR data NASA has produced an atlas of Antarctic ice conditions over the four-year period 1973-6 showing monthly averages (Zwally et al., 1983). Also using ESMR data Budd (1975) and Ackley and Keliher (1976) investigated ice-climate interactions. Their general conclusion, already well known, is that the Antarctic ice cover is everywhere divergent in its mean behavior, moving northward under the prevailing wind field and thus constantly opening new leads and polynyas. This divergence also affects the response of the ice to atmospheric lows. In winter a concentrated ice pack acts as a viscous material and responds to a low by converging. However, if the ice is divergent and loose it responds to a low by diverging further, allowing greater heat loss to the atmosphere which strengthens the low still more, providing a positive feedback mechanism to exacerbate atmospheric fluctuations. Another result of the net divergence of the Antarctic ice is the existence of a chain of semipermanent polynyas around the coast of the continent through the winter (Knapp, 1972) as mentioned in Section A.2.3.

Antarctic sea ice differs in many respects from Arctic ice. Buynitskiy (1967) and Lewis and Weeks (1971) reviewed the few studies made up to 1970, while recent more extensive work on the pack ice of the Weddell Sea and the fast ice of McMurdo Sound has been described by Gow et al. (1981). The structure of the ice usually includes a layer of "infiltration ice" near the surface, formed either from the flooding of the ice surface as it sinks under the weight of a snow cover, or else from an initial growth from the slushy snow-frazil mixture reported by Martin (1979). Such a layer is uncommon in Arctic ice because precipitation is much lower in the Arctic so that an ice cover seldom acquires enough snow to make it sink below sea level. Two other features once thought peculiar to Antarctic ice have also been observed in the Arctic. One is "underwater ice" or "anchor ice" which is scattered in plates through the normal columnar structure of the lower part of the ice sheet. This is believed to form in the water column just under the ice sheet and to float up against the growing ice-water interface. It has been observed in McMurdo Sound and is common along the Bering Sea coast of Alaska. An associated phenomenon is that of ice stalactites, long features enclosing the exit tubes of brine drainage channels. These were first discovered in McMurdo Sound (Dayton and Martin, 1971) and a theory of their formation was given by Martin (1974). They too have been observed in the Arctic (Lewis and Milne, 1977).

Perhaps the most surprising recent discovery (Gow et al., 1982) is that in the Weddell Sea pack a large proportion of the ice is frazil in origin (see Section B.4.2). More than 50% of the ice appears to be generated as frazil, which is found as layers in ice cores at all depths between layers of congelation ice. Frazil ice is derived from small crystals nucleated in the water column, and the assumption is that this frazil originates as a result of brine convection induced by rapid freezing of sea water in leads and polynyas. Such a scale of frazil production leads to exceptional floe thicknesses -- up to 5 m, growing in less than 2 years. This reinforces the existing view that the mean undeformed ice thickness in the Antarctic is much greater than in the Arctic, although the mean ice thickness may not be as great because of the much more extensive pressure ridging (up to 40% of ice volume) in the Arctic. Undeformed first year ice, for instance, is estimated to reach a mean thickness of 2.75-3.35 m in the Antarctic (Paige, 1966) compared to 2.0-2.2 m in the Arctic (Thorndike et al., 1975). Antarctic ice is also more saline than Arctic ice of similar thickness and age; first year floes average 4.6°/oo compared with 3°/oo in the Arctic while multiyear floes average 3.7°/oo compared with 2°/oo. Two factors appear to cause this: the low top surface summer melt, allowing floes to retain brine which would otherwise be flushed out by percolating meltwater, and the greater entrapment of brine by the spongy structure of the frazil ice. The fast ice sampled in McMurdo Sound was similar to Arctic fast ice in its tendency to c-axis orientation, but again differed in having a greater salinity.

The geometry of Antarctic ice is scarcely known. Floe size distribution, ice thickness distribution, pressure ridge depth and spacing distributions, and lead width and frequency distributions are all unknown. This means that modeling efforts are hampered by a lack of information suitable to estimate air and water drag coefficients. Qualitatively it is known that pressure ridging is less common and that the rate of production of new leads is greater than in the Arctic. However, no submarines have yet profiled the Antarctic ice cover and the first laser profiles of the top surface have only recently been obtained and are awaiting processing and publication (W. F. Weeks, personal communication).

A.3.2 The Greenland Sea

Seasonal variations in east Greenland ice extent have been dealt with elsewhere in this volume, so here we consider only the composition and characteristics of the ice cover. The composition of the East Greenland Current (EGC) pack was first treated thoroughly by Koch (1945), based on many years of personal shipborne and airborne observations. He defined various ice types such as "paleocrystic ice", "North Pole ice", and "Siberian ice" which he believed existed in a laminar distribution across the EGC width, ice originating from furthest west in the Arctic Basin being found on the western side of the current. In fact, the tracks of ice islands, data buoys, and floes tracked from satellites all show that the flow in the EGC is turbulent, with much churning and meandering of the ice superimposed on the overall southward drift. The result is that ice of all types is mixed irretrievably together, so that any description of the ice cover must be a statistical one.

Today we would say that the "paleocrystic ice" is partly very old ice from the Beaufort Gyre in the Canada Basin, which has spent many years circulating therein before crossing into the Trans Polar Drift Stream and exiting from the Arctic Ocean in the East Greenland Current, and partly ice which has undergone heavy deformation in the Greenland offshore zone (Section A.2.2) before entering

Fram Strait via Nordostrundingen. "North Pole ice" is ice of slightly more recent vintage, from both the Beaufort Gyre and the more distant parts of the Eurasian Basin (e.g., the northern part of the East Siberian Sea and Chukchi Sea). "Siberian ice" is younger first- and second-year ice formed in the nearshore and shelf areas of the Soviet Arctic, or in the region immediately north of Fram Strait. In winter, younger ice also exists in the pack, forming continuously in leads and polynyas as they open, so that the winter pack at the latitude of Denmark Strait contains a significant portion of ice which has formed south of Fram Strait. Recent work on the nature and characteristics of Greenland Sea ice has been reported by Overgaard et al., (1983) and reviewed by Wadhams (1981b).

A quantitative description of the thickness distribution of the ice cover can come only from the results of submarine sonar profiling, and here there are three sources of information. First, the ice in Denmark Strait was profiled by a U.S. Navy submarine in March 1971 (Kozo and Diachok, 1973; Kozo and Tucker, 1974; Westhall and Li, 1976). A sonar profile was obtained in the region 67-70°N, extending from the ice edge to 229 km inside the pack. It was found that marginal ice zone (MIZ) conditions extended as far as 166 km from the ice edge, beyond which lay a regime in which the ridge frequency and maximum ridge depth approached values found for Arctic Ocean pack ice. In this inner regime multiyear floes of 100-400 m diameter were common, mean ice draft was 2.44 m, and there were an average of 2.5 pressure ridge keels per km exceeding 6 m in draft. The spectral characteristics of the ice bottom were similar to those of Arctic Ocean ice. The maximum keel draft, however, was only 19.8 m, and toward the ice margin it fell even lower (12 m at 153 km penetration, 9 m further out), whereas profile sections of similar length in the Eurasian Basin have maximum drafts of 26-43 m. It is clear, then, that deep keels suffer considerable erosion during passage down the EGC, an erosion which is even more rapid near the ice edge.

In the outermost 166 km, or MIZ, the sonar profile revealed a sequence of four regimes. Regime 1 (0-58 km from edge) comprised floes ranging from pancake ice 3 m across to small floes less than 50 m across, an intimate mixture of first and multiyear ice. In Regime 2 (58-94 km) floe diameters increased slightly to 50-75 m and shallow ridges less than 5 m deep appeared. In Regime 3 (108-153 m) floe size increased to 100 m and ridge drafts and frequencies increased. In Regime 4 (153-166 m) there was a rather abrupt transition to larger floes, keels up to 12 m deep and an increased degree of long-wavelength variance in the bottom profile.

Many seasons of personal observation of the EGC MIZ have demonstrated to the author that such a sequence is typical, although the distances required become less with increasing latitude. Under an on-ice wind the ice edge can be highly compact, an abrupt line where open sea ends and 9-10/10 ice begins. The outermost floes are usually 20-50 m across and exist in a matrix of brash ice produced by abrasion between floes as they heave and surge under wave action; some of the brash is rafted over the floe edges. The floes are a mixture of first-year and multiyear ice, and may bear the remains on their upper surface of pressure ridge sails, although it appears that the corresponding keels have been eroded away. The floe size then steadily increases with increasing penetration, although at times a complex pattern of banding or vortices can be seen from the air, like a Swiss roll where small floes have been folded back into the inner pack. The complex behavior of the MIZ under wind and waves is explored more fully in Sections B and C of this chapter.

A second submarine profile from much further north was that of HMS SOVEREIGN discussed in Section A.2.2. The last three sections of that profile (37, 38, and 39 in Figure A.8 and Table A.1) cover the EGC where it begins to enter Fram Strait, with 39 describing marginal ice zone conditions at that latitude. It can be seen from Table A.1 that Section 37 resembles central Arctic conditions quite closely, while Section 38 shows a considerably diminished mean draft (3.77 m against 4.63 m) which is largely generated by a great increase in the percentage of thin ice (16.3% less than 0.5 m thick against 4.7%). In Section 39 (the MIZ) the further large decrease in mean draft (to 3.07 m) is largely due to a decrease in the percentage of ice thicker than 5 m. It is clear, then, that a great deal of bottom melting goes on in the outermost 100 km of the ice cover. The following table shows how the pressure ridge characteristics vary in these sections.

Table A.3

	Keels deeper than 9 m draft				
Section	Number n, km^{-1}	Mean draft $\bar{h}$, m	Maximum draft, m	Ridging intensity $n\bar{h}^2$	Leads per kilometer (width > 5 m)
North Pole	2.6	12.1	30.8	381	0.8
37	2.3	12.9	26.6	383	1.3
38	1.9	12.2	32.7	283	0.9
39	0.9	11.9	22.0	127	1.4

Section 37 has only a slightly lower frequency of keels than ice at the North Pole, while the "ridging intensity" (number per km multiplied by square of mean draft, proportional to volume of deformed ice per unit area of ice cover) is identical. In Sections 38 and 39 all of the ridging parameters are progressively reduced; note especially the low maximum keel draft of 22 m in Section 39, which comprises the outermost 135 km of the ice cover.

Even in Section 39 the mean ice draft is greater than that in the inner pack in Denmark Strait, while for Section 38, which offers a truer basis for comparison with respect to distance from the ice edge, the mean draft is 1 m higher than in Denmark Strait. Thus in reaching 70°N from 80°N the East Greenland ice diminishes in mean draft by some 35%. The relative contributions of divergence and bottom melting to this thinning are uncertain.

The third data source is a later cruise of HMS SOVEREIGN to Fram Strait and the western Eurasian Basin (Wadhams, 1983b). The cruise took place in April 1979, and ice conditions over the deep part of Fram Strait (79°–80°N, 0°–4°W) were milder than those of Section 39 in October 1976. Figure A.15 is a contour map of mean drafts assembled from 50 km averages and shows some very low values

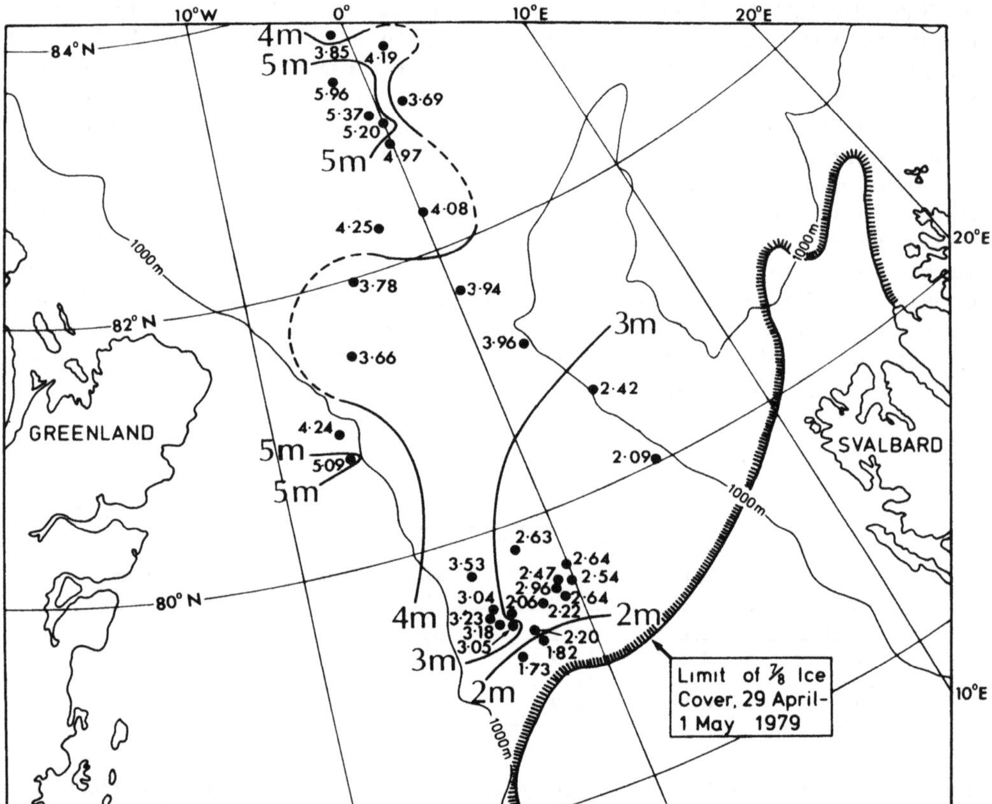

Fig. A.15. Contour map of mean drafts assembled from 50 km averages, showing some very low values near the ice edge.

near the ice edge. An analysis of undeformed ice in the near-edge zone (east of the 3 m contour in Fig. A.15) shows that it was mainly of first-year thickness (1.1-1.5 m) while in the Strait interior it was mainly multi-year (2.4-2.9 m). This suggests either that the extreme MIZ is really composed mainly of first-year ice, reviving the laminar flow theory of Koch, or that multi-year ice in this zone suffers exceptional thinning through bottom melt. Very thin multi-year ice was observed in this zone during the 1984 MIZEX experiment (A. J. Gow, personal communication). Pressure ridge frequencies and drafts agreed well with sections 38 and 39 from 1976. By combining the draft contours with drift rates it was possible to estimate the average ice volume flux through Fram Strait as 0.127 Sv.

A.3.3 The Barents Sea

The ice cover to the east and south of Svalbard in the Barents Sea differs in both composition and motion from the polar ice north of Svalbard that is moving west and southwest toward Fram Strait.

The Barents Sea is less than 500 m deep with a complex bottom topography which leads to a confused pattern of surface currents. According to Coachman and Aagaard (1974) there is little evidence of any strong net water transport through any of the channels which link the Barents Sea with the Kara Sea or the Arctic Ocean. This does not rule out wind-driven surface drifts, however. Near the coast of Svalbard itself there is a net southwestward drift known as the East Spitsbergen Current which transports ice around Sørkapp in winter. The subsequent progress, or survival, of this ice on the west coast of Svalbard is severely hampered by the powerful West Spitsbergen Current, which sweeps up the west Svalbard coast and keeps it clear of ice in many winters and all summers. From the drift tracks of Nimbus-6 buoys Loeng and Vinje (1979) estimate a mean southwestward velocity of 0.08-0.14 m s^{-1} for the ice in the East Spitsbergen Current. There are occasional flow reversals and also occasional episodes of rapid wind-driven ice flow, as in April 1976 (Vinje, 1977a) when ice drifted through the passage between Kvitøya and Nordaustlandet at such a speed that the passage was acting as a throttle; the ice accelerated from an initial speed of 0.15-0.20 m s^{-1} to a rate of 0.20-0.54 m s^{-1} within the strait, and the strait set up an arching fracture pattern in the approaching ice similar to that demonstrated by Sodhi (1977) for other restricted channels.

The major features of the seasonal ice variation have been given elsewhere in this volume. Mean conditions include a very narrow band of ice up the west Svalbard coast in midwinter and a "Whaler's Bay" (the bight of open water northwest of Svalbard) which is most strongly developed in summer. The largest seasonal variation occurs between Svalbard and Novaya Zemlya. The April ice limit develops a tongue near Bjørnøya; this was ascribed (Nansen, 1915) to the extremely slow circulation over the shallow bank area near the island. When the ice over the Barents Sea shelf breaks up in July it tends to drift onward in large coherent ice "massifs" similar to those found in the Bering Sea.

Ice composition and thickness in the Barents Sea are difficult to assess in the absence of unclassified submarine sonar profiles. From Project Birds Eye overflights, Wittmann and Schule (1966) estimated the winter composition as 58% multiyear ice, 23% thick winter ice, and 18% new ice (January-May) and the summer composition as 40% multiyear ice, 22% thick winter ice, and 38% new ice (August-October). The large fraction of multiyear ice implies a considerable ice transport into the Barents Sea from the Arctic Ocean, despite Coachman and Aagaard's (1974) estimate of negligible water transport. There is also considerable ridging; a Birds Eye flight in April 1967 (NAVOCEANO, 1967) counted an average of 18 ridges per kilometer in the Barents Sea (range 7-37) compared with 3-6 north of Svalbard. Evidently, the ice is very dynamic, the complex current system causing it to churn around and become heavily deformed. Another contribution to the heavy ridging could come from the fact that the "heavy winter ice" is probably fairly thin, less than 1 m thick, and therefore capable of undergoing rafting. Lunde (1963, 1965) found that the thickness reached by the fast ice off Hopen by the end of winter was only 0.7 m on average over the years 1946-1963, with an absolute maximum of 1.73 m in the unusual year of 1969 (Loeng and Vinje, 1979). Recent work on ice characteristics in the Barents Sea has been reported by Overgaard et al., (1983).

A.3.4 The Baltic Sea

The Baltic receives no surface ocean influx and is highly stratified in summer with a low-salinity surface layer which limits summer heat absorption. It therefore develops an ice cover in winter, but the extent and duration are very variable because of occasional warm air incursions from the Atlantic. In mild winters only the Gulf of Bothnia and eastern Gulf of Finland develop an ice cover, comprising less than 15% of the area of the Baltic, while in extreme winters (none since the 1940s) the entire Baltic is ice-covered. In a typical winter a cover extends for 5-7 months over the Gulf of Bothnia and Gulf of Finland down to 59°N, and the Gulf of Riga. Fast ice extends to the 10 m isobath and the rest of the ice cover is mainly wind-driven and therefore has a variable field of motion, except for a steady export from the Gulf of Finland to the Baltic proper. A typical ice thickness by the end of winter is 50 cm, with floe diameters of 0.2-2 km. Pressure ridges are usually loosely consolidated, with drafts of 3-15 m and sail heights of 0.5-2 m. A typical ridge density in the Gulf of

Bothnia in late winter is 7.5 km^{-1} with sail heights exceeding 30 cm (Leppäranta, 1981c). The ice structure, drift and deformation have been described and modeled by Leppäranta (1980, 1981ab).

A.3.5 Marginal Seas of the Soviet Union

The major aspects of ice distribution and drift in the Soviet Arctic are portrayed in Atlas Okeanov (1980).

The White Sea is similar to the Baltic in receiving little ocean influx and having a high stratification in summer. Its summer heat content is therefore soon lost. In a typical year the whole sea is ice-covered from January to May, but the ice season varies from 3 to 9 months (Doronin and Kheisin, 1977). Fast ice extends to the 10 m isobath and comprises 10% of the sea's area. The circulation is mainly wind-driven with tidal modification. There is a net ice drift out of the White Sea (Zubov, 1945), resulting in the occurrence of polynyas along the edge of the fast ice in the southern part of the sea, where rapid growth of young ice occurs.

The Kara Sea possesses a cyclonic gyral circulation on the east side of Novaya Zemlya and a northeastward trend of ice drift in the remainder of the Sea. There is a net transport of locally grown ice out towards Franz Josef Land, the Trans Polar Drift Stream and the eastern Barents Sea. Most of the Kara Sea ice is therefore first-year with a net divergence, although orientation of the c-axes of ice crystals over large areas (Cherepanov, 1971) suggests that the whole icefield tends to move as a body in winter and is resistant to relative rotation. In an average year the Kara Sea is icebound from October to late June. The fast ice reaches north from the mainland coast to encompass Severnaya Zemlya. The mechanism of breakup is similar to the Canadian Beaufort Sea in that a wide flaw lead opens between the fast ice and the offshore pack. It widens quickly under the effects of ice dynamics (northward drift) and thermodynamics (radiation absorption in the open water), until the fast ice itself breaks up. In August and September most of the Kara Sea is ice-free; the remaining ice tends to concentrate in a number of massifs (as in the Bering Sea in spring), of which the two major ones lie northeast of Dickson Island and between Novaya Zemlya and the Yamal Peninsula (Zubov, 1945).

The Laptev Sea also experiences a large net flux of ice north- and northeastwards out into the Arctic Ocean. During spring breakup much of this ice is river ice from the Lena and other Siberian rivers, or ice of low salinity originating as fast ice from the very brackish coastal waters. These distinctive ice types are recognizable in the ice which emerges into the East

Greenland Current. Breakup of the fast ice in the Lena Delta occurs in July, and by September the ice edge usually lies along a line from Vilkitskii Straits to the northern end of the New Siberian Islands. By October the sea is icebound again. During summer an ice massif--the Taimyr Massif--tends to remain off the east coast of Severnaya Zemlya and may move to block the shipping channel along the mainland coast. Zubov (1945) showed that the export of ice from the Kara Sea results in a lower mean ice thickness by the end of winter (1.0 m) than would be expected by pure thermodynamics (1.8 m).

The East Siberian Sea partakes of the general circulation of the Arctic Ocean, i.e., a strong net drift to the west and northwest of ice which has come from the Beaufort Sea and which will continue either within the Beaufort Gyre or into the Trans Polar Drift Stream. Local variations occur within this scheme, e.g., an eastward drift which sometimes develops along the coast and which transported the beset CHELYUSKIN during 1933-34 (Zubov, 1945). The East Siberian shelf is wide and gently sloping, and the fast ice extends far out, encompassing the northern fringes of the New Siberian Islands. It breaks up in August, creating a narrow band of open water which lasts until October. In winter the ice drift sometimes has an offshore tendency, creating a flaw lead, but more usually it creates a classic shear zone along the edge of the fast ice, similar to the Beaufort Sea. It was here that "stamukhi" were first observed and named (see Section A.2.1). Zubov (1945) noted that grounded pressure ridges in the region have drafts of 25 m or more, and submarine surveys have revealed scour marks on the seabed similar to those observed in the Beaufort Sea (A. S. McLaren, personal communication). The East Siberian Sea has the harshest thermal regime of the Soviet Arctic; undeformed first-year ice reaches a thickness of more than 1.8 m (Zubov 1945).

A.3.6 The Bering and Chukchi Seas

The ice found in the Bering Sea falls into two categories: ice which has come south through Bering Strait; and ice which is locally generated. Most of the ice is locally generated, since the net flow through Bering Strait is northward (amounting to 1-2 Sv) so that Chukchi Sea ice can pass southward only under favorable wind conditions. The Chukchi Sea itself is a seasonal ice zone in that in summer the ice retreats to the shelf break north of Bering Strait (Paquette and Bourke, 1981), but the ice in it resembles Arctic Ocean ice closely, differing only in having a greater proportion of first-year ice in the mix.

The locally generated ice is all first-year and, when undeformed, is quite thin, averaging 20-40 cm in thickness. Its properties have been reviewed by Bauer and Martin (1980), Pease (1980), McNutt (1979), Campbell et al. (1975) and Muench and Ahlnas

(1976). Much of the ice forms in the northern Bering Sea, especially in coastal polynyas in Norton Sound and to the south of St. Lawrence Island. It is then moved to the southwest by the strong northeast winds which accompany the anticyclonic circulation of the Siberian high-pressure system (Pease et al., 1982). The prevailing wind maintains the polynyas in existence as "factories" where rapid growth of new ice occurs to replace ice which is continuously advected away. The main pack of the Bering Sea is then a "conveyor belt" for the ice, transporting it out to the ice edge where it is destroyed by melting. Thus the ice edge, even in midwinter, is a melting ice edge where ice is continuously being lost. The winter ice edge maintains a constant position over the shelf break in the southeast Bering Sea from March until mid-May when it breaks up and rapidly retreats, leaving behind coherent islands of drifting ice called "ice massifs" which move about until they melt.

In the main pack of the Bering Sea the floe size is very large, but as the floes approach the ice edge their morphology changes because of the influence of the open sea manifested through ocean wave penetration. The mechanisms of interaction will be discussed in Section B, but the result of the wave action is that three ice zones can be distinguished in the Bering Sea MIZ (Bauer and Martin, 1980). These are:

1. The interior ice, comprising large, flat, relatively undeformed floes of order km in diameter and thickness 0.2-0.3 m. Ocean swell can be observed in this zone, but its amplitude is so low that it propagates through the ice as flexural-gravity waves (Section B.2.1) without fracturing the ice. The morphology of this zone is similar to that of the rest of the Bering Sea.

2. A transition zone, about 5 km wide, beginning 5-10 km from the ice edge. Here the floes are characterized by a rectangular pattern, the rectangles being approximately 20 m by 40 m with their long axes perpendicular to the direction of wave propagation. The floes are typically 0.3-0.6 m thick, consisting of two or three rafted layers. Here the incident wave energy is sufficient to fracture the ice in bending (Section B.8) and to cause a certain amount of rafting of the fractured pieces, due to the differential phases of the surge response.

3. The edge zone, the outermost 5-10 km. Here the ice meets the full force of the ocean wave spectrum, including the short steep waves which are rapidly attenuated and which therefore do not reach the transition zone. The ice consists of small broken floes 10-20 m in diameter. Each floe is heavily rafted and ridged, with sail heights (due to pieces rafted on top) of up to 1 m and keel depths of 2-4 m.

The three zones are illustrated schematically in Figure A.16.

The edge zone itself, especially under an off-ice wind, is likely to break up into long, linear bands of ice, typically 10 km in length and 1 km in width. The mechanism of formation is discussed in Section C.3.3. Since these bands advect downwind faster than the ice pack as a whole, they move into warmer water and melt. This is a very efficient way of melting ice at the edge, since the small rafted floes have a large surface/volume ratio, permitting enhanced melt rate, and since the turbulence associated with wave action around them (and waves breaking over them) also causes accelerated melting as discussed in Section B.6. Thus wave action increases the efficiency of the ice edge as a melting zone.

A.3.7 The Sea of Okhotsk

Sea ice conditions in the Sea of Okhotsk have been described by Watanabe (1972). Sea ice formation is promoted by low-salinity surface water which spreads from the northwestern part of the sea (where it receives outflow from the Amur River) towards the center. It is in the northwest that the first sea ice forms in November, and the ice spreads towards the center of the Sea, reaching the northern tip of Sakhalin Island at the beginning of December. It continues eastward off the east coast of Sakhalin and reaches the southern tip of the island by the beginning of January. By mid-January the drift ice, now a few tens of cm thick, reaches the northeast coast of Hokkaido to meet fast ice growing from the shore. By the end of January the whole coastline at 44-45°N is icebound, with heavy pressure ridging between the fast ice

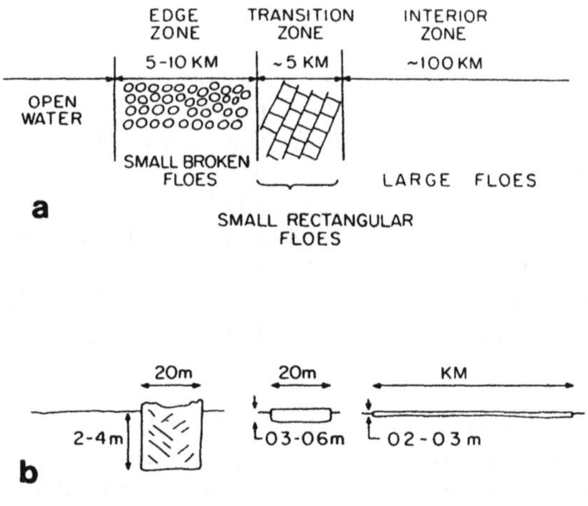

Fig. A.16. A schematic diagram of the three kinds of ice which occur near the ice edge in the Bering Sea, proceeding inward from open water. (After V. A. Squire).

and the pack. Maximum ice extent occurs in late March when the ice-free region, only about 25% of the sea, extends from 46°N to 55°N north of the northern Kurile Islands. Because of the net southward drift, the mean ice thickness off Hokkaido is greater than in the northern part of the sea. Some drift-ice runs through channels between the southern Kurile Islands into the Pacific and drifts south with the Oyashio Current as far as 41°N off southeast Hokkaido.

In spring the southward drift ceases, and under south and southwest winds a rapid melt and retreat begins. By April the southwest boundary of the icefield lying west of 150°E has retreated 200 km from the northeast coast of Hokkaido. The southern ice limit, located at about 45°N-146°E in mid-April, retreats to 48°N by mid-May. The last remnants of ice cover, which cling to the east coast of Sakhalin north of 52°N until mid-June, retreat to the western corner of the sea around Bolshoy Shantar Island, where they melt by late July.

Thus the Sea of Okhotsk shares some features with the Bering Sea in that the ice follows a wind-driven "conveyor belt" from source area to ablation area, except that here the coast of Hokkaido intervenes before a natural ice edge forms. The ice drift off Hokkaido has been observed by coastal radar (Tabata, 1972), which typically shows a southeastward drift under the prevailing NW wind. Tidal ice drift is also a non-negligible factor since the Sea is shallow. Tidal drift tends to build up coastal pressure ridging in the extreme west (55-57°N, west of 140°E) and northeast (east of 150°E) of the northern Sea of Okhotsk (Doronin and Kheisin, 1977).

A.3.8 The Canadian Arctic Archipelago

The ice of the Canadian Archipelago has been extensively described elsewhere in this chapter and volume. The fast ice and shear zone of the Beaufort Sea itself have been described in Sections A.1.1 and A.2.1, and the nature and variability of the brief open water season have been discussed by Milne (1978). Normally the breakup proceeds westward from Amundsen Gulf, from which a system of leads radiates into the southern Beaufort Sea. The open water can extend up to 200 km from the coast and the open water season may vary from a maximum of 120 days to a minimum of zero. The west coast of the Archipelago proper is flanked by a shear zone which is thought to be almost as heavily ridged as that of north Greenland (Section A.2.2). Recent observations from Prince Patrick Island northward suggest that massive hummock fields can be generated by a similar process to the stamukhi of Alaska (Section A.2.1) and that occasionally, when the shorefast ice fails to break up in summer, these can develop into multiyear hummock fields which, when they finally break out into the Beaufort Sea, pose great dangers to offshore structures.

The ice cover within the channels of the Archipelago is fast, and has been described by Lindsay (1969) and summarized in Section A.1.2. The thickness attained during a winter varies from 1.7 m at Coral Harbour (north end of Hudson Bay) to 2.3 m at Eureka (Ellesmere Island), and Bilello (1980) has described the way in which the ice decays during May-July under solar radiation until it finally breaks up. Breakup is not always complete, allowing multiyear fast ice to develop in certain areas and sometimes reach great thicknesses (Section A.1.3). After breakup, as Lindsay has shown, a generally eastward drift of the broken-out ice occurs, accompanied by incursions of polar ice through the western entrances to the Archipelago such as M'Clure Strait. The extent to which the channels are cleared out during the summer, and the extent of polar intrusion, determine the composition of the subsequent winter's ice and, to some extent, the nature of the following summer season.

A.3.9 The Labrador Sea and Baffin Bay

The Baffin Bay-Labrador Sea system forms a seasonal regime of mainly first-year ice which is almost completely isolated from the Arctic Ocean. A small amount of Arctic ice enters Baffin Bay via a southward current through Nares Strait, and in bad winters the ice of the East Greenland Current can reach around Kap Farvel into Baffin Bay, but most of the Baffin Bay ice is locally generated and thus is all first-year ice. Its seasonal distribution is treated elsewhere in this volume. The ice thickness and ridging distributions in the southern part of the region (Davis Strait between 61° and 65°N) were measured in February 1967 by the U.S.S. "Queenfish" (Wadhams et al., 1985). The profiles revealed a MIZ of width 50-100 km in which the mean ice draft was very low (0.25-0.51 m, mainly due to the large amount of open water), most ice was undeformed (93-100%) and the few ridges had drafts of less than 6 m. Beyond this the "interior zone" had higher mean ice drafts (0.77-1.66 m), a lower fraction of undeformed ice (67-91%) and deeper ridges (up to 19.6 m), but the ice was still clearly first-year in nature.

The Labrador Current forms a classic marginal ice zone exposed to the full force of North Atlantic wave action. Ice which forms in Davis Strait in October-November is carried south and joined by heavy floes from Foxe Channel coming out of Hudson Strait. The combined ice stream is carried south by the Labrador Current at a mean speed of some 16 km/day (Markham, 1969) and reaches the east coast of Newfoundland by January-February (Dinsmore, 1972). Its extreme limits, attained by March-April, tend to coincide with the edge of the Current, so that by the end of winter a wide seaward tongue of ice has formed off Newfoundland, corresponding to the eastward turn of the Labrador Current. Undeformed ice in this drift is of average thickness 1.3-1.5 m. In April the ice begins a swift retreat.

The morphology of the ice is similar to that of East Greenland ice except that the reworking by wave action is even more extensive, yielding smaller floe sizes (20 m or less near the edge) and a vast amount of brash. Structural analyses of the floes (Winsor and LeDrew, 1978) show that, as in the Antarctic, much of the ice is frazil in origin and the mean salinity is high. Obviously the turbulence and wind action to which the ice is exposed forces frazil and brash under the floes, causing a greater floe thickness than would be expected on thermodynamic considerations alone.

A.3.10 Icebergs and Ice Islands

The final ice type that requires a brief mention is ice of nonmarine origin, i.e., icebergs and ice islands, although these participate only minimally in the air-sea-ice interactive processes which affect hemispheric climate.

The greatest iceberg-producing region in the world is the Antarctic. Antarctic bergs calve from the floating ice shelves which form about 1/3 of the Antarctic coastline and the bergs therefore constitute the ice discharge from the Antarctic continental ice sheet. The estimated annual export is 10^3 km^3. Newly-calved bergs are typically tabular, about 200-250 m thick of which 35-55 m is freeboard, and up to a few km in linear dimensions, although occasionally giant bergs up to 180 km long are reported. A newly-calved berg normally begins by moving westward under the Antarctic coastal current then, if it does not ground again, it eventually breaks out into the Antarctic Circumpolar Current and begins a long voyage in an eastward and slightly northward direction until it melts or breaks up into fragments which melt. Breakup can be accomplished by long ocean swell (see Section B.8.3), by enlargement of crevasses or by collapse of ice cliffs following sidewall melting. When they first calve, the bergs have a large cold reserve (-16°C to -24°C at the top surface, rising to freezing point at the bottom) but this disappears through the percolation of meltwater down from the surface; the uppermost 30-50 m of a berg is permeable to water and air as its density is less than the 0.8 mg m^{-3} necessary for the transition from "firn" to "ice." Proposals have been made to tow Antarctic icebergs to coastal destinations in the Southern Hemisphere as sources of fresh water and energy (Husseiny, 1978; Annals of Glaciology, 1980) but it must be remembered that although the quantity of fresh water locked up in the Antarctic is immense, the <u>flux</u> represented by icebergs is much less than the flow rate of the Amazon River.

The main source area for Arctic icebergs is the tidewater glaciers of West Greenland, although some bergs come from the Greenland Sea. Starting from furthest north, Zubov (1945) describes the following iceberg sources: (1) Kvitøya, which produces only a few small bergs, (2) Franz Josef Land, a major source, and (3) northern Novaya Zemlya. Most bergs produced by these sources appear to go aground in the Barents Sea or Kara Sea, and the presence of a true iceberg in the ice drift emerging from the Trans Polar Drift Stream into the EGC is said to be a rare occurrence. Nevertheless, on August 23, 1980, the author observed no less than six small icebergs moving in company along the east Greenland ice edge at about 79°10′N. Along the East Greenland coast there are many iceberg-producing glaciers, but they are free to discharge icebergs only during the brief season (July-September) when the coast is free of land ice. Most icebergs remain at or near their glaciers of origin, either because of grounding on fjord sills or because of adverse winds and currents. In an estuarine circulation the outward flowing surface water in a fjord is compensated by a slow deep current moving up the fjord; an iceberg responds to the current integrated over a great depth and therefore does not feel the same rapid outward current as the pack ice. Fjords which are wide enough to possess a gyral circulation can discharge icebergs more easily, and thus the most fertile iceberg-producing fjord in east Greenland is Scoresby Sund, with Kong Oscars Fjord as a lesser source. These icebergs move southward in the EGC to Kap Farvel, where they join the far more numerous icebergs of west Greenland waters.

Some 12,000 bergs calve annually from West Greenland glaciers; the most productive glaciers lie in the latitude range 69-72° (70% of bergs) and 73-76° (20%). The bergs move north in the West Greenland Current at 10-14 km/day, then eventually cross over Baffin Bay, sometimes at a high latitude, and return southward off the coast of Baffin Island. If they do not run aground <u>en route</u> they eventually enter the Labrador Current, and end up south of Newfoundland, where they enter shipping lanes. Their final fate is to ground on the Grand Banks and melt, or else to melt at sea when they encounter warm water. The number reaching 48°N has varied from zero per year to 1400 (Groen, 1969) with a mean of 386, and an average of 51 per year reach 43°N. Arctic bergs are smaller than Antarctic bergs but often have a greater maximum freeboard because their low width/thickness ratio is conducive to capsizing; repeated capsizes and melting combine to yield a wide variety of shapes. A typical berg weighs 10^7 tonnes at calving and a tenth of that value when it reaches Newfoundland.

The equivalent of the iceberg in the Arctic Ocean is the ice island. This is a large tabular ice sheet 20-50 m thick which has broken off the floating apron of a far northern ice shelf. The chief source is northern Ellesmere Island, where the ice "shelves" concerned (e.g., Ward Hunt Ice Shelf) are not actively fed by glaciers but are in fact cases of extremely old sikussak (Section A.1.3), their age being some hundreds of years. When first injected into the Arctic, an ice island moves west in the Beaufort Gyre, but as it crosses the north pole area after its first circuit of the gyre it has a roughly 50% chance (Crary, 1958, 1960) of crossing over into the Trans Polar Drift Stream and entering the EGC. Large ice islands which are known to have done this include T2 (700 km^2, 1951), Arlis-II (15 km^2, 1965) and WH2 (68 km^2, 1973). Arlis-II was manned as a scientific station for four years (Schindler, 1968) and was eventually abandoned at 66°26′N near Iceland. The most famous ice island is T3, which was manned almost continuously since 1952, during which time it made several circuits of the Beaufort Gyre. It eventually entered the EGC and in 1986-5 is believed to have broken up west of the Kup Farvel.

Northeast Greenland has a source of small tubular icebergs in the form of Flade Isblink, and Svalbard possesses ice fronts on the south coast of Nordaustlandet and the east coast of Edgeøya which are capable of producing ice islands.

PART B. WAVE-ICE INTERACTION: THE EFFECT OF ICE ON WAVES

B.1 SUMMARY OF OBSERVATIONS AND EXPERIMENTS

The first navigators to encounter the Arctic and Antarctic ice margins rapidly became aware that ocean waves penetrate icefields. Scoresby (1820), Parry (1824), and Ross (1847) remarked on the phenomenon, and noted that it is unwise to take shelter from a storm within an icefield unless one penetrates a large distance (35 km according to Parry), for otherwise the ship is placed in great danger from the pounding of floes that are responding to the waves and swell. Ross lost a rudder from the TERROR in this way in the Antarctic. The technique of "ice docking," developed by sealers and whalers, consisted of cutting a basin in a multiyear floe with long saws in order to protect the ship from ice impact.

The Antarctic offers the clearest example of wave penetration into ice, since the ice edge is fully exposed to the very long swells of the Southern Ocean. It was an observation of swell action by Worsley (1916; Robin, 1963) which led Shackleton's marooned ice party from the ENDURANCE to conclude that their ice floe was approaching the ice margin. Permanent stations established on the coast have reported the presence of long swell and of microseisms (Imbert, 1954; Murauchi and Yoshida, 1959; Hatherton, 1960; Pratt, 1960; Smirnov, 1966) but without attempts at spectral analysis; significant periods were usually greater than 20 s and amplitudes less than 50 mm. This indicated that even in winter very long swell can cross the entire width of the Antarctic ice pack and reach the continent. The first systematic wave recordings at sea in the Antarctic ice were by Robin (1963) from RRS JOHN BISCOE en route from Stanley to Halley Bay. A shipborne wave recorder (Tucker, 1956) was used to make 10 minute recordings every 6 hours. As soon as a serious ice cover was encountered the energy below 8 s period fell to the noise level within the space of a single pair of recordings (~50 km), but the longer components decayed more slowly. In regions of small floes Robin found that the minimum wavelength at which significant penetration occurred was roughly three times the floe diameter, indicative of a scattering process as the main cause of decay. In regions of large floes (>1 km) Robin found an empirical relationship for the energy E_x at a given penetration relative to the energy E_o at the ice edge:

$$\frac{E_x}{E_o} \propto \frac{\lambda^4}{h^3} \tag{14.1}$$

where λ is wavelength and h ice thickness.

Further wave recordings were made by Dean (1966) using a shipborne wave recorder on the KISTA DAN in the outermost 2 km of a sharply defined ice edge. His work was not fully reported, and Figure B.1, drawn from his original data (Wadhams, 1973b), shows that at a given frequency the energy decay appears exponential, but with a decay rate which is greater in the outermost 300 m than further in. The decay rate decreased with wave period. Dean was able to estimate floe sizes by oblique photography from the ship, and found a change in mean floe size at the position where the decay rate changed. Dean's results are discussed further in Section B.3.3.

In the Arctic the first observations of wave penetration were made from the ice itself. Zubov (1945, ch. 94) reported on 10-12 s oscillations causing tilts of up to 1′ experienced by the NP-1 station in the Greenland Sea at 77°N, and on 9-10 s oscillations observed by theodolite on ice near the "Sedov" at 81°N, 100 km from the ice edge. The first gravimeter observations made on sea ice (Crary et al., 1952; Crary and Goldstein, 1959) showed the almost universal presence of long-period oscillations of the ice cover, with significant periods of 20-56 s and equivalent amplitudes of 0.5 mm. To investigate this phenomenon further, Hunkins (1962) used two gravimeters 400 m apart to show that the oscillations were due to progressive waves with a phase velocity approximately that of a flexural-gravity wave (see Section B.2.1). He inferred from spectral analysis that the energy spectrum had a trend in which energy increases at T^4, but Wadhams (1973b) has suggested that this trend may be spurious, a result of a failure to

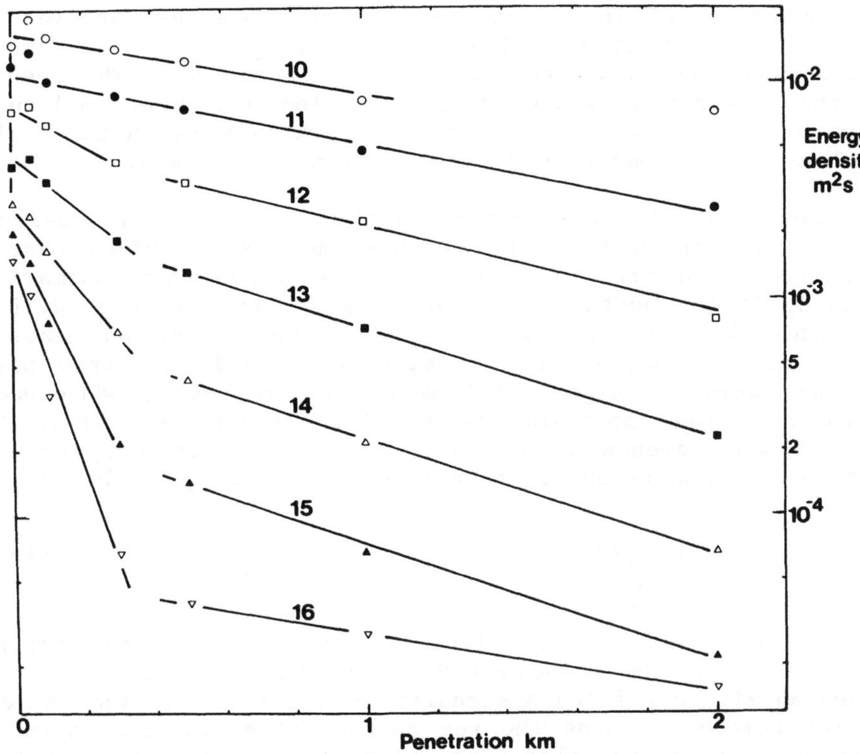

Fig. B.1. Energy decay with penetration, taken from spectra obtained in Dean's second series of wave recordings (designated 1963/4(ii) in Dean (1966)). The number n on each curve refers to a wave period of 96/n seconds. (After Wadhams, 1973b).

remove a mean value from acceleration recordings. A similar experiment was carried out by Sytinskiy and Tripol'nikov (1964) with a tripartite array of seismographs from ice stations NP-13 and NP-14. They found progressive waves of amplitude 0.2-0.25 mm and with phase velocities of 20-60 m s^{-1}. Next LeSchack and Haubrich (1964; LeSchack, 1965) made directional recordings with two gravimeters from ice island ARLIS-II in the Beaufort Sea. These revealed swell energy coming from the direction of the East Siberian Shelf, with three significant wave periods of 27.8 s, 38.5 s, and 62.5 s and an energy density of order 10^{-6} m^2s. The most recent measurements of this nature were carried out in spring 1981 by S.C. Moore of SPRI during the EUBEX experiment in the central Arctic Ocean (Manley et al., 1982). He deployed nine strainmeters in a triangular array, each vertex of the triangle consisting of a rosette of three strainmeters (Squire, 1978). Each rosette measures the principal axes of flexure of the ice surface with an ambiguity of 180°; the use of three rosettes resolves the ambiguity. A distant second station established by helicopter permitted decay rates of the waves to be measured. The results showed the presence of waves of 30 s period and 3 mm amplitude, and also of longer (10 min) oscillations which were in phase with internal wave measurements made from a thermistor chain.

Measurements of wave penetration and decay at and near the Arctic ice edge began in 1971 when Wadhams (1972, 1978a) carried out a series of observations by sonar from a submerged submarine, HMS ORACLE, in the northern Greenland Sea. The submarine drifted slowly under the ice edge, and a series of recordings was taken at different penetrations while floe sizes were estimated from periscope observation. Exponential decay was observed, with decay rates which varied approximately as T^{-2}. Figure B.2 shows the results of measurements on separate days (with different ice conditions) of the amplitude attenuation coefficient α_x, such that

$$E_x = E_o \exp(-2\alpha_x x) \quad (14.2)$$

Figure B.2 also shows results from measurements with an airborne laser profilometer by Wadhams (1975) off Newfoundland. In this technique an aircraft flies a straight course into the ice while a laser profilometer records the sea surface elevation and an infrared line scanner records floe sizes along the track. A correction must be made for the assumed directional properties of the sea relative to the flight direction. Again exponential decay was observed, with results that showed the same trend as the submarine data.

The most complete series of ice edge wave observations began in 1978 and has involved four field operations; one off Newfoundland, one in the Bering Sea, and two off east Greenland (Wadhams, 1979; Wadhams and Squire, 1980; Squire and Moore, 1980; Squire and Martin, 1980; Moore and Wadhams, 1980; Goodman et al., 1980). The principle is to combine direct wave decay measurements, made by inserting a wave buoy into the open water between floes at different distances from the ice edge, with concurrent observations of the flexural, heave, and surge responses of the ice floes. In this way it is hoped that a complete understanding of the mutual interaction of ice floes and waves can be obtained. Some results are discussed in Section B.3.3. The most recent extension of these experiments involved the use of directional wave buoys during the MIZEX-84 experiment in the Greenland Sea (Wadhams et al., 1986). The two major discoveries that emerged were that the width of the spectrum increases with increasing penetration of the icefield, becoming almost isotropic within a few km (due to the scattered wave vectors); and that significant reflection of wave energy (3-13%) occurs from the ice front back out to sea.

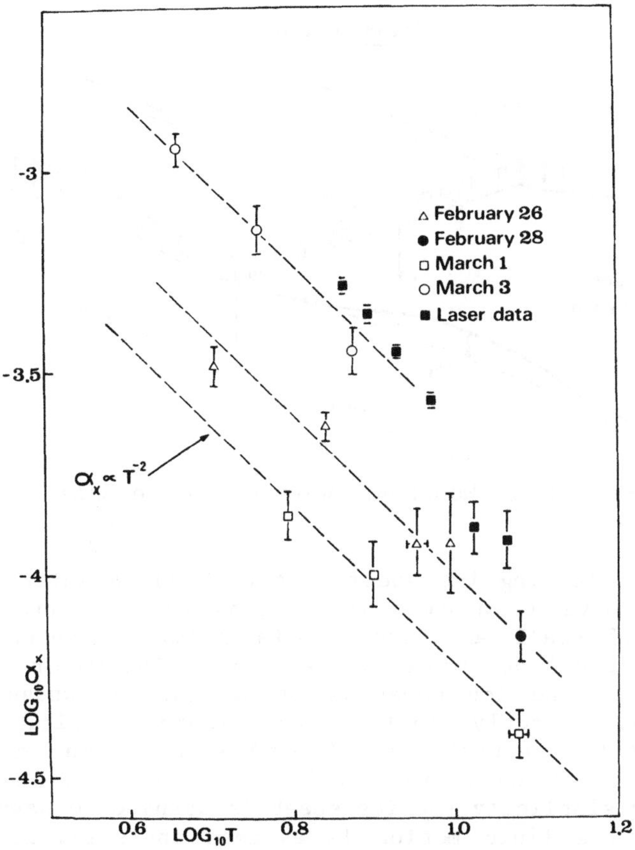

Fig. B.2. Log-log plot of α_x against wave period T, with laser data from Wadhams (1975) and submarine data from Wadhams (1978a).

For completeness it should be noted that a number of laboratory tank experiments have been carried out, but that no definite conclusions can be drawn from them, mainly because of scaling and constitutive problems. The experiments have been done by Ofuya and Reynolds (1967), Szczesny and LeMehaute (1965), Hendrickson (1966), Henry (1968) and Wadhams (1973b). More recent tank experiments by Martin and Kauffman (1981) have concentrated on the only type of sea ice -- frazil -- which can be represented in the full scale in the tank, so that valid conclusions can be drawn directly. This work is reported in Section B.4.2.

B.2 WAVE PROPAGATION THROUGH A UNIFORM FLOATING ICE SHEET

B.2.1 <u>The Flexural-gravity Wave</u>

We begin by considering the simplest case, that of a monotonic wave incident normally on the straight edge of a uniform

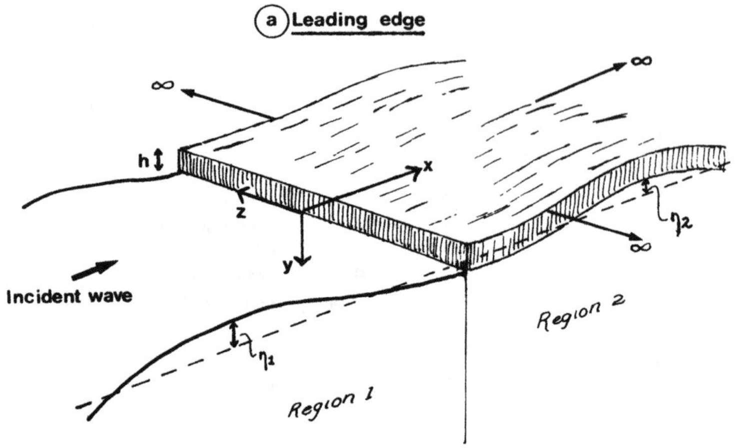

Fig. B.3. Monotonic wave on an ice edge.

semi-infinite floating ice sheet (Figure B.3) in water of uniform depth. The wave is assumed to be sinusoidal in profile, long-crested and of small amplitude. Its radian frequency is ω (= $2\pi/T$) and radian wave number k (=$\pi 2/\lambda$). The water depth is D, where $D \gg \lambda$. The ice sheet is of uniform thickness h and is assumed to be perfectly elastic with uniform constitutive properties, so that the neutral axis of bending is at the mid-point of its thickness. Young's modulus for the ice is E and Poisson's ratio ν. For simplicity the ice sheet is assumed to have zero submergence and the fluid motion is assumed inviscid, irrotational, and incompressible. Zero submergence allows us to ignore the effect of the submerged leading edge of the ice sheet on the wave.

The water column has two regions with differing surface boundary conditions. There exists a velocity potential ϕ_n in each region obeying the equation of continuity (Laplace's equation)

$$\nabla 2\phi_n = 0 \tag{14.3}$$

with the boundary condition

$$\frac{\partial \phi_n}{\partial y} = 0 \Big|_{y = D} \tag{14.4}$$

In region 1, let $\eta_1(x,t)$ be the instantaneous elevation of the water surface. Then, by the linearized Bernoulli equation for a free surface

$$\eta_1 = -\frac{1}{g} \frac{\partial \phi_1}{\partial t} \Big|_{y = 0} \tag{14.5}$$

with the kinematic surface boundary condition for potential

$$-\frac{\partial \eta_1}{\partial t} = \frac{\partial \phi_1}{\partial y}\Big|_{y=0} \tag{14.6}$$

A solution satisfying (14.3-14.4) is

$$\phi_1 = [Be^{ikx} + Re^{-ikx}]\, e^{-ky} e^{-i\omega t} \tag{14.7}$$

where $k = \omega^2/g$ if $D >> \lambda$. $|B|$ is the potential amplitude of the incident wave and $|R|$ of a wave reflected normally from the ice edge.

In region 2, let $\eta 2(x,t)$ be the instantaneous elevation of the water surface; assuming that the water remains in contact with the ice, this is also the vertical displacement of the ice. The equation of motion of the ice is

$$p(x, t) - L\frac{\partial^4 \eta_2}{\partial x^4} = \rho_i h \frac{\partial^2 \eta_2}{\partial t^2} \tag{14.8}$$

where $p(x,t)$ is the pressure just below the water/ice interface, ρ_i is the density of the ice, and L is its flexural rigidity. For a uniform ice sheet

$$L = Eh^3/12(1 - \nu^2) \tag{14.9}$$

Applying the linearized Bernoulli equation to the water/ice interface gives

$$p(x, t) = -\rho_w \left[g\eta_2 + \frac{\partial \phi_2}{\partial t}\Big|_{y=0} \right] \tag{14.10}$$

where ρ_w is the water density, with the boundary condition for potential

$$-\frac{\partial \eta_2}{\partial t} = \frac{\partial \phi_2}{\partial y}\Big|_{y=0} \tag{14.11}$$

Consider a solution for ϕ_2 as a sum of potentials

$$\phi_2 = \sum_n \left[A_n e^{ik_n x} + B_n e^{-ik_n x} \right] e^{-k_n y} e^{-i\omega t} \qquad (14.12)$$

This satisfies the boundary conditions provided

$$L\, k_n^5 + (\rho_w g - \rho_i h \omega^2) k_n - \rho_w \omega^2 = 0 \qquad (14.13)$$

Equation (14.13) is then the dispersion equation for periodic waves in the ice sheet region. It is a quintic, but two of the solutions for k_n are physically unfeasible since they lead to potentials that increase indefinitely with depth. The three physically feasible roots are

$$k_o = k_o$$
$$k_1 = a + bi \qquad (14.14)$$
$$k_2 = a - bi$$

where k_o, a, and b are real and positive.

The real solution k_o represents a conservative wave propagating through the ice sheet in the x-direction, the energy of the wave being coupled between the ice and the water. It is known as as a flexural-gravity wave or ice-coupled wave. Its properties were first deduced, almost correctly, by Greenhill (1887) and the first experimental observations of its dispersive behavior were made by Ewing and Crary (1934), who set off small charges in the ice of a frozen canal and measured the transit time of waves along a straight section. Its properties show remarkable agreement with the dispersion behavior of waves observed in the central Arctic Ocean by Hunkins (1962) and Sytinskiy and Tripol'nikov (1964) and with strainmeter observations in fast ice by Squire and Allan (1980).

There is no explicit solution for k_o in terms of ω, but (14.13) can be easily solved by iteration to obtain the dispersion relation. Figures B.4 and B.5 illustrate the dispersion relations; Figure B.4 shows $\lambda_o (= 2\pi/k_o)$, the wavelength of the flexural-gravity waves, as a function of period T for three different ice thicknesses, while Figure B.5 shows the phase velocity c and group velocity U as a function of T. The asymptotic behavior of c and U can be seen by rearranging (14.13) to give

$$c^2 = \frac{g\lambda o}{2\pi} \left[\frac{1 + (2\pi F/g\lambda_o^{\ 4})}{1 + G/\lambda_o} \right] \qquad (14.15)$$

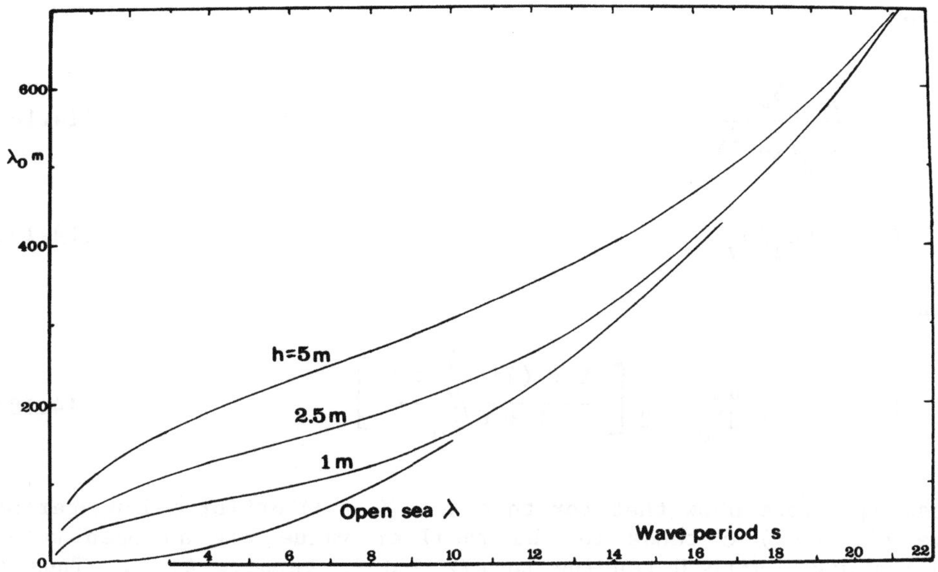

Fig. B.4. Ice-coupled wavelengths at three ice thicknesses.

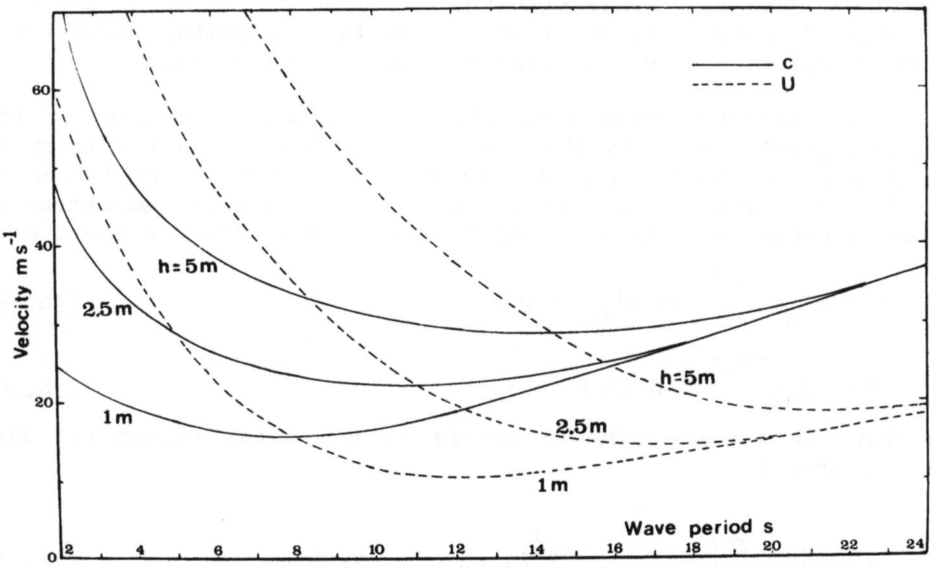

Fig. B.5. Dispersion relations for phase and group velocities of an ice-coupled wave.

where

$$F = \frac{2\pi^3 h^3 E}{3\rho_w(1-\nu^2)} \qquad (14.16)$$

$$G = 2\pi h\rho_i/\rho_w \qquad (14.16)$$

Then

$$U = c - \lambda_o \frac{dc}{d\lambda_o} = \frac{c}{2}\left[\frac{1 + (4F/\lambda_o^3\ c^2)}{1 + G/\lambda_o}\right] \qquad (14.18)$$

These equations show that for thin ice (L → 0) or for a long-period wave (λ_o → ∞) c tends to the familiar value for an open water gravity wave, with U tending to c/2 and λ_o tending to λ. Thus a flexural-gravity wave's properties tend to those of a free water wave at long periods or for thin ice; at shorter periods, however, as shown in Figure B.5, they tend to those of a pure flexural wave in a thin elastic sheet. Anomalous dispersion occurs, with c and U increasing as T decreases (c = U at the point where c takes its minimum value). The flexural arm of the dispersion is familiar to anyone who throws a stone onto an ice sheet in a pond, or who rams the edge of a thin ice sheet with a boat; a rippling disturbance in the sheet can be seen to radiate away at high speed.

Apart from its dispersion, the most important property of the flexural-gravity wave is the way in which the energy transport is shared between the ice and the water. If we let the amplitude of the flexural-gravity wave be A_i, then the x- and y- components of water particle velocity at a point (x,y,z) under the ice sheet are

$$U = \omega A_i e^{-k_o y} \cos(k_o x - \omega t) \qquad (14.19)$$

$$V = \omega A_i e^{-k_o y} \sin(k_o x - \omega t) \qquad (14.20)$$

so that the average kinetic energy in the water column per unit surface area is

$$E_i = \frac{1}{\lambda_o} \int_o^\infty \int_o^{\lambda_o} \rho_w \left(\frac{u^2+v^2}{2}\right) dxdy = \rho_w \omega^2 A_i^{\ 2}/4\ k_o \qquad (14.21)$$

The potential energy stored in the water column per unit surface area is

$$E_2 = \frac{1}{\lambda_o} \int_o^{\lambda_o} \frac{\rho_w g \eta_2^2}{2} dx = \rho_w g A_i^2/4 \qquad (14.22)$$

Since the ice moves only vertically with velocity $\dot{\eta}_2$, the average kinetic energy in the ice per unit surface area is

$$E_3 = \frac{h\rho_i}{2\lambda_o} \int_o^{\lambda_o} \dot{\eta}_2^2 dx = \omega^2 h\rho_i A_i^2/4 \qquad (14.23)$$

When a plate is flexed to a profile $\eta_2(x,t)$ it exerts a vertical stress $s(\eta_2)$ on the underlying water of

$$s(\eta_2) = - L\, d^4\eta_2/dx^4 \qquad (14.24)$$

The work done per wavelength in establishing this profile is

$$-\int_o^{\lambda_o}\int_o^{\eta_2} s(\eta)d\eta dx = L\, k_o^4 A_i^2\, \lambda_o/4 \qquad (14.25)$$

so that the mean potential energy in the ice per unit surface area is

$$E_4 = L\, k_o^4 A_i^2/4 \qquad (14.26)$$

It can be seen by comparing (14.21), (14.22), (14.23), and (14.26) with (14.13) that

$$E_1 + E_3 = E_2 + E_4 \qquad (14.27)$$

i.e., the mean kinetic and potential energies of the ice/water system are equal, a result expected for a progressive wave. However, the energy is unequally divided between ice and water, and both ice and water have unequal time-averaged kinetic and potential energies. For short waves most of the energy is transported by the ice, while for long waves most of the energy transport occurs in the water.

Now a wave of amplitude A_ω in open water of infinite depth has an energy per unit surface area of $\rho_w g A_w^2/2$, and the rate of energy transport per unit width of wavefront is $\rho_w g^2 A_w^2/4\omega$, since $g/2\omega$ is the group velocity. If we suppose that such a wave, incident normally on the ice as in Figure B.3, is transmitted perfectly into the ice and converted into a flexural-gravity wave, the amplitude A_i acquired by the wave is given by the equality of energy flux

$$\rho_w g^2 A_w^2 \; g/4\omega = \left(\sum_{i=1}^{4} E_i \right) . U \tag{14.28}$$

where U is given by (14.18). This yields the result

$$\frac{A_w^2}{A_i^2} = \frac{2\,\omega . U}{g} \left(1 + \frac{Lk_o^4}{\rho_w g}\right) \tag{14.29}$$

Figure B.6 shows A_w/A_i plotted against wave period for ice of three different thicknesses. At all ocean wave periods except those of long swell, the wave transmitted perfectly from water into ice displays a much lower amplitude within the ice. Only at

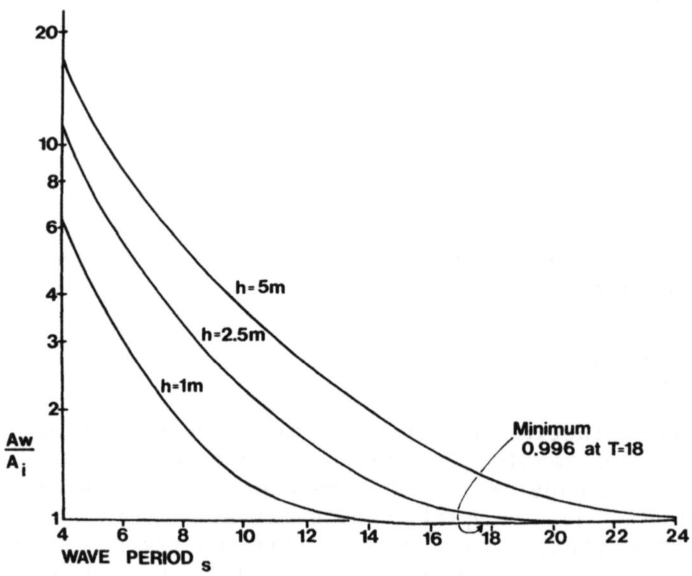

Fig. B.6. Ratio of the amplitude of a wave in water A_w, to its amplitude of flexure in ice A_i, assuming perfect transmission.

very long swell periods does the flexural-gravity wave have an amplitude tending to that of its open-water parent. It is noteworthy that periods observed in the central Arctic Ocean are extremely long--30 s or more--so that the energies of such waves are similar to those of open water waves of the same period.

An interesting property of flexural-gravity waves has been pointed out by Mollo-Christensen (1983). This is that under a compressive stress in the ice sheet the group and phase velocities of the wave are reduced, and may reach zero. This would lead to a concentration of energy and a buckling of the ice. The required compressive stress could come from the effect of wind, surface current or even the radiation pressure of the wave itself against the upwind edge of the ice sheet.

B.2.2 Evanescent Waves Near Ice Boundaries

We now consider the solutions k_1 and k_2 of (14.14). Substituting in (14.12) and neglecting the flexural-gravity terms A_o and B_o, we have

$$\phi_2' = \left(A_2 e^{ik_1 x} e^{-k_1 y} + B_2 e^{-ik_2 x} e^{-k_2 y}\right) e^{-i\omega t} \tag{14.30}$$

because A_2 and B_1 are not physically feasible terms in the sum, since they are exponentially unbounded at $x = \infty$. Expanding in terms of a and b we obtain

$$\phi_2' = \left[2B_2 \cos(ax-by) + (A_1-B_2)e^{i(ax-by)}\right] e^{-(bx+ay)} e^{-i\omega t} \tag{14.31}$$

Actual values of a and b must again be derived by solving (14.13) iteratively for k_o, then dividing (14.13) by $(k_1 - k_o)$ and equating real and imaginary parts (Appendix 1).

The first term in (14.31) is a standing wave. The nodes are defined by the planes

$$by = ax + (2n-1)\pi/2 \tag{14.32}$$

and the antinodes by

$$by = ax + n\pi \tag{14.33}$$

where $x \geq 0$, $y \geq 0$, and n is any integer including zero. The potential amplitude is $2B_2$ at (0,0,z), decaying rapidly with increasing x and y. Physically the term represents a diagonal oscillation of energy in the x-y plane with no net transmission, the wave being confined to the region near the ice edge -- an evanescent wave. We shall see later that when a wave is incident obliquely, propagation along the ice edge is possible.

The second term in (14.31) is a decaying progressive wave with a propagation vector $\underline{k}$ of (a,-b,0). Physically, this represents energy which is taken from the region 1/region 2 boundary plane and which propagates diagonally upwards, orthogonally to the nodal planes of the standing wave to be given to the ice as increased kinetic energy near the ice edge.

Both of these evanescent waves have decay lengths of 1/b in the x-direction, and 1/a in the y-direction. Figure B.7 shows plots of a and b against wave period for three different ice thicknesses. b is always close to, or greater than, k_o at normal wave periods, so that ϕ_2^i effectively disappears within one ice-coupled wavelength of the ice edge. At long wave periods a and b tend to asymptotic values, and it can be shown that

$$a \rightarrow \left(\frac{\rho_w g}{4L}\right)^{1/4}, \quad b \rightarrow a \text{ as } T \rightarrow \infty \tag{14.34}$$

The significant physical effect of the evanescent terms is to combine with A_1 to produce a peak value of bending stress in the ice at some characteristic distance ($<\lambda_o$) from the ice edge. A wave of sufficient amplitude entering the ice from the open sea will therefore fracture the ice along a line (x_{crit},0,z) where $x_{crit} < \lambda_o$. After the fractured ice fragments have broken away, a second fracture line at the same x_{crit} value will be produced inside the new ice edge. This is a mechanism for the progressive breakup of an ice edge under wave action, into long narrow floes of equal width x_{crit}. Figure B.8 shows what is believed to be this process in action in a fast ice sheet off Newfoundland. Within four hours the entire sheet within the field of view of the photograph had broken out. The spacing of the fracture lines is about 20 m.

B.2.3 Reflection and Transmission at an Ice Edge

Figure B.6 shows the amplitude adopted by an ocean wave when it enters a uniform ice sheet and continues to propagate as a flexural-gravity wave, assuming perfect transmission. We now consider what the transmission and reflection coefficients actually are.

The problem of evaluating |R|/|B| in (14.7), the amplitude reflection coefficient, is that of fitting two transition

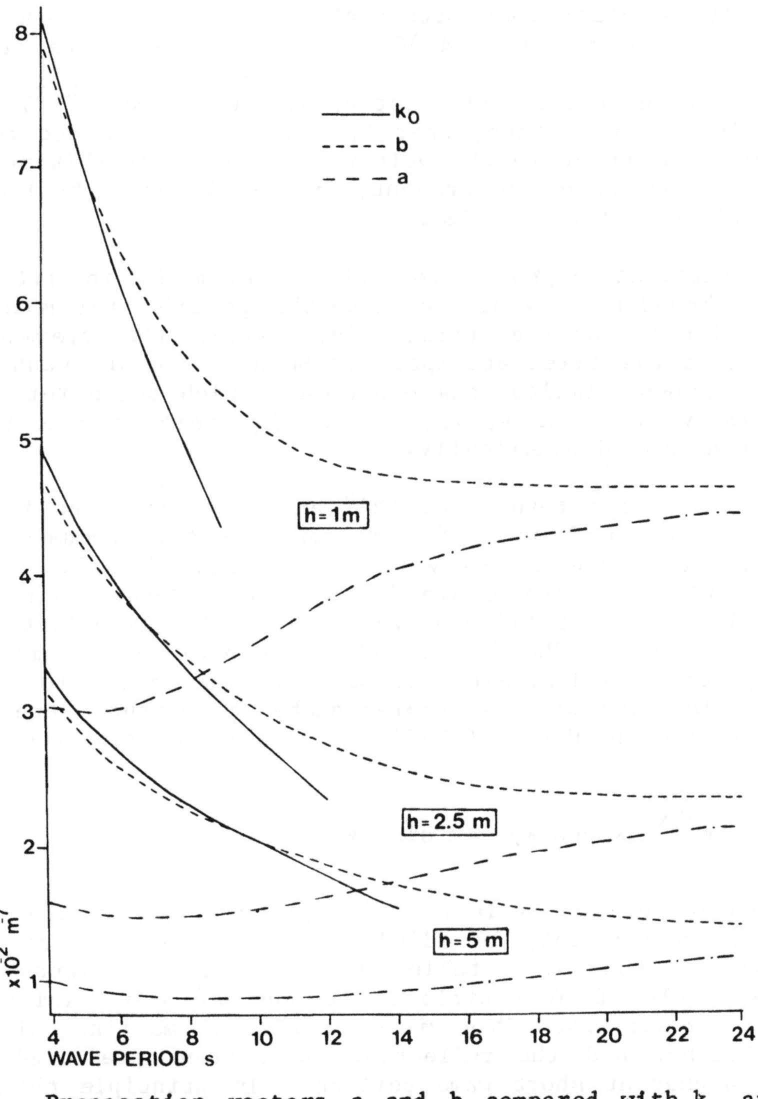

Fig. B.7. Propagation vectors a and b compared with k_0 at three ice thicknesses.

conditions across the region 1/region 2 boundary plane, i.e., the x = 0 plane in Figure B.3. The first condition is that there should be no bending moment or shear at the ice edge. So

$$\frac{\partial^2 \eta_2}{\partial x^2} = \frac{\partial^3 \eta_2}{\partial x^3} = 0\Big|_{x=0} \tag{14.35}$$

Now it is clear that potentials of form (14.7) and (14.12) cannot in themselves fit (14.36) at all depths, since the depth dependence of the two velocity potentials (e^{-ky} and $e^{-k_o y}$) are different. It is clear, then, that ϕ_1 and ϕ_2 still do not represent the complete solution to the velocity potential in this situation, but that further terms are present, presumably again having significance only near the ice edge.

The simplest approach to this problem is to fit (14.36) across the boundary only at $y = 0$, on the grounds that most of the energy is found near the surface (99.8% within the uppermost $\lambda/2$). This solution was first attempted by Hendrickson and Webb (1963), and the relevant simultaneous equations, which solve for the real and imaginary parts of R, A_o, A_1 and B_2, form an 8 x 8 matrix which can be solved numerically.

A graph of the result for $|R|^2$ at three different ice thicknesses is shown in Figure B.9. We can test the adequacy of this approximation by seeing how well it fulfills (14.37). In Figure B.10 the continuous line marked "true energy balance" is the square root of the r.h.s. of (14.37), while the dotted line is the square root of the l.h.s. The fit is not ideal and can be improved by searching for an additional form of potential to add to ϕ_1 and ϕ_2 so as to allow fitting at a greater number of depths. One type is a form discussed by Ursell (1947) for wave propagation past a fixed vertical barrier:

$$\phi_a = Ce^{-Kx} (K \cos Ky - k \sin Ky)e^{-i\omega t} \tag{14.38}$$

where K takes any real value. ϕ_a obeys the free surface boundary condition, so can only be added to ϕ_1. However there is no *a priori* way of choosing suitable values of K. In Figure B.10 we show the result of an empirical fit done at $y = 0$ and $y = \lambda/4$ using $K = k/2$ and $k/4$. This yields a 12 x 12 matrix. The energy fit is better, and the reflection coefficient predicted (Figure B.11) is higher at short wave periods. In principle the fitting could be extended to n depths, allowing $2(n-1)$ values of K to be chosen and producing a $4(n+1)$-dimensional matrix to be solved. An elaborate optimization technique based on approximating the energy balance could be used. This seems unrealistically difficult, and a better solution to the problem of wave transmission across an ice edge remains a current research problem.

B.2.4 Oblique Incidence

So far we have considered waves incident normally on the ice edge. If instead the wave is incident at angle θ to the normal it will suffer *refraction*. By matching potentials it can be shown

Fig. B.8. The breakup of fast ice in St. Anthony Bight, Newfoundland. (After Squire, 1978b).

where η_2 is defined in (14.8). The second condition is that the velocity potentials and the horizontal component of water particle velocity should match across the boundary plane, so

$$\phi_1 = \phi_2 , \quad \left. \frac{\partial \phi_1}{\partial x} = \frac{\partial \phi_2}{\partial x} \right|_{x=0} \tag{14.36}$$

Since the evanescent waves of Section B.2.2 transmit no energy, the far-field energy fluxes, due to the incident and reflected waves in open water, and the transmitted flexural-gravity wave in ice, must match. Setting B in (14.7) to unity, $|R|$ is then the amplitude reflection coefficient and $|R|^2$ the energy reflection coefficient. The only term in ϕ_2 (14.12) at deep penetration is A_o so, expressed in terms of potentials, (14.29) becomes

$$\frac{k^2}{k_o^{\,2}} \; \frac{(1-|R|^2)}{|A_o|^2} = \frac{2\omega U}{g} \left(1 + \frac{Lk_o^{\,4}}{\rho_w g}\right) \tag{14.37}$$

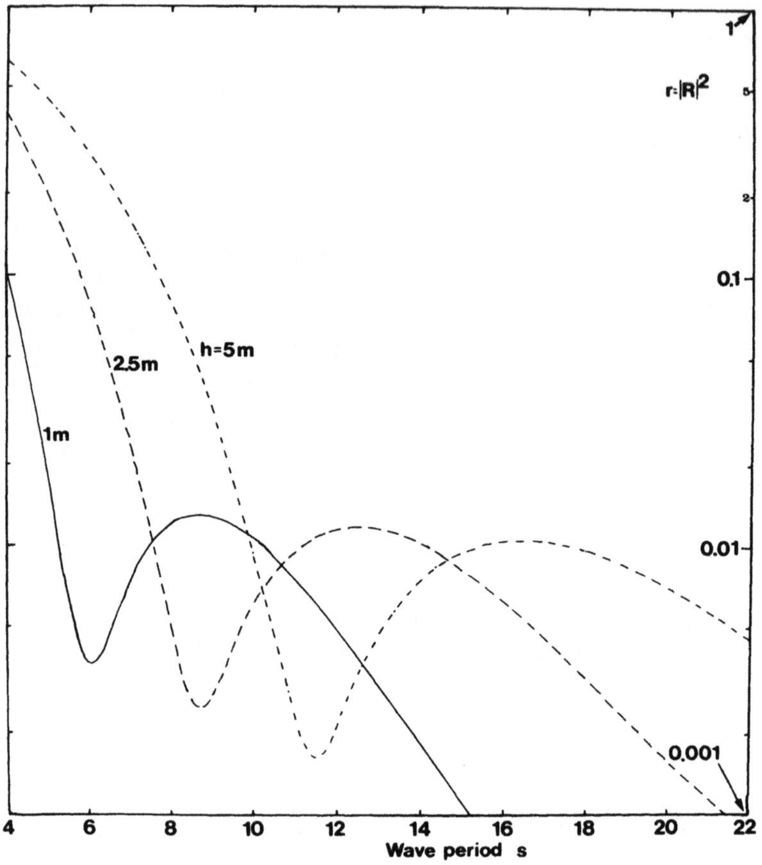

Fig. B.9. Energy reflection coefficient at the boundary of a semi-infinite floe, according to the surface-fitting Hendrickson theory.

that the flexural-gravity wave propagates into the ice with an angle of refraction θ_1, where

$$\sin \theta_1 = \frac{k}{k_0} \sin \theta \tag{14.39}$$

This implies $\theta_1 > \theta$ and hence the possibility of total reflection of the incident wave. In any case the reflection coefficient will be quite different from that predicted in Section B.2.3, especially since the evanescent components will also suffer a form of refraction. For these components there is a complex Snell's law and the refracted waves are inhomogeneous (Budden, 1961), in that surfaces of constant amplitude and phase do not coincide. The same would apply to the refracted flexural-gravity wave if one allowed

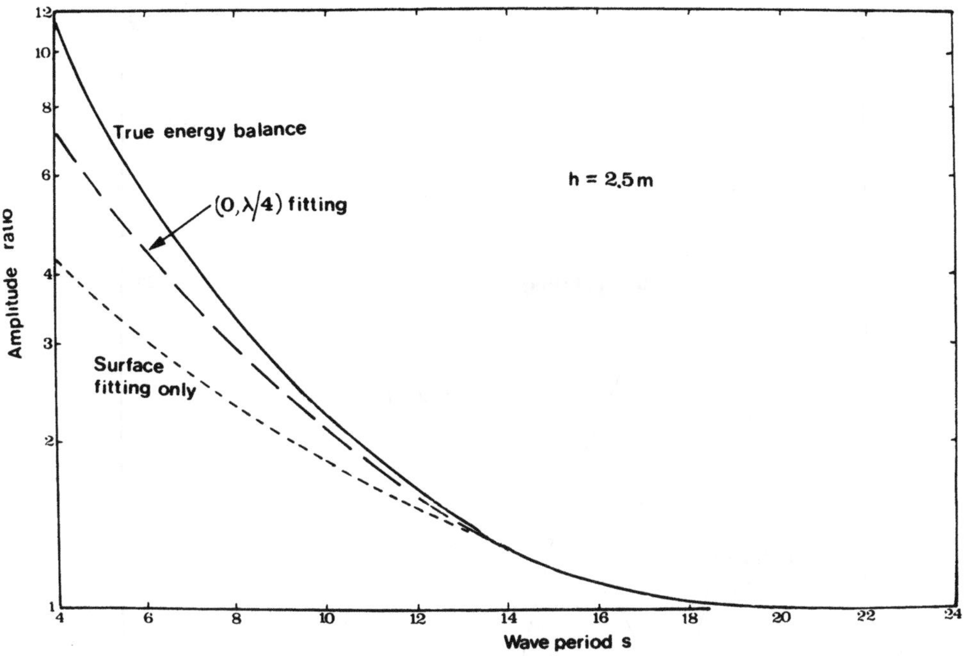

Fig. B.10. Fit of theory to the energy balance condition. Continuous line is $\sqrt{2WR_eU/g}$: dashed lines are square roots of the left-hand side of (14.37) and an attempt to fit at two depths.

the ice to have viscoelastic properties (as in Section B.2.6) which would introduce a mild attenuation term to the propagation.

If we assume perfect elasticity, (14.39) predicts a critical angle θ_c such that

$$\sin \theta_c = \frac{k_0}{k} = \frac{\lambda}{\lambda_0} \tag{14.40}$$

and it can be seen qualitatively from Figure B.4 that θ_c will be small for short-period waves, implying that refraction is a mechanism favoring the propagation of long waves into the ice, since the ice "accepts" long waves from a wider arc of incident angles.

To calculate the modified reflection coefficient we carry out the same analysis as in Sections B.2.1 to B.2.3. The z-axis runs parallel to the ice edge. For simplicity we consider only surface potentials. In open water this is

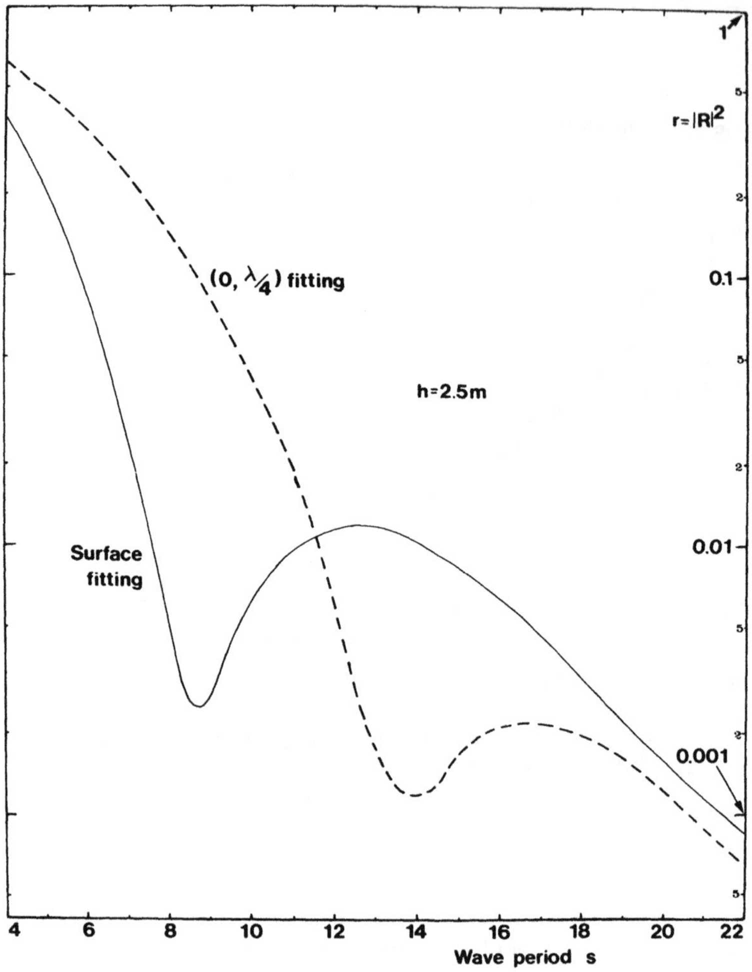

Fig. B.11. Comparison of energy reflection coefficients for the surface-fitting and the (0,λ/4)-fitting approximations. Ice thickness 2.5 m.

$$\phi_1^s = [e^{ik(x\cos\theta + z\sin\theta)} + Re^{-ik(x\cos\theta - z\sin\theta)}]e^{-i\omega t} \tag{14.41}$$

while within the ice cover

$$\phi_2^s = [A_1\, e^{ik_0(x\cos\theta_1 + z\sin\theta_1)} + A_2\, e^{ik_1(x\cos\theta_2 + z\sin\theta_2)}$$

$$+ B_3\, e^{-ik_2(x\cos\theta_3 - z\sin\theta_3)}]e^{-i\omega t} \tag{14.42}$$

where θ_2, θ_3 are angles of refraction (complex) for the evanescent modes k_1 and k_2. Matching surface potentials across the ice edge yields (14.39) and also

$$k \sin \theta = k_0 \sin \theta_1 = k_1 \sin \theta_2 = k_2 \sin \theta_3 \tag{14.43}$$

and

$$1 + R = A_1 + A_2 + B_3 \tag{14.44}$$

Matching first derivatives yields

$$k \cos \theta(1-R) = A_1 k_0 \cos \theta_1 + A_2 k_1 \cos \theta_2 - B_3 k_2 \cos \theta_3 \tag{14.45}$$

The zero bending moment condition (14.35) must now be extended to that for a thin plate which is (Graff, 1975, Section 4.2.2)

$$\frac{\partial^2 \eta_2}{\partial x^2} + \nu \frac{\partial^2 \eta_2}{\partial y^2} = 0 \tag{14.46}$$

where η_2 is defined as in (14.11). The zero shear condition likewise becomes

$$\frac{\partial^3 \eta_2}{\partial x^3} + (2-\nu) \frac{\partial^3 \eta_2}{\partial x \partial y^2} = 0 \tag{14.47}$$

Equations (14.43) to (14.47) can be solved numerically as before for R, A_1, A_2, B_3 as functions of θ and T.

Figure B.12 shows a computation (Squire, 1978b) of the magnitude of the amplitude reflection coefficient as a function of θ and T for ice 1 m thick. It can be seen that long waves cross the ice edge with little loss at all angles of incidence (except very close to 90°) while short waves can only penetrate if they are incident close to the normal. A further factor favoring long waves is that at long periods the division of energy among A_1, A_2 and B_3 favors A_1, the flexural-gravity component, so that little energy is diverted into evanescent waves. Further, the large refraction suffered by short waves which do penetrate (from 14.39) means that they have to travel much further in order to achieve a given penetration into the icefield, allowing dissipative mechanisms a greater distance to act upon them.

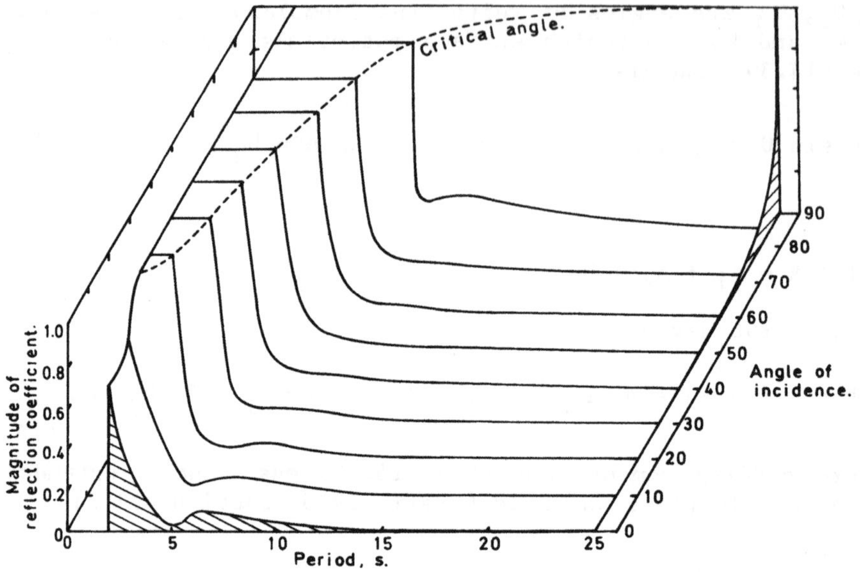

Fig. B.12. Three-dimensional graph showing dependence of reflection coefficient on both wave period and angle of incidence. The critical angle is shown. (After Squire, 1978b).

Thus deep within the icefield we expect to find only long period waves propagating almost normally to the ice edge, through the combined effects of:

1. Acceptance of incident long waves from a wider arc than incident short waves;

2. Smaller transmission coefficient for short waves even below the critical angle;

3. Large refraction away from the normal for transmitted short waves.

Squire (1978b) also went on to consider the more complex situation of refraction in viscoelastic ice.

B.2.5 Propagation Through a Field of Large Floes

So far we have considered the ideal case of a uniform continuous ice sheet. In practice the Arctic ice cover, even where it approximates to a continuous sheet (in the central Arctic Basin) contains irregularities of two basic types:

i) leads and polynyas, i.e., channels, generally long and narrow, of open water and young ice,

ii) pressure ridges and hummocks, i.e., extreme variations in ice thickness.

At present there is no theoretical or experimental description of wave reflection by a ridge or hummock, although this must undoubtedly occur. We can, however, estimate the effect of leads from the results of Section B.2.3.

Figure B.13 shows an open lead between two very large floes, 1 and 2. A flexural-gravity wave with unit energy density is incident normally on the lead from Floe 1. Assume that multiple reflections can occur, but that the edges of the lead are irregular enough for the reflections to be incoherent. Then the total energy transmitted into Floe 2 is

$$E_f = (1-r_1)\ (1-r_2)\ (1+r_1r_2+r_1^2r_2^2\ldots) = \frac{(1-r_1)(1-r_2)}{1-r_1r_2} \qquad (14.48)$$

where r_1, r_2 are energy reflection coefficients ($= |R|^2$) for the water/Floe 1 and water/Floe 2 boundaries respectively (it can be easily shown by an extension of the method used in B.2.1 and B.2.3 that the energy reflection coefficient at a trailing edge, i.e., for wave energy propagating out of ice into open water, is identical with the coefficient at a leading edge). If n is the mean number of leads encountered per unit distance, then the repeated partial reflection of (14.48) is equivalent to an exponential decay of forward-going wave energy of form

$$E_x = E_o \exp(-2\alpha\ x) \qquad (14.49)$$

where the amplitude attenuation coefficient α is given by

$$\alpha = \frac{-n}{2} \log_e \left[\frac{(1-r_1)(1-r_2)}{1-r_1r_2}\right] \qquad (14.50)$$

If the floes are of the same thickness so that $r_1 = r_2 = r$,

$$\alpha = nr + O(r^3) \qquad (14.51)$$

The simpler assumption that only a single reflection occurs at each boundary yields much the same result:

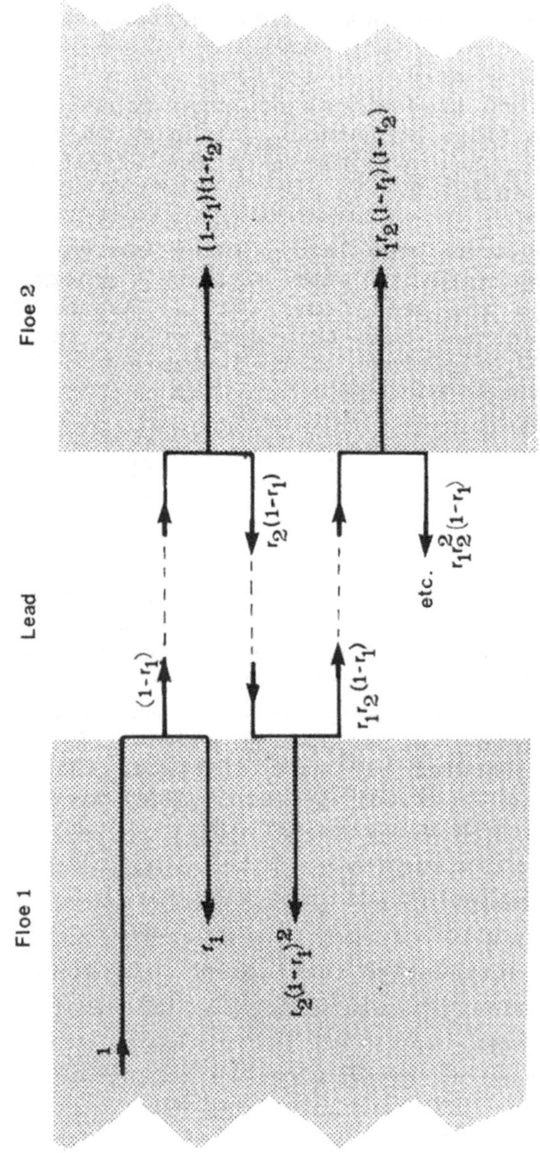

Fig. B.13. Schematic diagram of incoherent multiple reflection in a lead.

$$\alpha = -\, n\log_e (1-r) = nr + O(r^2) \qquad (14.52)$$

This model is suitable for describing the attenuation of flexural-gravity waves in the central Arctic due to leads--although they must be long leads, so that the opposite edges are dynamically independent--and it can also be applied to ice fields of very large floes, where (1/n) is the mean floe diameter. The minimum floe diameter for this model is of the order of 1 km, since smaller floes can heave in response to a wave so that the flexural wave energy is reduced. Small floes therefore tend to behave more as rigid floating rafts, attenuating wave energy by scattering, and will be treated in Section B.3.

It should be noted that the forward-going wave in Figure B.13 is split from a coherent sinusoidal wave into a sum of waves at different phases.

As a test of the model we can examine some wave measurements made by shipborne wave recorder in the Antarctic by Robin (1963). Figure B.14 shows the energy density of the 16 s swell component

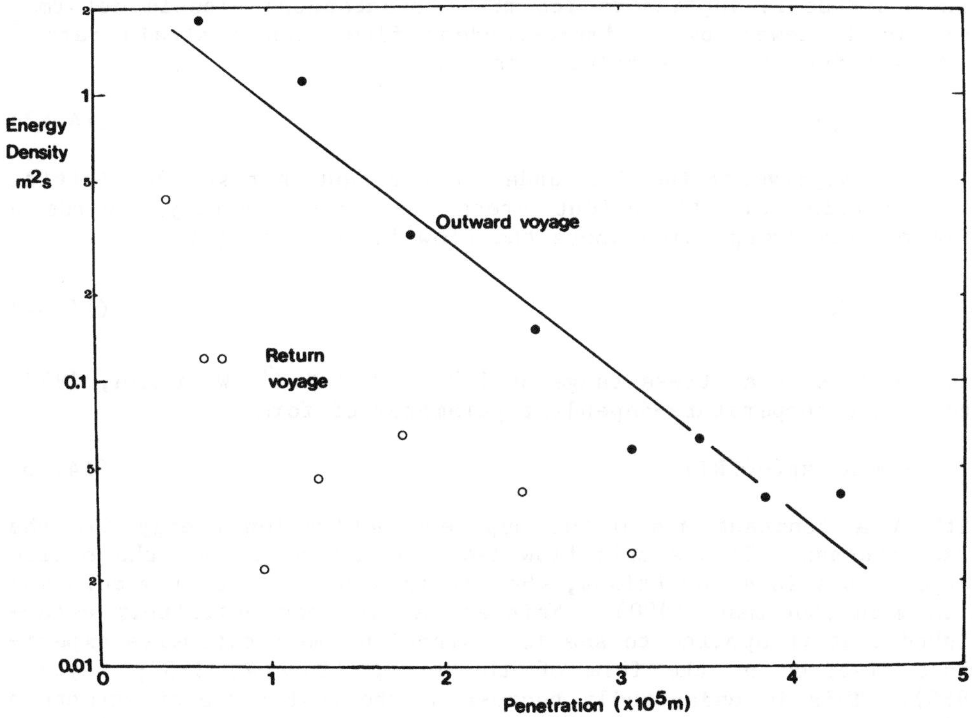

Fig. B.14. Decay of 16 s swell in the Antarctic ice cover as measured by Robin. (After Wadhams, 1973b).

plotted against penetration into the Antarctic ice cover. The results for the outward voyage approximately fit a negative exponential with $\alpha = (5.6 \pm 0.8) \times 10^{-6}\ m^{-1}$. Floe diameter varied from 1 km to more than 2 km. Taking the mean as 2 km and the results plotted in Figures B.9 and B.11 for 2.5 m thick ice we obtain

$$\alpha = 6.2 \times 10^{-6}\ m^{-1}, \ y = 0 \text{ fitting;}$$

$$\alpha = 2.1 \times 10^{-6}\ m^{-1}, \ y = 0, \ \lambda/4 \text{ fitting.}$$

The results are of the correct order of magnitude, which is remarkable considering the roughness of the estimates involved.

B.2.6 Attenuation Due to the Creep Properties of Ice

A further source of energy loss in a real ice field is the fact that ice is not a perfectly elastic material, but is viscoelastic, or in other words, possesses creep properties. A basic assumption of B.2.1 is therefore not strictly valid. The minimum stress required to initiate flow in ice (the elastic limit) is so small that it can be ignored (Pounder, 1965). This implies that when a stress is applied to ice the instantaneous elastic deformation is followed by a time-dependent flow with a strain rate $\dot{\varepsilon}$ that is a function of absolute stress, i.e.,

$$\dot{\varepsilon} = f(\sigma) \tag{14.53}$$

In polycrystalline ice under a constant stress the initial large strain rate (transient creep) decreases rapidly towards a steady-state creep which obeys the flow law of Glen (1955)

$$\dot{\varepsilon}_s = K\sigma^n \tag{14.54}$$

where n ~3 over a stress range of 10^5 to 10^6 Nm^{-2} (Weertman, 1972) and K is a temperature-dependent parameter of form

$$K = A \exp(-Q/RT) \tag{14.55}$$

with A a constant and Q the apparent activation energy for the creep process. This simple flow law does not hold over the entire range of possible conditions, when it is necessary to draw deformation maps (Goodman, 1980). Neither has it been definitely established that it applies to sea ice, since the most extensive experiments disagree on the form of the creep (Tabata, 1958; Peyton, 1966). This is undoubtedly because of the vast range of structure which sea ice can take up, on account of varying brine and air content and crystal size, shape and orientation in the fabric. However, Tabata's experiments showed that during a 5 s application of a stress of 10^6 Nm^{-2} the absolute creep strain was more than two orders of magnitude less than the elastic strain. This implies that our preceding analysis, based on the assumption that it is the elastic properties of the ice that determine the nature of wave propagation through it, is essentially valid. Creep needs to be considered only when we come to think of possible sources of energy loss.

Determination of the energy loss caused by creep to each volume element of the ice, through the alternating cycle of tension and compression as a wave passes through, is done by a dynamic strain analysis of a flexing sheet. The formal analysis is given in Wadhams (1973a,b) and will only be summarized here. We assume the validity of a Glen flow law of exponent n for sea ice, and we assume steady-state creep to occur throughout the oscillatory stress cycle (in fact the creep will be transient). We calculate the stress in the ice by linear elastic theory, then apply the flow law to calculate an energy dissipation rate, interpreted as a decay rate of the flexural-gravity wave with increasing penetration. Creep is thus seen as a small perturbation to a stress-strain situation dominated by elastic forces.

The flow law, as formalized by Nye (1953) for a three-dimensional state of stress, is

$$\dot{\varepsilon}_{ij} = \frac{\tau^{\eta-1}}{B^{\eta}} \sigma'_{ij} \tag{14.56}$$

where $\dot{\varepsilon}_{ij}$ = strain rate tensor, τ = effective shear stress = $|\sigma'_{ij}\sigma'_{ij}/2|^{1/2}$, σ'_{ij} = deviatoric stress tensor = $\sigma_{ij} - p$, σ_{ij} = applied stress, p = hydrostatic pressure, and i,j = 1 to 3. B, n are the flow law parameters with n the Glen exponent and B a function of temperature. If we use the geometry of Figure B.3 and neglect shear stresses due to bending and stresses which cause the ice sheet to spread under its own weight (Weertman, 1957), we are left with σ_{xx}, σ_{yy} and σ_{zz}. $\sigma_{yy} = 0$ since we assume elastic bending, and we have already assumed that $\sigma'_{zz} = 0$. For zero dilation it is necessary that

$$\sigma'_{xx} = -\sigma'_{yy} \tag{14.57}$$

Identifying x, y, and z as the principal axes, we have

$$\tau^2 = \frac{1}{2} (\sigma'^{\,2}_{xx} + \sigma'^{\,2}_{yy} + \sigma'^{\,2}_{zz}) \tag{14.58}$$

so combining (14.56) and (14.58) we are left with

$$\dot{\varepsilon}_{xx} = \left(\frac{\sigma_{xx}}{2B}\right)^n \tag{14.59}$$

as the applicable form of the flow law.

The rate of energy loss $\dot{W}$ per unit volume in the ice due to creep is

$$\dot{W} = \sigma_{xx}\dot{\varepsilon}_{xx} = \frac{\sigma_{xx}^{n+1}}{(2B)^n} \tag{14.60}$$

and if the ice profile at penetration x is

$$\eta_2 = {}_xA_i \cos (k_o x - \omega t) \tag{14.61}$$

then the tensile stress σ_{xx} at a vertical distance y from the center plane of the ice sheet (assuming that this is the neutral axis of bending) is

$$\sigma_{xx}(y) = \frac{2y}{h} \frac{Eh}{2(1-\nu^2)} \frac{\partial^2 \eta_2}{\partial x^2} = \frac{-yk_o^2 \eta_2 E}{(1-\nu^2)} \tag{14.62}$$

by the theory of elastic bending. This yields

$$\dot{W}(x,y,t) = 2B \left| \frac{k_o^2 E\eta_2 y}{2B(1-\nu^2)} \right|^{n+1} \tag{14.63}$$

The rate of energy loss per unit surface area is then

$$\dot{Q}(x,t) = 2\int_o^{h/2} \dot{W}(x,y,t)\, dy \tag{14.64}$$

But $\overline{\dot{Q}(x,t)}$, the rate of loss averaged over a wave period, is identical with $-(\partial E_t/\partial x)$ where E_t is the average rate of energy transmission per unit width of wavefront, given by (14.28) and (14.29) as

$$E_t = \frac{1}{2} \rho_w g \, {}_xA_i^2 \, U \left(1 + \frac{L k_o^4}{\rho_w g}\right) \quad (14.65)$$

(14.63-65) yield the final result

$$\frac{d\,{}_xA_i}{dx} = -S\,{}_xA_i^n \quad (14.66)$$

where

$$S = \frac{Kh^{n+2} k_o^{2n+2} I_n}{\rho_w g R_e U} \quad (14.67)$$

$$R_e = \left(1 + \frac{L k_o^4}{\rho_w g}\right) \quad (14.68)$$

$$K = \frac{(2B)^{-n}}{(n+2)} \left(\frac{E}{2(1-\nu^2)}\right)^{n+1} \quad (14.69)$$

$$I_n = \frac{1}{\pi} \int_o^{\pi} \sin^{n+1} \beta d\beta = \frac{1}{2} \text{ for } n = 1, < \frac{1}{2} \text{ for } n>1 \quad (14.70)$$

The solution is

$${}_xA_i^{n-1} = \frac{1}{(n-1)\, S_x + {}_oA_i^{-(n-1)}} \quad (14.71)$$

except if n = 1, when

$${}_xA_i = {}_oA_i \exp(-Sx) \quad (14.72)$$

Thus if sea ice creep were Newtonian ($n = 1$), it would cause exponential wave decay indistinguishable in form from the effect of progressive reflections or any other linear process. Field measurements of wave decay rate in a uniform ice sheet will therefore tell us the value of n for sea ice, from the form of the decay. If $n = 3$, for instance, (14.71) shows that the amplitude ${}_xA_i$ at a penetration relative to some arbitrary starting point $x = a$, is

$$\frac{1}{{}_x A_i^{\,2}} - \frac{1}{{}_o A_i^{\,2}} = 2S(x-a) \tag{14.73}$$

Measurements of wave decay in ice sheets are sparse: Wadhams (1973a) has discussed how this theory can be applied to existing observations, including Robin's (1963) results discussed in Section B.2.5, and has shown that the predictions of the theory are not incompatible with observations made to date, using an $n = 3$ flow law and a value for B slightly lower than that measured experimentally for polycrystalline ice.

A different approach is to assume Newtonian creep as in (14.72), and use a viscoelastic constitutive equation for ice from the start, instead of separating the analyses into an elastic analysis for propagation and a creep analysis for decay. This was done by Squire (1978b; Squire and Allan, 1980), who used a Maxwell-Voigt spring-dashpot model to represent sea ice behavior under tension. Figure B.15 shows the model, for which suitable values were suggested by Tabata (1958). On this model the stress-strain relationship is

$$\left(\frac{\partial^2}{\partial t^2} + a_1 \frac{\partial}{\partial t} + a_2\right) \sigma'_{xx} = \left(b_o \frac{\partial^2}{\partial t^2} + b_1 \frac{\partial}{\partial t}\right) \varepsilon'_{xx} \tag{14.74}$$

where σ'_{xx}, ε'_{xx} are deviatoric strain and stress, and a_i, b_i are related to the elastic moduli E_1, E_2 and the viscosities η_1, η_2 by

$$a_1 = \frac{E_1}{\eta_1} + \frac{E_1}{\eta_2} + \frac{E_2}{\eta_2} = 1.4 \times 10^{-2} s^{-1} \quad \text{(Tabata)} \tag{14.75}$$

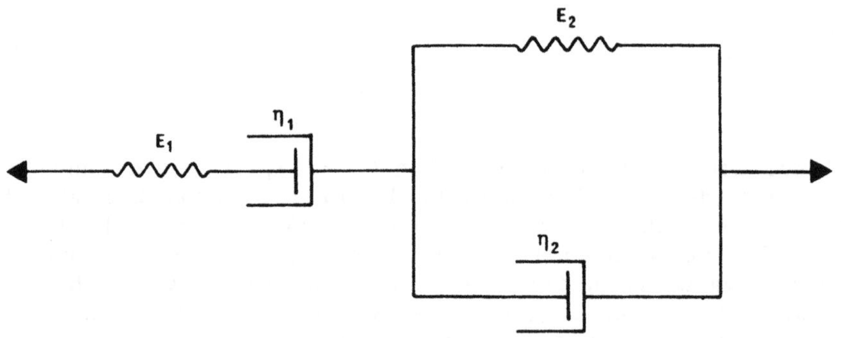

Fig. B.15. Maxwell-Voigt spring-dashpot model.

$$a_2 = \frac{E_1 E_2}{\eta_1 \eta_2} = 0.07 \times 10^{-4} s^{-2} \tag{14.76}$$

$$b_o = E_1 = 1.2 \times 10^9 \text{ N m}^{-2} \tag{14.77}$$

$$b_1 = \frac{E_1 E_2}{\eta_2} = 1.02 \times 10^7 \text{ Nm}^{-2} s^{-1} \tag{14.78}$$

Equation (14.8) is still valid as the equation of motion of the plate, except that L is replaced by a viscoelastic parameter which yields the equation

$$\frac{h^3}{6}\left(b_o \frac{\partial^2}{\partial t^2} + b_1 \frac{\partial}{\partial t}\right)\frac{\partial^4 \eta_2}{\partial x^4} + \rho_i h \left(\frac{\partial^2}{\partial t^2} + a_i \frac{\partial}{\partial t} + a_2\right)\frac{\partial^4 \eta_2}{\partial t^2}$$

$$= \left(\frac{\partial^2}{\partial t^2} + a_1 \frac{\partial}{\partial t} + a_2\right) p(x,t) \tag{14.79}$$

$p(x,t)$ derived in this way is applied to (14.10) and solutions of the form ϕ_2 are obtained as before, except that the quintic polynomial in k_n (equivalent of 14.13) is now

$$\frac{h^3}{6}(b_o\omega + ib_1)\, k_n^{\ 5} + (\omega - \frac{a_2}{\omega} + ia_1)(\rho_w g - \rho_i h\omega^2)\, k_n$$

$$- \rho_w\, (\omega - \frac{a_2}{\omega} + ia_1)\, \omega^2 = 0 \tag{14.80}$$

Again, three physically feasible solutions exist for k_n, of the form

$$\begin{aligned} k_o &= a_o + ib_o \\ k_1 &= a_1 + ib_1 \\ k_2 &= a_2 - ib_2 \end{aligned} \tag{14.81}$$

In other words, the two strongly evanescent waves do not form a pair with $a_1 = a_2$, $b_1 = b_2$, and the formerly conservative wave k_o now becomes a decaying wave, since $b_o \ll a_o$. In fact, b_o is several orders of magnitude less than a_o, and a_o is almost identical to k_o in (14.14). We therefore retain a situation almost like the elastic case, with two evanescent waves and a flexural-gravity wave that, instead of propagating with no energy loss, decays gradually in an exponential way with distance into the ice. Figure B.16

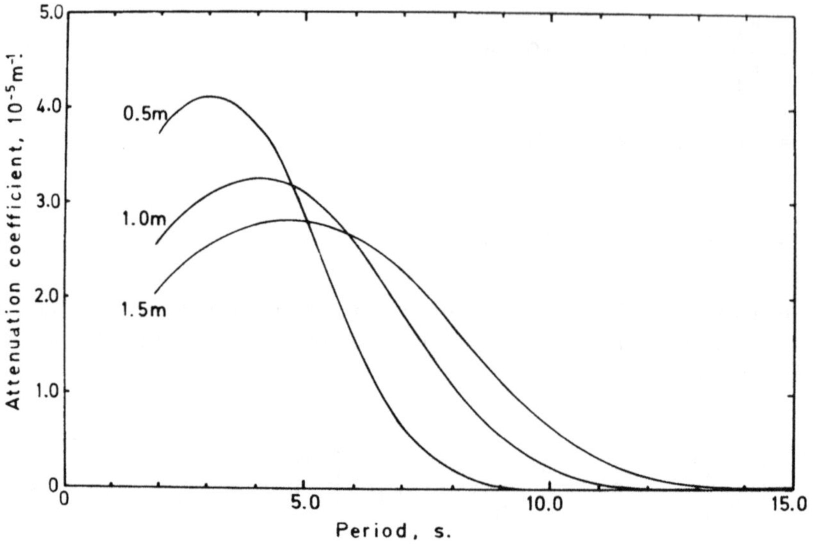

Fig. B.16. The attenuation coefficient versus period for a variety of ice thicknesses. (After Squire and Allan, 1980).

shows the values derived by Squire and Allan (1980) for b_0, which is an amplitude attenuation coefficient, for three different ice thicknesses. Squire and Allan also showed that at periods below about 5 s, the "amplitude" (i.e., magnitude of the velocity potential) at the ice edge of the strongly evanescent waves is greater than that of the flexural-gravity wave, but that above 10 s the strongly evanescent waves are quite negligible. Clearly this is because the flexural-gravity wave matches the open water wave in its properties much more closely at long periods. This implies that short-period waves should be most effective at breaking up an ice edge, since the evanescent components yield a peak of flexure near the edge which is relatively much greater than the "steady-state" flexure due to the flexural-gravity wave.

It is instructive to estimate the effect of the creep, or viscoelastic properties of ice upon the evolution over great distances of a wave spectrum incident from the open sea. Figure B.17 shows a typical spectrum for a wind sea, a Bretschneider (1963) spectrum chosen to have a peak period of 10 s and a spectral density of 2 m^2s at the peak. Entry into the ice immediately reduces the apparent spectral density (though not the energies) because of the amplitude ratio effect of Figure B.6. The progressive attenuation through creep (calculated using 14.73) reduces the energy preferentially at the short-period end of the spectrum, pushing the spectral peak to long periods until after 1000 km it reaches

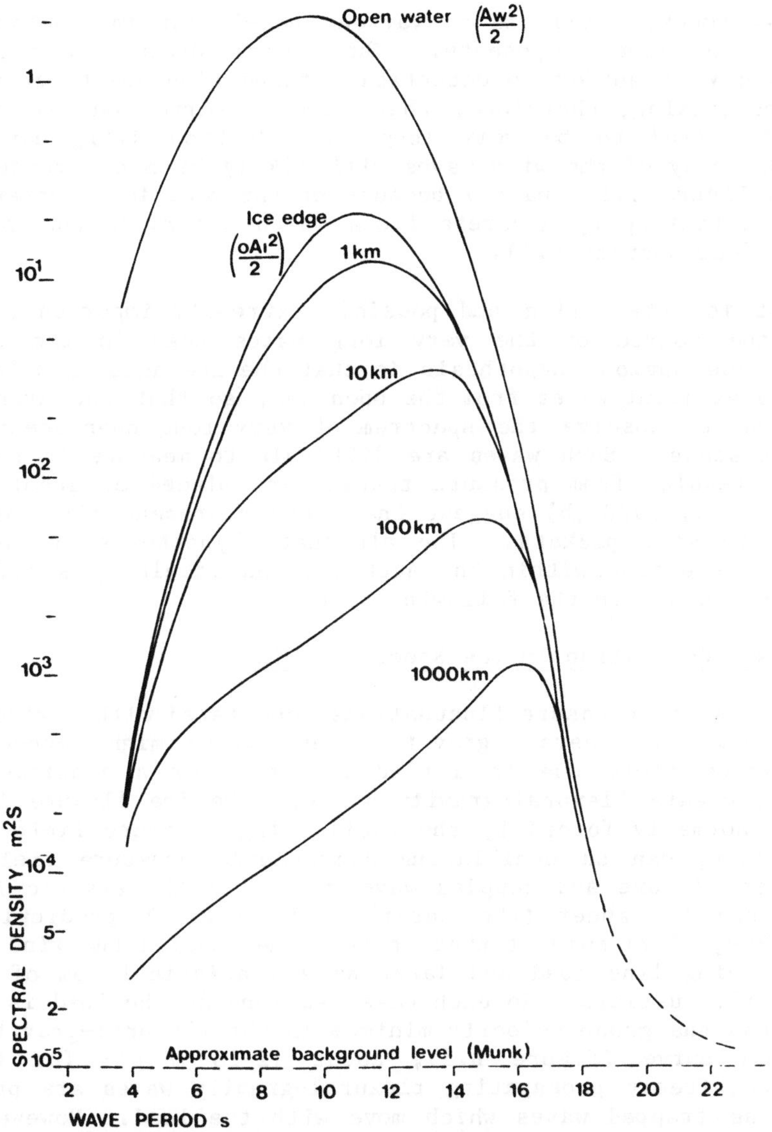

Fig. B.17. Evolution of a Bretschneider spectrum as it passes into a uniform ice cover. Ice thickness 5 m.

17 s. (The 5 m ice thickness mimics the mean thickness of the Arctic ice cover.) It can be seen that beyond about T = 18 s there is little energy loss even after 1000 km of propagation. Above 20 s, however, the Bretschneider spectrum fails to represent the real sea, since it falls off to minuscule values where in nature there is a background level of 10^{-5} - 10^{-6} m^2s ascribed to

surf beat (Munk, 1962) and to various non-linear interactions between shorter wave components. The theory shows that such very long waves will suffer no detectable attenuation due to creep. It is not surprising, therefore, that waves observed on the central Arctic ice tend to be very long (see Section B.1), especially since the decay of the wind waves will likely be much greater than shown in Figure B.17, partly because of the additional attenuation through scattering by discrete floes which occurs in the marginal ice zone (see Section B.3).

What is interesting and possibly extremely important, however, is the source of the very long waves seen in the central Arctic. The obvious hypothesis is that the ice acts as a low-pass filter to existing waves from the open sea, so that the Arctic ice permits us to observe the spectrum of very long open-ocean waves in a pure state. Such waves are difficult to measure in the open sea, and results from pressure transducers placed at great depths (e.g., Larsen, 1978a,b) suggest that they represent the envelopes of solitary wave packets. The alternate hypothesis is that the waves are generated within the Arctic Ocean itself by wind. This idea is discussed in the following section.

B.2.7 Wave Generation In Ice Sheets By Wind

Just as the pressure fluctuations associated with a wind blowing over the sea generate gravity waves, so we might expect that the pressure field due to a wind blowing over a continuous ice cover can create flexural-gravity waves. The ice flexure in such waves is normally forced by the oscillating pressure field in the water column; can an oscillating atmospheric pressure field also create ice flexure and coupled wave motion in the elastic foundation of the ice sheet (the water)? Theoretical predictions by Mills (1969, 1972) suggest that it can. He modeled the wind firstly as a moving line load and later as a statistical sum of micro-pressure fluctuations. In each case, so long as the load is moving faster than the group velocity minimum in the flexural-gravity wave dispersion curve (Figure B.5), about 12-15 m s^{-1} for typical Arctic ice, freely propagating flexural-gravity waves are produced as well as trapped waves which move with the load. However, the theory did not predict the magnitude of the coupling, and it did not predict the generation of the very long (~30 s) waves observed in the central Arctic.

The process is, of course, analogous to the generation of waves by a moving vehicle, and considerable theoretical and practical work has been done on this problem by Nevel (1970) and by earlier authors going back to Greenhill (1916). A critical combination of load and velocity can cause sufficient flexure to fracture the ice, and this has caused accidents to aircraft, vehicles, and skaters throughout history.

Whether the wind can act effectively like a moving load is a problem which is still unresolved. Hunkins (1962) found some correlation between wind speed and surface oscillation amplitude in the Arctic, so long as the wind speed exceeded 12 m s^{-1}. Sytinskiy and Tripol'nikov (1964) found that a gust of wind over a large multiyear floe produced firstly a short energy burst at 3 s period, ascribed to bobbing, then a 12-15 s wave continuing for a few minutes. As described in the next section, a lake experiment suggests that wind-ice coupling does indeed occur.

B.2.8 Current Research Problems

A large number of problems remain in this field. A fully satisfactory solution for the reflection coefficient at the ice boundary has not yet been obtained, nor has an analysis been done of reflection by variations in ice thickness, e.g., by pressure ridges. Pressure ridges can be 40 m or more deep, but a more typical depth is 10-20 m, and near the ice margin most ridges are eroded so that their depth is even less. They are typically linear features with slope angles of 30-35°, so the problem of a flexural-gravity wave crossing a pressure ridge is analogous to the problem of a wave in shallow water encountering a sea bottom feature. A better analysis is also required of the contribution made by evanescent waves to the flexure peak near the ice edge, i.e., the determination of x_{crit} and the peak flexure amplitude as a function of ice thickness and wave period.

Experimentally, a current problem is the measurement of directional spectrum, propagation velocity and attenuation rates of long flexural-gravity waves in the Arctic Ocean. The directional spectrum will determine the source area and the nature of the waves (locally wind-generated or coming from the open ocean), the velocity measurements will test the theory of Section B.2.1, and the attenuation rates will test the theory of Section B.2.6 and help establish the effective parameters of the creep process at these frequencies. As mentioned in Section B.1, an experiment has been carried out to make these critical measurements. Direct observations of the generation of flexural-gravity waves by wind are also required, in a situation which precludes generation by any other source. This implies a frozen lake as the site, and an experiment took place in March 1982 on a Norwegian lake in which arrays of strainmeters measured wave generation while an anemograph monitored the wind field, and a pressure sensor measured the micropressure variations at a point. Wave generation by wind has been observed, but results are still being analyzed by the present author and colleagues.

B.3 WAVE PROPAGATION IN A FIELD OF DISCRETE FLOES

B.3.1 A Model for Reflection and Transmission Due To a Single Floe

The effect of a floating body on an incident wave train is a problem that has received attention because of its relevance to ship motion. The major effect is one of wave drag, where the incident wave produces a forced response in the body, thus generating a scattered wave field. If the body is long compared with its beam, and if it faces a beam sea, the problem reduces to two dimensions and the scattering is simply equivalent to a partial transmission and partial reflection of wave energy. If we neglect viscosity and surface tension as well as quadratic terms in the body's displacement we are left with a two-dimensional problem in linear potential theory which is easy to formulate but difficult to solve.

Most studies have concentrated on the response of the body to the wave rather than the inverse effect. John (1950) showed that for an arbitrary cross-section the problem can be reduced to an integral equation, but its solution is difficult. MacCamy (1961) was able to obtain solutions for a flattened cylinder of shallow draft and a circular disc. A second approach is that of Ursell (1949, 1954), Bolton and Ursell (1973), and Havelock (1955), who considered the wave-making effect of an object undergoing forced vertical heave in still water. They found that the wave generation is equivalent to point- and dipole-sources distributed over the submerged surface of the object. The sources are most effective when the surface is horizontal, so a flattened body has a greater wave-making effect than one with steep sides.

Solutions to the problem of wave reflection and transmission by a freely floating body have been obtained only in the case of shallow water, where depth $\ll$ wavelength: John (1949) solved for a cylinder, and Stoker (1957) for a finite or semi-infinite raft. In deep water there are solutions only for a fixed surface obstacle (Ursell, 1961) or a fixed, thin vertical barrier (Ursell, 1947), together with some experimental studies reported in Wiegel (1964). We shall consider an approximate solution based on the same approach as for Section B.2.3.

Figure B.18 defines the geometry of the situation. The ice floe is of uniform thickness h, of length 2ℓ, and with an infinite lateral dimension so that the problem becomes two-dimensional; we assume that it has zero submergence. It floats in water of depth D, where $D \gg \lambda$, and faces a monochromatic wave of small amplitude incident from abeam.

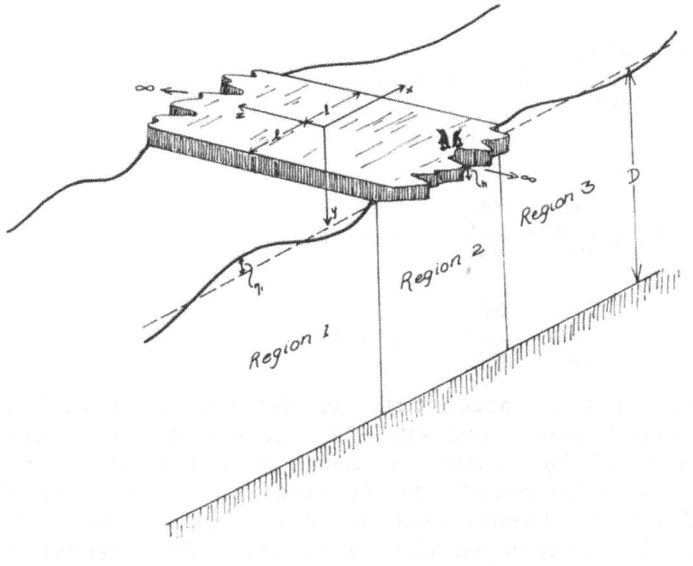

Fig. B.18. A long-crested monochromatic wave incident normally on an ice floe of infinite width.

There are now three regions to consider. Equations (14.3) to (14.6) are valid for all three regions. In equation (14.8) we should replace η_2 in the term containing $(\partial^4\eta_2/\partial x^4)$ by η_2', where η_2' is that part of η_2 which manifests itself as ice flexure (the remainder being heave and tilt). Since the heave and tilt must be a linear function of x we have $\partial^4\eta_2'/\partial x^4 = \partial^4\eta_2/\partial x^4$, so we can retain η_2 and leave (14.6) unaltered, so that (14.12) is still the solution for ϕ_2.

In region 3 the same boundary conditions prevail as in region 1, and are satisfied by a potential of the form

$$\phi_3 = Te^{ikx}\, e^{-ky}\, e^{-i\omega t} \tag{14.82}$$

where $|T|/|B|$ is the amplitude transmission coefficient.

If we set B equal to unity we need to derive R, A_n, B_n and T, i.e., eight complex quantities, for complete solutions to (14.7), (14.12), and (14.82). Since there is no dissipation the solution should satisfy

$$|R|^2 + |T|^2 = 1 \tag{14.83}$$

The necessary equations must come from the transition conditions between regions, which, by analogy with Section B.2.3 are now

$$\frac{\partial^2 \eta_2}{\partial x^2} = \frac{\partial^3 \eta_2}{\partial x^3} = 0 \ , \quad x = \pm \ell \tag{14.84}$$

and

$$\phi_1 = \phi_2 \ , \ \frac{\partial \phi_1}{\partial x} = \frac{\partial \phi_2}{\partial x} \qquad x = -\ell \tag{14.85}$$

$$\phi_2 = \phi_3 \ , \ \frac{\partial \phi_2}{\partial x} = \frac{\partial \phi_3}{\partial x} \qquad x = \ell \tag{14.86}$$

for all y. Again, potentials as defined in (14.7), (14.12), and (14.82) cannot alone satisfy the transition conditions since the y-dependence of ϕ_2 cannot be made to match that of ϕ_1 and ϕ_3. Once again we approximate by fitting only at $y = 0$; this solution was developed by Hendrickson et al., (1962), although the paper contained sign errors in the equations. The correct equations are presented in Appendix 2 in a form suitable for solution by a matrix-solving computer subroutine.

The solutions for the energy reflection coefficient r ($= |R|^2$) and transmission coefficient t ($= |T|^2$) by a single floe have to be interpreted in terms of an attenuation due to successive floes encountering the wavefront. A real ice field consists of an array of floes of differing sizes and shapes. Each floe scatters energy in all directions in the x-z plane, but, since its induced motion under the long-crested wave of Figure B.14 is all in the x-y plane, most of the scattered energy will be in the x-direction. Our two-dimensional assumption is still reasonably valid provided we picture our infinitely long ice floe as equivalent to a row of floes, each of diameter d ($= 2\ell$), situated with their centers at the same value of x so that they all respond in phase to the incoming wave. Energy scattered by individual floes in the positive and negative x-directions is then reinforced, whereas energy scattered to the side is not.

If floes of diameter d occupy a fraction p of sea surface area, there will be p/d "equivalent rows" per unit penetration in the x-direction. The simplest possible assumption is that only single scattering occurs, i.e., each row transmits a fraction t of the energy incident upon it, while the reflected fraction r is dissipated in an unspecified way before it can get back to the preceding row. Let unit energy per unit area be incident on the ice edge. Then after passing n rows of floes the forward energy is reduced to $E_f = t^n$. Now

$$\frac{\partial E_f}{\partial x} = \left(\frac{p}{d}\right) \frac{\partial E_f}{\partial n} = \frac{p}{d}\, t^n \log_e t \tag{14.87}$$

The amplitude attenuation coefficient for exponential decay α_x has been defined by

$$-\frac{\partial E}{\partial x} = 2\alpha_x E \qquad (14.88)$$

so we have

$$\alpha_x = -\frac{p}{2d}\log_e t = \frac{pr}{2d} + o(r^2) \qquad (14.89)$$

For a composite ice field, where floes of size class d_i occupy a fraction p_i of the sea surface area,

$$\alpha_x = \frac{1}{2}\sum_i \frac{p_i r_i}{d_i} + o(r_i^2) \qquad (14.90)$$

Figure B.19 shows some typical results of computations of α_x using (14.89) with p = 1. The following observations can be made:

i) At long wave periods α_x falls off following approximately a power law which is nearly the same for all sizes of floe. A rough guide is

$$\alpha_x \propto \left(\frac{1}{T}\right)^{6-8} \qquad (14.91)$$

where the exponent increases with T. The magnitude of α_x in this range is also similar for all size classes, though somewhat larger for big floes.

ii) As T decreases, α_x rises to a peak value at approximately $\lambda = 2d$, then falls steeply to a resonance at which the reflection is zero. For ice floes less than about 150 m in diameter there are two types of resonances. One occurs when $\lambda \doteqdot d/m$ with m an integer. The other is related to λ_o (= $2\pi/k_o$), the wavelength of conservative propagation under the floe, and takes place when $\lambda_o \doteqdot 2d/m$. These may be termed "water resonances" and "ice resonances" respectively. For larger floes the position is more complex as the resonances tend to become superimposed. λ_o is a function of ice thickness and at high frequencies is much greater than λ. Thus small floes ($d < 40$ m) experience a water resonance first with $\lambda \doteqdot d$, since λ_o is too great at that frequency for an ice resonance to occur. For floes with $d > 60$ m an ice resonance occurs first. There is a transition region at about d = 50 m where resonances of the two types occur at similar frequencies, so that the frequency band over which α_x is anomalously low is

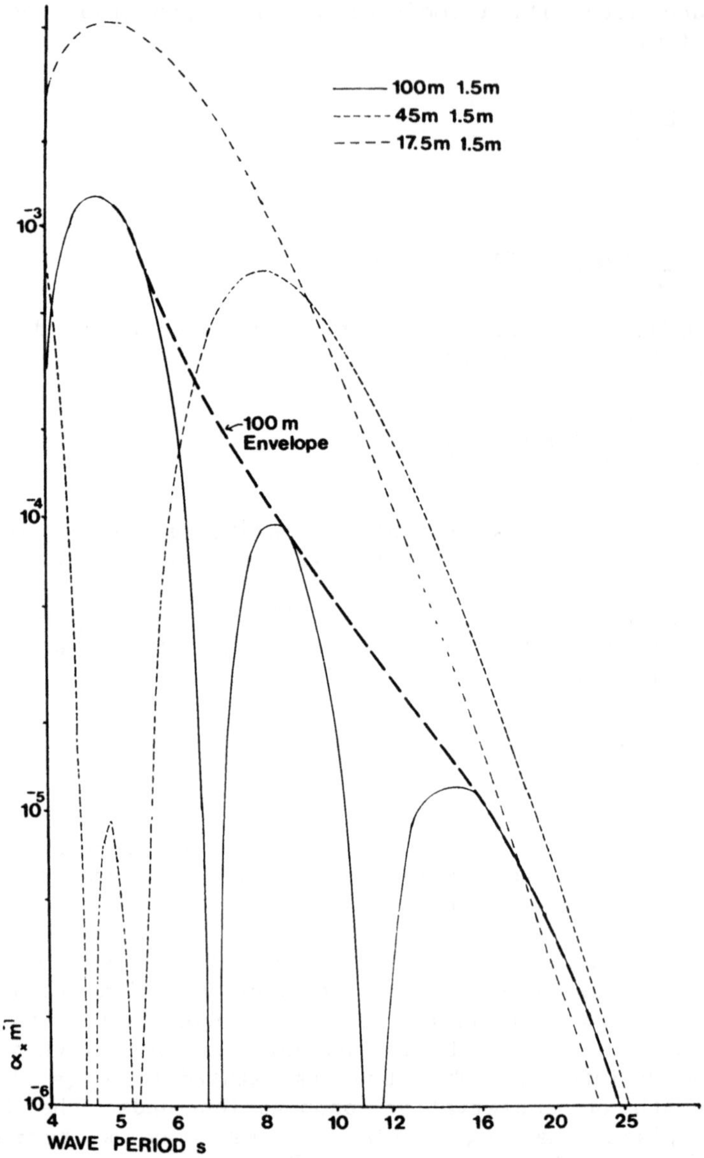

Fig. B.19. Computation of amplitude attenuation coefficient α_x for floes of different diameters.

considerably broadened. An example of this is the 45 m floe of Figure B.19.

iii) Between resonances α_x rises to touch an envelope which in real ice fields is a better measure of the true attenuation.

The resonance bands are normally narrow compared to the distance between them and are shifted considerably by a small change in d, so the range of floe sizes in a real ice field causes the anomalies to be smoothed out in the summing process of (14.90). We note also that the decay of α_x with increasing period is more gentle along the envelope than along the low-frequency tail beyond the first resonance.

iv) As the period decreases, the attenuation becomes dominated by successively smaller classes of ice floe. A change in the floe size distribution of an ice field will therefore alter the character of the attenuation rather than simply increase or decrease its overall efficacy.

v) The effect of varying ice thickness is shown in Figure B.20, using a 25 m floe as the example. At short periods $\alpha_x \propto h^{1/2}$ approximately, while at longer periods the relationship is almost linear. An additional effect occurring in floes more than 40 m across is that the frequency of their first (ice) resonance is reduced, since λ_o increases with h. Water resonances are unaltered.

vi) Small floes, in which the flexure is of minor importance, should obey Froude scaling. This suggests that two floes $_1$ and $_2$ with the same d/h ratio should have the same reflection coefficient at wavelengths λ_1 and λ_2 such that

$$\lambda_1/\lambda_2 = d_1/d_2 = h_1/h_2 \tag{14.92}$$

Figure B.21 shows that this is valid for floes less than about 30 m in diameter, provided the frequency is not close to that of a resonance. Of course, when the floe diameter becomes very small the entire analysis becomes invalid, since we can in no way think of the floe as being an elastic raft. Instead a theory based purely on inertia may be more valid; such a theory is presented in Section B.4.

B.3.2 Simple and Multiple Scattering

The single scattering approximation is valid if the ice cover is sparse and if the wave-measuring technique enables one to distinguish between the forward and returning waves; most ice fields encountered in nature are sufficiently concentrated for multiple scattering to be important. A simple treatment of this makes the following assumptions:

a) A reflected wave vector incident on the far side of a floe experiences the same reflection and transmission coefficients as the main forward wave vector incident on the front.

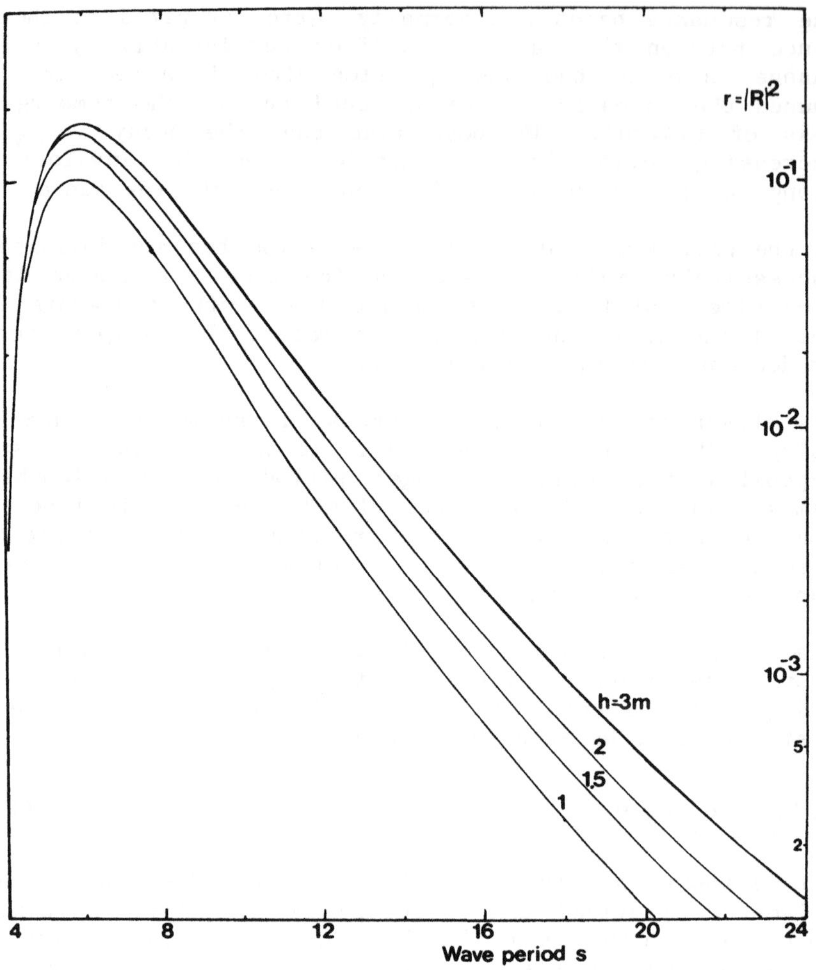

Fig. B.20. Effect of ice thickness on reflection coefficient, for a floe of 25 m diameter.

b) A wave vector is allowed to suffer up to two reflections only.

c) Energy contributions from individual wave vectors can be added arithmetically, i.e., the scattering is incoherent.

The forward energy after n rows of floes is now modified by contributions from waves which have been reflected twice, the second reflection occurring at the kth floe such that $1 \leq k \leq n$. Some typical contributing vectors are shown in Figure B.22. The total forward energy becomes

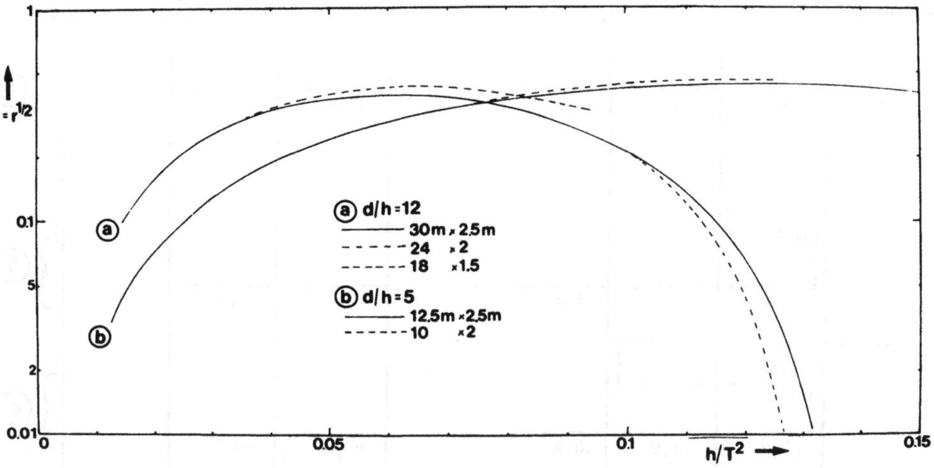

Fig. B.21. Froude scaling in small floes. For each d/h ratio the left-hand parts of the curves merge, showing true Froude scaling; the curves diverge from each other when they begin to approach their individual resonances, which depend on absolute diameter.

$$E_f = t^n + nt^n r^2 (1 + t^2 + t^4 + \ldots) = \frac{t^n}{1+t} [2 + (n-1)r] \tag{14.93}$$

The returning energy E_r is the sum of single reflections from $n+1 \leq k < \infty$ (for an ice field of finite length X, ∞ is replaced by $\overline{X}p/d$). These give a total contribution

$$E_r = t^n/(1 + t) \tag{14.94}$$

A wave-measuring technique which records only the forward wave yields, from (14.88) and (14.93), an attenuation coefficient of

$$\alpha_x = \left[r - \frac{r}{2+(n-1)r} + o(r^2)\right] \cdot \frac{p}{2d} \tag{14.95}$$

while one such as the shipborne wave recorder, which records the total energy E (= $E_f + E_r$), yields

$$\alpha_x = \left[r - \frac{r}{3+(n-1)r} + o(r^2)\right] \frac{p}{2d} \tag{14.96}$$

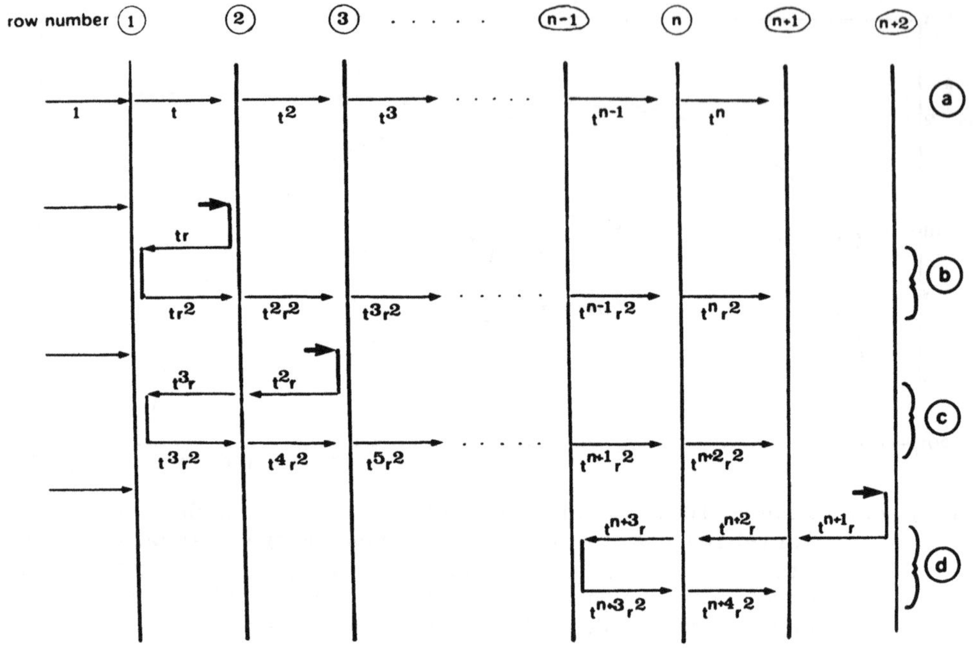

Fig. B.22. Some ray paths through a series of rows of ice floes. (a) is the primary wave vector; (b), (c) and (d) are some vectors that have each suffered two reflections when they reach the nth row. Vertical lines symbolize each row. Quantities shown are energies relative to unit incident energy.

Thus the effect of multiple scattering is to reduce the attenuation coefficient, especially in the outermost parts of the ice field. At the ice edge itself the factor is 1/2 (E_f only) or 2/3 ($E_f + E_r$) and it does not begin to increase significantly until we reach a penetration such that $(n-1)r \sim 2\text{-}3$. This corresponds to

$$x \sim 2 - 3 \left(\frac{d}{rp}\right) \tag{14.97}$$

which in most cases is of the order of km or tens of km. At very great penetrations, α_x asymptotically approaches its value for single scattering, going as $\frac{p}{2d}(r - d/x)$.

B.3.3 Inferred Attenuation Rates and Comparison with Observations

To compare this model with field observations we must assume that the reflection coefficient is the same for a spectral component of a sea as for a monochromatic wave of the same frequency; this assumption is basic to any linear theory.

Laser observations. The laser observations of Wadhams (1975) provide good data for testing this model, since the infrared line scanner provided precise information on floe size distributions. A theoretical curve for α_x was constructed using the floe sizes observed, and as can be seen from Figure B.23, it provides an excellent fit to the data so long as multiple scattering is assumed to occur. As we expect, the use of a sum of various floe sizes smoothes out the resonances described in Section B.3.1.

Submarine observations. It was more difficult to test the submarine observations of Wadhams (1978a) since floe sizes could be only imperfectly estimated from periscope measurements. However, a plausible floe size distribution was chosen and α_x computed; the results (Figure B.24) do not fit so well as for the laser, the theory predicting a higher decay than is actually observed. A possible cause is that the icefield was very concentrated, so that considerable interaction between floes was taking place. As well as causing multiple scattering, this may in some circumstances cause groups of floes to behave dynamically as a single floe. The effective floe sizes may therefore be much greater than the observed floe sizes; the effect of this would be to reduce the attenuation since the reflection coefficient r usually increases less than linearly with d, allowing the increased diameter to more than offset the effect of the increased reflection coefficient.

Shipborne wave recorder observations. The SBWR observations of Dean (1966) from the Antarctic can be tested against this model because Dean deduced floe sizes at different penetrations into the ice from oblique photographs taken from the ship's bridge. Dean found that at a given frequency the attenuation rate decreased with increasing penetration; on our theory this is due to a change in the floe size distribution. From Dean's spectra we have therefore obtained attenuation rates as a function both of period and penetration, and tested them against theoretical curves constructed from the floe size distributions found by Dean. The results are shown in Figure B.25 for the second series of observations by Dean. The fit is generally fairly good, especially if we assume multiple scattering and thereby reduce the attenuation coefficients shown. At low periods in the outer part of the icefield the attenuation was greater than predicted, but here there was large amount of brash ice between the floes, not recorded by Dean but observable on his photographs. This may have increased the attenuation rate; the case of propagation through an icefield composed purely of brash ice is considered in Section B.4.

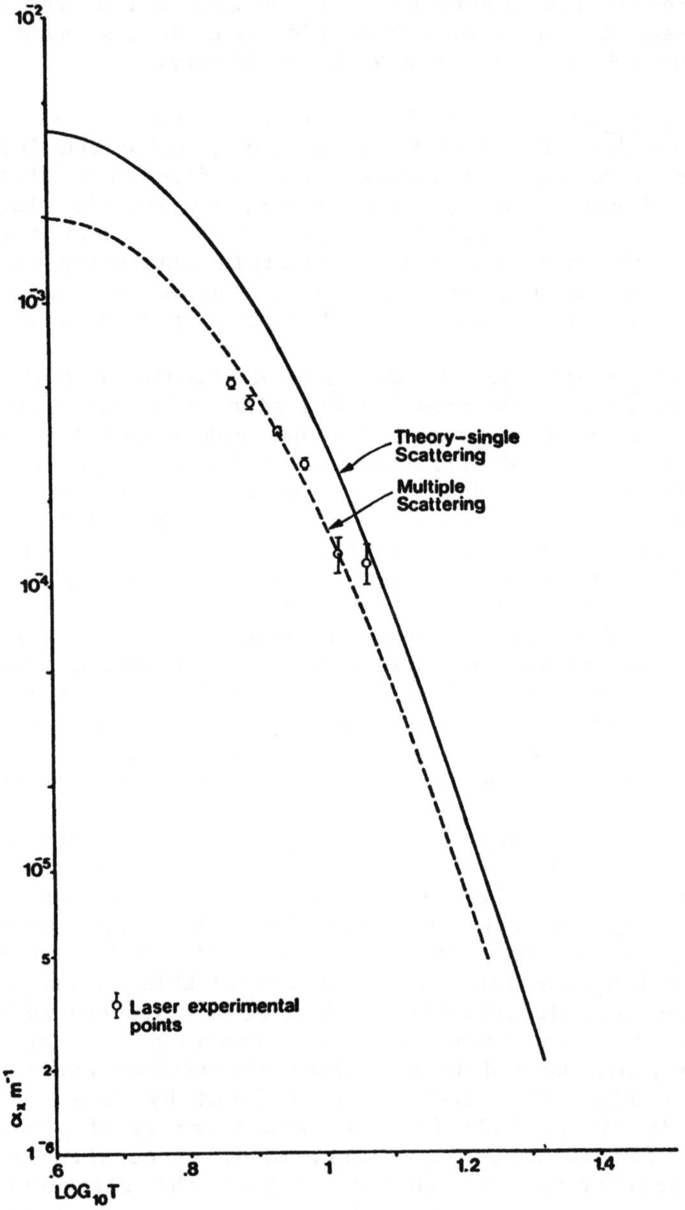

Fig. B.23. Fit of theory to laser observations. (After Wadhams, 1975).

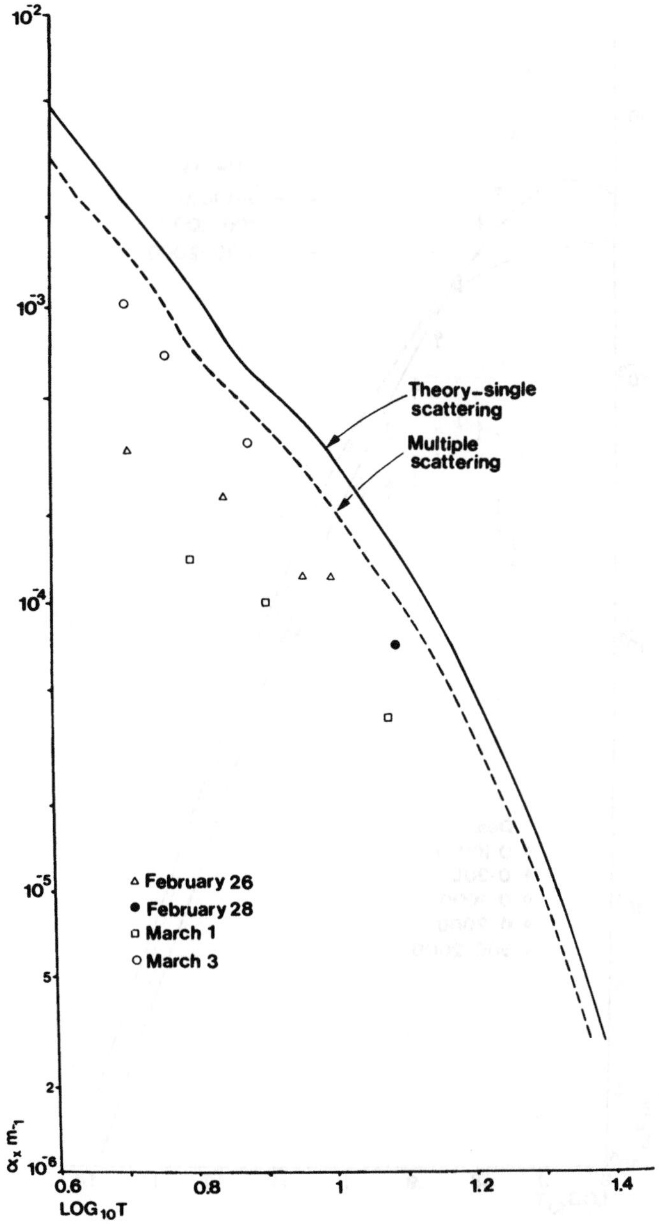

Fig. B.24. Fit of theory to submarine observations.

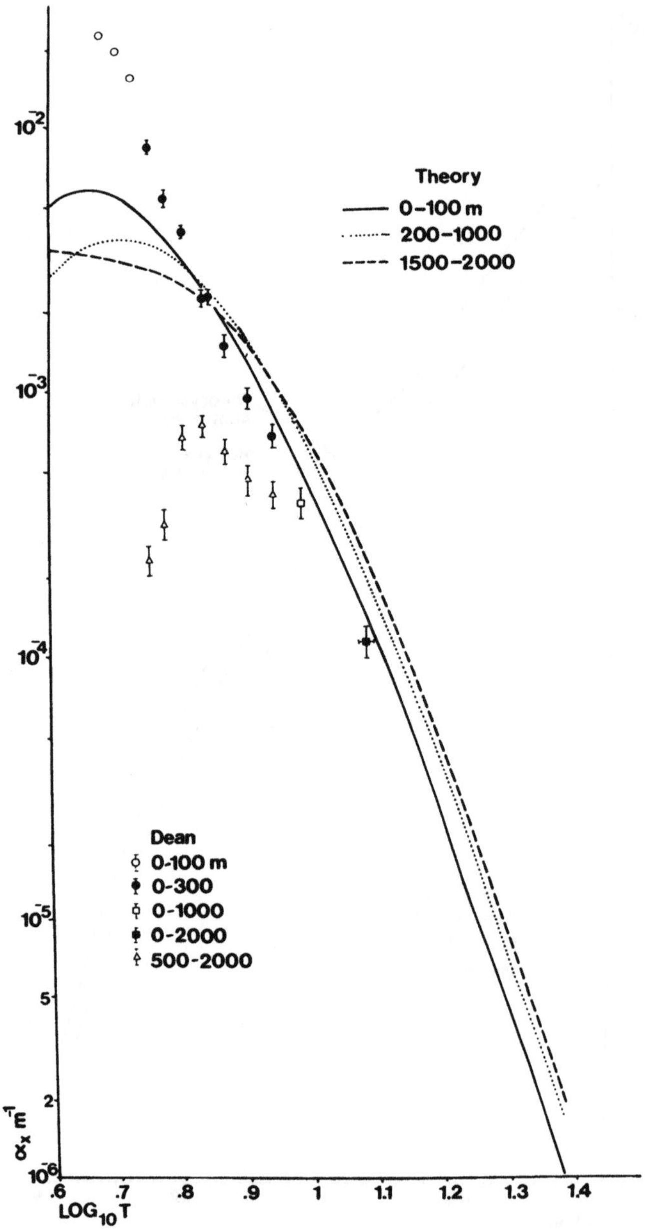

Fig. B.25. Fit of theory to Dean's second series of observations (designated 1963/4(ii) in Dean (1966)).

Accelerometer observations. The series of observations in east Greenland and the Bering Sea (Wadhams, 1979; Wadhams and Squire, 1980; Squire and Moore, 1980; Moore and Wadhams, 1980) provides excellent data to test this theory. Wave energy was measured using either a wave buoy inserted in the water between floes or an accelerometer on floes, and concurrent aerial photography and coring provided good control for floe sizes and thicknesses. Most of the data analysis has not yet reached the literature, but in general there is good agreement with the theory. For instance, some decay measurements from the Bering Sea (Squire and Moore, 1980) yielded exponential decay with the following energy attenuation coefficients:

Table B.1

Period s	$2\alpha_x$ (observed) $\times 10^{-4}\ m^{-1}$	$2\alpha_x$ (theory) $\times 10^{-4}\ m^{-1}$
12.2	0.27	--
12		0.26
10		0.86
9.4	0.44	
7.6	0.86	
6.4	1.09	
6	--	6.0
5.5	1.21	

The theoretical values on the right of the table were obtained using a 25 m diameter floe of 0.3 m thickness and assuming 50% ice cover. Actual floe diameters ranged from 10-20 m near the ice edge to 100 m some 70 km into the ice, with thicknesses of 0.3 m normally and 0.5 m with rafting. Considering the range of conditions compared with the single floe size used in the computation, the agreement between theory and observation is excellent, except at very short wave periods. Excellent agreement was also obtained in the case of the MIZEX-86 measurements made with a directional wave buoy (Wadhams et al., 1986).

B.3.4 Current Research Problems

The presently available theory of reflection from a floe, described in this section, is inadequate. Apart from the fact that it is not an analytical solution to the problem, it also ignores the wave-making effect of the heave and pitch of the floe, i.e., it assumes that only the mismatch between the free sea surface and the elastic floe surface causes the energy reflection. It also ignores diffraction and the effect of the finite submergence of the floe edges. Further, to convert a reflection into an attenuation by assuming that a typical irregular floe scatters energy only directly backwards, is also inadequate. A better theory must begin with a finite floe of some typical shape (round or square) and attempt to determine the entire scattered wave field due to such a floe. We do not expect that an analytical solution of this problem will be obtained, but there is much promise in the numerical techniques used to analyze the wave-making resistance of ships, where the immersed surface of the object is approximated by polygonal mapping using a grid, each grid point being an independent oscillating point source. An attack on the problem in this way is at present being carried out in a cooperative project between SPRI and C.M. Lee of David Taylor Model Basin. Considerable progress has been made using this technique to interpret measurements on sea ice in East Greenland (Squire, 1981,1983) and a series of papers will soon appear in the literature.

Plentiful well-controlled experimental evidence now exists of wave decay rates in different types of icefield. It appears from the evidence that in a highly concentrated ice cover the wave decay rate is reduced relative to an open icefield which is otherwise similar in floe size distribution. It seems that a dynamic interaction between floes occurs, such that aggregations of floes behave as single entities from the point of view of wave scattering. More research into this problem needs to be done.

Finally, there comes a point where any theory of the above type will become inapplicable because of small floe size. We lack measurements and a suitable theory for wave decay in fields of pancake, brash, or frazil ice. A first approach is given in the next section, but it completely lacks experimental verification as yet.

B.4 WAVE PROPAGATION THROUGH FRAZIL AND BRASH ICE

B.4.1 The NYU Analysis

When the floe size is very small we enter a new regime of wave-ice interaction in which we can no longer consider the ice as being made up of elastic rafts. Instead the individual floes are no longer important and we consider the properties of a continuous medium. Such a medium has two properties:

density - there is a mass per unit area of floating material covering the sea surface, which can still be thought of as equivalent to a thickness h of density ρ_i, and which contributes the $(\rho_i h \partial^2 \eta_2 / \partial t^2)$ term to the equation of motion of the ice (14.8).

coherence - the material interacts laterally in some way to form a floating mat. The interaction may approximate to an elastic sheet, or it may be simply one of friction between contiguous pieces of brash, shuga or pancake ice, or contiguous crystals of frazil ice.

If we assume the coherence to be zero, we can obtain a solution for wave propagation setting L = 0 in (14.8) and thus in (14.13) we obtain the solution

$$k_n = \frac{\rho_w \omega^2}{\rho_w g - \rho_i h \omega^2} \tag{14.98}$$

The ice field is now a continuous medium composed of a dense cover of small non-interacting mass points. A wave in such a medium propagates with the reduced wavelength $2\pi/k_n$, so it has the following properties:

i) The phase and group velocities in the ice are less than in open water, so a wave incident obliquely is refracted towards the normal.

ii) There is a frequency limit given by

$$\omega_o^2 = \frac{\rho_w g}{\rho_i h} \tag{14.99}$$

above which propagation is not physically possible; an incident wave at a higher frequency will suffer total reflection at the ice edge.

The wave of (14.98) may be termed the "NYU" wave because its properties were determined by colleagues of Stoker at New York University. The transmission into the ice field was studied by Weitz and Keller (1950) and by Peters (1950). Weitz and Keller considered a wave incident on the ice edge at an angle θ to the normal such that

$$k_y = \frac{2\pi \sin\theta}{\lambda} \tag{14.100}$$

and solved for ϕ under the ice cover using a Green's function. In a later paper (Keller and Weitz, 1953) they were able to simplify their solution to:

$$|R| = \left| \frac{\left(k^2-k_y^2\right)^{1/2} - \left(k_n^2-k_y^2\right)^{1/2}}{\left(k^2-k_y^2\right)^{1/2} + \left(k_n^2-k_y^2\right)^{1/2}} \right|_{\omega<\omega_c} \tag{14.101}$$

where $|R|$ is the amplitude reflection coefficient. The amplitude transmission coefficient $|T|$ is obtained by equating energy fluxes at the ice edge:

$$(1 - |R|^2) = |T|^2 U_i/U_w = |T|^2 k^2/k_n^2 \quad (14.102)$$

where U_w (= $\partial\omega/\partial k$) and U_i (= $\partial\omega/\partial k_n$) are group velocities in water and ice respectively. The energy transmission coefficient

$$t = |T|^2 U_i/U_w = \frac{4k k_n}{(k + k_n)^2} \quad (14.103)$$

These results refer to deep water: Keller and Weitz also derived $|R|$ and $|T|$ for water of depth $\ll \lambda$, obtaining the same values as Keller and Goldstein (1953); such results are relevant to an ice-filled bay or estuary.

Figure B.26 shows $|R|^2$ and $|T|^2$ for different ice thicknesses, assuming normal incidence. Note that $|T|^2 > 1$ at all frequencies. A remote sensing wave instrument, therefore, would record an increase in wave height inside the ice edge; it would also record an increased wave number.

An SBWR recording, however, involves placing the ship in an open pool, so the transmitted wave must pass an ice → water boundary before being recorded as a free water wave. Assuming that the multiple internal reflections occurring in a medium of finite length are incoherent, we obtain for the overall energy transmission coefficient:

$$t_2 = |T|^2 |T_{iw}|^2 (1 + |R|^4 + |R|^8 \ldots) = \frac{2k k_n}{k^2 + k_n^2} \quad (14.104)$$

where R is the same for ice → water as for water → ice, and T_{iw} is the amplitude transmission coefficient for ice → water, given by

$$T_{iw}^2 = \frac{4k^3}{k_n (k + k_n)^2} \quad (14.105)$$

A wave buoy placed in the icefield would record an amplitude intermediate between T_{iw} and $|T|$, depending on how closely it was surrounded by the material of the ice cover.

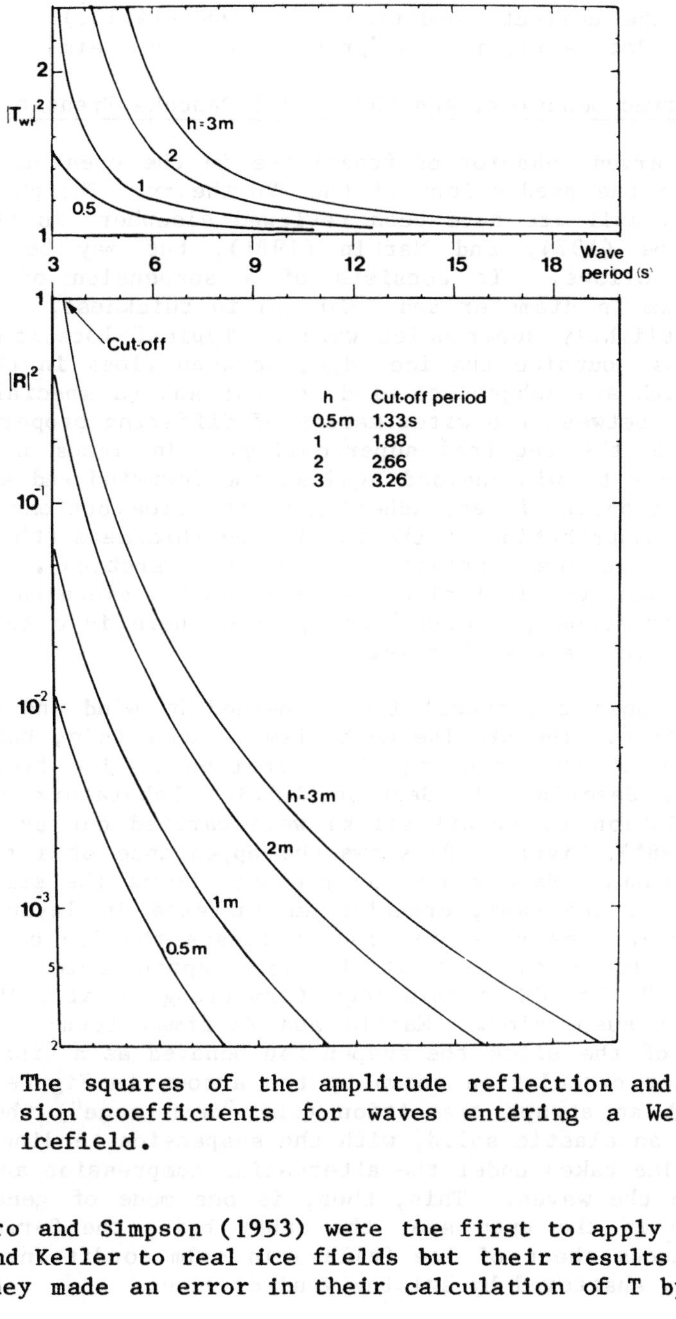

Fig. B.26. The squares of the amplitude reflection and transmission coefficients for waves entering a Weitz-Keller icefield.

Shapiro and Simpson (1953) were the first to apply the theory of Weitz and Keller to real ice fields but their results are erroneous. They made an error in their calculation of T by assuming that

$$1 - |R|^2 = |T|^2\, k/k_n$$

instead of the correct condition given by (14.102). This forced them to postulate a fictitious "pressure energy" term.

B.4.2 Observed Behavior, and the Frazil-Pancake Transition

The observed behavior of frazil ice in the open sea resembles in some ways the predictions of the NYU theory. The physical properties of frazil ice have been reviewed elsewhere in this volume and by Hobbs (1974) and Martin (1981), but may be summarized briefly as follows. It consists of a suspension of small ice discs, 1-4 mm in diameter and 1-100 μm in thickness, that form in turbulent, slightly supercooled water. Typical locations are the open sea just outside the ice edge; between floes in the MIZ; in polynyas which are subject to wind action; and (a special case) at the boundary between two water masses of differing properties whose mixing yields the required supercooling. In leads newly-formed frazil ice may be wind-herded against the downwind end and advected under neighboring floes, adhering to the floe bottoms to make a substantial contribution to the total floe thickness (this is especially common in the Labrador Sea and in Antarctica). In frazil observed at sea, the initial discs have usually sintered into clusters, 3-10 mm across, called "flocs," and there is a tendency for floc size to increase with time.

In the open sea frazil ice is herded by wind and waves into bands or slicks. The precise mechanism is uncertain, but is probably similar to that causing the formation of ice floe bands at the ice edge described in Section C.3.3. Laboratory experiments on the evolution of frazil slicks were carried out by Martin and Kauffman (1981); Figure B.27 shows the appearance of a slick grown in a wave tank. Wave radiation pressure herds the slick against the far end of the tank, creating an increase in layer thickness with distance. The pressure also increases the ice concentration within the slurry so that it is more appropriate to call it "grease ice," the WMO terminology (Armstrong et al., 1966) for a dense frazil suspension. Martin and Kauffman found that in the outer part of the slick the suspension behaved as a viscous liquid with an inner circulation shown by the arrows in Figure B.27, but that beyond an abrupt transition (the "dead zone") the behavior was that of an elastic solid, with the suspension tending to cohere into large ice cakes under the alternating compression and rarefaction due to the waves. This, then, is one mode of generation of pancake ice in the open sea, the other being the formation of a thin continuous sheet of ice (nilas) in calm conditions, which is subsequently shattered by arising wind or waves.

Martin and Kauffman found that in the zone of liquid behavior waves decayed linearly with distance with a slope β of

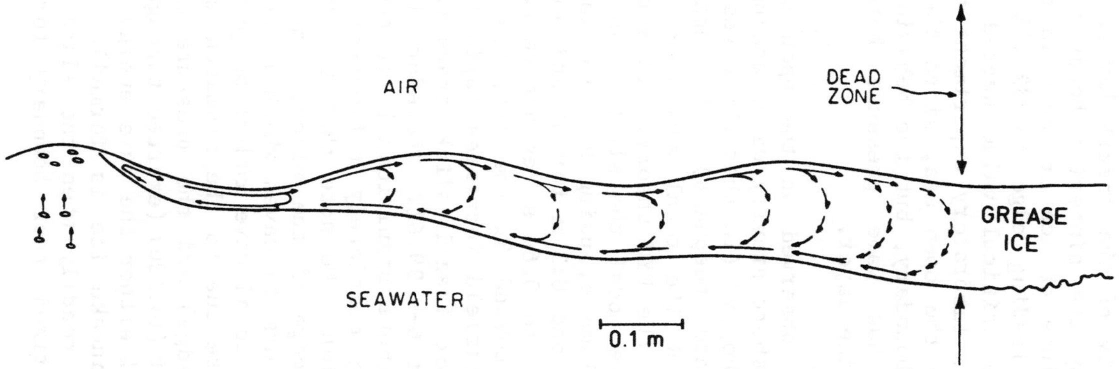

Fig. B.27. Circulation in a frazil ice slick subject to wave action. (After Martin and Kauffman, 1981.)

$$\beta \approx \frac{1}{4}(a_o\ k)^2 \qquad (14.106)$$

where a_o is the amplitude of incident waves and k is the wave number in open water. They explained this rate in terms of a shear-dependent viscosity of the frazil layer, the nonlinear viscosity being created by the sintering bonds between frazil crystals. The relative volume of ice crystals in the layer increased from 18-22% near the leading edge to 32-44% at the dead zone, which is apparently the critical value needed for the transition to pancake ice. In the laboratory experiment the herding is much exaggerated relative to the open sea, since the rear wall of the tank provides a fixed boundary, but the behavior is probably qualitatively correct. Thus wave pressure herds, thickens, and intensifies the frazil ice layer.

Frazil ice slicks observed on the open sea appear "smooth" (in that capillary and short gravity waves are unable to propagate) and at the same time "choppy" (in that the waves which do propagate appear to have an enhanced steepness). The lack of short waves is explained by (14.106) but the total appearance is also consistent with the predictions of the NYU theory. To attempt to apply this theory to frazil ice we model the slick as of constant thickness h′ and relative ice volume c, using h = h′c in equation (14.99). Typical values of 1 m and 0.3 m for h′ and 40% for c yield cut-off periods T_c of 1.20 and 0.66 s respectively. The behavior of (14.103) is such that beyond T_c the fraction of incident energy transmitted into the icefield increases rapidly to near unity; at 3 s period a 1-m thick frazil slick transmits 99% of incident energy (R = 0.09) and at 6-s 99.6%. Therefore it is only the very short waves that contribute significantly to the total wave radiation pressure by being completely reflected; longer waves make only a small contribution. We note that the wave force is developed only at the very edge of the slick, since on this model no further attenuation occurs to waves which succeed in entering the slick. In practice the total wave radiation pressure is likely to be a sum of the pressure due to the mismatching of NYU waves (exerted at the extreme edge) and the pressure due to the viscous attenuation mechanism of (14.106) (exerted throughout the outer portions of the slick until either the wave energy decays to a negligible value or else pancake ice is formed). It is likely that pancake ice forms most readily when the slick is herded by wind and waves against a fixed rear boundary formed by normal ice floes.

B.5 SUMMARY OF OBSERVATIONS

So far we have considered the effect of an icefield on waves. Of equal importance, especially since it determines the morphology of the marginal ice zone (MIZ), is the effect of waves on an icefield. Wave effects on ice can be divided into three categories:

i) Thermodynamic effects. This is principally an enhanced melt rate due to the oscillating shear current produced by the wave. The magnitude of the effect is estimated in Section B.6.

ii) Dynamic effects. A wave that suffers any kind of reflection, scattering, or loss of energy when it encounters an ice floe generates a force on that floe in the direction of wave incidence, known at the wave radiation pressure (Longuet-Higgins, 1977). The magnitude and importance of this pressure have already been discussed in connection with the wave force on the edge of a frazil ice streamer (Section B.4.2), and wave radiation pressure may also be an important factor in the generation of ice bands by an off-ice wind (Section C.3.3). In general, wave radiation pressure always acts as a compacting force on an icefield. In fields of floes where the wave decay rate is exponential, i.e., given by (14.2), the wave radiation pressure arising from the wave decay is always a compressional body force in the x-direction on any given area of icefield. A wave field may also cause a floe to move in the direction of wave propagation due to the Stokes drift (Stokes, 1847), or second-order streaming motion at the surface due to the wave. A contemporary estimate by Kenyon (1969) is that for a fully-developed sea the Stokes drift at the surface is between 1.6 and 3.6% of the wind speed. This falls off rapidly with depth, but especially for thin floes it may be a major factor causing motion down-wave.

iii) Structural effects. A typical MIZ icefield contains floes of varying sizes, and as we shall see in Section B.8 these respond bodily with surge, heave, and roll motions to the passage of waves. These responses are not in phase -- and in fact there can be very large differences betwen the phases of surge response for floes of differing sizes -- and the resulting differential motion can lead to frequent collisions between floes. These can break up floes, although a more usual effect of a collision is to break off relatively small lumps of ice from the edges of the floes. This enhances the ablation rate in summer still further beyond the effect described in Section B.6, since the broken-off brash, having a large surface-to-volume ratio, melts very rapidly. The collisions are also an important source of underwater noise (Diachok, 1980). More crucial from the point of view of overall ice morphology is the fracture of floes due to the flexure induced by wave action. This forms the chief substance of Sections B.7 and B.8. It is a major effect, observed throughout the MIZs of the world, and is the factor that determines the maximum floe size that can exist in a marginal icefield. This maximum is a function of the wave climatology in the fringing ocean, and of the distance from the ice edge, since the outer ice shields the interior ice from waves, so that larger floes can survive in the interior than at the extreme edge.

B.6 WAVE-INDUCED MELTING

Waves in the marginal ice zone produce an oscillating, horizontal velocity component in the water relative to the ice bottom that permits turbulent diffusion of heat into the ice provided the water is warmer than the ice (Wadhams, et al., 1979). According to Csanady (1972) the heat flux into a horizontal boundary surface due to a steady shear current V of water at a temperature ΔT above that of the solid surface is

$$H_v = c_H \rho_w c_w \Delta T V \quad (14.107)$$

where ρ_w is water density, c_w is its specific heat, and c_H is a dimensionless parameter estimated by Gill (1973) as being of the order of 3×10^{-3} for the bottoms of Antarctic ice shelves. The lower roughness length of ice floe bottoms, plus the likelihood that the turbulence in an oscillating shear flow is less than in a steady flow, suggest that a lower value of c_H, say 10^{-3}, is more appropriate for our estimates. H_v produces a melt rate of $(H_v/\rho_i L)$, where ρ_i is ice density, and L is the latent heat of fusion, provided the ice bottom is at the melting point.

The mean horizontal velocity at the surface due to a wave of amplitude A and angular frequency ω is $(2A\omega/\pi)$; if the initial amplitude A is reduced through exponential decay along a distance s at a decay rate α, then the velocity is

$$V = \left(\frac{2\,A\,\omega}{\pi}\right) \exp\,(-\alpha s) + V_0 \quad (14.108)$$

where V_0 is the steady shear current in the absence of waves. This assumes that the floe has no surge response to the wave, and is therefore likely to be an overestimate. Putting c_w = 4.0 kJ kg^{-1} $^oC^{-1}$, L = 335 kJ kg^{-1}, and ρ_i/ρ_w = 0.9 yields bottom melt rate M in ms^{-1} of

$$M = 8.4 \times 10^{-3}\, c_H\, \Delta T[A\omega\exp(-\alpha s) + V_0] \quad (14.109)$$

We can now estimate the time required to melt a 3 m floe, typical of East Greenland ice. In Figure B.28 we consider two wave periods, 10 s and 13 s (typical of spectral energy peaks for

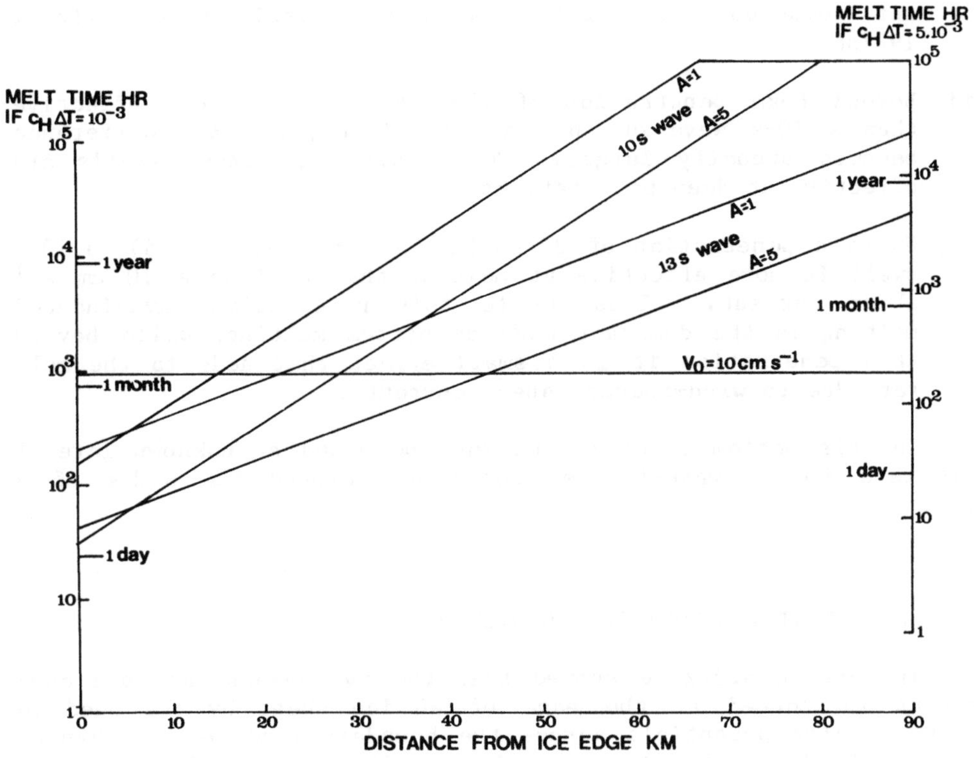

Fig. B.28. Melt times for a 3 m thick floe due to wave action at different distances inside the ice margin. (After Wadhams et al., 1979).

the Greenland Sea), and put α = 1.2 and 0.7 x 10^{-4} m^{-1} respectively, using the results of Wadhams (1978a). The figure then shows the melting time at different distances inside the ice edge for two values of $c_H\Delta T$ (10^{-3} and 5 x 10^{-3}) and of A (1 and 5 m). For comparison we also plot the melt time for V_0 = 10 cm s^{-1}, typical of wind-induced drift. Our conclusions are:

i) At the ice edge itself wave-induced melting is very effective. The waves with 5 m amplitude will melt a floe in 1 to 2 days, and in 6 to 8 days with 1 m amplitude even when $c_H\Delta T = 10^{-3}$ (i.e., approximately a 1°C temperature difference). With $c_H\Delta T$ = 5 x 10^{-3} (a 5°C temperature difference between ice and water) the times are only 6 to 8 h (5-m waves) and 1 to 2 days (1-m amplitude). It is clear that wave-induced bottom melting is constantly in progress at the ice edge, eating away at the floes and eroding back the edge. In winter this process is opposed by freezing, but it may account for the frequent observation that floes near the ice margin have a net component of drift towards the margin (e.g., Vinje, 1977ab).

ii) Beyond 6-km penetration of the 13-s wave is more effective than a 10-s wave of the same amplitude, and the difference becomes steadily larger. Thus only the longer swells are effective at deep penetrations.

iii) Up to a penetration of 22 km (A = 1) or 44 km (A = 5), a 13 s swell is more effective at melting the ice than a 10 cm s^{-1} shear current. Thus up to this penetration wave-induced melting is the dominant mode of bottom melting, while beyond this penetration it is a small effect that adds to the melt rate due to wind-induced shear currents.

To this bottom melting rate must be added an unknown lateral melt rate due to wave-induced turbulence around the sides of a floe.

B.7 BREAKUP AT THE EDGE OF A CONTINUOUS ICE SHEET

In Section B.2.2 we showed that the two evanescent wave components introduced at the edge of an ice sheet by the need to match velocity potentials across the boundary lead to an enhanced flexure of the sheet at some critical distance inside the edge. If the incident wave energy is almost great enough to cause fracture of the sheet due to the flexural-gravity wave alone, then these evanescent components make the sheet fracture preferentially at this critical distance from the edge.

A second mechanism for wave-induced breakup of ice near the edge of a large floe, discussed by Assur (1963) and Milne (1972), is that of standing waves. We have shown in Section B.2.3 that a flexural-gravity wave incident on the trailing edge of a large floe suffers partial reflection back into the floe and partial transmission out into the open water. The incident and reflected wave together create a partial standing wave, at the antinodes of which the amplitude of flexure is greater than that due to the propagating flexural-gravity wave alone. This again will lead to breakup at the antinodes if the propagating wave is almost energetic enough in itself to fracture the ice. The fracture due to standing waves will occur as a series of parallel cracks spaced half an ice-coupled wavelength (π/k_o in our notation) apart, with the first crack occurring just on the seaward side of a lead.

The critical amplitude at which a flexural-gravity wave alone will break up the ice is an asymptotic case of the breakup of a finite ice floe, and will be treated in Sections B.8.1 and B.8.2.

In practice, apart from exceptionally clear cases such as that illustrated in Figure B.8, it is difficult to determine the governing mechanism for any observed pattern of fracture in a continuous ice sheet, since wind-induced tilt (Weber and Erdelyi, 1976), internal ice stress and many other factors can cause breakup. For instance, parallel leads spaced widely apart have been observed in the Beaufort Sea (Section A.2.4) and north of Ellesmere Island, with no convincing explanation of their origin. Whether the calving of ice shelves and ice tongues is a wave-induced phenomenon is also an open question, and much work remains to be done in this field (see, for example, Holdsworth and Glynn, 1981).

B.8 BREAKUP OF ICE FLOES DUE TO FLEXURE

The flexural response of a finite ice floe to wave action is more complex than that of an infinite floating sheet. A floe possesses a rigid body response to waves that has six degrees of freedom. Three are rotations (pitch, roll, and yaw) while three are bodily translations (surge, heave, and sway). These responses help to relieve the oscillating pressure field produced by the waves on the floe underside, and hence the flexural response is in general less than that of an infinite sheet of the same thickness. Thus a casual conclusion we can immediately draw is that a large floe, in which the body response is lower, will flex more in a given wave field than a small floe, which readily moves in phase with the wave. In the marginal ice zone we therefore expect large floes to be broken up by waves until a floe size is attained such that the maximum flexure no longer exceeds the failure strain of sea ice. The attenuation of waves by ice in the MIZ is therefore accompanied by floe fracture and reworking of the floe size distribution such that near the ice edge the maximum floe size is small, while the maximum increases with increasing penetration until the distance from the ice edge is too great for wave action to break up floes at all. Beyond this point other factors (such as large-scale response to wind stress, inertial or tidal forces) take over as the determinants of maximum floe sizes.

B.8.1 Theory of the Flexural Response of an Elastic Raft

There is no analytical solution for the complete response of a floating elastic raft to a wave. To estimate the flexural response there are two possible approaches:

1) to obtain an analytical solution to a problem in which several simplifying assumptions have been made;

2) to make fewer simplifying assumptions and to adopt the finite element techniques employed by naval architects to solve problems of ship motion numerically.

(2) is the currently favored method of solution, but we will mainly describe method (1), since this gives greater insight into the physical principles involved.

The simplest approach (Goodman, et al., 1980) requires the following assumptions:

1. The incident wave is sinusoidal, long-crested, and of small amplitude.

2. Flow is irrotational and inviscid.

3. The added mass of the floe is zero. Added mass is a frequency-dependent hydrodynamic effect of the sea that makes the body appear to have a greater mass when it is undergoing motions of translation (surge, heave, sway) and an additional moment of inertia when undergoing rotations.

4. The damping coefficient of the floe is zero. The motion of a floating body is damped because part of the energy that is given to it by a wave is used by the body to generate a train of progressive outgoing waves.

5. The potential due to wave diffaction by the body may be ignored (the Froude-Krylov hypothesis).

6. The rigid body motion may be decoupled from the flexural response. In other words, we assume that the flexure is so small that the wave field does not "see" the body bending, but only sees the rigid body response. This enables us to calculate the pressure field on the basis of a rigid body response only, then to relax the condition and calculate the flexural response from the pressure field.

7. The hydrodynamic effect of the vertical sidewalls of the floe may be ignored.

Consider a floe of length ℓ, thickness h with an incident wave of amplitude A′, radian wave number k, and radian frequency ω. Measuring y positively upwards relative to mean sea level, and x horizontally relative to the upstream end of the floe, we have by Bernoulli's theorem that the pressure P(x,t) along the underside of the floe is given by

$$P(x,t) = -\rho_w g\xi + A\, \rho_w\, g\, \mathrm{Re}[e^{i(kx-\omega t)}] \tag{14.110}$$

Here $\xi(x,t)$ is the profile of flexure of the floe, and A is the amplitude of water particle motion at the bottom of the floe, which is related to A′ by

$$A = A' \exp(-\, k\, h\, \rho_I/\rho_w) \qquad (14.111)$$

ρ_I, ρ_w are ice and water densities. For a floe A and A′ are almost identical, but A is significantly lower for the case of ice islands or tabular icebergs.

We now assume that the flexural response of the floe to this pressure field is wholly elastic. This is justified because the period of the wave (6–20 s) is too small to permit significant creep to occur in order to relieve the stress (see Section B.2.6). The dynamic equation for a homogeneous elastic beam is

$$D \frac{\partial^4 \xi}{\partial x^4} + \rho_I\, h \frac{\partial^2 \xi}{\partial t^2} = P(x,t) \qquad (14.112)$$

where

$$D = E\, h^3/12(1 - \nu^2) \qquad (14.113)$$

Here, E is Young′s modulus and ν is Poisson′s ratio. The resulting equation is thus (e.g., Nevel, 1970):

$$D \frac{\partial^4 \xi}{\partial x^4} + \rho_I\, h \frac{\partial^2 \xi}{\partial t^2} + \rho_w\, g\xi = A\, \rho_w\, g\, e^{i(kx-\omega t)} \qquad (14.114)$$

where Re( ) is implicit.

We assume that the function $\xi(x,t)$ is separable and is periodic in t with radian frequency ω, i.e.,

$$\xi(x,t) = X(x)\, e^{-i\omega t} \qquad (14.115)$$

Then the equation of motion becomes

$$D \frac{d^4 X}{dx^4} + (\rho_w\, g - \rho_I\, h\omega^2)X = A\, \rho_w\, g\, e^{i\,k\,x} \qquad (14.116)$$

This can be solved easily using the boundary conditions that there is no bending moment or shear at the ends of the floe, i.e.,

$$\frac{d^2 X}{dx^2} = \frac{d^3 X}{dx^3} = 0 \;\Big|\, x = 0, \ell \qquad (14.117)$$

For simplicity, we present only the solution for the strain at the center of the floe, i.e., $x = \frac{\ell}{2}$, which is calculated from

$$\varepsilon(\tfrac{\ell}{2},t) = -\frac{h}{2}\frac{\partial^2 \xi}{\partial x^2}\Big|_{x=\frac{\ell}{2}} \qquad (14.118)$$

The solution is

$$\varepsilon(\tfrac{\ell}{2},t) = \frac{A\,\rho_w\, g\, k^2\, h \cos \omega t}{2\, D\Delta(k^4 + 4\,k'^4)}\left[\Delta - \Psi\cos\frac{k\ell}{2} - \frac{k}{k'}\sin\frac{k\ell}{2}\sin\frac{k'\ell}{2}\sinh\frac{k'\ell}{2}\right] \qquad (14.119)$$

where

$$k' = \left[\frac{\rho_w\, g - \rho_I\, h\omega^2}{4\,D}\right]^{1/4} \qquad (14.120)$$

$$\Delta = \frac{1}{2}(\sin k'\ell + \sinh k'\ell) \qquad (14.121)$$

$$\Psi = \sin\frac{k'\ell}{2}\cosh\frac{k'\ell}{2} + \cos\frac{k'\ell}{2}\sinh\frac{k'\ell}{2} \qquad (14.122)$$

Figure B.29 shows $\varepsilon(\frac{\ell}{2},t)/A'\cos\omega t$ plotted against the length of the body for some typical swell periods and for three types of floating ice raft, i.e., an ice island of 35 m thickness, a thin tabular iceberg of 100 m thickness, and an ice floe of 3 m thickness. Thus the left hand vertical scale represents the maximum surface strain per unit amplitude of incident wave. In each case it can be seen that the solution is initially oscillatory, but eventually becomes asymptotic at large body lengths, the asymptote representing the flexure of an infinite floating ice sheet. The iceberg takes much longer to approach the asymptote. A negative excursion of the function implies that the strain at the center is in antiphase with the wave.

There is a second approach to this problem that uses the same assumptions but employs the beam theory of Timoshenko et al. (1974, p. 435). This derives the strain response to a general sea forcing P(x,t) as a sum of overlap integrals of P(x,t) and the normal modes of vribration of the raft. In Figure B.29, a) and c), the contributions from the first and third normal modes are shown as dotted lines; when the length of the raft is much less than the wavelength of the forcing function, only the first mode is excited. This approach also identifies the radian frequencies p_n at which there is a bending resonance between the sea and the ice raft. These are

$$p_n^2 = (k^4 D + \rho_w g)/(\rho_I h) \tag{14.123}$$

where

$$k_n = (n + \frac{1}{2}) \pi, \; n = 1,3,5 \text{ etc. (Timoshenko et al., 1974)}$$

These frequencies are very high ($\sim 10^5$Hz for an ice island) and therefore such resonances are irrelevant except for thick icebergs.

Finally, the numerical approach (Squire, 1981) uses finite element analysis. The raft is replaced by a "patchwork" of smaller bodies with exactly the same physical properties. Equations are written down for each element and the elements linked together subject to necessary topological constraints so as to form a so-called stiffness matrix. The matrix equations are then solved

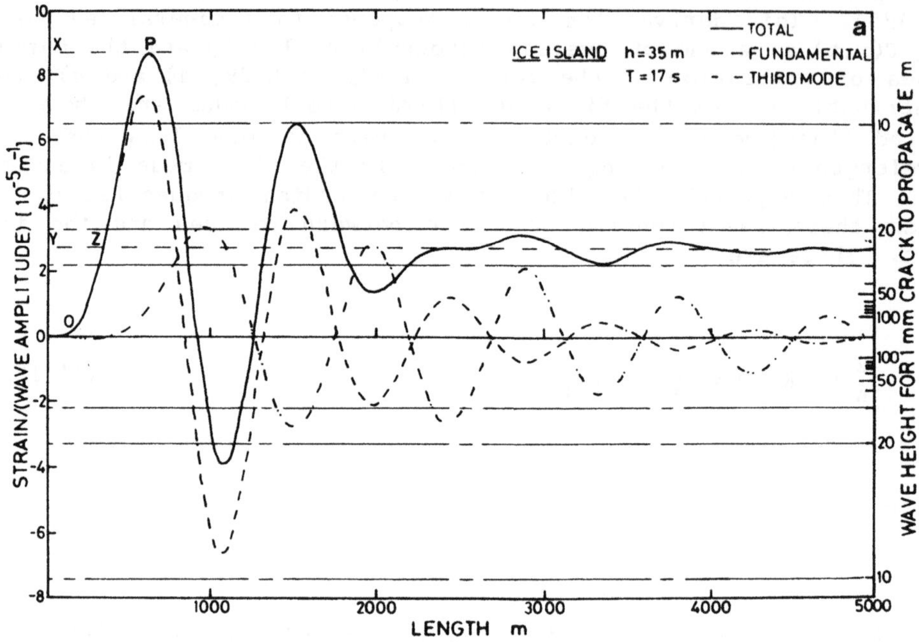
TOTAL
FUNDAMENTAL
THIRD MODE
a
ICE ISLAND h = 35 m
T = 17 s
X
P
Y
Z
O
STRAIN/(WAVE AMPLITUDE) [10^{-5}m^{-1}]
WAVE HEIGHT FOR 1 mm CRACK TO PROPAGATE m
LENGTH m

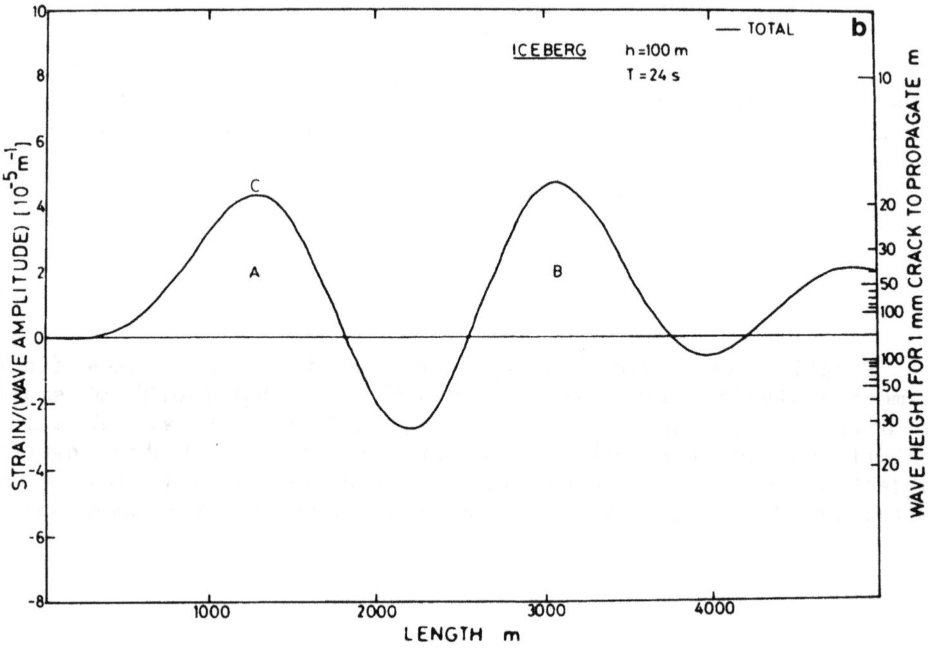
TOTAL
b
ICEBERG h = 100 m
T = 24 s
C
A
B
STRAIN/(WAVE AMPLITUDE) [10^{-5}m^{-1}]
WAVE HEIGHT FOR 1 mm CRACK TO PROPAGATE m
LENGTH m

Fig. B.29a-b.

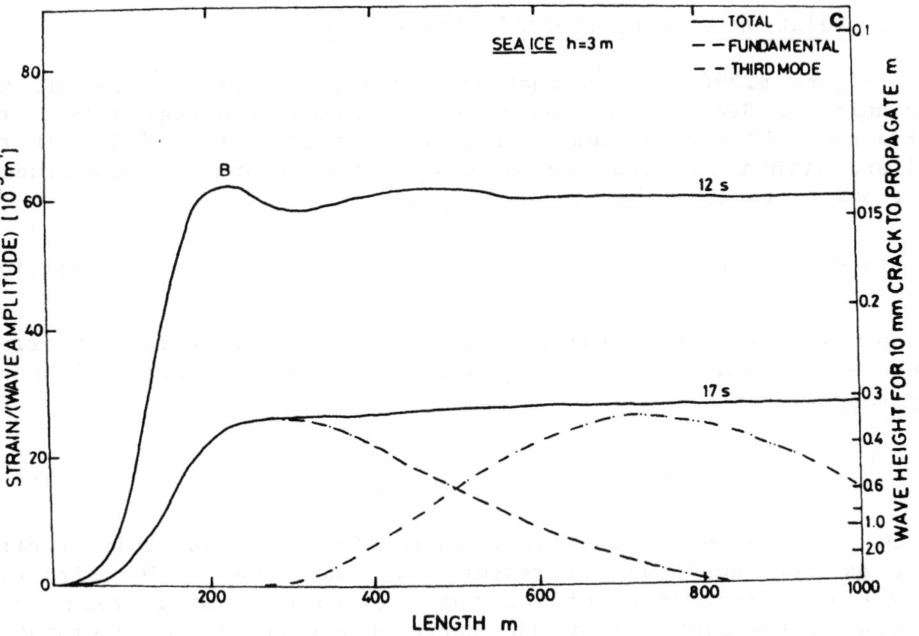

Fig. B.29. Theoretical predictions of the flexural response of an ice island (a), an iceberg (b), and a sea-ice floe (c) as a function of length in the swell direction for a given thickness and wave period. The wave height (peak to trough distance) to make a crack of a given size propagate is shown on the right-hand axis. (After Goodman et al., 1980).

numerically. The reader is referred to Desai and Abel (1972) or Zienkiewicz (1972) for further information on this technique. Figure B.30, from Squire (1981), is one result derived from this method, and shows schematically how the strain amplitude varies within the floe as a function both of distance along it and of position through its thickness. Clearly the maximum strain occurs at the top and bottom surfaces near the longitudinal center of the floe. It is here, then, that the floe is most likely to fracture under wave action.

B.8.2 Relation to Experimental Observations

Figure B.29(c) shows that the casual conclusion drawn at the beginning of Section B.8 was valid. Consider the region OB of the curve for a 12 s wave, and let it be represented by $y = f(\ell)$. If the failure strain for sea ice is ε_{crit}, then a wave of amplitude a will never fracture the ice so long as

$$a < \varepsilon_{crit}/y_B \qquad (14.124)$$

Suppose a wave has amplitude a_1, greater than this criterion. Then the maximum floe length ℓ_{max} that can survive without fracturing is given by

$$\ell_{max} = f^{-1}(\varepsilon_{crit}/a_1) \qquad (14.125)$$

Since $f(\ell)$ is a monotonically increasing function this implies that as wave amplitude increases, ε_{max} decreases. The situation is not so straightforward for the ice island and iceberg cases because of the overshoot of the curve before it becomes asymptotic. In Figure B.29(a), for instance, the region YX of the ordinate (ZP of the curve) represents strain per unit amplitude which is greater than that for an infinite ice sheet. It is therefore possible to find a wave amplitude which is not great enough to break up an infinite sheet (in this case, to cause an ice shelf to calve) but is great enough to fracture an ice island which lies within certain size limits. This is even more important in the case of icebergs, which are discussed more fully in Section B.8.3.

It is significant that the curve OB which defines maximum floe sizes in Figure B.29(c) lies within the size range 0 to 200 m, since this brackets the typical floe sizes found in the marginal ice zone. This implies that typical wave amplitudes are, indeed, great enough to exceed the criterion of (14.124). However, to obtain valid quantitative predictions to test against observation,

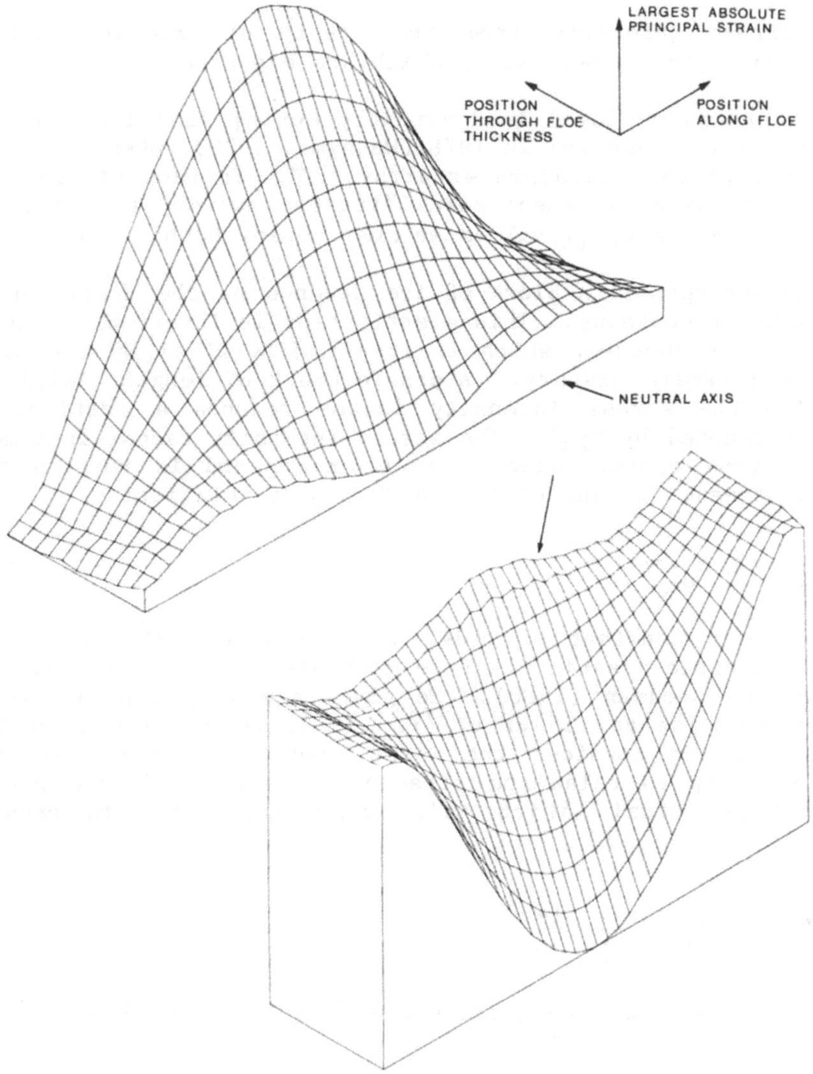

Fig. B.30. Three-dimensional schematic showing the largest absolute principal strain through the ice floe. (After Squire, 1981).

we need a value of ε_{crit} for sea ice. There are two ways of obtaining this:

1. Direct measurement of strain at the time of fracture;

2. Indirect inference from the theory of fracture mechanics, using typical crack sizes which are expected in sea ice.

Method (1) was employed inadvertently during floe flexure experiments off East Greenland in 1978 (Wadhams, 1979), when a floe broke directly beneath a strainmeter array. The failure strain was 3 x 10^{-5}. A single measurement means little, however; to estimate the range of possible ε_{crit} values we must employ Method (2).

The strength of a block of ice depends on the distribution of the cracks it contains. This distribution is invariably controlled by the microstructure, which is itself determined by the rate of growth and strain history. A single crack of length c will propagate when the stress intensity factor reaches a critical value (usually denoted by K_{Ic}). The stress intensity factor is a measure of the stress concentration near a crack, and is related to the applied stress, σ_o, and crack length by the relation:

$$K_{Ic} = \Omega \, \sigma_o \sqrt{\pi c} \tag{14.126}$$

where Ω is a constant of order 1. A critical discussion of how and when elastic-plastic fracture mechanisms can be applied to ice is given by Goodman (1980). He gives a representative value for K_{Ic} for pure ice as 115 kN $m^{-3/2}$. Then, if we take c to be the grain size, and consider the worst case where there has been no time for creep to relax the stresses, a purely elastic response, the critical strain which will cause the crack to propagate, ε_{crit}, is

$$\varepsilon_{crit} = K_{Ic} \frac{1}{\Omega\sqrt{\pi c}} \frac{1-\nu^2}{E} \tag{14.127}$$

which, for c = 1 mm (the grain size), Ω = 1, E = 8.72 GN m^{-2}, and ν = 0.3, is:

$$\varepsilon_{crit} = 2.14 \times 10^{-4} \tag{14.128}$$

For sea ice, because the cracks are water filled, K_{Ic} will be reduced. Since no good measurements yet exist for K_{Ic} for sea ice, we will take K_{Ic} = 50 kN $m^{-3/2}$, and $E/(1-\nu^2)$ = 6 MN m^{-2} (this will depend on brine volume), and suppose c = 10 mm, then

$$\varepsilon_{crit}^{sea\ ice} = 4.3 \times 10^{-5} \tag{14.129}$$

It is important, however, to recognize that ε_{crit} is inversely proportional to the square root of the crack length, and therefore it is not possible to identify a universal "strength" of ice or a universal ε_{crit}. Fortunately, in many situations the microstructure of ice is very uniform, and consequently so is the crack size distribution, and so a pseudostrength will exist.

In Figure B.29 these results have been incorporated in the right-hand ordinate, which shows the wave height (crest to trough) in meters necessary to propagate a crack of length 1 mm (iceberg and ice island) or 10 mm (ice floe). These are minimum crack lengths which can be expected to exist in the material, based on the crystal grain size which is small for glacial ice but much larger for the columnar grains found in sea ice. In practice, larger cracks probably exist, so that the required wave height will be less. Figure B.29(c) shows that a 12 s wave requires a height of 14 cm before it will fracture ice at all. When the height exceeds this value it will break large floes into floes of 200 m diameter; when the height reaches 30 cm the floes will be fractured further down to 100 m maximum size, while at 2 m wave height the maximum floe size will be about 60 m. These values tally very well with what is seen in the Greenland Sea marginal ice zone. We can also use these results to estimate the "active thickness" of the MIZ, i.e., the penetration to which breakup can occur. Again, using the 12 s period and a value of 0.8 x 10^{-4} m^{-1} for the amplitude decay rate, based on the results of Wadhams (1978a), we find that a wave of 5 m height at the ice edge causes breakup to a penetration of 45 km. A 17 s swell, however, although requiring a greater height of 31 cm to fracture ice (on account of the lower curvature of the surface), decays more slowly at a rate of about 0.4 x 10^{-4} m^{-1} (Wadhams, 1978a). This results in a 17 s swell of 5 m initial height being able to break up ice as far as 70 km into the MIZ.

The actual physical mechanism of failure will be a type of fatigue process. During part of each cycle, cracks will grow, reducing ε_{crit}. For each successive cycle the cracks will grow for a larger part of the cycle until eventually one crack will propagate unstably. The increasing hydrostatic pressure with depth will tend to arrest downward-growing cracks, but water filled cracks at bottom do not suffer this disadvantage and have a lower K_{Ic}. Consequently, the failure will probably originate at the bottom.

It is impossible to say which period will be most effective in floe fracture without a knowledge of the wave climatology of the area. Figure B.31 shows the strain per unit wave amplitude as a function of wave period for a floe of given dimensions (in fact, an ice island instrumented off east Greenland, Goodman et al., 1980). Firstly, this shows good agreement between theory and experiment. It also suggests that a 15 s wave would be most effective in breaking up this particular body. However, the ambient wave spectra may tend to have peaks at other frequencies; further,

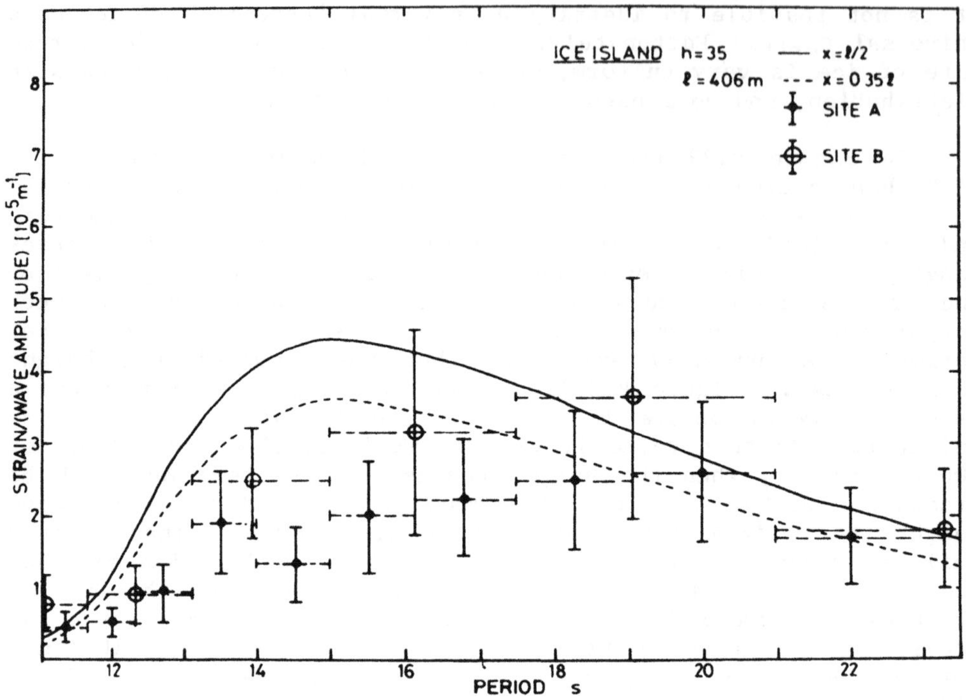

Fig. B.31. Strain per unit wave amplitude at two sites on an ice island. Solid and dotted lines are theoretical predictions. (After Goodman et al., 1980).

in pack ice, one must consider the decay of the wave spectrum within the ice, which is greatest at short periods. At a given penetration, then, the determination of the most effective wave component is a matter of convolving curves such as Figure B.31 with the curve of frequency dependence of attenuation rate and the curve of typical incident energy spectra.

B.8.3 Application to Icebergs

When applying this theory to ice islands and icebergs, we have to consider the oscillatory nature of the strain response as a function of body length. Consider first the 35 m thick ice island of Figure B.29(a). If ice islands with a wide distribution of sizes pass through a region where the maximum wave height is 10 m, those with any dimension between 490 and 760 m will not survive the passage. If the wave height is 20 m, the critical range where failure might occur is extended to 380-840, 1020-1140, and 1350-1760 m. Finally, if there is a likelihood of 30 m waves, only ice islands with dimensions within the range 0 to 340 m and

1840 to 2160 m will survive. Then, if a calving ice shelf generates a wide distribution of ice-island sizes, after a short exposure to sea swell the distribution would be biased in favor of those diameters which require a very high wave height for fracture. In practice, of course, an ice island will rotate so that it presents axes of varying length to an incident swell. Therefore, even if one axis is "stable" by the criterion of Figure B.29(a), the other probably will not be, so that the island will fracture until the fragments lie in the minimum "stable" size range on the limb OZ of the curve.

In the case of tabular icebergs of greater thickness, the Froude-Krylov hypothesis introduces large errors, as does the assumption that the sidewalls can be ignored. Figure B.29(b) should therefore be treated with caution, but it does give a qualitative idea of the type of behavior expected. Firstly, only the longest swell has enough energy to cause an oscillating pressure field at the base of the iceberg. Secondly, the large oscillations in the curve have a similar effect as for an ice island, in that ranges of alternating stable and unstable sizes will occur in a given wave field. Again in practice, no "stable" range of sizes is really stable except the lowest limb OC, where most icebergs will end after exposure to a few storms. This is in agreement with the observations by Landsat imagery in the open water of the Southern Ocean that most icebergs have dimensions no greater than 1.3 km (Hult and Ostrander 1974, Weeks and Mellor 1978). Another significant fact is that the maximum excursions of the strain response in Figure B.29(b) are much greater than the asymptote (which is approached at the right hand end of the curve). This implies that a very large iceberg, if free of major cracks, can survive storms which would break up medium sized bergs into smaller fragments. This may explain why vast bergs, up to 180 km in diameter, are sometimes observed with long lifetimes (e.g., "Trolltunga," Vinje, 1979).

The wave heights needed in Figure B.29(b) are exceptionally large, but the maximum crack size of 1 mm is probably unrealistically small. Therefore, the question can be reversed, and instead the crack size that a three-meter wave will propagate calculated. This is given by the expression:

$$\varepsilon_{crit} = \frac{1}{\pi}\left[\frac{K_{Ic}}{A'}\,\frac{1}{\Omega}\,\frac{(1-\nu^2)}{E}\,\left(\frac{A}{\varepsilon}\right)\right]^2 \qquad (14.130)$$

If 2 A' is 3 m, and ℓ for the iceberg is 1 km, ε_{crit} has the value of 66 mm. Cracks of this size would not be uncommon in a large iceberg.

Records of strain, heave, pitch, roll, and sway were obtained on the top surface of icebergs over 200 m thick in the South Atlantic in 1981 and 1982 (Kristensen et al., 1981; Kristensen and Squire, 1983; Orheim et al., 1982; Wadhams et al., 1983) using strainmeters, accelerometers, and tiltmeters. Measurable flexure occurred at wave periods as low as 10 s, where the pressure field at 150 m draft has only 0.2% of its surface amplitude. The strain data also showed three distinct, statistically independent spectral peaks occurring at about 10, 15, and 50 s, which demonstrate the iceberg's ability to tune to certain periods in the ocean. Such resonance would clearly be critical in determining the ultimate demise of tabular icebergs. Resonance in roll was also observed, with a typical period of about 100 s. Since oscillations at this period cannot be explained by ocean surface wave forcing alone, we seek an alternative explanation. One possibility is that the berg was set rolling by some impulsive event such as localized calving from its edges, and that the iceberg/water system has a high "Q-factor," so that it takes a very long time for the rolling to cease. This would appear reasonable since calving events are commonplace along the sides of an iceberg in a seaway. Another mechanism which has been suggested is forcing by wave packets at the period of the modulation envelope of surface waves in the open ocean (Larsen, 1978ab). If this is the case, it would be an interesting example of an otherwise rarely observed process. The 100 s roll is particularly significant since it implies that the visited berg was close to being unstable, and hence to capsizing. A concurrent program of iceberg modeling using the numerical techniques discussed earlier has enabled theoretical predications of iceberg motions and flexure to be made. Preliminary comparisons between data and theory are encouraging (Kristensen and Squire, 1982).

B.8.4 Current Research Problems

The problem of breakup of floating ice by waves is one of major practical importance, for a number of reasons. Proposals to tow Antarctic tabular icebergs to coastal desert locations for use as a fresh water source depend for their success on the berg not fracturing en route in a major storm. The floe size distribution in the MIZ, determined by wave action, is now known to be an important parameter in understanding ice dynamics and thermodynamics in the MIZ, and some way of predicting it in regions where it cannot be measured (e.g., the Antarctic ice margin) is highly desirable.

The theory outlined in Section B.8 has been tested against a large amount of data obtained from floes in East Greenland since 1978 (Wadhams, 1979; Wadhams and Squire, 1980) and works well in predicting flexural response to a given wave field. The associated problem of determining maximum floe size, however, still requires a better knowledge of failure strain, and thus of crack sizes in ice floes. A program of in situ measurements of crack distributiona and propagation in floes is therefore necessary. Further, it is not clear that fracture mechanics is indeed the best way of treating ice fracture problems under repeated cyclic loading, since fatigue may be the dominant mechanism for materials like ice which strain plastically. Little work has been done on the fatigue of sea ice or pure ice (some early results are reported by Nixon and Smith, 1984), but most materials become more brittle as fatigue progresses and so are more liable to fracture prematurely. The theoretical prediction of the onset of brittle failure for sea ice is therefore an extremely difficult and unresolved problem.

PART C. OCEAN FRONTS AND EDDIES

C.1 A BRIEF SURVEY OF POLAR AND ICE EDGE FRONTS

In this section we briefly survey areas where permanent polar fronts are found, and we also consider transient local fronts associated with ice edges. The term "polar front" has different meanings in meteorology and oceanography, and even in oceanography the same term is used for a number of different types of feature. Since our interest in this chapter is in the seasonal sea ice zone and in the air-sea-ice interaction which occurs within it, we shall consider a "polar front" (at least in its surface manifestation) to be a water mass boundary separating water which can possess an ice cover in winter from water which does not normally possess an ice cover. Thus a polar front is a feature with which an ice cover can interact, and which in its turn affects the ice dynamics and thermodynamics of the MIZ.

C.1.1 The Antarctic

The Antarctic Polar Front was first observed by the German South Polar Expedition of 1901-2 and the classic description of it is by Mackintosh (1946). A more recent treatment emphasizing its variability is by Taylor et al., (1978). It is a water mass boundary delineating the northernmost extent of near-surface Antarctic water. In winter this is visible as an intense sea surface temperature (SST) gradient of about 5°C in 10 km, the 2°C isotherm being conventionally used to define the front. In summer the surface water warms up and the front is less distinct at the surface, but a layer of minimum temperature remains near the surface, and now the subsurface 2° isotherm in this minimum defines the front. The front is circumpolar, occurring at about 50°S in the Atlantic and Indian Ocean sectors and nearer 60°S in Drake Passage, the Pacific sector and north of the Ross Sea. It corresponds to the north edge of the Antarctic Circumpolar Current and so there is a current shear across it. Its existence thus arises as the result of a planetary forcing. The front marks the extreme limit of winter pack ice, and also the limit of frequent occurrence of icebergs; in practice the main body of pack ice seldom reaches the latitude of the polar front.

The Polar Front in Drake Passage has been extensively monitored by very high resolution radiometer (VHRR) thermal infrared from NOAA satellites. An image taken in October 1976, i.e., early spring (Legickis, 1978) showed very large amplitude loops in the front, and successive images taken over a month (Legeckis, 1977) showed a progression of these loops along the front, implying that the loops represent wave-like instabilities in the front with wavelengths of about 250 km. At this time the ice edge lay 100-200 km south of the front. An image taken in September 1976 is shown in uncorrected digital form in Figure C.1. In this case a warm core eddy of diameter 200 km can be seen in the front representing a wave-like instability which has broken backwards ("forwards" being the downstream direction, i.e., eastward). The ice edge is now much nearer to the front, although

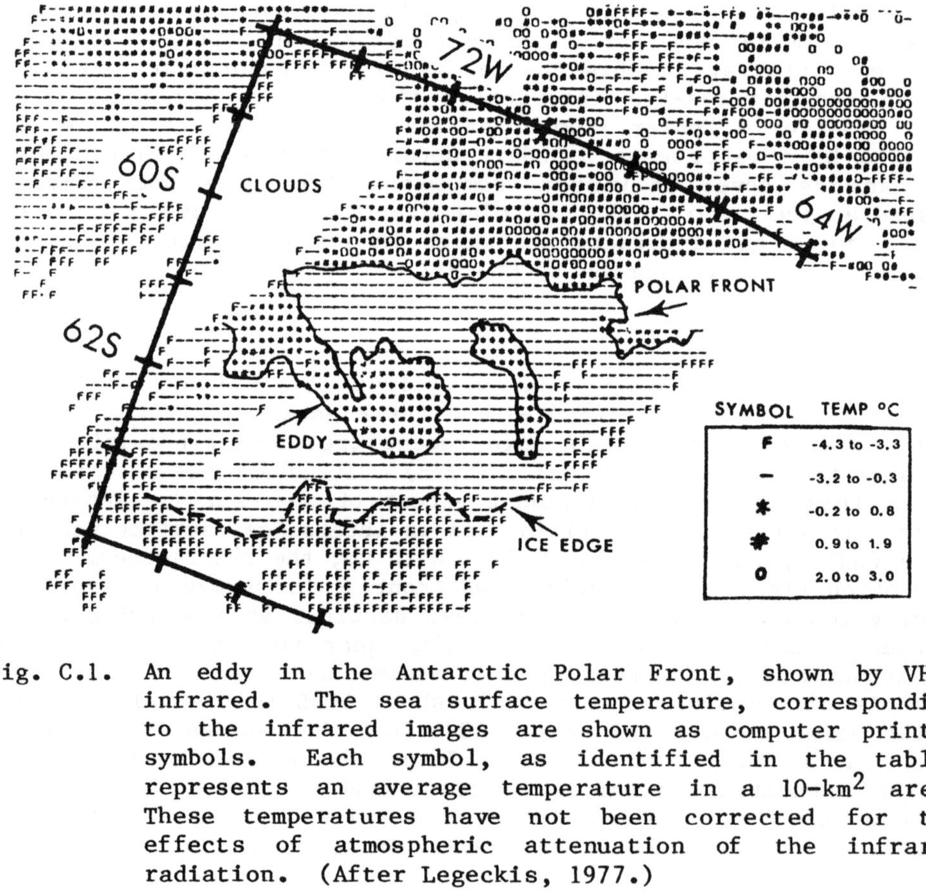

Fig. C.1. An eddy in the Antarctic Polar Front, shown by VHRR infrared. The sea surface temperature, corresponding to the infrared images are shown as computer printed symbols. Each symbol, as identified in the table, represents an average temperature in a 10-km^2 area. These temperatures have not been corrected for the effects of atmospheric attenuation of the infrared radiation. (After Legeckis, 1977.)

apparently not affected directly by the eddy. We shall see in Sections C.1.4 and C.2 that this type of eddy has also been observed in the East Greenland Current and in laboratory experiments, and can be explained in terms of baroclinic instability.

Warm core eddies have also been observed by VHRR in the zone 140-150°E, 54-58°S, south of New Zealand. The eddies were elliptical with a major axis of 150 km and SST gradients of 2°C in 2 km at the boundary. A cold core eddy of diameter 100 km was observed at 170°E, 53°S by the same means.

C.1.2 The Bering and Chukchi Seas

Kinder and Coachman (1978) have described a diffuse haline front lying over the shelf edge in the southeast Bering Sea, separating the waters of the tidally dominated regime above the broad continental shelf from the oceanic regime in the Aleutian basin. Such a front is greatly enhanced in strength during winter, since at this time it coincides with the ice edge. As we have shown in Section A.3.2, the ice edge is a melting edge, and the surface lens of meltwater pushed out just ahead of the ice margin creates a very strong near-surface salinity and temperature front (Pease, 1980).

In summer the ice edge retreats to the shelf break in the northern Chukchi Sea where its shape forms a series of bays which correspond to the channeling of northward-flowing water from Bering Strait by undersea canyons. A complex variety of fronts occurs at and near the ice edge, which have been described by Paquette and Bourke (1979, 1981). The water masses which define the fronts are "northern water," i.e., Arctic Ocean water, which is cold (-1.6° to -1.7°C) and saline (32.8 to 33.6°/oo) and "southern water" from the Bering Sea which forms a two-layer system, the uppermost 5-15 m being at 5-10°C and 30-32% while the lower layer is at -1.3° to +2°C and 32.6°/oo. The simplest form of front (Figure C.2) is a single wall between the northern water (far right of section) and the two layers of southern water, corresponding approximately to the ice edge. However, current shear can cause this wall to split into an upper level and lower level front which may be widely separated, and a number of even more complex frontal structures can occur.

C.1.3 The Labrador Sea

In the northern Labrador Sea the East Greenland Current extension around Kap Farvel has a front with the Atlantic water lying 50-100 km offshore with an SST change of 3°C across 10 km. The Labrador Current itself has a strong permanent front along its edge with the Atlantic water, lying 200-400 km offshore and roughly coincident with the 2000 m depth contour (outer edge of shelf break). The SST change is 4-5°C. Frequently wave-like meanders

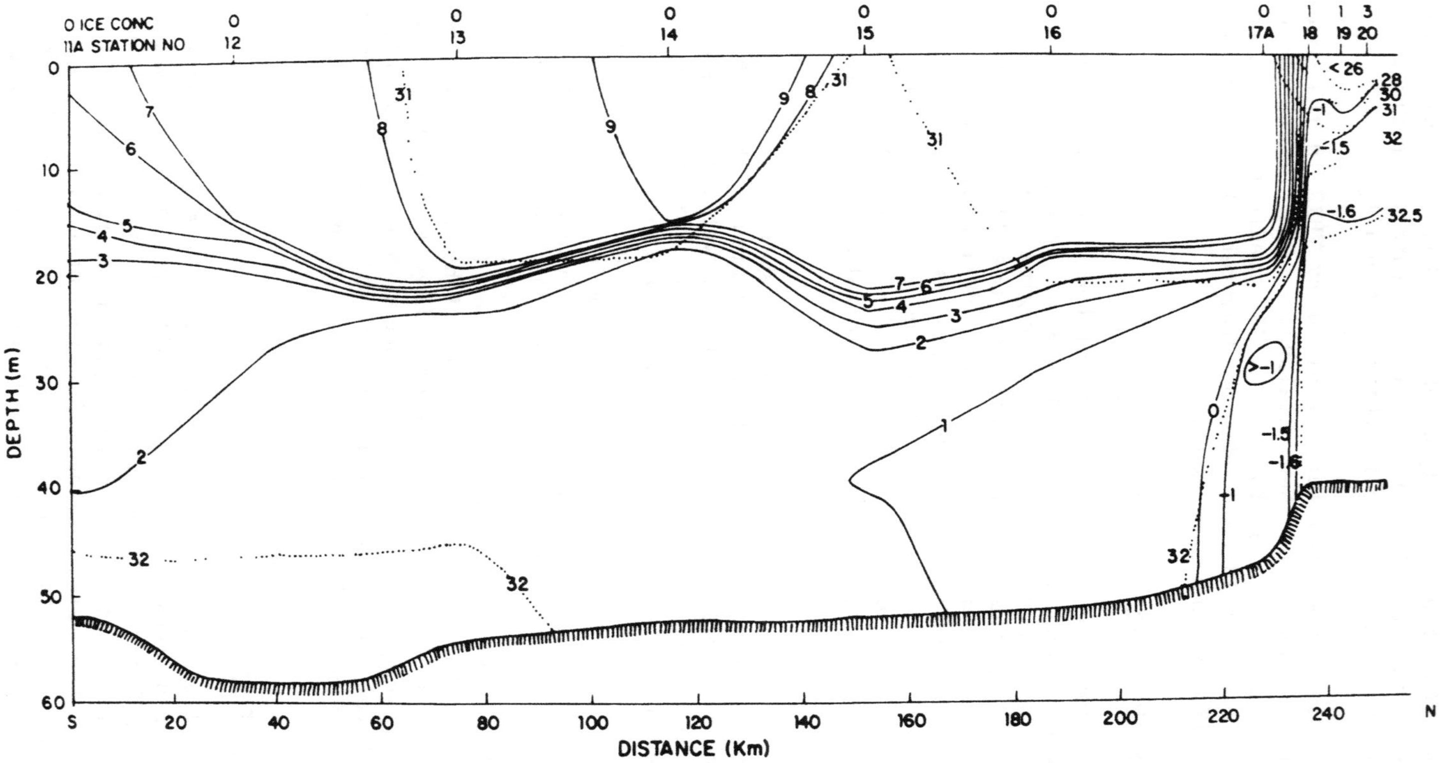

Fig. C.2. Temperature and salinity cross section from the westernmost line of stations in MIZPAC 1978 oriented approximately S-N. Station 18 marks the approximate boundary between warm southern water to the left (south) and cold northern water to the right (north). Close isotherms indicate regions of imbedded fine structure. The ice concentration is in oktas (eighths). Note that the salinity field is closely representative also of the density field at these low temperatures and relatively large salinity gradients. (After Paquette and Bourke, 1981.)

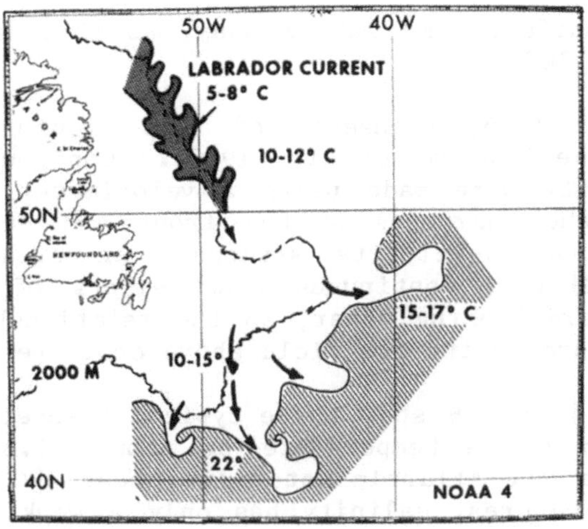

Fig. C.3. The SST boundaries of the Labrador Current and the North Atlantic Polar Front are shown on August 26, 1975, based on the interpretation of a VHRR infrared image. The arrows indicate the narrow filaments of cooler water that appear to be entrained in the eddies at the front. The surface temperatures represent the values recorded at the satellite and are not corrected for atmospheric effects. (After Legeckis, 1978.)

are observed on this front, which appear to be breaking upstream (e.g., Figure C.3, from Legeckis, 1978). The strongest front associated with the Labrador Current is the "cold wall" east and southeast of Newfoundland, where the Labrador Current meets Gulf Stream water at the North Atlantic Polar Front. The Labrador Current water appears to move over the Grand Banks in relatively narrow (30–50 km wide) filaments (arrows in Figure C.3) which become entrained in cyclonic eddies along the North Atlantic Polar Front. SST gradients here can exceed 10°C per km.

C.1.4 The Greenland Sea

The cold, fresh Polar Water that occupies the upper layers of the Arctic Ocean flows out along the east coast of Greenland, forming the upper part of the East Greenland Current. The eastern edge of this Polar Water, where it is found off East Greenland, can be referred to as the East Greenland Polar Front. Hydrographic sections across the East Greenland Current were made as long ago as 1905 and 1906 (Kiilerich, 1945), and this and more recent work were discussed by Aagaard and Coachman (1968). The sections show that the Polar Water occupies a wedge-shaped region bounded by the surface above, a sloping interface below, and the Greenland coast on the side. The interface is at a depth of 200 m or more where it strikes bottom on the shelf and slopes upward to the east to reach the surface some 200 km away, so that its average slope is roughly 1 in 1000.

In October 1976, transects of this sloping interface were made by submarine (Wadhams et al., 1979). Continuous measurements of sound velocity were made using a velocimeter, thus giving an indication of the sharpness of the front and of the horizontal scales in the sound-velocity structure on either side of the front. In addition, continuous measurements of ice draft were made using upward-looking sonar, so the relationship of the sound velocity structure to the ice field above could be seen.

Figures C.4 to C.6 show three typical transects of the front. The polar water has a temperature of about -1.5°C and salinity near 34%, while the Atlantic water lies near 4°C and 35°/oo. At these low temperatures, salinity has only a weak effect on sound velocity, so the relationship between velocity and temperature can be estimated from the approximation that a velocity change of 4.7 m s^{-1} is equivalent to a temperature change of 1°C. The ice draft profiles display "mean" and "extreme" ice drafts. The mean value is the mean draft of undeformed floes; the extreme value is the maximum draft of pressure ridges which occur with reasonable frequency. Isolated floes in open water areas are shown by single vertical bars.

The first transect (Figure C.4) is from the Fram Strait area between 80° and 81°N. In the top part of the transect the front is crossed between 1810 and 1900 hrs. At 85 m depth the front begins some 20 km inside the ice edge, neglecting a belt of isolated floes lying outside the main edge. Within the front there are large-amplitude temperature fluctuations of short wavelength, of the order of 100 m. There are two possible causes of these: isopycnal distortion due to internal waves, in which the temperature fluctuations are greatest in regions of high temperature gradient, i.e., the front; or double diffusive convection due to warm saline water underlying cold fresh water. In the lower transect the submarine has dived deeper to 122 m and encountered the front again some 44 km nearer the Greenland coast. This implies that the slope of the front is 1 in 1200, assuming that the submarine traveled orthogonally to the front. The front at the greater depth is the same width (15 km) as that at 85 m, but the temperature step is only 2.4°C instead of 4°C.

The second transect (Figure C.5) is from 76°N and shows a somewhat different structure. The front occurs in two extremely sharp steps, of which the first, at 2015 hrs, is of more than 3°C in less than 1 km. This front occurs about 6 km to seaward of the ice edge, implying that the surface front lies even further to seaward. Therefore at this latitude the ice edge lies several km back from the polar front, instead of coinciding with it as it did

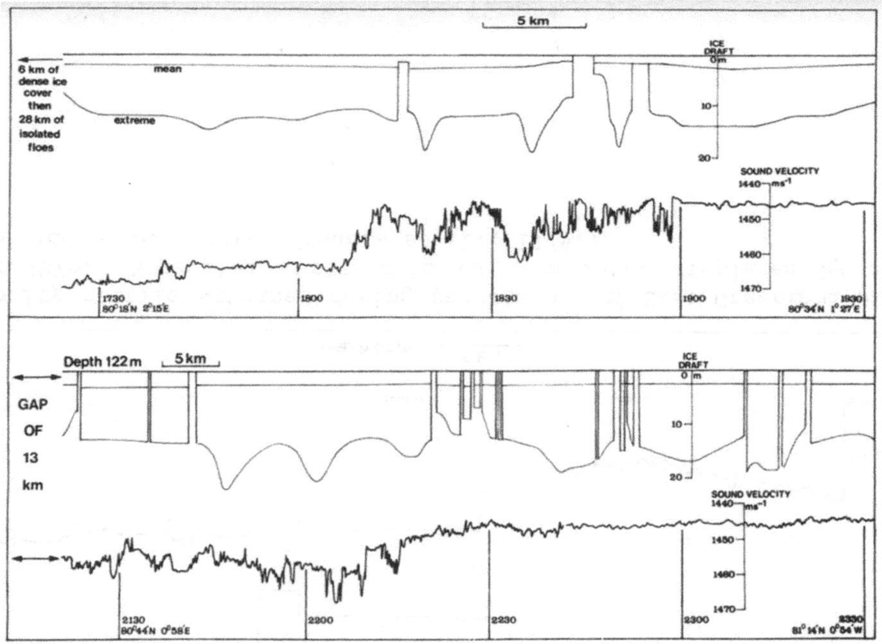

Fig. C.4. Two sound velocity profiles from passage into ice at 80-81°N, 18 October 1976. At a depth of 85 m (top profile) the Polar Front begins some 20 km inside the ice edge, neglecting a belt of isolated floes lying outside the main edge. At 19:30 GMT the submarine dived deeper and continued on her course at a depth of 122 m. At this depth (bottom profile) a front with a smaller and gentler velocity step was encountered 44 km further into the ice than the earlier front. This implies that the Polar Front slopes inwards towards the cold water zone with increasing depth, and that by diving deeper the submarine crossed the Front twice. If the direction of greatest slope lies parallel to the submarine's course (333° true), the slope of the front is 1 in 1200; otherwise this is a lower limit. (After Wadhams et al., 1979.)

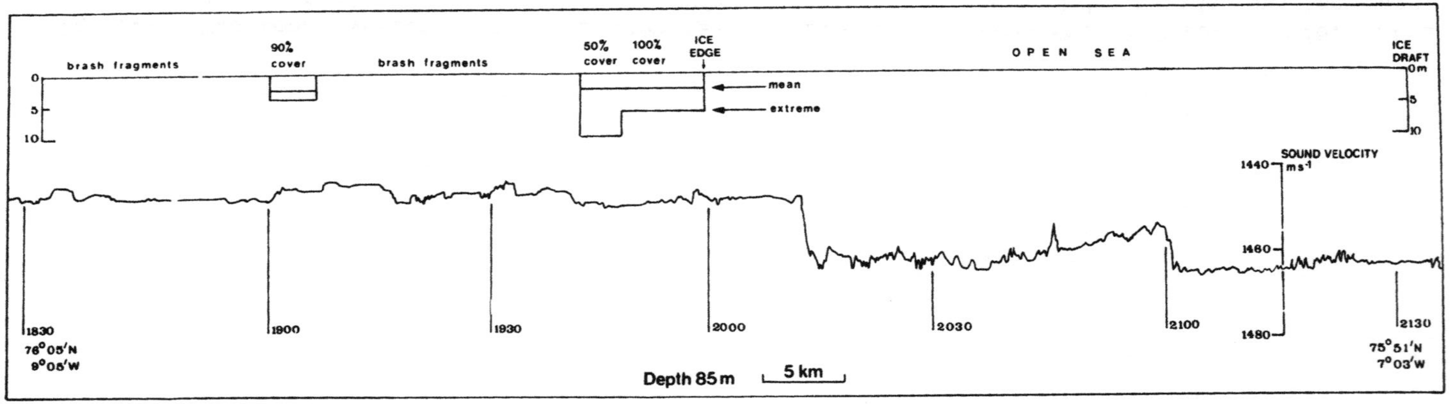

Fig. C.5. Sound velocity profile obtained during passage out of East Greenland pack ice at 76°N, 12 October 1976. Note sharp transition to warm water displaced by about 6 km from position of ice edge. (After Wadhams et al., 1979.)

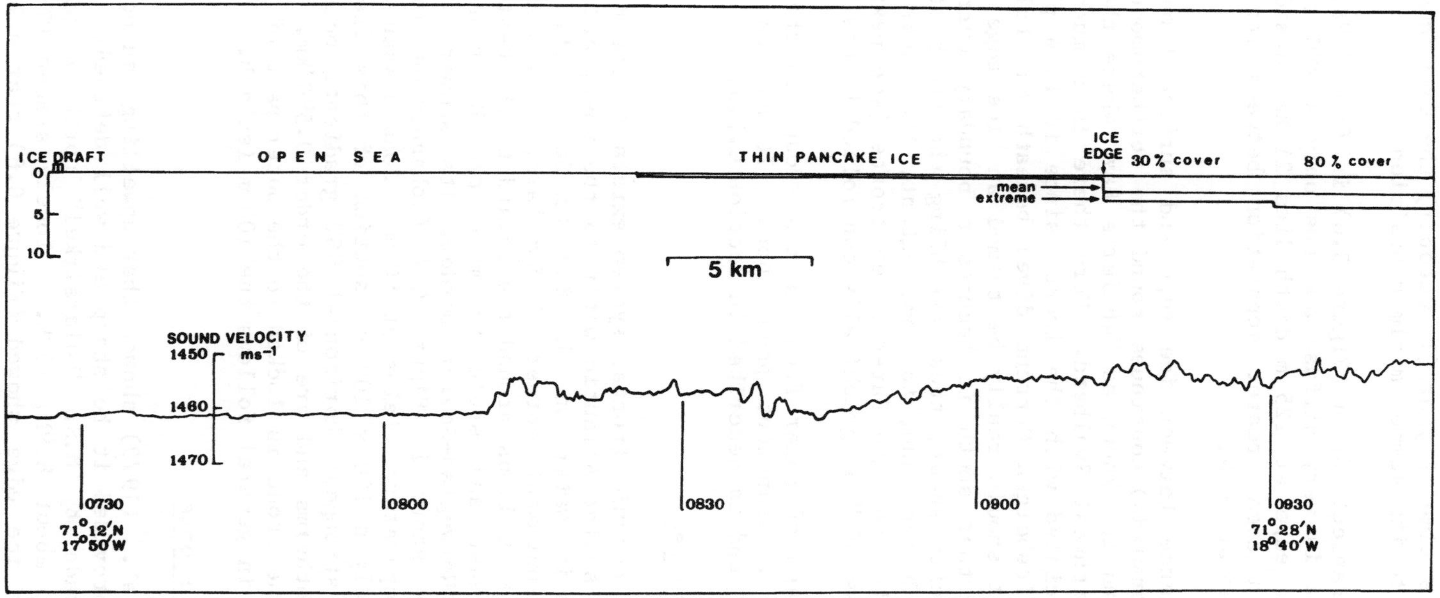

Fig. C.6. Sound velocity profile from passage into East Greenland ice at latitude of Jan Mayen, 7 October 1976. At a depth of 125 m, colder water begins 21 km outside the "ice edge," defined as the first point at which floes were seen. In this case there are a further 6 km of scattered floes before the denser 80% cover begins. (After Wadhams et al., 1979.)

at 80°N. At this time of year (12 October) the East Greenland ice still lies close to its summer minimum position.

The final transect shown (Figure C.6) is from 71°N, near Jan Mayen. The front is very diffuse and has only a 2°C temperature step. Its outer edge at 125 m depth lies 21 km outside the ice edge, implying an even greater separation between ice edge and surface front than at 76°N.

The coincidence between ice edge and surface front at high latitudes (80°N onwards) continues round the northernmost bight in the East Greenland ice known as "Whaler's Bay" where the ice edge turns towards northwest Svalbard. Here there is a much stronger convergence associated with the front, since it is a place where the warm West Spitsbergen Current dives beneath the Polar water. In this region it should really be termed an "ice edge controlled front," while further south it becomes a boundary current front governed by velocity shear, thus resembling the front at the edge of the Labrador Current and, to some extent, the Antarctic Polar Front. The Bering Sea and Barents Sea fronts (see next section) constitute a class of "topographically controlled fronts."

A fourth class of polar front is also found in the Greenland Sea but is only a transient phenomenon. It may be termed an "upwelling front" and is described in Section C.1.6.

C.1.5 The Barents Sea

A permanent oceanic frontal system exists in the Barents Sea between the warm saline Atlantic water to the west, and the cold, less saline Arctic water on the Barents Sea shelf. The front begins off the southwest corner of Svalbard, runs southward to Bear Island where it loops around the island to the west and south before running east and south to meet the European mainland roughly at the Norwegian-Soviet border. Its summer position in various years is shown in Figure C.7 (Johannessen and Foster, 1978) by the temperature fields at 50 m. Johannessen and Foster studied intensively a 100 x 100 km section of this front. They found that the strongest horizontal SST gradients occur between the 3° and 4° isotherms and are of the order 0.5°C/km. They also concluded that the front is locked to the outer part of the continental shelf and in general follows the 100 m isobath.

C.1.6 Upwelling Fronts

Buckley et al., (1979) showed that upwelling can occur at the pack ice edge, provided it be sharp and well defined. The measurements were made by M.S. "Polarsirkel" north of Svalbard in December 1977 at about 81°N, 13°E. Closely spaced CTD stations taken across the ice edge showed (Figure C.8) that a homogeneous region existed just outside the ice edge between the surface and a

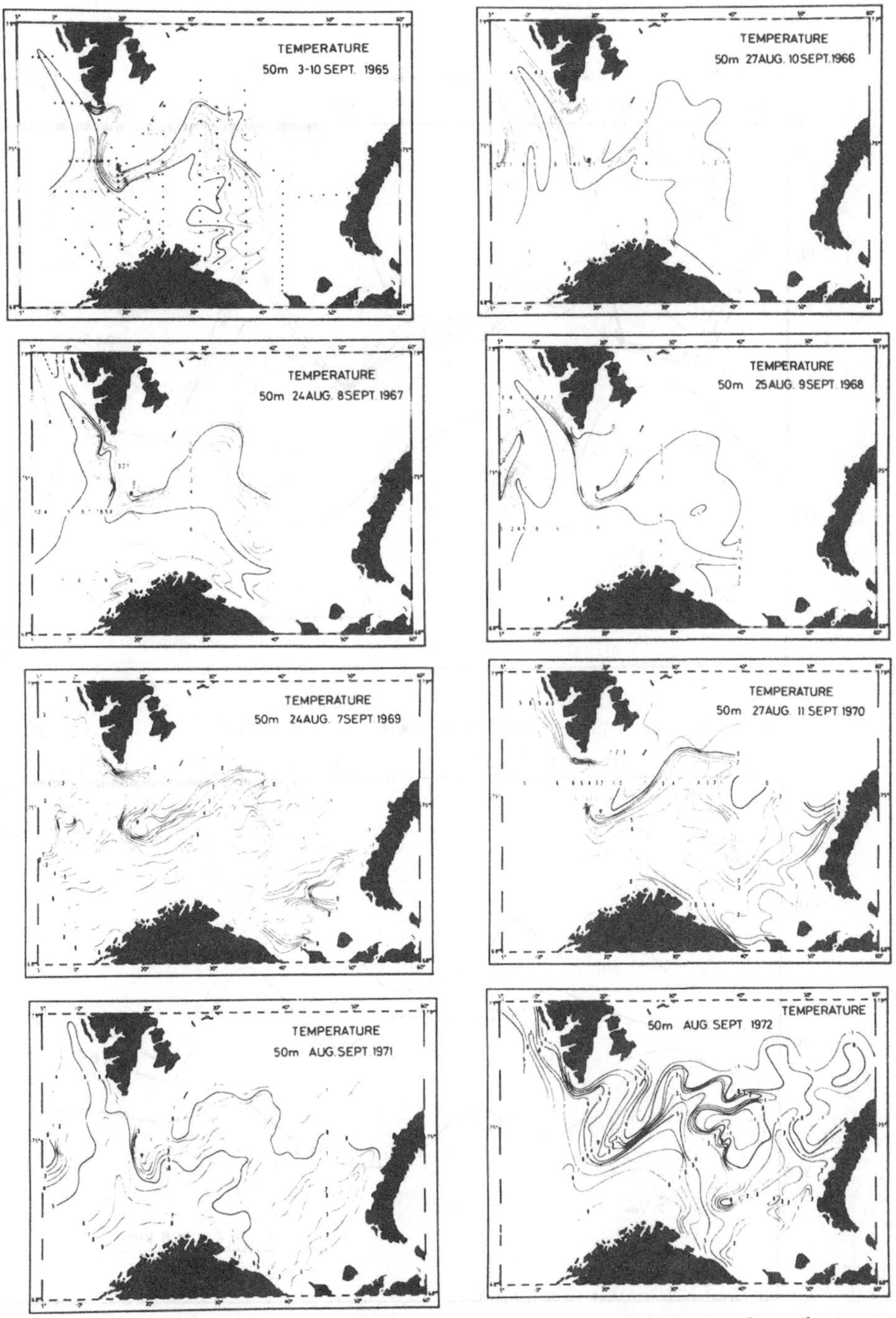

Fig. C.7. Horizontal temperature fields at 50 m based on the International 0-Group Fish Survey in the Barents Sea. Typical station spacing is indicated in the upper left panel. (After Johannessen and Foster, 1978.)

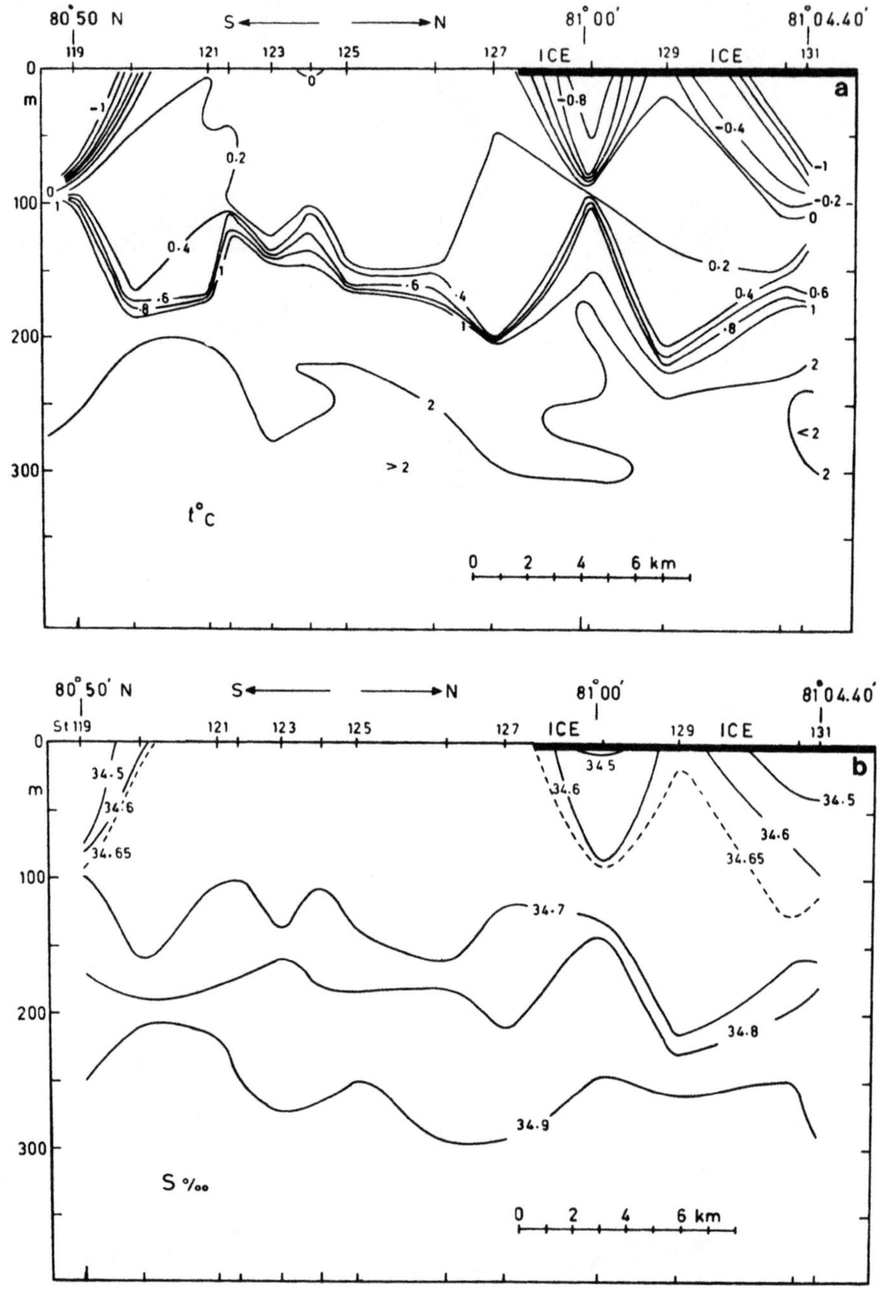
80°50 N
S ← → N
81°00′
81°04.40′
119
121
123
125
127
ICE
129
ICE
131
m
a
t°C
> 2
0 2 4 6 km
80°50′ N
St 119
b
S ‰
34.5
34.6
34.65
34.7
34.8
34.9

Fig. C.8 (a and b).

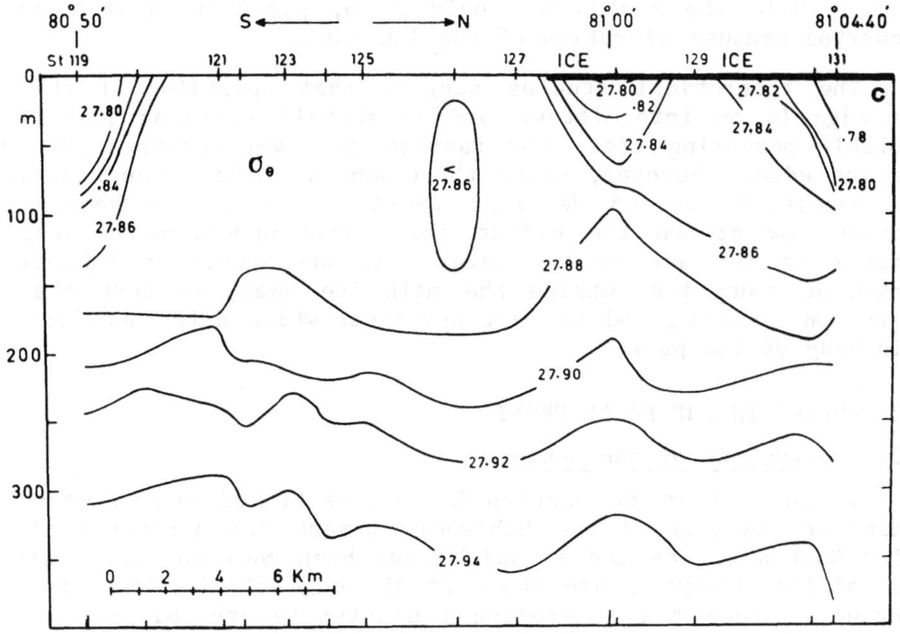

Fig. C.8. Sections drawn from CTD casts made at 13°E on 3 December 1977, showing upwelling near the ice edge. (After Buckley et al., 1979.) (a) Temperature °C; (b) salinity per mil; (c) density in units of σ_θ.

depth of about 140 m. Normally, the water column consists of cold Arctic water between the surface and 100-150 m, underlain by warmer, more saline Atlantic water. In a zone extending for 12 km seaward of the ice edge, however, there were distinct anomalies in temperature, salinity, and density, with the surface water being warmer and more saline than the water to seaward and within the ice zone. Water with these properties is found nearby only at depths of 150 m, implying that upwelling was occurring from that depth or greater. This upwelling split the polar front into two parts, of which the outer may be termed the "upwelling front."

Theoretical studies (Clarke, 1978) of coastal upwelling predict that a wind blowing parallel to the land boundary (or, in this case, to the ice edge, which is presumed to be stationary) will cause upwelling by inducing a divergence in the wind-driven flow field of the surface layer. The boundary must be sharp, much less than the baroclinic radius of deformation, and to cause upwelling from 150 m, a strong wind is required to blow in the correct direction for at least 2 days. The horizontal scale of the upwelling is then predicted to be the baroclinic radius, but, in fact, the observations showed the width of the upwelling to be

12 km, while the radius was only 3 km; possibly, distortion was occurring because of motion of the ice edge.

The theoretical studies suggest that upwelling at the pack ice edge is an intermittent and relatively uncommon phenomenon, probably occurring after the passage of low-pressure cells along the ice edge. However, since there are no other measurements at the necessary station density, there is no way of knowing for certain how common the effect is. When upwelling occurs, the effects on the ice at the margin are presumably to inhibit the growth of young ice outside the main ice edge, to keep the edge sharp and compact, and to melt any floes which drift away from the main body of the pack.

C.2 EDDIES IN THE POLAR FRONT

C.2.1 Structure of the Eddies

In our survey of Section C.1 we mentioned the existence of eddies at the edge of the Labrador Current and Antarctic Circumpolar Current. The eddies which have been most carefully studied and modeled, however, are those at the edge of the East Greenland Current. Since the generating mechanism is probably the same at all three locations, we shall use the EGC eddies as our example in exploring the nature of polar front ice-water eddies in general.

Figure C.9, from Vinje (1977), shows two views of an ice-water eddy observed in Fram Strait. The photograph is a Landsat image showing small floes in the MIZ (less than 80 m across and thus not individually resolved by Landsat) being drawn out into a cyclonic warm-core eddy of diameter about 60 km. The contoured diagram is a sea surface temperature distribution obtained from a Nimbus-tracked drift buoy two months later, which became entrapped in the eddy. It is interesting to note that during its entrapment the net drift of the buoy was ENE along the ice edge, undergoing several loops in one or more eddies. The eddy resembles in shape the Antarctic eddy shown in Figure C.1, but the great advantage of the "ice edge controlled front" north of 79°N in the Greenland Sea for eddy study is that the ice participates in the surface motion of water in the eddy, and thus enables the eddy to be visualized from Landsat and other visual imagery instead of only from thermal infrared.

Figure C.10 shows the temperature structure of the same or a similar eddy profiled during the submarine experiment of Section C.1.4 (Wadhams et al., 1979). Two lenses of warm water can be seen on the cold side of the polar front. Their widths are 20 km and 12 km, they have sharp boundaries and much high-frequency structure within them, and they are situated 40 km and 80 km from the beginning of the polar front proper. Each of the warm lenses has a reduced ice concentration over it. Since the profiling was

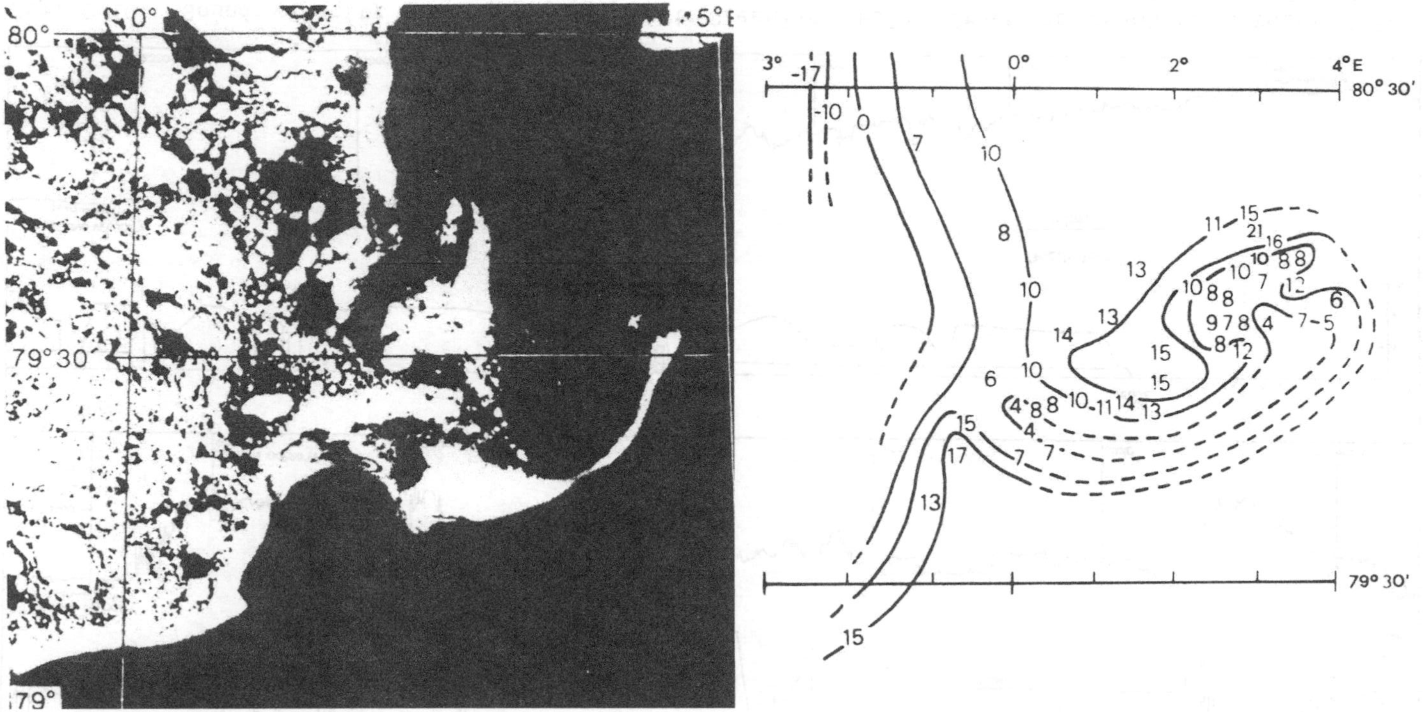

Fig. C.9. Left: Landsat image obtained on 9 May 1976 showing ice-water eddy. Right: Distribution of sea surface temperatures contoured in tenths of a degree as observed from Nimbus drift buoy, July 1976 (Vinje, 1977).

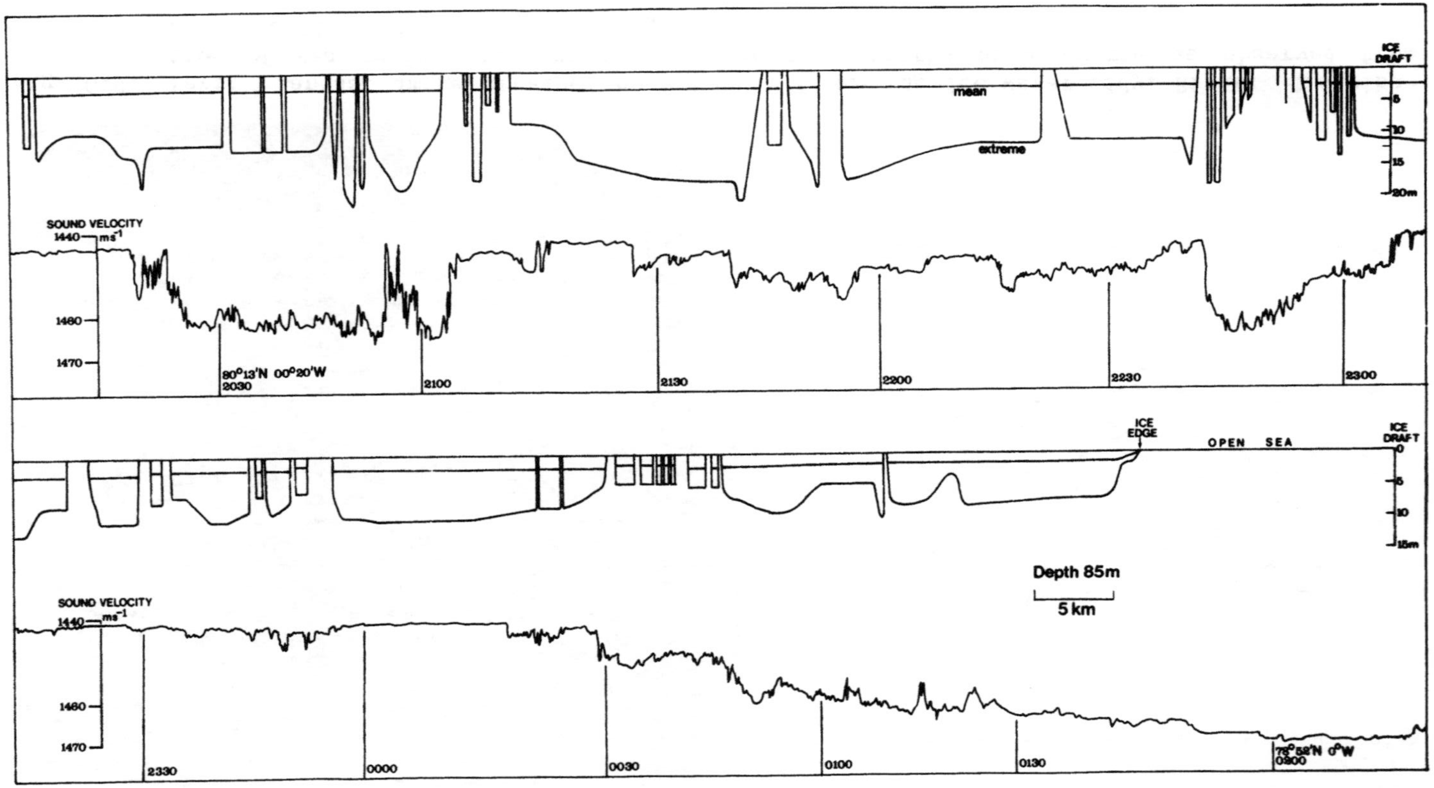

Fig. C.10. Sound velocity profile across East Greenland Polar Front at 79-80°N, October 1976. Two lenses of warm water with reduced ice concentration occur inside the Polar Front (Wahdams et al., 1979).

done at 85 m depth it is clear that the warm component of the eddy extends from the surface to at least this depth.

An eddy at the same location (centered at 79°15′N, 00°38′E) was studied during the YMER-80 cruise in August 1980 (Wadhams and Squire, 1983). The ice drift was followed using radar transponders, the ice structure was mapped by aerial photography and the water structure by CTD transects. The water structure showed a lens of warm (up to 4.3°C) water some 60 km inside the polar front; the lens was centered at 40 m depth, just below the polar surface water, and its effect extended to beyond 600 m. The temperature anomalies were accompanied by salinity anomalies so that there was little net effect on the density profile. Figure C.11 shows a map of the ice distribution in the eddy, with its characteristic "breaking wave" shape. Flight lines are shown, and two simplified temperature sections for the uppermost 100 m. In the sections, water below 0°C is lightly shaded, and it can be seen how the polar surface water changes from 10 m thickness in ice-free areas to 40-60 m thickness under ice at sharp ice edge fronts. Warm lenses are shown hatched, and clearly correspond well with the ice-free (less than 5/10 cover) core of the eddy.

In the region north of Svalbard, synthetic aperture radar imagery and summer ship observations (NORSEX, 1981) indicated the presence of smaller eddies with diameters of 5 to 10 km, approximately equal to the baroclinic radius of deformation. The vertical scale was 100-150 m while the mixed layer depth was about 20 m. These eddies appear to represent a different class from the larger eddies in Fram Strait and further south.

C.2.2 Mechanism of Generation

The repeated observation of a large eddy at the same location in the EGC (between 79° and 80°N) suggests bottom steering. A search for earlier references reveals large-scale oceanographic data from the 1960's (Gladfelter, 1964; Palfrey, 1967) which can be reinterpreted to show such an eddy, and anecdotal reports in Scoresby (1820, Vol. 1, Ch. 4, Sect. VIII). However, the ocean in this region is deep (about 3500 m on average) although it does possess a steep circular depression of 30 km diameter, the Molloy Deep, centered at 79°10′N, 2°50′E, and 2300 m deeper than the surrounding seabed. Theoretical studies of the effect of bottom topography on unstable boundary flows (P.F. Linden, personal comm.) suggest that at such great depths a feature such as this can have only a minor effect, perhaps acting as a trigger to a marginally unstable flow to break up into an eddy. What is equally likely is that the region around 79°N represents the beginning of the EGC as a narrow boundary current, and that this is therefore the first place where large eddies can form. Eddies should exist further downstream, and in fact are seen on satellite images, but are rarer because the ice edge retreats from the polar

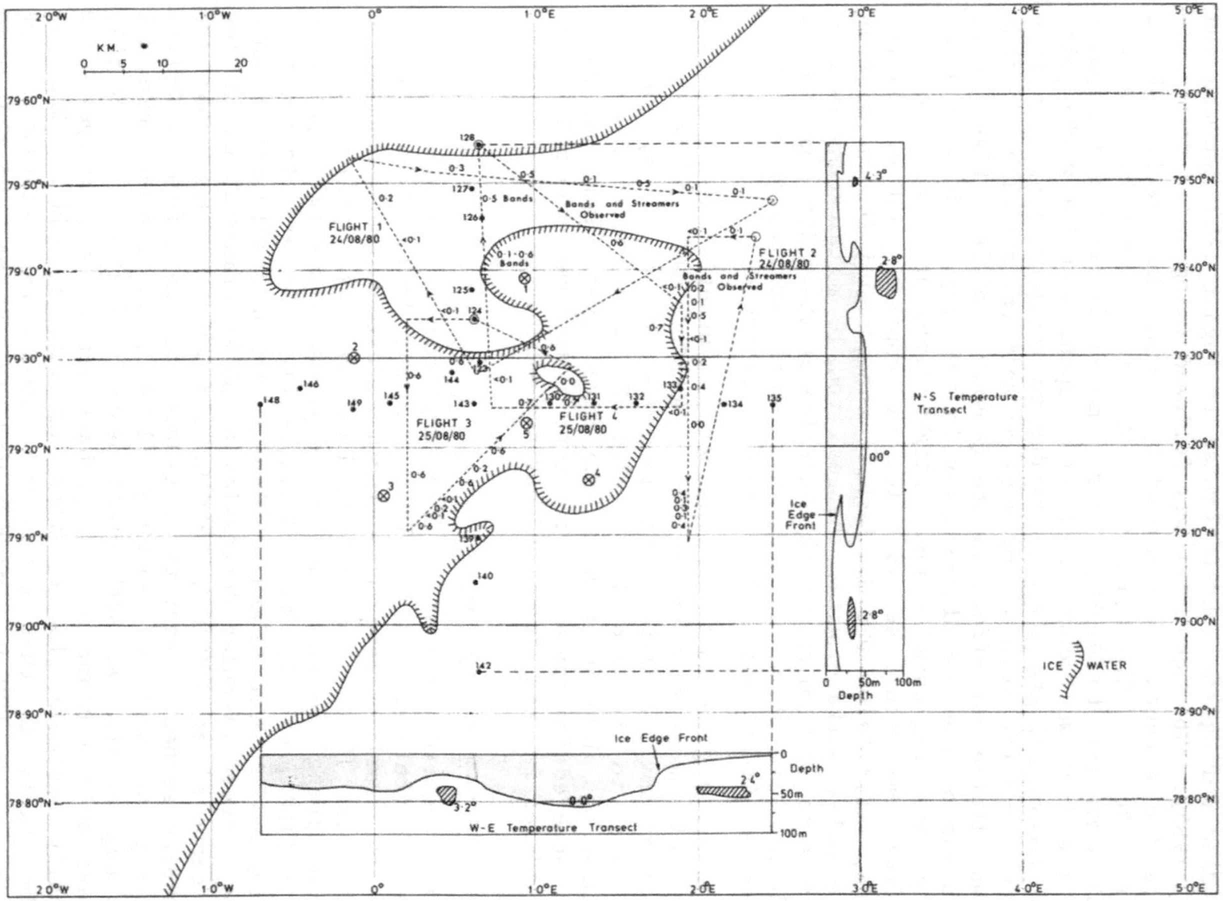

Fig. C.11. Shape of vortex observed in Fram Strait, showing main ice edge and simplified NS and EW temperature sections. (After Wadhams and Squire 1983.)

front so that the ice does not partake so readily in the water motion of the eddy.

Therefore the most likely mechanism for eddy generation is an instability in the polar front. In this connection, a study of boundary current instabilities by Jones (1977) is relevant. He considered a two-layer system where the upper layer forms a wedge of constant slope adjacent to a vertical boundary in an ocean of uniform depth H. If L' is the width of the wedge of less dense water, and D is its depth at the vertical wall (so that the interfacial slope is D/L', with $D \ll H$), Jones found that the criterion for instability is approximately given by $L' > L'_{crit}$, where

$$L'_{crit} = g'^{1/2}(HD)^{1/4}/f \ . \qquad (14.131)$$

Here f is the Coriolis parameter, and g' is the reduced gravity defined as $g' = g\Delta\rho/\rho$. L'_{crit} is about 25 km for the East Greenland Current. The Current is wider than L'_{crit} at all latitudes, suggesting that the East Greenland Polar Front is unstable along its whole length. It seams reasonable to conclude, therefore, that the observed eddies are the result of instability of the front.

The questions that remain are:

(i) can we explain why the eddies grow to their observed size and form?
(ii) is the instability mainly baroclinic or barotropic?

Griffiths and Linden (1981a) investigated these questions by means of theoretical analysis and laboratory experiments. The experiments were based on the fact that the East Greenland Current is basically a buoyancy-driven coastal current. The primary driving force for the current appears to be the density difference between the Polar water and the water of the Norwegian Sea. Where land masses are not present, the southward spreading of the less dense Polar water is inhibited by the effects of the earth's rotation. Southward motion induces an east-west flow in order to conserve angular momentum, and this flow, in turn, produces a north-south Coriolis force which opposes the buoyancy force. Consequently, further southward motion is inhibited. However, at a meridional barrier such as the coast of Greenland, the east-west flow and the north-south Coriolis force are removed. Southward motion is therefore possible near the coast, and this flow is accompanied by a Coriolis force directed normal to the boundary. The resulting current hugs the coastline. This type of current can be modeled by placing a circular (line) source of buoyant fluid around a cylindrical wall (Fig. C.12). As fluid was continuously supplied to the source, a stable zonal current developed and slowly grew in width until wave-like disturbances appeared. The waves grew in amplitude, broke on the upstream side and formed eddies which had widths comparable with the width of the original current. The

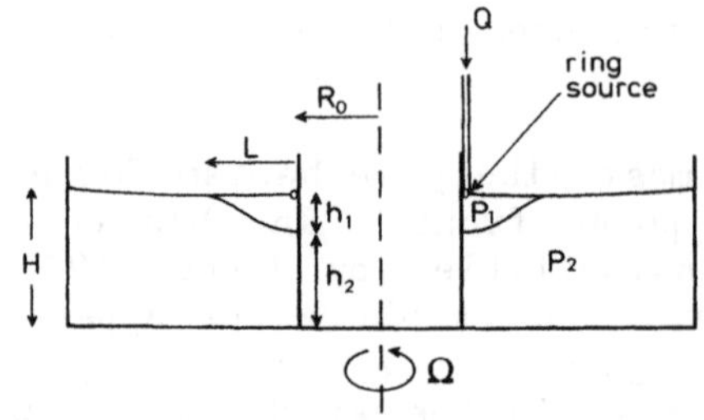

Fig. C.12. The apparatus of Griffiths and Linden (1981a). R = 12.7 cm.

width of the current before instability was always several times larger than the Rossby radius of deformation (about 8 km for the EGC) and Griffiths and Linden suggested, on the basis of this criterion and from further theoretical work (Griffiths and Linden, 1982), that the vortex is primarily baroclinic. When instability first occurred, it began as a small wave-like irregularity. Cyclonic vorticity then began to accumulate near the crests of the waves as they continued to grow and regions of cyclonic vorticity were eventually produced behind the crests, which entrained fluid from the waves so that they seemed to be breaking backwards. This corresponds well with full scale observations, and the surface motions associated with the vorticity fields correspond well with motions measured from YMER. The final form of the broken wave resembled that of isolated vortices studied by Griffiths and Linden (1981b), i.e., a closed anticyclonic circulation within the wave crest and a cyclonic region behind each crest.

The wavelength of the vortices can be predicted from the two layer model of baroclinic instability originally conceived by Phillips (1954) and discussed more fully by Pedlosky (1979). The model assumes the upper (subscript 1) and lower (subscript 2) layers of inviscid fluid to be of thickness h_1, h_2 and to be moving with uniform velocities U_1, U_2 in an infinitely long channel of rectangular section. The x and y axes are taken to be along and across the channel respectively, and the interfacial slope is assumed small and to intersect both sides of the channel. Since U_1 and U_2 do not vary across the width of the channel, the shear $(U_1 - U_2)$ is also independent of the crosschannel coordinate y. Phillips showed that perturbations on the basic flow are stable for any vertical shear if the Froude number is less than a critical value F_c which depends on the layer depths h_1 and h_2. In the East Greenland Current, where the critical value $F_c \sim 2$ and the Froude number is approximately 10^3, we expect baroclinically unstable modes to grow and the source of energy to be predominantly the potential energy in the mean flow. For a perturbation stream function of the form $e^{ik(x-ct)}\sin(\ell y)$, with $K^2=k^2+\ell^2$, the wave speed is given by:

$$c = c_r + c_i = \frac{1}{2(K^2+F_1+F_2)} [\{U_1(K^2+2F_2) + U_2 (K^2+2F_1)\} \pm \{-(U_1-U_2)^2 (4F_1F_2-K^4)\}^{1/2}]. \tag{14.132}$$

Hence, when $K^2 \ll F_1,F_2$ phase speed reduces to

$$c_r = \frac{U_1h_1 + U_2h_2'}{h_1 + h_2} \tag{14.133}$$

which indicates that the waves travel slowly downstream along the ice edge. The most unstable wavelength, λ (= $2\pi/k$) under these supercritical conditions is the wavelength with the largest growth rate kc_i. This is found to be $\lambda \sim 2\pi\hat{R}$ where R is the geometric mean of the Rossby radii for the two layers $[\hat{R}=(\bar{R}_1\bar{R}_2) = \bar{R}_1(\frac{h_2}{h_1})^{1/4}]$. This yields a wavelength of approximately 50 km for disturbances on the East Greenland Current which is in good agreement with observations. Recently, the eddy in Fram Strait has received further intensive study during the MIZEX experiments. Its apparent absence of downstream motion is strongly suggestive of a typographic steering mechanism, such as that proposed by Smith et al., (1984). Analysis of the relevant data is still in progress.

C.2.3 Effect on the Icefield

Eddies of this type may cause loss of ice from the marginal ice zone in two ways: (1) directly, by drawing a large expanse of ice into the warm water zone, where it melts, and (2) indirectly, by drawing large floes from deep within the ice pack out toward the margin, where they are broken up by waves (i.e., by increasing the rate of production of the pulverization factory). Figure C.9 shows how, in the vicinity of the eddy, the large floes are being drawn further out toward the margin.

The development and decay of the eddy involves mixing of water masses between the polar and Greenland Sea surface waters, which causes an unknown but possibly significant heat transfer into the East Greenland Current. Eddies are thus a way in which the barrier which the Polar Front normally presents to lateral heat flux may be broken down. The magnitude of this effect can only be determined by surveying the occurrence and life histories of eddies along the ice margin, both from satellite imagery and by direct shipborne measurements.

C.3 MESOSCALE AND SMALL-SCALE PHENOMENA AT THE ICE EDGE

C.3.1 Vortices Generated by Wind, Tide, and Inertial Forces

The ice in the MIZ drifts almost freely and has a free boundary on one side. Thus every kind of forcing function which, in the interior ice would produce only a modest effect on the ice drift averaged over a large area, produces in the MIZ a major effect on ice displacement, showing itself as a band, vortex, or oriented lead system. Aerial observations of the MIZ reveal a vast variety of such complex features at the ice edge (e.g., Fig. C.13) and it is quite impossible, in general, to unequivocally identify a generating mechanism for any given feature. We can only consider the type of forces that may be significant.

It has been known from the time of Nansen that tidal motion in the pack produces a periodic convergence and divergence of the ice. This was investigated further by Zubov (1945), and in the MIZ the effect may be especially intense where the shelf break (baroclinic tidal generating zone) and MIZ are nearly coincident. The effect on the morphology of the ice cover is, however, unknown. Similarly, floes drifting freely under a transient wind will undergo inertial oscillations which may produce whirl-shaped ice patterns in the MIZ (Hunkins, 1967). Possibly most important of all is the response of the MIZ to the passage of a cyclone along the ice edge, a common feature of the east Greenland coast. Using free drift equations it is possible to see qualitatively (Zubov, 1945) that the ice will respond in such a way as to generate a vortex-shaped pattern, but at present there has not yet been a simulation of the effect of a "typical" cyclone on a freely-drifting icefield, so the scale of the hypothetical cyclone-generated vortex pattern is unknown.

Fig. C.13. Bands and streamers outside ice edge. Photographed from 1300 m, Labrador Sea, February (by courtesy of A. M. Cowan).

Formalization of the forces on an isolated floe undergoing free drift was done by McPhee (1980), and is shown in Figure C.14.

Here V is the steady state ice drift relative to the undisturbed ocean surface current; m, a, h are the mass, area, and thickness of the floe, and f is the Coriolis parameter. The wind force term, which uses the wind velocity U_{10} at anemometer height, has a drag coefficient C which is given by:

$$C = C_{10} + \frac{1}{2}C_h\ H\ N \tag{14.134}$$

(Banke et al., 1976), where H is the mean height of ridges on the floe, N is the number of ridges per unit downwind distance, C_{10} is a flat plate drag coefficient describing the drag due to undeformed ice, and C_F is a form drag coefficient. C must include an important form drag component, since for a small floe the leading edge and any surface protuberances may yield drags far greater than those due to the flat surface. The term F is the wave radiation force, given by (14.139). The water drag term includes the stress term τ_w, which is given in the AIDJEX model (McPhee, 1975) by:

$$\tau_w = -c_w V\underline{V}\ e^{i\beta} \tag{14.135}$$

where c_w and β are experimentally determined boundary layer values. In the case of an isolated floe, c_w must include an addition for form drag, again a function of h/a and likely to be important if not dominant, and β may well be small, since an ice-water boundary layer cannot develop fully under a small floe. Therefore even steady state drift is difficult to model, and such an apparently simple problem as deriving the relative drift speed of floes of different sizes in the same wind is impossible to solve accurately.

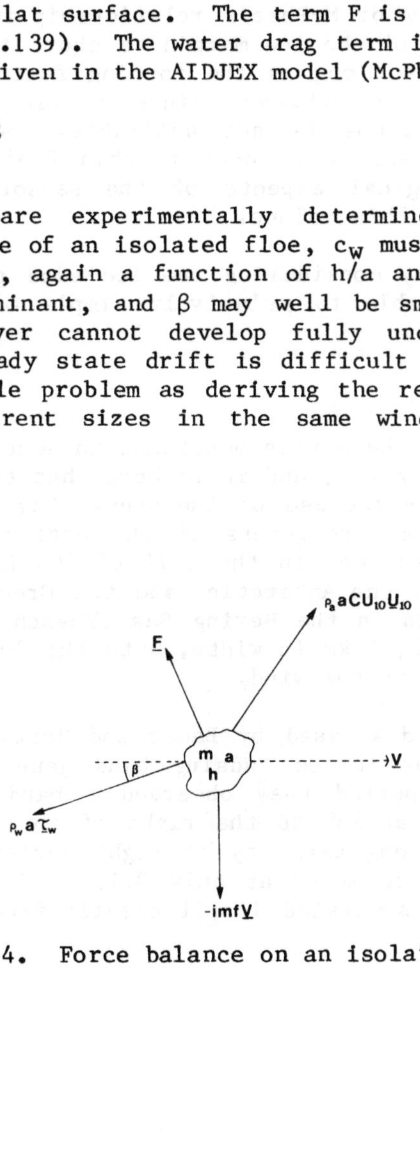

Fig. C.14. Force balance on an isolated floe.

A very rough guide to short-term ice response to wind is provided by the Zubov (1945) Law, which can be expressed in two forms:

1. Ice drifts on average at 28° to the right of the wind as measured at anemometer height and at 2% of the measured wind speed.

2. Ice drifts on average along the isobars, that is, in the direction of the geostrophic wind, at a speed V, which is given, in terms of the pressure gradient, by:

$$V = \frac{13{,}000}{\sin\phi}\frac{dp}{dx} \tag{14.136}$$

where V is in kilometers per month, ϕ is the latitude, and dp/dx is the pressure gradient in millibars per kilometer.

Using this law or McPhee's relationship it may be possible to model the response of the ice margin to short term wind, tidal, or inertial forcing, so long as the forcing function is realistic -- this is difficult to achieve, since a suitably dense grid of meteorological stations is not available. At present, however, there is a great deal to be done in this field. Further discussion on meteorological aspects of the seasonal sea ice zone is given elsewhere in this volume.

We conclude by considering one ice edge phenomenon that does appear to be amenable to relatively simple modeling -- ice edge bands.

C.3.2 Ice Bands

Ice edge bands have been mentioned in Section A.3.6 in connection with the Bering Sea, and it is here that they are most commonly observed, mainly because of the prevailing off-ice wind and the increased aerodynamic roughness of the edge zone ice. However, they have also been seen in the Gulf of St. Lawrence (Campbell et al., 1977, Fig. 5), the Antarctic, and the Greenland Sea. Typical dimensions of bands in the Bering Sea (Muench and Charnell, 1977) are 10 km in length, 1 km in width, with the long axis oriented at 40-90° to the left of the wind.

A case study discussed by Bauer and Martin (1980) enables us to assess the major forces causing band generation and movement. During a 14-hour period they observed a band move at an average speed of 0.4 m s^{-1} at 25° to the right of an off-ice wind of 10 m s^{-1}. The band's long axis lay at right angles to the wind. The remainder of the pack moved at only 0.15 m s^{-1} in the same direction, so the band separated itself clearly from the pack and moved approximately 20 km out into the open water zone. Its subsequent deterioration was rapid. The major forces on the band are assumed to be aerodynamic wind drag, hydrodynamic water drag, and wave radiation pressure; tidal and inertial forces may also be important, but probably act to alter the shape, orientation, and trajectory of the band rather than accounting for its existence as such. The total wind drag F_a on a floe is

$$F_a = \rho_a C_a A |U_a - U_i| (U_a - U_i) \quad (14.137)$$

and the water drag is

$$F_w = - \rho_w C_w A |U_i - U_w| (U_i - U_w) \quad (14.138)$$

where ρ is density, C is drag coefficient, A is surface area, U is velocity, and a, i, w denote air, ice, and water, respectively. For the small, yet exceptionally thick and rough floes of the "edge zone," discussed in Section A.3.6, both C_a and C_w are large and form drag may make as important a contribution as skin-friction drag to the overall drag. In the case study given, Bauer and Martin estimated F_a as of order 10^2N, and F_w as of order 10^3N, using reasonable estimates of C based on measured values for other types of ice. This implies that the major driving force on the band cannot be the wind but instead must be the third factor, wave radiation pressure.

Longuet-Higgins (1977) showed that when a wave is incident on a floating body, in deep water the force on the body per unit width of wavefront is

$$F_r = \frac{1}{4} \rho_w g(a^2 + a'^2 - b^2) \quad (14.139)$$

where a, a′, and b are amplitudes of the incident, reflected, and transmitted waves respectively. The force is in the direction of wave propagation. As we have seen in Section B.3, the reflection coefficient (a′/a) of a floe is large for short-period waves, and approaches unity when the floe diameter is greater than half the wavelength. For very thick floes an additional factor increasing the reflectivity beyond that calculated in Section B.3 is the floe submergence, which has an effect similar to that of a fixed vertical barrier and adds a large element to the reflection if the floe draft exceeds about half a wavelength. For floes 4 m thick by 20 m wide, then, we expect a large reflection coefficient for $T < 4.9$ s and an approach to unity for $T < 2.5$ s. These very short periods are not dominant in an open ocean spectrum, but they do dominate the fetch-limited spectrum that we would expect to see when an off-ice wind blows across a short expanse of open water between leads. Assuming zero absorption, (14.139) simplifies to

$$F_r = \frac{1}{2} \rho_w g a^2 R^2 \tag{14.140}$$

where $R = a'/a$ is the amplitude reflection coefficient. For a floe of 20 m diameter, assuming $R = 1$, F_r is of order 10^4N for a = 0.3 m and 10^3N for a = 0.1 m. This is clearly adequate to overcome the effect of water drag and propel the band at the observed velocity.

A feasible qualitative model for band formation, then, is as follows (Wadhams, 1983a). In initially wind-free conditions the incident waves and swell from the open sea (including long swell coming from the Pacific through gaps between the Aleutians) work upon the MIZ to produce zones of differing ice morphology as described in Section A.3.6. When an off-ice wind begins to blow, the "edge zone" ice, being more aerodynamically and hydrodynamically rough than the "transition zone" ice, is blown seaward slightly faster than the transition zone ice. This general diffusion of the ice edge opens up leads and polynyas of random sizes between the floes. In the bigger openings the off-ice wind can generate a significant fetch- limited spectrum of steep short-period waves. These exert a radiation pressure on the floes at the downwind end of the polynya which is large on account of the high reflectivity of floes to short waves, and hence the floes increase their downwind velocity. This in turn increases the size of the open water area, increases the fetch, increases the wave energy, and still further increases the radiation pressure. As the floes at the downwind edge of the polynya catch up on floes further downwind which are shielded from the wave energy, they herd them together into a compact band of high ice concentration.

Once established, the band keeps its integrity because the wave radiation pressure on both sides of it exerts a compressive force throughout the width of the band. The reason is as follows. We have shown in Section B.3 that progressive partial reflection from floes is equivalent to an exponential decay of wave energy with increasing penetration into an icefield, so that an initial amplitude a_o is reduced to a_x in distance x such that

$$a_x = a_o \exp(-\alpha x) \tag{14.141}$$

where α, the attenuation coefficient, is given by

$$\alpha = \frac{p\, R^2}{2\, d} \tag{14.142}$$

with d the floe diameter and p the fractional ice cover. Thus a wave radiation force F_r per unit wavefront transforms itself, in two dimensions, into a compressive force of $(-dF_r/dx)$ per unit wavefront on any element of the icefield. From (14.141) and (14.142) this is

$$- \frac{dF_r}{dx} = \rho_w \, g \, R^2 \, a_x^2 \, \alpha = \frac{\rho_w \, g \, p \, R^4 \, a_x^2}{2 \, d} \tag{14.143}$$

For $R = 0.5$, $p = 1$, $d = 20$ m, $a_x = 0.1$ m, this force is 0.2 N m^{-1}, sufficient to maintain the integrity of the band, especially near the windward end (the force decreases with distance from the windward end, since a_x is decreasing). An additional force maintaining the band integrity near the leeward, or seaward, end is the radiation pressure from the open ocean. This will be somewhat less than the pressure from upwind since under an off-ice wind there is very little wave generation in the upwind direction so that most of the incoming waves from the open sea are long-period swell from distant generating areas with low R offsetting higher a.

We can summarize the mechanism as follows:

Force originating band	= Off-ice wind drag on rough "edge zone" floes making them move seaward slightly faster than transition zone floes, opening leads within edge zone pack.
Force consolidating band	= Radiation pressure on floes at downwind end of a polynya due to fetch-limited wind wave spectrum.
Forces maintaining band integrity	= Compressive forces on band due to exponential decay of fetch-limited wave spectrum incident from windward and open water swell spectrum incident from leeward (seaward).
Forces separating band from pack	= Fetch-limited wave radiation pressure plus excess wind drag minus wave radiation pressure from open sea.

So far this mechanism describes the formation of a single band. To account for multiple bands we note that the initial band, for example Band 1 in Figure C.15, slows down as it grows wider since the excess force of the wave radiation pressure is being distributed over a greater number of floes. Its speed U_1 therefore tends towards U_o, the speed of an isolated floe not subject to wave action. Eventually the band encounters a second random polynya X_2; floes at the downwind end of X_2 are subject to

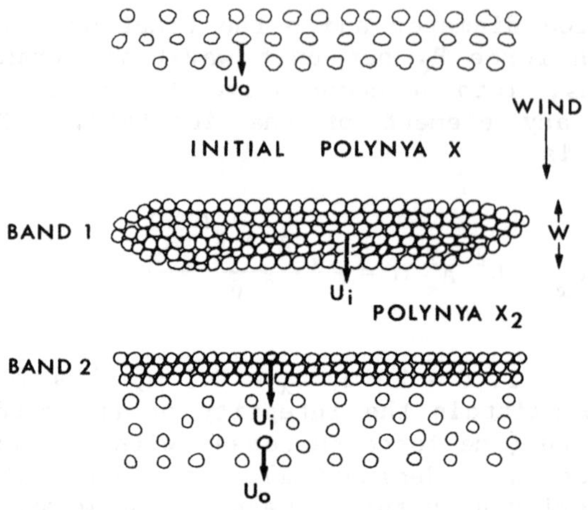

Fig. C.15. Multiple band formation. (After Wadhams, 1983a).

wave pressure from X_2 which gives them a speed of U_i. Band 1 cannot now close the gap; instead, floes at the downwind end of X_2 now herd together to form an independent Band 2. In this way the while MIZ region organizes itself into "immature bands" of varying width and spacing from an initially random state. This is followed by a second organization process by which the immature bands merge into a smaller number of "mature bands" with a tendency to uniform width and spacing. This is because their motion is governed by two laws:

Law 1. Other things being equal, a wide band moves more slowly than a narrow band, so narrow bands will catch up with wide bands and merge with them.

Law 2. Other things being equal, a band with a small upwind fetch moves more slowly than a band with a large upwind fetch, so bands driven by large fetches will catch up with bands driven by small fetches and merge with them.

Other mechanisms have been proposed for band separation and movement, including internal waves and boundary layer modification. These are described elsewhere in this volume.

ACKNOWLEDGMENTS

This chapter was written while the author occupied the Chair of Arctic Marine Science at the Department of Oceanography, Naval Postgraduate School, Monterey, California, sponsored by the Office of Naval Research. I am grateful to Anne Howe and Rob Massom for diagrams.

REFERENCES

Aagaard, K. and L. K. Coachman (1968) The East Greenland current north of Denmark Strait. Arctic, 21: 181-200, 267-290.

Ackley, S. F. and T. E. Keliher (1976) Antarctic sea ice dynamics and its possible climatic effects. AIDJEX Bull., 33: 54-76.

Annals of Glaciology (1980) Volume 1: Proceedings of the Conference on Use of Icebergs: Scientific and Practical Feasibility, Cambridge, United Kingdom.

Armstrong, T., B. B. Roberts, and C. W. M. Swithinbank (1966) Illustrated glossary of snow and ice. Scott Polar Research Institute, Special Publ. 4, 59 p.

Assur, A. (1963) Breakup of pack-ice floes. In Ice and Snow: Properties, Processes and Applications (W. D. Kingery, Ed.), MIT Press, Cambridge, Mass., p. 335-347.

Atlas Antarktiki (1966) Gidrometeoizdat, Leningrad.

Atlas Okeanov (1980) Severnyy Ledovityy Okean. Ministerstvo Oborony SSSR, Moscow, 184 p.

Banke, E. G., S. D. Smith, and R. J. Anderson (1976) Recent measurements of wind stress on Arctic sea ice. J. Fish. Res. Bd. Can., 33: 2307-2317.

Bauer, J. and S. Martin (1980) Field observations of the Bering Sea ice edge properties during March 1979. Mon. Wea. Rev., 108: 2045-2056.

Bentham, R. (1937) The ice-foot. In Arctic Journeys: The Story of the Oxford University Ellesmere Land Expedition 1934-5 (E. Shackleton, Ed.), Appendix III, Hodder and Stoughton, London.

Bilello, M. A. (1980) Decay patterns of fast sea ice in Canada and Alaska. In Sea Ice Processes and Models (R. S. Pritchard, Ed.), University of Washington Press, Seattle, Washington, p. 313-326.

Bolton, W. E. and F. Ursell (1973) The wave force on an infinitely long circular cylinder in an oblique sea. J. Fluid Mech. 57: 241-256.

Bretschneider, C. L. (1963) A one-dimensional gravity wave spectrum. In Ocean Wave Spectra, Prentice-Hall, Englewood Cliffs, New Jersey, p. 41-65.

Buckley, J. R., T. Gammelsrod, A. Johannessen, O. M. Johannessen, and L. P. Røed (1979) Upwelling: Oceanic structure at the edge of the arctic ice pack in winter. Science, 203: 165-167.

Budd, W. F. (1975) Antarctic sea-ice variation from satellite sensing in relation to climate. J. Glaciol., 15: 417-428.

Budden, K. G. (1961) The Wave-guide Mode Theory of Wave Propagation. Prentice-Hall, Englewood Cliffs, New Jersey.

Buynitskiy, V. Kh. (1967) Structure, principal properties and strength of Antarctic sea ice. Soviet Antarctic Expedition Information Bulletin, 65, p. 90-104.

Campbell, W. J., P. Gloersen, and R. O. Ramseier (1975) Synoptic ice dynamics and atmospheric circulation during the Bering Sea experiment. In USSR/US Bering Sea Experiment (K. Ya. Kondratyev, Ed.), Gidrometeoizdat, Leningrad, p. 164-185.

Campbell, W. J., R. O. Ramseier, R. J. Weaver, and W. F. Weeks (1977) Skylab floating ice experiment. Department of Fisheries and Environment, Ottawa, Canada. Miscellaneous Special Publication 34.

Cherepanov, N. V. (1964) Structure of very thick sea ice. Trudy Arkt. Antarkt. N.I. Inst., 267: 13-18.

Cherepanov, N. V. (1971) Spatial arrangement of sea ice crystal structure. Probl. Arkt. Antarkt., 38: 176-181. (Translation Amerind Publishing Co., New Delhi, India, 1973).

Clarke, A. J. (1978) On wind-driven quasi-geostrophic water movements near fast-ice edges. Deep-Sea Res., 25: 41-51.

Coachman, L. K. and K. Aagaard (1974) Physical oceanography of arctic and subarctic seas. In Marine Geology and Oceanography of the Arctic Seas (Y. Herman, Ed.), Springer, New York, p. 1-72.

Cooper, P. F. (1974) Landfast ice in the southeastern part of the Beaufort Sea. In The Coast and Shelf of the Beaufort Sea (J. C. Reed and J. F. Sater, Ed.), Arctic Institute of North America, p. 235-242.

Crary, A. P. (1958) Arctic ice island and ice shelf studies: I. Arctic, 11: 3-42.

Crary, A. P. (1985) Arctic ice island and ice shelf studies: II. Arctic, 13: 32-50.

Crary, A. P., R. D. Cotell, and J. Oliver (1952) Geophysical studies in the Beaufort Sea, 1951. Trans. Am. Geophys. U., 33: 211-216.

Crary, A. P. and N. Goldstein (1959) Geophysical studies in the Arctic Ocean. In Scientific Studies at Fletcher's Ice Island T-3 1952-1955 (V. Bushnell, Ed.), USAF Cambridge Research Center, Bedford, Mass., Vol. 1, p. 7-30, 183-219.

Csanady, G. T. (1972) Geostrophic drag, heat and mass coefficients for the diabatic Ekman layer. J. Atmos. Sci., 29: 488-496.

Dayton, P. K. and S. Martin (1971) Observations of ice stalactites in McMurdo Sound, Antarctica. J. Geophys. Res., 76: 1595-1599.

Dean, C. H. (1966) The attenuation of ocean waves near the open ocean/pack-ice boundary (abstract). In Symposium on Antarctic Oceanography, Scott Polar Research Institute, p. 221-222.

Desai, C. S. and J. F. Abel (1972) Introduction to the Finite Element Method. Van Nostrand, New York.

Diachok, O. (1980) Arctic hydroacoustics. Cold Reg. Sci. Tech., 2: 186-201.

Dinsmore, R. P. (1972) Ice and its drift into the North Atlantic Ocean. In Symposium on Environmental Conditions in the Northwest Atlantic, 1960-1969. International Commission for Northwest Atlantic Fisheries, Dartmouth, Nova Scotia, Special Publication 8, p. 89-128.

Doronin, Yu. P. and D. E. Kheisin (1977) Sea Ice. Gidrometeoizdat, Leningrad. Translation, Amerind Publishing Co., New Delhi, India, 323 pp.

Ewing, M. and A. P. Crary (1934) Propagation of elastic waves in ice: 2. Physics, 5: 181-184.

Feyling-Hanssen, R. W. (1953) Brief account of the ice-foot. Saetr. Nor. Geogr. Tidsskr., 14: 1-4, 45-52.

Fletcher, J. O. (1969) Ice extent in the Southern Ocean and its relation to world climate. Rand Corporation, Santa Monica, California, Report RM-S793-NSF.

Gill, A. E. (1973) Circulation and bottom water production in the Weddell Sea. Deep-Sea Res., 20: 111-140.

Gladfelter, W. H. (1964) Oceanography of the Greenland Sea: USS Atka (AGB-3) survey, summer 1968. U.S. Naval Oceanogr. Off., Washington, D.C., Informal Manuscript Report 0-64-63, 154 p.

Glen, J. W. (1955) The creep of polycrystalline ice. Proc. Roy. Soc., A(228): 519-538.

Goodman, D. J. (1980) Critical stress intensity factor (K_{Ic}) measurements at high loading rates for polycrystalline ice. In IUTAM Meeting on Physics and Mechanics of Ice, Copenhagen (P. Tryde, Ed.).

Goodman, D. J., P. Wadhams, and V. A. Squire (1980) The flexural response of a tabular ice island to ocean swell. Ann. Glaciol., 1: 23-28.

Gow, A. J. and S. Epstein (1972) On the use of stable isotopes to trace the origins of ice in a floating ice tongue. J. Geophys. Res., 77: 6552-6557.

Gow, A. J., S. F. Ackley, W. F. Weeks, and J. W. Govoni (1982) Physical and structural characteristics of Antarctic sea ice. Presented at 3rd International Symposium on Antarctic Glaciology, Columbus, Ohio. Ann. Glaciol., 3: 113-117.

Graff, K. F. (1975) Wave Motion in Elastic Solids. Ohio State University Press.

Greenhill, A. G. (1887) Wave motion in hydrodynamics. Am. J. Math., 9: 62-212.

Greenhill, A. G. (1916) Skating on thin ice. Phil. Mag., 31: 1-22.

Griffiths, R. W. and P. F. Linden (1981a) The stability of buoyancy-driven coastal currents. Dyn. Atmos. Oceans, 5: 281-306.

Griffiths, R. W. and P. F. Linden (1981b) The stability of vortices in a rotating, stratified fluid. J. Fluid Mech., 105: 283-316.

Griffiths, R. W. and P. F. Linden (1982) Laboratory experiments on fronts. Part 1: Density-driven boundary currents. Geophys. Astrophys. Fluid Dyn., 19: 159-187.

Groen, P. (1969) The Waters of the Sea. Van Nostrand, New York.

Hatherton, T. (1960) Microseisms at Scott Base. Geophys. J. Roy. Astr. Soc., 3: 381-405.

Havelock, T. H. (1955) Waves due to a floating sphere making periodic heaving oscillations. Proc. Roy. Soc., A231: 1-7.

Hendrickson, J. A. (1966) Interaction theory for a floating elastic sheet of finite length with gravity waves in water of finite depth (including a comparison with experimental data). Science Engineering Associates, San Marino, California, Report NBy-32215, 178 p.

Hendrickson, J. A., L. M. Webb, and R. J. Quigley (1962) Study of natural forces acting on floating ice fields. National Engineering Science Co., Pasadena, California, Report NBy-32215, 71 p.

Henry, C. J. (1968) Wave-ice interaction: Model experiments. Davidson Laboratory, Stevens Institute of Technology, Hoboken, New Jersey, Report 1314, 34 p.

Hibler, W. D. III (1979) A dynamic thermodynamic sea ice model. J. Phys. Oceanogr., 9: 815-846.

Hibler, W. D. III and S. F. Ackley (1983) Numerical simulation of the Weddell Sea pack ice. J. Geophys. Res., 88: 2873-2887.

Hibler, W. D. III, S. F. Ackley, W. K. Crowder, H. L. McKim, and D. M. Anderson (1974) Analysis of shear zone ice deformation in the Beaufort Sea using satellite imagery. In The Coast and Shelf of the Beaufort Sea (J. C. Reed and J. F. Sater, Ed.), Arctic Institute of North America, p. 285-296.

Hoare, R. D., B. W. Danielewicz, G. R. Pilkington, J. C. O'Rourke and R. D. Wards (1980) An upward-looking sonar system to profile ice keels for one year. Oceans '80, Conf. on Ocean Engng. in the '80's, Seattle, 8-10 Sept. 1980. Inst. Elec. & Electronic Engrs., New York, 123-126.

Hobbs, P. (1974) The Physics of Ice. Oxford University Press.

Holdsworth, G. and J. E. Glynn (1981) A mechanism for the formation of large icebergs. J. Geophys. Res., 86(C4): 3210-3222.

Hult, J. L. and N. C. Ostrander (1974) Applicability of ERTS to Antarctic iceberg resources. In Third Earth Resources Technology Satellite-1 Symposium (S. C. Freden and M. A. Becker, Ed.), NASA Publication SP-351, 1B, p. 1467-1490.

Hunkins, K. (1962) Waves on the Arctic Ocean. J. Geophys. Res., 67: 2477-2489.

Hunkins, K. (1967) Inertial oscillations of Fletcher's ice island (T3). J. Geophys. Res., 72: 1165-1174.

Husseiny, A. A. ed. (1978) Iceberg Utilization. Pergamon, London.

Imbert, B. (1954) Microséismes et houle dans l'Océan Indien Sud. Ann. Geophys., 10: 175-181.

Johannessen, O. M. and L. A. Foster (1978) A note on the topographically controlled oceanic polar front in the Barents Sea. J. Geophys. Res., 83: 4567-4571.

John, F. (1949) On the motion of floating bodies. Comm. Pure Appl. Math., 2: 13-57.

John, F. (1950) On the motion of floating bodies: II. Comm. Pure Appl. Math., 3: 45-101.

Jones, S. (1977) Instabilities and wave interactions in a rotating two-layer fluid. University of Cambridge, Ph.D. thesis.

Keller, J. B. and E. Goldstein (1953) Water wave reflection due to surface tension and floating ice. Trans. Am. Geophys. U., 34: 43-48.

Keller, J. B. and M. Weitz (1953) Reflection and transmission coefficients for waves entering or leaving an icefield. Comm. Pure Appl. Math., 6: 415-417.

Kenyon, K. E. (1969) Stokes drift for random gravity waves. J. Geophys. Res., 74: 6991-6994.

Kiilerich, A. (1945) On the hydrography of the Greenland Sea. Medd. om Gron., 144, part 2, 63 p.

Kinder, T. H. and L. K. Coachman (1978) The front overlying the continental shelf in the eastern Bering Sea. J. Geophys. Res., 83: 4551-4559.

Knapp, W. W. (1972) Satellite observations of large polynyas in polar waters. In Sea Ice (T. Karlsson, Ed.), National Research Council of Iceland, Reykjavik, p. 201-212.

Koch, L. (1945) The east Greenland ice. Medd. om Gron., 130, part 3.

Kovacs, A. and M. Mellor (1974) Sea ice morphology and ice as a geological agent in the southern Beaufort Sea. In The Coast and Shelf of the Beaufort Sea (J. C. Reed and J. F. Sater, Ed)., Arctic Institute of North America, p. 113-161.

Kovacs, A. and D. S. Sodhi (1980) Shore ice pile-up and ride-up: Field observations, models, theoretical analyses. Cold Reg. Sci. Tech., 2: 209-288.

Kozo, T. L. and O. I. Diachok (1973) Spatial variability of topside and bottomside ice roughness and its relevance to underside acoustic reflection loss. AIDJEX Bull., 19: 113-121.

Kozo, T. L. and W. B. Tucker (1974) Sea ice bottomside features in the Denmark Strait. J. Geophys. Res., 79: 4505-4511.

Kristensen, M., O. Orheim, and P. Wadhams (1981) Field experiments on Antarctic tabular icebergs. Polar Record, 20(128): 445-448.

Kristensen, M. and V. A. Squire (1983) Modelling of Antarctic tabular icebergs in ocean waves. In Proceedings, Second Symposium on Applied Glaciology, Hanover, New Hampshire, Ann. Glaciol., 4: 152-157.

Kulakov, I. Yu., M. I. Maslovsky, and L. A. Timokhov (1979) Seasonal variability of Antarctic sea ice extent: Its numerical modelling. In Sea Ice Processes and Models (R. S. Pritchard, Ed.), University of Washington Press, Seattle, Washington (Presented but not published.)

Larsen, L. H. (1978a) Surface waves and low frequency noise in the deep ocean. Geophys. Res. Lett., 5: 499-501.

Larsen, L. H. (1978b) The effects of finite depth on the propagation of non-linear wave packets. J. Phys. Oceanogr., 8: 923-925.

Legeckis, R. (1977) Oceanic polar front in the Drake Passage - satellite observations during 1976. Deep-Sea Res., 24: 701-704.

Legeckis, R. (1978) A survey of worldwide sea surface temperature fronts detected by environmental satellites. J. Geophys. Res., 83: 4501-4522.

Leppäranta, M. (1980) On the drift and deformation of sea ice fields in the Bothnian Bay. Winter Navigation Research Board, Helsinki, Research Report 29, 79 p.

Leppäranta, M. (1981a) An ice drift model for the Baltic Sea. Tellus, 33: 583-596.

Leppäranta, M. (1981b) On the structure and mechanics of pack ice in the Bothnian Bay. Finn. Mar. Res., 248: 3-86.

Leppäranta, M. (1981c) Statistical features of sea ice ridging in the Gulf of Bothnia. Winter Navigation Research Board, Helsinki, Research Report 32, 46 p.

Leschack, L. A. (1965) On the generation and directional recording of waves in the Arctic Ocean. U.S. Naval Oceanographic Office, Washington, D.C., Technical Report TR-179, 44 p.

Leschack, L. A. and R. A. Haubrich (1964) Observations of waves on an ice-covered ocean. J. Geophys. Res., 69: 3815-3821.

Lewis, E. L. and A. R. Milne (1977) Underwater sea ice formations. In Polar Oceans (M. J. Dunbar, Ed.), Arctic Institute of North America, p. 239-245.

Lewis, E. L. and W. F. Weeks (1971) Sea ice: Some polar contrasts. In Symposium on Antarctic Ice and Water Masses (G. Deacon, Ed.), Scientific Committee on Antarctic Research, Cambridge, p. 23-34.

Lindsay, D. G. (1969) Ice distribution in the Queen Elizabeth Islands. In Ice Seminar, Canadian Institute of Mining and Metallurgy, Special Volume 10, p. 45-60.

Loeng, H. and T. E. Vinje (1979) On the sea ice conditions in the Greenland and Barents Seas. In POAC 79: Proceedings, Fifth International Conference on Port and Ocean Engineering Under Arctic Conditions, University of Trondheim, Trondheim, Norway, p. 163-176.

Longuet-Higgins, M. S. (1977) The mean forces exerted by waves on floating or submerged bodies with applications to sand bars and wave power machines. Proc. Roy. Soc., A352: 463-480.

Lunde, T. (1963) Sea ice in the Svalbard region 1957-1962. Norsk Polarinstitutt Arbok 1962, p. 24-34.

Lunde, T. (1965) Ice conditions at Svalbard 1946-1963. Norsk Polarinstitutt Arbok 1963, p. 61-80.

Maccamy, R. C. (1961) On the heaving motion of cylinders of shallow draft. J. Ship Res., 5: 34-43.

McCann, S. B. and R. J. Carlisle (1972) The nature of the ice-foot on the beaches of Radstock Bay, south-west Devon Island, N.W.T., Canada. Institute of British Geographers, Special Publication 4, p. 175-186.

Mackintosh, N. A. (1946) The antarctic convergence and the distribution of surface temperature in Antarctic waters. Discovery Reports, 23: 177-212.
Mackintosh, N. A. (1972) Life cycle of Antarctic drill in relation to ice and water conditions. Discovery Reports, 36: 1-94.
McLaren, A. S., P. Wadhams, and R. Weintraub (1984) The sea ice topography of M'Clure Strait in winter and summer of 1960 from submarine profiles. Arctic, 37(2): 110-120.
McNutt, S. L. (1979) Remote sensing analysis of ice growth and distribution in the Eastern Bering Sea Shelf: Oceanography and resources (D. W. Hood and J. R. Calder, Eds.), Univ. Washington Press, Seattle, 1: 141-165.
McPhee, M. G. (1975) Ice-ocean momentum transfer for the AIDJEX model. AIDJEX Bulletin, 29: 93-111.
McPhee, M. G. (1980) An analysis of pack ice drift in summer. In Sea Ice Processes and Models (R. S. Pritchard, Ed.), University of Washington Press, Seattle, Washington, p. 62-75.
Manley, T. O., L. A. Codispoti, K. L. Hunkins, H. R. Jackson, E. P. Jones, V. Lee, S. Moore, J. Morison, T. T. Packard, and P. Wadhams (1982) The "Fram 3" Expedition. EOS, 63: 627-636.
Markham, W. E. (1969) Growth, break-up and movement of ice in Canadian coastal waters. In Ice Seminar, Canadian Institute of Mining and Metallurgy, Special Volume 10, p. 31-35.
Marko, J. R. and R. E. Thomson (1975) Spatially periodic lead patterns in the Canada Basin sea ice: A possible relationship to planetary waves. Geophys. Res. Lett., 2: 431-434.
Marko, J. R. and R. E. Thomson (1976) Rectilinear leads and internal motions in the pack ice of the western Arctic Ocean. J. Geophys. Res., 82: 979-987.
Martin, S. (1974) Ice stalactites: Comparison of a laminar flow theory with experiment. J. Fluid Mech., 63: 51-79.
Martin, S. (1979) A field study of brine drainage and oil entrapment in first year sea ice. J. Glaciol., 22: 473-501.
Martin, S. (1981) Frazil ice in rivers and oceans. Ann. Rev. Fluid Mech., 13: 379-397.
Martin, S. and P. Kauffman (1981) A field and laboratory study of wave damping by grease ice. J. Glaciol., 27(96): 283-314.
Maykut, G. A. and N. Untersteiner (1971) Some results from a time-dependent, thermodynamic model of sea ice. J. Geophys. Res., 76: 1550-1575.
Mills, D. A. (1969) Waves, pressure disturbances and floating ice fields. Horace Lamb Centre for Oceanographic Research, Flinders University, Bedford Park, Research Paper 33, 27 p.
Mills, D. A. (1972) On waves in a sea ice cover. Horace Lamb Centre for Oceanographic Research, Flinders University, Bedford Park, Research Paper 53, 64 p.
Milne, A. R. (1972) Thermal tension cracking in sea ice: A source of under-ice noise. J. Geophys. Res., 77: 2177-2192.

Milne, A. R. (1978) Oil, ice and climate change. Environment Canada, Victoria, B.C., Beaufort Sea Project Overview Report M5, 103 p.

Mollo-Christensen, E. (1983) Interactions between waves and mean drift in an ice pack. J. Geophys. Res., 88: 2971-2972.

Moore, S. C. and P. Wadhams (1980) Recent developments in strainmeter design. In Proceedings, Workshop on Stress and Strain Measurement in Ice, Memorial University, St. John's, Newfoundland, 97-125.

Muench, R. D. and K. Ahlnas (1976) Ice movement and distribution in the Bering Sea from March to June 1974. J. Geophys. Res., 81: 4467-4476.

Muench, R. D. and R. L. Charnell (1977) Observations of medium-scale features along the seasonal ice edge in the Bering Sea. J. Phys. Oceanogr., 7: 602-606.

Munk, W. H. (1962) Long ocean waves. In The Sea (M. N. Hill, Ed.), Wiley, New York, Vol. 1, p. 647-663.

Murauchi, S. and Y. Yoshida (1959) Reports of sea ice observations of the Japanese Antarctic Research Expedition 2, 1957-1958. Antarctic Record, 8: 22-39.

Nansen, F. (1915) Spitsbergen waters. Videnskapsselsk. Skr. Kl. I. Mat. Naturvidensk. Kl., no. 2.

Nansen, F. (1922) The strandflat and isostasy. Videnskapsselsk. Skr. Kl. I. Mat. Naturvidensk. Kl., no. 11.

NAVOCEANO (1967) Birds Eye 2-67, 15th March-19th April 1967. U.S. Naval Oceanographic Office, Washington, D.C., Informal Report 67-81.

Nevel, D. (1970) Moving loads on a floating ice sheet. U.S.A. Cold Regions Research and Engineering Laboratory, Research Report 261.

Newton, J. L. and L. E. Piper (1981) Oceanographic data from northwest Greenland Sea: Arctic East 1979 survey of USGS "Westwind." Science Applications, Inc., San Diego, California, Rept. SAI-202-81-003-LJ.

Nixon, W. A. and R. A. Smith (1984) Preliminary results on the fatigue behaviour of polycrystalline freshwater ice. Cold Regions Sci. & Technol., 9: 267-270.

NORSEX (1981) NORSEX marginal ice zone experiment 1979 by NORSEX Arctic working group. Geophysical Institute, University of Bergen, Reports 1-6.

NORSEX (1983) The Norwegian remote sensing experiment (NORSEX) in a marginal ice zone, by NORSEX Arctic working group. Science, 220: 781-787.

Nye, J. F. (1953) The flow law of ice from measurements in glacier tunnels, laboratory experiments and the Jungfraufirn borehole experiment. Proc. Roy. Soc., A219: 477-489.

Ofuya, A. O. and A. J. Reynolds (1967) Laboratory simulation of waves in an ice floe. J. Geophys. Res., 72: 3567-3583.

Orheim, O., P. Wadhams, and M. Kristensen (1982) Iceberg response to sea state. Ann. Glaciol., 3: 357-8; Iceberg Research, 1: 10-15.

Overgaard, S., P. Wadhams, and M. Leppäranta (1983) Ice properties in the Greenland and Barents Seas during summer. J. Glaciol., 29(101): 142-164.

Paige, R. A. (1966) Crystallographic studies of sea ice in McMurdo Sound, Antarctica. U.S. Naval Civil Engineering Laboratory, Technical Report R494, p. 1-31

Palfrey, K. M. (1967) Physical oceanography of the northern part of the Greenland Sea in summer 1964. University of Washington, Masters thesis, 52 p.

Paquette, R. G. and R. H. Bourke (1979) Temperature fine structure near the sea-ice margin of the Chukchi Sea. J. Geophys. Res., 84: 1155-1164.

Paquette, R. G. and R. H. Bourke (1981) Ocean circulation and fronts as related to ice melt-back in the Chukchi Sea. J. Geophys. Res., 86: 4215-4230.

Parkinson, C. L. and W. M. Washington (1980) Summary of a large-scale sea ice model. In Sea Ice Processes and Models (R. S. Pritchard, Ed.), University Washington Press, Seattle, Washington, p. 224-233.

Parry, W. E. (1824) Journal of a Second Voyage for the Discovery of a Northwest Passage. Performed in the Years 1821-2-3. John Murray, London.

Peary, R. E. (1910) The North Pole. Hodder and Stoughton, London.

Pease, C. H. (1980) Eastern Bering Sea ice processes. Mon. Wea. Rev., 108: 2015-2023.

Pease, C. H., S. A. Schoenberg, and J. E. Overland (1982) A climatology of the Bering Sea and its relation to sea ice extent. NOAA Environmental Research Laboratories, Boulder, Colorado, Technical Report ERL 419-PMEL 36.

Pedlosky, J. (1979) Geophysical Fluid Dynamics, Springer Verlag, New York, 624 p.

Peters, A. S. (1950) The effect of a floating mat on water waves. Comm. Pure Appl. Math., 3: 319-354.

Peyton, H. R. (1966) Sea ice strength. Geophysical Institute, University of Alaska, Report UAG R-182, 276 p.

Phillips, N. A. (1954) Energy transformations and meridional circulations associated with simple baroclinic waves in a two level quasigeostrophic model. Tellus, 6: 273-286.

Pounder, E. R. (1965) The Physics of Ice. Pergamon, Oxford.

Pratt, J. G. D. (1960) A gravity traverse of Antarctica. Trans-Antarctic Expedition 1955-1958, Science Report 2, 27 p.

Ramseier, R. O., W. J. Campbell, O. M. Johannessen, P. Gloersen, and J. Zwally (1979) Variation of the sea ice pack in the Norwegian, Greenland and Barents Sea via satellite passive microwave observations: September 1973-December 1974 (Film). Crawley Films, Ottawa, Canada.

Reimnitz, E. and K. F. Bruder (1972) River discharge into an ice-covered ocean and related sediment dispersal, Beaufort Sea, coast of Alaska. Geol. Soc. Am. Bull., 83: 861-866.

Reimnitz, E., C. A. Rodeick, and S. C. Wolf (1974) Strudel scour: A unique Arctic marine geologic phenomenon. J. Sed. Petrol. 44(2): 409-420.

Reimnitz, E., L. Toimil, and P. Barnes (1978) Arctic continental shelf morphology related to sea ice zonation, Beaufort Sea, Alaska. Marine Geology, 28: 179-210.

Robin, G. de Q. (1963) Wave propagation through fields of park ice. Phil. Trans. Roy. Soc. A.255: 313-339.

Ross, Sir James Clarke (1847) A Voyage of Discovery and Research in the Southern and Antarctic Regions During the Years 1839-1843. John Murray, London. (Reprinted 1969, David and Charles, Newton Abbot.)

Schindler, J. E. (1968) Impact of ice islands: The story of Arlis II and Fletcher's Ice Island T-3 since 1962. In Arctic Drifting Stations, Arctic Institute of North America, p. 49-80.

Scoresby, W. (1820) An Account of the Arctic Regions, With a History and Description of the Northern Whale-Fishery, Constable, Edinburgh. (Reprinted 1969, David and Charles, Newton Abbot.)

Serson, H. V. (1972) Investigation of a plug of multi-year ice in the mouth of Nansen Sound. Defence Research Establishment, Ottawa, Canada, Technical Note 72-6.

Serson, H. V. (1974) Sverdrup Channel. Defence Research Establishment, Ottawa, Canada, Technical Note 74-10.

Shapiro, A. and L. S. Simpson (1953) The effect of a broken ice-field on water waves. Trans. Am. Geophys. U., 34: 36-42.

Smirnov, V. N. (1966) Flexural and gravity oscillations of the ice cover of the Davis Sea. Inf. Byull. Sov. Antarkt. Eksped., 61: 61-65.

Smith, D. C., IV, J. H. Morison, J. A. Johannessen and N. Untersteiner (1984) Topographic generation of an eddy at the edge of the East Greenland Current. J. Geophys. Res. 89(5):8205-8208.

Sodhi, D. S. (1977) Ice arching and the drift of pack ice through restricted channels. U.S.A. Cold Regions Research and Engineering Laboratories, CRREL Report 77-18.

Squire, V. A. (1978a) An investigation into the use of strain rosettes for the measurement of propagating cyclic strains. J. Glaciol., 20: 425-431.

Squire, V. A. (1978b) Dynamics of ocean waves in a continuous sea ice cover. University of Cambridge, Ph.D. thesis.

Squire, V. A. (1981) Numerical simulation of ice floes in waves. Scott Polar Research Institute, Technical Report 81-1, 57 p.

Squire, V. A. (1983) Numerical modelling of realistic ice floes in ocean waves. In Proceedings, 2nd Symposium on Applied Glaciology, Hanover, New Hampshire, Ann. Glaciol., 4: 277-282.

Squire, V. A. and A. J. Allan (1980) Propagation of flexural gravity waves in sea ice. In Sea Ice Processes and Models (R. S. Pritchard, Ed.), University of Washington Press, Seattle, Washington, p. 327-338.

Squire, V. A. and S. Martin (1980) A field study of the physical properties, response to swell, and subsequent fracture of a single ice floe in the winter Bering Sea. University of Washington, Departments of Atmospheric Sciences and Oceanography, Scientific Report 18, 56 p.

Squire, V. A. and S. C. Moore (1980) Direct measurement of the attenuation of ocean waves by pack ice. Nature, 283: 365-368.

Stoker, J. J. (1957) Water Waves. Interscience, New York.

Stokes, G. G. (1847) On the theory of oscillatory waves. Trans. Camb. Phil. Soc., 8: 441-455.

Sytinskiy, A. D. and V. P. Tripol′nikov (1964) Some results of investigations of the natural vibrations of ice fields of the central Arctic. Izv. Akad, Nauk SSSR, Geofiz., 4: 615-621.

Szczesny, C. H. and B. Lemehaute (1965) Experimental investigation of ice floe under wave action. National Engineering Science Co., Pasadena, California, Report NBy-32256, 34 p.

Tabata, T. (1958) Studies on visco-elastic properties of sea ice. In Arctic Sea Ice, National Academy of Science - National Research Council, Publication 598, p. 139-147.

Tabata, T. (1972) Observations of deformation and movement of ice field with the sea ice radar network. In Sea Ice (T. Karlsson, Ed.), National Research Council, Reykjavik, p. 72-79.

Taylor, H. W., A. L. Gordon, and E. Molinelli (1978) Climatic characteristics of the Antarctic polar front zone. J. Geophys. Res., 83: 4572-4578.

Thorndike, A. S., D. A. Rothrock, G. A. Maykut, and R. Colony (1975) The thickness distribution of sea ice. J. Geophys. Res., 80: 4501-4513.

Timoshenko, S., D. H. Young, and W. Weaver, Jr. (1974) Vibration Problems in Engineering. Wiley, New York, 4th Edition.

Treshnikov, A. F. (1966) The ice of the Southern Ocean. In Proceedings, Symposium on Pacific-Antarctic Science, Japanese Antarctic Research Expedition Scientific Reports, Special Issue no. 1.

Tucker, M. J. (1956) A shipborne wave recorder. Trans. Inst. Nav. Archit., 98: 236-259.

Tucker, W. B., W. F. Weeks, and M. Frank (1979) Sea ice ridging over the Alaskan continental shelf. J. Geophys. Res., 84: 4885-4897.

Tucker, W. B., W. F. Weeks, A. Kovacs, and A. J. Gow (1980) Nearshore ice motion at Prudhoe Bay, Alaska. In Sea Ice Processes and Models (R. S. Pritchard, Ed.), University of Washington Press, Seattle, Washington, p. 261-272.

Ursell, F. (1947) The effect of a fixed vertical barrier on surface waves in deep water. Proc. Camb. Phil. Soc. Math. Phys. Sci., 43: 374-382.

Ursell, F. (1949) On the heaving motion of a uniform cylinder on the surface of a fluid. Quart. J. Mech. Appl. Math., 2: 218-231.

Ursell, F. (1954) Water waves generated by oscillating bodies. Quart. J. Mech. Appl. Math., 7: 427-437.

Ursell, F. (1961) The transmission of surface waves under surface obstacles. Proc. Camb. Phil. Soc. Math. Phys. Sci., 57: 638-668.

Vinje, T. E. (1969) The sea ice condition in Svalbard in 1967. Norsk Polar Institutt Arbok 1967, p. 194-196.

Vinje, T. E. (1977a) Sea ice studies in the Spitzbergen-Greenland area. NTIS, Springfield, Va., Landsat Report E77-10206.

Vinje, T. E. (1977b) Some observations from Nimbus-6 data collecting platforms in polar areas. In Proceedings, IAGA/IAMAP Assembly, Seattle, Washington, National Center for Atmospheric Research, Boulder Co., p. 124-132.

Vinje, T. E. (1979) On the drift ice conditions in the Atlantic sector of the Antarctic. In Proceedings, Fifth International Conference on Port and Ocean Engineering Under Arctic Conditions, University of Trondheim, Norway, Vol. 3, p. 75-82.

Vinje, T. E. and P. Steinbakke (1977) Nimbus-6 located automatic stations in the Svalbard waters in 1975. Norsk Polarinstitutt Arbok 1975, p. 109-117.

Wadhams, P. (1972) Measurement of wave attenuation in pack ice by inverted echo sounding. In Sea Ice (T. Karlsson, Ed.), National Research Council of Iceland, Reykjavik, p. 255-260.

Wadhams, P. (1973a) Attenuation of swell by sea ice. J. Geophys. Res., 78: 3552-3563.

Wadhams, P. (1973b) The effect of a sea ice cover on ocean surface waves. University of Cambridge, Ph.D. thesis.

Wadhams, P. (1975) Airborne laser profiling of swell in an open ice field. J. Geophys. Res., 80: 4520-4528.

Wadhams, P. (1978a) Wave decay in the marginal ice zone measured from a submarine. Deep-Sea Res., 25: 23-40.

Wadhams, P. (1978b) Characteristics of deep pressure ridges in the Arctic Ocean. In Proceedings, Fourth International Conference on Port and Ocean Engineering Under Arctic Conditions (D. B. Muggeridge, Ed.), Memorial University of Newfoundland, St. John's, Newfoundland, Canada, Vol. 1, p. 544-555.

Wadhams, P. (1979) Field experiments on wave-ice interaction in the Labrador and East Greenland currents, 1978. Polar Record, 19: 373-376.

Wadhams, P. (1980) A comparison of sonar and laser profiles along corresponding tracks in the Arctic Ocean. In Sea Ice Processes and Models (R. S. Pritchard, Ed.), University of Washington Press, Seattle, Washington, p. 283-299.

Wadhams, P. (1981a) Sea-ice topography of the Arctic Ocean in the region 70°W to 25°E. Phil. Trans. Roy. Soc. Lond., A302: 45-85.

Wadhams, P. (1981b) The ice cover in the Greenland and Norwegian Seas. Rev. Geophys. Space Phys., 19(3): 345-393.
Wadhams, P. (1983a) A mechanism for the formation of ice edge bands. J. Geophys. Res., 88(C5):2813-2818.
Wadhams, P. (1983b) Sea ice thickness distribution in Fram Strait. Nature, Lond., 305: 108-111.
Wadhams, P., and R. T. Lowry (1977) A joint topside-bottomside remote sensing experiment on Arctic sea ice. In Proceedings, Fourth Canadian Symposium on Remote Sensing, Quebec, Canadian Remote Sensing Society, p. 407-423.
Wadhams, P., A. E. Gill and P. F. Linden (1979) Transects by submarine of the East Greenland polar front. Deep-Sea Res., 26: 1311-1328.
Wadhams, P. and R. J. Horne (1980) An analysis of ice profiles obtained by submarine sonar in the Beaufort Sea. J. Glaciol., 25: 401-424.
Wadhams, P. and V. A. Squire (1980) Field experiments on wave-ice interaction in the Bering Sea and Greenland waters, 1979. Polar Record, 20: 147-158.
Wadhams, P. and V. A. Squire (1983) An ice-water vortex at the edge of the East Greenland Current. J. Geophys. Res., 88(C5): 2770-2780.
Wadhams, P., M. Kristensen and O. Orheim (1983) The response of Antarctic icebergs to ocean waves. J. Geophys. Res., 88: 6053-6065.
Wadhams, P., A. S. McLaren and R. Weintraub (1985) Ice thickness distribution in Davis Strait in February from submarine sonar profiles. J. Geophys. Res. 90(C1):1069-1077.
Wadhams, P., V. A. Squire, J. A. Ewing and R. W. Pascal (1986) The effect of the marginal ice zone on the directional wave spectrum of the ocean. J. Phys. Oceanogr., Feb. 1986.
Walker, E. R. and P. Wadhams (1979) On thick sea-ice floes. Arctic, 32: 140-147.
Watanabe, K. (1972) An average pattern of extending and retreating of an ice area in the Sea of Okhotsk. In Sea Ice (T. Karlsson, Ed.), National Research Council, Reykjavik, p. 152-153.
Weber, J. R. and M. Erdelyi (1976) Ice and ocean tilt measurements in the Beaufort Sea. J. Glaciol., 17: 61-71.
Weeks, W. F., A. Kovacs, and W. D. Hibler III (1971) Pressure ridge characteristics in the Arctic coastal environment. In Proceedings, First International Conference on Port and Ocean Engineering Under Arctic Conditions, Technical University of Norway, Trondheim, Norway, Vol. 1, p. 152-182.
Weeks, W. F. and A. J. Gow (1978) Preferred crystal orientations in the fast ice along the margins of the Arctic Ocean. J. Geophys. Res., 83: 5105-5121.
Weeks, W. F. and M. Mellor (1978) Some elements of iceberg technology. In Iceberg Utilization (A. A. Husseiny, Ed.), Pergamon, p. 45-98.

Weeks, W. F., W. B. Tucker, M. Frank, and S. Fungcharoen (1980) Characterization of the surface roughness and floe geometry of the sea ice over the continental shelves of the Beaufort and Chukchi Seas. In Sea Ice Processes and Models (R. S. Pritchard, Ed.), University of Washington Press, Seattle, Washington, p. 300-312.

Weertman, J. (1957) Deformation of floating ice shelves. J. Glaciol., 3: 38-42.

Weertman, J. (1972) Creep of ice. In Proceedings, Conference on Physics and Chemistry of Ice, Ottawa, National Research Council of Canada.

Weitz, M. and J. B. Keller (1950) Reflection of water waves from floating ice in water of finite depth. Comm. Pure Appl. Math., 3: 305-318.

Westhall, V. H. and H. Li (1976) An analysis of sea ice bottomside data in the Denmark Strait. AIDJEX Bull., 31: 101-114.

Wetzel, V. F., R. K. Atwater, and T. E. Huta (1974) Arctic ice movement and environmental data stations. In The Coast and Shelf of the Beaufort Sea (J. C. Reed and J. F. Sater, Ed.), Arctic Institute of North America, p. 269-284.

Wiegel, R. L. (1964) Oceanographical Engineering. Prentice-Hall, Englewood Cliffs, N.J.

Winsor, W. D. and B. R. Ledrew (1978) Ice feature characterization: Labrador offshore. Centre for Cold Ocean Resources and Engineering, Memorial University of Newfoundland, St. John's, Nfld., Project SAR'77 Field Data Report 14.

Wittmann, W. I. and J. J. Schule (1966) Comments on the mass budget of Arctic pack ice. In Proceedings, Symposium on Arctic Heat Budget and Atmospheric Circulation, Report RM-5233-NSF, p. 215-246, RAND Corp., Santa Monica, California.

Worsley, F. A. (1916) Journal kept during the Imperial Trans-Antarctic Expedition 1914-1916, entry for 9 March 1916. Scott Polar Research Institute, Cat. No. MS296, unpublished ms.

Zienkiewicz, O. C. (1977) The Finite Element Method. McGraw-Hill, New York, 3rd edition.

Zubov, N. N. (1945) L'dy Arktiki (Arctic Ice.) Izdatel'stvo Glavsevmorputi, Moscow, 360 p. (U.S. Navy Hydrographic Office, Translation 217, 1963; available as AD426972 from NTIS, Springfield, Va.)

Zwally, H. J. and P. Gloersen (1977) Passive microwave images of the polar regions and research applications. Polar Record, 18: 431-450.

Zwally, H. J., J. C. Comiso, C. L. Parkinson, W. J. Campbell, F. D. Carsey, and P. Gloersen (1983) Antarctic sea ice cover 1973-1976: satellite passive-microwave observations. NASA Scientific and Technical Information Branch, Report SP-459, 206 pp.

APPENDICES

1. Propagation Vectors

Firstly, solve for k_0 by Newton's method or a similar iteration. Then divide by $(k_n - k_0)$, giving

$$k_n^4 + k_n^3 k_0 + k_n^2 k_0^2 + k_n k_0^3 + \frac{\omega^2 \rho_w}{Lk_o} = 0 \tag{14.144}$$

Substitute for k_1 or k_2 in (1): The imaginary part gives

$$b^2 = \frac{(4 a^3 + 3 k_0 a^2 + 2 k_0^2 a + k_0^3)}{(k_0 + 4 a)} \tag{14.145}$$

and the real part

$$b^2 = 1/2 \left\{ (k_0^2 + 3k_0 a + 6a^2) - [(k_0^2 + 3k_0 a + 6a^2) - 4 (a^4 + k_0 a^3 + k_0^2 a^2 + k_0^3 a + \frac{\omega^2 \rho_w}{LK_o})]^{1/2} \right\} \tag{14.146}$$

a is obtained iteratively by equating (14.145) and (14.146), and then b is obtained by substitution in (14.145).

2. Reflection and Transmission Coefficients for a Discrete Floe

The eight equations are:

$$\Sigma_0^2 (A_n e^{-ik_n \ell} + B_n e^{ik_n \ell}) = Be^{-ik\ell} + Re^{ik\ell} \tag{14.147}$$

$$\Sigma_0^2 k_n (A_n e^{-ik_n \ell} - B_n e^{ik_n \ell}) = kB e^{-ik\ell} - k R e^{ik\ell} \tag{14.148}$$

$$\Sigma_0^2 k_n^3 (A_n e^{-ik_n \ell} + B_n e^{ik_n \ell}) = 0 \tag{14.149}$$

$$\Sigma_0^2 k_n^4 (A_n e^{-ik_n \ell} - B_n e^{ik_n \ell}) = 0 \tag{14.150}$$

$$\Sigma_0^2 (A_n\ e^{ik_n \ell} + B_n\ e^{-ik_n \ell}) = T\ e^{ik\ell} \tag{14.151}$$

$$\Sigma_0^2\ k_n\ (A_n\ e^{ik_n \ell} - B_n\ e^{-ik_n \ell}) = k\ T\ e^{ik\ell} \tag{14.152}$$

$$\, _0^2\Sigma k_n^3\ (A_n\ e^{ik_n \ell} + B_n\ e^{-ik_n \ell}) = 0 \tag{14.153}$$

$$\, _0^2\Sigma k_n^4\ (A_n\ e^{ik_n \ell} - B_n\ e^{-ik_n \ell}) = 0 \tag{14.154}$$

We define the factors

$$\alpha_0 = k_0\ell,\ \alpha = a\ell,\ \beta = b\ell,\ \gamma = k/k_0,\ \gamma_o = 1,\ \delta_0 = 0,$$

$$\gamma_m + i\delta_m = \left(\frac{a + b\ i}{k_0}\right)^m \quad B = 1 \tag{14.155}$$

and the functions

$$\begin{aligned} F_m &= 2\ (\gamma_m \cos\alpha\ \cosh\beta + \delta_m \sin\alpha\ \sinh\beta) \\ G_m &= 2\ (\gamma_m \cos\alpha\ \sinh\beta + \delta_m \sin\alpha\ \cosh\beta) \\ H_m &= 2\ (-\gamma_m \sin\alpha\ \sinh\beta + \delta_m \cos\alpha\ \cosh\beta) \\ I_m &= 2\ (\gamma_m \sin\alpha\ \cosh\beta - \delta_m \cos\alpha\ \sinh\beta) \end{aligned} \tag{14.156}$$

and finally, letting the suffices $_1$ and $_2$ represent real and imaginary parts, we set

$$\begin{aligned} P_1 &= R_1 + T_1 & P_5 &= A_{01} + B_{01} \\ P_2 &= R_2 + T_2 & P_6 &= A_{02} - B_{02} \\ P_3 &= R_1 - T_1 & P_7 &= A_{01} - B_{01} \\ P_4 &= R_2 - T_2 & P_8 &= A_{02} + B_{02} \\ Q_1 &= A_{11} + B_{21} + A_{21} + B_{11} & Q_5 &= A_{11} + B_{21} - A_{21} - B_{11} \\ Q_2 &= A_{11} - B_{21} + A_{21} - B_{11} & Q_6 &= A_{11} - B_{21} - A_{21} + B_{11} \\ Q_3 &= A_{12} + B_{22} + A_{22} + B_{12} & Q_7 &= A_{12} + B_{22} - A_{22} - B_{12} \\ Q_4 &= A_{12} - B_{22} + A_{22} - B_{12} & Q_8 &= A_{12} - B_{22} - A_{22} + B_{12} \end{aligned} \tag{14.157}$$

The sixteen linear equations for P and Q follow in Table 1.

Table 1. Linear equations for P and Q.

$$
\begin{pmatrix}
-\cos k\ell & +\sin k\ell & 0 & 0 & 2\cos\alpha_o & 0 & 0 & 0 & F_0 & 0 & 0 & 0 & 0 & 0 & 0 & -H_0 \\
0 & 0 & +\cos k\ell & -\sin k\ell & 0 & -2\sin\alpha_o & 0 & 0 & 0 & 0 & 0 & -I_0 & -G_0 & 0 & 0 & 0 \\
-\sin k\ell & -\cos k\ell & 0 & 0 & 0 & 0 & 0 & 2\cos\alpha_o & 0 & 0 & F_0 & 0 & 0 & H_0 & 0 & 0 \\
0 & 0 & \sin k\ell & +\cos k\ell & 0 & 0 & 2\sin\alpha_o & 0 & 0 & I_0 & 0 & 0 & 0 & 0 & -G_0 & 0 \\
0 & 0 & \gamma\cos k\ell & -\gamma\sin k\ell & 0 & 0 & 2\cos\alpha_o & 0 & 0 & F_1 & 0 & 0 & 0 & 0 & -H_1 & 0 \\
-\gamma\cos k\ell & \gamma\sin k\ell & 0 & 0 & 0 & 0 & 0 & -2\sin\alpha_o & 0 & 0 & -I_1 & 0 & 0 & -G_1 & 0 & 0 \\
0 & 0 & +\gamma\sin k\ell & +\gamma\cos k\ell & 0 & 2\cos\alpha_o & 0 & 0 & 0 & 0 & 0 & F_1 & H_1 & 0 & 0 & 0 \\
-\gamma\sin k\ell & -\gamma\cos k\ell & 0 & 0 & 2\sin\alpha_o & 0 & 0 & 0 & I_1 & 0 & 0 & 0 & 0 & 0 & 0 & -G_1 \\
0 & 0 & 0 & 0 & 2\cos\alpha_o & 0 & 0 & 0 & F_3 & 0 & 0 & 0 & 0 & 0 & 0 & -H_3 \\
0 & 0 & 0 & 0 & 0 & -2\sin\alpha_o & 0 & 0 & 0 & 0 & 0 & -I_3 & -G_3 & 0 & 0 & 0 \\
0 & 0 & 0 & 0 & 0 & 0 & 0 & 2\cos\alpha_o & 0 & 0 & F_3 & 0 & 0 & H_3 & 0 & 0 \\
0 & 0 & 0 & 0 & 0 & 0 & 2\sin\alpha_o & 0 & 0 & I_3 & 0 & 0 & 0 & 0 & -G_3 & 0 \\
0 & 0 & 0 & 0 & 0 & 0 & 2\cos\alpha_o & 0 & 0 & F_4 & 0 & 0 & 0 & 0 & -H_4 & 0 \\
0 & 0 & 0 & 0 & 0 & 0 & 0 & -2\sin\alpha_o & 0 & 0 & -I_4 & 0 & 0 & -G_4 & 0 & 0 \\
0 & 0 & 0 & 0 & 0 & 2\cos\alpha_o & 0 & 0 & 0 & 0 & 0 & F_4 & H_4 & 0 & 0 & 0 \\
0 & 0 & 0 & 0 & 2\sin\alpha_o & 0 & 0 & 0 & I_4 & 0 & 0 & 0 & 0 & 0 & 0 & -G_4
\end{pmatrix}
\begin{pmatrix}
P_1 \\ P_2 \\ P_3 \\ P_4 \\ P_5 \\ P_6 \\ P_7 \\ P_8 \\ Q_1 \\ Q_2 \\ Q_3 \\ Q_4 \\ Q_5 \\ Q_6 \\ Q_7 \\ Q_8
\end{pmatrix}
=
\begin{pmatrix}
\cos kl \\ -\cos kl \\ -\sin kl \\ \sin kl \\ \gamma\cos kl \\ -\gamma\cos kl \\ -\gamma\sin kl \\ \gamma\sin kl \\ 0 \\ 0 \\ 0 \\ 0 \\ 0 \\ 0 \\ 0 \\ 0
\end{pmatrix}
$$

CHAPTER 15

ASPECTS OF THE METEOROLOGY OF THE SEASONAL SEA ICE ZONE

R. G. Barry

Cooperative Institute for Research in Environmental Sciences
University of Colorado
Boulder, Colorado 80309 U.S.A.

1. INTRODUCTION

The hemispheric seasonal sea ice zones (SSIZ) are delineated in Figures 1 and 2 showing the range between the average seasonal maximum and seasonal minimum limits for $\geq$ 1/8 ice concentration. A marginal ice zone (MIZ) contiguous with the open ocean represents one general type of SSIZ and in Antarctica this is circumglobal, extending more or less concentrically around the continent between latitudes 60° and 70°S. The area of ice-covered waters varies between September and March from about 4 x 10^6 to 20 x 10^6 km^2. In the northern hemisphere, in contrast, there is a marked longitudinal asymmetry relating to the land-sea distribution and the seasonal range is much less, from 7.8 x 10^6 to 14.8 x 10^6 km^2 (Parkinson et al., in press), due to the land-locked Arctic Ocean. While the maximum limit of the MIZ extends to south of 50°N in the Sea of Okhotsk and the Labrador Sea, there is open water to the north of 75°N in the Norwegian-Barents Sea. From a meteorological standpoint, these MIZs are not subject to an unlimited fetch of wind and currents such as occurs around Antarctica and should, therefore, be regarded as a second type. A third type of SSIZ is represented in the essentially closed Arctic Ocean where the winter extension of the pack ice is constrained by land and shorefast ice, creating shear zones. These are prominent features from the Queen Elizabeth Islands to the Beaufort Sea coast of Alaska. The most extensive SSIZs in the Arctic occur in the Barents-Laptev and Chukchi-Bering "shelf seas."

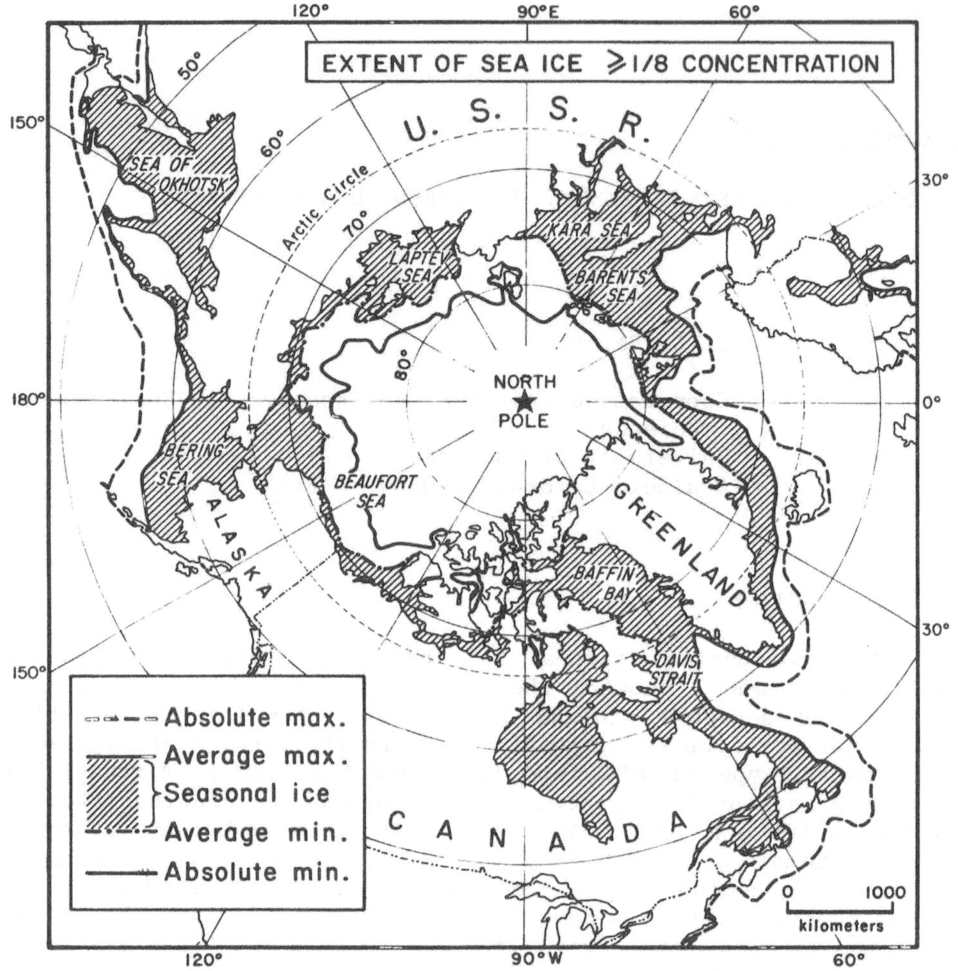

Fig. 1. The seasonal sea ice zone in the northern hemisphere; the average and extreme limits of seasonal ice, based on a concentration $\geq$ 1/8, are shown (after Central Intelligence Agency, 1978).

Since the SSIZ is defined by the seasonal range of ice limits, its meteorological characterization is not readily described in a systematic manner. Moreover, the ice limits are not solely determined by meteorological factors, nor does the SSIZ set up a unique climatological regime. Accordingly, this survey reviews selected aspects of large-scale circulation processes and examples of specific regional interactions.

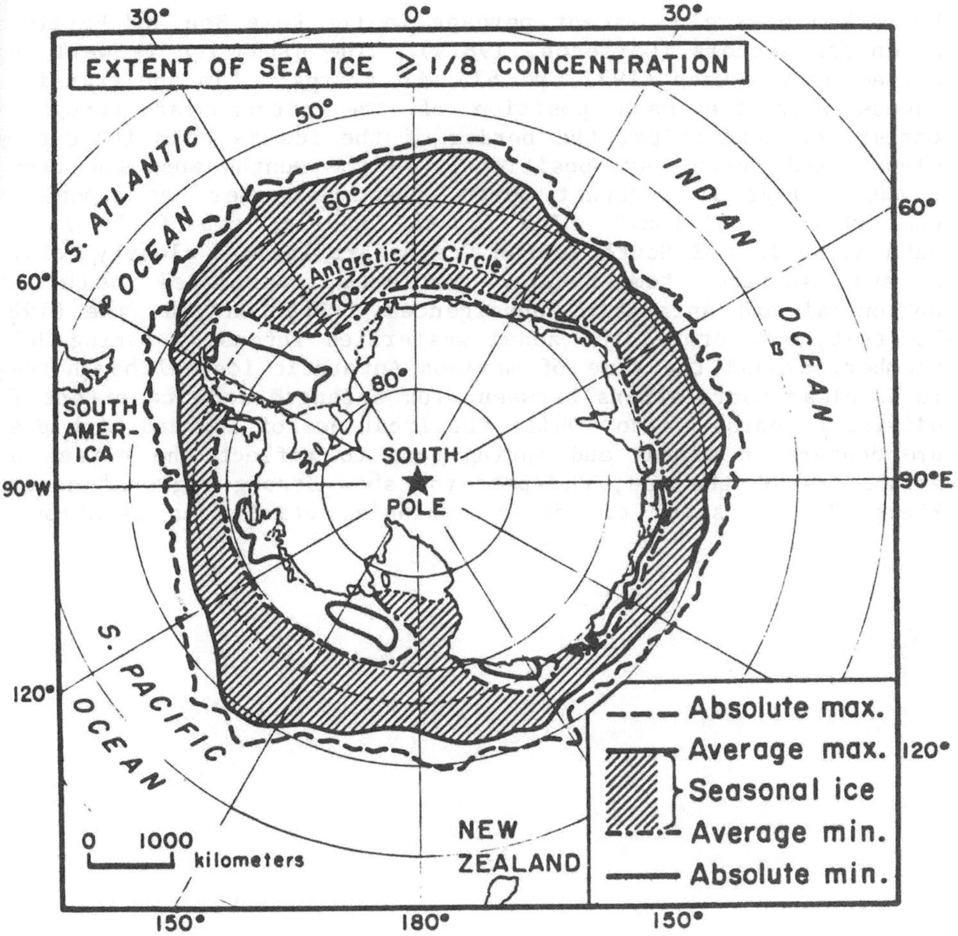

Fig. 2. As Figure 1 for the southern hemisphere; ice limits are for 1971-76.

2. ATMOSPHERIC CIRCULATION CHARACTERISTICS

Comparison between the summer ice limit in the Antarctic and atmospheric circulation parameters shows that the polar front is located about 45°S and although there is a subsidiary maximum of 1000-500 mbar thickness gradient at 70°S there is no strong evidence for a climatic front (Taljaard, 1972, Figure 8.9). In addition, in winter the frontal zones are spatially separated from the pack ice margin, except perhaps in the Ross Sea and Bellingshausen Sea sectors (Carleton, 1981a). The situation is well summarized by Van Loon (1972, p. 67) who states: "In the transition seasons when the mean position of the circumpolar trough is nearest the Antarctica, the border of the ice is near its extreme poleward and equatorward positions. In the months when the trough on the average is farthest north, the ice border has reached no such extreme position." The frontal maxima, especially in the South Atlantic and South Indian Oceans are instead closely related to sea surface temperature gradients associated with the Subtropical and Antarctic Convergences. Streten and Pike (1980) show that, on average, the zonal westerlies intensify during July-November, around the time of maximum Antarctic ice, although there are no clear correlations between wind strength and ice extent for individual years. Also, while the locations of low and high pressure centers in winter and spring tend to reflect the extent and variability of the ice, the patterns show strong regional control (Figure 3). A study of winter cyclonic activity in relation to

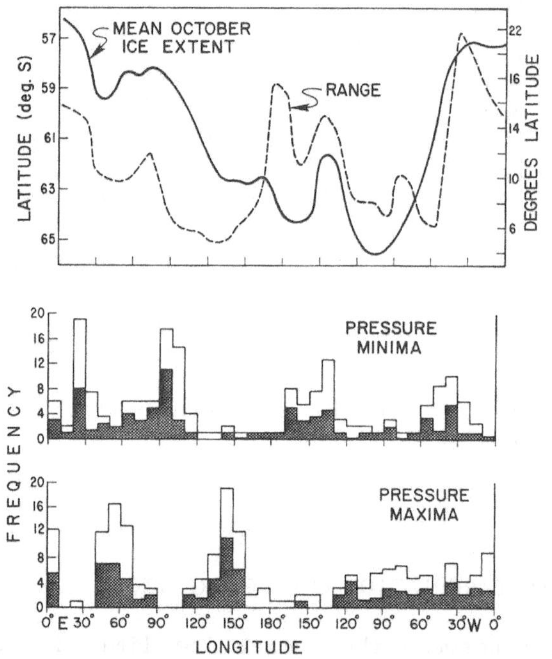

Fig. 3. Longitudinal distributions of pressure systems in relation to Antarctic sea ice extent. Above. Mean October ice extent and absolute range of ice extent, 1972-1977, in degrees latitude. Below. The frequency of pressure minima in the sub-Antarctic trough and pressure maxima (ridges) impinging on the same zone from north or south; open columns = annual, solid columns = winter plus spring (after Streten and Pike, 1980).

southern hemisphere ice by Carleton (1981a; 1981b; 1983) suggests that more extensive ice acts to increase cyclogenetic tendencies in the ice edge zone, at least regionally. He also notes instances where cyclonic activity may have enhanced ice advection. A statistical relationship between the ice margin and the latitudinal band of maximum cyclone frequency is reported for the transition seasons by Schwerdtfeger and Kachelhoffer (1973), but the types of cyclone and the location of cyclogenesis must be analyzed carefully to establish such relationships more precisely.

In the northern hemisphere, the three areas of extensive winter ice, in the Bering-Okhotsk Sea, the eastern Canadian Arctic and Subarctic, and the Barents Sea, are all located close to mean January and April 700 mbar pressure troughs. The mean contours indicate advection of cold northwesterly air streams on the western side of these troughs. In contrast, the longitudes of Scandinavia, Alaska, and to a lesser, extent Greenland, are sectors where, particularly in spring, there are maxima of blocking ridges (Rex, 1950). In winter the mean polar frontal zones are well south of the ice margins but there is a secondary "arctic" front in the northeastern North Atlantic. A major depression track from Newfoundland to Iceland parallels the ice margin (Klein, 1957), although it is displaced southward 8°-9° latitude. Petterssen (1950) was the first to propose a linkage between winter cyclonic activity and north polar ice based on the vorticity theorem. In summer, the arctic front lies over the forest-tundra ecotone on land (Reed, 1960; Bryson, 1966; Krebs and Barry, 1970), but there is a secondary cyclone track along the Arctic coasts of Siberia and Alaska both in summer and in fall (Klein, 1957).

Composite synoptic models of cyclonic systems occurring within the Arctic Basin were developed by Reed (1962) and a more complete characterization of cloud vortex patterns using satellite imagery and Navy Spot Data for middle and high latitudes has recently been compiled by Carleton (1985). This study also examines in detail the relationships between cyclonic activity, surface properties, latitude and season. Thermal effects of discontinuities in surface type on the atmosphere are discussed below.

3. ATMOSPHERIC CIRCULATION/ICE FEEDBACKS

The problem of interactions between atmosphere and ice has been approached by descriptive synoptic case studies, by simulation experiments with circulation models, and statistically. In the last category, Lemke et al. (1980) indicate from an 11-year data set for the Arctic, and a 6-year Antarctic set, that the space-time structure of monthly sea ice anomalies can be accounted for by a combination of white-noise atmospheric forcing (from weather disturbances), seasonal forcing (representing oceanic effects in both polar regions, plus solar radiation in the Arctic), and lateral diffusion and advection of ice, particularly in the Antarctic. The seasonal cycles are also shown to be different between the two polar regions. The growth stage is slower than the melting in the Antarctic whereas in the Arctic the seasonal cycle is nearly symmetrical (Comiso and Zwally, 1984). This undoubtedly reflects the large-scale geographical contrasts between the two regions.

The dramatic seasonal expansion and contraction of the Antarctic sea ice has been explained by Gordon and Taylor (1975; Gordon, 1981) in terms of: 1) the Ekman divergence due to the curl of the wind stress, and 2) the surface heat balance. The first factor is associated with divergence generating open water areas within the ice. In early winter, open water will tend to freeze over and the ice therefore expands northward. In summer, open water will provide a heat source and encourage lateral ice decay, although Gordon's (1981) calculations suggest that only half of the required heat-flux can be provided from the atmosphere while the remainder must come from deep water below the relatively weak pycnocline. Ackley (1981) states that the evidence relating to the ice advance is supportive of the mechanism proposed by Gordon and Taylor, but a modeling simulation by Hibler and Ackley (1983) indicates greater complexity. Their results suggest that the large seasonal fluctuation in ice extent in the Weddell Sea is associated with the ice dynamics. However, it is the rapid decay that depends critically on the role of leads and lateral ice advection, whereas the ice advance is primarily associated with freezing in situ, except in the western section of the Weddell.

The extent to which storm activity may cause a positive feedback between atmosphere and sea ice has been examined synoptically by Ackley and Keliher (1976). A rapid decrease in sea ice in the Bellingshausen Sea in winter 1973 is thought to have caused a relative gain in atmospheric heat, thereby augmenting cyclone activity which in turn helped to maintain less ice cover. Figure 4 illustrates the anomaly of atmospheric heat loss in 1973, which may have been up to an order of magnitude greater than the average total energy of a midlatitude cyclone (2.1×10^{16}kJ). Ackley and Keliher propose that usually the ice concentration is sufficient for storm activity to cause its convergence, reducing the sensible heat flux and thereby increasing the net atmospheric heat loss. However, poleward flow of warmer air, with meridional blocking patterns, may decrease the ice extent and its concentration. Under such conditions, enhanced storm activity could cause divergence of the pack.

From averaged data, Budd (1975) indicates that a 1°C increase in mean annual temperature corresponds to a 2.5° latitude change in sea ice extent at latitude 60°S, suggesting a stable feedback relationship. A 1° latitude change in the ice limit represents a change of about 15 percent in hemisphere sea ice cover according

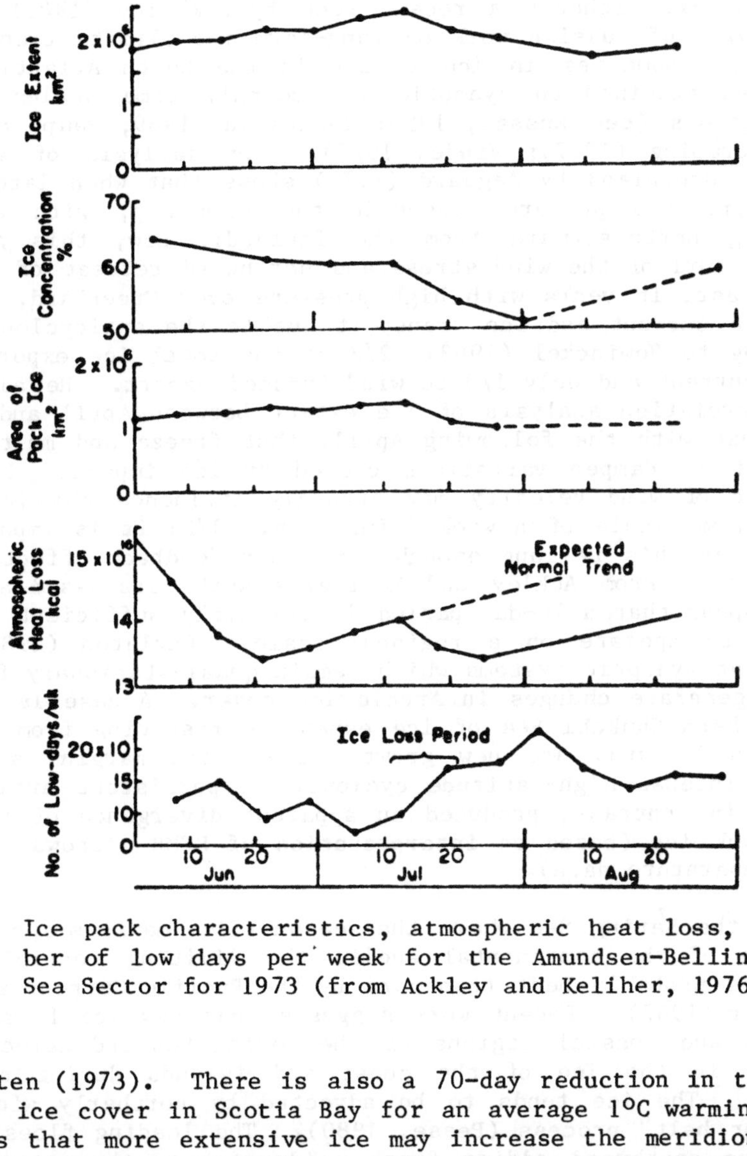

Fig. 4. Ice pack characteristics, atmospheric heat loss, and number of low days per week for the Amundsen-Bellingshausen Sea Sector for 1973 (from Ackley and Keliher, 1976).

to Streten (1973). There is also a 70-day reduction in the duration of ice cover in Scotia Bay for an average 1°C warming. Budd suggests that more extensive ice may increase the meridional temperature and pressure gradients and, therefore, the zonal (westerly) circulation. However, Miles (1975) demonstrates for the North Atlantic that such a relationship does not hold on the 5-30 year time scale. Indeed, there may have been a slight increase in poleward sensible heat flux associated with the post-1930 decrease in zonal circulation (Miles, 1976).

In the northern hemisphere, where the overall pattern of ice extent is strongly influenced by the interactive oceanic and atmospheric circulations, most investigations have been regional in character, although a recent book by Zakharov (1981) provides an overview of Russian work on long-term ice-climate trends. The causes of anomalies in ice extent in the North Atlantic sector have been examined on synoptic and monthly time scales by many investigators (see Nusser, 1958; Bjornsson, 1969; Haupt and Kant, 1976; Strübing (1967); Vinje, 1977).[1] An analysis of anomalies off East Greenland by Aagaard (1972) shows that when late-winter/spring pressure patterns resemble the mean map, with a trough extending northeastward from the Icelandic Low, this generates cyclonic curl of the wind stress and northward retreat of the ice. In contrast, in years with high pressure over Greenland, there is southward current and ice transport due to the anticyclonic curl. According to Vowinckel (1963), 2/3 of the total ice export is due to the current and only 1/3 to wind-induced export. He also shows from correlation analysis of ice extent between April and August, and August with the following April, that freeze and melt processes tend to dampen variations caused by ice import. Since the influence of wind velocity may strongly influence the ice margin on the time scale of a week (Einarsson, 1972) it is important to ask whether this is long enough for major feedback effects on the atmosphere. From Ackley and Keliher's work, discussed above, it would appear that a 10-day period is certainly sufficient for such effects to operate on a regional scale. Carleton (1984) also finds that synoptic systems which remain quasi-stationary for about a week generate changes in Arctic ice cover. A case is noted in the northern Chukchi Sea of ice expansion resulting from both ice advection by wind and new growth at the ice margin, associated with an intense high-latitude cyclone. A persistent anticyclonic regime, in contrast, produced an apparent divergence of the near-shore pack ice (based on interpretation of ESMR microwave brightness temperature data).

In the Bering Sea where the seasonal ice maximum approximates the edge of the continental shelf, the limiting control on ice extent is probably heat transport by the Pacific Current according to Dunbar (1967). Recent work suggests that new ice forms in the northern and coastal regions of the Bering Sea and especially in polynyas in the lee of the coast and islands during northerly airflow. The ice tends to be advected by northerly winds in a "conveyor belt" process (Pease, 1980). The leading floes melt as they move southward adding fresh cold water to the water column, thereby prolonging the residence time of new leading floes. A synoptic climatology of winter storm tracks in the region indicates that cyclonic activity is concentrated in the eastern (western) Bering Sea in heavy (light) ice years, which will tend to

---

[1]Processes operating in the Baltic Sea and Hudson Bay are rather special and are not discussed here. Schell (1964) has performed comparable investigations for the Sea of Okhotsk.

increase (decrease) the frequency of northerly airflow (Overland and Pease, 1982). These results accord with the hypothesis of Niebauer (1980) that the strength of the winter atmospheric circulation determines the interannual variations in ice extent in the Bering Sea.

The effects of sea ice anomalies on the atmosphere have been examined empirically by Wiese (1924) and Johnson (1980) and in a GCM experiment by Herman and Johnson (1978). Wiese showed that, in the North Atlantic, cyclone tracks are further south in years with heavy ice than in years with light ice. This result has also been found by Herman and Johnson using the GISS GCM. They compare a mean of two mid-winter simulations for "maximum" ice in the northern oceans with the mean of six simulations given minimum mid-winter ice extent. They demonstrate both local and hemispheric effects on the atmospheric circulation. With maximum sea ice there is an increase of sea-level pressure over the ice and a southward shift of cyclones in the North Atlantic, giving increased precipitation over northern Europe. On the hemispheric scale, deeper Icelandic and Gulf of Alaska lows at 700 mbar are coupled with stronger subtropical anticyclones and enhanced poleward energy flux in mid-latitudes. Herman and Johnson note, however, that the model omits feedback processes, including changes in sea surface temperatures. Moreover, observed ice anomalies do not tend to occur simultaneously in all sectors (Walsh and Johnson, 1979a; Lemke et al., 1980; Smirnov, 1980; Walsh and Sater, 1981). In a further data study, Johnson (1980) compares five years each of heavy and light Arctic ice with the concurrent winter circulation and finds similarity with the GISS GCM study for the Atlantic, but less agreement for the Pacific. She notes the importance of regional links between the atmosphere and ice extent, with northerly airflows causing ice advection, in Davis Strait and the Bering Sea, but indeterminate relationships for the sector east of Greenland.

A detailed analysis of northern hemisphere synoptic-scale ice-atmosphere interactions has recently been completed by Carleton (1985). Based on satellite imagery analysis of cloud vortex systems for two years of mid-season months he shows that, for the hemisphere as a whole, cyclones either are displaced equatorward and/or are more numerous in years when sea ice is more extensive in January and April. A reverse pattern is observed with the sea ice margin in July and October, however, and cryosphere-atmosphere interactions in October principally involve the continental snow cover. These results confirm the earlier findings of Wiese, although analysis of more years is needed to establish the variability of such interactions.

The extent to which the atmosphere forces the ice, or vice versa, is a question that is now beginning to yield to analytical studies, although many shortcomings in the atmospheric, oceanic and ice data place a limit on the confidence that can be attached to the statistical results (Walsh and Lemke, this volume). Nevertheless, eigenvector analysis of monthly ice anomalies and atmospheric variables over the Arctic for 1953-1977 suggests that during the autumn ice-growth period the response of the atmosphere to the ice anomalies is comparable to the response of the ice to atmospheric anomalies (Walsh and Johnson, 1979b). During summer, however, the ice responds to atmospheric forcing, up to four months earlier. A synoptic-scale analysis by Crane (1983) using ESMR data indicates similar relationships in the Beaufort Sea, but in the Kara-Barents Sea a more complex situation occurs. Here the direction of leads and lags varies not only seasonally, but also between the area of permanent Arctic pack and the SSIZ. In winter, over the permanent ice area, the ice tends to lead the atmosphere (~7-12 day interval) whereas in the SSIZ the atmosphere shows a greater control over ice conditions. In summer, in contrast, the ice generally leads the atmosphere except over the SSIZ where the air temperature distribution has some control on ice concentration. The synoptic analyses of Carleton (1985), discussed above, demonstrate that in January and April the sea ice margin exerts a predominantly cyclogenetic effect on the atmosphere although, since cyclones are also shown to be steered by the ice boundary, some consequent atmosphere-surface forcing is to be expected. The relative roles of dynamic and thermodynamic effects have not yet been resolved but air temperature rather than atmospheric pressure shows the strongest statistical relationship to sea ice extent in European and Alaskan sectors of the Arctic on a synoptic timescale (Crane, 1984; Crane et al., 1982).

4. ENERGY BUDGETS

The large-scale energy budget regimes of the Arctic Ocean and adjacent seas are well known in broad terms from observational and analytical studies. In particular, Vowinckel and Orvig (1966, 1970) and Vowinckel and Taylor (1965) have mapped calculated values for the whole area and distinguished several distinctive seasonal energy regimes. They identify the pack ice of the central Arctic Ocean as a transitional type, resembling oceanic areas in summer and continental areas in winter and the transition seasons. In the context of the SSIZ, the map delimiting regions investigated shows that only the Kara-Laptev Sea area is essentially located within the SSIZ. Their analysis of the Norwegian-Barents Sea includes all the area north of 65°N, for example. Table 1 summarizes their results for the central Arctic Ocean and the Kara-Laptev seas for annual totals and the extreme months. Interestingly, the results for the latter are consistently much closer to the central Arctic Ocean than to their calculated values for the Norwegian-Barents Sea. Their calculations indicate that the evaporative heat-loss from open pack ice is very similar to that from an open ocean, except in summer months at 65°/70°N and 70°/75°N when there is evaporative heat loss over open water but a small heat gain to open pack ice (Figure 5).

Further calculations by Vowinckel and Orvig (1973), using daily meteorological data for 1957-58 from the Soviet station NP7, illustrate the importance of synoptic conditions for the turbulent exchanges. In February, 10 percent of the month contributes 29 percent of the monthly sensible heat flux over ice, and 22 percent over open water. This synoptic control operates through variations in cloud cover for an ice surface, whereas over water there are additional effects due to wind and air mass changes. For the total surface energy loss, including long wave radiation, in January-February only 0.1 percent of the surface need be open water in order to equal the loss from the remaining 99.9 percent of the area that is ice covered. Subsequently, more detailed study has been made of the role of open leads and young ice. Maykut (1978; this volume) calculates energy fluxes for open water, thin ice and thick ice in mid-winter as shown in Table 2. The large fluxes for thin ice are maintained by latent heat released during ice growth, not by ocean transport. Maykut shows that up to one-half of the regionally-averaged turbulent heat losses may be contributed from the area of ice in the 20-80 cm thickness category. Estimated fluxes of sensible heat in the transition seasons in particular are affected by this young ice area. Wadhams and Horne (1980) estimate from sonar data that thin ice ($\leq$ 60 cm) occurred over 0.9

Table 1. Energy fluxes at the surface in the Arctic (W m^{-2}).

	ANNUAL		JANUARY		JULY	
	Central Polar Ocean	Kara-Laptev Sea	Central Polar Ocean	Kara-Laptev Sea	Central Polar Ocean	Kara-Laptev Sea
S	39	44	---	---	123	137
I↓	216	230	161	175	305	311
I↑	-255	-267	-198	-208	-337	-331
LE	-3	-6	1	---	-3	-2
H	-5	-7	2	3	1	2
O	8	4	36	31	-92	-116

S = absorbed short wave radiation
I = long wave radiation (↓ = down, ↑ = up)
LE = latent heat flux (positive upward)
H = sensible heat flux (positive upward)
O = oceanic heat exchange with deeper layers of the ocean

(After Vowinckel and Orvig, 1966.)

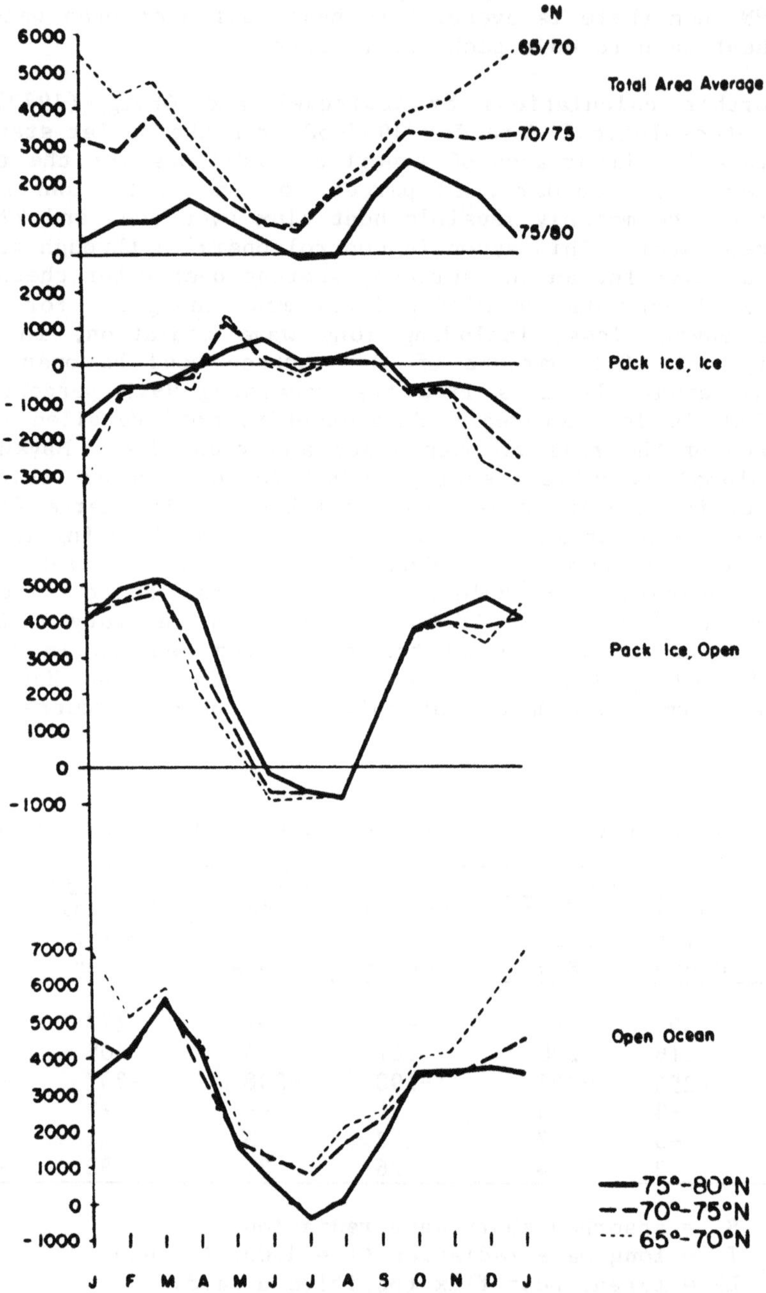

Fig. 5. Evaporation, Norwegian-Barents Sea (1,000 cal./cm^2 month $\simeq$ 16.1 W m^{-2}). (From Vowinckel and Taylor, 1965).

Table 2. Calculated energy fluxes over water and ice on 1 February (W m^{-2}).

	Open water	0.05 m ice	.4 m ice	3 m ice
Net radiation	-147	-93	-45	-29
Sensible heat	-575	-334	-77	-16
Latent heat	-147	-45	-6	0
Oceanic/conductive heat flux	869	472	128	13

(From Maykut, 1978.)

percent of the southern Beaufort Sea in April 1976, but was highly variable in extent along different tracks. Earlier measurement in the TransPolar Drift Stream in March 1971 gave an average of 2.7 percent thin ice.

There are few meteorological observations over open leads or young ice. Major practical difficulties hinder such field experiments, especially the rapid closure of leads and the irregular drift of steam fog, forming over the lead, across the instrument site. Radiation measurements over a freezing lead near Barrow in March 1974 show a daily net radiation value of -71 W m^{-2} over open water and -78 W m^{-2} on the next day over the frozen lead (Holmgren and Weller, 1974). This results from the marked decrease in absorbed short-wave radiation, but the maintenance of large long wave losses due to the relatively high surface temperature of the young ice. Related turbulent flux measurements by Andreas et al. (1979; Andreas, 1980) show that the sensible heat flux is approximately three times the latent heat flux in winter over Arctic leads and that the fluxes depend on the fetch. Using new turbulent transfer equations, the latent heat fluxes obtained by Andreas (1980) are close to those of Maykut (1978), but the sensible heat fluxes are substantially smaller, except over narrow ($\leq$ 10 m) leads. Almost all leads are < 50 m wide according to Wadhams and Horne (1980) but no classes of smaller size range are distinguished.

The surface energy balance in the Southern Ocean has received much less attention. Recently, Weller (1980) has made some order-of-magnitude estimates which suggest that the earlier figure of Fletcher (1969) for a midwinter turbulent heat flux to the atmosphere from the pack ice of 24 W m^{-2} is perhaps x 6 too small. Weller's figures shown in Table 3 assume that in the outer pack ice zone (with ice concentrations between 15 and 85 percent), 40 percent is open water, 20 percent ice < 40 cm thick, and 40 percent ice > 40 cm thick. In the inner pack, 10 percent is open water, 10 percent ice < 40 cm and 80 percent ice > 40 cm. Accordingly, Weller stresses the uncertainty, and potential variability

Table 3. Calculated energy budget components in midwinter in Antarctica (W m^{-2}).

Zone	Net radiation	Sensible heat flux	Latent heat flux	Subsurface heat flux
Coastal ice	-56	5	9	42
Inner pack (≥85% ice)		-89	-28	
Outer pack		-152	-37	
Open ocean*	-12	-29	-24	+65

*Zillman, 1972.
(From Weller, 1968, 1980.)

that is introduced into these estimates by the extent and occurrence of the polynyas (see below).

Little work has been performed relating to atmospheric energetics in the SSIZ zone. LeDrew (1980) illustrates a diagnostic approach to synoptic-scale processes; using the omega equation, he evaluated components of the vertical velocity field over the eastern Canadian Arctic for January-August, 1973. A detailed case-study of a synoptic event in July was also made relating to significant ice melt along the eastern coast of Baffin Island. LeDrew showed the role of warm air advection from the south and west, augmented by adiabatic compression due to descent in the lee of the coastal mountains. In this region, early ice retreat has been shown to occur in summers with frequent southerly flow components when the upper trough is 4°-5° further west (Crane, 1978a). Such airflows give warm air advection and high turbulent energy fluxes accelerating fast ice melt in western Davis Strait (Barry and Jacobs, 1978; Crane, 1978b). In further diagnostic studies, LeDrew (1984) considers the role of surface diabatic heat sources in synoptic systems over the Arctic Ocean. In both summer and winter cases, the effects of sensible heat flux from open water areas are found to play a minor role in 850 mb level vertical motion and heating rates. Large-scale vorticity and thickness advection play major roles, surface friction is an additional factor in the Beaufort Sea sector and, locally, orographic effects are also important.

Inter-annual and longer term variations in energy fluxes are essentially unknown, except via modelling estimates. The variables involved include cloudiness and advection (implying synoptic activity), fractional open water area, and current and ice transport effects. For example, Vowinckel (1963) estimates that in a given year the heat gain for the Polar Ocean could depart 15-20 percent from the long-term mean value due to ice and current export.

5. MESOSCALE PHENOMENA

5.1 Polynyas

The availability of satellite imagery has drawn attention to the extensive recurrent polynyas around Antarctic (Figure 6a), especially in the Ross Sea and the Weddell Sea (Knapp, 1972; Streten, 1975). In the Amundsen Sea (110°W), for example, two polynyas typically form in early November and join with the open ocean in late January-early February. The large Weddell Sea polynya, which forms by the extension of an ice tongue encircling an open water area, is not present in all years, but when present is remarkably persistent (Carsey, 1980). In the Northern Hemisphere, polynyas are mainly of local significance. Those along the Siberian coastline have received some study from Soviet scientists (see Barry, 1983). Apparent polynyas tentatively identified within the central Arctic pack ice in late summer using microwave data (Campbell et al., 1980) are considered by Crane (1981; Crane et al., 1982) to be more likely to comprise some combination of extensive areas of melt-water puddling with some

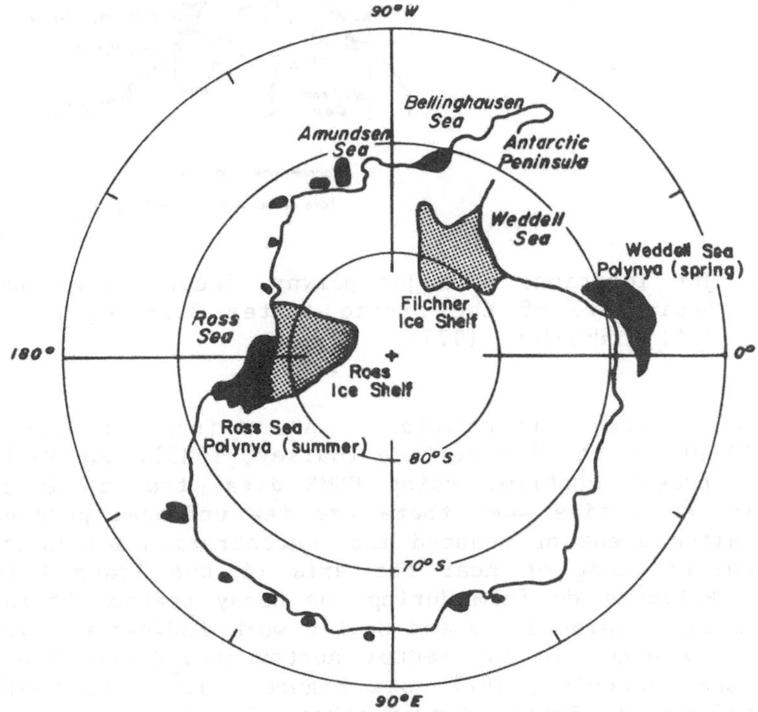

Fig. 6a. The locations of major polynya locations in Antarctica; ice shelves are cross-hatched (modified after Streten, 1973).

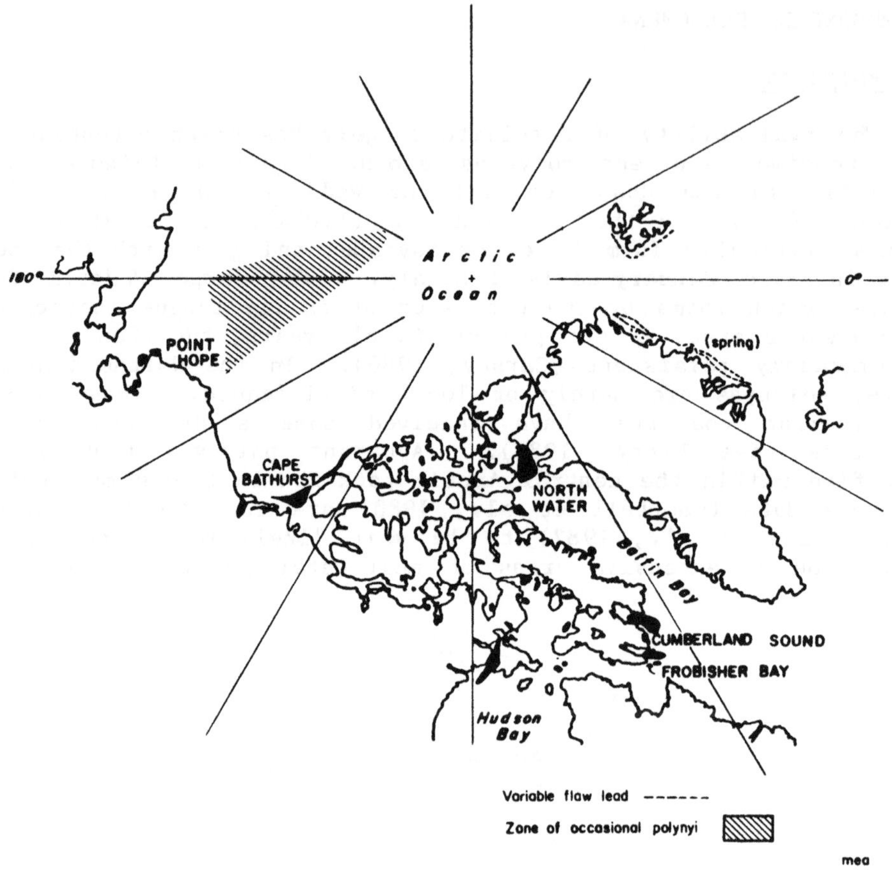

Fig. 6b. The locations of major polynya locations in the western hemisphere of the Arctic (after Stirling and Cleator, 1981; Yakovlev, 1977).

reduction in ice concentration. At maximum in July, puddles occupy 20-30% of the ice surface (Carsey, 1985), but Parkinson et al., (in press) confirm, using ESMR data, that in late August-September, at a time when there are few unfrozen puddles, there are extensive areas of reduced ice concentration across the Arctic Basin, mainly along or near the axis of the Trans Polar Drift Stream. Polynyas do form during the decay season within the ice in the northern Barents Sea and Soviet work indicates a zone where polynyas may occur in the sector northwestward from the northern Chukchi Sea (Yakovlev, 1977, see Figure 6b).[2] The small "North Water" polynya of Smith Sound-northern Bay has been most studied

[2]A polynya has been identified at about 80°N, 150°E during winter 1978-79 from SMMR data (D. Cavalieri, pers. commun., June 1982).

(Dunbar, 1973; Müller et al., 1976; Dey, 1980; Ito, 1982), but others occur in the Canadian Arctic Archipelago in spring (Haupt and Kant, 1976), off the Bering and Chukchi Sea coasts of Alaska and adjacent islands (Carleton, 1980; Pease, 1980), and along the flaw lead off east Greenland (Wadhams, 1982). They are related to coastal orientation in relation to wind and current directions and to the occurrence of local shoals. Other polynyas form within the ice cover, as in the Sea of Okhotsk (Parkinson and Gratz, 1983), perhaps in relation to warm upwelling water; in contrast to the North Water these do not open and close repeatedly during the winter. In the case of the North Water, Müller et al. (1976) show that precipitation on the adjacent land areas may be increased about 33 percent over that otherwise to be expected as the regional background amount. Its role in the surface energy balance and effects on cyclonic activity has not yet received any detailed study.

5.2 The Fast Ice and Shear Zones

From an ice-oriented viewpoint, there are striking differences between the landfast ice zone of eastern Baffin Island (Jacobs et al., 1975) and that of the Beaufort-Chukchi Sea off Alaska (Barry et al., 1979; Stringer, 1978). In the former the ice is not grounded and extends out over deep water in some areas such as Home Bay, whereas off northern Alaska it is severely ridged and anchored, particularly over coastal shoals and by the barrier islands (Reimnitz et al., 1978). As a result, much of the seasonal pack ice to seaward may become attached for significant periods in winter. From a meteorological standpoint, the boundary-layer windflow will exhibit greater mechanical turbulence as a result of the surface roughness (Banke et al., 1980), but since the average surface winds off Alaska are typically 5 m s^{-1} in winter months this factor is probably of little significance. More important is the occurrence of "flaw leads" in the shear zone. The role of such local heat sources has been discussed above.

Boundary-layer effects are important over other marginal ice zones. Off-ice winds tend to be decreased over the rough MIZ in the Bering Sea, while the horizontal temperature gradient across the ice edge gives a flow acceleration. Overland et al. (1983) estimate a net 10 percent increase in wind speed across the ice edge causing ice divergence and the formation of ice bands.

In late summer and fall, the decay of the fast ice allows encroachment of the pack into the coastal zone if wind conditions are suitable. On average, the limit of 4/8 ice concentration retreats 150 km off Point Barrow by 15 September (Barnett, 1976). Light-ice summers are associated with more frequent surface winds with a SE-S'ly component, while onshore winds predominate in heavy-ice summers (Rogers, 1978; Johnson, 1980). These patterns are associated with an accumulated thawing degree-day (TDD) total above (below) about 310°C in light- (heavy-) ice years and in fact the parameter most highly correlated with ice retreat is this TDD total. The observed biomodal distribution of total accumulated TDDs at Barrow is accounted for by the atmosphere-ocean-ice interaction which favors mild autumns to follow mild summers and vice versa. A case study of the severe ice conditions of 1975 along the Beaufort Sea coast has been made by Wendler and Jayaweera (1976). They draw attention to the role of persistent northwesterly wind components in keeping the pack ice close to the coast once the fast ice cleared. Such events and their frequency can now be examined via catalogs of synoptic pressure patterns such as those developed for the area by Barry (1977) and Moritz (1979).

5.3 Arctic Stratus

In summer, most of the Arctic Basin surface is melting ice or water with a temperature near 0°C. Low cloud averages 40-50 percent over most of the SSIZ and 60-70 percent over the central Polar Ocean according to conventional observations. However, satellite imagery analysis for summer 1979 by Kukla (1984) indicates a lower frequency of thick stratus over the Arctic Ocean than earlier studies. There remain considerable uncertainties in our knowledge even of mean cloud amounts and their characteristics (Crane and Barry, 1984). The predominant summer cloud type is stratus and stratocumulus. It is concentrated in the lowest few hundred meters and is often characterized by a multilayer structure (Reed and Kunkel, 1960; Tsay and Jayaweera, 1984). This stratus forms as a result of warm air advection which leads to condensation due to radiative cooling and to diffusive cooling at the surface (Herman and Goody, 1978). Summertime stratus appears to form in the cold mixing layer next to the surface, from observations near Svalbard (Holmgren, 1981); in contrast, wintertime stratus forms in the overlying relatively warm air. Stratus has a widely uniform distribution and is highly persistent, indicating that dissipative processes -- absorption of solar radiation, evaporation through convective heating, precipitation and synoptic scale vertical motion -- are all weak, although the first of these seems to be the main cause of the layering. There is insufficient information available at present to determine whether there are any major differences in stratus characteristics and their seasonal regime, between the SSIZ and the central Polar Ocean. Vowinckel and Orvig (1970) suggest that the mean water content of the clouds decreases poleward from the arctic coasts.

Empirical studies of cloudiness in relation to the sea ice margin suggest that the mean total amount and its variability differ in character across the ice margin, but additional work to discriminate reliably between cloud cover (and type) and ice/snow from remote sensing data is needed (Henderson-Sellers et al., 1982; Barry et al. 1984). A sensitivity study of the effects of cloudiness changes on a one-dimensional thermodynamic sea ice model (Shine and Crane, 1984) suggests that in the regions where the presence of ice in summer is variable from year-to-year, the predicted ice thickness is strongly dependent on cloud cover. However, there is little effect for regions of permanent ice or in the SSIZ that is always ice-free for part of the year.

6. OUTSTANDING PROBLEMS

A basic concern in meteorological studies in high latitudes is the reliability of the basic data on pressure and other parameters. Studies by Albright (1978) and Wohl (1978) using AIDJEX data for 1975 compared with NMC grid-point values illustrates this point. Wohl found that for July and August 1975, major synoptic features were missing over the Beaufort Sea on 7 days and poorly represented on a further 25 days. At 80°N,130°W, the MSL pressure agreed well only on 14 days, it was too high on 14 days and too low on 34 days. This problem has long been recognized but adjustments are seldom attempted (Reed and Kunkel, 1960). Certainly, studies of climatic trends and eigenvector analyses of meteorological fields and ice concentration (Walsh, 1978; 1980) must take careful account of the uncertainties in both sets of data although Walsh notes good agreement between monthly pressure values determined from NMC analyses and ice station measurements which became available later. A good historical series of daily pressure maps is urgently needed. For the future, an improved synoptic climatology of high-latitude disturbances should be obtained through interpretation of satellite imagery of cloud systems and diagnostic analysis of synoptic systems. Likewise, a reliable data base of ice limits and concentrations is required. The availability these sources should enable a much clearer understanding of local, regional, and large-scale ice-climate feedbacks to be obtained (Kellogg, 1975).

In light of the considerable data uncertainties and strong regional differences in ice conditions it seems unwise at this time to try and correlate long-term trends in ice extent with circulation trends within a hemisphere or between hemispheres (Fletcher, 1969; Kukla and Gavin, 1981). However, this is ultimately an important aspect of the possible global significance of the conditions in the SSIZs in the two hemispheres. Lemke et al. (1980) find no significant Arctic-Antarctic correlations in ice extent for the limited period of data overlap (1973-76).

The nature of sea ice-atmosphere interactions can only be determined if the large-scale investigations are complemented by selected studies of micrometeorological processes, especially ice ablation and the energy sinks, and of the characteristics of the planetary boundary layer over the SSIZ. The latter will involve a combination of aircraft profiling plus the now-available technology for boundary-layer investigations (Lenschow, 1975). As has been shown by Vowinckel and Orvig, and by Maykut, information on energy exchange processes can very usefully be incorporated into, or compared with, model calculations. From their studies, and from other work (Egorov, et al., 1974), it is clear that surface albedo and the fraction of open water are probably the two key factors. A promising technique to detect snow melt onset in the SSIZ using SMMR data (Anderson et al., 1985) can contribute to determining the time of albedo decrease in spring, while Carsey's (1985) work illustrates the feasibility of mapping melt-pond coverage and surface albedo.

A major area which has not been touched on this survey is the question of the long-term stability of the Arctic pack ice. Modelling studies (Parkinson, 1978) promise to address this question. Parkinson's results indicate that with a 5°C warming, arctic ice would essentially be reduced to a seasonal cover, reforming each winter. However, it must be noted that in Parkinson's model, the ice limit is forced by the monthly mean air temperature. A less inter-dependent coupling is necessary to examine this question adequately. An energy balance model incorporating changes in sea ice extent and ocean heat flux has been used to examine feedback processes, but further development and testing against observational data is needed (Saltzman and Moritz, 1980). From an empirical standpoint it seems reasonable to assume that a better understanding of the processes operating in the SSIZ could contribute greatly to this overall problem.

Finally, it is apparent that our basic knowledge of meteorological conditions over the SSIZs is very limited. This is, of course, due to the very unstable nature of the ice surface. On practical grounds, the problem cannot readily be resolved, which implies the need for more indirect observations (remote sensing and data buoys) coupled with attempts to estimate likely effects of various ice/water concentrations on heat, moisture and momentum exchanges through modelling studies.

7. CONCLUDING REMARKS

The results discussed in this review make it clear that certain typical meteorological "signatures" relating to the SSIZ can be identified at the mesoscale in terms of ocean-ice-atmosphere energy fluxes. It is these processes that are being studied via the Marginal Ice Zone Experiment (MIZEX) in progress off East Greenland in the Bering Sea (Wadhams et al., 1981). Over large regions, the effects of the sea ice margin and its seasonal and short-term shifts certainly modulate the occurrence of cyclonic activity on a synoptic time scale, although the net climatological consequences of this remain to be determined. Large-scale atmospheric circulation anomalies are associated with anomalies in sea ice extent in the marginal zones (see Walsh, Chapter 12), but in most cases so far examined the atmosphere is forcing the ice, apart from feedbacks on short time scales.

ACKNOWLEDGMENT

The work reported in this paper has been supported in part by the National Science Foundation under grant DPP-7920853 to R. G. Barry and H. J. Zwally (NASA-GSFC) and under DPP-8217265.

REFERENCES

Aagaard, K. (1972) On the drift of the Greenland pack ice. In Sea Ice (T. Karlsson, Ed.), National Research Council, Reykjavik, p. 17-22.

Ackley, S. F. (1981) A review of sea-ice weather relationships in the Southern Hemisphere. In Sea Level, Ice and Climatic Change (I. Allison, Ed.), International Association of Hydrologic Science, Publication 131, p. 127-60.

Ackley, S. F. and T. E. Keliher (1976) Antarctic sea ice dynamics and its possible climatic effects. AIDJEX Bulletin 33: 53-76.

Albright, M. (1978) Construction of atmospheric surface pressure maps from the AIDJEX data set. AIDJEX Bulletin 39: 111-119.

Anderson, M. R., R. G. Crane and R. G. Barry (1985) Characteristics of Arctic Ocean ice determined from SMMR data for 1979: Case studies in the seasonal sea ice zone. In: Space Observations for Climate Studies (G. Ohring and H.-J. Bolle, eds.), Advances in Space Research 5(b):257-261.

Andreas, E. L. (1980) Estimation of heat and mass fluxes over arctic leads. Mon. Wea. Rev., 108: 2057-2063.

Banke, E., S. D. Smith and R. J. Anderson (1980) Drag coefficients at AIDJEX from sonic anemometer measurements. In Sea Ice Processes and Models (R. S. Pritchard, Ed.), International Association of Hydrologic Science, Publication 124, University of Washington Press, Seattle, Washington, p. 430-442.

Barnett, D. G. (1976) A practical method of long range ice forecasting for the north coast of Alaska: Part I. Fleet Weather Facility, Suitland, Maryland, Technical Report 1, 16 p.

Barry, R. G. (1977) Study of climatic effects on fast ice extent and its seasonal decay along the Beaufort-Chukchi coasts. In Environmental Assessment of the Alaskan Continental Shelf, Vol. XIV, Transport, Environmental Research Laboratories, NOAA, Boulder, Colorado, p. 574-743.

Barry, R. G. (1983) Arctic Ocean ice and climate: Perspectives on a century of polar research. Annals Assoc. Amer. Geogr., 73: 485-501.

Barry, R. G. and J. D. Jacobs, with R. G. Crane, R. A. Keen, R. E. Moritz, E. F. LeDrew and R. L. Weaver (1978) Energy budget studies in relation to fast-ice breakup processes in Davis

Strait: Climatological overview. INSTAAR, University of Colorado, Boulder, Occasional Papers, No. 26, 284 p.
Barry, R. G., R. E. Moritz and J. C. Rogers (1979) The fast ice regime of the Beaufort and Chukchi Sea coasts, Alaska. Cold Reg. Sci. Tech., 1: 129-152.
Barry, R. G., A. Henderson-Sellers and K. Shine (1984) Climate sensitivity and the marginal cryosphere. In Climate Processes and Climate Sensitivity, (J. E. Hansen and T. Takahashi, Eds.)., Geophysical Monograph 29, American Geophysical Union, Washington, D.C., p. 221-237.
Bjornsson, H. (1969) Sea ice conditions and the atmospheric circulation north of Iceland. Jokull, 19: 11-17.
Bryson, R. A. (1966) Air masses, streamlines and the boreal forest. Geographical Bulletin, 8: 228-269.
Budd, W. R. (1975) Antarctic sea-ice variations from satellite remote sensing in relation to climate. J. Glaciol., 15(73): 417-427.
Campbell, W. J., et al. (1978) Arctic sea ice variations from time-lapse passive microwave imagery. Bound.-Layer Met., 18: 19-106.
Carleton, A. M. (1980) Polynya development in the Cape Thompson-Point Hope region, Alaska. Arct. Alp. Res., 12: 205-214.
Carleton, A. M. (1981a) Monthly variability of satellite-derived cyclonic activity for the Southern Hemisphere winter. J. Climatol., 1: 21-28.
Carleton, A. M. (1981b) Ice-ocean-atmosphere interactions at high southern latitudes in winter from satellite observation. Austral. Met. Mag., 29: 183-195.
Carleton, A. M. (1983) Variations in Antarctic sea conditions and relationships with Southern Hemisphere cyclonic activity, winters 1973-77. Arch. Meteorol. Geophys. Bioclim., B32: 1-22.
Carleton, A. M. (1984) Synoptic sea-ice atmosphere interactions in the Chukchi and Beaufort Seas from Nimbus 5 ESMR data. J. Geophys. Res., 89(D5): 7245-7258.
Carleton, A. M. (1985) Synoptic cryosphere-atmosphere interactions in the northern hemisphere from DMSP image analysis. Int. J. Remote Sensing, 6: 239-261.
Carsey, F. D. (1980) Microwave observation of the Weddell Polynya. Mon. Wea. Rev., 108: 2032-2044.
Carsey, F. D. (1985) Summer Arctic sea ice character from satellite microwave data. J. Geophys. Res., 90: 5015-5036.
Central Intelligence Agency (1978) Polar regions atlas. National Foreign Assessment Center, C.I.A., Washington, D.C., 66 p.
Comiso, J. C. and H. J. Zwally (1984) Concentration gradients and growth/decay characteristics of the seasonal sea ice cover. J. Geophys. Res., 89(C5): 8081-8083.
Crane, R. G. (1978a) Seasonal variations of sea ice extent in the Davis Strait-Labrador Sea area and relationships with synoptic-scale atmospheric circulation. Arctic, 31: 434-447.

Crane, R. G. (1978b) Synoptic controls on the energy budget regime of an ablating fast ice surface. Arch. Met. Geophys. Bioklim., A(28): 53-70.

Crane, R. G. (1983) Atmosphere sea-ice interactions in the Beaufort/Chuckchi Sea and in the European sector of the Arctic. J. Geophys. Res. 88: 4505-4523.

Crane, R. G., R. G. Barry and H. J. Zwally (1982) Analysis of atmosphere-sea ice interactions in the Arctic Basin using ESMR microwave data. Int. J. Rem. Sens. 3: 259-276.

Crane, R. G. and R. G. Barry (1984) The influence of clouds on cli-climate with a focus on high latitude interactions. J. Climatol. 4: 71-93.

Dey, B. (1980) Applications of satellite thermal infrared images for monitoring north water during the periods of polar darkness. J. Glaciol., 25(93): 425-438.

Dunbar, M. (1967) The monthly and extreme limits of ice in the Bering Sea. In Physics of Snow and Ice (H. Oura, Ed.), Institute of Low Temperature Science, University of Hokkaido, Sapporo, Japan, Vol. 1, p. 687-703.

Dunbar, M. (1973) Winter regime of the North Water. Trans. Roy. Soc. Can., Ser. IV(11): 275-281.

Egorov, K. L., D. L. Laikhtman and V. M. Radikevich (1974) Ice cover as a link in the ocean-atmosphere system. Bound.-Layer Met., 6: 169-182, 425-438.

Einarsson, T. (1972) Sea currents, ice drift and ice composition in the East Greenland current. In Sea Ice (T. Karlsson, Ed.), National Research Council, Reykjavik, p. 23-32.

Fletcher, J. O. (1969) Ice extent on the Southern Ocean and its relation to world climate. Rand Corp., Santa Monica, California, Memo RM-5793-NSF, 111 p.

Gordon, A. and H. Taylor (1975) Seasonal change of Antarctic sea ice cover. Science, 187: 346-347.

Gordon, A. L. (1981) Seasonality of Southern Ocean sea ice. J. Geophys. Res., 86(C5): 4193-4197.

Haupt, I. and V. Kant (1976) Satellite ice surveillance studies in the Arctic in relationship to the general circulation. In Proceedings, Symposium on Meteorological Observations from Space: Their Contribution to the First GARP Global Experiment, COSPAR, International Council of Scientific Unions, p. 179-187.

Henderson-Sellers, A., R. G. Barry, R. G. Crane, N. A. Hughes and K. Shine (1982) Cloud-cryosphere interactions near the coastal sea-ice boundary: Data retrieval and climate sensitivity. Preprint Volume. In Preprint Volume, First International Conference on Meteorology and Air/Sea Interaction of the Coastal Zone, The Hague, American Meteorological Society, Boston, Mass., p. 378-380.

Herman, G. F. and R. Goody (1976) Formation and persistence of summertime arctic stratus clouds. J. Atmos. Sci., 33: 1537-1553.

Herman, G. F. and W. T. Johnson (1978) The sensitivity of the general circulation to arctic sea ice boundaries: A numerical experiment. Mon. Wea. Rev., 106: 1649-1664.

Hibler, W. D., III and S. F. Ackley (1983) Numerical simulation of the Weddell Sea pack ice. J. Geophys. Res., 88: 2873-87.

Holmgren, B. (1981) Stratus and the atmospheric boundary layer over the Arctic Ocean (abstract). EOS, 62: 800.

Holmgren, B. and G. Weller (1974) Local radiation fluxes over open and freezing leads in the polar pack ice. AIDJEX Bulletin, 17: 149-166.

Ito, H. (1982) Sea ice atlas of northern Baffin Bay, Zürcher Geographische Schirften, No. 7, Dept. Geography, Swiss Federal Institute of Technology, 142 pp.

Jacobs, J. D., R. G. Barry and R. L. Weaver (1975) Fast ice characteristics, with special reference to the eastern Canadian Arctic. Polar Record, 17(110): 521-536.

Johnson, C. M. (1980) Wintertime arctic ice extremes and the simultaneous atmospheric circulation. Mon. Wea. Rev., 108: 1782-1791.

Kellogg, W. (1975) Climatic feedback mechanisms involving the polar regions. In Climate of the Arctic (G. Weller and S. A. Bowling, Eds.), University of Alaska, p. 111-116.

Klein, W. H. (1957) Principal tracks and frequencies of cyclones and anticyclones in the Northern Hemisphere. U.S. Weather Bureau, Washington, D.C., Research Paper 40, 60 p.

Knapp, W. W. (1972) Satellite observations of large polynyas in polar waters. In Sea Ice (T. Karlsson, Ed.), National Research Council, Reykjavik, p. 201-212.

Krebs, S. J. and R. G. Barry (1970) The arctic front and the tundra-taiga boundary in Eurasia. Geographical Review 60: 548-554.

Kukla, G. J. (1984) Variation of Arctic cloud cover during summer 1979. Lamont-Doherty Geological Observatory of Columbia University, Palisades, NY, Tech. Rep. LDGO-84-2, 67 p.

Kukla, G. and Gavin, J. (1981) Summer ice and carbon dioxide. Science, 214: 497-503.

LeDrew, E. F. (1980) Eigenvector analysis of the vertical velocity field over the eastern Canadian Arctic. Mon. Wea. Rev., 108: 1992-2005.

LeDrew, E. F. (1984) The role of local heat sources in synoptic activity within the Polar basin. Atmos.-Ocean 22: 309-327.

Lemke, P., E. W. Trinkl and K. Hasselmann (1980) Stochastic dynamic analysis of polar sea ice variability. J. Phys. Oceanogr., 10: 2100-2120.

Lenschow, D. (ed.) (1975) Instruments and techniques for probing the atmospheric boundary layer. Facilities for Atmospheric Technology No. 7, Facilities Laboratory, NCAR, Boulder, Colorado, 91 p.

Maykut, G. A. (1978) Energy exchange over young sea ice in the central Arctic. J. Geophys. Res., 83: 3646-3658.

Miles, M. K. (1975) North Atlantic circulation and associated temperature gradients. In Proceedings of the WMO/IAMAP Symposium on Long-Term Climatic Fluctuations. Geneva, WMO No. 421, p. 189-196.

Miles, M. K. (1976) The relation of the poleward flux of sensible heat during the winter half year to some zonal and meridional indices of the circulation. Quart. J. Roy. Met. Soc., 102: 763-768.

Moritz, R. E. (1979) Synoptic climatology of the Beaufort Sea coast of Alaska. INSTAAR, University of Colorado, Boulder, Occasional Papers, No. 30, 176 p.

Muench, R. D. and K. Ahlnas (1976) Ice movement and distribution in the Bering Sea from March to June 1974. J. Geophys. Res., 81: 4467-4476.

Müller, F., A. Ohmura and R. Braithwaite (1976) On the climatic influence of North Water. In International Geographical Congress: Symposium on the Geography of Polar Countries (Extended Summaries), Gidrometeoizdat, Leningrad, p. 55-58.

Niebauer, H. J. (1980) Sea ice and temperature variability in the eastern Bering Sea and the relation to atmospheric fluctuations. J. Geophys. Res., 85: 7507-7515.

Nusser, R. (1958) Distribution and character of sea ice in the European Arctic. In Arctic Sea Ice, National Academy of Sciences - National Research Council, Publication 598, p. 1-10.

Overland, J. E. and C. H. Pease (1982) Cyclone climatology of the Bering Sea and its relation to sea ice extent. Mon. Wea. Rev., 110: 5-13.

Overland, J. E., R. M. Reynolds, and C. H. Pease (1983) A model of the atmospheric boundary layer over the marginal ice zone. J. Geophys. Res., 88: 2836-40.

Parkinson, D. (1978) A numerical simulation of the annual cycle of sea ice in the Arctic and Antarctic. NCAR, Boulder, Colorado, Cooperative Thesis, No. 46, 191 p.

Parkinson, C. L. and A. J. Gratz (1983) On the seasonal sea ice cover of the Sea of Okhotsk. J. Geophys. Res., 88: 2793-2802.

Parkinson, C. L., J. C. Comiso, H. J. Zwally, D. J. Cavalieri, P. Gloersen, and W.J. Campbell (In press) Arctic Sea Ice, 1973-76: Satellite passive-microwave observations. NASA-SP.

Pease, C. H. (1980) Eastern Bering Sea ice processes. Mon. Wea. Rev., 108: 2015-2023.

Petterssen, S. (1950) Some aspects of the general circulation of the atmosphere. Centenary Proc. Roy. Met. Soc.: 120-155.

Reed, R. J. (1960) Principal frontal zones of the northern hemisphere in winter and summer. Bull. Amer. Met. Soc., 41: 591-598.

Reed, R. J. (1962) Arctic forecast guide. U.S. Navy Weather Research Facility, Norfolk, Va., NWRF 16-0462-058, 107 p.

Reed, R. J. and B. A. Kunkel (1960) The arctic circulation in summer. J. Met., 17: 489-506.
Reimnitz, E., L. Toimil and P. Barnes (1978) Arctic continental shelf morphology related to sea ice zonation, Beaufort Sea, Alaska. Marine Geology, 28: 179-210.
Rex, D. F. (1950) The effect of Atlantic blocking action upon European climate. Tellus, 2: 196-211, 275-301.
Rogers, J. C. (1978) Meteorological factors affecting interannual variability of summertime ice extent in the Beaufort Sea. Mon. Wea. Rev., 106: 890-897.
Saltzman, B. and R. E. Moritz (1980) A time-dependent climatic feedback system involving sea-ice extent, ocean temperature, and CO_2. Tellus, 32: 93-118.
Schell, I. I. (1964) Interrelations of the ice off northern Japan and the weather. J. Met. Soc. Japan, 42: 174-185.
Schwerdtfeger, W. and S. J. Kachelhoffer (1973) The frequency of cyclonic vortices over the Southern Ocean in relation to the extension of the pack ice belt. Ant. J. U.S., 8: 234.
Shine, K. P. and R. G. Crane (1984) The sensitivity of a one-dimensional thermodynamic sea ice model to changes in cloudiness. J. Geophys. Res., 89: 10, 615-10, 622.
Smirnov, V. I. (1980) Opposition in ice redistribution in the waters of the foreign Arctic. Sov. Met. Hydrol., 3: 73-7.
Stirling, I. and H. Cleator (Eds.) (1981) Polynyas in the Canadian Arctic. Canadian Wildlife Service, Ottawa, Occasional Papers No. 45, 75 p.
Streten, N. A. (1973) Satellite observations of the summer decay of the Antarctic sea ice. Arch. Met. Geophys. Bioklim. A(22): 119-134.
Streten, N. A. and D. J. Pike (1980) Characteristics of the broad-scale Antarctic sea ice extent and the associated atmospheric circulation 1972-1977. Arch. Met. Geophys. Bioklim., A(29): 279-299.
Stringer, W. (1978) Nearshore ice conditions by means of satellite and aerial remote sensing. In Environmental Assessment of the Alaskan Continental Shelf, Vol. X, Transport, Environmental Research Laboratories, NOAA, Boulder, Colorado, p. 1-220.
Strübing, K. (1967) On the relation between the ice transport in the East Greenland current and the atmospheric circulation over the Arctic Ocean. Deuts. Hydrogr. Zeit., 20(6). [Draft Translation 752, U.S.A., Cold Regions Research and Engineering Laboratory, 11p., 1980.]
Taljaard, J. J. (1972) Synoptic meteorology of the Southern Hemisphere. In Meteorology of the Southern Hemisphere (C. W. Newton, Ed.), Met. Monogr., 13(35): 25-58.
Tsay, S. C. and K. O. L. F. Jayaweera (1984) Physical characteristics of Arctic stratus clouds. J. Clim. Appl. Met., 23: 584-596.

Van Loon, H. (1972) Temperature in the southern hemisphere. In Meteorology of the Southern Hemisphere (C. W. Newton, Ed.), Met. Monogr., 13(35): 25-58.

Vinje, T. E., (1977) Sea ice conditions in the European sector of the marginal seas of the Arctic, 1966-1975. Norsk Polar-institutt Arbok 1975, p. 163-74.

Vowinckel, E. (1963) Ice transport between Greenland and Spitzbergen, and its causes. Arctic Meteorology Research Group, McGill University, Publications in Meteorology, No. 59, 23 p.

Vowinckel, E. and S. Orvig (1966) Energy balance of the Arctic. V: The heat budget over the Arctic Ocean. Arch. Met. Geophys. Bioklim., B(14): 303-325.

Vowinckel, E. and S. Orvig (1970) The climate of the North Polar Basin. In Climates of the Polar Regions, World Survey of Climatology (S. Orvig, Ed.), Vol. 14, Elsevier, New York, p. 129-252.

Vowinckel, E. and S. Orvig (1973) Synoptic energy budgets from the Beaufort Sea. In Energy Fluxes Over Polar Surfaces, World Meteorological Organization, Geneva, WMO Technical Note 129, p. 145-166.

Vowinckel, E. and B. Taylor (1965) Energy balance of the Arctic. IV: Evaporation and sensible heat flux over the Arctic Ocean. Arch. Met. Geophys. Bioklim., B(14): 36-52.

Wadhams, P. (1981) The ice cover in the Greenland and Norwegian Seas. Rev. Geophys. Space Phys., 19(3): 345-393.

Wadhams, P. and R. J. Horne (1980) An analysis of ice profiles obtained by submarine sonar in the Beaufort Sea. J. Glaciol., 25(93): 401-424.

Wadhams, P., S. Martin, O. M. Johannessen, W. D. Hibler, III and W. J. Campbell (1981) MIZEX: A program for mesoscale air-ice-ocean interaction experiments in arctic marginal ice zones. USA Cold Regions Research and Engineering Laboratory, Special Report 81-19, 20 p.

Walsh, J. E. (1978) Temporal and spatial scales of the arctic circulation. Mon. Wea. Rev., 106: 1532-1544.

Walsh, J. E. (1980) Empirical orthogonal functions and the statistical predictability of sea ice extent. In Sea Ice Processes and Models (R. S. Pritchard, Ed.), IAHS Pub. No. 124, University of Washington Press, Seattle, Wa., p. 373-384.

Walsh, J. E. and C. M. Johnson (1979a) An analysis of Arctic sea ice fluctuation, 1953-77. J. Phys. Oceanogr. 9: 580-591.

Walsh, J. E. and C. M. Johnson (1979b) Interannual atmospheric variability and associated fluctuations in arctic sea ice extent. J. Geophys. Res., 84: 6915-6928.

Walsh, J. E. and J. E. Sater (1981) Monthly and seasonal variability in the ocean-ice-atmosphere systems of the North Pacific and the North Atlantic. J. Geophys. Res., 86(C8): 7425-7445.

Weller, G. (1968) Heat energy transfer through a four-layer system: air-snow-sea ice-water. J. Geophys. Res., 73: 1209-1220.

Weller, G. (1980) Spatial and temporal variations in the South Polar surface energy balance. Mon. Wea. Rev., 108: 2006-2014.

Wendler, G. and K. O. L. F, Jayaweera (1976) Some remarks on the unusual Beaufort-Sea ice conditions in the summer 1975. University of Alaska, Geophysical Institute, Report UAG R-246, 28 p.

Wiese, W. (1924) Polareis and atmosphärische Schwankungen. Geografiska Annaler, 6: 273-299.

Wohl, G. M. (1978) A study of sea ice conditions and synoptic climatology for the Beaufort Sea coast of Alaska. University of Colorado, Boulder, M. A. thesis, 166 p.

Yakovlev, G. N. (1977) Ice routes of the Arctic, National Defence Headquarters, Ottawa, Canada, DGIS Translation Library No. 1653, 342 p.

Zakharov, V. F. (1981) L'dy Arktiki i Sovremenye Prirodnye Protsessy. Gidrometeoizdat, Leningrad, 136 p.

Zillman, J. W. (1972) Solar radiation and sea-air interaction south of Australia. American Geophysical Union, Antarctic Research Monograph 19, p. 11-40.

CHAPTER 16

REMOTE SENSING AS A RESEARCH TOOL

F. D. Carsey

Jet Propulsion Laboratory, Pasadena, CA 91109

and

H. J. Zwally

GSFC, Greenbelt, MD 20771

1. INTRODUCTION

Remote sensing of the polar oceans ice has revolutionized our view and thinking of sea ice as a global scale phenomenon. The record of pack ice extent and character from isolated ships and stations is woefully sparse even after careful compilations are made. Today the costs and hazards of in situ observations are great, and the time and space scales on which information is required create formidable requirements. Thus, remotely acquired data satisfy a major observational requirement in ice surveillance. Historically, isolated aircraft photographic flights over ice were made nearly throughout the century, and there have been flights with other observing instruments since the 1960's. The first microwave scanning or image producing instrument in space was the Electrically Scanning Microwave Radiometer (ESMR) on Nimbus 5 launched late in 1972; it provided the first global year-round record of the ice-covered seas. Spaceborne visible light and infrared sensors have flown steadily for more than a decade. The future holds promise of instruments of improved capability for application to sea ice science and operations.

There are a number of research problems involving sea-air-ice interaction which the other chapters in this volume deal with in detail. In order to examine the application of spacecraft and aircraft observing systems to such problems, it is necessary to summarize the problems and construct a list of basic observables - or variables which are candidate observables. The basic scientific issues of air-sea-ice interaction can be summarized as

(a) circulation of the atmosphere and ocean, (b) climatology, and (c) material response. Clearly, these are not truly separable. Within these broad areas of inquiry, it is possible to indicate the specific effects of ice in the system and the measurement parameters established by each effect. These measurement needs - basically ice features to be monitored on their intrinsic time and space scales - constitute a challenge to the observational community including, prominently, remote sensing.

There are several methods for obtaining information from aircraft and spacecraft about the ice cover and the adjacent open sea. The principal methods are (a) ambient visible light, (b) thermal (emitted) infrared (IR), (c) passive (emitted) microwave, (d) active (reflected with source and receiver both remote) microwave, and (e) altimetry. In general, the data acquired by any system is different from the input variable whose measurement is sought, and such issues as the cost, difficulty and error involved in the transformation from measurement to input variable are significant. In general, remote sensing systems are steadily becoming more complex and more capable. The number of channels of data is increasing while the surface resolution is improving (becoming finer or of smaller actual dimension). The newest Landsat (5), for example, consists of the standard 4 band Landsat Multi-Spectral Scanner (MSS) plus a 7 channel Thematic Mapper, but the 3 channel Return Beam Vidicon has been dropped. The total change is from 7 to 11 channels with broader spectral coverage and generally improved resolution. At this time, Landsat-5, the NOAA satellites, the Navy Remote Ocean Sensing System (NROSS), to be launched in 1990, a microwave scanner to be part of the U.S. Defense Meteorological Satellite System and deployment by Japan, Canada and the European Space Agency of an earth-sensing satellite are the future spacecraft observation systems of use for sea ice.

The overall purpose of this chapter is to describe measurement goals, the physical basis of observation systems on aircraft and spacecraft, and the methods from recent literature and ongoing work. The potential of future systems will be assessed; and remaining problems and information gaps will be discussed. The goal of the authors is to give an overview of remote sensing of sea ice from the standpoint of applications including what can be done, what needs to be done, what the future may hold, and what probably cannot be expected.

Every effort has been made to make this chapter truly current. There is rapid growth in remote sensing technology. This is especially true in the area of interpretation of data which is a most important consideration in applications. Thus the chapter contains much material from preprints and even from results not yet written up. The chapter is doubtless better for this inclusion and the contributing investigators are to be thanked for their generosity. Among those who assisted were AES Ice Centre of Canada for a SLAR photograph, Dr. R.O. Ramseier, Dr. B. Dey, Dr. D. Cavalieri, Dr. T. Grenfell, Prof. S. Martin, Prof. R.K. Moore, Dr. E.P. McClain, and Dr. T. Vinje. The work of assembing this review was supported by NASA through a contract to the Jet Propulsion Laboratory of the California Institute of Technology and by NASA Goddard Space Flight Center.

2. MEASUREMENT GOALS

The effects of ice in the air-sea-ice system can be listed briefly with considerable completeness.

1. Insulation. The ice cover, particularly when snow covered, insulates the ocean from the overlying air which is as much as 50 K cooler (Badgley, 1961; Maykut and Untersteiner, 1971; Maykut, 1978; Parkinson and Washington, 1979; Andreas et al. 1979). Heat and mass fluxes are both affected.

2. Albedo Modification. Ice fragments and floes have albedos higher than open ocean, and the snow-covered ice albedo is much higher. The ice margin, typically a large abrupt change in ice areal concentration, is thus a boundary of sharp insolation difference. Albedos in the polar seas range from about 0.15 for water to 0.58 for bare ice to as large as 0.84 for snow covered ice (Wendler, 1973; Marshunova and Chernigovsky, 1966; Langleben, 1968 and 1971; Grenfell and Perovich, 1984).

3. Latent Heat-of-Fusion Export. At the ice margin, ice floes are commonly advected equatorward to melt and thereby cool the upper ocean. This heat deficit term is substantial, amounting to about 3 x 10^6 watts per meter of margin for 1 m thick ice moving 1 cm s^{-1}. This heat transport is comparable to that in the atmospheric boundary layer under storm conditions (Pease, 1980).

4. Surface Stress Modification. Pack ice fragments typically move relative to both the atmosphere and the supporting oceanic mixed layer with the consequence that the ocean surface stress for open sea is different from that for ice covered sea. This generally meridional surface stress gradient produces at the margin an upwelling extending well below the undisturbed mixed layer. The overall significance of this flow is not fully assessed (Norsex, 1981), but it is known to lead to instabilities (Hakkinen, 1984).

5. Ocean Mixed Layer Modification. The ice cover not only effects the interaction between air and sea, it has its own effects. In particular, brine is excluded from forming ice to produce increased salinity in the mixed layer or even the penetration of cold saline water to depth. In melting, a lens of freshwater forms at the surface increasing the stability of the upper ocean

and providing something of a meridional salt transport (McPhee, 1981; Killworth, 1979; Carsey, 1980).

6. Radiatively and Sensibly Cool Surface. The snow covered or bare ice surface is typically the coldest point in the vertical temperature profile of the boundary layers. Thus, it is a heat sink both radiatively and sensibly with respect to the atmospheric boundary layer. This mechanism contributes to the cooling of air masses traveling over ice and also weakens the turbulent communication between the cooler surface layer and the remainder of the lower atmosphere (Herman and Goody, 1976; Curry, 1983), influencing other transport.

Of course, the roles of sea ice are viewed in other ways, i.e., as an operational hazard or as a climate indicator. Also, the above effects taken together may produce an identifiable net effect such as the often projected steering of migrating storms by the ice margin. If such steering exists, it is probably due to the ice pack modification of available heat both at the surface and advectively through the boundary layer, and this modification can be understood in the context of the effects listed above.

The material response of pack ice to stress has not been mentioned. Although it is clearly an important property of ice, it plays only a small direct part in air-sea interaction. An area in which ice rheology might influence air-sea-ice interaction is in determining the spatial distrubtions and orientations of features such as leads and ridges. At present it is not known if one large lead produces the same net air/sea modifications as a member of small leads with the same area, but a rough assessment of processes at work would indicate not. However, knowledge of ice rheology is a requirement for predicting its properties which in turn effect the interaction.

Each effect depends on a subset of the total set of possible ice variables. For the effects listed above, a table can be constructed of first-order dependencies, Table 1. The list of variables and the exact boxes to be checked are arguable, but with scent profit. There are inconsistencies such as having ice thickness, a point variable or a distribution, listed with ice concentration, an areal percentage, but the list will serve.

Before going any further, it is vital to point out that the variables listed are not observables in any practical sense. Snow cover and ice thickness have variation in almost all spatial scales, making even variable definition nearly impossible. Ocean velocity profiles are difficult to acquire, and, more so than wind speed, presumably, also difficult to smooth or to extrapolate. All of the variables are expensive to measure. These circumstances have produced a situation in which considerable modelling of what

Table 1. Effects and variables.

Insulation	Albedo	Heat export	Stress gradient	Ocean stability	Cool surface	Symbol	Variable
X	X				X	h_s	Snow cover
X		X				H	Ice thickness
	X		X			S	Ice type
X	X					C_I	Ice concentration
		X	X	X		$\underset{\sim}{W}$	Ice velocity field
		X	X	X		$\underset{\sim}{V}$	Ocean velocity profile
X			X		X	$\underset{\sim}{U}_{10}$	Surface wind vector field
X						T_{10}	Air temperature
	X				X	T_s	Surface temperature
					X	T_o	Ice surface temperature
		X		X		T_w	Ocean temperature

are actually input variables is required. Thus, in what follows, actual measureables will be merged with seasonal means, modelled quantities or analyses of extreme values which might reasonably occur. Strictly speaking, much of what is readily calculated is sensitivity; actual values are subsequently cadged.

Another distinction is also useful, that between monitoring and intensive study. In what follows, some measurements will be labor intensive requiring instrumented aircraft flights, surface parties and highly skilled data interpretation. From that extreme, the supervisory task can be relaxed to numerical modelling requiring a circulation model and adequate initial and/or boundary conditions. Somewhere in the middle is the observational goal of monitoring -- suitable description using operational platforms. To define suitable, the geophysical problems must be better defined than they now are; to define operational, platforms such as buoys and satellites with attendant data processing and dissemination must be properly planned.

3. OBSERVATIONAL ISSUES IN THE SEASONAL SEA ICE ZONE

While there is extensive history of in situ and remote sensing study of the central Arctic ice pack and the commercially important marginal seas, there has been much less learned about the ice margins at the land and open water transitions. Processes at these margins are receiving much current attention.

Ice at the margins is significantly different from ice within the pack, and the meteorology and oceanography pertinent to both the geophysics and the measurement process are also different. At the margin, taken to be a zone extending some 100 km on either side of the ice pack edge, there are more types of ice than are common in the central pack. These include grease, pancake, and brash ice between the major floes, and these floes, usually from the central pack, have been modified by flooding, wave wetting, repeated melt-thaw cycles, precipitation, severe deformation, and other events. Even if one is willing to forego specific information on these factors, they produce signals perceived as error by measurement systems, and they must be dealt with.

The driving terms in the atmosphere and ocean are different at the margin and create more stringent observational requirements. At the margin, storms influence the ice in a basically non-linear way so that long term averages, except of storm statistics, are not useful. Compared to those in the central pack storms at the margin are more intense, and they interfere more, and more often, with the remote sensing propagation path. The sea-ice interaction at the margin is strong and includes waves and swell not encountered in the central pack. In the adjacent open water ice fragments may interfere with aircraft and spacecraft methods for measuring temperature, surface stress and gravity wave properties. Finally, the gradients of all variables are very sharp within the margin so that typical resolution distances of the passive microwave remote sensors in the historical archive may be too large for the measurements to be directly interpreted there. This is usually not a problem when those sensors are applied deep within the pack.

4. PHYSICAL PRINCIPLES OF ICE REMOTE SENSING

The remote sensing methods for aircraft and spacecraft surveillance of sea ice require interpretation of electromagnetic waves either scattered by or emitted from the surface and propagating through the atmosphere. A summary of the measurement principles, characteristics, and systems follows.

4.1 Ambient Visible Light

Ambient visible light systems from handheld cameras to multichannel scanners utilize solar illumination scattered or reflected from the surface. The success of the method stems from the very high contrast or albedo difference between ice and ocean and from low sun angles which enhance the limited relief of sea ice features. Limitation of the method stems from the difficulty of identification of grey tones as either thin ice or mist-covered open water (Hall, 1980), from seasonal and weather problems, and from low coverage. Figure 1 shows sample measured spectral albedos of Grenfell (1984), taken near Barrow, Alaska, and the receiving bands of some satellite sensors. Other albedo measurements have been published by Wendler (1973) and Langleben (1968, 1971); results from these studies for Beaufort Sea ice and Amundsen Gulf fast-ice are shown in Figure 2.

Weather and season pose the most significant problems to visible light ice surveillance in the polar regions. Use of visible light is virtually nil in the winter when no solar illumination is available and in mid-summer when cloud cover is almost total. While many investigators have found clear skies in the summertime Arctic, data acquired at such times would be difficult to generalize, and methods dependent on unseasonal weather are not appropriate to remote sensing as a tool as reliability is called for. Although the global ice margin cannot be accurately generalized with regard to weather, it typically is found in regions of high mean surface-air convergence which indicates greater likelihood that storm interference will be significant (Overland and Pease, 1982).

The interpretation of grey areas in images is a problem for some investigations as thin ice, transition species ice, cloud-or mist-covered ocean, and sparse ice fragments all look grey. Also, thin ice or thick ice covered with much or little snow are similar. While data exist for albedo of ice types of the pack, no such measurements have been made for the wet ice, pancake ice, grease ice and other features common to the margin. Thus, quantitative interpretation is difficult even if all propagation path and lighting factors are well-behaved.

Finally coverage may be a problem. Aircraft surveillance is expensive and spatially limited. Satellite coverage is of two sorts, the NOAA/TIROS/DMSP type systems which have resolution rather coarse (about 1 km pixels), and the Landsat system in which coverage is less, but resolution is finer. The Landsat coverage improves with increasing latitude so that McNutt (1981) could track many Bering Sea floes for 2 consecutive days and for 3 days out of 5 with Landsat imagery compared to use of NOAA images in which tracking on 5 days out of 5 was not unusual. The Landsat

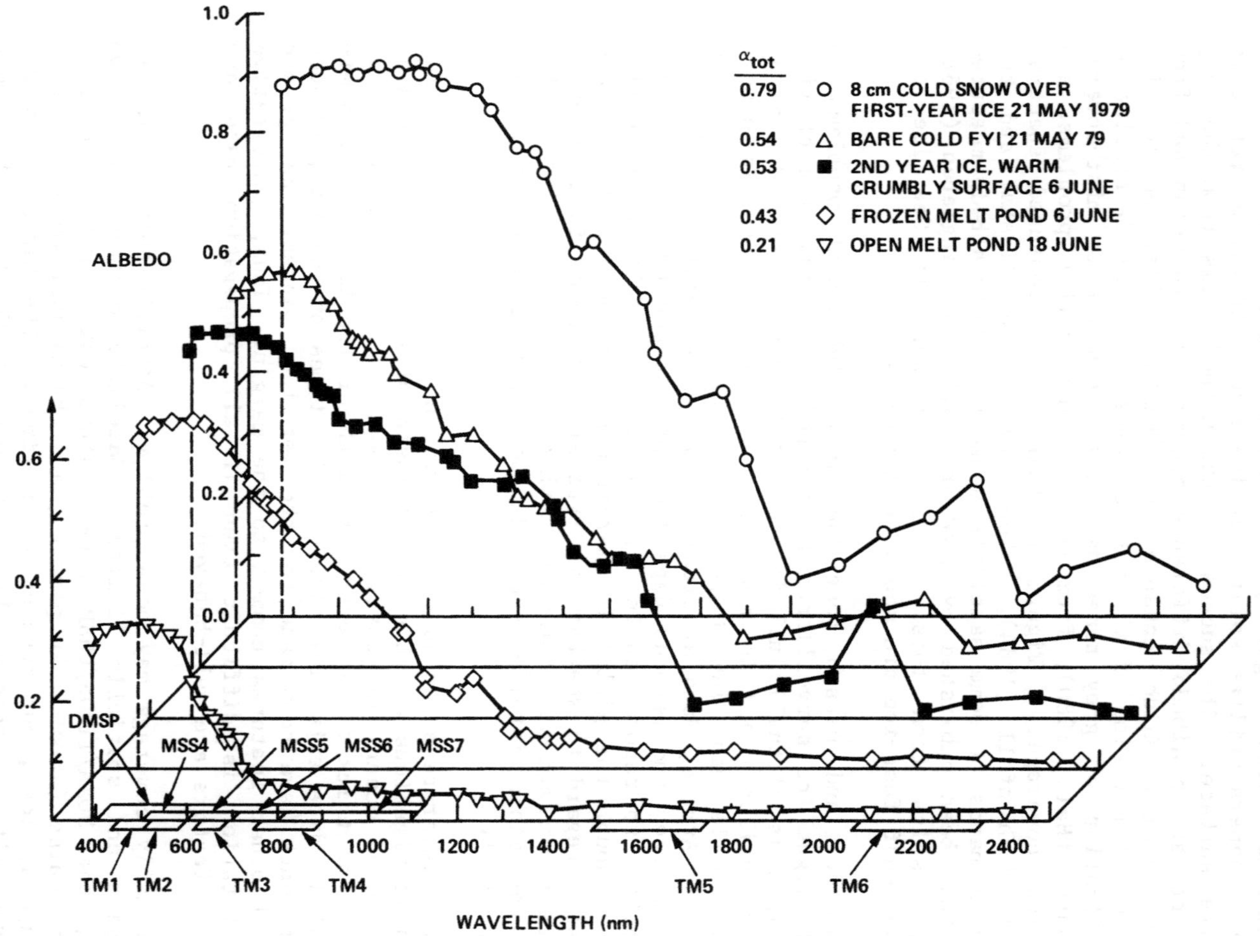

Fig. 1. Measured ice albedos taken by Grenfell (private communication) near Barrow, Alaska. These values should be regarded as samples as no effort was made to generalize in region or season. Also shown are some satellite sensor bands.

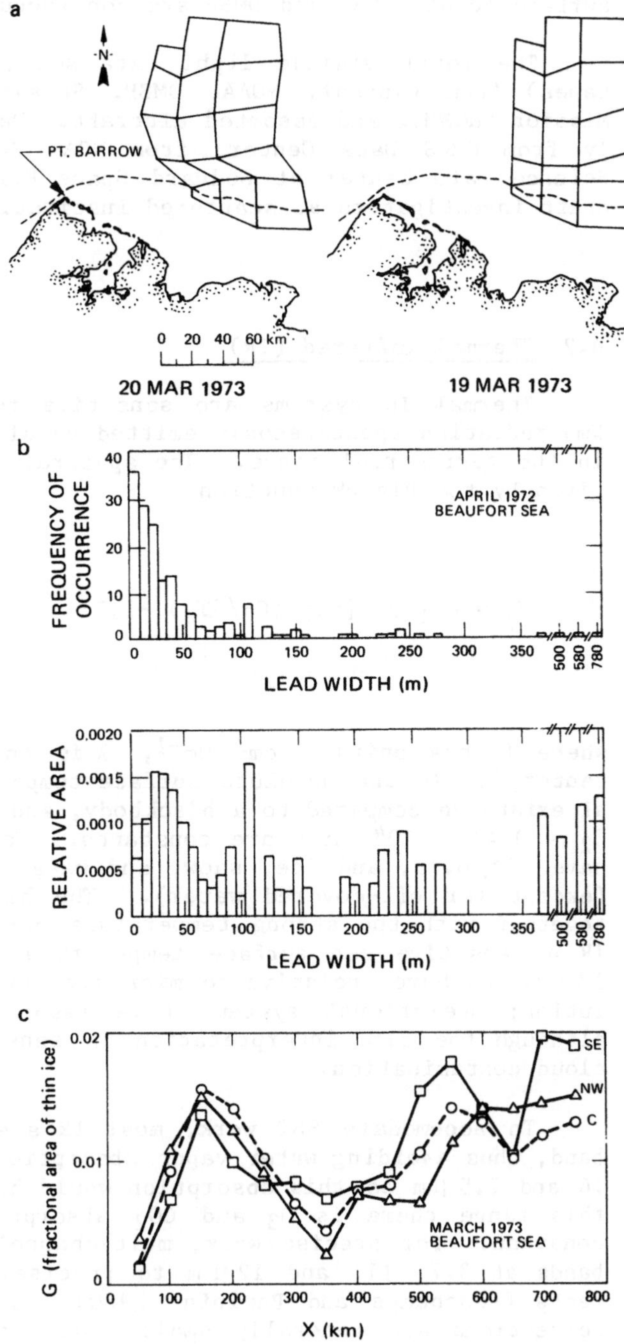

Fig. 2. Ice albedo results (a) from Landsat and previous literature for spring-summer-fall pack from Wendler (1973) and (b) from Langleben (1971) for summer fast-ice from air-craft. Fast ice is quite flat and can support very high melt pond fractional coverage.

cycle consists of 3 days of overlap and 15 days of gap for a given surface point; NOAA and DMSP are continuous.

The total visible light data set contains images (and some tapes) from Landsat, NOAA, DMSP, Seasat, Heat Capacity Mapping Mission (HCMM), and assorted aircraft. Data are available variously from EROS Data Center, from NOAA EDIS, from National Space Science Data Center at Goddard Space Flight Center and from aircraft investigators at scattered institutions.

4.2 Thermal Infrared (IR)

Thermal IR systems are sensitive to the long-wave (order 10 μm) radiation spontaneously emitted by all objects with temperature in the terrestrial range. The spectral emittance of a material is given by the Planck function

$$M_\lambda = \varepsilon C_1\ \lambda^{-5}[\exp\ (C_2/\lambda T_s)\ -1]^{-1} \qquad (16.1)$$

where M_λ has units W cm^2 μm^{-1}, λ is the wavelength of the band center, T_s is the absolute surface temperature, ε is the fractional existance compared to a blackbody, and C_1 = 3.74 W cm^{-2} μm^4 and C_2 = 1.44 x 10^4 m K are constants. Most terrestrial substances have $\varepsilon > 0.75$, and ice, snow, and water values are all above .98 (except for oil-covered water). The high open water value of ε together with the strong temperature dependence of eq. (1) makes IR a sensitive sea surface temperature (SST) tool. Furthermore, IR systems have, relative to microwave systems, fine surface resolution; operational systems have resolutions smaller than 1 km, although the data interpretation is sensitive to water vapor and cloud contamination.

To accomodate SST work, most IR systems work in the 8-14 μm band, thus avoiding water vapor absorption bands centered at about 16 and 7.5 μm as this absorption would be highly variable. Within this range there is O_3 and CO_2 absorption which is essentially constant. For precise work, multichannel IR work is optimum using bands at 3.7, 11, and 12 μm to precisely remove atmospheric effects (Deschamps and Phulpin, 1980). In the polar regions these corrections are generally small, but significant weather related errors near the ice margin can be expected. There has been work in the probably profitable simultaneous examination of IR and microwave radiances (Susskind et al., 1982).

IR imagery has found use in sea ice research primarily because of its availability during the polar winter (Dey, 1980a, b, c, 1981). The imagery has been used to identify the ice margins and to track features. Use of IR data to identify ice type or thickness has been proposed with some success (Jayaweera, 1970) but the unmonitored variability of air temperature has been an impediment (Gloersen et al. 1974; Poulin, 1975). The use of IR during the warm season is reduced by both weather and the breakdown of the thermal contrast, especially between open water and the smaller ice pieces. The use of IR data in calculation of radiative and sensible heat fluxes has not been attempted.

4.3 Passive Microwave

Passive microwave systems are sensitive to the thermally emitted radiation of wavelength from a few millimeters to tens of centimeters. These wavelengths are much longer than those useful for IR sensing. In the microwave part of the spectrum, the emitted intensity is usually called the brightness temperature T_B because it obeys the Rayleigh-Jeans Law which results in the relationship

$$T_b = \varepsilon T \qquad (16.2)$$

where ε is the emissivity and T the physical temperature. The utility of passive microwave methods in sea ice research is due to the difference in emissivity for different ice types and for open water (Zwally and Gloersen, 1977). Table 2 shows emissivities for ice at the commonly used wavelengths.

From space or from aircraft altitudes, the effect of the atmosphere must be taken into account. Atmospheric layers (including clouds) absorb microwave energy proportional to the upwelling electromagnetic intensity and emit radiation proportional to their temperature. The intensity T_B received by a spacecraft is given by

$$T_B = \varepsilon T_o e^{-\tau} + T_1 + (1 - \varepsilon)\, T_2 e^{-\tau} + (1 - \varepsilon)\, T_{sp} e^{-2\tau} \qquad (16.3)$$

where T_o is the ice surface temperature, τ is the total atmospheric opacity, T_1 is the upward emitted radiance of the atmosphere, T_2 is the total (downward emitted, to be reflected) atmospheric emission, and T_{sp} is the average celestial radiance which is twice attenuated and reflected. In the central pack τ is small enough to rewrite (Gloersen et al. 1978) for each wavelength and polarization:

Table 2. MICROWAVE EMISSIVITIES measured <u>in situ</u> at the ice margin (Fram Strait), by satellite over the central Arctic, and by aircraft over the eastern Arctic.

In situ (50°) (Norsex Arctic Working Group, 1981)							
f (GH_z)		4.9	10.4	21	36	94	Comment
Water	V	.505 ± 9	.545 ± 10	.633 ± 15	.685 ± 14	.75 ± 3	Calm
	H	.256 ± 9	.281 ± 10	.354 ± 10	.387 ± 12	.49 ± 5	
FY	V	.92 ± 1	.94 ± 1	.953 ± 3	.945 ± 1	.93 ± 2	Pancakes
	H	.84 ± 6	.88 ± 5	.91 ± 3	.90 ± 2	.90 ± 1	
MY	V	.93 ± 4	.89 ± 6	.79 ± 8	.68 ± 8	.57 ± 5	Complex snow/water
	H	.86 ± 5	.71 ± 6	.64 ± 12	.58 ± 7	.54 ± 6	Ice burden
Nimbus 7 (50°) (Cavalieri et al. 1981)							
f (GH_z)		6.5	11	18	21	37	
FY	V	.99	.99	.99	.99	.99	
	H	.97	.96	.94	.94	.91	
MY	V	.99	.95	.89	.85	.76	
	H	.91	.86	.79	.76	.66	
Aircraft (NADIR) (Troy et al. 1981)							
			14	19	31	90	
FY			.99 ± 1	.97 ± 1	.98 ± 1	.95 ± 2	
MY			.90 ± 5	.86 ± 6	.80 ± 5	.81 ± 4	
Young			.91 ± 2	.91 ± 2	.93 ± 2	.94 ± 2	
New			.85 ± 3	.84 ± 2	.87 ± 2	.92 ± 1	
Nimbus 5 (NADIR) (Carsey 1982)							
MY				19.35			
				.82 ± 2			

$$T_B = \varepsilon T_o + \tau[(1 + k)\ T_A - \ \varepsilon(kT_A + T_o)] + (1 - \varepsilon)\ T_{sp} \qquad (16.4)$$

where T_A is the density weighed atmospheric temperature such that $T_1 \simeq \tau T_A$, and k, usually taken to be 1, is an estimate of the diffusiveness of the surface reflection and the vertical distribution of absorption. In the MIZ τ may too large for the linearization process in which case it must be determined carefully. Little work has been done for this case although the data (Gloersen and Barath, 1972) is theoretically capable of this correction. In the above derivation, the snow cover has been assumed to play no part in microwave emission. This is not entirely accurate, and the error grows with increasing frequency; but there is now no way to parameterize the snow contribution as little data have been taken. This problem is appreciable for wet snow conditions common to late spring and ice margin conditions. In general, an ice layer covered by wet snow can be treated as a surface with an increased emissivity (Carsey, 1985).

Both emissivity and opacity are functions of wavelength: $\varepsilon = \varepsilon(\lambda)$ and $\tau = \tau(\lambda)$. (In fact, the complicated profiles and wavelength dependence of dielectric constant and temperature within the ice probably mean that the emitter temperature is no longer the surface temperature, or equivalently, $T_o = T_o(\lambda)$.) Thus, for multichannel systems (4) becomes a set of simultaneous linear equations in which

$$\varepsilon(\lambda)\ T_o = \sum_{j=1}^{n} C_j \varepsilon_j(\lambda)\ T_{oj} \qquad (16.5)$$

and

$$\sum_{j=1}^{n} C_j = 1 \qquad (16.6)$$

where for each jth ice type there is a proportional areal concentration, (wavelength and polarization dependent) emissivity, and surface temperature.

An alternative scheme that has been developed by Cavalieri et al. (1982) takes advantage of the large contrast between open water and ice at longer wavelengths, the smaller differences in signals of different polarizations for ice than water, the opposite trends of emissivity with wavelength for ice and water, and the increase in contrast with decreasing wavelength between different ice types. The method computes total ice concentration by parameterizing the polarizaion of the surface through the wavelength-dependent polarization (PR)

$$PR = \frac{T_{BV} - T_{BH}}{.5(T_{BV} + T_{BH})} \tag{16.7}$$

where subscripts V and H refer to vertical and horizontal polarization. The multiyear fraction is then computed by defining a spectral Gradient R (GR).

$$GR = \frac{T_{BV}(.81 \text{ cm}) - T_{BV}(1.7 \text{ cm})}{.5[T_{BV}(.81 \text{ cm}) + T_{BV}(1.7 \text{ cm})]} \tag{16.8}$$

a parameter used to identify first-year ice, multiyear ice and open water through their respective spectral signatures. Both the polarization and gradient ratios are used in an iterative scheme to obtain estimates of ice concentration, multiyear fraction and ice temperature from the multifrequency passive observations.

Much work has been done with the single wavelength horizontally polarized radiometer on Nimbus 5 called the Electrically Scanning Microwave Radiometer, ESMR, (Wilheit, 1972; Wilheit et al., 1972; Cavalieri and Parkinson, 1981; Zwally and Gloersen, 1977; Campbell et al., 1978; Comiso and Zwally, 1982). The single channel permits accurate global ice extent determination including surveillance of substantial polynyas. These data have also been further used in special cases such as in the Southern Ocean where there is little multiyear ice and in the Arctic summer when all ice is of one "type" and temperature (Carsey, 1982; Zwally et al., 1985). Examples will be shown and discussed in a later section.

The more advanced Scanning Multichannel Microwave Radiometer (SMMR) (Gloersen and Barath, 1977) on Nimbus 7 and Seasat permits calculations based on emissivity differences of different ice types (Gloersen et al. 1978; Cavalieri et al., 1984) and provides a method for measuring sea surface temperature and windspeed (Wilheit and Chang, 1980; Lipes, 1982).

The variation of emissivity with ice type which gives passive microwave surveillance its value is due to the different properties of the topmost few tens of centimeters of ice. In the Arctic, where much _in situ_ and aircraft work has been done, ice typically forms on the sea by the sintering of frazil ice to make up a matrix of millimeter size pure ice fragments separated by random pockets of brine. Growth of the ice usually takes the form of penetration of congelation ice into the sea (Weeks and Ackley, this volume). This ice has a generally high emissivity controlled entirely by surface properties. Ice which has survived a summer melt season has undergone drainage of brine resulting in the formation of approximately cylindrical voids in the freeboard portion and a decrease in salinity throughout. This ice has lower salinity and density than first-year ice as well as an inhomogeneous index of refraction. These properties affect the emissivity as discussed by Vant et al. (1974, 1978). In short, the emissivity is decreased for decreased temperature, decreased salinity, increased density, and brine channel scattering. In terms of measured T_B for the 1.55 cm ESMR data, first-year ice has a T_B near 240 K; multiyear ice near 220 K and open water near 140 K. The emissivity of Arctic multiyear ice has been shown to have a spatial variation producing a regional T_B variation comparable to the variation produced by typical open water or first-year/multiyear ice mixtures (Carsey, 1982). The ice of the Southern Ocean is presumed to behave as Arctic first-year ice but with microwave variability (Comiso et al., 1984). Electromagnetic aspects of ice will be discussed below.

4.4 Active Microwave

The active microwave observation of sea ice is conveniently broken down into 4 categories according to instrument and geometry. These are radar altimetry, airborne scatterometry, imaging radar, and land-based radar. In usual usage a radar is a ranging instrument, and a scatterometer is a backscatter-coefficient measuring instrument.

A. Radar Altimetry. The microwave radar altimeter is an aircraft or spacecraft instrument that measures the range from surface to altimeter, and may also measure the backscattered energy received by a narrow vertically oriented antenna beam. It could be thought of as a near-vertical scatterometer. The reflection from sea ice is much stronger than that from water, and the form of the return pulse is different. When short-pulse radar altimeter signals are reflected and scattered from water, the return has a characteristic rapid rise, plateau, and gradual fall as the illuminated area on the ocean surface grows from a point to a circle and thence to an annular ring of nearly constant area, where the incidence angle dependence of the backscatter controls the signal fall-off. When ice is below the altimeter, the returned signal is stronger than for typical water roughness and drops off with angle much more rapidly to a reduced value which persists for an unusually long time. Altimeter returns from extremely calm water are strongly peaked but show no persistent higher-angle return. The effect is tentatively accounted for (Brown, 1982) by scattering from a distribution of variously sized horizontal flat faces in the ice surface. Further examination of altimeter scattering signatures show that this could be a valuable tool for classifying ice, particularly when used in conjunction with other remote sensing methods.

B. Scatterometry. While near vertical radar altimetry has a return signal primarily sensitive to ice flatness characteristics, other radar methods are sensitive to ice roughness on a variety of spatial scales. The difference of course is due to the geometry of incidence angles, nadir for the altimeter and 20° - 60° for other radar. The spacecraft scatterometer for oceanographic remote sensing is designed for observation of ocean capillary waves with characteristics trough to trough distances of a centimeter or so. The scattering from a "slightly rough" planar surface is Bragg-like requiring not truly periodic structure but merely sufficient energy in the appropriate region of the roughness power spectrum (Barrick and Peake, 1968; Brown, 1978). The oceanographic motivation for scatterometry is due to the response of the ocean surface roughness to wind stress (Moore and Pierson, 1966; Moore and Young, 1977; Jones et al. 1979; Brown, 1979) making the scatterometer a useful device for measuring ocean surface wind (stress).

One scatterometer design works by transmitting a "fan beam" - a beam which is wide in one direction and narrow in the other. The fan is oriented with the long direction of the footprint aligned about 45° from with the direction of flight. In operation, the Doppler shift from ocean areas depends on their position along the flight line relative to the transmitter and is used to resolve spatial location (Johnson et al. 1980). The instrument samples the surface at a wide variety of incidence angles and thus requires good knowledge of the scattering function. Another scatterometer design usually applied in aircraft instruments relies on steerable or fixed orientation "pencil" beams. These systems produce a data set which is more simply analyzed and may be more accurately calibrated.

It is well known to any ice observer that the various ice types have different roughness characteristics and should thus at least have different backscatter characteristics. However, the acquisition of precise backscatter values at different incidence angles for different ice types in different situations in the different seasons is a task requiring a great investment in skill and time. While gathering of this information is indeed successfully underway (Onstott et al. 1981; Onstott and Gogineni, 1985), and the results are promising, there is as yet not enough information to spell out the capabilities and weaknesses of scatterometry as a tool in sea ice research.

Examination of the Seasat Scatterometer data set has been undertaken by Peteherych (private communication) who has pointed out that the ice and water backscatter coefficients are very different, ice having a larger signal by 10 - 15 dB, making the ice margin readily resolvable. Also Carsey (1985) noted strong seasonal dependence as scattering increased when the intersticial water froze. Thus the scatterometer data may contain information to estimate the area of these low-albedo features of Arctic summer ice.

Finally the data set seemed to indicate a backscatter difference associated with the Beaufort Gyre giving encouragement to the idea that the scatterometer may provide an estimate of total ridged-ice area. It will be interesting to see how this develops.

C. Imaging Radar. The data product of imaging radar systems is a picture-format array in which each point has an intensity or hue according to its backscatter strength σ^o or some function of σ^o. These images typically look a great deal like visible light or IR images. There are two common methods for production of radar images, Real-Aperture Radar (RAR, also previously called Side-Looking-Airbore Radar or SLAR) and Synthetic-Aperture Radar or SAR. It is useful to start by describing RAR, a well established technology (Jensen et al. 1977).

The RAR approach also relies on a fan beam. In this case the beam is usually oriented to place the long axis of the footprint on the surface normal to the flight line on one or both sides of the aircraft. An image of backscatter is swept out line by line as on a TV screen by transmitting pulses and receiving a return signal whose strength is a measure of surface roughness and whose time of flight is a simple function of distance along the surface. Clearly RAR is instrumentally simple, and it has advantages of low cost and immediate data processing (Ulaby et al. 1982; Reeves et al. 1975, p. 443 f). It has the disadvantage that incidence angle, range, and footprint vary along the image from near nadir to near horizontal with attendant changes in feature resolution and signal to noise ratio.

The RAR (SLAR) method is valuable in the monitoring of ice conditions (Dunbar, 1975; Dunbar and Weeks, 1975), and it continues in operational use at this time. While data have not been published from the ice margin, the method may be well suited to measurements there. Measurements requiring quantitative backscatter data are more difficult with RAR observation than with instruments designed to be scatterometers, as the incidence angle variation, aircraft motion and range differences render accurate calibration difficult. In general, RAR images have coarser resolution than (aircraft) SAR images.

The principle of the SAR method is also to transmit a narrower beam normal to track than along track. The returned echo is spatially located normal to track (or in range) by time delay of the signal and along track (or azimuth) by Doppler shift analysis as shown in Figure 3 (Elachi, 1981). To produce an image the scattering strength from each surface resolution element is observed for a number of pulses as the platform moves past and this Doppler/intensity/range history is used to synthesize the signal which would be received from a far larger antenna with a correspondingly small surface resolution.

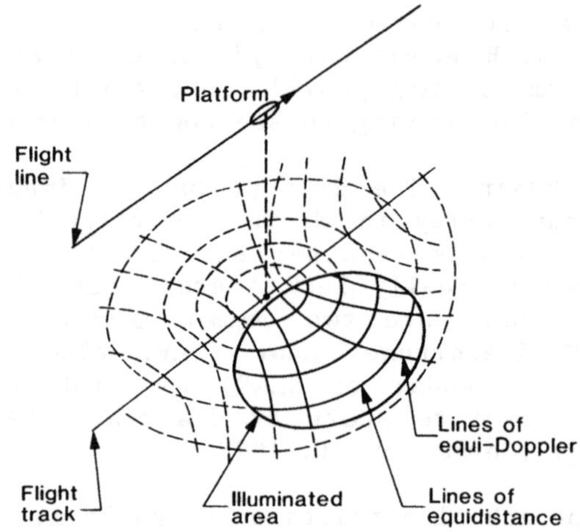

Fig. 3. A schematic showing SAR operaion principles. The platform is an aircraft or spacecraft with the SAR antenna on its right, in this case (Elachi, 1980).

The SAR method has proved applicable to a wide variety of problems (Elachi, 1980). In ice work images have been interpreted for location of recognizable features, e.g. ridges, to track relative and absolute positions of the features (Hall and Rothrock, 1981) and to describe ice conditions for specific purposes. Use of backscatter signal strength (or its contrast) to estimate areas of different ice types shows promise. Since SAR operates in a narrow range of incidence angles, smaller for spacecraft than aircraft, the interpretation of backscatter involves knowledge of ice-type backscatter at the incidence angle used; also needed is a knowledge of system relative calibration. Time, frequency shift and antenna location are recorded, but antenna orientation is not a first order consideration, although illumination distribution is affected. At present, SAR systems are calibratable to perhaps 3 dB absolute and 1 dB relative. For ice velocity and deformation determination high absolute and relative registration accuracy is called for (Hall and Rothrock, 1981). This can be obtained from Seasat data under certain conditions and can be computed in general with accuracy of a few resolution elements (Brown et al. 1982; Curlander, 1982). Resolution is limited by data handling capability, but the Seasat radar with 25 m resolution is probably adequate to most applications. Aircraft systems may have more severe registration problems compared to spacecraft (Leberl et al., 1979).

At the ice margin near Fram Strait data were acquired as part of the NORSEX program in November, 1979. These data will be discussed further below. In the L-band NORSEX SAR images, the clear distinction between floes, ridges and open water noted in images from the central pack (Bryan et al., 1977; Carsey, 1982) is absent. It appears that the leads have become "bright" (high σ^o) at the margin although they are "dark" (low σ^o) within the pack. The cause is likely that the NORSEX leads were filled with pancake ice which is effectively rough to the radar. This may be a blessing if the presence of pancake ice is a required determination, or it may be a problem as leads will resemble ridges. The capability to track features and determine velocities will probably be impaired but not crippled.

Data both in situ and from aircraft have been obtained at a variety of wavelengths from L-band (about 25 cm) to X-band (about 3 cm). In general, ice features are more distinguishable at shorter wavelength, but this is still under study. At present higher frequency radars pose technological problems for work in space and are more expensive to construct and operate. The SAR can be used in a variety of incidence angles over ice, but best resolution of ice type requires rather large angles of 40° - 70° (Larsen et al. 1981).

D. Land-Based Radar. For smaller-range studies (less than 100 km) the azimuthally sweeping radar has proved a useful method for tracking floes and monitoring conditions. This method uses a transmit-receive antenna on land with the consequence that incidence angles near 90° are used; these angles are very efficient for locating floating ice, especially icebergs. Surveillance of ice in Japanese harbors has been successfully done using such a radar (Tabata, 1972), and detailed motion of ice floes has been measured (Tabata et al., 1980).

4.5 Laser Altimeter and Sonar

In a fashion analogous to satellite radar altimetry, the upper and lower surface contours of sea ice may be probed by the aircraft laser altimeter and the upward-looking submarine sonar. In both cases, pressure ridge statistics and mass balance information are derived. The observation utilized is the measurement of range from aircraft (submarine) to the ice surface by recording time-of-flight of the light (sound) pulse from the vehicle to reflection from the discontinuity at the ice surface and back to the vehicle. For both platforms, occasional open water in leads produces an identifiable backscatter signal, serving as a check of aircraft (submarine) absolute height (depth) from the water surface. In one case near-simultaneous colocated laser/sonar results were acquired (Wadhams, 1981) and strong correspondence was found between numbers and sizes of sails and keels. While laser altimeters have not flown in space, they show great promise for ice and snow work.

Interpretation of both aircraft laser altimeter and submarine sonar data is well established, if complex. Complexities arise because vertical aircraft motions must be removed with care (Hibler et al., 1972; Lowry and Brocher, 1978). Good results have been obtained for much of the Arctic through much of the seasonal cycle (Wadhams, 1981; Weeks et al., 1980; Wadhams and Horne, 1980; Williams et al., 1975; Hibler et al., 1974).

Laser altimetry from space has been suggested for ice sheet profiling and may be applicable to sea ice mass balance study. The likelihood of such a system giving good ridge sail distribution, however, seems scant.

5. DEMONSTRATION OF METHODS

5.1 Visible and Infrared

The best established use of satellite visible and infrared imagery has been the description of sea ice cover on basin-wide and synoptic scales. The utility of this technique is due to the usually abrupt nature of the ice edge and the strong ice-water contrast which combine to provide an easily measured line of ice pack advance/retreat. Taken together the visible and IR systems provide coverage over the spring, fall and winter (the low-cloud period) with a mix of fine and coarse resolution. These observations provide good support for seasonal processes and large scale dynamics studies (McClain, 1974, 1975; Dey, 1980a,b,c, 1981; McNutt, 1981; Crane, 1978; Vinje, 1976; Ahlnas and Wendler, 1976, 1977, 1980). Samples of figures of Dey are shown in Figure 4, of McNutt in Figure 5. Regional short-wave energetics have been examined by Wendler (1973) who also constructed seasonal albedo charts as well as the means in Figure 2 for the Southern Beaufort Sea. McNutt (1980) recorded registration differences between Landsat and NOAA 3 of the order of 10 km.

Individual features, notably large leads and polynyas, have also been monitored by these types of imagery. Examples are Aber and Vowinkel (1972) and Dey (1980a) who studied the North Water polynya; Barry and co-workers (1979) who examined the climatology of fast ice; Muench and Charnell (1977) who examined the form of grease ice plumes at the Bering ice edge; and Knapp (1969) who studied polynyas in the southern ocean.

Quantitative use of visible and IR images has enjoyed most success in the tracking, with much higher spatial density of observations than buoy data permits, of ice features in time. Notable examples are the ice velocity fields of Figure 6 generated by Vinje (1976; by private communication) for the East Greenland drift stream, the Bering Sea ice motion and deformation by Shapiro and co-workers (1975, 1976 and 1977), and the velocities of McNutt (1981) in the Bering Sea, Hall (1980a) in the central Arctic, and DeRycke (1973) in the Southern Ocean. Except for Hall, all made use

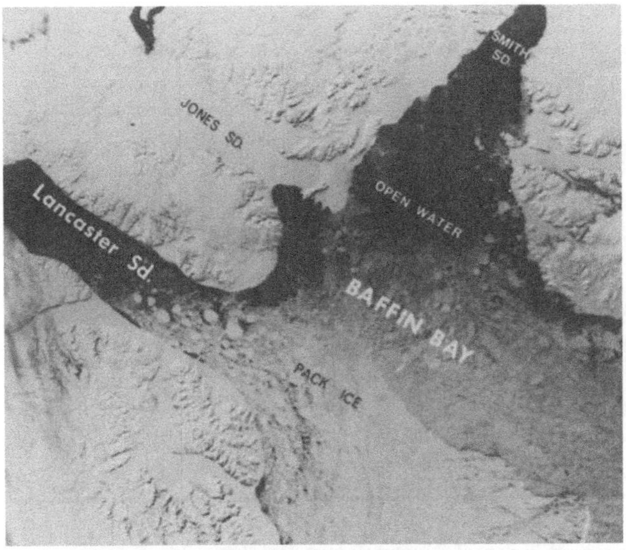

28 MAY 1977 NOAA VIS

30 NOV 1976 NOAA TIR

Fig. 4. Samples of visible and IR (TIR) data from the NOAA series of satellites. Note that ice coverage and some floe structure are discernible in both about equally well. (From Dey, 1981, 1980a).

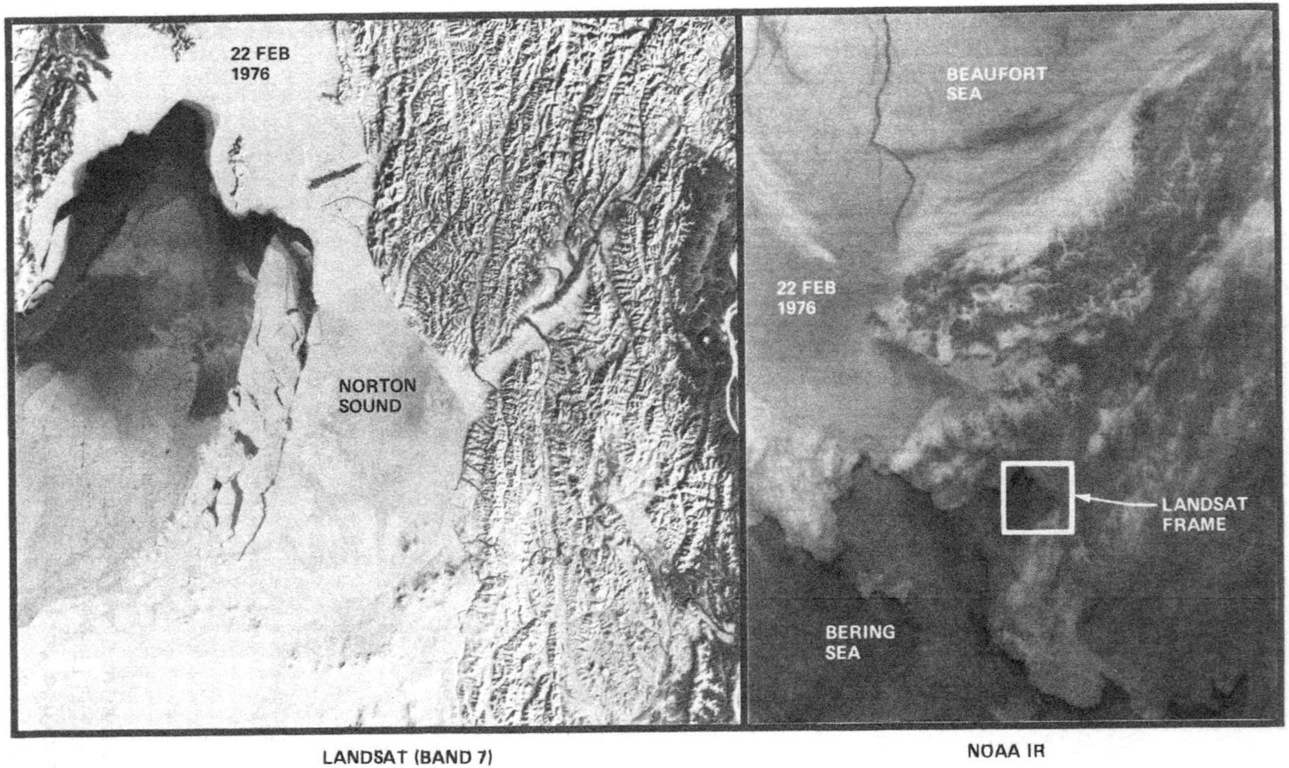

Fig. 5. Landsat visible-light and NOAA IR images for the Bering Sea area. The comparison shows strikingly the consequences of the finer resolution of Landsat and the regional coverage of NOAA.

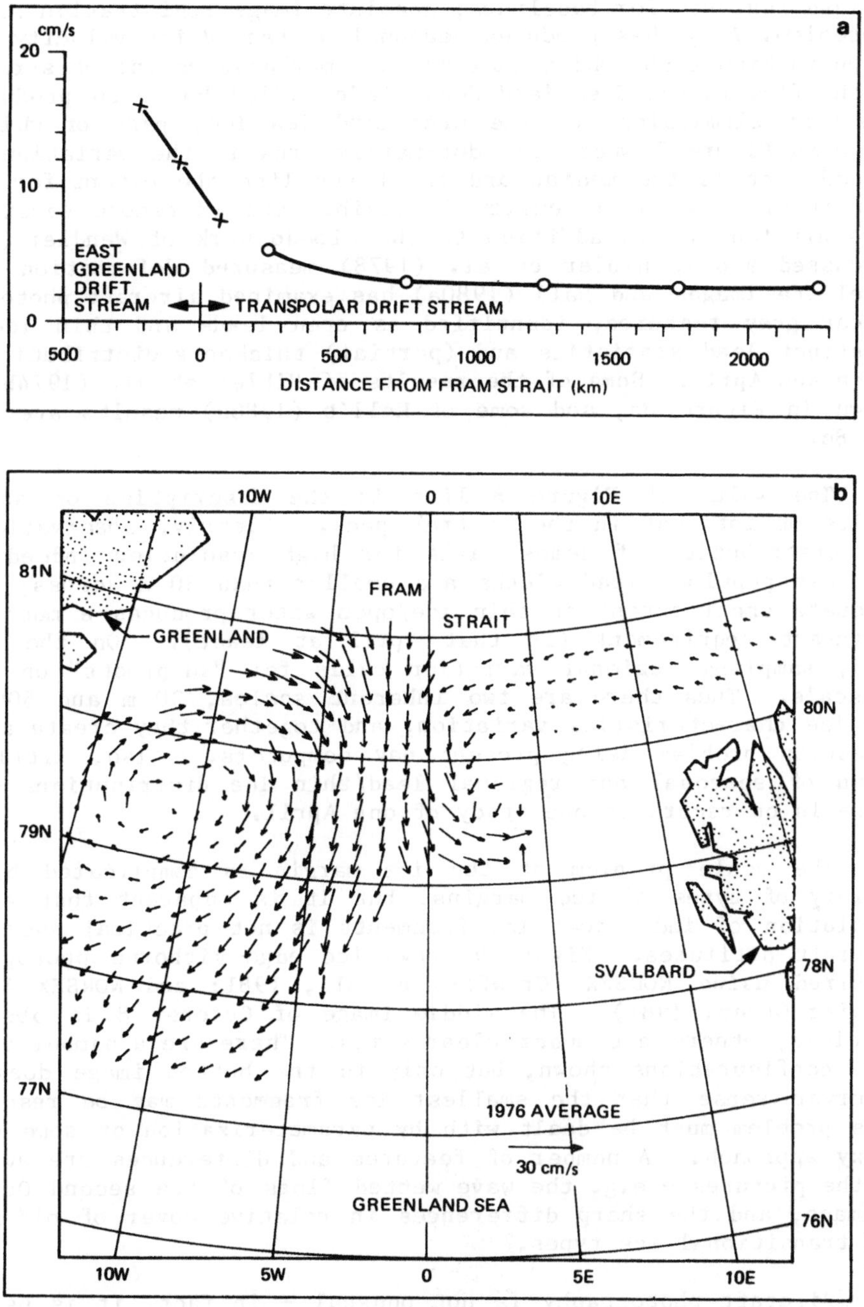

Fig. 6. Ice velocities determined by analysis of Landsat data for the Eastern Arctic. Part a shows mean values of the whole stream (Vinje, 1970), and part b shows long-term mean trajectories (Vinje, private communication).

of land features for baselines; absolute image registration remains a problem. Vinje has produced seasonal average drift velocity plots shown in Figure 6b, which indicate a remarkable amount of structure in the flow in the Greenland Sea. Vinje (1978) has also produced a ten year climatology of the Greenland Sea ice, part of which is shown in Figure 7 where the dot-matrix area is the variation, the beaded line is the median and the heavy line the extent for 1978. Quantitative use of intensity in visible and IR remote sensing is more difficult. In addition to the albedo work of Wendler (1973) discussed above, Hibler et al. (1978) measured deformation using satellite images and Hall (1980a) has examined aircraft photos for linear grey features, identified as open leads and thin ice, to construct lead statistics and (partial) thickness distributions in March and April. Some of the results of Hibler et al. (1974b) are shown in Figure 8a, and some of Hall's (1980a) results are in 8b and 8c.

The value of Figure 8 lies in the description of spatial scales of interest in the central pack. Clearly, observation of the distribution of leads calls for high resolution indeed, for the most populous lead widths are smaller than 30 m across; while adequate areal extent of thin ice/open water produces a much less stringent requirement (in that April at least). On the other hand, sampling regional variation calls for "snapshots" on a 500 km scale. Thus there are two inherent scales, 30 m and 500 km, for ice characteristics variation, and together they create a data digestion problem using present day computers. Also, little is known of seasonal and regional lead/thin ice distributions - all there is on record is one study of one April.

The scale problem at the ice margin is complicated by the variety of types of ice margins, but it is apparent that actual resolution of individual ice fragments is not practical even from aircraft altitudes. Figure 9 shows ice edge airborne photography acquired using NORSEX (Crawford et al., 1981; and NORSEX Arctic Working Group, 1981). The middle image of October 8 is obscured by cloud; others are under clear skies. There are a number of ice edge configurations shown, but only in the bottom image does the observer sense that the smallest ice fragments may be resolved. This problem must be dealt with by parameterization or some other proxy approach. A number of features and differences are evident in the pictures - e.g. the wave wetted floes of the second October 1 image, and the sharp differences in relative cover of old floes and transitional ice types.

Aircraft photography is not unusual - in fact, it is usually acquired on research flights and operational ice surveillance flights. But the use of the photography for geophysical means is not common. In addition to the work of Hall (1980a) discussed above, Hall (1980b) used mosiac images (Hall, 1978) in the central

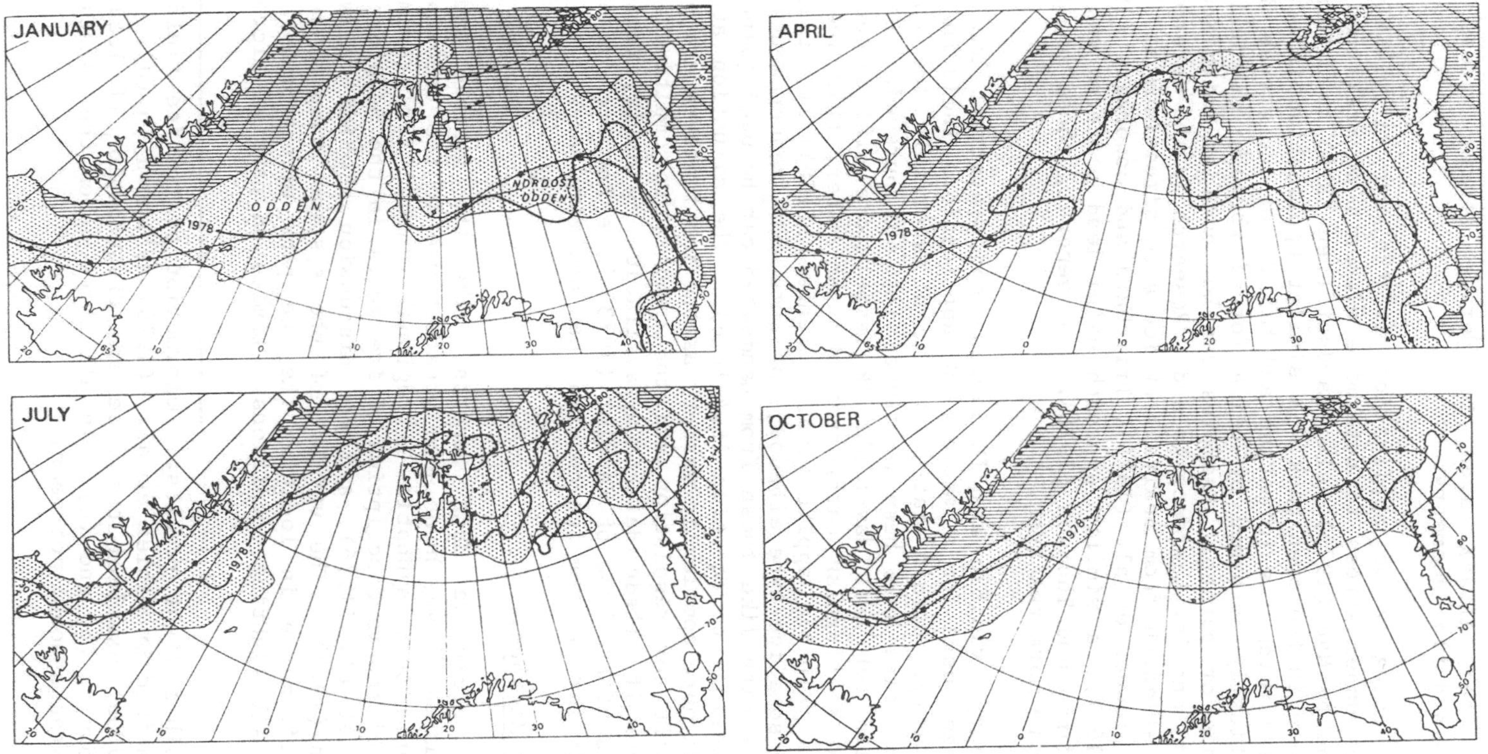

Fig. 7. Ice climatology from Landsat. The line-stippled area was always ice-covered, the dot-stippled area was ice-covered some years, and the clear area was never ice-covered. The beaded line is the mean ice edge, and the heavy line is the ice edge for 1978 (Vinje, 1978).

Arctic over the AIDJEX area to derive strain rates, ice concentration and floe geometry.

Another category of application is the use of visible/IR images of ice under stress to determine elements of ice strength. This was done by Nye (1975) using Landsat (nee ERTS) to determine a failure curve in the open pack. A similar application is accomplished by examining the formation and failure of ice arches where ice is plastically deformed by being forced through channels or around islands or points. Examples can be found at Smith Sound, Bering Strait, and St. Lawrence Island. Research in the determination of ice strength from analysis of ice arch situations has been reported by Crowder et al. (1974), Pritchard and coworkers (1980) and Sodhi (1977); a similar approach using lead patterns was described by Marco and Thomson (1976). In general, these analysis methods require more research.

In summary, visible and IR images from spacecraft and aircraft have been used for a variety of applications including ice extent, velocity, strain rate, and open water/thin ice cover. Other possibilities have been mentioned but not pursued. Barnes (1972) pointed out that day-night IR temperature differences may be usable to distinguish categories of open water, thin ice and heavy ice, a valuable application. Grenfell (1984) in Figure 1 has measured some ice type albedos, but more such measurements are needed before data like those from Landsat-5 can be used quantitatively (in analogy to equation 16.4-16.6). The situation at the ice margin is more complex in variety as the ice features are probably too small for resolution and too complex for multi-channel analysis even if albedo data were available. Finally, regional albedo charts after those of Wendler (1973) have not been generated seasonally for all the ice covered seas and margins.

5.2 Passive Microwave

Passive microwave data records, like the visible and IR images discussed above, have seen greatest use in the description of ice extents on a synoptic to regional scale. Early examples include the microwave description of ice zones in the Beaufort Sea by Campbell et al. (1976), and the discussion of ESMR by Zwally and Gloersen (1977). The utility of the systems stems from its adequate performance in locating the ice pack extent under most weather and all lighting conditions. Limits on passive microwave

→

Fig. 8. Visible light results. Deformation of the ice field was examined by Hibler et al., (1978) as shown in (a). Hall (1980a) examined the size and distribution of leads as shown in (b) and (c). In all three cases the values are instantaneous, not mean.

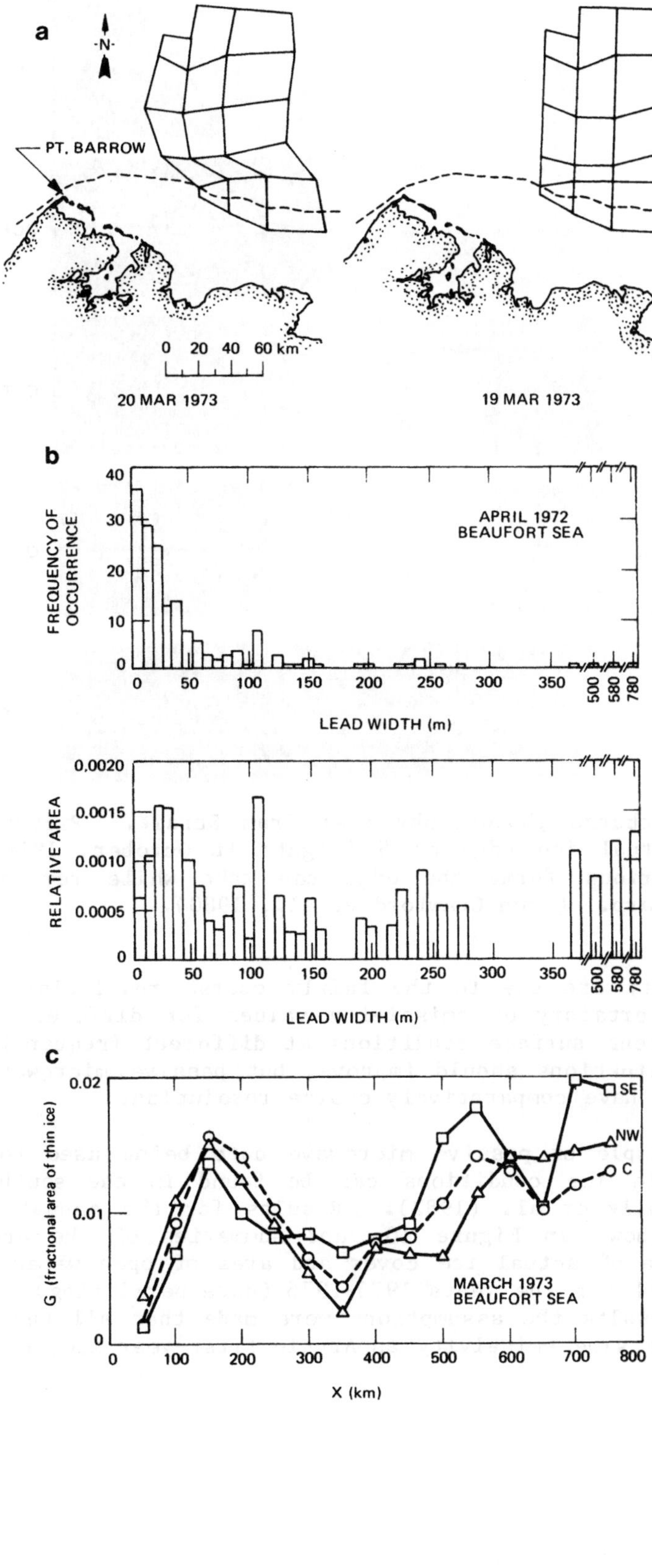
a
N
PT. BARROW
0 20 40 60 km
20 MAR 1973
19 MAR 1973
b
FREQUENCY OF OCCURRENCE
APRIL 1972
BEAUFORT SEA
LEAD WIDTH (m)
RELATIVE AREA
LEAD WIDTH (m)
c
G (fractional area of thin ice)
SE
NW
C
MARCH 1973
BEAUFORT SEA
X (km)

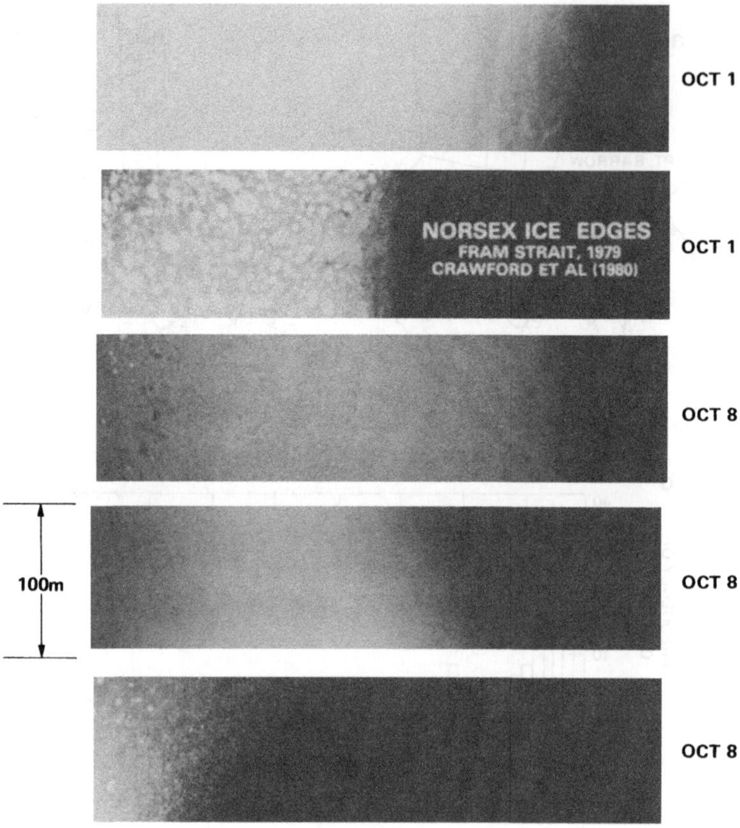

Fig. 9. Airborne photography over Fram Strait. Pictures of the actual ice edge on 5 flights in October, 1979 show the various forms the edge can take while remaining quite sharp. (From Crawford et al., 1980).

applicability are due to the fairly coarse resolution (> 10 km) and the uncertainty of emissivity values for different ice types with different surface conditions at different frequencies. Both of these situations should improve, but passive microwave sensors will always have comparatively coarse resolution.

An example of passive microwave data being used to describe larger scale ice conditions can be found in the southern ocean work of Zwally et al. (1983). Results for the annual cycle for 1974 are shown in Figure 10, and summaries of the area of ice extent, area of actual ice cover and area of open water are shown in Figure 11 for the years 1973-1976 (data permitting). To arrive at these results the assumptions were made that all the ice present had the same emissivity as Arctic first-year ice, that the ice

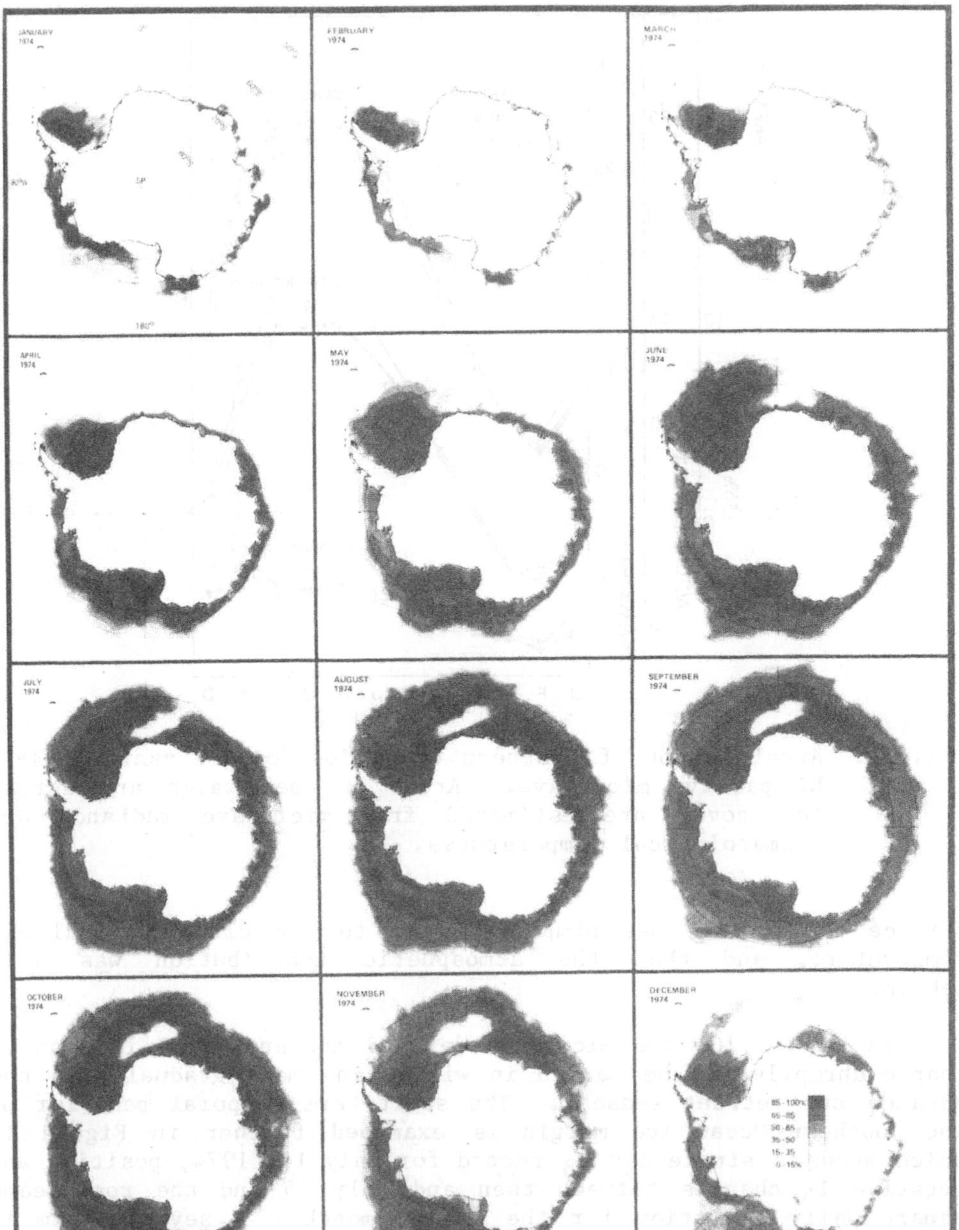

Fig. 10. Maps of the annual cycle of southern ocean ice for 1974 by passive microwave. Data from Nimbus 5 was analyzed to produce ice concentration maps such as these for the years 1973-1976.

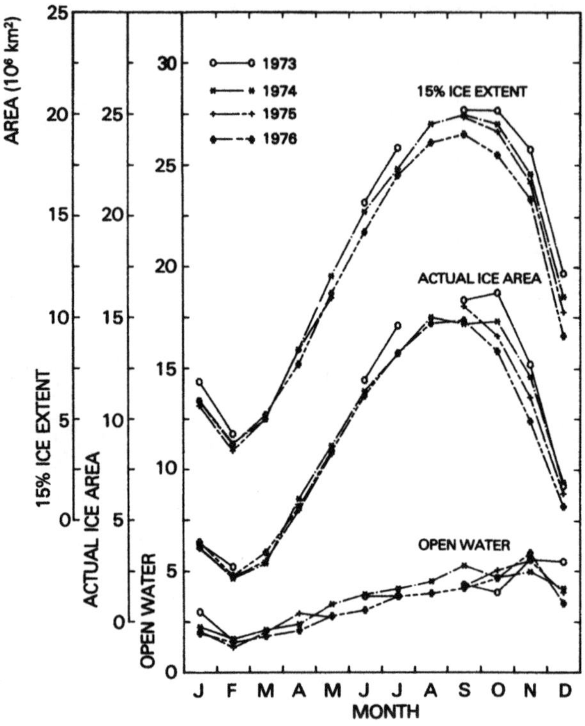

Fig. 11. Areal extent of southern ocean ice for the years 1973-76 by passive microwave. Areas of open water and actual ice cover are estimated from microwave radiance and climatological temperatures.

surface temperature was simply related to the climatological air temperature, and that the atmospheric contribution was well behaved.

In Figure 10, the microwave-derived concentration is seen to change abruptly at the margin in winter and more gradually in the advance and retreat seasons. The short-term temporal behavior of the Southern Ocean ice margin is examined further in Figure 12 which shows a single-day T_B record for July 14, 1974, positive and negative T_B changes between then and July 15 and the root-mean-square daily deviation for the entire month. A severe storm is visible in the South Atlantic and the result T_B changes are substantial. In the RMS variation plot, the variations at ice margins at the continent and the open ocean are seen to be 3-4 times greater than they are within the pack.

The change of the pack extent over a range of time and space scales as seen by passive microwave has been further studied in

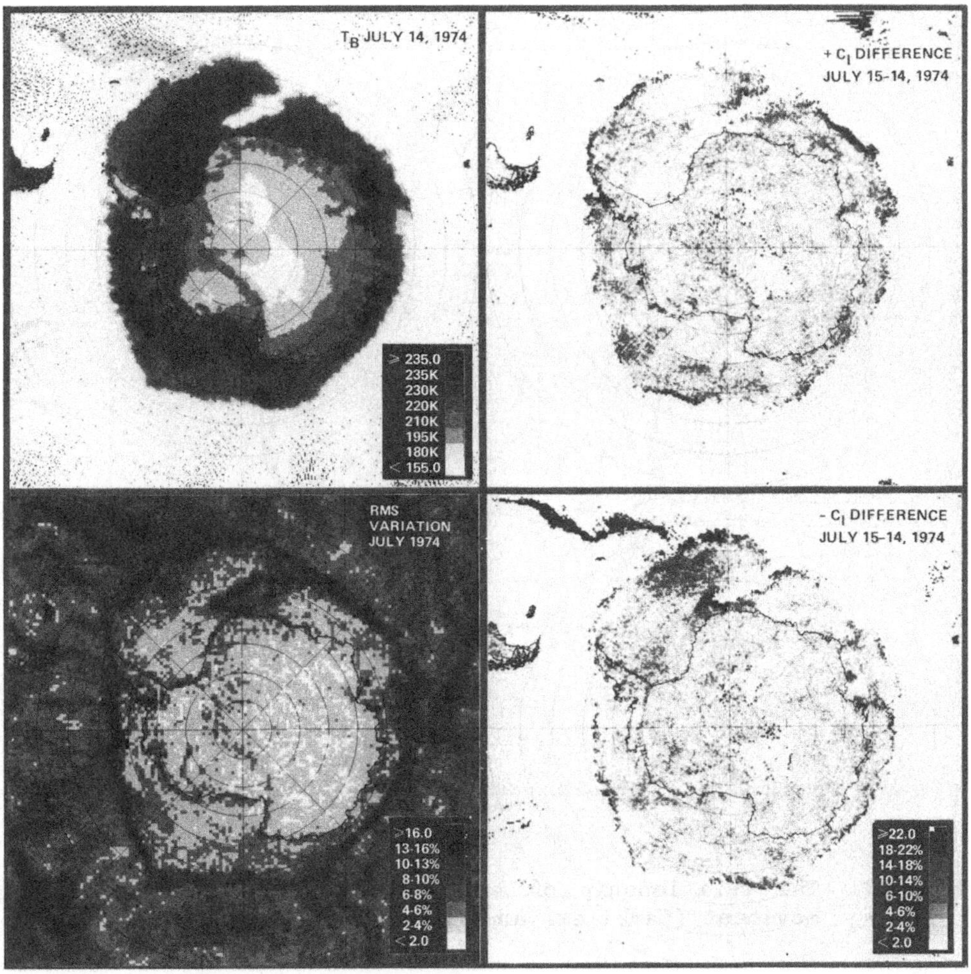

Fig. 12. Variation in microwave radiance of Southern Ocean at 19.35 GHz. In upper left the daily mean, on the right the positive and negative ice concentration changes over 1 day and at lower left the square root of the average squared daily concentration change are shown. A storm is visible in the upper left of the negative C_I image and in the mean image.

Cavalieri and Parkinson (1981) using Australian Weather Service Analyses (obtained from NCAR) and Nimbus 5 ESMR data. Figure 13 is their Figure 9, in which the advance and retreat of the ice edge over a month of late fall, 1974, is compared with the mean surface pressure for the same period. Clearly the amount of edge motion is large enough that the satellite microwave record is of adequate

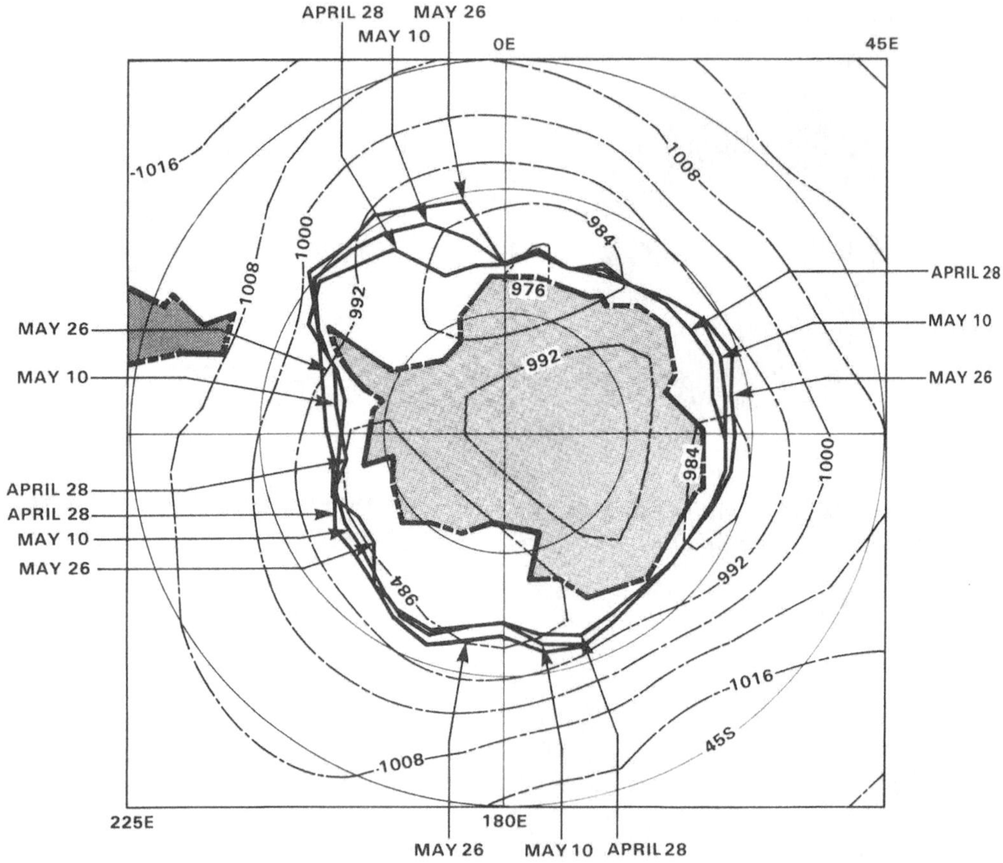

Fig. 13. The relationship of mean weather patterns and ice edge movement (Cavalieri and Parkinson, 1982).

resolution to monitor this interaction. A time series examination using principal component analysis for the Arctic has been reported by Crane et al. (1981). They found much stronger correlation of T_B changes with air temperature (~0.8) than with surface atmospheric pressure (~0.4) although the meteorlogical record is known to be weak.

In satellite IR work images of composite brightness have been found useful and are routinely produced (McClain, 1973). Figure 14 shows a group of demonstration composite brightness images from ESMR on Nimbus 5, the composite maximum from day 256 (near the day of minimum ice extent, see Carsey, 1982) to day 315, the composite minimum days 227-315, and the composite maximum day 227-256. The beginning and ending days, 227 and 315, are chosen because they border the period of minimum Arctic Ocean radiance. In a maximum

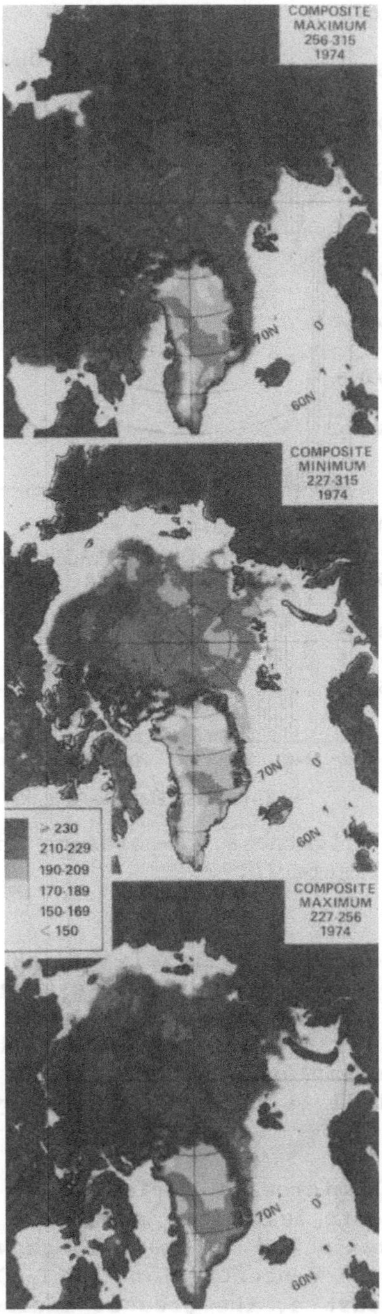

Fig. 14. Composite microwave radiance images for the Arctic region surrounding the period of least ice.

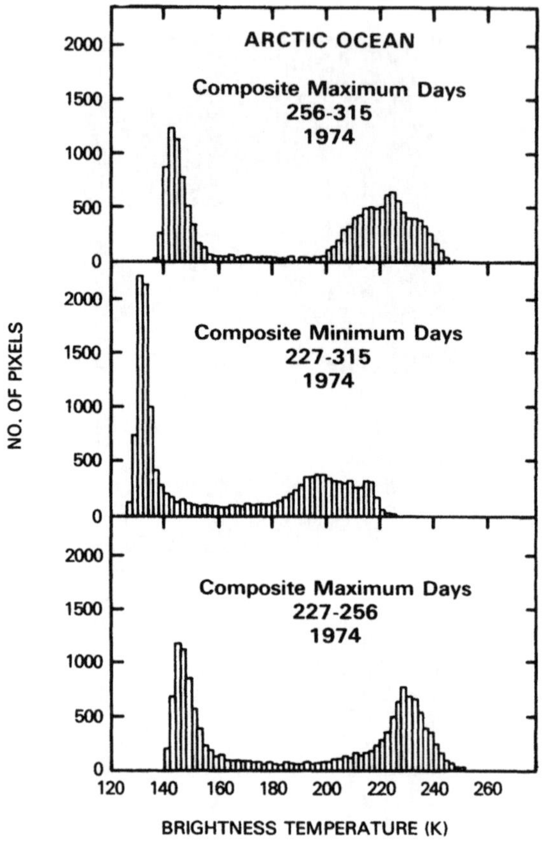

Fig. 15. Microwave brightness temperature distributions for the images in Figure 14.

(minimum) composite image each pixel retains the greatest (least) brightness recorded in any of the 3-day average maps at that pixel location during the interval. Figure 15 shows brightness frequency in number of pixels versus T_B for the three images. Similar distributions have been employed by Ramseier et al. (1974), Gloersen et al. (1974), Campbell et al. (1978), Carsey (1982), and Zwally and Comiso (1981).

The summer-fall microwave variation of the Arctic pack has been further examined to learn more about the amount and distribution of ice at summer's end. Figure 16 shows the areal coverage of ice of at least 15% concentration (or $T_B > 150$ K) and the map of its extent. In part a, the area at minimum extent is surprisingly constant from year to year considering the regional variation shown in part b. The seasonal change in T_B values of the ice pack was further analyzed by Carsey (1982) as shown in Figure 17 where

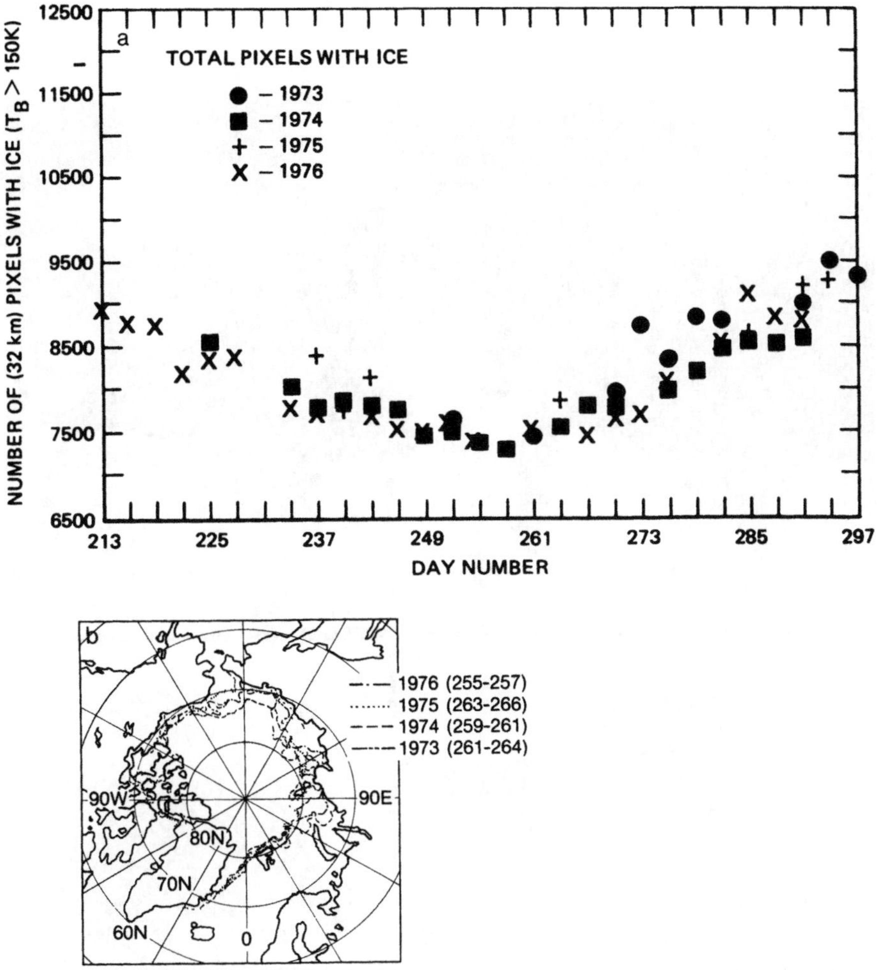

Fig. 16. Area and distribution of ice at the summer ice-extent minimum. (From Carsey, 1982.)

T_B images in mid-summer and early fall are shown with resulting multiyear ice emissivities and concentration. The approach used was the application of eqs. 4 to 6. It was necessary to assume that there was no first-year ice in the Arctic at the time of minimum ice extent (about day 260), that the ice present was substantially free of surface water, and that there was no open water at a period in the fall when the T_B reached a nearly stationary value (near day 300).

To illustrate further some methods of examining seasonal change, Figure 18 shows T_B distributions for a 64 x 10^4 km^2 area

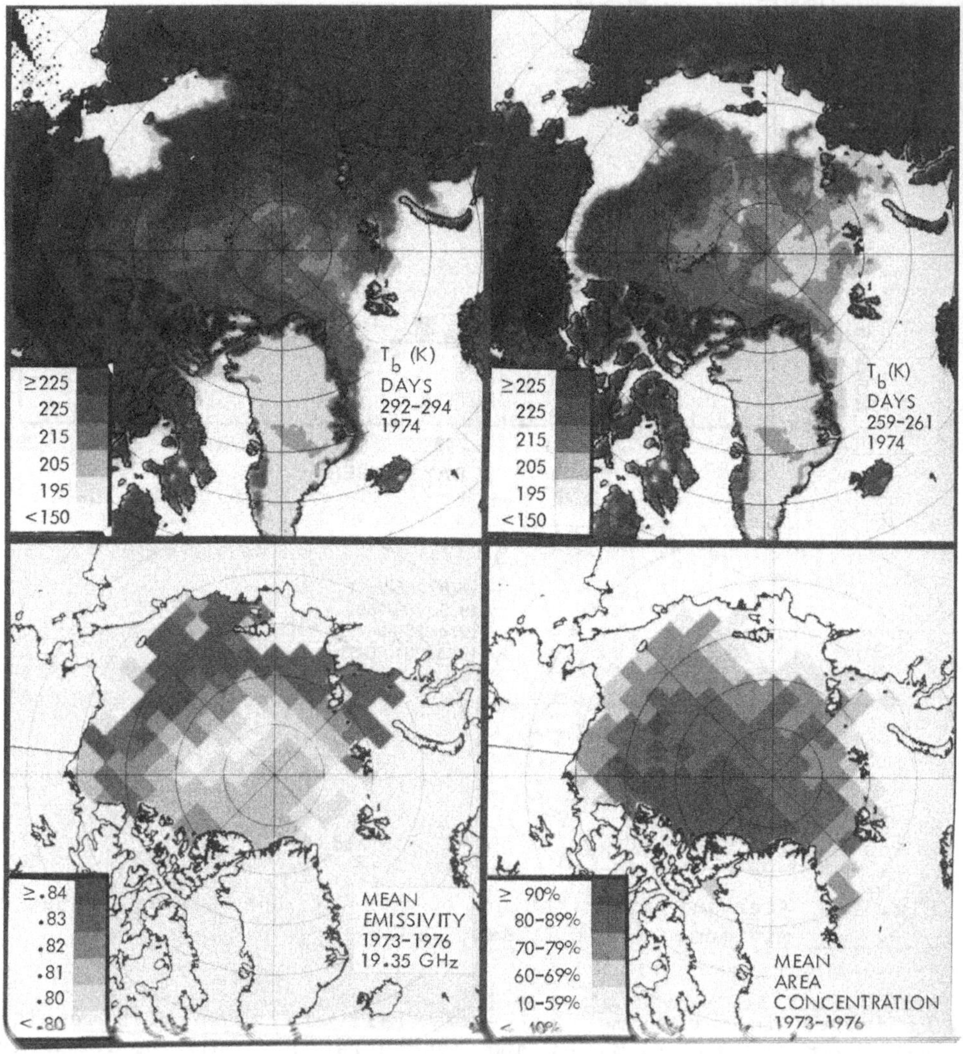

Fig. 17. Analysis of seasonal change in microwave radiance over late-summer to early fall. (From Carsey, 1982.)

in the central Arctic. The radiance distribution is seen to go through 2 changes, from highly varied behavior in the summer, days 241-249, (possibly also observed by Gloersen et al. 1978b, and interpreted as concentration changes) to a sharper and more stable distribution (256-264) to higher T_B values and an asymmetric high T_B shoulder. When the specific emissivities of various ice types are better known such distribution histories will be valuable in monitoring changing conditions; now it is only possible to infer, for example, that the first changes are due to the freezing of surface water, and the second changes are due to freezing of leads. Figure 18 also shows the large change in areal emissivity, defined as T_B divided by the multiyear ice surface temperature, over the summer-fall transition for the years 1973-1976. To develop Figure 18, T_B values were averaged to arrive at areal emissivity for the central pack. The results for multiyear ice concentration of Cavalieri et al. are shown in Figure 19. These values were determined by applying eq. 16.7 and 16.8. The overall behavior of the ice pack is similar to that shown in Figure 18.

There has been limited application of passive microwave data to ice margin problems (Svendren et al., 1983; Cavalieri et al., 1983). Figure 20 shows Landsat images with superimposed ESMR data closest in time for images of the summer ice margins in the Southern Ocean off the Queen Maud Coast (Comiso and Zwally, 1982). In the Landsat figure (a) the open margin shows the complex eddy structure noted by Vinje (private communication) and the sharp fall-off of T_B with ice concentration. Two sets of numbers are shown - those for the day before the Landsat image was taken and those for the day after. In Figure 20(b) the coastal margin is shown and the difficulty created by coarse resolution in the passive microwave is clearly apparent.

The property of passive microwave data to delineate the ice edge and thus make calculation of ice extent possible is useful for some kinds of ice margin research. In Figure 21, the ice edges of the Weddell Polynya and its associated open ocean margin are shown (Carsey, 1980). The most striking result is the stability of form of the polynya compared to that of the open margin, but even the general persistence of form of the open ice edge is unexpected. Interpretation of this behavior yields clues as to the origin and maintenance of the polynya.

For the north polar region, similar data are shown in Figures 22 and 23. In Figure 22, the microwave marginal band, defined as that part of the ocean with T_B between 160 K and 200 K is shown for a selection of days throughout 1974. Also shown is the T_B

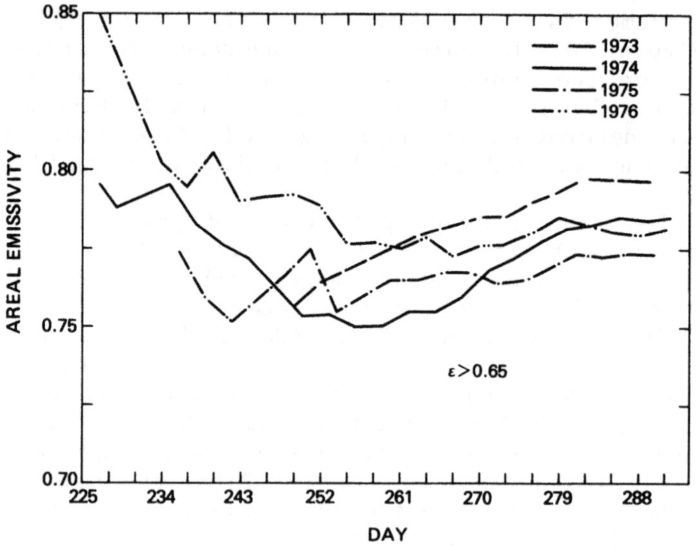

Fig. 18. Microwave brightness and areal emissivity changes over the summer-fall transition. The minimum in areal emissivity is interpreted to be due to a drop in radiance as the free water on the ice surface is frozen followed by an increase as leads freeze.

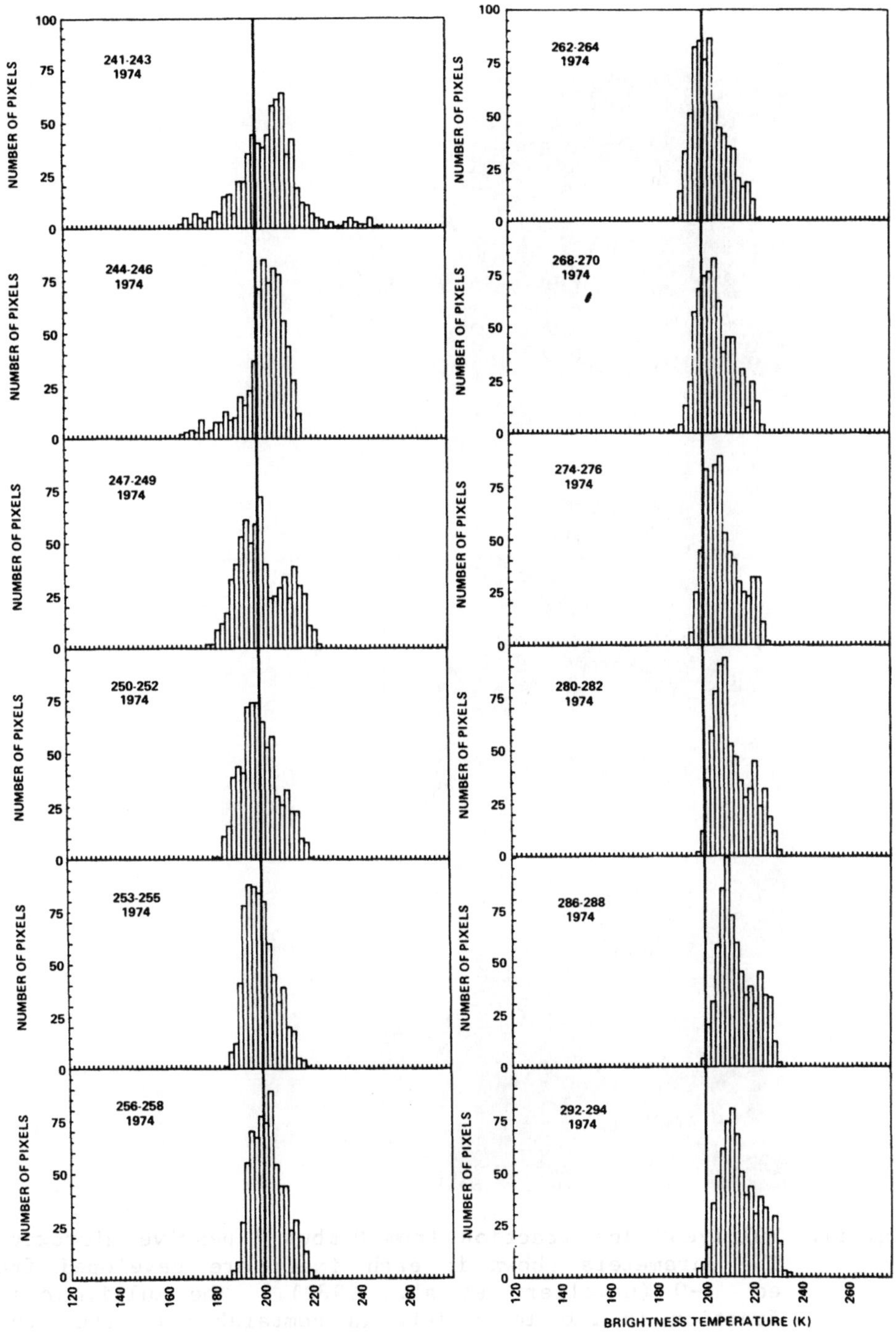
241-243
1974
244-246
1974
247-249
1974
250-252
1974
253-255
1974
256-258
1974
262-264
1974
268-270
1974
274-276
1974
280-282
1974
286-288
1974
292-294
1974
NUMBER OF PIXELS
BRIGHTNESS TEMPERATURE (K)

Fig. 18 - cont'd.

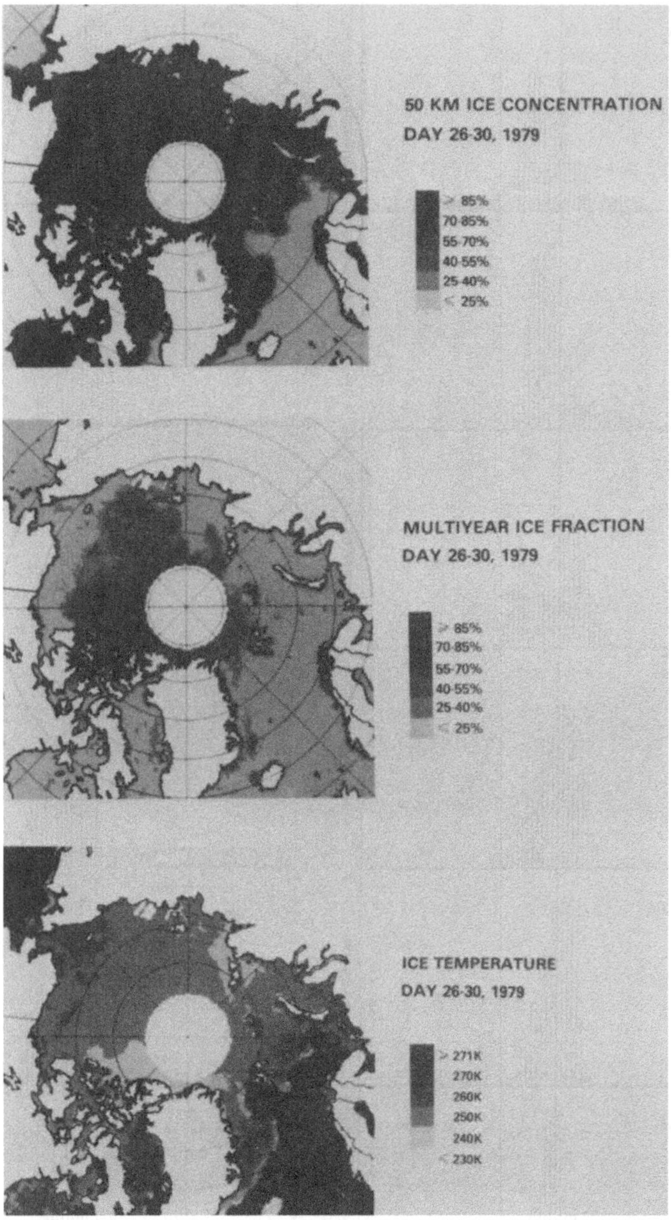

Fig. 19. Multiyear ice fraction from Nimbus 7 passive microwave. The parameters shown in each image are developed from eq. 7-9 (Cavalieri et al., 1982). The multiyear ice fraction in the lower left is comparable to the areal concentration of Figure 17.

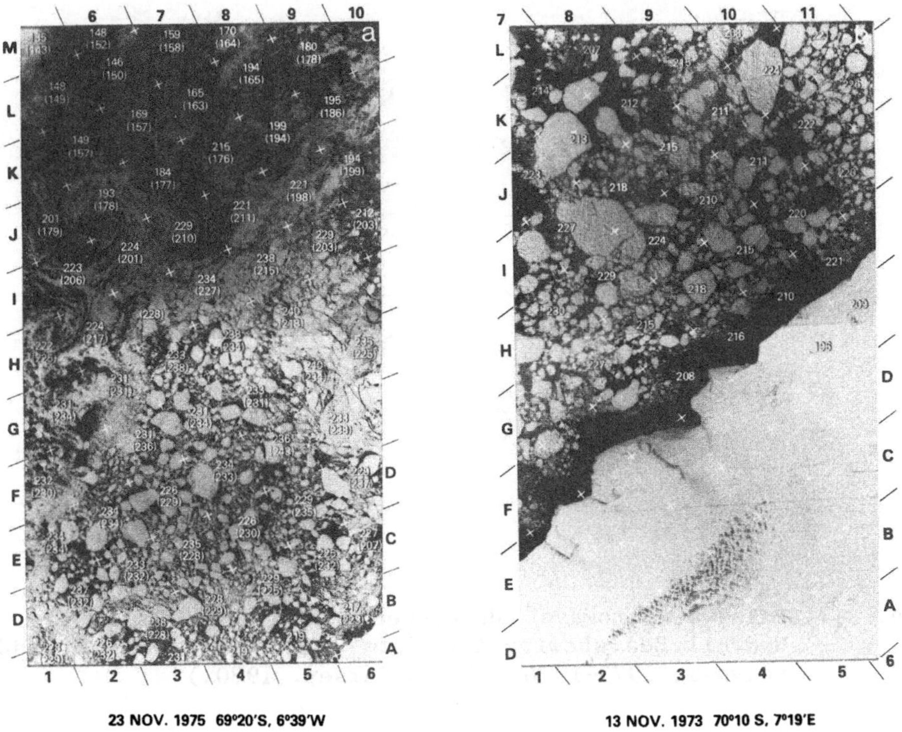

Fig. 20. Landsat and ESMR passive microwave overlay (Comiso and Zwally, 1982). In (a) the brightness temperatures in parentheses are taken the day before, the rest the day after. In (b) the data are taken on the same day.

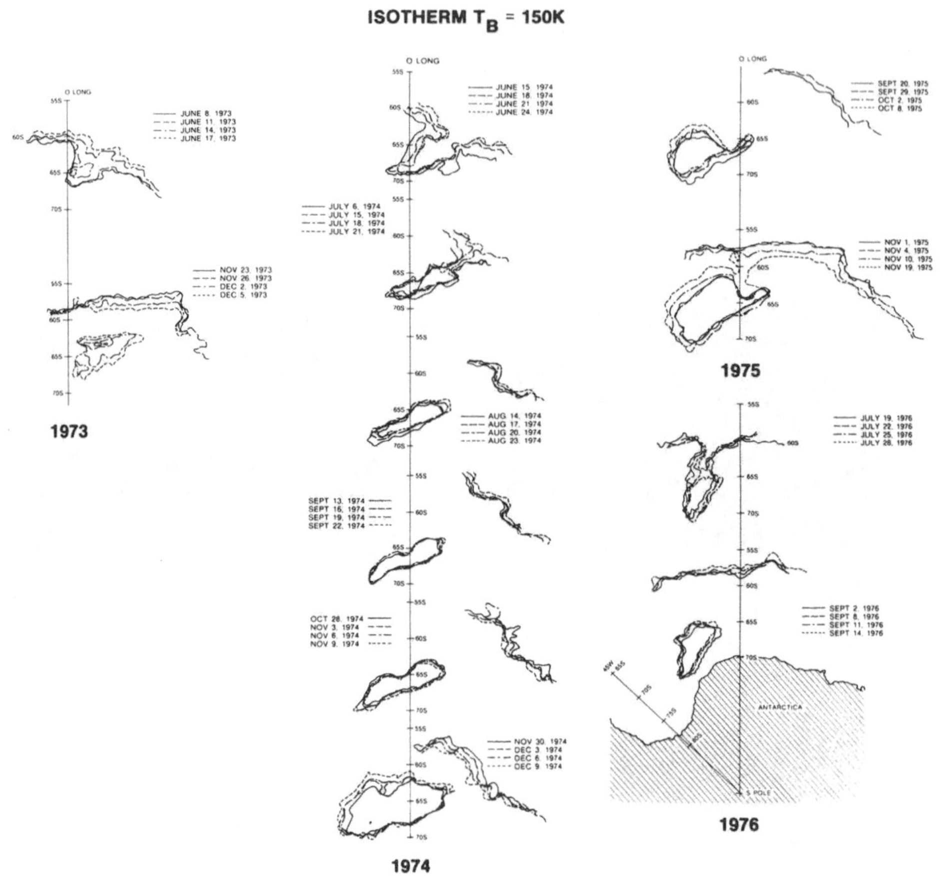

Fig. 21. Passive-microwave developed ice edge data for the Weddell Sea showing the form and change of the Weddell Polynya, 1973-1976. (From Carsey, 1980.)

distribution for selected days in which the number of pixels in the marginal band is seen to vary greatly, depending presumably on the thermal state (melting, forming or stable) within the band. The ocean/weather peak near 140 K and the ice peak near 210 K also change during the year. Finally, the T_B distribution for Fram Strait for one satellite orbit on day 195, 1974 is shown at the bottom of Figure 22. Greenland and Svalbard are shown at passive microwave resolution, and T_B values are placed over the sea areas. The reader is invited to draw isotherms at 20 K intervals and analyze the T_B map.

Similar data are shown for the Bering Sea in Figure 23 showing 160 K and 200 K isotherm off the Alaska Coast. Once again the number of pixels in the microwave marginal band increase in the

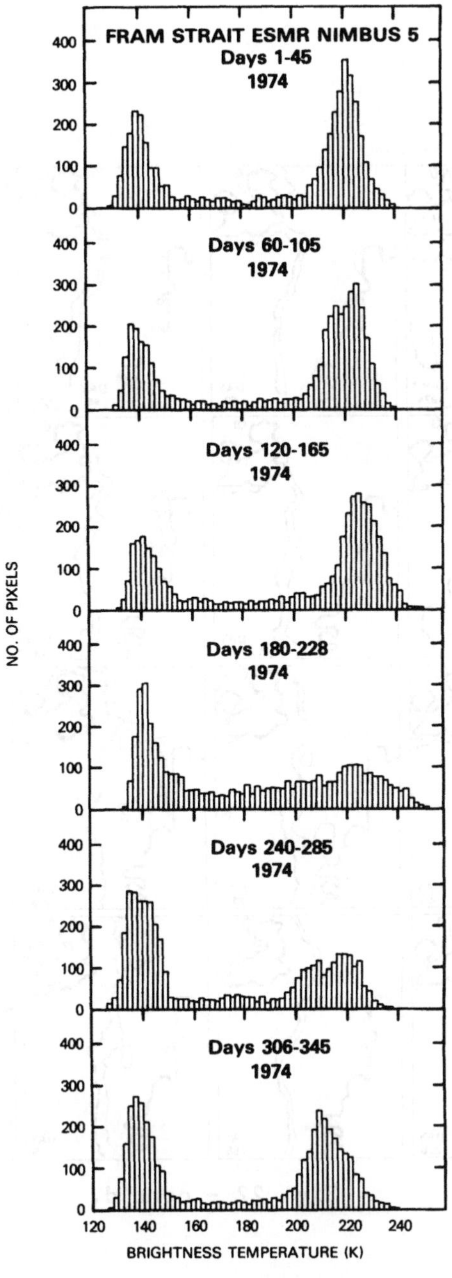

Fig. 22. Passive microwave data from Fram Strait. Shown are ice edge locations through 1974, some sample brightness distributions, and a single orbit swath in which the T_B value are placed in their geographical location. Isotherms at lower brightness values are difficult to construct. (Fig. 22 - cont'd.)

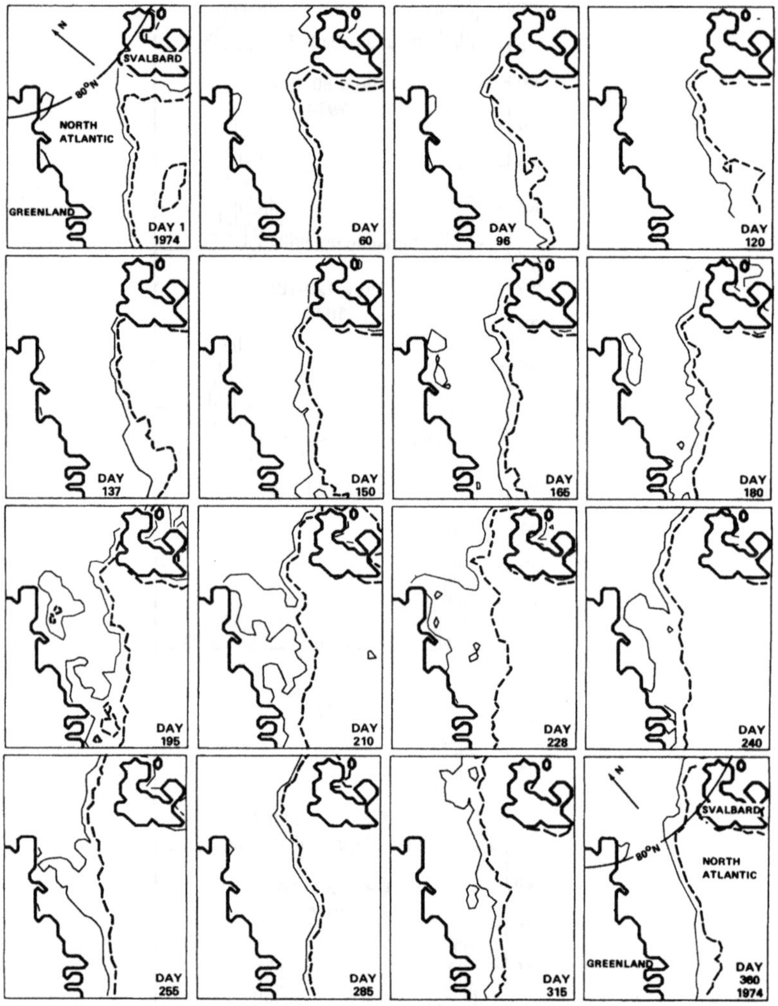
N
80°N
SVALBARD
NORTH ATLANTIC
GREENLAND
DAY 1 1974
DAY 60
DAY 96
DAY 120
DAY 137
DAY 150
DAY 165
DAY 180
DAY 195
DAY 210
DAY 228
DAY 240
DAY 255
DAY 285
DAY 315
DAY 360 1974

Fig. 22 - cont'd.

FRAM STRAIT DAY 195, 1974 ESMR NIMBUS 5

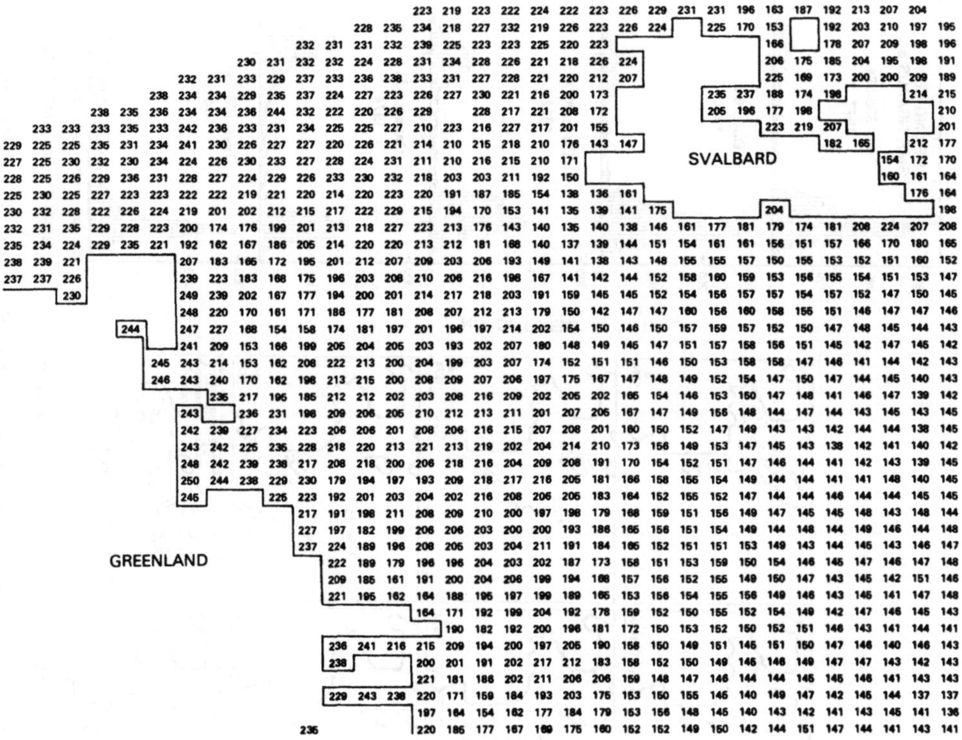

Fig. 22 - cont'd.

retreat season, but in general little can be determined except for the advance-retreat cycle itself.

Figures 24 and 25 show the results of on-going work by Martin, Cavalieri, McNutt, and Gloesen (private communication) in the Bering Sea using SMMR data from Nimbus 7 (still in operation at this writing). In Figure 24, is the SMMR determined ice concentration with research aircraft flight lines (solid lines) and ship stations (marked by C's) superimposed. Much structure in the ice pack is apparent. In Figure 25, Landsat-derived ice conditions are shown for the same period (McNutt, private communication).

The North-Water Polynya has been studied with visibile and IR methods for part of the freezing season (Aber and Vowinkel, 1972; Dunbar, 1973) and with passive microwave by Crawford and Parkinson (1981) whose results are shown in Figure 26. The dark heavy line on the abcissa indicates when the ESMR was operational; the

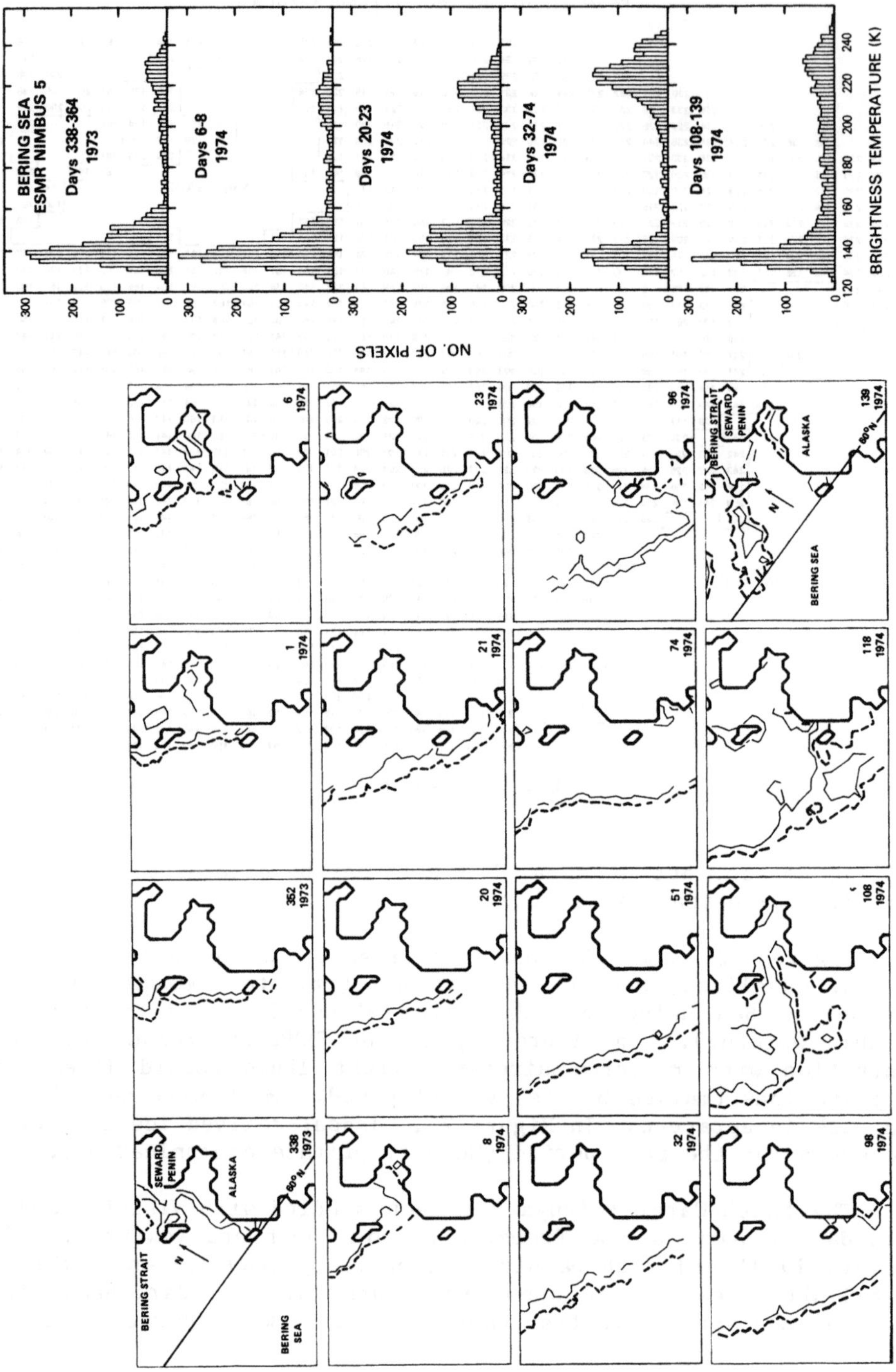

Fig. 23. Passive-microwave determined ice edge locations and sample brightness distributions for the Bering Sea in 1974 from ESMR, 19.35 GHz.

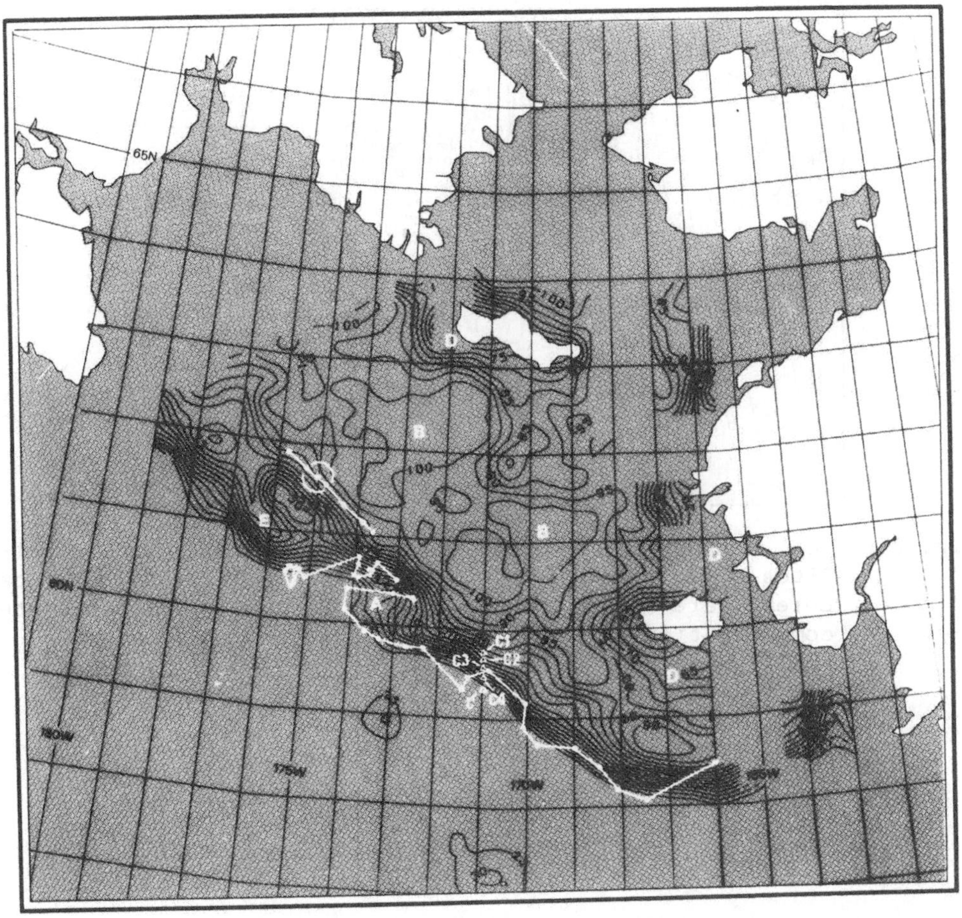

Fig. 24. Ice concentration results from Nimbus 7 passive microwave shown overlaid with aircraft tracks and ship stations. (Martin et al., 1982, private communication).

histogram show open water area. A single pixel has about 1000 km^2 -- this polynya is clearly resolvable by passive microwave sensor.

The connection between passive microwave characteristics and mechanical properties of sea ice is clearly important in both air-sea-ice interaction studies and engineering problems. Most mechanical properties depend strongly on brine volume (Schwarz and Weeks, 1977), which also influences microwave behavior (Vant et al. 1978), although it may be that different actual brine distributions are involved. In addition to scale problems, in which the ice features of importance are far smaller (< 1 cm) than observation resolution,

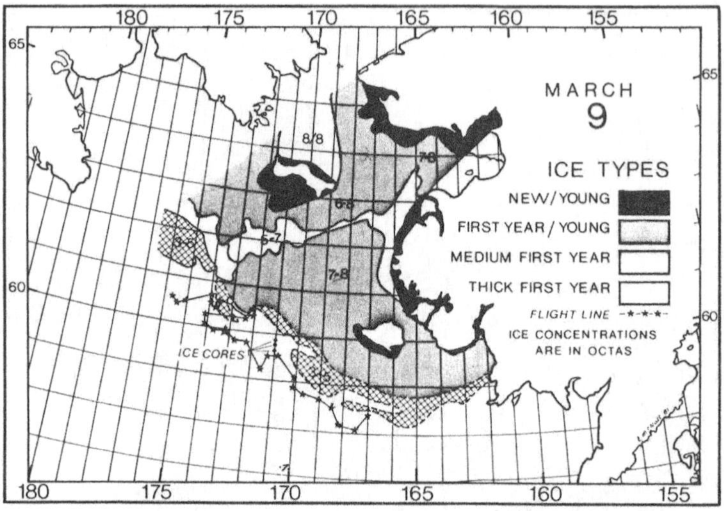

Fig. 25. Ice concentration results from Landsat and NOAA satellites. Aircraft flight lines and ship stations (ice cores) are also shown. (L. McNutt, 1982, private communication.)

there is a significant gap in theory and data on microwave characteristics of sea ice. Of course this situation is aggravated at the ice margin.

5.3 Imaging Radar. The imaging radar systems, Synthetic Aperture Radar (SAR) and Real-Aperture Radar (RAR/SLAR), because of data rate and field of view considerations, are generally not used for regional and synoptic ice-condition descriptions as was the case for visible/IR and passive microwave sensors, but instead are used for detailed location of ice features and description of pack morphology. Operational use of imaging radar, usually RAR but leaning toward SAR for resolution, consists of location and tracking of ice hazards and features. Figure 27 shows an aircraft SLAR (RAR) image (courtesy of AES ICE Centre of Canada) taken in Lancaster Sound, February 11, 1981, using an X-band SLAR manufactured commercially. A host of ice features is visible including new or young ice, the flaw lead, older floes, leads, ridges, small icebergs and, perhaps, grease ice forming in the eastern end of

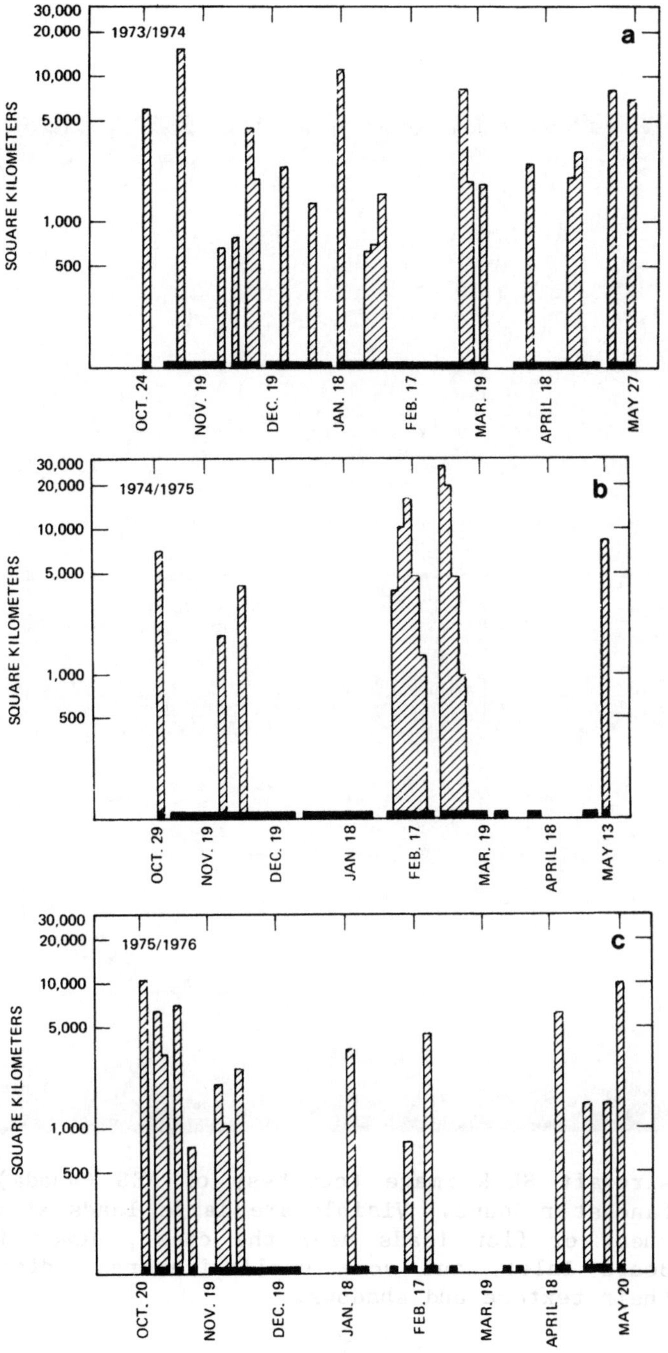

Fig. 26. North-water polynya area. The dark line indicates data availability. Data used were 19.35 GHz radiance recorded by ESMR on Nimbus 5. (Crawford and Parkinson, 1981.)

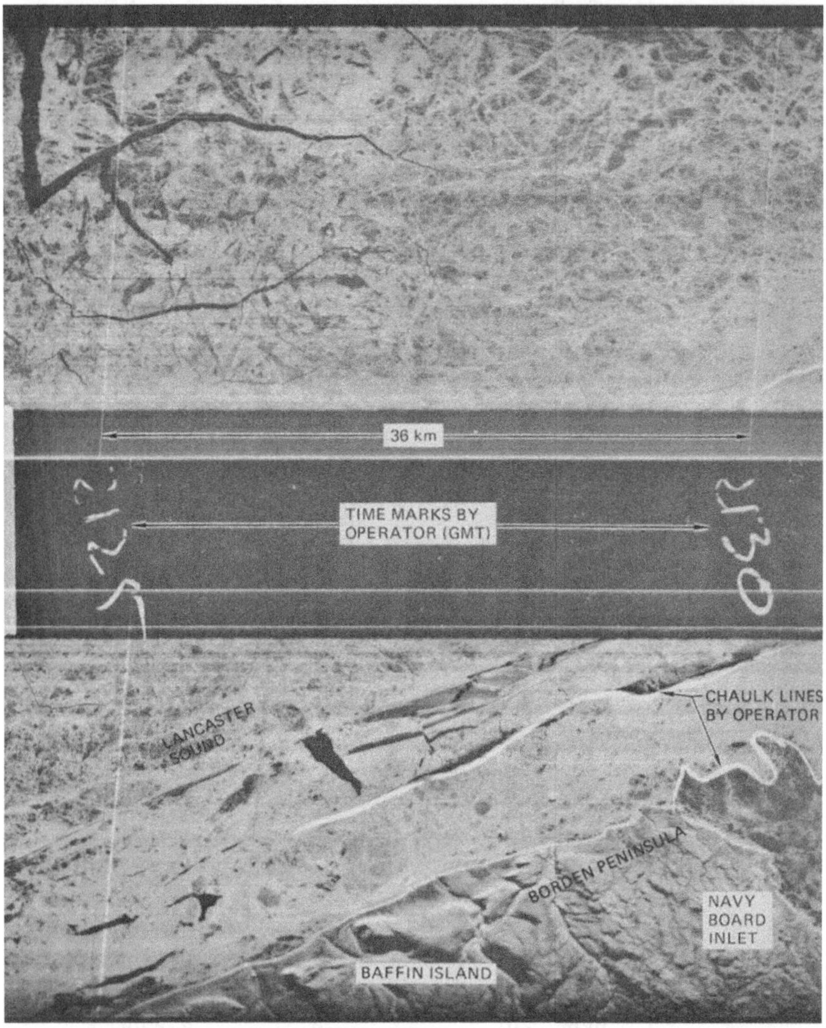

Fig. 27. Aircraft SLAR image (courtesy of AES Canada) of ice in Lancaster Sound. Visible are large leads at upper left, shear or flow leads near the coast, newer ice in Navy Board Inlet, and some small icebergs discernible by their texture and shadows.

the flaw lead. In operational ice work features which are problems for stationary platforms are bergs, large multiyear floes, and large ridges. In addition, large leads or areas of thin ice offer opportunities for navigation (Barnes and Bowley, 1981).

Although the promise of SAR and RAR for ice research is clearly great, geophysical application to problem solving has been limited. This is no doubt due to the high cost of data acquisition and manipulation, and the limited coverage of archived data. Demonstration calculations have been reported by Leberl et al. (1981) and a significant scientific contribution from analysis of SAR data from Seasat has been made by Hall and Rothrock (1981) examining high density ice motion data in the Beaufort Sea. It is known that significant deformation occurs on scales well below 50 km (Thorndike and Colony, 1977) but the exact structure of a motion field has not previously been established. Sample results are shown in Figure 28.

Hall and Rothrock did not extend their work to the ice margins because ice drift rates there were too large to permit high-density ice feature tracking. Thus, there is probably a coverage gap for ice-margin ice velocity work. There are, however, more important problems. Figures 29 and 30 are aircraft L-band SAR images from the Bering and Beaufort Seas (Bryan, 1976) and Figure 31 shows L-band data from the NORSEX area of Fram Strait (NORSEX Arctic Working Group, 1981). It is convenient to compare these with Figure 32 (Carsey, 1981) which shows Seasat images from the eastern Beaufort. Although the Bering and Fram Strait images are both of ice edges, they are very different. What they have in common is a lack of customary ice features clearly seen in Figures 31 and 32. The tracking of such features will be quite difficult. This is an area requiring much work.

Quantitative work with radar images will require calibrated radar systems and a knowledge of backscatter properties. This is particularly important at the ice margins when the ice pack is complex and rapidly changing.

6. MICROWAVE CHARACTERISTICS

It has been about 10 years since the first airborne passive microwave data from sea ice were published (Wilheit et al. 1972) and the important emissivity differences for Arctic first-year and multiyear ice discussed (Gloersen et al. 1973). Since that time, it has been widely hoped that differences in emissivity and radar backscatter among the various ice types and features would permit quantitative descriptions and estimates of occurrence. Although this capability remains the goal of microwave remote sensing of sea ice, the tabulation and explanation of specific emissivities and backscatter coefficients for various ice types and features as

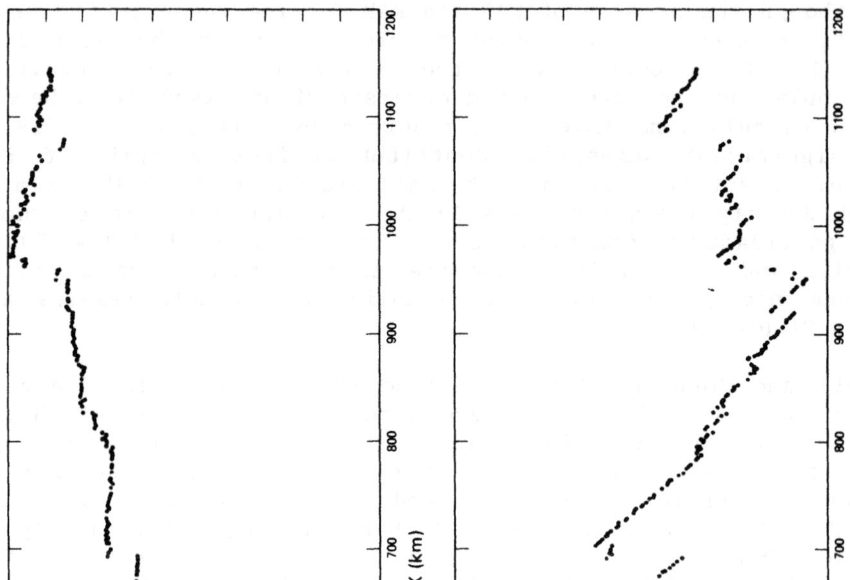

Fig. 28. Ice motion data from tracking of ice features in Seasat SAR data (Hall and Rothrock, 1981).

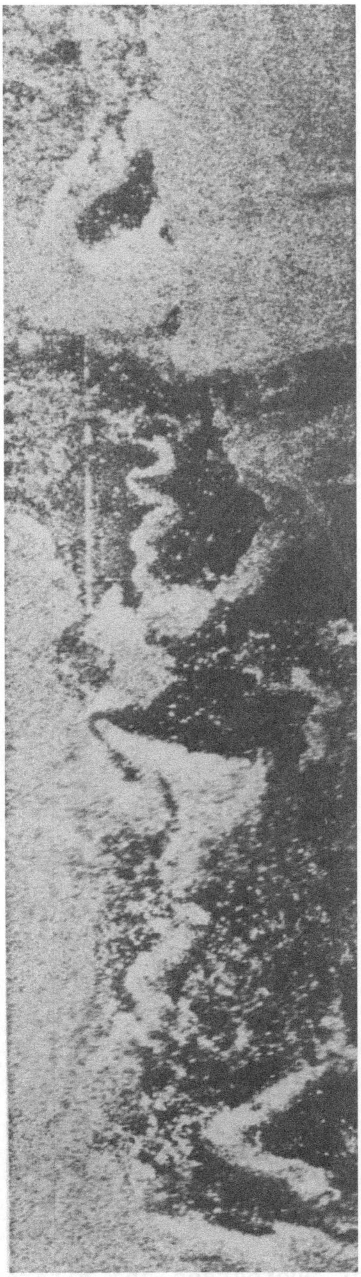

Fig. 29. Aircraft SAR image of the ice margin of the Bering Sea. The tendency of this ice margin to decay by production of snake-like bands is apparent. The open sea is dark. (From Bryan, 1976.)

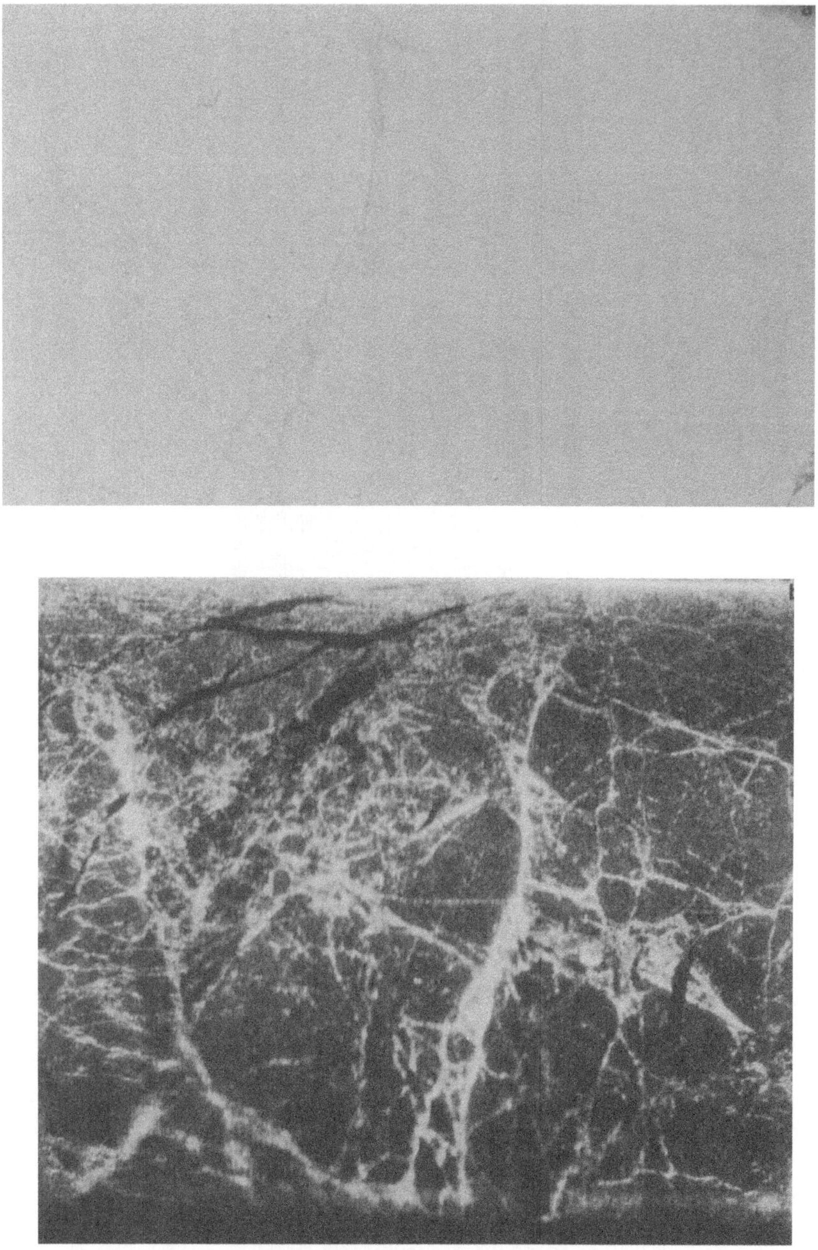

Fig. 30. Visible light photograph and aircraft SAR image of floes and ridges in the Beaufort Sea. More structure can be seen in (b), the SAR image.

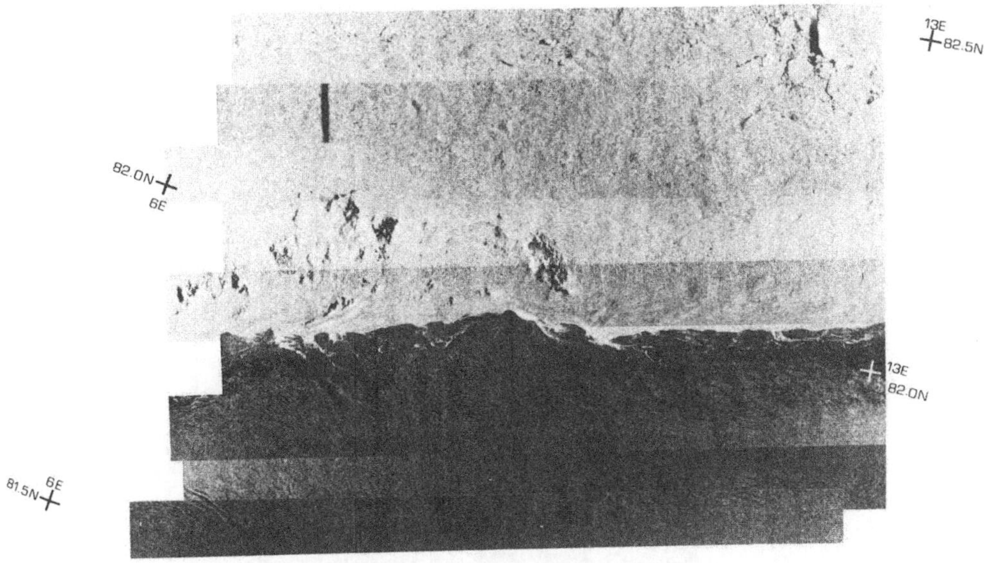

Fig. 31. Aircraft L band SAR image of the ice edge in Fram Strait. Open water is dark, and a strong wind is blowing right-to-left (~80°).

function of wavelength, incidence angle, season, snow cover and environment have turned out to be time consuming and difficult tasks. For example, as of this writing, only tentative values have been published (Cavalieri et al. 1984) of emissivity for the wavelenghts and incidence angles of the Nimbus 7 SMMR for "average" first-year and multiyear ice. Region-specific measurements, such as those of NORSEX (1981), are still needed. Finally, emissivity variations for given ice type have been observed to be large (NORSEX, 1981; Carsey, 1982, 1984; Troy et al. 1981), but there is no way at present to accomodate this problem or profit from this opportunity.

In an isothermal material the emissivity depends on 2 variables, the dielectric constant and the opacity, or absorption. The dielectric constant determines the amount of energy reflected at the material surface - the impedance match or efficiency of escape of the electromagnetic energy; the opacity determines the coupling between the total thermal energy in the various electronic and molecular states and the portion of the microwave spectrum involved. Thus, a substance such as freshwater ice has a total dielectric constant not too different from air or vacuum and thus has a high emissivity, but ice has a low absorption so that is is bright in the microwave only when it is thick enough to have a strong net interaction with the electromagnetic field (Swift, 1980). Otherwise, it is essentially transparent, and a radiometer

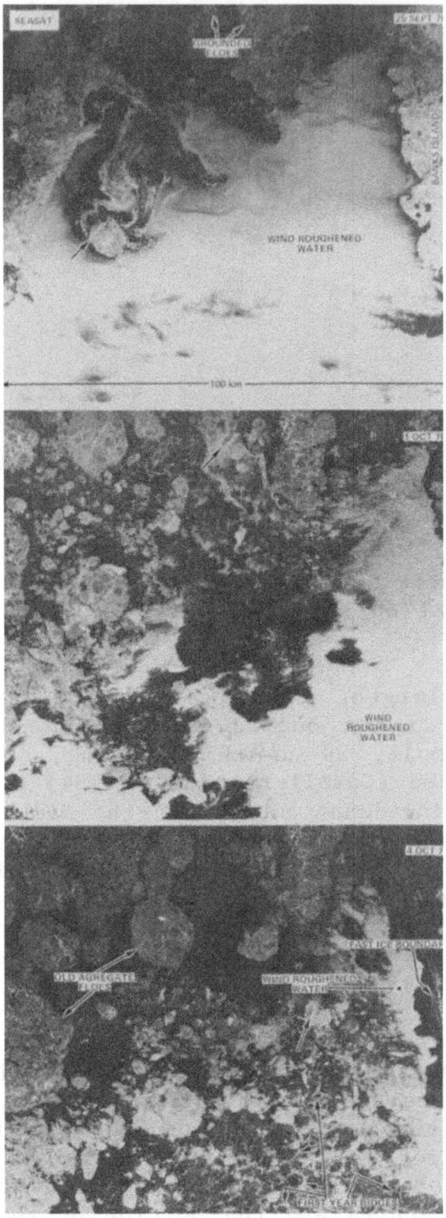

Fig. 32. Seasat L band SAR images from the Eastern Beaufort Sea. The arrow locates a floe visible in all three images. The distinct appearance of floes, ridges, open water, and new ice in these images is in puzzling contrast to the textured look of images in Figures 29, 30 and 31.

"sees" the microwaves incident on the lower surface of the ice with appropriate reflections at the fresh-ice surfaces.

The emissivity of sea ice cannot be convincingly calculated, and a discussion of the trial methods is beyond the scope of this review. The critical aspect of sea ice is its basic upper-layer structure, which consists of a matrix of fairly pure ice imbedded with roughly ellipsoidal brine pockets for first-year ice, and of air inclusions where brine has largely drained out for Arctic multiyear ice. The brine/air elements tend to be of sub-millimeter scale (Weeks and Ackley, this volume) on sub-centimeter spacing-scales similar to the microwave wavelengths of remote sensing. Emissivity computation consists of describing the exchange of microwave energy between the matrix and the inclusions until it emerges through the somewhat rough upper surface. First-year ice has a very large absorption so that only the ice within the top-most part-wavelength of depth emits microwaves, but multiyear ice has a freeboard portion with low salinity and thus, low absorption so that microwaves originate throughout, are emitted according to the local dielectric constant gradient, which is temperature dependent, and are reflected (essentially downward) by air inclusions (Gloersen et al., 1973; Vant et al., 1974). Snow, when dry and cold, has a very low absorption and an emissivity intermediate between sea ice and air. Thus, it acts to weakly match impedance between sea ice and air and increase the effective emissivity at longer wavelengths and as a scatterer at short wavelengths.

The emissivity of first-year ice can be modelled by assuming that the index of refraction at the surface can be approximated by a linear combination of that for brine and fresh ice. This method, essentially an assumption that all emission occurs in a thin skin at the surface, appears to give a satisfactory result. The emissivity of multiyear and first-year ice is under investigation by Fung and Kong and their co-workers employing radiative transfer and modified radiative transfer methods (Tsang and Kong, 1981; Kong et al. 1979) in which wave phase is carefully modeled. An approach that models the dielectric constant by mixing that of fresh ice and brine has been utilized by Vant and co-workers (Vant et al., 1978; Vant et al., 1974) for both ice types with some success. However, the result is very sensitive to the orientation of the brine elipsoids, and this is a bit troublesome since the actual radiance does not vary strongly with incidence angle.

The backscatter case is similar, but the emphasis is shifted, ice surface roughness being more important than volume dielectric constant variation. However, the backscatter is again related to the surface roughness, the average dielectric properties, and a measure of the fluctuations of the dielectric constant both vertically and horizontally. This area is under investigation by Fung, Kong and Parsahar among others.

This discussion points up the weakness of the current morphological description of ice as being either first-year or multiyear ice approximating structure as from Vant et al. (1978) shown in Figure 33. Clearly the brine content and (near-surface) distribution in first year ice is variable according to water surface conditions at freezing (see Pounder, 1965; Lake and Lewis, 1970; Eide and Martin, 1975). The structure of the freeboard portion of multiyear ice will depend on the warmth of the summers survived, the amount of freeboard, the amount of ice and snow which has melted and percolated and other factors. Ridges of various kinds will be far more complicated and different from frozen melt ponds. Clearly the exact formulation of microwave behavior of sea ice will be complex even before the complex environmental factors of the ice margin are included (NORSEX, 1981). Much is left to be done in the area of sea ice microwave properties.

Two promising avenues are now being explored in order to increase our knowledge of sea ice microwave properties. One of these is the comparison of various types of satellite data with each other using arguments of internal consistency and the known elements of seasonal change. This approach has led to important insights into emissivity variation among a given overall ice type by Comiso (1983) and Comiso et al. (1984). A similar method was used by Carsey (1985) comparing SAR and SMMR data, and by Gray et al. (1982) using aircraft active and passive data. Another approach has been that of comparing aircraft, satellite, and in situ data for given seasons and locales. This approach was used by Carsey (1985), Onstott and Gogineni (1985), Grenfell and Perovich (1985), Onstott et al. (1982), Hollinger et al. (1984), and others. At present the summer and some elements of the fall season have been examined with a combination of in situ and aircraft/spacecraft data. The remainder of the year needs examination as does other significant geographical regions, notably Antarctica.

7. ICE THICKNESS MEASUREMENT

In almost every interaction involving sea ice, there is a strong dependence on ice thickness. In fact, in the absence of snow cover even the short-wave energy budget is thickness dependent. At the same time the thickness of a given ice feature is very difficult to define although of course the pointwise thickness is simply measured. In effect, the definition of thickness is a bit different for every problem. This has been previously recognized, and for detailed examination a thickness distribution function has been defined (Thorndike et al. 1975; Thorndike and Maykut, 1973).

Actual auger measurements of ice thickness of a flat multiyear floe were carried out by Hibler et al. (1972). They found that a floe some 200 m x 300 m could be characterized in thickness

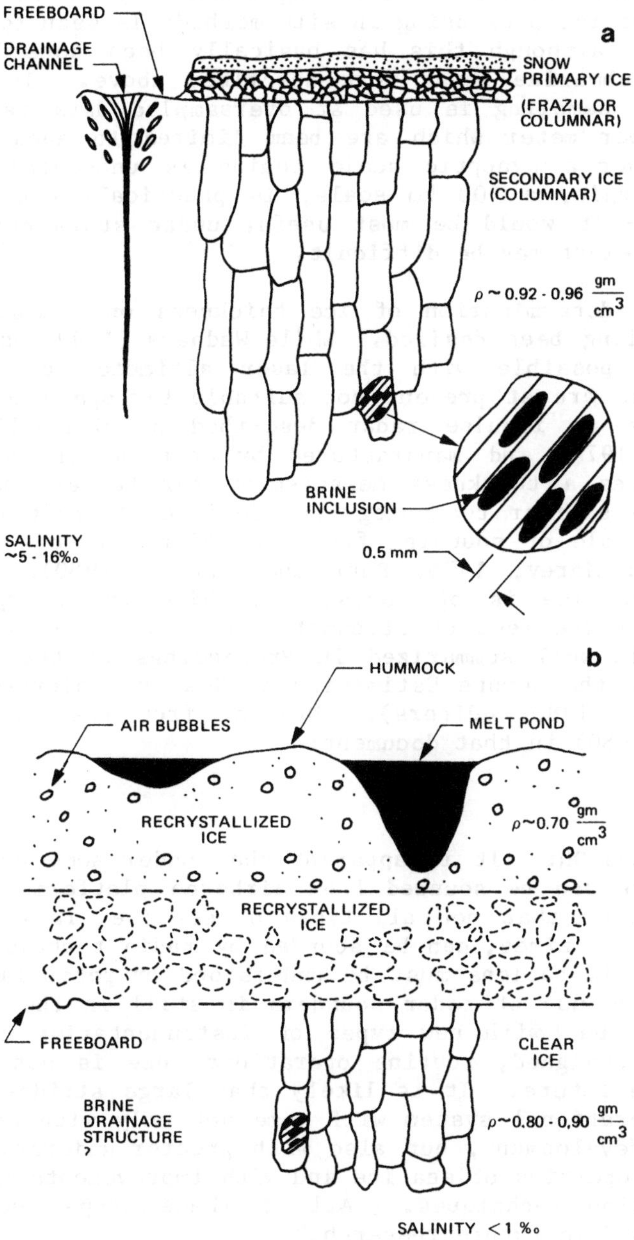

Fig. 33. Simplified geometry of (a) first year and (b) multi-year sea ice (Vant et al., 1978). This picture has now been somewhat modified (see Weeks and Ackley, this volume).

to $\pm$ 1/2 meter with 6 measurements. When ridges, leads, and new ice are added, the problem of designing a sampling scheme to adequately describe the ice pack using in situ methods is seen to be challenging indeed, although this has basically been done by the laser altimeter/submarine sonar work described above. In that work, 2 dimensional sampling is used at oversampled data rates of order 0.6 to 1 per meter which are beam limited to about 1 per 20 m. Sampling over a synoptic scale region is uneconomical; sampling the ice margin, a 100 km scale, is practical, but has not been attempted - it would be most useful under storm conditions when good measurement may be difficult.

Remote determination of ice thickness on almost any spatial scale has long been desired. While Wadhams (1980) and others have shown this possible with the laser altimeter and/or submarine sonar, these are at present not suitable for operational use. For surface use an impulse radar described by Campbell and Orange (1973 and 1974) and manufactured by Geophysical Survey Systems, Inc. produces a thickness measurement far faster than the auger method, and similar technology has lead to aircraft and helicopter systems capable of sounding from some 40 m altitude (Rossiter and Butt, 1979; Morey, 1975; Butt and Gamber, 1980). The complex structure of ice is of course a problem in interpreting radio echoes under the best of circumstances. The overall situation in this area is well summarized in Proceedings of the International Workshop on the Remote Estimation of Sea Ice Thickness, (Rossiter and Bazeley, 1980, editors). Drawing from the review paper by Rossiter (1980) in that document:

> "CONCLUSION: It is apparent that under some conditions sea ice can be sounded from airborne platforms. It is also true that not all types of sea ice, in all areas, or at all times, can be sounded by radio techniques, and these limitations must be understood by potential users. Airborne use of radar sounders is still in the research stage, but with new types of instrumentation currently being designed, routine operational use is not too far in the future. It is likely that large strides towards an operational system will come not only with new hardware development, but also with greater understanding of the properties of sea ice and with improvements in interpretation techniques. All of these steps need to be followed in future research."

Of course, a spacecraft thickness sounder cannot be conceived of before high altitude aircraft systems are operational.

8. SUMMARY

At the outset of the chapter (in Table 1) a fairly complete list of variables required to describe sea-air-ice interaction was compiled. Subsequent sections described the physical basis of remote sensing methods and discussed specific applications of remote sensing data to problems. It is appropriate but not satisfying to re-examine the variable list in the light of proven applications.

Snow Cover. There is no method known for snow cover estimation over ice. Some multi-spectral methods are partially successful for snow over land (Rango et ál. 1979; Hofer and Matzler, 1980), and some speculation suggest that higher frequencies, 90 GHz and above, will be useful. Because the ice surface temperature figures into the passive microwave calculations (eq. 3-6) some rough estimate of snow cover may be obtainable by comparing T_B spatial behavior with some kind of air temperature. The snow cover signal in early fall has been discussed by Carsey (1985).

Ice Thickness. There is no method for measurement of ice thickness from space, but from aircraft a laser altimeter recipe has been established by Wadhams (1980). There is little chance of a direct measure of ice thickness or any sort of spatial average ice thickness from aircraft cruise altitude or space.

Ice Type. Ice type (first-year, multiyear and first-year-thin) measurement in the Arctic is apparently obtainable by both active and passive microwave methods on various space scales. Although methods now in use have large errors, that situation should improve steadily. Variations of ice property within general type are now of interest, and research on how remote sensing could be an investigative tool should be initiated. Also the measurement of areal coverage of transitionphase ice types such as pancake ice should be attacked.

Ice Concentration. Ice concentration is really a special consideration under ice thickness, and both passive and active microwave systems appear to measure it, but suitable winter open-water coverages, of order 1%, are not resolvable. The measurement of temporal changes in fractional coverage of open water, thin ice, first-year ice, ridges and multiyear ice might be one way of developing a post-hoc thickness distribution - a bookkeeping job keeping track of the thin ice to ridge + open water process would be needed.

Ice Velocity Field. Ice velocity measurements can be made on a several-points-per-10 km basis, and the essential discontinuity of the velocity field is apparent for ice floes within the pack. Some problems may weaken the method at the margin and, under any circumstances, the technique is very labor intensive. Automation of data extraction is not foreseeable. Larger scale ice velocity is obtainable from buoy tracking by satellite in a straightforward manner.

Ocean Velocity Profile. The ocean velocity profile will not be obtainable via remote sensing in the even distant future. Buoys can and do record this information for satellite communication.

Surface Wind Vector Field. The surface wind vector within the pack is adequately obtained via analysis of surface pressure from environmental buoys. At the ice margin this method is untried, but the results from the Bering Sea are promising (Pease 1980). Precise direct wind measurements from space via laser radar are possible, but no data should be expected before the end of the century.

Air, Snow, and Ice Surface Temperatures. Air temperature is now obtained from buoys and by circulation modelling. The multichannel microwave sensors such as SMMR may be capable of an ice surface temperature, and IR instruments can measure snow surface temperature in winter (Comiso, 1983) thereby allowing heat flux calculations without resorting to air temperature. The newer visible/IR/microwave satellites may permit this calculation in all seasons.

Ocean Temperature. The sea surface temperature is measured by both IR and microwave methods, but both are severely contaminated when ice is in the field-of-view. Ocean temperature profiles must be acquired by buoy or other in situ method.

Data Bases. For all investigators access to data sets themselves and to some measure of data quality will remain a problem. For future field studies satellite and aircraft availability will continue to be troublesome. For those working with archived data the inevitable incompleteness of the record is a difficulty. The future of microwave remote sensing of ocean and ice is uncertain. However, some things are certain, much work with *in situ* instruments in the visible, IR, and microwave bands is needed; the historical record still can contribute to physical understanding; and only persistent, exacting work will prevail.

REFERENCES

Aber, P. G. and E. Vowinckel (1972) Evaluation of north water spring ice cover from satellite photographs. Arctic, 25: 263-271.

Ahlnas, K. and G. Wendler (1977) Arctic sea-ice conditions in early spring viewed by satellite. Arct. Alp. Res., 9: 61-72.

Andreas, E. L., C. A. Paulson, R. M. Lindsay and J. A. Businger 1979) The turbulent heat flux from arctic leads. Bound.-Layer Met., 17: 57-91.

Badgley, F. I. (1961) Heat balance at the surface of the Arctic Ocean. In Proceedings, Western Snow Conference, Spokane, Washington, pp. 101-104.

Barnes, J. C., D. T. Chang and J. H. Willand (1972) Application of ITOS and Nimbus infrared measurements to mapping sea ice. NOAA/NESS report.

Barnes, J. C. and C. J. Bowley (1981) Comparative analysis of sea ice features using side-looking radar (SLAR) and Landsat imagery. NASA report. (Available from Lewis Research Center, Cleveland, Ohio 44135.)

Barrick, D. E. and W. H. Peake (1968) A review of scattering from surfaces with different roughness scales. Radio Science, 3: 865-868.

Barry, R. G., R. E. Moritz and J. C. Rogers (1979) The fast-ice regimes of the Beaufort and Chukchi sea coasts, Alaska. Cold Reg. Sci. Tech., 1: 129-152.

Brown, G. S. (1978) Backscattering from a Gaussian-distributed perfectly conducting rough surface. IEEE Trans. Ant. Prop., AP-26: 472-482.

Brown, G. S. (1979) The estimation of wind speed using a space-borne radar altimeter. J. Geophys. Res., 84: 3974-3978.

Brown, G. S. (1982) A theory for near-normal incidence microwave scattering from first-year sea ice. Radio Science, 17: 233-243.

Bryan, M. L. (1977) Interpretation key for SAR (L-band) imagery of sea ice. In Proceedings, ASCM-ASP Convention, 1976, Seattle, Washington.

Bryan, L., T. Farr, F. Leberl and C. Elachi (1977) Synthetic aperture radar imagery of the AIDJEX triangle. AIDJEX Bulletin, 37: 161-end.

Butt, K. A. and J. B. Gamberg (1980) Technology of an airborne impulse radar. In Proceedings, International Workshop on Remote Estimates of Sea Ice Thickness. (Rossiter and Bazeley, Ed.), C-CORE, St. Johns, Newfoundland, p. 383-412.

Campbell, K. J. and A. S. Orange (1974) The electrical anisotropy of sea ice in the horizontal plane. J. Geophys. Res., 79: 5059-5063.

Campbell, W. J., P. Gloersen, R. O. Ramseier, W. J. Webster and T. T. Wilheit (1976) Beaufort Sea ice zones as delineated by microwave imagery. J. Geophys. Res., 81: 1103-1110.

Campbell, W. J., J. Wayenberg, J. B. Ramseyer, R. O. Ramseier, M. R. Vant, R. Weaver, A. Redmond, L. Arsenault, P. Gloersen, H. J. Zwally, T. T. Wilheit, T. C. Chang, D. Hall, L. Gray, D. C. Meeks, M. L. Bryan, F. T. Barath, C. Elachi, F. Leberl and T. Farr (1978) Microwave remote sensing of sea ice in the AIDJEX main experiment. Bound.-Layer Met., 13: 309-337.

Carsey, F. D. (1980) Microwave observation of the Weddell polynya. Mon. Wea. Rev., 108: 2032-2044.

Carsey, F. D. (1982) Arctic sea ice distribution at end of summer 1973-1976 from satellite microwave data. J. Geophys, Res., 87: 5809-5835.

Carsey, F. D. (1985) Summer arctic sea ice character from satellite microwave data. J. Geophys. Res., 90, 5015-5034.

Cavalieri, D. J., P. Gloersen and W. J. Campbell (1984) Determination of sea ice parameters with the Nimbus-7 SMMR. J. Geophys. Res., 89: 5355-5369.

Cavalieri, D. J. and C. L. Parkinson (1981) Large scale variations in observed Antarctic sea ice extent and associated atmospheric circulation. Mon. Wea. Rev., 109: 2323-2336.

Cavalieri, D. J., S. Martin, and P. Gloersen (1983) Nimbus 7 SMMR observations of the Bering Sea ice cover during March 1979. J. Geophys. Res., 88: 2743-2754.

Chernigovskii, N. T. (1976) Radiational properties of the ice cover of the central Ārctic. In Hydrometeorology of the Polar Regions (G. Y. Vangengeim and A. F. Laktionoff, Ed.). (Available from NSF, Washington, D.C.)

Comiso, J. C. and H. J. Zwally (1982) Antarctic sea ice concentrations inferred from Nimbus 5 ESMR and Landsat imagery. J. Geophys. Res., 87: 5836-5844.

Comiso, J. C. (1983) Sea ice effective emissivities from satellite passive microwave and infrared observations. J. Geophys. Rev., 80: 7686-7704.

Comiso, J. C., S. F. Ackley and A. L. Gordon (1984) Antarctic sea ice microwave signatures and their correlation with in situ ice observations. J. Geophys. Res., 89: 662-672.

Crane, R. G. (1978) Seasonal variations of sea ice extent in the Davis Strait - Labrador Sea area and relationships with synoptic-scale atmospheric circulation. Arctic 31: 434-447.

Crane, R. G., R. G. Barry and H. J. Zwally (1982) Analysis of atmosphere - sea ice interactions in the Arctic Basin using ESMR microwave data. Int. J. Rem. Sens., 3, 259-276.

Crawford, J. P., S. E. Kerpelman, P. R. Schaffner and H. A. Thompson (1979) NORSEX microwave remote sensing summary report. NASA Langley report.

Crawford, J. P. and C. L. Parksinson (1981) Wintertime and microwave observations of the north water polynya. In Oceanography from Space (J. F. R. Grower, Ed.), Plenum, New York, p. 839-844.

Curlander, J. C. (1982) Location of spaceborne SAR imagery. IEEE Trans. Geosci. & Remote Sens., GE-20: 359-364.

Curry, J. (1983) On the formation of continental polar air. J. Atmos. Sci., 40: 2278-2292.

DeRycke, R. J. (1973) Sea ice motions off Antarctica in the vicinity of the Eastern Ross Sea as observed by satellite. J. Geophys. Res., 78: 8873-8879.

Deschamps, P. Y. and T. Phulpin (1980) Atmospheric correction of infrared measurements of sea surface temperature using channels at 3.7, 11 and 12 m. In Passive Radiometry of the Ocean, Proceedings, 6th IUCRM, Bound.-Layer Met., 18: 131-144.

Dey, B. (1980a) Applications of satellite thermal infrared images for monitoring north water during the periods of polar darkness. J. Glaciol., 25: 425-438.

Dey, B. (1980b) Seasonal and annual variations in ice cover in Baffin Bay and Northern Davis Strait. Canadian Geographer, 24: 308-384.

Dey, B. (1980c) Sea ice cover and related atmospheric conditions in arctic Canada during the summer of 1978. Mon. Wea. Rev., 108: 2092-2097.

Dey, B. (1981) Monitoring winter sea ice dynamics in the Canadian Arctic with NOAA-TIR images. J. Geophys. Res., 86: 3223-3235.

Dunbar, M. (1975) Interpretation of SLAR imagery of sea ice in Nares Strait and the Arctic Ocean. J. Glaciol., 15: 193-213.

Dunbar, M. and W. F. Weeks (1975) Interpretation of young ice forms in the Gulf of St. Lawrence using side-looking airborne radar and infrared imagery. U.S.A. Cold Regions Research and Engineering Laboratory, Research Report 337.

Elachi, C. (1980) Spaceborne imagery radar: Geologic and oceanographic applications. Science, 209: 1073-1082.

Eide, L. I. and S. Martin (1975) The formation of brine drainage features in young sea ice. J. Glaciol., 14.

Gloersen, P., W. Nordberg, T. J. Schmugge, T. T. Wilheit and W. J. Campbell (1973) Microwave signatures of first-year and multiyear sea ice. J. Geophys. Res., 78: 3564-3572.

Gloersen, P., R. Ramseier, W. J. Campbell, T. C. Chang and T. T. Wilheit (1974a) Variation of ice morphology of selected mesoscale test areas during the Bering Sea experiment. In Results of the US Contribution to the Joint US/USSR Bering Sea Experiment, Goddard Space Flight Center, NTIS X-910-74-141.

Gloersen, P., T. T. Wilheit, T. C. Chang, W. Nordberg and W. J. Campbell (1974b) Microwave maps of the polar ice of the earth. Bull. Am. Met. Soc., 55: 1442-1448.

Gloersen, P. and F. Barath (1977) A scanning multichannel microwave radiometer for Nimbus G and Seasat. IEEE J. Oceanic Eng., OE-2: 172-178.

Gloersen, P., H. J. Zwally, T. C. Chang, D. K. Hall, W. J. Campbell and R. O. Ramseier (1978) Time dependence of sea ice concentration and multiyear ice fraction in the Arctic Basin. Bound.-Layer Met., 13: 339-359.

Gray, A. L., R. K. Hawkins, C. E. Livingstone, L. D. Arsenault and W. M. Johnstone (1982) Simultaneous scatterometer and radiometer measurements of sea ice microwave signatures. IEEE Trans. Oceanic Engr. OE-7: 20-33.

Grenfell, T. C. and D. K. Perovich (1984) Spectral albedos and incident spectral irradiance over arctic sea ice. J. Geophys. Res., 89: 3573-3580.

Grenfell, T. C. and A. W. Lohaneck (1985) Temporal variations of the microwave signatures of sea ice during the late spring and early summer near Mould Bay NWT. J. Geophys. Res., 90, 5035-5044.

Hakkenen, S. (1984) Dynamics of the coupled ice ocean system in the marginal ice zone: study of the mesoscale processes and constitutive equations for sea ice. Ph.D. dissertation, Florida State University, unpublished.

Hall, R. T. (1978) Seasonal photo mosaics of the AIDJEX triangle. AIDJEX Bulletin 39: 79-84.

Hall, R. T. (1980a) A test of the AIDJEX ice model using Landsat images. In Sea Ice Processes and Models (R. S. Pritchard, Ed.), University of Washington Press, Seattle, Washington, p. 89-101.

Hall, R. T. (1980b) AIDJEX modelling group studies involving remote sensing data. In Sea Ice Processes and Models (R. S. Pritchard, Ed.), University of Washington Press, Seattle, Washington, p. 151-162.

Hall, R. T. and D. A. Rothrock (1981) Sea ice displacement from Seasat synthetic aperture radar. J. Geophys. Res., 86: 11078-11082.

Herman, G. and R. Goody (1976) Formation and persistance of summertime arctic stratus clouds. J. Atmos. Sci., 33: 1537-1553.

Hibler, W. D., W. F. Weeks and S. J. Mock (1972a) Statistical aspects of sea ice ridge distributions. J. Geophys. Res., 77: 5954-5970.

Hibler, W. D., S. Ackley, W. F. Weeks and A. Kovacs (1972b) Top and bottom roughness of a multiyear ice floe. AIDJEX Bulletin, 13: 77-92.

Hibler, W. D., S. J. Mock and W. B. Tucker (1974a) Classsification and variation of sea ice ridging in the Western Arctic Basin. J. Geophys. Res., 79: 2735-2743.

Hibler, W. D., S. F. Ackley, W. K. Crowder, H. L. McKim and D. M. Anderson (1974b) Analysis of shear zone ice deformation in the Beaufort Sea using satellite imagery, Coast and Shelf of the Beaufort Sea (J. C. Reed and J. E. Sater, Ed.), Arctic Institute of North America, p. 285-300.

Hofer, R. and C. Matzler (1980) Investigations on snow parameters by radiometry in the 3 to 60 mm wavelength region. J. Geophys. Res., 85: 453-460.

Hollinger, J. P., B. E. Troy, R. O. Ramseier, K. W. Asmus, M. F. Hartman, and C. A. Luther (1984) Microwave emission from high Arctic sea ice during freeze-up. J. Geophy. Res., 89: 8104-8122.

Jayaweera, K. O. L. F., (1976) Use of enhanced infrared satellite imagery for sea ice and oceanographic studies. Ocean Engineering, 3: 293-298.

Jensen, H., L. C. Graham, L. J. Porcello and E. N. Leith (1977) Side looking airborne radar. Scientific American, 237: 84-95.

Jones, W. L., P. G. Black, D. M. Boggs, E. M. Bracalente, R. A. Brown, G. Dome, J. A. Ernst, I. M. Halberstam, J. E. Overland, S. Peteherych, W. J. Pierson, F. J. Wentz, P. M. Woiceshyn and M. G. Wurtele (1979) Seasat scatterometer results of the Gulf of Alaska Workshop. Science, 204: 4131-4915.

Killworth, P. D. (1979) On "chimney" formations in the ocean. J. Phys. Oceanogr., 9: 531-554.

Knapp, W. W. (1969) A satellite study of large stationary polynyas in Antarctic coastal waters. University of Wisconsin, Ph.D. thesis.

Kong, J. A., R. Shin, J. C. Shine and L. Tsang (1979) Theory and experiment for passive microwave remote sensing of snowpacks. J. Geophys. Res., 84: 5669-5673.

Lake, R. A. and E. L. Lewis (1970) Salt rejection by sea ice during growth. J. Geophys. Res., 75: 583-597.

Langleben, M. P. (1968) Albedo measurements of an arctic ice cover from high towers. J. Glaciol., 7: 289-297.

Langleben, M. P. (1971) Albedo of melting sea ice in the southern Beaufort Sea. J. Glaciol., 10: 101-104.

Larsen, P. and S. Orwig (1962) Albedo of arctic surfaces. McGill University, Report AFERL-62-1069, 33 pp.

Larsen, R. W., J. D. Lyden, R. A. Shuchman and R. T. Lowry (1981) Determination of backscatter characteristics of sea ice using synthetic aperture radar data. Environmental Research Institute of Michigan, Ann Arbor, Michigan, ERIM Report 142600-1-F.

Leberl, F., M. L. Bryan, C. Elachi, T. Farr and W. J. Campbell (1979) Mapping of sea ice and measurement of its drift using aircraft synthetic aperture radar images. J. Geophys. Res., 84: 1827-1835.

Leberl, F., J. Roggam, C. Elachi and W. J. Campbell (1983) Sea ice motion measurements from Seasat SAR images. J. Geophys. Res., 88(C3): 1915-1928.

Lipes, R. G. (1982) Description of seasat radiometer status and results. J. Geophys. Res., 87: 3385-3396.

Lowry, R. T. and C. J. Brocher (1978) An interactive correction and analysis system for airborne laser profiles of sea ice. Can. J. Rem. Sens., 4: 149-160.

Marko, J. R. and R. E. Thomson (1977) Rectilinear leads and internal motions in the ice pack of the western Arctic Ocean. J. Geophys. Res., 82: 979-987.

Marshunova, M. S. and N. T. Chernigovsky (1966) Numerical characteristics of the radiation regime in the Soviet Arctic. In Proceedings, Symposium on the Arctic Heat Budget and Atmospheric Circulation (J. O. Fletcher, Ed.), Rand Corp., Santa Monica, California.

Martin, S., D. Cavalieri, P. Gloersen and L. McNutt (1982) Bering Sea ice cover during March, 1979: Comparison of surface and satellite data with Nimbus 7 SMMR. University of Washington, Technical Report.

Maykut, G. A. (1978) Energy exchange over young sea ice in the central Arctic. J. Geophys. Res., 83: 3646-3658.

Maykut, G. A. and N. Untersteiner (1971) Some results from a time dependent thermodynamic model of sea ice. J. Geophys. Res., 76: 1550-1575.

McClain, E. P. (1973) Quantitative use of satellite vidicon data for delimiting sea ice conditions. Arctic, 26: 44-57.

McClain, E. P. (1974) Earth satellite measurements as applied to sea ice problems. In Approaches to Earth Survey Problems Through Use of Space Techniques (P. Bock, Ed.), COSPAR, Constance, F.R.G. May 1973, Akademie-Verlag, Berlin, p. 149-163.

McClain, E. P. (1975) Potential value of earth satellite measurement to oceanographic research in the Southern Ocean. NOAA TM NESS 61.

McNutt, L. (1981) Ice conditions in the Eastern Bering Sea from NOAA and Landsat imagery: Winter conditions 1974, 1976, 1977, 1979. NOAA TM-ERL, PMEL-24, NOAA/PMEL, Seattle, Washington 98105.

McPhee, M. G. (1982) Sea ice drag laws and simple boundary layer concepts, including applications to rapid melting. CRREL, CR 82-4.

Moore, R. K. and W. J. Pierson (1966) Measuring sea state and estimating surface winds from a polar orbiting satellite. In Proceedings, International Symposium on Electromagnetic Sensing of Earth from Satellites, R1-R18.

Moore, R. K. and J. D. Young (1977) Active microwave measurement from space of sea-surface winds. IEEE J. Oceanic Eng., OE-2: 309-317.

Morey, R. M. (1975) Airborne sea ice thickness profiling using an impulse radar. USCG Technical Report CG-D-178-75.

Muench, R. D. and R. L. Charnell (1977) Observations of medium-scale features along the seasonal ice edge in the Bering Sea. J. Phys. Oceanogr., 7: 602-606.

NORSEX Arctic Working Group (1983) The Norwegian Remote Sensing Experiment (NORSEX) in a marginal ice zone. Science 220(4559): 781-787.

Nye, J. F. (1975) A test of the ice thickness redistribution by measurements of ERTS pictures. AIDJEX Bulletin, 28: 141-150.
Onstott, R. G., R. K. Moore, S. Gogineni and C. Delker (1981) Four years of low altitude sea ice broadband backscatter measurements. IEEE J. Oceanic Engr. OE-7: 44-50.
Onstott, R. G. and S. P. Gogineni (1985) Active microwave measurements of Arctic sea ice under summer conditions. J. Geophys. Res., 90, 5035-5044.
Overland, J. E. and C. H. Pease (1982) Cyclone climatology of the Bering Sea and its relation to sea ice extent. Mon. Wea. Rev. 110(1): 5-13.
Parkinson, C. L. and W. M. Washington (1979) A large-scale numerical model of sea ice. J. Geophys. Res., 84: 311-337.
Poulin, A. O. (1975) Significance of surface temperature in the thermal infrared sensing of sea and lake ice. J. Glaciol., 15: 277-283.
Pounder, E. R. (1965) The Physics of Ice. Pergamon, New York.
Pease, C. H. (1980) Eastern Bering Sea ice processes. Mon. Wea. Rev. 108: 2015-2033.
Pritchard, R. S., R. W. Reimer and M. D. Coon (1980) Ice flow through straits. In POAC 79: Proceedings, Fifth International Conference on Port and Ocean Engineering Under Arctic Conditions, University of Trondheim, Norway.
Ramseier, R. O., P. Gloersen, W. J. Campbell and T. C. Chang (1974) Mesoscale description for the principal Bering Sea experiment. In Results of the U.S. Contribution to the Joint US/USSR Bering Sea Experiment.
Rango, A., T. C. Chang and J. L. Foster (1979) The utilization of spaceborne microwave radiometers for monitoring snowpack properties. Nordic Hydrology, 10: 25-40.
Reeves, R. G., A. Anson and D. Landen (1975) Manual of Remote Sensing. American Society of Photogrammetry, Fall Church, Virginia.
Rossiter, J. R. and K. A. Butt (1979) Remote estimation of the properties of sea ice, Beaufort Sea field trip report. In C-Core Data Report.
Rossiter, J. R. (1980) Review of impulse radar sounding of sea ice. In Proceedings, International Workshop on the Remote Estimation of Sea Ice Thickness (Rossiter and Bazeley, Ed.), C-Core, St. Johns, Newfoundland, Canada.
Schwarz, J. and W. F. Weeks (1977) Engineering properties of sea ice. J. Glaciol., 19: 499-513.
Shapiro, L. H. (1976) A preliminary study of ridging in landfast ice at Barrow, Alaska using radar data. In Proceedings, Third International Conference on Port and Ocean Engineering Under Arctic Conditions, University of Alaska, Fairbanks, Alaska, U.S.A., p. 417-425.

Shapiro, L. H. and J. J. Burns (1975) Satellite observations of sea ice movement in the Bering Strait region. In Climate of the Arctic (G. Weller and S. Bowling, Ed.), University of Alaska, p. 379-386.

Sodhi, D. S. (1977) Ice arching and the drift of pack ice through restricted channels. U.S.A. Cold Regions Research and Engineering Laboratory, CRREL Report 77-18.

Susskind, J. J. Rosenfield, D. Renter and M. T. Chahine (1982) The GLAS physical inversion method for analysis of HIRS/MSU sounding data. NASA Tech. Memo 84936, avail. GSFC, Greenbelt, MD 20771.

Svendren, E., K. Kloster, B. Farrelly, O. M. Johannessen, J. A. Johannessen, W. J. Campbell, P. Gloersen, D. Cavalieri and C. Maetzler (1983) Norwegian Remote Sensing Experiment: Evaluation of the Nimbus 7 Scanning Multichannel Microwave Radiometer for sea ice research. J. Geophys. Res., 88: 2781-2792.

Tabata, T. (1972) Radar network for drift ice observation on Hokkaido. In Sea Ice (T. Karlson, Ed.), National Research Council of Iceland.

Tabata, T., T. Kawamura and M. Aoto (1980) Divergence and rotation of an ice field off Okhotsk sea coast of Hokkaido. In Sea Ice Processes and Models (R. S. Pritchard, Ed.), University of Washington Press, Seattle, Washington 98195.

Tsang and Kong (1981) Energy conservation for reflectivity and transmissivity at a very rough surface. J. Appl. Phys., 51: 673-680.

Thorndike, A. S. and G. A. Maykut (1973) On the thickness distribution of sea ice. AIDJEX Bulletin, 21: 20-21.

Thorndike, A. S., D. A. Rothrock, G. A. Maykut and R. Colony (1975) The thickness distribution of sea ice. J. Geophys. Res., 80: 4501-4513.

Thorndike, A. J. and R. Colony (1977) Estimating the deformation of sea ice. AIDJEX Bulletin, 37: 25-36.

Troy, B. E., J. P. Hollinger, R. M. Lerner and M. M. Wisler (1981) Measurement of the microwave properties of sea ice at 90 GHz and lower frequencies. J. Geophys. Res., 06: 4283-4289.

Ulaby, F. T., R. K. Moore and A. K. Fung (1982) Microwave Remote Sensing, Vol. I and II. Addison Wesley.

Vant, M. R., R. B. Gray, R. O. Ramseier and V. Makios (1974) Dielectric properties of fresh and sea ice at 10 and 35 GHz. J. Appl. Phys., 45: 4712-4718.

Vant, M. R., R. O. and Ramseier, V. Makios (1978) The complex dielectric constant of sea ice at frequencies in the range 0.1 - 40 GHz. J. Appl. Phys. 49: 1264-1280.

Vinje, T. E. (1976) Sea ice conditions in the European sector of the marginal seas of the Arctic, 1966-1975. In Norsk Polarinstitutt Arbok 1975, Oslo, p. 163-174.

Vinje, T. E. (1978) Sea ice conditions and drift of Nimbus-6 buoys in 1978. In Norsk Polarinstitutt Arbok 1978, Oslo, p. 57-66.

Vinje, T. E. (1981) Private communication. Presented at NATO ASI, Maratea, Italy.

Weeks, W. F., W. B. Tucker, M. Frank and S. Fungcharoen (1980) Characterization of surface roughness and floe geometry of sea ice over the continental shelves of the Beaufort and Chukeki Seas. In Sea Ice Processes and Models (R. S. Pritchard, Ed.), University of Washington Press, Seattle, Washington, p. 300-312.

Weeks, W. F. and S. F. Ackley (1984). The growth, structure, and properties of sea ice. This volume.

Wadhams, P. (1980) A comparison of sonar and laser profiles along corresponding tracks in the Arctic Ocean. In Sea Ice Processes and Models (R. S. Pritchard, Ed.), University of Washington Press, Seattle, Washington.

Wadhams, P. (1981) Sea ice topography of the Arctic Ocean in the region 70°W to 25°E. Phil. Trans. Roy. Soc., London, 203: 45-85.

Wadhams, P. and R. H. Horne (1980) An analysis of ice profiles obtained by submarine sonar in the Beaufort Sea. J. Glaciol., 25: 401-424.

Wendler, G. (1973) Sea ice observation by means of satellite. J. Geophys. Res., 78: 1427-1448.

Wilheit, T. T. (1972) The electrically scanning microwave radiometer (ESMR) experiment. In Nimbus 5 User's Guide (R. R. Sabatini, Ed.).

Wilheit, T. T., W. Nordberg, J. Blinn, W. J. Campbell and A. Edgerton (1972) Aircraft measurements of microwave emission from arctic sea ice. Rem. Sens. Envir., 2: 129-239.

Wilheit, T. T. and A. T. C. Chang (1980) An algorithm for retrieval of ocean surface an atmospheric parameters from the observations of SMMR. Radio Science, 15: 525-544.

Williams, E., C. W. M. Swithinbank and G. de Q. Robin (1975) A submarine sonar study of arctic pack ice. J. Glaciol., 15: 349-362.

Zwally, H. J. and P. Gloersen (1977) Passive microwave images of the polar regions and research applications. Polar Record, 18: 431-450.

Zwally, H. J., F. D. Carsey, P. Gloersen, W. J. Campbell and R. O. Ramseier (1979) Antarctic sea ice variations 1973-1975. In Fourth NASA Weather and Climate Review (E. R. Kreins, Ed.), p. 335-340.

Zwally, H. J. and J. C. Comiso (1982) Antarctic sea ice concentrations inferred from Nimbus 5 ESMR and Landsat imagery. J. Geophys. Res., 87(C8): 5836-5844.

Zwally, H. J., J. C. Comiso, C. L. Parkinson, W. J. Campbell, F. D. Carsey and P. Gloersen (1983) Antarctica Sea Ice 1973-76, NASA, Washington, D.C., 206 p. (Supt. of Doc., USGPO, Washington, D.C. 20402).

APPENDIX. DATA COLLECTION AND LOCATION PLATFORMS

Surface measurements that are complimentary or comparative with remote sensing measurements are obtainable by *in situ* platforms. Generally, the platforms perform two functions. Data is collected from nearby sensors such as thermistors and is relayed to interrogating satellites along with information to determine the location of the platform. Successive measurements of location provide velocities of the platform. Platforms may be located on land or ships or may be buoys that free-drift with ocean currents or sea ice floes. In this appendix the capabilities and limitations of two existing systems and a proposed Advanced Location and Data Collection System (ALDCS) are described. The ALDCS design allows lower-cost platforms, which is desirable since many platforms are intended to be nonrecoverable.

Two satellites are currently in operation with capabilities for in-situ surface data acquisition: Geostationary Operational Environmental Satellite (GOES) Data Collection System (DCS), operated by the National Oceanic and Atmospheric Administration (NOAA), and the TIROS/ARGOS system, operated by the Service Argos for the French Centre National d'Etudes Spatiale.

Geostationary Operational Environmental Satellite (GOES) Data Collection System (DCS)

The GOES DCS is included on two geostationary satellites located over the equator with continuous view of most of the Atlantic and Pacific Oceans. The far Western Pacific and Indian Oceans are not covered. Coverage at the high latitudes (above 75°) is also poor due to the low angle above the horizon presented by the satellites. Presently, no position determination of drifting buoys is provided by the system. On-board radionavigation positions could be transmitted with the data stream, but this technique would require the addition of radionavigation receiver/processors at a cost of several thousand dollars per buoy. Surface platforms may transmit on a timed basis or be commanded to transmit the data to the satellite. GOES-DCS platforms with omnidirectional transmitting antennas have been used on buoys to avoid motion-related antenna pointing problems. The reduced gain provided by these antennas requires compensating high power (75 watt) transmitters. Use of these high power transmitters significantly increases buoy costs and requires relatively expensive battery packs.

TIROS/ARGOS System

The TIROS/ARGOS system is included on the NOAA low-latitude weather satellites which are in 860 km altitude near-polar sun-synchronous orbits. They view the entire earth twice per day (more frequently at high latitudes), once during the day and once during the night. Two satellites located in different orbits planes are always in operation, increasing the available coverage. At the equator, the elapsed time between successive viewing of a surface platform by the two-satellite ARGOS system average six hours. At 40° latitude, this reduces to about four hours. Typically, the time delay between remote sensor observations by another sun-synchronous satellite and surface data acquisition by TIROS/ARGOS would average two hours at 40° latitude. These delays exceed some needs for concurrent surface measurements of, for example, winds and waves. Alternatively, data could be acquired by a buoy during the remote sensing satellite overpasses and stored on the buoy until the TIROS comes along; however, this process increases data delivery delay time and increase buoy cost. Therefore, a data collection and location system on the remote sensing satellite is desirable to obtain concurrent surface and remote sensing measurements.

The ARGOS cannot provide concurrent position locations, especially in the case of research ship platforms and balloons, because the velocity estimate requires coverage on two successive orbits and assumes constant speed and direction of the platform between passes. ARGOS employs the traditional Doppler frequency measurement techniques to estimate platform position. Buoy platform motion effects contribute position errors which can only be removed with knowledge of platform velocity.

The ARGOS platform transmitter is a relatively simple, low power device which transmits four sensor data samples per frame and as many as eight frames per transmission. The system can accommodate approximately 200 platforms in view of the single frame transmission mode, fewer in the eight frame mode. The ARGOS platform transmitter requires an expensive, highly stable oscillator and closely-controlled transmitter power.

Advanced Location and Data Collection System

The ALDCS concept comprises an advanced on-board data collection system with a one-pass *in situ* data acquisition and platform location capability. It is a proposed system having innovations that overcome all described shortcomings of the present alternatives and could be implemented on future satellites. *In situ* platform sensor data are transmitted to the spacecraft where they are detected and combined with on-board measurements of the platform transmitter frequency and phase. The combined data are relayed to a surface data processing facility where *in situ* platform position locations are calculated from the frequency and phase measurements. The *in situ* measurements and their point of origin are then distributed to various users.

The ALDCS design employs a combination Doppler-frequency measurement and a radiofrequency-interferometer phase-difference measurement technique for position location to allow these functions to be performed during a single overpass and to provide concurrency of the remote sensor data with the corresponding surface measurements and th surface position of data platforms. The combined system has less stringent requirements on transmitter frequency stability due to the insensitivity of the interferometer technique to variations in signal frequency. System capacity is optimized by the use of narrow band modulation techniques and a fast signal searching device based upon fast Fourier transform techniques.

The ALDCS estimation of platform position and velocity during an overpass is derived from a series of two distinct measurements of each signal received from a platform. One of these measurements is the received frequency. The other is the relative RF phase received at two antennas separated by approximately three meters. The line between these two antennas is perpendicular to the orbit plane of the satellite.

The phase and frequency measurements could be used independently to locate surface platforms. However, it is advantageous to combine their complimentary characteristics as indicated by Figure A1. Two conditions are presented: the upper condition corresponds to a platform significantly removed from the satellite's subtrack. Assuming that several signals are received from the platforms, the frequency measurements enable Doppler lines-of-position (LOP) for the platform to be drawn on the earth's surface. The intersection of these LOP's is the estimated position of the platform. Similarly, the relative phase of interferometric measurements also generate the LOP's from which platform position can also be estimated. The upper sketch of Figure A1 illustrates that, for a platform located some distance from the satellite subtrack, the lines of position cross at an angle approaching the orthogonal for both the Doppler and interferometer techniques. In this case, the accuracy of the derived position is good, even in the event of slight line-placement errors. Both techniques, therefore, provide accurate positions for the off-track platforms. The lower sketch shows that the lines of position become more nearly parallel near the satellite subtrack. The crossovers become ambiguous, yielding low accuracy position estimates for either technique alone.

It is apparent from Figure A1 that the increased errors are in the along-track dimension for the interferometer and in the cross-track direction for the Doppler method. When used together, these techniques provide high quality positions near the subtrack and everywhere in the satellite field of view. Figure A2 compares expected ALDCS performance with the Doppler-only technique. The difference in performance near the subtrack as previously

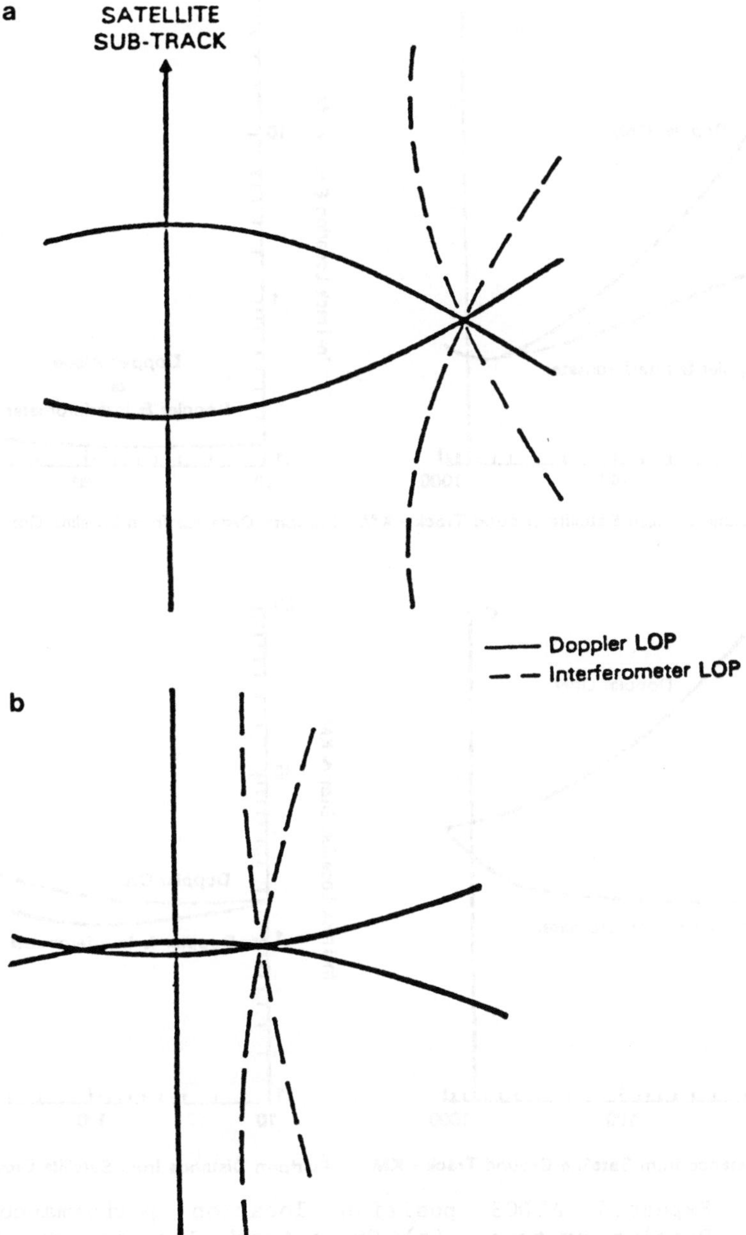

Fig. A1. Doppler versus interferometer performance: (a) Platform removed from sub-track; and (b) platform near subtrack.

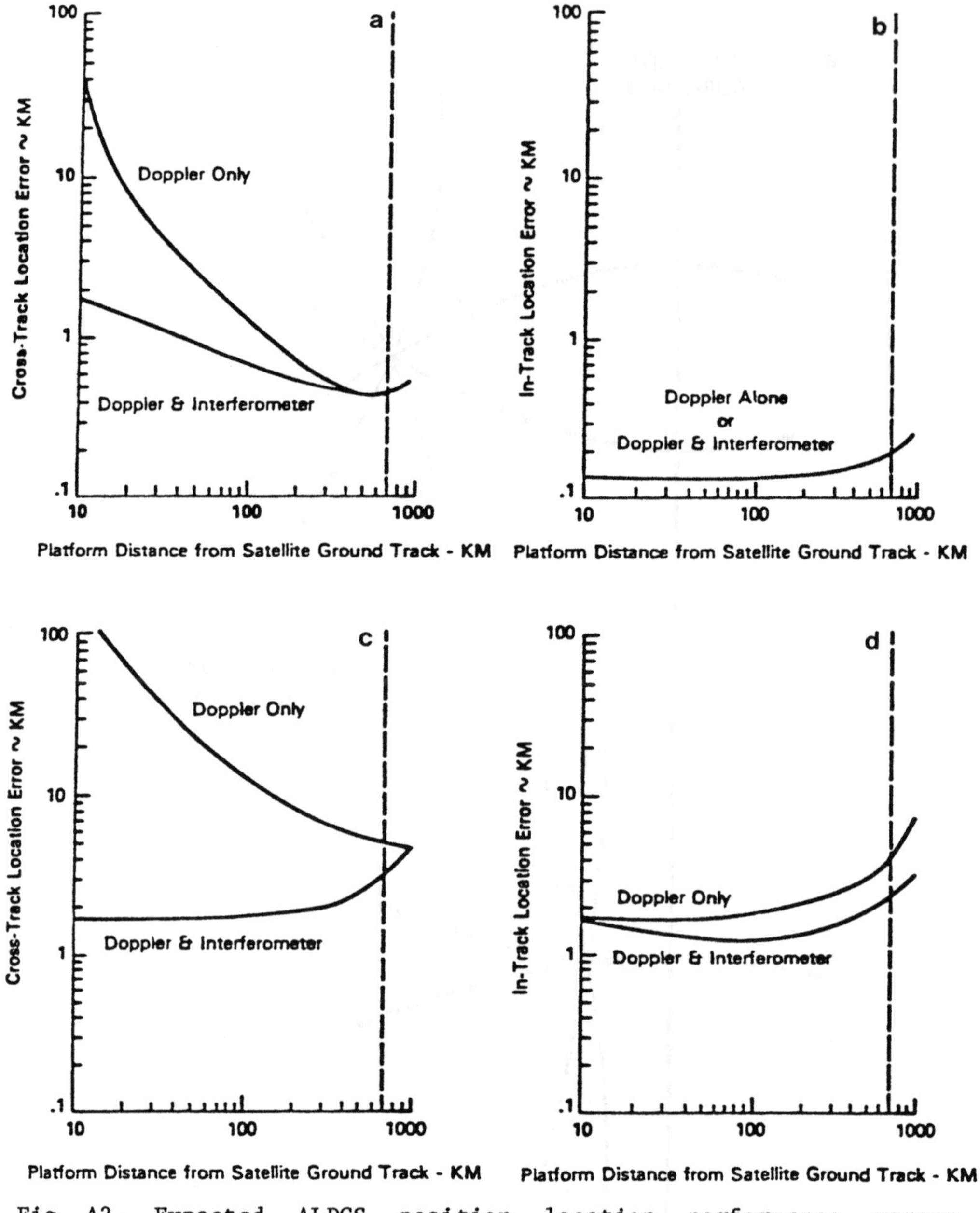

Fig. A2. Expected ALDCS position location performance versus Doppler system: (a) Cross-track location errors (stationary platforms); (b) in-track location errors (stationary platforms); (c) cross-track location errors (moving platform - 20 knots NE).

described is clearly indicated. It should be noted that errors shown in Figure A2 are for single measurements during one pass of the satellite. Since the errors are expected to be random, the position errors can be reduced by averaging of multiple measurements.

Table A1 presents a comparison of the various in situ surface data collection systems. Relative platform costs are included only for comparison purposes.

Appendix Table 1. Alternative data collection and location systems.

System	Position fixing	Data delivery delivery (minimum)	Global coverage	Moving surface platforms (ships)	Buoy terminal cost (relative)	Buoy capacity (in field of view)	NADIR position fixing
GOES	No	30 min	No	No	3.0 k	1,000	No
TIROS/ ARGOS	Yes	7 hrs[1]	Yes	No	1.8 k[2]	200	No
ALDCS	Yes	None	Yes	Yes	0.8 k	600	Yes

[1] Average delay for surface platform at mid-latitude.

[2] Includes 0.2 k for data storage.

CHAPTER 17

THE SEA ICE DATA BASE

R. G. Barry

Director, World Data Center-A for Glaciology [Snow and Ice]
University of Colorado
Boulder, Colorado

1. INTRODUCTION

Sailing vessels and shore stations have long recorded sea ice conditions in many parts of the world oceans and recently a more complete global picture has become available via aerial surveillance and satellite remote sensing. The types of measurements that may be made vary considerably in terms of the parameters recorded, their frequency, resolution, reliability, and geographical coverage. Also, whereas standard terminology, practices for ice observations, and codes for their reporting have been documented (e.g., U.S. Naval Oceanographic Office, 1956; Heap, 1972) and procedures for display of ice information on charts have recently been standardized, international standards for data archiving are only now beginning to be formulated (WDC-A for Glaciology, 1979; Thompson, 1981; Krutskikh, 1984). As a result, the state of sea ice data bases remains primitive by comparison with those for meteorological information. Accordingly, this survey indicates what is required in these areas as well as the current status.

2. THE HISTORICAL DATA BASE

There are few long-term historical records of sea-ice conditions. The longest and best-known is that compiled by Koch (1945) from an historical survey of Iceland's climate by P. Thoroddsen in 1917. This describes the incidence of East Greenland ice on the coasts of Iceland since AD 860. Lamb (1977, p. 583) tabulates the yearly record of weeks/year with drift ice at the coast, after

Koch with some additions, from AD 1600 to 1975; the annual series is believed to be complete from 1780. A related record of the extent of the Storis[1] drift along the west coast of Greenland exists from 1821 (Speerschneider, 1931; Koch, 1945; Lamb, 1977, p. 582). These types of record are shown in Figure 17.1 which illustrates the different ranges of variability according to the parameter selected. Recently, new attempts have been made to cross-check the reliability and completeness of the early Icelandic sources used by Koch and others. Vilmundarson (1972) illustrates discrepancies in 17th century sources concerning the ice extent in particular years and notes the potential value of official letters on conditions in Iceland, especially for the 18th century. Ogilvie (1984) has identified further sources showing that sea ice was common off the north coast in the 1590s, whereas the Koch graph suggests little or none. For the waters around Denmark, Speerschneider (1915) has drawn attention to the problems of early records: unreliable ancient reports of ice conditions, the defects of old maps, the changing designations used to refer to segments of Danish coastal waters, and the misidentification of the year in which a severe winter season occurred.

Other long series are available for the Baltic Sea and adjacent waters. Two major ones provide an annual winter maximum extent of ice from 1720 to 1956 and the spring breakup, indicated by the final opening of the port of Riga, annually from 1710 with less continuous data from 1530. Tabulations of these records, compiled by Speerschneider (1915, 1927), and Betin and Preobazhensky (1959), are provided by Lamb (1977, pp. 586-9). Jurva (1952) has analyzed changes in ice area in the Baltic for 1830-1951. For Antarctica, the only long series is the record from 1903 for Scotia Bay, South Orkneys (Fletcher, 1969) and the representativeness of this locality is uncertain. Rogers (1983) shows that ice duration there is inversely related to the strength of the westerly winds and that a long ice season is favored by a deep trough in the lee of the Andes. Ship reports of iceberg drift have been compiled for the southern oceans by Burrows (1976) dating back to 1770. However, the causes of ice breakup include tsunami, glacier surges, and subglacial volcanic activity as well as higher temperatures and it is not known whether the observations of icebergs in low latitudes (to 30°-40°S) represent normal or exceptional events.

The problems of historical data are illustrated by the work of Hunt and Naske (1979) for the Alaskan coasts. They used the log reports of whaling and trading ships operating in the Bering, Chukchi and Beaufort Seas from the late nineteenth century. Charts of the average ice edge are shown for the summer months for

---

[1]Ice from the East Greenland current drift carried around Kap Farvel.

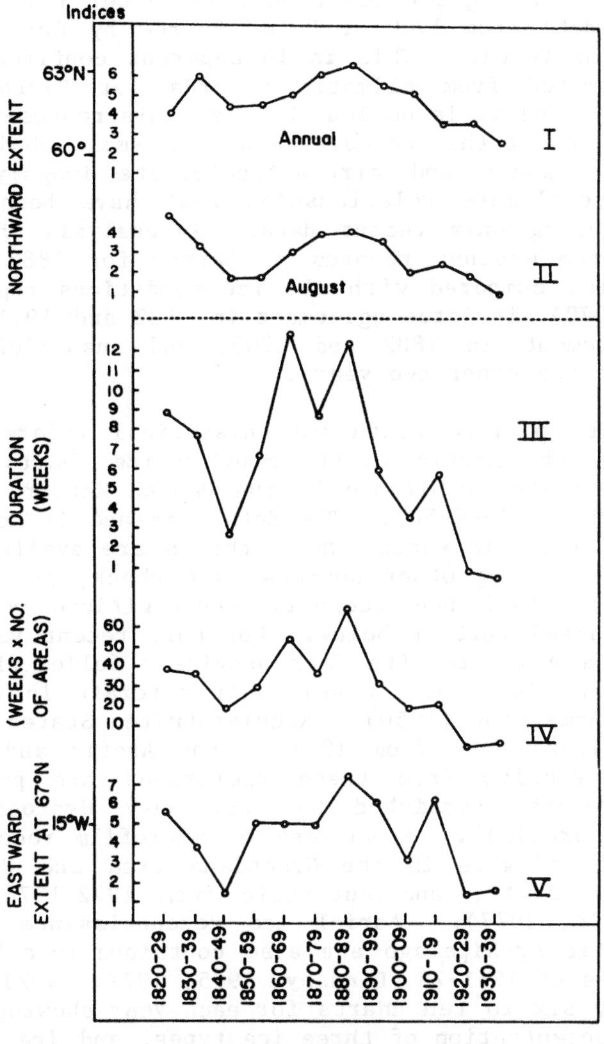

Fig. 17.1. Examples of long records of ice conditions around Greenland and Iceland, 1820-1939. I = West Greenland annual ice (after Speerschneider), II = West Greenland August ice (after Koch), III = weeks with East Greenland ice off Iceland (after Thoroddsen), IV = weeks x number of areas with ice off Iceland (Koch), V = east-west width of the ice belt north of Iceland (Koch). Note that the vertical scale for curve II was undefined (after Koch, 1945).

1860-79, 1880-1919, 1920-39 and 1940-70, but no details are given as to the procedures used or number of data points available. These charts show lighter ice conditions after 1940 and a tabulation of ice data from 1900 at Point Barrow by Swithinbank (1960) gives the same result. This is in apparent conflict with the ice retreat estimated from climatic records for Barrow since 1926 (Barry, 1979), and it is unclear how far this represents differences in definition of the ice edge seen from small whaling ships and from modern vessels and aircraft/satellite imagery, or whether there are ice-climate relationships that have been inadequately interpreted using only recent data. An analysis of inferred ice conditions from weather records at Barrow for 1882, 1883, 1902, 1911 and 1916, compared with the ice conditions reported by Hunt and Naske (1979), indictes agreement in 1902 and 1911 (light ice), some disagreement in 1882 and 1903, and insufficient data for comparison in the other two years.

The most complete ocean-wide historical information is that available for the Arctic in the monthly ice charts in April to August or September published by the Danish Meteorological Institute (1902-39 and 1946-50). The data coverage is moderately good in the North Atlantic sector where charts are available from 1877 (Ryder, 1896). Using other sources as a check, the ice limits in this area since 1901 have recently been digitized by Kelly (1979). This is discussed further below. For more recent years, aircraft reconnaissance and satellite data provide a fuller picture of both polar regions. Early Soviet work with aircraft in the Arctic is reported by Armstrong (1950). Regular United States aerial reconnaissance flights date from 1952 in the Arctic and 1962 in the Antarctic. Results from these operations are presented in a series of reports published by U.S. Navy Hydrographic Office (1954-69 (Antarctic)). A summary of microfilm records of Danish reconnaissance flights in the Greenland area and U.S. satellite data for the Arctic and Antarctic from 1972-1976 is given by Mitchell (1976, 1977). Aerial ice reconnaissance data for the Canadian Arctic Archipelago are also contained in atlasses for the summer seasons of 1961-78 (Lindsay, 1975, 1977, 1982). These give approximately six to ten charts for each year showing the existing fractional concentration of three ice types, and ice forms such as ridging. A complementary analysis for the same area for 1900-1958 is provided in an earlier atlas by Swithinbank (1960). This contains maps of the frequency of three ice concentration classes at 324 locations for each season in the Canadian arctic islands and adjacent seas. Swithinbank uses both ship and shore observations and some U.S. Navy Hydrographic Office aerial reconnaissance information. The categories of ice concentration that are assumed equivalent are as follows:

Traversed by ship:	without difficulty	Aircraft:	1-5/10 ice
	with difficulty		5-8/10

All but icebreakers brought to standstill	8-10/10
Unnavigable	10/10

There is much more information available in Swithinbank's study for the 1950s than for the preceding 50 years, however.

The problem of bias in earlier 'historical' records should be carefully considered in any studies of trends in ice conditions (e.g. Ogilvie, 1984). Koch (1945, p. 167), for example, draws attention to some specific inadequacies in the Danish ice charts for Greenland waters. He also points out that a monthly chart must obscure important short-term variability in ice limits; moreover, what is depicted need not be representative of the average state for that month. Recent chart series from other sources tend to depict conditions at a particular point in time (e.g. end of month) rather than portraying monthly average conditions.

3. PROXY DATA

By comparison with paleoclimatic information, there are few sources of proxy data indicative of sea ice conditions in the past. Three categories can be mentioned.

The first involves the use of faunal assemblages in marine sediments. In the Antarctic Ocean, for example, a decrease in diatom input during the last glaciation is related to an increase in summertime sea ice cover according to Hays (1978).

A second, more direct and closely related approach, uses ocean sedimentary evidence of the rafting of detritus and volcanic ash by sea ice and/or icebergs. Ash zones are especially important in this regard since they can be closely dated. The wintertime maximum extent of Antarctic sea ice can be estimated from the equatorward limit of volcanic ash, since ash may fall on the ice at any time of year and is transported generally northward until the ice melts. On the basis of ash and diatom evidence, Hays (1978) suggests that the sea ice 18,000 years ago was about twice as extensive as present in winter and roughly 10 times so in summer (ignoring any differences in ice concentration). Cooke (1978) has reconstructed seasonal ice limits around Antarctica for selected time intervals over the last 140,000 years using these approaches. More recently, Burckle et al. (1982) have reinterpreted these data and suggested that the observed sediment boundary represents the approximate springtime limit of sea ice. The occurrence and extent of ice-rafted detritus in the North Atlantic has also been used as an indicator of ocean circulation conditions during the last glaciation and of the volume and pattern of ice-sheet buildup in North America by Ruddiman (1977; Ruddiman et al., 1980).

The third category of possible proxy data is ocean-bottom scour by icebergs. This has been investigated particularly in the coastal waters of eastern Canada where the topic is of significance for seabed installations. (As a result, some of the information that may exist is proprietary). The existence of linear gouges several meters deep in waters 300 m in depth off Labrador is attributable to large icebergs (>50 m in height, with an average draft/height ratio = 3.5:1). Gustajtis (1979) shows that annually ca. 100 such icebergs reach latitude 56°N off Labrador and ca. 25 reach latitude 52°N. Approximately 10 percent of these icebergs may scour the seabed in waters 300 m deep. The interpretation of scour marks in deep waters as relict Pleistocene features (King, 1976) remains uncertain since the rates of postglacial isostatic uplift of the coast following deglaciation and of eustatic sea level rise due to the ice sheet melt are not well known. Hence, the estimation of past limits of iceberg extent from such information seems unlikely in the near future. Seabed scour by ice ridges in the Arctic shelf seas presents similar problems of interpretation, although there the time-scale of reworking of the bottom sediments is quite short (Reimnitz et al., 1977).

4. CURRENT DATA PRODUCTS

4.1. Data Streams

Sea ice data on a worldwide basis are reported by synoptic observations on ships, by aerial ice reconnaissance patrols, and by satellite and occasionally submarine remote sensing. The ship observations are coded and reported in digital form; while the aircraft observations include coded observer reports, photography and processed imagery. The satellite data are both vidicon and multispectral scanning imagery. However, while the latter are received in digital form, they are processed to give imagery and then manually transferred to chart format (Figure 17.2). Currently, much attention is being given to the digitization of such charts. Digitization of charts, such as shown in Figure 17.2, is not a trivial problem given the variety of types of data they contain.

4.2. Data Products

The variety of regularly-produced ice charts is summarized for hemispheric and regional products in Tables 17.1 and 17.2. A listing of national sea-ice information services and publications has recently been published by the World Meteorological Organization (1981). The most substantial ongoing analysis programs are those of the U.S. Navy-NOAA Joint Ice Center for both polar regions (U.S. Fleet Weather Facility, 1976a, b, and subsequent annual publications by the Naval Polar Oceanography Center - now the Joint Ice Center). These weekly analyses illustrated in

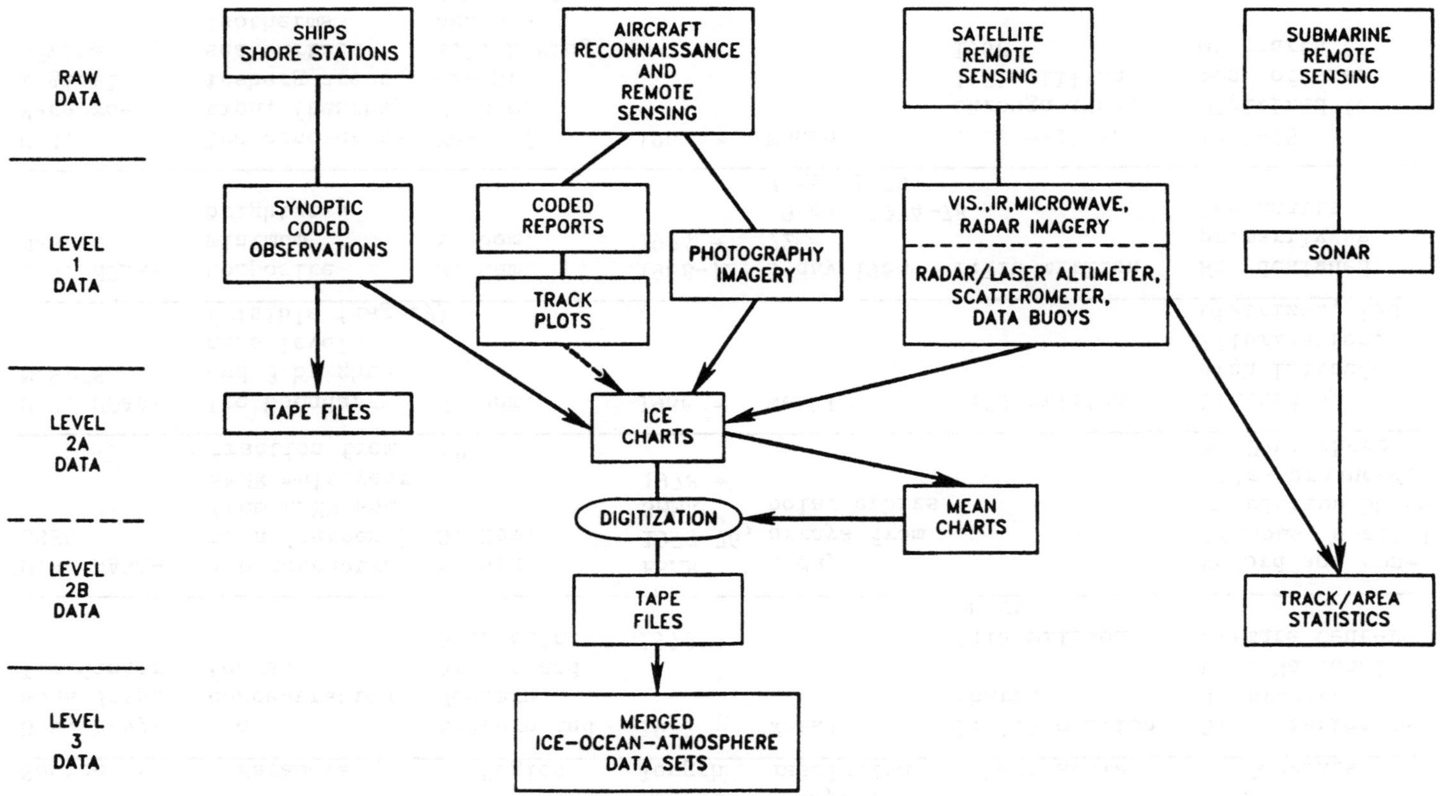

Fig. 17.2. A schematic diagram of sea-ice data streams. The data levels in the left column refer to GARP categories of data processing.

Table 17.1. Hemispheric charts.

Source	Parameter	Region	Record length	Temporal resolution	Chart scale	Comments
U.S. Navy/ NOAA Joint Ice Center	Ice concentration (oktas)	Eastern and Western Arctic and Antarctic	1972 → 1973 →	Weekly	1:11.6 million charts 1:16 million chart	Digitization is in progress at U.S. National Climate Center
U.S. NASA-GSFC	Ice concentration (percent) from ESMR and SMMR multiyear fraction from SMMR	N. Hem. S. Hem.	ESMR 1973-76, SMMR 1978 →	3-day arrays from polar orbits		Record not continuous, spatial resolution 50 km. Atlas produced, S. Hemisphere
U.S. NOAA-NESDIS	Ice boundary and 3 brightness levels (visible imagery)	N. Hem.	1966 →	Weekly	1:50 million	Limited by high latitude illumination. Digitized, 1981
U.S. NOAA-NESDIS	Composite minimum brightness	N. Hem. S. Hem.	1968-72; 1974 →	5-day 1968-72 10-day 1974-78 7-day 1979	1:110 million	Not designed primarily as ice charts
U.K. Meteorological Office	Ice concentration, (tenths) iceberg areas, sea surface isotherms	Most of N. Hem. except S.E. Bering and S'n Othotsk Seas	1962 →	Monthly	1:22 million through 1967; 1:21 million 1968	1966-76 Digitized for most of area of charts

Table 17.2. Regional charts.

Source	Parameter	Region	Record length	Temporal resolution	Chart scale	Comments
Canada, Atmos. Environ. Service	(a) Composite ice charts Ice concentration (tenths)	Eastern seaboard, Hudson Bay	1972 →	2-7 days Dec.-June 7 days June-Nov.	1:4 million	
		E'n Arctic W'n Arctic	1972 →	7 days June-Oct.	1:4 million	
		Great Lakes	1972 →	7 days	1:5 million	
Canada Atmos. Environ. Service	(b) Historical Series Ice concentration	North of 65°N	1960-74	7 days May Oct.	1:6 million	
	Ice concentration	Eastern seaboard, Hudson Bay, Foxe Basin	1958-74	7 days Dec.-May	1:6 million	Few data N. of 55°N
Danish Meteorol. Institute	Ice concentration (tenths)	Greenland waters	1957-64	Weekly (April-Aug.)	1:13 million	Tabulations 1959-74 data to be published in 1982
Norwegian Meteorol. Institute	Ice concentration (oktas); sea surface isotherms	North Atlantic 30°W-69°E	1970 →	3-4 days	1:10 million	

Table 17.2. Regional charts - continued.

Source	Parameter	Region	Record length	Temporal resolution	Chart scale	Comments
Swedish Met. Hydrol. Inst.	Ice conditions (9 categories of type, form, concentration	Baltic Sea and Danish waters	1953 →	3-4 days October - May/June	1:4 million	A 1:2 million facsimile chart is prepared daily
Institute of Marine Research, Finland	Each series now uses the Baltic Sea code.	Baltic Sea	1920 →	Oct. 15- May 31	1:4 million	Digitized records of chart data from 1963
Deutsches Hydrograph. Inst.		Baltic Sea and Gulf of Finland	1939/40→	3-4 days Oct-May/ June	1:3 million	
		Western Baltic-German Bight		2-3 days	1:2,200,00	
U.S. NOAA-GLERL	Ice concentration (tenths), types (4 classes)	Great Lakes, North America	1962 →	2-3 days	1:400,000 to 1:3 million	Digitized
U.S. NOAA-NESDIS	Ice concentration (oktas) type (6 classes)	Beaufort-Chukchi Seas Chukchi-Bering Seas	1973-79	3-4 days	1:10 million	Discontinued

Source	Parameter	Region	Record length	Temporal resolution	Chart scale	Comments
Argentina Navy Hydrographic Service	Ice edge, concentration	Bellinghausen Sea-Antarctic Peninsula-Weddell Sea sector (20°-100°W		3-4 days	1:5 million	
Japanese Weather Association	Ice concentration (okta classes); sea surface isotherms	Sea of Okhotsk	Dec. 1970 -1983	3-4 days	1:9.5 million	
Japanese Meteorological Agency Maritime Meteorol. Division	Ice Condition Chart (tenths); sea-surface isotherms	Sea of Okhotsk	Dec. 1938 →	3-4 days Dec.-May	1:12 million	

Figure 17.3 draw on a wide variety of sources, including the microwave imagery collected for research and development purposes by NASA. It is worth noting that nowadays most national agencies draw heavily on United States satellite data. Hence, close similarity in most of the information depicted is to be anticipated.

The principles of remote sensing in relation to sea ice are discussed by Carsey and Zwally (Chapter 16). From the mid-1960s through the 1970s most sea ice analyses depended on visible imagery (McClain, 1978). The weekly charts of the northern hemisphere snow and ice boundary produced by NOAA-NESDIS since November 1966 used visible imagery until January 1971 and infrared scanning radiometer data in conjunction with the visible channels thereafter, except for intervals of system failure (see Crane, 1979, p. 64, for details). The analyses based only on visible data are necessarily incomplete in high latitudes during the period of polar darkness. The composite minimum brightness/composite maximum temperature products provide coverage with no polar cut-off in the case of the thermal channel data from December 1974. The reflectivities are normalized by a cosine function to take account of most solar altitude effects but this data source, which has a resolution of 50-100 km, is not specifically an ice chart. McClain (1973) has shown that five categories of ice condition can be distinguished from a compressed and calibrated brightness scale, but little routine analysis of CMB products appears to have been carried out.

Essentially complete coverage of the polar regions with microwave data has become available since the launch of the Electrically Scanning Microwave Radiometer (ESMR) on Nimbus 5 in December 1972 and the Scanning Multichannel Microwave Radiometer on Nimbus 7 in 1978. Orbital data using an operational algorithm are available to the Navy-NOAA Joint Ice Center in near-real time. Research products have been produced from ESMR (see 4.5) and have recently become available for 1979-81 from SMMR (Cavalieri et al., 1984). The use of the dual-polarized multispectral SMMR data has permitted the mapping of ice concentration, multiyear ice fraction and ice surface temperature with a 50 km resolution (see Carsey and Zwally, Chapter 16). Beginning in 1986 the Special Sensor Microwave Imager (SSM/I) on Defense Meteorological Satellite Program (DMSP) satellites will continue to provide similar data with 25 km resolution (and 12.5 km for ice edge delimitation) into the 1990s.

The regional analyses, such as those for the Baltic Sea and Canadian Arctic (Table 17.2) provide much more information in ice types, the occurrence of ridging, etc. than is possible on the hemispheric analyses. The problem of how this information could be integrated into more spatially-extensive digital data sets has not yet been addressed.

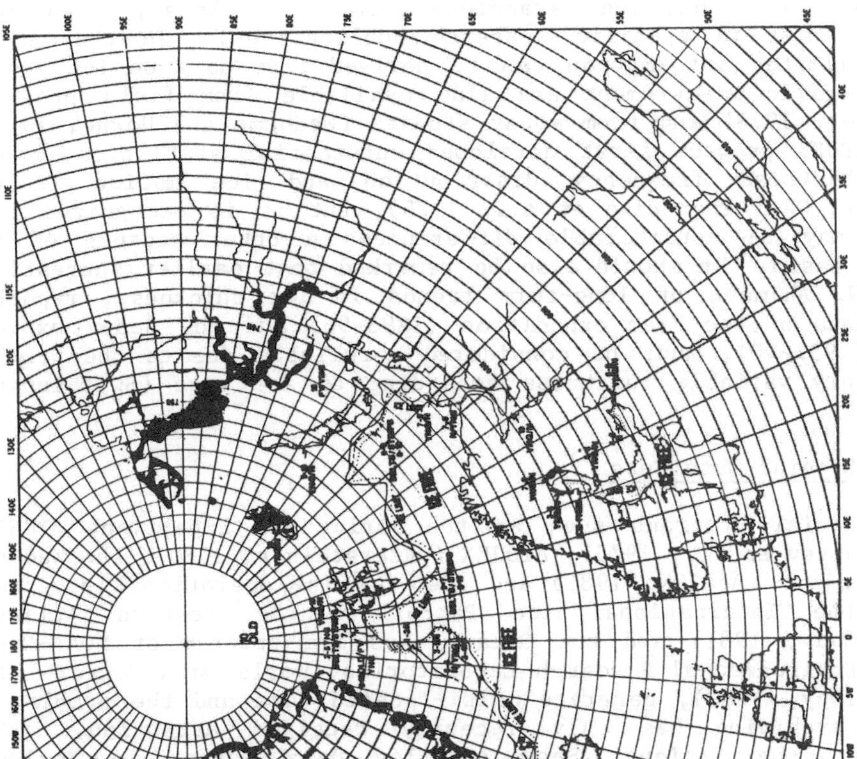

Fig. 17.3. The southern ice limit for 30 December 1980 as depicted by the Navy-NOAA Joint Ice Center Chart.

Data on other ice parameters such as thickness, or motion, mostly exist in specific research publications. In the former case, limited information can be derived from data on ice types, but specific measurement are also available from field programs (Koerner, 1973) and from sonar studies (Wadhams and Horne, 1980). The AIDJEX project first developed substantial statistics on the ice thickness frequency distribution and ice motion in the Beaufort Sea (Thorndike et al., 1975), but it is recognized that there may be considerable differences in other sectors of the Arctic Basin. It should also be remarked that earlier suggestions by H.W. Ahlmann of long-term trends in ice thickness, reported from data of the Fram expedition (1893-96) and the Sedov expedition (1937-40), must be considered unreliable due to the problem of highly variable thickness according to ice type (Swithinbank, 1972).

4.3. Special Data Sets

Special programs such as the Bering Sea Experiment (BESEX) (Kondratyev et al., 1975), AIDJEX (Untersteiner, 1980), the Weddell Sea Project (Ackley, 1981) and even ongoing operational programs, like the International Ice Patrol in the western Atlantic (Anonymous, 1981) and the Great Lakes ice program of NOAA-Great Lakes Environmental Research Laboratories (World Data Center-A for Glaciology, 1980), generate multi-faceted data and the problem of their archiving has only recently received attention. For example, the National Snow and Ice Data Center in Boulder, Colorado, is currently involved in managing the multidisciplinary data being collected in the East Greenland Sea-Fram Strait area through the Marginal Ice Zone Experiment (MIZEX).

The summarizing of display information produced by coastal radar stations on ice distribution or ice drift, such as that of for Hokkaido (Tabata et al., 1980) also presents new problems, especially for comparisons with earlier data (Hakodate Marine Observatory, 1955). In general, it is preferable to archive 'raw' data (the lowest level of data useful to other prospective users), since the most appropriate calibrations of, say, sensor systems may not be immediately known, or improved data treatment becomes possible at a subsequent date. However, the computer software necessary to read special tapes (or other media) sometimes becomes unavailable after a certain time interval and the original data are then irretrievable. This is the case, for example, with airborne radar data collected by NASA Lewis over the Great Lakes. In other instances, routine satellite data received with extremely high resolution cannot all be archived. The GOES (geostationary satellite) system has a data rate of 1 million bits per second, for example, necessitating time and space sampling (Jenne, 1981). At the United States Air Force Global Weather Central (AFGWC), the tapes of Defense Meteorological Satellite Program (DMSP) data are 'cleaned' on a daily basis and only imagery/chart products are archived. These images (visible and infrared channels) are available from 1975 in orbital swath and mosaic format at the National Snow and Ice Data Center. Many instances of similar problems exist in smaller magnitude in most research projects.

4.4. Ice Indices

Many of the longer series of ice data have been compiled in the form of indices, usually as annual values. Table 17.3 (after Kelly, 1979) summarizes most of the regional ones and indicates the variety of parameters selected. A few examples using different bases for the indices are discussed.

Several indices describe the maximum ice extent including those for Newfoundland (Miles, 1974) and West Greenland (Koch, 1945, Valeur, 1976, and Speerschneider, 1931). As recognized by Koch, however, ice drift is a complex variable which particularly reflects wind velocity. Other studies define ice "severity" by a relative scale. Some of these are not readily interpreted since no examples of the selected scale are given. This is true of the historical series for the Kara Sea (Nazarov, 1964), and the indices of Walker (1947), Meinardus (1905, 1906), and Brooks and Quennell (1928). The last authors develop a weighted index (I) of ice conditions in the Greenland (G), Barents (B) and Kara (K) Seas, partly according to partial regression coefficients relating the ice to atmospheric parameters. The index is:

$$I = 5(G_{5-8}) + 7\,(B_{5-8}) + 2\,[3K_8 + K_{5-8}]$$

where the subscripts denote the months considered.

i.e., G5-8 denotes the ice conditions in the Greenland Sea in May through August.

Several problems need to be recognized with regard to ice indices. First, there is the question of the reliability and representativeness of the data used in historical sources. This point has already been discussed. Second, many of the indices are seasonal averages which may obscure crucial differences between different years. Also, such averages may be biased due to the equal weighting of the monthly data that they incorporate. A third difficulty with indices concerns their spatial representativeness if, for example, the ice anomalies are of opposite sign in different sectors of a large region such as that used by Brooks and Quennel. A fourth problem may arise when a single date is used to define some stage of the seasonal ice cycle. This is illustrated by the identification of the "final ice clearance" in Baffin Bay (Dunbar, 1967; Keen, 1980). Its timing may be difficult to determine in the first place and its physical significance

Table 17.3. Sea ice indices.

Region	Parameters	Time interval	Period of record	Source(s)
Geographic sectors: (see text)	Ice area (km^2) $\geq$ 1/8 concentration and per- cent open water in pack			Kukla and Gavin (1979)
N. Hemisphere		Weekly	1972 →	U.S. Navy charts
S. Hemisphere			1973 →	
N. Hemisphere		Monthly	1960 →	British Met. Office
Arctic (sectors)	Area (km^2) between limit of 7/10 ice and minimum extent; graphs	Monthly; Feb.-Apr. Aug.-Sep.	1966-1974 (Atlantic sectors) 1969-1974 (Pacific sectors)	Sanderson (1975) British Met. Office charts
Greenland-Barents-Kara seas	Ice occurrence weighted for each area (see text)	Annual for Apr.-Aug.	1895-1924	Brooks & Quennell (1928) from Danish data mainly
Greenland-Barents-seas and Spitsbergen area	Severity index (-7 to +7) estimated for each of three sectors and mean tab- ulated (2 estimates)	Annual	1897-1938	Walker (1947)
Greenland Sea	a) Area	Apr.Aug.	1877-1915	Speerschneider (1917)
	b) Area (44°W-15°E) deviations from average for 1898-1913; graph	Apr.-May Apr.-Aug.	1900-1939; 1946-1957	Skov (1970)
	c) Percent of area (monthly and seasonal)	Apr.-Aug.	1924-1939; 1945-1968	Kirillov & Khromtsova (1972) from Danish data and "Marine Observer"

Table 17.3. Sea ice indices - continued.

Region	Parameters	Time interval	Period of record	Source(s)
West Greenland (Davis Strait)	a) N'wd extent of Storis	Annual	1821-1939	Kock (1945)
			1900-1972	Valeur (1976)
	b) N.most extent; W.most extent of Storis; index 0-10 from extent for each mon and duration summarised as annual mean	Annual for Jan.-Aug.	1820-1930	Speerschneider (1931) from ships logs and Danish annals from 1897.
	c) Ice area W. of 44°W. (weight of cut-out paper traced from maps)	Apr.-Aug.	1901-1937	Hansen & Svistrup (1943) from Danish maps
Newfoundland	a) Index of sea ice area; (Lat. of E.most extent + Long. of S.most extent), 87 (max. to 103 (min. ice)	Apr.	1920-1973 (except 1940-1943)	Miles (1974) from British Met. Office, U.S. Coast Guard and "Marine Observer"
	b) Severity index for sea ice (-2 to +2)	Annual (? spring)	1860-1902	Meinardus (1905) from British, German and North American data
West Atlantic	a) Iceberg number S. of 48°N	Annual	1900-1970	Bailey (1972)
		Sep.-Aug.		U.S. Coast Guard data
	b) Iceberg severity index (N/24-13/2, where N=number of icebergs	Mar.-July	1886-1939	Walker (1947) from Meinardus and U.S. Coast Guard

Table 17.3. Sea ice indices - continued.

Region	Parameters	Time interval	Period of record	Source(s)
Iceland	a) Duration of drift ice at coasts with 3 categories of width of ice belt E-W; monthly ice maps	Monthly	1877-1939 1940-1971	Kock (1945) from Thoroddsen, Sigtrygsson (1972) updated by Lamb (1977)
Baffin Bay	Date of total clearing of sea ice	Annual	1952-1970 1952-1976	Dunbar (1972) updated by Keen (1980)
Barents Sea	a) Ice area	Apr.-Aug.	1895-1915	Danish Met. Inst. (1917)
	b) Ice area as % of Barents Sea	May-Aug.	1900-1960	Maksimov et al. (1964)
Kara Sea	a) Ice area	Apr.-Aug.	1895-1915	Danish Met. Inst. (1917)
	b) Severity index (-2 to +2)	Unspecified	1868-1946 except 1885, 1891-1892 (some data back to 1530)	Nazarov (1947)
Beaufort Sea	Severity index; ranking of years based on 6 features--distance to pack-ice off Barrow, conditions along ship route to Prudhoe Bay, duration of navigation season	Aug.-Sep.	1953-1975	Barnett (1976) from U.S. Navy data

Region	Parameters	Time interval	Period of record	Source(s)
Okhotsk Sea	Anomaly of drift ice (days) at Abshira and at Shana, Hokkaido	Dec.-Apr.	1892-1945	Sawada (1957)
	Dates of appearance/ disappearance of fast ice and pack ice at 7 coastal stations	Dec.-Apr.	1959 → (and less complete earlier data)	Bulletin of the Hakodate Marine Observatory (No. 20, 1981) (annual reports)

is complex since this stage marks the culmination of a protracted process involving thermodynamic and dynamic processes (Dey, 1980). The typical decay stages for fast ice in the same region, which can be related to meteorological parameters, have been defined by Weaver et al. (1975).

The most comprehensive indices are those of Kukla and Gavin (1979). Digital files of ice area and the percentage of open water within the pack have been developed for geographical sectors in both polar regions. The weekly U.S. Navy ice maps are used since 1972 (Arctic) or 1973 (Antarctic) while the British Meteorological Office maps are used to provide a monthly file of Arctic ice area since 1960. A limitation of these indices is their use of geographical sectors ca. 10° latitude meridionally by a variable longitudinal span.

4.5. Digitization

The need for cryospheric data to be readily accessible in digital form is now widely acknowledged (WDC-A for Glaciology, 1979; Kukla et al., 1981), but only limited progress has been made in this direction as indicated in Tables 17.1 and 17.2. On the practical level the digitization process involves prior decisions as to the projection or grid to be used, and the element size and resolution appropriate to the data. Commonly-used grid arrays for meteorological archives are the U.S. National Meteorological Center (NMC) grids, but these are not equal area. It is recommended by the Snow Watch Workshop that projection (x,y) coordinates be used for archiving digital information with appropriate documentation provided to transform the data into geographical coordinates (Kukla et al., 1981).

Several sets of digitized monthly ice data are now available. Walsh (1978) has extracted northern hemisphere ice concentration values for 1953-1977 on a 110 km grid using a variety of sources (Crane, 1979, p. 163). Corresponding meteorological data grids are also available. The data quality is recognized to be spatially and temporally variable, especially in the 1950s and early 1960s. A preliminary data set of monthly ice extent (> 4/10 concentration) for the northern hemisphere for 1901-1956, also on a 110 km grid, has been prepared by Kelly (1979) using the Danish Meteorological Institute charts. Large segments of "unreliable" data are present.

For Arctic and Antarctic ice conditions (on a common base) there are now several digital data sets. One has been extracted by Lemke et al. (1980) from monthly charts of the British Meteorological Office for the Arctic, for 1966-1976, and weekly U.S. Navy Charts for Antarctica, 1973-79. The parameter is the latitude where ice cover is > 5/8 for 5° longitudinal sectors on the

Antarctic charts. In the Arctic it is the area of ice per 10° longitude x 1° latitude sector; each 1° rectangle is counted as ice-covered where $\geq$ 7/10 ice is shown. Another set comprises the digital 3-day average microwave brightness temperature data obtained by NASA from the Electrically Scanning Microwave Radiometer (ESMR) system on Nimbus 5 and 6 satellites. The data array is a 293 x 293 grid with an approximate 50 km resolution. These data are available for about 85 percent of the time from December 1972 through May 1977 (except for March-May 1973 and June-August 1975) from ESMR 5 and for parts of the period August 1975-July 1977 from ESMR-6. The Antarctic data for 1973-76 are presented in a detailed atlas (Zwally et al., 1983) and an Arctic atlas is in preparation.

Several sets of digitized ice data have recently been constructed for Antarctica, all based on the Navy-NOAA Joint Ice Center (JIC) chart series that started in 1972/73 (Chiu, 1983; Jacka, 1983; Ropelewski, 1983) and a further digital product using the international sea ice grid (SIGRID) format described by Thompson (1981) with 0.25^o latitude resolution for both hemispheres is also available (Knight, 1984). Despite the common source, estimates for the mean monthly and mean annual ice areas in the Antarctic differ between these studies. Sturman and Anderson (1985) show that this arises as a result of different analysis techniques and the use of weekly charts from the beginning, middle or end of the month to characterize the monthly averages. This can cause important differences in the transition seasons especially. The mean annual ice area given by Chiu (1983) is significantly larger than the other sources for reasons that could not be determined.

A little known set of digital data has recently been prepared by the U.S. Naval Oceanographic Office (Westhall, 1975). Visual observations from the BIRDSEYE ice reconnaissance flight and other missions in the Arctic during 1964-72 have been coded and filed in sectors 2° latitude by 10° longitude. The files include location, time, ice concentration, the primary ice form (floe size), stage of development and fractional concentration, topography, extent of ridging, opening, and number of bergs in three size categories.

4.6. Ice Atlases

In addition to time series and spatial patterns for individual months, it is also necessary to have long-term average distributions of ice extent. Published atlases[2] of sea ice include ones for the northern hemisphere (U.S. Navy Hydrographic Office, 1946),

---

[2]'Atlases' that display a series of daily or weekly conditions are not included in this context. Examples have already been referred to above.

Arctic Canada (Swithinbank, 1960; and Markham, 1981), Alaska (LaBelle et al., 1983), the eastern Canadian seaboard (Markham, 1980), seas north of the USSR (Armstrong, 1958), the Baltic (Swedish Meteorological and Hydrological Institute, 1982), and Antarctica (Heap, 1963; Sovetskaya Antarkticheskaya Expeditsia, 1966). In addition, general atlases of the polar regions usually depict at least the mean seasonal ice limits. A new Soviet Atlas of World Snow and Ice Resources including sea ice maps will be published in a few years' time (Kotlyakov and Dreyer, 1978).

Two principal types of map display of ice information have been used up to now (Armstrong, 1964). Isopleths of average and extreme limits at specified time intervals are the most common (Sowden and Geddes, 1980), but sector ('pie') diagrams of relative frequency located at points, or for box areas, have been used by Armstrong (1958) and Swithinbank (1960). The ice limit may be for > 1/8 or some other concentration. In the case of the new Antarctic atlas (Zwally et al., 1983), the 15 percent concentration is approximated from microwave brightness temperatures. As noted earlier, these new data sources are likely to cause considerable revision of older atlases and charts.

Other regional presentations of mean monthly ice limits are available in yearbooks, journals, and reports. Some examples includes studies for the Bering Sea (Dunbar, 1967; Webster, 1981), the Okhotsk Sea (Watanabe, 1967), the North Atlantic-Barents Sea sector (Vinje, 1977), the Baltic (Palosuo, 1966), and Hudson Bay (Danielson, 1971). For icebergs off Labrador, mean seasonal maps of relative densities have recently been published (Gustajtis and Buckley, n.d.).

5. DATA NEEDS

As a basis for evaluating existing sea ice data, or for assessing future requirements, it is essential to consider the types of use to which these data are put. For example, detailed local data on ice extent, concentration, thickness, and motion are required in "real-time" for operational purposes, such as navigation. These and associated meteorological/oceanographic data are also used in short-term forecasting of ice conditions for shipping. Similar types of data are needed over a number of years in particular localities for operational or commercial purposes to assess the probability of certain ice conditions occurring at a given time of year. Such assessments may also require data on the frequency of open water, ridged ice, etc.

From the research standpoint, sea ice is an important component of the oceans and of the climate system. Most general circulation models (GCMS) of the atmosphere now incorporate sea ice cover effects at least as a fixed surface boundary condition. The ice is expected to influence the occurrence of cyclone activity in the models, although the nature of such interactions is inadequately known (see Barry, Chapter 15). For atmospheric GCMs, gridded data are required on monthly mean ice concentrations with a resolution of at least 250 km. To adapt such models for extended-range (7 day) weather-forecasting purposes would necessitate the availability of at least weekly ice data on a routine-basis. Recently, coupled models incorporating sea ice - atmosphere interactive systems have been developed (Washington et al., 1980) building on present knowledge of the operative thermodynamic and dynamic processes. Sea ice models using climatological mean forcing by the atmosphere have also been tested (Parkinson and Washington, 1979; Hibler, 1980). In the case of the mean ice conditions, adequate verification data exist for ice extent, although not for ice thickness, and studies attempting model simulations for individual years encounter inadequacies in the available data (Walsh et al., 1984). Also, there are major problems of adequate parameterization of sub-grid scale processes, such as heat fluxes from leads or thin ice, in the lower boundary region of the models.

The parameters that characterize sea ice conditions are listed in Table 17.4 with an indication of the typical levels of accuracy, spatial and temporal resolution required for purposes of basic research and climatological studies. However, these levels may be inadequate for other purposes such as shipping operations. For assessments of ice forces on structures, and other hazards relating to ice motion, information on ice mechanical properties and on meteorological/oceanographic factors is necessary.

For many of the parameters listed in Table 17.4, the "data base" is limited to the published results of specific research investigations and data sets, as such, are non-existent. Most of the requirements shown in Table 17.4 are within the technical capabilities of present measurement systems, although many of these systems are not yet operational and indeed may not become so within the next several years.

A major problem with most existing ice data is that they are not readily meshed with concurrent meteorological and oceanographic information. Few "Level 3" data sets (see Figure 17.2) have been compiled, in part because most research projects tend to be unidisciplinary, and in part because of the multi-organizational nature of data collection and archiving relating to sea ice and the associated atmosphere/ocean conditions. The Arctic Ice Dynamics Joint Experiment (AIDJEX) provided a rare exception to this statement and the data deriving from this project are well documented and archived (Stateman, 1978). Nevertheless, the time span of AIDJEX was limited.

Table 17.4. Basic requirements for sea ice observations (after NASA, 1979, with modifications).

Parameter	Accuracy[1]	Resolution[1] Space	Time	Sensor type(s)	Comments
ICE					
Boundary	5-20 km	5-20 km	1-3 day	Passive microwave, multi-frequency	Imaging radar for high resolution
Concentration	2-5%	25-100 km	1-3 day	Passive microwave, multi-frequency	Imaging radar for high resolution
Type (fractional area, by type)	5-10%	25-100 km	7-30 day	Passive microwave, multi-frequency	Gives some thickness information
Leads: frac. area orientation	5-10% 10°-30°	5-25 km N.A.	1-3 day 1-3 day	Imaging radar	Occasional measurement
Ice thickness	0.2-1 m	25-100 km	7-30 day	Sonar	Occasional measurement
Motion (point displacement	0.1-1 km per day	5-100 km	1-7 day	Data buoy; imaging radar	Occasional measurement
Ridging: density height orientation	10-50% 1 m 10°-30°	5-25 km N.A. N.A.	7-30 day 1-30 day 7-30 day	Imaging radar	Scatterometer; Sonar
Albedo (areal mean)	0.02-0.04 (5-10%)	25-100 km	3-10 day	Scanning multiband radiometer	
Surface melting (frac. area)	wet/dry	100 km	1-3 day	Passive microwave	
Surface temperature	1-3 K	25-100 km	1-3 day	Scanning multiband radiometer	

Table 17.4. Basic requirements for sea ice observations (after NASA, 1979, with modifications) - continued.

Parameter	Accuracy[1]	Resolution[1] Space	Time	Sensor type(s)	Comments
OTHER PARAMETERS					
Icebergs:					
Location	100 m	100 m	6-48 hr	Imaging radar	Location for operational needs around offshore platforms
Volume discharge (ocean current system)	5-10%	(area average)	3-12 mo		
Meteorological data:					
Surface pressure (point)	0.5-1 mb				
Wind velocity (point)	10°-20° 1-2 m s^{-1}	25-50 km	3 hr	Data buoy	
Air temperature	0.2-1 K				
Sea surface temp.	0.2-2 K	10-25 km	1-3 day	Scanning multiband radiometer	

[1]Smaller value is desired standard; larger value is minimum for basic/climate studies. One or other may be more appropriate also according to location and season.

Coupled ice-atmosphere data sets have been prepared on a monthly time scale by Walsh, but meteorological observations in the polar regions are inadequate for detailed synoptic scale studies. Even for many of the remote sensing analyses of ice conditions, there are insufficient ground-truth data to verify the existing interpretations (e.g. Crane et al., 1982).

6. DATA PROBLEMS AND FUTURE PROSPECTS

Some of the complexities involved in securing useful data bases on sea ice are readily apparent from the foregoing. A major difficulty is posed by the continual improvement of the sensor system. Historically, the observational records have changed emphases from shore station, to ship observations, to aerial reconnaissance support, and finally to global coverage via satellite remote sensing. The 'matching' of different types of data is rarely considered, yet this presents serious problems in attempting to define any changes in 'typical' or mean conditions. This was noted by Gloersen et al. (1974) in comparisons of early microwave results with ice atlas depictions of ice extent. However, the higher spatial and temporal resolution of each succeeding generation of instrumentation will cause this problem to keep recurring. Indeed, there may be a case for 'degrading' high resolution data for purposes of historical comparisons. An alternative may be to devise or adopt suitable regional indices of ice conditions of the type described in Section 17.4.4.

Another general problem is that the standard parameters characterising sea ice may be inadequate for process-oriented studies at the frontiers of basic or applied research problems. It is, nevertheless, impossible to develop all-purpose data bank structures to take account of all possible future requirements. New data files must be developed as new data types are generated.

The procedures involved in establishing and maintaining data bases are outlined in Figure 17.4. The steps are as follows:

(1) Data need identified (e.g. by the user community).

(2) Existing data sources determined (e.g. by a workshop and an inventory).

(3) Data files received at data center.

(4) Documentation/data formats checked or set-up, if not available.

(5) Archived data advertised (data fliers, catalogs) and distributed to meet requests.

(6) Data files updated periodically.

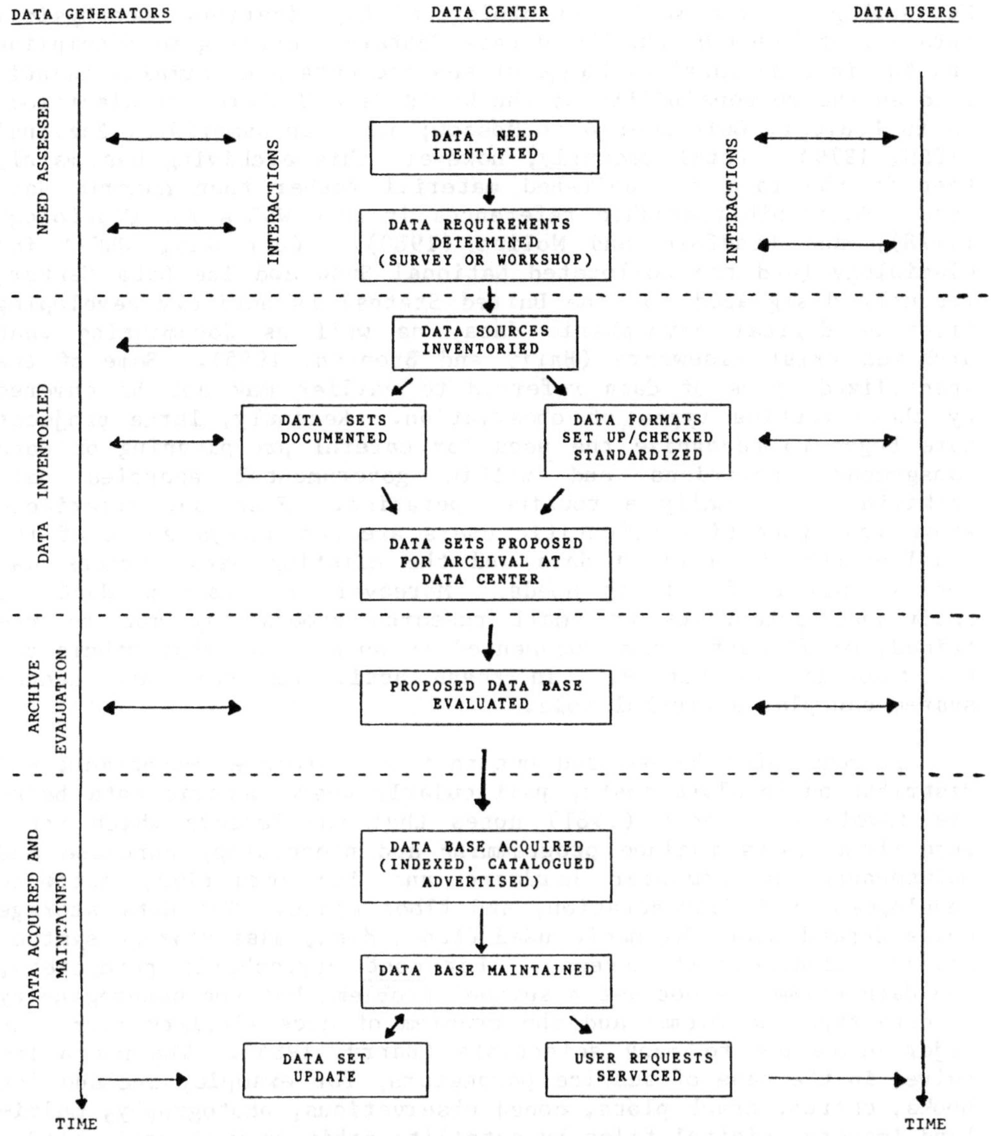

Fig. 17.4. The functions of a data center and its interaction with the generators and users of data.

The focal point for such activities is normally a data center. In many countries, sea ice observations are the responsibility of the national meteorological or hydrographic service but most of these data are available only in chart form, as noted earlier. Standards for ice observing codes and chart displays are prepared and reviewed by a Working Group of the Commission for Maritime

Meteorology of the World Meteorological Organization. Geophysical data are archived by the World Data Centers according to discipline and the international exchange of sea ice data are formally identified as the responsibility of the World Data Centers for Glaciology (A in Boulder, Colorado; B in Moscow; and C in Cambridge, England) (ICSU, 1979). Until recently, however, this archiving has mainly been in the form of published material rather than numeric data sets. Major bibliographic references include WDC-A for Glaciology (1978), and Bradford and Moline (1980). Currently, WDC-A for Glaciology (and the co-located National Snow and Ice Data Center, recently designated for the United States) is actively developing files of digital cryospheric data, as well as documenting what archives exist elsewhere (Barry and Brennan, 1985). Some of the specialized types of data referred to earlier may not be covered by these routine types of observation. Recently, large projects have begun to recognize the need for careful pre-planning of data management operations and within governmental agencies data archiving is normally a routine operation. Even so, experience shows that potential scientific users are not always aware of the availability of archived data and the existing data formats may not be optimal for their needs. Moreover, the data products of individual scientists or small research groups may not be retained, or formatted and documented in such a way that others can use them in the future. In these activities the data center system can play a crucial role.

It must also be pointed out that data storage, management and distribution involves costs, particularly where numeric data banks are involved. Jenne (1981) notes that the factors which enter into these costs include programming and processing, purchase and maintenance of computer hardware and its operation, software development and documentation, and floor space. The data storage costs depend upon the media used (tape, disk, mass storage system, and the frequency of access). For <u>most</u> cryospheric parameters, the data volume is not yet a serious problem, but the heterogeneity of data type and format and the problem of accessibility represent major obstacles to many scientists (Barry, 1984). The media involved in the case of sea ice parameters, for example, include log books, charts, track plots, coded observations, photography, multiband imagery, digital files by satellite orbit as well as in grids, and analog sonar traces. Recognition of the potential future problems that may arise, if appropriate data management strategies are not devised, is the essential first step to well-structured data bases that can adequately serve the needs of users.

ACKNOWLEDGMENTS

I wish to thank W. E. Markham, Director of the Ice Branch, Atmospheric Environment Service, Canada, for his comments on a first draft of this chapter. The work was supported in part by

DOE-NOAA Contract NA84RAH05116 from the National Geophysical Data Center, NOAA/NESDIS. I am indebted to M. Strauch for her word processing assistance.

REFERENCES

Ackley, S. F. (1981) Sea-ice atmosphere interactions in the Weddell Sea using drifting buoys. In Sea Level, Ice and Climatic Change (I. Allison, Ed.), IAHS Publication 131, p. 177-191.

Anonymous (1981) The International Ice Patrol. Marine Observer 51: 20-31.

Armstrong, T. (1950) Study of sea ice in the Soviet Arctic, 1920-45. Polar Record, 5(39): 468-473.

Armstrong, T. (Compiler) (1958) Sea ice north of the USSR. Part 1: Frequency charts (19 p.); Part 2: Sources of information and their evaluation (39 p.). Hydrographic Department of the Admiralty, U.K., H.D. 511.

Armstrong, T. (1964) Ice atlases. Polar Record, 12(77): 161-163.

Bailey, W. B. (1972) A review of Coast Guard published data on icebergs in the North Atlantic. In Proceedings, Canadian Seminar on Icebergs, Department of National Defence, Halifax, N.S., Canada, p. 30-47.

Barnett, D. G. (1976) A practical method of long-range ice forecasting for the north coast of Alaska. Part 1. Fleet Weather Facility, Suitland, Md., Technical Report, 16 p.

Barry, R. G. (1979) Study of climatic effects on fast ice extent and its seasonal decay along the Beaufort-Chukchi coasts. In Environmental Assessment of the Alaska Continental Shelf: Final Reports of the Principal Investigators, Vol. 2., Physical Science Studies. National Oceanic and Atmospheric Administration, Boulder, Colorado, p. 272-375.

Barry, R. G. (ed.) (1984) Workshop on Antarctic Climate Data. Glaciological Data, Report GD-15, World Data Center-A for Glaciology, University of Colorado, Boulder, 69 p.

Barry, R. G. and A. M. Brennan (1985) World Data Center-A for Glaciology: Functions and services. Antarctic J. United States, in press.

Betin, V. V. and Yu. V. Preobazhensky (1959) Variations in the state of the ice on the Baltic Sea and in the Danish Sound. Trudy Gos. Okean. Inst. (Moscow), 37: 3-13. (Translation 102, U.S. Navy Hydrographic Office, 1961.)

Bradford, J. D. and M. Moline (Compilers) (1980) Bibliography on northern sea ice and related subjects. Canadian Special Publication of Fisheries and Aquatic Sciences, No. 45, Ottawa, 171 p.

Brooks, C. E. P. and W. A. Quennell (1928) The influence of arctic ice on the subsequent distribution of pressure over the eastern North Atlantic and Western Europe. Geophysical Memoirs (London), 4(41), 36 p.

Burckle, L. H., P. Robinson and D. Cooke (1982) Reappraisal of sea-ice distribution in Atlantic and Pacific sectors in the Southern Ocean at 18,000 yr. BP. Nature 299: 435-437.

Burrows, C. J. (1976) Icebergs in the Southern Ocean. New Zealand Geogr. 32: 127-138.

Canada, Department of Transport (1965) MANICE (Manual of standard procedures for ice reconnaissance). Meteorology Branch, CIR-4130, ICE-18, Toronto, 53 p.

Cavalieri, D. J., P. Gloersen and W. J. Campbell (1984) Determination of sea ice parameters with the Nimbus 7 SMMR. J. Geophys. Res. 89(D4): 5355-5369.

Chiu, L. S. (1983) Variation of Antarctic sea ice: an update. Mon. Wea. Rev. 111: 578-580.

Cooke, D. W. (1978) Variations in the seasonal extent of sea ice in the Antarctic during the last 140,000 years. Columbia University, Ph.D. thesis, 287 pp.

Crane, R. G. (Compiler) (1979) Inventory of snow cover and sea ice data. World Data Center-A for Glaciology, Boulder, Colorado, Glaciological Data Report GD-7, 171 pp.

Crane, R. G., R. G. Barry and J. Zwally (1982) Analysis of atmosphere-sea interactions in the Arctic Basin using ESMR microwave data. Int. J. Rem. Sens. 3: 259-276.

Danielson, E. W., Jr. (1971) Hudson Bay ice conditions. Arctic, 25: 90-107.

Danish Meteorological Institute (1901-1956) The State of the Ice in the Arctic Seas. Charlottenlund, Copenhagen, Annual Appendix to Nautical-Meteorological Annuals.

Danish Meteorological Institute (1960-1968) The Ice Conditions of the Greenland Waters. Annual. Charlottenlund, Copenhagen.

Dunbar, M. (1967) The monthly and extreme limits of ice in the Bering Sea. In Physics of Snow and Ice, (H. Oura, Ed.), Vol. 1, Institute of Low Temperature Science, Hokkaido University, Sapporo, Japan, p. 687-703.

Gloersen, P., T. T. Wilheit, T. C. Chang, W. Nordberg and W. J. Campbell (1974) Microwave maps of the polar ice of the earth. Bull. Am. Met. Soc., 55: 1442-1448.

Gustajtis, K. A. (1979) Iceberg scouring on the Labrador Shelf, Saglek Bank. Memorial University of Newfoundland, St. John's, Newfoundland, Canada, C-CORe Publication 79-13, 89 p.

Gustajtis, K. A. and T. J. Buckley (n.d.) Iceberg density distribution along the Labrador coast. C-CORE, Memorial University of Newfoundland, St. John's, Newfoundland, Canada. (4 seasonal maps, 1:200,000).

Hakodate Marine Observatory (1955) Sea-ice distribution maps in the southern Okhotsk Sea (1937-1944). Japan, 43 p.

Hansen, G. J. A. and P. P. Sveistrup (1943) Arctic ice fluctuations in Julianehaab Bay, 1901-1937. Medd. Gron., 131(13), 32 p.

Hays, J. D. (1978) A review of the Late Quaternary climatic history of Antarctic seas. In Antarctic Glacial History and World Palaeoenvironments (E.M. Van Zinderen Bakker, Ed.), A.A. Balkema, Rotterdam, p. 57-71.

Heap, J. A. (1963) Sea ice distribution in Antarctica between longitudes 7°W and 92°W. Admiralty Hydrographic Department, London, H.D. 542, 5 p. and 144 sheets.

Heap, J. A. (1972) International cooperation in sea ice observing, recording and reporting. In Sea Ice (T. Karlsson, Ed.), Proceedings, International Conference, Reykjavik, National Research Council of Iceland, p. 80-83.

Hibler, W. D. III (1980) Modeling a variable thickness sea ice cover. Mon. Wea. Rev., 108: 1943-1973.

Hunt, W. R. and C. M. Naske (1979) A baseline study of historic ice conditions in the Beaufort Sea, Chukchi Sea, and Bering Strait. In Environmental Assessment of the Alaska Continental Shelf: Final Reports of the Principal Investigators, Vol. 1, Physical Science Studies, National Oceanic and Atmospheric Administration, Boulder, Colorado, p. 122-534.

ICSU (1979) Fourth Consolidated Guide to International Data Exchange Through the World Data Centres, Secretariat, International of Council Scientific Unions, Panel on World Data Centres, Washington, D.C., 113 p.

Jacka, T. H. (1983) Computer data base for Antarctic sea ice extent. Austral. Nat. Antarct. Res. Exped., ANARE Res. Notes 13.

Jenne, R. L. (1981) Strategies to develop and access large sets of scientific data. In Frontiers in Data Storage, Retrieval and Display. NOAA and Environmental Data and Information Service, National Geophysical and Solar-Terrestrial Data Center, Boulder, Colorado, p. 5-11.

Jurva, R. (1952) About the variations and changes of freezing in the Baltic during the last 120 years. Fennia, 75: 17-24.

Keen, R. A. (1980) Temperature and circulation anomalies in the eastern Canadian Arctic, summer 1946-76. Institute of Arctic and Alpine Research, University of Colorado, Occasional Papers, 34, 159 p.

Kelly, P. M (1979) An Arctic sea ice data set, 1901-1956. In Workshop on Snow Cover and Sea Ice Data, Glaciological Data Report GD-5, World Data Center-A for Glaciology, Boulder, Colorado, p. 101-106.

King, L. H. (1976) Relict iceberg furrows on the Laurentian Channel and western Grand Banks. Can. J. Earth Sci., 13: 1082-1092.

Kirillov, A. A. and M. S. Khromtsova (1972) Many-year variations of the ice coverage of the Greenland Sea and methods of forecasting it. AIDJEX Bulletin, 16: 46-55.

Koch, L. (1945) The East Greenland ice. Medd. Gron., 130(3), 374 p.

Koerner, R. M. (1973) The mass balance of the sea ice of the Arctic Ocean. J. Glaciol., 12(65): 173-186.

Kondratyev, K. Ya., Yu. I. Rabinovich and W. Nordberg (1975) USSR/USA Bering Sea Experiment. Gidrometeoizdat, Leningrad, 316 p.

Kotlyakov, V. M., and N. N. Dreyer (1978) The World Atlas of Snow and Ice Resources. Compiled in the Soviet Union. World Data Center-A for Glaciology, Boulder, Colorado, Glaciological Data Report GD-3, p. 53-57.

Knight, R. W. (1984) Introduction to a new sea-ice database. Ann. Glaciol. 5: 81-84.

Krutskikh, B. A. (ed.) (1984) Mezhdunarodnaya simbolika dlya morskikh ledovykh kart i nomenklatura morskikh l'dov. Gidrometeoizdat, Leningrad, 56 p.

Kukla, G. and J. Gavin (1979) Snow and pack ice indices. World Data Center-A for Glaciology, Boulder, Colorado, Glaciological Data Report GD-6, Snow Cover, p. 9-14.

Kukla, G., A. Hecht and D. Wiesnet (Eds.) (1981) Snow Watch Workshop. World Data Center-A for Glaciology, Boulder, Colorado, Glaciological Data Report GD-11, 148 p.

LaBelle, J. C., J. L. Wise, R. P. Voelker, R. H. Schulze and G. M. Wohl (1983) Alaska Marine Ice Atlas, Arctic Environmental Information and Data Service, University of Alaska, Anchorage, 302 p.

Lamb, H. H. (1977) Climate: Present, Past and Future, Vol. 2. Methuen, London, p. 515-520, 580-589.

Lindsay, D. G., ed. (1975,1977,1982) Sea ice atlas of arctic Canada, 1961-1968; Sea ice atlas of arctic Canada, 1969-1974; Sea ice atlas of Arctic Canada, 1975-1978. Department of Energy, Mines and Resources, Canada, 213 p., 219 p., 139 p.

Maksimov, I. V., N. P. Smirnov and V. N. Vorobiev (1964) Long-range forecasting of secular changes of the ice in the Barents Sea. Materialy Ryboxozyaistvennyx Issledovanii Severnogo Basseina, Murmansk, 4, p. 75-83.

Markham, W. E. (1980) Ice atlas: Eastern Canadian seaboard. Environment Canada, Atmospheric Environment Service, Toronto, 96 p.

Markham, W. E. (1981) Ice atlas. Canadian arctic waterways. Environment Canada, Atmospheric Environment Service, Toronto, 198 p.

McClain, E. P. (1973) Quantitative use of satellite vidicon data for delimiting sea ice conditions. Arctic, 26: 44-57.

McClain, E. P. (1978) Sea ice observations by NOAA's National Environmental Satellite Service. World Data Center-A for Glaciology, Boulder, Colorado, Glaciological Data Report GD-2, p. 33-42.

Meinardus, W. (1905) Uber Schwankungen der nordatlantischen Zirkulation und damit zusammenhangende Erscheinungen. Met. Zeit., 22: 398-412.

Meinardus, W. (1906) Periodische Schwankungen der Eisdrift bei Island. Ann. Hydrogr. Marit. Met., 34: 148-162, 227-239, 278-285.

Miles, M. K. (1974) An index of pack-ice severity off Newfoundland and its secular variation. Met. Mag., 103: 121-125.

Mitchell, P. A. (1976) Aerial ice reconnaissance and satellite ice information microfilm file. Naval Oceanography Office, Ref. Publ. 17, 33 p.

Mitchell, P. A. (1977) Aerial ice reconnaissance and satellite ice information microfilm file--1976. Naval Oceanography Office, Ref. Publ. 17(76), 8 p.

NASA (1979) Ice and Climate Experiment (ICEX). Goddard Space Flight Center, Greenbelt, Md.

Nazarov, V. S. (1947) Historical variation of ice conditions in the Kara Sea. Izv. Vses. Geog. Obsch. (Leningrad), 79(6): 653-655. (Translation T523R, Defence Research Board of Canada, Ottawa, 1969.)

Ogilvie, A. E. J. (1984) The past climate and sea-ice record from Iceland, Part I: Data to A.D. 1970. Climatic Change 6: 131-152.

Palusuo, E. (1966) Ice in the Baltic. Ann. Rev. Ocean Mar. Biol., 4:79-90.

Parkinson, C. L. and W. M. Washington (1979) A large-scale numerical model of sea ice. J. Geophys. Res., 84: 311-337.

Reimnitz, E., P. W. Barnes, L. J. Toimil, and J. Melchior (1977) Ice gouge recurrence and rates of sediment reworking, Beaufort Sea, Alaska. Geology, 5: 405-408.

Rogers, J. C. (1983) Spatial variability of Antarctic temperature anomalies and their association with the southern hemisphere atmospheric circulation. Ann. Ass. Amer. Geogr., 73: 502-518.

Ropelewski, C. F. (1983) Spatial and temporal variations in Antarctic sea-ice (1973-1982). J. Clim. Appl. Met., 22(3): 470-473.

Ruddiman, W. F. (1977) Late Quaternary deposition of ice-rafted sand in the subpolar North Atlantic (lat. 40° to 65°N). Geol. Soc. Am. Bull., 88: 1813-1827.

Ruddiman, W. F., A. McIntyre, V. Niebler-Hunt and J. T. Jurazzi (1980) Oceanic evidence for the mechanism of rapid Northern Hemisphere Glaciation. Quat. Res. 13: 33-64.

Ryder, C. (1896) Isforholdene i Nordhavet, 1877-1892. Tidsskr. f. Sovaesen, Kobenhaven, 28 p.

Sanderson, R. M. (1975) Changes in the area of arctic sea ice, 1966 to 1974. Met. Mag., 104: 313-323.

Sawada, T. (1957) A study of the statistic forecast of summer temperature over northern Japan using the amount of sea-ice over Okhotsk Sea (I). J. Met. Soc. Japan, 35: 60-66.

Skov, N. A. (1970) The ice cover of the Greenland Sea. Medd Gron., 188(2), 56 p.

Sigtryggsson, H. (1972) An outline of sea ice conditions in the vicinity of Iceland. Jokull, 22: 1-11.
Sovietskaya Antarkticheskaya Expeditsia (1966) Atlas Antarktiki. Glavnoe Upravlenie Geodesii i Kartografii, Moscow. Plates 124-126.
Sowden, W. J. and F. E. Geddes (1980) Weekly median and extreme ice edges for eastern Canadian seaboard and Hudson Bay. Environment Canada, Ice Climatology and Applications Division, Ottawa.
Speerschneider, C. I. H. (1915) Om Isforholdene i danske Farvande i aeldre og nyere Tid: Aarene 690-1860. Medd. Danske Met. Inst., Nr. 2 (Copenhagen), 123 p.
Speerschneider, C. I. H. (1917) Summary of the state of the ice in arctic seas. In Nautisk Meteorologisk Aarbog, 1916, Danske Met. Inst., Copenhagen, xxiii-xlvii.
Speerschneider, C. I. H. (1927) Om Isforholdene i danske Farvande: Aarene 1861-1906. Med. Danske Met. Inst., Nr. 6 (Copenhagen), 84 p.
Speerschneider, C. I. H. (1931) The state of the ice in Davis Strait, 1820-1930. Medd. Danske Met. Inst., Nr. 8 (Copenhagen), 53 p.
Stateman, M. (1978) Final disposition of AIDJEX data bank files. AIDJEX Bulletin, 40: 190-201.
Sturman, A. P. and M. R. Anderson (1985) A comparison of Antarctic sea ice data sets and inferred trends in ice data. J. Clim. Appl. Met., 24:275-280.
Swedish Meteorological and Hydrological Institute (1982) Climatological Ice Atlas for the Baltic Sea, Kattegat, Skagerrak and Lake Vänen (1963-1979). SMHI, Norköpping, Sweden, 220 p.
Swithinbank, C. W. M. (1960) Ice Atlas of Arctic Canada. Queen's Printer, Ottawa, 67 pp.
Swithinbank, C. (1972) Arctic pack ice from below. In: Sea Ice (T. Karlsoon, Ed.), Proceedings, International Conference, Reykjavik, National Research Council of Iceland, p. 246-254.
Tabata, T., M. Ishikawa, M. Oi, H. Fukushi, M. Aota and T. Kawamura (1980) Distribution of pack ice off Okhotsk Sea coast of Hokkaido observed with sea ice radar network, January-April 1980. Low Temp. Sci., Ser. A, 39, Data Rep.: 35-61.
Thompson, T. (1981) Proposed format for gridded sea ice information (SIGRID). World Meteorological Organization, Geneva, Report, World Climate Programme, 28 p.
Thorndike, A. S., D. A. Rothrock, G. A. Maykut and R. Colony (1975) The thickness distribution of sea ice. J. Geophys. Res., 80: 4501-4513.
Untersteiner, N. (1980) AIDJEX review. In Sea Ice Processes and Models, (R. S. Pritchard, Ed.), IAHS Publication 124, p. 3-11.
U.S. Fleet Weather Facility (1976a) Eastern Arctic sea ice analyses, 1972-75. U.S. Navy, Suitland, Maryland, 212 p. (AD-A033 345).

U.S. Fleet Weather Facility (1976b) Weatern Arctic Sea ice analyses, 1972-75. U.S. Navy, Suitland, Maryland, 212 p. (AD-A033 344).

U.S. Naval Oceanographic Office (1956) Ice observations. Washington, D.C., 2nd ed., H.O. Pub. No. 606-d, 42 p.

U.S. Naval Oceanographic Office (1962-74) Report(s) of the arctic ice observing and forecasting program - 1960-71. Washington, D.C., Special Publication SP-70 (60-71).

U.S. Naval Oceanographic Office (1963) Antarctic ice observations October 1962-March 1963. Washington, D.C., Special Publication SP-80(62), 58 p.

U.S. Naval Oceanographic Office (1964-69) Report(s) of the Antarctic ice observing and forecasting program. Special Publication 80(62-69), 1966-69.

U.S. Navy Hydrographic Office (1946) Ice atlas of the Northern Hemisphere. Washington, D.C., H.O. Publ. 550, 106 p. + charts.

U.S. Navy Hydrographic Office (1954) Report of ice observations and forecasting for the season 1953. Washington, D.C., H.O. Misc. 15869, 65 p.

U.S. Navy Hydrographic Office (1954-1959) Report of ice observations and forecasting for the season 1954 annually to 1959. Washington, D.C., Technical Reports TR-49-52, 66, 69.

Valeur, H. H. (1976) Polar ice variations off the Greenland west coast 1900-72. International Commission for the Northwest Atlantic Fisheries, Dartmouth, Nova Scotia, Canada, Special Publication No. 9, p. 7-28.

Vilmundarson, T. (1972) Evaluation of historical sources on sea ice near Iceland. In Sea Ice (T. Karlsson, Ed.), Proceedings, International Conference, Reykjavik, National Research Council of Iceland, p. 159-169.

Vinje, T. E. (1977) Sea ice conditions in the European sector of the marginal seas of the Arctic, 1966-1975. In Norsk Polarinstitutt Arbok 1975, Oslo, Norway, p. 163-174.

Wadhams, P. and R. J. Horne (1980) An analysis of ice profiles obtained by submarine sonar in the Beaufort Sea. J. Glaciol., 25(93): 401-426.

Walker, G. (1947) Arctic conditions and world weather. Quart. J. Roy. Met. Soc., 73: 226-256.

Walsh, J. E. (1978) A data set on Northern Hemisphere sea ice extent, 1953-76. World Data Center-A for Glaciology, Boulder, Colorado, Glaciological Data Report GD-2, p. 49-51.

Walsh, J. E., W. D. Hibler, III and B. Ross (1984) A model simulation of 20 years of northern hemisphere sea ice fluctuations. Ann. Glaciol., 5: 170-176.

Washington, W. M., A. J. Semtner, Jr., G. A. Meehl, D. J. Knight and T. A. Mayer (1980) A general circulation experiment with a coupled atmosphere, ocean and sea ice model. J. Phys. Oceanogr. 10: 1887-1908.

Watanabe, K. (1967) Summary of drift ice in the Okhotsk Sea. In Physics of Snow and Ice (H. Oura, Ed.), Vol. 1, Institute of Low Temperature Science, Hokkaido University, Sapporo, Japan, p. 667-686.

Weaver, R. L., R. G. Barry and J. D. Jacobs (1976) Fast ice studies in western Davis Strait. In Proceedings, Third International Conference on Ports and Ocean Engineering under Arctic Conditions, Vol. 1, University of Alaska, Fairbanks, USA, p. 455-466.

Webster, B. D. (1981) A climatology of ice extent in the Bering Sea. Anchorage, Alaska, NOAA Tech. Memorandum NWS, A-33.

Westhall, V. H. (1975) An arctic sea ice data bank from airborne observations. Naval Oceanographic Office, Washington, TN 6140-2-75, 10 p.

World Data Center-A for Glaciology (1978) Arctic sea ice. Boulder, Colorado, Part 1 and Part 2, Glaciological Data Report GD-2, Colorado, 262 p.

World Data Center-A for Glaciology (1980) Great Lakes ice, Boulder, Colorado, Glaciological Data Report GD-9, 79 p.

World Meteorological Organization (1970) WMO sea-ice nomenclature. WMO, Geneva, WMO No. 259, TP 145, 147 p.

World Meteorological Organization (1981) Sea-ice information services in the world. WMO, Geneva, WMO No. 574, 108 p.

Zwally, H. J., J. C. Comiso, C. L. Parkinson, W. J. Campbell, F. Carsey and P. Gloersen (1983) Antarctic sea ice cover, 1973-76, from satellite passive microwave imagery. NASA Goddard Space Flight Facility, Greenbelt, Md.

SPECIAL TOPIC 1

ACCURACY OF SURFACE GEOSTROPHIC WIND FORECASTS IN THE CENTRAL ARCTIC

Richard E. Moritz

Polar Science Center
University of Washington
Seattle, WA 98195, U.S.A.

1. INTRODUCTION

Surface geostrophic winds are related simply to observed surface winds (Figure 1) and pack ice motion in the central Arctic (Albright, 1980; Thorndike and Colony, 1982). Their prediction, therefore, assumes special importance for sea ice forecasting, air and marine navigation and offshore resource exploration and development.

Systematic assessments of geostrophic wind forecast accuracy in the central Arctic have hitherto been hampered by the paucity of synoptic pressure measurements; neither forecast initializations nor verification fields were reliable. This situation improved when, in early 1979, an array of drifting data buoys was deployed on the arctic pack ice, measuring surface pressure as part of the First GARP Global Experiment (FGGE) (Untersteiner and Thorndike, 1982). In this article we evaluate arctic geostrophic wind forecasts produced operationally at the U.S. National Meteorological Center (NMC) during 1979 and 1980, when good coverage of surface pressure was provided by the buoys.

The data comprise about 18 months of daily geostrophic winds. We restrict our analysis to prediction time intervals of one, two and three days. Readers interested in climatological estimates of average arctic pressure distribution are referred to Keegan (1958), Reed and Kunkel (1960) and Walsh (1978). Longer term predictability of large scale arctic pressure patterns was investigated by Walsh (1980). Previous works on short term arctic wind forecasts

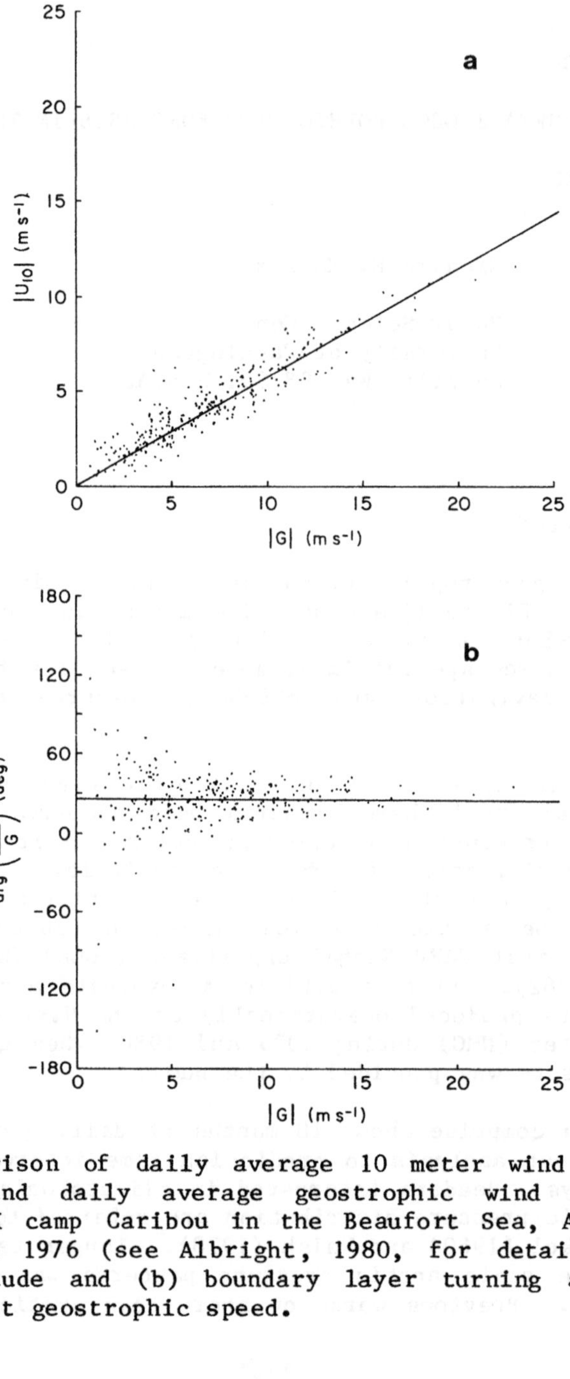

Fig. 1. Comparison of daily average 10 meter wind measurements U_{10} and daily average geostrophic wind estimates at AIDJEX camp Caribou in the Beaufort Sea, April, 1975 - April, 1976 (see Albright, 1980, for details). (a) U_{10} magnitude and (b) boundary layer turning angle plotted against geostrophic speed.

include those by Walsh (1977) and by Denner and Ashim (1980). These studies suffer from insufficient arctic basin pressure measurements in the forecast initializations and verification fields. Space and time statistics of arctic surface pressure fluctuations are analyzed by Thorndike (1982).

One application for this work is in the design and evaluation of prediction models for sea ice motion. With this application in mind, we study statistical properties of the forecasts, computed from many daily realizations, in contrast to a case study approach that focuses on particular features of given weather situations, such as the motion of individual high and low pressure centers. The main objectives are: 1) to estimate statistical properties of the wind forecasts and compare them with observed parameter estimates, and 2) to find linear models that make optimal use of the NMC forecasts by minimizing the error variances. It is not the intention here to trace each forecast statistic back to its origins in the NMC prediction model. However, some of the results raise questions about the ability of the model to reproduce atmospheric statistics analyzed from data.

2. STATISTICAL CHARACTERIZATION OF THE WINDS

The geostrophic wind data are based on observations of the surface pressure field p(x,y,t) where x and y are Cartesian space coordinates and t is time. Suppose p is regarded as a time-dependent, random surface, whose statistics are homogeneous, isotropic, stationary, and Gaussian. Then any statistical property involving p or its derivatives can, in principle, be derived from the covariance function

$$R(r,t) = \overline{p(0,0,0)\, p(\alpha_1, \alpha_2,t)} \qquad (1)$$

where $r = \sqrt{\alpha_1^2 + \alpha_2^2}$, the overbar denotes statistical expectation (see below), and the origin in space and time is arbitrary. Thorndike (1982) showed that several observed properties of the arctic surface pressure field agree with predictions based solely on knowledge of R (estimated from the buoy measurements) and the Gaussian model. These properties include the covariance functions for surface geostrophic wind and vorticity and the average areal density of highs, lows, and saddles on the pressure surface. This agreement suggests that a second moment description of analysis and forecast statistics should capture much of the information contained in the data sets.

Vectors in the horizontal plane are treated as complex numbers, whose real and imaginary parts correspond to x and y components. Our coordinate system has its origin at the pole, the positive x-axis along the Greenwich meridian, and the positive y-axis along 90°E. Cartesian coordinates (x,y) are related to latitude ϕ and longitude λ by

$$x = a_o\ (90^\circ - \phi)\cos(\lambda)$$

$$y = a_o\ (90^\circ - \phi)\sin(\lambda) \tag{2}$$

where a_o = 110.949 km. We assume that sea level and 1000 mb geostrophic winds may be compared directly. It follows that we may express the geostrophic wind as the complex number

$$G = \frac{\nabla p}{\rho_a f_o}\, e^{i\frac{\pi}{2}} = \frac{g\nabla Z}{f_o}\, e^{i\frac{\pi}{2}} \tag{3}$$

where $\nabla = \frac{\partial}{\partial x} + \frac{\partial}{\partial y}$, ρ_a is the air density, f_o is the Coriolis parameter, g is earth's gravitational acceleration, Z is the height of the 1000 mb pressure surface, and $i^2 = -1$. For notational convenience, we distinguish different geostrophic wind data sets with a subscript system (Table 1). The symbol G_b represents the surface geostrophic wind analyzed at the Polar Science Center using all available buoy data. G_N stands for the geostrophic wind computed from NMC's 1000 mb height analysis. The NMC 1000 mb height forecasts over 24, 48, and 72 hours are designated by G_{24}, G_{48}, and G_{72}. Individual measurements or forecasts are thought of as functions of discrete space and time. For example the symbol $G_k(j,t)$ represents the geostrophic wind computed from data set k, at the jth grid point, at time t. Expectation values are estimated from time series at a single grid point

$$\bar{G}_k(j) = \frac{1}{N}\sum_{t=1}^{N} G_k(j,t) \tag{4}$$

Table 1. Geostrophic wind data sets.

Location	Dates	Variables*
Arctic, 28 points	Mar-Dec, 1979 May-Dec, 1980	G_b, G_N, G_τ
40°N, 86°W (ML)	Jan-Dec, 1979 Jan-Dec, 1980	G_N, G_τ

*G_b - Surface geostrophic wind, buoy analysis
G_N - 1000 mb geostrophic wind, NMC analysis
G_τ - 1000 mb geostrophic wind, τ - hour NMC forecast

or over space and time together

$$\bar{G}_k = \frac{1}{NM} \sum_{j=1}^{M} \sum_{t=1}^{N} G_k(j, t) \ . \tag{5}$$

The averages (eqs. 4 and 5) are referred to as "grid point" and "overall" parameters. When a given function or statistic appears with only a single argument, it represents the time-dependence, and the spatial argument (j, as in eq. 4) is suppressed. Averages over the M grid points at fixed time are called "spatial" statistics. Fluctuations about an average value are denoted by primed quantities, such as $G_k'(t) = G_k(t) - \overline{G_k}$.

Statistical parameters encountered in second moment descriptions of stationary, real valued time series have counterparts in the complex case. Denoting complex conjugation with an asterisk, we define the variance

$$\mathrm{var}(G_1) = \overline{G_1 G_1^*} - \overline{G_1}\ \overline{G_1^*} \tag{6}$$

covariance

$$\mathrm{cov}(G_1, G_2) = \overline{G_1 G_2^*} - \overline{G_1}\ \overline{G_2^*} \tag{7}$$

and squared correlation

$$\rho^2(G_1, G_2) = \frac{[\mathrm{cov}(G_1, G_2)]^2}{[\mathrm{var}(G_1)\mathrm{var}(G_2)]} \tag{8}$$

Here G_1 and G_2 could represent winds taken from different data sets, or winds from a single data set, lagged in space or time, and so forth.

We shall also consider the variances and covariances of forecast and analysis winds in different frequency bands, as an aid to understanding the accuracy of predictions on various time scales. Techniques described in Jenkins and Watts (1968, sec. A5.3, A7.1, and A9.2), and modified for complex time series are used to estimate the cross spectrum, at a single grid point

$$\Gamma_{12}(f) = \int_{-\infty}^{\infty} \gamma_{12}(\tau) e^{-2\pi i f \tau} d\tau = \alpha_{12}(f) e^{i\phi_{12}(f)} \tag{9}$$

where f is the frequency, $\gamma_{12}(\tau) = \mathrm{cov}(G_1(0), G_2(\tau))$, $\alpha_{12}(f)$ is the real cross amplitude spectrum and $\phi_{12}(f)$ is the phase spectrum. The squared coherency

$$\kappa_{12}^2(f) = \frac{\alpha_{12}^2(f)}{\Gamma_{11}(f)\Gamma_{22}(f)} \tag{10}$$

measures the variance at the frequency f that can be explained by a linear model relating G_1 and G_2. Spectra of complex time series vary on both positive and negative frequencies corresponding to anticlockwise and clockwise rotation of the velocity vector as time increases. The integral of $\Gamma_{12}(f)$ over the interval $f_1 \leq f \leq f_2$ gives the portion of $\mathrm{cov}(G_1, G_2)$, due to fluctuations with frequencies in this range. For positive frequencies the phase $\phi_{12}(f)$ is positive (negative) accordingly as fluctuations in G_1 lag (lead) those in G_2.

3. DATA

The buoy array covered a substantial portion of the Arctic Ocean. Figure 2 shows the configurations in mid-July, 1979 and mid August, 1980. Typically neighboring buoys were separated by distances on the order of 500 km. Spatial variations in pressure on smaller scales, and their contributions to the geostrophic wind are not resolved by these measurements. A characteristic diameter for surface high and low pressure systems in the Arctic is 1100 km (Thorndike, 1982) so that these circulation features are usually

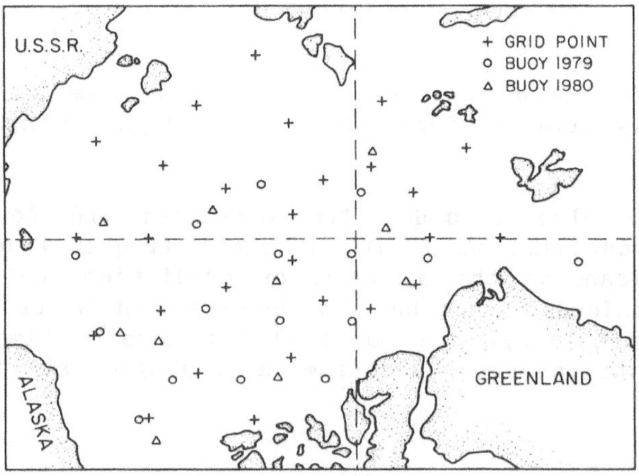

Fig. 2. Arctic base map showing 28 grid points (+) and summer buoy positions in 1979 (o) and 1980 (Δ).

resolved. The time periods for which sufficiently dense buoy coverage was available are March to December, 1979 and May to December, 1980.

Buoy pressure and position data are determined at Service Argos in France. All relevant reports are then archived on magnetic tape and forwarded to the Polar Science Center for further analysis. These tapes provide the data base for the buoy winds described below.

Pressure and positions are also authorized for real time transmission to participating weather services, including NMC, via the Global Telecommunications System (GTS). The accuracy of NMC's arctic analyses and forecasts depends, therefore, on timely receipt and assimilation of these data.

NMC data available for this research include three forecast fields and a final analysis field. Unfortunately, the operational data used to initialize the forecasts are not archived by NMC. The data base for the final analysis is updated three times subsequent to the operational initialization, raising questions about the amount of buoy data incorporated in the actual forecast initial fields. We obtained printouts of buoy reports available for the NMC final analyses in each of three 7-day periods during March, June and September, 1979. Virtually all relevant buoy reports had been received by NMC. Moreover, selected samples of NMC operational bulletins showed that essentially all buoy data are received there between one and five hours after the pressure observations are made. The data cutoff time for 0000 GMT forecasts is +3.5 hours and, since 1979, arctic buoy measurements made as early as -3.0 hours have been utilized in the 0000 GMT operational runs. Thus we can assume that most of the buoy reports enter directly into the operational initial fields and into the NMC final analysis. The use of some non-synoptic data introduces only small errors here, as illustrated by the three hour lag autocorrelation coefficient r = 0.99 for arctic buoy pressure time series (Thorndike, 1982). It is also worth noting that buoy information received after the cutoff is incorporated indirectly, because the first-guess field used in the 0000 GMT operational update is a 12-hour forecast initialized with all data available 9 hours after the nominal initial time (-1200 GMT). In short, the NMC final analysis should correspond closely to the operational data base used to initialize the forecasts.

Table 1 summarizes the wind data sets, all valid daily at 0000 GMT. In 1979 and 1980, surface geostrophic winds G_b, hereafter called the buoy winds, are calculated at 28 grid points (Figure 2). Data editing and analysis procedures are described in Thorndike and Colony (1980, 1981). For present purposes it suffices to note that all relevant buoy reports are utilized to estimate

G_b by optimal interpolation (Gandin, 1965). The variance of errors due to measurement noise and interpolation is estimated not to exceed $(2\ ms^{-1})^2$. The Polar Science Center analysis insures that all buoy data are included, and that buoy reports are given full weight in the procedure. In the NMC final analysis algorithm buoy measurements get half as much weight as regular land station reports (McPherson et al., 1979).

The buoy position measurements provide data for independent estimation of sea ice velocity time series at the 28 grid points. We found the daily ice velocities to be more highly correlated with G_b than with NMC wind analyses and on this basis we regard G_b as the better estimate.

Routine analyses and forecasts of 1000 mb geopotential height are produced daily at NMC. The predictions are valid at 0000 GMT, 24, 48, and 72 hours after initialization and are computed by numerical integration of a multi-level, primitive equation model (Shuman and Hovermale, 1968; Stackpole, 1978). Previous forecast fields, valid at the initial time, are updated with current observations to provide the initial fields (Bergman, 1979; McPherson et al., 1979). For this study we use 1000 mb heights archived on the NMC octagonal grid with a spacing of about 400 km in the Arctic. A 15 by 15 subset of each daily grid, centered on the North Pole, is fitted to a bicubic spline (deBoor, 1962). The spline coefficients are used to find height gradient components at each of the 28 grid points, and geostrophic winds are then computed according to eq. (3).

In order to provide background for evaluating arctic statistics, NMC 1000 mb winds were also estimated, by finite difference, at an arbitrary middle latitude site "ML" (40°N, 86°W), in the central United States.

NMC replaced their hemispheric, finite difference model with a global, spectral model that became operational on 12 August 1980. We estimated arctic forecast statistics for periods before and after this change and found no differences that could be attributed unambiguously to the change. Therefore, no distinction between these models is made in the following discussion.

4. RESULTS

Time averages include data from all seasons. Significant interseasonal differences in the mean and variance of arctic sea level pressure were reported by Thorndike (1982), and surface geostrophic winds are also subject to these variations. However, the statistics describing forecast performance do not exhibit large seasonal changes, so all seasons are grouped together here to obtain larger samples.

4.1 Properties of the Buoy Analysis

Consider first the properties of G_b. Overall estimates of the mean, variance and squared autocorrelation at 24 hour lag are shown in Table 2, along with maximum and minimum grid point values for the study periods in 1979 and 1980. If we think of the mean square wind $\overline{G_b G_b^*}$ as measuring the signal strength for our predictand, then it is evident that most of the signal results from fluctuations about the mean, as given by $var(G_b) \approx (8\ ms^{-1})^2$. The square of the mean wind is small by comparison.

The geographical distribution of $\overline{G_b}$ can be visualized with the aid of mean pressure charts (Figure 3). In both study periods a high pressure center dominates the Arctic Basin. The location of maximum pressure is 82°N, 160°E in 1979 and 75°N, 150°W in 1980. These patterns are not greatly different from estimates of longer term annual mean patterns (Walsh, 1978).

The most striking features of Table 2 are the large squared autocorrelations. The statistic $\rho^2(G_1,G_2)$ gives the percentage of $var(G_1)$ explained by the generalized linear model

$$G_1'(t) = A_{12}G_2'(t) + \varepsilon_{12}(t) \tag{11}$$

and is therefore a measure of the accuracy with which G_1 can be forecast from G_2. Here the complex constant

Table 2. Parameter estimates for arctic winds.

		Mar-Dec, 1979					May-Dec, 1980				
	X →	G_b	G_N	G_{24}	G_{48}	G_{72}	G_b	G_N	G_{24}	G_{48}	G_{72}
$\lvert\bar{X}\rvert$ (ms^{-1})	Min*	0.4	0.6	0.3	0.4	0.4	0.6	0.6	0.5	0.3	0.3
	Ovr	0.4	0.5	0.1	0.2	0.3	1.2	1.1	0.8	0.7	0.6
	Max	3.4	2.9	3.4	3.4	3.7	3.4	2.9	3.3	2.7	2.2
$var(X)$ $(ms^{-1})^2$	Min	44	45	36	34	33	50	49	51	50	49
	Ovr	60	59	45	46	44	79	75	65	65	60
	Max	67	68	59	59	59	92	90	93	95	83
$\rho^2[X(0),X(24)]$	Min	0.40	0.39	0.38	0.33	0.30	0.37	0.38	0.35	0.31	0.30
	Ovr	0.52	0.47	0.49	0.46	0.42	0.47	0.51	0.46	0.41	0.40
	Max	0.56	0.52	0.51	0.50	0.45	0.53	0.58	0.53	0.49	0.48

*Min, Max - extremes of time averages at 28 grid points. Ovr - overall space and time averages.

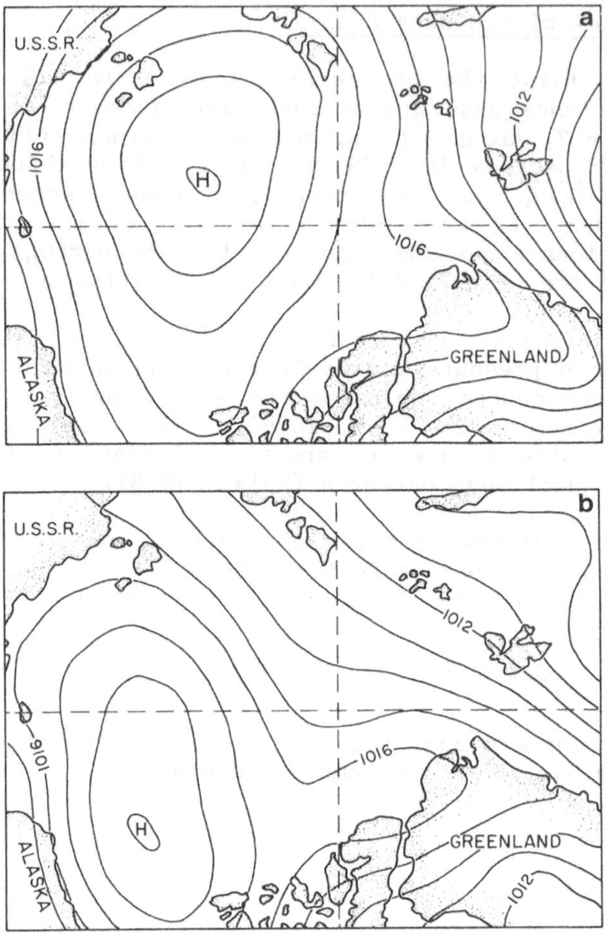

Fig. 3. Average sea level pressure from the buoy data set. (a) March-December, 1979 and (b) May-December, 1980. Contour interval 1 mb.

$$A_{12} = |A_{12}|e^{i\theta_{12}} = \frac{\text{cov}\ (G_1, G_2)}{\text{var}\ (G_2)}\ ,$$

is estimated by imposing the constraints

$$\overline{\varepsilon_{12}} = 0$$

$$\text{cov}(G_2, \varepsilon_{12}) = 0 \qquad (12)$$

on the errors ε_{12}. These constraints yield the minimum error variance, given by

$$\mathrm{var}(\varepsilon_{12}) = (1-\rho^2(G_1,G_2))\mathrm{var}(G_1) \tag{13}$$

If we replace G_1 with $G_b(t)$ and G_2 with $G_b(t-24)$ in eq. (11), we obtain a damped persistence forecast $\hat{G}_b(t) = A_{12}G_b'(t-24)$ that accounts for 50% of the variance of G_b, according to $\rho^2(G_b(0),G_b(24))$ in Table 2. Such a forecast is "damped" in the sense that predicted fluctuations G_1' tend to zero as $|A|$ decreases, no matter how large the particular fluctuation G_2 might be. In the limit as $\mathrm{cov}(G_1,G_2)$ vanishes, the model, eq. (11), implies an optimum prediction $\hat{G}_1(t) = \overline{G_1}$, since $\hat{G}_1(t) = \overline{G_1} + \hat{G}_1'(t)$, thereby utilizing climatological information (the mean). The damped persistence model performs significantly better than the forecast $\hat{G}_b(t) = G_b(t-\tau)$ at lags τ = 24, 48 and 72 hours. Of course, at sufficiently long forecast intervals, the covariance between forecasts and observations vanishes and in this limit the damped model has an error variance equal to $\mathrm{var}(G_1)$, while the undamped error variance is twice as large.

Overall squared autocorrelations at two and three day lags are about 0.2 and 0.1 (Table 3). It follows that large amplitude fluctuations of G_b have characteristic periods longer than a few days. This notion is reinforced by the estimates $\hat{\Gamma}_{bb}(f)$ of the spectrum, a typical example of which is presented in Figure 4.

Table 3. Parameter estimates for arctic winds.

		Mar-Dec, 1979				May-Dec, 1980			
	k—>	N	24	48	72	N	24	48	72
	Min	0.75	0.64	0.49	0.32	0.83	0.59	0.42	0.29
$\rho^2(G_b,G_k)$	Ovr	0.85	0.74	0.58	0.42	0.90	0.74	0.55	0.38
	Max	0.93	0.80	0.64	0.50	0.95	0.81	0.62	0.45
	Min	0.85	0.86	0.71	0.58	0.91	0.75	0.61	0.59
$\|A_{bk}\|$	Ovr	0.93	0.99	0.87	0.76	0.97	0.95	0.82	0.71
	Max	1.02	1.10	0.98	0.83	1.06	1.06	0.95	0.84
	Min	-4	-4	-6	-5	-4	-4	-4	-7
$\theta_{bk}(^o)$	Ovr	0	0	0	2	0	1	3	4
	Max	4	12	17	19	6	14	18	26
	Min	0.2	0.1	0.1	0.2	0.1	0.2	0.2	0.2
$B(G_b,G_k)(ms^{-1})$	Ovr	0.3	0.3	0.4	0.4	0.6	0.0	0.4	0.7
	Max	2.2	1.1	1.4	1.7	0.7	1.1	1.4	1.7
$[\mathrm{var}(\varepsilon_{bk})]^{\frac{1}{2}}(ms^{-1})$	Ovr	3.0	4.0	5.1	5.9	2.8	4.5	60.	7.0
$\rho^2[G_b(0),G_b(k)]$	Ovr	-	0.52	0.21	0.10	-	0.47	0.21	0.12
$S(G_b,G_k)$	Ovr	-	0.45	0.47	0.36	-	0.51	0.43	0.30
$\hat{r}(G_b,G_k)(ms^{-1})$	Ovr	3.1	4.0	5.2	6.2	2.8	4.6	6.2	7.4

Approximately 80% of var(G_b) is associated with fluctuations on periods longer than six days. In terms of forecasts over one, two and three days, damped persistence effects a substantial reduction in error variance that must be exceeded by the NMC forecasts if they are to be useful.

4.2 Comparison Between Buoy Analysis and NMC Analysis

The NMC forecasts cannot be expected to correlate better with G_b than do the initial conditions from which they derive. In order to estimate this baseline accuracy, we compared time series of the two analyses G_N and G_b. Table 2 shows that the mean, variance and autocorrelation of the G_N series are similar to corresponding G_b parameters. The spectral estimates $\hat{\Gamma}_{NN}(f)$ (not shown) are likewise similar to Figure 4. The analysis bias $B(G_b,G_N) = |\overline{G_b} - \overline{G_N}|$ measures systematic differences between the two data sets. Since eq. (11) concerns only the deviations G′ from the mean values, it follows that bias will not be represented by ρ^2. Table 3 shows that overall biases are smaller than the overall mean wind. At individual grid points, the analysis bias is typically less than half as large as $|\overline{G_b}|$.

Fluctuations in $G_N(t)$ and $G_b(t)$ are highly correlated over time, as expected for two different analyses derived from essentially the same data (Table 3). The fact that the series are not identical reflects differences in the analysis procedures,

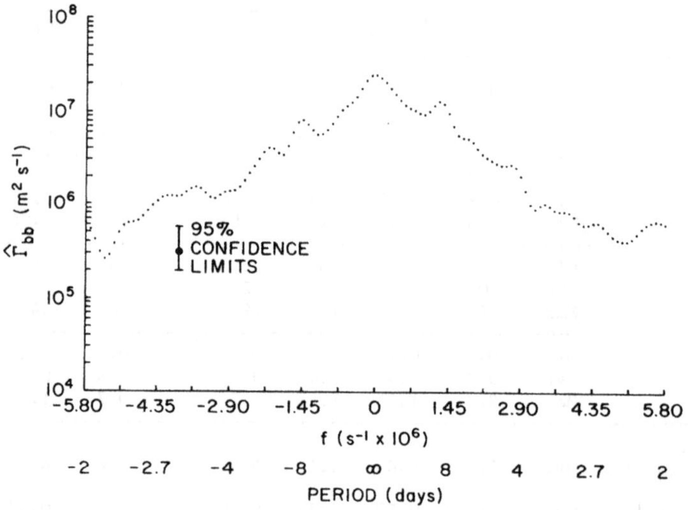

Fig. 4. Spectral estimate for G_b, March-December, 1979 at grid point 17 (82°N, 170°W). Vertical bars show the interval bracketing the estimates wherein the true spectrum lies, with 95% probability.

including the initial guess field and the reduced weight given to buoy measurements by NMC. Roughly 10% of var(G_b) is not linearly related to G_N, corresponding to a typical wind deviation of 2.5 ms^{-1}.

The grid point estimates of $\rho^2(G_b,G_N)$ are not uniform over the basin, having smallest values north of Greenland in both 1979 and 1980 (Figure 5). The cross-basin differences in correlation depicted on these maps differ significantly from zero at the 95% confidence level, assuming Gaussian distributions for G_b and G_N. Similar maps (not shown) of the grid point bias $B(G_b,G_N)$ have

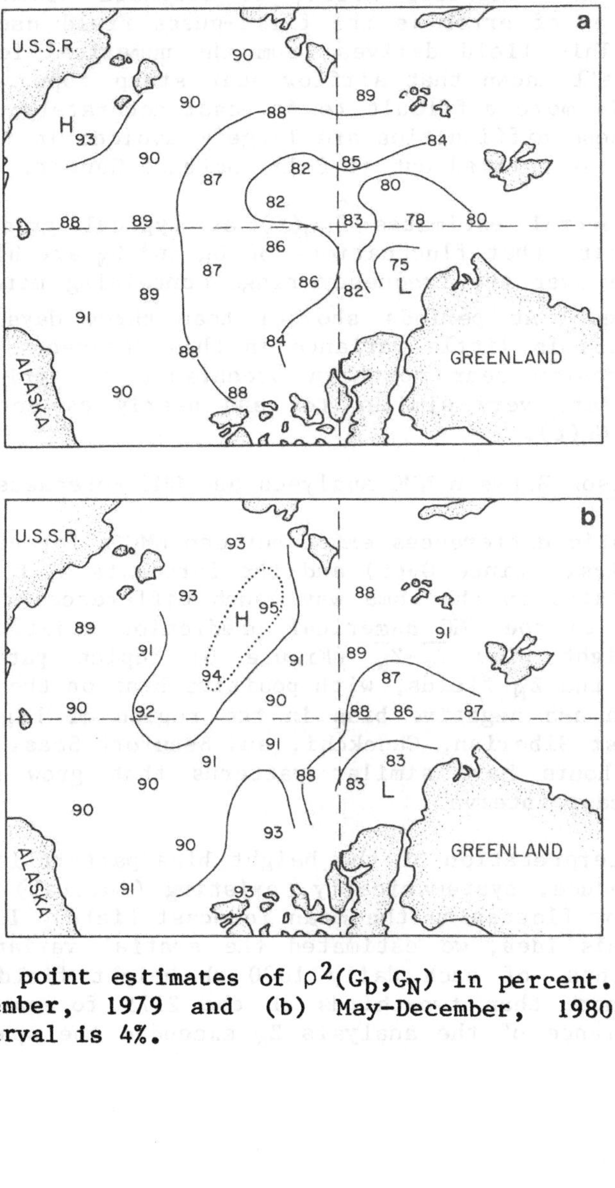

Fig. 5. Grid point estimates of $\rho^2(G_b,G_N)$ in percent. (a) March-December, 1979 and (b) May-December, 1980. Contour interval is 4%.

maxima north of Greenland in 1979, but no pronounced pattern in 1980. The smallest correlation $\rho^2(G_b,G_N)$ occurred at grid point 27 (86°N, 40°W) in 1979 and in 1980. At this same point, the two wind analyses were correlated independently with the estimates of daily sea ice motion in 1979, with the result that G_N accounted for 52% of the ice velocity variance, while G_b accounted for 65%. On this basis G_b is taken to be the better estimate. The 13% difference in explained ice motion here is the largest out of the 28 grid points, a more typical value being 5%.

The occurrence of small $\rho^2(G_b,G_N)$ north of Greenland, in both years, may be caused by the procedure used at NMC to interpolate updated variables from the terrain-following sigma coordinate system, of the numerical model, to standard isobaric surfaces. Another source of error is the first-guess field used in the NMC analysis. This field derives from the numerical forecast model, and it is well known that airflow near steep topography, such as Greenland, is more difficult to forecast accurately than over flat terrain. These difficulties are largely avoided in the two dimensional analyses carried out at Polar Science Center.

The spectral estimates $\hat{\Gamma}_{bN}(f)$ at typical grid points (not shown) indicate that fluctuations of G_b and G_N are highly coherent and in phase over the frequency range containing more than 90% of the variance. At periods shorter than three days $\hat{k}^2_{bN}(f)$ drops off, but there is little variance in this frequency range. Aside from the problem near northern Greenland, the grid point $G_N(t)$ time series are very similar to, and nearly as accurate as the time series $G_b(t)$.

4.3 Comparison Between NMC Analyses and NMC Forecasts

Systematic differences exist between NMC's forecast and analysis statistics. Since $G_N(t)$ and the forecasts $G_\tau(t)$ are analyzed and interpolated in the same way, such differences can be attributed solely to the NMC numerical prediction model. Maps of the 1000 mb height bias $\overline{Z_\tau}-\overline{Z_N}$ (Figure 6) depict patterns roughly opposite to the $\overline{Z_N}$ fields, with positive bias on the European side of the basin and negative bias in the region of higher pressures over the East Siberian, Chuckchi, and Beaufort Seas. Bias maps at 24 and 48 hours have similar patterns that grow stronger with longer forecast interval.

One interpretation of the height bias pattern is that the NMC model may reduce, systematically, existing (initial) height gradients, thereby flattening the mean forecast field. In order to investigate this idea, we estimated the spatial variance, over the 28 grid points, of each daily 1000 mb height field in 1979 and 1980. In more than two-thirds of the 2175 forecast grids, the spatial variance of the analysis Z_N exceeded the spatial variance

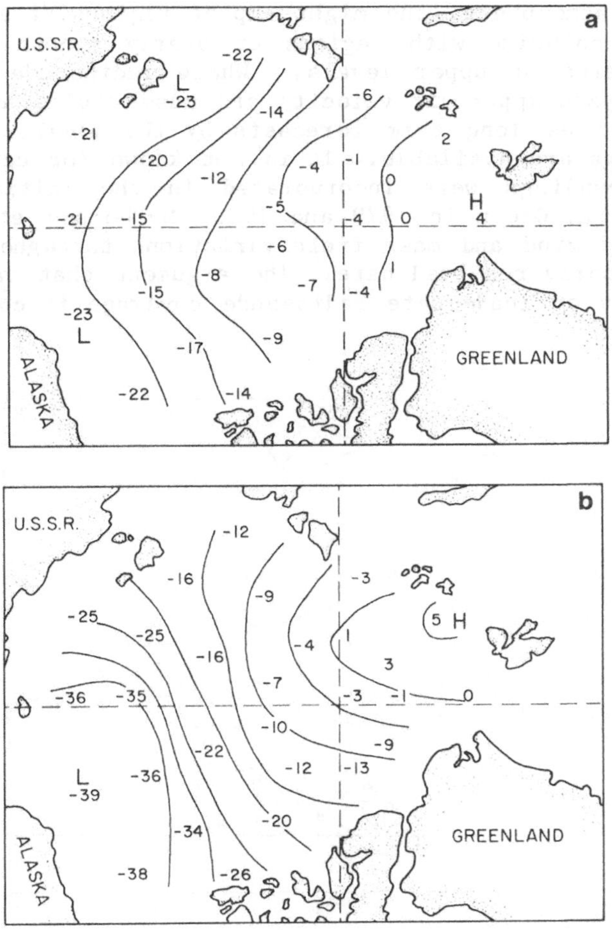

Fig. 6. Grid point estimates of $\overline{Z_{72}-Z_N}$ in meters. (a) March-December, 1979 and (b) May-December, 1980. Contour interval is 5 m.

of the forecast Z_τ. Two-year time averages of these spatial variances, for Z_N, Z_{24}, Z_{48}, and Z_{72} are 301 m^2, 243 m^2, 244 m^2, and 235 m^2. There is indeed a tendency for the forecast model to smooth existing geopotential height gradients.

Another discrepancy is that the overall geostrophic wind forecast variances are 13 to 25% smaller than var(G_N) in the Arctic (Table 2). The geographical distribution of the difference var (G_τ) - var(G_N) (Figure 7) has maxima over the central polar ocean, far removed from dense radiosonde coverage. Since surface pressure tendencies and accompanying geostrophic wind fluctuations derive ultimately from mass flux divergence integrated vertically through the atmosphere, one might expect regions of small forecast variance to coincide with regions of overly smooth initial wind and mass fields at upper levels. Where radiosonde stations are sparse, analyzed upper air velocity and mass fields are essentially determined as long term forecasts by the model, unless satellite soundings are available. It is not known for certain whether satellite soundings were incorporated in the initialized fields over the Arctic Ocean in 1979 and 1980 (McPherson et al., 1979). Thus analyzed wind and mass field variations throughout the troposphere are poorly resolved here. The argument that var(G_T) is too small because of inadequate radiosonde coverage is consistent with

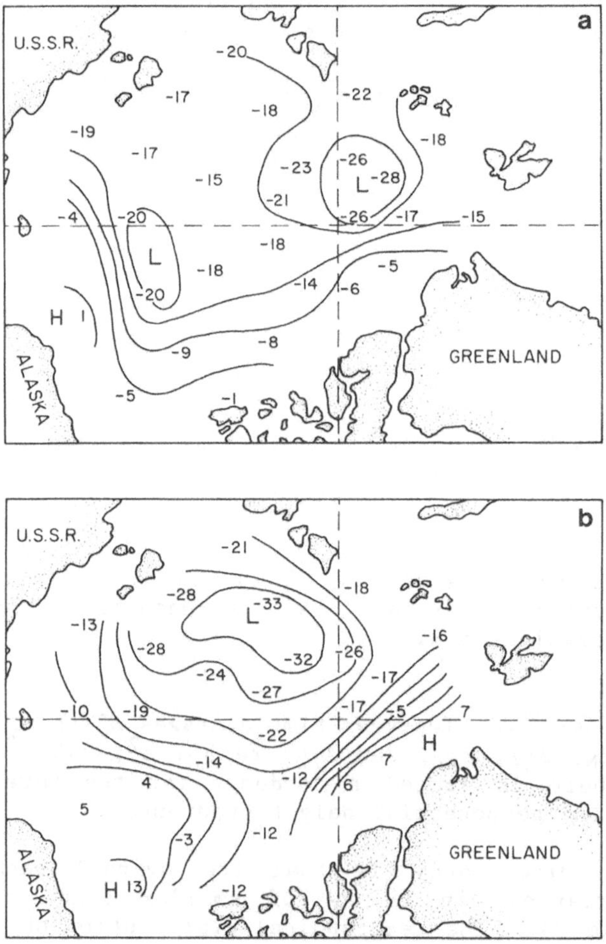

Fig. 7. Grid point estimates of var(G_{72}) - var(G_N) in $(ms^{-1})^2$. (a) March-December, 1979 and (b) May-December, 1980. Contour interval is 5 $(ms^{-1})^2$.

the fact that the forecast variances actually exceed var(G_N) at the mid latitude site (Table 5) where dense coverage is available.

4.4 Forecast Skill

We turn now to the performance of NMC's forecasts as predictors of $G_b(t)$ based on the linear model eq. (11). Table 3 shows that all three forecasts correlate better with G_b than damped persistence, measured by $\rho^2(G_b(0),G_b(\tau))$. It is customary meteorological practice to express forecast skill as the ratio of forecast accuracy achieved, beyond a zero level, to total achievable forecast accuracy. With ρ^2 as a measure of accuracy, the skill over persistence becomes

$$S(G_b,G_\tau) = \frac{\rho^2(G_b(0),G_\tau(0)) - \rho^2(G_b(0),G_b(\tau))}{1-\rho^2(G_b(0),G_b(\tau))} \qquad (14)$$

When forecasts and observations correlate perfectly, eq. (16) shows that S = 1. If forecast correlations are just equal to damped persistence correlations, S = 0. Negative S-values indicate that persistence is better than the numerical forecasts. Table 3 shows positive skill for the NMC forecasts at 24, 48 and 72 hours. S-values for the first two intervals are similar, and skill begins to fall off at 72 hours. The 74% of var(G_b) explained by G_{24} in the Arctic corresponds to a typical forecast error of $(\mathrm{var}(\varepsilon_{b24}))^{\frac{1}{2}} = 4.2\ \mathrm{ms}^{-1}$. From a practical standpoint, the 72 hour error (> 6 ms^{-1}) limits substantially the operational utility of forecasts over this interval.

The geographical patterns of grid point accuracy $\rho^2(G_b,G_{24})$ (Figure 8) are strikingly similar to the analysis accuracy patterns shown in Figure 5, with maxima of 0.75 to 0.80 in the Beaufort Sea and minima about 0.15 smaller than this near northern Greenland. One might speculate from this similarity that the forecast correlation pattern is due to patterns in the correlation between initial winds G_N and G_b, and does not indicate poorer performance of the numerical model increment $\Delta_\tau G_\tau(t) \equiv G_\tau(t) - G_N(t-\tau)$ off northern Greenland. In that case, however, the patterns of correlation between G_N and the forecasts G_τ should <u>not</u> exhibit the structure shown in Figures 5 and 8. In fact, grid point patterns of $\rho^2(G_N,G_{24})$ (not shown) are essentially identical to Figures 5 and 8, so the smaller forecast accuracy near Greenland can be ascribed to larger initial errors in G_N <u>and</u> to larger errors in the model increment $\Delta_\tau G_{24}$ here. Correlation patterns for the forecasts G_{48} and G_{72} (not shown) are similar to the G_{24} patterns discussed above.

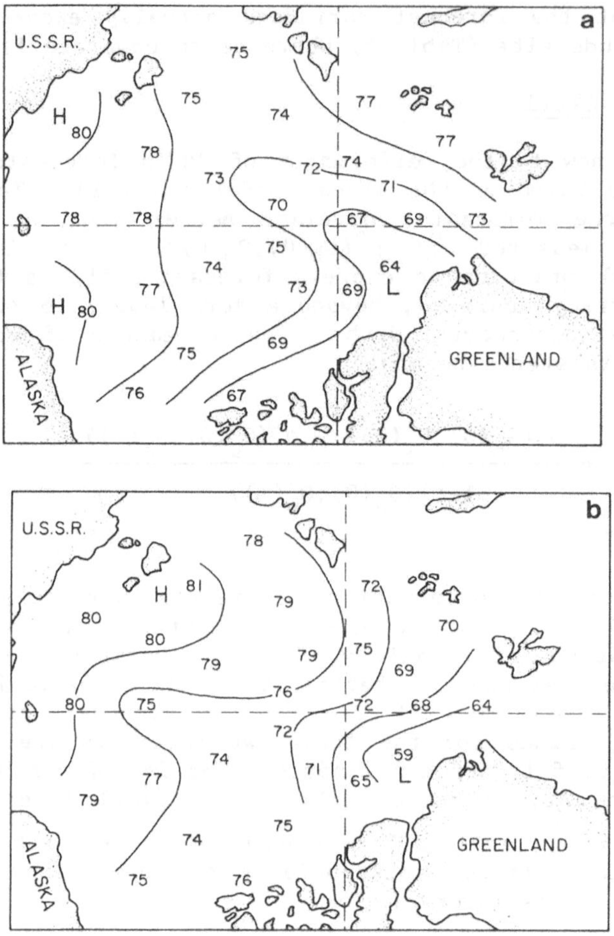

Fig. 8. Grid point estimates of $\rho^2(G_b,G_{24})$, in percent. (a) March-December, 1979 and (b) May-December, 1980. Contour interval is 4%.

In an operational setting the NMC numerical forecast $G_\tau(t+\tau)$ is available near time t, and one seeks the optimal prediction for $G_b(t+\tau)$. The linear model eq. (11) involves G'_τ and $A_{b\tau}$, yielding the deviation $\hat{G}'_b$ as a predictand. Therefore, application of eq. (11) requires estimates of the three parameters $\overline{G_\tau}$, $\overline{G_b}$, and $A_{b\tau}$.

The latter quantity can be described by its magnitude $|A_{b\tau}|$ and its argument $\theta_{b\tau}$. The magnitude provides for damping out the influence of the numerical forecast fluctuations as $cov(G_b, G_\tau)$ decreases with increasing τ, giving more weight to the long-term mean $\overline{G_b}$. $|A|$ and θ are presented in Table 3. As expected, $|A|$ decreases as the forecast interval lengthens. The angular deviations $\theta_{b\tau}$ are small, with the exception of grid point 4 (74°N, 140°W) where θ_{b72} reached 19° and 26° in 1979 and 1980. We have no explanation for these large angular deviations. More typically, the grid point estimates of $\theta_{b\tau}$ lie between -6° and 6°.

In view of the fact that $|A_{b\tau}|$ is fairly close to 1.0 and θ is almost zero, one might ask how much additional accuracy is achieved by estimating the linear model, eq. (11), as opposed to the "face value" forecast $\hat{G}_b(t) = G_\tau(t)$ that assumes A = 1 and $\overline{G_b} = \overline{G_\tau}$? This question is answered by comparing

$$(var(\varepsilon_{b\tau}))^{\frac{1}{2}}$$

with the root mean square error

$$\hat{r}(G_b, G_\tau) = \left[\overline{(G_\tau - G_b)^2}\right]^{\frac{1}{2}} .$$

Table 3 indicates that the improvement obtained in using eq. (11) is small in the central Arctic. To some extent, though, this circumstance is fortuitous, because the NMC model underestimates the variance of the 1000 mb geostrophic wind. If the correlation between forecasts and observations is fixed, any increase in var(G) will lead to reductions in $|A|$ because, using eq. (8),

$$|A_{b\tau}| = \rho(G_b, G_\tau) \left[\frac{var(G_b)}{var(G_\tau)}\right]^{1/2} \qquad (15)$$

If, as argued above, $var(G_\tau)$ is too small because of sparse radiosonde information in the initial fields, we may find changes in $A_{b\tau}$ if the coverage improves in the future, perhaps with the aid of satellite soundings. For the present, we conclude that the NMC forecasts G_τ, taken at face value, provide arctic forecasts almost as good as those obtained from the linear model, eq. (11), using G_τ as the input.

4.5 Cross Spectra for Analysis and Forecasts

Typical cross-spectra comparing arctic forecasts with analyses are shown in Figure 9. For each forecast lag τ we have strong

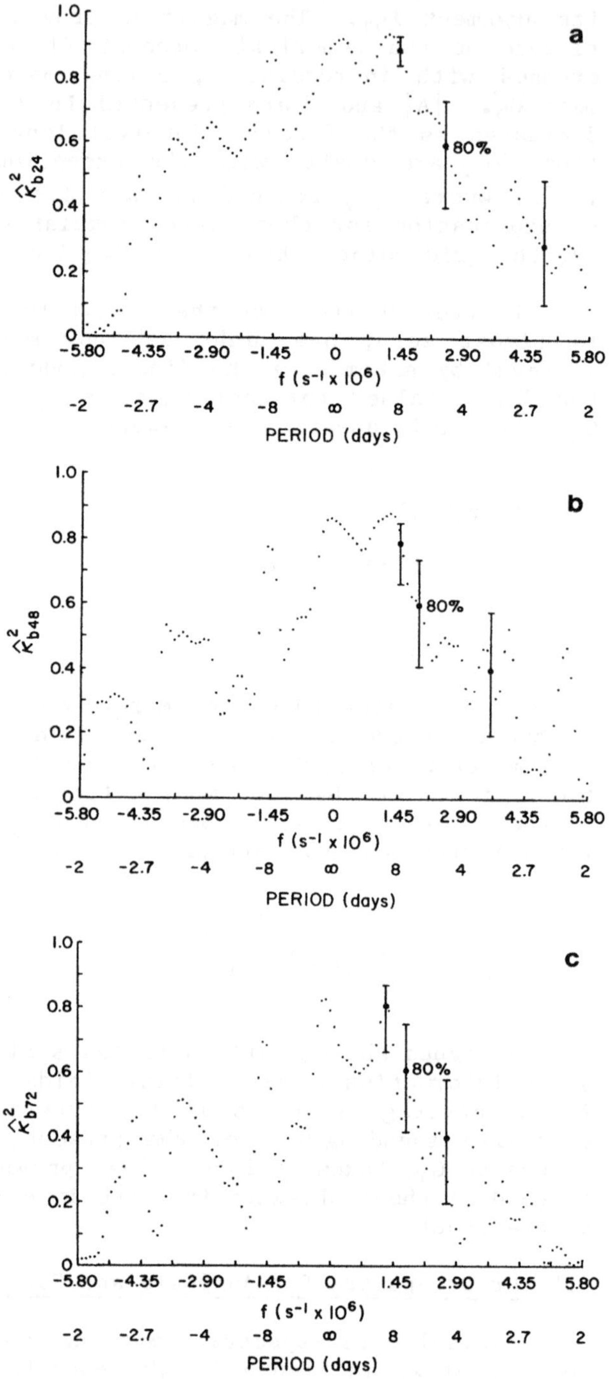

Fig. 9. Cross spectral estimates for buoy and NMC forecast winds at grid point 17, March-December, 1979. (a-c) squared coherency and (d-f) phase, for 24, 48 and 72-hour forecasts. Vertical bars show the intervals bracketing the estimates wherein the true parameters lie with the probabilities indicated.

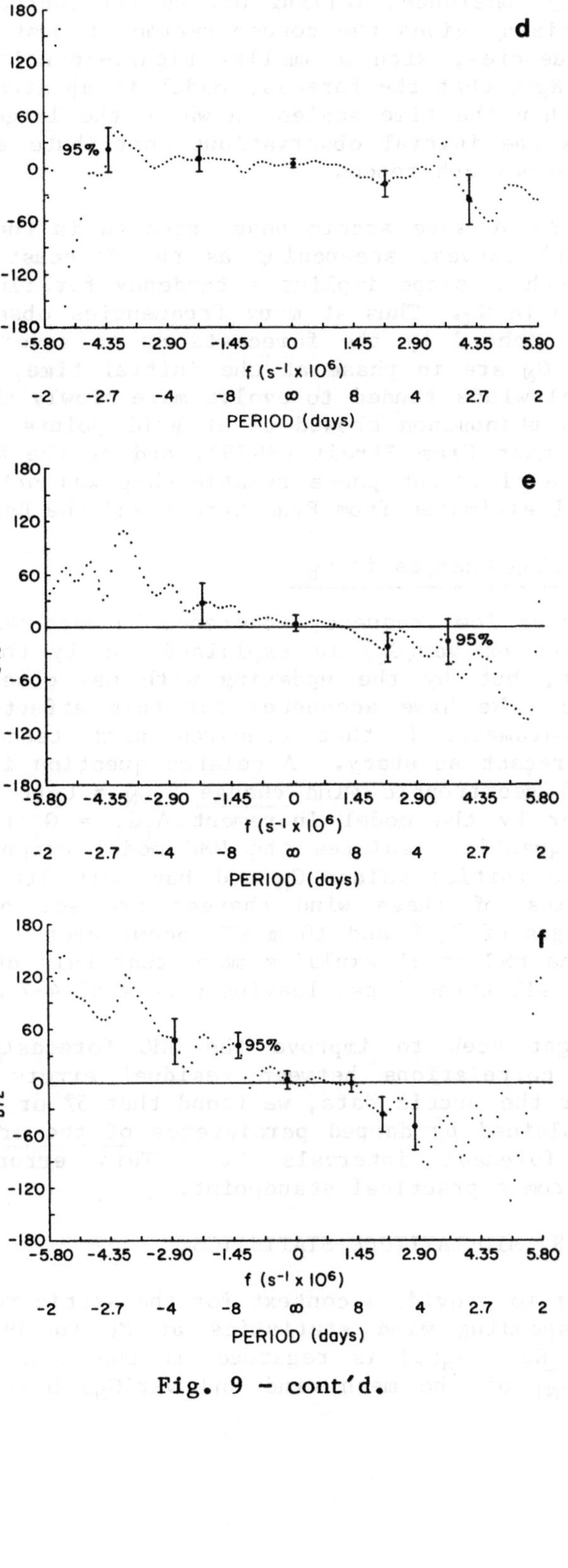
d
95%
φ̂b24 (deg)
f (s⁻¹ x 10⁶)
PERIOD (days)
e
95%
φ̂b48 (deg)
f (s⁻¹ x 10⁶)
PERIOD (days)
f
95%
φ̂b72 (deg)
f (s⁻¹ x 10⁶)
PERIOD (days)

Fig. 9 – cont'd.

low frequency coherence, falling off as $|f|$ increases. This fact is not surprising given the concentration of most of the variance at low frequencies, with a smaller signal-to-noise ratio at high frequency. Note that the forecast model is updated more frequently (daily) than the time scales on which the largest fluctuations occur. Thus the initial observations contribute substantially to the low frequency coherence.

A feature of some arctic phase spectra is the negative slope of the $\hat{\phi}_{b\tau}(f)$ curves, steepening as the forecast interval τ increases. Such a slope implies a tendency for fluctuations in G_b to lead those in G_τ. Thus at many frequencies observed wind fluctuations are echoed by the forecasts up to several hours later. Since G_b and G_N are in phase at the initial time, it follows that the NMC model winds tended to evolve more slowly than the observations. This phenomenon showed up at grid points in the Beaufort Sea (1979), near Fram Strait (1979), and at the North Pole (1979 and 1980). A clear-cut phase relationship was not present in the 1980 spectral estimates from Fram Strait and the Beaufort Sea.

4.6 Forecasting Changes in G_b

The strong low frequency coherence between G_b and G_τ implies that a portion of var(G_b) is explained not by the NMC numerical model output, but by the updating with new observations at the initial time. We have accounted for this effect by introducing the skill parameter S that compares persistence accuracy with numerical forecast accuracy. A related question is: how much of the observed geostrophic wind change $\Delta_\tau G_N \equiv G_N(t + \tau) - G_N(t)$ is accounted for by the model increment $\Delta_\tau G_\tau = G_\tau(t + \tau) - G_N(t)$? The latter quantity isolates the NMC model output from additive errors in the initial values G_N and has very little persistence. The statistics of these wind changes are set out in Table 4. Typical changes of 7, 9 and 10 m s^{-1} occur over 1-, 2-, and 3-day periods. The NMC model explains more than half of the respective variances at all three lags, leaving errors of 4-7 ms^{-1}.

One might seek to improve the NMC forecasts by exploiting time lagged correlations between residual errors $\varepsilon_{b\tau}(t + \tau)$ and $\varepsilon_{b\tau}(t)$. For the arctic data, we found that 5% or less of var($\varepsilon_{b\tau}$) could be explained by damped persistence of the error over any of the three forecast intervals τ. This error reduction is negligible from a practical standpoint.

5. ARCTIC VS. MID-LATITUDE STATISTICS

In order to provide a context for the arctic results, we estimated corresponding wind statistics at ML in 1979 and in 1980 (Table 5). Here $G_N(t)$ is regarded as the best estimate. The magnitude $|\overline{G_N}|$ of the mean wind and var(G_N) here are similar to

Table 4. Parameter estimates for time changes of arctic winds.

		1979			1980		
	$\tau \rightarrow$	24	48	72	24	48	72
$var(\Delta_\tau G_N)(ms^{-1})^2$	Min	25	53	68	33	51	60
	Ovr	34	68	86	50	85	103
	Max	45	78	101	62	106	123
$var(\Delta_\tau G_\tau)(ms^{-1})^2$	Min	27	49	61	33	50	60
	Ovr	33	60	75	44	77	91
	Max	44	73	90	53	94	107
$\rho^2(\Delta_\tau G_N, \Delta_\tau G_\tau)$	Min	.30	.38	.43	.34	.33	.33
	Ovr	.50	.54	.52	.51	.53	.47
	Max	.65	.66	.62	.63	.64	.56
$\lvert A_{N\tau} \rvert$	Min	.56	.67	.69	.54	.55	.55
	Ovr	.72	.78	.77	.76	.76	.73
	Max	.83	.87	.89	.87	.87	.84
$\theta_{N\tau}(^o)$	Min	-4	-3	-2	-6	-5	-6
	Ovr	2	2	2	2	2	2
	Max	8	9	9	10	9	9
$[var(\varepsilon_{N\tau})]^{\frac{1}{2}}(ms^{-1})$	Ovr	4.1	5.6	6.4	5.0	6.3	7.4

arctic grid point values. Again, most of the signal $\overline{G_N G_N^*}$ contained in the fluctuations $[var(G_N)]^{\frac{1}{2}} = 9\ ms^{-1}$. These ML winds are less persistent than those at any arctic grid point. Only 5 to 10% of $var(G_N)$ is explained at ML by 24-hour damped persistence, with negligible autocorrelations at 48 and 72 hours.

The ML forecast statistics are similar to their ML analysis counterparts, with the exception of variance. All three forecast variances exceed $var(G_N)$ in both years, in marked contrast to the situation at most arctic grid points. This fact is consistent with the idea that the NMC model does not damp excessively the temporal fluctuations in G_τ if adequate upper air and surface data are available locally in the initial values.

Table 5. Parameter estimates for mid-latitude winds.

	1979				1980			
X→	G_N	G_{24}	G_{48}	G_{72}	G_N	G_{24}	G_{48}	G_{72}
$\lvert\bar{X}\rvert (ms^{-1})$	2.2	1.8	2.4	3.2	1.7	2.1	2.8	2.5
$var(X)(ms^{-1})^2$	82	86	103	96	76	94	110	105
$\rho^2[X(0),X(24)]$	0.11	0.10	0.09	0.11	0.05	0.02	0.05	0.10
τ→	N	24	48	72	N	24	48	72
$\rho^2(G_N,G_\tau)$	1.0	0.80	0.59	0.34	1.0	0.75	0.53	0.31
$\lvert A_{N\tau}\rvert$	1.0	0.88	0.69	0.54	1.0	0.77	0.61	0.47
$\theta_{N\tau}(^o)$	0	1	3	-4	0	0	3	4
$B(G_N,G_\tau)(ms^{-1})$	0.0	0.4	1.4	1.7	0.0	0.7	1.7	1.2
$[var(\varepsilon_{N\tau})]^{\frac{1}{2}}(ms^{-1})$	0.0	4.0	5.8	7.4	0.0	4.4	6.0	7.2
$\rho^2[G_N(0),G_N(\tau)]$	-	0.11	0.00	0.00	-	0.05	0.00	0.00
$S(G_N,G_\tau)$	1.0	0.78	0.59	0.34	1.0	0.74	0.53	0.31
$\hat{r}(G_N,G_\tau)(ms^{-1})$	-	4.2	6.8	8.8	-	5.0	7.4	10.3

Despite the tougher prediction problem posed by the meagre time autocorrelation of G_N at ML, the NMC model performs as well here as in the Arctic, as shown by values of $\rho^2(G_N,G_\tau)$ exceeding 70%, 50%, and 30% at the three forecast lags. The skill is correspondingly larger than in the Arctic because of the reduced persistence at ML. The coefficients $|A_{N\tau}|$ are smaller at ML because the variance in G_τ is not underestimated here (cf. eq. 15). Therefore the least-squares model, eq. (11), represents a greater improvement over the "face value" forecast $\hat{G}_N(t) = G_\tau(t)$ at ML. For example, at $\tau = 72$ hours the standard deviation of the errors (using eq. 11) is smaller by about 2 ms^{-1} than the root mean square error $\hat{r}$. It seems remarkable that so simple an application of climatological information (in the form of $A_{b\tau}$ and $\overline{G_N}$) as eq. (11) can make such significant improvements on the "raw" numerical output from the NMC model. As noted previously, these improvements are smaller in the Arctic basin.

Variance spectra at ML (not shown) reflect the weaker persistence with only 50% of $var(G_N)$ due to fluctuations on periods greater than six days. Twenty-five percent of var (G_N) occurs on time scales shorter than 3.5 days, compared to only 7% in the Arctic. None of the ML phase functions $\phi_{N\tau}(f)$ showed a clear sloping tendency in 1979 or 1980.

6. CONCLUSIONS

We have estimated statistical properties of surface geostrophic winds and wind forecasts over the central Arctic basin. The results are based on 18 months of daily winds, derived from pressure measurements at an array of drifting buoys. The main points follow:

1) Arctic surface geostrophic winds have a variance of about $(8\ ms^{-1})^2$, and 50% of this signal can be predicted with a 24-hour damped persistence forecast.

2) Geostrophic winds G_N derived from NMC's 1000 mb analysis are nearly as accurate as the buoy winds G_b, based on their correlations with independently measured ice velocity time series. Except near northern Greenland, about 87% of var(G_b) is explained by G_N.

3) NMC's 1000 mb height forecasts produce geostrophic winds that are clearly superior to persistence forecasts at lags of one, two and three days. Typical NMC forecast errors are 4, 6, and 7 ms^{-1} at these lags.

4) The NMC primitive equation model produces arctic 1000 mb fields that are spatially smoother than the buoy and NMC final analysis fields. The smoothing may result from differences between final and operational analysis procedures, from the numerical prediction model, or from a combination of these. Temporal fluctuations are also underestimated by up to 25% in the model output. It is suggested that these effects may be caused, in part, by the paucity of radiosonde data over the Arctic Ocean.

5) At least 50% of the variance in the time changes of G_N over one, two and three day intervals is accounted for by the NMC forecasts of wind changes.

6) The NMC forecast accuracy and residual errors at an arbitrary mid-latitude site are similar to corresponding arctic statistics. The persistence of G_N is smaller at ML, and forecast skill, over 24 and 48 hours, is correspondingly larger here than in the Arctic. The time variances were not underestimated by the NMC model at ML.

7) Phase spectra at several arctic grid points show a tendency for NMC forecasts to evolve more slowly than the observed atmosphere, from the same initial conditions. This tendency was absent in the ML data.

8) By 72 hours, the NMC forecasts have little practical value as predictors of fluctuations in daily sea ice velocity, despite

the existence of statistically significant skill, because of the large errors (7 ms^{-1}). The 24- and 48-hour forecasts should be useful in this application.

7. ACKNOWLEDGMENTS

This work was supported by the National Oceanic and Atmospheric Administration Grant NA80-AA-D-00015, which was funded in part by the Global Atmospheric Research Program and the Office of Climate Dynamics, Division of Atmospheric Sciences and the Meteorology Program, Division of Polar Programs, of the National Science Foundation, and the Office of Naval Research, Arctic Programs.

The author thanks Alan Thorndike for reviewing the manuscript, and Roger Colony for helpful discussions.

8. REFERENCES

Albright, M. (1980) Geostrophic wind calculations for AIDJEX. In Sea Ice Processes and Models (R. S. Pritchard, Ed.), University of Washington Press, Seattle, Washington, ISBN 0-295-95658-5 (see p. 402-409).

Bergman, K. (1979) A multivariate optimum interpolation analysis system for temperature and wind fields. Mon. Wea. Rev., 107: 1423-1444.

de Boor, C. (1962) Bicubic spline interpolation. J. Math. Phys., 41(3): 212-218.

Denner, W. W. and L. D. Ashim (1980) Operational determination of wind stress on the arctic ice pack. In Sea Ice Processes and Models (R. S. Pritchard, Ed.), University of Washington Press, Seattle, Washington, p. 76-88.

Gandin, L. S. (1963) The Objective Analysis of Meteorological Fields. Leningrad. (English Translation by Israel Program for Scientific Translations, Jerusalem, 1965, 242 pp.)

Jenkins, G. M. and D. G. Watts (1968) Spectral Analysis and its Applications. Holden-Day, 525 p.

Keegan, T. J. (1958) Arctic synoptic activity in winter. J. Met. 15: 513-521.

McPherson, R. D., K. H. Bergman, R. E. Kistler, G. E. Rasch and D. S. Gordon (1979) The NMC operational global data assimilation system. Mon. Wea. Rev. 107: 1445-1461.

Reed, R. J. and B. J. Kunkel (1960) The arctic circulation in summer. J. Met., 17, 489-506.

Shuman, F. G. and J. B. Hovermale (1968) An operational six-layer primitive equation model. J. Appl. Met., 7(4): 525-547.

Stackpole, J. D. (1978) The NMC 9-layer global primitive equation model on a latitude-longitude grid. NMC, Suitland, Md., Office Note 178, 31 p.

Thorndike, A. S. (1982) Statistical properties of the atmospheric pressure field over the Arctic Ocean. J. Atmos. Sci. 39(10): 2229-2238.

Thorndike, A. S. and R. Colony (1982) Sea ice motion in response to geostrophic wind. J. Geophys. Res., 87: 5845-5852.

Untersteiner, N. and A. S. Thorndike (1982) Arctic data buoy program. Polar Record, 21(131): 127-135.

Walsh, J. E. (1977) Ice forecasting limitations imposed by the accuracy of atmospheric prediction models. AIDJEX Bulletin 36: 1-13.

Walsh, J. E. (1978) Temporal and spatial scales of the arctic circulation. Mon. Wea. Rev. 106(11): 1532-1544.

Walsh, J. E. (1980) Empirical orthogonal functions and the statistical predictability of sea ice extent. In Sea Ice Processes and Models (R. S. Pritchard, Ed.), University of Washington Press, Seattle, Washington, Vol 2, p. 373-384.

SPECIAL TOPIC 2

INTERNAL WAVES IN THE ARCTIC OCEAN: A REVIEW

James Morison

Polar Science Center
Applied Physics Laboratory
University of Washington Seattle, WA 98105

INTRODUCTION

One of the first modern scientific observations of internal waves in the ocean was made in the Arctic by Nansen (1902). Since that time, the bulk of observations of internal waves have been made in temperate oceans. There are a number of reviews of this work, among them are those given in a 1975 collection of internal wave papers in Volume 80 of the Journal of Geophysical Research and by Munk (1981). After three decades of intensive research in temperate waters there are still major questions regarding internal waves: How are they generated? Why do spectral properties of internal waves have the universal properties they appear to have? How do the waves decay and interact with each other? How do they affect the small-scale ocean structure?

In this review, work on internal waves which has been done in the Arctic will be discussed. The unique ways in which internal waves are forced by the surface ice cover will be described. Observations made in the Arctic will be compared to the current general ideas about internal waves and arguments will be presented for further and more comprehensive internal wave research in the Arctic.

GENERAL OBSERVATIONS

There are a number of observations of internal waves in the Arctic which have come as offshoots of other studies. Nansen (1902) suggests large variations observed in temperatures at a thermocline in the Eastern Arctic may be due to internal waves. Morse and Smith (1972) observed internal waves with a magnitude on the order of a few meters. Arnold Hanson (personal communication, 1981) has found small (order 2 mm) fluctuations in the draft of ice island Arlis II with periods of about eight minutes. Similarly, Wadhams (personal communication) has observed strains in the ice with a ten-minute period which appear to be due to the small surface displacements of internal waves.

Concerted efforts to study internal waves in the Arctic have centered on the generation of waves and the spectral character of the free wave field. This review is therefore broken into one section dealing with wave generation and one section dealing with free wave spectra.

FORCING OF INTERNAL WAVES IN THE ARCTIC

The earliest and most famous study of internal wave forcing is that of Ekman (1906). During his expedition in 1883 Nansen had encountered the phenomenon of "dead water" while sailing the Fram north of Taimur Island (approximately 76° 20′, 96°E). He observed that when the vessel came into a region where brackish water was laying over salt water, the Fram lost speed to the point where she could make only one knot at full steam. Water at the stern appeared to move with the vessel and transverse slicks appeared at the water surface astern.

On hearing of these observations, V. Bjerkenes hypothesized that the phenomenon was related to the generation of internal waves on the interface between the brackish surface layer and the salt water beneath. He also suggested that the subject might be studied by V. J. Ekman. Ekman took up the project and pursued two lines of research. First, by taking out newspaper advertisements all over the world, he acquired a large collection of interesting personal accounts of encounters with "dead water". Second, he performed laboratory experiments with model ships in a stratified tow tank. Figure 1 from Ekman (1906) shows a sequence of photographs of a model of the Fram being towed through the stratified tank. The first three photographs show the model moving "in dead water, with a towing force, increased in successive experiments." The internal wake behind the model is obvious. The fourth photograph shows the model "at a high speed without dead water." In this case no wake is formed. Ekman found that when a vessel moves through a two-layer body of water in which the upper layer is of a thickness on the order of the vessel draft, the towing or propulsion force and the internal wake magnitude increase with increasing speed until the long wave-length internal wave speed,

$$c = \sqrt{\frac{gD\Delta\rho}{\rho}} , \qquad (1)$$

Fig. 1. A model of the Fram in a stratified tow tank from Ekman (1906). The first three photographs (a,b,c) show the model moving in "dead water with a towing force increased in successive experiments." The internal wake behind the model is obvious. The fourth photograph shows the model "at high speed without dead water". Towing force is decreased and the wake is absent.

is nearly reached. In this equation g is the acceleration of gravity, D is the surface layer depth, $\Delta\rho$ is the difference in density between the two layers, and ρ is the water density. At speeds well below the critical velocity (Figures 1a and 1b) the waves in the internal wake are short and the group velocity is less than the phase velocity (ship's speed). So, some energy is left behind. Near the shallow water wave velocity (Figure 1c), the group velocity approaches the phase (ship) velocity and little energy is left behind. The wake grows due to resonant forcing and the drag reaches a maximum at a ship speed slightly less than the long-wave internal wave velocity. Above this critical speed (Figure 1d) the wake cannot form, and consequently the wave drag decreases. The situation is analogous to that of a vessel in shallow water accelerating from operation as a displacement hull to a planing hull.

The maximum in drag near the critical speed is viewed by Ekman as an explanation for the apparent capriciousness of the "dead water" phenomenon described in personal accounts. For example, a sailing ship would be cruising at several knots, come about, fall below the critical speed and then be limited to a maximum speed of one knot. Ekman was also able to explain other observed characteristics of "dead water" (loss of steerage, appearance of slicks at the water surface, etc.) in terms of internal wakes.

A more recent study by Rigby (1974, 1976) has also examined the generation of internal wakes, in this case by pressure ridge keels underneath Arctic pack ice. Rigby modified a solution by Lamb (1932) to yield expressions describing the internal wake produced at the pycnocline of a two-layer ocean by a semi-elliptical pressure ridge keel as illustrated in Figure 2. He shows that a wake will be generated with a form

$$\eta = A \sin Kx \tag{2}$$

in a reference frame moving with the pressure ridge keel. For a relative water velocity C_o, the wavenumber is given by

$$K = \frac{\Delta\rho g}{2\rho C_o^2}, \tag{3}$$

which is the dispersion relation for an interfacial wave in deep water. For major and minor ridge axes a and b, he finds the wave amplitude to be

$$A = \pi K b(a+b)\ e^{-KD}. \tag{4}$$

Rigby (1976) performed an experiment to see if he could measure such wakes. He deployed thermistors in the pycnocline along a line parallel to the prevailing current direction and, given the

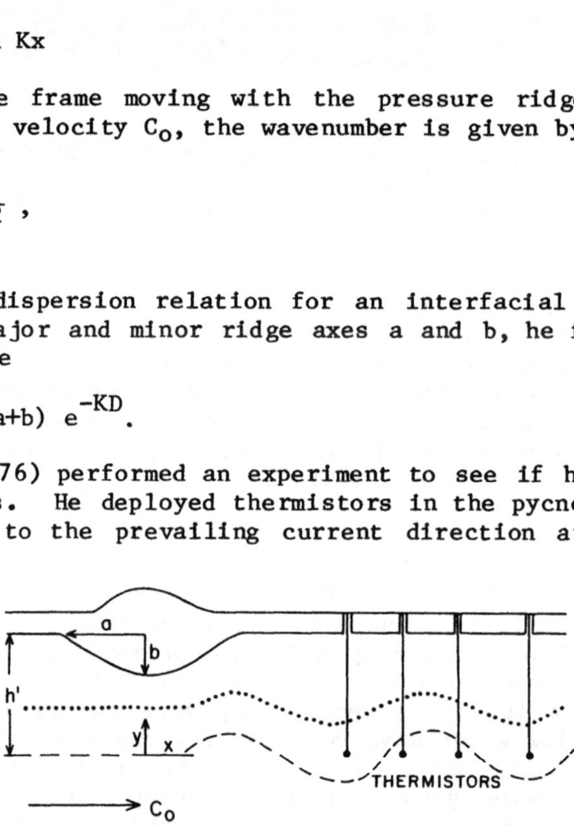

Fig. 2. Schematic of an internal wave generated at a pycnocline (depth D) by a semi-elliptical pressure ridge keel (from Rigby, 1976). In his experiment Rigby used a linear array of thermistors to determine the wavelength and magnitude of waves which were stationary with respect to the ice.

observed temperature gradient in the pycnocline, was able to determine the pycnocline displacement as a function of horizontal position. He examined waves stationary with respect to the ice by fitting sine waves of arbitrary length and amplitude to the isopycnal displacements averaged over ten or fifteen minutes. Stationary waves with amplitudes from 0.1 m to 1.0 m were found. Figure 3 from Rigby (1976) shows measured wavelengths from these waves plotted against relative water velocity. The solid line is the wavenumber from (3). The agreement between (3) and the data is reasonably good, which indicates that the waves were indeed nearly stationary and moving parallel to the current. In fact, the data seem to indicate a slightly lower wavelength than that predicted by the simple single wake theory. This is apparently due to the fact that the observations are not of a single wake from a single ridge perpendicular to the flow but, rather, are the cumulative effect of a great number of ridges at a variety of angles to the flow.

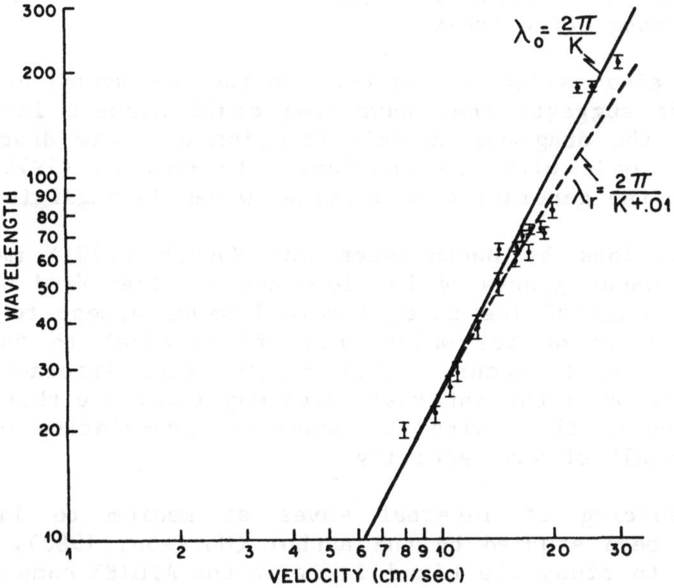

Fig. 3. A log-log plot of the wavelength against the total thirty meter current for Rigby's experiment (from Rigby, 1976). The solid line is from Equation (3) for the wavelength of a wake behind a single keel. In addition, the dashed line shows the wavelength which Rigby finds would be observed in a field of wakes generated by a hypothetical random pressure ridge keel distribution.

In order to investigate the effect of a random-ridge field, Rigby first shows that the wake behind the ridge at an angle to the flow will produce a wake with a wavenumber

$$k_{wave} = \frac{\Delta\rho g}{2\rho c_o^2 \cos^2\theta} . \quad (5a)$$

Observed on a line parallel to the flow the wavenumber is

$$k_{obs} = k_{wave} \cos\theta = \frac{K}{\cos\theta} . \quad (5b)$$

He numerically simulated the field of wakes from a random distribution of ridges. With these results he also simulated the type of wake observations he had made in the field and found that the observed wavenumber for the aggregate field of wakes was k_{total} = K + 0.01 $\pm$ 0.02. The plot of this as a dashed line added in Figure 3 shows improved agreement with the experimental data. The fact the k_{total} is only slightly greater than K indicates that a random ridge field generates wakes running nearly parallel to the flow. This is so because ridges perpendicular to the flow generate waves move effectively.

Rigby also estimated the drag on the ice caused by wake generation. He suggests that wave-drag could account for up to 30 percent of the drag due to skin friction and form drag. This is somewhat at odds with the conclusion by Hunkins (1974) that the drag due to the generation of internal waves is negligible.

Observations by Hachmeister and Muench (1981) may also be related to wakes generated by ridge keels. They find that, in the Bering Sea marginal ice zone, internal waves appear to be associated with bands of ice which move off parallel to the ice edge when off-ice winds occur. They suggest that interaction of the bands of ice with the internal waves may involve either generation of the waves by the moving ice bands or accumulation of bands of ice as a result of wave activity.

The forcing of internal waves at medium to large length scales has been studied by the author (Morison, 1980). During an experiment to study the mixed layer at the AIDJEX camp Big Bear in the Beaufort Sea in the spring of 1976, a response in the upper pycnocline was observed which seemed to be related to the onset of a storm. After a prolonged calm, a storm system with southeasterly winds moved from west to east through the area of the AIDJEX camp array. At the time of the storm the isopycnals displayed the response shown in Figure 4. Generally, the isopycnal surfaces dropped and then rebounded with oscillations. The response seemed to precede the local wind and inertial oscillations were excited both above and below the pycnocline. A model was developed to see

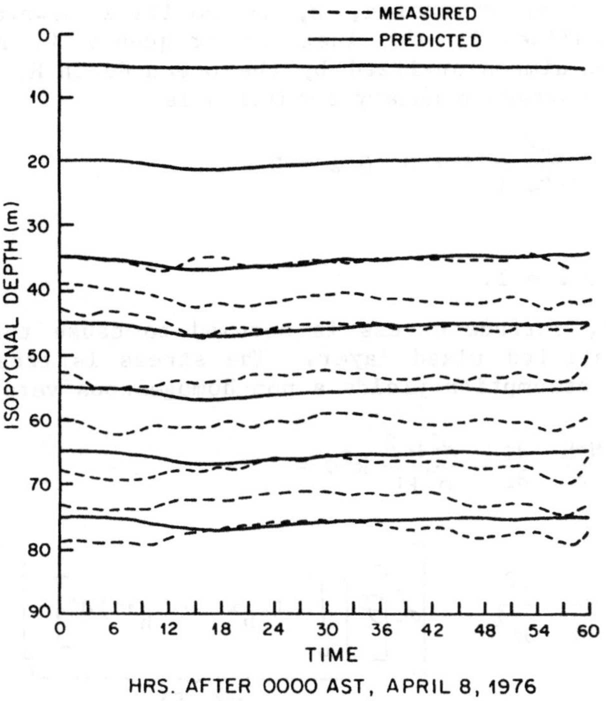

Fig. 4. Measured and predicted isopycnal displacements during an Arctic storm lasting from 0 to 24 hours (from Morison, 1980). The measured responses show initial deepening with rebound and oscillations. The predicted displacements are the combined forced internal wave responses to surface stress and buoyancy flux.

if the observed motions could be explained as forced waves driven by surface stress or buoyancy flux.

The equations for free internal waves in a continuously stratified ocean can be written as a differential equation for ϕ, the Laplace transform in time, and Fourier transform in the horizontal, of the vertical displacement. The equation of motion is

$$\frac{\partial^2\phi}{\partial z^2} + \frac{N^2H}{g}\,\frac{\partial\phi}{\partial z} - \frac{N^2+\sigma^2}{\sigma^2+1}\,k^2\phi = 0, \qquad (6)$$

were N is the Brunt-Vaisala frequency

$$N^2 = \frac{g}{\rho}\,\frac{d\rho}{dz}\,. \qquad (7)$$

The vertical dimension, z, is positive downward. Time is non-dimensionalized by the inertial frequency f, spatial dimensions are non-dimensionalized by the ocean depth H, and $\mathcal{N}$ is equal to N/f. The surface boundary condition is

$$\frac{\partial\phi}{\partial z} - \frac{gk^2}{Hf^2(\sigma^2+1)}\phi = 0 \quad @\ z = 0 \tag{8}$$

$$\phi = 0 \quad @\ z = 1. \tag{9}$$

In the model, surface stress is assumed to cause turbulent stress in an unstratified mixed layer. The stress is treated as a body force. This assumption yields a non-homogeneous version of (6).

$$\frac{\partial^2\phi}{\partial z^2} + \frac{N^2H}{g}\frac{\partial\phi}{\partial z} - \frac{\mathcal{N}^2+\sigma^2}{\sigma^2+1}k^2\phi =$$

$$\frac{1}{r_o\,(\sigma^2+1)}\frac{\partial^2}{\partial z^2}\left[\underbrace{\mathcal{L}\left[\mathcal{F}\left[\int_o^t (\nabla_h x\tau - \nabla_h\cdot\tau)\,d\xi\right]\right]}_{T(\tau,\bar{k})}\right]. \tag{10}$$

$\mathcal{L}$ denotes the Laplace transform and $\mathcal{F}$ denotes the Fourier transform. If surface pressure is non-zero, the surface boundary condition (8) becomes non-homogeneous,

$$\frac{\partial\phi}{\partial z} - \frac{g\ \ k^2}{Hf^2(\sigma^2+1)}\phi = \frac{-k^2P_c}{(\sigma^2+1)}, \tag{11}$$

where $P_c(\sigma,\bar{k})$ is the transformed surface pressure.

Solutions for both pressure forcing and stress forcing are found as sums of the eigenfunction solutions ψ_n, of the free-wave equations (6), (8), and (9) for eigen values,

$$\lambda_n = \frac{-k^2}{\sigma_n^2+1} = \frac{k^2}{\omega_n^2-1}, \tag{12}$$

where ω_n is the eigen frequency.

One form of the solution for stress, obtained using an inverse Laplace transform technique similar to that of Leonov and Miropolskiy (1973), is

$$\psi_\tau(t,z,k) = \sum_{n=1}^{\infty} -\frac{\partial\psi n}{\partial z}\Big|_{z=o} \frac{\psi n(z)}{\lambda_n\omega_n} \int_o^t T(\alpha)\Big|_{z=o} \sin\omega_n(t-\alpha)d\alpha. \quad (13)$$

T is the fourier transform of surface stress forcing as indicated in (10). Note that the response depends only on surface stress curl (convergence of Ekman transport) and stress divergence, and is expressed as a convolution integral in time, weighting the eigenfunctions. For pressure, the solution has a similar form

$$\psi_p(t,z,k) = \sum_{n=1}^{\infty} R_p(\psi_n)\frac{\psi_n(z)}{\lambda_n\omega_n}\int_o^t k^2 P_i(\alpha)\sin\omega_n(t-\alpha)d\alpha. \quad (14)$$

The pressure response factor, R_p, is a function of the eigenfunction shape and is determined by a change of variables and use of the orthogonality condition. Alternately, it can be obtained from the results of Leonov and Miropolskiy (1973). $P_i(t,\bar{k})$ is the surface pressure Fourier transformed in space.

Of course, in order to estimate an actual wave response it is necessary to obtain the forcing functions over a wide range of length scales. In this study the surface stress forcing was estimated at long wavelengths (300 km to 1000 km) using the stress of ice on water, calculated with a quadratic drag law from ice velocities obtained from the AIDJEX buoy array. Short wavelength (less than 22 km) forcing was estimated on the basis of the total large scale forcing at the camp location and lead patterns observed in satellite photographs of the Beaufort Sea. Intermediate scale forcing was arbitrarily set equal to the 22 km wavelength forcing. The forcings from the three length scales were combined in a phase relation chosen to produce a maximum response. The resulting displacements and velocities agree qualitatively with the observations, but the maximum magnitude of the theoretical responses is only on the order of one-third that of the observations.

The response to large scale atmospheric pressure fluctuations is always negligible. However, an effective pressure due to buoyancy flux in leads was derived by Morison (1980) from results of Magaard (1973). This effective pressure, derived for conditions prevailing at the time of the experiment, actually produced an internal wave response twice as great as that due to surface stress forcing. (This is in contrast to the results of Magaard for temperate oceans where the buoyancy flux response is typically less than the stress response.) If the waves forced by stress and buoyancy flux are combined in such a way as to most nearly duplicate the observed motions of the pycnocline, displacements on the order of 2 m, as shown by the solid lines in Figure 4, result. The combined forced wave response can be made to agree with the observations above 50 m but the match is poor below that depth. The observations show a complicated depth dependence but the theoretical solution, in spite of being a weighted sum of 50 modes with a great deal of vertical structure at each of about 20 wavenumbers, add up to a response which shows a simple monotonic decrease with depth. The velocity predictions show inertial motions with an amplitude of about 2 cm/sec both above and below the pycnocline. These agree qualitatively with the velocity observations made during the experiment.

The conclusions drawn from this study are that surface forcing may produce displacements in the pucnocline on the order of one meter and that velocity disturbances with a magnitude of several centimeters per second and a high intertial frequency content may also be produced. The observed responses differ from the observations probably largely as a result of the advection of previous disturbances in the density surface past the measurement site. In this sense, isopycnal displacements observed from a moving ice camp will always show a combination of immediate forced internal waves responses and the remnants of previous disturbances.

In summary, studies of internal wave forcing in the Arctic indicate that short internal waves with amplitudes on the order of a meter are excited by the motion of pressure ridge keels at the surface. At longer length scales the forcing is due to buoyancy flux. Large scale surface stress also excites waves of about one meter amplitude in the upper pycnocline. With time, the energy in these waves generated at the surface must propagate downward into the main pycnocline and may supply the energy manifested in the ambient internal wave field. This energy may appear as larger oscillations in the less densely stratified lower portions of the main pycnocline. Also, as pointed out by Mork (1972), waves near the coast can be amplified as internal Kelvin waves so that amplitudes there may be two orders of magnitude greater than in the Central Basin.

In general, the mechanisms by which internal waves are forced at the surface of the Arctic Ocean are reasonably well understood. In order to estimate the total amount of energy going into the internal wave field it still remains to integrate the effects of surface forcing with other mechanisms of internal wave generation.

Free-waves, Spectra

The spectral models of internal waves by Garrett and Munk (1972, 1975, 1979) have been especially useful as standards for comparing various studies of free internal waves. A complete description of the internal wave field requires a knowledge of the nergy spectrum in horizontal wavenumber, vertical wavenumber or mode number, and temporal frequency. Garrett and Munk have used measured spectra in the individual variables and the dispersion relation for internal waves, which relates any one of the variables above to the other two, to produce models of internal wave spectra in all three dimensions. For the temperate oceans, internal wave measurements display very similar spectra in good agreement with the Garrett-Munk spectra. Even energy magnitudes are remarkably similar except near obvious points of generation (Munk, 1981; Wunsch, 1976; and Wunsch and Webb, 1979). It has been suggested that this is due to a universiality of source mechanisms or the presence of dynamic processes which cause the internal wave energy to be spread out in frequency in the same way at all locations (Munk, 1981; Briscoe, 1975). However, as will be shown, the available data indicate that internal wave spectra in the Arctic differ significantly from the usual temperate ocean spectra.

Perhaps the most commonly used internal wave spectrum is the frequency spectrum of vertical displacement at a single point. Figure 5 shows the shape of a spectrum such as given by Garrett and Munk (1975). Internal wave energy is confined to frequencies greater than the interial frequency, f, and less than the local Brunt-Vaisala frequency.

Desaubies' (1976) representation of the Garrett-Munk model is convenient:

$$S(\omega) = \frac{2\,f}{\pi\,N}\, r\, \omega^{-2}\, (1-f^2/\omega^2)^{1/2} \qquad f < \omega < N \,. \tag{15}$$

The last term results in a smooth roll-off as ω decreases toward f but, over most of the frequency band, energy drops off as ω^{-2}. The parameter r is a measure of internal wave energy scaled by the local value of N. Observed spectra commonly show a slight peak in the energy just below the Brunt-Vaisala frequency (dashed curve in Figure 5) not present in the Garrett-Munk model. This peak can be attributed to waves which are reflected at the depth where the Brunt-Vaisala frequency drops below the propagation frequency (Munk, 1981). Similarly, a peak is found near the inertial period in horizontal velocity spectra because waves propagating poleward encounter a turning point at latitudes where f drops below the wave frequency.

An example of a Brunt-Vaisala frequency profile in the Arctic, taken from Morison (1980) is shown in Figure 6. The profile is constructed on the basis of hydrographic data gathered during the AIDJEX experiment in the Beaufort Sea. The high values of N in the upper 300 m are characteristic of the seasonal and main pycnocline. The processes controlling stratification in the Arctic Ocean are different from those in most temperate oceans; the Arctic Ocean being stratified by salinity rather than temperature. However, the overall magnitude of stratification is not

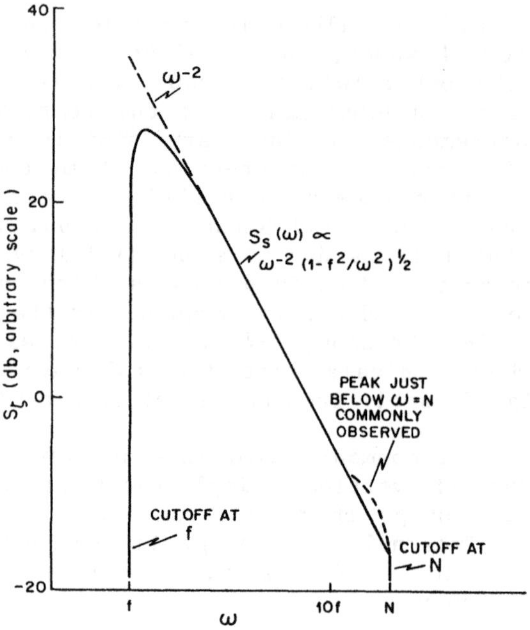

Fig. 5. The shape of the vertical displacement spectrum of internal waves as described by Desaubies (1976). The model is based on that of Garrett and Munk (1972, 1975, 1979) and is characterized by what is basically an ω^{-2} fall-off.

unusually high or low. For example, one way of assessing the overall stratification is to examine $\bar{N}H$ where $\bar{N}$ is N averaged over the ocean depth, H. For the profile of Figure 6 $\bar{N}H$ is about 6 m sec^{-1} and corresponds to an $\bar{N}$ of about 1.1 cph. Briscoe (1983) has estimated an $\bar{N}H$ of 4 m sec^{-1} for the data from Project Ski Jump in the Beaufort Sea, and data from Yearsley (1966), gathered at T-3 in the Beaufort Sea, suggests an $\bar{N}H$ of 5 m sec^{-1}. By comparison, Briscoe (1983) has examined $\bar{N}H$ values for a large number of locations in temperate oceans and has found values ranging from 1.6 m sec^{-1} in the Mediterranian to 8.9 m sec^{-1} in the North Pacific. Thus the $\bar{N}H$ values for the Arctic Ocean fall roughly in the middle of the range of values found in other oceans.

Three examples of internal wave spectra from the Arctic are available. The first is from Yearsley (1966). Yearsley suspended thermistors at 60 and 125 meters below the ice island T-3. He observed temperature fluctuations and converted these to vertical displacements using temperature gradients determined from hydrocasts. Figure 7 shows a composite of eight spectra obtained from

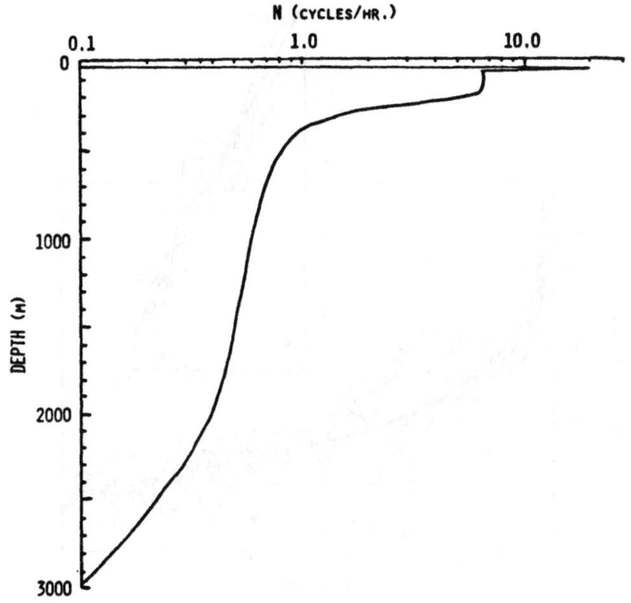

Fig. 6. Brunt-Vaisala frequency, N, vs. depth in the Beaufort Sea from Morison (1980).

the displacement records by Yearsley. An ω^{-2} curve has been added. The observed spectra appear to fit the curve fairly well. Two possible values of local Brunt-Vaisala frequency are shown, one from an exponential curve fit to the hydrographic data collected several weeks prior to the experiment and one estimated directly from the hydrographic data. In either event the characteristic peak in energy just below the Brunt-Vaisala frequency mentioned above is apparent here.

The internal wave amplitudes shown by Yearsley's data are lower than those usually found in the pycnocline of temperate oceans. In terms of Desaubies' spectral model, (15), and using N = 14.2 cph, the magnitude of the variance parameter r, is 3.9 m^2 cph for Yearsley's spectra, the rms amplitude of the internal waves is 0.4 m and the total energy (equal to rN) is 0.17 J m^{-3}. Typical values of r for temperate oceans lie between 190 and 341 m^2 cph (Roth et al, 1981, and Desaubies, 1976) and a corresponding rms amplitude for N = 14.2 cph would be 3.7 m. The significance of this low level of internal wave energy is difficult to assess from a single experiment. It may be the result of an unusual condition during Yearsley's experiment, but other results described below suggest it is not. Rather, it appears to be an important characteristic of the internal wave field in the Arctic.

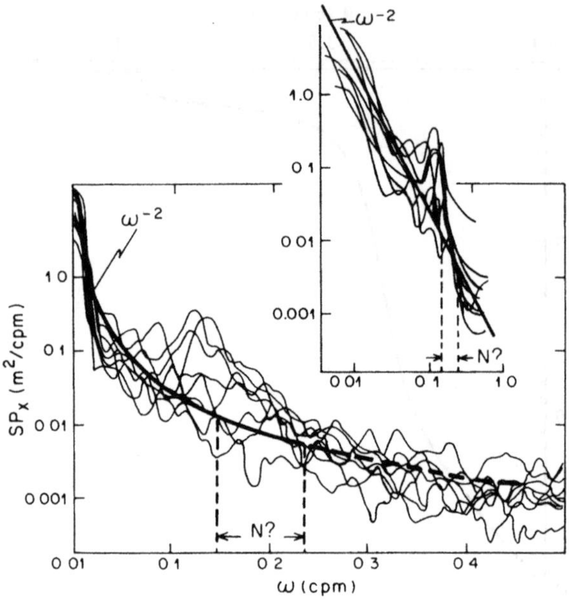

Fig. 7. A composite of spectra of vertical displacement measured 60 m below the ice island T-3 by Yearsley (1966). The data shows an ω^{-2} fall-off. The value of N is somewhere between the values shown. There is energy at $\omega > N$ perhaps due to finestructure contamination. The supplemental figure shows the same data replotted on a log-log scale by Murray Levine.

Yearsley's spectra from 125 meters depth are given over a range of frequencies from well below the lowest frequency he should be able to resolve to less than one-tenth the Brunt-Vaisala frequency. Thus, the results cannot be interpreted with a great deal of confidence, but they suggest there is even less internal wave energy at 125 m than at 60 m.

Yearsley's use of a linear frequency scale obscures some details at low frequencies (examples of spectra replotted by Levine [personal communication, 1981] on a log-log scale are also shown in Figure 7). However, the linear scale does show the high frequency behavior quite well. Yearsley's spectra continue to fall off as ω^{-2} rather than drop off completely above the Brunt-Vaisala frequency as internal wave theory would predict. This is indicative of the presence of fine structure. When internal waves move in a density structure which consists of layers of constant density separated by thin, highly-stratified sheets, a discontinuous temperature record is produced. When one transforms this temperature record into a record of vertical displacement using the mean gradient, a discontinuous displacement record results. It was shown by Phillips (1971) that spectra from such records decay as ω^2.

The problem of fine structure contamination of internal wave records, and one solution to the problem, has been described quite clearly by Neshyba et al. (1972). They used a single thermistor lowered by a winch at T-3 (a wonderfully stable platform) to measure detailed temperature profiles in the thermocline of the Arctic Ocean every four minutes. A sequence of their profiles is shown in Figure 8. The uniform layers of about 2.5 meters thickness, separated by thin, highly stratified sheets, are obvious. The layering is due to double diffusive convection, which is an important means of heat transfer from the Atlantic Water in this region of the Arctic.

Neshyba et al. (1972), determined internal wave amplitudes in two ways. They measured the displacement of individual isotherms, selected by eye from successive temperature profiles, to obtain measurements uncontaminated by fine structure. They also computed

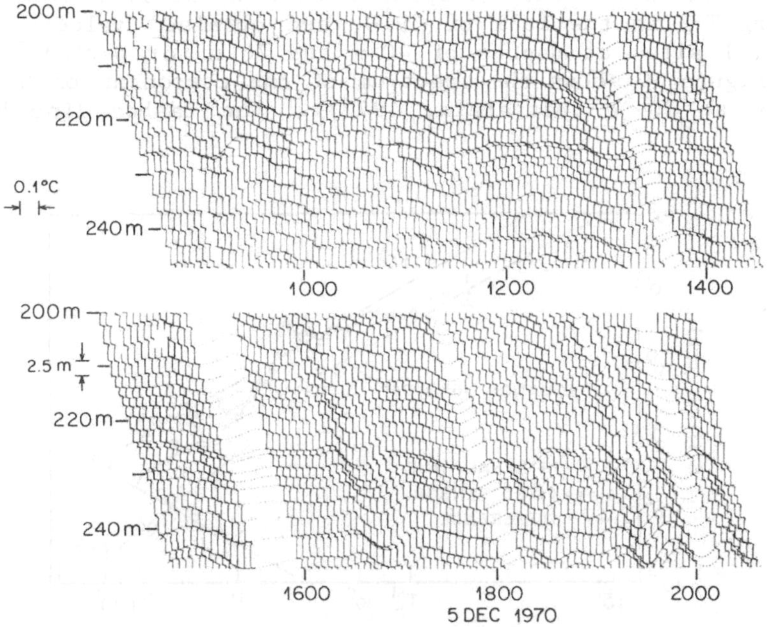

Fig. 8. A sequence of temperature profiles at T-3 (from Neshyba, et al., 1972). The profiles show uniform layers of about 2.5 m thickness separated by thin highly stratified sheets. The layering is due to double diffusive convection. Fluctuations in layer depth have been traced by Neshyba et al. to determine internal wave spectra uncontaminated by finestructure.

the effective moored spectra for a fixed sensor by using the temperature at a given depth and the mean temperature gradient to estimate an effective vertical displacement. Figure 9, taken from their paper, shows spectra computed for the two types of displacement record. The effective moored spectrum varies as ω^{-2} even at frequencies above the Brunt-Vaisala frequency while the true internal wave spectrum from tracing layer depths shows a dramatic drop in amplitude near the Brunt-Vaisala frequency (the amplitude varies as ω^{-7} to ω^{-8} from 2 cycles/hour to N).

The magnitude of r in (15) computed for the spectra of Neshyba, et al. (1972), is 94 m^2 cph. This is for the given value of N, 3 cph, and corresponds to an rms amplitude of 4.0 m and an energy of 0.89 J m^{-3}. While this value of r is not as low as that from Yearsley's data, it is still lower than that usually encountered in temperate oceans. Thus, these data also suggest that the internal wave energy is low in the Arctic.

The final set of historical data to be discussed are the horizontal velocity spectra of Bernstein and Hunkins (1971) and Bernstein (1972). They suspended current sensors from the ice adjoining T-3 in the fall of 1970, and obtained velocity records at several sites at depths of 40 m and 70 m for periods up to 70 days. Figure 10 shows the spectrum of one component of horizontal velocity at 70 m (Bernstein, 1972). The dashed line has been

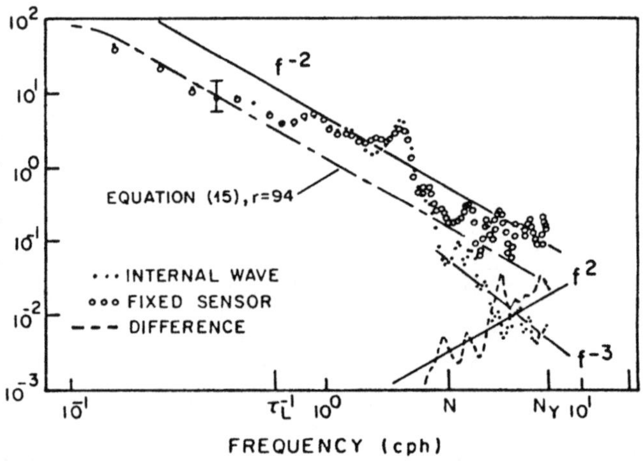

Fig. 9. Vertical displacement spectra at T-3 computed by tracing layer depths (internal wave) and by using temperature at a fixed depth (fixed sensor) (from Neshyba et al. 1972). The former spectrum is uncontaminated by spikes due to finestructure and shows a rapid roll-off above N. The later spectrum rolls of as ω^{-2} even above N.

added to represent a Garret-Munk type spectrum for one component of horizontal velocity given by an equation analogous to (15):

$$S_v(\omega) = 4\pi r \frac{fN(\omega^2 + f^2)}{\omega^3(\omega^2 - f^2)^{1/2}} \quad . \tag{16}$$

This equation is derived from the results of Levine et al. (1982) and yields the curve in Figure 10 for $r = 12.8\ m^2$ cph and N = 10 cph. This value of N is used because it is suggested by Bernstein (1972) and because the spectrum shows a slight increase in energy at frequencies just below 10 cph. These values of r and N correspond to an internal wave energy of 0.40 J m^{-3}. Aside from the "bump" just below 10 cph, the spectral model fits the data fairly well in the internal wave frequency band. Again, the value of r is substantially lower than that found in temperate oceans. It might be added that the results for 40 m are essentially the same as those for 70 m. Also, by comparing measurements at different sites, Bernstein and Hunkins (1971) find the velocities to be coherent over 400 m of horizontal separation for frequencies below one cycle per hour and coherent over 80 m at higher frequencies.

Aside from the studies described above, there are other indications that internal wave energies are relatively low in the Arctic. Measurements of sound propagation over 300 km in the Arctic Ocean show an extremely stable sound channel as compared with that found in temperate oceans (Mikhalevsky, 1981; Dyer,

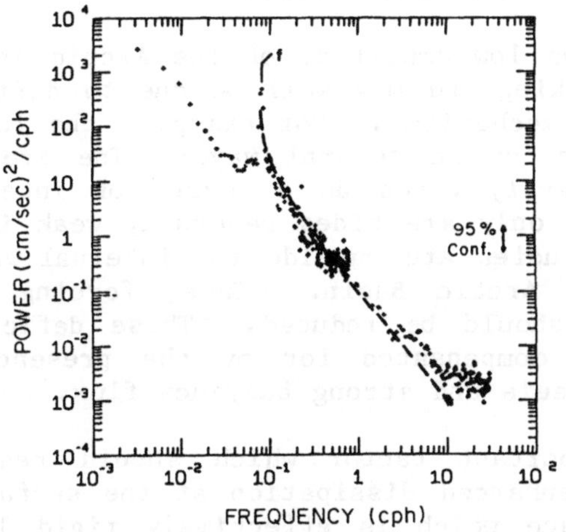

Fig. 10. The spectrum of one component of horizontal velocity at 70 m from Bernstein (1972). The Garret/Munk spectrum, as represented by equation (16) for $r = 12.8\ m^2$ cph, is shown by the dashed line.

personal communication, 1982). This may be due to lower levels of internal wave activity. Also, Hunkins (personal communication, 1982) reports time series of temperature and salinity at fixed depths, made during the 1975-76 AIDJEX experiment, show rms vertical displacement amplitudes on the order of 1 m. Finally, results of an internal wave study, undertaken at the Fram III ice camp at 82°N 9°E in 1981 by C. Paulson, M. Levine, and the author, yield values of r on the order of 50 m^2.

All of the measurements described here for the Arctic come from relatively shallow depths (<300 m), while most of the data which have been referred to regarding temperate oceans pertain to deep ocean (>300 m) measurements. Roth et al (1981) have analyzed shallow water internal wave data gathered from various sources and have compared them to appropriate deep water spectra. Taking the deep water energy levels as a base level they find that spectra near the surface have the same general shape but sometimes have elevated energies, presumably due to surface generation. Following their results, we would expect near surface internal wave energies, measured in the Arctic, to be high rather than low.

The consistency of the spectral shapes obtained by Yearsley (1966), Neshyba et al (1972), and Bernstein (1972) with the Garrett-Munk model is reassuring. In view of the differences in surface forcing between the Arctic Ocean and temperate oceans the agreement gives some support to the idea that the spectral shapes are controlled by dynamic processes, such as internal wave interaction and breaking, and not by some common source mechanism. These dynamic processes would appear to act in the same way in the Arctic Ocean as in temperate oceans.

However, the low amplitude of the Arctic internal wave measurements is striking and may well be due to differences in source and dissipation mechanisms. For example, the forcing of internal waves may be reduced in several ways: The ice cover eliminates most surface gravity waves as a source of internal wave excitation. Also, not only are tides generally weak in the Arctic, but the tidal frequencies are outside the internal wave frequency band for most of the Arctic Basin. Thus, forcing of free internal waves by tides should be reduced. These deficiencies may be at least partially compensated for by the presence of forcing by pressure ridge keels and strong buoyancy flux.

Another important factor which should reduce the internal wave energy is enhanced dissipation at the surface. The ice pack provides a surface which is effectively rigid longitudinally for the frequencies and wavelengths of most internal waves. Horizontal motions at the surface due to the waves must generate oscillatory boundary layers imbedded in and interacting with whatever turbulent boundary layer exists without the waves. Such boundary layers may provide important sinks for internal wave energy. In fact, the horizontal velocities due to internal waves are higher at the surface than at the ocean bottom because, for typical profiles of N, the eigenfunction derivatives at the surface are greater than those at the bottom. Therefore, dissipation under the ice cover is more important than dissipation at the bottom. The added dissipation should be roughly proportional to the internal wave energy to the three halves power, so the energy level required to dissipate a given input of energy is less if the surface is ice covered.

SUMMARY

Similarities between the internal wave fields in the Arctic and temperate oceans suggest the physics controlling the frequency distribution of internal wave energy are universal. The differences, both in the magnitude of the internal wave spectra and in the types of forces affecting internal waves at the surface, are intriguing because they suggest the level of internal wave energy may be controlled, in part by factors which are different in the Arctic than they are in temperate oceans. More internal wave experiments are needed in the Arctic in order to examine these important differences. This is necessary for two reasons: First, our general understanding of internal waves would be improved by examining conditions in a region where forcing and dissipation are different from those found in temperate oceans. Second, internal waves may be important to the dynamics and thermodynamics of the Arctic Ocean. They transfer momentum through wave propagation and interaction and they transfer heat and momentum through nonlinear processes. For example, Perkin and Lewis (1978) have shown breaking internal waves to be a crucial factor in the mixing of Cambridge Bay. Also, the manner in which heat is diffused from the Atlantic Water in the Arctic Basin is of crucial importance to the Arctic Ocean heat budget and may involve the enhancement of mixing by internal waves in a thermocline.

To resolve these questions it will be necessary to stage a major internal wave experiment in the Arctic. Such an experiment might focus on the determination of moored spectra, horizontal and vertical coherence, and the interaction of internal waves with the surface boundary layer. The experiments would be facilitated by the relatively rigid surface platform the ice provides. It simplifies micro-structure profiling, the construction of horizontal and vertical arrays, and the measurement of turbulence and boundary layer velocities at the surface. Also, the fact that, during quiet periods, the ice tends to move with the underlying ocean, may reduce the Doppler shifting problem over the situation in which arrays moored to the bottom are used.

ACKNOWLEDGMENTS

The author would like to thank Murray Levine for his helpful advice and discussions and Ken Hunkins for providing the reference, Bernstein (1972). The preparation of this review was supported by the Office of Naval Research Contract N-00014-79-C-0024.

REFERENCES

Bernstein, R. L. and K. Hunkins (1971) Inertial currents from a three-dimensional array in the Arctic Ocean. EOS. 52(4): 255 (abstract).

Bernstein, R. L. (1972) Observations of currents in the Arctic Ocean. Lamont-Doherty Geological Observatory, Columbia University, Technical Report 7, Contract N-00014-67-A-0108-0016.

Briscoe, M. G. (1975) Internal waves in the ocean. Rev. Geophys. Space Phys. 13: 591-598, 636-645.

Briscoe, M. G. (1983) Observations on the energy balance of internal waves during JASIN. Phil. Trans. R. Soc. London, A308, p. 427-444.

Desaubies, Y. I. F. (1976) Analytical representation of internal wave spectra. J. Phys. Oceanogr. 6: 976-981.

Ekman, V. W. (1906) On dead-water. In The Norwegian North-Polar Expedition, 1893-96, Scientific Results. (F. Nansen, Ed.), Christiania, Norway, Vol. 5, p. 1-152.

Garrett, C. J. R. and W. H. Munk (1972) Space-time scales of internal waves. Geophys. Fluid Dyn. 3: 225-264.

Garrett, C. J. R. and W. H. Munk (1975) Space-time scales of internal waves: A progress report. J. Geophys. Res. 80: 291-297.

Garrett, C. J. R. and W. H. Munk (1979) Internal waves in the ocean. Ann. Rev. Fluid Mech. 11: 339-369.

Hachmeister, L. E. and R. D. Muench (1981) Interaction between internal waves and ice bands in the marginal ice zone. EOS 62(45): 895 (abstract).

Hunkins, K. (1974) An estimate of internal wave drag on pack ice. AIDJEX Bulletin, 26: 141-152.

Lamb, H. (1932) Hydrodynamics. Dover, New York, p. 410-413.

Leonov, A. I. and Yu. Z. Miropolskiy (1973) Resonant excitation of internal gravity waves in the ocean by atmospheric-pressure fluctuations. Bull. (IZC.) Acad. Sci. USSR Atmospheric and Oceanic Physics 9(8) English translation, 480-485.

Levine, M. D., R. A. deSzoeke, and P. O. Niiler (1983) Internal waves in the upper ocean during MILE. J. Phys. Oceanogr. 13: 240-257.

Magaard, L. (1973) On the generation of internal gravity waves by a fluctuating buoyancy flux at the sea surface. Geophys. Fluid Dyn. 5: 101-111.

Mikhalevsky, P. N. (1981) Characteristics of cw signals propagated under the ice in the Arctic. J. of the Acoust. Soc. of Am. 70(6): 1717-1722.

Morison, J. H. (1980) Forced internal waves in the Arctic Ocean. University of Washington, Ph.D. thesis.

Mork, M. (1972) On oceanic response to atmospheric forces. Extrait Rapp. Proces. Verb. Cons. Cons. Int. Explor. Mer. 162: 184-190.

Morse, B. A. and J. D. Smith (1972) Density structure of the mixed layer under multi-year ice. EOS, 53(11): 1011 (abstract).

Munk, W. (1981). Internal waves and small scale processes. In Evolution of Physical Oceanography, B. Warren and C. Wunch, Ed.), MIT Press Cambridge, Mass., p. 264-291.

Nansen, F. (1902) The oceanography of the North Pole Basin. In Scientific Results of the Norwegian North Polar Expedition, 1893-1896 3(9): 427.

Neshyba, S. J., B. T. Neal and W. W. Denner (1972) Spectra of internal waves: In situ measurements in a multiple-layered structure. J. Phys. Oceanogr. 2: 91-95.

Perkin, R. G. and E. L. Lewis (1978) Mixing in an arctic fjord. J. Phys. Oceanogr. 8: 873-880.

Phillips, O. M. (1971). On spectra measured in an undulating layered medium. J. Phys. Oceanogr. 1: 1-6.

Rigby, F. A. (1974) Theoretical calculations of internal wave drag on sea ice. AIDJEX Bulletin, 26: 129-140.

Rigby, F. A. (1976) Pressure ridge generated internal wave wakes at the base of the mixed layer in the arctic Ocean. University of Washington, Master's thesis.

Roth, M. W., M. G. Briscoe and C. H. McComas III (1981) Internal waves in the upper ocean. J. Phys. Oceanogr. 11: 1234-1247.

Wunsch, C. (1976) Geographical variability of the internal wave field: A search for sources and sinks. J. Phys. Oceanogr. 6: 471-485.

Wunsch, C. and S. Webb (1979) The climatology of deep ocean internal waves. J. Phys. Oceanogr. 9: 235-243.

Yearsley, J. R. (1966) Internal waves in the Arctic Ocean. University of Washington, Master's thesis.

INDEX

9781489953544VOL02